Fundamental constants

Constant	Symbol	Value (SI units)	cgs-esu or other units
Gas constant	R	8.31447 J K^{-1} mol^{-1}	8.31447 × 10^7 erg K^{-1} mol^{-1}
		8.31447 m^3 Pa K^{-1} mol^{-1}	0.0820575 L atm K^{-1} mol^{-1}
			1.98721 cal K^{-1} mol^{-1}
Avogadro's number	N_A	6.02214 × 10^{23} mol^{-1}	6.02214 × 10^{23} mol^{-1}
			6.02214 × 10^{23} amu g^{-1}
Faraday constant	F	96485.3 C mol^{-1}	23.06055 kcal V^{-1} mol^{-1}
		96.4853 kJ V^{-1} mol^{-1}	
Speed of light	c	2.99792458 × 10^8 m s^{-1}	2.99792458 × 10^{10} cm s^{-1}
Planck's constant	h	6.62607 × 10^{-34} J s	6.62607 × 10^{-27} erg s
Boltzmann's constant	k_B	1.38065 × 10^{-23} J K^{-1}	1.38065 × 10^{-16} erg K^{-1}
Electronic charge	e	1.602176 × 10^{-19} C	4.803204 × 10^{-10} esu (statC)
Coulomb's law constant	k	8.98755 × 10^9 J m C^{-2}	1 erg cm esu^{-2}
Electron mass	m_e	9.10938 × 10^{-31} kg	9.10938 × 10^{-28} g
			0.0005485799 amu
Proton mass	m_p	1.672622 × 10^{-27} kg	1.672622 × 10^{-24} g
			1.0072765 amu
Neutron mass	m_n	1.674927 × 10^{-27} kg	1.674927 × 10^{-24} g
			1.0086649 amu
Gravitational constant	G	6.673 × 10^{-11} J m kg^{-2}	6.673 × 10^{-8} erg cm g^{-2}
Acceleration of gravity	g	9.80665 m s^{-2}	980.665 cm s^{-2}

Source: P. J. Mohr and B. N. Taylor, *Rev. Mod. Phys.* **72,** 351 (2000).

A LEARNING PACKAGE BUILT TO ENSURE YOUR SUCCESS, THIS TEXTBOOK IS INTEGRATED WITH THE MOST ADVANCED CHEMISTRY TUTORIAL MEDIA AVAILABLE.

Mastering GENERAL CHEMISTRY

For instructor-assigned homework, MasteringGeneralChemistry™ (www.masteringgeneralchemistry.com) provides the first adaptive-learning online tutorial and assessment system. Based on extensive research of precise concepts students struggle with, the system is able to coach you with feedback specific to your needs, and simpler subproblems and help when you get stuck. The result is targeted tutorial help to optimize your study time and maximize your learning.

If your professor requires MasteringGeneralChemistry as a component of your course, your purchase of a new copy of Peter Siska's *University Chemistry* already includes a free Student Access Kit. You will need this Access Kit to register.

If you did not purchase a new textbook and your professor requires you to enroll in MasteringGeneralChemistry, you may purchase online access with a major credit card. Go to www.masteringgeneralchemistry.com and follow the links to purchasing online.

Minimum System Requirements
System requirements are subject to change. See website for the latest requirements.
Windows: 250 MHz CPU; OS Windows 98, NT, 2000, XP
Macintosh: 233 MHz CPU; OS 9.2, 10.2, 10.3
RedHat Linux 8.0, 9.0
All:
- 64 MB RAM
- 1024 x 768 screen resolution
- Browsers (OS dependent): Firefox 1.0, Internet Explorer 6.0, Mozilla 1.7, Netscape 7.2, Safari 1.3
- Flash 7.0

MasteringGeneralChemistry™ is powered by MyCyberTutor by Effective Educational Technologies

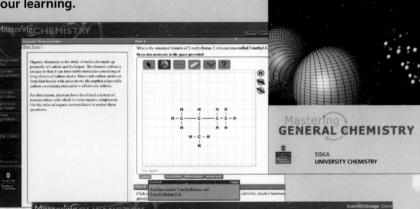

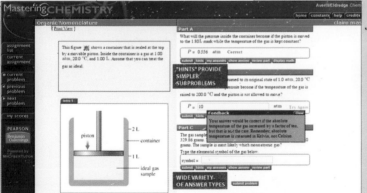

For self study, the open-access website The Chemistry Place™ (www.aw-bc.com/chemplace) provides interactive graphs, multiple-choice quizzes, animated molecular structures, flashcards, InterAct Math, and many other interactive resources to enhance your learning.

Minimum System Requirements
System requirements are subject to change. See website for the latest requirements.
Windows: 266 MHz CPU; OS Windows 98, NT, 2000, XP
Macintosh: 266 MHz CPU; OS 9.2, 10.2, 10.3
Both:
- 64 MB RAM
- 1024 x 768 screen resolution
- Browsers (OS dependent): Internet Explorer 6.0, Netscape 7.2
- Flash, Shockwave, QuickTime, and Chime

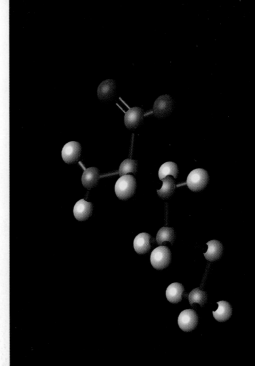

University Chemistry

Peter E. Siska
University of Pittsburgh

PEARSON
Benjamin Cummings

San Francisco Boston New York
Cape Town Hong Kong London Madrid Mexico City
Montreal Munich Paris Singapore Sydney Tokyo Toronto

Publisher: Jim Smith
Marketing Manager: Scott Dustan
Project Editors: Katie Conley, Lisa Leung
Assistant Editor: Cinnamon Hearst
Editorial Assistant: Kristin Rose
Media Producer: Claire Masson
Production Supervisor: Shannon Tozier
Production Editor: Lori Dalberg, Carlisle Editorial Services
Composition: Carlisle Publishing Services
Illustrators: Imagineering Media Services, Inc.
Manufacturing Buyer: Pam Augspurger
Text Designer: Patrick Devine, Patrick Devine Design
Cover Designer: Jeff Puda
Text Printer and Binder: Courier Kendallville
Cover Printer: Coral Graphics

Library of Congress Cataloging-in-Publication Data
Siska, Peter E.
   University chemistry / Peter E. Siska.
      p. cm.
   Includes bibliographical references and index.
   ISBN 0-8053-9349-8 (alk. paper)
   1. Chemistry—Textbooks. 2. Chemistry—History—Textbooks. I. Title.
   QD31.3.S57 2006
   540—dc22

                                                                 2005025051

ISBN 0-8053-9349-8

ISBN 0-8053-3076-3 (Instructor)

Copyright © 2006 Pearson Education, Inc., publishing as Benjamin Cummings, 1301 Sansome St., San Francisco, CA 94111. All rights reserved. Manufactured in the United States of America. This publication is protected by Copyright and permission should be obtained from the publisher prior to any prohibited reproduction, storage in a retrieval system, or transmission in any form or by any means, electronic, mechanical, photocopying, recording, or likewise. To obtain permission(s) to use material from this work, please submit a written request to Pearson Education, Inc., Permissions Department, 1900 E. Lake Ave., Glenview, IL 60025. For information regarding permissions, call (847) 486-2635.

Many of the designations used by manufacturers and sellers to distinguish their products are claimed as trademarks. Where those designations appear in this book, and the publisher was aware of a trademark claim, the designations have been printed in initial caps or all caps.

1 2 3 4 5 6 7 8 9 10-CRK-08 07 06 05
www.aw-bc.com

DEDICATION
To Jeanne, David, and Sarah

Brief Table of Contents

Chapter 1
Physical Principles Underlying Chemistry 1

Chapter 2
The Quantum Revolution: The Failure of Everyday Notions to Apply to Atoms 41

Chapter 3
Wave Mechanics and the Hydrogen Atom: Quantum Numbers, Energy Levels, and Orbitals 68

Chapter 4
Atoms with Many Electrons and the Periodic Table 98

Chapter 5
Valence Electron Configurations, Periodicity, and Chemical Behavior 130

Chapter 6
Orbitals and Chemical Bonding I: The Valence Bond Model and Molecular Geometry 177

Chapter 7
Orbitals and Chemical Bonding II: The Molecular Orbital Model and Molecular Energy Levels 205

Chapter 8
Molecular Motion and Spectroscopy 237

Chapter 9
Properties of Gases and the Kinetic Molecular Theory 267

Chapter 10
Energy Changes in Chemical Reactions 301

Chapter 11
Spontaneity of Chemical Reactions 342

Chapter 12
Free Energy and Chemical Equilibrium 375

Chapter 13
Electrochemistry 435

Chapter 14
States of Matter and Intermolecular Forces 472

Chapter 15
Rates and Mechanisms of Chemical Reactions 538

Chapter 16
The Nucleus 598

Chapter 17
The Transition Metals 643

Chapter 18
The Chemistry of Carbon 678

Appendix A: Blackbody Radiation: The Origin of the Quantum Theory 732
Appendix B: A Particle in a Box 739
Appendix C: Selected Values of Thermodynamic Properties at 298.15K 746
Appendix D: A Brief Look at Statistical Thermodynamics 774
Appendix E: Answers to Exercises 782
Photo Credits 804
Index 805

Table of Contents

Chapter 1
Physical Principles Underlying Chemistry 1

- 1.1 The Province of Chemistry 2
- 1.2 What We Want to Know: Objectives of Modern Chemistry 3
  - Atomic Theory 4
  - The Periodic Law 7
  - Modern Questions 9
- 1.3 Units of Measure 10
  - Errors in Measurement 11
- 1.4 Force, Work, and Energy 13
  - Mechanical Work 14
  - Conservation of Energy 16
  - Temperature and Heat Energy 19
- 1.5 The Electrical Nature of Matter 20
  - The Charge on the Electron 22
- 1.6 The Nuclear Atom 23
  - Isotopes 26
  - Avogadro's Number and the Mole 28
- 1.7 Coulomb Force and Potential Energy 30
  - Summary 34
  - Exercises 35

Chapter 2
The Quantum Revolution: The Failure of Everyday Notions to Apply to Atoms 41

- 2.1 The Wave Theory of Light 43
  - Diffraction and Interference of Light Waves 45
  - The Electromagnetic Spectrum 46
- 2.2 The Line Spectra of Atoms 46
  - The Hydrogen Atom Spectrum 47
- 2.3 The Ultraviolet Catastrophe and Planck's Quantum Hypothesis 49
  - The Quantum of Energy 49
- 2.4 The Photoelectric Effect: Particles of Light 51
- 2.5 The Nuclear Atom and the Quantum: Bohr's Explanation of Line Spectra 53
- 2.6 de Broglie's Matter Waves: The Beginning of a New Mechanics 56
  - The Bohr–de Broglie Model of the Hydrogen Atom 58
- 2.7 The Schrödinger Equation: A Wave Equation for Particles 62
  - Summary 63
  - Exercises 64

Chapter 3
Wave Mechanics and the Hydrogen Atom: Quantum Numbers, Energy Levels, and Orbitals 68

- 3.1 Solving the Schrödinger Equation for the Hydrogen Atom: A Brief Description 68
  New Mechanics but Old Mathematics 70
  Boundary Conditions 71
- 3.2 The Quantum Numbers n, l, and m 72
  Specification of the Energy: Ionization 73
  Specification of the Wave Functions: Orbitals 74
- 3.3 Characteristics of the Orbitals: Three-Dimensional Waves 76
- 3.4 Born's Interpretation of Orbitals 83
  Radial Distribution Functions 84
- 3.5 Heisenberg's Uncertainty Principle 87
  A Thought Experiment 90
- 3.6 Orbitals and Orbits: An Analogy 91
- 3.7 A Qualitative Description of Electronic Transitions 93
  Summary 94
  Exercises 95

Chapter 4
Atoms with Many Electrons and the Periodic Table 98

- 4.1 Hydrogen-Like Orbitals for Many-Electron Atoms 99
- 4.2 Electron Spin: Evidence for a Fourth Quantum Number m_s 101
  The Stern-Gerlach Experiment 103
  Dirac's Relativistic Electron 104
- 4.3 Pauli's Exclusion Principle 105
- 4.4 The Aufbau Principle and Electron Configurations of the Elements 105
  The Energies of Orbitals: Screening 106
  Hund's Rule 108
  Reading Configurations from the Periodic Table 109
  Anomalous Configurations 112
- 4.5 Periodic Properties of the Atoms of the Elements 113
  Ionization Energy 113
  Electron Affinity 116
  Atomic Radius 118
  Ionic Configurations and Isoelectronic Sequences 120
- 4.6 Computers in Chemistry: The Many-Electron Problem 120
- 4.7 "Seeing" Atoms 122
  Summary 126
  Exercises 127

Chapter 5
Valence Electron Configurations, Periodicity, and Chemical Behavior 130

5.1 **Metals, Nonmetals, and Binary Salts: The Noble Gas Rule of Lewis** 131
Simple Oxidation–Reduction (Redox) Reactions 133
Ionic Bonding 137

5.2 **Electron Sharing: Lewis's Covalent Bond** 138
Electronegativity 140
Resonance Forms and Expanded Octets 141
A Lewis Structure Description of a Reaction 144

5.3 **Reactions of Metals and Nonmetals with Water: Acidic and Basic Behavior** 145
Metal–Acid Reactions 149
Nonmetals in Water and Base 150

5.4 **Binary Oxides and Hydrides: Chemical Periodicity** 150
Hydrides 154

5.5 **Acid–Base Reactions** 156
Brønsted–Lowry Acids and Bases 157
The Lewis Concept 158
Solvent Dependence of Acidity and Basicity 159

5.6 **Oxidation–Reduction Reactions** 160
Addition and Displacement 160
Combustion Reactions 162
Redox in Aqueous Solution 162
Oxidation Numbers 165

5.7 **Precipitation Reactions** 167
Reaction Classification 168

Summary 169

Exercises 171

Chapter 6
Orbitals and Chemical Bonding I: The Valence Bond Model and Molecular Geometry 177

6.1 **The Hydrogen Molecule: The Simplest Electron-Pair Bond** 178
The Heitler–London Treatment of H_2 178
The Importance of Orbital Overlap 181

6.2 **Polyatomic Molecules: Pauling's Valence Bond Model** 181
Atomic Hybridization 183

6.3 **Double and Triple Bonds: σ and π Bond Types** 187

6.4 **Electron-Pair Repulsion: Rationalizing and Predicting Molecular Shapes** 190
σ-Bonded Molecules 190
Molecules with π Bonds 193
The Effect of Heteroatoms 193
An Overall Assessment of VSEPR 195

- 6.5 Asymmetric Electron Sharing: The Electric Dipole Moment 195
  Valence Bond Description of Partial Ionic Character 199
- 6.6 Shortcomings of the Valence Bond Model 200
  Paramagnetism of Molecular Oxygen 200
  Reaction Mechanisms and Molecular Energy Levels 201

  Summary 201

  Exercises 202

Chapter 7
Orbitals and Chemical Bonding II: The Molecular Orbital Model and Molecular Energy Levels 205

- 7.1 The Hydrogen Molecular Ion: The Prototype Molecular Orbital System 206
  Linear Combinations of Atomic Orbitals 207
  Why Covalent Bonds Form 209
- 7.2 The Molecular Orbital Description of Diatomic Molecules 211
  Homonuclear Diatomic Molecules from Second-Row Atoms 212
  Heteronuclear Diatomic Molecules 217
  Acidity, Basicity, and the Frontier Orbitals 221
  Frontier Orbitals and Redox Reactions 223
- 7.3 Molecular Orbitals of the Water Molecule 224
- 7.4 Delocalized Bonding: A Solution to the Resonance Problem 227
- 7.5 Shortcomings of Molecular Orbitals 230
- 7.6 Valence Bonds, Molecular Orbitals, and Quantum Chemistry 231

  Summary 232

  Exercises 233

Chapter 8
Molecular Motion and Spectroscopy 237

- 8.1 Degrees of Freedom: Translation, Rotation, and Vibration 238
- 8.2 Rotational Motion: Defining a Molecular Day 240
  Quantization of Rotational Energy 243
  Rotational Spectroscopy 246
- 8.3 Vibrational Motion: Molecular Calisthenics 250
  Quantization of Vibrational Energy 253
  Vibrational Spectroscopy 255
- 8.4 Electronic Transitions in Molecules 259
  Photoelectron Spectroscopy 261
- 8.5 Looking Back and Peeking Ahead 262

  Summary 263

  Exercises 264

Chapter 9
Properties of Gases and the Kinetic Molecular Theory 267

9.1 Pressure of a Gas: Barometric Principles 268
9.2 The Ideal Gas Law 270
9.3 Gas Mixtures: Dalton's Law of Partial Pressures 276
9.4 The Kinetic Molecular Theory 278
   Predictions of the Kinetic Theory of Gases 282
   The Maxwell–Boltzmann Velocity Distribution 285
   Gas Dynamics 288
9.5 The Behavior of Real Gases 291
   Summary 295
   Exercises 296

Chapter 10
Energy Changes in Chemical Reactions 301

10.1 Conceptual Developments 302
10.2 Heat and Work in a Chemical System 304
10.3 The First Law of Thermodynamics:
   The Energy of the Universe Is Constant 309
   State Variables 309
   State Changes for an Ideal Gas 311
10.4 Measuring the Heat of a Chemical Reaction: Calorimetry 314
10.5 Systematics of Heats of Reaction 317
   Standard Heats of Formation 320
10.6 Origins of the Heat of Reaction: Bond Energies 324
   Resonance Energy 328
10.7 Heat Capacities: A Molecular Interpretation 329
   Molecular Energy Storage: The Boltzmann Distribution 331
   Temperature Dependence of the Enthalpy 333
   Summary 334
   Exercises 337

Chapter 11
Spontaneity of Chemical Reactions 342

11.1 Berthelot's Hypothesis 343
11.2 Carnot's Ideal Engine and the Entropy S 345
   Spontaneity and Reversibility 345
   The Carnot Cycle 347
   A New State Variable: The Entropy S 351

11.3 The Second Law of Thermodynamics: The Entropy of the Universe Tends to a Maximum 353

11.4 Entropy and Probability: Boltzmann's Postulate $S = k \ln W$ 354

11.5 The Third Law of Thermodynamics: The Entropy of a Perfect Crystal at $T = 0$ K Is Zero 357
Absolute Entropies 358
Entropy Changes in Chemical Reactions 361

11.6 Focusing on the System: The Gibbs Free Energy $G = H - TS$ 363
Temperature Control of Spontaneity 365

11.7 Use and Misuse of the Second Law 366
Entropy and Time 367
Entropy and Evolution 367
Entropy and Society 368
Where It All Leads 368

Summary 369

Exercises 371

Chapter 12
Free Energy and Chemical Equilibrium 375

12.1 The Attainment of Equilibrium 376
Dependence of Free Energy on Pressure 376
The Equilibrium Constant $K = \exp[-\Delta G°/RT]$ 380
The Position of Equilibrium in Gas-Phase Reactions 381

12.2 LeChâtelier's Principle: Controlling the Position of Equilibrium 384

12.3 Equilibria in Aqueous Acid–Base Chemistry 387
The Arrhenius Acid–Base Reaction 388
Ionization of Weak Acids and Bases 392
Brønsted–Lowry Acid–Base Reactions 400
Common Ions and Buffers 403
Ionization of Polyprotic Acids 406
Acid–Base Titrations 411
Aqueous Complexes 416

12.4 Equilibria in Multiphase Systems 418
Solubility and Precipitation 419
Qualitative Metal Ion Analysis 421

12.5 Deviations from Ideal Behavior 423

Summary 426

Exercises 429

Chapter 13
Electrochemistry 435

13.1 Voltage from the Free Energy of a Redox Reaction 436

13.2 Electrochemical Cells: Principles and Practice 440
The Daniell Cell 441
Leclanché's Dry Cell 443

Planté's Lead-Acid Storage Battery 445
Fuel Cells 447

13.3 **Electrode Potentials** 449
The Standard Hydrogen Electrode 450

13.4 **Concentration Cells** 454
The pH Meter 455

13.5 **Electrolysis: Redox Chemistry in Reverse** 458
Faraday's Laws of Electrolysis 460
Applications of Electrolysis 461

13.6 **The Electrochemistry of Corrosion** 464

Summary 465

Exercises 466

Chapter 14
States of Matter and Intermolecular Forces 472

14.1 **It's All a Matter of Temperature** 473

14.2 **Heating Curves** 475

14.3 **The Free Energy of Phase Changes** 479
The Phase Diagram: Water 485
Phase Behavior in General 487

14.4 **Phase Transitions, Molecular Structure, and Intermolecular Forces** 495
Attractive Intermolecular Forces 496
Hydrogen Bonding 500
Repulsive Intermolecular Forces and the Intermolecular Potential 502

14.5 **Structure and Bonding in Solids** 503
Metal Crystals 504
Nonmetallic Network Solids 510
Allotropy in Elemental Solids 513
Ionic Solids 514
Molecular Solids 519

14.6 **Liquids and Solutions, Featuring Water** 521
Solutions 522
Colligative Properties of Solutions 524

Summary 529

Exercises 532

Chapter 15
Rates and Mechanisms of Chemical Reactions 538

15.1 **On the Way to Equilibrium** 539

15.2 **Rate Laws and Rate Constants** 541
Determining the Rate Law: The Method of Initial Rates 547
Determining the Rate Law: The Isolation Method 549

15.3 **Reaction Mechanisms** 551
Elementary Reactions 551
Rate-Limiting Step 552
The Steady-State Approximation 555
The Principle of Detailed Balance 557
The Temperature Dependence of Reaction Rates 559

15.4 **Molecular Theories of Elementary Reactions** 562
Collision Theory 562
Transition State Theory 566
Molecular Reaction Dynamics 569
Reactions in Solution 573

15.5 **Chemical Catalysis and Complex Reactions** 577
Catalysis: Homogeneous and Heterogeneous 578
Enzyme Catalysis 580
Surface Catalysis 581
Chain Reactions, Explosions, and Flames 583
Oscillating Reactions 586

Summary 589

Exercises 591

Chapter 16
The Nucleus 598

16.1 **A Chemical Microcosm: Nuclear Structure and Bonding** 599
16.2 **Nuclear Stability and Radioactivity** 604
16.3 **Kinetics of Nuclear Decay** 611
16.4 **Nuclear Reactions** 616
Stellar Nucleosynthesis 616
Nuclear Reactions in the Laboratory 618

16.5 **Applying our Nuclear Knowledge** 623
Isotopic Chemistry 624
Nuclear Medicine 625
Nuclear Magnetic Resonance (NMR) 626
Radioactive Dating 630
Nuclear Energy Sources 632
The Genie Is Out of the Bottle 635

Summary 636

Exercises 638

Chapter 17
The Transition Metals 643

17.1 **The *d*-Block Metals: Energies, Charge States, and Ionic Radii** 645
17.2 **Chemistry of the Early Transition Metals: Oxyions** 648
Spectroscopy and Structure of Oxyanions 648

17.3 **Chemistry of the Late Transition Metals: Coordination Complexes** 651
   Stoichiometry, Isomerism, and Geometry of Complexes 653
   Nomenclature 654
   The Nature of Ligands 654
   Magnetism 657
   Lability 657

17.4 **The Spectrochemical Series and Bonding in Complexes** 658
   The Crystal Field Model 661
   Tetrahedral and Square-Planar Complexes 664
   The Ligand Field Model 665

17.5 **Chemical Kinetics of Complexes** 668
   Ligand Substitution 668
   Oxidation–Reduction 669

17.6 **The Lanthanides and Actinides** 672

   Summary 674

   Exercises 675

Chapter 18
The Chemistry of Carbon 678

18.1 **What Makes Carbon Special?** 680
   Boron: Making Do with Too Few 681
   Nitrogen: Three's a Crowd 683
   Silicon: More Than Just Computer Chips? 683

18.2 **Building Blocks of Organic Chemistry: The Hydrocarbons** 684
   Structural Isomers 686
   Alkane Reactions 688
   Unsaturated Hydrocarbons: Alkenes and Alkynes 689
   Addition Reactions: Carbocations and Formal Charge 691

18.3 **Derivatives of the Hydrocarbons: Functional Groups** 692
   Substitution Reactions 694
   The Asymmetric Carbon Atom: Enantiomers 695

18.4 **Benzene and Its Derivatives: The Aromatics** 697

18.5 **Orbitals in Organic Reactions: The Diels–Alder Reaction** 699

18.6 **Polymers** 702

18.7 **The Molecules of Life: Organic Motifs in Biology** 704
   Carbohydrates and Fats 704
   Amino Acids and Proteins 707
   Cyclic Bases, Nucleotides, and Nucleic Acids 711

18.8 **The Chemical Dynamics of Life** 715
   Photosynthesis: The Light and Dark Reactions 715
   Enzyme Catalysis and Carbohydrate Metabolism 717
   How Enzymes Work: The Hydrolysis of a Peptide Bond
   by Chymotrypsin 720
   Making the Right Enzymes: DNA, RNA, and the Code of Life 722

The Molecular Roots of Life 727
Where Are We Heading? 728

Summary 728

Exercises 730

Appendix A: Blackbody Radiation: The Origin of the Quantum Theory 732
Appendix B: A Particle in a Box 739
Appendix C: Selected Values of Thermodynamic Properties at 298.15 K and 1 atm 746
Appendix D: A Brief Look at Statistical Thermodynamics 774
Appendix E: Answers to Exercises 782

Photo Credits 804

Index 805

About the Author

Peter Siska is a professor of chemistry at the University of Pittsburgh. He received his undergraduate education at DePaul University, his PhD from Harvard University, and postdoctoral experience at the University of Chicago, before arriving at Pitt in 1971, where he has been ever since. His research interests lie in the romantic field of molecular beams and their use in elucidating intermolecular forces and reaction dynamics; in 1984 he chaired a Gordon Research Conference on atomic and molecular interactions. At Harvard he held a National Science Foundation predoctoral fellowship, and at Pitt he was an Alfred P. Sloan Fellow. He has been an instructor in the University Honors College since its inception at Pitt in 1987. In addition to teaching physical chemistry at both the undergraduate and graduate levels, he has taught Pitt's honors general chemistry courses since 1981. In 1987 he received a Chancellor's Distinguished Teaching Award, in 1998 a visiting professorship at Harvard, in 2001 an Innovations in Education Award, and in 2003 a Bellet Undergraduate Teaching Excellence Award. In addition to pursuing beam studies of excited atom chemistry, he is currently working on a physical chemistry text. What he relishes most about academic life is the chance to interact with young scholars at every level.

Foreword

It is a jubilation to welcome this superb text by Peter Siska, developed over many years in response to his students. A much wider audience can now enjoy his remarkably lucid and engaging expositions, seasoned with verve and humor, offering an ample menu of chemical delectations. Equally important in my view is the motivating cultural perspective of his text. It presents a compelling saga of human adventure, replete with foibles as well as feats, in exploring a fabulous molecular world.

Siska's approach fosters the chief aim of a liberal arts education: to instill the habit of self-generated questioning and thinking, of actively scrutinizing evidence and puzzling out answers. That is also the essence of a genuine scientific literacy. It accords with a favorite definition, ironic but applicable to both science and the humanities: "Education is what's left after all you've learned has been forgotten." This defines the aim to be understanding rather than ritualistic training; cultural perspective and self-reliant thinking rather than conventional knowing. For earnest students, whether novice scientists or destined for other careers, these "what's left" aspects transcend any technical particulars.

Teachers or textbooks best serve as catalysts, accelerating efforts of students to take ownership of a subject. In Siska's approach, this catalytic role is enhanced by bringing out at each stage reinforcing links, both to topics already met and to those ahead. Students thereby gain a broadening view of both cultural and technical aspects as they progress through the text. Siska also aptly points out vistas of current research frontiers.

When students ask how to study science, I urge that they approach it like a foreign language: "Make it your own! Once you get it in your ear, it gets easier and easier; otherwise, harder and harder!" Actual counts have shown that in introductory science textbooks, the number of new words or ordinary words used with special meanings exceeds the vocabulary of a typical one-year language course. Likewise, the array of interlocking concepts met in a science course functions much like grammatical rules.

This perspective emphasizes the kinship of students with research scientists. Nature speaks to us abundantly, but in many alien tongues. In frontier research, we try to add to knowledge of the vocabulary and grammar of some strange dialect. To the extent that we succeed, we gain the ability to decipher many messages that Nature has left for us, blithely or coyly. No matter how much human effort and resources we devote to solving a practical problem in science or technology, progress will come only when we can read the answers that Nature is willing to give us. That is why curiosity-driven research is an essential and practical investment, as well as an exhilarating intellectual adventure.

Siska's text celebrates and elucidates for students a bountiful legacy of chemical linguistics. However, those eager to acquire this legacy should recognize that neophytes and veteran scientists alike will often be quite confused! That makes many students uneasy. Yet puzzlement is welcomed by scientists, as it is usually prerequisite for any exciting new insight. Realizing this, and relying on the sure and cheerful guidance of Siska's book, students can persevere with sangfroid.

DUDLEY R. HERSCHBACH
Harvard University

Preface

This work arose out of an honors-level introductory chemistry course I have been teaching at the University of Pittsburgh for more than two decades. It was begun in a gap period in the 1980s, when older texts, such as Busch, Shull, and Conley's *Chemistry* and Dickerson, Gray, and Haight's *Chemical Principles,* were going out of print, and newer offerings did not seem suitable to the course as it gradually developed. It is as much a result of my students teaching me what they were capable of learning as neophyte chemists, as of my own efforts to knit the course into a whole. The thought of publishing was then remote, but I was desperate for something to get my course off to a good start. The beginnings of this project also coincided with another I was urging on my physical chemistry colleagues at the University of Pittsburgh, to reverse the order of topics in our physical chemistry courses. We now begin with quantum mechanics and molecular structure, then weave molecular statistics into classical thermodynamics and kinetics. At a lower level, this text attempts to do the same, while incorporating the topics traditionally included in general chemistry. Aside from this, one's creations are inevitably a product of one's experience, and those old enough to have used either of the two texts cited above will recognize the great debt owed them for any virtues that may be found here.

Just as important as presentation and topic sequence is placing knowledge in an historical context. In her research essay *They're Not Dumb, They're Different,* Sheila Tobias finds that much of what drives bright students away from the study of science lies in introductory courses that provide little motivation for the humanistically inclined to "get interested," by failing to include historical developments and the human face behind every scientific idea. This is not a history book, but it does attempt to use historical evolution of a topic to maintain a story line, hoping to carry the students' thinking from one stepping stone to the next as they cross the stream of science. No one has done a better job with chemical history than Dickerson, Gray, and Haight in *Chemical Principles;* I am still in love with that book and its *postscripts,* and could not hope to improve on them. There are no postscripts here; the history of a topic in briefer form is integrated into its presentation, serving a dual role of providing context and aiding concept development.

The text is deliberately very connective, the topics flowing into one another, hinging back and projecting forward, with the big picture always in mind. This resonates with another of Tobias's findings—that treating science as a series of problem types to be mastered loses the students who desire perspective, who are motivated by knowing why they should strive for mastery. Connectivity has the major disadvantage of lacking flexibility as to topic sequence. Applications from later chapters can be brought in to enrich a topic, but skipping around is predicted to be difficult.

The number of chapters, 18, is smaller by 10 or more than in most contemporary texts; this reflects a realism as to how much of twenty-first century chemistry can be covered, or better "uncovered," in a two-semester or three-trimester academic year. In particular, I have found that entirely too much time and space is generally devoted to chemical equilibrium, the study of "dead" reactions. While equilibrium concepts pervade the text, the presentation of this topic is confined to Chapter 12. The text naturally divides into three sections: Chapters 1 through 8 are microscopic, 9 through 15 macroscopic, and 16 through 18 applications. (The original title was *Foundations of Chemistry;* hence it is, remembering Isaac Asimov, a "foundation trilogy.") In our two-term sequence of 40 lectures per term at the University of Pittsburgh, I cover Chapters 1 through 8 in the first term and 9 through 17 in the second, with suitable omission of certain topics in the latter. Chapter 18, an introduction to organic and biochemistry, has gotten squeezed out in my courses, though it has provided enrichment for topics in molecular structure, thermodynamics, and kinetics. The majority of my students go on to enroll in organic chemistry in any event.

It has become common for incoming students, especially those wanting a more rigorous science track, to have had not only physics and chemistry but mathematics through calculus in secondary school. Modest use of calculus is made periodically throughout the text, beginning in Chapter 1, not only because it makes certain problems easier to state and solve, but because it is an essential part of the language of science. The sooner students appreciate this, the better. Physical science is also unique among branches of knowledge in that its concepts all come from numbers, whether measured or theoretically imposed; this "triumph of numbers" is the ultimate rationale for the use and utility of numerical exercises in mastering the subject. That mathematics is so unreasonably effective in describing nature is wondrous but true. To encourage problem solving, answers to *all* exercises are included; this ensures that no problems need become dead letters (or numbers).

At first glance the organization of topics does not seem to lend itself to a concurrent introductory laboratory program. Laboratory being of critical importance to learning chemistry—to learn piano you have to touch the keys—an honors lab program has been developed at Pittsburgh that meshes to the extent possible with the text. More detail cannot be given here; please contact me at pes@pitt.edu for further information.

Writing this book has actually been a joy, made so by being able to put on paper explanations and pictures that my students found useful in learning chemistry. Finished since 1999, it has also made teaching the honors courses a near-celestial experience. I would like to acknowledge, in addition to the authors of the wonderful books cited above, the following honors students for their particularly extensive efforts to provide feedback that improved the end result: Robert Hackenberg, John Ernsthausen, Erin Dougherty (who suggested changes that required upheaving the index, and then redid it herself), and John Gansner (actually a Harvard student who happened to take a one-term version of the course taught while I was on sabbatical there, and who saw things through to the finish). A great debt is also owed to the honors chemistry teaching fellows who assisted me while the text was evolving, particularly David W. Martin, who early on helped with exercise solutions, and Joseph H. Noroski, who has found numerous errors and inconsistencies, and who is now coauthor of the solutions manual. Referees from three different potential publishers have assiduously ferreted out myriad mistakes and unhappy turns of phrase. The hospitality of Harvard University Department of Chemistry and Chemical Biology in the fall of 1998, where Chapter 16 was created and the earlier ones tested by fire, was of inestimable value. Among other authors and teachers of chemistry I gratefully acknowledge my own early teachers, Alvin Metzger, Robert Clay Miller, and Avrom Blumberg, the late Walter J. Moore for the most erudite physical chemistry texts around, each edition a new treasure, and Dudley R. Herschbach, without whose vital push no one would be reading this now, nor would this preface exist at all. The 1500+ students who have passed through my honors sequence, many helpful and encouraging colleagues, and my wife and children all have my perpetual thanks.

Turning a manuscript into a book requires, as I now know, a combination of stubborn faith and subtle shepherding of wrangling forces. My appreciation to Ben Roberts, who became the latest in a line to think my work publishable, to Jim Smith, who made sure it finally happened, to Katie Conley, editorial shepherdess, and to the ever-patient Lori Dalberg. A special word of thanks also to the artists at Imagineering Media Services for their patience with my continual insistence on accuracy and for producing some truly lovely art, and to Richard Megna and Kim Peticolas of Fundamental Photographs for collaborating on some unique photos.

The publication of this book makes me more than a little apprehensive. In the words of Pooh Bear, "When you are a Bear of Very Little Brain, and you Think of Things, you find sometimes that a Thing which seemed very Thingish inside you is quite different when it gets out into the open and has other people looking at it." But the buck stops here. Full responsibility for remaining errors and other shortcomings is mine. Please notify me electronically at pes@pitt.edu or by snail mail, Department of Chemistry, University of Pittsburgh, Pittsburgh, Pennsylvania 15260, USA, of anything that needs correction or improvement.

PETER E. SISKA
University of Pittsburgh

Reviewers

Warren Beck
Michigan State University

Philip C. Bevilacqua
Pennsylvania State University

Malcolm Chisholm
Ohio State University

Ronald C. Cohen
University of California, Berkeley

Terry L. Gustafson
Ohio State University

Dale Hawley
Kansas State University

Cindy Harwood
University of Illinois, Chicago

Michael Heinekey
University of Washington

Carl Hoeger
University of California, San Diego

Joseph T. Hupp
Northwestern University

Paul Kiprof
University of Minnesota, Duluth

Jeffrey Kovac
University of Tennessee

Jeffrey L. Krause
University of Florida

John Krenos
Rutgers University

Chi H. Mak
University of Southern California

John Nelson
University of Nevada, Reno

Liz Ottinger
Swarthmore College

Lee Y. Park
Williams College

Robert Parson
University of Colorado

Bill Pennington
Clemson University

Joe Perry
University of Arizona

Seth Rasmussen
North Dakota State University

Michael P. Rosynek
Texas A&M University

George C. Schatz
Northwestern University

Mike Shen
Colgate University

Michael Sommer
University of Wyoming

Richard Thompson
University of Missouri, Columbia

Professor Dino S. Tinti
University of California, Davis

John Todd
Dublin, Ohio

Michael Topp
University of Pennsylvania

John H. Weare
University of California, San Diego

Thomas Webb
Auburn University

Brian Woodfield
Brigham Young University

Zhiping Zheng
University of Arizona

Physical Principles Underlying Chemistry

CHAPTER 1

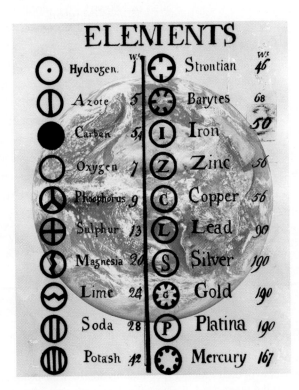

The atoms of the known elements and their relative masses according to John Dalton (1808). Dalton's cryptic symbols were eventually replaced by the capital letters of J Berzelius, but his atomic theory provided the first, and still the accepted, description of the microscopic makeup of our world based on observation.

The atoms move in the void and, catching each other up, jostle together, and some recoil in any direction that may chance, and others become entangled with one another in various degrees according to the symmetry of their shapes and sizes and positions and order, and they remain together, and thus the coming into being of composite things is effected.

Simplicius (6th century A.D.)

CHAPTER OUTLINE

1.1 The Province of Chemistry

1.2 What We Want to Know: Objectives of Modern Chemistry

1.3 Units of Measure

1.4 Force, Work, and Energy

1.5 The Electrical Nature of Matter

1.6 The Nuclear Atom

1.7 Coulomb Force and Potential Energy

Chemistry is concerned with the material world—the everyday world around us. Everything, from the most ordinary rock or block of wood to a silicon computer chip, from the air we breathe to the exhaust of a jet engine, from a crystal-clear glass of water to the protoplasm of a living cell, is of interest in chemistry. As we begin the

chem•is•try (kem′-is-trē) *n.*
the science dealing with the composition and properties of substances, and with the reactions by which substances are produced by or converted into other substances
WEBSTER'S

Early and modern chemists at work: "The Alchymist in Search of the Philosopher's Stone Discovers Phosphorus," by Thomas Wright of Derby (1734–1797) (top); an organic chemist in an academic laboratory performs an aqueous workup of a reaction using a separatory funnel (bottom). A modern look at phosphorus chemistry may be found in Chapter 5, while Chapter 18 introduces organic chemistry.

21st century, an appreciation of underlying chemistry is becoming increasingly important for our lives and the future of our small planet. Grasping the technological advances—lasers, microelectronics, genomics and proteomics, the green revolution, . . . —and coping with the attendant problems—energy supply, changes in global climate, the pitfalls and perils of cloning, . . . —often boil down to questions of chemistry. Perhaps you, dear reader, may one day provide fresh insight into one of these areas, or a new one we aren't yet aware of. New ideas aside, just being a well-informed citizen, capable of choosing wisely, will often oblige you to draw on chemical knowledge.

At another level, chemistry is a discipline, a field of study with ever-advancing frontiers. The nature of the discipline is defined by the people who "practice" chemistry, the chemists. Be it a chemist in an industrial laboratory attempting to perfect a new high-strength fiber, or a faculty member in a college chemistry department (your current college chemistry instructor, for example) researching the details of new kinds of chemical reactions, the problems that are tackled, the goals that are set, and the new ideas that evolve make chemistry what it is. Because these facets are constantly changing, so is the nature of chemistry. As a student of chemistry, you can follow the signposts of knowledge planted by chemists in the past in the early stages, but eventually you must examine where chemistry is going. And in order to grasp where it's going, you've got to know where it's been.

1.1 The Province of Chemistry

The science we now call chemistry has two distinct roots in the ancient past. The early Greek philosophers with their rational speculations are the wellspring of modern scientific thought. On the other hand, the useful discoveries of metals and alloys in the ancient Middle East led through the centuries to a practice called **alchemy**. Alchemy was based more on experience as collected in chemical recipes than on reasoning, with strong religious and magical overtones in its later incarnations in medieval Europe. Modern chemistry is a remarkable blend of these disparate branches. In principle it proceeds from large numbers of observations to a grand rational scheme that explains them, but in practice it is often hard to tell whether the data or the scheme came first.

Given this backdrop, you should not be surprised to learn that modern equivalents of the two ancient schools have evolved. The philosophers these days are called "theoretical chemists," who try to derive everything chemical from first principles (but often are secretly guided by the experimental facts); whereas the alchemists are simply experimenters in chemistry (but seldom do any experiment without a preconceived idea of what will happen). You may also find in your college experience that certain chemists seem to possess an unusual fervor about their subject—shades of their alchemical forebears!

During the 19th century, the forenoon of the modern scientific era, the various scientific disciplines gradually distilled out of what was formerly called *natural philosophy*. These disciplines form a natural hierarchy:

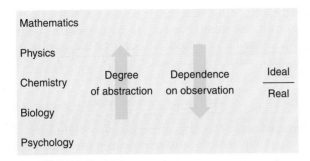

It is no coincidence that mathematics was the first to emerge as a distinct discipline, followed in order by the rest. Progress in the "lower" sciences was possible only through accurate observations, which required the development of increasingly elaborate experimental methods. Nonetheless, though observation is paramount in the experimental sciences, really significant advances are made when abstract *models*—descriptive systems of postulates and inferences—are available to guide experiment. The location of chemistry in the middle of this list signifies a strong role for both experiment and theory: chemistry is a truly "real-world" science, but one strongly influenced by idealized models.

It has been said that physics is applied mathematics, chemistry is applied physics, biology is applied chemistry, etc., and there is more than a little truth in this. Many have attempted to reduce the world with all its beauty and wonder to an abstract mathematical structure. This may be a rational goal (known as *reductionism*), but from a human standpoint it is hopelessly impractical. Nature simply has too many layers; in the words of the renowned physicist Julian Schwinger, "The role of a truly fundamental theory should not be to confront the raw data." The sorts of models used in chemistry are designed to do just that and are, therefore, not fundamental in the purest sense, though they do organize and rationalize experimental results.

In the 20th century, chemistry itself divided into four areas of study: analytical, inorganic, organic, and physical chemistry. There are signs, however, of a return to a less specialized approach to chemical research. Many chemists have become engaged in "interdisciplinary" research projects in which the traditional boundaries within chemistry and between chemistry and physics or biology are being crossed. Modern chemistry can also be viewed as consisting of two main branches, one concerned with the *synthesis and structure* of matter, the other with the *kinetics and dynamics* involved in its transformation. But no matter how chemistry is subdivided, its unifying focus is the atomic theory of matter.

1.2 What We Want to Know: Objectives of Modern Chemistry

Chemistry was launched on its modern course by Robert Boyle in the 17th century, when he proposed the definition of an **element** as a substance that could not be broken down into other substances. Anything that could be so decomposed was therefore a **compound** or perhaps just a **mixture.** Mixtures could be separated by purely physical means, for instance, the evaporation of water from brine, whereas chemical transformations were necessary to decompose a compound. In 1766 Henry Cavendish isolated what would turn out to be the simplest and lightest element, hydrogen gas, and was undoubtedly the first person to run the explosive chemical reaction

$$\text{hydrogen} + \text{air} \longrightarrow \text{water}$$

under controlled conditions. This reaction is exemplary of the sorts of developments in experimenting and hypothesizing that have brought chemistry to its present stage. Cavendish had proven that water, regarded as an element since the time of the ancient Greeks, was actually a compound.

Joseph Priestley and Antoine Lavoisier established within the next two decades that the air itself was a mixture of gases, and that the burning of hydrogen was a reaction with the gaseous element oxygen:

$$\text{hydrogen} + \text{oxygen} \longrightarrow \text{water}$$

Lavoisier further showed that in the course of a chemical transformation there was no detectable change in the mass of material present, and formulated the **law of conservation of mass:**

In any physical or chemical change, mass is neither created nor destroyed.

Essential to Lavoisier's success was his development of techniques for observing reactions in enclosed containers (vessels), preventing the often-unnoticed escape of reagents and/or products. Figure 1.1 shows two of Lavoisier's vessels. But, equally important, Lavoisier was open to new ways of thinking. Priestley had carried out several similar experiments prior to Lavoisier, but his strong belief in an earlier, incorrect theory of combustion, called the *phlogiston theory,** caused him to misinterpret his results.

We will see at several points in the development of different areas in chemistry that Priestley's failure to see the "truth" contains a general take-home lesson. Albert Szent-Györgi put it this way: "Scientific research consists in seeing what everyone else has seen, but thinking what no one else has thought." Reasoning must be guided by observations; observations must not be bent to fit preconceived ideas.

Atomic Theory

By the wee 1800s it had been (somewhat tenuously) established that not only was mass conserved in chemical reactions, but in compounds such as water, formed from two elements, the masses of the elements that would combine always stood in the same ratio. This was called the **law of definite proportions.** For water the ratio of oxygen mass to hydrogen mass m_O/m_H was invariably (within the limited accuracy of the available experiments) 8:1. If one added oxygen in excess of this ratio, it would

---

* The phlogiston theory, which held sway in chemistry for about a century before the work of Lavoisier, held that a combustible substance possessed a fluid called phlogiston that was *given off* in the flame when the substance burned in air. Lavoisier disproved this notion by demonstrating that calxes (oxides) of metals actually *gained* mass relative to the pure metal.

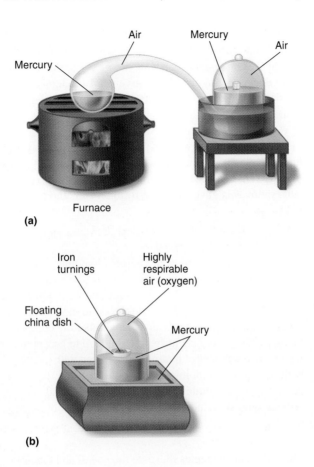

Figure 1.1
Lavoisier's vessels for studying calcination (oxidation) of metals. In vessel (**a**) mercury was heated for 12 days in a fixed amount of air to form the red calx (oxide) of mercury (HgO). The level of mercury in the bell jar rose, showing a decrease in the volume of the air by 1/6; the remaining mercury plus oxide increased in mass. The air remaining in the bell jar was not fit to breathe. In vessel (**b**) iron turnings were burned in pure oxygen. In this case the mercury rose almost to the top; Lavoisier warned against using too much iron, thus removing so much oxygen that the hot china dish would touch the top of the bell jar and crack it. "In which case, the sudden fall of the column of mercury, which happens the moment the least flaw is produced in the glass, causes such a wave, as throws a great part of the quicksilver [mercury] from the basin." The iron and oxygen were carefully weighed beforehand, and the iron oxide afterward, establishing the conservation of mass.

refuse to combine. Moreover, when elements combined to form more than one compound, as in the "carbonic oxides" (compounds containing the elements carbon and oxygen), the ratio of combining masses of one of the elements, for a given mass of the other, would always be in the ratio of small whole numbers. This was the **law of multiple proportions.** In one oxide (today known as carbon monoxide) $m_O/m_C = 1.33$, and in the other (today carbon dioxide) $m_O/m_C = 2.67$, giving an oxygen combining mass ratio of 2.0. This invariability and discrete character of the combining masses led an English schoolteacher named John Dalton to propose the first truly *chemical* **atomic theory** in 1803.

The notion that matter is composed of tiny indestructible **atoms** (the word *atom* derives from the Greek *atomos,* meaning "indivisible") dates back to the Greek philosopher Democritus (*ca.* 420 B.C.), as reported by the Roman poet Lucretius in the first century B.C. After a lapse of some 1500 years, the notion was revived by such notables as René Descartes, John Locke, and Sir Isaac Newton. A corpuscular theory of matter was nothing new, but until Dalton's work it had been pure speculation. Dalton's hypothesis was soundly based on observation, and it explained many experimental results, including Dalton's own, in great detail.

Dalton proposed that each element consists of tiny atoms, each with the same mass. The masses of the atoms of different elements are different. The law of definite proportions is then a statement that a given compound substance is made of these atoms combined in a definite atomic ratio. In a reaction in which that substance is made from its elements (a formation reaction), the atoms are merely reshuffled into this definite ratio without being altered in any essential way. The overall conservation of mass is a consquence of the indestructibility of the atoms. The existence of more than one compound of a given pair of elements corresponds to more than one possible atomic ratio; the ratio of combining weights then reflects the different atomic ratios and must necessarily be in the ratio of small whole numbers—the law of multiple proportions.

Unfortunately, it was not possible in Dalton's day to find the atomic ratios in any known compounds; the data were simply inadequate. So Dalton was forced to assume that, lacking evidence to the contrary, the atoms always combine in the simplest possible ratio. This gave water the formula HO, and our example reaction is, according to Dalton,

$$H + O \longrightarrow HO$$

In the decades that followed Dalton's hypothesis, the work of Joseph Gay-Lussac, Amedeo Avogadro, and Stanislao Cannizzaro established the **law of combining volumes** for gas reactions. This law states that the volumes of reacting gases are in the ratio of small whole numbers, as described in more detail in Chapter 9, and that a given volume of any gas contains an equal number of "compound atoms" or **molecules**. Although 19th-century chemists found it hard to accept, reconciling Gay-Lussac's finding that two volumes of hydrogen combined with one of oxygen to make two of steam with the known combining mass ratio required that hydrogen and oxygen gases be composed of **diatomic** molecules denoted H_2 and O_2. By the early 1860s, a century after Cavendish's studies, a modern version of our reaction could finally be written:

$$2H_2 + O_2 \longrightarrow 2H_2O \tag{1.1}$$

The following examples illustrate the power and the pitfalls of Dalton's thinking.

EXAMPLE 1.1

For Reaction 1.1, show that the ratio of combining weights of oxygen and hydrogen is 8:1 if relative atomic masses of 16.0 for O and 1.0 for H are assumed.

Solution:

$$\text{Mass of hydrogen} = 2(1.0 + 1.0) = 4.0$$
$$\text{Mass of oxygen} = 1(16.0 + 16.0) = 32.0$$
$$\text{Ratio } m_O/m_H = 32.0/4.0 = 8:1$$

By assuming the simplest atomic ratio HO for water, Dalton's O atom was only 8 times as heavy as hydrogen. Later comparison with ratios from other oxides and hydrides turned up many inconsistencies, casting doubt on Dalton's hypotheses.

EXAMPLE 1.2

A second compound of H and O you are probably familiar with is hydrogen peroxide, with the molecular formula H_2O_2. Write a reaction similar to Reaction 1.1 for the formation of H_2O_2, and determine the ratio of combining masses as in Example 1.1. What law does this illustrate? Given Dalton's formula for water, HO, what formula would he assign to hydrogen peroxide?

Solution:
The reaction is $H_2 + O_2 \longrightarrow H_2O_2$. The procedure for finding the combining mass ratio is the same as in Example 1.1, except that now $m_H = 2.0$, so that $m_O/m_H = 32.0/2.0 = 16.0$. The ratio of these ratios is $16.0/8.0 = 2.0$ or 2:1, a ratio of small whole numbers. This result illustrates the law of multiple proportions. Dalton would be obliged to give this compound the formula HO_2.

EXAMPLE 1.3

By use of the (experimental) combining weights for water, "nitrous gas" (nitric oxide, NO), and ammonia (NH for Dalton; actually NH_3), along with Dalton's simplest ratio assumption, show that inconsistent atomic masses of nitrogen are obtained.

Solution:
As mentioned in Example 1.1, Dalton's assumption of HO as water's formula leads to $m_O = 8.0$. In NO, the combining mass ratio m_N/m_O is actually $14.0/16.0 = 0.875$, implying $m_N = (0.875)(8.0) = 7.0$. However, in ammonia, $m_N/m_H = 4.7$, implying $m_N = (4.7)(1.0) = 4.7$, an incompatible result.

The long interval between Dalton's enunciation of the atomic theory and the correctly stated chemical equation caused many scientists to doubt Dalton's reasoning and to disregard the atomic hypothesis. As late as 1905 a chemistry text was published that made no mention of atoms. But Reaction 1.1 as written has sacrificed none of the logic and simplicity of the atomic theory; it is still both a statement of atomic ratios and a statement about combining masses and the conservation of mass. And Dalton's concept of a chemical reaction as a reshuffling of atoms has stood the test of time; today we are beginning to appreciate the intricacies of that reshuffling.

The Periodic Law

By the time Reaction 1.1 had finally been correctly represented, *ca.* 1860–1865, about 60 elements, conforming to Boyle's definition, had been isolated; and from combining weights and atomic ratios, most of the relative masses of the elements, based on a mass of one for a hydrogen atom, were known to two figures or a little better. Much of this information was organized and systematized as the result of an historic meeting of the world's leading chemists at Karlsruhe, Germany, in 1860. As a result, chemists could now confidently write formulas for chemical compounds, such as SiO_2 or NH_3. Aside from the grand division of the elements into metals and nonmetals, however, chemistry remained a hodgepodge of strange names, unusual properties, and hundreds of different reactions. (A similar situation existed in high-energy physics precisely a century later, when the number of subatomic "elementary particles" exceeded 200 with no order in sight, except for a grand division into particles and antiparticles.) Experiment was in the ascendancy; chemical philosophy was weak or nonexistent.

A few chemists still clung to the idea that chemistry ought to conform to some rational order, that it ought to make sense. In 1864 John Newlands noted that when the elements were arranged in order of increasing atomic mass, every eighth element fell into a group with similar chemical and physical properties. Unfortunately he

Figure 1.2
An early periodic table published by Dmitri Mendeleyev in 1872. Note the gaps for atomic masses 44, 68, and 72. Mendeleyev made detailed predictions of the physical and chemical properties of these three elements based on his periodic law. The later discovery of scandium, gallium, and germanium bore out his predictions remarkably well. Superscripts instead of subscripts were then used to indicate chemical formulas, and many metals, such as the sequence Ti to Cu, are now placed in extra "transition" columns in modern tables.

TABELLE II

REIHEN	GRUPPE I. — R^2O	GRUPPE II. — RO	GRUPPE III. — R^2O^3	GRUPPE IV. RH^4 RO^2	GRUPPE V. RH^3 R^2O^5	GRUPPE VI. RH^2 RO^3	GRUPPE VII. RH R^2O^7	GRUPPE VIII. — RO^4
1	H=1							
2	Li=7	Be=9,4	B=11	C=12	N=14	O=16	F=19	
3	Na=23	Mg=24	Al=27,3	Si=28	P=31	S=32	Cl=35,5	
4	K=39	Ca=40	—=44	Ti=48	V=51	Cr=52	Mn=55	Fe=56, Co=59, Ni=59, Cu=63.
5	(Cu=63)	Zn=65	—=68	—=72	As=75	Se=78	Br=80	
6	Rb=85	Sr=87	?Yt=88	Zr=90	Nb=94	Mo=96	—=100	Ru=104, Rh=104, Pd=106, Ag=108.
7	(Ag=108)	Cd=112	In=113	Sn=118	Sb=122	Te=125	J=127	
8	Cs=133	Ba=137	?Di=138	?Ce=140	—	—	—	— — —
9	(—)	—	—	—	—	—	—	
10	—	—	?Er=178	?La=180	Ta=182	W=184	—	Os=195, Ir=197, Pt=198, Au=199.
11	(Au=199)	Hg=200	Tl=204	Pb=207	Bi=208	—	—	
12	—	—	—	Th=231	—	U=240	—	— — —

called this the "law of octaves"; the musical analogy laid his ideas open to ridicule by chemical buffoons, and denied them a fair hearing. Five years later a Russian chemist named Dmitri Mendeleyev presented what he called the **periodic law** to the Russian Chemical Society. He had organized all known elements into a table arranged by increasing mass (unknowingly using Newlands' idea) whose columns defined chemical families with similar chemical properties. Figure 1.2 shows one of the early tables Mendeleyev constructed.

Mendeleyev's method of classification left vacancies in his table, and he boldly claimed that there were as-yet-undiscovered elements that would eventually occupy these spaces. He even predicted the chemical and physical properties of three of these unknown elements. Within the next decade the elements gallium, scandium, and germanium were isolated, and proved to have properties remarkably close to Mendeleyev's predictions. Mendeleyev also pronounced in error all measurements of atomic masses that did not conform to his ordering scheme. This prompted more precise and careful measurements, and some discrepancies still remained; for example, the atomic mass of Te was found to be higher than that of I, and, when the noble gases had been discovered, Ar was found to weigh more than K; while in each case chemical properties dictated a reverse order.

Meanwhile Lothar Meyer, unaware of Mendeleyev's work, had constructed a similar table based on the physical properties of the elements alone. However, he made no predictions and did not exploit the extra information contained in the chemistry of the elements. So most of the "fame and glory" went to Mendeleyev.

The new organization, regularity, and predictive powers of the periodic table accelerated greatly the search for and discovery of new elements, so that at the dawn of the 20th century nearly all of the 90 naturally occurring elements had been isolated and their chemistry explored. The mass discrepancies alluded to earlier were acknowledged to be real; chemists therefore simply numbered the elements from 1 to 92—two "missing" elements, technetium and promethium, have now been made by us—and called the integer assigned to each element the atomic number Z. In 1914 Henry Moseley published measurements of X-ray emission (emission of light of very short wavelength; see Chapter 2) from 39 elements from aluminum to gold, which confirmed, from the regular variation of X-ray wavelength, that the atomic number Z, and not the atomic mass, is the natural ordering parameter for the elements. Other advances in our understanding of Dalton's atoms, which we'll consider in turn, eventually led to the modern form of the periodic table that appears inside the front cover of this book.

Modern Questions

In the last century chemistry has largely passed from consideration of the identity and composition of the molecular participants in a chemical reaction (though in synthetic chemistry, product identification is still of central importance) to more subtle questions aimed at elucidating the structure of atoms and molecules and how this relates to and determines their reactivity. This sort of knowledge can enable us to choose certain reagents to produce a desired set of products, and otherwise to control to the extent possible the outcome of a chemical event. It can also help us in the search for new kinds of chemical reactions, and to understand on a fundamental level reactions between exceedingly complex molecules, such as those occurring in living systems.

Concerning Reaction 1.1, a few more modern questions (which are already partially or completely answered at this writing, at least for this reaction) might be framed as follows:

1. Why are the reagents H_2 and O_2 gases under normal conditions while H_2O is a liquid?
2. The reaction needs a spark, but then proceeds explosively with the evolution of a great deal of heat. Why the spark? Where does the heat come from? Where does it go?
3. What is the nature of the bonding force that holds H_2O together and, more baffling, that causes two atoms of H or O to cling together to form the diatomic gases?
4. Are the atoms in these species bonded together in a definite geometry (atom–atom distances and, in the case of H_2O, a "bond angle"), and, if so, what are the geometrical parameters?
5. How do the nature and geometry of bonding affect the natural states of the molecules? The course of the reaction?
6. What are the details of the reshuffling of atoms necessary to cause reaction? What motions and geometrical contortions must the atoms and molecules undergo, and what is the nature of the forces that govern them?

Answers to these and other similar questions probe to the very heart of chemistry and form the basis for much of modern chemical research. It is our hope that, as you work your way through this book, you will learn the answers to some of these questions, at least for "simple" cases, and, more importantly, how chemists think about such problems. This is where models come into play. Models help us to frame questions about chemistry, just as Dalton's atomic model is the basis for the preceding six questions. When a model is "right" (note, however, that models are never right all the time), it can lead to rapid chemical advances, as in the case of Mendeleyev's periodic law. The exceptions to the rule (masses out of order) stimulate more careful experimental work, and in the past have always led to even deeper understanding. In any case, good models are always inspired by observations, as you will see again and again as we explore the terrain of chemistry.

1.3 Units of Measure

When you can measure what you are speaking about, and express it in numbers, you know something about it; but when you cannot express it in numbers your knowledge is of a meagre and unsatisfactory kind; it may be the beginning of knowledge, but you have scarcely, in your thoughts, advanced to the stage of science.

William Thomson (Lord Kelvin)

Answers to chemical questions almost invariably are stated in or are based on numbers derived from experimental measurement or calculation. Some of these numbers are "pure," like atomic ratios, but most have mechanical or electrical units of some sort. The fundamental physical quantities of interest in chemistry are **length, mass, time,** and **electric charge**.* In the future, books such as this will not have to refer to the English units of feet and pounds or inches and ounces, but at least everyone agrees on the second as a basic unit of time. **Metric units** (based on the meter) have been nearly uniformly used in science for more than a century, but have been adopted slowly in the world at large, especially in the United States. Until 1960, chemists and atomic/subatomic physicists had universally used the **centimeter-gram-second** (cgs) **electrostatic-unit** (esu) metric system, while the rest of physics used the **meter-kilogram-second** (mks) system. Except for electric charge, to be discussed later, these two systems are interconvertible through powers of 10: 100 centimeters (cm) = 1 meter (m) and 1000 grams (g) = 1 kilogram (kg). In 1960 the International Bureau of Weights and Measures recommended the universal adoption of the so-called SI (Système International) system of units, essentially the mks units. While the SI system has not been ideal for chemistry, chemists have been gradually switching to the new system. But because of the depth to which the cgs-esu system is ingrained in chemistry, changing chemistry textbooks to the new system has generally given rise to a plethora of errors and inconsistencies. What is done in this book is to employ the SI system as an overall rule, but to use cgs-esu units when particularly apropos for certain problems, followed by units conversion. This presents few difficulties, because both cgs-esu and SI are metric systems, and also carries the virtues of convenience of calculation and a tie with chemistry's past.

Table 1.1 lists the fundamental quantities of the SI and cgs-esu systems. The conversion factors multiply cgs quantities to give SI units. Any of the units without prefixes (m, g, s, C) may be given a prefix to denote a power of 10. This enables scientists working in different regimes to communicate in the same basic units. These prefixes and their abbreviations are listed in Table 1.2.

All of the other quantities we will use in this book, such as force, energy, volume, and dipole moment, are derived from combinations of these basic units and will be de-

---

* In the SI system to be described, *current* rather than charge is considered fundamental, where charge = current × time. For our purpose, the distinction is academic.

TABLE 1.1
Basic SI and cgs-esu units

	SI unit	cgs-esu unit	SI/cgs conversion factor
Length	Meter (m)	Centimeter (cm)	10^{-2}
Mass	Kilogram (kg)	Gram (g)	10^{-3}
Time	Second (s)	Second (sec)	1
Charge	Coulomb (C)	Electrostatic unit (esu)	$1/2.9979 \times 10^9$

TABLE 1.2
SI unit prefixes

Prefix	Power of 10	Prefix	Power of 10
deci (d)	10^{-1}	deka (da)	10^{1}
centi (c)	10^{-2}	hecto (h)	10^{2}
milli (m)	10^{-3}	kilo (k)	10^{3}
micro (μ)	10^{-6}	mega (M)	10^{6}
nano (n)	10^{-9}	giga (G)	10^{9}
pico (p)	10^{-12}	tera (T)	10^{12}
femto (f)	10^{-15}	peta (P)	10^{15}
atto (a)	10^{-18}	exa (E)	10^{18}
zepto (z)	10^{-21}	zetta (Z)	10^{21}
yocto (y)	10^{-24}	yotta (Y)	10^{24}

fined in their contexts. Other "special" units, such as the angstrom, the electron-volt, the wavenumber, the kelvin, and the calorie, will be introduced where appropriate. A complete conversion table may be found on the inside back cover of this book.

Errors in Measurement

Every number derived from experiment has an uncertainty associated with it that arises from limitations on the **precision** and the **accuracy** of the measuring instrument. (It is presumed that the person using the instrument is skilled in its use—not always a safe presumption!) For example, a chemical sample may be weighed to the nearest 0.01 g on a crude beam balance, or to the nearest 0.0001 g (0.1 mg) on an "analytical" balance. In the first case, the mass might be stated as 3.26 ± 0.01 g or simply 3.26 g, whereas in the second it might be 3.2637 ± 0.0001 g or 3.2637 g. We say the second number has a higher *precision* than the first, as indicated by the larger number of **significant figures.** Generally speaking, the precision of a measurement is assumed to be one unit in the least significant (rightmost) digit.

Even though the measurement with the analytical balance appears to be better as far as precision goes, how accurate the measurement is—how close it is to the true mass—depends on how well the balance has been calibrated (usually done with a standard weight set). For example, a student in a general chemistry laboratory checked his weighings on two balances with the same sample, and found masses of 0.5735 and 0.5748 g. This disagreement, which is an order of magnitude (factor of 10) greater than the precision of the balances, indicates that one (or possibly both) of them is not properly calibrated. This is an example of an error introduced by lack of *accuracy* in the balance. In many cases errors of this sort go undetected; in the present case the student should report a mass of 0.574 ± 0.001 or just 0.574 g.

To summarize, *precision* refers to the inherent capability of a measuring device to yield a certain number of significant figures in a measured quantity, whereas *accuracy* refers to the closeness of the measurement to the "true" value of the quantity. Errors of precision are *randomly* distributed about a mean value; errors of accuracy cause a *systematic* shift of that mean above or below the true value.

Significant figures, then, are a direct reflection of the quality of a measurement; when measured values are used to carry out other calculations, the answers are often limited in the number of "good" digits by the input. For example, when a length measured with an ordinary ruler (good to ±0.05 cm) is *added* to a length measured with a vernier caliper (±0.002 cm), the sum can only be quoted to ±0.05 cm; the extra precision of the vernier is rendered useless. When one or more measured quantities enter as *factors* in a product or quotient, the result can only be given to a number of significant figures equal to that in the least precise input. Often much time and effort is wasted carrying out careful measurements of one type, when the accuracy of the desired result is actually limited by some other poorly known variable. For example, weighing a solid sample on an analytical balance in order to make a solution of known concentration (see later discussion) is inconsistent with using a beaker with volume markings (typically accurate to only 10%) to measure the solution volume. A calibrated volumetric flask should be used instead.

At present, electronic calculators are standard equipment for all of you, most of which carry 8 to 13 digits in all calculations. All too many students we have known simply copy everything in the calculator display as the answer to a problem in which the most precise number given has only three or four significant figures. Such answers are arithmetically correct but scientifically wrong. It is quite all right to carry many digits through a calculation, since truncation errors are thereby avoided, but the answer must be truncated with rounding to the appropriate number of significant figures. When numbers very large or very small compared to one are encountered, either the units are adjusted using one of the prefixes given earlier, or so-called **scientific notation** is used. For example, a time of 83,400 s, where only the first three digits are significant, could be written as 83.4 ks or, more commonly, as 8.34×10^4 s. As the example shows, in scientific notation a decimal point is placed to the right of the first significant digit, and the number of places the decimal has thereby been moved is indicated as a power of 10. This notation makes it easy to indicate the number of significant figures in a quantity.

At various places in this book the fundamental constants of nature will be presented and used. These constants all result from experimental measurements, and therefore have intrinsic errors; in addition, they are continually being revised to reflect new and more precise experiments. This can involve changes in both the value of the constant (due to inaccuracy in previous results) and its error (due mainly to errors of precision, but sometimes to discrepancies among two or more independent measurements). The most recent update was completed in 1998; these values are collected inside the front cover of this book. This variability is exemplary of the difference between mathematics and the observational sciences: the number 3 will never change!

1.4 Force, Work, and Energy

Many modern chemical questions such as those we introduced in Section 1.2 involve mechanical concepts: forces and motions. Indeed, as we shall find, the atom itself is a tiny mechanism with moving parts. So it is important for you to know a few concepts and relationships of so-called **Newtonian mechanics,** the system for describing material (massive) bodies moving under the influence of forces devised by Sir Isaac Newton in the short span of 18 months in the years 1665–1667. In this same brief period Newton also invented the branch of mathematics called **calculus,** in order to ease the task of stating and solving his equations describing the motions of bodies. (It is a bit unfortunate that "the calculus" these days is part of the mathematics curriculum, and students often don't get to see how beautifully it blends with and underpins the natural sciences until they enroll in advanced science courses.) In this book we will make modest use of calculus for the same reasons Sir Isaac did, as well as to acquaint you with this unique language for expressing scientific ideas.

Perhaps the central idea in Newton's mechanics is what he called *quantity of motion,* today called the **momentum** p, defined by

$$p = mv = m\frac{dx}{dt} \tag{1.2}$$

where m is the mass of a material object and v its velocity. Momentum therefore has units of kg m s^{-1} (SI) or g cm s^{-1} (cgs). The *time derivative* dx/dt defines the velocity in terms of the *rate* at which the distance x is covered per unit time t, at a given instant t. The notation dx means a tiny change in x, and likewise for dt; the ratio of these two *differentials* is the derivative, or instantaneous rate of change, of the variable in the numerator (x) with respect to that in the denominator (t). If an object is moving with a constant velocity, the distance traveled is increasing linearly with time, and the velocity is the *slope* of the straight line plot of x versus t. If the velocity is not constant, the x versus t graph will be curved and the instantaneous velocity dx/dt is the slope of the line tangent to the curve at time t. So Newton's momentum involves **differential calculus** right off the bat.

Newton's idea of **force** was any external influence that changes the momentum of an object. So a force F is defined in terms of its *effect* on motion, rather than its cause, whatever that might be. **Newton's second law of motion** describes this relationship:

$$F = \frac{dp}{dt} = m\frac{dv}{dt} = ma \tag{1.3}$$

where the leftmost and rightmost quantities in this multiple definition are probably familiar to you: force equals mass times acceleration. Equation 1.3 says that a force of a fixed magnitude, like the force of gravity on the earth's surface, will cause a constant rate of change of the momentum with time. Because the mass m is a constant (as long as v is not too near the speed of light) this is a constant change in v

with t, or a constant **acceleration** a. The acceleration is also "the rate of change of the rate of change of" x with t, or

$$a = \frac{d}{dt}\left(\frac{dx}{dt}\right) = \frac{d^2x}{dt^2} \tag{1.4}$$

where the rightmost quantity is called a *second derivative*. Equation 1.3 also works for forces that are not constant as the displacement x of the object changes, or for time-dependent acceleration, a virtue enabled by the use of differential calculus to define velocity and acceleration.

From Equation 1.3 the units of force are kg m s^{-2} (SI) or g cm s^{-2} (cgs). Because forces are so common in mechanics, their unit of measure is given a special label; in SI units it is the **newton** (N), 1 N = 1 kg m s^{-2}; in cgs units it is the *dyne*, 1 dyne = 1 g cm s^{-2}. From the basic units you can see that 1 N = 10^5 dyne.

A complication we will invoke only when necessary in this book is the fact that all the length-based quantities we are using must in general be represented by *vectors* in three-dimensional space (a *direction* as well as a *magnitude* must be specified); this is an important consideration in many types of mechanical problems. We note that Equation 1.3 requires that the change in momentum be in the direction of the force that causes it. If you hold a string attached to a ball and pull, the ball will move toward you. If you begin to swing the ball in a circle around you, your hand still must exert an inward force (holding the string tightly) to keep the ball on a circular path; there is still an inward change in momentum that overcomes the tendency of the ball to fly off tangent to its path. The acceleration in this case is called *centripetal* acceleration.

It is easy to ponder forces that arise, like the ball-on-a-string example, from a mechanical connection between the source of the force and the object that is moved by it. A pitcher throws a baseball, a batter hits it, a sprinter pushes off the starting blocks, a gasoline engine propels an automobile. In each case there is a tangible link between force and object. What is harder to grasp is the notion of an "invisible" force that acts at a distance with no strings attached, although anyone who has played with magnets has experienced this. Newton also introduced this idea in his theory of gravity, formulated in that same flash of brilliance we mentioned earlier. In the classic example, the apple falls to the ground in response to the force of gravity, and so does the moon orbit the Earth, though no sensible connection exists between them. Newton's gravity is an **inverse-square law of force** between two masses m_1 and m_2:

$$F = -G\frac{m_1 m_2}{r^2} \tag{1.5}$$

where r is the distance between the centers of the massive objects and G is a proportionality constant known as the gravitational constant. It is always, as far as we know, an *attractive* force that tends to pull the masses closer together. The electrical forces that, as we will see, are important in chemistry also follow an inverse-square law, though they are not always attractive.

Mechanical Work

Everyone is acquainted with the everyday concept of **work,** that is, manual labor. Work also has a more precise mechanical definition, but one consistent with its common usage. We say that work is done when an applied force results in the movement of a material (massive) object through some distance. If there is little or no opposing force, like friction or gravity, to slow the object down after the force has been applied, the object has acquired *energy of motion* as a result, which is called **kinetic energy.** If we analyze this relationship in terms of a very small change in distance dx, then a very small amount of work dw will be done as a result of the applied force F given by

$$dw = F\,dx \tag{1.6}$$

If the object moves from a point x_1 to a point x_2 *while the force is acting on it,* then the total work w is given by the *sum* of all the little bits of work done between x_1 and x_2—the *definite integral*—

$$w = \int_{x_1}^{x_2} F\,dx \tag{1.7}$$

where the symbol $\int$ (a special kind of S, standing for the sum of all the differential bits of the quantity appearing to its right) is called an *integral sign,* and x_1 and x_2 are the *limits* or bounds on the integral. If we regard the force F as a y coordinate (ordinate), then finding the integral in Equation 1.7 means finding the *area* under the graph of force versus displacement, as illustrated in Figure 1.3.

Let's suppose you were to give this book a shove. In doing that you've done work on the book by the repulsive force of your hand, although due to friction the book stops moving after you're done pushing it. (The existence of friction confused early attempts to formulate laws of motion.) So let's imagine a frictionless, gravity-free world, like that in a space station, where the pushed book drifts away with a constant momentum. We say that the book has *kinetic energy* as a result of the work of pushing it, as follows. In the work formula of Equation 1.7, we see from Equation 1.2 that $dx = v\,dt$, and further that $F = m\,dv/dt$ from Newton's law (Equation 1.3). Then Equation 1.7 becomes

$$w = \int_0^{t_2} \left(m\frac{dv}{dt}\right) v\,dt = \int_0^{v_2} mv\,dv \tag{1.8}$$

where the second equality results from canceling the dt's. In making these substitutions and in canceling, the *integration variable* has been changed from distance x to time t to velocity v. Whenever the integration variable is changed, the limits on the integral must also change to reflect this. The time can be taken to be zero when the force is first applied, then t_2 when it is removed (the book leaves your hand). Correspondingly, the book is taken to be at rest before you push it, $v = 0$, and finally its velocity is v_2. As Figure 1.3 shows, the last integral in Equation 1.8 is particularly easy to do by finding the area of the triangle formed by plotting the momentum $p = mv$ against v for $v = 0$ to v_2, with the result

$$w = \tfrac{1}{2} m v_2^2 \tag{1.9}$$

The quantity $\tfrac{1}{2} mv^2$ is called the *kinetic energy* K of the object (book). This formula says that you can tell how much work was done by knowing the mass of the object worked on and measuring its velocity. Recall that this is only true in the absence of friction.

Because the kinetic energy gained is equal to the work done, we say that both are **energy** in different forms. The units of energy may be found either from Equation 1.7

Figure 1.3
The case of the pushed book. Illustration of work done pushing a book in frictionless, gravity-free surroundings, and its relation to kinetic energy. Before the book is pushed ($t = 0$) it has potential energy due to the force about to be applied, equal to the final kinetic energy.

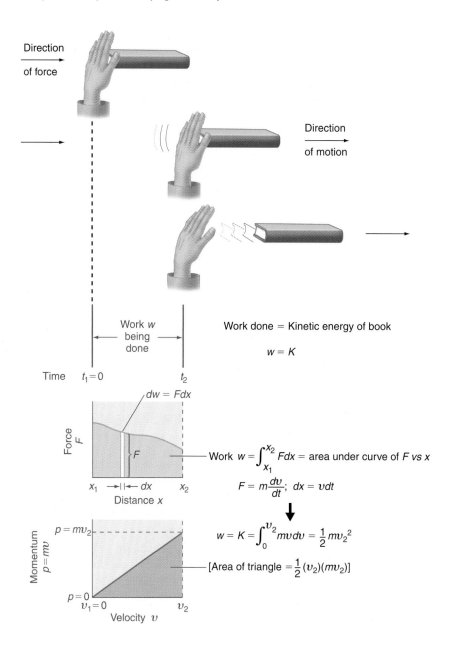

or 1.9 to be kg m² s⁻² (SI, known as a **joule** J) or g cm² s⁻² (cgs, an *erg*). A pound (0.454 kg) of butter lobbed across a room has about 1 J of kinetic energy; a beetle crawling quickly down the sidewalk has about an erg. From the basic units, $1\text{ J} = 10^7$ erg.

Conservation of Energy

Coming back down to earth with its force of gravity, if instead of pushing the book you lift it up, you have clearly done some work, but where is the kinetic energy? You can answer this by dropping the book; it falls quickly to the nearest landing. In falling it has gained kinetic energy, energy that could not have been acquired unless you had first lifted it. The difference here is that you have worked against an opposing force, the force of gravity, and *your work has become stored energy* or **potential energy** V. In falling, the book converted its potential energy V into kinetic energy K. Near the earth's surface, the force of gravity is nearly constant, equal to the mass (of the book) m times g, the acceleration of gravity (9.8 m s^{-2}). If you raise the book a distance h above a table, the work you've done is just

$$w = \int_0^h F\,dx = -mg\int_0^h dx = -mgh \qquad (1.10)$$

where the constant force can be factored out of the integral, because every differential dx is multiplied by the same factor, leaving the area of a rectangular block of unit height and length h. The minus sign arises because gravity is a *downward* force, while you are lifting the book *upward,* exerting a force in the opposite direction. From our discussion above, $-w = V_1$, the initial potential energy before the fall. So we have

$$\begin{aligned}\text{Before fall:} \quad & V_1 = +mgh \quad K_1 = 0 & K_1 + V_1 &= mgh \\ \text{After fall:} \quad & V_2 = 0 \quad K_2 = \tfrac{1}{2}mv_2^2 = mgh & K_2 + V_2 &= mgh\end{aligned} \qquad (1.11)$$

On the way down, some of the potential energy has been converted into kinetic energy, but it can be reasoned that at every point in the fall, $K + V = mgh =$ constant. We call this constant the **total energy** E and write

$$E = K + V = \text{constant} \qquad (1.12)$$

that is, the total energy, the sum of kinetic and potential energies, is conserved for any object that is subject only to conservative (nonfrictional) forces. In the present example, air resistance would slow the book down slightly; however, that amount of energy lost by the book would be carried away by the molecules of air, so no energy would be destroyed. The principle of the conservation of energy is quite general, applying as well to atoms, molecules, and chemical reactions as to falling books, and we shall use it so often that in many instances it will simply be assumed without being explicitly appealed to.

The potential energy V plays a central role in rationalizing a variety of chemical phenomena, and there are three important, related properties of V with which you should be familiar:

1. *Potential energy exists only when there is a force acting on the object of interest.* It is related to the force by

$$V = -\int F\,dx \quad \text{or} \quad F = -\frac{dV}{dx} \qquad (1.13)$$

The second equality follows by *differentiating* (taking the derivative of) both sides of the first equality. Because derivatives and integrals are inverse operations (the so-called "fundamental theorem of the calculus"), the integral is undone, leaving just the force F. In our gravitational example, $F = -mg$, and the potential energy for an arbitrary height x is $V = +mgx$; the relations of Equations 1.13 are easily verified. It is often useful to think of potential energy in terms of work done, as we did in our falling book example. The derivative relation means that the force is the negative of the *slope* of the potential, thought of as a curve $V(x)$ plotted against x.

2. *An object will naturally tend to minimize its potential energy.* Books fall, balls roll downhill in response to the attraction of gravity. Magnets push each other away when like poles are brought together, in response to magnetic repulsion; helium-filled balloons rise in response to the repulsion of atmospheric pressure. In each case events occur so as to minimize V. In view of the conservation of energy, Equation 1.12, this implies conversion of V into K, into actual motion.

3. *After an object has minimized its potential energy as far as possible, it has reached a state called equilibrium.* After the book has dropped, after the explosive reaction (Reaction 1.1) is over, things quiet down, with no further apparent changes. Reaching such a state requires nonconservative (frictional) outside influence; the object or reaction system communicates its energy to its surroundings. If the book couldn't do this, it would simply bounce like a rubber ball! Generally, without friction the object would remain in perpetual motion. The **equilibrium** concept will be particularly important in chemical thermodynamics (see Chapter 12).

EXAMPLE 1.4

A common example of the application of Newton's mechanics is the action of a spring. Suppose a ball of mass m is attached to an immovable wall by a spring:

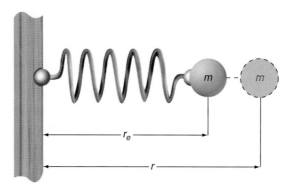

Hooke's law of springs states that the restoring force of the spring is proportional to the displacement from its rest position r_e:

$$F = -k(r - r_e)$$

where k is called the *force constant* of the spring. The minus sign comes from the opposition of the force to the direction of motion. From Hooke's law find the potential energy created when the spring is stretched (or compressed) from r_e to an arbitrary distance r. Plot the potential energy versus r.

Solution:
Using Equation 1.13,

$$V = -\int_{r_e}^{r} F\, dr = +k\int_{r_e}^{r} (r - r_e)\, dr = \tfrac{1}{2}k(r - r_e)^2$$

The result of the integration can be obtained either geometrically or by using the simple rule for integrating powers $\int x^n\, dx = x^{n+1}/(n + 1)$, $n \neq -1$, where in this case

$x = r - r_e$. A rule like this is usually given in the form of an *indefinite* integral, an integral without limits. To get a physical result out of such a rule, you simply plug the upper and lower limits into the right-hand side, and subtract the lower from the upper. The resulting plot of V versus r is an upward-opening parabola:

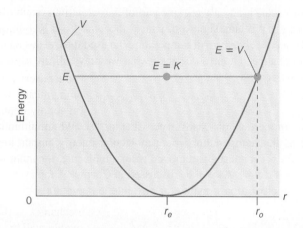

If the spring is released after being extended or compressed, the ball will vibrate to and fro in accord with Newton's second law (Equation 1.3). In the absence of air resistance and imperfections in the spring, the conservation of energy $E = K + V$ applies. At $r = r_0$, $K = 0$ and $E = V = \frac{1}{2}k(r_0 - r_e)^2$, that is, the energy is present entirely as potential energy and the ball is at rest. At $r = r_e$, $V = 0$ and $E = K$, the energy is all kinetic. The potential in effect confines the ball to moving between well-defined limits on r, much like a marble rolling to and fro on a parabolic track subject to gravity. Air and spring friction eventually allow the ball to return to rest at r_e, where V is a minimum. (The ball is said to undergo *simple harmonic motion,* and the potential is called the *harmonic oscillator* potential. This potential is frequently used to describe vibrations of molecules, because a chemical bond is much like a spring.)

As we examine the various facets of chemistry, it is hoped you will gain a deeper understanding and appreciation of the potential energy concept, through seeing it used in different situations. For now, if the preceding discussion is not quite clear to you, be patient. Few of us have the mental capacity of Newton!

Temperature and Heat Energy

The qualities of hotness and coldness in everyday experience are measured quantitatively with thermometers, which give us the temperature of the air outside, of a person's body, of a refrigerator or oven. *The temperature of an object is an indicator of the amount of mechanical energy its constituent molecules possess.* We call this energy **heat energy.** Heat energy will be discussed in more detail when the subdiscipline known as **thermodynamics** is introduced in Chapters 10 and 11.

Temperature scales all have some temperature defined to be "zero degrees." For example, the Fahrenheit scale (°F) was set up somewhat arbitrarily by Gabriel Fahrenheit in the early 1700s so that 0°F corresponded to a cold day in winter and 100°F to his wife's body temperature (she must have had a fever!). On this scale, water freezes at 32°F and boils at 212°F. A few years later Anders Celsius developed the **centigrade,** now called **Celsius,** scale, °C, based on 0°C as the freezing point of water and 100°C as its boiling point at standard pressure. Note that, because the number of degrees between freezing and boiling points is greater for the Fahrenheit scale (180°F), a Celsius degree is 180/100 = 1.8 times as large as a Fahrenheit degree.

The Celsius scale gradually became part of the scientific system of units, so that when in the 1850s William Thomson, later to become Lord Kelvin, proposed that there should be an *absolute zero* of temperature, he used the Celsius degree as his unit of temperature change. Thomson defined what we now call the **absolute** or **kelvin (K) temperature** scale by

$$T(K) = t(°C) + 273.15 \tag{1.14}$$

based on the properties of gases (see Chapter 9). We now understand 0 K ($-273.15°C$) as that temperature where molecular energy has its lowest possible value. When all excess energy is removed from a molecule, it cannot lose any more and is as cold as it can be. As will be discussed in Chapter 9, the average energy $\bar{\varepsilon}$ of a molecule is directly proportional to the absolute temperature:

$$\bar{\varepsilon} \approx k_B T \tag{1.15}$$

where k_B is a proportionality constant known as **Boltzmann's constant;** its modern value is $k_B = 1.3807 \times 10^{-23}$ J K^{-1}. We use an "approximately equal" sign ($\approx$) because there is usually a numerical factor, such as $\frac{3}{2}$, that must be included to give a precise relationship. The amount of energy represented by $k_B T$ is called **thermal energy,** and we shall frequently be comparing other energies, such as that needed to sever a chemical bond between two atoms, to $k_B T$. Because k_B is so small, astronomical numbers of hot molecules are needed before a person can feel their heat energy.

For more than a century, heat energy has been measured in units of **calories,** with 1 calorie defined as the energy necessary to raise the Celsius temperature of 1 g of water by 1 degree. We now use the **defined calorie,**

$$1 \text{ cal} = 4.184 \text{ J} \tag{1.16}$$

with the 4.184 factor taken as an exact number. Calories are not a part of the SI system, but again the unit is strongly ingrained in chemistry and is still widely used in the current literature.

1.5 The Electrical Nature of Matter

Figure 1.4 illustrates an electrical discharge tube, a glass tube with metal disk electrodes sealed into the ends. As early as 1748 it had been found that when the pressure of air in the tube was reduced to less than about 1% of atmospheric pressure and opposite charges placed on the metal disks, the circuit was closed and the gas began to glow a purplish-orange color. This is called *electrical breakdown* of the gas. According to the electricity theory of Ben Franklin and others, this meant that the air had somehow become a carrier of charges through the tube. Progress in this field was hampered by the lack of good vacuum pumps. But by the late 19th century, a series of careful experiments had been done by several scientists, most notably William Crookes and Heinrich Hertz, that established the existence of **cathode rays** that em-

1.5 The Electrical Nature of Matter

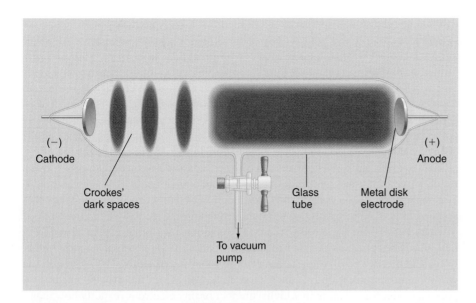

Figure 1.4
An electrical discharge tube ("Crookes' tube") containing air at about 0.01% of atmospheric pressure. The tube glows due to emission of light by excited atoms and molecules. The energy to excite them comes from the invisible cathode rays traveling from cathode to anode.

anated from the negatively charged electrode (the cathode). These rays were then found to be composed of *massive, negatively charged particles of electricity,* mainly through the experiments of Joseph John Thomson and his students in the 1890s at the Cavendish Laboratory in England (the site of the first controlled hydrogen–air reaction some 130 years earlier). Thomson contrived to measure the charge-to-mass ratio of the cathode rays by an ingenious combination of electric and magnetic fields; his apparatus is illustrated in Figure 1.5. Thomson identified these rays with the **electrons** proposed 20 years earlier by G. Johnstone Stoney as the carriers of electric current in wires. He determined that the nature of the cathode rays was independent of the metal used for the cathode and the gas used to fill the tube, with a universally valid **charge-to-mass** (q/m) **ratio** of

$$\frac{q}{m} = -5.2728 \times 10^{17} \text{esu g}^{-1} = -1.7588 \times 10^{11} \text{C kg}^{-1}$$

(modern value). With the discharge tube technique, neither the charge nor the mass of the electrons could be determined separately.

Earlier experiments had suggested that the gas in the tube must be undergoing **ionization,** separation into electrons and **positive ions.** Thomson and his group used

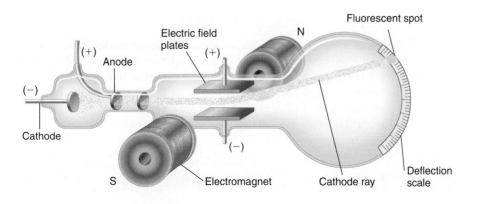

Figure 1.5
J. J. Thomson's elaborate Crookes' tube for measuring the charge-to-mass ratio of the cathode rays (electrons) using crossed electric and magnetic fields. The cathode rays are actually invisible except for the fluorescent spot they produce on the far end. The size of the spot is determined by the slot in the anode. Note that the rays are deflected toward the positive field plate, indicating their negative charge. The same electric deflection technique is used in modern television picture tubes. First measuring the electric deflection and then balancing it with an opposite magnetic deflection allows q/m to be obtained.

a discharge tube similar to that shown in Figure 1.5, but with the cathode and anode reversed, in order to see whether any positive rays would shoot through. Although these were much weaker, their existence confirmed the earlier work. Thomson performed similar measurements of the charge-to-mass ratios of the positive ions, finding the ratios to be much smaller than that for the electron and highly dependent on the identity of the gas fill. The largest value occurred for the lightest element, hydrogen, but was still 1836 times smaller than that for the electron. Thomson correctly hypothesized that, to maintain electrical neutrality, the **hydrogen ion** must carry a charge equal and opposite to that of the electron, but be 1836 times heavier.

The experiments of Thomson and his predecessors had profound implications for chemistry. Dalton's once indestructible atom could now be torn asunder at will and made to reveal its constituents. Although there had been much evidence that certain kinds of compounds (salts) contained charges, especially from the work of Michael Faraday in the 1830s, these new results showed that it was quite likely that all elements are composed of charged particles, lightweight negative electrons, and heavier positive particles, later to be called **protons.** The hydrogen ion was thought to consist of a single proton. Based on these findings, Thomson proposed a model for the atom in which a number of tiny electrons (exactly how many was unknown then) was embedded in a uniform sphere composed of a like number of protons. This was later dubbed the "raisin pudding" model by Ernest Rutherford (to Thomson's chagrin). In addition, one of Thomson's students, Francis Aston, went on to develop and refine the positive ion analysis into what is known today as **mass spectrometry.**

The Charge on the Electron

Thomson also was the first to obtain a sound estimate of the charge on the electron, by observing water droplets bearing static charges. Robert A. Millikan, an American physicist, refined Thomson's method by substituting charged oil droplets (so they wouldn't evaporate during measurement) and adding an electric field. Through the use of a reticulated telescope, he was able to make measurements on *individual droplets,* arresting their free fall in gravity with the opposing electric field. Millikan's apparatus is diagrammed in Figure 1.6. The balance condition on the electric field allowed Millikan to calculate the charge on each droplet; he always found it to be an integer multiple of a minimum value, which he correctly took to be the charge of a single electron. In 1909 Millikan published a value for $|q_e|$, usually just denoted e, of 4.774×10^{-10} esu, very close to the modern value of

$$e = 4.803204 \times 10^{-10} \text{ esu} = 1.602176 \times 10^{-19} \text{ C}$$

Unlike Thomson's experiments on water droplets and earlier work, Millikan's oil drop method established beyond doubt that electrons each carry the *same* charge, that the static charges of nature come in multiples of a fundamental, discrete charge unit. Dividing this result by Thomson's charge-to-mass ratio, the mass of the electron could be obtained. Its modern value is

$$m_e = 9.10938 \times 10^{-28} \text{ g}$$

Because the charge had been shown to be discrete, the mass had to be also; there could not be "light" and "heavy" electrons, with m_e being an average mass. Then, as-

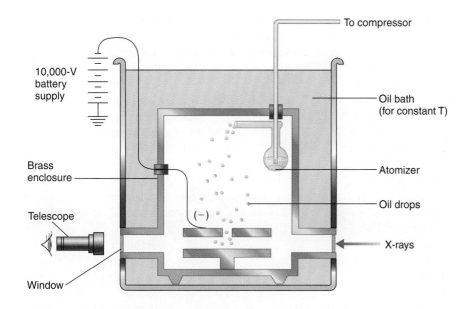

Figure 1.6
Millikan's oil drop apparatus for measuring the charge on the electron. Individual droplets could be followed and "played with" for hours. If a drop lost or acquired a charge, it would suddenly "jump," altering its velocity by discrete amounts. Millikan went to great lengths to eliminate systematic errors. He also declared emphatically that there are no "subelectrons" with fractional charge. (The modern theory of the nucleus posits the existence of "quarks" with ±1/3 and ±2/3 of the electron's charge.)

suming every proton possesses the same positive charge of magnitude e, Thomson's positive ion analysis leads to (again a modern value of) the proton mass

$$m_p = 1.672622 \times 10^{-24} \text{ g}$$

The sum of the electron and proton masses should yield the mass of a neutral hydrogen atom, and allow us to relate the relative atomic masses of the periodic chart, then based on a value of one for H to the actual masses of the atoms; more on this later.

1.6 The Nuclear Atom

In the work on "atomic electricity" we have already described, as well as in what follows, the discovery of X-rays, emanations from the anodes and walls of the discharge tubes, and natural **radioactivity** from heavy elements such as uranium and its daughters, polonium and radium, were closely intertwined with the elaboration of the atomic theory. We have omitted the details of these developments in order to focus on the results that bear directly on this elaboration. Moseley's experiments on X-ray emission have already been mentioned, and X-rays will be of importance again in Chapters 2 and 14; radioactive elements will form a starting point for the material in Chapter 16. But the understanding of both of these phenomena was greatly enhanced by the remarkable experiments of Ernest Rutherford and his students Geiger and Marsden.

Rutherford's early work (1895–1910) had elucidated the transmutation of uranium into lighter elements largely through the radioactive emission of **α particles,** which appeared (and were later proven) to be helium atoms stripped of their electrons. He then began to make beams of α's in order to scatter them from thin metal foils, using a sample of the strong α-emitter radium, kindly given him by its

Figure 1.7
Rutherford's α-particle scattering apparatus. The experiments were carried out in vacuum, with the viewing microscope sealed into a rotatable hollow cylinder surrounding the stationary source and foil target. Marsden's modification was to rotate the cylinder around the foil target in order to look for backscattering. The angle-dependence of the scattering was found to be $1/\sin^4(\theta/2)$, where θ is the angle of deflection with respect to the α-particle beam.

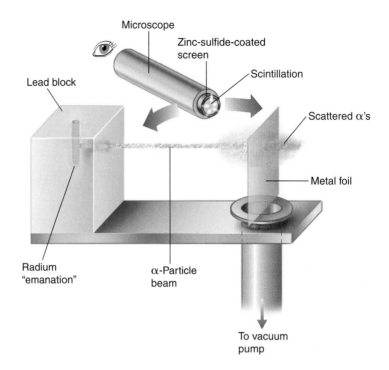

discoverer, Marie Curie. As a scattering detector, a zinc sulfide screen, which scintillated each time it was struck by an α, was used. Figure 1.7 diagrams the scattering setup. In Rutherford's own words:

> In the early days I had observed the scattering of α-particles, and Dr. Geiger in my laboratory had examined it in detail. He found in thin pieces of heavy metal that the scattering was usually small, of the order of one degree. One day Geiger came to me and said, 'Don't you think that young Marsden, whom I am training in radioactive methods, ought to begin a small research?' Now I had thought that too, so I said, 'Why not let him see if any α-particles can be scattered through a large angle?' I may tell you in confidence that I did not believe they would be, since we knew that the α-particle was a very fast massive particle, with a great deal of energy, and you could show that if the scattering was due to the accumulated effect of a number of small scatterings, the chance of an α-particle's being scattered backwards was very small. Then I remember two or three days later Geiger coming to me in great excitement and saying, 'We have been able to get some of the α-particles coming backwards.' ... It was quite the most incredible event that has ever happened to me in my life. It was almost as incredible as if you fired a 15-inch shell at a piece of tissue paper and it came back and hit you.
>
> On consideration I realized that this scattering backwards must be the result of a single collision and when I made calculations it was impossible to get anything of that order of magnitude unless you took a system in which the greater part of the mass of the atom was concentrated in a minute nucleus.*

---

* Ernest Rutherford, in *Background to Modern Science,* J. Needham, ed. (MacMillan, 1938), p. 61.

Rutherford chose the word **nucleus** to describe the tightly packed center of an atom by analogy with the nucleus of a living cell. The Thomson model was clearly incorrect, and Rutherford, in a paper published in 1911, proposed instead a planetary model, in which the light electrons orbit the massive nucleus much as planets orbit the sun. From many scattering experiments of this sort, he was able to show that the number of positive charges in the nucleus is roughly half its atomic mass, and suggested that this number was identical to the atomic number Z that orders the elements in the periodic table. In a neutral atom of an element, the number of electrons in orbit would also be Z, whereas in positive ions one or more of these electrons was missing, leaving a net positive charge. The nucleus proved to be extremely small, about 10^{-13} cm in diameter, while the atoms themselves were about 10^{-8} cm across. The electrons in the atom, to use Rutherford's metaphor, were like "a few flies in a cathedral," while the nucleus would be no bigger than a pea in a pew. In other words, instead of Thomson's pudding, the atom was actually mostly empty space.

Insofar as the constitution of the atom was concerned, the major question that remained was the extra atomic mass not accounted for by the protons in the nucleus. At first it was believed that there were extra protons that had combined with electrons within the nucleus. It was not until 1932 that James Chadwick, one of Rutherford's students, discovered an electrically *neutral* particle with about the same mass as a proton, as a product of a nuclear reaction between α particles and the element beryllium. He called this particle the **neutron.** The neutron is actually slightly heavier than the proton, with a mass

$$m_n = 1.674927 \times 10^{-24} \text{ g}$$

In our present-day picture the nucleus is composed of a mixture of protons and neutrons, with the number of protons determining the number of electrons in the neutral atom, Z, and hence the identity of the element, while the extra mass is due to the neutrons. (Whether the nucleus can be accurately depicted in this way is currently a matter of debate among nuclear physicists, but for the most part only the charge and mass of the nucleus are of consequence in chemistry, and not the details of its structure.)

Though Rutherford's planetary model seemed on the face of it a pleasing way to view the internal structure of an atom, it raised two troubling questions that in some respects still occupy chemists and physicists today. The first regarded the *stability* of the planetary atom. Unlike the solar system, the electrons and nucleus are charged particles, and an electron orbiting the nucleus forms an *oscillating electric dipole* when the orbit is viewed from its edge. James Maxwell had developed a highly successful theory of electricity and magnetism 50 years before Rutherford's work, in which oscillating dipoles must gradually "run down" from radiating their energy into space. *A planetary atom should collapse with a twinkle in a few trillionths of a second!* The other question concerned the nucleus itself. Protons, being positively charged particles, ought to *repel* each other according to Coulomb's law of electrostatics (which we will examine in some detail in Section 1.7). The 79 protons in a gold nucleus, packed into a sphere of only 10^{-13} cm in diameter, ought to explode! Unfortunately, humankind has succeeded in releasing this awesome power in destructive nuclear bombs, but, luckily, most nuclei somehow are stable as rocks despite the Coulomb repulsion of their protons. This nuclear stability chemists ordinarily take for granted (see Chapter 16), but dealing with the planetary stability problem was to be central to the development of models for and answers to the modern chemical questions we have posed.

Figure 1.8
Aston's 1920 mass spectrum of pure neon gas, showing the three isotopes $^{20}_{10}\text{Ne}$, $^{21}_{10}\text{Ne}$, and $^{22}_{10}\text{Ne}$. The relative peak heights are proportional to the natural abundances of the isotopes. Although the *x* axis is labeled by mass, actually the positive ions (Ne$^+$) are separated according to their mass-to-charge ratios. It is usually the case that the ions are singly charged, having lost one electron apiece. Typical ion currents are 10^{-9} amp or less.

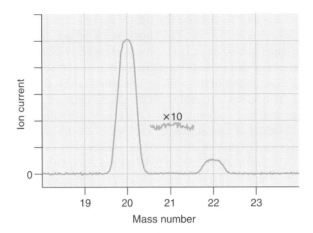

Isotopes

Chadwick's neutron provided a natural explanation for the discovery of **isotopes** of the elements in the rapidly developing field of mass spectrometry in the early 1920s. Figure 1.8 shows an early **mass spectrum** of neon gas. The mass axis is labeled by the so-called **mass number *A*,** which could now be understood as the sum of the number of protons and neutrons in the nucleus. Because the neon gas used was chemically pure, the three peaks had to belong to atoms with different numbers of neutrons in the nucleus, but all with the same **atomic number *Z*;** these were called **isotopes** of neon, written as $^{20}_{10}\text{Ne}$, $^{21}_{10}\text{Ne}$, and $^{22}_{10}\text{Ne}$ and containing respectively 10, 11, and 12 neutrons. In general isotopes are denoted $^{A}_{Z}\text{X}$, where X is the chemical symbol for the element. The early measurements indicated that the masses of different isotopes were all integer multiples of the hydrogen atom mass, but as mass spectrometric techniques improved it became clear that all the heavier atoms weighed *less* than the sum of their protons, neutrons, and electrons. This **mass defect** is due to conversion of a small part of the nuclear mass into the **binding energy** of the nucleus in accord with Einstein's famous formula $E = mc^2$, where *E* is the binding energy and *m* the mass defect. A thorough discussion and calculations are reserved for Chapter 16. Since 1960, scientists have agreed to use the so-called **unified mass scale** for expressing relative atomic masses, in which the carbon-12 isotope is assigned a mass of exactly 12, that is,

$$\text{mass}(^{12}_{6}\text{C}) = 12.\overline{0} \textbf{ atomic mass units (amu)}$$

where the bar over the zero in $12.\overline{0}$ indicates that the zero is repeated *ad infinitum*. This definition sets the scale for all mass spectra, and makes it independent of the relative abundances of the carbon isotopes $^{12}_{6}\text{C}$ and $^{13}_{6}\text{C}$. Table 1.3 lists the masses on the unified scale for the electron, proton, neutron, and selected isotopes, along with natural isotopic abundances. The natural abundances of the isotopes of an element allow the frequently nonintegral atomic masses of the elements to be understood. For example, the two stable isotopes of chlorine are $^{35}_{17}\text{Cl}$ and $^{37}_{17}\text{Cl}$, and the atomic mass (AM) is a *weighted average* of the isotope masses. Using the data of Table 1.3,

$$\text{AM(Cl)} = (0.7577)(34.969) + (0.2423)(36.966) = 35.45$$

Even elements that appear to have nearly integral atomic masses may consist of substantial fractions of two or more isotopes. Bromine, for example, with an atomic

TABLE 1.3

Masses and abundances of particles and selected isotopes of the elements

Particle	Symbol	Mass (amu)	Mass (g)
Electron	$_{-1}^{0}e$	0.00054858	9.10938×10^{-28}
Proton	$_{1}^{1}p$	1.0072765	1.672622×10^{-24}
Neutron	$_{0}^{1}n$	1.0086649	1.674927×10^{-24}

Isotope	Mass (amu)	% Abundance	Isotope	Mass (amu)	% Abundance
$_{1}^{1}H$	1.007825	99.985	$_{11}^{23}Na$	22.989767	100.00
$_{1}^{2}H$	2.014102	0.015	$_{17}^{35}Cl$	34.968852	75.77
$_{2}^{3}He$	3.01603	0.0000014	$_{17}^{37}Cl$	36.965903	24.23
$_{2}^{4}He$	4.002602	100.000	$_{26}^{54}Fe$	53.939612	5.8
$_{3}^{6}Li$	6.015121	7.5	$_{26}^{56}Fe$	55.934939	91.8
$_{3}^{7}Li$	7.016003	92.5	$_{26}^{57}Fe$	56.935396	2.1
$_{6}^{12}C$	12.0̄	98.90	$_{26}^{58}Fe$	57.933277	0.3
$_{6}^{13}C$	13.003355	1.10	$_{29}^{63}Cu$	62.929598	69.2
$_{7}^{14}N$	14.003074	99.63	$_{29}^{65}Cu$	64.927793	30.8
$_{7}^{15}N$	15.000108	0.37	$_{35}^{79}Br$	78.918336	50.69
$_{8}^{16}O$	15.994915	99.76	$_{35}^{81}Br$	80.916289	49.31
$_{8}^{17}O$	16.999131	0.04	$_{47}^{107}Ag$	106.905092	51.839
$_{8}^{18}O$	17.999160	0.20	$_{47}^{109}Ag$	108.904757	48.161
$_{9}^{19}F$	18.998403	100.00	$_{79}^{197}Au$	196.966543	100.00
$_{10}^{20}Ne$	19.992435	90.48	$_{92}^{234}U$	234.040946	0.005
$_{10}^{21}Ne$	20.993843	0.27	$_{92}^{235}U$	235.043924	0.720
$_{10}^{22}Ne$	21.991383	9.25	$_{92}^{238}U$	238.050784	99.275

mass of 79.904, does not possess a stable $A = 80$ isotope, being composed instead of nearly equal amounts of ^{79}Br and ^{81}Br.

EXAMPLE 1.5

From more precise mass spectra than that shown in Figure 1.8, accurate isotope abundances can be obtained from the ratios of peak heights in the spectrum. Use the data of Table 1.3 to confirm the atomic mass of Ne given in the periodic table.

Solution:
The atomic mass is the average of the three isotopes:

$$\text{AM(Ne)} = (0.9048)(19.9924) + (0.0027)(20.9938) + (0.0925)(21.9914)$$
$$= 20.180$$

Note that the isotope masses are more precisely known than the abundances; the mass spectral peak positions are much easier to measure than the relative heights.

Dalton's concept of a unique mass for all atoms of a given element is thus true only on the average; because this average depends on isotopic abundances, the atomic mass of an element can change with time. In the years 1945–1963, during atmospheric testing of nuclear weapons, the proportion of ^{18}O in the atmosphere increased slightly, thereby increasing by a tiny amount the atomic mass of oxygen.

Avogadro's Number and the Mole

By combining isotope masses in amu with the actual masses in grams of the electron, proton, and neutron, we can put an *absolute* number to each of the masses listed in Table 1.3. The mass of the proton in amu, 1.007276, is equivalent to an actual mass of 1.67262×10^{-24} g, so that 1 amu = 1.66054×10^{-24} g. The mass in grams of each of the isotopes is the product of this number and its mass in amu. A more enlightening way to express the conversion is through its inverse: the number of amu in a gram of matter is 6.0221×10^{23} amu g^{-1}. This is called **Avogadro's number** N_A. Because masses weighed in the chemical laboratory are normally on the order of a gram, Avogadro's number allows you to translate an ordinary mass quickly and directly into microscopic terms.

If you contrive to weigh out exactly 26.98 g of aluminum metal, Avogadro's number tells you that the number of Al atoms in the sample is

$$(26.98 \text{ g Al})(6.0221 \times 10^{23} \text{ amu g}^{-1})(1 \text{ atom Al}/26.98 \text{ amu})$$
$$= 6.0221 \times 10^{23} \text{ atoms of Al}$$

Notice that we deliberately chose a mass in grams numerically equal to the atomic mass of Al in amu listed in the periodic table; this made the first and last factors in our calculation cancel exactly. The result would therefore be the same if a mass in grams equal to the atomic mass in amu of *any* element were weighed out. Any such sample will contain 6.0221×10^{23} atoms, that is, an Avogadro's number of atoms. Chemists call this a **mole** of atoms. Although the mole is based on the gram (cgs units), it is still an "approved" SI unit as well. In the SI system it must be referred to as a pure number, 6.0221×10^{23} mol^{-1}, where **mol** is the approved abbreviation.

It is also useful to think of *moles of molecules* when interpreting a chemical reaction such as Reaction 1.1, $2H_2 + O_2 \longrightarrow 2H_2O$. Instead of dealing with the reaction a few molecules at a time, we can interpret the chemical equation as the reaction of 2 *moles* H_2 with 1 *mole* O_2 to yield 2 *moles* H_2O. This relates the equation directly to laboratory-scale quantities: 2 moles H_2 weighs 2(2.016) = 4.03 g; 1 mole O_2, 32.00 g; and 2 moles H_2O, 2(18.015) = 36.03 g. The study and use of relations between moles in a chemical equation and laboratory-scale amounts of compounds is called **stoichiometry**. The masses just calculated are obtained by adding atomic masses in amu, multiplying by the stoichiometric coefficients (the numbers of each of the molecules involved in the reaction), and scaling the entire equation up by N_A, the number of molecules per mole. Notice that you do not need to know the actual value of N_A to analyze a reaction in this way; the molecular mass of H_2O can be thought of as 18.015 g mol^{-1} (the mass of 1 mole) as well as 18.015 amu, and only the mass ratios need be known to do quantitative mass (stoichiometric) calculations for reactions.

We can usually calculate the number of moles of a reagent or product in one of three ways:

1. If the compound is a pure solid or liquid, moles are obtained by dividing its weighed mass by its molecular mass (grams divided by grams/mole).

2. If it is a gas, the ideal gas law (Chapter 9) can be used to find the number of moles from the pressure, volume, and temperature of the gas. At 0°C (273 K) and atmospheric pressure (standard temperature and pressure or STP), 1 mole of gas occupies a volume of 22.4 liters. The **liter** (L) is defined as 1000 cm^3 (10^{-3} m^3). So at STP, moles of a gas are obtained by dividing its volume by the molar volume (liters divided by 22.4 liters/mole).

3. If it is dissolved in a liquid solvent such as water, that is, if it is a **solute** in a **solution,** its concentration in the solution is usually given in **moles/liter** of solution (or **molarity** M). In this instance moles are calculated from the product of the molarity and the measured volume of solution (moles/liter times liters).

All of the simpler stoichiometric problems, in which a single unknown quantity is wanted, can be worked out in the following sequence:

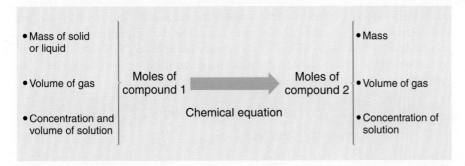

The last step usually involves reversing the methods of (1), (2), or (3). The example illustrates these stoichiometric methods.

EXAMPLE 1.6

We add 4.86 g of metallic zinc (Zn) to 75.0 mL of 1.49 M hydrochloric acid, HCl, solution, and the reaction

$$Zn + 2HCl \longrightarrow ZnCl_2 + H_2$$

proceeds to completion. Find the number of liters of hydrogen gas produced at STP, and the molar concentration of zinc chloride.

Solution:
We do not know at the start whether the amounts of reagents conform to the balanced chemical equation, so we calculate

$$\text{mol Zn} = (4.86 \text{ g})(1 \text{ mol}/65.39 \text{ g}) = 0.0743 \text{ mol Zn}$$
$$\text{mol HCl} = (0.0750 \text{ L})(1.49 \text{ mol/L}) = 0.1118 \text{ mol HCl}$$

Because the ratio mol HCl/mol Zn < 2, where 2 is the ratio of stoichiometric coefficients, the amount of HCl present is insufficient to consume all of the Zn, and HCl is the limiting reagent. (We could have reached this same conclusion by using the balanced chemical equation to compute the ratio of combining masses, and comparing that to the mass ratio actually present.) The amounts of ZnCl$_2$ and H$_2$ formed are thus determined by the amount of HCl. For H$_2$,

$$\text{L H}_2 = 0.1118 \text{ mol HCl } (1 \text{ mol H}_2/2 \text{ mol HCl})(22.4 \text{ L/mol H}_2) = 1.25 \text{ L}$$

and for $ZnCl_2$,

$$\text{mol } ZnCl_2 = 0.1118 \text{ mol HCl } (1 \text{ mol } ZnCl_2/2 \text{ mol HCl}) = 0.0559 \text{ mol}$$

If we assume negligible change in solution volume during reaction, then M $ZnCl_2$ = 0.0559 mol/0.0750 L = 0.745 M. If the reaction had been the result of mixing two solutions, the volumes would be (approximately) additive, and the molarity of the products would be obtained using the total solution volume.

Two features of Avogadro's number deserve mention. The first is the sheer enormity of it: 602 *sextillion* objects per mole. There are more molecules in a mole than there are stars in the known universe. A mole of ice cubes stacked one on another would reach to the farthest edge of our Milky Way Galaxy—and back again! The second is the window N_A gives us on the world of the very small. The **density** (mass per unit volume) of liquid water is 1.00 g cm^{-3}, and, like most liquids and solids, water is quite incompressible. So we can safely assume that in water the H_2O molecules are packed closely together. The density of the liquid can then be combined with N_A to yield the volume occupied by one *molecule* of water:

$$V(H_2O) = (1 \text{ cm}^3/1.00 \text{ g})(18.0 \text{ g mol}^{-1})(1 \text{ mol}/6.022 \times 10^{23} \text{ molecules})$$
$$= 2.99 \times 10^{-23} \text{ cm}^3/\text{molecule}$$

The cube root of this molecular volume gives an estimate of the diameter of a water molecule, about 3×10^{-8} cm. Distances on the order of 10^{-8} cm (10^{-10} m) are so typical of molecular dimensions that the **angstrom** (Å) length unit (1 Å = 10^{-8} cm = 10^{-10} m) was adopted to express them in small whole numbers. Although the angstrom is not an official SI unit, it is a metric unit, and its use is so widespread in the chemical and physical literature that we would be remiss not to use it in this book.

1.7 Coulomb Force and Potential Energy

The experiments we have described in Sections 1.5 and 1.6 established beyond doubt that Dalton's atoms are composite entities, tiny mechanisms of moving electrically charged particles. If we are to understand the formation of bonds between atoms and the course of chemical reactions, the nature of electrical forces must be explored. In experiments in the late 18th century, little noted in most accounts of the history of science, Charles Augustin de Coulomb established the **inverse-square law of force** for the interaction of charged bodies, through the study of statically charged pith balls suspended by rods and silk threads. **Coulomb's law** of force is

$$F(r) = k\frac{q_1 q_2}{r^2} \tag{1.17}$$

where q_1 and q_2 are the charges on two bodies, r is the distance between them, and k is a proportionality constant. Equation 1.17 is remarkably similar to Newton's gravitational law (Equation 1.5), but there are important differences. Newton's law applies to all massive bodies, whereas Coulomb's affects only charged objects. The gravitational force is (so far as we know) always attractive, whereas the existence of both positive and negative charges means that the Coulomb force (also called the *electromagnetic* force) can be either *attractive,* pulling the charges q_1 and q_2 together when they are of opposite sign, or *repulsive,* pushing charges apart when their signs are the same. The rule is *unlike charges attract, like charges repel.* In the case of unlike charges, the Coulomb force is negative, corresponding to inward acceleration and tending to decrease the distance r. For like charges, F is positive, and the charges are accelerated away, toward increasing r. In either situation, the force weakens as the particles move apart, approaching zero at very large r. On the other hand, the force becomes arbitrarily huge as r approaches zero, but then the "point charge" picture implicit in defining r breaks down.

According to Equation 1.13, we can get the mutual potential energy of a pair of charges by integrating the force over some distance (calculating the work). Moving two charges from r_1 to r_2 apart changes the potential energy by an amount

$$V = -\int_{r_1}^{r_2} \frac{kq_1q_2}{r^2} dr = -kq_1q_2 \int_{r_1}^{r_2} \frac{dr}{r^2}$$

The integral is a particular case of the integral of a power; see Example 1.4 for the general rule for evaluating it. Using this rule, the potential energy is

$$V = kq_1q_2 \left(\frac{1}{r_2} - \frac{1}{r_1} \right)$$

With an electron and a nucleus in mind, we choose r_1 to be very large, so that the force is zero. The potential is then the negative of the work required to pull the electron completely away from the nucleus, or

$$V(r) = k \frac{q_1 q_2}{r} \tag{1.18}$$

where r_2 is replaced by an arbitrary distance r.

Calculation of V can be done in either cgs-esu or SI units. In the cgs-esu system, the charge q is *defined* so that the proportionality constant k is unity; the resulting unit of charge is the **electrostatic unit** (esu), and the potential energy comes out in ergs. In SI units, $k = 8.98755 \times 10^9$, the charge is in coulombs (C) and the potential in joules. If a proton and an electron are at a typical separation of 1 Å, then $q_1 = +e$ and $q_2 = -e$ and we have

$$V = \frac{-(1)(4.803 \times 10^{-10} \text{ esu})^2}{1 \times 10^{-8} \text{ cm}} \cdot \frac{1 \text{ J}}{10^7 \text{ erg}} = \frac{-(8.988 \times 10^9)(1.602 \times 10^{-19} \text{ C})^2}{1 \times 10^{-10} \text{ m}}$$
$$\text{(cgs-esu)} \qquad\qquad\qquad \text{(SI)}$$
$$= -2.307 \times 10^{-18} \text{ J}$$

As you might expect, the potential energy is minuscule by everyday standards because of the small charges involved. It is negative, however, reflecting the attractive force and the consequent lowering of potential energy as the particles come together. A convenient unit for expressing these microscopic energies is the **electron-volt** (eV), defined as the kinetic energy gained by an electron in falling through an electric potential difference (voltage) of 1 volt (V). Because 1 V is defined to be the voltage that bestows one joule of energy on one coulomb of charge, that is, $1 \text{ V} = 1 \text{ J C}^{-1}$, we have

$$1 \text{ eV} = 1 \text{ V} (1.602 \times 10^{-19} \text{ C}) = 1.602 \times 10^{-19} \text{ J} \tag{1.19}$$

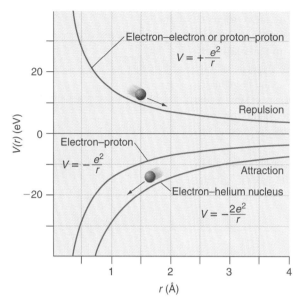

Figure 1.9
Potential energy curves for the interaction of pairs of charged particles. Motion in these potentials is analogous to a ball rolling on a track or slide in the shape of the curve subject to gravity, as shown. In each case the ball rolls toward lower potential. The formulas are given in cgs-esu. All potentials involving neutrons are exactly zero, except at nuclear distances (10^{-13} cm) (see Chapter 16).

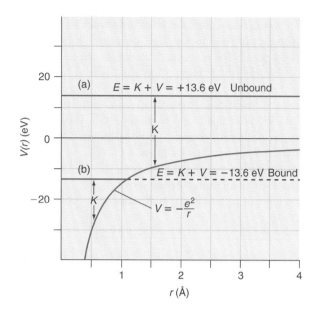

Figure 1.10
Potential, kinetic, and total energies for an electron interacting with a proton. In case (a), $E > 0$, the electron has positive kinetic energy at all distances and can "fly away" from the proton; whereas in case (b), $E < 0$, the kinetic energy would be negative for distances beyond $r = 1.06$ Å, impossible in Newton's mechanics, and the electron is confined (bound) to distances less than this. Note how the kinetic energy increases as the particles approach, characteristic of an attractive force. Case (a) occurs in the discharge tubes of Figures 1.4 and 1.5, while case (b) is found in a normal hydrogen (or other) atom.

So our electron–proton potential energy is

$$V = (-2.307 \times 10^{-18} \text{ J})/(1.602 \times 10^{-19} \text{ J eV}^{-1}) = -14.40 \text{ eV}$$

This turns out to be a typical magnitude for the potential energy of chemically active electrons in an atom. This calculation also yields a useful "engineering formula" for the quick calculation of V: $V (\text{eV}) = 14.40 q_1 q_2 / r (\text{Å})$, where the charges are expressed in units of e.

Plots of the potential energy for the interaction of various charged atomic particles are shown in Figure 1.9. Because the slopes of these curves give the forces between the particles, motion subject to these forces can be represented by a ball rolling or sliding frictionlessly on the potential curves subject to gravity. Imagine you are an electron sliding on one of the attractive potentials; as you approach the nucleus your "slide" gets steeper and steeper and you gain more and more kinetic energy. An electron in an atom has a very exciting time of it!

Disregarding for now Maxwell's requirement that the atom should collapse due to radiation (it can't be true!), the total energy $E = K + V$ is constant for any of the pairs of particles in Figure 1.9, and can be plotted as a horizontal line on the same graph as the potential. Figure 1.10 illustrates this for the electron–proton potential energy. In one case the total energy lies below the r axis; since the kinetic energy must be zero or positive (at least in Newton's mechanics), the electron is confined to a region "inside" the potential curve. Thus, *negative total energy indicates a "bound"*

1.7 Coulomb Force and Potential Energy

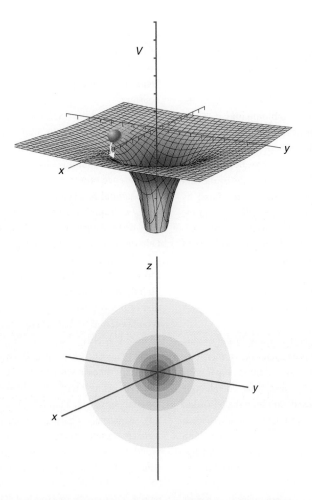

Figure 1.11
Two- and three-dimensional representations of the Coulomb potential between an electron and a proton. The 2-D surface (top) is generated by rotating the $-e^2/r$ curve about the vertical axis. The edges of the surface actually extend to infinite r. A ball rolling on this surface can execute planet-like orbits. The 3-D cloud (bottom) represents concentric spheres of constant potential energy. Each smaller sphere represents a drop in V of 5 eV, and the darkness indicates the total magnitude of V.

electron, one that cannot leave the vicinity of its nucleus. The other case corresponds to an *unbound* electron, since with a *positive* total energy it will still possess kinetic energy even at distances far from the nucleus, leaving behind a positive ion. Note that these rules result from our choice of zero potential energy at large distances.

Electrons within an atom are bound by their attractive Coulomb interaction with the nucleus, whereas in a cathode-ray tube the electrons are freely moving, not attached to any nucleus. The energy to free the electrons is supplied by the source of the tube voltage. By analogy the orbiting planets are bound to the sun, but the *Voyager* spacecrafts are capable of leaving the solar system.

Figures 1.9 and 1.10 are one-dimensional plots of the potential energy, but the particles that compose atoms exist in (at least) a three-dimensional space. (According to the latest "unified" theories of the forces of nature, there may be 10 or more dimensions!) In two dimensions the attractive Coulomb potential becomes a tapered cone, as illustrated in Figure 1.11. Rolling a ball round the edges of the cone gives an impression of the circular and elliptical planet-like orbits of Rutherford. In three dimensions the Coulomb potential cannot be properly graphed, because the value of the potential would have to be plotted in a fourth dimension. But because this potential is *spherically symmetric,* it can be represented as a series of concentric spheres of constant potential energy, with the inner spheres being successively lower in potential. In this way you can see that the Coulomb potential is something like a hole in space.

SUMMARY

Chemistry is a science squarely in the center of the study of nature. Springing from twin roots of Greek philosophy and eastern **alchemy,** modern chemistry is based on the **atomic theory** of matter. Hypothesized by John Dalton in 1803 to explain the **conservation of mass** and the definite combining masses of the **elements** in forming simple **compounds,** the atomic theory led to systematic investigations of chemical reactions, to the concept of **molecules,** and eventually to the proposal of the **periodic law,** the regular recurrence of similar chemical properties of the elements when arranged in order of increasing atomic mass, by Dmitri Mendeleyev in 1869. Mendeleyev's periodic table of the elements showed gaps that stimulated the discovery of additional elements. Finally, Henry Moseley demonstrated in 1914 that the atomic number Z, rather than the atomic mass, is the natural ordering parameter for the elements, through his studies of X-ray emission from metals. Modern chemistry seeks to probe the structure of molecules and the course of reactions in great detail.

Chemical conclusions such as those leading to the atomic theory and the periodic law are based on quantitative measurement, that is, experimental numbers. These numbers have units composed of the basic quantities **length, mass, time,** and **electric charge.** In science **metric units** are used, and both the **SI (mks)** and **cgs-esu** metric systems are introduced. Unlike mathematical numbers, numbers from observations have errors associated with them, usually expressed by a limited number of **significant figures.** These errors are of two basic types: errors of **precision** and errors of **accuracy.** Precision refers to the span of numbers resulting from repeated measurements of a given quantity, whereas accuracy is the deviation of the average of those measurements from the "true" value of the quantity.

Answers to modern chemical questions involve forces within and among molecules and their motions, particularly in the course of chemical reactions. These are mechanical concepts, and their appreciation starts with Isaac Newton's mechanical laws. A moving object has **momentum,** the product of mass and velocity $mv = m\, dx/dt$, that can be changed by the application of a **force** F. **Newton's second law of motion** states that the applied force is equal to the rate of change of the momentum with time, $F = d(mv)/dt = m\, dv/dt$, where $dv/dt = d^2x/dt^2$ is the **acceleration** a, the rate of change of the velocity, or the "rate of change of the rate of change" of distance x with time. When the object moves through some distance while under the influence of a force, **work** w is done, given by $w = \int F\, dx$. As a result of the work the object, if initially at rest, acquires **kinetic energy** $K = \frac{1}{2} mv^2$. This means that an object about to be moved by a force has the potential to acquire kinetic energy; this is called **potential energy** V. The law of the conservation of energy states that, in the absence of frictional forces, the sum of the kinetic and potential energy of an object is a constant, called the **total energy** E; that is, $E = K + V =$ constant.

The potential energy has three important properties: (1) It only exists when there is a force acting on the object. It is related to the force by $V = -\int F\, dx$ or $F = -dV/dx$. (2) An object naturally tends to minimize its potential energy, that is, it responds to the presence of a force by accelerating toward lower V. (3) After the potential has been minimized, the object may reach a state called **equilibrium.** To reach this state requires friction; the energy of the object must be communicated to its surroundings. Otherwise the object remains in a state of perpetual motion.

A macroscopic object possesses **heat energy** at any finite **absolute (kelvin) temperature** T. The amount of heat energy is directly proportional to the absolute temperature and is a reflection of the average energy $\bar{\varepsilon}$ of the molecule of which the object is composed: $\bar{\varepsilon} \approx k_B T$ where $\bar{\varepsilon}$ is a total mechanical (kinetic plus potential) energy of an individual molecule, and k_B is a proportionality constant known as **Boltzmann's constant.**

Experiments with gases in electrical discharge tubes in the late 19th century led to the discovery of **cathode rays.** Through the analysis of J. J. Thomson these rays were shown to consist of tiny, negatively charged particles called **electrons.** Electrons were shown to be universal constituents of all matter, whereas the properties of weaker, positively charged rays were found to depend on the identity of the gas in the tube. These rays proved to be positively charged **ions,** gaseous atoms and molecules with one or more electrons removed. Thomson's experiments yeilded the **charge-to-mass ratios** of the electron and ions. Ions of the lightest element, hydrogen, possessed the largest charge-to-mass ratio, but this was still 1836 times smaller than that of the electron. This **hydrogen ion** became known as the **proton,** with a positive charge equal in magnitude to the electron's charge and a mass 1836 times as large. The charge on the electron was measured accurately by Robert Millikan in 1909 by observing statically charged oil droplets. When combined with Thomson's charge-to-mass ratio the masses of the electron and proton could then be obtained.

The arrangement of the charged particles within the atom was clarified by the α-particle-on-metal-foil scattering experiments of Ernest Rutherford in 1911, which showed that all of the positive charge (the protons) and nearly all of the mass of the atom were concentrated in a minute **nucleus** at the center of the atom, with the electrons in comparatively distant orbits about it. The nucleus was found to contain roughly twice the mass of its constituent protons; this was explained by the discovery of the **neutron** in 1932 by James Chadwick as the product of a nuclear reaction. The neutrons, with a mass slightly larger than a proton's but no charge, make up the remainder of an atom's mass. The **atomic number Z** was identified with the number of protons in the nucleus, while the **mass number A** was the sum of the protons and neutrons. Though the interpretation of Rutherford's experiments was unquestionably correct, problems of stability of the orbiting electrons and of the nucleus arose.

The existence of **isotopes** of the elements—atoms with the same number of protons but different numbers of neutrons in their nuclei, that is, different mass numbers—was established by **mass spectrometry** in the early 1920s. The atomic mass of an element was then understood to be the average of the masses of its isotopes, weighted by their natural abundances. In 1960 the **unified scale of atomic masses** was defined so that the mass of the isotope of carbon with six protons and six neutrons in its nucleus is exactly 12, that is, $m(^{12}_{6}C) = 12.\overline{0}$ amu, **atomic mass units.**

The combination of accurate absolute masses of the elementary particles and the accurate relative mass scale allows the number of amu in a gram of matter to be obtained. This is called **Avogadro's number** N_A, 6.0221×10^{23} amu g^{-1}. If a mass in grams of any element or compound numerically equal to its atomic or molecular mass in amu is weighed out, the sample will contain N_A atoms or molecules, called a **mole** of atoms or molecules. Because a mole of chemical reagent or product is a laboratory-scale quantity, moles are normally used in **stoichiometric** calculations (involving measured amounts of compounds). Procedures for finding moles of solids, liquids, gases, and solutions are given. Avogadro's number also allows estimation of molecular sizes from the density of liquids or solids; typical molecular diameters are a few **angstroms**, Å, where 1 Å = 10^{-8} cm = 10^{-10} m.

Coulomb's law governs the interactions of the charged particles within atoms and molecules. Like the force of gravity, it is an **inverse-square law of force,** but may be attractive (for unlike charges) or repulsive (for like charges). The **Coulomb potential** $V = kq_1q_2/r$ contributes to the total energy of a pair of charges by an amount that vanishes at very large distance r but becomes arbitrarily large at small r. Total energies for unlike charges may be either negative, corresponding to bound motion over a restricted range of r, or positive, corresponding to unbound motion.

EXERCISES

Combining weights, combining volumes, and atomic theory

1. Lavoisier gave a detailed account of his work in his pioneering 1789 monograph "Traité élémentaire de Chimie" (*Elements of Chemistry*), from which the diagrams and description in Figure 1.1 were taken. From the experiment depicted in Figure 1.1(a), Lavoisier was able to recover 45.0 grains (2.39 grams) of the red calx of mercury. When the calx was later heated to red heat, it decomposed back into 41.5 grains (2.20 grams) mercury liquid, liberating a new gas, which Lavoisier later named oxygen. Using the modern values of the atomic masses of mercury (Hg) and oxygen (O), what formula for the red calx is predicted by Lavoisier's masses?

2. In Lavoisier's study of the burning of phosphorus, he found that "100 parts of phosphorus require 154 parts of oxygen for saturation." Use Lavoisier's finding along with the modern values for the atomic masses of P and O to find a formula for phosphorus oxide, assuming that "parts" refer to masses. Your result will not agree with the modern formula P_2O_5; what change in the "parts of oxygen" would be required to bring Lavoisier's measurement into agreement?

3. In an appendix to his book, Lavoisier tabulated carefully measured masses of "a cubical inch" of various gases. The values were given in "grains," but their ratios are independent of unit system:

Gas	Mass of a cubic inch	Modern formula
Azotic gas	0.30064	N_2
Oxygen gas	0.34211	O_2
Hydrogen gas	0.02394	H_2
Carbonic acid gas	0.44108	CO_2

    Compare the ratios of these masses to that of O_2 with predictions based on Avogadro's hypothesis (equal volumes contain equal numbers of molecules) using modern molecular mass values.

4. For carrying out the iron calcination experiment depicted in Figure 1.1(b), Lavoisier recommended limiting the amount of iron to 1.50 gros (1 gros =

72 grains; 1 grain = 0.05313 grams), to avoid using all the oxygen in the bell jar and cracking it. Assuming that the iron calx has the (modern) formula Fe_2O_3, and using modern atomic masses and the data in Exercise 3, estimate the volume of Lavoisier's bell jar in "cubical inches" and in liters (1 inch = 2.54 cm).

5. Dalton appears to have been led to his law of multiple proportions by experiments with "carburetted hydrogen" (now methane, CH_4) and "olefiant gas" (now ethylene, C_2H_4). (Dalton's original notebooks were destroyed during the bombing of Manchester, England, during World War II.) Using modern atomic masses, find the ratio of combining weights m_C/m_H for each of these two compounds, and take their ratio. How does this illustrate the law?

6. Give balanced modern chemical equations that reflect Gay-Lussac's 1808 summary of his and others' results on combining volumes, in the light of Avogadro's hypothesis, for the following gaseous systems:

   a. 2 vol. carbon monoxide + 1 vol. oxygen ⟶
      2 vol. carbon dioxide
   b. 3 vol. hydrogen + 1 vol. nitrogen ⟶
      2 vol. ammonia
   c. 1 vol. nitrogen + 1 vol. oxygen ⟶
      2 vol. nitric oxide
   d. 1 vol. nitrogen + 2 vol. oxygen ⟶
      2 vol. nitrogen dioxide

   Choosing any one of the preceding examples, show that incorrect combining volumes are obtained if the elemental gases are assumed to be monatomic.

Errors and significant figures

7. Many of Lavoisier's measurements were quite precise, even by modern standards, but often were inaccurate due to systematic errors both known and unknown to him. In Exercise 3, for example, the mass ratios do not agree to the number of significant figures given with those computed from the molecular masses. What amount of contamination by air (expressed as a fraction of molecules or moles) in Lavoisier's hydrogen sample would account for the difference between his measured H_2/O_2 mass ratio and that computed from molecular masses? Take the molecular mass of air to be 29.0.

8. A gas fills a container of volume 1.033×10^3 cm^3. When filled with the gas, the container weighs 386.3 g, while evacuated it weighs 383.0 g. Find the density of the gas in mg cm^{-3}, being careful of significant figures. Find the molecular mass of the gas assuming STP.

Units and mechanics

9. A typical highway speed is 55 mph. Convert this to SI units (1 mi = 5280 ft; 1 in. = 2.54 cm). Calculate the kinetic energy in kJ of a 2600-lb car at this speed (1 lb = 453.6 g). The burning of 1 mol of hydrogen gas (Reaction 1.1) liberates 237 kJ of useful energy. How many grams of hydrogen are required to accelerate a frictionless car to this speed? How much work is done on this ideal car?

10. A 45-kg high diver is about to execute a 10-m dive. What is her potential energy in joules before diving, with respect to pool level ($g = 9.8$ m s^{-2})? What is her kinetic energy (neglecting air resistance) when she enters the water? What is her velocity? How much work did she do in climbing the diving tower?

11. For the case of the dropped book discussed in this chapter, sketch and label an energy diagram illustrating the conservation of total energy E by plotting E and V versus x, where x is the distance above the table top where the book lands, assuming a book of mass $m = 1.02$ kg dropped from a height $x = h = 100.$ cm. Referring to your diagram, compute the following:

    a. The kinetic energy of the book in ergs and joules at $x = 50.$ cm and $x = 0.$ cm. Indicate these results on your energy diagram.
    b. The velocity of the book in centimeters per second at these two points.

12. When bonded to oxygen atoms, hydrogen atoms H are well described by Hooke's law of springs (see Example 1.4), with equilibrium distance $r_e = 0.96$ Å and force constant $k = 7.8 \times 10^5$ dyne/cm (780 J m^{-2}).

    a. Convert k into units of eV Å^{-2}. Compute the potential energy in electron-volts when the O–H bond is stretched by 0.10 Å.
    b. How much kinetic energy will the H atom have when it returns to $r = r_e$ after being stretched by 0.10 Å?
    c. Illustrate your calculations with an energy diagram showing E and V plotted against r.

13. If you swing a ball of mass m attached to a string in a circle, the ball is said to possess **angular momentum,** because the *angle* of the ball and string with respect to some fixed direction is constantly changing. The mechanical definition of angular momentum is $L = mr^2 \, d\theta/dt$, where r is the length of the string (radius of the ball's orbit) and θ is the angle of rotation. If the orbital velocity v is constant in magnitude, show that the ball's angular momentum is $L = mvr$. (*Hint:* If v is constant, then $d\theta/dt$, often denoted ω, is also constant, and equal to $2\pi/\tau$ where τ is the rotational period, $2\pi r/v$.) Give the units of L in two forms, one involving joules. Express the kinetic energy of the ball in terms of L. (Angular momentum is useful in thinking about electrons orbiting nuclei and spinning on their axis, and about atoms in a molecule tumbling around their center of gravity.)

Temperature and heat

14. Give the Celsius and kelvin equivalents of the following familiar Fahrenheit temperatures: 0°F, 65°F, 98.6°F, and 451°F, which is the combustion point of books, as in the novel *Fahrenheit 451* by Ray Bradbury. Can you give a metric title for Mr. Bradbury's book? At what temperature are the Celsius and Fahrenheit scales numerically equal? What is absolute zero (0 K) in °F? (The absolute Fahrenheit scale is called the *Rankine* scale, °R.)

15. Estimate the average mechanical energy in joules of a hydrogen molecule at 25°C. What is the corresponding heat energy of 6.02×10^{23} molecules (one mole), in joules and calories?

16. A baseball pitcher throws a 92-mph fastball at a 10.0-L plastic bag filled with water. Assuming the water completely absorbs the kinetic energy of the baseball (mass = 140 g), how much will its temperature rise?

Isotopes and atomic mass

17. From inspection of the periodic table and Table 1.3, what can you say about Rutherford's conclusion that an atom's mass is roughly twice the mass of its protons?

18. Compute the mass defect in amu and grams of the isotope $^{12}_{6}C$. Explain why the atomic mass of carbon is slightly larger than 12.

19. You may have noticed that the isotope masses are very nearly integers equal to the respective mass numbers A. Use this fact to estimate the abundances of the 85 and 87 isotopes of rubidium (Rb) from its atomic mass, assuming these are the only stable ones.

20. Using the data in Table 1.3, find the atomic mass of iron (Fe) and compare it with the value on the periodic chart.

21. The mass spectrum of chlorine gas (Cl_2) shows three peaks at mass numbers 70, 72, and 74. Assign these to combinations of Cl atomic isotopes. Use the abundances in Table 1.3 to predict the relative heights of the three peaks, assuming that the bonding of two Cl atoms is not influenced by the mass number. (*Hint:* If the isotopes were of equal abundance, the ratio of peaks would be 1:2:1.)

22. Marie Curie in 1907 determined the atomic mass of radium (Ra) by converting 403.3 mg $RaCl_2$ to silver chloride AgCl. If the mass of Ag required was 292.6 mg, find the atomic mass of Ra.

Moles and chemical reaction stoichiometry

23. How many carbon atoms are there in 6 g of pure carbon? In 6 g of carbon dioxide, CO_2? In 6 g of sugar, $C_{12}H_{22}O_{11}$?

24. Samples of Zn and S of equal mass are combined to make ZnS, zinc sulfide. Which sample contains more atoms? Which sample limits the amount of ZnS formed? What percentage by mass of the element in excess remains after reaction? (A sometimes useful analogy is the attempt to marry a given mass of fat men to an equal mass of thin women—the fair s-ex will be in ex-s!)

25. Try Exercise 24 again, but using Al, O, and Al_2O_3.

26. Iron is recovered from its ore by heating the ore with charcoal in a process known as smelting. The principal reaction is

$$2Fe_2O_3 + 3C \longrightarrow 4Fe + 3CO_2$$
$$\text{(ore)} \quad \text{(charcoal)}$$

What is the percent iron by mass in the ore? What mass of carbon will be required to smelt a metric ton (1000 kg) of iron ore? How many moles of carbon is this? How many kilograms of iron will result? How many liters of CO_2 at STP?

27. In his landmark work leading to the discovery of the noble gases in the 1890s, W. Ramsay used a reaction he had discovered in earlier work to remove nitrogen from the air. When deoxygenated dry air, obtained by removing

the O_2 by reaction with P (see Exercise 2), is passed over hot magnesium (Mg), the reaction

$$3Mg + N_2 \longrightarrow Mg_3N_2$$

occurs quantitatively, leaving behind a residual gas that proved to be a new element argon Ar (the lazy one).

a. What volume of N_2 in liters at STP can be removed with 18.2 g Mg?

b. Given that air is 78.084 mol % N_2 and 0.934 mol % Ar, what volume of Ar in milliliters at STP can be recovered from the deoxygenated air sample of part (a)?

Formulas of compounds

28. Once atomic masses were known, stoichiometric measurements could be used to determine chemical formulas for new compounds. Because mass ratios alone enable us to carry out stoichiometric calculations, only the *atomic ratios* in the compound (the so-called **empirical** or **simplest formula**) can be established without additional data. Exercises 1 and 2 are early examples. For compounds that do not consist of molecular units, such as metal oxides and salts, one need go no further. When 1.000 g of an oxide of manganese (Mn) was heated in the presence of excess hydrogen gas, 0.632 g Mn metal and an unmeasured amount of water vapor were produced. What is the simplest formula of the oxide? Write a chemical equation that describes the reaction. What mass of water vapor was formed?

29. Most compounds not containing metals consist of molecular units; this is always true of compounds that are gases. A certain volume of a gaseous compound containing only carbon and hydrogen was burned in excess oxygen to yield 1.637 g CO_2 and 0.335 g H_2O. What is the simplest formula of the compound? If the volume of gas used was 0.417 L at STP, what is the **molecular formula,** giving the number of carbon and hydrogen atoms in one molecule of the compound? (The molecular formula must always be an integer multiple of the simplest formula.)

Solution stoichiometry

30. Find the molarity M (mol L^{-1}) of the following solutions:

a. 12.30 g $KClO_3$, potassium chlorate, is dissolved in water to a final volume of 250 mL.

b. 5.78 STP liters of HCl, hydrogen chloride gas, is dissolved in water to make a 500.0-mL solution, a hydrochloric acid solution.

c. 15.0 mL of concentrated HNO_3, nitric acid (65% by mass HNO_3, density 1.40 g mL^{-1}), is diluted with water to 1.000 L.

d. 52.8 mL liquid ethanol, C_2H_5OH, density 0.789 g/mL, is dissolved in enough CCl_4, carbon tetrachloride, solvent to make 750. mL of solution.

31. Give the amounts of the following solutes that will yield 300 mL of 0.500 M aqueous solution:

a. Mass of NaOH, sodium hydroxide

b. Volume of 12.0 M HCl, concentrated hydrochloric acid

c. Volume of NH_3, ammonia gas, at STP

d. Volume of liquid acetone, C_3H_6O, density 0.790 g/mL

32. A common type of aqueous solution reaction is that between hydroxides of metals, such as $Ba(OH)_2$, barium hydroxide, and compounds of hydrogen and nonmetals, such as HCl:

$$2HCl + Ba(OH)_2 \longrightarrow BaCl_2 + 2H_2O$$

This is an example of an *acid–base neutralization reaction;* these will be discussed more fully in Chapter 5. Suppose you mix equal volumes of 0.030 M $Ba(OH)_2$ and 0.075 M HCl solutions. Find the molar concentrations of $BaCl_2$, barium chloride, and of the excess reagent. If the total volume of the mixed solutions is 400 mL, what mass of $BaCl_2$ can be recovered?

33. To ensure *stoichiometric equivalence* (amounts of reagent that exactly satisfy the mole ratio specified by the chemical equation), a technique called **titration** is used, in which one of the reagent solutions is slowly delivered out the bottom of a volume-calibrated tube called a **buret,** and the equivalence point is shown by an **indicator,** a compound that undergoes a color change when the added reagent is ever so slightly in excess. An unknown acid HA reacts with NaOH according to

$$HA + NaOH \longrightarrow NaA + H_2O$$

Find the moles of HA present in a solution if 27.85 mL of 0.1094 M NaOH is required to titrate it.

What is the molecular mass of the unknown acid if the pure acid sample weighed 177 mg prior to being dissolved?

34. The precise molarity of an NaOH solution cannot be determined by directly weighing the solid NaOH because of its tendency to accumulate moisture from the air [NaOH(s) is *hygroscopic*]. Instead, an approximate molarity is prepared and standardized by titrating a known acid, usually potassium hydrogen phthalate $KHC_8H_4O_4$ (or KHP for short). What mass of KHP should be weighed out so that 30.0 mL of 0.100 M NaOH will be needed to titrate it? The reaction is

$$NaOH + KHC_8H_4O_4 \longrightarrow NaKC_8H_4O_4 + H_2O$$

If 32.73 mL were actually required, what is the molarity of NaOH?

35. In the *Kjeldahl method* for analysis of nitrogen in protein, the sample is digested in concentrated sulfuric acid, converting the nitrogen to ammonium sulfate $(NH_4)_2SO_4$. NaOH is then added, converting $(NH_4)_2SO_4$ to ammonia gas NH_3, which is distilled into a standard HCl solution, where the neutralization reaction $NH_3 + HCl \longrightarrow NH_4Cl$ occurs. The unneutralized acid is determined by titration. Find the percent nitrogen by mass in a 5.503-g protein sample if the HCl concentration in 500.0 mL of acid solution is reduced from 0.2000 M to 0.0742 M in a Kjeldahl analysis.

Avogadro's number

36. The density of liquid CCl_4 is 1.594 g/mL. Use Avogadro's number to estimate the diameter in angstroms of a CCl_4 molecule.

37. The diameter of a silicon atom (Si) is 2.72 Å. Estimate the edge length in centimeters of a 1.000-kg cube of silicon, and the density in grams per cubic centimeter of silicon.

38. It takes 1 calorie of heat energy to raise the temperature of 1 g of water by 1°C. How much energy, in joules and electron-volts, is gained on the average by one molecule of water when this occurs?

39. When a mole of water is formed in Reaction 1.1, 68.3 kcal of heat energy is released. How much energy, in joules and electron-volts, is released for each molecule of water formed?

Coulomb's law

40. Compute the Coulomb potential energy (Equation 1.18) between two protons at distances of 0.9 and 1.1 Å in both SI and cgs-esu systems. Find $(V_2 - V_1)/(r_2 - r_1) \equiv \Delta V/\Delta r$ and compare with the inverse-square law of force (Equation 1.17) evaluated at $r = 1.0$ Å. What is the significance of the algebraic signs of V and F?

41. Compute the Coulomb potential energy in electron-volts between the following pairs of charges:

    a. An electron and a helium nucleus at a distance $r = 0.31$ Å

    b. Two protons at $r = 0.74$ Å

    c. A helium nucleus and a gold nucleus at $r = 1.00 \times 10^{-12}$ cm

    d. An electron and a particle of charge $+2.52e$ at $r = 0.84$ Å

42. For case (a) of Figure 1.10, compute the kinetic energy at a few distances in the angstrom range, and plot K versus r. What simple alteration of your plot would enable it to represent case (b) of Figure 1.10?

43. In the Rutherford–Geiger–Marsden experiment scattering α particles ($_2^4$He nuclei) from gold foil, α's of kinetic energy $K = 4.87$ MeV (1 MeV = 10^6 eV) were used.

    a. Compute the distance r between the α and the gold nucleus, assuming a head-on collision, where the α must come to rest, that is, where $K = V$.

    b. Use your result for part (a), along with the general form of the Coulomb potential energy curve, to construct an energy diagram showing E and V versus r for this pair of particles. On your diagram, indicate the point where K has decreased to half its value before collision, and calculate the value of r where this occurs.

44. In Rutherford's model, the electron in a hydrogen atom is in orbit around the proton. Use the Coulomb potential and the result of Exercise 13 for a circular orbit to write the total energy E in terms of the distance r and the angular momentum L. For a distance $r = 2.12$ Å and total energy $E = -3.40$ eV, find V, K, L, and the orbital velocity v and orbital period (an electron's year) $2\pi r/v$. Make an energy diagram showing E and V versus r.

45. In atoms with several electrons, each electron feels forces from the other electrons as well as from the nucleus. The potential energy of an electron is simply the sum of the potential energy terms from each of the other charged particles. Suppose two electrons and a helium nucleus are arranged in an equilateral triangle that is 1.00 Å on each side. What is the potential energy of either electron? What is the total potential energy of the three particles? (Be careful to count each unique potential term only once.) Can you think of a better way (a way of lower total potential energy) of arranging the electrons that preserves the electron–nucleus distances?

46. Compute the charge, in both coulombs and esu, carried by 1 mol of electrons. (The number of coulombs per mole is called *Faraday's constant;* see Chapter 13.)

The Quantum Revolution: The Failure of Everyday Notions to Apply to Atoms

CHAPTER 2

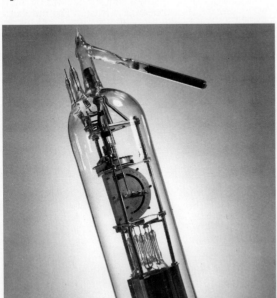

Electron tube used by Davisson and Germer (1927) to discover electron diffraction by metal crystals.

Things on a small scale behave like nothing that you have any direct experience about.

Richard Feynman (1963)

CHAPTER OUTLINE

2.1 The Wave Theory of Light
2.2 The Line Spectra of Atoms
2.3 The Ultraviolet Catastrophe and Planck's Quantum Hypothesis
2.4 The Photoelectric Effect: Particles of Light
2.5 The Nuclear Atom and the Quantum: Bohr's Explanation of Line Spectra
2.6 de Broglie's Matter Waves: The Beginning of a New Mechanics
2.7 The Schrödinger Equation: A Wave Equation for Particles

What today we call science is a relatively young discipline; even so, it has not had a very tumultuous history. Scientific revolutions that completely overturn old ways of thinking have been a rare occurrence—certainly less common than political upheavals—and even the most profound of these never made headline news. Revolutions result when contradictions are brought into a collision course. Scientific contradictions, like other sorts, are not always apparent as they accumulate, owing to limited, perhaps inaccurate, information and to the parochial views of the scientists who gather it. These gathering inconsistencies arise not from the facts themselves, but from the interpretations, dictated by the conventional wisdom of the time, laid on them.

2 The Quantum Revolution: The Failure of Everyday Notions to Apply to Atoms

By the end of the 19th century, an enormous quantity of chemical information had been gathered and systematized, based on the new periodic table of Mendeleyev and Meyer, and the chemical principles described in Chapter 1. The young science of thermodynamics also enabled the analysis of the energetics of chemical reactions and compounds. The atomic theory of Dalton, nearly a century old, was by then widely accepted, and provided a unifying theme for chemistry. On the other hand, the separately evolving discipline of physics had welded the mechanics of Newton, which governed the motions of material objects, with the newer wave theory of light developed by Maxwell, into an apparently sound description of the material world and our own experience with it.

The discovery of the internal "parts" of the atom, that is, that the atom is a tiny mechanism, naturally led to the attempt to describe it physically. This brought previously hidden or ignored inconsistencies to light, and fomented a revolution of scientific thought so profound that both chemistry and physics today still tremble with its reverberations. The famous Danish physicist Niels Bohr, about whom we shall hear more in this chapter, was once asked by a despairing colleague, "Will we ever understand the atom?" Bohr paused in his characteristically thoughtful way, then replied, "Yes, we will, but only if we change what we mean by the word *understand*."

Revolutionary council of physicists, 1927. They carried no placards, and were armed only with new ideas. Max Planck is second from the left in the front row, Niels Bohr is in the second row, far right, Louis de Broglie is two seats to his left, and Erwin Schrödinger is in the center of the third row. Can you pick out Einstein?

In this chapter we attempt a brief description of the basis of the revolution that would bring a man of Bohr's insight and intelligence to make such a statement. We begin when the revolution began, at precisely the turn of the 20th century, with a look at what was known and believed at that critical juncture.

2.1 The Wave Theory of Light

By the end of the 19th century, it was widely accepted that what we call *light* consisted of traveling electromagnetic waves known as **electromagnetic radiation.** These waves all travel or radiate with the same speed c—in a vacuum, $c = 2.99792 \times 10^{10}$ cm s^{-1} (186,000 mi s^{-1})—but have different **wavelength λ.** The wavelength of a wave is the distance between successive crests or troughs, as illustrated in Figure 2.1. The reciprocal of the wavelength is called the **wave number $\bar{\nu}$**, usually given in inverse centimeters, and represents the number of waves that fit into a 1-cm length. If the wave is traveling with a speed c, then there will be a **frequency ν** (cycles s^{-1} or hertz, Hz) with which successive crests pass by, given by $\nu = c/\lambda$. The frequency is the number of crests passing a point in 1 s. In Maxwell's model, the frequency ν is that of a vibrating charge in the light-emitting object, and the wavelength λ is determined by $\lambda = c/\nu$.

The sort of mechanical motion we call *vibration* corresponds to standing or stationary waves, such as those set up when a bow is drawn across a cello string. Normally light or electromagnetic radiation is always traveling, but if two mirrors are set up to form a cavity (as in a **laser**), light can also be trapped to form standing waves.

EXAMPLE 2.1

A laser is a source of light that emits the purest, brightest color of any device known. The word *laser* is an acronym standing for *l*ight *a*mplification by *s*timulated *e*mission of *r*adiation. A gas laser is just a discharge tube with mirrors on each end. One of the mirrors is only partially reflecting, to allow some light to escape. The earliest gas laser (invented in 1961) is the helium-neon laser, which emits red light of wavelength $\lambda = 6.328 \times 10^{-5}$ cm (632.8 nm); it came into widespread use in barcode scanning. Calculate the frequency and wave number of the laser light.

Solution:

$$\nu = c/\lambda = (2.998 \times 10^{10} \text{ cm s}^{-1})/(6.328 \times 10^{-5} \text{ cm})$$
$$= 4.738 \times 10^{14} \text{ cycles s}^{-1} = 4.738 \times 10^{14} \text{ Hz}$$
$$\bar{\nu} = 1/\lambda = 15{,}800 \text{ cm}^{-1}$$

Note that the units of Hz (cycles s^{-1}) arise because the wavelength is implicitly centimeters per cycle. As we will see in Section 2.2, the energy of light is inversely proportional to the wavelength, and thus directly proportional to the wavenumber. Because the wavenumber involves no fundamental constants whose values might change, it has become the practice among spectroscopists to express energies in terms of wavenumbers. **Spectroscopy** is the name given to the study of the interaction of light and matter; both chemists and physicists engage in this study.

Figure 2.1
Wave characteristics. (a) Definition of wavelength λ and wave amplitude A and examples. (b) Illustration of a traveling wave. The colored portion moves to the right; when its position has changed by λ, the period τ, frequency ν, and speed of travel c can be defined. (c) The electric (**E**) and magnetic (**B**) waves that comprise light. According to Maxwell, these are broadcast by a vibrating charge at the point of origin.

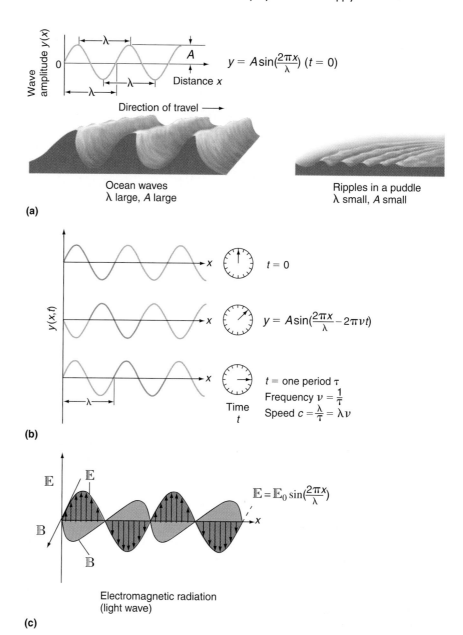

The mathematical function describing a traveling wave is a **sine function**,

$$y(x, t) = A \sin\left(\frac{2\pi x}{\lambda} - 2\pi\nu t\right) \tag{2.1}$$

where t is the time, and the wave is propagating along the x axis with an **amplitude** A, the height of a crest above the axis. Note that the wave oscillates between positive and negative values. Light has both an **oscillating electric field E** and **magnetic field B**. For the interaction of light with atoms and molecules, composed as they are of charged particles, it is mainly the electric field with which we are concerned. The

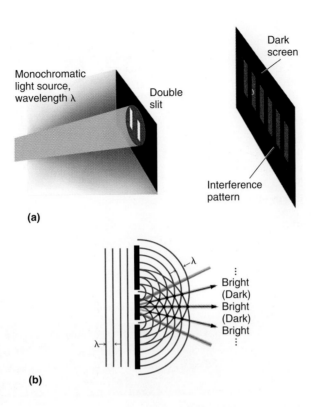

Figure 2.2
Diffraction interference of light waves. (**a**) In an experiment first performed by Thomas Young in 1801, light of selected wavelength shines through a pair of closely spaced slits, producing alternating light and dark bands on a screen. (**b**) Huygens' wave construction interprets the pattern as a result of interference between waves diffracted through each slit. The concentric semicircles represent the crests of diffracted wavelets; where these intersect the waves interfere constructively, producing the bright bands.

radiant intensity (light energy falling on a given area in unit time) is, according to the wave theory, proportional to $\mathbf{E}_0^2$, the square of the wave amplitude; that is, brighter light sources have larger amplitudes.

Diffraction and Interference of Light Waves

The main evidence for the validity of the wave theory of light came from **diffraction experiments** of the sort illustrated in Figure 2.2. When light of a particular wavelength λ shines through a pair of closely spaced slits (or is reflected from the two jaws of a single slit), a **diffraction interference pattern,** consisting of alternate bright and dark bands, may be observed on a screen. If light is a wave, then the crests and troughs of the waves coming from one slit will alternately reinforce or cancel those from the other slit where they intersect, and the interference pattern results. The cancellation of the two waves occurs when the trough of one wave coincides with the crest of the other. We say that the waves are "out of phase" and call the phenomenon **destructive interference.** The reinforcement of waves, which produces the bright places on the screen, is called **constructive interference.**

Note that the wave picture, along with the mathematical function describing the wave, is very much a part of our daily experience. Even the interference effects can be simulated with water-filled ripple tanks, or just by two stones dropped side by side in a puddle. Recreational "wave pools" always have regions of complete calm, due to destructive interference between two wave sources.

Figure 2.3
The spectrum of electromagnetic radiation. The slanted lines in the upper panel indicate overlap between spectral regions, and the lower panel details the visible region. The energies are assigned in accord with Planck's quantum hypothesis $E = h\nu$, where $h = 6.63 \times 10^{-34}$ J s and 1 eV = 1.602×10^{-19} J.

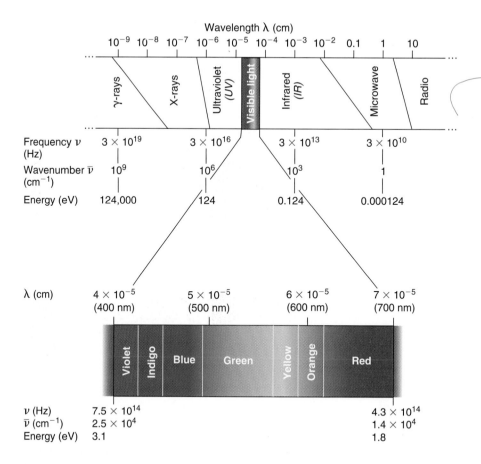

The Electromagnetic Spectrum

By the turn of the century we were also coming to realize that visible light, the electromagnetic radiation that our eyes can respond to, possesses a very small range of wavelengths, between $\lambda = 4 \times 10^{-5}$ cm and 7×10^{-5} cm, compared to the seemingly limitless range of wavelengths, from meters to 10^{-9} cm, that was then being discovered. This enormous span became known as the **electromagnetic spectrum** (the word *spectrum* comes from Newton), and is diagrammed in Figure 2.3. In the X-ray region of the spectrum, $\lambda \sim 1$ Å, or of the same order as the spacing between atoms in a crystalline solid. As we will see in Chapter 14, and will refer to later in this chapter, the rows of atoms in a crystal can act like a set of slits, and produce **X-ray diffraction** when a beam of X-rays shines on a crystal.

2.2 The Line Spectra of Atoms

Although the wave theory of light appeared at that time to be a very satisfactory description, the seeds of its destruction had already been sprouting for more that 80 years! As early as 1814, Josef Fraunhofer, a German lensmaker, had observed a series of dark lines in the spectrum of sunlight (see Figure 2.4). Using a narrow slit and a **diffraction grating** (glass ruled with a series of parallel scratches) he was able to index the wavelengths of about 700 of these lines. Their origin, however, could not be fathomed at the time.

2.2 The Line Spectra of Atoms

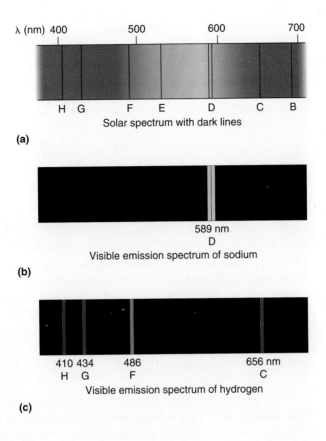

Figure 2.4
Fraunhofer's solar spectrum with dark lines (**a**) is compared to the flame spectrum of a sodium salt (**b**) and the gas-discharge spectrum of hydrogen (**c**). The fainter lines in (**a**) are not shown. The sodium "D-line" is actually a closely spaced doublet and will be discussed further in Chapter 4. The lines in (**a**) that do not correspond to those in (**b**) and (**c**) are due to helium.

Forty-seven years later Gustav Kirchhoff and Robert Bunsen undertook an extensive series of studies of the colored glows produced by various salts vaporized in a natural gas burner (now known as a Bunsen burner). These **flame spectra** each consisted of a series of bright lines characteristic of the elements present in the salt, with some of the wavelengths coinciding exactly with Fraunhofer's dark lines. Kirchhoff concluded that his bright lines were *emitted* by atoms of the elements, while those same elements, present in the sun's outer atmosphere, were *absorbing* sunlight at the same wavelengths to produce the dark lines. As illustrated in Figure 2.4, a flame of sodium chloride NaCl shows a strong **emission line** at $\lambda = 589$ nm (5.89×10^{-5} cm) due to the sodium atom (any sodium salt shows the same line), which coincides with Fraunhofer's "D-line," the corresponding **absorption line** of Na.

The uniqueness of the spectrum of an element led to the discovery of many new elements, as well as to the identification of elements in the sun and more distant stars. The element helium was discovered in the solar atmosphere before it was found here on earth! Other weaker and more numerous lines were also observed that turned out to be due to molecules.

The Hydrogen Atom Spectrum

The lightest element, hydrogen, turned out to have a particularly pleasing and intriguing spectrum, illustrated in Figure 2.5. The appearance of distinct groups (**series**) of lines, each with wavelengths smoothly converging to a limiting value, suggested a mathematical relationship of some sort. A high school math teacher named Balmer

Figure 2.5
The spectrum of atomic hydrogen. Each series of lines in the atomic hydrogen spectrum occurs in separate regions. As drawn, the line spacings are wider for the series at longer wavelength, but the wavelength scales for the different series are not identical.

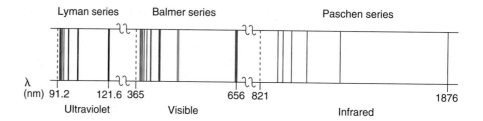

discovered in 1885 that the wavelengths of the visible series of hydrogen, which now bears his name, could be represented to uncanny accuracy by the formula

$$\frac{1}{\lambda} = R_H \left(\frac{1}{4} - \frac{1}{n^2} \right), \qquad n = 3, 4, 5, \ldots \qquad (2.2)$$

where R_H is a constant (the **Rydberg constant;** modern value 109,677.58 cm^{-1}) and n is an index for the lines in the Balmer series, beginning with the longest wavelength, $\lambda = 656.469$ nm, assigned to $n = 3$, the next to $n = 4$, etc. Later Rydberg showed that all the series could be represented by the following formula (the **Rydberg formula,** of course!):

$$\frac{1}{\lambda} = R_H \left(\frac{1}{n_1^2} - \frac{1}{n_2^2} \right), \qquad n_1 = 1, 2, 3, \ldots; \quad n_2 = n_1 + 1, n_1 + 2, n_1 + 3, \ldots \qquad (2.3)$$

Different values of n_1 label different series—setting $n_1 = 1$ yields the ultraviolet **Lyman series,** $n_1 = 2$ the **Balmer series,** etc.—while, as in the Balmer case, n_2 generates the various lines within a series. Rydberg predicted the existence of further series (larger n_1) in the infrared; when these were discovered, their wavelengths were found to be given "exactly" by the Rydberg formula.

Why such "magic formulas" existed was quite a mystery at the time; the line spectra of heavier elements followed no such simple pattern. However, the special case of hydrogen provided a crucial piece of the puzzle of atomic structure, as we will see later.

EXAMPLE 2.2

The line of longest wavelength in the ultraviolet Lyman series, the so-called **Lyman-α** line, has been used extensively in photoionization experiments and in detection of hydrogen atoms in chemical reactions. Compute the wavenumber, wavelength, and frequency of Lyman-α to six significant figures from the Rydberg formula, Equation 2.3.

Solution:
The Lyman series corresponds to $n_1 = 1$, and the longest wavelength in the series occurs for the smallest n_2, $n_2 = 2$. So

$$\frac{1}{\lambda} = \bar{\nu} = 109{,}677.6 \text{ cm}^{-1} \left(\frac{1}{1} - \frac{1}{4} \right) = 82{,}258.2 \text{ cm}^{-1}$$

$$\lambda = 1.21568 \times 10^{-5} \text{ cm} = 121.568 \text{ nm}$$

$$\nu = c/\lambda = c\bar{\nu} = (2.99792 \times 10^{10} \text{ cm s}^{-1})(82{,}258.2 \text{ cm}^{-1})$$

$$= 2.46604 \times 10^{15} \text{ Hz}$$

Note that the limiting short wavelength of each series is obtained by taking the limit $n_2 \rightarrow \infty$; for Lyman this yields $\lambda \rightarrow 1/R_H = 91.1763$ nm. This example could also be worked in the opposite direction: a given wavelength can be used to find the quantum numbers n_1 and n_2. The Rydberg formula is strictly valid only for the hydrogen atom. Although analogous formulas can be derived for some of the shorter wavelengths in the spectra of other atoms, these are beyond the scope of our introduction. You should *not* at present try to use Equation 2.3 for atoms other than H.

Given the evidence then accumulated from dozens of experiments, such as those we described in Chapter 1, that atoms are composed of charged particles, it was naturally thought that these line spectra must have to do with some type of oscillations of the charges, but nothing more detailed than this could be derived. The concept that would eventually solve the riddle of atomic spectra (and simultaneously undermine the wave theory of light) was to come from observations of a different sort.

2.3 The Ultraviolet Catastrophe and Planck's Quantum Hypothesis

James Clerk Maxwell in the 1860s had given a complete mathematical theory for electromagnetic radiation, and physicists over the next 30 years were calmly working out its consequences. It came as quite a shock to the scientific community when it was found that *the wave theory failed to explain why a red-hot iron bar is red!* Metals and other solids with high melting points glow when they are heated, a familiar sight in electric ranges, toasters, and space heaters. The ideal radiation from a perfect emitter and absorber of radiation is called **blackbody radiation.** As an object is heated to higher and higher temperatures, its glow changes from red to orange to yellow and finally to "white heat." In the late 1890s the German physicists Lummer and Pringsheim had precisely measured the spectrum of blackbody radiation and showed that this changing color reflects a *continuous* spectrum with a peak wavelength that becomes shorter and shorter as the temperature increases (see Figure 2.6).

The problem was not with this quantitative measurement of a common observation, but with the wave theory's prediction that *all heated objects should glow with a violet color,* which should become more intense with rising temperature! Quantitatively, as shown in Figure 2.6, the theory predicts that the spectral intensity should increase without bound as the wavelength goes to zero. This result was regarded with such dismay that it was called the *ultraviolet catastrophe.*

The Quantum of Energy

Max Planck, a young professor of physics at Berlin, set out to explain the data of Lummer and Pringsheim in the late 1890s. He had become a physicist (instead of a musician) despite warnings that this was a closed subject in which new discoveries were scarcely to be expected. At a meeting of the Berlin Physical Society on October 19, 1900, Planck presented a new theory of blackbody radiation. His theoretical expression for the blackbody spectrum fit the data of Lummer and Pringsheim very closely (see Figure 2.6; Appendix A gives a detailed account of formulas arising from both the classical wave theory and Planck's new theory that

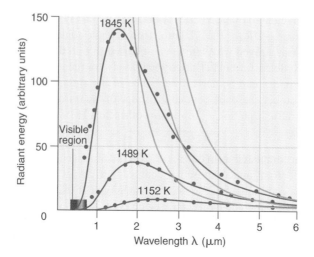

Figure 2.6
Blackbody radiation spectra at three temperatures of Lummer and Pringsheim (points) compared with the predictions of the wave theory (green curves) and Planck's quantum theory (blue curves). Appendix A presents the actual functions used to plot the curves. The wavelengths of the peaks in the spectra are inversely proportional to the absolute temperature. The wave theory curves have been attenuated to agree with experiment at the right-hand end of the graph.

yield the curves shown in the figure). But in order to achieve this, Planck had to discard a hallowed tenet of classical physics: *Natura non facit saltum*—"Nature does not make jumps."

According to the electromagnetic theory of Maxwell, the frequency of the light given off by the hot object is dictated by the allowed frequencies of standing light waves within the object. The *amount* of radiant energy emitted at a particular wavelength then depends only on the *number* of waves with that frequency. The high-frequency waves could easily be shown, using the wave theory, to outnumber greatly those of lower frequency (see Appendix A). Planck's new theory implied that the radiant energy of the light comes in discrete packets or bundles called **quanta,** with the light energy ε proportional to the frequency:

$$\varepsilon = h\nu \tag{2.4}$$

The proportionality constant $h = 6.6261 \times 10^{-34}$ J s (modern value) is now known as **Planck's constant.** Figure 2.3 also indicates the energies in electron-volts corresponding to the electromagnetic spectrum.

EXAMPLE 2.3

Use Planck's hypothesis, Equation 2.4, to compute the energy in joules and electron-volts of the Lyman-α spectral line of Example 2.2.

Solution:

$$\varepsilon = h\nu = hc/\lambda$$
$$\lambda = 121.568 \text{ nm}$$
$$\varepsilon = (6.6261 \times 10^{-34} \text{ J s})(2.9979 \times 10^{8} \text{ m s}^{-1})/(121.57 \times 10^{-9} \text{ m})$$
$$= (1.9865 \times 10^{-25} \text{ J m})/(121.57 \times 10^{-9} \text{ m})$$
$$\varepsilon \text{ (J)} = 1.6340 \times 10^{-18} \text{ J}$$
$$\varepsilon \text{ (eV)} = 1.6340 \times 10^{-18} \text{ J}/(1.6022 \times 10^{-19} \text{ J eV}^{-1}) = 10.199 \text{ eV}$$

The important clue that this simple yet profound calculation gives us is that the energy of a spectral line "quantum" or "photon" (see the description of the photoelectric effect further on) is comparable to the mechanical energy of an electron and proton at atomic distances, as shown in Section 1.7.

This quantum postulate averted the ultraviolet catastrophe as may be seen using the following analogy. The energy packets or quanta may be thought of as stones of various sizes, which must be thrown out of the solid. The low-frequency waves correspond to small pebbles of light that are easily cast away, but according to Equation 2.4 the higher frequencies must heave out huge boulders of light. The temperature of the solid dictates the average energy available for radiation, $\bar{\varepsilon} \approx k_B T$ (see Chapter 1), and the chance that energy could be gathered into chunks much larger than this is slight. Hence, the higher frequency waves cannot radiate as effectively, and the spectrum falls to zero at short wavelength. Appendix A puts these ideas into mathematical terms.

The notion that the energy of electromagnetic radiation is present in the form of discrete bundles became known as the **quantum hypothesis,** and later the **quantum theory of radiation.** When a hot solid emits light of a certain frequency, it must also lose a well-defined amount of energy, it being impossible to lose any other intermediate amount. Thus, in Planck's own words, "Nature indeed makes jumps, and very extraordinary ones!"

It's important to realize that Planck made his hypothesis in response to precise experimental observation—it was not "a bolt out of the blue." Steeped as he was in the theories and traditions of classical physics, Planck was uncomfortable with the quantum idea for several years thereafter. Only after corroboration of his hypothesis through the interpretation of other types of experiments was he able to accept it fully. His work began an avalanche of new theories and experiments that was to bury 19th-century physical notions concerning light and the atom and pave the way for modern physics, chemistry, and biology.

2.4 The Photoelectric Effect: Particles of Light

The year 1905 was a remarkable one for a young patent examiner in Switzerland named Albert Einstein. In that year Einstein published three landmark papers in the German periodical *Annalen der Physik* (*Annals of Physics*) on three distinct topics: the theory of relativity, the phenomenon of Brownian motion, and the **photoelectric effect.** This last showed how Planck's quantum hypothesis provided a natural explanation for the phenomenon in which electrons are ejected from a metal surface when light shines on it.

Einstein's interpretation was based mainly on the experiments of Heinrich Hertz in the 1880s. The main features of the photoelectric effect were as follows:

1. Red light, or other wavelengths on the long-wavelength end of the spectrum, could not cause electron emission, no matter how intense, whereas blue light or other short-wavelength radiation could, no matter how weak.
2. The kinetic energy of the "photoelectrons" depended on the wavelength of the light, but not its intensity.

Figure 2.7
Measurements of the kinetic energy of photoelectrons versus frequency of incident light for a sodium metal surface made by Robert Millikan. The slope of the line is, according to Equation 2.5, Planck's constant h, and the best line through the data gave a value of h close to the modern one.

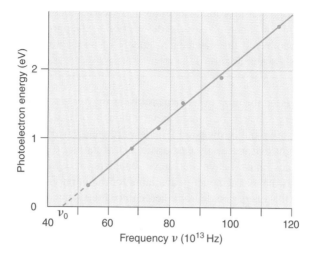

Both of these features were in conflict with the electromagnetic wave theory, in which the energy of the light was proportional to its intensity only. Sufficiently intense red light should have caused electron ejection, and the kinetic energy of the photoelectrons should have increased with the light intensity. Einstein invoked Planck's energy quanta, arguing that if one thought of the quanta as little *particles* of pure energy, later to be dubbed **photons,** then the electron emission process could be pictured as a collision between a photon and an electron embedded in the metal. In the collision the photon was destroyed, its energy, given by the Planck formula $\varepsilon = h\nu$, being imparted to the electron to free it from the metal. Einstein proposed the **photoelectric equation:**

$$K = \begin{cases} h(\nu - \nu_0), & \nu > \nu_0 \\ 0, & \nu \leq \nu_0 \end{cases} \tag{2.5}$$

where K is the final kinetic energy of the free electron, h is Planck's constant, ν is the frequency of the incident light ($\nu = c/\lambda$), and ν_0 is a "threshold" frequency below which no electrons can be ejected. The energy $h\nu_0$ is the minimum needed to free an electron from the metal, and is known as the **work function** W. (Actually the electrons emerge with a distribution of kinetic energies for $\nu > \nu_0$, and Equation 2.5 represents the maximum possible K.) A decade later Robert Millikan, the discoverer of the charge on the electron (see Chapter 1), published extensive measurements that confirmed Equation 2.5. Some of his data are shown in Figure 2.7.

EXAMPLE 2.4

Equation 2.5 predicts, and Figure 2.7 illustrates, that a plot of the kinetic energy of photoelectrons versus frequency ν is a straight line with slope h and ν-intercept ν_0. Use two data points from Millikan's results [($\lambda = 546.1$ nm, $K = 0.47$ eV) and ($\lambda = 312.6$ nm, $K = 2.13$ eV)] to obtain a value for Planck's constant h and the work function W for sodium metal.

Solution:
First convert the K's into SI (or cgs) units:

$$K_1 = (0.47 \text{ eV})(1.60 \times 10^{-19} \text{ J eV}^{-1}) = 7.52 \times 10^{-20} \text{ J}$$
$$K_2 = (2.13 \text{ eV})(1.60 \times 10^{-19} \text{ J eV}^{-1}) = 34.1 \times 10^{-20} \text{ J}$$

and the λ's into ν's:

$$\nu_1 = c/\lambda_1 = (2.998 \times 10^8 \text{ m s}^{-1})/(546.1 \times 10^{-9} \text{ m}) = 5.490 \times 10^{14} \text{ Hz}$$
$$\nu_2 = c/\lambda_2 = (2.998 \times 10^8 \text{ m s}^{-1})/(312.6 \times 10^{-9} \text{ m}) = 9.591 \times 10^{14} \text{ Hz}$$

The slope of the line is

$$h = \frac{K_2 - K_1}{\nu_2 - \nu_1} = \frac{34.1 - 7.5}{9.591 - 5.490} \times 10^{-34} \text{ J s} = 6.49 \times 10^{-34} \text{ J s}$$

and $W = h\nu_0$ is

$$W = h\nu_2 - K_2 = (6.49 \times 10^{-34} \text{ J s})(9.591 \times 10^{14} \text{ Hz}) - 3.41 \times 10^{-19} \text{ J}$$
$$= 2.81 \times 10^{-19} \text{ J} = 1.76 \text{ eV}$$

Using all his data Millikan found $h = 6.56 \times 10^{-34}$ J s (he expressed it as 6.56×10^{-27} erg sec) and $W = 1.80$ eV.

Now the quantum concept had been shown to be a natural explanation for two separate phenomena related to the interaction of light with matter. While some scientists of the day had regarded Planck's hypothesis as somewhat *ad hoc*—perhaps a different modification of the wave theory of light might work as well—Einstein's analysis of the photoelectric effect left little doubt as to the reality of the quantum. In fact, now one had to imagine "particles of light," *photons*. But the wave theory of light could not simply be discarded; the phenomena we have described could only be understood if light possesses wave-like properties as well. In diffraction experiments light behaves as a wave, whereas in blackbody radiation and photoelectric emission light behaves as a particle. Is a beam of light a wave or a stream of particles? It is neither and it is both; the question presupposes that light can be understood in everyday terms. This paradox became known as **wave-particle duality.**

Here we have the first inkling that nature on a small scale does not behave "sensibly," and that, like Niels Bohr, we are going to have to "change what we mean by the word *understand*" if we want to appreciate the world of atoms.

2.5 The Nuclear Atom and the Quantum: Bohr's Explanation of Line Spectra

The next step in the unfolding of the nature of atoms was to connect the antics of individual gas-phase atoms with the quanta of light. The breakthrough was made by a young postdoctoral student, a Dane by the name of Niels Bohr, working in Ernest Rutherford's laboratory in 1913. Rutherford had published his nuclear model for the atom 2 years earlier (see Chapter 1), and Bohr saw the essence of what was needed to reconcile the nuclear atom with the old puzzle of line spectra, again using Planck's quantum hypothesis.

Bohr had to postulate that *an atom could only exist in certain allowed states of specific total energy E* (kinetic plus potential energy). He called these **stationary states,** because the atom could not gradually lose energy through radiation as required by Maxwell's theory (see Chapter 1). If an atom happened not to be in its state of lowest total energy, its **ground state,** then it could make a downward jump, a **transition,** to a state of lower energy, and in the process liberate a quantum of light, a photon. If we call the energy of the upper state E_u, and the lower state E_l, then

$$E_u - E_l = \Delta E = h\nu \tag{2.6}$$

where Δ means "change in" and ν is the frequency of the newly created photon. Planck's ubiquitous constant h appears again in Equation 2.6 to connect the frequency of light, a wave property, with its role as an energy quantum. Bohr's postulate of discrete stationary states with fixed energy levels leads naturally to lines in the spectrum, as well as to the coincidence of the dark lines of an absorption spectrum for a particular element with the bright emission lines. In the case of absorption, light of the appropriate frequency can cause an upward transition between the same two energy levels, and Equation 2.6 again applies, leading to a dark line at the same wavelength as in emission. Figure 2.8 illustrates Bohr's idea for the D-line of sodium.

EXAMPLE 2.5

Bohr's hypothesis implies that in Example 2.3 on the Lyman-α line of hydrogen, what was calculated was not only the energy of Lyman-α photons, but also the difference in energy of two stationary states of the H atom. Helium has an intense spectral line lying even further into the ultraviolet at 58.435 nm. This so-called He I resonance line has been widely used in the technique of *photoelectron spectroscopy,*

Figure 2.8
Bohr's stationary state hypothesis for explaining absorption and emission line spectra in the case of the D-line of sodium. As noted in Figure 2.4, the D-line is actually a closely spaced doublet. The labels 3s and 3p on the stationary states are taken from the orbital model for many-electron atoms discussed in Chapter 4. In this case $\lambda = 589$ nm and $\Delta E = 2.11$ eV.

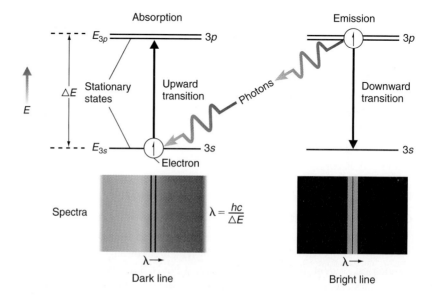

the generalization of Millikan's photoelectric experiment. Calculate the energy-level spacing in joules and electron-volts in helium corresponding to this line.

Solution:
The calculation is the same as in Example 2.3, but the interpretation of the result is embellished.

$$\Delta E = hc/\lambda = (1.9865 \times 10^{-25} \text{ J m})/(58.435 \times 10^{-9} \text{ m})$$
$$= 3.3995 \times 10^{-18} \text{ J}$$
$$= 21.218 \text{ eV}$$

Note that this line is of shorter wavelength and thus higher energy than any line in the hydrogen spectrum. This is consistent with the greater magnitude of the potential energy of the electrons in helium due to the doubling of the nuclear charge (see Figure 1.9). The ΔE you have calculated corresponds to the energy gap between $1s$ and $2p$ energy levels (see Chapters 3 and 4).

This was a giant step beyond Planck's quantum of light. Now the atom itself was "quantized." That is, the rules that govern the motions of the electrons within the atom must lead in some way to Bohr's stationary states. The trouble with this was that Newton's laws of motion could never constrain the atom, as a mechanism, to possess only certain discrete energies, any more than they could restrict the motions of the planets and moons in such a way. In the next dozen years, Bohr and an older colleague, Arnold Sommerfeld, searched for the new rules obeyed by atoms, knowing that, if and when such rules were found, the world of Newton as well as of Maxwell would be overthrown.

Bohr proposed a theory of the hydrogen atom spectrum in which the stationary states for the simplest atom correspond to circular orbits of various fixed radii of the single electron.* We will present a hybrid version of Bohr's theory later because it contains a number of important ideas; for now we note that he was able to derive the Rydberg formula, Equation 2.3, and showed that the Rydberg constant R_H was actually a composite of other constants; in cgs-esu

$$R_H = \frac{2\pi^2 m_e e^4}{h^3 c} \tag{2.7}$$

where m_e is the mass of the electron, e the charge on the electron in esu, h is again Planck's constant, and c the speed of light. This formula gave R_H to better than 0.1%, using the best values of the constants e, h, and c then known, and Bohr's theory won him wide acclaim. Equation 2.7 turns out to be a nearly exact formula if the mass of the electron m_e is replaced by the "reduced mass" μ of the electron–proton pair:

$$\mu = \frac{m_e}{1 + \dfrac{m_e}{m_p}} = 0.999456\, m_e \tag{2.8}$$

---

* Bohr postulated the *quantization of angular momentum L* as a criterion for fixing the radii of the orbits. Although the idea that L is quantized was correct, the values Bohr obtained for L were not. See Exercise 1.13 for a discussion of angular momentum; also see Exercise 29 at the end of this chapter.

Unfortunately, Bohr's model could not be made to work for atoms with more than one electron, although it could be extended to one-electron ions such as He^+ and Li^{2+}, as we will see later.

2.6 de Broglie's Matter Waves: The Beginning of a New Mechanics

Maxwell's wave theory of light was a relatively new theory in 1913, and chinks in its armor were to be expected. But Newton's mechanics had stood the test of almost 250 years of application without failure, and the possibility that it might be invalid for the tiny mechanical system of the atom was unthinkable to most scientists of the time. Those physicists working on atomic structure knew that the atom was "strange," but were unable to see how to modify Newton's laws to explain it. What was required was a step back, away from particles in orbits, in order to replace rather than modify the classical laws. That refreshing yet perplexing step was taken by a French nobleman-turned-physicist, the duke Louis de Broglie, in 1924.

de Broglie was struck by the appearance of integers in the Rydberg formula (Equation 2.3). The only place where integers entered naturally in physical problems was in constrained wave motion, as in a vibrating violin string. As illustrated in Figure 2.9, a string tied down at its ends can only vibrate with certain wavelengths λ, which satisfy $n(\lambda/2) = L$, where λ is the wavelength of a standing-wave vibration of the string, L the length of the string, and n is an integer, where $n = 1$ yields the fundamental tone of the string, and $n = 2, 3, \ldots$ corresponds to overtone vibrations.

This suggested to de Broglie that perhaps some sort of wave motion was occurring within the atom. Could the electron, widely acknowledged until then as a particle, have wave-like properties? It would certainly ascribe a pleasing if baffling symmetry to nature: light, formerly thought of as a wave, could behave as a particle, and electrons, formerly thought of as particles, were now behaving as waves! de Broglie sought a way of defining a wavelength for a particle. For light he had Planck's formula $E = hc/\lambda$ and the now-famous Einstein relation $E = mc^2$, where,

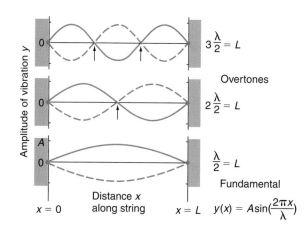

Figure 2.9
The three simplest allowed standing wave vibrations of a plucked string of length L. Each waveform obeys the equation $n(\lambda/2) = L$, with $n = 1, 2, 3, \ldots$, and λ the wavelength of the vibration. The arrows indicate the nodes of the waves, places where the waves change sign. The constraint on λ results from tying down the ends of the string.

2.6 de Broglie's Matter Waves: The Beginning of a New Mechanics

if applied to a photon, m is the "equivalent" (not actual) mass of the photon. Setting these two E's equal, de Broglie found

$$\lambda_{\text{photon}} = \frac{h}{mc}$$

So the obvious choice for the wavelength of a material particle would be

$$\lambda_{\text{particle}} = \frac{h}{mv} \qquad (2.9)$$

where now m is the actual mass of the particle and v its velocity. de Broglie had electrons in mind, but he realized that Equation 2.9 had to apply equally well to electrons, nuclei, dust motes, baseballs, boulders, and planets. We now call the wavelength calculated from Equation 2.9 the **de Broglie wavelength.** Table 2.1 lists de Broglie wavelengths for a range of particles. Because of the small size of Planck's constant, Equation 2.9 predicts minuscule wavelengths for all but the very lightest objects.

de Broglie proposed these "matter waves" in his Ph.D. thesis in 1924. Just 3 years later, a striking confirmation of de Broglie's hypothesis was provided by the nearly simultaneous discovery by Davisson and Germer in the United States and by G. P. Thomson (son of J. J.) in England of the phenomenon of **electron diffraction.** When a beam of electrons of appropriate velocity impinges on a crystalline powder or foil, bright and dark rings of scattered electron intensity are found. Figure 2.10 compares photographic exposures of X-rays and electrons scattered from aluminum foil. The bright rings of electrons can only be interpreted (in everyday terms!) by assuming wave-like constructive interference, and the spacings between the rings are precisely predicted from de Broglie's formula, Equation 2.9.

TABLE 2.1

de Broglie wavelengths for various moving objects

Object	Mass (g)	Wavelength (Å)
1-Volt electron	9.11×10^{-28}	12.3
10-Volt electron	9.11×10^{-28}	3.88
100-Volt electron	9.11×10^{-28}	1.23
Helium atom at room temperature	6.65×10^{-24}	0.73
α particle from radium	6.65×10^{-24}	0.000066
Average protein molecule	6.64×10^{-20}	0.0073
Floating chalk dust	$\sim 10^{-6}$	$\sim 6.6 \times 10^{-13}$
Driven golf ball	45	4.9×10^{-24}
Pitched baseball	140	1.9×10^{-24}
Chemistry professor, pacing	8×10^{4}	8.3×10^{-26}
Car at 30 mph	9×10^{5}	5.5×10^{-28}
Earth in orbit around sun	6×10^{27}	3.7×10^{-53}

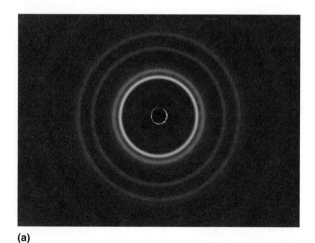

(a) (b)

Figure 2.10
Comparison of the scattering of (**a**) X-rays and (**b**) electrons by aluminum foil. The striking similarity of the photographs, whose bright rings can be explained using a wave model, is persuasive evidence of the correctness of de Broglie's postulate of matter waves.

EXAMPLE 2.6

In Davisson and Germer's discovery of electron diffraction (which occurred accidentally when a shattered glass vacuum tube was too hastily repaired), they generated a beam of electrons from a wire cathode heated to incandescence. The accelerating voltage was 54 V, giving an electron energy of 54 eV. Calculate the de Broglie wavelength of the electrons in the beam.

Solution:
Because $K = \frac{1}{2} mv^2$, $v = (2K/m)^{1/2}$ and

$$\lambda = \frac{h}{m\sqrt{2K/m}} = \frac{h}{\sqrt{2mK}}$$

In SI units,

$$K = (54 \text{ eV})(1.60 \times 10^{-19} \text{ J eV}^{-1})$$
$$= 8.64 \times 10^{-18} \text{ J}$$

$$\lambda = \frac{6.626 \times 10^{-34} \text{ J s}}{[2(9.109 \times 10^{-31} \text{ kg})(8.64 \times 10^{-18} \text{ J})]^{1/2}}$$
$$= 1.67 \times 10^{-10} \text{ m} = 1.67 \text{ Å}$$

Because this value is comparable to the spacing between nickel atoms in the metal foil target Davisson and Germer used, a strong diffraction effect was seen, with pronounced maxima and minima in scattered electron current detected as the angle with respect to the beam was increased. Thomson originated the use of photographic detection of the scattered electrons, as depicted in Figure 2.10.

The Bohr–de Broglie Model of the Hydrogen Atom

Whereas Bohr had to make an arbitrary assumption of quantized circular orbits to arrive at a workable model for the H atom, de Broglie could offer a more logical derivation. Each stationary state, according to de Broglie, corresponds to a **standing-wave**

2.6 de Broglie's Matter Waves: The Beginning of a New Mechanics

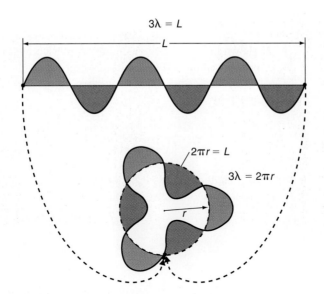

Figure 2.11
The similarity between the vibration of a string and the ring that results when the ends of the string are tied together. The ring represents the path of an orbiting electron in Bohr's atomic model. The standing-wave condition imposed by tying the ends requires the wave to meet itself "in phase" after 360° is swept out. This is a special case of Equation 2.10 with $n = 3$.

vibration of the electron wave. Bohr's orbits become the paths around which the wave may vibrate, and only certain orbits fulfill the standing-wave condition, that an integral number (n) of wavelengths "fit" into the circumference of the orbit:

$$n\lambda_e = 2\pi r, \quad n = 1, 2, 3, \ldots \quad (2.10)$$

where r is the radius of the orbit. This is quite analogous to the case of the vibrating string, except that here the ends of the string are tied together. The analogy is illustrated in Figure 2.11. Of course, λ_e is given by de Broglie's Equation 2.9, and using Newton's second law of motion for a circular orbit governed by the Coulomb force, the allowed energies of the H atom can be obtained. Newton's $F = ma$ becomes, in cgs-esu,

$$-\frac{e^2}{r^2} = m_e\left(-\frac{v^2}{r}\right) \quad (2.11)$$

where all quantities have been defined earlier, and $-v^2/r$ is the centripetal acceleration. Equations 2.9, 2.10, and 2.11 form a system of three equations in the unknowns λ_e, v, and r and are easily solved. When these are substituted into the total energy

$$E = K + V = \tfrac{1}{2}m_e v^2 - \frac{e^2}{r} \quad (2.12)$$

where we have used the result for the Coulomb potential energy from Chapter 1, a formula for E is obtained:

$$E_n = -\frac{2\pi^2 m_e e^4}{h^2}\left(\frac{1}{n^2}\right), \quad n = 1, 2, 3, \ldots \quad (2.13)$$

where the subscript n has been added to indicate the **quantization of E,** and n is called a **quantum number.** Because we choose our zero of energy to correspond to the proton and electron at very large distance, the **energy levels** of Equation 2.13 are all negative, indicating **bound states** of the electron–proton system. These energy levels are shown in Figure 2.12. A slightly more accurate formula is obtained by substituting the reduced mass μ (Equation 2.8) for m_e.

Corresponding to the allowed energies E_n are the radii r_n that satisfy Equation 2.10, given by

$$r_n = \left(\frac{h^2}{4\pi^2 m_e e^2}\right) n^2, \quad n = 1, 2, 3, \ldots \tag{2.14}$$

and also illustrated in Figure 2.12. The constant factor multiplying n^2 is a characteristic distance in the atom called the **Bohr radius** a_0:

$$a_0 = \frac{h^2}{4\pi^2 m_e e^2} = 0.5292 \text{ Å} \tag{2.15}$$

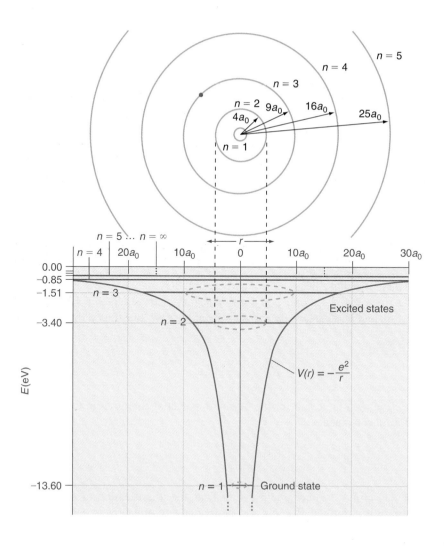

Figure 2.12
Bohr orbits (top) and energy levels (bottom) for the hydrogen atom. The energy levels are shown superimposed on a two-dimensional "funnel" representation of the Coulomb potential (see Chapter 1) to illustrate the confinement of the orbit by the potential. The orbits would be visible when viewing the funnel from the top.

In terms of a_0, the formulas for E_n and r_n are written simply:

$$E_n = -\frac{e^2}{2a_0 n^2}, \qquad r_n = n^2 a_0 \tag{2.16}$$

Newton's law (Equation 2.11), and the total energy (Equation 2.12), can easily be written for an electron interacting with a nucleus of arbitrary charge Ze (see Chapter 1). The formulas of Equations 2.16 then become

$$E_n = -\frac{e^2 Z^2}{2a_0 n^2}, \qquad r_n = \frac{n^2 a_0}{Z} \tag{2.17}$$

These are appropriate for the so-called hydrogen-like ions He^+, Li^{2+}, etc.

With the energy level formula of Equation 2.13, Bohr was able to give a complete interpretation of the Rydberg formula of Equation 2.3, as mentioned in Section 2.5. If we take Equation 2.3, multiply both sides by hc, and expand the parentheses, we get

$$\frac{hc}{\lambda} = \frac{hcR_H}{n_1^2} - \frac{hcR_H}{n_2^2} \tag{2.18}$$

According to Bohr's hypothesis (see Equation 2.6), the left-hand side of Equation 2.18 is an energy change ΔE, and the right-hand side is the difference of two energy levels, spectroscopically designated as **term values,** and identical to those given by the energy level formula of Equation 2.13. The identification of lines in the Balmer series with Bohr's transitions is illustrated in Figure 2.13. From this line of reasoning, Bohr was able to identify the Rydberg constant R_H as given in Equation 2.7; and using the general formulas of Equations 2.17 he was the first to assign correctly some

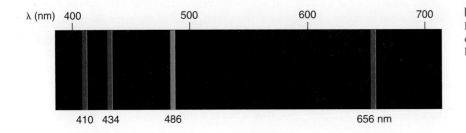
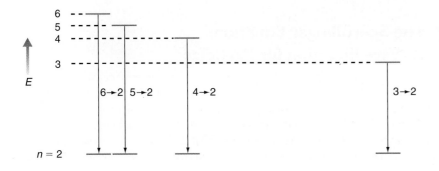

Figure 2.13
Bohr's assignment of the Balmer series of atomic hydrogen to its energy levels.

previously unidentified spectral lines as belonging to He$^+$. Of equal importance, Bohr's characteristic atomic length a_0 was in agreement with what was known about the sizes of atoms at that time.

EXAMPLE 2.7

Use Equation 2.17 to calculate E (eV) and r (Å) for $n = 1$ and 2 for the helium ion He$^+$, and find the wavelength of a photon emitted during a transition between these two levels.

Solution:
For He$^+$ $Z = 2$, so (in cgs-esu)

$$E_n = -\frac{(4.803 \times 10^{-10} \text{ esu})^2(2)^2}{2(0.5292 \times 10^{-8} \text{cm})n^2} = -\frac{8.718 \times 10^{-11} \text{ erg}}{n^2} = -\frac{54.42 \text{ eV}}{n^2}$$

$$r_n = \frac{0.5292 \text{ Å}}{2} n^2 = 0.2646 \, n^2 \text{ Å}$$

For $n = 1$, $E_1 = -54.42$ eV and $r_1 = 0.2646$ Å

For $n = 2$, $E_2 = -13.61$ eV and $r_2 = 1.0584$ Å

$$\Delta E = E_2 - E_1 = -13.61 - (-54.42) = 40.81 \text{ eV}$$

$$\lambda = hc/\Delta E = (1.2399 \times 10^{-6} \text{ eV m})/(40.81 \text{ eV})$$

$$= 3.038 \times 10^{-8} \text{ m} = 30.38 \text{ nm}$$

Consulting Figure 2.3, this is in the far-ultraviolet region of the spectrum. With calculations such as this, Bohr was able to make assignments of spectral lines.

Although this model was eventually supplanted by a more general one, it serves to illustrate how the postulate of wave-like properties for the heretofore particulate electron can lead in a natural way to the quantization of the mechanical energy of the atom. Conceptually, however, we are stuck with a wave-particle paradox for matter complementary to that for light. Whether an electron (or other small "particle") is a wave or a particle depends on the experiment performed on it. Once again, our macroscopic world has not prepared us to picture such behavior. Scientists have nonetheless continued to refer to these small manifestations of matter as "particles," though perhaps "wavicles" would be a better term.

2.7 The Schrödinger Equation: A Wave Equation for Particles

There is a perhaps apocryphal story about how the required new mechanics—to be called **quantum mechanics** or **wave mechanics**—was born. The winter of 1925 was approaching. A middle-aged German physics professor named Erwin Schrödinger had elected to give a lecture on the new postulate de Broglie had put forth concerning the wave nature of matter. At the conclusion of the lecture, the famous physical chemist Peter Debye stood up and thundered words to the effect, "What is this foolishness? If there are waves, they must obey a **wave equation**!" Schrödinger, himself an acknowledged expert in mathematical physics, reacted to these words as if he had been struck a blow. He at once realized what was needed, and quickly ran off to begin laboring feverishly to produce the wave equation Debye had demanded!

What Debye was referring to was a *differential equation* obeyed by all standing or traveling waves like the sine wave of Equation 2.1, which was (and is) well known to most physicists, probably including de Broglie himself. Schrödinger's task, which he successfully completed in a few months, was to put de Broglie's wavelength (Equation 2.9) into a standard wave equation. Here we will settle for this brief description; Exercise 30 outlines a one-dimensional analysis, and in a later course you may examine the question in more detail. The result that Schrödinger finally arrived at (after a few "almost" tries) for the standing-wave motion of a particle of mass m under the influence of a potential energy $V(x, y, z)$ can be written

$$-\frac{h^2}{8\pi^2 m}\left(\frac{\partial^2 \psi}{\partial x^2} + \frac{\partial^2 \psi}{\partial y^2} + \frac{\partial^2 \psi}{\partial z^2}\right) + V(x, y, z)\psi = E\psi \qquad (2.19)$$

This is a formidable-looking equation, and, make no mistake, it *is*, even to people called "quantum chemists" who devote their energies to solving it for chemical problems. You will *not* be expected to understand the meaning of the **Schrödinger equation**—recall Niels Bohr's remark—but it helps to allay the mystery to see the equation, and to be able to identify the quantities that enter it. The (x, y, z) are the independent variables, the Cartesian coordinates of points in three-dimensional space, the terms with curly ∂'s are second-order partial derivatives ($\partial^2 \psi / \partial x^2$ means "take the second derivative of ψ with respect to x while holding y and z constant"), E is the total energy of the particle, and ψ is the **wave function** for the particle.

Most of the mystery of the Schrödinger equation centers on the meaning of ψ. In the absence of a potential energy V, ψ becomes a 3-D generalization of the sine wave of Equation 2.1, being in any case a function of the point in space under consideration, that is, $\psi = \psi(x, y, z)$. Because there is nothing in Equation 2.19 that specifies a path or trajectory of the particle through space, ψ may be thought of as replacing in some sense the orbit of the particle, and in the chapters to follow we will refer to ψ, when it applies to an individual electron, as an **orbital**. We shall have much more to say on the subject of orbitals and their interpretation in Chapters 3 and 4. But lest you think that everything you read here or elsewhere is "gospel," bear in mind that Schrödinger himself puzzled over the meaning of ψ for several years after he published his equation, and Albert Einstein rejected the whole business! The reason you are going to learn something about wave mechanics, or quantum mechanics, is that, whatever it may mean in everyday terms, it *works! We have yet to find a case where wave mechanics fails to describe experimental measurements correctly.*

SUMMARY

Our modern view of the structure of the atoms of which our world is made is based on the **wave-particle duality** of light and subatomic particles; that is, that both light, or **electromagnetic radiation,** and electrons, which make up the outer reaches of an atom, can behave as either waves or particles. These notions, which are nonsensical based on everyday experience, arose first from the analysis of **blackbody radiation** by Planck and of the **photoelectric effect** by Einstein, which showed that light, despite its wave-like behavior in the phenomenon of **diffraction,** must also be thought of as a quantum of energy or **photon.** **Planck's constant** h, which defines the particle-like aspects of light and the wave-like aspects of matter, was first introduced as the proportionality constant between the energy of the quantum and the **frequency** ν of the light wave: $\varepsilon = h\nu$.

Bohr then demonstrated that Planck's **quantum hypothesis** could explain the **line spectra** of gas-phase atoms, provided that the atom could only exist in certain allowed **stationary states** of fixed total energy. The existence of these stationary states was a mystery until de Broglie proposed that matter could display wave-like properties with a **de Broglie wavelength** $\lambda = h/mv$. The stationary states of an atom then correspond to electron waves that fulfill standing-wave conditions, in analogy to a vibrating string. Confirmation of de Broglie's hypothesis was provided by the discovery of **electron diffraction** by Davisson and Germer and by Thomson. We presented a simple model, based on the work of Bohr and de Broglie, for the hydrogen atom that explains its line spectrum as described by the **Rydberg formula**.

Finally, Schrödinger postulated a differential equation now called the **Schrödinger equation** that was to be the foundation for a new mechanics, **wave mechanics** or **quantum mechanics**, that supplants Newton's laws of motion in the atomic world.

EXERCISES

Light waves and photons

1. When bombarded by high-voltage cathode rays, an iron plate emits X-rays of wavelength 1.932 Å. What is the frequency of these X-rays? According to Planck's quantum hypothesis, what is the energy (in joules and electron-volts)?

2. A popular FM (frequency-modulation) radio station broadcasts at a mean frequency of 93.7 MHz. What is the wavelength of the radio waves? According to Planck's hypothesis, what is the energy (in joules and electron-volts)?

3. Verify the range of energies in electron-volts given in Figure 2.3 for the wavelengths spanned by visible light (400–700 nm). In addition to joules, give these energies in kJ/mol of photons and kcal/mol of photons. (One mole of photons is called an **einstein**.)

4. Light sources such as lasers are rated according to their **power** in **watts** W, where 1 W = 1 J/s, the energy radiated per second. An *argon ion laser* (a gas laser like HeNe) can be made to "lase" on nine different spectral lines of Ar^+; one with high power is a green line at 514.5 nm. If 5 W of power is obtained on the green line, how many photons s^{-1} are emitted? How many einsteins (moles of photons) s^{-1}?

5. Wien's law of blackbody radiation states that the product of the wavelength of maximum intensity and the temperature of the hot body is a constant (see Figure 2.6), $\lambda_{max} T$ = constant. The data of Lummer and Pringsheim, as well as Planck's quantum theory, give a value for the constant of 0.288 cm K. The sun's spectrum (aside from Fraunhofer's dark lines) is roughly that of a blackbody radiator, and is found to peak at λ = 530 nm. Find the temperature of the sun's surface. (The sun is much hotter than this in its interior.) For further exercises on blackbody radiation, see Appendix A.

Photoelectric effect

6. Certain metals and alloys (mixtures or compounds of two or more metals) have unusually low work functions and are useful as photocathodes (surfaces yielding photoelectrons). Light of wavelength 389 nm shines on a particular "trialkali" photocathode, and electrons of maximum kinetic energy 1.45 eV are given off. What is the work function W (in electron-volts and joules) of the surface? What maximum wavelength of light will cause photoemission from this cathode?

7. The "electric eyes" used to open doors automatically are based on the photoelectric effect. The metal surface inside the "eye" [an evacuated glass envelope containing the surface (cathode) and a photocurrent collector (anode)] continuously provides current until the light beam that actuates it is interrupted. What is the maximum kinetic energy of photoelectrons produced by a mercury vapor lamp, emitting 436-nm violet light, actuating an electric eye of work function 2.1 eV?

8. Modern lasers are such intense light sources that some of the "rules of the game" in the photoelectric effect now have been revised. For example, when light from a focused, high-power carbon dioxide (CO_2) laser, which emits infrared light of wavelength 10 μm, strikes a metal surface with a work function of a few electron-volts, electrons are in fact ejected. This phenomenon is a *multiphoton absorption* effect; so many photons impinge on the surface in such a short time

that a single electron can collide with and absorb the energy of several photons. How many CO_2 laser photons must an electron absorb to be emitted from a surface of work function 4.5 eV?

Atomic spectra and stationary states

9. In addition to the far-ultraviolet 58.4-nm line (see Example 2.5) the same upper, excited stationary state of the helium atom also emits an infrared line of wavelength 2.06 μm. From the energies of these two wavelengths, construct an energy-level diagram (like the bottom of Figure 2.12, but without the Coulomb potential) showing by downward arrows the energy levels involved in the two transitions. Find the energy difference between the two lowest stationary states.

10. As a rule both absorption and emission of light by gas-phase atoms of an element occur at the same set of wavelengths. Yet, when infrared light is passed through a sample of helium gas, no absorption is observed at 2.06 μm, although emission at this wavelength in helium discharges is seen. Based on the results of Exercise 9, speculate on why this is. Can you imagine an experiment that would show absorption at 2.06 μm by helium?

11. Metallic elements often show visible emission lines when their salts are vaporized in a Bunsen burner flame, and the distinctive colors produced are the basis for the visual **flame tests** for the presence of various elements. Lithium salts show a bright red flame ($\lambda = 671$ nm), sodium yellow-orange (589 nm), and potassium a faint red (764 nm). Calculate ΔE in electron-volts and inverse centimeters for the three elements.

12. The "D-line" in the solar dark line spectrum is assigned to absorption by sodium (Na) atoms, and actually consists of two closely spaced lines (a *fine-structure doublet*) of wavelengths $\lambda = 589.158$ and 589.755 nm, as illustrated in Figures 2.4 and 2.8.
    a. Calculate the energy-level spacings in electron-volts and inverse centimeters between the stationary states involved. Assuming the transitions occur out of a common lower state, sketch an energy-level diagram, indicating the observed transitions by upward arrows.
    b. Find the spacing between the two upper levels in electron-volts and inverse centimeters. Predict the wavelength λ in microns of a transition between these levels and give its spectral region.

13. Calcium (Ca) yields a red-orange flame test, due to transitions between excited energy levels. The brightest emission line actually consists of three closely spaced lines, a *fine-structure triplet*, of wavelengths $\lambda = 610.441$, 612.391, and 616.387 nm.
    a. Calculate the energy-level spacings in electron-volts and inverse centimeters between the stationary states involved. Assuming the transitions occur out of a common upper state, sketch an energy-level diagram, indicating the observed transitions by downward arrows.
    b. Find the spacings between adjacent levels for the three lower levels in electron-volts and inverse centimeters. Predict the wavelengths λ in microns of transitions between adjacent levels and give their spectral region.

14. When a gas-phase atom is irradiated with light of sufficiently short wavelength λ, it will ionize. This is analogous to the photoelectric effect and is called *photoionization*; Einstein's photoelectric equation applies, with the work function W replaced by the atom's ionization energy (IE).
    a. When Na vapor is irradiated by Lyman-α ($\lambda = 121.568$ nm), electrons of kinetic energy 5.060 eV are produced. Find the ionization energy IE of Na in electron-volts.
    b. Assuming the lower state of Exercise 12 (a) is the ground state, and that the energy E_1 of the ground state of Na is the negative of its IE, find the energies of the upper states of the D-line doublet of Exercise 12. Redraw the energy-level diagram to include the ionization continuum above $E = 0$.

The Bohr theory of the hydrogen atom

15. a. Using only the Rydberg formula and Bohr's hypothesis $\Delta E = hc/\lambda$, find a formula for $\Delta E_{n_1 n_2}$, the energy-level spacing between any two stationary states n_1 and n_2 in electron-volts. Use your formula to compute ΔE for the longest wavelength line in the Balmer series (Balmer-α).
    b. Your formula from part (a) also implies a formula for the energy E_n of a stationary state. Use this formula to verify the energies of the excited states given in Figure 2.12.

16. Use the Rydberg formula to compute the wavelengths in centimeters of the 4⟶3 (Paschen series) and the 4⟶1 and 3⟶1 (Lyman series) transitions. Show numerically, using inverse centimeter units for the reciprocal wavelengths, that

$$\frac{1}{\lambda_{41}} = \frac{1}{\lambda_{43}} + \frac{1}{\lambda_{31}}$$

Also show using the Rydberg formula that in general

$$\frac{1}{\lambda_{ki}} = \frac{1}{\lambda_{kj}} + \frac{1}{\lambda_{ji}}$$

where i, j, k are Bohr's quantum numbers with $i < j < k$. What statement does this imply about the energy levels involved?

17. Use the values of the fundamental constants inside the front cover of this book to calculate the Rydberg constant using Bohr's formulas (Equations 2.7 and 2.8) and compare with the best experimental value, being careful of significant figures. (R_H has been used to improve the precision of the less well-known constants.)

18. A branch of atomic spectroscopy is concerned with the properties of so-called "Rydberg atoms"; the prototype is a hydrogen atom in a highly excited state with quantum number $n \approx 20$–100. Compute the radius of the Bohr orbit in angstroms and the corresponding energy in electron-volts for $n = 100$.

19. The limiting wavelength of each Rydberg series can be computed by allowing n_2 to approach ∞ for fixed n_1. Give a formula for these wavelengths. If a hydrogen atom in its ground state absorbs a photon of this limiting wavelength, what will happen?

20. Find a spectral line in the spectrum of He^+ that lies in the visible and give its quantum numbers. Give a Rydberg-like formula for computing the spectrum of He^+.

21. The helium ion He^+ in its first excited state has the same total energy as the neutral hydrogen atom in its ground state, but it has a Bohr orbit radius twice as large. Draw superimposed semiquantitative energy diagrams for H and He^+ for these two coincident states, indicating their kinetic and potential energies at the respective Bohr orbit radii. Discuss briefly.

22. X-rays, such as those used by Moseley and Millikan in their groundbreaking experiments and by every doctor and dentist today, are short-wavelength radiation produced by a special high-voltage Crookes tube. The X-rays are emitted by the excited metal atoms produced when cathode rays knock electrons out of a metal anode. X-ray wavelengths λ are found to follow the Bohr model for nuclear charge Z, with $n_1 = 1$ and $n_2 = 2$. Use the Bohr model to predict the X-ray λ in angstroms produced by a Mo anode.

Matter waves

23. Top off the list of de Broglie wavelengths in Table 2.1 by calculating the wavelength of a so-called "thermal" electron at a temperature of 1500 K. (Such an electron can be emitted from a hot metal wire, a *filament*.)

24. Recently it has become possible to alter the velocity of gaseous atoms with radiation pressure from laser beams. In one experiment an atomic beam of sodium atoms was brought almost to a standstill by shining lasers from six directions tuned to the D-line of Na. In this "optical molasses" the atoms move sluggishly about with an average velocity of 40. cm s^{-1}. Find the de Broglie wavelength of these atoms. (Using this method, a so-called *Bose–Einstein condensate* has been made, an ultracold cluster of atoms that behaves like one giant atom.)

25. Because the de Broglie wavelength is inversely proportional to momentum $p = mv$, a particle subject to forces will have a continuously changing wavelength. For a particle of fixed total energy $E = K + V$, derive a more general formula for the de Broglie wavelength that accounts for the presence of potential energy.

26. Fill in the algebraic steps necessary to derive Bohr's energy-level formula of Equation 2.13, from Equations 2.9 through 2.12. In the process, obtain formulas for the velocity v_n and the de Broglie wavelength λ_n of the nth Bohr orbit. Compute these latter quantities for $n = 1, 2,$ and 100.

27. In the course of completing Exercise 26, you may have noted that $-V = 2K$ and $-E = K$. These relations are a consequence of the **virial theorem**, which relates kinetic and potential energies of a particle in an inverse-square force field. They hold whether the energy is quantized or not. Use these relations to simplify the generalized de Broglie wavelength of Exercise 25, and show that the formula found in Exercise 26 for λ_n is recovered.

28. Using the results of Exercises 25 and 27, and assigning the hydrogen electron a total energy corresponding to the $n = 3$ state, compute its de Broglie wavelength as a multiple of a_0 for $r = 9a_0$ (the radius of the $n = 3$ Bohr orbit), and also for $r = 0, 4.5a_0$, and $13.5a_0$. Make a rough plot of λ versus r. What happens to $\lambda(r)$ as r approaches $18a_0$? (The other r's are accessed by the electron in Schrödinger's generalization of de Broglie's idea.)

29. When Bohr derived Equation 2.13, he had no inkling of wave-particle duality for the electron. Instead of using a standing-wave condition, Bohr postulated the *quantization of angular momentum*:

$$L = mvr = \frac{nh}{2\pi}$$

as a starting point. Show that this requirement is equivalent to de Broglie's standing-wave condition of Equation 2.10.

30. The equation referred to by Debye in response to Schrödinger's lecture on de Broglie waves is the *classical wave equation*. For describing a standing wave ψ in one dimension, it is

$$\frac{d^2\psi}{dx^2} + \frac{4\pi^2}{\lambda^2}\psi = 0$$

where λ is the wavelength of the standing wave. (For a fixed λ, a solution of this equation is the sine wave $\psi = A\sin(2\pi x/\lambda)$ of Figure 2.1.) Show that this equation becomes a one-dimensional form of Schrödinger's wave equation

$$-\frac{h^2}{8\pi^2 m}\frac{d^2\psi}{dx^2} + V\psi = E\psi$$

when the generalized form of the de Broglie wavelength λ from Exercise 25 is substituted into the classical equation and some algebraic rearrangement is made. You can see how to solve this equation for a simple case in Appendix B.

31. Apply de Broglie's standing-wave condition (Equation 2.10) to the earth in orbit (radius = 1.5×10^{13} cm) around the sun, using the data in Table 2.1. Find the quantum number n for the earth–sun system. What would be the effect of a $\Delta n = 1$ transition?

32. In Example 2.6, a 54-eV beam of electrons was found to have a (de Broglie) wavelength of 1.67 Å.

    a. What is the energy of X-ray photons of this same wavelength?

    b. Find a formula relating the energies of particles of light and particles of matter of the same wavelength. Your result should not contain Planck's constant h.

3 Wave Mechanics and the Hydrogen Atom: Quantum Numbers, Energy Levels, and Orbitals

Hydrogen is hydrogen, no matter where you find it. Spectra from three stars in the belt of the constellation Orion and a star in the handle of the Big Dipper (Ursa Major) all show the Balmer series of the hydrogen atom.

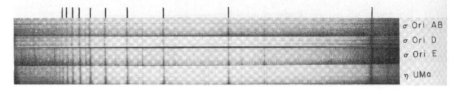

All we know of nature is in the same boat, to sink or swim together. The constructions of science are merely expositions of the characters of things perceived.

Alfred North Whitehead (1920)

In three short years after Schrödinger proposed his equation in 1926, the foundations for understanding atomic structure and spectra and the nature of chemical bonding had been laid. In many areas we are still working out the consequences of the theory of quantum mechanics developed in that unbelievably fruitful period. If the means exist to understand the workings of a thing, then it is part of our nature to want to investigate, to pick the thing apart. That is the way science progresses. For chemistry this has meant learning to speak the language of quantum mechanics, for in that strange tongue we can find ways to express the mysteries of atomic and molecular structure, even approaching the molecular workings of life itself, of our own brains. In this chapter we explore the consequences of wave-particle duality for the description of the simplest atom, hydrogen, hoping to learn the rules of behavior in the quantum world inside an atom.

CHAPTER OUTLINE

3.1 Solving the Schrödinger Equation for the Hydrogen Atom: A Brief Description

3.2 The Quantum Numbers n, l, and m

3.3 Characteristics of the Orbitals: Three-Dimensional Waves

3.4 Born's Interpretation of Orbitals

3.5 Heisenberg's Uncertainty Principle

3.6 Orbitals and Orbits: An Analogy

3.7 A Qualitative Description of Electronic Transitions

3.1 Solving the Schrödinger Equation for the Hydrogen Atom: A Brief Description

As complicated as the **Schrödinger equation** (Equation 2.19) appears, it can be written in a very compact form that suggests how it is related to the Newtonian ideas discussed in Chapter 1. So that you can appreciate this, we will introduce the concept of an **operator** by way of example. When you find the first derivative of a function $f(x)$ you write df/dx, and it is normal to regard the d/dx part as having its own existence, that is,

3.1 Solving the Schrödinger Equation for the Hydrogen Atom: A Brief Description

$$\frac{df(x)}{dx} = \left(\frac{d}{dx}\right)f(x) \equiv \hat{D}f(x)$$

where $\hat{D}$ is a *differential operator*,* an entity that will automatically "operate" on the factor to its right by taking its derivative. (Medical analogies are appropriate here!) An operator is distinguished from an ordinary factor by the possibility of its modifying the other factors. We can apply the same formal factorization to the Schrödinger equation (Equation 2.19), writing it as

$$\left[-\frac{h^2}{8\pi^2 m}\left(\frac{\partial^2}{\partial x^2} + \frac{\partial^2}{\partial y^2} + \frac{\partial^2}{\partial z^2}\right) + V(x,y,z)\right]\psi(x,y,z) = E\psi(x,y,z) \quad (3.1)$$

The quantity in square brackets is called the **Hamiltonian operator** $\hat{H}$, and the Schrödinger equation is often written in the elegant (and therefore less useful) form

$$\hat{H}\psi = E\psi \quad (3.2)$$

Note that in Equation 3.2 you can't simply cancel the ψ's because $\hat{H}$ is an operator that may modify the wave function ψ. In Newton's mechanics, the Hamiltonian (invented by William Hamilton in the 19th century) is just the sum of the kinetic and potential energies, considered as functions of the coordinates and momenta: $H = K + V$. If we identify the collection of partial differential operators on the left of Equation 3.1 as a **kinetic energy operator** $\hat{K}$, then Equation 3.1 can be put into the suggestive form

$$[\hat{K} + V]\psi = E\psi \quad (3.3)$$

In this form the Schrödinger equation seems to embody mechanical energy conservation, $K + V = E$.

Solving the Schrödinger equation is a bit unlike solving algebraic equations such as a quadratic equation. When you apply the quadratic formula, you find a pair of numbers that "satisfy" the equation. But solving a wave equation requires you to find the wave function, an *infinite set* of numbers that form a curve when plotted against one of the coordinates x,y,z. Finding unknown *functions* instead of a few numbers is a general characteristic of solving *differential* equations. Solving Newton's second law of motion (Equation 1.3) is a similar sort of mathematical exercise, in which you find the position of a particle as a function of time, $x(t)$.

The solutions to the Schrödinger equation, the wave function ψ and the energy E, depend on what potential energy function $V(x,y,z)$ is used, and the equation has been solved for several choices of V. Because this is a book about chemistry, not about quantum mechanics, we will not explore different potentials, but specialize at once to the Coulomb potential. An especially simple potential to deal with, however, is called a *box potential*, somewhat like a prison cell for confining a particle. Appendix B contains the details of solving the Schrödinger equation for a **particle in a box.** You may be surprised to learn that you *can* handle the mathematics for this simple case.

Erwin Schrödinger (1887–1961). He began his research career in color theory, and, enlisting in the artillery in World War I in 1914, continued to write research papers while in the trenches. See W. Moore, *Schrödinger: Life and Thought* (Cambridge, 1989).

---

* All good operators are normally found wearing little hats.

Figure 3.1
Spherical polar coordinates (r,θ,ϕ) in relation to Cartesian coordinates (x,y,z).

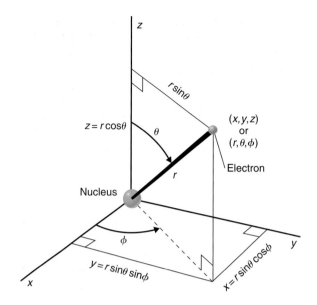

Because the Coulomb potential is *central* or *radial* (that is, it depends only on the distance r between charged particles), it is mathematically convenient to switch from the Cartesian coordinates (x,y,z) to **spherical polar coordinates** (r,θ,ϕ) when solving the Schrödinger equation for the hydrogen atom. Figure 3.1 illustrates the relationship between these coordinate systems. Polar coordinates are natural for describing motion in radial forces, and they are commonly used in Newton's mechanics to represent planetary and satellite motion. As you sit reading this book, your present location on the surface of the earth (a nearly spherical body) is conveniently given by your distance from the earth's center (r), your latitude (θ), and your longitude (ϕ). Switching or *transforming* coordinates like this makes the kinetic energy operator $\hat{K}$ more complicated than its Cartesian form in Equation 3.1,* but allows the Coulomb potential V to be written just as in Equation 1.18, in cgs-esu units, $V = -e^2/r$. (Otherwise $V = -e^2/[x^2 + y^2 + z^2]^{1/2}$ would have to be used.) Because the proton is 1836 times heavier than the electron, these coordinates are essentially the coordinates of an electron moving around a proton fixed at the origin of the coordinate system, as Figure 3.1 suggests.

New Mechanics but Old Mathematics

When Schrödinger invented his equation, one of the first things he did with it was to solve it for the hydrogen atom, that is, a particle in a Coulomb potential. He was able to do this very quickly because he could just look up the solutions! The French math-

---

*In spherical polar coordinates,

$$\hat{K} = -\frac{h^2}{8\pi^2 m}\left[\frac{1}{r^2}\frac{\partial}{\partial r}\left(r^2\frac{\partial}{\partial r}\right) + \frac{1}{r^2 \sin\theta}\frac{\partial}{\partial \theta}\left(\sin\theta\frac{\partial}{\partial \theta}\right) + \frac{1}{r^2 \sin^2\theta}\frac{\partial^2}{\partial \phi^2}\right]$$

ematicians Laguerre and Legendre had already found the solutions to Schrödinger's equation for a Coulomb potential in the early part of the 19th century. Of course, they were interested only in solving different types of differential equations, and had little inkling of the use to which their work would be put. This is one of the ever-increasing number of examples of pure mathematics eventually finding an application in the real world. There may be times when studying your math does not seem to be leading anywhere—but you never know when some esoteric topic might turn out to be more than a strain on the brain.

Schrödinger began the solution of the spherical polar form of Equation 3.1 by assuming

$$\psi(r,\theta,\phi) = R(r)\, Y(\theta,\phi) \tag{3.4}$$

that is, that the wave function can be factored into **radial** (R) and **angular** (Y) functions.* When he substituted Equation 3.4 into Equation 3.1, he found that the original equation broke up into separate differential equations for R and Y. These equations are the ones that Laguerre and Legendre had already solved; that is, they had already found the functions $R(r)$ and $Y(\theta,\phi)$—which is exactly why Schrödinger was able to make such quick progress.

Boundary Conditions

At this point it was time for the physics (and chemistry) of the problem to come into play. Whenever you are solving an equation that relates to a physics or chemistry problem, you have to keep thinking about the realities you are getting at, to avoid going down mathematical blind alleys. The constraints that reality places on the solutions to a physically relevant equation are called **boundary conditions.** For example, sometimes a quadratic equation has both positive and negative solutions, but only one of these is physically admissible because of the way the problem has been set up; if the quantity desired is the concentration of a product of a chemical reaction, it can only be positive. So what are we trying to describe here? An electron in a hydrogen atom. The **wave function** ψ is just the embodiment of de Broglie's hypothesis of matter waves (Equation 2.9), and ψ is somehow going to be determined by a proper three-dimensional generalization of de Broglie's electron standing-wave condition (Equation 2.10). This implies that ψ must be a well-behaved function: smooth, single-valued, and finite everywhere in space. In addition, ψ must become small at large distances r from the nucleus (proton) in order to depict a bound stationary state where the electron is held close to the proton. These requirements on ψ are the boundary conditions for the hydrogen atom solutions to the Schrödinger equation.

Just as you sometimes cannot "use" both solutions to a quadratic equation, Schrödinger found that he could only use certain solutions among those found by

---

* This is a common procedure for partial differential equations. Equation 3.4 is valid for any central potential.

Laguerre and Legendre, only those solutions that satisfied the boundary conditions on ψ. Once Schrödinger had selected the proper solutions, the total energy E followed naturally from Equation 3.1. *Operating on a proper wave function with the Hamiltonian operator "automatically" gives the energy times the wave function.* In addition, the well-behaved wave functions were found to be characterized by a set of three integers, denoted by n, l, and m; these became known as the **quantum numbers**. The integers actually represented the degrees of polynomials appearing in ψ, as we will see later when we inspect the solutions.

3.2 The Quantum Numbers n, l, and m

Why *three* quantum numbers? Wouldn't one be enough? Three quantum numbers are needed to impose standing-wave conditions on motions in each of the three spherical polar coordinates r, θ, and ϕ. In other words, *one quantum number for each dimension is required.* The reason only a single quantum number occurred in the Bohr–de Broglie model is that it was formulated as a one-dimensional (1-D) problem. As the electron circles the proton in the x-y plane, the only coordinate that changes is ϕ; for a given stationary state, r and θ are fixed. But there is no reason to expect the electron to do its wavy business in only one dimension, and solving the 3-D Schrödinger equation takes this fact into account.

The significance of the quantum numbers will become clearer as we go along; first we establish nomenclature and definitions:

> The quantum number n is called the **principal quantum number**; it is the closest thing to a successor to Bohr's quantum number n. The energy E that Schrödinger obtained by solving Equation 3.1 depended only on n, and was given by a formula identical to that obtained by Bohr (Equation 2.13). (If Schrödinger had obtained a different formula, he would have had to throw out his equation and start over! Bohr's formula agreed with experiment.) Just as in Bohr's model, n can take on the values $n = 1, 2, 3, \ldots$
>
> The quantum number l is called the **azimuthal quantum number**. In our later discussions we will also refer to l as the **angular momentum quantum number**. Both of its names suggest an association with variation of an angle, θ in this case. The allowed values of l are limited by n: $l = 0, 1, 2, \ldots, n-1$. The reason for this limitation is discussed in Section 3.3.
>
> The quantum number m is called the **magnetic quantum number**. We will also refer to m as a **projection quantum number**. The magnetic label comes about because m explains how the hydrogen atom behaves in a magnetic field. The allowed values of m are in turn limited by the value of l: $m = -l, -l+1, \ldots, -1, 0, 1, \ldots, l-1, l$. In words, m goes from $-l$ to l in integral steps.

It is important to realize that these rules for specifying the quantum numbers are not arbitrary; they result from imposing boundary conditions on the wave functions. The rules do depend on the choice of coordinates; the spherical polar coordinates allow the simplest forms for the energy and wave functions for a Coulomb potential, so the quantum numbers go along with these forms.

EXAMPLE 3.1

List all possible values for the unspecified quantum number or numbers for the cases: $n = 2$; $n = 3$; $n = 4, l = 2$.

Solution:

Using the preceding l-rule, for $n = 2$, l can be 0 or 1; using the m-rule for each l, for $l = 0$, m can only be 0, while for $l = 1$, m can be -1, 0, or 1. Writing the possible combinations as a string of three digits, $nlm = 200, 211, 210,$ and $21-1$. For $n = 3$, l may be 0, 1, or 2, and we have $nlm = 300, 311, 310, 31-1, 322, 321, 320, 32-1,$ and $32-2$. When n and l are specified, then only m may vary, so for $n = 4, l = 2$ we have $nlm = 422, 421, 420, 42-1,$ and $42-2$. Note that for $n = 2$ four combinations are allowed, while for $n = 3$, nine are allowed. In general there are n^2 combinations for a given n. For a given n and l, m can take on $2l + 1$ possible values, and the number of combinations is therefore $2l + 1$. Right now this seems like just "playing with numbers," but this numerology will prove to be pivotal in understanding the structure of atoms and the periodic table.

Specification of the Energy: Ionization

As we noted when defining the principal quantum number n, the total energy depends only on n and not on l and m. This makes the hydrogen atom problem peculiar, because "normally" the energy should depend on all three quantum numbers (see Appendix B). (If this peculiarity did not exist, there would be no simple Rydberg formula, Bohr would not have been led to develop his model, de Broglie would not have been struck by the appearance of integers in describing atoms, . . .) The hydrogen atom therefore has a lot of "hidden" structure; for each energy level n there are n^2 possible combinations of quantum numbers to go with it. Bohr's "stationary states" are much more complicated than the simple circular orbits suggest. The situation where two or more sets of quantum numbers correspond to the same energy is referred to as *degenerate* (no moral implications here), and the number of such sets is called the **degeneracy**.

The conclusions that Bohr drew about the relation between the allowed energies E_n and the H atom's spectrum (see Section 2.5) still stand after the dust raised by Schrödinger has settled. Another aspect of atomic behavior that can be interpreted with the aid of the E_n is the **ionization energy** IE. The ionization energy is defined as the work necessary to remove the electron from the H atom, that is, to cause the reaction

$$H \longrightarrow H^+ + e^- \tag{3.5}$$

Figure 3.2 recaps the hydrogenic energy levels and illustrates the so-called **ionization continuum,** which corresponds to an ion and a free electron. In this continuum the electron's energy is no longer quantized; that is, it can take on any value from $E = 0$ on up. Recall from our discussion in Section 1.7 that positive total energies correspond to unbound or free states. To reach the continuum $E = 0$ from its ground state, a hydrogen atom requires energy exactly equal to the negative of its bound-state energy. If we regard the IE as the energy change ΔE in the ionization reaction of Reaction 3.5, then IE $= 0 - E_1 = +13.6$ eV. For a hydrogen atom in excited state n, IE $= -E_n$. These ionization energies correspond to the limiting short wavelengths in each of the spectral series of hydrogen (see Section 2.2).

Figure 3.2
Energy levels of the hydrogen atom with the ionization continuum. Each arrow represents the ionization energy out of a particular energy level.

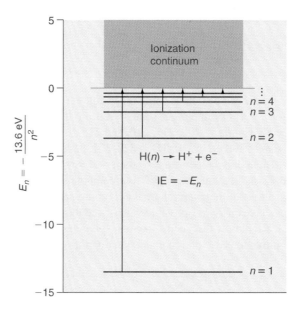

Specification of the Wave Functions: Orbitals

The wave functions, unlike the energies, were found by Schrödinger to depend on all three quantum numbers as follows:

$$\psi_{nlm}(r,\theta,\phi) = R_{nl}(r)\, Y_{lm}(\theta,\phi) \tag{3.6}$$

This is just Equation 3.4 restated with quantum number subscripts. Every permissible combination of quantum numbers corresponds to a different wave function. This means that, for example, for $n = 3$, instead of a single circular orbit, as in the Bohr–de Broglie model, there are *nine* different ways the electron can "vibrate," to form a standing wave, within the atom. As indicated by the subscripts, the radial part of the wave R changes its form as n and l change, whereas the angular part Y alters with l and m.

Because these wave functions are evidently replacements for Bohr's orbits, Robert S. Mulliken (not to be confused with Robert A. Millikan) suggested the term **orbital** to refer to these functions. As mentioned earlier, the functions have polynomial parts to them. The radial function $R_{nl}(r)$ is a *polynomial* in r of degree $n - 1$ (highest power r^{n-1}, called a *Laguerre polynomial*) multiplied by an exponential *function* of the form $e^{-r/(na_0)}$, where a_0 is the Bohr radius. This exponential factor ensures that $R(r)$ will become small as r gets large, in order to conform with the boundary condition for a bound state. Similarly the angular function $Y_{lm}(\theta,\phi)$ consists of *products of polynomials* in $\sin\theta$ and $\cos\theta$ (called a *Legendre function*) multiplied by a *complex exponential function* $e^{im\phi}$ where $i = \sqrt{-1}$. The forms of the θ polynomials and the complex exponential ensure that ψ will be single valued and continuous as a function of θ and ϕ; that is, that the wave will "meet itself" as θ and ϕ cycle around the origin. Table 3.1 collects the radial and angular parts of the hydrogen atom orbitals for $n \leq 4$ and $l \leq 3$. In accord with Equation 3.6, the total wave function is a product of R_{nl} and Y_{lm}, with the proviso that only those functions

TABLE 3.1
Wave functions for the hydrogen atom*

$R_{nl}(r)$ ($\rho = r/a_0$)	$Y_{lm}(\theta,\phi)$
n = 1	**l = 0**
$l = 0$, $1s$, $R_{10}(r) = a_0^{-3/2}\, 2\, e^{-\rho}$	$m = 0$, $Y_{00}(\theta,\phi) = \dfrac{1}{2\sqrt{\pi}}$
n = 2	
$l = 0$, $2s$, $R_{20}(r) = \dfrac{1}{2\sqrt{2}} a_0^{-3/2}(2-\rho)\, e^{-\rho/2}$	($l = 0$ as above)
	l = 1
$l = 1$, $2p$, $R_{21}(r) = \dfrac{1}{2\sqrt{6}} a_0^{-3/2}\, \rho\, e^{-\rho/2}$	$m = 0$, $Y_{10}(\theta,\phi) = \dfrac{\sqrt{3}}{2\sqrt{\pi}}\cos\theta$
	$m = \pm 1$, $Y_{1\pm 1}(\theta,\phi) = \dfrac{\sqrt{3}}{2\sqrt{2\pi}}\sin\theta\, e^{\pm i\phi}$
n = 3	
$l = 0$, $3s$, $R_{30}(r) = \dfrac{2}{81\sqrt{3}} a_0^{-3/2}(27 - 18\rho + 2\rho^2)\, e^{-\rho/3}$	($l = 0$ as above)
	($l = 1$ as above)
$l = 1$, $3p$, $R_{31}(r) = \dfrac{4}{81\sqrt{6}} a_0^{-3/2}(6 - \rho)\rho\, e^{-\rho/3}$	**l = 2**
$l = 2$, $3d$, $R_{32}(r) = \dfrac{4}{81\sqrt{30}} a_0^{-3/2}\rho^2\, e^{-\rho/3}$	$m = 0$, $Y_{20}(\theta,\phi) = \dfrac{\sqrt{5}}{4\sqrt{\pi}}(3\cos^2\theta - 1)$
	$m = \pm 1$, $Y_{2\pm 1}(\theta,\phi) = \dfrac{\sqrt{15}}{2\sqrt{2\pi}}\sin\theta\cos\theta\, e^{\pm i\phi}$
	$m = \pm 2$, $Y_{2\pm 2}(\theta,\phi) = \dfrac{\sqrt{15}}{4\sqrt{2\pi}}\sin^2\theta\, e^{\pm i2\phi}$
n = 4	
$l = 0$, $4s$, $R_{40}(r) = \dfrac{1}{768} a_0^{-3/2}(24 - 18\rho + 3\rho^2 - \rho^3)\, e^{-\rho/4}$	($l = 0$ as above)
	($l = 1$ as above)
$l = 1$, $4p$, $R_{41}(r) = \dfrac{1}{256\sqrt{15}} a_0^{-3/2}(80 - 20\rho + \rho^2)\rho\, e^{-\rho/4}$	($l = 2$ as above)
	l = 3
$l = 2$, $4d$, $R_{42}(r) = \dfrac{1}{768\sqrt{5}} a_0^{-3/2}(12 - \rho)\rho^2\, e^{-\rho/4}$	$m = 0$, $Y_{30}(\theta,\phi) = \dfrac{3\sqrt{7}}{4\sqrt{\pi}}\left(\dfrac{5}{3}\cos^3\theta - \cos\theta\right)$
$l = 3$, $4f$, $R_{43}(r) = \dfrac{1}{768\sqrt{35}} a_0^{-3/2}\rho^3\, e^{-\rho/4}$	$m = \pm 1$, $Y_{3\pm 1}(\theta,\phi) = \dfrac{\sqrt{21}}{8\sqrt{\pi}}\sin\theta(5\cos^2\theta - 1)\, e^{\pm i\phi}$
	$m = \pm 2$, $Y_{3\pm 2}(\theta,\phi) = \dfrac{\sqrt{105}}{4\sqrt{2\pi}}\sin^2\theta\cos\theta\, e^{\pm i2\phi}$
	$m = \pm 3$, $Y_{3\pm 3}(\theta,\phi) = \dfrac{\sqrt{35}}{8\sqrt{\pi}}\sin^3\theta\, e^{\pm i3\phi}$

* For hydrogen-like ions such as He^+, Li^{2+}, etc., the radial (R) wave functions may be obtained by replacing a_0 by a_0/Z, where Z is the nuclear charge; that is, ρ becomes Zr/a_0 and the factor $a_0^{-3/2}$ becomes $(Z/a_0)^{3/2}$. The Y functions are independent of Z.

with the same *l*-value can be combined. The premultiplying constants in Table 3.1 are **normalization constants;** we discuss their significance later.

3.3 Characteristics of the Orbitals: Three-Dimensional Waves

All of the waveforms we have looked at so far (see Figures 2.1, 2.9, and 2.11) have been one-dimensional functions of a single coordinate. The hydrogen atom orbitals, however, are three-dimensional functions and, like the 3-D Coulomb potential, cannot be plotted all at once (see Section 1.7). In this section we develop various ways of plotting and picturing the orbitals. Aside from its intrinsic importance, the main reason for taking the time to do this is that the hydrogenic orbitals will serve as models for the structure of bigger, many-electron atoms, and for a wave-mechanical description of forming chemical bonds. The hydrogen atom is unique in that *it is the only atom for which the Schrödinger equation can be solved "exactly" to produce a table of wave functions* such as Table 3.1. In chemistry, as in the other experimental sciences, there are very few problems for which we can find the exact answers. When such things happen, we must try to extract as much useful information as possible.

Let's start with the simplest of the orbitals, that which represents the ground state of the H atom. From Table 3.1 we extract the $n = 1$, $l = 0$, $m = 0$ functions $R_{10}(r)$ and $Y_{00}(\theta,\phi)$, and multiply them together to get

$$\psi_{100}(r,\theta,\phi) = \frac{1}{\sqrt{\pi}} a_0^{-3/2} e^{-r/a_0} \tag{3.7}$$

According to the quantum number rules given earlier, this is the *only* orbital of energy E_1. Note that θ and ϕ do not appear in the formula; the orbital is a function of r only. This means that the orbital is *spherically symmetric;* at any particular distance r from the nucleus, the electron wave has the same value regardless of angles. Instead of ψ_{100}, this orbital is usually denoted 1s, where the 1 stands for the principal quantum number, and the *s* for $l = 0$. This notation arose from spectroscopy, where spectral lines originating from $l = 0$ orbitals were "sharp" (well resolved). Figure 3.3 shows various ways of plotting the 1s orbital: a 1-D graph of ψ versus r, a 2-D surface obtained by rotating the curve about the vertical (z) axis, a 2-D "contour map" of the orbital amplitude in the *x-y* plane, a 3-D cloud with density proportional to the amplitude of the wave, and finally a 3-D **boundary surface diagram,** in which a surface of constant ψ is chosen to enclose most of the region of significant orbital amplitude.

The 1s orbital is a "simple decaying exponential" in r; it is worth examining the shape in some detail, because this sort of function crops up in a number of chemical problems. The orbital has fallen to half of its $r = 0$ value at $r = a_0 \ln 2$, where ln2, the natural logarithm (log to the base e) of 2, is 0.693..., giving $r =$ (0.529 Å)(0.693) = 0.367 Å. If this r-value is doubled, the orbital falls again by half, to a quarter of its value at the origin. This continued halving as r is doubled is a characteristic of *exponential decay*. The decay of the orbital as r increases is required if it is to represent a bound state.

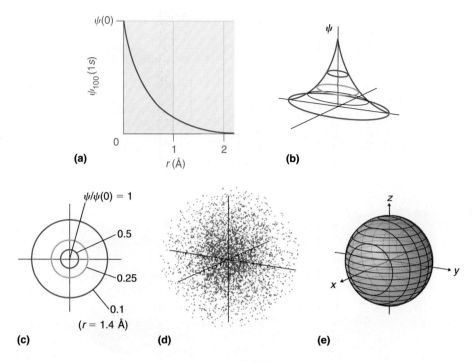

Figure 3.3
Graphic representations of the 1s orbital of the hydrogen atom. (a) 1-D ψ versus r. (b) 2-D "mountain" obtained by rotating the graph in (a) about the vertical axis. This represents the orbital amplitude in the x-y plane. (c) 2-D "contour map" of the orbital amplitude in the x-y plane obtained from the rings of constant altitude shown in (b). The contour values are fractions of the peak value at $r = 0$. This is what you would see looking down on (b) from above. (d) 3-D cloud whose density is proportional to orbital amplitude. (e) 3-D boundary surface diagram enclosing most of the cloud of (d).

You should have noticed that the 1s orbital does not have any **nodes**—places where the wave passes through zero amplitude. This is a general property of the lowest energy wave function for a quantum mechanical system and is analogous to the primary-tone vibration of a string (see Figure 2.9). Nodes begin to appear in the excited states, which brings us to $n = 2$. In Example 3.1 we found the possible sets of quantum numbers nlm to be 200, 211, 210, and 21−1. The 200 orbital is referred to as 2s, while the three $n = 2$, $l = 1$ combinations are 2p orbitals, where p stands for *principal*, the most intense lines in the spectrum. As before, we take $(R_{20})(Y_{00})$ to get the 2s function:

$$\psi_{200}(r,\theta,\phi) = \frac{1}{4\sqrt{2\pi}} a_0^{-3/2}\left(2 - \frac{r}{a_0}\right) e^{-r/2a_0} \qquad (3.8)$$

Again θ and ϕ do not enter; this is a general characteristic of the $l = 0$ (ns) orbitals. They are all spherically symmetric because Y_{00} is a constant. The main difference between this orbital and the 1s is the factor $2 - r/a_0$, which becomes zero at $r = 2a_0$. Because the orbital is spherical, this node occurs on the surface of a sphere of radius $2a_0 = 1.06$ Å, and is called a **nodal sphere** or **radial node.** If the sign of ψ_{200} is taken as positive for $r < 2a_0$, as in Equation 3.8, then the wave function is negative for $r > 2a_0$; that is, the node implies an oscillation in the wave that we have seen to be a normal characteristic of waves. Equation 3.8, however, does not have the form of a sine wave; as mentioned in Section 2.7, only when $V = 0$ (or a constant) do we obtain the ordinary sine wave. Figure 3.4 illustrates the 2s orbital, among others. Note that in the boundary surface diagram for 2s, the radial node is in the interior and does not "show."

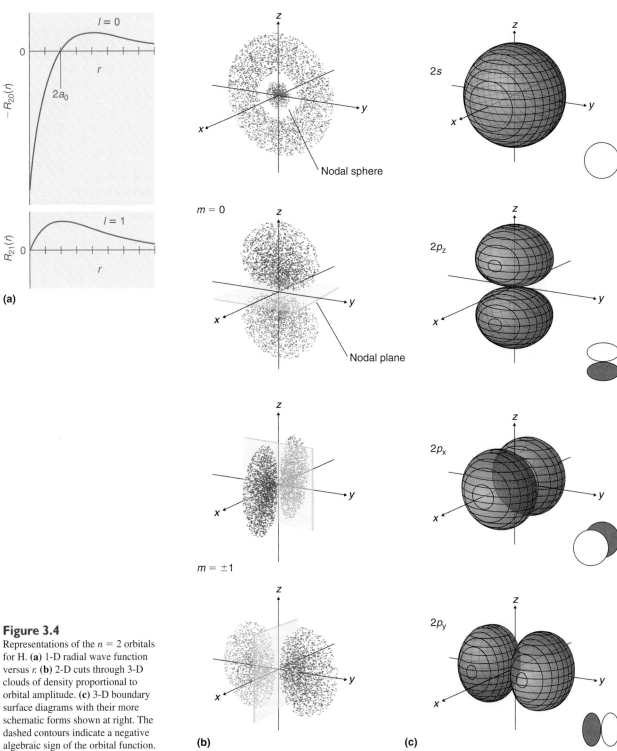

Figure 3.4
Representations of the $n = 2$ orbitals for H. **(a)** 1-D radial wave function versus r. **(b)** 2-D cuts through 3-D clouds of density proportional to orbital amplitude. **(c)** 3-D boundary surface diagrams with their more schematic forms shown at right. The dashed contours indicate a negative algebraic sign of the orbital function. In the schematic forms, shading is conventionally used to indicate the negative lobes.

3.3 Characteristics of the Orbitals: Three-Dimensional Waves

The first angle dependence is encountered in the $2p$ orbitals. The simplest of these is the $m = 0$ orbital, $2p_0$, the product of R_{21} and Y_{10} from Table 3.1:

$$\psi_{210}(r,\theta,\phi) = \frac{1}{4\sqrt{2\pi}} \, a_0^{-3/2} \frac{r}{a_0} e^{-r/2a_0} \cos\theta \tag{3.9}$$

Because ϕ does not appear in Equation 3.9, the $2p_0$ orbital function is the same in any plane passing through the z axis, regardless of the angle ϕ this plane makes with the x axis. We call this **cylindrical symmetry** about the z axis; this is a lower (less general) symmetry than the spherical symmetry of the s orbitals. The polynomial factor in $R_{21}(r)$ is simplified to r/a_0, implying no radial nodes except at the origin. The $\cos\theta$ factor, on the other hand (which you can regard as a first-degree polynomial in $\cos\theta$), passes through zero at $\theta = 90°$, as θ goes from $0°$ to $180°$ (0 to π radians), giving rise to an **angular node,** in this case a **nodal plane** coincident with the x-y plane. Above the x-y plane, $2p_0$ is positive, whereas below it $2p_0$ is negative. Thus, the oscillatory behavior transfers from $R(r)$ to $Y(\theta,\phi)$ as l goes from 0 to 1, but there is still only one **nodal surface** for $2s$ or $2p_0$. This constancy in the number of nodes is characteristic of degenerate orbitals, orbitals with the same energy. The product $r\cos\theta$ obtained by reordering the factors in Equation 3.9 is just the z coordinate as given in Figure 3.1, so the orbital is usually labeled $2p_z$ instead of $2p_0$. Figure 3.4 also illustrates this orbital. Unlike a radial node, the angular node in $2p_z$ is readily visible in the boundary surface diagram. The fact that the orbital "points" along the z axis (has a preferred direction in space) and has two lobes of opposite algebraic sign will be of great importance in describing chemical bonding.

When we encounter the $2p_{+1}$ and $2p_{-1}$ orbitals, we come up against a problem of graphing because of the complex exponential factors $e^{\pm i\phi}$ coming from $Y_{11}(\theta,\phi)$. According to **Euler's formula,** $e^{\pm i\phi} = \cos\phi \pm i\sin\phi$, that is, the ϕ dependence of $2p_{\pm 1}$ has *real* and *imaginary* parts. One way around this is to graph the modulus* of the orbital; we refer to this approach again later in Section 3.6. The method usually used, especially in describing chemical bonding, is to construct two *real* orbitals, $2p_x$ and $2p_y$, from the complex ones by taking **linear combinations** as follows:

$$2p_x = \frac{1}{\sqrt{2}}(2p_{+1} + 2p_{-1})$$
$$2p_y = \frac{1}{i\sqrt{2}}(2p_{+1} - 2p_{-1}) \tag{3.10}$$

You can easily verify using Euler's formula that $2p_x$ and $2p_y$ are strictly real functions, given by

$$2p_x = \frac{1}{4\sqrt{2\pi}} a_0^{-3/2} \frac{r}{a_0} e^{-r/2a_0} \sin\theta \cos\phi \tag{3.11}$$

---

* For a complex number $z = a+ib$, the modulus of z, written $|z|$, is defined as $[a^2 + b^2]^{1/2}$. For $2p_{\pm 1}$, this gives two identical, ϕ-independent orbitals, since $\cos^2\phi + \sin^2\phi = 1$.

$$2p_y = \frac{1}{4\sqrt{2\pi}} a_0^{-3/2} \frac{r}{a_0} e^{-r/2a_0} \sin\theta \sin\phi \qquad (3.12)$$

These differ from $2p_z$ of Equation 3.9 only in the angular factors. The labels x and y arise from identifying the Cartesian coordinates from their spherical polar definitions $x = r\sin\theta\cos\phi$ and $y = r\sin\theta\sin\phi$, respectively, as they appear in Equations 3.11 and 3.12. These real orbitals can no longer be associated with unique m-values, since $m = +1$ and $m = -1$ were mixed in forming them. When they are plotted, they are found to differ from $2p_z$ only in orientation, as shown in Figure 3.4; $2p_x$ points along x, $2p_y$ along y.

It is important for you to remind yourself, as we go through the algebra of the wave functions, that we are trying to represent the hydrogen atom in all of its possible stationary states, and that these orbitals are the 3-D realization of de Broglie's standing waves for the electron. What these waves really mean is impossible to say in terms of our everyday experience; in Sections 3.4, 3.5, and 3.6 we will grapple with this question.

For principal quantum number $n = 3$, l may be 0(3s), 1(3p), or 2(3d, where d stands for *d*iffuse, describing some weak, fuzzy lines in the spectra of atoms). Using the procedures we have detailed for $n = 1$ and 2, we form products $R_{3l}(r)\,Y_{lm}(\theta,\phi)$ from Table 3.1 to obtain the orbital functions, then take linear combinations of these for $\pm m$ as in Equation 3.10 to get real orbitals. Each possible combination of quantum numbers (see Example 3.1) gives a unique orbital, so there are nine in all. For easy reference all of the real orbital functions for $n = 1$, 2, and 3 are collected in Table 3.2, and those for $n = 3$ are illustrated in Figure 3.5.

Comparing the $n = 3$ orbitals with those for $n = 1$ and 2 allows us to begin to see generalizations concerning the structure and appearance of these electron waves. The 3s orbital has two radial nodes, compared to one for 2s and none for 1s. In general, an ns orbital has $n - 1$ radial nodes, the same as the degree of the $R(r)$ polynomial, which has $n-1$ real roots. As l increases from 0 to $n - 1$, the number of nodes stays the same, but l of the radial nodes are exchanged for angular nodes, in the form of nodal planes or **nodal cones.** Algebraically, $R(r)$ loses one root for an increase in l by one, and the degree of the $Y(\theta,\phi)$ polynomial in $\cos\theta$, $\sin\theta$ increases by one. The restriction on the possible values of l is just right to preserve the total number of nodes at $n-1$ and to ensure the same energy for each orbital of a given n.

Figure 3.6 compares the boundary surfaces of the $nl0$ orbitals for $n = 1$, 2, and 3 on the same distance scale. These make it clear that n controls the spatial extent, the size of the orbital. This is in accord with the Bohr orbits, whose radii are proportional to n^2. On the other hand l controls the angular shape, the lobal structure of the orbital; note the similarity between 2p and 3p. And from Figures 3.4 and 3.5, m controls the orientation in space, as well as the details of the nodal structure, in each orbital.

Since n may range to as high a value as you care to choose, l may do so as well. Table 3.1 gives the $Y_{3m}(\theta,\phi)$ functions, where $l = 3$ is labeled an f orbital with f (for *f*undamental) again taken from atomic spectroscopy. Beyond $l = 3$, the orbital labels are alphabetical, $l = 4$ being g, $l = 5$ h, etc. The preceding rules allow you to predict the size, shape, and nodal structure of any orbital.

TABLE 3.2

Hydrogen atom orbitals*

n	l	m	Orbital label	Orbital function ($\rho = r/a_0$)
1	0	0	$1s$	$\dfrac{1}{\sqrt{\pi}} a_0^{-3/2} e^{-\rho}$
2	0	0	$2s$	$\dfrac{1}{4\sqrt{2\pi}} a_0^{-3/2} (2-\rho) e^{-\rho/2}$
2	1	0	$2p_z$	$\dfrac{1}{4\sqrt{2\pi}} a_0^{-3/2} \rho\, e^{-\rho/2} \cos\theta$
2	1	±1	$2p_x$	$\dfrac{1}{4\sqrt{2\pi}} a_0^{-3/2} \rho\, e^{-\rho/2} \sin\theta \cos\phi$
			$2p_y$	$\dfrac{1}{4\sqrt{2\pi}} a_0^{-3/2} \rho\, e^{-\rho/2} \sin\theta \sin\phi$
3	0	0	$3s$	$\dfrac{1}{81\sqrt{3\pi}} a_0^{-3/2} (27 - 18\rho + 2\rho^2) e^{-\rho/3}$
3	1	0	$3p_z$	$\dfrac{\sqrt{2}}{81\sqrt{\pi}} a_0^{-3/2} (6 - \rho) \rho\, e^{-\rho/3} \cos\theta$
3	1	±1	$3p_x$	$\dfrac{\sqrt{2}}{81\sqrt{\pi}} a_0^{-3/2} (6 - \rho) \rho\, e^{-\rho/3} \sin\theta \cos\phi$
			$3p_y$	$\dfrac{\sqrt{2}}{81\sqrt{\pi}} a_0^{-3/2} (6 - \rho) \rho\, e^{-\rho/3} \sin\theta \sin\phi$
3	2	0	$3d_{z^2}$	$\dfrac{1}{81\sqrt{6\pi}} a_0^{-3/2} \rho^2 e^{-\rho/3} (3\cos^2\theta - 1)$
3	2	±1	$3d_{xz}$	$\dfrac{\sqrt{2}}{81\sqrt{\pi}} a_0^{-3/2} \rho^2 e^{-\rho/3} \sin\theta \cos\theta \cos\phi$
			$3d_{yz}$	$\dfrac{\sqrt{2}}{81\sqrt{\pi}} a_0^{-3/2} \rho^2 e^{-\rho/3} \sin\theta \cos\theta \sin\phi$
3	2	±2	$3d_{xy}$	$\dfrac{\sqrt{2}}{81\sqrt{\pi}} a_0^{-3/2} \rho^2 e^{-\rho/3} \sin^2\theta \cos\phi \sin\phi$
			$3d_{x^2-y^2}$	$\dfrac{1}{81\sqrt{2\pi}} a_0^{-3/2} \rho^2 e^{-\rho/3} \sin^2\theta (\cos^2\phi - \sin^2\phi)$

*As in Table 3.1, orbitals for hydrogen-like ions may be obtained by replacing a_0 with a_0/Z.

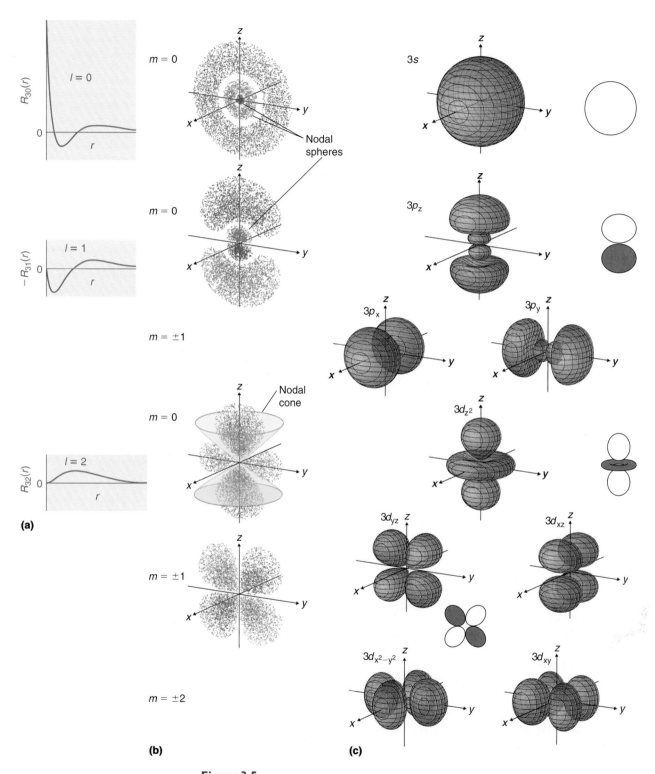

Figure 3.5
Representation of the $n = 3$ orbitals for H. As in Figure 3.4, **(a)** shows the radial parts of the waves, **(b)** 2-D cuts through 3-D amplitude clouds, and **(c)** boundary surface diagrams, with schematic forms. Like $\pm m$ values are grouped in pairs.

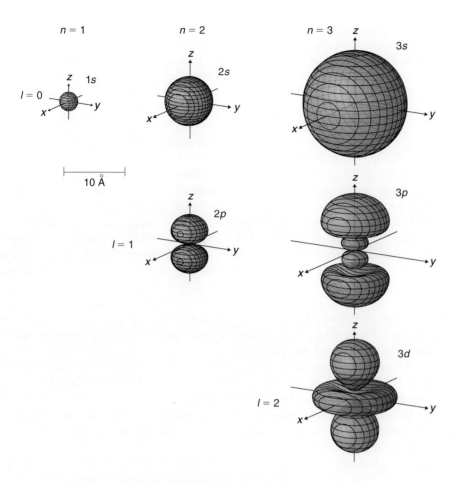

Figure 3.6
Boundary surfaces on the same distance scale for the *nl*0 orbitals of H.

3.4 Born's Interpretation of Orbitals

Is there anything more we can say about the orbitals of the hydrogen atom, other than that they supplant the planet-like orbits of the Bohr model? Schrödinger preferred that not much physical significance be attached to the orbitals themselves; they should be regarded as tools for predicting atomic properties. The more philosophical Bohr was not satisfied with this. By this time devices had been built that could detect electrons *one at a time,* based on the work of Geiger (of α-scattering fame). Bohr reasoned that anything that could create a pulse in one of Geiger's detector tubes couldn't be a smooth, continuous wave. Bohr, the man who first pulled the atom out of the mold of classical physics, had become the philosophical leader of the quantum revolution.

Max Born, who had just been involved with Werner Heisenberg in developing quantum mechanics along a somewhat different path, as discussed further in the next section, was inspired by Bohr to invent a more satisfying interpretation of the Schrödinger waves. Born presumed, as we have mentioned at several junctures, that the wave function must in some way be the quantum version of the orbit of the electron. In those regions of space where the wave function is large, the electron is *more likely* to be found in orbit. This was, after all, the physical basis for requiring the H atom orbitals to decay away at large r, so as to represent a bound state where, as we saw in Section 1.7, the electron is confined close to the nucleus by the Coulomb potential. To get a more quantitative relationship, Born appealed to the wave theory of light, in which the intensity of the light at any point in space is proportional to the *square* of the wave amplitude. In quantum language, the larger this squared amplitude is, the more likely it is that a *photon* will be found at that point. It is entirely consistent with de Broglie's symmetry between light and matter, then, to suppose that *the absolute square of an orbital is proportional to the probability of finding an electron at some point in space*. In mathematical terms,

$$\text{Prob}(r,\theta,\phi) \propto |\psi(r,\theta,\phi)|^2 \tag{3.13}$$

In probability theory $|\psi|^2$ is called a **probability density,** and ψ itself is called a **probability amplitude.** When $|\psi|^2$ is multiplied by a **differential volume element** $dx\,dy\,dz$ (the volume of a tiny cube; in spherical polar coordinates it is $r^2 \sin\theta\, dr\, d\theta\, d\phi$), the result is proportional to the probability of finding an electron within that tiny volume. To make the proportionality sign in Equation 3.13 an equality, a proportionality constant is needed. This is the **normalization constant,** the collection of factors premultiplying each of the functions in Tables 3.1 and 3.2. These constants ensure that the net probability of finding the electron somewhere is exactly 1. To achieve this, the point probabilities must add up to 1 when summed (integrated) over all possible locations in space. These constants are obtained by carrying out this integration over space for the square of each of the orbital functions and then taking the square root of the reciprocal of the result.

With Born's interpretation one can think of a "probability cloud," a strange sort of mist made up of a single electron. This makes $|\psi|^2$ an **electron density** and $-e|\psi|^2$ a **charge density.** Figure 3.7 compares cloud representations of ψ and $|\psi|^2$ for 1s, 2p, and 3d orbitals. You should take note that, although nodal surfaces are left unchanged (zero squared is still zero), any negative parts of the orbitals become positive. The boundary surfaces for ψ and $|\psi|^2$ are identical—a surface of constant ψ is the same as one of constant $|\psi|^2$—and the same schematic forms, omitting shading, suffice to represent $|\psi|^2$. Born's picture (part of what is called the "Copenhagen interpretation" of quantum mechanics) is not the only possibility for ascribing meaning to ψ, but it is the most useful one yet devised. It embodies wave-particle duality in a somewhat less mysterious way, by replacing the certainty of the electron's orbit predicted by Newtonian mechanics by a probability. One way of building up a probability distribution is by taking repeated measurements of a quantity, such as the position of the electron; $|\psi|^2$ could be the end result of thousands of such measurements on a single atom, or individual measurements on thousands of atoms. Either way, quantum mechanics predicts, instead of the trace of a circle, a smeared-out cloud surrounding the nucleus.

Radial Distribution Functions

Armed with the probability idea, we can ask what the probability is that an electron will be found at a certain distance r from the nucleus irrespective of direction. As illustrated in Figure 3.8, we are asking for the probability that an electron is within a thin spherical shell of radius r surrounding the nucleus. We calculate the probability by multiplying the square of the wave function by the volume of the spherical shell, $4\pi r^2\, dr$, and then averaging (integrating) over angles. The angle average exactly cancels the 4π factor, leaving us with

$$\text{Prob}(r) = r^2[R(r)]^2$$

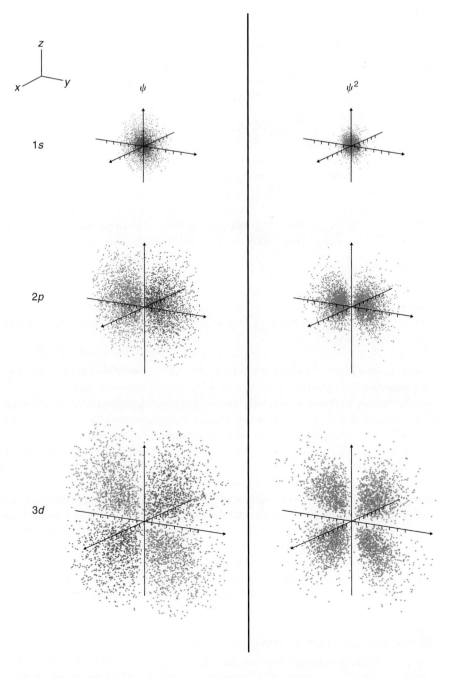

Figure 3.7
Comparison of densities for orbitals and their squares.

The result is called the **radial distribution function** (RDF). It differs from just the square of $R(r)$ by the r^2 factor, which drives the RDF to zero at $r = 0$. With the RDFs we can get a clearer idea of the relation between Schrödinger's waves and Bohr's orbits. Each Bohr orbit has a fixed radius $r_n = n^2 a_0$; the electron in this case is *always* found at $r = r_n$, and Bohr's RDF is a sharp spike located at r_n. The Schrödinger RDF, however, is a smooth curve with one or more peaks (radii of maximum probability),

Figure 3.8
Concept of the radial distribution function. The RDF is obtained by summing the electron density $|\psi|^2$ in the spherical shell.

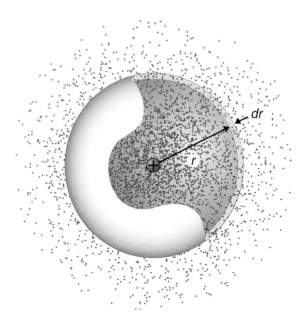

as illustrated in Figure 3.9. As expected from our earlier consideration of nodal structure, the distributions for $l = n-1$ (1s, 2p, 3d, ...) show one peak, and Figure 3.9 indicates that this peak occurs precisely at $r = r_n$. The other distributions oscillate due to the nodal structure in $R(r)$; the nodes are now interpreted as radii of zero probability density for finding the electron. In addition, each curve falls smoothly to zero at large r, with no abrupt cutoff beyond a certain radius. These features raise two questions, which we now consider in turn.

Using the 2s RDF as an example (see Figure 3.9), we can say that the electron "spends most of its time" in the range of r surrounding the outer peak, and much less around the inner peak. But how does the electron "get" from the outer to the inner region, since there is a node in between where the RDF, the probability of finding the electron at r, is zero? Why doesn't the nodal structure "trap" the electron in a particular peak in the RDF? The answer to this riddle lies in the mistake of trying to think of the electron as a particle that can be found at a particular r. If instead the electron is regarded as a wave within the atom, the questions we are asking lose their meaning; the standing wave has amplitude everywhere simultaneously. (These same considerations apply to an electron in, say, a 2p orbital; being a standing wave, the electron doesn't have to travel from one lobe to another.)

In Newton's mechanics, when a particle is confined in a bound orbit by a potential, it cannot stray beyond a certain fixed radius, called the **outer turning point**, as we saw in Sections 1.4 and 1.7. As shown in Figure 3.9, however, an electron in the 1s orbital (or any other orbital) has a small but finite probability of being found beyond the outer turning point! This peculiarity is referred to as **quantum tunneling**; it's somewhat akin to a person being able to walk through walls. We won't refer to tunneling again until Chapter 15, but you should be aware that many chemical reactions involving transfer of electrons (so-called oxidation–reduction reactions) proceed by way of electrons tunneling through potential hills

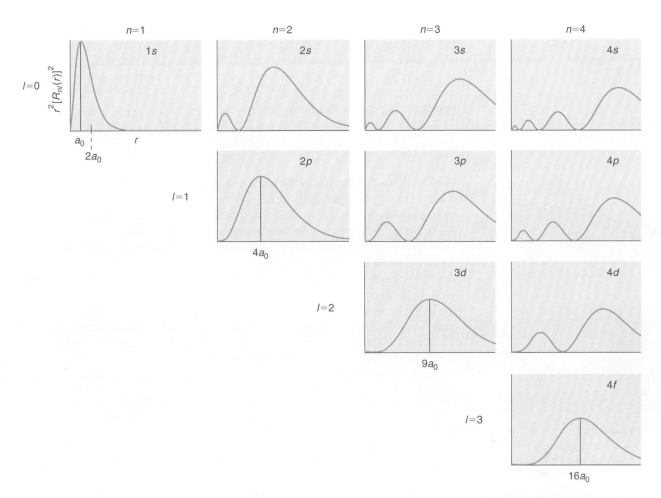

Figure 3.9
Radial distribution functions for the lowest four energy levels of H. The curves for $n = 2$, 3, and 4 have been amplified, and the distance scales for 3 and 4 have been compressed. Each location where the curves touch the r axis corresponds to a nodal sphere in the specified set of orbitals. The $n = 3$ and 4 curves decay smoothly to zero beyond the domain of the plots. For the 1s RDF, $r = 2a_0$ is the outer turning point.

or barriers, and virtually all solid-state microelectronics components operate through a tunneling mechanism.

We will use RDFs again in Chapter 4 when we discuss many-electron atoms. A way of thinking about these puzzling aspects of RDFs is coming up next.

3.5 Heisenberg's Uncertainty Principle

During the same year that Schrödinger developed his wave equation, 1925–1926, Werner Heisenberg, Max Born, and Pascual Jordan hit upon a completely independent approach to a quantum theory of matter, which came to be known as **matrix mechanics.** Although Schrödinger was quick to show that matrix mechanics and wave mechanics were actually two different ways of formulating precisely the same theory, the matrix formulation provided some important insights. It was based on *what we can measure* about the atom rather than a physical picture of the atom itself. Because of the finite size of Planck's constant h, Heisenberg found that attempting to measure the position and momentum of an electron in an atom, the quantities essential to the Newtonian description of an orbit, gives different results depending on the *order* in which the measurements are made. Heisenberg interpreted this to mean that *measuring the position of the electron disturbs its momentum* and, vice versa, *measuring the momentum disturbs its position*. This led to **Heisenberg's uncertainty principle** for position x and momentum along x, p_x:

$$\Delta x \Delta p_x \geq \frac{h}{4\pi} \tag{3.14}$$

Here the Δ's are regarded as *errors* in the measurements of x and p_x; the *product* of these errors can never be smaller than $h/4\pi = 5.27 \times 10^{-35}$ J s. This means that the more precisely we know (or specify) the position of the electron (or some other particle), the less well we can know the momentum, and vice versa.

Equation 3.14 is really another very explicit way of expressing wave-particle duality. As described earlier (see Section 2.1), when a wave passes through a slit, it diffracts, or spreads out, forming circular wavelets on the other side. You can replace the wave by an electron—perfectly legal according to de Broglie—and regard the passing through the slit as determining the electron's position perpendicular to its flight path within a certain error, Δx, the width of the slit. The diffraction spreading on the far side of the slit, which also occurs perpendicular to the initial direction of the electron, reflects the *induced* uncertainty in the momentum Δp_x. This is illustrated in Figure 3.10. A more quantitative analysis using wave propagation for a de Broglie wave, $\lambda = h/p$, leads to Equation 3.14.

Figure 3.10
Two views of Heisenberg's uncertainty principle. In **(a)** a beam of electrons passing through a slit will, according to de Broglie's postulate of matter waves, be diffracted, forming hemispherical wavelets on the outer side of the slit. The spreading corresponds to acquiring a velocity component v_x perpendicular to the beam path, with Δv_x given by the uncertainty formula. In **(b)** light of wavelength short enough to resolve the position of the electron in an atom is so energetic that it destroys the atom. This illustrates the disturbance aspect of the uncertainty principle.

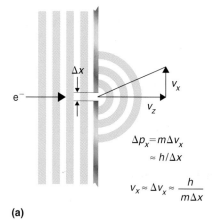

(a)

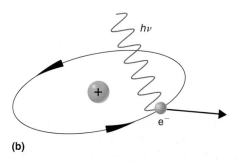

(b)

Just as de Broglie wavelengths are very short for ordinary macroscopic objects (see Section 2.6), Heisenberg's uncertainty principle is a significant constraint on our ability to measure both position and momentum only for very small particles at very small distances. This is brought out by the following examples.

EXAMPLE 3.2

Suppose that by some means an electron in a hydrogen atom with a total energy of -3.40 eV is determined to be between 2.0 and 2.2 Å away from its proton. By using Equation 3.14, estimate the resulting uncertainty in radial momentum of the electron, and compare this to the root-mean-square average total momentum obtained from $K = p^2/2m = -E$.

Solution:

$$\Delta p_r = h/(4\pi \Delta r) = 6.63 \times 10^{-34} \text{ J s}/[(4\pi)(0.2 \times 10^{-10} \text{ m})]$$
$$= 2.64 \times 10^{-24} \text{ kg m s}^{-1}$$
$$p = [2mK]^{1/2} = [(2)(9.11 \times 10^{-31} \text{ kg})(3.40 \text{ eV})(1.60 \times 10^{-19} \text{ J/eV})]^{1/2}$$
$$= 9.96 \times 10^{-25} \text{ kg m s}^{-1}$$

So Δp is more than twice as large as p itself. This is another way of explaining the impossibility of the Bohr orbits, with their fixed total momentum, and of understanding the breadth of the radial distribution functions.

EXAMPLE 3.3

To hit a baseball squarely, a batter must locate the 140-g ball (by eye) within about 1 mm. Use Equation 3.14 to find the uncertainty in momentum of the ball, and compare this to the total momentum of a 92-mph fastball.

Solution:

$$\Delta p = h/(4\pi \Delta x) = 6.63 \times 10^{-34} \text{ J s}/[(4\pi)(1.0 \times 10^{-3} \text{ m})]$$
$$= 5.28 \times 10^{-32} \text{ kg m s}^{-1}$$
$$p = (0.140 \text{ kg})(92 \text{ mph})(1609 \text{ m/mi})(1 \text{ hr}/3600 \text{ s})$$
$$= 5.76 \text{ kg m s}^{-1}$$

The uncertainty in momentum in this case influences only the 32nd decimal place of the total! This completely unmeasurable error means that ballplayers and you in your everyday life don't have to cope with intrinsic uncertainties. The situation would be quite different, however, if h were 34 orders-of-magnitude larger. For a description of the amusing consequences, read George Gamov's *Mr. Tompkins in Wonderland*.

The uncertainty principle also provides interesting answers to these questions: Why does a hydrogen atom have a ground state, and what determines the energy of that state? Just to require the electron to be roughly a distance a_0 from the nucleus implies an uncertainty in momentum $\Delta p \approx h/a_0$.* Whatever the actual momentum of the electron is, it cannot be much less than its own uncertainty, so $p \approx h/a_0$ and $K \approx p^2/m \approx h^2/ma_0^2$. Using the definition of a_0 (Equation 2.15), $a_0 \approx h^2/me^2$, we get $K \approx (h^2/m)(me^2/h^2)^2 \approx me^4/h^2$, which agrees, aside from a factor of $2\pi^2$, with Equation 2.13 for the total energy $E_1 = -K$. The generalization of this sort of argument can be stated as follows: *confining a particle more tightly increases its kinetic energy.* As the example shows, the kinetic energy increases as the inverse square of the confining distance. The result of this tendency is that, in the case of the hydrogen atom, the electron cannot collapse into the proton even though the Coulomb attraction is drawing it in. To do so would rapidly increase the kinetic energy to the point where the electron could not remain bound. It is a general characteristic for small particles in bound states to exhibit a lowest state where the kinetic energy is nonzero; the particle is said to possess **zero point energy,** being constrained by the uncertainty principle from going any lower in total energy.

A Thought Experiment

Heisenberg suggested a *gedanken* (thought) experiment to illustrate how the uncertainty principle works, and what the role of the *observer* (you!) is in all of this. Let's suppose you want to make a movie of the electron in motion in the hydrogen atom. To get sharp pictures, the wavelength of the spotlight you shine on the electron must be short compared to its size or, failing that, compared to the size of its orbit. This implies wavelengths of the order of 0.1 Å or less. But according to Planck's hypothesis, the shorter the wavelength, the higher the energy; $\lambda = 0.1$ Å corresponds to a photon with 10^5 eV of energy! As illustrated in Figure 3.10, when such an energetic photon strikes the electron, the electron will be knocked out of the atom completely—recall that the IE of a hydrogen atom is only 13.6 eV. In attempting to observe the detailed workings of an atom, you destroy it! The uncertainty principle says that it is impossible to make a movie of electrons in atoms. And it's not just a matter of building a better microscope; this is an *intrinsic,* natural limitation on our ability to know the details of small things in motion. Furthermore, the act of observation inevitably *disturbs* the system being observed; in a sense *you* are determining what the atom will do by trying to see it. This is not really as bizarre as it sounds. A child often behaves differently if his parents are watching. A spy who infiltrates higher echelons of a foreign government to observe its workings inevitably influences its policies.

The uncertainty principle does not prevent us from measuring, say, the momentum of the electron ejected by our attempt to observe its orbit, since such a measurement does not require us to know its position precisely. In the same way, we can observe the spectrum of an atom, which reveals its energy levels, because again the positions of its electrons are not being measured simultaneously, nor are we asking exactly *how* an electron hops between levels. We discuss atomic transitions further in the last section of this chapter.

---

* Here we are making a rough estimate, and hence it is customary to omit factors close to unity, such as π and 4. The treatment given here becomes exact (equivalent to the Bohr–de Broglie model) if $\Delta r = a_0/2$, $\Delta p = p$, and Equation 3.14 is taken as an equality.

3.6 Orbitals and Orbits: An Analogy

Is the uncertainty principle, or equivalently, wave-particle duality, going to prevent us from really knowing the details of atomic structure? At the least, we are forced to develop new ways of thinking about the atom. But in point of fact, we have not thrown out all notions of Newtonian mechanics. In particular, the notion of a potential energy function that depends on well-defined coordinates of *particles* survives the translation into Schrödinger wave mechanics. In a sense, quantum mechanics cannot stand on its own, but needs a classical underpinning to make it a complete theory; that is, it is built on a Newtonian foundation. So it is not a meaningless question to ask for the classical analogues of the orbitals—what the electron "would be" doing if it could be treated as a Newtonian corpuscle. The answer is not always a Bohr orbit!

Analyzing the analogous classical motion of the electron gives fuller meaning to the quantum numbers nlm. The motion of the electron may be broken down into in-and-out, *vibrational* motion in which the distance r changes, and round-and-round, *rotational* motion in which the angles θ and ϕ change. The amount of vibrational motion is determined by $n-l-1$, the same quantity that specifies the number of nodes in the radial wave function. The smaller l is for a given n, the more vibration. The amount of rotational motion is determined directly by l; for this reason l is often referred to as the *angular momentum quantum number*. As in vibration, the more nodes in the angular wave function, the more energy that is tied up in rotation. Finally, m, often called the *projection quantum number* because it specifies the projection of the angular momentum on the z axis, determines the possible orientations of the plane of rotation. For fixed l, $m = 0$ requires the plane of rotation to contain the z axis, while $m = l$ fixes the plane of rotation in the x-y plane, perpendicular to the z axis. If m lies between these extremes, the plane is "tilted." The sign of m dictates the sense of rotation; for example, for $m = +l$, the rotation is counterclockwise as you look down on the x-y plane, whereas for $m = -l$ it is clockwise.

What are things like for an electron in a $1s$ orbital? The angular momentum quantum number $l = 0$, so there is no rotation, and the motion is entirely radial or vibrational. An orbit of this type is unknown in classical mechanics with an inverse-square force, but can be approached if the rotational motion is made extremely small. In this limit the orbit becomes highly *elliptical*, with one of the foci being the nucleus, and the minor axis of the ellipse approaching zero. Figure 3.11 illustrates this analogy. This "pencil" orbit is similar to cometary orbits in our solar system, and a far cry from the circular Bohr orbit for $n = 1$. The orbit may point in any direction, and the spherically symmetric $1s$ orbital may be thought of as an average over all possible angles. Please bear in mind that this is only an analogy; as we saw in the last section, no observation we could make would ever reveal an electron in such an orbit.

So as not to overdraw the classical analogy, we will only go as far as $n = 2$ to find orbits with nonzero angular momentum. The $2s$ orbital again has $l = 0$, so the orbit is again a pencil, but longer than $1s$, as illustrated in Figure 3.11. The $2p$ orbitals, $l = 1$, finally display rotational motion and are the closest to having circular Bohr orbits as their classical counterparts. For $m = 0$, $2p_0$ or $2p_z$, the plane of the orbit passes through the z axis, whereas for $m = \pm 1$, $2p_{\pm 1}$, rotation occurs in the x-y plane. The squares of the $2p_{\pm 1}$ orbitals are identical (recall they are not the same as $2p_x$, $2p_y$) doughnut-shaped figures, with the orbit residing within (again see Figure 3.11). As noted earlier, the sign of m reflects the sense of rotation, clockwise or counterclock-

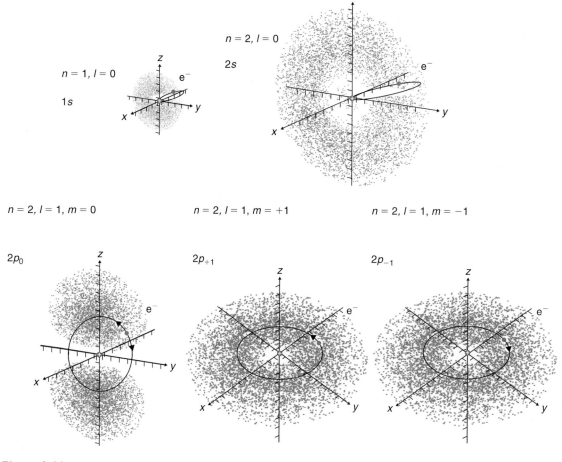

Figure 3.11
Classically analogous orbits of the electron in 1s, 2s, and 2p orbitals. For 1s and 2s, the orbit may point in any direction, and for $2p_0$ the plane of rotation may have any orientation so long as it passes through the z axis. The dot densities represent $|r\psi|^2$ in the plane of the orbit.

wise. To maintain the same energy as 2s, this 2p rotational motion comes at the expense of vibration. The vibrational component cannot be reduced to zero, however, for this would violate the uncertainty principle for the radial coordinate.

As the quantum numbers get larger and larger, the classical analogy is found to get better and better. This is called **Bohr's correspondence principle:** *in the limit of large quantum numbers, the behavior of any mechanical system becomes classical.* Another way of stating this principle is that as the de Broglie wavelength diminishes relative to the dimensions of the system, Newton's laws become valid. This means that quantum mechanics contains Newtonian mechanics as a special case.* For the hydrogen atom, both n and l must be large such that both the radial and angular motion become classical. What is "large"? Roughly $n \approx l \geq 10$. For the states we will be considering in Chapters 4 and 5, the classical analogy is a bit strained, but nevertheless useful.

---

* Perhaps quantum mechanics is also a special case of some other mechanics nobody knows about yet!

3.7 A Qualitative Description of Electronic Transitions

Up until now, we've considered only the standing-wave solutions, the orbitals, of the hydrogen atom. But let's imagine a hydrogen atom, off by itself in its ground state, suddenly bathed in a ray of light (a stream of photons). Recall from Section 2.1 that light waves are in part an oscillating electric field **E** and will therefore exert an electrical force on the electron, given by $-e\mathbf{E}$, the product of the electron's charge and the field strength. The electron now feels not only the pull of the nucleus, but also the pull of the passing light wave. The result is that the electron-wave, the orbital, is temporarily jarred out of its stationary state.* If the frequency of the light satisfies the Bohr condition $\nu = (E_n - E_1)/h$, the electron may hop into (actually become) a new orbital associated with the energy level E_n. Light waves that can excite atoms, by being *in resonance* with the energy level spacing, are, as we've seen, thousands of angstroms in wavelength, much larger than the diameter of an atom. As a result, in the vicinity of an atom the electric field is nearly constant in space at any particular time, and the potential energy of interaction is accurately given by $eE_x x$ if, say, the light wave oscillates along the x axis. The product $e \cdot x$, charge times length, is called an **electric dipole**, and the electronic transitions that normally occur under these conditions are called **electric dipole transitions.** The dipole nature of the interaction gives rise to **selection rules** that dictate the *orbitals* between which transitions can occur. For the hydrogen atom, for example, in the Lyman-α absorption the electron can jump between $1s$ and $2p$ orbitals, but not between $1s$ and $2s$; the general selection rule for H is $\Delta l \equiv l_2 - l_1 = \pm 1$.

Once in an excited state, an atom will not remain there for long; in a matter of a few nanoseconds (10^{-9} s) it falls to a lower state, emitting a photon that again satisfies the Bohr condition.† Because of the quantized energy levels, the electron cannot slowly dribble away its energy, but must give it off in large chunks, as we've already discussed in Chapter 2. Once again the electric dipole selection rule applies, due again to the large wavelength of the emitted light. For example $1s \longrightarrow 2p$ absorption implies $2p \longrightarrow 1s$ emission. The average time that it takes the atom to radiate is called the *radiative lifetime* τ. As you might imagine, the uncertainty principle has something to say about all this. The units of Planck's constant are joule-seconds, energy times time. You can take the uncertainty principle product and manipulate it like this:

$$\Delta x \Delta p_x = \Delta x \Delta p_x \frac{\Delta t}{\Delta t} = \left(\frac{\Delta x}{\Delta t} m \Delta v \right) \Delta t = [mv \Delta v] \Delta t$$

The quantity in brackets is, by differential calculus, a ΔK or ΔE, an *energy uncertainty*, while Δt is a time uncertainty, roughly equal to τ. So the uncertainty principle becomes

$$\Delta E \tau \geq \frac{h}{2\pi} \qquad (3.15)$$

---

* The generalization of the Schrödinger equation that can describe a particle wave that changes with time is $\hat{H}\psi = \frac{ih}{2\pi} \frac{\partial \psi}{\partial t}$, the so-called *time-dependent* Schrödinger equation.

† The natural tendency to fall to lower levels may be thought of as the best the atom can do to emulate the atomic collapse required by Maxwell's theory (see Chapters 1 and 2).

where ΔE is not to be confused with the energy level difference; rather it is the *uncertainty in the energy of the excited state* owing to its finite lifetime τ. If the energy is uncertain, then the transition wavelength is also uncertain, giving rise to a "natural" broadening of the emission line. This natural broadening is very slight, typically one part in 10^9 or so, and is difficult to observe, but it limits the precision with which we can measure the energy-level spacings.

A decade before quantum mechanics was formulated, Albert Einstein analyzed atomic absorption and emission of light and showed that the absorption and emission line intensities are proportional to each other. He further predicted a phenomenon known as **stimulated emission:** an atom in an excited state will be stimulated to emit its photon by the presence of other photons of the same frequency. It is this principle on which **lasers** are based (see Example 2.1). In a gas laser, a few atoms begin by emitting light in all directions. The photons that happen to be emitted along the mirror axis get reflected back into the tube, causing more atoms to become excited. Finally, when the photon density reaches a certain threshold value, atoms already excited are stimulated to emit, leading to laser action.

SUMMARY

Solving the **Schrödinger equation** for the hydrogen atom, one of a small set of real-world problems that can be solved exactly, subject to physically motivated **boundary conditions**, yields **wave functions** or **orbitals** and corresponding quantized energy levels. In **spherical polar coordinates** (r,θ,ϕ) the orbitals are found to have the form $R(r)Y(\theta,\phi)$, the product of a **radial** function $R(r)$ and an **angular** function $Y(\theta,\phi)$. The form of these functions, as well as the value of the energy, is specified by three **quantum numbers**, one for each dimension, n, l, and m, where n is called the **principal quantum number**; for the hydrogen atom, n alone determines the energy, just as in the Bohr model, and can take on the values $1, 2, 3, \ldots$. The number l is called the **azimuthal** or **angular momentum quantum number,** and can be $0, 1, 2, \ldots, n-1$, and m is called the **magnetic** or **projection quantum number**, whose possible values are $-l, -l+1, \ldots, 0, \ldots, l-1, l$. The quantum numbers and their allowed values are a direct result of imposing physical boundary conditions on the orbitals.

The fact that the energy levels depend only on n, but that all three quantum numbers n, l, and m are needed to specify a unique orbital, implies that a number (n^2) of orbitals corresponds to each level. The number of sets of quantum numbers that yield the same energy is called the **degeneracy** of the level. The energy needed to remove the electron from its ground state is called the **ionization energy** (IE), and is the negative of the lowest allowed energy. Energies above $E = 0$ are in the **ionization continuum** and are no longer quantized.

The radial parts of the orbitals are specified by n and l and denoted $R_{nl}(r)$, whereas the angular parts depend on l and m, $Y_{lm}(\theta,\phi)$. Their explicit forms are displayed and discussed, with each being described as a polynomial times an exponential function with the degree of the polynomial determined by the quantum numbers. The notation $s, p, d, f, g, \ldots$, corresponding to $l = 0, 1, 2, 3, 4, \ldots$, for labeling the orbitals is introduced. Various graphical representations of the orbitals are presented; the most useful of these will later prove to be the **boundary surface diagram,** a three-dimensional surface that encloses most of the region of large orbital amplitude. The orbitals display 3-D **nodal surfaces** (surfaces on which the orbital amplitude is zero) of two types: **radial nodes,** which take the form of **nodal spheres,** and **angular nodes,** which are **nodal planes** or **cones.** For an orbital with a given n, there are $n-1$ total nodes, l of them being angular and the rest $(n-l-1)$ radial. Forming real orbitals from the complex ones that arise for $m \neq 0$ is illustrated for $2p$ orbitals. In general, n controls the size of the orbital, l controls its lobal structure, and m its orientation.

The most useful interpretation of the orbitals, due to Born, is as **probability amplitudes,** whose absolute squares give the **probability density** of finding the electron in a small volume surrounding a particular point in space. With this interpretation, the square of the orbital $|\psi|^2$ gives the **electron density** or **charge density.** The orbitals are therefore normalized so that, when $|\psi|^2$ is integrated over all space, the result is unity. The probability that an electron will be found at a certain distance from the nucleus irrespective of angles is called the **radial distribution function** (RDF), and is given by $r^2[R(r)]^2$. The RDF is used to discuss the relation between the orbitals and the Bohr orbits, the wave nature of the electron, and to introduce **quantum tunneling** into regions of space forbidden by Newton's laws.

Heisenberg's uncertainty principle helps us to understand the lack of a well-defined orbit for the electron in terms of our intrinsic inability to measure position and momentum simultaneously to arbitrary precision. The mathematical inequality $\Delta x \Delta p_x \geq h/(4\pi)$ embodies this natural limitation, where the Δ's are errors or uncertainties, and h is again Planck's constant. The more accurately the position is known, the less well can we know the momentum, and vice versa. This means we cannot photograph the electron in orbit without disturbing the orbit. Confining an electron (or other small particle) to a small range of distances Δx induces a momentum uncertainty or a nonzero kinetic energy, leading to a lowest energy state where the particle is still in motion. In this situation the particle is said to possess **zero point energy**.

Despite the uncertainty principle and wave-particle duality, we are free to imagine what the electron would be doing if h were zero. The orbitals are analyzed in terms of radial and angular motion, and the meaning of the quantum numbers is thereby clarified. According to **Bohr's correspondence principle,** as the quantum numbers get larger, the classical description improves.

Finally, the mechanism by which an atom interacts with light to produce a line spectrum is described qualitatively. In the hydrogen atom, only transitions between orbitals differing by one in the l quantum number are allowed, according to **selection rules** for **electric dipole transitions.** The finite radiative lifetime τ of an atom in an excited state leads to a natural broadening of emission lines dictated by the uncertainty principle in the form $\Delta E \tau \geq h/2\pi$, where ΔE is an uncertainty in the energy of the excited state. Einstein showed that the emission and absorption line intensities are proportional to each other, and predicted the phenomenon of **stimulated emission,** the principle on which lasers operate.

EXERCISES

Quantum numbers, energy levels, and orbital labels

1. Give all possible values for the missing hydrogen atom quantum number in each case:
   a. $n = 3, l = 2, m = ?$
   b. $n = 5, l = ?, m = 0$
   c. $n = 4, l = ?, m = -2$
   d. $n = ?, l = 2, m = 1$
   e. $n = 10, l = 0, m = ?$

2. For the cases of Exercise 1, give the corresponding orbital labels.

3. Which of the following sets of hydrogen atom quantum numbers are degenerate (have the same energy)?

   $nlm = 200, 522, 322, 31-1, 210, 100$

4. What is the ionization energy in electron-volts of a hydrogen atom in a $4p$ state? A $100s$ state?

5. Describe the ionization energy of a hydrogen atom in its ground state in terms of work and in terms of potential energy.

6. Give the number of orbitals and their labels for all orbitals with
   a. $n = 3$
   b. $n = 4, l = 2$
   c. $n = 5, l = 1$
   d. $n = 2, l = 1, m = \pm 1$
   e. $n = 3, l = 2, m = \pm 1$

7. Use Euler's formula to verify the $3p_x$ and $3p_y$ orbital functions of Table 3.2 using the appropriate functions from Table 3.1 (see Equation 3.10).

8. State the possible quantum numbers nlm that match the following orbitals:
   a. $2p_z$
   b. $4p_x$
   c. $3s$
   d. $3d_{x^2-y^2}$
   e. $5d_{yz}$

Orbital properties and shapes

9. Use Table 3.1 to construct explicitly a normalized orbital with $n = 4, l = 3$, and $m = 0$. Suggest a label for this orbital, and sketch its boundary surface.

10. Give the number of radial nodes (nodal spheres) and the number of angular nodes (nodal planes or cones) for each of the following orbitals:
    a. 1s
    b. 6s
    c. 3d
    d. 5d
    e. 4f
    f. 2p

11. Sketch radial wave functions $R_{nl}(r)$ on a single set of labeled axes for 1s, 2s, and 3s orbitals, explicitly calculating and indicating the values of r (in units of a_0 and in angstroms) at the nodal positions. How are these nodes manifested in the 3-D orbitals?

12. Make a plot on a common set of labeled axes of the (unnormalized) angular functions $Y_{10}(\theta) = \cos\theta$ and $Y_{20}(\theta) = \frac{3}{2}\cos^2\theta - \frac{1}{2}$ versus θ for $0 < \theta < \pi$. Explicitly find the nodes in these functions, and indicate them on your plot. How are these nodes manifested in the corresponding 3-D orbitals $2p_z$ and $3d_{z^2}$?

13. Using Table 3.2, show that the radial node in a $3p_z$ orbital occurs at $r = 6a_0$ and the angular node in the x-y plane, $\theta = 90°$.

14. Without referring to the figures, sketch signed boundary surfaces on appropriate axes of the following orbitals: 1s, 3s, $3p_x$, $2p_y$, $3d_{z^2}$, $3d_{xy}$.

15. Without referring to the figures, sketch signed boundary surfaces on appropriate axes of the following orbitals: 2s, 4s, $4p_z$, $3p_y$, $3d_{x^2-y^2}$, $3d_{yz}$.

16. The orbitals have units embedded in their normalization constants given in Table 3.1 or 3.2. Determine the units of the orbitals and their squares. What is the meaning of these units according to Born's interpretation?

17. At what point(s) in space (r, θ, ϕ) is an electron in a $2p_z$ state most likely to be found? You may approach this problem graphically, by plotting $[R(r)]^2$ and $[Y(\theta,\phi)]^2$ for the $2p_z$ orbital, or by using calculus. Using calculus, maximum values of a function are located by taking first derivatives, for example, $d(R^2)/dr$, setting them to zero, and solving for the coordinate. Note that when finding the extrema of a function, premultiplying constants may be omitted.

Radial distribution functions

18. Using the radial function from Table 3.1, form the radial distribution function (RDF) for the 1s orbital, expressing it in terms of $\rho = r/a_0$. Setting $a_0 = 1$ in the normalization constant for convenience, compute the RDF for $\rho = 0.2, 0.5, 1.0, 1.5, 2.0, 2.5,$ and 3.0 ($r = 0.2a_0, 0.5a_0, a_0, 1.5a_0, 2a_0, 2.5a_0,$ and $3.0a_0$) and plot it on suitable axes. Verify graphically and by setting $d(\text{RDF})/dr = 0$ that the 1s RDF peaks at $r = a_0$. Estimate the area under your curve by counting blocks on your graph paper.

19. Using your graph from Exercise 18, estimate the total probability that an electron in a 1s state will be found beyond the classical outer turning point $r = 2a_0$. (This is just the area under the curve for $r \geq 2a_0$; this probability could also be obtained by finding the integral $\int_{2a_0}^{\infty} r^2 [R(r)]^2 \, dr$).

20. Find the radii in units of a_0 and in angstroms at which the radial distribution functions for 2s and 2p states have local maxima. Using Table 3.1, form $\rho^2 R^2$ for each state, take its first derivative with respect to ρ, set the resulting expression to zero, and solve for ρ and hence $r = \rho a_0$. Does the radius you obtained for 2p agree with that from Exercise 17? Briefly explain.

21. The orbitals for which $l = n-1$ (1s, 2p, 3d, ...) have radial wave functions $R(\rho) \propto \rho^{n-1} e^{\rho/n}$, where $\rho = r/a_0$. Show that the corresponding radial distribution functions have single maxima at $\rho = n^2$, identical to the radii of the corresponding Bohr orbits.

Uncertainty principle

22. A 5-kg watermelon is packed in a crate so that it can shift its position by only 1 cm in any direction. Estimate the uncertainty in the watermelon's momentum in a given direction and its zero point kinetic energy. Compare this kinetic energy with that arising from "thermal agitation," $k_B T$ at 298 K. Repeat the calculation for a beryllium atom packed in a beryllium metal crystal, where the constraint on motion is 0.05 Å.

23. An electron is launched from earth orbit destined for the moon 230,000 mi away. If the electron emerges from its launcher through a 1-mm-diam opening with an energy of 10 eV, give a rough estimate of the uncertainty in its position when it hits the moon. [See Figure 3.10(a).]

24. From the $1s$ radial distribution function plotted in Figure 3.9, or from the graph created in Exercise 18, estimate the uncertainty in radius Δr from the width of the curve at half-maximum. Then use the uncertainty principle to estimate the uncertainty in momentum Δp_r and the zero point kinetic energy K (eV).

25. The kinetic energy of rotation of an electron about the nucleus is given by $K_{\text{rot}} = l(l+1)h^2/(8\pi^2 m_e r^2)$, where $l(l+1)h^2/4\pi^2$ is the square of the angular momentum. For the case of the $2p$ orbital discussed in Section 3.6, find the (average) kinetic energy remaining in vibration for $r = 4a_0$. Assuming this vibrational energy is "zero point energy" for the radial motion, estimate the corresponding uncertainty in position Δr (Å).

Correspondence principle

26. A general underpinning of Bohr's correspondence principle is provided by the shortening of the deBroglie wavelength λ relative to the size of the system as the quantum numbers increase. Find an expression for the ratio λ_n/r_n for the Bohr model of the H atom, and show that it becomes vanishingly small as $n \to \infty$. You may use the results of Exercise 2.26.

27. According to the correspondence principle as applied to the hydrogen atom, when the quantum numbers are large, the frequency $\nu_{n,\,n-1}$ of a transition from state n to $n-1$ approaches the classical frequency of motion of the electron's orbit $\nu_n = v_n/(2\pi r_n)$, as required by Maxwell's wave theory of light. Show that this holds, if Bohr's circular orbits are used to find ν_n. You may use the results of Exercise 2.26.

Atomic transitions

28. An electron in a $4d$ state cannot fall directly to the ground $1s$ state because l would have to change by two units, in violation of the selection rule $\Delta l = \pm 1$. Suggest two paths by which the electron can eventually reach the ground state.

29. The radiative lifetime of the $2p$ state of hydrogen is 2.13 ns. Find the uncertainty in the transition energy in J, eV, and cm^{-1} and also the corresponding spread in transition wavelength $\Delta \lambda$ in nm. Compute $\Delta\lambda/\lambda$ for the Lyman-α line.

30. Recently techniques have been developed for producing ultrashort laser pulses in the picosecond to femtosecond (10^{-12} to 10^{-15} s) regime. This means that the "lasing medium" must emit its photons in this twinkling. Use the energy-time uncertainty principle to find the intrinsic spread in energy (in J, eV, and cm^{-1}) and wavelength (in nm) for 532-nm laser pulses of widths 1 ps and 1 fs.

CHAPTER 4

Atoms with Many Electrons and the Periodic Table

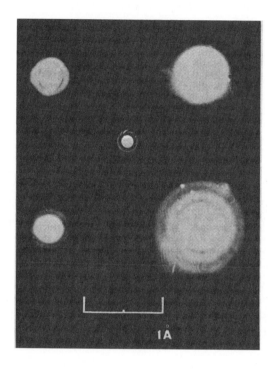

Holograms of electron densities in atomic neon (top) and argon (bottom) at 260,000,000 power. Short (left) and long (right) exposures are reproduced. The diffuse rings in the atomic photographs are instrumental diffraction effects, and are *not* due to shell structure. See Section 4.7 for a brief description of the experiment.

The underlying physical laws necessary for . . . the whole of chemistry are thus completely known, and the difficulty is only that the exact application of these laws leads to equations much too complicated to be soluble.

Paul Dirac (1929)

CHAPTER OUTLINE

4.1 Hydrogen-Like Orbitals for Many-Electron Atoms

4.2 Electron Spin: Evidence for a Fourth Quantum Number m_s

4.3 Pauli's Exclusion Principle

4.4 The Aufbau Principle and Electron Configurations of the Elements

4.5 Periodic Properties of the Atoms of the Elements

4.6 Computers in Chemistry: The Many-Electron Problem

4.7 "Seeing" Atoms

It is fair to say that the discovery of the laws of quantum mechanics alluded to by Dirac has changed the face of chemistry. As we will see, it is possible to generalize the Schrödinger equation to apply to any number of interacting particles, with the way made particularly facile by the pairwise interaction property of Coulomb's law. The complication arises from the addition of three new coordinates for each particle (electron) added to an atom or molecule, making the wave function a multidimensional entity impossible to extract exactly—to say nothing of comprehending it. Nonetheless, in the same year Dirac made this oft-quoted statement, the atomic physicists D. R. Hartree and J. C. Slater were working out approximate but practical methods for dealing with the complexities of many-electron atoms. These reduced the impossible to merely extremely difficult and tedious calculations, whose results, although obviously approximate, proved to be unexpectedly accurate. The advent of high-speed computers in the late 1950s began the routine and rapid application and extension of these early methods. (We discuss the impact of computers on chemistry further at the end of this chapter.)

The key idea that Hartree and Slater demonstrated to be useful is that, to good accuracy, *each electron in a many-electron atom can be regarded as occupying, or being represented by, a hydrogen-like orbital*. This is an example of a model, the **orbital model**. Although this model is not exact, nor even always well defined, describing chemical phenomena in any detail is impossible without it.

Paul A. M. Dirac (1902–1984) lecturing on chemical bonding. Dirac held the Lucasian professorship at Cambridge University, the same chair once held by Newton. His relativistic quantum theory of 1928 simultaneously validated the notion of electron spin and predicted the existence of antimatter.

4.1 Hydrogen-Like Orbitals for Many-Electron Atoms

Let's consider the simplest atom next to hydrogen, the helium atom, $Z = 2$, with two electrons. If we label the electrons 1 and 2 and their coordinates $\mathbf{r}_1 \equiv (r_1, \theta_1, \phi_1)$, $\mathbf{r}_2 \equiv (r_2, \theta_2, \phi_2)$, then the Hamiltonian operator $\hat{H}$ is given by

$$\hat{H} = \hat{K}_1 + \hat{K}_2 - \frac{2e^2}{r_1} - \frac{2e^2}{r_2} + \frac{e^2}{r_{12}} \tag{4.1}$$

where each electron contributes a kinetic energy term $\hat{K}_i$ (an operator of the same sort as in the hydrogenic Schrödinger equation, Equation 3.1, except with subscripts on the coordinates), $r_{12} \equiv |\mathbf{r}_1 - \mathbf{r}_2|$ is the interelectron distance, and the Coulomb interactions are pairwise, as illustrated in Figure 4.1. This generalization of the Hamiltonian is clear if you remember that it represents the total mechanical energy of the system. The nucleus is fixed at the origin of our coordinate system, so it does not contribute a kinetic energy term.* The $\hat{K}$'s and the electron–nucleus attractions are familiar from the hydrogen atom, but the last term in Equation 4.1, the electron–electron repulsion, is new. It is this term, and many more like it for bigger atoms, that make the Schrödinger equation impossible to solve exactly. This same repulsion is what caused the Bohr model to fail for He and other atoms.

The problem is that both electrons are mobile, so that the interelectron distance r_{12} does not bear any fixed relation (length or orientation) to the electron–nucleus distances. One statement that can be made at the outset is that, because of the natural tendency of a mechanical system to minimize its potential energy (see Sections 1.4 and 1.7), the electrons will try to avoid each other. For example, if one electron is on one side of the nucleus at a particular instant, the other electron will move to the far side, in order to remain close to the nucleus (lower electron–nucleus potential energy) while keeping away from the other electron (lower electron–electron potential energy). This sort of *correlated motion* is expressed in classical terms, but all we have to do is to speak in probabilities to become more quantum mechanical. The quantum way of

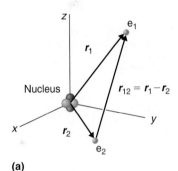

(a)

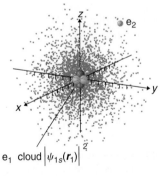

e_1 cloud $|\psi_{1s}(r_1)|^2$

(b)

Figure 4.1
(a) Interparticle distances in the helium atom. The sizes of the nucleus and electrons are greatly exaggerated. Coulomb forces act between pairs of particles. (b) Screening of one electron by another in the orbital model. Electron 1 is screened in the same way by the orbital of electron 2.

---

* This doesn't mean that helium atoms don't move! In the absence of external forces, the motion of an atom as a unit through space is independent of the relative motion of its parts. At ordinary temperatures, however, an atom moves quite sluggishly compared to its dancing electrons (see Chapter 9).

saying this is that the wave function for the atom will have larger amplitude when $\mathbf{r}_1$ and $\mathbf{r}_2$ are small but $\mathbf{r}_{12}$ is large.

Speaking of the wave function, it is what we would get if we could solve $\hat{H}\Psi = E\Psi$ with $\hat{H}$ given by Equation 4.1. For He Ψ is a function of six coordinates, $\Psi(r_1,\theta_1,\phi_1,r_2,\theta_2,\phi_2)$ or $\Psi(\mathbf{r}_1,\mathbf{r}_2)$. After our travails with 3-D wave functions, it's no fun to think of how to represent this 6-D one, even if we could find it. But a shrewd look at Equation 4.1 suggests what Hartree and Slater showed in the early 1930s: that the last repulsive term cancels only about one-quarter of the total potential energy even if all the interparticle distances were comparable, and that we should not be too badly off if each electron were treated as if it were *independent*. Without going into the mathematical details, if the last term in Equation 4.1 is (temporarily) neglected, then the total wave function can be written as a *product* of functions each depending on the coordinates of only one electron, and each a hydrogen-like orbital,

$$\Psi(\mathbf{r}_1,\mathbf{r}_2) \approx \psi_a(\mathbf{r}_1)\,\psi_b(\mathbf{r}_2) \tag{4.2}$$

where the lowercase ψ's represent the one-electron orbitals.* This is an independent-electron approximation. The electron–electron repulsion can then be taken into account to some extent once the electrons are assumed to be independent waves. The electron density arising from electron 1, represented by $|\psi_a(\mathbf{r}_1)|^2$, will **screen** the nuclear charge to a small extent from electron 2, and vice versa, as illustrated in Figure 4.1. This **screening effect** can be incorporated into the orbitals by replacing the actual nuclear charge $Z = 2$ by an **"effective" nuclear charge** Z_{eff} wherever Z appears in the orbital functions. See Tables 3.1 and 3.2 and their footnotes for writing hydrogen-like orbital expressions. The best value for Z_{eff} for He is 1.6875, when both electrons are assumed to be $1s$ orbitals. The wave function in this case should represent the ground state of helium, and is easily found from Table 3.2 to be

$$\Psi(\mathbf{r}_1,\mathbf{r}_2) \approx \frac{Z_{\text{eff}}^3}{\pi a_0^3}\left(e^{-Z_{\text{eff}}r_1/a_0}\right)\left(e^{-Z_{\text{eff}}r_2/a_0}\right) \tag{4.3}$$

with $Z_{\text{eff}} = 1.6875$ for He. With each electron "in" an identical $1s$ orbital, the **electron configuration** of the ground state of He is written $1s^2$, where the 2 superscript indicates the "occupancy" of the $1s$ level. Just how good is the orbital approximation? The easiest comparison is with the total energy of the atom. The ionization energy (IE) of helium derived from its ultraviolet spectrum is 24.59 eV. After ionization the one-electron He$^+$ ion remains, and its IE is easily calculated from the Bohr formula of Equation 2.17 to be 54.40 eV. Note that the second electron is harder to remove than the first, due to the loss of screening. The total energy of He is thus $-24.59 - 54.40 = -78.99$ eV. When the orbital approximation, Approximation 4.3, is substituted into the full Schrödinger equation with the Hamiltonian of Equation 4.1, the total energy is found to be -77.46 eV; the percent error is $(-78.99 + 77.46)/(-78.99) \times 100\% = 1.9\%$, accurate enough for many purposes. The absolute error is 1.53 eV,

---

* Without electron–electron repulsion, the Hamiltonian of Equation 4.1 splits into two hydrogenic (He$^+$) Hamiltonians, $\hat{H} = \hat{H}_1 + \hat{H}_2$, with each $\hat{H}_i$ containing only the coordinates of one electron. When Equation 4.2 is substituted into the approximate version of $\hat{H}\Psi = E\Psi$, it is found to satisfy the equation exactly if the ψ's are hydrogen-like orbitals.

with the orbital approximation being higher*; as we'll see later, this error is comparable to the chemical energies commonly encountered in reactions. This 1930 result for helium nonetheless rightfully encouraged chemists to regard orbitals as a useful concept, while more elaborate wave functions than Equation 4.3 were found to give energies within 0.01 eV of the experimental value, bolstering the then-fledgling theory of quantum mechanics. Even though the Schrödinger equation for He cannot be solved exactly, we can come very close.

Generating hydrogen-like orbitals with effective nuclear charges is easy given Tables 3.1 and 3.2. Only the radial parts of the orbitals are modified from that for hydrogen, by simply inserting Z_{eff}; the angular parts remain the same. Thus all the orbital plots and diagrams of Chapter 3 carry over to many-electron atoms, with only a change in scale necessary. Based on a series of calculations of total atomic energies, Slater in 1930 gave a set of rules for estimating Z_{eff} for the orbitals of any atom.

Knowing how to construct atomic orbitals, we can now ask how they are "occupied" in atoms. Bear in mind that when we speak of an electron being "in" an orbital, we really should say that the electron *is* an orbital. But if an orbital is thought of in a mathematical sense, as a function taken from the tables, then the notion of orbital occupation has some meaning. An orbital has a *mathematical* existence independent of whether it is used in a particular atom, but takes on a *physical* existence when an electron is actually being described by it. Thus, although in Equation 4.3 each electron is its own 1s orbital, we speak of a *doubly occupied* 1s orbital, or a 1s orbital holding two electrons, since the two orbital functions are identical.

If we examine the next atom in Mendeleyev's table, lithium, with $Z = 3$, we find its spectrum to resemble that of hydrogen more than that of helium. From its spectrum the IE of Li is 5.39 eV, much lower than that of either H or He. In addition, one is hard-pressed to find three more chemically distinct elements, as we will discuss in the following chapters. You might expect the ground state of Li to result from adding a third electron to the 1s level, giving $1s^3$. But in that case lithium's IE would be even higher than helium's, and it would not be the highly chemically reactive metal that it is, but rather more inert than He, which is already the most noble of the noble gases. So the ground state of Li is not $1s^3$, but, it turns out, $1s^2 2s^1$. Why? The answer literally lies in another dimension, in an intrinsic property of the electron called **spin.**

4.2 Electron Spin: Evidence for a Fourth Quantum Number m_s

In the dozen years between the Bohr theory and the advent of wave mechanics, Bohr and others worked extensively on correlating the location of an element in the periodic chart and the electronic structure of its atoms. It was Bohr who first gave us the concept of electron shells as an explanation for the periodic behavior of the elements; Newlands' law of octaves resulted from exhausting the capacity, eight, of a given shell for electrons, forcing additional electrons into higher orbits. But this was pretty

---

* According to the *variation theorem,* due to Eckart, the total energy obtained from any approximate wave function must be higher than the true energy. In the present case, it is easy to understand why the orbital-model energy is higher—the electrons, being in separate orbitals, can't cooperate (or *correlate*) in order to lower the potential energy.

much an after-the-fact rationalization, and could not have been predicted from the Bohr orbits. Nonetheless, Bohr deserves much credit for taking Mendeleyev's table very nearly to its modern form.

As often happens, the answer to the electron configuration enigma was to come from the resolution of a seemingly unrelated problem. Recall from Chapter 2 how the D-line of sodium is actually a closely spaced doublet (see Figures 2.4 and 2.8). This implies that either the upper or the lower energy level is *split* into two closely spaced levels. The orbitals involved in the D-line transition are 3s and 3p as Figure 2.8 indicates. Why 3p is higher in energy than 3s is a question we consider in the next section (recall that in hydrogen these orbitals are degenerate). The 3p orbital, as we discussed in Section 3.6, involves rotation of the electron around the Na^+ ion core, while 3s involves only vibration and no rotation. Now a circulating charge will generate a magnetic field; this is the principle on which electromagnets are based. In 1925, on the eve of quantum mechanics, Uhlenbeck and Goudsmit showed that the only plausible explanation for the D-line doublet (as well as other "fine structure" in atomic spectra) was to ascribe to the electron an *intrinsic angular momentum* or **spin** *s*, and an accompanying magnetism due to the charged electron's spinning on its axis. The **spin quantum number** *s* can only have the value ½ and its **projection** m_s on any axis can only be +½ or −½, referred to as "spin up" and "spin down," respectively. In everyday terms, each electron is itself a tiny magnet with north and south poles, with the north pole only able to point "up" or "down," and not anywhere in between. Then, as illustrated in Figure 4.2, the D-line doublet is readily explained as being due to the splitting of the 3p orbital by favorable and unfavorable alignment of the orbital and spin magnetic fields. As you have undoubtedly experienced, if two bar magnets

Figure 4.2
Uhlenbeck and Goudsmit's explanation for the doubling of the sodium D-line using the concept of electron spin. The **B**'s are magnetic fields due to orbit and spin. The splitting of magnetic sublevels is 17.2 cm^{-1}.

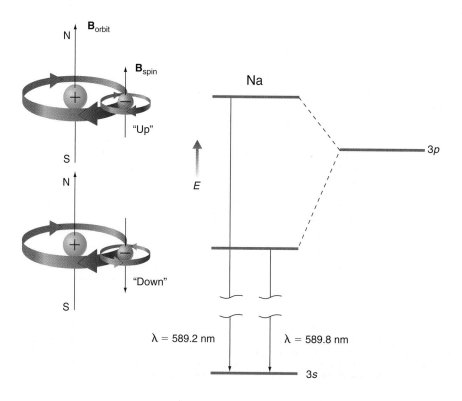

are brought side by side with their north poles adjacent, they repel; whereas if north is adjacent to south, they attract. Translating this observation into potential energy terms, the potential energy is higher when like poles are adjacent. By analogy, the energy of the $3p$ orbital is slightly higher when m and m_s have the same sign ($m = +1$, $m_s = +\frac{1}{2}$) than when their signs are opposed, for the signs indicate the sense of rotation, and hence the direction of the magnetic fields. Because $3s$ can only have $m = 0$, no splitting of this sort can occur.* As Figure 4.2 shows, this yields two closely spaced spectral lines instead of a single line.

The postulate of electron spin clarified so many mysteries in atomic spectroscopy that it was rapidly accepted; the spin of the electron is now regarded as just as fundamental a property as its mass and charge. The idea was easy to accept again because of an analogy with the solar system, where the earth, as well as the other planets, spins on its axis as well as revolving about the sun. The spin of the electron, however, is **quantized in space,** with the directions "up" or "down" depending on the frame of reference. In the case of the spin-orbit magnetic interaction described earlier, a preferred spatial direction within the atom is defined by the direction perpendicular to the plane of rotation of the electron, and the spin may only point along or against this direction. Another peculiarity is that the spinning electron generates *twice* the magnetic field that it should, when compared to the orbital magnetism. These quantum properties imply that the word "spin" is really just a convenient label for a certain property the electron possesses, just as an "orbit" for the electron has no everyday meaning.

At this point (1925) electron spin was a convenient postulate that aided in interpreting atomic spectra, but, just as in the case of Planck's quantum hypothesis, further developments during the next few years were to firmly entrench this feature of nature.

The Stern-Gerlach Experiment

Spectral lines are always a reflection of the properties of two *different* states of an atom, and inferences drawn from multiple lines concerning a particular state can always be questioned. In 1926 Stern and Gerlach reported the results of an experiment that cleanly measured the magnetic properties of a *single state* of an atom. As illustrated in Figure 4.3, they passed a beam of sodium atoms in their ground states through a strong, inhomogeneous (changing rapidly in strength over a small distance) magnetic field. As we'll see shortly, Na in its ground state has one electron in a $3s$ orbital, and all other electrons in inner orbitals arranged in such a way that all spin magnetism is cancelled (like a horseshoe magnet with a keeper). Because the $3s$ orbital has no rotational motion, the only way the atom can respond to the external field is through any intrinsic magnetism of the $3s$ electron. As shown in the figure, the result of the experiment was to split the original beam into two!

What had happened was that the big magnet established a preferred direction in space, and the $3s$ electrons were obliged to line up with their north poles either

---

* Nuclei can also have spin, but due to their larger mass, they spin more slowly and the associated magnetic moment is much smaller. When this nuclear magnetism interacts with the electron's, it gives rise to so-called "hyperfine" structure in atomic spectra, which has been observed even for s orbitals. The phenomenon of *nuclear magnetic resonance* (see Chapter 16) is based on nuclear spin.

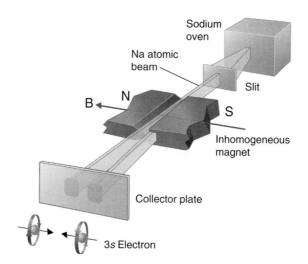

Figure 4.3
Stern and Gerlach's 1926 magnetic deflection experiment on a beam of sodium vapor. The oven produces vapor by heating a Na metal sample. The experiment was carried out in vacuum, and run continuously until measurable deposits of metal were formed on the collector plate. The deposits were spread out due to the distribution in speeds of the Na atoms. (Stern and Gerlach had been doing this type of experiment for the previous 5 years; the first success was found with a silver atom beam. At this point they did not yet realize that electron spin was responsible for splitting the beam in two.)

along or against the north pole of the magnet. Those aligned north-to-north were repelled, and moved to where the field was weaker, while those favorably disposed were drawn into stronger field. The lack of a "straight-through" beam with the magnet on was clear evidence for the existence and two-valued nature of the electron's magnetism.

Electron spin (or whatever the quantum version of "spin" is) was a reality, but how did it arise? There didn't seem to be any room in the Schrödinger equation for another quantum number. But that was because the Schrödinger equation is cast in only three dimensions.

Dirac's Relativistic Electron

In 1928 a young British scientist named Paul A. M. Dirac succeeded in using Schrödinger's idea to find a quantum mechanical equation consistent with Einstein's 1905 special theory of relativity. The result, now known as the **Dirac equation,** resolves itself into four coupled Schrödinger-like equations. When these equations were solved for the hydrogen atom (using the same techniques Schrödinger had employed), a *fourth quantum number* arose that could be identified with the spin quantum number m_s. You can appreciate this development by relating it to Einstein's conclusion that our world is actually four dimensional, three spatial dimensions and the time. To satisfy the de Broglie condition in 4-D demands four quantum numbers. This means that electrons are naturally required to have spin; it's a law of nature, not just a postulate.

Dirac's theory was "the clincher"; no one after this point doubted the spin as a physical attribute. But, as in most other groundbreaking developments, new and puzzling aspects arose. Besides the ordinary spin-up and spin-down electrons, Dirac found two additional, strange solutions to his equation. Instead of throwing them out as physically unreasonable, Dirac suggested that they might represent some new particles or states. He had predicted what we now call **antimatter;** the two strange particles were antielectrons or **positrons** with spin up and down. Positrons are like electrons in every way except for their positive charge and a property called helicity or "handedness." When a positron and electron meet, they annihilate each other in a burst of gamma rays. Fortunately, all the matter in our immediate vicinity here on earth seems to be "ordinary"; you and your friends can shake hands or hug without being annihilated! But beware of the stranger who extends his *left* hand to you for a handshake.*

4.3 Pauli's Exclusion Principle

It was 1925, and the young professor Wolfgang Pauli had just been asked to lecture on the periodic law. Uhlenbeck and Goudsmit had published their electron spin postulate a few months earlier. Desperate to make sense out of the relation between atomic structure and the chemical periods, Pauli invented what we now call the **Pauli exclusion principle:** *no two electrons in a given atom can have the same set of four quantum numbers* $nlmm_s$. The two electrons in helium therefore must have the quantum numbers $100 + \frac{1}{2}$ and $100 - \frac{1}{2}$. Lithium cannot have the configuration $1s^3$ because the third electron would have to be assigned an m_s identical to one of the first two $1s$ electrons. Given that each set nlm specifies a unique orbital, it follows that each orbital can "hold" at most two electrons with quantum numbers $nlm + \frac{1}{2}$ and $nlm - \frac{1}{2}$. There are magnetic implications as well: an orbital fully occupied with two electrons cannot contribute to the magnetism of the atom because the electron spins are paired, causing their magnetism to cancel. This underpinning is needed to interpret fully the results of the Stern–Gerlach experiment on sodium.

Pauli had now equipped us to examine the structure of many-electron atoms, and to understand the relation between that structure and the location of an atom in the periodic table. What we are about to convey to you in the next sections is just what Pauli is likely to have told his students, in what must have been an exciting series of lectures. As with the electron spin, the exclusion principle began as a convenient postulate, but in a short time Dirac and Enrico Fermi showed that it is a necessary consequence of the **indistinguishability** of electrons—we can't hang tags on them because of the uncertainty principle—and the symmetry requirements consequently imposed on the total wave function.

4.4 The Aufbau Principle and Electron Configurations of the Elements

The **Aufbau** or **building-up principle,** a rule for finding electron configurations of atoms also developed by Pauli, is a natural consequence of the exclusion principle. It states that *to find the most stable or ground-state electron configuration of an atom, you add the electrons (Z in number) one by one into the lowest energy orbitals available, while obeying the Pauli exclusion principle.* The Aufbau principle

---

* The particles of chemistry all have their antiparticle counterparts. An antihydrogen atom is a positron bound to an antiproton. It is conceivable that entire regions of our universe are composed of antimatter, with antiplanets, populated with antipeople, orbiting antisuns. The interaction of positrons with ordinary matter is an active field of chemical research.

combines the requirements of the Pauli principle with the natural tendency of electrons to reduce their potential energy. Literally, if you take a bare nucleus and a bunch of electrons and put them together in a small region of space, they will arrange themselves (after perhaps radiating a number of photons) into a ground state given by this principle.

The Energies of Orbitals: Screening

We stated earlier that the ground state of the lithium atom is given by the electron configuration $1s^2 2s^1$. However, in view of the degeneracy of the $2s$ and $2p$ hydrogenic orbitals, an equally stable configuration might be $1s^2 2p^1$, where $2p$ can be any of the three $2p$ orbitals. Why choose $2s$? The answer lies in the fact that, unlike hydrogen, lithium has a *filled* $1s$ orbital between the $2s$ or $2p$ orbital and the nucleus. The $1s^2$ electron density partially *screens* the outer electron from the nucleus; if this electron is far enough away from the nucleus, it will see a charge of $+1$ instead of $+3$ (in units of e, the charge on the electron) due to the *cancellation* of two positive charges by the two $1s$ electrons. The distribution of electron–nucleus distances is given by the radial distribution function (RDF) as presented in Chapter 3. As Figure 4.4 illustrates, the RDF for the $2s$ orbital, with its inner "hump," allows an electron in the $2s$ to be in a region close to the nucleus where the $1s^2$ screening of the nucleus is less complete. A $2s$ electron will thus see a larger effective nuclear charge Z_{eff} than a $2p$ electron, since the $2p$ RDF has no inner maximum. This lowers the potential and total energy of the $2s$ orbital relative to $2p$. The ground state of Li is therefore $1s^2 2s^1$, and the configuration $1s^2 2p^1$ represents an excited state. Although there is a sizable $2s$–$2p$ energy difference, 1.8 eV in Li, it is very much smaller than the $1s$–$2s$ gap of 70 eV.

Now we can begin to look at the periodic table and observe how the $1s$, $2s$, $2p$ energy-level ordering is reflected there. Figure 4.5 is a standard table with orbital assignments added. In the second row or period, beyond lithium are the elements Be through Ne with atomic numbers $Z = 4$–10. At neon, 10 electrons total have

Figure 4.4
Illustration of orbital screening effects due to core electrons using radial distribution functions. Owing to its inner maximum, the $2s$ orbital possesses more electron density in the region of the $1s^2$ RDF than $2p$, as indicated by the shaded area, and is therefore less effectively screened by the $1s^2$ density from the nuclear charge.

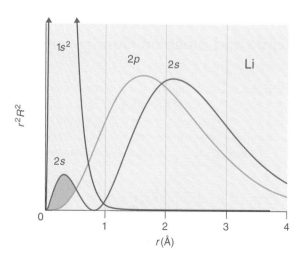

4.4 The Aufbau Principle and Electron Configurations of the Elements

been added to the atom, 8 of these in the second period, and the period is complete in a chemical sense. At the end of the period, the $2s$ and $2p$ orbitals that comprise the $n = 2$ shell have also been filled: with beryllium, $1s^2 2s^2$, the $2s$ is at capacity, and the next six atoms, B through Ne, exhaust the capacity of the $2p$ orbitals. The electron configurations and spin-orbital quantum numbers of each added electron for the second period elements are shown in Table 4.1. Note how the $1s^2$ part of the configuration remains invariant for the entire period, a necessary consequence of the Aufbau principle. The $1s$ electrons are deep inside these atoms, very close to the nucleus, with binding energies of hundreds of electron-volts. They are known as **core electrons** and play little if any role in the structure and reactions of these atoms and their molecules. The $2s$ and $2p$ electrons, on the other hand, have more ordinary energies, and form the periphery of the second-row atoms. These outer electrons endow the elements with their chemical powers, and are known as **valence electrons.** The $n = 1$ and $n = 2$ orbital capacities and energies thus provide a natural correspondence with the lengths of the first and second periods. Recall that the number of orbitals of a given n is n^2. Then, because of the two-valued m_s, the capacity for electrons of a given n is $2n^2$; that is, 2 for $n = 1$, 8 for $n = 2$, 18 for $n = 3$, etc. The slightly different energies of $2s$ and $2p$ create subshells within the $n = 2$ shell that are often chemically significant. *The electron capacity of a subshell of a given l is $2(2l + 1)$.*

Figure 4.5
Periodic table of the elements, with orbitals filled in various regions of the table indicated. The period index can be identified with the principal quantum number n. Except for the anomalies listed in Table 4.2, the electron configuration of any element may be read from this table. The table has been modified slightly by placing He in group II.

	IA	IIA											IIIA	IVA	VA	VIA	VIIA	0
$n=1$	1 H	2 He																
2	3 Li	4 Be					$(n-1)d$						5 B	6 C	7 N	8 O	9 F	10 Ne
3	11 Na	12 Mg	IIIB	IVB	VB	VIB	VIIB		VIIIB		IB	IIB	13 Al	14 Si	15 P	16 S	17 Cl	18 Ar
4	19 K	20 Ca	21 Sc	22 Ti	23 V	24 Cr	25 Mn	26 Fe	27 Co	28 Ni	29 Cu	30 Zn	31 Ga	32 Ge	33 As	34 Se	35 Br	36 Kr
5	37 Rb	38 Sr	39 Y	40 Zr	41 Nb	42 Mo	43 Tc	44 Ru	45 Rh	46 Pd	47 Ag	48 Cd	49 In	50 Sn	51 Sb	52 Te	53 I	54 Xe
6	55 Cs	56 Ba	57 *La	72 Hf	73 Ta	74 W	75 Re	76 Os	77 Ir	78 Pt	79 Au	80 Hg	81 Tl	82 Pb	83 Bi	84 Po	85 At	86 Rn
7	87 Fr	88 Ra	89 †Ac	104 Rf	105 Db	106 Sg	107 Bh	108 Hs	109 Mt	110 Uun	111 Uuu	112 Uub	113	114	115	116	117	118

$(n-2)f$

*Lanthanides	58 Ce	59 Pr	60 Nd	61 Pm	62 Sm	63 Eu	64 Gd	65 Tb	66 Dy	67 Ho	68 Er	69 Tm	70 Yb	71 Lu
†Actinides	90 Th	91 Pa	92 U	93 Np	94 Pu	95 Am	96 Cm	97 Bk	98 Cf	99 Es	100 Fm	101 Md	102 No	103 Lr

TABLE 4.1

Electron configurations and spin-orbital quantum numbers for the second period elements

Z	Element	Configuration	n l m m_s (for new electron)
3	Li	$1s^2 2s^1$	2 0 0 $+\frac{1}{2}$
4	Be	$1s^2 2s^2$	2 0 0 $-\frac{1}{2}$
5	B	$1s^2 2s^2 2p^1$	2 1 +1 $+\frac{1}{2}$
6	C	$1s^2 2s^2 2p^2$	2 1 0 $+\frac{1}{2}$
7	N	$1s^2 2s^2 2p^3$	2 1 −1 $+\frac{1}{2}$
8	O	$1s^2 2s^2 2p^4$	2 1 +1 $-\frac{1}{2}$
9	F	$1s^2 2s^2 2p^5$	2 1 0 $-\frac{1}{2}$
10	Ne	$1s^2 2s^2 2p^6$	2 1 −1 $-\frac{1}{2}$

Hund's Rule

As we mentioned earlier in discussing the $1s^2 2p^1$ excited state of Li, the $2p$ energy level is degenerate, and the first electron to occupy it has a total of $2(2l + 1) = 6$ sets of quantum numbers from which to choose. The choice given earlier for boron is therefore arbitrary, but from carbon onward the arbitrariness is reduced. As you can see from Table 4.1, the added electrons are assigned new m_s quantum numbers with the same spin until the possibilities are exhausted for spin up $(+\frac{1}{2})$, and then similarly for spin down $(-\frac{1}{2})$. This may be visualized by use of **orbital occupation diagrams**, as given below for B through Ne:

	$2s, l=0$ $m=0$	$2p, l=1$ $m=+1$	0	-1
B	↑↓	↑	—	—
C	↑↓	↑	↑	—
N	↑↓	↑	↑	↑
O	↑↓	↑↓	↑	↑
F	↑↓	↑↓	↑↓	↑
Ne	↑↓	↑↓	↑↓	↑↓

The electrons with their spins are represented by the up and down arrows, and the underlinings (in other books perhaps boxes or circles) represent the orbitals.

In understanding why the $2p$ orbitals fill in this particular way, you will learn what is known as **Hund's rule**. In an analogy that is unfortunately a commentary on our times, the orbitals may be likened to two-passenger seats on a bus. If the bus is less than half full, generally speaking only one passenger will occupy each seat, due to the alienation of strangers. Like strangers, electrons make poor seatmates, owing to the Coulombic repulsion between them. Their potential energy is lower if they can be far from each other while each has a window seat looking in toward the nucleus. Thus, *degenerate orbitals are always filled with single electrons before any of them are doubly occupied.* This is one aspect of Hund's rule.*

The fact that electrons prefer to fill the separate orbitals *with their spins aligned,* another part of Hund's rule, has a somewhat more subtle cause. Having the same value of m_s prevents the electrons from wandering into each other's orbitals, since a violation of the Pauli principle would result. Because of this "spin avoidance" the electron–electron repulsion is further reduced, lowering the potential and hence total energy. The electrons sitting by themselves are referred to as **unpaired.** Any other electron arrangement permitted by the Pauli principle is an excited state of the atom.

In oxygen, the fourth p electron is forced to sit with a disagreeable fellow passenger. As we'll see later in this chapter, the discomfiture this electron feels is apparent in the oxygen atom's properties.

Reading Configurations from the Periodic Table

After knowing how the first and second periods correlate with the orbitals and their capacities, you should not be surprised to learn that the entire periodic table has an orbital description. Although, due to magnetic effects of the sort we have discussed for Na, orbitals for the heavier atoms are less well defined, the lengths of the periods and subperiods are precisely given by orbital capacities.

As we pass to the third and higher (higher n, lower on the table) periods, the two most important effects to bear in mind are those of screening, which splits an n shell into $n-1$ l subshells, with $l = 0$ lowest in energy, $l = 1$ next, etc; and the drawing together in energy of successive n shells. The former can be understood by extending the arguments we have used for $2s$ and $2p$. The latter is in strict analogy with the hydrogenic energies. Beginning with $n = 3$ the energetic effects of screening become comparable to the energy differences between shells. This means that a high-l orbital of a given n may be higher in energy than a low-l orbital of the next n.

Inspecting the third period, Na to Ar, in Figure 4.5, we find only 8 elements instead of the 18 we might have expected for $n = 3$. The 10 "missing" electrons would have filled the $3d$ orbitals [$2(2 \cdot 2 + 1) = 10$], but, due to screening of the nucleus by the $n = 1$ and $n = 2$ shells, the $3d$ level is higher in energy than the $4s$. The fourth period thus commences before the $3d$ orbitals can be occupied. As indicated in the periodic table, the $3d$ orbitals are used immediately after the $4s$, and are followed by $4p$. In turn, the fourth period, K to Kr, completes before either the $4d$ or $4f$ orbitals can be occupied, again due to intervening $n = 5$ levels that are lower in energy. This sort of intrusion simplifies both the fourth and fifth periods, in that for each the ns, $(n - 1)d$ and np orbitals are filled, giving 18 elements. None of this is contrary to either the Pauli principle or the Aufbau principle, since the lowest energy orbitals are used first in each case.

---

* Hund actually proposed three rules, of which we are giving the first. The second concerns the ordering of excited states for a given configuration, and the third the influence of magnetic spin-orbit "fine structure."

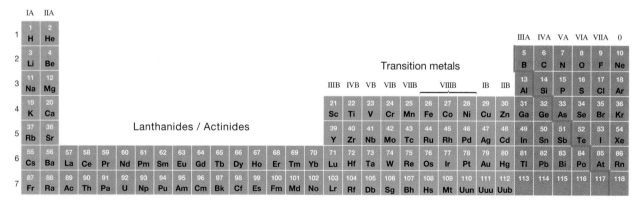

Figure 4.6
The "long form" of the periodic table.

The 4f orbitals finally make their appearance in the sixth period, giving rise to a separate row of 14 elements [2(2 · 3 + 1)] usually demoted to the bottom of the periodic table for the sake of compactness. Actually the sixth period is a record 2 + 14 + 10 + 6 = 32 elements long. Figure 4.6 depicts schematically the so-called extended form of the table, in which the f orbitals are put where they belong. By the end of the sixth period at radon (Rn), Z = 86, we have neared the edge of nature's supply of elements, defined not by any problem with the electrons, but by increasingly unstable nuclei (see Chapter 16). The seventh period thus starts out like the sixth, but is incomplete as yet.

Figure 4.7 illustrates the splitting and interleaving of the hydrogenic energy levels that occurs in many-electron atoms. The level ordering is obtained from atomic spectra, but the amazing and reassuring fact is that it accounts perfectly for the form of the periodic table, which was originally fixed by empirical chemical observation.

We would like to encourage you to use the periodic table to deduce the electron configurations of the elements, because this method will stay with you far longer than any ordering rules you may have learned previously. Simply counting over from the left of the period containing the element of interest allows orbital occupancies to be deduced in nearly all cases.

EXAMPLE 4.1

Give the ground-state electron configurations of the following elements: Si, S, Ca, Ga, As, Nb, Ir.

Solutions:

Si is in $n = 3$, four elements over from the left, so its configuration is $1s^2 2s^2 2p^6 3s^2 3p^2$. The core electrons ($n = 1$ and 2) are in a Ne-like configuration, so as a shorthand we can write the configuration as $[Ne]3s^2 3p^2$, which emphasizes the chemically important valence electrons. Similarly, the configurations of S and Ca are $[Ne]3s^2 3p^4$ and $[Ar]4s^2$, respectively. For Ga and As, the 3d level is filled as well, so the configurations are $[Ar]4s^2 3d^{10} 4p^1$ for Ga and $[Ar]4s^2 3d^{10} 4p^3$ for As. Niobium, Nb, is in the 4d series of $n = 5$, and its configuration is $[Kr]5s^2 4d^3$, while Ir, iridium, lies in $n = 6$ beyond the 4f series, so we have $[Xe]6s^2 4f^{14} 5d^7$.

4.4 The Aufbau Principle and Electron Configurations of the Elements

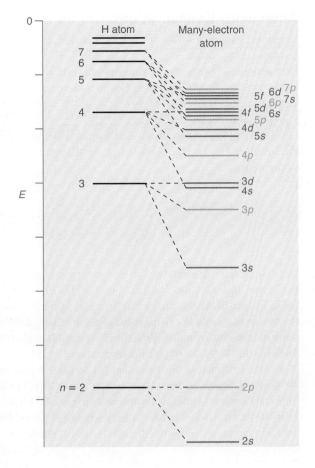

Figure 4.7
Splitting and interleaving of orbital energies due to screening. The 1s level (not shown to allow an expanded view of the rest) is not split.

EXAMPLE 4.2

Find the number of unpaired electrons for the elements of Example 4.1 by applying Hund's rule.

Solutions:
Using orbital occupancy diagrams for the highest energy orbital "buses," we get

Si	$3p^2$	↑ ↑ __	2 unpaired electrons
S	$3p^4$	↑↓ ↑ ↑	2 unpaired electrons
Ca	$4s^2$	↑↓	0 unpaired electrons
Ga	$4p^1$	↑ __ __	1 unpaired electron
As	$4p^3$	↑ ↑ ↑	3 unpaired electrons
Nb	$4d^3$	↑ ↑ ↑ __ __	3 unpaired electrons
Ir	$5d^7$	↑↓ ↑↓ ↑ ↑ ↑	3 unpaired electrons

Anomalous Configurations

Beginning in the $3d$ series of the fourth period, anomalies (exceptions to the Aufbau principle) in the orbital occupancy are observed. These occur in larger numbers in the later periods, as the listing of **anomalous configurations** in Table 4.2 indicates. The earlier ones, such as Cr and Cu, can be understood to some extent by combining the closeness in energy (near-degeneracy) of the valence orbitals for the later periods with orbital occupancy diagrams.

For Cr, the Aufbau prediction for the valence configuration is $4s^2 3d^4$, while that observed is $4s^1 3d^5$. Comparing the diagrams for the predicted and observed occupancies,

Cr, predicted: $4s^2 3d^4$ ↑↓ ↑ ↑ ↑ ↑ __

Cr, observed: $4s^1 3d^5$ ↑ ↑ ↑ ↑ ↑ ↑

you can see that the electron–electron repulsion is reduced in the observed configuration, since each electron has its own seat. This more than compensates for having to board a slightly higher energy bus. For Cu, $4s^1 4d^{10}$ is found instead of $4s^2 3d^9$. Here the bus analogy doesn't help, because the number of unpaired electrons (one) is the same in each configuration. What has happened is a reversal of the level ordering from $4s<3d$ to $3d<4s$ by the time Cu is reached at $Z = 29$.

In the fifth period the anomalies for Mo and Ag are the counterparts of those for the previous period, but new ones appear owing to s–d level reversal and increasingly important magnetic effects. By the sixth period the levels $6s$, $4f$, $5d$ are so nearly degenerate that the delighted electrons early in the period find it hard to choose among the large, empty buses. Nonetheless, the configuration for Gd shows a half-filled $4f$ level, analogous to the d^5 configurations for periods 4 and 5.

TABLE 4.2
Anomalous electron configurations[*]

Period	Z	Element	Configuration	Period	Z	Element	Configuration
4	24	Cr	$[Ar]4s^1 3d^5$	6	57	La	$[Xe]6s^2 5d^1$
4	29	Cu	$[Ar]4s^1 3d^{10}$	6	58	Ce	$[Xe]6s^2 4f^1 5d^1$
5	41	Nb	$[Kr]5s^1 4d^4$	6	64	Gd	$[Xe]6s^2 4f^7 5d^1$
5	42	Mo	$[Kr]5s^1 4d^5$	6	78	Pt	$[Xe]6s^1 4f^{14} 5d^9$
5	44	Ru	$[Kr]5s^1 4d^7$	6	79	Au	$[Xe]6s^1 4f^{14} 5d^{10}$
5	45	Rh	$[Kr]5s^1 4d^8$	7	89	Ac	$[Rn]7s^2 6d^1$
5	46	Pd	$[Kr]4d^{10}$	7	90	Th	$[Rn]7s^2 6d^2$
5	47	Ag	$[Kr]5s^1 4d^{10}$	7	91	Pa	$[Rn]7s^2 5f^2 6d^1$
				7	92	U	$[Rn]7s^2 5f^3 6d^1$

[*] These configurations cannot be deduced by following the Aufbau ordering indicated in Figure 4.5, with the possible exception of La and Ac, where these elements are retained in the d series.

4.5 Periodic Properties of the Atoms of the Elements

Although, as depicted in Figure 4.7, the energies of the orbitals have a more or less fixed ordering, the energies of individual levels plunge rapidly as Z increases. The energy of an orbital is roughly given by the Bohr formula of Equation 2.17:

$$E_{nl} = -\frac{Z_{\text{eff}}^2 e^2}{2a_0 n^2} \quad (4.4)$$

Figure 4.8 illustrates the dependence of these energies on n and Z. Each passage of a chemical period corresponds to a jump from n to $n + 1$, and Equation 4.4 and Figure 4.8 indicate that the energy of the outer orbitals will also show a discontinuity, the newly occupied orbitals of the next period being considerably higher in energy. This sets us on our way to understanding the atomic properties of the elements and, more importantly, the relationship between the chemistry of an element and its position in the periodic table.

Ionization Energy

More than any other atomic property, the **ionization energy** (IE) ties together the orbital model and periodicity. Recall from Section 3.2 that the IE is the energy (work) required to remove an electron completely from an atom, leaving a positive ion:

$$A \longrightarrow A^+ + e^-, \quad \Delta E = E(A^+) - E(A) = \text{IE} \quad (4.5)$$

The definition is ambiguous for a many-electron atom unless we specify which orbital the ionized electron is coming out of. The so-called **first IE** is defined to be *the energy needed to remove the most weakly bound electron*, that is, an electron from the highest occupied (valence) atomic orbital HOAO. The second, third, and continuing IEs correspond to removing the easiest electron from A^+, A^{2+}, etc., and we expect them to increase steadily. For example, the element boron, $Z = 5$, has five ionization energies: 8.30, 25.15, 37.93, 259.4, and 340.2 eV; the element iron has 26 IEs. The last IE (the "Zth") corresponds to creation of a bare nucleus and can be computed accurately from the Bohr formula. But it is the first IE that is of most significance with respect to chemical properties.

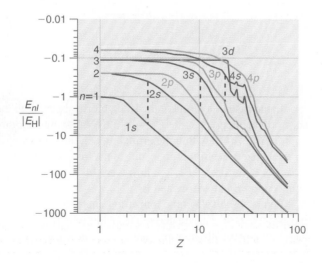

Figure 4.8
Orbital energies as a function of atomic number for the first four periods. The vertical dashed lines indicate the jumps in energy occurring at the start of new periods. Note the crossover between $4s$ and $3d$. The spikes in $3d$ and $4s$ occur for anomalous configurations (Table 4.2).

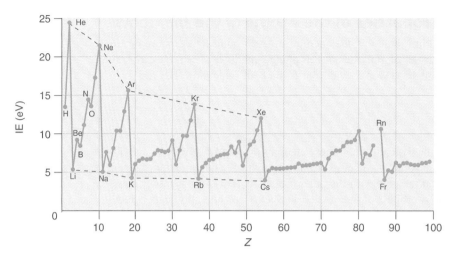

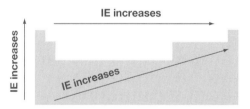

Figure 4.9
(**Top**) Ionization energies of the elements plotted against atomic number. The dashed lines connect elements of two families (noble gases and alkalis) to emphasize the decline with increasing Z within a family. (**Middle**) A periodic table of IEs (eV). (**Bottom**) Overall trends in IE with position in the table.
Source: Natl. Inst. Stand. Tech. (2003).

In the case of hydrogen, the IE is rigorously equal to the negative of the ground-state orbital energy: $IE = -E_1 = +13.6 \text{ eV}$. For bigger atoms this relationship is only approximate; even within the orbital model, the remaining electrons "relax" a bit after the ionized electron is gone, due to the loss of electron–electron repulsion (Z_{eff} increases). Thus, $IE < -E_{\text{HOAO}}$; however, this inequality is not so strong as to preclude connecting orbital energy and IE.

Figure 4.9 displays the ionization energies of the elements in graphic form. In most cases, these data are derived from ultraviolet spectroscopy on the gaseous atoms and are known with great precision. Note how each period builds to a climax at the noble gas, the last element of a period, followed by an abrupt fall to begin the next. This is easily understood if you identify the IE with orbital energy as given by Equation 4.4. As a period is traversed, n is held fixed while Z_{eff} increases, thereby lowering the orbital energy and increasing the IE. The atoms are thus becoming increasingly stable toward loss of electrons, an important chemical factor, as we will see in the next chapter.

A closer inspection of Figure 4.9 reveals, however, that a voyage across a period is not entirely smooth sailing. Little "glitches" in the curve for the second period, for example, occur at boron, which has a lower IE than beryllium (9.32 eV for Be, 8.30 eV for B) and oxygen, which has a lower IE than nitrogen (14.53 eV for N, 13.62 eV for O). This is not due to experimental error, but to the effects of subshell structure. Beryllium has a filled $2s$ subshell and the next electron (in B) must board the more effectively screened $2p$. The increase in Z_{eff} is then insufficient to bridge the $2s - 2p$ energy gap, and the IE falls. Why the IE should drop between N and O is a more subtle question, but we've already seen the explanation for it in Section 4.4. Nitrogen has a half-filled p subshell, and the fourth p electron (in O) is the first to be obliged to share a bus seat with another electron. The repulsion between these two electrons is not offset by the increase in Z_{eff}, and again the IE falls. Thus, in addition to the more obvious stability of filled shells, we may also identify filled and (for $l > 0$) half-filled subshells as possessing special stability. You can view the small drops in IE within a period as being due to effects similar to those which cause the large drop between periods, but of lesser degree.

As indicated in Figure 4.9, as we descend a given column (chemical family) of the periodic table, the IE generally falls slowly. Equation 4.4 does not make this trend immediately transparent, because both Z_{eff} and n are increasing down a column and tend to cancel each other's effect. The screening phenomenon, however, allows only a modest increase in Z_{eff}; if screening by the core electrons were perfect, Z_{eff} would have to start at unity for each new period. Using the first column of the table, group IA, the **alkali metals,** as an example, the first IEs are Li, 5.39 eV; Na, 5.14 eV; K, 4.34 eV; Rb, 4.18 eV; and Cs, 3.89 eV. Equation 4.4 can be used to estimate Z_{eff}: $Z_{eff} \approx n\left[\text{IE}/13.6 \text{ eV}\right]^{1/2}$, giving the values Z_{eff} = 1.26, 1.84, 2.26, 2.77, and 3.21, for lithium through cesium. These are all greater than 1.00, indicating imperfect screening, but the tremendous effect of screening may be noted by comparing Z_{eff} with the bare nuclear charge Z. The n dependence of the orbital energies thus dominates, and the IEs of the heavier elements of a given family are generally lower.

As also illustrated in Figure 4.9, the increase in IE from left to right and the decrease from top to bottom of the periodic table imply a general trend in IE, increasing from lower left to upper right. That is, any element lying to the right and at or above a given element will have a higher IE, subject to the "glitches" we discussed earlier. Just as you can use the periodic table to derive electron configurations, with a few exceptions, you can also deduce relative IEs with it.

The intervention of the d and f orbitals manifests itself in Figure 4.9 in the relatively flat portions of the curve with IEs generally between 6 and 8 eV. In these so-called **transition elements,** the **transition metals,** the **lanthanides** and the **actinides,** the electrons are filling inner shells, and the electrons that are lost in ionization come from the outer s orbitals instead. The result is that Z_{eff} for the s electrons is relatively constant across a d or f series, the increase in nuclear charge being offset by the screening provided by the added inner-shell electron. Only when the d subshell is full does a significant increase in IE occur, for the elements Zn (9.39 eV), Cd (8.99), and Hg (10.44). Although the heaviest element in its family, mercury has the highest IE, generally the transition metals of the sixth period share this property. This owes to the prior filling of the 4f orbitals; being of higher *l*, these do not screen the nucleus quite as effectively as *d* orbitals, and as a result Z_{eff} is a bit higher than in previous periods as the 5*d* orbitals become occupied. This provides a significant exception to the general trend for IE.

Electron Affinity

Although certain atomic negative ions, atoms with one or more excess electrons, are quite common in solutions and crystals (e.g., table salt, NaCl, containing the Cl$^-$ ion), such species are rare in the gas phase. It is only within the last few decades that such ions have been extensively studied spectroscopically in an isolated state. The **electron affinity** (EA) of a singly charged negative ion is defined as *the energy required to remove the least tightly bound electron* to leave a neutral atom:

$$A^- \longrightarrow A + e^-, \qquad \Delta E = E(A) - E(A^-) = EA \qquad (4.6)$$

This definition is analogous to that for IE, except that the electron being removed is an "extra." W. C. Lineberger and coworkers have measured the EAs for a large number of elements and small molecules using laser photodetachment, in which a beam of negative ions passes through the cavity of a laser, and the kinetic energy of the electrons knocked out by the laser light is measured.

The EAs are always smaller than IEs, because the extra electron of a negative ion does not feel any Coulomb attraction to the neutral atom when at a large distance from it. As Figure 4.10 illustrates, the potential energy is much weaker for an electron interacting with a neutral atom than for the electron-positive ion case, making it easier to detach. The formation of negative ions, the reverse of Equation 4.6, depends on both the stability and the availability of the orbital that holds the electron. Although a noble gas has very stable valence orbitals, they are filled, and an additional electron

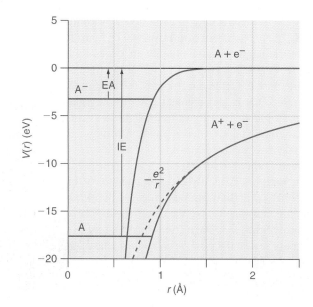

Figure 4.10
Potential energies for an electron e^- interacting with a neutral atom A and a positive ion A$^+$. The deviation of the A$^+$ + e^- curve from the $-e^2/r$ dotted curve, and the A + e^- curve from near zero, indicates the presence of the valence orbitals. Inside this distance the nucleus is imperfectly screened from the electron. The A + e^- curve is not exactly zero at large distance due to "polarization" of the atom by the electron's charge. The curves and energies refer to F and F$^-$.

cannot be accommodated because this would violate the Pauli principle. By far the most common atomic negative ions are the halide ions F^-, Cl^-, Br^-, and I^-, in which the extra electron has filled a "hole" in the p subshell. These ions exist only because the new electron can share the nuclear charge due to imperfect screening. In a rationale that is the reverse of that employed for the IE, the added electron provides greatly increased $e^-\cdots e^-$ repulsion, particularly within the valence shell, and hence $EA \ll -E_{HOAO}$, where the orbital energy refers to the neutral atom.

Figure 4.11 graphs the known electron affinities of the elements. Its appearance is choppier than that of Figure 4.9 for the IEs because of the requirement of an orbital vacancy. Those elements with filled shells or subshells simply have no EA to speak of; trying to add an electron to some of these atoms actually costs energy. Based on Equation 4.4, we might expect the same trends in EA as for IE: increasing from lower left to upper right of the periodic table. As the figure indicates, however, EAs for the third period are actually somewhat larger than those for the second. This is testimony to the leading role played by electron repulsion in determining the stability of the negative ion. The $n = 3$ orbitals are larger than those for $n = 2$, allowing the electrons to distance themselves from one another to a greater extent. This size effect brings us to another central atomic property: its radius.

Figure 4.11
Electron affinities of the elements plotted against atomic number, and a periodic table of EAs (eV). Those elements that require energy to add an electron are assigned an EA of zero.
Source: Hotop & Lineberger, *J. Chem. Phys. Ref. Data* (1985).

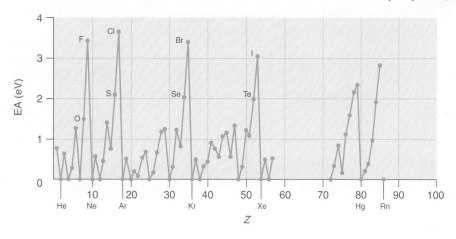

1 H 0.754																	2 He 0
3 Li 0.62	4 Be 0											5 B 0.28	6 C 1.263	7 N 0	8 O 1.461	9 F 3.399	10 Ne 0
11 Na 0.55	12 Mg 0											13 Al 0.44	14 Si 1.39	15 P 0.746	16 S 2.077	17 Cl 3.617	18 Ar 0
19 K 0.50	20 Ca 0	21 Sc 0.19	22 Ti 0.08	23 V 0.53	24 Cr 0.67	25 Mn 0	26 Fe 0.16	27 Co 0.66	28 Ni 1.16	29 Cu 1.23	30 Zn 0	31 Ga 0.3	32 Ge 1.2	33 As 0.81	34 Se 2.02	35 Br 3.365	36 Kr 0
37 Rb 0.49	38 Sr 0	39 Y 0.31	40 Zr 0.43	41 Nb 0.89	42 Mo 0.75	43 Tc 0.55	44 Ru 1.05	45 Rh 1.14	46 Pd 0.56	47 Ag 1.30	48 Cd 0	49 In 0.3	50 Sn 1.2	51 Sb 1.07	52 Te 1.97	53 I 3.059	54 Xe 0
55 Cs 0.47	56 Ba 0	57 La 0.5	72 Hf 0	73 Ta 0.32	74 W 0.82	75 Re 0.15	76 Os 1.1	77 Ir 1.57	78 Pt 2.13	79 Au 2.31	80 Hg 0	81 Tl 0.2	82 Pb 0.36	83 Bi 0.95	84 Po 1.9	85 At	86 Rn 0

Atomic Radius

The size of an atom directly determines the lengths and strengths of the bonds it will make with other atoms. The **atomic radius** varies in a systematic way with location in the periodic table just as IE and EA do, and it is again interpretable in terms of orbital properties. Equation 4.4 can be rewritten as

$$E_{nl} = -\frac{Z_{\text{eff}} e^2}{2 r_{nl}}; \quad r_{nl} = \frac{n^2 a_0}{Z_{\text{eff}}} \qquad (4.7)$$

where r_{nl}, the radius of peak probability in the radial distribution function, may be identified with the radius of the atom if n and l refer to the highest occupied atomic orbital (HOAO).

Figure 4.12 graphs the atomic radii against Z in the same way we've seen for IE and EA, and Figure 4.13 illustrates the relative sizes of atoms and atomic ions. You

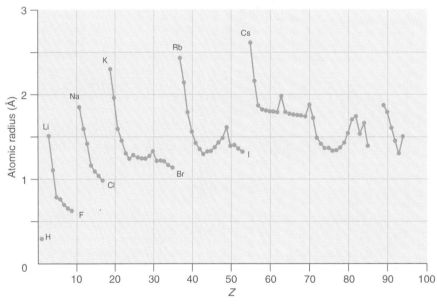

Figure 4.12
Atomic radius plotted against atomic number, and a periodic table of atomic radii (Å). Gaps in the graph appear at the positions of the noble gases.
Source: L. Pauling, *The Nature of the Chemical Bond,* and *Handbook Chem. Phys.*

should be struck by the similarity between Figures 4.9 and 4.12. The difference is that the spikes in Figure 4.12 occur at the alkali atoms, which show by far the largest radii. These data are obtained by analyzing not isolated gaseous atoms, but elemental solids and molecules to obtain atom–atom separations, where the atoms are chemically bonded. The noble gases, which do not bond to one another, are not included in the graph. The atomic radius falls as a period is traversed, and grows when traveling down a chemical family, just the opposite of the trends for IE and EA. This is easily understood by noting the inverse relationship between r_{nl} and $-E_{nl}$ in Equation 4.7. The radius does not show the irregularities due to subshell structure that were seen for IE and EA. This is due in part to the use of chemical bond distances to obtain the radius; as we'll see in Chapter 6, in bonding situations the atomic orbitals are often strongly "blended," and the subshell distinctions no longer dominant. However, the anomalously high IEs for the sixth period (5d) transition metals do correlate with unusually small atomic radii. As explained earlier, the filling of the 4f subshell prior to 5d causes an increase in Z_{eff} and correspondingly shrinks the valence orbitals in a phenomenon known as the **lanthanide contraction.**

The relative sizes of atoms can thus be deduced from the periodic table just as the IEs can. In addition, these radii can be applied directly to the estimation of atom–atom separations (bond lengths) in molecules.

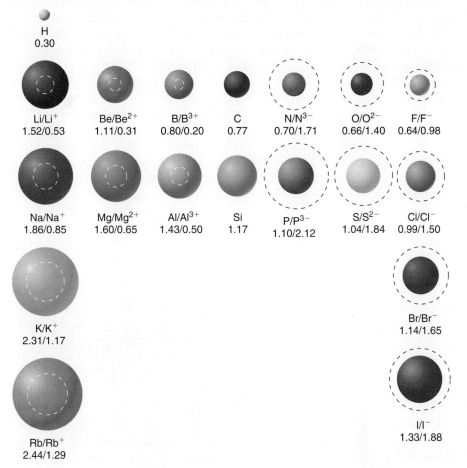

Figure 4.13
Atomic (shaded spheres) and ionic (dashed circles) sizes in pictorial form for some representative elements. The numbers beneath each picture give the covalent and ionic radii in angstroms.

Ionic Configurations and Isoelectronic Sequences

In addition to neutral atomic radii, Figure 4.13 also shows a number of **ionic radii.** These correspond to atoms that have gained or lost electrons to achieve the electron configuration of the nearest noble gas. Such ions occur in many chemical compounds, and reflect a natural tendency to attain what we've seen is the most stable atomic electronic structure possible. In Chapter 5 we'll explore the chemical implications of this tendency.

Note first that the positive ions are smaller than their neutral parent atoms, while the negative ions are larger. The positive ions have lost their entire valence shell, leaving only a noble gas-like core, so a sharp decrease in size is expected. The size increase for the negative ions is relatively smaller, reflecting the fact that electrons are being added to the same shell. These radii differences also reflect the different types of bonding (covalent versus ionic) that atoms and their ions engage in. Again, Chapter 5 contains more details of this.

Consider the sequence of ions

$$N^{3-} \quad O^{2-} \quad F^{-} \quad Na^{+} \quad Mg^{2+} \quad Al^{3+}$$

Each of these ions has achieved an identical $1s^2 2s^2 2p^6$ electron configuration, the negative ions by filling holes in their valence shells, the positive by losing theirs. They are called **isoelectronic ions,** and each is also isoelectronic with neutral Ne. As Figure 4.13 indicates, their radii smoothly decrease from N^{3-} to Al^{3+}, due to the steadily increasing nuclear charge. N^{3-} and O^{2-} are unstable as isolated ions (even N^{-} is unstable) but can exist in ionic compounds surrounded by positive ions. The energy needed to remove an electron steadily increases in the sequence, as expected from the relation between orbital energy and radius in Equation 4.7. For F^{-} through Al^{3+} these energies are 3.40 eV (EA of F), 47.3 eV (2nd IE of Na), 80.1 eV (3rd IE of Mg), and 120 eV (4th IE of Al). The IE of Ne, 21.56 eV, falls neatly between F^{-} and Na^{+}.

Within the **main group** (s- and p-orbital) chemical families, the ionic radii follow the same trends as the neutral radii do; for ions of a given charge, the radii increase down a column of the periodic table, and decrease from left to right. Peculiarities arise for the transition metals, as we've discussed for IE and neutral atom radius. Transition metal ions nearly always are multiply charged (exceptions are Cu and Ag, which also have anomalous neutral configurations), having lost their outer s electrons and perhaps one or two d electrons. For example, neutral iron has the electron configuration $[Ar]4s^2 3d^6$, and its ions are predominantly Fe^{2+} and Fe^{3+} with $[Ar]3d^6$ and $[Ar]3d^5$ configurations, respectively. Transition metal ions all tend to have similar radii, but can differ greatly in their chemistry. Examples of their reactions will appear throughout the remainder of the book, but particularly in Chapter 17.

4.6 Computers in Chemistry: The Many-Electron Problem

The increasing availability of high-speed electronic computing machines—computers to you and me—by the mid-1960s had enabled the development of highly accurate orbitals, and has in fact allowed us to go beyond the notion of electrons in well-defined orbitals in quantitatively describing atoms. The basis for most of these computer calculations is the **self-consistent-field** (SCF) method developed and applied by Hartree in 1928–1936. In the beginning of this chapter we touched on the complexities of many-electron atoms; for the most part we've simply assumed the existence of the orbital functions with appropriate effective nuclear charges. But how would you go about finding the best orbitals? This is the problem that Hartree solved—with accommodation for the exclusion principle provided by V. Fock and J. C. Slater—at least in principle, and that computers have made practical, even routine.

To get an idea of the difficulty of the problem, consider the Hamiltonian operator $\hat{H}$ for an atom with N electrons ($N = Z$ for a neutral atom, but not for an ion):

$$\hat{H} = \underbrace{\sum_{i=1}^{N} \hat{K}_i}_{\substack{\text{kinetic} \\ \text{energies} \\ N \text{ terms}}} - \underbrace{\sum_{i=1}^{N} \frac{Ze^2}{r_i}}_{\substack{\text{electron–nucleus} \\ \text{attractions} \\ N \text{ terms}}} + \underbrace{\sum_{i<j} \frac{e^2}{r_{ij}}}_{\substack{\text{electron–electron} \\ \text{repulsions} \\ N(N-1)/2 \text{ terms}}} \qquad (4.8)$$

Here the r_{ij} are the interelectron distances, and taking only terms with $i < j$ ensures that no term is counted twice. For sodium, $Z = 11$, for example, there are 77 terms in the Hamiltonian, 55 of them electron–electron repulsions, and the wave function is 33 dimensional.

Hartree began by assuming, as in Equation 4.2 for He, that the total wave function is an *orbital product*:

$$\Psi = \psi_a(\mathbf{r}_1)\,\psi_b(\mathbf{r}_2)\,\psi_c(\mathbf{r}_3)\cdots\psi_k(\mathbf{r}_N) \qquad (4.9)$$

where the subscripts *a, b, c,* etc. represent different sets of quantum numbers *nlm*. Then the electron repulsion terms in Equation 4.8 were replaced by an average over the electron density; for a particular electron *i* interacting with another electron *j*,

$$\frac{e^2}{r_{ij}} \longrightarrow e^2 \int \frac{|\psi_\alpha(r_j)|^2}{r_{ij}}\,d\tau_j \qquad (4.10)$$

where the orbital ψ_α is the one occupied by electron *j*, and the integral occurs over the volume τ_j of that orbital. This is an approximation that *neglects correlated motion*, but it takes conceptual advantage of Born's interpretation of orbitals: the integral includes all possible positions of electron *j*, weighted by the probability of finding the electron at each position. The average over electron *j* means that *the repulsion term now depends only on the position of electron i*. When the wave function of Equation 4.9 is substituted into the Schrödinger equation $\hat{H}\Psi = E\Psi$, with Equation 4.10 substituting for each term of the last sum in $\hat{H}$ in Equation 4.8, the equation breaks up into N one-electron equations, each a modified form of the hydrogenic Schrödinger equation. The difference is the terms of Equation 4.10, which already contain the orbitals for the other electrons. So you need to know the orbitals already just to set up wave equations to solve for the orbitals! This *Catch-22* situation was remedied by introducing an **iteration**: guess the orbitals, solve the one-electron equations for new orbitals, substitute them into Equation 4.10 to generate new potential energies, solve the equations again, and test to see whether the orbitals have changed.

If they are unchanged, then a self-consistent set of orbitals has been obtained.* The procedure can be diagrammed like this:

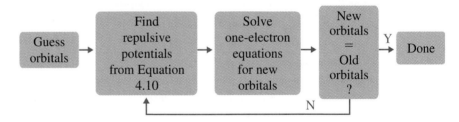

Even for this approximate theory (neglecting correlated motion), it took Hartree and his workers several months to complete a calculation for a single third-period atom. This is a task that a modern computer, even the ones commonly found on desktops or in laps, can complete in a few seconds.† The computer, after careful programming, is an ideal tool for the mindless repetition Hartree's iterative method entails.

The combination of a proper theory for describing atoms and molecules along with computing power for producing numbers from the theory has spawned a class of chemical specialists, called **quantum chemists,** who are adept at implementing Hartree's method and its descendants on computers. The ground-state energies of atoms well into the second period may now be obtained with high accuracy, but only for He can we match the spectroscopic results. Caution is advised, however, when reading statements like this. New developments in hardware, software (the computer codes, usually written in FORTRAN, but sometimes in machine language to improve speed), and the theory of implementation beyond SCF, can rapidly make statements of limitation obsolete. Currently the *ab initio* (first-principles) prediction of the spectrum of any atom beyond H (i.e., most of the electronically excited states) to spectroscopic accuracy is beyond reach. But for many chemical applications such accuracy is unnecessary, especially those for which an orbital approximation is acceptable. Often quantum theory can now be carried through for molecular systems too complicated to be understood using experimental data alone. Future conceptual advances in chemistry are likely to be made based on a combination of *ab initio* theory and experiment. Quantum mechanics on computers has become an integral part of today's chemistry. And there's a chance that, for certain chemical problems, Dirac may yet be proven wrong.

4.7 "Seeing" Atoms

Now that we're able to write down electron configurations for atoms, e.g., Si or Ar, can we say what an atom "looks like"? Our quantum mechanical description yields a "cloud" of electron density surrounding the nucleus. This cloud, in the orbital approximation, is the *sum* of individual electron densities, the squares of orbitals, for each electron in the atom. For Ar, this gives rise to an electron cloud of spherical shape, having concentric shells of high electron density within the cloud, a reflection

---

* If the starting guess is too far off, the procedure may not converge. The validity of the SCF method, however, has been verified by other wave-mechanical means.
† The word *computer* actually appears in the early articles describing these types of calculations, but it refers to a person skilled in calculation, not a machine.

of the $n = 1, 2,$ and 3 orbitals. Using Hartree's method as outlined in the previous section, the optimum electron cloud within the orbital approximation can be found. But can this cloud be observed?

First, we can say with some certainty that our feeble eyes cannot see an atom. The human eye is sensitive to a limited range of wavelengths of light (Chapter 2), all much larger than the diameter of an atom. We are limited to seeing, even with a powerful optical microscope, living cells and the larger aggregates (organelles) within them. As discussed in Chapter 3, shortening the wavelength into the X-ray regime only results in destruction of the atom, although X-rays can be photographed, and a limited number "scatter" from the atom to give a diffraction pattern that does reflect the arrangement of the electrons (recall G. P. Thomson's experiment, Chapter 2, Figure 2.11; see also Chapter 14). But now we have to believe the photograph; we cannot see what made it.

In 1953 L. Bartell and L. Brockway reported a pioneering *electron* diffraction experiment on a *gas*. The experiment was similar to that of G. P. Thomson, except that electrons of very high energy (40 keV) were passed through an atomic beam of Ar atoms, like the beam used by Stern and Gerlach described in Section 4.2. The high energy gave the electrons a very short de Broglie wavelength, much shorter than an X-ray, enabling the details of the electron cloud to be probed. By analyzing the photographic image of the scattered electrons—this image did not contain sharp rings, but rather more gradual variations in exposure—using the known de Broglie wavelength, the radial electron distribution D as a function of the distance r from the nucleus could be found; $D(r)$ is like a radial distribution function, except all of the electrons are included. The solid curve in Figure 4.14 shows the resulting $D(r)$; the electron shells of Ar are clearly visible. The dashed curve is the prediction of the orbital model; the agreement is remarkable, and lends credence to the atomic model we have built in this chapter. Bartell later developed a holographic method for converting the scattered electron pattern directly into an image of Ar's electron cloud (see the chapter-opening image).

Bartell's method, like Thomson's, requires millions of individual encounters between atoms and the high-energy electrons; it takes advantage of the fact that all Ar atoms are alike, a consequence of their quantized nature. In 1982 a new type of microscope was developed by G. Binnig and H. Rohrer for examining *solid surfaces* that eventually proved capable of imaging *individual* atoms. This new microscope is called

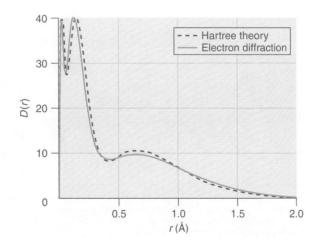

Figure 4.14

Radial electron distribution for Ar. The solid curve is the result of an electron diffraction experiment, while the dashed curve is the orbital approximation. The three maxima in the curves arise from the $n = 1, 2,$ and 3 shells of Ar.
Source: Bartell & Brockway, *Phys. Rev.* (1953).

Figure 4.15
(a) STM image of the surface of a silicon crystal. The smallest Si–Si distance is 2.72 Å. (b) A repeat of the scan at the left at a higher tip bias; the electrons being imaged here are engaged in forming Si–Si bonds.

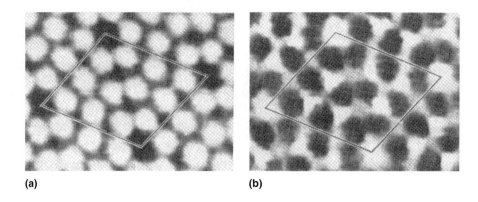

(a)　　　　　　　　　　(b)

a **scanning tunneling microscope** (STM), and consists of a metal electrode so sharply pointed that its tip consists of a single atom, located within a few angstroms of a solid surface of a metal or a semiconductor. Applying a small voltage (a "bias") between the tip and the surface causes electrons to "tunnel" (see Chapter 3); if the tip is more positive than the surface, electrons are "sucked" out of the surface into the tip. The tiny electrode spacing makes the STM the world's smallest Crookes' tube.

Individual atoms can be "seen" by scanning the STM tip over the surface; when a surface atom is directly beneath the tip, many electrons (*ca.* 10^{10}/s or so) jump out, yielding a measurable current I, while the current drops when the tip is "between" atoms. (In recent instruments, the height z above the surface is feedback-controlled to maintain constant I.) The resulting $I(x,y)$ is a map of the atoms on the surface. Figure 4.15 is one of the first examples of an atomically resolved surface map of a single crystal of solid silicon; the bright regions are those of higher I, and the regular pattern of atoms is clearly visible. Because the probe is also an atom, the resolution cannot be much better than an angstrom or so, which is not as fine as electron diffraction; thus each Si atom appears as a somewhat fuzzy ball of electron density. But a significant advantage of STM is the ability to vary the bias of the tip; at larger biases, electrons from lower-energy surface states are enabled to tunnel into the tip. Figure 4.15 also shows a high-bias scan that reveals electrons located "between" the atoms; these are "bonding" electrons that hold the crystal together (see Chapters 6, 7, and 14).

Figure 4.16
(a) STM image of Si crystal surface, similar to Figure 4.15. (b) STM image after surface has been exposed to ammonia NH_3. The darkened areas have been hydrogenated.

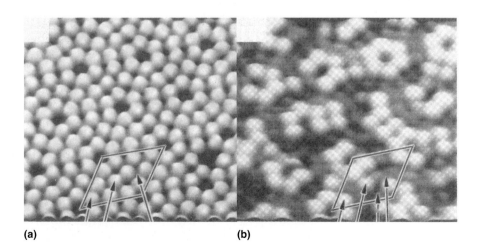

(a)　　　　　　　　　　(b)

The chemical capabilities of the STM are illustrated in Figure 4.16. On the left is again a pure Si surface, while on the right is a scan taken after the surface was exposed to ammonia NH_3, which hydrogenates the surface, making Si–H bonds. Hydrogenated regions are nonconducting, so few electrons are available, and the image remains dark. Improvements in scan rate have now allowed surface reactions to be observed in progress. Figure 4.17 shows the synthetic potential of the STM. When the tip is brought closer to the surface, attractive forces begin to act; this allows atoms and molecules to be "dragged" along the surface to form any sort of arrangement. In Figure 4.17, 48 Fe atoms have been arranged in a circle on a Cu surface; the shadowing is added by computer to give a landscape effect. The enclosed region has been dubbed a "quantum corral"; electrons on the surface of the Cu are forced to obey boundary conditions and form a standing wave within the corral.

The STM only works on conducting and semiconducting surfaces, as the hydrogenated silicon example illustrated. But a slight modification of the apparatus allowed a new mode of operation, wherein the tip itself is brought very close to the surface and dragged over it, and deflections of the tip arm (a "cantilever") are detected by laser reflection. When the tip encounters a bulge (an atom), it is repelled and deflected very slightly by some angle θ, and $\theta(x,y)$ becomes a map of the surface. This **atomic force microscope** (AFM) can be used on any type of surface; Figure 4.18 presents scans of graphite (an allotrope of carbon) and sodium chloride. The tip-to-surface gap is controlled to avoid damaging the surface.

Together with other variations, these methods comprise **scanning probe microscopy** (SPM). They are all enabled by the *piezoelectric effect*, whereby certain crystals expand or contract on an angstrom scale when voltage is applied to opposite faces; the tip mount generally contains a piezoelectric crystal. SPMs have already

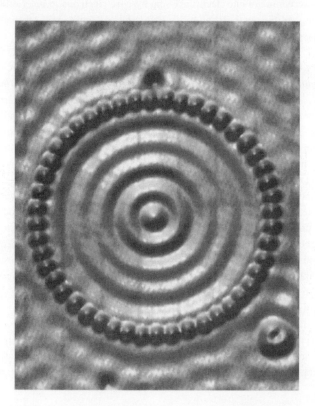

Figure 4.17
A circle of 48 Fe atoms on the surface of a Cu crystal, assembled by a STM tip, and then scanned by the same tip. The wave pattern in the center of this "quantum corral" is formed by surface electrons trapped within the circle.

Figure 4.18
Atomic force microscope AFM images of the surface of graphite (**left**) and a sodium chloride crystal (**right**).

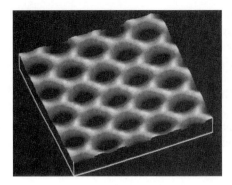
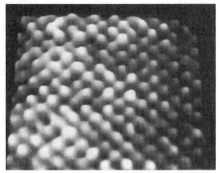

been commercially available for some time, and are being employed in advanced undergraduate labs at many schools. It is important to bear in mind that these SPM images are not actual photographs, pictures of actual atoms, but rather a measured response to the presence of atoms. But there is no doubt that individual atoms are responsible for the forms of these images, and it is in that sense that we are now finally able to "see" an atom.

SUMMARY

The orbital approximation for many-electron atoms is introduced. A detailed description of the helium atom, with two electrons, gives rise to the concept of **screening** of one electron by another from the full positive charge of the nucleus, and to the concept of **effective nuclear charge** Z_{eff}, which is used to describe an orbital and its energy. The hydrogenic orbital model for He yields a ground-state $1s^2$ configuration with a total energy within 2% of the experimental value derived from its spectrum.

To understand the structure of atoms beyond He, two further concepts are introduced, the **electron spin,** with quantum number $m_s = +1/2$ (spin up) or $-1/2$ (spin down), and the **Pauli exclusion principle,** according to which *no two electrons in an atom may have the same four quantum numbers $nlmm_s$*. Electron spin was necessary for the understanding of fine structure in atomic spectra and the result of the Stern–Gerlach magnetic deflection experiment, while the Pauli principle produced a complete interpretation of periodic properties of the elemental atoms in terms of their spin-orbital electron configurations. Both of these postulates were later reinforced by detailed theoretical treatments, the spin in Dirac's relativistic quantum theory and the exclusion principle in the analysis of indistinguishability and symmetry by Dirac and Fermi.

With these ideas in hand, the electron configuration of any atom can be obtained by adding electrons one by one into the lowest energy orbitals while observing the exclusion principle.

This is called the **Aufbau, or building-up, principle,** and follows from the natural tendency to minimize potential energy. Screening by **core electrons** in inner orbitals causes the **valence electron** orbitals to split into l subshells of slightly different energy, with s ($l = 0$) orbitals lowest in energy, p ($l = 1$) next, then d ($l = 2$), etc. Combining this effect with the capacity of one orbital for two electrons allows the entire periodic chart to be correlated with the ground-state energy-level and orbital configurations of the atoms. **Hund's rule** predicts the arrangement of electrons in partially filled $l \neq 0$ subshells, stating that *the electrons will tend to occupy separate orbitals with their spins aligned* or **unpaired.**

The inner-shell screening effect combined with the closing ranks of the energy levels as n increases causes the n shells to interleave their l subshells beginning at $n = 4$, where $4s$ orbitals are lower in energy than $3d$. This gives rise to a lag of one period in filling the d orbitals and two periods for f orbitals. With these features in mind, you can use the periodic table to read off electron configurations with a few exceptions. These exceptions, or **anomalous configurations,** can usually be traced to half-filled or filled subshells, or to reversal of level ordering for closely spaced levels.

Three atomic properties that show periodic variation correlated to the electron configuration are **ionization energy** (IE), **electron affinity** (EA), and **atomic radius.** IE and EA are strictly properties of isolated elemental atoms, defined as the energy or work needed to remove the least tightly held electron from a neutral atom (IE) or a negative ion (EA), while atomic radii are derived from lengths of chemical bonds an atom makes with its neighbors. IE and EA both generally increase from left to right in a given period and decrease from top to bottom of a chemical family, while the atomic radius shows just the opposite trends. This systematic behavior is analyzed in terms of orbital energies and radii. Irregularities in these trends with position in the periodic table are traced to the influence of filled and half-filled subshells, and to the filling of 4f orbitals, which do not as effectively screen the nucleus.

Positive and negative ions display properties that are related sensibly to those of their parent atoms and other atoms or ions with isoelectronic configurations. The commonly occurring ions (found in metal–nonmetal compounds called salts) have configurations isoelectronic with the noble gas nearest in Z. **Ionic radii**, for ions of a given charge, vary with position in the periodic table in the same way as atomic radii, and also decrease uniformly with increasing positive charge along an isoelectronic sequence, such as Na^+, Mg^{2+}, and Al^{3+}. The positive ions of transition metals have similar radii, owing to the fact that the atoms all lose their outer s electrons first when forming ions.

Finally, Hartree's **self-consistent-field** (SCF) method for obtaining the best orbitals for atoms is described. The modern high-speed computer is capable of solving the SCF equations very quickly to give the optimum orbital structures for atoms and many molecules. To get quantitative agreement with atomic energies derived from spectroscopy requires going beyond the orbital approximation. This has been done to spectroscopic accuracy only for He. Nonetheless such calculations are useful in a wide variety of chemical applications. SCF calculations are in good agreement with a high-resolution electron diffraction analysis of the charge distribution in the argon atom. "Pictures" of atoms on surfaces from newly developed **scanning probe microscopes** are surveyed.

EXERCISES

The orbital model

1. Must the Bohr model fail for atoms bigger than H? Consider the following "choreographed" model for He:

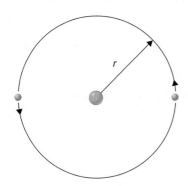

   The electrons always stay on opposite sides of the nucleus, so that $r_{12} = 2r$. Applying the de Broglie standing-wave condition (Equation 2.10) independently to each electron, find the radius r and use the Hamiltonian of Equation 4.1 to find the total energy of the system. (Note that for the Bohr model $K_i = \frac{1}{2}m_e v_i^2$.) Compare with the experimental value given in Section 4.1.

2. Suppose the electron configuration of Li were $1s^3$, that is, that the Pauli principle did not hold. Using the result for He that an electron screens 0.31 of a unit nuclear charge, estimate Z_{eff} for $1s^3$ Li, and find the orbital radius and energy using Equation 4.7.

3. An excited state of helium has the configuration $1s^1 2p^1$. In this state the $1s$ electron ought to screen a full unit of nuclear charge from the $2p$ electron. Estimate the $2p$ orbital energy based on this assumption. (The experimental value is -3.36 eV.)

4. Using a plot of the radial distribution functions for $3s$, $3p$, and $3d$ orbitals, make an argument based on screening to establish the sublevel energy ordering.

5. If electron–electron repulsion were completely neglected, what would be the total energy of a helium atom?

Spin, the exclusion principle, and electron configurations

6. Using the periodic table inside the front cover of this book, give the full ground-state electron configurations of the following atoms or ions: C, Si, Mg, S^{2-}, S^{6+}, C^-, Cl^+, Cl^-, Mg^{2+}, P^{3-}, Kr.

7. Using noble gas notation for the core electrons, predict the electron configurations of the following atoms or ions: Be, Ba, Al, Mn, Mn^{2+}, Mn^{7+}, Ga, As, Sn^{2+}, Sn^{4+}, Cu^+, Nd (neodymium), Ra, U.

8. Write out valence orbital occupation diagrams and give the number of unpaired electrons in the following

species: O^-, Sc, Ni^{2+}, Se, Sm, Bi^{3+}. Give a possible set of quantum numbers for an electron in the highest occupied atomic orbital.

9. The following are putative valence orbital occupation diagrams for the sulfur atom. Decide which of them is the ground configuration, and whether the others are excited states or impossible.

	3s	3p	4s
a.	↑↓	↑ ↓ ↑↓	—
b.	↑↓	↑↓ ↑ ↑	—
c.	↑↓	↑↑ ↑ ↑	—
d.	↑↓	↑↓ ↑↓ —	—
e.	↑↓	↑ ↑ ↑	↑

10. Fill in the missing items:

Configuration	Charge	Unpaired e^-'s	Element
[Ne]	−3	0	?
?	0	?	Ca
$[Ar]4s^2 3d^{10} 4p^2$	0	?	?
?	+2	4	Fe
[Kr]	?	?	Rb
?	+?	3	Cr
$[Xe]4f^{14}5d^{10}$	?	?	Hg

11. If screening did not split the l sublevels of a given n shell, how many elements would occur in the fifth period? Give the configuration of the noble gas of this period. Would you expect this period to be completed by the naturally occurring elements?

Atomic properties

12. Arrange the following elements in order of increasing ionization energy IE:
    a. Ne   Ba   Ga   F   S
    b. Cl   I   Br   Te   Ar
    c. Mg   Cs   N   Si   Al

13. Elements lying along a downward-sloping diagonal of the periodic table tend to have similar IEs. For example, the elements C, P, and I have IEs of 11.26, 10.49, and 10.45 eV. Rationalize this occurrence. Explain why the IE of Se, 9.75 eV, is a bit out of line.

14. Which one of the following pairs of atoms has the higher IE? In the case of an exception to the general trend, state the cause.
    a. Ba,Sr   f. Cu,Zn
    b. Mg,K   g. Ga,Ca
    c. Hg,Cd   h. Cl,I
    d. P,S   i. Ar,Br
    e. Si,N   j. Kr,Rb

15. The elements that deviate from the general trend in IE with anomalously large IEs tend to have anomalously small electron affinities. Explain.

16. For chemical interactions among atoms, both the ionization energy and the electron affinity contribute to the electron attracting power, or *electronegativity*, of an atom's valence shell. We can define an index $\chi = 0.192(IE + EA)$ as a measure of this power, where the coefficient is chosen to give F a "perfect" 4.0 rating. Using data from Figures 4.9 and 4.11, find χ for B, C, N, O, and F. Note that the anomaly in IE is not present in χ. In the next chapter, we'll give a more "chemical" definition of χ, and use it in discussing chemical bonding.

17. Arrange the sequences of Exercise 12 in order of increasing atomic radius.

18. For the pairs of atoms given in Exercise 14, choose the atom with the larger atomic radius.

Energy and radius calculations

19. Two ions not illustrated in Figure 4.13 are Ca^{2+} and Sc^{3+}. With which ions shown are these ions isoelectronic? Using the general trends for radii evident in that figure, give rough estimates of the ionic radii for these ions.

20. In the orbital approximation for He described in Section 4.1, the total energy E of a helium atom He is predicted to be the sum of the energies ε of the individual screened electrons

$$E(\text{He}) = \varepsilon_1 + \varepsilon_2 = -2\left(\frac{Z_{\text{eff}}^2}{1^2}\right)E_H$$

where $E_H = e^2/2a_0 = 13.60$ eV, and, in accord with the Bohr model of Equation 4.4, the two 1s electrons contribute identical energies. From Equation 4.5, the first ionization energy IE = $E(He^+) - E(He)$, where $E(He^+)$ is also given by the Bohr model.

  a. Use these relations with the given Z_{eff} for He to confirm the total energy of He in eV in the orbital model quoted in Section 4.1, and predict the IE in eV of He. Compare your IE with the value given in Figure 4.9, and compute a % error.

  b. Use Equation 4.7 to predict the orbital radius of He in Å.

21. In 1930 Slater gave a set of empirical rules for estimating Z_{eff} based on the electron configuration. For s and p electrons with $n = 2$ or 3, Slater's rules (with an update from more recent calculations) yield

$$Z_{eff} = Z - [0.35(N_n - 1) + 0.86 N_{n-1} + 1.00 N_{n-2}]$$

where N_n is the number of electrons in the same n shell as the electron of interest, N_{n-1} is the number of electrons in the $(n - 1)$ shell, and N_{n-2} the number in shells $n - 2$ and lower. For any atom A in periods 2 or 3, the Bohr model can then be used to predict the orbital radius from Equation 4.7, and IE = $E(A^+) - E(A)$ as in Exercise 20.

  a. Find Z_{eff} from Slater's rule for an Ar $3p$ electron, and use it to predict the $3p$ orbital radius.

  b. When an atom A ionizes, Slater's rule predicts that only its remaining valence electrons will relax, due to the loss of the screening effect of the ionized electron. Using the Bohr energy formula, the first IE is then given by the difference in valence electron energies

$$\frac{IE}{E_H} = -(N_n - 1)\frac{[Z_{eff}(A^+)]^2}{n^2} - \left(-N_n \frac{[Z_{eff}(A)]^2}{n^2}\right)$$

where Z_{eff} refers to the valence shell, $E_H = 13.60$ eV, and the energies of the core electrons cancel in the difference. If A has more than one valence electron, $Z_{eff}(A^+) = Z_{eff}(A) + 0.35$; the loss of the ionized electron increases Z_{eff} for *all* the remaining valence electrons. (Note that this yields a simple result for the alkali metals, as used in Section 4.5.) Calculate the IE of Ar in eV, and compare with the value in Figure 4.9.

22. Use Slater's rule as given in Exercise 21 to compare Z_{eff} and orbital radius for the $2s$ electron in Li $1s^2 2s^1$ and excited He $1s^1 2s^1$. Based on Z_{eff}, which species is expected to have the lower IE?

23. Use information given in Exercise 21 to compute Z_{eff} for a $2p$ electron in ground-state C, N, and O atoms, and to estimate orbital radii and IEs for these atoms. Compare with values in Figures 4.12 and 4.9, and discuss briefly. (Note that Slater's rule does not distinguish between an electron in the same orbital as the one considered and those in other orbitals in the same shell.)

24. Use information given in Exercise 21 to compute Z_{eff} for a $3p$ electron in Cl and Cl^-, and estimate orbital radii and EA(Cl). Compare with data in Figures 4.11 and 4.13, and discuss briefly.

25. Use information given in Exercise 21 to compute Z_{eff} for the $3s$ electron in Na and a $2p$ electron in Na^+. Use these to estimate the radii of these two orbitals and IE(Na), and compare with data in Figure 4.13. Discuss briefly.

26. Draw a boron nucleus with five electron-dots around it. Draw lines between the dots until all dots are connected to all other dots. Count the number of distinct line segments and verify the formula $N(N-1)/2$ for the number of electron–electron repulsion terms given in Equation 4.8. How many such terms are there in a bromine atom? In a uranium atom?

CHAPTER 5

Valence Electron Configurations, Periodicity, and Chemical Behavior

The reaction between phosphorus and pure oxygen. The phosphorus oxide product dissolves in the column of water below. See Figure 5.9 and surrounding text for a description.

CHAPTER OUTLINE

5.1 Metals, Nonmetals, and Binary Salts: The Noble Gas Rule of Lewis

5.2 Electron Sharing: Lewis's Covalent Bond

5.3 Reactions of Metals and Nonmetals with Water: Acidic and Basic Behavior

5.4 Binary Oxides and Hydrides: Chemical Periodicity

5.5 Acid–Base Reactions

5.6 Oxidation–Reduction Reactions

5.7 Precipitation Reactions

The traditional physics professor is supposed to point to chemistry as a monumental example of what can be done with electrons, a few nuclei, and Coulomb's Law. He might be surprised to learn what fun can be had . . .

Dudley Herschbach (1966)

The orbital model of atomic structure outlined in the preceding chapter allowed the chemist not only to understand the form of the periodic table, but also to begin to probe the nature of chemical transformations and how an element's position in the table affects its role in them. This put the vast array of known chemical data into a new conceptual framework and endowed the chemist with new predictive powers, the likes of which had not been seen since the days of Mendeleyev. In this chapter we take a broad-brush tour across the periodic table, examining representative elemental properties and reactions, the hard data of chemistry, in the light of our knowledge of atomic structure and the fundamental role played by the interaction of charges.

But the fascination of chemistry is not just dry analysis. It is witnessing chemical reactions as they happen, enjoying the often spectacular evolution of heat and light, and the wondrous changes in color and state as products are formed. Combining such lively action with the glimmerings of understanding provided by the atomic model can give a sensory and intellectual thrill that's hard to match in any area of study.

Within a chemical family, a column of the periodic table, the valence orbital occupation is analogous. In group I, the alkali metals, for example, all of the valence configurations can be represented as ns^1, while in group VII, the halogens, they are all ns^2np^5, where n is both the period number and the principal quantum number. Even though the elements are seldom found as isolated atoms with these orbital occupancies, the electron configurations will take you a long way toward understanding chemical reactions and compounds.

Gilbert Newton Lewis (1875–1946) at work in his laboratory at the University of California, Berkeley in the 1930s. In 1902 Lewis began his speculations about atomic structure and valence that were to revolutionize chemical thought, though his abiding interest was chemical thermodynamics. He later worked in spectroscopy, and coined the word "photon."

5.1 Metals, Nonmetals, and Binary Salts: The Noble Gas Rule of Lewis

Ordinary table salt, sodium chloride (NaCl), is a common example of a large class of chemical compounds composed of **metals** and **nonmetals.** We call these compounds **salts,** though many of them would be quite disagreeable or even poisonous sprinkled over food.

Metals in their elemental states are solids under normal conditions (except for mercury, Hg, a liquid), can take on a high luster when polished, are more or less malleable, and are good electrical conductors. Figure 5.1 pictures some common metals in their elemental forms. In the periodic table they all occur to the left of a stair-step dividing line, the heavy line visible on most periodic tables (see Figure 5.2), descending to the right from just to the left of boron to between polonium and astatine. Along this line the ionization energies (IEs) are all about 8 eV, with metals being lower in accord with the general trend in IE. The dividing line steps down instead of being exactly vertical because of the competing effects on the IE of increasing Z_{eff} and increasing n, as you've seen in Chapter 4. The low IEs characteristic of metals are consistent with their generic properties. Electrical conductivity, for example, results from electrons being able to move freely through a metal wire (occasionally *very* freely as in a superconductor). This is only possible if the metal atoms can give up their valence electrons easily to form a maelstrom of flowing electrons. Since the IE measures the work necessary to remove these electrons, metallic conduction correlates with position in the periodic table.

Figure 5.1
Eight common metals. MP is the melting point. Only mercury is a liquid at room temperature.

Na sodium
MP 98°C

Ca calcium
MP 839°C

Al aluminum
MP 660°C

Fe iron
MP 1535°C

Cu copper
MP 1083°C

Zn zinc
MP 420°C

Hg mercury
MP −39°C

Pb lead
MP 328°C

Due to the filling of d and f orbitals immediately after the s level (instead of after the p level), metals greatly outnumber the nonmetals—68 to 22 among the naturally occurring elements, 23 to 0 for the man-made (mainly transuranic) elements. Unlike the metals, 11 of the 21 nonmetallic elements are *gases* under normal conditions. However, they also count one liquid, bromine, among their number, with the rest being solids. Some common nonmetals are pictured in Figure 5.3. With the exception of carbon in the form of graphite, they are either poor electrical conductors (near the divide) or completely insulating. Again using electrical conductivity as an example, the high IEs of the nonmetals prevent electrons from migrating out of their orbital homes into the sea. Once again this property can be understood by referring to the periodic table.

The elements either on or directly under a step of the metal–nonmetal divide form a special class known as **metalloids** or semimetals. These elements often behave ambiguously both as to elemental properties and chemical reactions, and, by no coincidence, they are the cornerstones of the microelectronics industry. For the purposes of our discussion in this section, this exceptional behavior will not be highlighted.

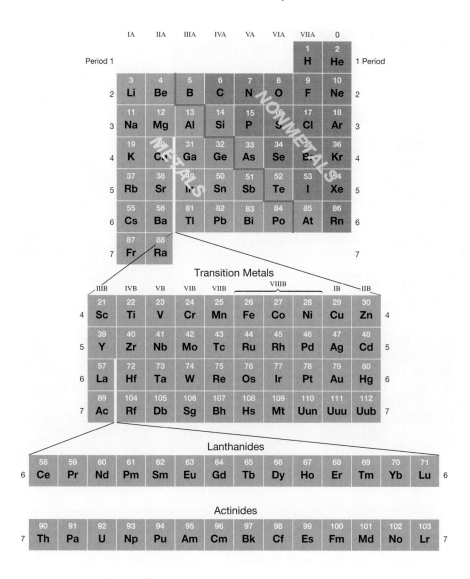

Figure 5.2
A so-called short form of the periodic table, illustrating the division of the known elements into metals and nonmetals. The blue elements (the metalloids) have properties that allow them to act as either metals or nonmetals.

Simple Oxidation–Reduction (Redox) Reactions

When a metal and nonmetal are brought together, a special sort of chemical reaction (formation of a new substance with properties different from either reagent) takes place. In the reaction *electrons are transferred from the metal to the nonmetal* to make a salt, composed of positive metal ions and negative nonmetal ions. The metal, which loses electrons, is said to be **oxidized,** and the nonmetal, which gains them, is **reduced.**

Such reactions are most often observed today in the slow combination of metals with oxygen from the air, but more spectacular reactions may have occurred on the hot young planet earth, because the earth's crust abounds with salts in the form of the minerals fluorspar (CaF_2), bauxite (Al_2O_3), sphalerite (ZnS), and the like. As

Figure 5.3
Eight common nonmetals. N_2 and O_2 are shown in liquid form. Carbon and phosphorus may exist in different forms, or allotropes. BP, normal boiling point; MP, melting point.

N_2 nitrogen
78% of atmosphere
BP −196°C

O_2 oxygen
21% of atmosphere
BP −183°C

Cl_2 chlorine gas
BP −3°C

Br_2 bromine liquid and vapor
BP 59°C

I_2 iodine solid and vapor
MP 113°C

C carbon solid
MP 3550°C

red   white
P (P_4) phosphorus solid
MP 44°C

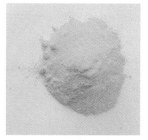

S (S_8) sulfur solid
MP 119°C

an example, ordinary table salt can be obtained from the the glowing combination of sodium and chlorine (*s* stands for solid, *g* for gas) on modest heating:

$$2Na(s) + Cl_2(g) \longrightarrow 2NaCl(s) \tag{5.1}$$

as Figure 5.4 illustrates. There is much evidence to suggest that the salt product NaCl is actually composed of ions Na^+ and Cl^-. If the salt is melted by heating to a very high temperature, it becomes an electrical conductor, and dissolving it in water (H_2O), a nonmetallic compound and poor electrical conductor, produces a solution that conducts electricity readily. The total energy of the salt crystal can be accurately calculated by assuming Coulombic attractions between ions. From these studies the magnitudes of the charges are found to be unity on both Na and Cl. Reaction 5.1 can therefore be thought of as the loss of an electron by Na:

$$Na([Ne]3s^1) \longrightarrow Na^+ ([Ne]) + e^- \tag{5.2}$$

and the gaining of an electron by Cl:

$$Cl([Ne]3s^2 3p^5) + e^- \longrightarrow Cl^-([Ar]) \tag{5.3}$$

5.1 Metals, Nonmetals, and Binary Salts: The Noble Gas Rule of Lewis

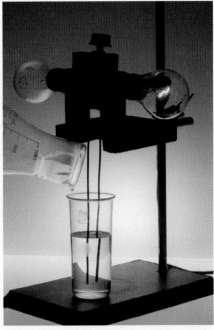

Figure 5.4
The reaction of sodium metal with chlorine gas. In the left panel, the reaction flask gives off a bright orange-yellow chemiluminescence (the sodium D-line) arising from the energy liberated by the reaction. The NaCl product coats the walls of the cooled reaction flask. In the right panel, the ionic nature of NaCl is established by rinsing the product into a beaker of distilled water, which becomes at once an electrical conductor.

where the parentheses enclose the electron configurations of the atoms and ions. Here we have ignored the bonds between atoms in Na(s) and $Cl_2(g)$, focusing instead on the electron configurations of the atoms in these nonionic elemental substances. The important point is that *the charges on the ions result from the attainment of a noble gas electron configuration.*

When a host of other binary salts is examined, this **noble gas rule,** first formulated in 1916 independently by W. Kossel and by Gilbert N. Lewis, is found to hold remarkably well, especially if compounds involving only the so-called **representative elements** (*s* and *p* valence shells, excluding the transition metals, lanthanides, and actinides) are considered. For example, in salts such as AlF_3, Mg_3N_2, BaO, Na_2S, and KBr, the metal has lost electrons to become like the noble gas of the previous period, while the nonmetal has gained them to attain the noble gas configuration of its own period. As we have seen, the closed-shell noble gas structure is by far the most stable way to arrange electrons. The stoichiometry of these compounds then results from charge conservation; that is, electrically neutral reagents must yield electrically neutral products. No electrons are lost or created during the reaction.

EXAMPLE 5.1

For the compound AlF_3, write a balanced chemical equation like Equation 5.1, using the naturally occurring form of fluorine, $F_2(g)$. Then write equations analogous to Equations 5.2 and 5.3 for the oxidation and reduction, showing the electron configurations. Finally, add these latter two equations so as to give a set of ions that is electrically neutral overall.

Solution:
The unbalanced equation is

$$Al(s) + F_2(g) \longrightarrow AlF_3(s)$$

This redox reaction may be balanced by inspection; more complex reactions to be considered later in this chapter require explicit accounting of the electrons transferred. In this case, first consider the element with the larger number of atoms in the participating compounds, F. Balancing F atoms with integer coefficients requires $3F_2$ and $2AlF_3$; Al is then balanced by 2Al to yield

$$2Al(s) + 3F_2(g) \longrightarrow 2AlF_3(s)$$

Using the noble gas rule, the oxidation and reduction equations are

$$Al([Ne]3s^23p^1) \longrightarrow Al^{3+}([Ne]) + 3e^-$$
$$F([He]2s^22p^5) + e^- \longrightarrow F^-([Ne])$$

Note that both atoms attain Ne-like configurations. Because no net change in the number of electrons can occur, three F atoms must acquire electrons for each Al atom that becomes Al^{3+}. So if we multiply the F reduction equation by 3 and add it to the Al equation, we get

$$Al + 3F \longrightarrow Al^{3+} + 3F^-$$

In adding the equations you should think of the reagents and products as algebraic quantities, and the factor of 3 as being necessary to make the electrons "cancel." The new feature of the atomic-ionic equation is that it shows explicitly not only the conservation of mass, but also of charge. (If you regard the atoms and ions in this last equation as isolated gas-phase entities, the reaction is energetically impossible. It "goes" owing to the bonding differences in the reagents and products.)

When Lewis proposed the noble gas rule, 10 years before the advent of quantum mechanics but 3 years after the Bohr atom, he could not write electron configurations as we have done. Though he drew on the work of Bohr in advancing the modern form of the periodic table, he did not wholly subscribe to the planetary model, preferring at first to think of atoms as tiny cubes with electrons at the vertices. But this picture of the atom did not affect the ability of the noble gas rule, combined with conservation of charge, to rationalize as well as to predict the stoichiometry of ionic compounds.

Not knowing anything of orbitals, Lewis diagrammed his atoms by using the chemical symbol surrounded by valence electron dots. If an imaginary square were drawn around the atomic symbol, each side could accommodate up to a pair of electrons (just as an orbital can). The combination of a sodium atom and a chlorine atom to make salt would be diagrammed

$$Na\cdot + \cdot \ddot{\underset{..}{Cl}}: \longrightarrow Na^+ + [:\ddot{\underset{..}{Cl}}:]^- \tag{5.4}$$

Thus was born what you already know as **Lewis electron-dot structures.** Because of the great usefulness of these structures in depicting the noble gas rule, they are still widely used today.* Of course, you should now regard Equation 5.4 and the like as shorthand notation for the orbital electron configurations actually involved.

---

* It is interesting to speculate on what chemistry today would have been like if Lewis had been born 10 years later. He probably could not have gotten away with such simplistic diagrams under the glare of quantum mechanics!

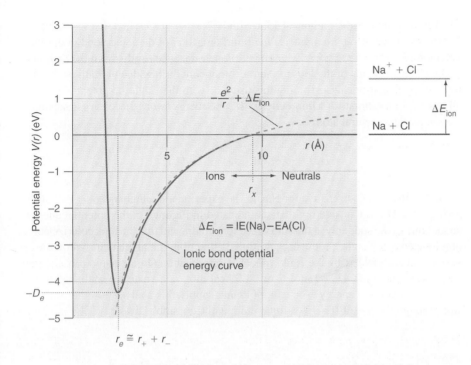

Figure 5.5
Energetics of forming an ionic bond between Na and Cl. As the isolated atoms are brought together, they remain neutral until the energy of Coulomb attraction crosses the energy of the neutral atom pair; an electron is then transferred from metal to nonmetal. The ions are then drawn toward each other by the Coulomb force until they "touch." The dashed curve is the Coulomb potential. The solid curve deviates from this because of polarization attraction. D_e, bond dissociation energy; r_e, bond length of the ionic bond.

If the d and f inner-shell electrons are ignored for the heavier elements, Lewis's noble gas rule is a rule of eight (the capacity of the valence s and p orbitals) or an **octet rule**.

Ionic Bonding

The attainment of noble gas shells is an important driving force behind the chemistry of salt formation, but it is by no means the only one. Of equal influence is the Coulombic attraction between the ions formed, which binds them together in a strong **ionic bond**. As illustrated in Figure 5.5, if a neutral Na atom and Cl atom are brought together, the Coulomb potential energy of the ion pair must eventually become lower than the combined energy of the neutrals. At this critical atom–atom distance r_x an electron is transferred from Na to Cl to make the ion pair, since this produces lower potential energy. Because the Coulomb potential must make up the energy shortfall for ion formation, given by the difference IE(Na) − EA(Cl), the electron-transfer distance can easily be estimated from

$$\frac{e^2}{r_x} = \text{IE(Na)} - \text{EA(Cl)} \tag{5.5}$$

Using IE(Na) = 5.14 eV, EA(Cl) = 3.62 eV, r_x is found to be 9.5 Å, a rather large distance.*

---

* As will become clear from later discussion, the orbitals of the atoms are overlapping very weakly at such distances, making the actual "jump" of the electron a rather breathtaking event, much like a ballet dancer's *grande jeté*.

From the ionic radii of Figure 4.13, we find $r(Na^+) = 0.85$ Å and $r(Cl^-) = 1.50$ Å. The sum of these is (approximately) the **bond length** r_e of the ionic bond in the isolated ion pair Na^+Cl^-, 2.35 Å. This finite ion-pair distance results from the inability of the electron clouds of the ions to interpenetrate any further, due to Pauli exclusion for closed-shell species. Combining this distance with the Coulomb potential energy allows our first estimate of a **bond energy** D_e, the energy (work) necessary to separate the ion pair into neutral atoms, as

$$D_e \approx \frac{e^2}{r_e} - [IE(Na) - EA(Cl)] \qquad (5.6)$$

giving a value of $D_e = 4.61$ eV for NaCl. The experimental numbers are $r_e = 2.36$ Å and $D_e = 4.27$ eV. Considering that, as indicated in Figure 5.5, the potential energy cannot turn up so suddenly at the sum of the ionic radii, and that the ions polarize each other's electron clouds, producing additional attraction, the agreement with the experimental D_e yielded by the ionic model is remarkable. On a per-mole basis, the NaCl bond energy is 412 kJ mol^{-1} or 98.5 kcal mol^{-1}, a value we shall find to be quite a typical bond energy. You can, of course, apply this analysis to other salts as well through the use of IEs, EAs, ionic radii, and Coulomb's law.

5.2 Electron Sharing: Lewis's Covalent Bond

Nineteenth-century chemists found it easy to accept the existence of salts, the result of mating radically unlike atoms. But an anthropomorphic view of chemistry left little room for the right-thinking person of that time to contemplate molecules like H_2, O_2, or Cl_2, homoatomic species with no obvious ionic attraction to bind them together. Such molecules, as well as a good many heteroatomic molecules like H_2O or CO_2, composed of nonmetals alone, do not show evidence of being ionic in character. Lewis found, however, that he could extend the noble gas rule to these species as well, as long as the atoms were made to *share* rather than transfer their electrons. For the Cl_2 molecule, for example, Lewis could depict its formation from Cl atoms as

$$:\ddot{C}l\cdot + \cdot\ddot{C}l: \longrightarrow (:\ddot{C}l(:)\ddot{C}l:)$$

Each Cl atom requires one electron to attain an Ar-like configuration, but neither would enjoy giving up an electron to satisfy the other. So each shares an electron with the other, producing, as the circles around each Cl indicate, at least the illusion of a completed noble gas shell. With the sharing idea, Lewis could rationalize the stoichiometry of nearly all known nonmetal compounds, including the elemental diatomic molecules H_2, N_2, O_2, F_2, Cl_2, Br_2, and I_2, and a myriad of organic compounds. What Lewis had done was provide the first firm footing for the concept of **valence,** the capacity of an atom to bind with only certain fixed numbers of other atoms. The shared-electron case was later dubbed the **covalent bond.** This amazingly successful concept took the pre-quantum world of chemistry by storm in the late teens and early twenties, revolutionizing the subject both in the laboratory and in the classroom.

While Lewis's sharing idea cannot provide the quantitative sort of analysis we've seen for ionic bonds, it does yield qualitative correlations and trends consistent with known molecular properties. For example, the molecules N_2, O_2, and

F_2 can be formed by allowing the atoms to share three, two, and one electrons apiece, respectively:

$$:\!\dot{N}\cdot + \cdot\dot{N}\!: \longrightarrow \;:\!N\!::\!:\!N\!: \quad (or \;:\!N\!\equiv\!N\!:)$$

$$:\!\ddot{O}\cdot + \cdot\ddot{O}\!: \longrightarrow \;:\!\ddot{O}\!::\!\ddot{O}\!: \quad (or \;:\!\ddot{O}\!=\!\ddot{O}\!:)$$

$$:\!\ddot{\underline{F}}\cdot + \cdot\ddot{\underline{F}}\!: \longrightarrow \;:\!\ddot{\underline{F}}\!:\!\ddot{\underline{F}}\!: \quad (or \;:\!\ddot{\underline{F}}\!-\!\ddot{\underline{F}}\!:)$$

The differing valences or bond-forming powers of the elements result from the differing number of electrons required to form noble gas shells. N_2 is said to contain a **triple bond,** due to the N atom's valence of 3, and consequent sharing of three pairs of electrons, while O_2 possesses a **double bond** and F_2 a **single bond.** The electrons not shared, which always occur in pairs, became known as **lone pair electrons.** Lewis coined the term **bond order** (3, 2, or 1) to denote the number of electron pairs shared between two given atoms, and reasoned that a triple bond ought to be stronger than double, and double stronger than single. The modern values of r_e and D_e for these molecules are shown in Table 5.1. As expected from the Lewis structures, the bond energy decreases in the sequence N_2, O_2, F_2, while the bond length increases. The increase in r_e normally accompanies weakening of a chemical bond, and can be thought of as due to the less intimate contact of the atoms required when fewer electrons are shared.

In Lewis's picture H_2 is H—H, since this makes each H feel like He. Other possible molecules like H_3 and H_4 are ruled out because it is not possible to construct covalent bonds that satisfy the noble gas rule. (Try it!) Similarly, the stoichiometry of the nonmetal hydrides of C, N, O, and F is strictly dictated by Lewis's rule. These molecules have the following Lewis structures:

$$CH_4, \; H\!-\!\underset{H}{\overset{H}{\underset{|}{\overset{|}{C}}}}\!-\!H \qquad NH_3, \; H\!-\!\underset{H}{\overset{\cdot\cdot}{\underset{|}{N}}}\!-\!H \qquad H_2O, \; H\!-\!\underset{H}{\overset{\cdot\cdot}{\underset{|}{\ddot{O}}}}\!: \qquad HF, \; H\!-\!\ddot{\underline{F}}\!:$$

Methane Ammonia Water Hydrogen fluoride

Each hydrogen must share one electron out of the valence shell of the nonmetal to become He-like, while the number of H's in each molecule is sufficient to satisfy the noble gas requirement of the central atom. Because of their unit valence, the H atoms must always be located on the periphery of hydride molecules.

TABLE 5.1

Modern values for bond length and bond dissociation energy for three representative molecules

	Bond order	r_e(Å)	D_e
N_2	3	1.10	9.76 eV (942 kJ mol^{-1}, 225 kcal mol^{-1})
O_2	2	1.21	5.12 eV (494 kJ mol^{-1}, 118 kcal mol^{-1})
F_2	1	1.41	1.60 eV (155 kJ mol^{-1}, 37 kcal mol^{-1})

In most cases it is possible to formulate molecular Lewis structures that satisfy the normal valences of the atoms involved, but occasionally exceptions are encountered. For the oxides of carbon, CO_2 and CO, the only acceptable Lewis structures are

$$:\ddot{O}=C=\ddot{O}: \qquad :C\equiv O:$$

In CO_2, the normal valences, four for C and two for O, are found, but in CO each atom is forced out of its normal valence to satisfy the Lewis rule. The bond energy of CO is 11.09 eV, the strongest bond known in all of chemistry, and certainly consistent with the triple bond of the Lewis structure.

Significant deviations from normal valences also occur for so-called odd-electron molecules such as the air pollutant nitric oxide (NO). If shared pairs of electrons comprise the only mode of bonding, then one of the atoms is left with an incomplete octet. On the other hand, if "half-bonds" are allowed, then octets may be realized. So NO can be written as

$$:\dot{N}=\ddot{O}: \quad \text{or} \quad :\dot{N}\doteq\dot{O}:$$

The first structure was preferred by Lewis, but the second, with a bond order of 2½, is in better accord with the bond energy of NO, 6.52 eV, which falls between those of O_2 and N_2. Note that in CO and NO, the bond order is the average of the normal valences of the bonded atoms (if half-bonds are permitted). Half-bonds certainly must exist; the well-known molecular ion H_2^+, $D_e = 2.77$ eV, observed in abundance in mass spectrometry, has no choice but to "half-bond," with only one electron to share. In Chapter 7 we will start with this molecule to construct an orbital theory of molecules.

Although Lewis's dots and lines would appear to have nothing to do with orbitals, Lewis's model for molecules works so well most of the time that chemists who began to use it before quantum mechanics continued to do so afterward, and we still do today. As you can easily surmise, there's no comparison between the difficulty of drawing a Lewis structure and solving the Schrödinger equation for a molecule!

Electronegativity

The ionic salts and the homoatomic covalent molecules we have just examined clearly represent two extremes of chemical bonding. The bonding in compounds of two different nonmetals cannot be purely covalent (of equal power) because the atoms and their orbitals are different. In 1932 Linus Pauling, of whom we'll speak at greater length in the next chapter, developed the concept of **electronegativity,** a property of an atom that is a relative measure of its electron-attracting power when engaged in bonding.

Pauling noted that bond energies for heteroatomic molecules AB were always greater than the average (arithmetic or geometric) of the bond energies for each atom bonding to its own kind, that is,

$$D_e(AB) > [D_e(A_2)D_e(B_2)]^{1/2} \qquad \text{or} \qquad (5.7)$$

$$\Delta \equiv D_e(AB) - [D_e(A_2)D_e(B_2)]^{1/2} > 0$$

where the square root represents a geometric mean. Taking this as an indication of an ionic, Coulomb contribution to D_e, Pauling defined the **electronegativity difference** between A and B as

$$\chi_A - \chi_B = 0.21 \Delta^{1/2} \qquad (5.8)$$

where χ symbolizes the electronegativity, the D_e's in Δ are expressed in kcal mol^{-1}, and 0.21 converts to electron-volt units. The square root of Δ was used because this yields a more consistent set of χ's when many chemical bonds are compared. Equation 5.8 determines only the difference of two χ's, so the **electronegativity scale** is set by defining that of flourine atoms, χ_F, to be 4.00, and then computing average χ's for other atoms from Equation 5.8 by averaging all available data on D_e's.

For example, the bond energy of ClF, chlorine monofluoride, is 60.4 kcal mol^{-1}, (253 kJ mol^{-1}, 2.62 eV), whereas $D_e(Cl_2) = 57.2$ kcal and $D_e(F_2) = 36.9$ kcal. From Equation 5.7, $\Delta = 60.4 - [(57.2)(36.9)]^{1/2} = 14.5$ kcal, and therefore $\chi_F - \chi_{Cl} = 0.21[14.5]^{1/2} = 0.80$ from Equation 5.8. Using $\chi_F = 4.00$ yields $\chi_{Cl} = 3.20$, a contributing value to the overall $\chi_{Cl} = 3.2$. The bond-energy difference and the electronegativity difference point to a **partial ionic character** of the bond in ClF, with F more strongly attracting the shared electron pair to itself. We indicate this by a **partial charge separation** δ, a fraction of a full electronic charge ($0 < \delta < 1$), in the Lewis structure,

$$\overset{+\delta}{:\ddot{\underset{..}{Cl}}} - \overset{-\delta}{\ddot{\underset{..}{F}}:}$$

with the more electronegative atom bearing the partial negative charge. This type of bonding occupies the middle ground between the covalent and ionic extremes, and is referred to as **polar covalent.** The word *polar* describes the partial charge separation and indicates that the molecule possesses an intrinsic imbalance in its charge distribution. This imbalance gives rise to an **electric dipole moment,** and will be discussed at greater length in Chapter 6.

Equation 5.8 is a strictly empirical definition of χ based on bond energies, but 2 years after Pauling's proposal, Robert Mulliken (who coined the word *orbital*) showed that values of χ very similar to Pauling's could be obtained from the average of the IE and EA of the atoms, that is, the valence orbital energy. Figure 5.6 collects the IE, EA, and electronegativity in periodic form. Note that, in accord with Mulliken's correlation, χ follows the same trend with position in the table that IE and EA do. You can use the χ values to assess the extent of ionic character of a bond. Roughly speaking, an electronegativity difference $\Delta\chi$ greater than 1.7 indicates an ionic bond, 0.4 to 1.7 polar covalent, and less than 0.4 essentially **pure covalent.**

Resonance Forms and Expanded Octets

Lewis's model does not always provide a satisfactory description of molecular structure for polyatomic molecules. For example, the molecule ozone, O_3, an **allotrope** (different elemental form) of oxygen formed in lightning storms and electric arcs through air, may be represented by the Lewis structure

$$:\ddot{O}=\ddot{O}-\ddot{O}:$$

Figure 5.6

Ionization energies IE (eV), electron affinities EA (eV), and electronegativities χ (Pauling scale) of the elements.

	1	2	3	4	5	6	7	8	9	10	11	12	13	14	15	16	17	18
	H																	**He**
IE	13.60																	24.59
EA	0.75																	0
χ	2.2																	–
	Li	**Be**											**B**	**C**	**N**	**O**	**F**	**Ne**
IE	5.39	9.32											8.30	11.26	14.53	13.62	17.42	21.56
EA	0.62	0											0.28	1.26	0	1.46	3.40	0
χ	1.0	1.6											2.0	2.5	3.0	3.5	4.0	–
	Na	**Mg**											**Al**	**Si**	**P**	**S**	**Cl**	**Ar**
IE	5.14	7.65											5.99	8.15	10.49	10.36	12.97	15.76
EA	0.55	0											0.44	1.39	0.75	2.08	3.62	0
χ	0.9	1.3											1.6	1.9	2.2	2.6	3.2	–
	K	**Ca**	**Sc**	**Ti**	**V**	**Cr**	**Mn**	**Fe**	**Co**	**Ni**	**Cu**	**Zn**	**Ga**	**Ge**	**As**	**Se**	**Br**	**Kr**
IE	4.34	6.11	6.56	6.83	6.75	6.77	7.43	7.90	7.88	7.64	7.73	9.39	6.00	7.90	9.79	9.75	11.81	14.00
EA	0.50	0	0.19	0.08	0.53	0.67	0	0.16	0.66	1.16	1.23	0	0.3	1.2	0.81	2.02	3.37	0
χ	0.8	1.0	1.4	1.5	1.6	1.7	1.6	1.8	1.9	1.9	1.9	1.7	1.8	2.0	2.2	2.5	3.0	–
	Rb	**Sr**	**Y**	**Zr**	**Nb**	**Mo**	**Tc**	**Ru**	**Rh**	**Pd**	**Ag**	**Cd**	**In**	**Sn**	**Sb**	**Te**	**I**	**Xe**
IE	4.18	5.69	6.22	6.63	6.76	7.09	7.28	7.36	7.46	8.34	7.58	8.99	5.79	7.34	8.61	9.01	10.45	12.13
EA	0.49	0	0.31	0.43	0.89	0.75	0.55	1.05	1.14	0.56	1.30	0	0.3	1.2	1.07	1.97	3.06	0
χ	0.8	1.0	1.2	1.3	1.6	2.2	1.9	2.3	2.2	2.2	1.9	1.7	1.8	2.0	2.1	2.1	2.7	–
	Cs	**Ba**	**La**	**Hf**	**Ta**	**W**	**Re**	**Os**	**Ir**	**Pt**	**Au**	**Hg**	**Tl**	**Pb**	**Bi**	**Po**	**At**	**Rn**
IE	3.89	5.20	5.58	6.83	7.55	7.86	7.83	8.43	8.97	8.96	9.23	10.44	6.11	7.42	7.29	8.41	–	10.75
EA	0.47	0	0.5	0	0.32	0.82	0.15	1.1	1.57	2.13	2.31	0	0.2	0.36	0.95	1.9	2.8	0
χ	0.8	0.9	1.1	1.3	1.5	2.4	1.9	2.2	2.2	2.3	2.5	2.0	2.0	2.3	2.0	2.0	2.2	–

According to the bond-order–bond-length-and-strength correlation, O_3 should have one shorter, stronger bond and one longer, weaker one. Spectroscopic examination of the molecule (see Chapter 8), however, reveals that the two bonds are identical. To rescue the Lewis idea, Pauling suggested that the structure of O_3 above was in **resonance** with a structure in which the double bond has shifted to the right-hand side:

$$:\ddot{O}=\ddot{O}-\ddot{O}: \longleftrightarrow :\ddot{O}-\ddot{O}=\ddot{O}:$$

In this way the observation of two identical bonds could be considered an average of the two resonance forms of O_3. At first Pauling conceived of an actual mechanical resonance, a rapid shifting in the electron arrangement. But when experiments failed to reveal such a phenomenon, this idea gave way to an admission that the "true" O_3 structure was neither of the resonance forms, but something in between, such as

$$O-O-O$$

where the dashed line indicates an extra bond that somehow spans both O—O linkages. You should not attempt to insert the electrons not used in bonding, the lone pair electrons, into the hybrid structure shown, since they must shift around as well, as the resonance forms show.

The use of resonance forms must be regarded as a patch on the Lewis quilt, though in its favor is the fact that in this and other cases the bond order correlations work out quite well. In the case of O_3 the bond characteristics are intermediate between a single and full double bond. The bond order is fractional, 1½ for each bond, just as in the odd-electron case.

5.2 Electron Sharing: Lewis's Covalent Bond

In the similar molecule, sulfur dioxide (SO_2), you might expect to find the same problem; however, some chemists patch things up with a structure that violates or expands the octet rule in the case of the third-row atom S,

$$:\ddot{O}=\ddot{S}=\ddot{O}:$$

While we prefer not to violate the Lewis rule except when absolutely necessary, there is justification for this: the availability of d orbitals for $n = 3$ and greater, allowing atoms like S to accommodate more than eight electrons. These orbitals, however, carry a price tag, being higher in energy than the $3s$ and $3p$ orbitals, and the "dominant" structures remain the resonance forms

$$:\ddot{O}=\ddot{S}-\ddot{O}: \longleftrightarrow :\ddot{O}-\ddot{S}=\ddot{O}:$$

which are analogues of the O_3 structures.

EXAMPLE 5.2

Give a proper Lewis structure along with any resonance forms for the carbonate ion, CO_3^{2-}, where the carbon atom is central.

Solution:
In more complicated cases with multiple bonding, as in CO_3^{2-}, it is useful to have a systematic procedure for deciding how many bonds and lone pairs there will be. A procedure that includes the case where an atom is forced out of its normal valence is as follows:

1. Find the number of valence electrons required to satisfy every atom's noble gas quota.
2. Find the actual number of valence electrons present, accounting for the overall charge.
3. Subtract (2) from (1) to find the number of electrons that must be shared. The number of bonds is half this number.
4. Finally, after bonds are drawn, add enough lone pairs to each atom until its octet is supplied.

For CO_3^{2-},

$$e^- \text{ required} = 8 + 3(8) \qquad\qquad = 32\ e^-$$
$$e^- \text{ available} = \underset{(C)}{4} + \underset{(3O)}{3(6)} + \underset{(charge)}{2} = 24\ e^-$$
$$\overline{\qquad\qquad\qquad 8\ e^- \text{ shared}}$$

Therefore, four bonds are formed. For a central carbon atom, the structures are

$$\left[\begin{array}{c}:\ddot{O}:\\|\\:\ddot{O}-C=\ddot{O}:\end{array}\right]^{2-} \longleftrightarrow \left[\begin{array}{c}:\ddot{O}\\\|\\:\ddot{O}-C-\ddot{O}:\end{array}\right]^{2-} \longleftrightarrow \left[\begin{array}{c}:\ddot{O}:\\|\\:\ddot{O}=C-\ddot{O}:\end{array}\right]^{2-}$$

that is, three resonance forms. For CO_3^{2-}, three identical C—O bonds are predicted, each with a bond order of $1\frac{1}{3}$, in agreement with spectroscopic evidence.

In the cases of the phosphate ion PO_4^{3-} and the sulfate ion SO_4^{2-}, the central P or S is forced to share more electrons than its normal valence would dictate. In these cases the normal, nonexpanded Lewis structures are resonance free and isoelectronic:

$$\left[\begin{array}{c} :\ddot{O}: \\ | \\ :\ddot{O}-P-\ddot{O}: \\ | \\ :\ddot{O}: \end{array} \right]^{3-} \text{and} \left[\begin{array}{c} :\ddot{O}: \\ | \\ :\ddot{O}-S-\ddot{O}: \\ | \\ :\ddot{O}: \end{array} \right]^{2-}$$

The bonds are found to be equivalent in each ion, as the Lewis structures imply, but the bonds are shorter and stronger than "ordinary" single bonds. One way of accounting for this is to violate the octet rule and allow the oxygen lone pairs to become involved in double bonds. Then the problem is to account for all of the resonance forms thereby generated. But you can also understand the increased bond strength in terms of an increased ionic contribution to each bond. Being bonded to four electronegative oxygen atoms makes the central atom quite poor in electron density, since all bond pairs are drawn out to the O's. The central atoms in these ions are thus more electropositive than usual, giving an added Coulomb attraction in bonding to the electronegative O's. Expansion of the octet is absolutely essential for a few compounds, chlorides and fluorides of nonmetals such as PCl_5 and SF_6, to allow bonding of all peripheral atoms.

We recommend that you draw structures that obey Lewis's octet rule when possible, recognizing that abnormal valences may alter bond characteristics, and expanding the octet only when necessary.

A Lewis Structure Description of a Reaction

As introduced in Chapter 1, a major theme of modern chemistry is the relationship between structure and reactivity. Now equipped with a short course in the Lewis school of molecular structure, we return briefly to a simplified version of our salt-forming reaction,

$$Na(g) + Cl_2(g) \longrightarrow NaCl(g) + Cl(g)$$

to describe the mechanism (microscopic pathway) of this reaction. Placing all reagents and products in the gas phase is not unrealistic, since a substantial fraction of the reaction actually occurs in the way just shown, owing to the great heat liberated by the reaction. The heat vaporizes solid sodium, giving mainly atomic vapor.

In terms of Lewis structures, the above reaction is

$$Na\cdot + :\ddot{Cl}-\ddot{Cl}: \longrightarrow Na^+[:\ddot{Cl}:]^- + \cdot\ddot{Cl}:$$

Compared to the simple atomic reaction of Equation 5.4, it is harder to perceive how the products can be formed, because here the chlorines are bonded so that no vacancies occur in their valence shells. Part of the answer lies in recognizing that in Cl_2, the satisfaction of Cl's noble gas tendency is less complete than it would be in a Cl^- ion, and so one Cl gives up sharing an electron with another stingy Cl in exchange for a complete electron transfer:

where the curly fish-hook arrows indicate the migration of single electrons, and the charges in double parentheses are incipient, developing in the course of forming a new ionic bond. This type of picture is not entirely satisfactory for electron transfer reactions; as we will see in Chapter 7, an orbital model can provide a more reasonable description. However, it is a general rule that *the acceptor molecule in an electron transfer reaction suffers some bond rupture if it began as a species satisfying Lewis's noble gas rule.*

5.3 Reactions of Metals and Nonmetals with Water: Acidic and Basic Behavior

By far the largest part of chemistry is concerned with reactions occurring in a liquid solvent, and the plentiful compound water (H_2O) is the most frequently employed solvent, both naturally and in the chemical laboratory. Often the solvent is not included in the balanced chemical equation (though it is always an important ingredient in the reaction), but in this section we consider reactions directly with water as a way of launching a description of the chemistry of the elements.

Water is a compound of nonmetals, and has the structural geometry shown here:

$$\overset{+\delta}{H} \xrightarrow{0.96\ \text{Å}} \overset{-2\delta}{O}$$

$$104.5° \quad 0.96\ \text{Å}$$

$$H\ +\delta$$

A critically important facet of the water molecule is that it is bent, with the HOH bond angle as indicated. In the next chapter we will consider why H_2O is bent, but for now we simply state that water's being a liquid for a wide range of temperatures (thus providing a versatile reaction medium) and many aspects of its chemical behavior can only be interpreted in terms of a bent structure. A feature of great significance to the present discussion is the separation of charge, denoted by the partial charges $+\delta$ and -2δ, owing to the substantial electronegativity difference between O ($\chi = 3.5$) and H ($\chi = 2.2$).

As mentioned earlier, when salts, as well as a variety of purely nonmetallic compounds, are dissolved in water, positive and negative ions—**cations** and **anions**—are produced. An important part of learning aqueous chemistry is the proper identification of the ions present (i.e., using the accepted nomenclature for these species). Aside from H^+ and positive metal ions, which are referred to by their elemental names, for example, barium ions, the most commonly occurring positive ion is NH_4^+, ammonium ion. Negative ions, particularly oxyanions are, as we will see later, much more diverse, and naming them is generally a stumbling block for fledgling chemists. We have already examined the anions phosphate (PO_4^{3-}) and sulfate (SO_4^{2-}). Table 5.2 collects the formulas and names of the more common anions, including in appropriate cases the names of the acids arising from protonating (adding H^+ to) these ions. No, we won't ask you to memorize this table; rather refer to it to learn the names as you encounter their formulas in laboratory and homework.

TABLE 5.2
Common anions and their protonated forms

Formula	Name	Protonated form(s)	Name(s)
H^-	Hydride	H_2	Hydrogen (g)
F^-	Fluoride	HF	Hydrogen fluoride (g); hydrofluoric acid (aq)
Cl^-	Chloride	HCl	Hydrogen chloride (g); hydrochloric acid (aq)
Br^-	Bromide	HBr	Hydrogen bromide (g); hydrobromic acid (aq)
I^-	Iodide	HI	Hydrogen iodide (g); hydroiodic acid (aq)
O^{2-}	Oxide	OH^-	Hydroxide
		H_2O	Water
S^{2-}	Sulfide	HS^-	Hydrogen sulfide (bisulfide)
		H_2S	Hydrogen sulfide (g); hydrosulfuric acid (aq)
N^{3-}	Nitride	NH_2^-	Amide
		NH_3	Ammonia (g or aq)
P^{3-}	Phosphide	PH_3	Phosphine (g or aq)
OH^-	Hydroxide	H_2O	Water
BO_3^{3-}	Borate	H_3BO_3	Boric acid
CO_3^{2-}	Carbonate	HCO_3^-	Hydrogen carbonate (bicarbonate)
		H_2CO_3	Carbonic acid [also $CO_2(aq)$]
CN^-	Cyanide	HCN	Cyanic acid
CNO^-	Cyanate	...	...
SCN^-	Thiocyanate	...	...
$C_2H_3O_2^-$	Acetate	$HC_2H_3O_2$	Acetic acid
$C_2O_4^{2-}$	Oxalate	$H_2C_2O_4$	Oxalic acid
SiO_3^{2-}	Silicate	H_4SiO_4	Silicic acid ($H_2SiO_3 \cdot H_2O$)
NO_3^-	Nitrate	HNO_3	Nitric acid
NO_2^-	Nitrite	HNO_2	Nitrous acid
PO_4^{3-}	Phosphate	HPO_4^{2-}	Monohydrogen phosphate
		$H_2PO_4^-$	Dihydrogen phosphate
		H_3PO_4	Phosphoric acid
PO_3^{3-}	Phosphite	H_3PO_3	Phosphorous acid
O_2^{2-}	Peroxide	H_2O_2	Hydrogen peroxide
SO_4^{2-}	Sulfate	HSO_4^-	Hydrogen sulfate (bisulfate)
		H_2SO_4	Sulfuric acid
SO_3^{2-}	Sulfite	HSO_3^-	Hydrogen sulfite (bisulfite)
		H_2SO_3	Sulfurous acid [also $SO_2(aq)$]
$S_2O_3^{2-}$	Thiosulfate	...	...
ClO_4^-	Perchlorate	$HClO_4$	Perchloric acid

(continued)

5.3 Reactions of Metals and Nonmetals with Water: Acidic and Basic Behavior

TABLE 5.2
Common anions and their protonated forms (continued)

Formula	Name	Protonated form(s)	Name(s)
ClO_3^-	Chlorate	$HClO_3$	Chloric acid
ClO_2^-	Chlorite	$HClO_2$	Chlorous acid
ClO^-	Hypochlorite	$HClO$	Hypochlorous acid
BrO_3^-	Bromate	$HBrO_3$	Bromic acid
BrO^-	Hypobromite	$HBrO$	Hypobromous acid
IO_3^-	Iodate	HIO_3	Iodic acid
IO^-	Hypoiodite	HIO	Hypoiodous acid
I_3^-	Triiodide	...	...
CrO_4^{2-}	Chromate	H_2CrO_4	Chromic acid
$Cr_2O_7^{2-}$	Dichromate	...	...
MnO_4^-	Permanganate	...	...

Being a nonmetallic compound, water, although usually quite stable against attack by other reagents, will react with some of the more active metals from **Group I**, the **alkali metals**, and **Group II**, the **alkaline earth metals**. For example, sodium metal will react with water (see Figure 5.7) as follows:

$$2Na(s) + 2H_2O(l) \longrightarrow 2NaOH(aq) + H_2(g) \qquad (5.9)$$

The reaction products are aqueous sodium hydroxide and hydrogen gas. NaOH(aq) is present in solution in the form of Na^+ and OH^-, where OH^-, hydroxide ion, is isoelectronic with HF. You can understand why the reaction goes like this by considering the Lewis–Pauling version of the mechanism. For the interaction of one Na atom with one H_2O molecule the picture is

$$Na \cdot \quad H \overset{+\delta}{-} \overset{-2\delta}{O} \overset{}{-} H^{+\delta} \longrightarrow Na^+ + H\cdot + \left[:\ddot{O}\!:\!H \right]^- \qquad (5.10)$$

where the full-headed arrow implies movement of a *pair* of electrons. The H·, an atomic radical, must combine with another H· from a similar reaction to give H:H or H_2. Note that, like Cl_2, H_2O is a "good" Lewis molecule, and therefore it must undergo bond breakage if it is to accept an electron. The electron from Na is naturally (à la Coulomb) attracted to the electropositive H end of H_2O; the more electronegative O then draws the bonding pair of electrons to itself. This *concerted movement* of electrons and electron pairs is frequently invoked in mechanistic explanations of this sort. Even a detailed picture such as that presented by Reaction 5.10 is necessarily schematic, however, because the reaction actually occurs at the surface of Na metal, and the newly formed hydrogen atoms are probably bonded to the metal prior to H_2 formation.

Figure 5.7
The sodium–water reaction. Phenolphthalein indicator added to the water turns deep pink, showing the production of strong base. The energy given off by the reaction melts the sodium metal; the molten ball skitters erratically on the surface until reaction is complete. Larger quantities of metal result in a flame produced by combustion of the hydrogen.

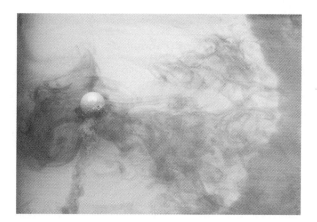

The reactions of water with all of the alkali atoms are analogous to Equation 5.9; just substitute Li, K, Rb, or Cs for Na. This is part of what we mean by referring to the alkalis as a chemical family; they react in similar ways to form similar products, owing to their common ns^1 valence electron configurations. The production of OH^- ions in solution makes the solution **basic.** The earliest systematic definition of a base is due to Svante Arrhenius, who gave us, as a part of his general theory of ions in solution, the classification **acid** as *a substance that produces H^+* (in solvated form H_3O^+, hydronium ion) *in solution* and **base** as *a substance that produces OH^-*. According to Arrhenius, an alkali metal acts as a base, since OH^- is produced as a result of reacting it with water.

The reactions of the alkaline earths, electron configuration ns^2, with water are stoichiometrically different, for example,

$$Ca(s) + 2H_2O(l) \longrightarrow Ca(OH)_2(aq) + H_2(g) \tag{5.11}$$

but mechanistically similar,

$$Ca\!:\quad H\overset{+\delta}{\underset{H^{+\delta}}{\rightleftarrows}}\overset{-2\delta}{\ddot{O}:} \longrightarrow Ca^+ + H\cdot + \left[:\!\ddot{O}\!:\!H\right]^- \tag{5.12}$$

with a second, similar step involving $Ca\cdot^+$ as reagent. One calcium atom, with two valence electrons, produces as much OH^- as two alkali atoms on its way to the nearest noble gas configuration, Ar.

Although all the alkali atoms react readily with water, only the heavier alkaline earths, Ca through Ra do so, with Be completely inert and Mg reacting very slightly. The driving forces behind these reactions are several, including the bond strength within the solid metal reactant and the strength of solvation (hydration) of the positive ion, but an important factor is the ionization energy of the metal. The IE measures the ease of removal of the electron to be transferred from the metal to water, and the unreactive character of Be and Mg reflects their high IEs. Be is adjacent to the metal–nonmetal divide, giving it near-metalloid properties. This correlation between reactivity and IE illustrates the chemical effect of both the left-to-right increase in IE (compare Na and Mg) and the top-to-bottom decrease (compare Mg and Ca).

Metal–Acid Reactions

The metals to the right of Group II, including the transition metals, will not react with water directly, but most of them, as well as Be and Mg, will react with an acidic solution (see Figure 5.8), for example,

$$\text{Mg}(s) + 2\text{H}^+(aq) \longrightarrow \text{Mg}^{2+}(aq) + \text{H}_2(g) \tag{5.13}$$

The H^+ originates from an acidic compound such as HCl(aq), hydrochloric acid, or $\text{H}_2\text{SO}_4(aq)$, sulfuric acid. Equation 5.13 is an example of an **ionic equation,** in which only the atoms and ions that are reacting are shown. If the acidic reagent were HCl, then the neutral form of Reaction 5.13 would be

$$\text{Mg}(s) + 2\text{HCl}(aq) \longrightarrow \text{MgCl}_2(aq) + \text{H}_2(g) \tag{5.14}$$

According to Arrhenius's ionic theory, the compounds HCl and MgCl_2 actually exist as independent positive and negative ions in aqueous solution, so Reaction 5.14 can be expanded into

$$\text{Mg}(s) + 2\text{H}^+(aq) + 2\text{Cl}^-(aq) \longrightarrow \text{Mg}^{2+}(aq) + 2\text{Cl}^-(aq) + \text{H}_2(g) \tag{5.15}$$

In Reaction 5.15, the chloride ions (Cl^-) appear on each side of the chemical equation in exactly the same form and are therefore not participating in the chemical transformation. These so-called **spectator ions** may therefore be omitted from the equation from a mechanistic (but not stoichiometric) viewpoint, giving Reaction 5.13. In ionic equations such as Reaction 5.13, the charge as well as the mass must balance.

Figure 5.8
Magnesium metal reacts only very slowly with cold water, but when concentrated HCl is added, hydrogen gas is evolved vigorously.

The (hydrated) proton H^+, bearing a full positive charge, makes a more attractive target for the metal's valence electrons than did the polar H on H_2O; hence, Mg and the other less electropositive metals react more readily in water laced with acid.

Nonmetals in Water and Base

Because water is itself a nonmetal compound, the nonmetallic elements, with the exception of fluorine gas, react either weakly or not at all with it. When chlorine gas is bubbled through water, the reaction

$$Cl_2(aq) + H_2O(l) \longrightarrow HClO(aq) + HCl(aq) \tag{5.16}$$

occurs to a slight extent. The reaction products are acidic, and their structures are

$$H-\ddot{O}-\ddot{Cl}: \qquad H-\ddot{Cl}:$$

Hypochlorous acid **Hydrochloric acid**

Of these two Arrhenius acids, HCl is by far the stronger, dissociating completely in water to give H^+ and Cl^- ions. Br_2 and I_2 react even more weakly, but, in analogy to the way the less active metals react with H^+ but not H_2O, the halogens all react readily with OH^-:

$$X_2(aq) + 2OH^- \longrightarrow OX^- + X^- + H_2O \tag{5.17}$$

where X is Cl, Br, or I. This is followed by **disproportionation,** a redox reaction that occurs increasingly readily for the heavier halogens,

$$3XO^- \longrightarrow 2X^- + XO_3^- \tag{5.18}$$

Reactions 5.16 and 5.17 must result from the attack of the halogen on the oxygen lone pair electrons on H_2O or OH^-, in contrast to the metal reactions, where the hydrogen-atom end is the center of interest. This is a consequence of the tendency of the halogens to acquire electrons, and they are called **electrophilic,** or electron loving. The active metals on the other hand seek out electropositive regions where they can donate their electrons, and are called **nucleophilic,** or nucleus loving. These "philicities" are traceable to the electronegativities of the atoms involved.

Nonmetals from the other groups (IV, V, and VI) are remarkably inert to water; since their electronegativities are near the average for the three atoms in H_2O, $[\chi(H_2O) \approx (2\chi_H + \chi_O)/3 = 2.6]$, they neither tend to oxidize nor reduce the water molecule.

5.4 Binary Oxides and Hydrides: Chemical Periodicity

If you look back to Figure 1.2, one of Mendeleyev's early periodic tables, you may note formulas for the oxides of the various groups. These empirical formulas provided Mendeleyev with the chemical guidance he needed to organize the elements. Consider the following sequence of binary compounds with oxygen (compounds containing two elements) of the third period:

$$Na_2O, MgO, Al_2O_3, SiO_2, P_2O_5, SO_3, Cl_2O_7$$

5.4 Binary Oxides and Hydrides: Chemical Periodicity

Oxygen is more electronegative than any of the third-period elements, so that, if all of these compounds were regarded as completely ionic, each element would have given up its entire valence shell. P, S, and Cl also form other oxides, but those listed are the most fully oxidized that can be made. The oxides Na_2O and MgO are actually well characterized as ionic compounds, but all the others are polar-covalent, decreasing in polarity to the right, as may be inferred from the electronegativities. This decrease is also reflected in the physical states of the oxides; the leftmost five are solids at room temperature, while SO_3 and Cl_2O_7 are liquids.

This trend in bonding character is evident in the reaction of these oxides with water. On the metallic side, Na_2O reacts rapidly with water to produce NaOH,

$$Na_2O(s) + H_2O(l) \longrightarrow 2NaOH(aq) \qquad (5.19)$$

yielding a strongly basic solution. This is not a redox reaction, but mechanistically it bears some relation to the reaction with Na metal of Equation 5.9. Na_2O is an ionic compound, $2Na^+ + O^{2-}$, and Reaction 5.19 can be represented as

$$:\ddot{O}:^{2-} \quad H\overset{+\delta}{}\!\!\!\!\!\!\!\!\!\!\overset{-2\delta}{\ddot{O}:} \longrightarrow [:\ddot{O}-H]^- + \left[\begin{array}{c}:\ddot{O}:\\|\\H\end{array}\right]^- \qquad (5.20)$$

In the oxide, the Na valence electrons have already been transferred to O, whose extra valence pair then attacks the electropositive H of H_2O. MgO undergoes a similar reaction,

$$MgO(s) + H_2O(l) \longrightarrow Mg(OH)_2(aq) \qquad (5.21)$$

except that $Mg(OH)_2$ is relatively *insoluble*, giving only a weakly basic solution. $Mg(OH)_2$ is the chief ingredient in *milk of magnesia*, a stomach antacid.

Al_2O_3 is a ceramic material, highly inert to attack by water or acid, and SiO_2, the main component of glass and quartz, is likewise inert, and useful of course for containers of solutions. These properties are indicative of the transition from ionic to polar covalent bonding that takes place as the metal–nonmetal divide is crossed. The oxygen lone pair electrons are now not as anxious to find a more positive center, since the more electronegative Al or Si draws electron density away from the oxygens. The mode of attack that might replace the preceding mechanism, namely, the attack by the oxygen of water on the relatively electropositive Al or Si centers, is inhibited by highly stable bonded networks in the Al and Si oxides.

In P_2O_5, there is no bonded network, although the oxide exists in P_4O_{10} subunits:

Phosphorus skeleton   **Phosphorus (V) oxide**

Figure 5.9
Red phosphorus burns readily in pure oxygen (here over water) to produce a bright flame. As O_2 is consumed, the water level rises. The resulting nonmetal oxide, which occurs in P_4O_{10} units, is an acid anhydride, reacting promptly with water to yield phosphoric acid. This causes the universal indicator, added to the water beforehand, to turn from green to red.

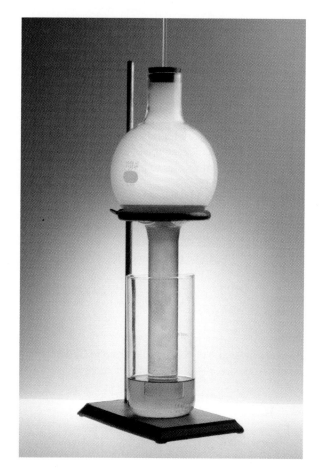

Here the reaction with water (see Figure 5.9) again leads to an addition product (an *adduct*, discussed later),

$$P_4O_{10}(s) + 6H_2O(l) \longrightarrow 4H_3PO_4(aq) \tag{5.22}$$

where the product is known as *phosphoric acid,* and has the structure

$$\begin{array}{c} \ddot{\text{O}}: \\ | \\ \text{H}-\ddot{\text{O}}-\text{P}-\ddot{\text{O}}-\text{H} \\ | \\ :\ddot{\text{O}}: \\ | \\ \text{H} \end{array}$$

This structure could be obtained by protonating three of the oxygens in the phosphate ion (PO_4^{3-}), whose structure was discussed earlier in this chapter. Reaction 5.22 typifies the production of acidic compounds from nonmetal oxides in water. The mechanism in this case involves nucleophilic attack on the more electropositive P atom by a water lone pair,

$$\text{(5.23)}$$

with six such interactions required in all for the full reaction.

Sulfur trioxide SO_3 reacts in a similar way,

$$SO_3(l) + H_2O(l) \longrightarrow H_2SO_4(aq) \quad (5.24)$$

The Lewis structure of SO_3 involves resonance forms very similar to those of CO_3^{2-}:

but the mechanism is insensitive to this:

$$\text{(5.25)}$$

Instead of severing an S—O bond, the addition converts the double bond to a single bond. Finally, the chlorine oxide, an explosive yellow oil, produces perchloric acid,

$$Cl_2O_7(l) + H_2O(l) \longrightarrow 2HClO_4(aq) \quad (5.26)$$

$HClO_4$ is the strongest inorganic (non-carbon-containing) acid among the oxyacids of the nonmetals; that is, it most readily produces protons in solution. In order of increasing acid strength, $H_3PO_4 < H_2SO_4 < HClO_4$, in concert with the increasing electronegativity of the central nonmetal atom *and* the increasing number of nonhydroxyl oxygens bonded to it.

H_3PO_4 and H_2SO_4, together with HNO_3 and HCl, are the so-called **mineral acids,** and are the most commonly used acids both in the laboratory and in industry. More than 3.5×10^7 metric tons (1 metric ton = 1000 kg) of H_2SO_4 are produced in the United States each year, making it the leading manufactured chemical.

Both the metallic and nonmetallic reaction products can be written as hydroxyl (OH) compounds, for example, $Mg(OH)_2$ and $PO(OH)_3$, and the fact that the first is basic and the second acidic can be understood on the basis of relative electronegativities. The bonding in each case is X—O—H, putting the X and H atoms in competition for the oxygen's electrons. If $\chi_X < \chi_H$ (e.g., X = Mg) then H wins the battle and the ionization

$$X-O-H \longrightarrow X^+ + OH^-$$

occurs. If on the other hand $\chi_X > \chi_H$ (e.g., X = P), then X wins, and acidic ionization

$$X—O—H \longrightarrow X—O^- + H^+$$

dominates.

Hydrides

As an element, hydrogen is really in a family by itself. By losing an electron, it becomes a "no-electron" ion, but with a single positive charge as in the alkalis. By gaining an electron it becomes noble-gas-like, as do the halide ions. But its properties place it between these extremes; its electronegativity lies near the metal–nonmetal border. Hence, the binary compounds H makes with a given period of elements might be expected to provide a more balanced illustration of periodic chemical variations than do the oxides. The second-period hydrides are

LiH(s),	BeH$_2$(s),	B$_2$H$_6$(g),	CH$_4$(g),	NH$_3$(g),	OH$_2$(l),	FH(g)
lithium hydride	**beryllium hydride**	**diborane**	**methane**	**ammonia**	**water**	**hydrogen flouride**

OH$_2$ and FH are usually written H$_2$O and HF. Only the first two and the last compound are named in a standard way. None of these compounds is completely ionic, although LiH and HF show considerable ionic character in their bonding, lithium hydride being polar in the direction Li$^{+\delta}$H$^{-\delta}$ and hydrogen fluoride F$^{-\delta}$H$^{+\delta}$. Among these hydrides, only LiH reacts rapidly and completely with water (see Figure 5.10):

Figure 5.10
Lithium hydride powder reacts violently with water to yield a strongly basic solution as indicated by the pink color of phenolphthalein. Some of the rising smoke is unreacted LiH blown away by the force of the reaction and the liberation of hydrogen gas.

$$\text{LiH}(s) + \text{H}_2\text{O}(l) \longrightarrow \text{LiOH}(aq) + \text{H}_2(g) \tag{5.27}$$

An unusual aspect of this reaction is the participation of three different "kinds" of hydrogen. The reaction mechanism involves the coming together of the negative and positive hydrogens,

$$\overset{+\delta}{\text{Li}}\overset{-\delta}{\text{H}:} \quad \overset{+\delta'}{\text{H}} \overset{..}{\text{O}:} \longrightarrow \text{Li}^+ + \text{H}-\text{H} + \left[:\overset{..}{\text{O}}\diagdown \text{H} \right]^- \tag{5.28}$$

to make neutral H_2, where the fractional charge $\delta' \neq \delta$. This reaction resembles both the reaction of the metal itself (Reaction 5.9) and the reaction of the oxide (Reaction 5.19) and is classified as a redox reaction because of the production of elemental hydrogen, with LiH acting as a reducing agent.

Metallic hydrides are often used as reducing agents in the synthesis of various organic (carbon-containing) compounds. Two common examples are lithium aluminum hydride (LiAlH_4) and sodium borohydride (NaBH_4). Each of these can be regarded as an **adduct** (**add**ition prod**uct**) of two binary hydrides (e.g., LiH and AlH_3) and reacts with water to form hydrogen and the corresponding hydroxides (e.g., LiOH and Al(OH)_3). Neither of these is as strong a reducing agent as LiH itself, because B and Al are not as metallic as Li, and their bonds to hydrogen are more covalent. Both $\text{BeH}_2(s)$ and $\text{B}_2\text{H}_6(g)$ react slowly with water but more rapidly with acid, for example, HCl(aq), to liberate H_2. (Beryllium and its compounds are extremely toxic.)

For the nonmetal hydrides the reactivity changes dramatically. The electronegativity difference between C and H is only 0.3, so that CH_4 (methane) is wholly unreactive toward water. Ammonia (NH_3) dissolves readily in water but reacts only weakly with it:

$$\text{NH}_3 + \text{H}_2\text{O} \longrightarrow \text{NH}_4^+ + \text{OH}^- \tag{5.29}$$

Reactions like that of Reaction 5.29 will be considered in more detail when acid–base chemistry (coming next) and equilibria (Chapter 12) are discussed. This reaction, like those of the metal hydrides, produces OH^-; the mechanism, however, does not involve the ammonia hydrogens, but rather the lone pair on the nitrogen:

$$\text{H}-\overset{\text{H}}{\underset{\text{H}}{\text{N}:}} \quad \overset{+\delta'}{\text{H}} \overset{..}{\text{O}:} \longrightarrow \left[\text{H}-\overset{\text{H}}{\underset{\text{H}}{\text{N}}}-\text{H} \right]^+ + \left[:\overset{..}{\text{O}}\diagdown \text{H} \right]^- \tag{5.30}$$

This reaction occurs rather than $\text{NH}_3 + \text{H}_2\text{O} \longrightarrow \text{NH}_2^- + \text{H}_3\text{O}^+$ because (1) N is less electronegative than O and more readily donates its electron pair and (2) the N—H bonds are less polar than O—H, making the ammonia hydrogen less susceptible to attack by nucleophiles.

Because water is our reference hydride (as well as oxide), you might suppose we should skip it, but in fact water does react to a very small extent with itself, in a reaction analogous to Reaction 5.29:

$$\text{H}_2\text{O} + \text{H}_2\text{O} \longrightarrow \text{H}_3\text{O}^+ + \text{OH}^- \tag{5.31}$$

In pure water this reaction produces only 10^{-7} M concentrations of **hydronium** (H_3O^+) and hydroxide ions, but these are of great importance, for they define the boundary between acidic and basic solutions. Again, this is a topic in aqueous equilibrium (Chapter 12).

The rightmost hydride in our sequence, HF(g), is the only member polar enough to protonate water to any appreciable extent,

$$HF + H_2O \longrightarrow H_3O^+ + F^- \tag{5.32}$$

and the only one aside from H_2O itself to require the participation of an oxygen lone pair:

$$\overset{-\delta}{:\!\ddot{F}\!:}\!-\!\overset{+\delta}{H}\quad\overset{-\delta'}{:\!\ddot{O}}\!-\!H \longrightarrow :\!\ddot{F}\!:^- + \left[H-\ddot{O}-H \atop H \right]^+ \tag{5.33}$$

As in the case of ammonia (Equation 5.30), this reaction occurs, rather than one producing $H_2F^+ + OH^-$, because now (1) F is more electronegative than O and less readily donates its electrons, and (2) the F—H bond is more polar than O—H, making HF more susceptible to nucleophilic attack.

For both the oxides and hydrides, the noble gas rule accounts for the stoichiometry, but only with the aid of the atomic orbital model can we make sense out of the bonding and chemical trends. However, we still have not utilized the spatial characteristics of the orbitals; this awaits a more detailed discussion of molecular geometry and energy levels, to be given in the next two chapters.

5.5 Acid–Base Reactions

In Sections 5.3 and 5.4 we have outlined the acidic and basic behavior, entailing the production of H^+ and OH^- ions in aqueous solution, of the elements and their simple compounds. When such solutions are mixed, according to Arrhenius's definition, an **acid–base** or **neutralization** reaction occurs, in which the H^+ and OH^- ions combine to make H_2O. For example, the acidic product of Reaction 5.26, $HClO_4$, will react with the basic product of Reaction 5.19, NaOH,

$$HClO_4 + NaOH \longrightarrow NaClO_4 + H_2O \tag{5.34}$$

but, according to Arrhenius's ionic theory, and to the best of our modern knowledge, Reaction 5.34 and the many other possible combinations of strong acids and bases all reduce to the simple $H^+ + OH^- \longrightarrow H_2O$. If we write the hydrated proton as H_3O^+, then the Arrhenius acid–base reaction is

$$H_3O^+ + OH^- \longrightarrow 2H_2O \tag{5.35}$$

In this form we see it as a proton-transfer reaction, exactly the reverse of Reaction 5.31.

That Reaction 5.31 occurs to such a slight extent implies that Reaction 5.35 takes place essentially stoichiometrically, leaving a negligible amount of the limiting reagent. This makes the acid–base reaction a useful tool in chemical analysis, in a technique known as **titration** (see Exercise 1.33 and Chapter 12).

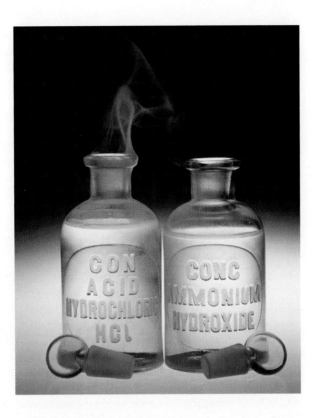

Figure 5.11
The non-aqueous reaction of NH_3 and HCl vapors to yield ammonium chloride in the form of small white particles. This is a proton transfer Brønsted-Lowry acid–base reaction occurring without the mediation of a solvent.

Brønsted–Lowry Acids and Bases

Although Arrhenius's definition of an acid–base reaction encompasses many reactions, from the point of view of the emerging chemical theories of the 1920s, it was too limited in scope. In 1923, Johannes Brønsted and T. M. Lowry independently introduced a generalization of the definition to include all **proton-transfer reactions.** This allowed reactions in nonaqueous solvents and even some gas-phase reactions to be categorized as acid–base. For example, consider the reaction

$$HCl(g) + NH_3(g) \longrightarrow NH_4Cl(s) \tag{5.36}$$

As a gas-phase reaction (which frequently occurs when reagent bottles of concentrated HCl and NH_3 are placed in proximity; see Figure 5.11), this is an **addition reaction** that yields the solid salt NH_4Cl. Ammonium chloride is an ionic compound, $NH_4^+Cl^-$, however, and there has thus been a proton transfer from HCl to NH_3. The reaction also occurs readily in aqueous solution to yield the solvated ions NH_4^+ and Cl^-:

$$HCl(aq) + NH_3(aq) \longrightarrow NH_4^+(aq) + Cl^-(aq) \tag{5.37}$$

In the aqueous reaction, Brønsted and Lowry recognized that what had been formed were a new acid and base, NH_4^+ and Cl^-, respectively, although these were weaker than the originals. The original acid, HCl, is the **proton donor** and the base NH_3 is the **proton acceptor,** while the protonated base NH_4^+ is called the

conjugate acid, and the acid less a proton, Cl^-, is the **conjugate base.** Using Brønsted and Lowry's definition, an acid–base reaction could now be written in a general form,

$$HA + B \longrightarrow BH^+ + A^- \qquad (5.38)$$
$$\text{Acid} \quad \text{Base} \quad \text{Conjugate acid} \quad \text{Conjugate base}$$

where HA is any protonic acid and B is any molecule (or ion) capable of protonation. In the form of Reaction 5.35, the Arrhenius acid–base reaction is just a special case of Reaction 5.38, where HA is H_3O^+ and B is OH^-. The concept of conjugate acids and bases will be useful to us in considering the degree of acidity of various aqueous solutions in Chapter 12.

The Lewis Concept

The oxidation–reduction chemistry by which ions were formed from neutrals, described at the start of this chapter, had been put on a firm quantitative footing by Lewis's noble gas rule. Now Lewis sought to expand the notion of acid–base reactions, in order to analyze and categorize the rich chemistry that fell into neither the redox nor the traditional acid–base category, some of which we've discussed in presenting the reactions of oxides and hydrides.

Lewis began with the recognition of the importance of the *lone electron pair* in the traditional definitions of acid–base reactions, the "leftover" electrons that play no role in bond formation (in Lewis's view). For example, NH_3 can act as a Brønsted–Lowry base, a proton acceptor, while CH_4 cannot,* because the lone pair on N can be shared with a proton, whereas in CH_4, C has no "spare" electrons. With the aid of his dot structures, Lewis claimed that the important factor in an acid–base reaction was the attainment of a new shared pair of electrons, a new polar covalent bond. Lewis's idea can be summarized in the general reaction type

$$A + :B \longrightarrow A:B \qquad (5.39)$$
$$\text{Acid} \quad \text{Base} \quad \text{Adduct}$$

It is straightforward to observe that the Arrhenius acid–base reaction in the form $H^+ + OH^- \longrightarrow H-OH$ and the more general Brønsted–Lowry reaction written as $H^+ + B \longrightarrow H-B^+$ are *both* special cases of Reaction 5.39. For Lewis an **acid is an electron pair acceptor** and a **base is an electron pair donor,** and their reaction results in a new species called the adduct. As in the Brønsted–Lowry definition, the species A and :B may be neutral or charged. H^+ is now just a particular example of an electron pair acceptor, and all of the oxide and hydride reactions we've described that did not come under the redox category can be classified as Lewis acid–base reactions. Consider, for example, the reaction between SO_3 and H_2O, Reaction 5.24. As the mechanism of Reaction 5.25 indicates, the adduct sulfuric acid H_2SO_4 results from the acceptance of an oxygen lone pair from H_2O by the sulfur in SO_3, and thus SO_3 is the Lewis acid and H_2O the Lewis base.

---

* At least not in aqueous solution. The ion CH_5^+ *is* known as a transient gas-phase species, however.

The Lewis definition allows the consideration of reactions that are both nonaqueous and nonprotonic. For example, a reaction widely used in industrial plants to "scrub" smoke from sulfur-containing fuels is

$$CaO + SO_2 \longrightarrow CaSO_3 \tag{5.40}$$

Mechanistically, an oxide ion lone pair from CaO attacks the electropositive sulfur center in SO_2, producing a new bond in the calcium sulfite adduct. Thus, CaO is the Lewis base and SO_2 the Lewis acid.

It is no accident that when CaO is dissolved in water it produces a conventional Arrhenius base, $Ca(OH)_2$, while SO_2 produces the weak Arrhenius acid H_2SO_3. That is, Lewis bases are usually compounds that will lead to the production of OH^- in water, Lewis acids to H^+. The reactions of the oxides with water, Reactions 5.19 to 5.26, can all be written in one of the two general forms

$$\underset{\text{Base}}{B{:}} + \underset{\text{Acid}}{H_2O} \longrightarrow \underset{\text{Adduct}}{B{:}H^+ {-}{-}{-}{-}{-} OH^-} \tag{5.41}$$

or

$$\underset{\text{Acid}}{A} + \underset{\text{Base}}{:\!\overset{\displaystyle |}{\underset{\displaystyle H}{O}}\!-\!H} \longrightarrow \underset{\text{Adduct}}{A{:}\overset{\displaystyle |}{\underset{\displaystyle H}{O}}^- {-}{-}{-}{-}{-} H^+} \tag{5.42}$$

Water can be either acidic or basic because it has both two electropositive hydrogens, which can act as e^- pair acceptors, and two donative lone pairs.

Lewis's notion has also led to the more recent concepts of electrophile and nucleophile, which we introduced earlier in connection with reaction mechanisms. **Lewis acids are electrophiles,** that is, they seek out regions of high electron density, while **Lewis bases are nucleophiles,** that is they seek out nuclei that have been partially or fully depleted of their electron density, or that have an affinity for more electrons. Such interactions can lead to either redox or acid–base reactions, and thus go beyond even Lewis's ideas. We will meet these "philicities," electro- and nucleo-, again in Chapter 18, when organic reactions are discussed.

Please bear in mind as you ponder these ideas that the "electron dot" is to be viewed, in light of our present ideas about atoms, as just a convenient representation for those outer regions in the atom where the valence orbitals are, and that the electron pair arises because of the Pauli principle, which forces each orbital to be (at most) doubly occupied. An orbital description of bonding will be given in the next two chapters; there we will find that one can picture Lewis acid–base interactions in terms of donor and acceptor *orbitals*.

Solvent Dependence of Acidity and Basicity

The *strength* of an Arrhenius acid or base, or even whether a given compound is classed as an acid or base, depends on the properties of water, the solvent. A protonic acid is not just a compound that produces H^+ in water, but one that produces more H^+ than water itself does in Reaction 5.31, written simply as $H_2O \longrightarrow H^+ + OH^-$. Since Reaction 5.31 yields 10^{-7} mol/liter H^+, a solution with greater concentration of H^+ than this is called *acidic,* and the compound that when dissolved yields such a concentration is an acid (of any stripe). The strength of an acid is defined in terms of the percentage yield of H^+ per mole of compound dissolved. In Chapter 12 we will learn how to compute this percentage quantitatively; for now, when an acid is described as a **weak acid,** it means this percentage is small ($\leq 10\%$), while **strong acids** are close to 100% ionized, yielding H^+ stoichiometrically.

While HF is a weak acid, HCl, HBr, HI, HNO_3, $HClO_4$, and so on, are all equally strong, essentially 100% ionized. Why should all of these compounds, which differ greatly in structure, produce the same acid strength? It is because they all are completely converted to hydrated protons, to H_3O^+, and it is that species that dictates the degree of acidity in water. All of these compounds are much stronger Brønsted–Lowry acids than protonated water is; therefore, they all donate their protons to water stoichiometrically. This phenomenon, by which different strong acids are reduced to the same degree of acidity in water, is known as the **leveling effect.** The phenomenon is capsulized by stating that *the hydronium ion H_3O^+ is the strongest acid that can exist in water.* All acids stronger than H_3O^+ are converted to it. A similar phenomenon occurs for bases stronger than OH^-. There are very few of these, the most common being O^{2-}, the oxide ion, and S^{2-}, the sulfide ion. Reactions like Reaction 5.19 represent the conversion of O^{2-} to OH^- by leveling. Species that might be considered even stronger bases, such as the elemental alkaline earth metals, cannot retain a share of the electrons they donate, and their reactions with water pass over into the redox category.

5.6 Oxidation–Reduction Reactions

At the beginning of this chapter you were introduced to the rudiments of **oxidation–reduction** or **redox reactions** in the chemistry of metals with nonmetals. In the cases we considered, there was a simple relationship among the valence electron configurations, the number of electrons transferred, and the stoichiometry of the reaction. In more complex redox reactions, such as those that occur in aqueous solution, however, reagent ions are often involved whose valence configurations are not those of the isolated elements. In addition, many reactions involve oxyanions, "complexed" metal ions, organic compounds, and other complex species for which a simple analysis of atomic electron configurations may be precluded.

Addition and Displacement

The direct combination of metals with nonmetals to make salts, such as $Mg(s)$ + $Br_2(l) \longrightarrow MgBr_2(s)$, are called **redox addition** reactions. There are also many redox reactions where a more redox-active species displaces a less active one already present in a combined form. For example, in the spectacular thermite reaction (see Figure 5.12, which shows this and other reactions)

$$2Al(s) + Fe_2O_3(s) \longrightarrow Al_2O_3(s) + 2Fe(s) \qquad (5.43)$$

aluminum, which is more easily oxidized than iron, displaces it in the oxide. This **redox displacement** reaction type encompasses all of the active-metal-plus-water re-

Figure 5.12
Three redox displacement reactions. On the left, in a reaction yielding enough heat to melt iron, aluminum displaces iron from its oxide. This reaction is known as the "thermite," and was formerly used to weld steel rails in the field. In the center, chlorine water added to a solution of potassium bromide yields a yellow color characteristic of dilute $Br_2(aq)$. On the right, zinc metal displaces copper ion in aqueous solution, with finely divided copper plating out on the zinc strip.

actions, as in Reactions 5.9 and 5.11, the metal–acid reactions exemplified by Reaction 5.14, and the active-metal-hydride-plus-water reaction, Reaction 5.27 (in which the hydride ion replaces hydroxide ion in bonding to H^+). Other examples in aqueous solution include halide displacement, for example,

$$Cl_2(aq) + 2KBr(aq) \longrightarrow 2KCl(aq) + Br_2(aq) \quad (5.44)$$

where the yellow of dilute $Br_2(aq)$ gives evidence of the reaction, and conversion directly on a metallic surface,

$$CuSO_4(aq) + Zn(s) \longrightarrow Cu(s) + ZnSO_4(aq) \quad (5.45)$$

in which a strip of zinc dipped in $CuSO_4$ solution immediately blackens from a finely divided copper metal deposit.

A study of many displacement reactions involving metals has led chemists to formulate the **redox activity series:**

$$Li>K>Ba>Ca>Na>Mg>Al>Zn>Cr>Fe> \\ Ni>C>Sn>Pb>H_2>Cu>Ag>Au>Pt \quad (5.46)$$

in which an element in uncombined form will displace another lying to its right in the series in a variety of salts and other compounds. The gross ordering of the activity series can be understood on the basis of ionization energy and, hence, position in the periodic table; but there are a few exceptions, such as Li and K, generated by differences in bond and/or solvation energies of the combined forms of the metals. The number of electrons transferred in displacement is determined by the valence shell electron configurations and the noble gas rule in most cases.

Combustion Reactions

A milestone in the ascent of early man was the discovery of fire. Now, of course, we understand fire to be a result of the combining of elements or compounds with oxygen from the air (see Chapter 1), the primordial oxidation process. A reaction that powers your modern-day laboratory Bunsen burners, and is also widely used in home and industrial heating, is

$$CH_4(g) + 2O_2(g) \longrightarrow CO_2(g) + 2H_2O(l) \tag{5.47}$$

where methane CH_4 is the major (typically 95% by volume) component of natural gas. This is one of a large class of reactions between carbonaceous fuels and oxygen known as **combustion reactions.** Other familiar examples are the burning of wood, coal, kerosene, gasoline, candles, butane in cigarette lighters, and acetylene in the oxyacetylene torch. These reactions are classed as redox reactions although no ions nor apparent changes in charge state are involved. The telling factor is the conversion of oxygen from an elemental to a combined state. Reaction 5.47 and others like it are easily balanced by "extended inspection" wherein the carbon atoms, hydrogen atoms, and oxygen atoms are balanced in succession.

Combustion reactions are mechanistically complex in the extreme, but a clue to their workings may be found in the production of soot—elemental carbon—and carbon monoxide in *incomplete* combustion. This often occurs with the higher molecular weight fuels, such as gasoline (C_8H_{18}) and paraffin (candle wax, roughly $C_{20}H_{42}$), and indicates that first the hydrogens are stripped off, followed by oxidation of the underlying carbon.

Redox in Aqueous Solution

By far the most numerous and important sorts of redox reactions in chemistry occur between solutes in aqueous solution. These range from "simple" electron transfer between metal ions to reactions of oxyanions to metabolic reactions in living organisms, with applications in all of the traditional areas of chemistry and biochemistry. Here you will learn to analyze such reactions by assessing the electron loss and gain in the same way we introduced at the beginning of this chapter. As in all of the aqueous chemistry we've already seen, the solvent water is always an essential, though sometimes unaccounted, factor in aqueous redox, as is the acidity or basicity (in the Arrhenius sense) of the solution. Thus there are always more or less explicit acid–base effects that accompany such chemistry, as you will see.

The oxyanions (see Table 5.2) are generally redox active; those with greater numbers of oxygens, or with more electronegative central elements, are typically moderate-to-strong oxidizing agents. Two oxyanions commonly employed in chemical analysis are MnO_4^- and $Cr_2O_7^{2-}$. For example, a sample containing iron can be analyzed by first converting all of the Fe to Fe^{2+} [iron(II)] in acid solution, usually H_2SO_4, and reacting (via a *titration;* see Exercise 1.33 and Chapter 13) with $KMnO_4$ according to

$$2KMnO_4 + 8H_2SO_4 + 10FeSO_4 \longrightarrow \\ 5Fe_2(SO_4)_3 + 2MnSO_4 + K_2SO_4 + 8H_2O \tag{5.48}$$

Figure 5.13 illustrates a titration utilizing this reaction. Note first that the stoichiometric coefficients are large, and the equation would be difficult to balance by trial and error. This is a redox reaction, since Fe changes its cationic charge from +2 to

5.6 Oxidation–Reduction Reactions

Figure 5.13
In a redox titration commonly used to assay ore samples for percent iron, the solution in the reaction flask turns pink when the endpoint has been reached.

+3, but the change in the state of Mn from being covalently bonded in MnO_4^- to the Mn^{2+} [manganese(II)] cation cannot be so easily assessed. Further, with such an extreme change in constituency, the mechanism is clearly complex.

Reaction 5.48 can be analyzed on a more phenomenological level by first assuming ions in solution and then omitting ions that do not participate, the so-called spectator ions (see Section 5.3), to obtain

$$MnO_4^- + 8H^+ + 5Fe^{2+} \longrightarrow 5Fe^{3+} + Mn^{2+} + 4H_2O \qquad (5.49)$$

Neither K^+ nor SO_4^{2-} metamorphoses during the reaction and can be ignored; note that the K_2SO_4 product, composed entirely of spectators, disappears completely in this ionic equation. In Equation 5.49 the total charge (+17 on each side) as well as the atoms must balance. Next we attempt to isolate the oxidation and reduction half-reactions as in the simpler example of Equations 5.2 and 5.3. For the oxidation step, we get without difficulty

$$5Fe^{2+} \longrightarrow 5Fe^{3+} + 5e^- \qquad (5.50)$$

just 5 times the simple $Fe^{2+} \longrightarrow Fe^{3+} + e^-$. Subtracting Equation 5.50 from Equation 5.49, we find the reduction step:

$$MnO_4^- + 8H^+ + 5e^- \longrightarrow Mn^{2+} + 4H_2O \tag{5.51}$$

According to Equation 5.51, both electrons and protons are required to reduce the MnO_4^- ion to Mn^{2+}, implying both acid–base and redox chemistry; but if we focus on the electrons transferred, it is as though Mn started as Mn^{7+}, that is, minus its entire $4s^2 3d^5$ valence shell. While Mn does not exist as Mn^{7+} in the polar covalently bonded MnO_4^-, it is sometimes convenient from a "bookkeeping" viewpoint to assign Mn an oxidation number of +7. We discuss oxidation numbers in more detail later.

Instead of proceeding from a balanced chemical equation to the oxidation and reduction half-reactions as was just done, it is more often useful to go the other way. Consider the oxidation of Sn^{2+} [tin(II)] ion by dichromate in acid solution, written schematically as

$$Cr_2O_7^{2-} + Sn^{2+} \xrightarrow{H^+} Cr^{3+} + Sn^{4+} \tag{5.52}$$

We can produce a balanced redox equation for Reaction 5.52 by deducing the proper half-reactions. The redox half-reaction **couples** are easily seen to be

$$Cr_2O_7^{2-} \longrightarrow Cr^{3+} \quad \text{reduction}$$
$$Sn^{2+} \longrightarrow Sn^{4+} \quad \text{oxidation}$$

These need to be balanced with respect to atoms and charge. To do so, we use a systematic procedure for balancing half-reactions, called the **ion-electron half-reaction method**. We illustrate this method with the dichromate-chromium(III) couple as follows:

1. Balance the equation with respect to all atoms other than H and O. This yields

$$Cr_2O_7^{2-} \longrightarrow 2Cr^{3+}$$

2. Balance the O atoms if necessary by adding H_2O to the side deficient in O:

$$Cr_2O_7^{2-} \longrightarrow 2Cr^{3+} + \mathbf{7H_2O}$$

3. Balance the H atoms if necessary by adding H^+ to the side deficient in H:

$$Cr_2O_7^{2-} + \mathbf{14H^+} \longrightarrow 2Cr^{3+} + 7H_2O$$

4. If the reaction occurs in *neutral* or *basic* solution, add enough OH^- to each side to neutralize the H^+ (making one H_2O for each H^+) and cancel the extra H_2O's if necessary. This does not apply to our present example.

5. Finally, add electrons e^- to the appropriate side so as to balance the charge on each side:

$$Cr_2O_7^{2-} + \mathbf{14H^+} + \mathbf{6e^-} \longrightarrow 2Cr^{3+} + 7H_2O \tag{5.53}$$

Note that it is not necessary to know beforehand whether reduction or oxidation is occurring; the electrons needed to balance the charge tell you this automatically. Equation 5.53 represents reduction.

Often one or more of steps 1 through 4 can be skipped. In our example the oxidation half-reaction is balanced simply by step 5:

$$Sn^{2+} \longrightarrow Sn^{4+} + 2e^- \qquad (5.54)$$

The balanced half-reactions of Equations 5.53 and 5.54 can now be added to yield the balanced ionic form of Equation 5.52, by multiplying each reaction by an integer factor to balance the number of electrons acquired and lost. If Equation 5.54 is multiplied by 3 and 5.53 left alone (multiplied by 1), the electrons will cancel when the equations are added:

$$Cr_2O_7^{2-} + 14H^+ + 6e^- \longrightarrow 2Cr^{3+} + 7H_2O$$

$$3Sn^{2+} \longrightarrow 3Sn^{4+} + 6e^-$$

Sum $\quad Cr_2O_7^{2-} + 14H^+ + 3Sn^{2+} \longrightarrow 2Cr^{3+} + 3Sn^{4+} + 7H_2O \qquad (5.55)$

We now have a balanced ionic equation that completely characterizes the redox process. To make it useful stoichiometrically, the spectator ions must be added. Suppose the dichromate salt is $Na_2Cr_2O_7$, the tin(II) salt is $SnSO_4$, and the acid is H_2SO_4; then Na^+ and SO_4^{2-} must be added until all ions are converted into neutral formulas:

$$Na_2Cr_2O_7 + 7H_2SO_4 + 3SnSO_4 \longrightarrow Cr_2(SO_4)_3 + 3Sn(SO_4)_2 + Na_2SO_4 + 7H_2O$$

Note how certain stoichiometric coefficients must change to accommodate the neutral formulas, and how the necessity of balancing the spectators produces the new term Na_2SO_4 on the right. Occasionally it is necessary when adding spectators to multiply the entire equation by a factor (usually 2 or 3) to ensure integer stoichiometric coefficients.

The mechanisms of reactions of this sort are quite complicated, but are likely to involve, in the acidic reactions, attack of a protonated form of the oxyanion on the reducing agent.

Oxidation Numbers

With one important class of exceptions, all of the redox reactions we have discussed thus far are easy to recognize as such by analysis of charges. But what allows us to classify combustion as a redox process? In addition, as we pointed out at the beginning of this section, analysis of charge states can sometimes be difficult. In reactions such as the oxidation of hypochlorite ion by permanganate in basic solution,

$$4KMnO_4(aq) + 3KClO(aq) + 2H_2O(l) \longrightarrow 4MnO_2(s) + 3KClO_3(aq) + 4KOH(aq)$$

written in ionic form as

$$4MnO_4^- + 3ClO^- + 2H_2O \longrightarrow 4MnO_2 + 3ClO_3^- + 4OH^- \qquad (5.56)$$

none of the species occurs atomically. To cover cases such as this, chemists of the Lewis school have invented the concept of **oxidation number.** The oxidation number of an atom or an atomic ion is simply equal to its charge, while for molecular reagents or products *the oxidation number is the charge on an atom that would result were the molecule containing it completely ionic.* The sign of the oxidation number is determined by the relative electronegativities of the bonded atoms. Thus, for the water molecule, the *fictitious* ionic form would be $2H^+ + O^{2-}$, and the oxidation number of O is therefore -2 and that of H $+1$. In the hypochlorite ion ClO^-, Cl has an oxidation number of $+1$ (not -1) and O -2, since O is more electronegative than Cl (Figure 5.6). In a similar way, you will find that the oxidation number of Mn in MnO_4^- is $+7$ (recall our earlier analysis of permanganate reduction), Mn in MnO_2 is $+4$, and Cl in ClO_3^- is $+5$. Note that the sum of the oxidation numbers in a molecule or ion is always the total charge on the species (zero for a neutral molecule).

The general definition of oxidation number just given can be formulated as a set of five rules for assigning them. In order of decreasing priority, these are as follows:

1. Uncombined elements are assigned an oxidation number of zero.
2. Alkali atoms are assigned $+1$, alkaline earths $+2$.
3. Hydrogen atoms are $+1$ except in metal hydrides, where they are -1.
4. Oxygen atoms are -2 except when bonded to F.
5. Halogens are -1 except in interhalogen compounds such as ICl.

The priority of the earlier rules over the later ones can be illustrated by the compound $Na_2O_2(s)$, sodium peroxide, in which O must be assigned -1 (not -2) owing to rule 2. Cases not covered explicitly by the preceding rules are decided on the basis of electronegativity, applying the noble gas rule to the negative atom. Thus in As_2S_3, S is -2 and As is $+3$. Redox reactions can now be generally defined as reactions in which *the oxidation numbers of at least two atoms change*. From our analysis of the reagents and products of Reaction 5.56, Mn may be found to change from $+7$ in MnO_4^- to $+4$ in MnO_2, and Cl from $+1$ in ClO^- to $+5$ in ClO_3^-. Equation 5.56 is thus classified as a redox reaction. Or consider the combustion of methane, Equation 5.47. From rules 1, 3, and 4, C goes from -4 in CH_4 to $+4$ in CO_2, while O changes from zero in O_2 to -2 in both CO_2 and H_2O. H does not change. It is clear that in these cases the oxidation numbers do not reflect ionic charges (except in MnO_2, perhaps), but they do represent a consistent way of keeping tabs on electron transfer.

Reactions that do not occur in aqueous solution (which may still be balanced by the half-reaction method, though this represents mechanistic fiction) are conveniently balanced by a method employing oxidation numbers. Simply, the number of electrons transferred must balance, and the stoichiometric coefficients of the reducing and oxidizing agents are adjusted to ensure this. For example, the mineral galena (PbS) is the chief source of lead for use in lead-acid storage batteries. A key reaction in lead production is, in unbalanced form,

$$PbS(s) + PbO(s) \xrightarrow{\Delta} Pb(l) + SO_2(g)$$

To balance this reaction, we note that Pb passes from $+2$ in both PbS and PbO to zero in Pb(l), a gain of two electrons per Pb, while S is -2 in PbS and $+4$ in SO_2, a loss of six electrons per sulfur atom. Since O is unchanged at -2, three Pb atoms must be reduced for every S oxidized. The coefficient of PbS cannot be changed without changing the S atom balance, so we increase that of PbO to 2. Then 3Pb(l) will balance Pb, giving the following balanced equation:

$$PbS(s) + 2PbO(s) \longrightarrow 3Pb(l) + SO_2(g) \qquad (5.57)$$

with O balanced automatically. This oxidation number method is one you may have learned in high school. It can, of course, also be used in solution redox, but for the purpose of electrochemical analysis (Chapter 13), we urge you to learn and practice the half-reaction method for aqueous reactions. Recall that half-reactions involve no assumptions about charge states. You should view oxidation numbers as a quick method of classification and assessment of the redox character of a reaction, without attaching Coulombic significance to the oxidation numbers you assign.

5.7 Precipitation Reactions

A test you can perform at home or in your dorm to discover whether your drinking water contains excessive lead Pb is to add table salt (NaCl) to a sample of tap water (about 10 g per cup of water). Clouding of the resulting solution, due to the formation of lead(II) chloride **precipitate,** indicates a possibly dangerous concentration of lead. Precipitation is the result of mixing anions and cations whose combination results in a **sparingly soluble salt,** which appears opaque due to the accretion of macroscopic particles containing salt crystals. Tap water contains a variety of anions and cations, but for the sake of definiteness the reaction could be written, for example, as

$$Pb(NO_3)_2 + 2NaCl \longrightarrow PbCl_2(s) + 2NaNO_3 \qquad (5.58)$$

where $PbCl_2(s)$ is the precipitate. This swapping of ionic partners is called **metathesis.** Of course, the reactants in aqueous ionic metathesis are separate and independent ions, so that, stripped of spectator ions, Reaction 5.58 is simply

$$Pb^{2+}(aq) + 2Cl^-(aq) \longrightarrow PbCl_2(s) \qquad (5.59)$$

The tendency for precipitation to occur results from the relatively greater stability of the salt crystal versus that of the hydrated ions.

The likelihood of a precipitation metathesis may be judged by the following empirical **salt solubility rules:**

1. All *alkali* ion (Li^+ through Cs^+) and *ammonium* ion (NH_4^+) salts are *soluble.*
2. All *nitrate* (NO_3^-), *acetate* ($C_2H_3O_2^-$), *chlorate* (ClO_3^-), and *perchlorate* (ClO_4^-) salts are *soluble.*
3. All *chlorides, bromides,* and *iodides* (Cl^-, Br^-, I^-) are *soluble except* those of Ag^+, Pb^{2+}, and Hg_2^{2+}.
4. All *sulfates* (SO_4^{2-}) are *soluble except* those of Ca^{2+}, Sr^{2+}, Ba^{2+}, Hg_2^{2+}, and Ag^+.
5. All metal *oxides* (O^{2-}) and *hydroxides* (OH^-) are *insoluble except* those of rule 1, Sr^{2+}, and Ba^{2+}.
6. All *carbonates* (CO_3^{2-}), *phosphates* (PO_4^{3-}), and *sulfides* (S^{2-}) are *insoluble except* those of rule 1.

Soluble in these rules means roughly 1 g or more of the salt dissolves in 100 mL of water. These rules are firmly based in periodic variation in elemental properties and Coulomb's law; two examples are shown in Figure 5.14. Ions of metals lying to the right in the periodic table, such as Ag^+ or Pb^{2+}, tend to form stronger bonds, with some covalent character, with many anions due to their more stable orbitals. In

Figure 5.14
Typical precipitation reactions.
(a) NaOH(*aq*) is added to NiCl$_2$(*aq*).
(b) Cd (NO$_3$)$_2$(*aq*) is added to Na$_2$S (*aq*). You can use the solubility rules to deduce the identity of the precipitate in each case, and to write balanced neutral and net ionic equations for these reactions.

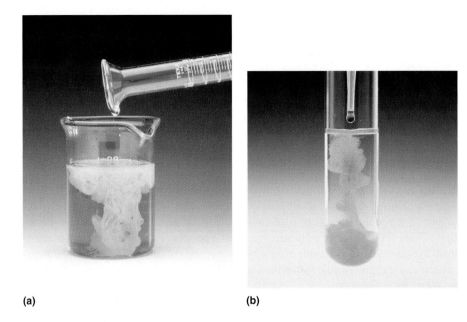

(a)   (b)

addition, a multiply charged ion often has more to gain energetically speaking by nestling among ions of opposite charge (often called **counter ions**) in the solid crystal than by remaining solvated. You should note that most of the precipitates involve multiple charges on both cation and anion. Many chemists regard all precipitation reactions as Lewis acid–base reactions at bottom, with the cation the electron-pair acceptor and the anion the donor.

Reaction Classification

Lewis's model for acids and bases allows much of the chemistry we have described in this chapter to be classified in only two categories: oxidation–reduction or acid–base reactions. Figure 5.15 is a flowchart that will help you to categorize many of the reactions you're encountering here and in chemistry laboratory. Though the diagram is relatively simple, classification is not always easy. For example, when salts are heated, particularly those containing oxyanions, **decomposition** often occurs.

Figure 5.15
Flowchart for categorizing oxidation–reduction and acid–base reactions.

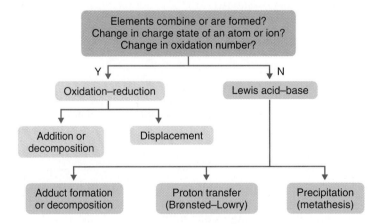

$$2KClO_3(s) \xrightarrow{\Delta} 2KCl(s) + 3O_2(g) \qquad (5.60)$$

$$CaCO_3(s) \xrightarrow{\Delta} CaO(s) + CO_2(g) \qquad (5.61)$$

These reactions, while superficially similar, fall into different categories. Reaction 5.60 forms elemental oxygen, establishing its redox character, whereas Reaction 5.61 can be analyzed using Lewis structures to show that it is an adduct decomposition, the reverse of the typical Lewis acid–base reaction.

Classification is a means of separation, but you should also appreciate the unifying theme of all of chemistry: the general seeking out by atoms and molecules, through their valence electrons, of ever lower potential energy and, hence, of greater chemical stability. If electrons must "jump" to accomplish this, then a redox reaction will result. If sharing electron pairs with new partners is favored, an acid–base reaction ensues. The relation of these tendencies to the structures of the atoms involved provides you with the beginnings of a system for *understanding* rather than merely classifying chemical behavior. What we will attempt in the next two chapters is to flesh out that system through an *orbital* description of bonding and reactions, that is, we will go a step beyond Lewis's dots.

SUMMARY

The division of the periodic table into **metals** on the lower left and **nonmetals** on the upper right correlates with the trend in atomic ionization energy and electron affinity. In compounds of metals and nonmetals known as **salts,** electrons are transferred from valence orbitals of metals to vacancies in the valence shells of nonmetals, resulting in positive and negative ions, **cations** and **anions. Lewis's noble gas rule** states that the charges on the ions and the stoichiometry of the neutral salt result from the attainment of noble-gas-shell electron configurations through gain or loss of electrons. All simple salts are crystalline solids, and their stability arises from the Coulombic attraction between ions of opposite charge. This is called **ionic bonding;** the **bond energy** D_e and **bond length** r_e of ionic bonds can be analyzed using IEs, EAs, Coulomb's law, and ionic radii.

Lewis extended his noble gas rule to nonionic molecules such as H_2, Cl_2, and HCl by postulating the sharing of electron pairs rather than the transfer of electrons. For nonmetal molecules this sharing occurred to the extent necessary to complete the noble gas shell of each atom, and became known as **covalent bonding.** When the noble gas rule required it, more than one electron pair might be shared between atoms, as in O_2 or N_2, giving **double** or **triple bonds,** corresponding to a **bond order** of 2 or 3, respectively, where the bond order is the number of shared pairs of electrons between a pair of atoms. Bonds of higher bond order were generally found to be stronger and shorter than single bonds. With this model, Lewis was able to rationalize the **valences** or bond-forming powers of the atoms, and thereby the atomic constituency of nearly all known molecules. Molecules such as NO, with an odd number of electrons, fall outside Lewis's rule unless "half-bond," or fractional bond orders, are allowed. Although Lewis's model does not invoke orbitals, it is so useful that it is still widely employed to describe molecules.

Linus Pauling recognized that in bonds between unlike atoms the electron sharing is not democratic; the atoms lying up and to the right in the periodic table tend to exert a stronger pull on the shared pairs. To quantify this tendency, Pauling invented the concept of **electronegativity** χ, the electron-attracting power of an atom when engaged in bonding. Pauling proposed a scale of atomic electronegativities that follows the trends in IE and EA, as given in Figure 5.6. This concept allows bonding to be viewed on a continuum from **pure covalent** bonding (difference $\Delta\chi < 0.4$) through **polar covalent** ($0.4 \leq \Delta\chi \leq 1.7$) to **ionic** ($\Delta\chi > 1.7$).

In polyatomic molecules or ions (three or more atoms) containing double bonds, such as ozone (O_3), often more than one Lewis dot structure satisfies the noble gas rule. The experimentally determined structure of such molecules is always the "average" of these different structures, dubbed **resonance forms,** and thus no one Lewis structure can fully represent such molecules. Other molecules, such as PCl_5 or SF_6, require the "expansion" of the noble gas shell of the central atoms to accommodate all the bonds; this normally occurs only for atoms of the third period or below, where d orbitals are available in the valence shell.

A consequence of the Lewis model is that covalent molecules satisfying the noble gas rule must always undergo bond rupture when participating in redox reactions.

An exploration of the chemistry of the elements, in light of the periodic properties and Lewis's model, is begun with consideration of reactions of **Group I**, the **alkali metals, Group II**, the **alkaline earth metals**, and **Group VII**, the **halogens**, with liquid water H_2O. Since water is composed of nonmetals, groups I and II undergo redox, transferring their valence electrons (one for the ns^1 alkalis, two for the heavier ns^2 alkaline earths) to water, rupturing the water molecule into OH^-, hydroxide ions, and H_2 gas, and becoming cations. The halogens, for example, Cl_2, on the other hand, react weakly with water to produce halide ions, Cl^-, aqueous H^+, hydrogen ions, and the polar covalent weak acid HClO, in which the halogen acquires electrons. Metal reactivity *increases* as the family column is *descended*, in accord with the decrease in IE, but *decreases* for adjacent elements in going from Groups I to II, where IE *increases*. In Group VII reactivity *decreases* on *descending*, as IE and EA decrease, because electrons are gained rather than lost. Other metals become reactive if the water contains aqueous H^+, or is **acidic**, while halogen activity increases greatly if a **basic** solution containing aqueous OH^- is used. Elements from Groups IV, V, and VI are wholly unreactive to H_2O, being unable either to yield or acquire electrons. But for those families that do react, a trend from production of strongly basic solutions from elements on the far left of the periodic table to weakly acidic on the far right is established.

Periodic variation is also strongly reflected in the properties of the binary oxides and hydrides of the elements. In the oxides the bonding ranges from fully ionic for Group I to polar covalent for Group VII, owing to the high electronegativity of O, whereas for the hydrides, polarity switches from negative on the H end in metallic hydrides to positive for the hydrogen halides. The reactivities and products in reactions of these compounds with water correlate well with those of the pure elements themselves: strongly basic solutions from the metallic oxides and hydrides giving way to inert behavior in the center of the periodic table, thence to a wide range of weak-to-strong acidic products for nonmetallic oxides and hydrides. For oxides the essential principle that determines acidity or basicity is the nature of the X—O—H linkage formed on reaction with water; for $\chi_X < \chi_H$, the X—O bond cleaves ionically to give OH^-, whereas for $\chi_X > \chi_H$ the O—H bond breaks instead to yield H^+.

In the nonmetal hydrides the critical factor is the relative activity (owing to polarity and strength) of the X—H bonds versus any lone pair electrons. Thus, while HCl ionizes strongly in water to yield H^+ and Cl^-, NH_3 does not yield H^+, but uses its lone pair to abstract H^+ from H_2O weakly to yield $NH_4^+ + OH^-$. Lone pairs are less available the further to the right the element lies in the periodic table. In discussing these ideas, the notions of a **nucleophilic** (nucleus-loving) molecule that seeks out positive centers in other molecules, and an **electrophilic** molecule looking for locales of rich electron density, are found useful.

When solutions containing acids and bases are mixed, the ionic reaction that always occurs is $H^+ + OH^- \longrightarrow H_2O$, or $H_3O^+ + OH^- \longrightarrow 2H_2O$, where H_3O^+ represents a hydrated proton $H^+ \cdot H_2O$ called the **hydronium ion.** This is a so-called **Arrhenius acid–base reaction;** in the form involving H_3O^+, it is a proton-transfer reaction. Brønsted and Lowry broadened the acid–base definition to include all proton-transfer reactions, even those not occurring in aqueous solution. This allowed classification of reactions in nonaqueous solvents or even in the gas phase. Brønsted–Lowry acids are **proton donors,** while bases are **proton acceptors.** Further, the species formed when the acid has donated its proton is to some extent a base, called the **conjugate base,** while upon accepting a proton the original base is converted to its **conjugate acid.** Lewis drew on his dot structures for covalent molecules to produce an even more general definition of acids and bases based on lone pair electrons: a **Lewis acid is an electron pair acceptor,** while a **Lewis base is an electron pair donor.** This definition encompasses both the Arrhenius and Brønsted–Lowry notions, and allows all simple inorganic reactions to be classified as either redox or acid–base. The three definitions correspond to the following reactions:

Arrhenius	H^+ +	OH^-	$\longrightarrow$	H_2O
	Acid	**Base**		**Water**
Brønsted–Lowry	HA +	B	$\longrightarrow$ BH^+ +	A^-
	Acid	**Base**	**Conjugate acid**	**Conjugate base**
Lewis	A +	:B	$\longrightarrow$	A:B
	Acid	**Base**		**Adduct**

Whether a compound is acidic or basic in the Arrhenius sense depends on the properties of water, the solvent. Pure water ionizes weakly to produce 10^{-7} M H^+; acidic compounds produce greater concentrations than this. **Weak acids** yield a small percentage of H^+ per mole of acid, while **strong acids** are 100% ionized. Equal concentrations of different strong acids yield solutions of the same acidity due to the **leveling effect,** by which any Brønsted–Lowry acid stronger than H_3O^+ is converted to H_3O^+ in water. Similarly, soluble metallic oxides or sulfides are converted quantitatively to OH^- in water because they are better proton acceptors than OH^- is.

Oxidation–reduction or **redox reactions** involing metals may be classified as **redox addition,** as in metal + nonmetal $\longrightarrow$ salt, or **redox displacement,** where a more active metal or nonmetal replaces a less active one in a compound or solution. For metals the **redox activity series** establishes the propensity for displacement of one metal by another. Nonmetal **combustion reactions,** the combining of carbonaceous fuels with oxygen to produce CO_2 and H_2O, do not involve charged species, but are classified as redox reactions because of the conversion of elemental oxygen into a combined form. Redox reactions in aqueous solution range from simple electron transfer between metal ions to metabolic redox in living systems. The reactions most useful in chemical analysis involve oxyanions such as MnO_4^- or $Cr_2O_7^{2-}$. Both the stoichiometry and the mechanism of such reactions are exceedingly complex, but the former may be tackled by the **ion-electron half-reaction method** of redox balancing, wherein the reaction is first stripped of its spectator ions and then separated into oxidation and reduction half-reactions. A systematic five-step procedure is given for balancing these, in which the transferred electrons appear explicitly. When the half-reactions are added so as to balance the electrons gained and lost, a balanced ionic equation results. Spectator ions must then be reintroduced to yield a stoichiometrically useful result. Finally, the concept of oxidation number is introduced to allow classification, assessment, and balancing of reactions not involving elements, atoms, or atomic ions.

The small solubility of certain salts leads to **precipitation** when two solutions of soluble salts containing, respectively, the cation and anion of a **sparingly soluble salt** are mixed. This exchange of ionic partners is called **metathesis,** but because of the existence of independent ions in solution, the reaction is of the Lewis acid + base $\longrightarrow$ adduct type. Empirical rules are given that allow you to determine whether and what sort of a precipitate may form in a given reaction. The solubility tendencies can be understood in broad terms using Coulomb's law, which causes salts of multiply charged ions to be more stable in crystalline form and hence less soluble, and periodic trends, dictating more covalent, less soluble salts of metals lying further to the right in the periodic table.

Based on Lewis's definition of acid–base reaction, a block diagram is given to help you classify inorganic reactions as either redox or acid–base, with a few subcategories for each. But you are also encouraged to see the common theme of these various reaction types, the seeking out by the valence electrons of their most stable configurations.

EXERCISES

Salts and ionic bonding

1. For each of the following metal–nonmetal pairs, write oxidation and reduction half-reactions as in Equations 5.2 and 5.3 and Example 5.1, giving electron configurations of the atoms and ions in parentheses. Also give the formula of the salt resulting from their reaction.

    a. Rb, Br
    b. Al, S
    c. Li, N
    d. Ca, O
    e. Sr, F

2. Predict the reaction products and balance the following metal–nonmetal reactions using the minimum integer coefficients:

    a. $Ca + P_4 \longrightarrow$
    b. $Li + O_2 \longrightarrow$
    c. $Be + Cl_2 \longrightarrow$
    d. $Ga + Se \longrightarrow$
    e. $Mg + N_2 \longrightarrow$

3. One family of recently discovered high-temperature superconductors is typified by the trimetallic oxide $YBa_2Cu_3O_7(s)$, a so-called "1–2–3 superconductor." Using the noble gas rule for Y, Ba, and O, what is the charge state of Cu in this compound?

4. 5.383 g of an unknown metal from Group I or II was reacted with excess chlorine gas to yield 8.163 g of solid metal chloride. Identify the metallic element from these data.

5. Use the ionic radius data of Figure 4.13 (either as given or estimated in the manner of Mendeleyev) to calculate the Coulomb potential energy in eV, kcal mol^{-1}, and kJ mol^{-1} for the following ionic combinations, where the ions are assumed to "touch":

    a. $Li^+ \cdots F^-$
    b. $Cs^+ \cdots I^-$
    c. $Na^+ \cdots S^{2-}$
    d. $Ba^{2+} \cdots O^{2-}$
    e. $F^- \cdots Ca^{2+} \cdots F^-$
    (collinear)

6. For the ionic molecule LiF, find the distance r_x where the ion pair becomes more stable than the neutral atoms as they are brought together, and estimate bond length r_e and bond energy D_e in the ionic model. Use data from Figure 4.13 and Figure 5.6. (The experimental values are $r_e = 1.51$ Å and $D_e = 6.02$ eV.)

7. Carry out the calculations requested in Exercise 6 for the ionic molecule CsCl. (Experimental values for r_e and D_e are 2.87 Å and 105.8 kcal mol^{-1}, respectively.)

Lewis structures and electronegativity

8. Give molecular formulas and draw proper Lewis structures for the third-row gaseous hydrides of Si, P, S, and Cl. In which molecule is the bonding the least polar (most covalent)? In which is it most polar (most nearly ionic)?

9. Draw Lewis structures for the following molecules or molecular ions. If you are uncertain how the atoms are bonded, first invoke the normal valences of the atoms; if this does not remove the ambiguity, then make the least electronegative atom the central one, valence permitting. Show resonance forms if they exist.

   HBr   CN$^-$ HCN   CS$_2$ BeH$_2$ N$_2$O   BF$_3$
   NO$_3^-$ PCl$_3$ C$_2$H$_2$ N$_2$H$_4$ CH$_2$O

10. The 10 most common chemicals produced in the United States are as follows: H$_2$SO$_4$, NH$_3$, CaO, O$_2$, N$_2$, C$_2$H$_4$ (ethylene), NaOH, Cl$_2$, H$_3$PO$_4$, and NH$_4$NO$_3$. Which of these may be classified as salts? Give the chemical name and draw a Lewis structure for each.

11. The following molecules or ions require an "expanded octet" for the central atom. Write Lewis structures by assigning unit valence to all peripheral atoms and surrounding the central atom with the remaining electrons.

   PCl$_5$ SF$_4$ SF$_6$ XeF$_2$ XeO$_3$ XeO$_4$
   I$_3^-$ BrF$_4^-$ IF$_7$

12. The following are molecular radicals containing an odd number of electrons. Draw Lewis structures by assigning the less electronegative atom (except H) one less than its required octet.

   NO   NO$_2$ NO$_3$ OH   HO$_2$ ClO   CH$_3$ HCO

13. The following simple compounds are sole or major ingredients in products common in many households. Give the chemical name and write a Lewis structure for each. Can you identify the household products that contain these molecules?

   NaHCO$_3$ H$_2$CO$_3$ CH$_3$CO$_2$H   NaOCl   H$_2$O$_2$
   NH$_3$ HCl   Na$_3$PO$_4$ NaNO$_2$ Mg(OH)$_2$ CH$_4$
   CH$_4$S   CH$_4$O   CH$_3$CH$_3$O   (CH$_3$)$_2$CH$_2$O
   CH$_3$(CH$_2$)$_2$CH$_3$

14. The following compounds are frequently used in chemical and biological laboratories. Give their chemical names if you can and also their Lewis structures.

   CH$_2$O   (CH$_3$)$_2$CO   HNO$_3$ H$_2$SO$_4$ H$_3$PO$_4$ HCl
   NaOH   "NH$_4$OH"   H$_2$S   Na$_2$S$_2$O$_3$ CrO$_3$

15. Use the electronegativity data of Figure 5.6 to arrange the following chemical bonds in order of increasing polarity, that is, with least polar or most covalent first. Use a + or − symbol to denote the sense of the bond polarity, and divide your sequence into the three segments: covalent, polar covalent, and ionic.

   Sb—Br   B—O   C—S   Ca—P   N—H   Al—I
   Rb—F   Si—Te   P—H   Ga—As

16. Although salts are nominally ionic compounds, many metal–nonmetal compounds exhibit some degree of polar covalent bonding. Based on the data in Figure 5.6, classify the following compounds as purely ionic or polar covalent:

   a. AlCl$_3$
   b. CsF
   c. CaCl$_2$
   d. SrS
   e. Na$_2$O
   f. LiBr

17. From the bond energy data $D(H_2) = 103$ kcal mol^{-1}, $D(F_2) = 37$ kcal mol^{-1}, and $D(HF) = 135$ kcal mol^{-1}, estimate the electronegativity of the hydrogen atom.

Reactions of the elements with water, acids, and bases

18. Based on analogy with reactions given in Section 5.3, complete and balance the following reactions (if no reaction, write NR after the arrow):

   a. Rb(s) + H$_2$O(l) ⟶
   b. Ba(s) + H$_2$O(l) ⟶
   c. Be(s) + H$^+$(aq) ⟶
   d. Br$_2$(l) + H$_2$O(l) ⟶
   e. Br$_2$(l) + OH$^-$(aq) ⟶
   f. Al(s) + H$_2$O(l) ⟶
   g. Al(s) + H$^+$(aq) ⟶
   h. I$_2$(s) + OH$^-$(aq) ⟶
   i. Li(s) + H$^+$(aq) ⟶
   j. Mg(s) + H$_2$O(l) ⟶

19. For those reactions involving $H_2O(l)$ in Exercise 18 that proceed to yield products, classify the resulting solutions as either acidic or basic.

20. Give Lewis-structure-based mechanisms for reactions (a) and (i) in Exercise 18.

21. Give Lewis-structure-based mechanisms for reactions (b) and (g) in Exercise 18.

22. For the ionic reactions (c), (g), and (i) in Exercise 18, give balanced neutral chemical equations when the source of $H^+(aq)$ is
    a. $HCl(aq)$
    b. $H_2SO_4(aq)$

23. Using electronegativity χ as a guide (Figure 5.6), arrange the following elements in order of decreasing reactivity toward water (i.e., most reactive first): Sr, Cs, Al, Ca, Rb, Ga. Discuss any apparent anomalies in your order.

Binary oxides and hydrides

24. Based on the reaction types discussed in Section 5.4, complete and balance the following oxide–water reactions, and classify the resulting solutions as acidic or basic (if no reaction, write NR after the arrow):
    a. $Li_2O(s) + H_2O(l) \longrightarrow$
    b. $SeO_3(s) + H_2O(l) \longrightarrow$
    c. $CO_2(g) + H_2O(l) \longrightarrow$
    d. $CaO(s) + H_2O(l) \longrightarrow$
       (called "slaking of lime")
    e. $SO_2(g) + H_2O(l) \longrightarrow$
    f. $As_4O_{10}(s) + H_2O(l) \longrightarrow$
    g. $SiO_2(s) + H_2O(l) \longrightarrow$
    h. $Cs_2O(s) + H_2O(l) \longrightarrow$
    i. $N_2O_5(s) + H_2O(l) \longrightarrow$
    j. $P_2O_3(s) + H_2O(l) \longrightarrow$

25. Give Lewis-structure-based mechanisms for reactions (c), (d), and (h) of Exercise 24.

26. Using Lewis structures and electronegativities, compare the acid–base character of $Ca(OH)_2$ and $SO_2(OH)_2$ (a.k.a. H_2SO_4).

27. As mentioned in Section 5.4, the nonmetals frequently show more than one binary oxide. Predict the empirical formulas of the oxides of the following elements based on sharing (a) all valence electrons and (b) only the p-valence subshell: C, P, N, S, As, Pb, Se. Which element has many more oxides than these two? List them and write their Lewis structures.

28. When burned in pure oxygen, the heavier alkali metals and alkaline earths form peroxides, such as Na_2O_2 and BaO_2, or superoxides, such as KO_2, where the oxygen atoms are bonded together. Give the charges on the peroxide and superoxide ions and draw their Lewis structures.

29. Based on reactions discussed in Section 5.4, give balanced chemical equations for the following hydride–water chemical systems:
    a. $NaH(s) + H_2O(l) \longrightarrow$
    b. $CaH_2(s) + H_2O(l) \longrightarrow$
    c. $AlH_3(s) + H_2O(l) \longrightarrow$
    d. $LiAlH_4(s) + H_2O(l) \longrightarrow$

30. Give a Lewis-structure-based mechanism for reaction (a) of Exercise 29, and point out the features it shares with the reactions of both the pure metal and its oxide with water.

31. Using electronegativities discuss the relative acid strengths in water of H_2S and HCl.

32. Based on the electronegativities of the halogens, what relationship would you predict among the hydrogen halides as to their relative acid strength? It is found experimentally that HI is the strongest of the hydrohalic acids. Does this agree with your prediction? If not, what other factors might dominate?

Acid–base reactions

33. Complete and balance the following acid–base reactions, and classify them according to the most specific definition that applies:
    a. $NaOH(aq) + H_2SO_4(aq) \longrightarrow$
    b. $NH_3(aq) + HClO_4(aq) \longrightarrow$
    c. $CaO(s) + CO_2(g) \longrightarrow$
    d. $NaHCO_3(s) + HCl(aq) \longrightarrow$
    e. $H_2O(l) + HCl(g) \longrightarrow$
    f. $Ca(OH)_2(aq) + H_3PO_4(aq) \longrightarrow$
    g. $H_2O(l) + NH_3(g) \longrightarrow$
    h. $Na_2S(s) + H_2O(l) \longrightarrow$
    i. $SO_2(g) + H_2O(l) \longrightarrow$
    j. $LiH(s) + AlH_3(s) \longrightarrow$

34. For those reactions of Exercise 33 involving water or aqueous solution, write the equations out in ionic form, and cancel spectator ions, if present. Into what common form do the reactions classified as Arrhenius acid–base fall?

35. For the reactions of Exercise 33 categorized as Brønsted–Lowry acid–base, identify the acid, base, conjugate acid, and conjugate base.

36. Acid rain contains mainly sulfuric acid (at concentrations as high as 10^{-2} M), and in many parts of the world it is destroying marble statuary at an alarming rate. Marble is mainly calcium carbonate, and the reaction is acid–base. Give a balanced chemical equation for the reaction and classify it.

37. For the reactions of Exercise 33 classified as Lewis acid–base, identify the acid, base, and adduct, and illustrate the mechanism of the reaction.

38. All of the oxide–water reactions of Exercise 24 are acid–base reactions. Classify each of them.

39. In Chapter 1, Exercises 1.33 and 1.34, you were introduced to the rudiments of acid–base analysis by titration. In analyzing solid samples, such as stomach antacids, whose solubility in pure water is limited, a technique known as *back-titration* can be used. In this method, a weighed amount of antacid is dissolved in a known volume of standard HCl solution, and the excess acid is then titrated with standard NaOH. Suppose a 0.3960-g sample of antacid is ground up and dissolved in 20.00 mL 0.5000 M HCl. The resulting solution is diluted to 50 mL and titrated with 0.2000 M NaOH, requiring 26.43 mL to reach the equivalence point. Use these results to find g HCl neutralized per g antacid. (Note that dilution of the analyte solution has no effect on the result.)

Redox reactions

40. Based on the activity series, which of the following can be expected to undergo displacement reactions? Possible transition metal ion products are shown in parentheses.
    a. Ni + ZnS (Ni^{2+})
    b. K + NaCl
    c. H_2 + CuO
    d. Fe + Al_2O_3 (Fe^{3+})
    e. Ba + Ag_2S
    f. Zn^{2+}(aq) + Cu(s) (Cu^{2+})
    g. Au^{3+}(aq) + Cr(s) (Cr^{3+})
    h. Li^+(aq) + Mg(s)
    i. Zn + PbO_2 (Zn^{2+})
    j. Al + $CaBr_2$

    For those that do react, give a balanced chemical equation, and deduce the number of electrons transferred per unit reaction.

41. Many combustion reactions are common in industry and in everyday life. Try your hand at balancing them.
    a. $C_3H_8 + O_2 \longrightarrow CO_2 + H_2O$ (burning propane fuel)
    b. $C_6H_{12}O_6 + O_2 \longrightarrow CO_2 + H_2O$ (burning wood, sugar metabolism)
    c. $C_2H_6O + O_2 \longrightarrow CO_2 + H_2O$ (flaming desserts)
    d. $C_4H_{10} + O_2 \longrightarrow CO_2 + H_2O$ (cigarette lighters)
    e. $C_2H_2 + O_2 \longrightarrow CO_2 + H_2O$ (oxyacetylene torches)
    f. $NH_3 + O_2 \xrightarrow{Pt} NO + H_2O$ (Ostwald reaction; step in making nitric acid)
    g. $SO_2 + O_2 \xrightarrow{V_2O_5} SO_3$ (step in the synthesis of sulfuric acid)
    h. $H_2N\text{—}CH_2\text{—}COOH + O_2 \longrightarrow N_2 + CO_2 + H_2O$ (burning beefsteak on the grill)

42. Even in a well-tuned gasoline engine, the combustion of octane $C_8H_{18}(l)$ produces about 1 mol of carbon monoxide, CO(g), per mole of C_8H_{18}. Give a balanced chemical equation that includes this molar ratio in addition to the usual combustion products. What is the percentage of CO in the exhaust gases? (Recall that air is 21% O_2 and regard H_2O as a gas.) For an automobile yielding 25 miles per gallon of gasoline, how many grams of CO are produced per mile? (1 gal = 3.76 L; density of gasoline = 0.85 g cm^{-3}).

43. Write a balanced chemical equation for the production of soot (elemental carbon) and water vapor from the burning of candle wax (paraffin, $C_{20}H_{42}$). (The glow of a candle comes mainly from soot heated to incandescence.)

44. Using the rules given in Section 5.6, assign oxidation numbers to all atoms in the following molecules or ions:

    H_2SO_4 HNO_3 H_3PO_4 H_2O_2 SO_3^{2-}
    ClO_3^- $CaCl_2$ $NaHCO_3$ NO_2^+ NO_2^-
    NH_3 NH_4^+ CH_2O CH_4 CH_3Cl

45. Balance the following redox reactions occurring in acidic aqueous solution by the half-reaction method. As in the example in Section 5.6, first write the equation in ionic form and eliminate spectator ions. When you have obtained a balanced ionic equation, add spectators to give a balanced neutral equation.

    a. $CuS(s) + HNO_3(aq) \longrightarrow CuSO_4(aq) + NO(g)$
    b. $Br_2(aq) + H_2C_2O_4(aq) \longrightarrow HBr(aq) + CO_2(g)$
    c. $H_2O_2(aq) + CH_2O(aq) \longrightarrow H_2O(l) + CO_2(g)$
    d. $NO_2(g) + H_2O(l) \longrightarrow HNO_3(aq) + NO(g)$
    e. $H_2S(aq) + K_2Cr_2O_7(aq) \xrightarrow{HCl}$
       $S(s) + Cr^{3+}(aq)$
    f. $KMnO_4(aq) + HBr(aq) \longrightarrow$
       $MnBr_2(aq) + Br_2(aq) + H_2O(l)$
    g. $KIO_3(aq) + KI(aq) \xrightarrow{H_2SO_4}$
       $I_3^-(aq) + H_2O(l)$

46. Balance the following redox reactions occurring in neutral or basic solution. Follow the prescribed procedure of Exercise 45.

    a. $I_2(aq) + S_2O_3^{2-} \xrightarrow{KOH} I^-(aq) + S_4O_6^{2-}$
    b. $CN^-(aq) + MnO_4^-(aq) \xrightarrow{NaOH}$
       $MnO_2(s) + CNO^-(aq)$
    c. $CrO_4^{2-} + HSnO_2^-(aq) \xrightarrow{KOH}$
       $HSnO_3^-(aq) + CrO_2^-(aq)$
    d. $H_2O_2(aq) + N_2H_4(aq) \longrightarrow N_2(g) + H_2O(l)$
       (In pure form, the reagents are useful as a rocket fuel.)
    e. $VO^{2+}(aq) + ClO_3^-(aq) \xrightarrow{NaOH}$
       $VO_2^+(aq) + Cl^-(aq)$

47. The titration illustrated in Figure 5.13 has long been used to analyze iron ore for percent Fe by mass. The ore is dissolved in acid and reduced to Fe^{2+} by $SnCl_2$ before titration with standard $KMnO_4$ solution. A 3.3111-g sample of ore, after dissolution and reduction, required 36.79 mL of 0.2000 M $KMnO_4$ to reach the equivalence point, indicated by a persistent pale pink color due to slight excess of MnO_4^-. Find the percent by mass Fe in the ore. (See the text for the balanced chemical equation.)

48. The technique of back-titration, introduced for acid–base analysis in Exercise 39, was developed for analysis of redox agents by Robert Bunsen in the 1850s. In this adaptation, a known mass of oxidizing agent is reacted with excess iodide ion (I^-) in acid solution, producing I_2. This is then titrated with standard sodium thiosulfate ($Na_2S_2O_3$), reducing the I_2 to I^- again. The equivalence point is indicated by the sudden disappearance of the deep blue color of a complex formed by I_2 and added starch. (This is called *iodometric analysis*.) For analysis of hypochlorite (ClO^-), the active ingredient in laundry bleach, the unbalanced ionic reactions are

    $$ClO^- + I^- \longrightarrow Cl^- + I_2$$

    and

    $$S_2O_3^{2-} + I_2 \longrightarrow S_4O_6^{2-} + I^-$$

    After balancing these reactions, determine the molarity of ClO^- in bleach if 1.000 mL of bleach required 17.32 mL of 0.1000 M $Na_2S_2O_3$ to titrate the I_2 produced by it.

49. Like NaOH (see Exercise 1.34), $Na_2S_2O_3$ is hygroscopic, and must be standardized by titration before use in the iodometric analysis of the previous exercise. In one method, a known mass of potassium dichromate ($K_2Cr_2O_7$) is reacted with excess I^- in acid solution, and the I_2 titrated with thiosulfate. The unbalanced ionic reaction is

    $$Cr_2O_7^{2-} + I^- \longrightarrow Cr^{3+} + I_2$$

    After balancing this reaction, determine the mass in grams of $K_2Cr_2O_7$ needed so that about 25 mL of about 0.1 M $Na_2S_2O_3$ will be required to titrate the I_2 it produces. (You will need the balanced reaction between $S_2O_3^{2-}$ and I_2 from the previous exercise.)

50. Nonmetallic elements in their free states are often prepared industrially or in the laboratory by suitable redox reactions. Balance the following preparation reactions using either the half-reaction method or,

for nonaqueous systems, the oxidation number method:

a. $HCl(aq) + MnO_2(s) \longrightarrow$
$\qquad Cl_2(g) + MnCl_2(aq) + H_2O(l)$
b. $NaI(s) + H_2SO_4(l) \longrightarrow$
$\qquad I_2(g) + Na_2SO_4(s) + SO_2(g) + H_2O(l)$
c. $Zn(s) + HCl(aq) \longrightarrow ZnCl_2(aq) + H_2(g)$
d. $NH_4^+(aq) + NO_2^-(aq) \longrightarrow N_2(g) + H_2O(l)$
e. $KClO_3(s) \longrightarrow KCl(s) + O_2(g)$
f. $NH_3(g) + Br_2(l) \longrightarrow N_2(g) + NH_4Br(s)$
g. $H_2S(g) + SO_2(g) \longrightarrow S_8(s) + H_2O(l)$
h. $Ca_3(PO_4)_2(s) + CO(g) \longrightarrow$
$\qquad CaO(s) + CO_2(g) + P_4(s)$
i. $KBr(s) + H_2SO_4(l) + MnO_2(s) \longrightarrow$
$\qquad Br_2(l) + K_2SO_4(s) + MnSO_4(s) + H_2O(l)$
j. $K_2SiF_6(s) + K(s) \longrightarrow Si(s) + KF(s)$
k. $H_2SeO_3(aq) + SO_2(aq) + H_2O(l) \longrightarrow$
$\qquad Se(s) + H_2SO_4(aq)$

51. The famous Comstock Lode, discovered in Virginia City, Nevada, in 1859, was "played out" in only a decade; but due to increases in the price of gold, mining the remaining low-grade ore became profitable in the 1980s. The process for extracting gold from the ore is known as *cyanide leaching*;

$$Au(s) + CN^-(aq) + O_2(g) \xrightarrow{NaOH} Au(CN)_4^-(aq)$$

followed by reduction with zinc,

$$Au(CN)_4^-(aq) + Zn(s) \longrightarrow Au(s) + Zn(CN)_4^{2-}(aq)$$

Complete and balance these two reactions. What mass (g) of Zn is required to produce one troy ounce (31.1 g) Au?

Precipitation

52. A test used for aqueous cadmium ion analysis is sulfide precipitation to give the yellow CdS(s):

$$Na_2S(aq) + CdCl_2(aq) \longrightarrow CdS(s) + 2NaCl(aq)$$

Write the equation in ionic form, and identify the Lewis acid and base.

53. When solutions of the following compounds are mixed, will precipitation occur? If so, give the precipitate and balance the chemical equation.

a. $KNO_3 + (NH_4)_3PO_4$
b. $Na_2CO_3 + BaCl_2$
c. $Pb(C_2H_3O_2)_2 + NaOH$
d. $FeCl_3 + KI$
e. $Ca(ClO_3)_2 + AgNO_3$
f. $Cr(NO_3)_3 + Na_2S$

54. What mass in grams of nickel(II) phosphate precipitate may be expected when 100. mL of 0.15 M $Ni(NO_3)_2$ solution is mixed with 200. mL of 0.65 M K_3PO_4 solution?

Classification

55. Using Figure 5.15, classify the following reactions as either redox or acid–base, and give the subcategory of each (see the blocks at the bottom of the figure) if applicable.

a. $NH_3(g) + SO_3(g) \longrightarrow HSO_3NH_2(s)$ (sulfamic acid)
b. $I_2(s) + 5HNO_3(aq) \longrightarrow$
$\qquad HIO_3(s) + 5NO_2(g) + 2H_2O(l)$
c. $4NaO_2(s) + 2CO_2(g) \longrightarrow 2Na_2CO_3(s) + 3O_2(g)$
d. $BaCl_2(aq) + (NH_4)_2SO_4(aq) \longrightarrow$
$\qquad BaSO_4(s) + 2NH_4Cl(aq)$
e. $Cr_2O_3(s) + 2Al(s) \longrightarrow Al_2O_3(s) + 2Cr(s)$
f. $H_2CO_3(aq) + N_2H_4(aq) \longrightarrow$
$\qquad HCO_3^-(aq) + N_2H_5^+(aq)$
g. $2NO_2(g) + 2N_2H_4(g) \longrightarrow 3N_2(g) + 4H_2O(g)$
h. $Al_2(SO_4)_3(aq) + 8NaOH(aq) \longrightarrow$
$\qquad 2NaAl(OH)_4(aq) + 3Na_2SO_4(aq)$
i. $NaHSO_3(aq) + HCl(aq) \longrightarrow$
$\qquad NaCl(aq) + SO_2(g) + H_2O(l)$
j. $NaOH(aq) + CO_2(g) \longrightarrow NaHCO_3(aq)$

56. Samples of pure $CO_2(g)$ may be obtained from a number of different reactions. Classify them.

a. $2NaHCO_3(s) \longrightarrow Na_2CO_3(s) + CO_2(g) + H_2O(l)$
b. $3C(s) + 2Fe_2O_3(s) \longrightarrow 3CO_2(g) + 4Fe(s)$
c. $NiCO_3(s) + H_2SO_4(aq) \longrightarrow$
$\qquad NiSO_4(s) + CO_2(g) + H_2O(l)$
d. $2CO(g) + O_2(g) \longrightarrow 2CO_2(g)$
e. $CO(g) + CuO(s) \longrightarrow CO_2(g) + Cu(s)$

Orbitals and Chemical Bonding I: The Valence Bond Model and Molecular Geometry

Liquid oxygen suspended between poles of a magnet as it boils away. See Section 6.6 for a discussion.

The ideas involved in modern structural chemistry are no more difficult . . . than the familiar concepts of chemistry. Some of them may seem strange at first, but with practice there can be developed an extended chemical intuition which permits the new concepts to be used just as confidently as the old ones. . . .

Linus Pauling (1938)

CHAPTER OUTLINE

- 6.1 The Hydrogen Molecule: The Simplest Electron-Pair Bond
- 6.2 Polyatomic Molecules: Pauling's Valence Bond Model
- 6.3 Double and Triple Bonds: σ and π Bond Types
- 6.4 Electron-Pair Repulsion: Rationalizing and Predicting Molecular Shapes
- 6.5 Asymmetric Electron Sharing: The Electric Dipole Moment
- 6.6 Shortcomings of the Valence Bond Model

The reassuring words of Linus Pauling refer to the impact of quantum mechanics on chemists' notions about molecular structure and the nature of the chemical bond. It was largely through the work of Pauling in the early 1930s that chemists began to incorporate the "strange" wave mechanical ideas, such as orbitals and energy levels, into their thinking. The problem chemistry faced in those days could be succinctly stated: how to make Lewis electron-dot structures somehow consistent with quantum mechanics. Lewis structures had provided chemists with such a powerful and predictive

6 Orbitals and Chemical Bonding I: The Valence Bond Model and Molecular Geometry

tool that abandoning them seemed (and still seems) unthinkable. Yet quantum theory was by that time widely accepted (with good reason) as the correct theory for describing the tiny mechanism we call an atom, and, in view of the uncertainty principle, Lewis electron dots were certainly a schematic picture at best. The young Pauling brought all of his great chemical genius to bear on this problem, with a particular emphasis on describing Lewis's covalent bond in wave mechanical terms.

Linus Pauling (1901–1994). His visit to Germany on a fellowship in the late 1920s led to his lifelong interest in describing molecular structure using the new wave mechanics, summarized in his book *The Nature of the Chemical Bond*. His hybridization and resonance models led to his discovery of the protein α-helix. Pauling won two unshared Nobel Prizes, for chemistry and for peace.

6.1 The Hydrogen Molecule: The Simplest Electron-Pair Bond

Does quantum mechanics work for molecules? You can imagine there were chemists in the early days who hoped it wouldn't, who hoped wave mechanics would remain a plaything for physicists. But it had been known for about 50 years that small molecules also have line spectra (although *many* more lines than atoms) and other peculiar physical properties that imply an underlying wave nature.

The Heitler–London Treatment of H$_2$

In 1927, during the infancy of quantum mechanics, W. Heitler and F. London (more on London later) published their work on the simplest of all covalent bonds, that of H$_2$. Figure 6.1 illustrates the mechanical problem they encountered—four charged particles, two of them nuclei, and six Coulomb terms in the Hamiltonian (see Chapters 3 and 4). There was clearly no hope of finding the exact wave function, so Heitler and London did what we've described for atoms: they *guessed* a wave func-

Figure 6.1
When two hydrogen atoms are brought together to form a H$_2$ molecule, four new Coulomb terms arise in the potential energy. (For N charged particles there are $N(N-1)/2$ Coulomb interactions; see Chapter 4.)

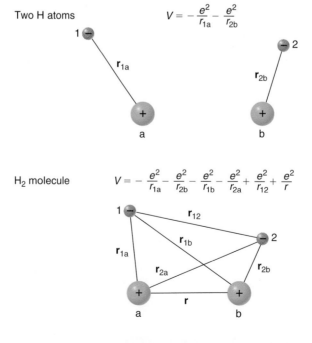

tion built from orbitals. To generate a guess, they asked what the wave function would reduce to if the neutral atoms were very far apart. From Figure 6.1 you can see that all but two electron–nucleus attractions become vanishingly small, and the system becomes two independent H atoms. The wave function for the ground state of this system is just the *product* of the 1s orbitals on the two atoms:

$$\Psi(\mathbf{r}_1, \mathbf{r}_2) = 1s_a(\mathbf{r}_1) \cdot 1s_b(\mathbf{r}_2) \tag{6.1}$$

However, when they substituted the guess of Equation 6.1 into the Schrödinger equation and calculated the energy for various nucleus–nucleus distances, they found the dashed curve shown in Figure 6.2, a potential energy curve with only a shallow minimum. What they had hoped to find was a curve with a deep "well" in it, a bond potential energy curve of the sort seen for ionic molecules in Chapter 5. To improve on the guess of Equation 6.1, they drew on Lewis's idea of sharing. Equation 6.1 always assigns electron 1 to nucleus a and electron 2 to nucleus b, but if the nuclei are to share their electrons, electron 1 must spend some time near nucleus b, while electron 2 samples nucleus a. This suggests

$$\Psi(\mathbf{r}_1, \mathbf{r}_2) = \frac{1}{\sqrt{2}}[1s_a(\mathbf{r}_1)1s_b(\mathbf{r}_2) + 1s_a(\mathbf{r}_2)1s_b(\mathbf{r}_1)] \tag{6.2}$$

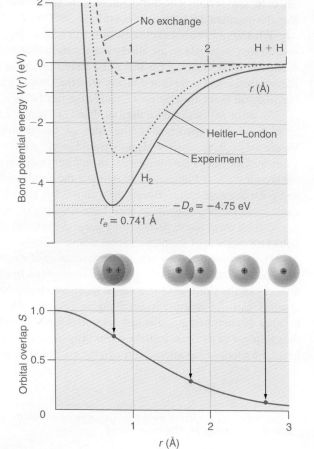

Figure 6.2
Bond potential energy curves and orbital overlaps for the H_2 molecule. In the upper panel, the dashed curve is obtained using the wave function of Equation 6.1, while the dotted curve comes from Equation 6.2. The solid experimental curve is determined mainly from spectroscopy. In the lower panel the overlap integral S (Equation 6.3) is plotted, along with pictures showing overlap of the s boundary surfaces at three internuclear distances. Note the correlation between the magnitude of S and the attractive part of the Heitler–London or experimental potential curves.

where the factor $1/\sqrt{2}$ keeps the wave function approximately normalized, and the second orbital product allows for swapping or **exchange** of the electrons. When this guess is substituted into the Schrödinger equation, the dotted energy curve shown in Figure 6.2 is obtained. This curve has a minimum corresponding to a bond energy D_e of 3.16 eV (72.9 kcal/mol) and a bond length r_e of 0.80 Å. While this is still far from the experimental curve, also shown in Figure 6.2, which shows a bond energy of 4.75 eV (109.5 kcal/mol) and length 0.74 Å, it is clear that this very approximate treatment contains the essential ingredients for describing the chemical bond. As discussed later, it is possible to improve on simple guesses like Equation 6.2 and carry through very complicated calculations to obtain a curve that is indistinguishable from the experimental one. But here we are interested in a *description* of bonding that may lead to what you can think of as understanding (taken in that peculiar quantum mechanical sense of the word).

When Heitler and London analyzed their calculation to find out what it was that gave rise to the deep minimum in the energy, they discovered a *nonclassical* contribution that "turned on" when the orbitals of the atoms got close enough to overlap each other. They referred to this term in the energy as the *exchange energy* because it arose from adding the extra orbital product in Equation 6.2. It was nonclassical because *it could not be described as an electrostatic interaction among charges*. That is, *formation of a covalent chemical bond is a fundamentally quantum-mechanical phenomenon*. As Pauling was among the first to recognize, chemists were after all going to have to reckon with the quantum view. In the next chapter we will discuss further the quantum description of bonding.

What about Pauli's exclusion principle? To satisfy Pauli's symmetry requirements (see Chapter 4) on the bond wave function (Equation 6.2), Heitler and London found that the electron spins had to be paired just as in an atomic orbital.* Lewis' ideas thus seemed to be fitting nicely with the laws of quantum mechanics, with the electron pair bond drawing new quantitative justification from the orbital analysis.

As we have discussed for the ionic bond in Chapter 5, the bond length r_e of a *covalent* bond is approximately the sum of the *neutral* atomic radii (given in Figure 4.12). In the case of H_2, this would imply $r(H) = \frac{1}{2} r_e(H_2) = 0.37$ Å. When many hydride bonds are considered, an optimum value is $r(H) = 0.30$ Å. You can use the atomic radii of Figure 4.12 to estimate the bond lengths of a variety of covalent and polar covalent bonds, usually to within 0.1 Å and often much more closely. Because covalent bonds seem to arise from merging of atomic orbitals, the radii obtained from them are expected to be generally smaller than the radii of the orbital boundary surfaces of the participating atomic orbitals.

The idea that bonding arises from exchange of electrons has much in common with modern theories of the fundamental interactions of matter. For example, there is a theory of the Coulomb force called *quantum electrodynamics* that postulates that charged particles interact by exchanging photons. The strong nuclear force (see Chapter 16) between a proton and neutron that binds nuclei together is mediated by the exchange of particles called *mesons*. And those *quarks* of which you've no doubt heard are supposed to interact through the exchange of *gluons*.

---

* Equation 6.2 is a two-electron wave function, not an orbital, and the electrons described by such functions need not always have their spins paired.

The Importance of Orbital Overlap

Electrons can only be exchanged between atoms if their orbitals are "in touch," that is, when the electron waves from the two nuclei have overlapping amplitudes. This **orbital overlap** is quantitatively expressed by the overlap integral,

$$S = \int \psi_a \psi_b \, d\tau \qquad (6.3)$$

where ψ_a and ψ_b are valence orbitals on atoms a and b, $d\tau$ is a differential volume element, and the integral extends over all space. The major contribution to S comes from that region of space where the atomic orbitals overlap, for only in that region is the orbital product, the integrand in Equation 6.3, appreciable. Figure 6.2 shows both pictorially and quantitatively how the overlap changes with internuclear distance for the H_2 case, where the ψ's are $1s$ orbitals.

You can see from Figure 6.2 that the overlap continues to increase as the H atom separation shrinks; that is, the exchange contribution to the energy continues to grow. The overall energy, however, turns up again at distances smaller than r_e, owing to internuclear Coulomb repulsion (Figure 6.1).

Pauling seized on the overlap idea as a *sine qua non* for chemical bonding, reflecting as it does the all-important **exchange energy.** He made it the basis for a model of chemical bonding known as the **valence bond** (VB) theory, in which orbitals are injected into Lewis structures.

6.2 Polyatomic Molecules: Pauling's Valence Bond Model

Pauling proposed that all electron sharing within a Lewis structure for a molecule is approximately accounted for by **valence bond wave functions** of the sort Heitler and London had proposed for H_2 (Equation 6.2). For example, in the HF molecule the single electron pair bond in the Lewis structure H—F̈: should be represented by*

$$\Psi_{VB}(\mathbf{r}_1, \mathbf{r}_2) = \frac{1}{\sqrt{2}} [1s_H(\mathbf{r}_1) 2p_F(\mathbf{r}_2) + 1s_H(\mathbf{r}_2) 2p_F(\mathbf{r}_1)] \qquad (6.4)$$

where fluorine uses its valence $2p$ orbital in making the overlap bond as illustrated in Figure 6.3. The other F electrons remain paired in their filled atomic orbitals (AOs), thus giving a perfect correspondence between valence bonds and filled AOs in VB theory and shared and lone pairs in the Lewis model.

The spirit in which Pauling then tackled the problem of polyatomic molecules was first to examine the Lewis structure, and then construct valence bond overlaps for each shared electron pair, assigning the remaining electrons to doubly occupied AOs on the atoms bearing lone pairs. While this approach succeeds in interpreting bonding in a more wave-mechanical manner, it is somewhat *ex post facto,* and lends the VB model the same shortcomings as well as advantages of Lewis structures.

---

* The VB wave function, strictly speaking, accounts only for the covalent contribution to the bond; for HF there is also a substantial ionic contribution to the overall polar covalent bond (see Section 6.5).

Figure 6.3
Orbitals and electron pairs (denoted by ↑↓) in the valence bond model for HF and H_2O. The axes selected for the H—F and H—O bonds are arbitrary. On the left are shown only the orbitals used in constructing valence bonds using realistic boundary surfaces, while on the right both bonding and lone pair orbitals are shown in a more schematic fashion. Pauling's rationale for the opening out of the HOH bond angle due to Coulombic repulsion of the electropositive hydrogens is illustrated at the bottom.

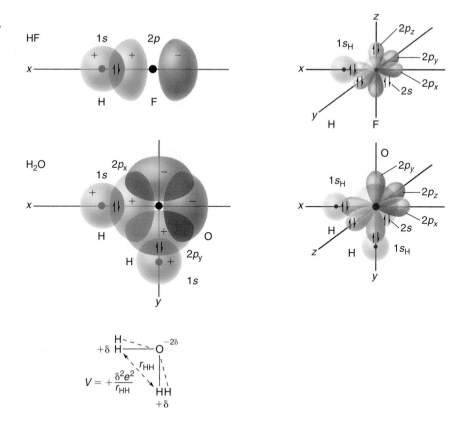

There was an aspect of molecular structure, however, that was enormously clarified by use of the orbital overlap idea: the **bond angles** of triatomic or larger molecules. We have already described the structure of the water molecule in Chapter 5. As with most known molecules and molecular ions surveyed in Chapter 5, H_2O possesses a well-defined equilibrium structure as to the angle between the O—H bonds as well as the lengths of those bonds. As you may recall, the HOH bond angle is 104.5°. A VB description of H_2O would naturally employ two mutually perpendicular $2p$ orbitals, $2p_x$ and $2p_y$, say, in making overlaps with the $1s$ orbitals on the hydrogen. There are then two identical VB wave functions, each of the form of Equation 6.4 with $2p_O$ replacing $2p_F$, corresponding to the two (polar) covalent bonds pointing at a right angle, 90°, with respect to each other, as illustrated in Figure 6.3. The lone pairs on O in the Lewis structure must occupy the remaining $2p$ orbital ($2p_z$) and the lower energy $2s$. The VB model thus predicts a *bent* structure for water, with a bond angle of 90°, only 14.5° away from experiment.

Once a result like the one for water is obtained, chemists often seek *qualitative* explanations for the remaining differences between the model and experiment, to establish whether the model pushes our thinking in the right direction. For H_2O Pauling noted that, due to the polar covalent bonding, each H bears a partial positive charge $+\delta$, and the two hydrogens should therefore repel each other by Coulomb's law. This repulsion would tend to open up the bond angle, but not so much as to disrupt the $1s_H – 2p_O$ overlap. Though the argument sounds plausible, you should exercise some caution in accepting such a line of reasoning if it is not backed up by numbers (recall the quotation by Lord Kelvin in Chapter 1). In the next section we examine another viewpoint concerning the H_2O bond angle. But this example should help you to understand why chemists now place so much stock in the orbital model.

Atomic Hybridization

Lewis structures have had by far their greatest impact on **organic chemistry,** the chemistry of carbon-containing molecules. The building block of organic chemistry is the methane molecule, CH_4. By 1930 a great deal of chemical and spectroscopic evidence had been amassed to show that methane has a **tetrahedral** structure, so named because each H atom nucleus is located at a vertex of a regular tetrahedron (a four-faced figure, with each face an equilateral triangle) as illustrated in Figure 6.4. (The ordinary way of drawing a Lewis structure for methane gives a square planar structure, probably a bias due to drawing on a flat 2-D piece of paper.) In a perfect tetrahedral structure all bond angles are equal, 109.5°; in addition, all the C—H bond lengths and energies are identical.

On the face of it, the valence bond method looks unworkable for CH_4. To begin with, the orbital configuration of C is $1s^2 2s^2 2p^2$; only the $2p$ electrons are unpaired and therefore ready to bond with H. (Lewis structures were not formulated to recognize subshells.) To make four bonds, carbon must suffer breakup of the paired $2s$ electrons in the VB model. Even given that, the four bonds should consist of three at 90° angles, built from the $2p$ orbitals on C, and one with no particular preferred direction arising from the spherically symmetric $2s$. In addition, the highly directional $2p$ orbitals are more effective at overlapping the hydrogen $1s$, and therefore should make stronger, shorter bonds than $2s$.

To overcome these formidable difficulties, Pauling in 1931 invented the concept of the **hybrid orbital.** To be consistent with the VB overlap bond concept, the four tetrahedral bonds in methane should be built from orbitals that overlap strongly; ideally the C orbitals should point along the C—H bond axes. Moreover, the equivalence of the bond properties requires that the four orbitals be identical except for spatial orientation. Pauling found he could construct a new set of orbitals fulfilling

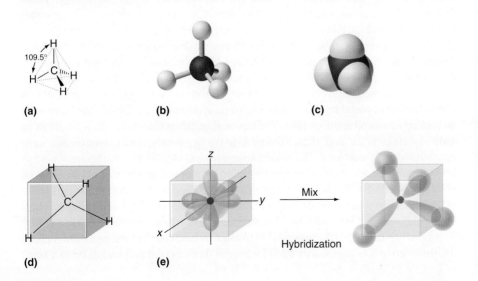

Figure 6.4
Geometrical structure and orbital hybridization in methane. In (a), (b), and (c) the conventional ways of depicting the tetrahedral bond arrangement in the CH_4 molecule are shown: **(a)** most common, a modification of the Lewis structure, in which the top and left bonds are in the plane of the paper; the lowest, solid bond protrudes toward you, and the dashed bond recedes; **(b)** ball-and-stick model frequently used to display geometry; and **(c)** space-filling model roughly representing the overall electron density boundary surface. In **(d)** the molecule is inscribed in a cube, the hydrogens occupying diagonally opposite vertices on each face. In **(e)** the four valence orbitals of carbon are schematically illustrated; when these are combined into sp^3 hybrids, indicated by the arrow, valence bond orbital overlaps may be formed with the H $1s$ orbitals. The hybrid orbital h_1 of Equation 6.5 points to the upper right. The actual form of the hybrids is much "squatter" and less pointed, and they each possess a small tail not shown on the right. See Figure 6.5 for more realistic hybrid boundary surfaces.

these requirements by forming **linear combinations** of the original 2s and 2p orbitals. This is an algebraic procedure involving multiplying the orbital functions of Table 3.2 by carefully chosen coefficients and adding them to yield an orbital with new spatial characteristics termed a *hybrid orbital*. One of these orbitals, call it h_1, is simply the normalized sum of all the valence orbitals:

$$h_1 = \tfrac{1}{2}(2s + 2p_x + 2p_y + 2p_z) \tag{6.5}$$

This orbital and the three others, h_2, h_3, and h_4, formed by other linear combinations obtained by inserting two minus signs in front of the three possible pairs of 2p orbitals in turn, are all identical in shape and point toward the vertices of a regular tetrahedron, as shown in Figure 6.4. Four equivalent valence bonds are then constructed by overlapping the h_i's with hydrogen 1s orbitals; the bond wave functions are given by expressions in the form of Equation 6.4, substituting an h_i for 2p, with each function corresponding to a bonding pair of electrons with their spins paired.

The tetrahedral hybrid atomic orbitals, as in Equation 6.5, are designated sp^3, being composed of one part 2s and three parts 2p. The number of hybrids obtained equals the number of orbitals originally combined: 4 sp^3 orbitals from the four $n = 2$ valence orbitals.

With the aid of the atomic hybridization model, chemists could now interpret, if not explain, the geometry not only of CH_4, but of a whole host of organic molecules. For example, the next higher hydrocarbon molecule beyond methane is ethane, C_2H_6, with the following Lewis structure:

$$\begin{array}{c} \text{H} \quad \text{H} \\ | \quad\; | \\ \text{H}-\text{C}-\text{C}-\text{H} \\ | \quad\; | \\ \text{H} \quad \text{H} \end{array}$$

We know from spectroscopy that the bonding in ethane is tetrahedral around each of the two carbon atoms, with bond angles of 109.5° for both HCH and HCC. The VB model then posits sp^3 hybridization for each carbon, with the C—C bond resulting from overlap of two sp^3 hybrids on different centers. Any length of carbon chain (C—C—C⋯) can be interpreted in this way. Chapter 18, Section 18.2, provides much more about carbon chains; here we need to pursue principles further.

The VB hybridization model for methane, while extremely useful for the tetrahedral geometry problem, is difficult to visualize, because of the need to combine four orbitals at a time to make the sp^3 hybrids. To get a better understanding of how hybrids are formed, let's consider a simpler case, BeH_2 in the gaseous state. BeH_2 is a polar covalent metal hydride known from its spectrum to be *linear*, that is, to have an equilibrium bond angle of 180°. The Lewis structure is simply H—Be—H. Be in its ground state is $1s^2 2s^2$ and, thus, in order to make two bonds, the 2s electron pair must be ruptured as it was for C. To build two *equivalent* Be—H bonds requires two hybrid orbitals that are identical *except* for pointing in opposite directions. These can be obtained by combining two Be atomic orbitals. As Figure 6.5 illustrates, the only two orbitals that can contribute are the 2s and the $2p_x$, if we take the x axis to be the common bond axis for the molecule, because only these two orbitals can successfully overlap with the H 1s. (As you may encounter in Exercise 2 at the end of this chapter, the overlap S is necessarily zero for p orbitals that point perpendicular to a bond

6.2 Polyatomic Molecules: Pauling's Valence Bond Model

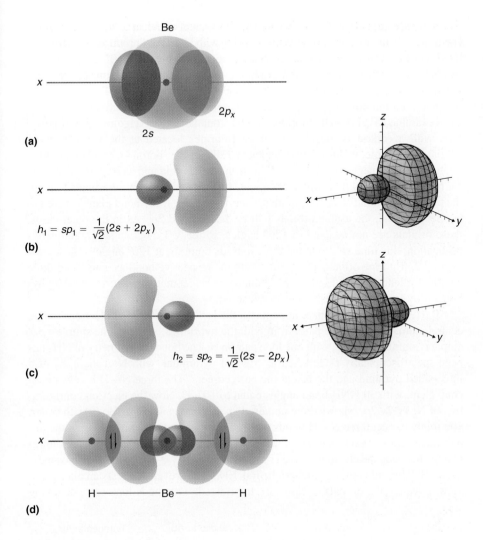

Figure 6.5
Valence bond model for hybridization and bonding in BeH$_2$, the simplest triatomic neutral molecule. In **(a)** the boundary surfaces for the $2s$ and $2p_x$ orbitals, those suited to making bonds with H atoms lying along the x axis, are depicted. (For this purpose, the $2s$ orbital is taken to be positive outside its nodal sphere.) From these are formed **(b)** and **(c)**, sp hybrids pointing toward the H's. (The sp^2 and sp^3 hybrids have shapes nearly identical to those shown.) In **(d)** the overlaps of the hybrids are made with H $1s$ orbitals to yield two identical valence bonds.

with a neighboring H atom.) The hybrids are obtained by simply adding or subtracting the $2s$ and $2p_x$ orbital functions:

$$h_1 = \frac{1}{\sqrt{2}}(2s + 2p_x) \tag{6.6}$$

$$h_2 = \frac{1}{\sqrt{2}}(2s - 2p_x)$$

These are called sp hybrids (one part s and one part p). By studying Figure 6.5 you can see how the hybrids come to have their characteristic lopsided shape, with one big lobe (used for bonding) and one small one. In the case of h_1 of Equation 6.6, adding the orbitals produces a buildup of amplitude (constructive interference) along the positive x axis, and a reduction along the negative x axis, where the orbitals have opposite sign and tend to cancel each other (destructive interference). For h_2, subtracting $2p_x$ from $2s$ is equivalent to changing the sign of $2p_x$; then buildup occurs along the negative axis and reduction along the positive, giving an orbital identical to h_1 except for orientation. Overlapping each of these sp hybrids with H $1s$ then produces the requisite bonding characteristics: two identical bonds and a bond angle of 180°.

Now that you've seen the four-bond tetrahedral and two-bond linear cases, you are clearly eager to hear about the three-bond systems in between. The hydrides of boron and aluminum, however, are dimeric and polymeric, respectively, so we instead examine BF_3, boron trifluoride, which is found to have a **trigonal-planar** structure, with a central boron and the three fluorines located at the vertices of an equilateral triangle, yielding bond angles of 120°. In BF_3 boron lacks an octet but is relatively electronegative, making BF_3 a strong Lewis acid and an important catalyst for many organic reactions. Since the molecule is planar, only those orbitals lying in the plane of the bonds can contribute to their formation. If the bond plane is taken to be the x-y plane, the usable orbitals will be $2s$, $2p_x$, and $2p_y$. As in the other cases, these orbitals are not adapted to 120° angles, so the VB model again requires hybridization, this time sp^2, to make three hybrids pointing at 120° angles. These now overlap $2p$ fluorine orbitals that point along the respective bond axes to make three identical valence bonds. Figure 6.6 summarizes the geometry and corresponding hybridization for the three cases considered so far.

BeH_2, BF_3, and CH_4 share in common the lack of lone pairs of electrons on the central atom. How do lone pairs fit in with the hybridization idea? The simplest example is ammonia (NH_3), with one lone pair on N. The bonding geometry of NH_3 has been spectroscopically established as trigonal pyramidal, with N at the apex of a three-sided pyramid and the H's at the base corners. The three N—H bonds are all equivalent, with all HNH bond angles equal to 107.3°. Now the electron configuration of N, $1s^2 2s^2 2p^3$, shows three unpaired $2p$ electrons and, by analogy with water, you might expect three N—H bonds at 90° angles from overlap of $1s_H$ with the unhybridized $2p$'s. Again as in H_2O, the bond angles must open out to reduce the repulsion between the electropositive H's in each polar $N^{-\delta}$—$H^{+\delta}$ bond. The closeness of the HNH bond angles to the tetrahedral 109.5°, however, suggests an alternative view, in which the orbitals of N are sp^3 hybridized, with the N lone pair occupying

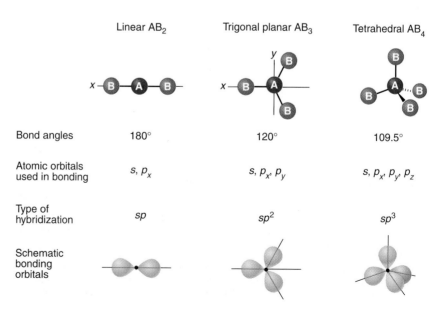

Figure 6.6
Atomic hybridization schemes for atoms A bonded to two, three, or four atoms B, with no lone pairs on A. The p orbitals used for the AB_2 and AB_3 cases depend on the choice of coordinates, as indicated at the top. The hybrids are directed along the bond axes in each case. The drawings of boundary surfaces are artificially "pointy" in order to show directionality clearly; they actually overlap in space, and the sp^3 case corresponds to a spherical overall electron density. The small negative "tails" of the hybrid orbitals are not shown. See Figure 6.5 for more realistic hybrid boundary surfaces. The hybridizations shown also apply approximately when one or more B atoms are replaced by lone pairs.

an sp^3 hybrid orbital. In this case the electron pair geometry would be roughly tetrahedral, and the overlap with $1s_H$ much stronger at the equilibrium bond angles than in the 90° case. Thus, under the general assumptions of the VB model, the modern view assigns sp^3 hybridization to N in NH_3.

Next in a sequence following CH_4 and NH_3 would be H_2O. The H_2O bond angle is only 5° smaller than tetrahedral, again encouraging one to think of an sp^3-hybridized O atom rather than the unhybridized $2p$ plus repulsion of the original VB treatment outlined earlier. The fact that the bond angles decrease slightly in the CH_4, NH_3, H_2O sequence suggests an alternate rationale based on *electron-pair repulsion;* in Section 6.4 you will be introduced to a general scheme for making geometrical predictions founded on this notion.

How do you choose between alternative explanations, like the two we have seen for the bond angle of H_2O? In the absence of quantitative calculations of bond energies and angles, the choice is not without uncertainty, but one way of judging is to ask which scheme produces a "model molecule" closer to reality. The hybrid model wins this one, because there is less bond-angle discrepancy to be explained away.

Before moving on, we add a clarifying note. Pictures of sp^3 hybrids such as those in Figure 6.6 tend to leave the impression that electron pairs poke out like rabbit ears from a hybridized atom. This is an incorrect impression (that apparently is held even by a few experienced chemists); just as a Ne atom with filled $2s$ and $2p$ orbitals is spherical in shape, so is a hybridized atom such as C in CH_4 still spherical in shape prior to the forming of C—H bonds. A nonspherical charge density cannot be obtained out of a spherical one unless one or more orbitals from the original set are omitted. In H_2O, for example, there are no articulated lobes of electron density on the lone pair end of the molecule, but rather a rounded shape characteristic of a spherical atom.

6.3 Double and Triple Bonds: σ and π Bond Types

The valence bond model also accounts neatly for multiple bonding. The prototypical organic molecule containing a double bond is ethylene (C_2H_4), with the planar structure

$$
\begin{array}{c}
H \quad\quad\quad H \\
\diagdown \; 121.5° \; \diagup \\
C = C \quad 117° \\
\diagup \quad\quad\quad \diagdown \\
H \quad\quad\quad H
\end{array}
$$

The CCH and HCH bond angles are each very close to the 120° trigonal-planar angles of BF_3, suggesting sp^2 hybridization of each C atom. As shown in Figure 6.7, this allows all of the atoms in C_2H_4 to be bonded together in the proper geometry. In addition, each carbon atom has one "leftover" $2p$ orbital that was not used to make the hybrids, and one electron occupying it. Although these orbitals do not "point" at each other in the usual VB manner, they can still overlap each other when the two carbon atoms are brought together, with regions of overlap *above* and *below* the C—C bond axis, but zero amplitude along the bond axis. Thus a *second valence bond* between C's can be formed.*

---

* This is not the scheme originally proposed by Pauling, who preferred to use sp^3 hybridization and "bent" orbital overlaps to make the double bond. Explaining bond angles is more difficult in this form.

Figure 6.7
Valence bond $\sigma-\pi$ picture of the bonding in ethylene. The σ-bonded framework is shown at the top in schematic form, while the π bond between the carbons is more realistically represented. To simplify the π-bond illustration, the σ bonds are shown in skeletal form. The schematic at the bottom is the usual way of representing the orbitals involved in a double bond. As the diagrams suggest, π orbital overlap is smaller than σ at a given interatomic distance.

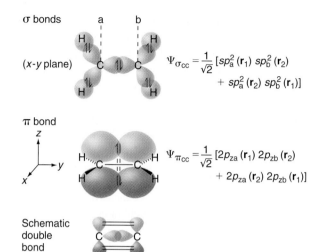

The new bond wave function is written in the same way as earlier (Equation 6.4), but it has a new type of *symmetry*. Until now, all the bonds we've described have been *cylindrically symmetric* about the bond axis; that is, if you rotate the overlapping orbitals about the axis, the appearance of the wave function is unchanged. Since cylindrical symmetry about an axis is analogous to spherical (s-orbital) symmetry about a point, the bond is called a **σ bond**, with the sigma being the Greek equivalent of s. The symmetry of our new ethylene wave function, however, is not cylindrical. As you can see by thinking about Figure 6.7, if the overlapped orbitals are *reflected* through the molecular plane, they are unchanged except for algebraic sign. This is a lower symmetry than σ, and corresponds to the symmetry of a *p*-orbital when the orbital is reflected through a plane passing through the origin, perpendicular to the pointing axis of the orbital. Hence it is labeled a **π bond**, with pi being the Greek equivalent of *p*.

Thus despite appearances the Lewis double bond consists of two nonequivalent types of bonds, one σ and one π. The exposure of the π-electron density above and below the plane of the molecule helps to explain the relatively great reactivity of C_2H_4 compared to C_2H_6. C_2H_4 is readily attacked by electrophiles; for example, $C_2H_4 + Br_2 \longrightarrow C_2H_4Br_2$ is a facile redox addition reaction in which Br_2 adds across the double bond. C_2H_6 will not react with Br_2 except under extreme conditions. This is an early clue that orbital-based models will bring you closer to an understanding of the relation between structure and reactivity than Lewis structures allow.

Acetylene (C_2H_2), which is even more reactive than C_2H_4, has the Lewis structure H—C≡C—H and is found to be linear (180° CCH bond angles). Earlier we associated linearity with *sp* hybridization as in BeH_2.* Then the σ-bonded framework H—C—C—H can be constructed from three valence bonds, the C—C arising from *sp–sp* overlap. Now there are two unused *p* orbitals on each C, each pointing perpendicular to the molecular axis. Each parallel pair of *p*'s can overlap to form a π bond, so that C≡C consists of one σ and two π bonds. The π-electron density merges to make a sort of "muff" for the molecule to sit in, as illustrated in

---

*Again, Pauling maintained sp^3 hybridization and used bent-overlap bonding for C≡C. This also naturally leads to a linear structure and has a few proponents in the current literature.

Figure 6.8. It is even more tempting for an electrophile to "eat π for lunch" when reacting with C_2H_2 than with C_2H_4.

The molecules C_2H_2, C_2H_4, and C_2H_6 have structures analogous to those of N_2, O_2, and F_2, which were discussed in Section 5.2. Each C—H bonding electron pair becomes a lone pair in the corresponding diatomic molecule. Thus the picture of σ and π bonds carries over to the diatomics, with N_2 having one σ and two π bonds, O_2 one σ and one π, and F_2 one σ bond. It is not necessary to hybridize the atomic orbitals here to account for the bonding, although this is often done. For example, in O_2 we can use unhybridized $2p_\sigma$ (pointing along the bond axis) and $2p_\pi$ orbitals on each atom to form the double bond, putting the lone pairs in the atomic $2s$ and the remaining $2p$ orbitals not used in bonding. Or you could consider sp^2 hybridizing the O orbitals just as in ethylene. Then the only difference between O_2 and C_2H_4 is the lone pairs on each O, replacing the C—H bonds. Either alternative is acceptable, since no bond angles are in need of rationalization, but in view of the ideas in the next section, the hybridized model may be preferable. For later reference, we note that in either picture all electrons are paired in doubly occupied orbitals or in VB wave functions.

As a general rule, for polyatomic molecules the VB prescription calls for hybridizing the central atoms as appropriate for the observed molecular geometry, in order to construct a σ-bonded framework. Then any p orbitals on the central atoms that have not been used in making hybrids may be combined with p orbitals on the bonded atoms to account for the π components of any multiple bonds present.

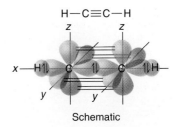

Schematic

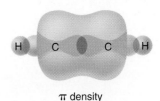

π density

Figure 6.8
Valence bond σ–π picture of the bonding in acetylene. The σ bonds are constructed with C sp hybrids blended from $2s$ and $2p_x$ atomic orbitals, while the $2p_y$ and $2p_z$ orbitals on each C overlap to form π bonds.

EXAMPLE 6.1

Give a valence bond description of and sketch the bonding in the nitrite ion NO_2^-, given an ONO bond angle of 118°, and two equal bond lengths.

Solution:
Because VB follows Lewis structures in its formulation, you should always start with the electron-dot structure:

$$[:\ddot{O}=\ddot{N}-\ddot{O}:]^- \longleftrightarrow [:\ddot{O}-\ddot{N}=\ddot{O}:]^-$$

(See Example 5.2 to review the rules for drawing Lewis structures.) You need consider only one of the resonance forms, because the other will have identical bonding characteristics. Both the bond angle and the presence of the double bond suggest sp^2 hybridization of the N atom. Then the bonding consists of a pair of N—O σ bonds involving sp^2 hybrids on N and either unhybridized $2p$ or sp^2/sp^3 hybrids on O. The N lone pair resides in an sp^2 N atom orbital, while the extra bond is a π bond formed from the unused $2p$ orbital on N pointing perpendicular to the molecular plane and the corresponding $2p_\pi$ orbital on one of the O's. A schematic sketch looks like:

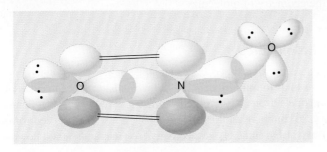

In this example, as in the isoelectronic O_3 molecule (see Section 5.2), the inability of Lewis structures to account in a satisfactory way for the two equivalent bonds carries over to the VB description. We can conclude from the resonance forms that each N—O bond in NO_2^- possesses some "π character"; the bond orders being 1½ suggests each bond consists of one σ and ½ π bonds. In the next chapter we will learn a more satisfying way of describing the resonance situation.

The Lewis structure of a molecule can be used in a shorthand way to predict the type of hybridization (and thereby the approximate bond angles) for many common molecules. If an atom from the second period is involved in only single bonds, its hybridization is dictated by the number of bonds plus the number of lone pairs. When this sum is two, the hybridization is sp, three gives sp^2, and four gives sp^3, while all the bonds are σ. Only when the sum is four, of course, is the octet rule satisfied, generally for carbon rightward, occasionally for boron. When the atom obeys the octet rule, its multiple bonds determine its hybridization. If it makes no multiple bonds, the atom is sp^3 hybridized and all bonds are σ; if one double bond is made, it is sp^2 hybridized and there is one π bond (in the double bond); if two double bonds or one triple bond is made, the hybridization is sp and the atom participates in two π bonds.

It is important for you to appreciate that the hybridization concept is a *rationale* concocted to put the observed facts of molecular geometry into the VB mold. Though hybridization can be used to predict bond angles by analogy with known examples, it does not provide an *explanation* for the angles observed. This leads us into an offshoot of VB theory that turns a simple notion into a powerful tool for predicting the shapes of molecules.

6.4 Electron-Pair Repulsion: Rationalizing and Predicting Molecular Shapes

Imagine two VB electron pairs, located in directional orbitals and mounted on a particular atom, that are free to swivel about the nucleus, as illustrated in Figure 6.9. We already know that electrons make poor bedfellows due to Coulomb repulsion, so the electron pairs will tend to repel each other. Making use of the principle of minimization of potential energy (natural response to a repulsive force; see Section 1.4 and Exercise 1.45), the orbitals will swing away from each other until the potential is smallest, which occurs for a bond angle of 180°. Thus the linear geometry of BeH_2, our example of sp hybridization, has a natural explanation. R. J. Gillespie in 1959 proposed that **valence-shell electron-pair repulsion** (VSEPR) is a general principle governing the geometry of molecules.

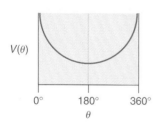

Figure 6.9
Potential energy minimization in the VSEPR model. The potential curve $V(\theta)$, called a *bending potential*, dictates the equilibrium angle θ.

σ-Bonded Molecules

VSEPR in general predicts that σ bonds and lone pairs will arrange themselves about an atomic core so as to minimize their mutual repulsion. For the cases of two, three, and four pairs of electrons, the required geometries are linear (180° bond angle), trigonal planar (120° angles), and tetrahedral (109°), just as observed for our example molecules BeH_2, BF_3, and CH_4, respectively. You can understand how these geome-

6.4 Electron-Pair Repulsion: Rationalizing and Predicting Molecular Shapes

tries are obtained at least qualitatively by imagining what happens when a third orbital lobe is inserted into the linear two-lobed arrangement. The 180° angle between the two unperturbed orbitals is forced to close up (to 120°) due to the interloper's repulsive effect. Then, if a fourth orbital approaches from above the trigonal plane, the three orbitals already present are forced to bend down, out of the plane, closing the bond angles to the tetrahedral 109.5°. Figure 6.10 illustrates this line of thinking.

The model can be extended further to cover the expanded-octet cases such as PCl_5 (**trigonal bipyramidal**) and SF_6 (**octahedral**). Figure 6.11 summarizes the geometries and orbital appearance for the five cases, in which two, three, four, five, or six bonds and lone pairs surround the central atom. While the octahedral geometry, with all bond angles equal to 90°, fits in nicely with the two-, three-, and four-lobe cases, which also have equal angles, the five-pair geometry is irregular. The three lobes that form an equilateral triangle, in the so-called *equatorial* position, are at 120° angles, while the two *axial* lobes make 90° angles with their three equatorial neighbors. The smaller angles for the axial positions cause the axial bonds to be somewhat longer. For example, in PCl_5 the equatorial P—Cl bonds are 2.05 Å long, while the axial bonds are 2.19 Å.

An appealing way of handling the situations where there are one or more lone pairs is also provided by VSEPR. A lone pair of electrons is intrinsically more repulsive than a bonding pair, because there is no bonded atom to draw electron density away from the atom under consideration. Lone pairs therefore cause a slight distortion from the symmetrical geometries of Figure 6.11. Consider again the sequence CH_4, NH_3, OH_2 in which the bond angles HXH decrease: 109°, 107°, 105°. In NH_3, the single lone pair, being slightly more repulsive than the N—H bonding pairs, causes the HNH angles to close slightly from the tetrahedral value. The arrangement of electron pairs about N is still very close to tetrahedral, while the atomic geometry is trigonal pyramidal, just as in the sp^3-hybridization scheme of the last section. For

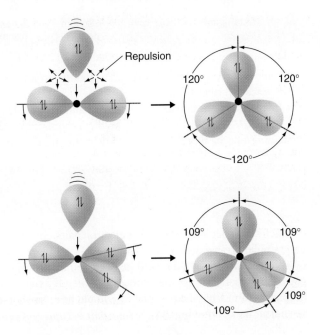

Figure 6.10
Building up electron-pair repulsion geometries by successively adding new electron pairs for the three- and four-pair cases. The five- and six-pair geometries to be described later can also be justified in this way. Performed in reverse, this scenario can be enacted by oval-shaped balloons tied tightly together, and pricked one by one.

Figure 6.11
Summary of bond angles and geometries for the five common cases of two through six electron pairs about a central atom, with lone pair substitutions. Beneath the schematic electron pair pictures are the corresponding VB hybridizations. The lone pairs cause the bond angles to be a few degrees smaller than the ideal values given on the left.

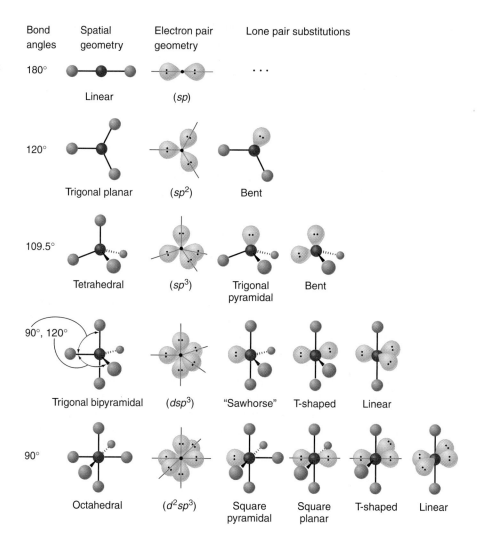

water, the two oxygen lone pairs exert a slightly stronger effect, causing the bond angle to deviate further from 109°.

In addition to its consistent (though qualitative) explanation of small bond-angle differences, VSEPR also accurately predicts geometries of expanded-octet molecules containing one or more lone pairs. For example, the SF_4 molecule has the Lewis structure

$$F-\ddot{S}-F$$
$$/\quad\backslash$$
$$F\quad\quad F$$

in which sulfur accommodates five electron pairs. SF_4 therefore must have a trigonal-bipyramidal arrangement of electron pairs (see Figure 6.11). The question is, which position, equatorial or axial, will be occupied by the lone pair? Where the lone pair goes dramatically affects the geometry of SF_4: if axial, SF_4 would be described as a trigonal pyramid, but if equatorial SF_4 would be a "sawhorse." Experiment favors the sawhorse shape, implying that *the lone pair occupies an equatorial position* as shown

6.4 Electron-Pair Repulsion: Rationalizing and Predicting Molecular Shapes

in Figure 6.11. This correlates with the bond-length distortion in the five-lobe case discussed earlier. In the isolobal molecules ClF_3 and XeF_2, the additional lone pairs also occupy equatorial sites, yielding T-shaped and linear geometries, respectively. As in the four-lobe cases, lone pairs cause a slight distortion of the ideal VSEPR geometry; for example, the bond angles in ClF_3 are 87° instead of 90°.

Molecules with π Bonds

As we saw in Section 6.3, the ethylene molecule (C_2H_4) is approximately trigonal planar (120° angle) about each C atom with σ and π valence bonds between the carbons. Chemical intuition, which always works best after the facts are established, hastens to add that this is entirely expected in the VSEPR model. The C═C π-electron density (see Figure 6.7) is concentrated above and below the molecular plane and, due to its consequent greater distance from the C—H σ-bond lobes, exerts a much smaller effect on the bond angles than the C═C σ bond does. Thus the HCH angle (117°) is only a bit smaller than the 120° angle that would prevail in the absence of π bonding. This means that multiple bonding can be very simply incorporated into the VSEPR scheme by regarding the π bonds as minor perturbations, and ignoring them in classifying the geometric type. The ethylene carbons then fall into the three-lobe class, where the idealized geometry is trigonal planar about each C. The effect of the π bond is then similar to that of a lone pair: the adjacent bond angles close slightly.

From the discussion of hybridization and bonding at the end of Section 6.3, you may begin to see that the bonding geometry goes hand in hand with hybridization, and that atoms engaged in one π bond (sp^2) always fall in the three-lobe trigonal planar class, while two π bonds (sp) (whether arising from two double bonds or one triple bond) yield the two-lobed linear case.

EXAMPLE 6.2

Compare the geometries of SO_2 and CO_2 using the VSEPR model.

Solution:
The Lewis structures are

$$:\ddot{O}=C=\ddot{O}: \quad \text{and} \quad :\ddot{O}-\ddot{S}=\ddot{O}:$$

Ignoring the components of the double bonds as well as the additional resonance form of SO_2, CO_2 has two σ bonds and no lone pairs on C and is therefore linear (180° OCO angle), while SO_2 has two σ bonds and one lone pair on S, yielding a bent SO_2 with an angle somewhat less than 120° due to lone pair repulsion. (The experimental angle is 119°.)

The Effect of Heteroatoms

Here we consider substituting different atoms (heteroatoms) on the periphery of a polyatomic molecule. As with lone pairs and multiple bonds, to a first approximation the ideal VSEPR structures of Figure 6.11 determine the orientation of electron pairs and the geometry of the molecule, while the heteroatom may induce small changes in the ideal bond angles. For example, consider the HCH bond angles in the sequence of monohalomethanes CH_3F (110°), CH_3Cl (110°), CH_3Br (111°), and CH_3I (111.4°). These are all larger than the ideal tetrahedral angle 109.5°, apparently owing to the greater electronegativity of the halogens, causing the electron pair in the C—X bond to be drawn away from the carbon atom, thereby reducing the repulsion with the other pairs. The angle also seems to be greater for the heavier halogens, although the experimental uncertainty in these angles is usually 1° or so. The HCX angles are large enough so that the differing "bulkiness" of the halogen atoms does not influence the angles appreciably. Many other examples of this sort show very little effect of heteroatom substitution; typically the VSEPR ideal geometries generally hold within 1° to 2° for second-row central atoms provided no lone pairs or multiple bonds are present. Even with lone pairs and double bonds, the ideal geometry is nearly always close to reality.

EXAMPLE 6.3

Make a geometrically realistic sketch of the amino acid *glycine* H_2NCH_2COOH (aminoacetic acid), whose Lewis structure is

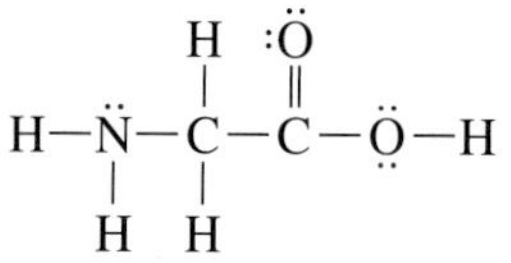

$$\begin{array}{c} \text{H} \quad\; :\ddot{\text{O}}: \\ | \quad\;\; \| \\ \text{H}-\ddot{\text{N}}-\text{C}-\text{C}-\ddot{\text{O}}-\text{H} \\ |\quad\; | \\ \text{H} \quad \text{H} \end{array}$$

Include in your structural sketch an indication of approximate bond angles and hybridizations.

Solution:

We may deduce the approximate bonding geometry of the entire molecule by making use of the VSEPR rules for each atom bonded to more than one other atom. From left to right, for N, the sum of lone pairs (1) and σ bonds (3) is 4; for C, also 4; for the next C, 3; and for the O, 4. Referring to Figure 6.11, these lead to approximate bond angles of ≅109°, 109°, 120°, and ≅109° around the respective atoms, making an allowance for the lone pairs on N and O. The corresponding hybridizations are also read from Figure 6.11. To draw the resulting structure, we use by convention a zigzag conformation, giving

$$\begin{array}{c} \quad\quad\quad \text{H} \;\; \text{H} \\ \quad\quad\quad\quad\; \diagdown \;\, /sp^3 \\ \text{H} \diagdown \;\; ≅109° \;\; \text{C} \quad\; 120° \;\;\; :\ddot{\text{O}}: \, sp^3 \\ \quad\; :\text{N} \diagup\!\!\diagdown\, 109° \diagup\!\!\diagdown\, \text{C} \diagup\!\!\diagdown ≅109° \diagdown \text{H} \\ \quad / sp^3 \quad\quad\; sp^2 \| \\ \;\;\text{H} \quad\quad\quad\quad\quad\; \ddot{\text{O}} \end{array}$$

Nearly free rotation around the σ bonds makes the above zigzag structure one of a number of possible *conformers;* see Chapter 18 for further discussion. Generally you do not have to worry about terminal atoms, such as the doubly bonded *carbonyl* O, since whether or not you hybridize it has no bearing on geometry. In aqueous solution, glycine may exist as a *zwitterion*, $^+H_3NCH_2COO^-$. Can you now deduce the geometry of this species?

An Overall Assessment of VSEPR

The significance of the VSEPR model is that it provides a rationale for Pauling's hybrids beyond the simple need to account for experimental facts, and the combination of these ideas forms a powerful method for predicting and analyzing the structure of polyatomic molecules.

For a few molecules, principally the hydrides of nonmetals of the third or higher periods from groups V and VI, the VSEPR predictions are substantially in error. For example, the bond angles in H_2S and PH_3 are 91° and 93°, respectively, much further from 109.5° than expected based on the extra repulsion due to the lone pairs. For central atoms of greater girth, the valence orbitals are more diffuse and the mean distance between them greater. We would therefore expect *less* difference between bonding and lone pairs, and more nearly ideal geometry. Instead, the bond angles are nearly those that would obtain in the *absence* of hybridization. We may qualitatively argue that it is electron pair repulsion that provides the impetus to hybridize, and when this repulsion is reduced, the central atom reverts to its usual valence orbitals in making bonds. But this argument is just a patch applied to an already approximate model of molecular structure, and perhaps it is time to shrug our shoulders with the observation that all models have their failings. Moreover, there are more serious problems with our model to be discussed in the last section.

6.5 Asymmetric Electron Sharing: The Electric Dipole Moment

The motivation for the concept of electronegativity χ, introduced in Chapter 5, was that an electron pair bond between atoms with different χ's is not purely covalent; that is, the electrons are shifted toward the more electronegative atom. Hence the partial charges $+\delta$ and $-\delta$ were introduced to indicate this, as in $H^{+\delta}F^{-\delta}$. Two unlike charges separated by a finite distance constitute an **electric dipole,** and a molecule that has such a charge separation is said to possess a **dipole moment.** As you saw in Chapter 5, the partial charges play an important role in defining molecular affinities for certain types of chemical reactions. Dipolar molecules also may liquefy or solidify more easily than ones without a dipole, while the rule "like dissolves like" refers to the polar or nonpolar character of the molecules comprising the solvent and solute (see Chapter 14). Thus it is of chemical significance to know whether a particular molecule has a nonzero dipole moment, as well as its magnitude.

If you consider a molecule or ion as simply a cloud of charge (negative electrons and pointlike positive nuclei), that cloud can be analyzed using classical electrostatics (applying Coulomb's law to find electric fields) to yield a set of quantities known as the *multipole moments.* Chemists ordinarily consider only the two simplest multipole moments, the *electric monopole moment*—simply the net charge on the species—and the **electric dipole moment.** In cgs-esu the dipole moment $\boldsymbol{\mu}$ is given by $\boldsymbol{\mu} = q\mathbf{r}$, where q is the magnitude of the charge separation and $\mathbf{r}$ is the distance between the charges $+q$ and $-q$:

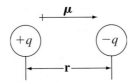

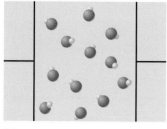

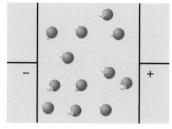

Figure 6.12
Classic method for measuring dipole moments, illustrated for gaseous HCl. The molecules, in gaseous or liquid form, are placed between the plates of a capacitor. When the plates are uncharged as in (a) the molecules are oriented randomly, but when + and − charges are imposed as in (b), dipolar molecules will on the average line up with the resulting electric field, and increase the capacitance over that with no sample present. If the capacitor is part of an oscillator circuit, the polar molecules will cause a change in the resonant frequency, which is readily measured on an oscilloscope. The ratio of capacitances with and without sample is related to the square of the dipole moment, as first shown by Debye. More modern methods for obtaining μ are based on rotational spectroscopy (see Chapter 8) in the presence of an electric field.

where $\boldsymbol{\mu}$ is a vector quantity, pointing conventionally toward the negative charge. For a polar molecule such as HF, the charge separation q is δe and r is taken to be the bond length r_e, so the dipole magnitude $\mu = \delta e r_e$. If the length of the bond were 1 Å and the bond were completely ionic ($\delta = 1$), then μ would be $(1)(4.80 \times 10^{-10}$ esu$)$ $(1 \times 10^{-8}$ cm$) = 4.8 \times 10^{-18}$ esu cm $= 4.8$ Debye (D). The last equality defines the **Debye unit** of dipole moment, $1\,\text{D} = 1 \times 10^{-18}$ esu cm, named after Peter Debye, who showed how μ is related to experimentally measurable properties. Measurements on HF, whose bond length is 0.917 Å, yield a dipole moment $\mu = 1.82$ D. This implies that the fractional charge separation δ is

$$\delta = \frac{\mu}{e r_e} = \frac{1.82\,\text{D}}{(4.8)(0.917)\text{D}} = 0.41$$

that is, 41% of a full electronic charge has been transferred from H to F, or the HF bond is 41% ionic. Note that in the calculation e is expressed in units of 10^{-10} esu and r_e in Å (10^{-8} cm), giving the ionic dipole $e r_e$ in Debyes.

Figure 6.12 illustrates the classic method for measuring dipole moments, and Table 6.1 collects experimental dipole moment data on diatomic and polyatomic molecules. As you might expect, μ is identically zero for **homonuclear** diatomic molecules (both atoms the same) since $\Delta\chi = 0$ in those cases, while the magnitude of μ correlates with $\Delta\chi$ in the **heteronuclear** systems. For example, μ declines monotonically in the HF, HCl, HBr, HI sequence despite the increasing bond length, due to the decreasing $\Delta\chi$. The largest dipoles occur (also as expected) for the alkali halides, where the ionic character approaches 100%.

To understand the data for polyatomic molecules, the VB model suggests we consider each bond as possessing its own **bond dipole**. The total dipole moment of the molecule is then the vector sum of the individual bond dipoles. This makes the geometry of the molecule a critical factor; if identical bond dipoles point in directions that produce overall cancellation, the net molecular dipole is zero. For example, the CO_2 molecule is linear, and the bond dipoles for each C=O bond point in opposite directions and therefore cancel:

$$\overset{\overset{\mu_a}{\leftarrow}\;\overset{\mu_b}{\rightarrow}}{\ddot{\text{O}}\!=\!\text{C}\!=\!\ddot{\text{O}}\!:}$$

$$\mu(CO_2) = \mu_a + \mu_b = 0$$

For all the symmetrical no-lone-pair VSEPR structures illustrated in Figure 6.11, μ is identically zero provided that all the bonded atoms or groups are identical. However, if lone pairs are present and unsymmetrically arranged, or a heteroatom is present, a net dipole will result due to the imbalance in bond dipoles. Thus, while methane has no dipole moment, with the (small) C—H bond moments canceling in the symmetrical tetrahedral geometry, NH_3 is dipolar, as is H_2O, due to the substitution of lone pairs for bonds to hydrogen:

$\mu = 0$ $\mu = 1.49$ D $\mu = 1.85$ D

TABLE 6.1
Selected molecular bond lengths r_e (Å) and dipole moments μ (D)

Diatomic molecules			Polyatomic molecules		
Molecule	r_e	μ	Molecule	r_e	μ
H_2	0.741	0	BeH_2	1.19	0
HF	0.917	1.82	H_2O	0.96	1.85
HCl	1.27	1.08	H_2S	1.35 (92°)*	0.97
HBr	1.41	0.82	HCN	1.06(C—H); 1.16(C≡N)	2.98
HI	1.60	0.44	CO_2	1.16	0
CO	1.13	0.11	NO_2	1.20 (134°)*	0.32
CS	1.53	1.98	N_2O	1.13(N≡N); 1.19(N—O)	0.17
N_2	1.10	0	SO_2	1.43	1.63
NO	1.15	0.15	O_3	1.28	0.53
O_2	1.21	0	CS_2	1.56	0
F_2	1.42	0	COS	1.19(C=O); 1.58(C=S)	0.71
SO	1.48	1.55	OF_2	1.42	0.30
ClF	1.63	0.88	ONF	1.13(N=O); 1.52(N—F)	1.81
Cl_2	1.99	0	HgI_2	2.77	0
Br_2	2.29	0	BF_3	1.44	0
I_2	2.66	0	NH_3	1.00	1.49
LiH	1.60	5.88	PH_3	1.43	0.58
LiF	1.56	6.33	SO_3	1.74	0
LiCl	2.02	7.13	PCl_3	2.10	0.78
LiBr	2.17	7.27	$SOCl_2$	1.56(S=O); 2.03(S—Cl)	1.45
LiI	2.39	7.43	NHF_2	1.00(N—H); 1.34(N—F)	1.92
NaF	1.93	8.16	CH_4	0.98	0
NaCl	2.36	9.00	CH_3F	1.41(C—F)	1.81
KF	2.17	8.60	CH_3Cl	1.76(C—Cl)	1.87
KCl	2.67	10.27	CH_2Cl_2		1.58
KBr	2.82	10.41	$CHCl_3$		1.01
RbF	2.27	8.55	CCl_4		0
CsF	2.35	7.88	CH_3I	2.10(C—I)	1.91
CsCl	2.57	10.42	SiF_4	1.81	0
SrO	1.92	8.90	SF_4	1.68	0.63
			PCl_5	2.05, 2.19	0
			IF_5	1.97	2.18
			SF_6	1.66	0
			C_2H_4	1.34(C=C); 0.99(C—H)	0
			N_2H_4	1.40(N—N); 1.00(N—H)	1.75
			H_2O_2	1.32(O—O); 0.96(O—H)	2.2

*Anomalous bond angle.

The bond dipoles grow larger in the CH_4, NH_3, H_2O sequence owing to the increasing $\Delta\chi$. You should note carefully that, despite the appearance of these structures, the lone pairs *do not* contribute to the net dipole. As we discussed at the end of Section 6.2, rearranging the orbitals to form hybrids does not alter the original spherical form of the overall electron density. The nonbonded electron density, although shown in the artificially lobular form of hybrids, is distributed spherically about the central atom, with only the bonds being articulated.

The effect of heteroatoms is illustrated by the series of **chloromethanes** CH_3Cl, CH_2Cl_2, $CHCl_3$, CCl_4:

$\mu = 1.87$ D $\quad\quad \mu = 1.58$ D $\quad\quad \mu = 1.01$ D $\quad\quad \mu = 0$

A single Cl atom converts the zero dipole of CH_4 to a large value, due mainly to the large bond dipole of the C—Cl bond. The influence of bond length on μ can be noted by comparing CH_3Cl ($\mu = 1.87$ D) and CH_3F (1.81 D), where the greater $\Delta\chi$ and hence δ of the C—F bond is offset by the greater length of the C—Cl bond.

As you may have surmised, most molecules do have dipole moments; only those where the atoms or bonds are symmetrically arranged about a central point have no dipole. To bring this point out, consider the ozone molecule:

$\mu = 0.53$ D

Even though all of the atoms are the same, the central oxygen is hypervalent, sharing three instead of the usual two electron pairs, and hence is electron poor and must bear a partial positive charge. VSEPR rules predict a bent molecule, making μ nonzero.

EXAMPLE 6.4

Predict whether or not the following molecules will have nonzero dipole moments:

$CS_2 \quad SCl_2 \quad CH_2O \quad$ (C=C with F, H on one carbon and H, F on the other)

Solution:
In the VB-VSEPR model, you always start with the Lewis structures:

$:\ddot{S}=C=\ddot{S}: \quad\quad :\ddot{C}l-\ddot{S}-\ddot{C}l: \quad\quad H-\overset{:\ddot{O}}{\underset{\|}{C}}-H$

with $C_2H_2F_2$ as above. CS_2 falls into the two-electron pair case and is therefore, like CO_2, linear. The C=S bond dipoles then cancel, and $\mu = 0$. SCl_2 on the other hand

6.5 Asymmetric Electron Sharing: The Electric Dipole Moment

is a four-pair case, bent, and will show a nonzero μ due to the noncanceling S—Cl dipoles. CH_2O falls into the three-pair category, roughly trigonal planar, and the C=O bond thus is the dominant factor that makes $\mu \neq 0$. The structure of $C_2H_2F_2$ given above is one of three **structural isomers** possible with this molecular formula. In this case the C—F bond dipoles (as well as the small C—H bond dipoles) point in opposite directions,

$$\begin{array}{c} F \searrow \quad \quad \nearrow H \\ C\!=\!C \\ H \nearrow \quad \quad \searrow F \end{array}$$

giving a zero resultant μ. The molecule, known as *trans*-1,2-difluoroethylene, remains in the planar configuration, without allowing the F atoms to flip up or down, due to strongly hindered rotation around the double bond. This will be discussed more thoroughly in Chapter 18.

Valence Bond Description of Partial Ionic Character

In a polar covalent bond, the bond energy is partially accounted for by Coulomb attraction between the partially charged ($+\delta e$ and $-\delta e$) atoms, thereby taking on a mixed covalent/ionic character. This is represented in VB theory by adding an **ionic term** to the VB bond wave function,

$$\Psi_{bond} = a\, \Psi_{covalent} + b\, \Psi_{ionic} \tag{6.7}$$

where $\Psi_{covalent}$ is a VB function like Equation 6.2 or 6.4, Ψ_{ionic} places the bonding pair entirely on the more electronegative atom, and a and b are coefficients, chosen so that $a^2 + b^2 = 1$, that determine the amount of ionic character. For the HF example of Section 6.2, the bond wave function would be

$$\Psi_{bond} = \frac{a}{\sqrt{2}} [1s_H(\mathbf{r}_1)2p_F(\mathbf{r}_2) + 1s_H(\mathbf{r}_2)2p_F(\mathbf{r}_1)] + b\, 2p_F(\mathbf{r}_1)2p_F(\mathbf{r}_2) \tag{6.8}$$

and the fractional ionic character is given by b^2. From our discussion of dipoles and the partial charge separation δ, we can identify $b^2 \approx \delta$; for HF, we then expect $b^2 = 0.41$ or $b = 0.64$.

As the coefficient b in Equation 6.7 increases, the bonding becomes more ionic; in the limit of a purely ionic bond ($a = 0$, $b = 1$), we revert to the Coulombic description given in Chapter 5. This means that the fraction of the bond energy that arises from exchange also shrinks, disappearing in the ionic limit, and thus the description of the nature of bonding passes from purely quantum mechanical to classical electrostatic. This diversity of molecular bonding is included in the bond function of Equation 6.7.

Both the charge separation δ and the coefficient b are closely related to the electronegativity difference $\Delta\chi$; an estimate of the fractional ionic character can be obtained from the completely empirical formula

$$\delta \approx 0.84\, \{1 - \exp[-0.27(\Delta\chi)^{2.4}]\} \tag{6.9}$$

For HF $\Delta\chi = 1.8$, giving $\delta = 0.56$; the formula does better with nonhydride molecules.

6.6 Shortcomings of the Valence Bond Model

As successful as the VB-VSEPR model is at describing molecular structure, it still falls short of being a comprehensive theory of bonding. As noted earlier, it necessarily shares the shortcomings of Lewis structures, namely, inadequate treatment of odd-electron molecules and molecules with resonance forms. The VB model is based on the electron pair, and VB bond wave functions such as Equation 6.2 or 6.4 cannot describe unpaired electrons. And although Lewis's multiple bonds are given a wave-mechanical σ–π character, when these bonds along with lone pairs are required to shift about in possible resonance forms, the VB wave functions must switch atomic allegiance as well. But an even more serious objection lies in a well-known property of a common molecule, O_2.

Paramagnetism of Molecular Oxygen

As early as 1848 Michael Faraday, whose major contributions to chemistry are detailed in Chapter 13, noted that clouds of smoke would tend to thicken around wires carrying large currents. Through various tests Faraday concluded that oxygen in the air was being drawn into the magnetic lines of force surrounding the wire. Figure 6.13 shows liquid oxygen suspended between the poles of a powerful permanent magnet; liquid nitrogen cannot be so suspended. Modern measurements show this to be due to *unpaired electron spins* that align themselves along the magnetic lines; the magnitude of the effect indicates two unpaired electrons in each O_2 molecule. Molecules or solids with unpaired electrons are referred to as **paramagnetic;** if no unpaired e's are present, the substance is weakly repelled by a magnetic field, a property called **diamagnetism.** O_2 is paramagnetic while N_2 is diamagnetic.

Figure 6.13
(a) Liquid oxygen is drawn into the pole gap of a powerful magnet, showing that O_2 is paramagnetic, (b) while liquid nitrogen passes through the gap without sticking.

(a)

(b)

We have already noted above that the VB picture, while accounting well for the bonding properties of O_2, inescapably predicts that O_2 (as well as N_2) should be diamagnetic; that is, O_2 should possess no unpaired electrons. Thus the details of the VB wave functions cannot be entirely accurate. This signal discrepancy, along with the more general failings discussed earlier, point to the need for a better, or at least complementary, bonding model.

Reaction Mechanisms and Molecular Energy Levels

In chemistry, knowing the structure of a molecule is not an end in itself, but a means for understanding its reactions. The VB model fails to move us beyond our Lewis structure description of a reaction (see Section 5.2 for example) in σ-bonded reagents, because it does not provide a mode for transferring electrons to a molecule that has all of its valence orbitals occupied and all of its electrons already paired. This is partly due to the reliance of VB on electron pairs, but also to its failure to provide a description of *the energy levels and associated orbitals* for molecules, particularly polyatomic molecules. Atoms, as you have learned, usually have partially filled or empty orbitals that are exploited to yield bonding between them. But a VB molecule is not generally endowed with similar properties that would allow it to react or form new bonds. In addition, transitions between molecular electronic energy levels, generally excited by ultraviolet light for most of the common molecules we have considered so far, cannot be easily described within our VB model.

As we will discuss at greater length in Chapter 7, it is possible to generalize the valence bond method to remedy these shortcomings, but at the expense of the simple bonding picture VB provides. Here again, an alternative view of molecular structure is desirable. That view is provided by the molecular orbital model presented in Chapter 7.

SUMMARY

In 1927 Heitler and London presented the first application of the then newly invented Schrödinger wave mechanics to the simplest example of a Lewis electron-pair bond, the H_2 molecule. They found that a simple product of atomic orbitals, taken as a guess for the molecular wave function, did not yield a strong bond between the atoms, but when an additional term, allowing for **exchange** of the electrons between the nuclei was added, a stable molecule was predicted. The calculations took the form of a potential energy curve for the covalent bond, in fair agreement with the curve obtained from spectroscopy. Two important conclusions Heitler and London drew from their calculations were (1) the covalent bond arises from a *nonclassical* term in the molecular energy, the so-called **exchange energy,** and hence covalent bonding is a strictly quantum mechanical phenomenon; and (2) this exchange energy becomes important when the atomic orbitals are **overlapping.**

Linus Pauling generalized the findings of Heitler and London into a model for chemical bonding called **valence bond** (VB) theory. The model assumes that each bond in a molecule is approximately represented by a VB electron-pair wave function of a form identical to that given by Heitler and London, but using the appropriate valence orbitals of the bonding atoms. Bonding is possible only when (1) each orbital contains only one electron or one orbital two, the other none, and (2) the orbitals possess nonzero overlap. Pauling found that in this way Lewis structures could be made consistent with wave-mechanical bonding principles and, in addition, the geometries of molecules, bond lengths, and particularly bond angles, now could be given an interpretation in terms of the need for orbital overlap combined with the directional character of the p valence orbitals.

The requirement of orbital overlap for strong chemical bonding led Pauling to invent the concept of atomic hybridization to interpret the structure of **tetrahedral** molecules such as methane CH_4. Using ordinary $2p_C$–$1s_H$ overlap leads to an HCH angle of $90°$, while the tetrahedral HCH angle is $109.5°$. Pauling discovered how to form **linear combinations** of carbon's $2s$ and

2p orbitals to yield four identical, highly directional orbitals called sp^3 **hybrid orbitals** that point toward the vertices of a tetrahedron. These hybrids are ideally adapted to form four identical valence bonds by overlap with H $1s$, as observed for CH_4. For linear molecules with no lone pairs on the central atom, such as BeH_2, two atomic orbitals (AOs), $2s$ and the $2p$ that points along the molecular axis, are mixed to form two sp hybrids suitable for bonding. Molecules, such as BF_3, having a **trigonal-planar** geometry, require three sp^2 hybrids formed from $2s$ and the two $2p$ orbitals pointing in the plane of the four atoms.

The $2p$ orbitals unused in the sp- and sp^2-hybridization schemes provide a natural extension to molecules with double and triple bonds. In the ethylene molecule (C_2H_4), the C=C bond is described by overlap between sp^2 hybrids to form one bond, called a **σ bond,** and also between the unused $2p$ orbitals perpendicular to the molecular plane to form a second bond called a **π bond.** The orbitals involved in σ bonds are cylindrically symmetric around the bond axis, while those in π bonds change sign when rotated 180° around the axis. In acetylene C_2H_2 the C≡C bond is composed of one σ bond from sp–sp overlap and two π bonds from the pair of $2p$'s on each C perpendicular to the axis.

In diatomic molecules such as N_2, O_2, and F_2, hybridization need not be invoked to account for the bonding, though there continues to be π bonding in N≡N and O=O, and all electrons are spin-paired in either bonding or lone pair orbitals. For molecules with more than two heavy (nonhydrogenic) atoms bonded together, such as CO_2 or SO_2, the observed molecular geometry and/or the number of multiple bonds in the Lewis structure determines the hybridization. Only one double bond (σ and π components) dictates sp^2 hybridization for the atom in the center of a three-atom linkage, while two double or one triple bond indicates sp hybridization.

The **valence-shell electron pair repulsion** (VSEPR) model allows a rationalization of the observed bond angles in polyatomic molecules. Mutual repulsion among the electron pairs surrounding a central atom induces geometrical arrangements that minimize that repulsion. The linear, trigonal-planar, and tetrahedral geometries are dictated by repulsion among two, three, and four electron pairs, respectively. For atoms with expanded octets, five and six electron pairs correspond to **trigonal-bipyramidal** and **octahedral** geometries, respectively. Lone pairs are slightly more repulsive than bonding pairs, while π bonds influence the bond angles only slightly, accounting in a qualitative way for the slight deviations from ideal VSEPR geometries in lone pair and π-bonded molecules. The case of one or two lone pairs is ambiguous only in the five-pair trigonal-bipyramidal case, for which the lone pairs both occupy the trigonal (equatorial) rather than pyramidal (axial) positions. The actual geometry of the molecule in lone pair cases is a subclass of the electron pair geometry.

When two unlike atoms bond, the electrons are not shared equally, and the result is a nonzero **electric dipole moment** $\mu = \delta e r_e$ pointing toward the more electronegative atom in the bond. All **heteronuclear** diatomic molecules possess a dipole moment, while $\mu = 0$ for all **homonuclear** ones. The only polyatomic molecules that do not show a dipole are those in which identical possibly dipolar bonds are arranged symmetrically about a center. Whether μ is nonzero can be determined by analyzing the geometry of the molecule using VSEPR. An **ionic term** is added to the VB wave function to account for the partial ionic character of polar covalent bonds. The square of the coefficient of this extra term is identified with the charge separation δ, and both can be estimated from the electronegativity difference $\Delta \chi$ using the empirical relation of Equation 6.9.

Being based on Lewis structures, the VB-VSEPR model, though powerful in predicting and pictorializing molecular structure, suffers from poor accounting of odd-electron molecules and molecules with resonance forms. It also fails to account for the paramagnetism of O_2, and does no better than electron dots in describing the relation between structure and reactivity except in π-bonded systems. This motivates the examination of an alternative model of molecular electronic structure in the next chapter.

EXERCISES

Valence bond wave functions and orbital overlap

1. Write valence bond wave functions for the bond between (a) two Li atoms; (b) two F atoms; (c) Li and H atoms. In which case does the covalent VB wave function omit an important term?

2. Successful overlap using p orbitals depends on their orientation with respect to the covalent bond axis. By using symmetry arguments, show that the overlap integral S is identically zero for overlapping an s orbital on one atom with a p orbital on a neighboring atom that points perpendicular to the internuclear axis. (That is to say, σ-symmetry orbitals have zero overlap with π-symmetry orbitals.) It helps to make a sketch.

3. Compare the bond potential energy curve for H_2 given in Figure 6.2 with that for NaCl from Figure 5.5, listing any differences between them. How does the bond energy D_e arise in each case?

4. Use the neutral atomic radii of Figure 4.12 to estimate the bond lengths in HF and H_2O. (Experimental values are given in Table 6.1.) Are these radii larger or smaller than the radii of the atomic orbital boundary surfaces? (See Figure 6.3.) Why?

Atomic hybridization

5. Using the orbital functions of Table 3.2, construct the hybrid orbital functions of Figure 6.5. Plot one of the functions along the x axis to examine its lobes and nodal structure. (As described in the caption to Figure 6.5, note that you should use the negative of the $2s$ function; otherwise, the hybrids will be reversed.)

6. The geometry of a molecule and the hybridization required to rationalize it are determined by the number of σ bonds plus lone pairs around a particular atom, with the hybrids accommodating them. Using the summary of Figure 6.6, give the approximate bond angles and hybridization for the central atom in each of the following:

   a. CH_3Cl
   b. SiF_4
   c. AlH_4^-
   d. MgI_2
   e. BCl_3
   f. PH_3
   g. PO_4^{3-}

7. Recalling that a double bond consists of one σ and one π component, repeat Exercise 6 for the following molecules:

   a. CO_2
   b. NO_2^-
   c. NO_3^-
   d. SO_3
   e. O_3
   f. ClO_3^-

8. For the molecule methyl fluoride (CH_3F, fluoromethane), give valence bond wave functions describing the C—H bonds and the C—F bond, and sketch the overlapping orbitals for each.

9. For the formaldehyde (methanal) molecule CH_2O, give valence bond wave functions for the C—H bonds and the C=O bond, and sketch the overlapping and lone pair orbitals.

10. The active ingredient in vinegar is acetic acid (CH_3COOH, ethanoic acid). Draw the Lewis structure of acetic acid, then deduce the hybridization and approximate bond angles for the two C's and the nonterminal O. Give a valence bond wave function for the bond between C and the nonterminal O.

11. Give Lewis structures and hybridizations for the mineral acids H_2SO_4, HNO_3, and H_3PO_4.

12. Using the valence bond model with hybridization, explain why BF_3 is a strong Lewis acid.

Valence shell electron-pair repulsion (VSEPR)

13. Use the VSEPR model (see Figure 6.11) to predict and sketch the bond angles and geometry of the following molecules or ions:

    a. SCl_2
    b. SO_4^{2-}
    c. NO_3^-
    d. N_2O
    e. HCN
    f. I_3^-
    g. SF_6
    h. PO_3^{3-}
    i. IF_5

14. Repeat Exercise 13 for the following species:

    a. GeH_3Cl
    b. CO_3^{2-}
    c. S_2O
    d. PCl_3
    e. PCl_5
    f. NHF_2
    g. SF_4
    h. HgI_2
    i. XeF_4

15. For each of the following molecules, use VSEPR to predict the bond angles (and hybridization) of each atom bonded to more than one other atom, and sketch the molecular shape.

```
   O  H              H                  H
   ‖  |              |                  |
H—C—N—H          H—C—O—H            H—C=C=O
                    |
                    H

Formamide         Methanol            Ketene
  (a)               (b)                 (c)
```

```
   O                 H
   ‖                 |
H—C—O—H          H—C=N—H

Formic acid       Methylenimine
   (d)               (e)
```

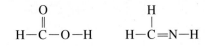

16. The molecule diimine (N_2H_2) exists in two geometrical isomers, in which the atoms are bonded in the same order but with different spatial arrangements. Use VSEPR to deduce the approximate geometries of the two isomers. What is the hybridization of each N atom?

17. Odd-electron molecules are difficult to treat using VSEPR, but often the molecular geometry may be estimated by finding the angles of the positive and negative ions of the molecule in question and interpolating. Use this method to estimate the bond angles of nitrogen dioxide NO_2 and the hydroperoxyl radical HO_2.

Dipole moments

18. Using the data of Table 6.1, calculate and compare the charge separation δ and the percent ionic character of NaCl and HCl. Check to see how well your results agree with values obtained from Equation 6.9, using data from Figure 5.6.

19. Repeat Exercise 18 for the molecules KBr and HBr. Draw a Lewis structure and indicate the direction of the dipole moment for each.

20. In Chapter 5 we gave a criterion for "pure ionic" bonding as an electronegativity difference $\Delta\chi$ of 1.7 or greater and "pure covalent" as $\Delta\chi = 0.4$ or less. Using Equation 6.9, find the charge separations corresponding to these two limits.

21. For a triatomic molecule with identical end atoms and a bond angle θ, use vector addition to show that the magnitude of the bond dipole μ_b is related to that of the overall molecular dipole μ by $\mu = 2\mu_b \cos(\theta/2)$. Use this relation along with the data of Table 6.1 to find the bond dipoles and partial charges δ for the following molecules. Give a Lewis structure and the directions of the bond dipoles and overall dipole in each case.

    a. H_2O
    b. H_2S
    c. NO_2
    d. SO_2

22. Deduce which of the following molecules have nonzero dipole moments:

    a. GaH_3
    b. PH_3
    c. SiF_4
    d. SF_4
    e. $SeCl_2$
    f. $BeCl_2$
    g. $SbCl_5$
    h. IF_5

23. Arrange the following molecules in order of increasing dipole moment: HOCl, H_2O, HOF, Cl_2O.

24. Find the partial charge δ on the halogen atom for CH_3F and CH_3I by assuming the entire molecular dipole is due to the C—X bond dipole. Use data from Table 6.1.

25. Give a bond wave function analogous to Equation 6.8 for the HCl molecule, and estimate the coefficient b quantifying the ionic contribution using any convenient method.

Orbitals and Chemical Bonding II: The Molecular Orbital Model and Molecular Energy Levels

CHAPTER 7

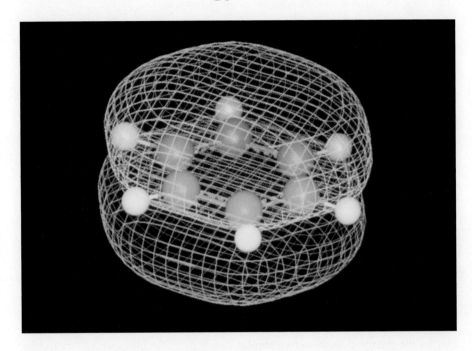

The lowest energy π orbital in benzene C_6H_6. See Section 7.4 for a discussion.

We can ask whether Graham Greene is in London, but we cannot ask whether electron A is at point B.

Walter J. Moore (1962)

Describing a chemical bond in terms of a Lewis electron-dot structure and its quantum-mechanical successor, valence bond (VB) theory, while very useful and predictive, does have its drawbacks, as you have seen. The problem of molecular radicals, the unreality of resonance forms, and the inexplicable paramagnetism of molecular oxygen, as well as the need for a better structure–reactivity picture, motivated the search for an alternative theory of bonding. **Molecular spectroscopy,** to be described in the next chapter, was in the 1920s and 1930s revealing new details of **molecular electronic energy levels,** between which *single* electrons were undergoing transitions, a process not easily described by a theory based on electron pairs. Robert Mulliken, whom we've mentioned earlier in connection with atomic orbitals and electronegativity, was a physicist by training, and the leading figure in the development of an orbital theory of molecular structure now known as **molecular orbital theory,** or MO theory for short. As we'll describe shortly, Mulliken found that Lewis structures are

CHAPTER OUTLINE

7.1 The Hydrogen Molecular Ion: The Prototype Molecular Orbital System

7.2 The Molecular Orbital Description of Diatomic Molecules

7.3 Molecular Orbitals of the Water Molecule

7.4 Delocalized Bonding: A Solution to the Resonance Problem

7.5 Shortcomings of Molecular Orbitals

7.6 Valence Bonds, Molecular Orbitals, and Quantum Chemistry

205

Robert S. Mulliken (1896–1986). Together with Friedrich Hund (of Hund's rule fame) he developed the molecular orbital theory in the late 1920s. His work fostered a new understanding of molecular electronic spectroscopy, but was slow to be taken up by chemists.

an inadequate starting point for developing such a theory, and in addition the theory rapidly grows in complexity when going beyond the simplest diatomic molecules. Thus, although MO theory developed nearly in parallel with VB theory, it didn't "catch on" with most chemists, who were interested in describing bigger molecules and complex chemistry, and were definitely *not* interested in giving up that redoubtable teaching and research tool, the Lewis electron dot. It is only in more recent times, since roughly 1970, that MO theory has found its way into freshman courses and textbooks such as this one, and even today there is some resistance in the chemical community to exposing you, the budding chemist (perchance!), to the "new" ideas contained in the theory. There has also been a tendency in modern texts as well as in research to combine certain aspects of MO theory with VB and Lewis structures, in order to patch up some of the difficulties of the latter without having to abandon it altogether. We will discuss this further at chapter's end. For now, it seems best that you see the logic behind this decidedly different yet already familiar approach to chemical theory. Then you will better appreciate the attempt to marry the VB and MO approaches.

7.1 The Hydrogen Molecular Ion: The Prototype Molecular Orbital System

Recall the great success of the atomic orbital (AO) theory in correlating the structure of the periodic table with the orbital structure of the atoms that comprise it. Half the atoms in the table have an odd number of electrons, yet this poses no particular problem for the theory. If there were a similar theory for molecules, perhaps molecular radicals would also be accommodated and, further, perhaps some of the other problems of the Lewis–VB model could be attacked.

Now, you can think of a molecule as an atom with a fragmented nucleus, or, in the other direction, that every molecule corresponds to an atom, the so-called *united atom*, that would result if the nuclei in the molecule were fused together. The task then is to construct the *one-electron orbitals* corresponding to the arrangement of nuclear charges presented by a particular molecule. Think of this problem as similar to a "pitch the ball and win a prize" booth at a carnival. The nuclei are our targets, and we will "throw" electrons at them and see where they stick. Such a program requires a model system as a guide, because, as with atoms, orbitals for many-electron molecules will be approximate. The H atom was the paradigm for atoms; correspondingly, we select the *simplest one-electron molecule* H_2^+ as the paradigm for MO theory. Figure 7.1 shows the geometry of the three-particle H_2^+ problem. Note that, unlike the case for H_2, the Coulomb potential does not contain the troublesome e^2/r_{12} electron–electron repulsion term, since there is only one electron. In 1927 O. Burrau showed that the Schrödinger equation for H_2^+ could be separated and solved exactly for any fixed distance r between the protons. In this case the French mathematicians had not already solved the separate differential equations, so Burrau solved them numerically (without the aid of a computer, of course). The results were in perfect agreement with spectroscopic evidence on the ground state of H_2^+: bond length $r_e = 1.06$ Å and bond energy (to dissociate the molecular ion to $H^+ + H$) $D_e = 2.77$ eV.

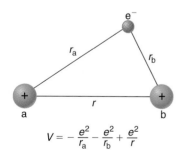

Figure 7.1
Coulomb interactions in the H_2^+ problem. Note the simplicity of this problem compared to H_2 (Figure 6.1).

7.1 The Hydrogen Molecular Ion: The Prototype Molecular Orbital System

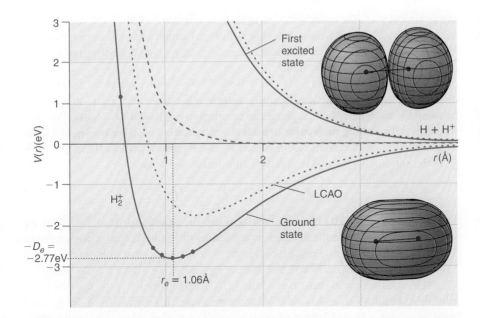

Figure 7.2
Potential energy curves and wave function boundary surfaces for the H_2^+ molecule. The solid curves are derived from a combination of experiment and exact solution of the Schrödinger equation for H_2^+. The points on the ground-state curve are the original 1927 calculations of Burrau. The short-dashed curves are the result of the simplest LCAO-MO approximation, and the long-dashed curve is the result of forcing the electron to remain localized on one of the H nuclei.

Figure 7.2 gives the bond potential energy curve for H_2^+. It is important to know that, for a case like this, in which the Schrödinger equation can be solved exactly, the results agree with experiment, but for our purposes it is also critical to know the wave functions, the molecular orbitals for this simplest of all molecules. Since there is only one electron, the molecular orbital here is a well-defined concept, but of course it depends on the internuclear distance r. Normally, though, we will only be interested in what it looks like when $r = r_e$, the equilibrium bond length.

Figure 7.2 also displays boundary surfaces for the two lowest MOs corresponding to the energies at $r = r_e$. Both of them are σ-type orbitals, *the lower energy orbital, corresponding to a stable H_2^+ molecule, being nodeless*, and *the higher one, the lowest excited state, corresponding to* the repulsive potential energy curve in Figure 7.2 and therefore *an unstable molecule, with a node midway between the nuclei*. The appearance of these MOs may be understood very simply by analogy with the 1-D particle-in-a-box wave functions (see Appendix B) for the two lowest energy states. The H_2^+ orbitals are of course three dimensional, but their gross structure along the bond axis can be thought of in a 1-D context. We will exploit this analogy further below.

Linear Combinations of Atomic Orbitals

Enter Robert Mulliken. Examining the orbitals of Figure 7.2, Mulliken saw how they could be readily approximated by a **linear combination of atomic orbitals**—LCAO—as follows. The lowest orbital, whose occupation leads to molecule formation and is therefore termed a **bonding molecular orbital**, may be emulated by simply *summing* the 1s orbitals on each H atom (labeled a and b):

$$\psi_1 \equiv \sigma_{1s} \approx 1s_a(\mathbf{r}) + 1s_b(\mathbf{r}) \tag{7.1}$$

where **r** represents the coordinates of the single electron.* Adding the orbitals produces *constructive interference*—buildup of orbital amplitude—between the nuclei, already familiar to you as *orbital overlap,* and we may say that an electron "in" this orbital is *shared* between the atoms.

Subtraction of the two $1s$ AOs, on the other hand, approximates the first excited orbital, with the orbitals *destructively interfering*—canceling each other out—between the nuclei. This nodal plane can be taken to imply that an electron "in" this so-called **antibonding molecular orbital** is *not* shared between the nuclei, and leads to dissociation of the atoms, as indicated by the repulsive potential energy curve. With the antibonding orbital denoted by an asterisk, we write

$$\psi_2 \equiv \sigma_{1s}^* \approx 1s_a(\mathbf{r}) - 1s_b(\mathbf{r}) \tag{7.2}$$

The resulting LCAO-MO boundary surfaces are illustrated in Figure 7.3. They are very similar, but not identical, to the exact results of Figure 7.2; as a consequence, they yield a bond potential energy curve, the short-dashed curve in Figure 7.2, with a somewhat smaller D_e, 1.77 eV, and a larger r_e, 1.32 Å.

Note that the approximate part of the LCAO-MO approach is the use of atomic orbitals to build MOs, and not the MOs themselves. This differs from the valence bond treatment of H_2 in Chapter 6 in two ways: (1) the H_2^+ problem is a simpler problem than H_2, allowing the exact solution of the Schrödinger equation; and (2) in the

Figure 7.3
Energy levels and molecular orbitals arising from linear combination of $1s$ atomic orbitals (LCAO). The energy-level diagram represents two points taken from the potential curves of Figure 7.2, the separate atomic orbitals at $r \to \infty$ and the equilibrium distance r_e.

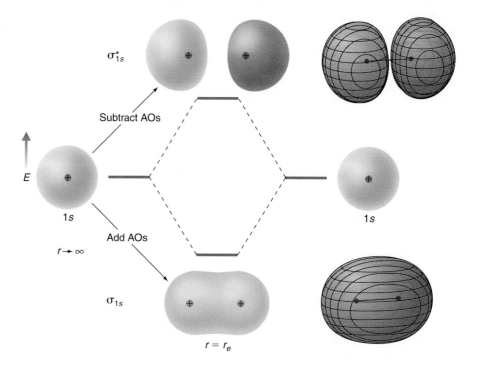

* For simplicity, we are using unnormalized LCAO-MOs.

VB method atomic orbitals are built in from the beginning. Of course, when we move on to many-electron molecules, the MOs themselves become approximations, just as AOs are for many-electron atoms.

Why Covalent Bonds Form

In Chapter 6 we noted that covalent bond formation is a quantum mechanical phenomenon, described in the Heitler–London valence bond method by the exchange energy and atomic orbital overlap. However, if we want a deeper understanding of covalent bonding, we must abandon any description that relies on atomic orbitals, since, according to the MO way of thinking, *atomic orbitals cease to exist* as well-defined quantities *once atoms bond together*. In the exact solution of the Schrödinger equation for H_2^+, there are only the whole-molecule wave functions, the molecular orbitals.

An explanation of the covalent bond must therefore be couched in more basic quantum-mechanical terms. Figure 7.4 illustrates the potential energy in two dimensions (see Figures 1.10 and 2.12) of an electron in the field of one proton and of two. The two-proton curve (H_2^+) is shifted upward to account for Coulomb repulsion between the protons, and its bicuspid appearance is due to the two "Coulomb holes." The ground-state or zero point energy (ZPE) is indicated in each case, as obtained from exact solution of the wave equation. The ZPE drops in response to the wider potential well in H_2^+. Recall our discussion in Chapter 3 of how the uncertainty principle leads to zero point kinetic energy due to the confinement of a particle. In H_2^+ the electron is less confined along the bond axis. Thus, for the most part, *chemical bonds are formed due to a drop in zero point energy;* a molecule is a "bigger box" than an atom, and electrons that are able to occupy its lowest energy level stabilize it against dissociation.

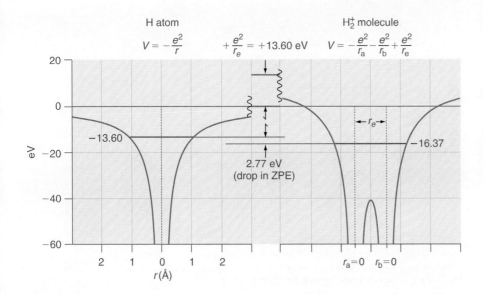

Figure 7.4

Two-dimensional Coulomb potential energies for an electron in the field of a single proton (H atom) and of two protons at fixed distance $r = r_e$ (H_2^+ molecule). In the molecule case the plot gives the potential energy along the bond axis, adjusted for proton–proton repulsion as indicated in the center. The wider Coulomb well for H_2^+ causes the zero point energy (ZPE) to drop and gives rise to bond formation. See Figures 1.10 and 2.12 for details of 2-D potentials and energies in the H atom.

In this view of bonding, the overlap and exchange ideas from the VB model are artifacts of assuming that the atomic orbitals still exist in the molecule. In so far as the LCAO approximation is valid, atomic overlap ensues only after the nuclei are brought together closely enough that the "Coulomb hill" between the nuclei (see Figure 7.4) is reduced to the point that the electron is free to migrate between them. Hence exchange, a concept founded on atomic orbitals, is just a way of describing the **delocalization** of the electron that causes the bond to form. It is this idea of delocalization of electrons into a larger space as a cause of bonding that supplants exchange in the MO model.

This doesn't mean that we must discard the overlap idea. In LCAO, the overlap signals the onset of delocalization. As Figure 7.5 illustrates, when the σ_{1s} wave function of Equation 7.1 is squared to find the electron density, the cross-term $2[1s_a(\mathbf{r})1s_b(\mathbf{r})]$ produces a buildup of electronic charge in the region between the nuclei. This buildup is over and above that which results from the squares of the individual 1s orbitals, and is sometimes called the *overlap charge*. If the σ_{1s}^* orbital of Equation 7.2 is analyzed in the same way, the complete depletion of the internuclear electron density expected from the boundary surface (Figures 7.2 and 7.3) is observed.

Within the LCAO approximation, we can also try an informative "computational experiment," by freezing the electron in one of the H atom orbitals, that is, by not allowing it to delocalize. If the energy is calculated with this restriction, the curve in Figure 7.2 that lies in the middle of the others, without a well in it, is obtained. As for the analogous VB trial function Equation 6.2, this indicates the essential role in bonding played by the electron's visiting both nuclei.

Figure 7.5
Electron density along the bond axis in H_2^+. As you may recall from Chapter 3, densities are obtained by squaring wave functions. Comparison of the bonding and antibonding densities with the sum of the atomic densities shows the buildup and depletion of interatomic electron density, respectively, between the nuclei. The amplitudes of the MOs have been scaled to emphasize these differences.

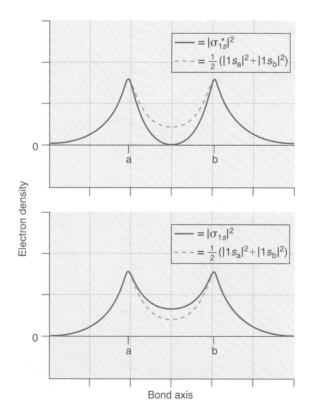

7.2 The Molecular Orbital Description of Diatomic Molecules

The lowest two MOs for H_2^+, along with the LCAO prescription for approximating them, allowed Mulliken (and allows us as well) to construct a general orbital theory for molecules. To begin, we examine the molecules that can be described using only $1s$ orbitals, that is, molecules that employ only σ_{1s} and σ_{1s}^* MOs. At first we restrict our discussion to homonuclear diatomic molecules, leaving heteronuclear diatomics until later in this section. Thus, without further discussion we can take a look at H_2^+, H_2, He_2^+, and He_2.

The following energy-level diagrams show the occupancy of the σ_{1s} MOs for these four diatomics, which contain 1, 2, 3, and 4 valence electrons respectively.

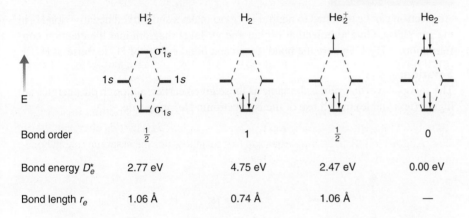

	H_2^+	H_2	He_2^+	He_2
Bond order	$\frac{1}{2}$	1	$\frac{1}{2}$	0
Bond energy D_e^*	2.77 eV	4.75 eV	2.47 eV	0.00 eV
Bond length r_e	1.06 Å	0.74 Å	1.06 Å	—

The MOs are filled in accordance with the Aufbau and Pauli principles just as for atoms (see Chapter 4); addition of a second electron into the bonding MO (giving a pair of "bonding" electrons) strengthens and shortens the H_2 bond compared to H_2^+. When a third electron is added, as in He_2^+, however, the bond energy and bond length revert to values close to those in H_2^+. This destabilization is clearly due to the antibonding electron, which effectively cancels the bonding influence of one of the bonding electrons. We therefore define a **bond order,** analogous to the Lewis idea, as

$$\text{Bond order} = \tfrac{1}{2}(n_{\text{bonding}} - n_{\text{antibonding}}) \tag{7.3}$$

where the n's are total numbers of bonding and antibonding electrons. MO theory thus predicts "half-bonds" for H_2^+ and He_2^+, as well as a full bond for H_2. The case of He_2 is of particular importance; He is already a noble gas, and Lewis would predict no bond between neutral heliums. In MO theory there are an equal number of bonding and antibonding electrons, giving a bond order of zero; the lack of bonding is the result of a *cancellation* of stabilizing and destabilizing effects.

---

* Experimental values from molecular electronic spectroscopy. Simple LCAO-MO theory predicts smaller D_e's and larger r_e's for H_2 and He_2^+, just as we've seen for H_2^+, but the relative values show the same relationships.

Just as in atoms, the actual molecular orbital energies are different for different molecules, and it is our knowledge of the *ordering* of the levels that allows us to "fill" them in the Aufbau order. As for atoms, we can then write molecular orbital electronic configurations as

$$\begin{aligned} H_2^+ &\quad (\sigma_{1s})^1 \\ H_2 &\quad (\sigma_{1s})^2 \\ He_2^+ &\quad (\sigma_{1s})^2(\sigma_{1s}^*)^1 \\ He_2 &\quad (\sigma_{1s})^2(\sigma_{1s}^*)^2 \end{aligned}$$

where, as before, the exponent denotes orbital "occupancy."

EXAMPLE 7.1

An electron can be attached to neutral $H_2(g)$ to make a short-lived negative ion H_2^- in the gas phase. Give a molecular orbital energy-level diagram and the electron configuration of H_2^-. Compare the bond length and bond energy of H_2^- to those of H_2.

Solution:
The energy-level diagram is the same as those given earlier, although the energies are higher (less stable) than those of the isoelectronic He_2^+ molecule:

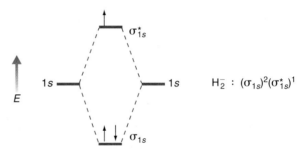

The bond order is then $\frac{1}{2}(2 - 1) = \frac{1}{2}$ from Equation 7.3. The bond is therefore longer $[r_e(H_2^-) > r_e(H_2)]$ and weaker $[D_e(H_2^-) < D_e(H_2)]$ than the H_2 bond. (Recent MO calculations give $D_e = 1.0$ eV and $r_e = 0.8$ Å; the rise in the σ and σ^* energy levels makes H_2^- unstable with respect to electron detachment.)

Homonuclear Diatomic Molecules from Second-Row Atoms

Of course, we want and need to go beyond the molecules of the first-row atoms. Mulliken's LCAO method gives us a facile way of doing this. For each pair of AOs we combine, a pair of MOs should result, one of bonding and one of antibonding character.

We don't expect the core electrons to contribute to bonding; for the second row of the periodic chart these electrons occupy very tight, tiny $1s$ AOs. These orbitals will be too small to overlap at normal bonding distances (1–2 Å), and even if they did, the fact that they are fully occupied means that bonding and antibonding contributions will cancel, as in He_2.

We can therefore focus (as you might have guessed!) on the valence AOs. In making combinations among s and p orbitals with different energies, we use two rules developed by Mulliken:

I. Only orbitals with the same symmetry along the bond axis (most often σ or π) can be combined. This arises because the overlap integral between a σ and a π AO is zero.

II. Orbitals of the same or similar energies will interact more strongly than those of widely different energy.

Using these rules, while recalling the screening-induced splitting between s and p orbitals (see Figure 4.7), suggests that $2s$ on one atom will combine mainly with $2s$ on another, $2p_\sigma$ with $2p_\sigma$, and $2p_\pi$ with $2p_\pi$, with each of these generating a bonding and antibonding MO. Figure 7.6 illustrates the MOs that result from these AO combinations. Note that eight MOs are generated from this pairwise combination of the four valence orbitals on each atom, and that the operation applies not only to all second-row homonuclear diatomics but also to molecules from the third-row representative elements and beyond.

To begin filling these orbitals with electrons, we need to know their (relative) energies. We do know that, for a given AO combination, the bonding orbital is always lower in energy because it represents constructive overlap, but how do we know the relative ordering of, say, the σ_{2s}^* and σ_{2p} MOs? The best answer to that question lies in the analysis of experimental electronic spectra—far beyond what we are aiming for here—but we can make a simple but powerful analogy with the particle-in-a-box problem of Appendix B. In Figure 7.7 the σ- and π-symmetry MOs are separated and compared with the "box" wave functions. A general property of these and other wave

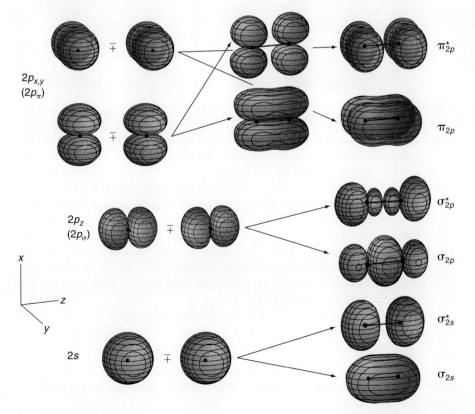

Figure 7.6
LCAO-MOs for the $n = 2$ valence shell homonuclear diatomic molecules. Each pair of atomic orbitals gives a bonding and antibonding MO. Note the nodes midway between the nuclei (heavy dots) for the antibonding orbitals (i.e., those denoted by an asterisk). Similar boundary surfaces are obtained for the later periods.

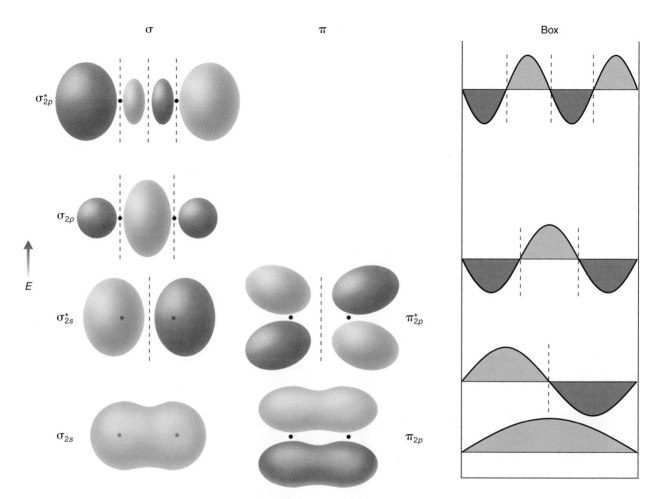

Figure 7.7
Determination of the energy ordering of $n = 2$ valence MOs by analogy with the particle-in-a-box wave functions (see Appendix B). Level order is given by the number of bond-axis nodes. This scheme does not establish the relative ordering of σ and π.

functions for bound states (see Chapter 3) is an increase in the number of nodes by one for each new energy level. Both σ and π MOs are expected to follow this rule, implying that the correct ordering of the orbitals is $\sigma_{2s}, \sigma_{2s}^*, \sigma_{2p}, \sigma_{2p}^*$ for the σ group and π_{2p}, π_{2p}^* (as expected) for the π group, with there being a degenerate pair of π and of π^* orbitals. For the ordering of σ and π with respect to each other, experimental results must be invoked. These show that π_{2p} lies lower than σ_{2p}, and π_{2p}^* lower than σ_{2p}^*, for all the second-row molecules lighter than O_2, whereas for O_2 and F_2, σ_{2p} lies lower than π_{2p}.* To avoid confusion, we present just one energy-level diagram, Figure 7.8, appropriate for the lighter diatomics. In O_2 and F_2 the σ–π orbital reversal does not affect the bonding properties, although you could not use Figure 7.8 to analyze electronic transitions in these species.

Armed with the orbitals and energy-level ordering, we are now in a position similar to that for many-electron atoms (see Chapter 4). Bear in mind that not only are

---

*Other diatomic molecules composed of atoms in Groups VI and VII also show this σ–π reversal.

7.2 The Molecular Orbital Description of Diatomic Molecules

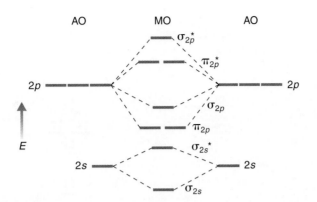

Figure 7.8
MO energy levels for second-row homonuclear diatomic molecules lighter than O_2. For O_2 and F_2, the ordering of the σ_{2p} and π_{2p} orbitals is reversed. To obtain electron configurations, valence electrons are placed in the lowest energy orbitals, consistent with the Pauli exclusion principle and Hund's rule. Molecules from later periods show similar valence MO levels.

we using LCAO to form MOs, but that the MOs themselves are only approximations to the true, many-electron wave functions in bigger molecules. We nonetheless expect much of them, since AOs allowed such a detailed understanding of atomic behavior. We now examine some of the second-row homonuclear molecules to see what advantages the MO view may offer.

In Li_2, there are only two valence electrons, and they both are placed, spins paired, into the σ_{2s} bonding MO, giving a single Li—Li bond (just as Lewis had predicted) and a valence electron configuration $(\sigma_{2s})^2$. Skipping over the less common (but nonetheless by now well studied) molecules Be_2, B_2, and C_2, molecular N_2 brings a total of 10 valence electrons, 5 from each atom, to fill the MOs. The Aufbau valence electron configuration for N_2 is therefore $(\sigma_{2s})^2(\sigma_{2s}^*)^2(\pi_{2p})^4(\sigma_{2p})^2$, and the bond order is $\frac{1}{2}(8-2) = 3$, again in agreement with the Lewis and valence bond pictures. There is one important difference, however. The Lewis and VB models for N_2 show a lone pair of electrons on each N, :N≡N:, while in the MO picture there are canceling bonding and antibonding σ_{2s} pairs. This means that these electrons, although not making a net contribution to the bond, are nonetheless delocalized; that is, they cannot be associated with just one of the N atoms. This seems a more natural state of affairs: electrons don't want to be "cooped up," because this would raise their energy according to the uncertainty principle. (In addition, spectroscopic evidence points to the existence of two distinct σ_{2s} energy levels; the VB lone pairs would presumably be degenerate.)

The first striking divergence between VB and MO predictions occurs for O_2. Recall our discussion of the paramagnetism of O_2 at the end of Chapter 6, and the failure of VB theory to account for it. O_2 has 12 valence electrons and therefore the electron configuration $(\sigma_{2s})^2(\sigma_{2s}^*)^2(\sigma_{2p})^2(\pi_{2p})^4(\pi_{2p}^*)^2$. Note that we have interchanged the σ_{2p} and π_{2p} levels (as required by experiment; see earlier discussion). Now the π_{2p}^* level consists of two degenerate orbitals and, as in degenerate atomic orbitals, electrons will tend to occupy separate MOs with their spins aligned according to Hund's rule. Therefore, O_2 has two unpaired electrons, one in $\pi_{2p_x}^*$ and the other in $\pi_{2p_y}^*$, and is predicted to be paramagnetic, in agreement with a long and indubitable succession of observations.

The publication of this result for O_2 by Mulliken in 1932 made the world of chemistry take notice; perhaps this success meant that some of the ideas behind MO theory might be useful—perhaps even correct. More on this facet later in this

Figure 7.9
Molecular orbital occupation diagrams and bonding properties for the four most common second-row homonuclear diatomic molecules. Note that the σ_{2p} and π_{2p} levels switch order for O_2 and F_2. The bond orders are the same as would be obtained from the Lewis structures (see Chapter 5), although they arise from partial cancellation of bonding and antibonding electrons except for Li_2. On an absolute energy scale, the orbital energies drop and the $2s$–$2p$ energy gap increases from left to right. Can you explain why the bond lengths of Li_2 and F_2 are so different despite their identical bond orders and similar bond energies?

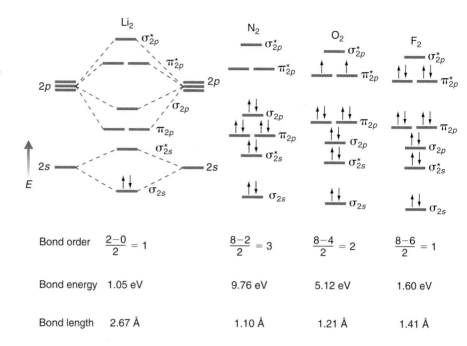

chapter. Agreement persists, nonetheless, between the MO and VB theories as to the bond order of O_2: both predict 2.

The MO configuration of molecular fluorine F_2, 14 valence electrons, is $(\sigma_{2s})^2(\sigma_{2s}^*)^2(\sigma_{2p})^2(\pi_{2p})^4(\pi_{2p}^*)^4$. Here the π^* orbital is completely filled, giving a diamagnetic molecule with a net single bond, in agreement with Lewis and VB theories and consistent with experiment. Delocalized nonbonding electrons continue to replace the lone pairs, however.

At the end of the second row of the periodic table, we encounter Ne_2. Under special experimental conditions Ne_2 can be formed, but it is found to have a "bond energy" of only 0.01 eV, and is known to be held together by forces other than covalent bonding (see Chapter 14). The MO electron configuration fills the $n = 2$ valence orbitals completely, $(\sigma_{2s})^2(\sigma_{2s}^*)^2(\sigma_{2p})^2(\pi_{2p})^4(\pi_{2p}^*)^4(\sigma_{2p}^*)^2$, and as in He_2, there is no net bonding.

Figure 7.9 summarizes the discussion of the Li_2, N_2, O_2, and F_2 molecules with orbital occupation diagrams.

EXAMPLE 7.2

The bonding properties of B_2, diboron, were first measured in electronic spectroscopy by Herzberg in 1940. Use MO theory to predict the bond order and magnetic properties of B_2.

Solution:
The Aufbau valence electron configuration is $(\sigma_{2s})^2(\sigma_{2s}^*)^2(\pi_{2p})^2$, accommodating the six valence electrons. Using the orbital occupation diagram of Figure 7.8 for the early molecules of the second period, we find two unpaired electrons:

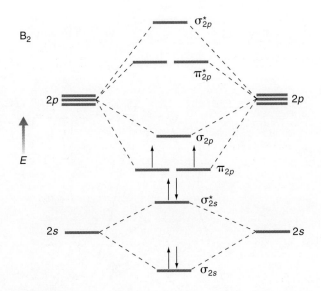

Thus, B_2 is predicted to have a bond order of 1 (a single π bond) and to be paramagnetic; both of these are in accord with experiment. (B_2 is electron deficient, so no valid Lewis structure can be drawn; a single-bonded structure, :B—B:, would appear to be diamagnetic, and VB theory would predict a σ bond, not π.)

Homonuclear molecules of the third and higher periods are easily accommodated by replacing the AOs $2s$, $2p$ with ns, np. The bonding description remains the same, although the bonds tend to be weaker and longer due to the more diffuse orbitals of the heavier atoms, leading to weaker overlap, and to their larger atomic radii. Similarities within a chemical family are thus maintained, as first established by the Lewis dot descriptions. Figure 7.10 illustrates how the bond energies of the homonuclear diatomics correlate with the bond order predicted by MO theory. Although Lewis–VB theory can also produce the Group I, V, VI, VII, and noble gas correlations, Groups II, III, and IV represent a notable advantage to the MO method. MO theory agrees with VB theory in cases where the latter is successful, and breaks with it to retain agreement with observation where it fails. It is easy to pinpoint the reason for the successful prediction of bond orders and magnetic properties in MO theory: it is *the existence of antibonding orbitals,* as required by quantum rules regarding nodal structure and energy ordering.

Heteronuclear Diatomic Molecules

Mulliken's LCAO method can be applied equally well to heteronuclear molecules. The atomic energy levels no longer match up in a case such as CO, say, but we can expect to be able to combine the $2s$ and $2p$ orbitals in the same way as the homonuclear cases, since the same nodal structure in the σ- and π-orbital manifolds must

Figure 7.10
The MO valence energy-level scheme of Figure 7.8 leads to definite predictions of bond orders for the diatomic molecules of all representative elements. This diagram shows how the predicted orders correlate with experimental bond energies.

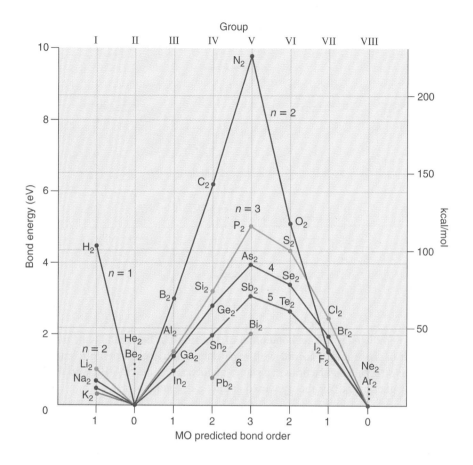

appear. Figure 7.11 illustrates the situation for a general AB molecule where B is more electronegative than A. Because the atomic orbitals have different energies, however, they do not enter the linear combination with equal weight. So we write, for the CO σ_{2s} orbitals, for example,

$$\sigma_{2s} = a\, 2s_C + b\, 2s_O$$
$$\sigma_{2s}^* = b\, 2s_C - a\, 2s_O \qquad (7.4)$$

where $b > a$, giving a greater contribution to the bonding MO from O, the more electronegative atom.*

Giving the greater weight to the orbital of lower energy ensures that an electron will "spend most of its time" in the "more comfortable" orbital. In combining atomic orbitals of different energy, the bonding MO energy will necessarily be closer to that of the lower energy AO, and the coefficients reflect this. The antibonding MO, on the other hand, gets stuck with the less stable AO as the major contributor, since its energy is closer to that of the less electronegative atom. (Note how the coefficients

---

* Coefficients a and b are obtained from self-consistent-field solution of the molecular Schrödinger equation (see Chapter 4 for the atomic case), to be discussed briefly at the end of this chapter.

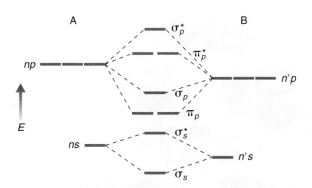

Figure 7.11
MO energy levels for heteronuclear diatomic molecules, given for a hypothetical AB molecule with $\chi_B > \chi_A$. A and B may be from different periods. The ordering of the levels is identical to that of Figure 7.8 despite the disparity in the atomic levels. (The ordering of the σ_p and π_p levels reverses when A and B are from Groups VI and/or VII.)

switch for the σ^* orbital.) As long as we restrict ourselves to pairs of orbitals combining, these results hold in general for all valence MOs. Figure 7.12 illustrates the "lopsided" MOs that result from the unequal orbital coefficients.

Despite the weird-looking orbitals, the MO electron configurations come out for the most part as in the homonuclear case, as Figure 7.11 indicates. Thus, for example, the configuration of CO is $(\sigma_{2s})^2(\sigma_{2s}^*)^2(\pi_{2p})^4(\sigma_{2p})^2$, the same as the isoelectronic N_2. Odd-electron molecules present no special difficulty, as the next example shows.

EXAMPLE 7.3

For the widespread air pollutant NO, nitric oxide, analyze the bonding using MO theory and sketch the orbital occupied by the unpaired electron.

Solution:

With 11 valence electrons, NO's configuration is $(\sigma_{2s})^2(\sigma_{2s}^*)^2(\sigma_{2p})^2(\pi_{2p})^4(\pi_{2p}^*)^1$. (Note the σ–π reversal as in O_2.) The bond order is therefore $\frac{1}{2}(8 - 3) = 2\frac{1}{2}$. (As mentioned in Chapter 5, this is consistent with the bond energy of NO, which falls between O_2, bond order 2, and N_2, bond order 3.) The highest occupied MO (HOMO) is π_{2p}^*, where the unpaired electron must reside by the Aufbau principle. An antibonding orbital always shows greater amplitude on the less electronegative atom, so

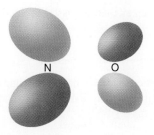

represents π_{2p}^* (see Figure 7.12).

Diatomic hydrides are a special case, since the H atom lacks p orbitals in its valence shell. The H 1s orbital can combine only with s or p_σ orbitals on the partner atom; the p_π orbitals are then nonbonding MOs, truly lone pair orbitals.

Figure 7.12
MO boundary surfaces for heteronuclear diatomic molecules AB with $\chi_B > \chi_A$. These correspond to the energy levels of Figure 7.11. The bonding orbitals are lopsided in favor of B, the antibonding in favor of A.

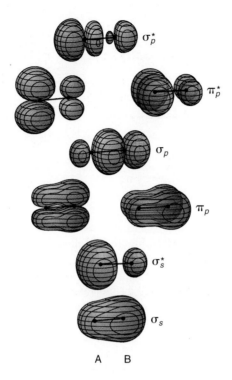

Figure 7.13 illustrates the LCAO scheme for the HF molecule. Fluorine is considerably more electronegative than hydrogen, so even its $2p$ orbitals lie much lower than H $1s$. The $1s - 2s$ energy mismatch is so great that Mulliken's rule applies, and these orbitals interact only weakly. Thus, for the most part, the HF bond arises from the $1s - 2p_\sigma$ overlap, which gives a σ bonding MO that looks like a slightly lopsided $2p$. The overall picture yields a single σ bond and three pairs of electrons localized on F, in accord with the Lewis–VB picture. The partial ionic character and the Coulomb contribution to the bond energy are already contained in the MO coefficients a and b,

$$\sigma_{2p} = a\ 1s_H + b\ 2p_F$$

with $b > a$; modifications such as that discussed in Section 6.5 are therefore not required in the MO picture.

EXAMPLE 7.4

The CH radical has been detected in interstellar clouds, and may be a precursor to larger organic molecules also found there. Analyze the bonding and magnetic properties of CH.

Solution:
The energy-level diagram is much like that of Figure 7.13, except that the $1s_H$ and $2p_C$ orbitals are more nearly equal in energy, since C is only slightly more electronegative than H. The orbital configuration can be written $(\sigma_{2s})^2 (\sigma_{2p})^2 (n\pi_{2p})^1$ to account for the five valence electrons. The single unpaired electron, which makes CH paramagnetic, is in the lone pair $n\pi$ orbital, and there is again a single σ bond. (A corresponding Lewis structure would be :C—H.)

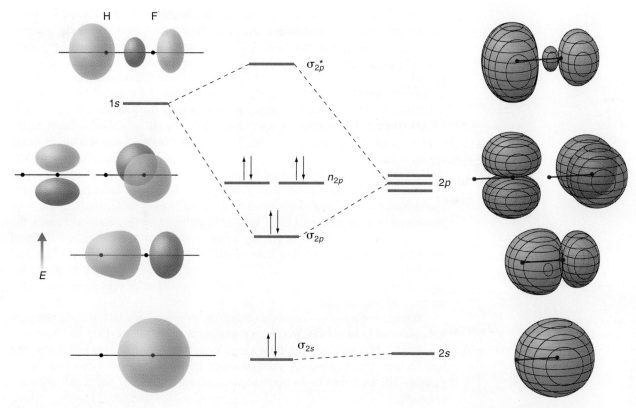

Figure 7.13
LCAO-MO energy levels and boundary surfaces for the HF molecule. This marks the first occurrence of true lone pairs, the nonbonding "n" $2p$ electrons on F, in our MO survey. A similar level scheme, with adjustment of the right-hand atomic energy levels, holds for all diatomic hydrides with representative elements.

In the limit of a nearly completely ionic bond such as in LiF, the valence energy levels on the two atoms are so widely separated that very little interaction can occur between them, and the bonding is entirely electrostatic; the electron originally on Li must be paired in the lower orbital ($2p$) associated with the F atom, effectively a redox electron transfer. Here again we find a concurrence between the MO and VB theories in the ionic case, as well as in the idea that the bonding passes from a purely quantum effect in covalent molecules to purely classical electrostatic in the ionic ones.

Acidity, Basicity, and the Frontier Orbitals

How does MO theory help us in our quest for the structural origins of chemical reactivity? We can already give the broad outlines of a structure–reactivity model based on the diatomic examples thus far treated. It was first suggested by Fukui in 1952, and later elaborated by Woodward and Hoffmann in the early 1960s, that we should focus on two orbitals in a molecule in particular, namely, the **highest energy occupied** MO (HOMO), and the **lowest energy unoccupied** MO (LUMO). These orbitals are generally out on the "edges" of the molecule, both energetically and spatially, and have been dubbed the **frontier orbitals.**

Let's use HF as an example of the frontier MO model. Refer to Figure 7.13 and pick out the HOMO and LUMO. The HOMO consists of the doubly degenerate $n(\pi)$ orbitals localized on F, and the LUMO is the two-node antibonding σ_{2p}^* orbital, with most of its amplitude on H. As we discussed in Chapter 5, HF is classified by its reaction with water as an acid. In the Lewis acid–base model, an acid is an electron pair acceptor. The problem we had in the Lewis–VB structure of HF was *where to put the electron pair,* seeing that HF is already a "good" Lewis molecule satisfying the noble gas rule. MO theory has a natural Aufbau-style answer to this question: the next available empty orbital, the LUMO! As illustrated in Figure 7.13, the σ^* HF MO has its maximum amplitude on the H atom, in accord with the general rules governing LCAO. Thus the electron pair that is to be accepted by HF is localized mainly on the H atom, in agreement with the notions introduced in Chapter 5, wherein electrons are drawn to the more electropositive atom in a molecule. We conclude that *the LUMO dictates the Lewis acid behavior of a molecule.*

Another principle we introduced in Chapter 5 was that bond rupture always accompanies acceptance of electrons by a "good" Lewis molecule. This can also be viewed in terms of the LUMO: since it is antibonding (generally the case), the H–F bond must be severed by placing a pair of electrons in σ^*_{2p}, for then the bond order becomes zero. One reaction of HF we examined in Chapter 5 (Equation 5.32) was $HF + H_2O \longrightarrow H_3O^+ + F^-$, in which the H–F bond breaks in the process of protonating water. Figure 7.14 shows a frontier orbital picture of the progress of the reaction.

What is the significance of the other frontier orbital, the HOMO? As you may have guessed from our discussion of the relation of the LUMO to acidic behavior, *the HOMO dictates the Lewis base behavior of a molecule.* As we will find in the next section, the oxygen lone pairs on water are HOMO electrons. In addition, in gaseous HF an appreciable concentration of *dimers,* $(HF)_2$, is found, with the geometry

$$\begin{array}{c} H \\ \diagdown \\ F \cdots\cdots\cdots H-F \end{array}$$

This association between HF monomers is not a chemical bond, but a weaker form of attraction known as *hydrogen bonding* (see Chapter 14). The two HFs are not

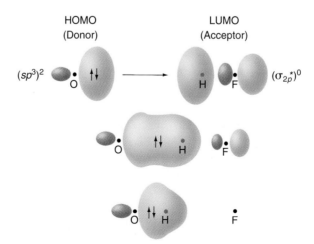

Figure 7.14
Frontier orbital picture of an acid–base proton transfer reaction. The lone pair orbital on O (shown hybridized here) interacts with the empty σ^*_{2p} orbital on HF. As the species approach, a new O—H bond forms while the old H—F bond breaks as a result of the HOMO–LUMO merger. Note that the contribution of the F atom to the new bond is zero (as you might expect!). Of course, F retains the other eight electrons, shown in Figure 7.13, to become the F^- product.

equivalent in this dimer, with the left-hand one acting as a base, through its HOMO lone pairs on F. Thus a given molecule's LUMO and HOMO properties can allow it to be either an electron donor (base) or acceptor (acid) depending on its chemical environment. This behavior is examined in the next section in the case of the world's most important molecule, water.

Strong bases, such as OH⁻, have energetically high-lying HOMOs, while strong acids, such as the hydrogen halides or BF_3, have relatively low-lying LUMOs. This induces electrons to migrate more readily out of the HOMO or into the LUMO.

You can regard these frontier orbital ideas as extensions of the concepts you have already become familiar with for atoms into the molecular realm. This parallels MO theory itself in its role as an extension of atomic structure theory.

Frontier Orbitals and Redox Reactions

In general, frontier MO theory predicts that both acid–base and redox reactions are mediated by the interaction between the HOMO of the eventual electron donor and the LUMO of the electron acceptor. For example, in the reaction between Na and Cl_2 (Chapter 5), the $3s$ valence electron of Na ($3s$ being the HOMO of Na) is transferred into the antibonding σ_{3p}^* orbital of Cl_2 (the LUMO of Cl_2) as shown in Figure 7.15. As before, the existence of antibonding orbitals provides a mechanism for electron transfer that was missing from the Lewis–VB theory. In the spirit of the LCAO approximation, the interacting orbitals must be of the proper symmetry to overlap in order to promote electron transfer; this requirement often suggests geometrical constraints on the "supermolecule" formed in the course of the reaction. In the Na + Cl_2 reaction, a linear geometry Na $\cdots$ Cl—Cl is probably preferred over a T-shaped

$$\text{Na} \cdots \begin{array}{c} \text{Cl} \\ \text{Cl} \end{array} \text{one,}$$

because the $3s - \sigma_{3p}^*$ overlap is zero for the latter.

In the frontier MO model, the breaking and making of chemical bonds in reactions is a natural by-product of the populating of antibonding LUMOs as the reaction proceeds. Once the newly added antibonding electrons weaken or break the old bond(s), new atoms or groups are liberated, along with their orbitals, for the formation of new bonds.

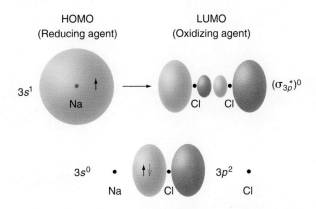

Figure 7.15

Frontier orbital picture of an electron-transfer (redox) reaction. The electron transfer from Na to Cl_2 occurs "suddenly" over a small range of distances near r_x (see Chapter 5). Acceptance of the electron by the empty σ_{3p}^* orbital of Cl_2, coupled with the Coulomb acceleration that ensues, severs the Cl_2 bond. The product Cl^- $3p$ orbital already has an electron as a result of pulling apart the σ_{3p} orbital, denoted by the dotted arrow.

7.3 Molecular Orbitals of the Water Molecule

Considering the importance of the water molecule in chemistry, it seems natural to examine the MOs of water using the qualitative survey of the MO ideas to which you have now been exposed. Dihydrides such as H_2O are the simplest triatomic molecules from a structural perspective, since the hydrogens bring only their $1s$ AOs to the LCAO problem. The MO principle of "throwing electrons at the [three] nuclei and seeing where they stick" applies, obliging us to consider orbitals that are in general delocalized over the entire molecule, rather than individual pair overlaps as in the VB model. Consistent with the uncertainty principle, the lowest possible energy will result when the electrons are allowed to "spread out" as much as possible.

We proceed by examining the symmetry and nodal structure of the orbitals of linear triatomic dihydrides with a second-period atom X in the center, and then ask what happens to the energies of these orbitals when the bond angle is changed. If we take the z axis as the common bond axis, then we find, as in HF, that only the $2s$ and $2p_z$ orbitals of X have nonzero overlap with $1s$ on the H atoms. The LCAO now involves three AOs; for example, the $1s - 2s - 1s$ combination is

$$\sigma_{2s} = a\ 1s_A + b\ 2s_X + c\ 1s_B \tag{7.5}$$

where we have labeled the H atoms A and B to distinguish the orbital centers. Based on the symmetry of the problem, we expect $|a| = |c|$; that is, the $1s$ orbitals should enter the combination with equal weight. The resulting MOs are predictable by again appealing to the particle-in-a-box problem (see Appendix B), which embodies the general principle that each energy level has one more node in its wave function than the one below it. Figure 7.16 shows the resulting triatomic MOs. Note how the nodal

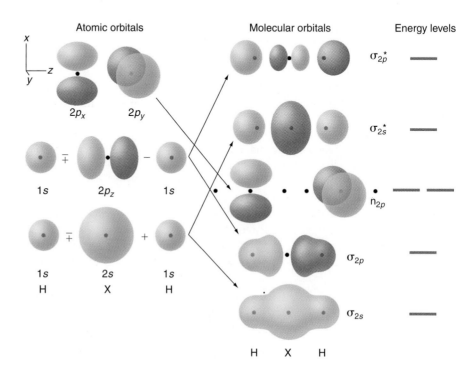

Figure 7.16
Molecular orbitals and energy levels of a linear triatomic dihydride HXH, where X is a second-row atom (e.g., BeH_2). The resulting orbitals are similar to those for second-row diatomics (see Figures 7.6 and 7.7), although the constituent AOs differ. Note the nodal structure and compare it to Figure 7.7.

requirement causes the σ_{2p} orbital to lie lower in energy than the σ_{2s}^*. The nonbonding character of the $2p_\pi$ orbitals is the same result we saw for HF.

If we now put the eight valence electrons of water into the lowest available levels, we find that the two lowest e$^-$ pairs fill the bonding MOs σ_{2s} and σ_{2p}, while the next two fill the degenerate $n\pi_{2p}$ levels, lone pair orbitals. This looks like what might be expected based on the Lewis structure and the VB model of H$_2$O. But H$_2$O, as you know only too well by now, is bent, not linear, and in order to find the MO prediction of its structure, we must ascertain (in a qualitative way) what happens to the orbitals and energy levels of Figure 7.16 when the H atoms are swung together, that is, when the bond angle closes up from 180°.

Taking the MOs of Figure 7.16 one by one, we can reason qualitatively whether the orbital energy will go up or down upon bending the molecule. The σ_{2s} orbital will be distorted in an energetically *favorable* way when the bond angle is closed, since this will allow constructive overlap between the end hydrogens, further delocalizing the electrons occupying it. The σ_{2p} MO, on the other hand, will be *destabilized* by bending, since the $1s$ orbitals have opposite signs in σ_{2p} and will destructively interfere. What happens to the nonbonding n_{2p} orbitals depends on their orientation relative to the plane of the bend (the plane of the bent molecule). Let's assume that the bending occurs in the xz plane. Then the $2p_y$ orbital on O is perpendicular to this plane, and its energy will change very little when the bond angle changes. The $2p_x$ orbital, however, points in the plane, and can begin to overlap with the H $1s$ orbitals when the bond angle closes, thus lowering its energy. This means that bending the molecule splits the formerly degenerate n_{2p} levels of Figure 7.16 into separate levels, and introduces a small amount of bonding character into the lower energy in-plane MO. To summarize, of the four occupied orbitals in the linear case, two go down in energy (σ_{2s} and n_{2p_x}), one goes up (σ_{2p}), and one remains the same (n_{2p_y}) as the bond angle is decreased. Because more levels drop than rise, we can conclude that H$_2$O is bent. (We'd be in trouble if we couldn't!) Figure 7.17 illustrates the changes in energy and the resulting orbitals for water. You should compare it carefully to Figure 7.16 to observe the changes in the orbitals wrought by bending.

This line of argument, originally due to A. D. Walsh, is more difficult, and the result less precise, than the VB–VSEPR model detailed in Chapter 6 regarding the bond angle of H$_2$O. To go any further requires an actual calculation of the atomic orbital coefficients and MO energies at various bond angles, which is certainly well beyond what our aim must be here. What we gain from this look at water's MOs is a knowledge of energy-level ordering and, more importantly for chemistry, a new picture of the relationship between water's structure and its chemical properties.

Based on our earlier discussion of the frontier MO model, we should focus on the LUMO and HOMO orbitals of water to understand its acid–base properties. Figure 7.17 indicates that the LUMO, the antibonding orbital labeled $a_1^*(\sigma)$, has most of its amplitude on the H atoms, and also that it is *stabilized* in the bent geometry, implying that a bent water molecule makes a better Lewis acid than a linear one. The HOMO, the orbital labeled b$_1$, is truly a lone pair orbital, the $2p$ oxygen orbital perpendicular to the molecular plane, and must account for the basicity of water. As we saw in our chemical explorations in Chapter 5, water behaves either acidically or basically depending on the reagent with which it interacts. For example in the (weakly occurring) reaction with ammonia (Equation 5.29)

$$H_2O + NH_3 \longrightarrow NH_4^+ + OH^-$$

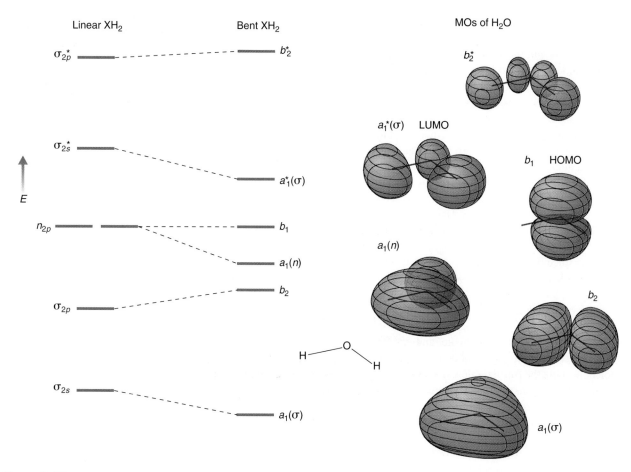

Figure 7.17
Molecular orbitals of the water molecule. Illustrated at the left is the change in energy of the various XH$_2$ energy levels of Figure 7.16 when the molecule is allowed to bend. The new MO labels a_1, b_1, and b_2 are necessary because the bent molecule is no longer cylindrically symmetric. The lowest four MOs on the right are occupied by electron pairs in H$_2$O. The most notable change from the linear case is in the $a_1(n)$ MO, which acquires some bonding character.

water is the acid, accepting the lone pair from ammonia. As in the case of HF, electron density placed in the antibonding LUMO of water causes bond breakage, with a proton from H$_2$O transferring to NH$_3$. When HCl is dissolved in water, on the other hand, nearly complete proton transfer to water occurs:

$$\text{HCl} + \text{H}_2\text{O} \longrightarrow \text{H}_3\text{O}^+ + \text{Cl}^-$$

Frontier MO theory has the HOMO accepting the proton, thereby donating its lone pair and behaving as a Lewis base. Figure 7.18 illustrates the H$_2$O frontier orbital role in these reactions.

To gain this advantage in describing reactions with orbitals, we have sacrificed much of the simplicity of the Lewis–Pauling conception of the electronic structure of a triatomic molecule such as H$_2$O. Instead of the two single, localized-electron-pair O—H bonds, we now have three MOs, all delocalized, with some bonding character. In the MO picture there is only one true lone pair, not two as in VB. However, if you consider the *overall electron density* in H$_2$O, the sum of the squares of the MOs yields

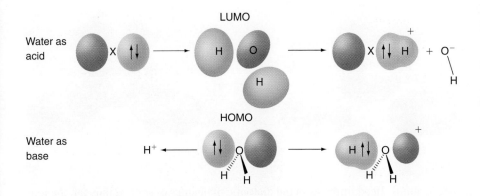

Figure 7.18
Frontier MO picture of the acid–base properties of water. Migration of the base pair (↓↑) at the top into water's antibonding LUMO causes dissociation, while acceptance of a proton by the HOMO causes changes in the other two O—H bonds to produce a trigonal pyramidal structure.

a buildup of electron density along the bond directions (from the three bonding MOs) that strongly suggests the existence of valence bonds.*

7.4 Delocalized Bonding: A Solution to the Resonance Problem

The first example of the need for resonance brought forward in Chapter 5 was the ozone molecule O_3. Recall that experimental evidence points to two equivalent O—O bonds in O_3, while the Lewis structure and its VB counterpart show one single and one double bond that "resonate":

The actual structure of the molecule must be taken as an "average" of the two resonance forms, as indicated by the dashed-line structure. In VB theory with hybridization, the double bond in either resonance form consists of a σ-type bond between sp^2 hybrids and a π bond between the leftover p orbitals perpendicular to the plane containing the nuclei.

MO theory offers a natural way of accounting for O_3's symmetrical structure. As illustrated in Figure 7.19, all three oxygen atoms can contribute perpendicular p orbitals to the formation of a π bond that extends over all three atoms. This delocalization is a natural consequence of the tendency to minimize electronic energy; that is, a lower energy can be achieved by extending the space the π electrons occupy, as implied by the uncertainty principle. There is no intrinsic requirement in the physics of the molecule that forces electrons to remain between a particular

---

* The MOs can even be recombined (with sacrifice of the energy-level structure) into so-called "natural bond orbitals" that point along the bond directions as in hybridized VB theory.

Figure 7.19
Delocalized π bonding in the ozone (O_3) molecule. The three $2p$ orbitals perpendicular to the molecular plane combine in phase to produce the three-atom delocalized π bond at the right. Two other π MOs can be formed from the three AOs; one of these holds a nonbonding pair of electrons, while the other is not occupied in O_3. Based on the particle-in-a-box analogy, can you deduce the shapes of these other two π MOs?

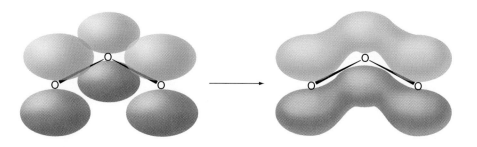

pair of atoms. In addition, as in the diatomic molecules examined thus far, the lone pairs of the Lewis structure are also actually delocalized, so that the shifting of lone pairs that occurs with changing resonance forms is also resolved in the MO picture.* Thus the end oxygens contribute simultaneously and equally to the π component of the bond, leading to the prediction of equal bond lengths, in agreement with observation.

All of the other examples of resonance among molecules discussed in Section 5.2 can be given a similar delocalized-π-bond description, as illustrated by the following example.

EXAMPLE 7.5

In Example 5.2 you examined the resonance forms of the carbonate ion CO_3^{2-}. Now interpret the experimentally observed symmetrical trigonal planar structure of CO_3^{2-} with the delocalized π MO picture.

Solution:
In this case all four atoms will have p orbitals perpendicular to the molecular plane, and the delocalization principle dictates that the π bond MO should consist of a simultaneous constructive overlap of all three O $2p$'s with the C $2p$, as shown here:

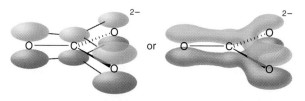

One-third of the π bond then binds each CO pair together, leading to the observed symmetrical structure.

The most famous case of delocalized π bonding occurs in the benzene molecule C_6H_6. In benzene the six carbons are bonded together in a planar *six-membered ring*. The Lewis structure for benzene shows two resonance forms,

---

* The σ bonding, however, is more complicated in MO theory. As in H_2O, O_3 has more than two σ-type MOs with some bonding character, making the overall bond order difficult to assess.

7.4 Delocalized Bonding: A Solution to the Resonance Problem

while experiment, both spectroscopic and chemical, demonstrates that all C—C bond lengths are identical. Since every carbon is involved in a π bond, each can contribute a $2p$ orbital perpendicular to the molecular plane toward the formation of delocalized π MOs, as illustrated in Figure 7.20. In this case there are three π electron pairs, so that

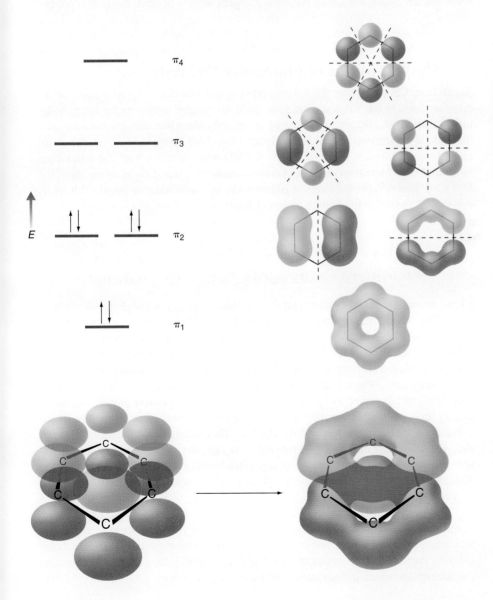

Figure 7.20
Delocalized π bonding in benzene (C_6H_6). The six carbon $2p$ orbitals perpendicular to the molecular plane combine to make six π MOs, as shown in detail for the lowest level and in a top view for the others, with the lowest three levels occupied by electrons. The number of nodal planes increases with energy just as in the particle-in-a-box problem (Appendix B); this pattern is shown by a so-called "particle-on-a-ring" model.

three π MOs are required to "hold" them. In the lowest energy MO, all six C 2p orbitals overlap constructively to form doughnut-shaped lobes above and below the plane, while the two degenerate orbitals each show a single perpendicular nodal plane. Because of the delocalization, the net result of occupying these three orbitals is to produce equal π-bonding character in each of the six C—C bonds, along with remarkable stability.

In general, you can expect delocalized π bonding in any molecule with *alternating single and double carbon–carbon bonds* in its Lewis structure, as well as in molecules with "resonating" double bonds. Not only does π delocalization allow a ready explanation of symmetric geometries in these cases, but it also helps us to interpret the higher stability (lower energy) of such molecules and their lower chemical activity. These two topics will be exposed in more detail in Chapters 10 and 18. There are also a few molecules with adjacent π bonds, such as CO_2, for which no resonance forms are necessary, but which show the energetic effects of delocalized π bonding.

7.5 Shortcomings of Molecular Orbitals

Recall from Chapter 6 how the valence bond model was constructed by first considering the form of the VB wave function when the bonded atoms are far apart. A significant merit of this approach is that it properly describes the dissociation of a diatomic molecule into neutral atoms. In the LCAO-MO picture, however, the dissociation (large r) limit is not "built in" to the MO wave function from the start (when molecules with more than one electron are considered). Using H_2 as an example, the MO wave function for the electron pair is given by the product of two σ_{1s} MOs (recall the discussion of the helium atom in Chapter 4),

$$\Psi(\mathbf{r}_1,\mathbf{r}_2) \approx \sigma_{1s}(\mathbf{r}_1)\sigma_{1s}(\mathbf{r}_2) \tag{7.6}$$
$$= [1s_a(\mathbf{r}_1) + 1s_b(\mathbf{r}_1)][1s_a(\mathbf{r}_2) + 1s_b(\mathbf{r}_2)]$$
$$= 1s_a(\mathbf{r}_1)1s_b(\mathbf{r}_2) + 1s_a(\mathbf{r}_2)1s_b(\mathbf{r}_1) + 1s_a(\mathbf{r}_1)1s_a(\mathbf{r}_2) + 1s_b(\mathbf{r}_1)1s_b(\mathbf{r}_2)$$

where, as earlier, $\mathbf{r}_1$ and $\mathbf{r}_2$ represent the coordinates (x,y,z) or (r,θ,ϕ) for electrons 1 and 2. The second line of Equation 7.6 is the LCAO approximation, and in the third line we have expanded the orbital product into its four terms. You should recognize the first two terms of the third line as the valence bond wave function of Equation 6.2. The last two terms each represent *ionic* contributions H^+H^- where both electrons are on the same atom. These latter terms are given the same weight as the VB or covalent terms, and do *not* go away when the molecule dissociates into atoms. The dissociation limit is then not just the energy of the neutral atoms, but an average of the energies of the neutral atoms and the ion pair. This means that *MO theory does not describe molecular dissociation properly*. This can only be remedied in the LCAO approximation by adding at least one new orbital configuration to Equation 7.6, and thereby losing the connection, invaluable as you have seen, between orbital and energy level.

Add to this rather serious drawback the difficulty we found, in our examination of the water molecule, of estimating bond angles, and you may begin to appreciate why many chemists have stood by the Lewis–VB model, especially for large molecules. In the next section, we examine an MO-VB "hybrid" description of molecules that attempts to retain the advantages of each.

7.6 Valence Bonds, Molecular Orbitals, and Quantum Chemistry

In Section 4.6 we briefly described the self-consistent field (SCF) method of Hartree for solving the Schrödinger equation to find the best orbitals for atoms. Because of the close relationship of MOs to AOs, MO theory was the first to be developed into a molecular SCF method for calculating the orbitals and energies of molecules. C. C. J. Roothaan and his coworkers developed this theory in the early 1950s. When computers arrived on the scene in the early 1960s, the MO SCF theory became a widely used tool for computational prediction of molecular properties. In the late 1960s and early 1970s, W. A. Goddard and his students pioneered the development of VB theory into an SCF computational tool of comparable power.

In principle, when each of these methods is carried to its ideal limit, by making the orbitals employed more and more flexible, they should give the same "answer" for the energy levels of a molecule. In practice, however, the answer obtained depends on how elaborate the calculation is, and the advantages and limitations of each theory that we have described in these two chapters often show up in the calculated results. As of 2005 the most widely used computer algorithm for molecular structure is the molecular orbital GAUSSIAN program of the late J. A. Pople and coworkers, which runs on a home computer! Quantum chemistry has now become a field in which both purely theoretical chemists and experimentalists can use available computer algorithms to calculate molecular features, often enabling the interpretation of complex experiments and the planning of new ones.

As valuable as it is, this approach of "letting the computer do it" has one major drawback: a loss of intuition. Understanding the results of experiment or calculation requires a backdrop based on the orbital model such as we have attempted to give you in these chapters. But in the VB and MO models we have two rather different views of molecular structure and reactivity, and within the confines of the simplest versions of these models, these views cannot be entirely reconciled. Even so, to provide a descriptive approach that can be readily extended to large molecules, most chemists subscribe to a hybrid molecular model that is essentially the Lewis–VB model with MOs employed where needed, that is, for delocalized bonding or for describing the frontier orbitals. A classic example of this occurs in the benzene molecule, in which the six-membered ring skeleton,

often called the σ-bonded framework, is given a localized-electron-pair sp^2 hybrid VB description, and only the π electrons are regarded as delocalized. In many cases this suffices to describe the important chemical properties of the molecule—for benzene, the π-orbital manifold provides both the HOMO and LUMO, and the energy levels of the σ framework are generally irrelevant. In this way the obvious ease and utility of describing a molecule "one bond at a time" is retained, along with its geometrical implications.

It is this author's opinion that the MO method for describing molecules, despite its drawbacks mentioned earlier, arises more naturally and with fewer assumptions from the attempt to use the Schrödinger equation—quantum mechanics—to compute molecular energies and orbitals and is, therefore closer to the "truth" of molecular structure at its simplest level. The author sides with the majority of chemists in this view, though a minority disagree. At a higher level of theory, though, where improvements are made to the orbital description, the best theory to use may depend on the case to which it is applied. And chemists, buoyed by our great progress in finding new reactions and building new molecules, will undoubtedly continue to use the "marriage of convenience" combination of MO and VB models.

SUMMARY

Molecular orbital (MO) theory is in essence an extension of the atomic orbital (AO) model to molecules. The paradigm of MO theory is the H_2^+ molecule, whose orbitals can be found exactly for fixed values of the interproton distance. The two lowest energy orbitals of H_2^+ can be approximated by adding or subtracting the $1s$ AOs on each atom,

$$\sigma_{1s} = 1s_a + 1s_b$$
$$\sigma_{1s}^* = 1s_a - 1s_b$$

where the sum corresponds to constructive interference of the AOs and a **bonding MO** σ_{1s}, with σ indicating the cylindrical symmetry of the orbital and the $1s$ subscript its AO parentage, and the difference to a destructively interfering **antibonding MO**, indicated by the asterisk. These σ orbitals are an example of a method devised by Robert Mulliken for constructing MOs called the **linear combination of atomic orbitals** (LCAO) approximation. The bonding combination is nodeless and the antibonding shows a node midway between the nuclei.

An examination of the potential energies involved in H_2^+ bonding shows that *covalent chemical bonding arises mainly (but not wholly) from the drop in zero point energy* due to the presence of the extra proton added to H. The extra proton widens the Coulomb well and allows the electron to delocalize over a larger spatial region. With less spatial constraint, the uncertainty principle allows a lower uncertainty in momentum and hence a lower ZPE. The delocalization concept replaces the electron-exchange idea of VB theory as the root cause of covalent bonding. In the LCAO approximation, orbital overlap, a central quantity in VB theory, signals the onset of **delocalization** and, hence, remains important in the MO picture.

MOs for diatomic molecules can be approximated by combining pairs of AOs of similar energy and the same symmetry (σ or π) into bonding and antibonding combinations. The first-row homonuclear diatomics, H_2^+, H_2, He_2^+, and He_2, are predicted to have bond energies and lengths in qualitative agreement with experiment provided electrons in the antibonding σ_{1s}^* orbital cancel the bonding effects of electrons in the σ_{1s} orbital. The MO definition of **bond order** is bond order = $\frac{1}{2}(n_{bonding} - n_{antibonding})$, where the n's are the total numbers of bonding and antibonding electrons. This leads to the conception that two helium atoms will not bond, not because each has a closed shell, but because $n_{bonding} = n_{antibonding}$. A line notation for MO occupancy analogous to that for AOs is introduced; He_2^+, for example, has the configuration $(\sigma_{1s})^2(\sigma_{1s}^*)^1$.

MOs for the second-period homonuclear diatomic molecules arise from combining $2s$ and $2p$ orbitals on each atom, subject to the rules of energetic similarity and symmetry. The result is four σ MOs and two doubly degenerate π MOs, with the energy ordering $\sigma_{2s}, \sigma_{2s}^*, \pi_{2p}, \sigma_{2p}, \pi_{2p}^*, \sigma_{2p}^*$. For O_2 and F_2 the σ_{2p} and π_{2p} are reversed, but this does not affect the bond order. Using the Aufbau principle in a manner analogous to that for atoms, it is found that N_2 has a triple bond, in agreement with the Lewis–VB prediction, but that the "lone pair" electrons are actually delocalized in the σ_{2s} and σ_{2s}^* orbitals. O_2 has two electrons in the doubly degenerate π_{2p}^* MO; these electrons occupy separate MOs with their spins parallel, in accordance with Hund's rule and the experimental observation of O_2's paramagnetism. The s–p MO picture is easily extended to the representative elements of later periods. Its success owes to the existence of antibonding orbitals.

In a heteronuclear molecule AB where $\chi_B > \chi_A$ and neither A nor B is a hydrogen atom, ns and np orbitals can be combined to give the same set of MO energy levels, but the orbitals become lopsided, with the bonding combinations favoring the more electronegative B atom, and the antibonding favoring A. In the case of diatomic hydrides, the $1s$ orbital of H can combine only with ns and np_σ orbitals of the larger atom; the np_π orbitals remain localized and are termed nonbonding orbitals. As the bonding changes from covalent to ionic, the weight given to the electropositive atom's orbitals goes to zero, giving localized MOs and an ion pair.

MOs bring us a step closer to the structure–reactivity relationship if we focus on the highest occupied MO (HOMO) and lowest unoccupied MO (LUMO) of a molecule, the so-called **frontier orbitals**. The HOMO of a molecule characterizes its Lewis base or electron-donor properties, while the LUMO embodies its electron-accepting, Lewis acid character. In "good" Lewis molecules, the LUMO is some type of antibonding orbital, and resolves the difficulty in Lewis acid–base theory of where to put the donated electron pair and how to picture the bond rupture that normally ensues. The general utility of the frontier orbitals in describing redox as well as acid–base reactions is discussed briefly.

The MOs of simple polyatomic molecules are exemplified by those of the water molecule. The analysis begins with a linear dihydride, and the effect of bending the molecule on the energies of the resulting orbitals is qualitatively assessed. Examination of water's frontier orbitals suggests how it can act as either an acid or base, but the structure of its bonding MOs does not show the two single localized-electron-pair O—H bonds expected from the Lewis structure. The overall electron density, however, is very similar to that of VB theory, since both theories use the same AOs.

The problem of "resonating" double bonds inherent in the Lewis–VB model is eliminated by the naturally delocalized bonding in MO theory. The classic case of the benzene molecule is discussed in particular.

MO theory is not without its drawbacks. In addition to its cumbersome description of larger molecules, good estimates of bond angles are difficult to obtain, and it does not describe bond breaking properly. These difficulties have prompted chemists to describe the structure of molecules and their reactions using a hybrid theory that employs the localized VB description for the σ bonds and delocalized π MOs where needed for the π bonding. The modern versions of VB and MO theories as implemented on 21st-century computers are both self-consistent field (SCF) theories that ideally lead to the same results for molecular electronic structure, while abandoning the simpler versions of these theories described in the preceding two chapters.

EXERCISES

The nature of covalent bonding

1. The Lewis structure for H_2^+ (see Chapter 5) would have to be $[H \cdot H]^+$, where the Lewis dot is located midway between the nuclei. Compute the potential energy in J and eV (see Chapter 1 and Figures 7.1 and 7.4) for an electron (a) midway between two protons 1.06 Å ($2a_0$) apart, and (b) along the interproton axis but 0.25 Å closer to one proton than the other. Which is more stable? What are the implications for the Lewis picture of H_2^+? (Recall that the origin of covalent bonding is nonclassical.)

2. The one-dimensional particle-in-a-box model (see Appendix B) yields insight into the nature of covalent bonding. The box dimension, of length a, lies along the bond axis. Apply this model to the covalent bond in H_2^+. The ground-state (zero point) kinetic energy is $K_1 = h^2/(8ma^2)$, where m is the mass of the electron. For the H atom, $K_1 = 13.60$ eV. Now suppose the box is expanded along the bond axis by an amount δa as a proton is added to form H_2^+. Use differentials to show that δK, the drop in zero point kinetic energy, is approximately given by $\delta K \approx -2K_1(\delta a/a)$. (This relation shows that lengthening the box indeed lowers the energy.) Find the fractional increase in box length $\delta a/a$ beyond that of the atom that will lower the kinetic energy by an amount equal to the bond energy of H_2^+, 2.77 eV. [Note that this same argument can be made starting from the uncertainty principle $\Delta x \Delta p_x \approx h$, where x is the bond axis (see Section 3.5), implying that the covalent bond owes its existence to the intrinsic uncertainties in the electron's motion.]

3. An improvement to the simple $1s$ LCAO approximation for H_2^+ can be made by allowing a free parameter Z_{eff} in the $1s$ orbitals, as in the He atom (see Chapter 4). As the two protons are brought together, the H_2^+

molecule shrinks into the united atom, He^+. Use this observation to place bounds on Z_{eff}. (Finding the Z_{eff} that gives the lowest possible energy yields $Z_{eff} = 1.23$ at $r = r_e$, with a bond length $r_e = 1.06$ Å and bond energy $D_e = 2.24$ eV.)

4. If an improvement analogous to that for H_2^+ in Exercise 3 is made for H_2, what bounds may be placed on Z_{eff}? (*Hint:* What is the united atom? See Section 4.1.) (The lowest possible energy for H_2 is obtained when $Z_{eff} = 1.19$, giving $r_e = 0.73$ Å and $D_e = 3.47$ eV.)

5. Atoms A and B are brought together along the z axis. Which of the following pairs of atomic orbitals will yield nonzero overlap?

   a. $p_z(A), p_x(B)$
   b. $p_y(A), s(B)$
   c. $p_x(A), p_y(B)$
   d. $s(A), p_z(B)$
   e. $s(A), s(B)$
   f. $p_y(A), p_y(B)$

Homonuclear diatomic molecules

6. The molecule C_2 has been observed in acetylene (C_2H_2)-oxygen flames. Use the MO energy-level diagram of Figure 7.8 to deduce the bond order of C_2, and give the electron configuration. Referring to Figure 7.6 or 7.7, sketch the orbital or orbitals occupied by the electrons responsible for bonding in C_2.

7. The species Be_2 has recently been observed spectroscopically to have a very small bond energy, $D_e \approx 0.02$ eV. What would you predict the bond order to be, based on the MO analysis of second-row diatomic molecules? Give the electron configuration of Be_2, and sketch the valence MOs.

8. The Al_2 molecule has been observed in the gas phase in electric arcs, and has been found to be paramagnetic, with two unpaired electrons. Explain the magnetism of Al_2 using MO theory. Give the electron configuration, bond order, and a sketch of the bonding orbital(s).

9. The atmospheric gases N_2 and O_2 undergo a process known as photoionization in the earth's upper atmosphere, for example,

$$O_2 + h\nu \longrightarrow O_2^+ + e^-$$

where the ultraviolet photons arise from the short-wavelength end of the solar spectrum. (See Chapter 2 and Exercise 2.5.) Use MO theory to find the bond orders of the molecular ions N_2^+ and O_2^+, and give their electron configurations. Compare each with regard to bond order, bond energy, and bond length with its parent neutral molecule.

10. Neon signs glow with the light emitted by excited Ne atoms. After a sign is switched off, these atoms combine to make a high concentration of Ne_2^+ ions that lasts for a few milliseconds. Use MO theory to show that, while neutral Ne_2 is not chemically bonded, Ne_2^+ should be stable.

11. Low-energy electrons in gas discharges can attach to neutral molecules to make anions according to $A_2 + e^- \longrightarrow A_2^-$. Which (if any) of the molecules N_2, O_2, or C_2 would be stabilized by the addition of an electron? Sketch a boundary surface for the newly occupied orbital in each case.

12. When ignited in pure oxygen, barium metal produces barium peroxide (BaO_2), while potassium yields potassium superoxide (KO_2). Assuming that the oxygen atoms in these compounds are covalently bonded, use MO theory to deduce the bond orders of the anions in these compounds. In which anion do you expect a shorter O—O distance?

13. Calcium carbide (CaC_2), through a violent reaction with water, is an important source of acetylene (C_2H_2). Assuming CaC_2 is an ionic compound, give the MO electron configuration and bond order of the dicarbide anion. With which neutral homonuclear diatomic molecule is this anion isoelectronic?

Heteronuclear diatomic molecules

14. Use MO theory to analyze the differences in bonding and magnetism in the cyanide ion (CN^-) and the cyanogen radical (CN). Give electron configurations for each, and give the expected relation between the bond energies and lengths of the neutral and ion. Sketch the highest energy occupied orbital in CN^-.

15. In the late 1980s an "ozone hole" was discovered over Antarctica. Chlorofluorocarbons (CFCs, also known as freons) such as CF_2Cl_2 are suspected of causing the hole through release of chlorine into the stratosphere. The implication of chlorine in ozone destruction has been verified through the detection of chlorine monoxide (ClO) in the upper atmosphere. Use MO theory to analyze the bonding in ClO, and determine its stability (bond energy) relative to a "single" bond involving Cl (such as Cl—Cl).

16. NO^+ is found to be the dominant molecular ion in the ionized region of the upper atmosphere known as the *ionosphere*. Compare the bond order of NO^+ to those of the cations of N_2 and O_2 (see Exercise 9). Which is predicted by MO theory to have the strongest bond? Sketch the boundary surface of the highest occupied MO of NO^+.

17. Assess the effect of adding an electron, $AB + e^- \longrightarrow AB^-$, on the bond lengths and strengths of the molecules BC, CO, and ClO. In each case, sketch a boundary surface for the newly occupied MO.

18. Arrange the following molecules or ions in order of increasing bond energy, that is, lowest first:

    BC OF$^-$ N$_2^-$ NO$^+$ BN

Diatomic hydrides and frontier orbitals

19. Analyze the HCl molecule using MO theory, giving an energy-level diagram showing orbital occupancy. How does your diagram differ from that for HF in Figure 7.13? Considering the radii of Cl and F, how will the MOs differ in appearance? Does the MO description of HCl differ from the VB picture in terms of bonding and lone pairs?

20. Ultraviolet laser photodissociation of ammonia (NH_3) yields the NH molecule along with H_2. Give a MO description of NH, and predict its magnetic properties.

21. The hydroxyl radical (OH) is a common intermediate in combustion reactions, as well as an important, almost central, species in stratospheric chemistry. Analyze OH using MO theory, and predict its magnetic properties.

22. Give a MO energy-level diagram and sketch the associated orbitals for the hydroxide ion (OH^-). Which orbital(s) is(are) important for describing the acid–base chemistry of OH^-? Identify the frontier orbitals involved in the simplest form of the Arrhenius acid–base reaction $H^+ + OH^- \longrightarrow H_2O$.

23. In Chapter 5 you saw that metal hydride bonds are polar in the direction $M^{+\delta}H^{-\delta}$. Give an MO description of the simplest metal hydride, LiH. (Note that the relative energies of the contributing AOs are much different than in HF.) How is the bond polarity reflected in the lowest bonding MO? Sketch the two lowest MOs for LiH. Which of these is important in its redox chemistry (see Section 5.4)?

24. For the redox reaction between K and ICl, identify the important frontier orbitals and sketch them. On the basis of your sketches, do you expect to see a greater yield of KI or of KCl?

Dihydrides

25. The methylene molecule (CH_2) is an important combustion intermediate and has served as a test case for quantum-chemical calculations. On the basis of the MO analysis of water given in Section 7.3, decide whether CH_2 is linear or bent. Sketch the HOMO of CH_2.

26. Hydrogen sulfide (H_2S) is a weak acid that is responsible for the "rotten egg" smell of some rural water supplies and of catalytic automobile mufflers. By analogy with the H_2O example, sketch the molecular orbitals of H_2S, indicating the contributing atomic orbitals and their occupancy by electrons. Identify the orbital responsible for the acidic character of H_2S.

27. The simplest neutral chemical reaction is $H_a + H_bH_c \longrightarrow H_aH_b + H_c$, where the hydrogen atoms are labeled to show the breaking of the old bond and the making of a new one. (Isotopes are usually used to study the reaction.) The reaction intermediate is "H_3," an unstable temporary "reaction complex." Analyze linear H_3 using LCAO-MO theory, giving the energy levels and sketches of the three lowest energy MOs and their occupancy.

28. Analogous to Exercise 27, the simplest gas-phase ionic reaction is $H_a^+ + H_bH_c \longrightarrow H_aH_b^+ + H_c$. Here the reaction intermediate H_3^+ is a well-known stable molecule. Analyze H_3^+ using LCAO-MO theory and, by following the discussion of orbital energy versus angle for H_2O, argue that H_3^+ should be a bent molecule. (It is in fact an equilateral triangle.)

Delocalized bonding

29. Describe the resonance forms of the NO_2^- and NO_3^- ions in terms of delocalized bonds, using sketches of the π MOs. What can be concluded about the relative N—O bond lengths within each ion? Which ion has shorter bonds?

30. Compare the delocalized π bonding in SO_2 and SO_3, giving sketches of the delocalized MOs, and relate it to their resonance forms. How does the MO picture explain the observed equal S—O bond lengths within each molecule? Which molecule has shorter bonds?

31. The principal class of organic acids is the carboxylic acids, containing the carboxyl group

$$-\underset{\underset{}{\overset{\overset{O}{\|}}{C}}}{}-O-H$$

the simplest example of which is formic acid (methanoic acid) HCOOH. When formic acid ionizes, the formate ion

$$\left[H-\underset{}{\overset{\overset{O}{\|}}{C}}-O \right]^{-}$$

is formed. Draw the resonance forms, with the appropriate VSEPR bonding geometry, of the formate ion, and interpret them in terms of delocalized π bonding, giving a sketch of the π MO. Describe the σ bonding using VB theory. Can the lone pairs be handled within VB? If yes, how? If not, why not?

32. Describe and compare the VB and MO pictures of the π bonding in CO_2. In what way (if any) will the predictions of CO_2's properties differ in the two pictures?

Molecular Motion and Spectroscopy

CHAPTER 8

In 1800 William Herschel discovered infrared radiation by passing the sun's light through this prism. Infrared light can excite molecular vibration, as described in Section 8.3.

Swift as a shadow, short as any dream,
Brief as the lightning in the collied night . . .
And ere a man hath power to say 'Behold!'
The jaws of darkness do devour it up:
So quick bright things come to confusion.

A Midsummer Night's Dream, W. Shakespeare (1595)

CHAPTER OUTLINE

- **8.1** Degrees of Freedom: Translation, Rotation, and Vibration
- **8.2** Rotational Motion: Defining a Molecular Day
- **8.3** Vibrational Motion: Molecular Calisthenics
- **8.4** Electronic Transitions in Molecules
- **8.5** Looking Back and Peeking Ahead

A molecule, a set of atoms bonded together for the mutual comfort of their electrons, is *not* a frozen geometrical object, though our discussion of the shapes of molecules might have led you to believe so. Recall from our introduction of the idea of temperature in Chapter 1 the interpretation of T (absolute or Kelvin scale) as a measure of the average amount of *mechanical energy* ε possessed by the molecules comprising a sample: $\varepsilon \approx k_B T$, where k_B is Boltzmann's constant. Mechanical energy is just

Gerhard Herzberg (1904–1999). In addition to his pioneering experiments in electronic spectroscopy of molecules and ions in the gas phase, over the 40-year period 1937–1977 Herzberg authored a series of books on atomic and molecular structure and spectra that have become classics in the field.

kinetic and potential energy, which means that, at any finite T, *all molecules are in constant, unceasing motion.* This in turn implies that, not only are molecules always moving about as a whole, but also they are "tumbling and jiggling," with their bond lengths and angles constantly changing in time. The consequences of this molecular agitation will be fleshed out in Chapters 9, 10, and 11. In this chapter we classify and describe these motions using both Newtonian and quantum pictures, and show how information about the *average* molecular geometry can be obtained through **molecular spectroscopy,** in which these motions are altered by the absorption or emission of photons.

8.1 Degrees of Freedom: Translation, Rotation, and Vibration

Suppose we've isolated a single argon atom in 3-D space. What is needed to analyze its motion, assuming it has some kinetic energy? In Newton's mechanics, we need to know where it is, that is, its 3-D coordinates (x, y, z), and also where it is headed, that is, the three components of its momentum (p_x, p_y, p_z). If we don't apply a force, then the atom will move according to Newton's first law, in a straight line along its initial direction of motion, and we can predict its position and momentum at any later time. Each coordinate-momentum pair, for example, (x, p_x), is referred to as a **degree of freedom;** an argon atom freely moving through 3-D space has three degrees of freedom. This motion through space is referred to as **translation.**

Now let's consider *two* argon atoms in motion. Keeping track of two atoms clearly requires *six* coordinates, three for each atom, and *six* momentum components. Hence two atoms have six degrees of freedom. It is easy to generalize this: *a collection of N atoms possesses 3N degrees of freedom* (referred to from here on as DFs). If these N atoms happen to be bonded together into an N-atom molecule, the number of DFs is still $3N$; that is, you still need to know where each atom of the molecule is and where it is going to predict the motion of the molecule. But, as illustrated in Figure 8.1, while N argon atoms have $3N$ translational DFs, the atoms in a molecule cannot "translate" independently of each other, but rather must contribute their DFs to *internal* as well as translational motion. These internal motions are of two types: *rotation* and *vibration.* We have already discussed rotational and vibrational motion in Section 3.6 with regard to the classical analogues of electron motion in an atom. You should reread Section 3.6 either now or after you finish this section to gain some appreciation of the similarities between electronic and atomic motion.

One of the properties by which we know a molecule from a disconnected bunch of atoms is the ability of a molecule to move about in space as a unit, without suffering any change in its structure. Spilled or boiled, water molecules remain intact. The molecule when it moves as a whole can be represented as a point mass situated at the **center of mass** (the balance point) of the molecule. Using a homonuclear diatomic molecule such as N_2 as an example, the center of mass lies midway between the nuclei. Motion of the center of mass then requires 3 DFs to describe it, just as for an atom. In general, regardless of its size or complexity, a molecule has three translational DFs. This in turn implies, from our earlier discussion, that $3N - 3$ DFs remain to describe the internal motions of rotation and vibration.

8.1 Degrees of Freedom: Translation, Rotation, and Vibration

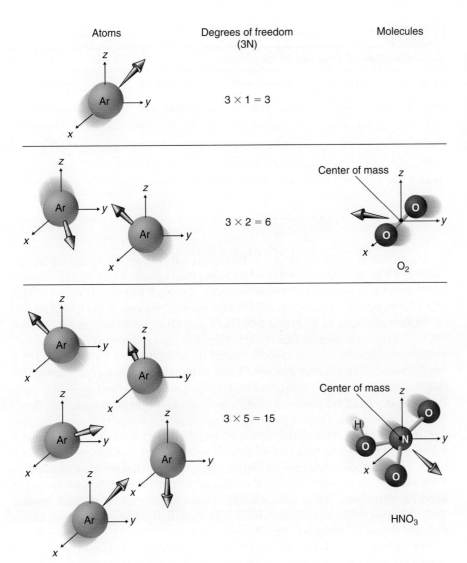

Figure 8.1
Molecular motion and degrees of freedom. The number of DFs is the same for a set of N independent atoms or an N-atom molecule, but in the molecule only three of these are translational, the rest being assigned to rotation and vibration within the molecule. The coordinate systems on each atom or molecule indicate the three translational DFs.

Just as planets revolve about the sun while the entire solar system hurtles through space, so atoms in a molecule revolve about their center of mass while the molecule translates. This tumbling motion is called **rotation.** Let's again use N_2 as an example. The isolated molecule will rotate, much like a twirling baton, in a fixed plane, owing to the conservation of angular momentum (see next section) in the absence of torques. To deduce the number of DFs this motion represents, we place the molecule with its center of mass at the origin of our coordinate system (which travels along with the hurtling molecule) and its bond axis along the z axis. Because the plane of rotation must then pass through the z axis, N_2's rotation can be represented by superimposing rotations in the xz and yz planes. The xy plane does not contribute, since rotation around the bond axis does not move the nuclei. There are therefore two rotational DFs arising from the two independent rotational planes. Without further development, we state the general result: *a linear molecule has 2 rotational DFs,*

TABLE 8.1
Classification of degrees of freedom

Molecule	Total	Translational	Rotational	Vibrational
N-atom molecule	$3N$	3	2 if linear 3 if nonlinear	$3N$–5 if linear $3N$–6 if nonlinear
N_2	6	3	2	1
CO_2	9	3	2	4
H_2O	9	3	3	3
CH_4	15	3	3	9

while a nonlinear molecule has 3. The former is easily extrapolated from the N_2 discussion, while the latter follows from including the xy plane, which must contribute a DF when the bonds of a polyatomic molecule do not all line up on the z axis. This rule, like translation, applies regardless of the number of atoms in the molecule.

After translation (3 DFs) and rotation (2 or 3 DFs) are accounted for, $3N - 6$ ($3N - 5$ if linear) molecular DFs remain. All of these are assigned to vibrational motion. A diatomic molecule (necessarily linear) therefore has only 1 vibrational DF, a linear triatomic has 4, a bent triatomic 3, and so on. As mentioned at the start of this chapter, the bond lengths and angles that describe a molecule's bonding arrangement are not rigidly fixed, but represent an *average* over all possible bond stretching, bending, and twisting. The single vibrational DF of a diatomic is simply a to-and-fro motion of its atoms about the equilibrium bond length r_e, and can be likened to a ball on a spring or a ball rolling back and forth on a parabolic track (see Chapter 1 and Section 8.3). Vibrations in which bond *lengths* change are called **stretching vibrations.** Triatomic and larger molecules can also change their bond *angles* in what is termed **bending vibrations.** These characteristic vibrations are called **normal modes.** Table 8.1 presents the overall classification of molecular motion with simple examples, and Figure 8.2 illustrates two of them.

In general, for an isolated molecule (which, as you may realize, is an entity found only in the gas phase), the translational DFs can be analyzed separately from the internal ones; that is, a molecule's internal motions are not influenced by its center-of-mass motion through space. These translational motions will be analyzed in more detail in Chapter 9, while here we emphasize the rotations and vibrations, which are susceptible to change through the action of electromagnetic radiation.

8.2 Rotational Motion: Defining a Molecular Day

In imagining molecular motion, it may be useful for you to think of yourself as a molecule, or, if that makes you uncomfortable, as a passenger riding a molecule. When a molecule rotates, or tumbles end over end (see Figure 8.2), it possesses **rotational kinetic energy.** We will confine our discussion to the simplest examples, so think of a diatomic molecule such as H_2 or N_2. The energy of internal motion is always defined with respect to the center of mass of the molecule. For a diatomic, this is the point

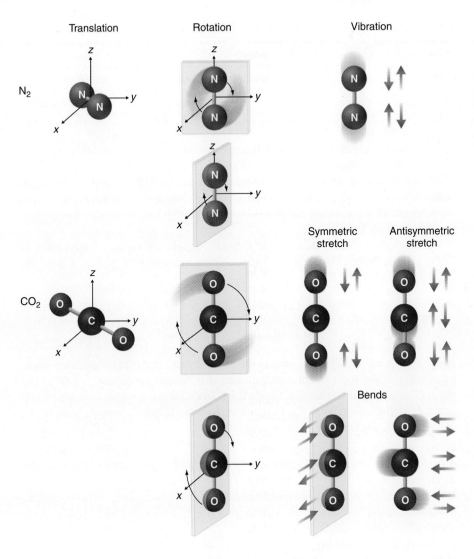

Figure 8.2
Classification of degrees of freedom for isolated N_2 and CO_2 molecules. The coordinate axes for translation represent 3 DFs, while each internal DF is illustrated separately. Absorption of infrared radiation by the bending vibration of CO_2 present in the atmosphere may contribute to the so-called greenhouse effect, in which global warming may occur due to trapping of radiation from the earth's surface.

between the atoms at which a fulcrum could be placed that would precisely balance the atoms if gravity were acting on them:

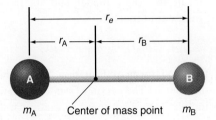

The balance condition, which determines the distances r_A and r_B, is that the torques on either side of the fulcrum must cancel, or that

$$m_A r_A = m_B r_B \tag{8.1}$$

where m_A and m_B are the atomic masses. In terms of the bond length $r_e = r_A + r_B$, we can find these distances as follows:

$$r_A = \frac{m_B}{m_A + m_B} r_e \qquad (8.2)$$

$$r_B = \frac{m_A}{m_A + m_B} r_e$$

These formulas say that if atom A is heavier than atom B, $m_A > m_B$, then $r_B > r_A$; that is, B must be placed farther from the center of mass than A. This is an observation that is familiar to anyone who has played on a seesaw with someone of much different mass: successful seesawing requires the heavier person (atom) move toward the fulcrum. This discussion should also recall your high school physics: just replace the weights with atoms!

When rotating, then, atoms A and B tumble around the center-of-mass point as we've now defined. Each atom undergoes circular or angular motion, so that the rotational kinetic energy is

$$K_{rot} = \tfrac{1}{2} m_A v_A^2 + \tfrac{1}{2} m_B v_B^2 \qquad (8.3)$$

$$= \tfrac{1}{2} m_A (r_A \omega)^2 + \tfrac{1}{2} m_B (r_B \omega)^2$$

where the v's are velocities *tangent* to the circular paths and ω is the *angular velocity* $d\theta/dt$, with the angle of rotation θ measured in radians (π radians = 180°). The angular velocity ω is the same for each atom, since (as we assume) they are rigidly bonded together. To express K_{rot} in terms of the bond length r_e, we substitute for r_A and r_B from Equation 8.2 to find

$$K_{rot} = \tfrac{1}{2} \frac{m_A m_B}{m_A + m_B} r_e^2 \omega^2 \qquad (8.4)$$

$$\equiv \tfrac{1}{2} \mu r_e^2 \omega^2$$

$$\equiv \tfrac{1}{2} I \omega^2$$

where μ is the **reduced mass**

$$\mu = \frac{m_A m_B}{m_A + m_B} \qquad (8.5)$$

of the rotating pair (recall the Bohr model and its reduced-mass correction, Chapter 2), and I is the **moment of inertia:**

$$I = m_A r_A^2 + m_B r_B^2 = \mu r_e^2 \qquad (8.6)$$

The resemblance of the final form of Equation 8.4 to the familiar $K = \tfrac{1}{2} m v^2$ should help you to remember it. The second form of Equation 8.4 evidently describes a single, hypothetical "atom" of mass μ rotating at a distance r_e from the origin.

The rotational kinetic energy can also be cast in terms of the **angular momentum**, usually denoted J, for molecular rotation. The angular momentum is actually a vector pointing perpendicular to the plane of rotation, but K_{rot} depends only on its magnitude. Just as I replaces m and ω replaces v in K_{rot}, we have $J = I\omega$ instead of

$p = mv$ for the definition of J. Then, using the last form of K_rot in Equation 8.4, we can write K_rot in terms of J as

$$K_\text{rot} = \frac{J^2}{2I} \tag{8.7}$$

analogous to $K = p^2/2m$ for linear momentum.

Once initiated, molecular rotation is *unforced*, that is, the potential energy V is independent of the angle θ. This simplifies the mechanics of rotation, and dictates a constant energy $\varepsilon_\text{rot} = K_\text{rot}$ and also a constant angular momentum J and angular velocity ω.

Now let's suppose a molecule has $k_\text{B}T$ of energy stored in rotation with $T = 300$ K (room temperature). We can then deduce how fast a molecule typically rotates, as follows. The **rotational period** τ_rot is given by

$$\tau_\text{rot} = \frac{2\pi \text{ radians/revolution}}{\omega \text{ (radians/sec)}} \tag{8.8}$$

with $\omega = [2K_\text{rot}/I]^{1/2}$ from Equation 8.4. For $K_\text{rot} = k_\text{B}T$, this gives us

$$\tau_\text{rot} = 2\pi \sqrt{\frac{I}{2k_\text{B}T}} = 2\pi \sqrt{\frac{\mu r_e^2}{2k_\text{B}T}} \tag{8.9}$$

Let's put numbers into Equation 8.9. For N_2, $r_e = 1.098$ Å, and

$$\mu = \frac{m_N m_N}{m_N + m_N} = \frac{m_N}{2}$$

$$= (7.003 \text{ amu}) \left(\frac{1 \text{ g}}{6.022 \times 10^{23} \text{amu}} \right) \left(\frac{1 \text{kg}}{10^3 \text{g}} \right)$$

$$= 1.163 \times 10^{-26} \text{ kg}$$

(Note that μ is always less than the mass of either atom of the diatomic.)

Then from Equation 8.9

$$\tau_\text{rot} = 2\pi \left[\frac{(1.163 \times 10^{-26} \text{ kg})(1.098 \times 10^{-10} \text{ m})^2}{2(1.381 \times 10^{-23} \text{ JK}^{-1})(300 \text{ K})} \right]^{1/2}$$

$$= 8.17 \times 10^{-13} \text{ s}$$

If you are sitting on one of the atoms, this is the time between "atom-rises" and thus defines what could be called a day in the life of a molecule. Of course, τ_rot depends on the mass and bond length of a particular molecule as well as on the energy; for many molecules at room temperature, $\tau_\text{rot} \approx 10^{-12}$ s, that is, one molecular day is one picosecond. In the twinkling of an eye, a molecule has rotated about 100 billion times.

Quantization of Rotational Energy

Electrons in molecules are not the only particles subject to the laws of quantum mechanics. Everything has a de Broglie wavelength λ, and whether quantum rules will have a noticeable effect depends on whether λ is large enough to be comparable to the dimensions of the system. For a molecule with $k_B T$ of (rotational) kinetic energy, λ is easily estimated:

$$\lambda = \frac{h}{\mu v} = \frac{h}{\sqrt{2\mu\varepsilon}} = \frac{h}{\sqrt{2\mu k_B T}} \tag{8.10}$$

Notice that we have used the reduced mass μ, by which the rotational (and vibrational, see next section) motion of a diatomic molecule is reduced to that of a single "atom." For N_2 at 300 K, Equation 8.10 yields $\lambda = 0.67$ Å, quite comparable to the bond length of N_2 (1.10 Å) and therefore to the circumference of a rotational orbit. We must conclude that quantum mechanics will exercise great influence over molecular rotation; that is, that *molecular rotational energy is quantized*. Quantization of the rotational energy then implies from Equation 8.7 that the rotational angular momentum J is also quantized.

The problem is to set up and solve the Schrödinger equation (Equation 3.1) for rotational motion, where r_e is assumed fixed (the so-called rigid-rotor approximation) and the potential V is zero. Before this prospect gives you conniptions, we point out that the problem is mathematically identical to that of an electron rotating around the nucleus! So Schrödinger and Legendre have already done the job for us! But just so you don't lose track of how the result we will give shortly came about, we recap a bit from Chapter 3. The **boundary condition** on the **rotational wave function** $Y(\theta,\phi)$ that gives rise to energy and angular momentum quantization is that the wave meet itself after a full rotation around the origin. As before, the wave functions replace the classical rotatory motion, and they are given by $Y_{jm}(\theta,\phi)$, the so-called spherical harmonics, with j the angular momentum quantum number (equivalent to l for the H atom) and m the projection quantum number (as in H). The Y's are the very same angular functions given in Table 3.1, replacing l by j.

The quantum formula for the rotational energy (see Exercise 3.25) turns out to be*

$$\varepsilon_j = \frac{h^2 j(j+1)}{8\pi^2 I}, \qquad j = 0, 1, 2, 3, \ldots \tag{8.11}$$

By comparison with Equation 8.7, the classical angular momentum becomes

$$J^2 \longrightarrow \frac{h^2}{4\pi^2} j(j+1) \tag{8.12}$$

The m quantum number, $-j \leq m \leq j$ in integer steps, specifies the direction of rotation by its sign, and the tilt of the rotational plane with respect to the z axis by its magnitude; for $m = \pm j$ the molecule is rotating in the xy plane, while for $m = 0$ the plane passes through the z axis. Each rotational energy level, specified by j, has $2j + 1$ m-values (rotational planes and direction) associated with it, and is therefore $(2j + 1)$-fold degenerate, just like the H atom energy levels. Spectroscopists normally state the rotational

---

* There is more than an accidental similarity between Equation 8.11 and the particle-in-a-box formula $E_n = h^2 n^2/(8ma^2)$ (see Appendix B). We have just turned the "box" boundary conditions into those of a closed ring; Figure 2.11 illustrates the analogy.

levels in wavenumber (inverse centimeters) units, obtained by dividing the energy by hc (see Chapter 2):

$$\frac{\varepsilon_j}{hc} = \frac{h}{8\pi^2 cI} j(j+1) \equiv Bj(j+1) \tag{8.13}$$

where $B = h/(8\pi^2 cI)$ is called the **rotational constant,** and has units of inverse centimeters, since j is a unitless integer. Figure 8.3 illustrates the approximately quadratic dependence of ε_j on j, and Table 8.2 lists some experimental rotational constants B for linear molecules. These vary by a factor of more than 1000 due to the wide range of reduced masses and bond lengths.

Equation 8.13 holds for linear molecules like CO_2 as well as diatomics. Although certain symmetrical polyatomic molecules (called *symmetric tops,* for example, CH_3Cl or NH_3) follow a more complicated formula with two quantum numbers, no one has as yet been able to derive a general energy-level formula for an arbitrary polyatomic molecule. Though such molecules still have energy levels, these do not obey a simple formula like that of Equation 8.13.

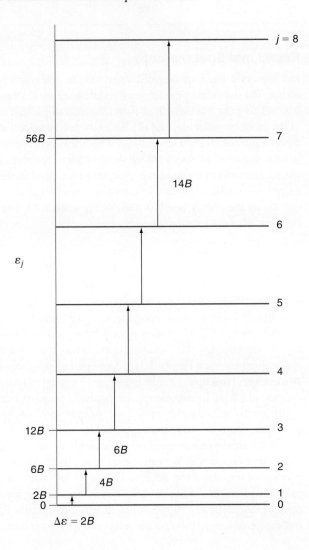

Figure 8.3
The rotational energy levels for linear molecules increase in spacing as the rotational quantum number j increases. The arrows indicate allowed transitions in an absorption spectrum, for which $\Delta j = +1$ and the molecule must have a dipole moment. The spectrum then consists of a series of equally spaced lines as in Figure 8.4.

TABLE 8.2

Rotational constants B for linear molecules

Molecule	B (cm^{-1})	Molecule	B (cm^{-1})
H_2	60.8	NaCl	0.218
D_2	30.4	CsI	0.0236
HF	21.0	LiH	7.51
HBr	8.46	CO_2	0.39
Li_2	0.67	CS_2	0.109
N_2	2.01	N_2O	0.418
O_2	1.45	HCN	1.48
F_2	0.89	C_2H_2	1.18
CO	1.93		
NO	1.70		
I_2	0.037		

Rotational Spectroscopy

In Chapters 2 and 3 we devoted some time to discussing the interaction of light with matter. The important principles we restate here are (1) the Bohr postulate $\Delta\varepsilon = h\nu = hc/\lambda$ relating the wavelength of light absorbed or emitted to the energy-level spacings in the atom or molecule, and (2) the existence of **selection rules** due to the electric-dipole nature of "optical transitions." The spacings between rotational levels determine in what region of the spectrum we should look to see rotational transitions. As we shall see later, rotational transitions *accompany* transitions in the vibrational and electronic degrees of freedom, but here let's confine ourselves to *pure* rotational spectroscopy.

Using the energy-level formula of Equation 8.13, you can show that the spacing between adjacent rotational levels j and $j-1$ is

$$\frac{\Delta\varepsilon_{j,j-1}}{hc} = 2Bj \qquad (8.14)$$

as illustrated in Figure 8.3. For the (typical) rotational constants in Table 8.2, we find spacings of a few inverse centimeters, or wavelengths λ in the range of millimeters to centimeters. Such λ's, which are much longer and of lower energy than the visible-to-UV wavelengths that atoms absorb, cover the **far-infrared** to the **microwave** spectral regions (see Figure 2.3), a large part of pure rotational spectroscopy is **microwave spectroscopy**. Thanks to the development of radar during World War II, and the ease of tuning frequencies in the gigahertz range (0.1 cm^{-1} × 3 × 10^{10} cm s^{-1} = 3 GHz) with ordinary electronics, microwave spectroscopy has attained a very high precision, down to ~1 Hz in frequency (10^{-10} cm^{-1}).

What are the requirements for a molecule to show a pure rotational spectrum (absorption or emission)? We can appeal to the hydrogen atom for help, since the electron's rotation about the nucleus is analogous. The most basic prerequisite is the presence of an *oscillating electric dipole*. An electron rotating (or vibrating) about a nucleus automatically supplies this, due to the opposite charges. But a rotating neutral molecule cannot, *unless the molecule has a permanent dipole moment μ*. This rules out homonuclear diatomic molecules such as H_2 or N_2, and centrosymmetric molecules such as CO_2 or CH_4 (see Chapter 6), but includes most other molecules such as HCl, H_2O, or SO_2. Selection rules supply a critical constraint; again by analogy with the H atom, where $\Delta l = \pm 1$ must be upheld for allowed transitions, the **rotational selection rule** for linear molecules is $\Delta j = \pm 1$. Because only the rotational energy is changing, absorption corresponds to $\Delta j = +1$, and emission to $\Delta j = -1$. And, of course, the Bohr frequency condition $\Delta \varepsilon_{j,j-1} = h\nu = hc\bar{\nu}$ must hold.*

Figure 8.4 diagrams a microwave spectrometer and shows the pure rotational absorption spectrum of HCl. According to Equation 8.14, transitions $j = 0 \longrightarrow j = 1$, $1 \longrightarrow 2$, $2 \longrightarrow 3$, etc., occur at $2B$, $4B$, $6B$, etc., respectively, giving a spectrum with lines equally spaced by $2B$ when plotted against $\bar{\nu}$. This is actually a line spectrum, except that the absorption of the microwave radiation is measured electronically by a detector instead of photographically while the frequency of the microwaves is swept. *The spectrum thus measures the rotational constant directly* through the line spacing, and thus the bond length may be obtained by a simple calculation, as the following example shows.

EXAMPLE 8.1

The line positions of the fourth, fifth, and sixth lines of the spectrum shown in Figure 8.4 are $\bar{\nu}_4 = 83.03$ cm^{-1}, $\bar{\nu}_5 = 103.8$ cm^{-1}, and $\bar{\nu}_6 = 124.3$ cm^{-1}. Find the equilibrium bond length r_e for HCl.

Solution:

First we use the $\bar{\nu}$'s to find B, and then B to find r_e. The differences $\bar{\nu}_5 - \bar{\nu}_4$ and $\bar{\nu}_6 - \bar{\nu}_5$ are 20.8 and 20.5 cm^{-1}, respectively, giving an average $B = \frac{1}{2}(20.65) = 10.32 \pm 0.2$ cm^{-1}. Then, using Equation 8.6 and 8.13, the moment of inertia is

$$I = \mu r_e^2 = \frac{h}{8\pi^2 cB} \quad \text{or} \quad r_e = \sqrt{\frac{h}{8\pi^2 cB\mu}}$$

The reduced mass μ, in kilograms, is

$$\mu = \frac{m_H m_{Cl}}{m_H + m_{Cl}} = \frac{(1.008)(35.00)}{(1.008 + 35.00)} \text{amu} \left(\frac{1 \text{ g}}{6.0221 \times 10^{23} \text{ amu}}\right)\left(\frac{1 \text{ kg}}{10^3 \text{ g}}\right)$$

$$= 1.627 \times 10^{-27} \text{ kg}$$

where the m's are atomic masses in atomic mass units. In this case we choose the more abundant ^{35}Cl isotope; the ^{37}Cl isotope will show a slightly different spectrum, which can be observed with better spectral resolution. Then

$$r_e = \left[\frac{6.6261 \times 10^{-34} \text{ J s}}{(8\pi^2)(2.9979 \times 10^8 \text{ m s}^{-1})(10.32 \times 10^2 \text{ m}^{-1})(1.627 \times 10^{-27} \text{ kg})}\right]^{1/2}$$

$$= 1.291 \times 10^{-10} \text{ m} = 1.291 \pm 0.01 \text{Å}$$

If the calculation is to be repeated for several molecules, then all of the constants and conversion factors can be gathered together and precalculated, resulting in the following "engineering" formula:

---

* The *classical* (Maxwellian) requirement for absorption or emission is that the frequency of oscillation of the dipole match the frequency of the light. It can be shown (see Exercise 11 at the end of this chapter) that in the limit of small $\Delta\varepsilon$, the Bohr condition reduces to this classical one.

$$r_e\,(\text{Å}) = \frac{4.106}{\sqrt{B\,(\text{cm}^{-1})\,\mu\,(\text{amu})}}$$

The frequency of the $0 \longrightarrow 1$ line of $H^{35}Cl$ has now been measured to nine significant figures, resulting in a bond length of $r_e = 1.274552$ Å, limited in precision by the uncertainty in Planck's constant. Our calculated value is slightly larger due to bond-stretching effects in the higher rotational states.

Pure rotational spectra of nearly all common polar diatomic and polyatomic molecules have now been measured to high precision, yielding much of the information we have quoted in earlier chapters on bond radii, bond lengths, and molecular shapes. Elec-

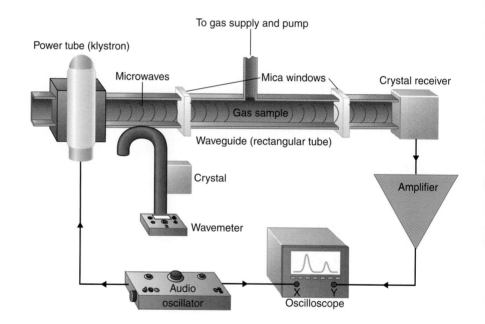
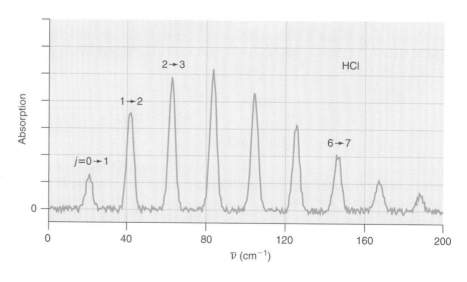

Figure 8.4
A waveguide microwave spectrometer and a pure rotational spectrum for HCl. The microwaves are attenuated by the sample when their frequency matches a rotational transition frequency, resulting in a reduced signal at the receiver. The audio oscillator allows sweeping of the microwave frequency to obtain the spectrum. The microwave technique is workable at wavelengths of about 0.05 cm or larger; hence, the HCl spectrum shown is a hybrid of microwave ($j = 0 \longrightarrow 1$) and far-infrared ($1 \longrightarrow 2$ and higher) absorption measurements. The equally spaced rotational lines reflect the increasing rotational spacing with increasing j and the $\Delta j = 1$ selection rule illustrated in Figure 8.3. The relative intensities of the spectral lines reflect a random (statistical) population of rotational levels of HCl at 300 K.

trodes inserted into microwave spectrometers have also allowed the determination of electric dipole moments, such as those referred to in Chapter 6, through the effect of the applied electric field on the rotational energy levels (known as the *Stark effect*). The spectral resolution achieved in the microwave region has become so high that coupling of nuclear spin with rotation is now routinely measured (so-called hyperfine splittings). Part of this enhanced resolution has come from using molecular beams (like those of Stern and Gerlach) instead of static gas samples. Microwave spectroscopy of free radicals combined with radioastronomy has told us that in many regions of interstellar space there are "clouds" containing dozens of different kinds of molecules.

Although nonpolar molecules cannot absorb or emit rotational quanta directly, they *can* undergo changes in rotational state while *scattering* light of much shorter wavelength. The scattered light then shows weak satellites due to the rotational transitions, as illustrated in Figure 8.5, in what is known as the **Raman effect.** The

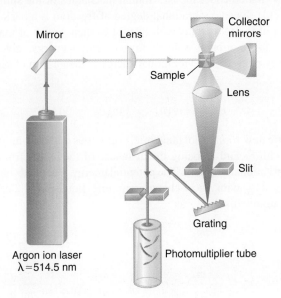

Figure 8.5
A schematic of a laser-based Raman spectrometer, and a Raman rotational spectrum of N_2O. The diffraction grating reflects different wavelengths of scattered light through different angles; it is rotated by a motor to scan the spectrum. Because the scattered light is all in the visible range, it is readily detected by the photomultiplier tube, which operates on the photoelectric effect (see Chapter 2). Note that the N_2O lines are more closely spaced than those for HCl (Figure 8.4) due to the heavier atoms and longer bonds, though the shape of the intensity envelope is similar. The largest frequency shift is less than 1% of that of the exciting line.

resolution of the experiment is lower than the microwave, but the use of a narrow-band laser as the exciting light source has greatly improved it. The type of information obtained on bond lengths and angles is the same as for conventional absorption spectroscopy.

8.3 Vibrational Motion: Molecular Calisthenics

How does a molecule occupy itself during a typical molecular day? As we mentioned at the opening of this chapter, molecules are always stretching and compressing their bonds and flexing their bond angles in a frenetic, ceaseless molecular exercise routine. The extent to which they do this depends on their (average) energy $\bar{\varepsilon}_{vib}$ (i.e., on the temperature) and on the strength of the forces that tend to keep them at their equilibrium geometry.

Again let's first consider the Newtonian mechanics of the simplest example, a diatomic molecule. Its sole vibrational degree of freedom is a back-and-forth or in-and-out motion of the two atoms with respect to their center of mass. Unlike rotational motion, this motion is subject to a potential energy, given by the **bond potential energy curve**.* The vibrational energy is then a sum of kinetic and potential energy terms,

$$\varepsilon_{vib} = \tfrac{1}{2} m_A v_A^2 + \tfrac{1}{2} m_B v_B^2 + V(r) \tag{8.15}$$

where the v's are now *radial* (not tangential) velocities of the atoms, $v_A = dr_A/dt$ and $v_B = dr_B/dt$, with r_A and r_B being the distances of the atoms from their center of mass as in the rotational problem. The potential energy V depends on the interatomic distance $r = r_A + r_B$, with r now changing in time. Equations 8.2 can be written for a general interatomic distance r as

$$r_A = \frac{m_B}{m_A + m_B} r; \quad v_A = \frac{dr_A}{dt} = \frac{m_B}{m_A + m_B} \frac{dr}{dt} \tag{8.16}$$

$$r_B = \frac{m_A}{m_A + m_B} r; \quad v_B = \frac{dr_B}{dt} = \frac{m_A}{m_A + m_B} \frac{dr}{dt}$$

where the equations on the right result from differentiating the original relations with respect to time. The derivative dr/dt represents the *relative* vibrational velocity v, and you can easily demonstrate that Equation 8.15 reduces to

$$\varepsilon_{vib} = \tfrac{1}{2} \mu v^2 + V(r) \tag{8.17}$$

when Equations 8.16 are substituted for v_A and v_B. Here μ is again the reduced mass of Equation 8.5; note that again, as in rotation, the problem reduces to that of a single "particle" of mass μ subject to V.

We would like now to know how the interatomic distance r depends on the time t. We did not bother to find how the rotational angle θ depended on t, since the motion was unforced and the t dependence was therefore linear: $\theta = \theta_0 + \omega t$. However,

---

* We assume that the nuclei vibrate slowly enough that the electrons can "keep up"; this is called the **Born–Oppenheimer approximation.** For most vibrational motions, this assumption holds well; in the course of certain electron transfer reactions it often does not.

8.3 Vibrational Motion: Molecular Calisthenics

$r(t)$ should be more interesting, since $V \neq 0$. In the absence of external forces, energy is conserved, $\varepsilon_{\text{vib}} = $ constant. This implies that

$$\frac{d\varepsilon_{\text{vib}}}{dt} = 0 = \frac{d}{dt}(\tfrac{1}{2}\mu v^2 + V(r)) \qquad (8.18)$$

$$0 = \tfrac{1}{2}\mu \cdot 2 \frac{dr}{dt}\frac{d^2r}{dt^2} + \frac{dV}{dr}$$

$$0 = \mu \frac{dr}{dt}\frac{d^2r}{dt^2} + \frac{dV}{dr}\frac{dr}{dt}$$

$$0 = \mu \frac{d^2r}{dt^2} + \frac{dV}{dr}$$

where we have carried through the time derivative using the chain rule and, in the last line, cancelled the common dr/dt factor. Because the force $F(r) = -dV/dr$, you should recognize the last line as Newton's second law of motion $F = ma$ (see Chapter 1), in the form

$$-\frac{dV}{dr} = \mu \frac{d^2r}{dt^2} \qquad (8.19)$$

In this case you can see that Newton's law is equivalent to, or implied by, conservation of energy, a result that is quite general for nonfrictional forces.

Solving the differential equation (Equation 8.19) will give us $r(t)$, but this requires that we specify the potential V. Figure 8.6 depicts the by-now-familiar bond potential energy curve. Although the qualitative form of this curve is well known, it in general follows no analytical force law because of its complicated dependence on orbital overlap.* However, as we will see shortly, the vibrational energy at ordinary temperatures is much less than the bond energy D_e ($\bar{\varepsilon}_{\text{vib}} \approx k_B T \ll D_e$), and thus a diatomic

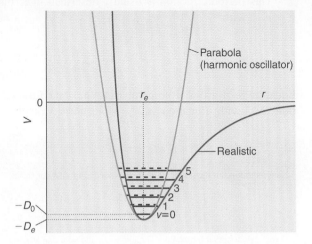

Figure 8.6
Comparison of the harmonic oscillator (HO) potential $V = -D_e + \tfrac{1}{2}k(r - r_e)^2$, green curve, with a realistic bond potential, blue curve. The lowest six vibrational levels are shown for each. While the HO levels remain equally spaced, the realistic level spacing shrinks as v grows. The energy required to break the bond is D_0 rather than D_e, due to the zero point vibration energy. The D_e term characterizes the well depth of the potential, and may be thought of as the classical dissociation energy.

---

* Several bond potential forms have been proposed. The most widely used is the *Morse potential*, $V(r) = D_e\{1 - \exp[-\beta(r - r_e)]\}^2 - D_e$, which has the correct qualitative form of Figure 8.6, but is by no means an exact bond potential.

molecule will ordinarily vibrate near the bottom of the potential well. As Figure 8.6 illustrates, near the bottom the potential may be well approximated by a parabola

$$V(r) = -D_e + \tfrac{1}{2} k(r - r_e)^2 \tag{8.20}$$

where k is a constant.* You may recall from Chapter 1, Example 1.4 (or your high school physics), that this is the "spring" potential reflecting Hooke's law of springs $F = -k(r - r_e)$. With this simplification Newton's law (Equation 8.19), becomes

$$\mu \frac{d^2 r}{dt^2} = -k(r - r_e) \tag{8.21}$$

By direct substitution you can verify that

$$r(t) = r_e + \sqrt{\frac{2\varepsilon_{\text{vib}}}{k}} \sin\left(\sqrt{\frac{k}{\mu}}\, t\right) \tag{8.22}$$

is a solution of Equation 8.21, where $r = r_e$ at $t = 0$. This sinusoidal oscillation of r about r_e is called **simple harmonic motion,** and the mechanical system subject to the potential of Equation 8.20 is called the **harmonic oscillator.** The motion is illustrated in several ways in Figure 8.7. Notice that the energy ε_{vib} of the oscillation goes as the *square* of the amplitude (the coefficient of the sine function in Equation 8.22), just as in the classical wave theory of light. The energy ε_{vib} is continuously being exchanged between kinetic and potential energy, becoming entirely kinetic when $r = r_e$, and entirely potential when r reaches its maximum or minimum possible values,

$$r_\pm = r_e \pm \sqrt{\frac{2\varepsilon_{\text{vib}}}{k}} \tag{8.23}$$

the so-called *turning points* of the classical motion.

This periodic motion is usually described in terms of frequency ν or period $\tau = \nu^{-1}$, with

$$\nu = \frac{1}{2\pi} \sqrt{\frac{k}{\mu}} \tag{8.24}$$

Equation 8.24 arises from comparing the standard sine wave of Chapter 2, Equation 2.1, with Equation 8.22. Molecules typically have force constants $k \approx$

---

* Near its minimum the potential may be expanded in a Taylor series

$$V(r) = V(r_e) + \left.\frac{dV}{dr}\right|_{r=r_e} (r - r_e) + \tfrac{1}{2} \left.\frac{d^2 V}{dr^2}\right|_{r=r_e} (r - r_e)^2 + \cdots$$

If we neglect higher terms, take $V(r_e) = -D_e$, and note that at $r = r_e$ $dV/dr = 0$, we arrive at the form of Equation 8.20, with

$$k = \left.\frac{d^2 V}{dr^2}\right|_{r=r_e}.$$

Thus, k gives the *curvature* of V at its minimum.

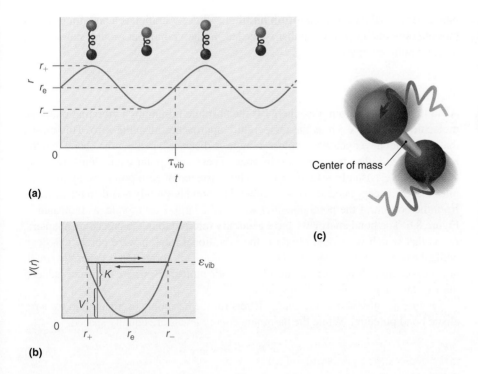

Figure 8.7
Graphic descriptions of a vibrating chemical bond. In (a) the sinusoidal variation of the interatomic distance about its equilibrium value with time is plotted. In (b) this in-and-out motion, represented by the arrows, is projected onto the harmonic oscillator potential to show the continual interconversion of kinetic and potential energy. At $r = r_e$ the vibrational energy is entirely kinetic, while at either turning point (r_+, r_-) it is entirely potential. The diagram in (c) illustrates the simultaneous rotation and vibration occurring in a diatomic molecule. Note that many vibrations typically occur even in a small fraction of a rotational period.

500 N m^{-1} (5 × 10^5 dyne cm^{-1}), and for a typical $\mu \approx$ 10 amu = 1.7 × 10^{-26} kg, Equation 8.24 yields $\nu \approx 3 \times 10^{13}$ Hz ($\bar{\nu} = \nu/c \approx 1000$ cm^{-1}) and $\tau_{vib} \approx 2 \times 10^{-14}$ s (20 fs). If this is compared to a typical rotational period $\tau_{rot} \approx 10^{-12}$ s, we can see that a molecule vibrates ~50 times during a typical molecular day. Unlike rotation, the vibrational period (in the harmonic oscillator model) is *independent* of the vibrational energy. As illustrated in Figure 8.7, a molecule actually vibrates and rotates simultaneously.

Quantization of Vibrational Energy

The same de Broglie wavelength argument we used to indicate the need for a quantum-mechanical treatment of rotational motion also applies to vibration. Here the quantum laws are more stringent, though, because the strong restoring force provided by a chemical bond allows only a small uncertainty in bond length r. According to the uncertainty principle $\Delta r \Delta p_r \geq h/4\pi$, this induces a sizable uncertainty Δp_r and hence a significant **zero point vibrational energy.** This is the same principle we invoked to explain the zero point energy of the hydrogen atom in Chapter 3.

In the harmonic oscillator model, the Schrödinger equation for vibration can be solved *exactly* to yield analytic wave functions and energy levels. The equation and boundary conditions are much like those of the 1-D particle-in-a-box of Appendix B, and were first solved by the French mathematician Hermite long before the modern era. The resulting energy levels are given by

$$\varepsilon_v = h\nu(v + \tfrac{1}{2}), \qquad v = 0, 1, 2, \ldots \tag{8.25}$$

where v is the **vibrational quantum number**.* Since vibration is a one-dimensional motion, only one quantum number is needed. In the vibrational ground state $v = 0$, the zero point energy is

$$\varepsilon_0 = \tfrac{1}{2}h\nu, \qquad (v = 0) \tag{8.26}$$

As in the hydrogen atom, this energy is the absolute minimum a vibrating diatomic molecule can have, even as the temperature approaches absolute zero. This means that the nuclei never cease to explore a range of distances, making the "bond length" r_e an *average* rather than a fixed distance. These zero point excursions are fairly small, however, usually ~0.2 Å or less. The existence of zero point energy also implies that the energy needed to break a chemical bond is slightly *less* than the distance from the bottom of the bond potential well to the large-r asymptote, as indicated in Figure 8.6. The bond energy we have generally referred to in the preceding chapters, D_e, is that which we would obtain if the vibrational motion were purely classical, while the actual energy needed is $D_0 = D_e - \tfrac{1}{2}h\nu$. This correction is generally small, never exceeding ~5%, and is greatest for hydride molecules, where the small reduced mass inflates ν according to Equation 8.24.

Figure 8.6 also shows the energy levels for the harmonic oscillator and for a realistic bond potential. While the harmonic energy levels are equally spaced,

$$\Delta\varepsilon_v = h\nu \qquad \text{for any } v \tag{8.27}$$

the spacings for the real-world case decrease slightly. (Note that the ν in Equation 8.27 is both the classical oscillator frequency and the frequency of the quantum of light corresponding to ν.) As in rotation, spectroscopists usually express ε_v and $\Delta\varepsilon_v$ in wavenumber units:

$$\frac{\varepsilon_v}{hc} = \bar{\nu}(v + \tfrac{1}{2}); \qquad \frac{\Delta\varepsilon_v}{hc} = \bar{\nu} \tag{8.28}$$

From Equation 8.24, the force constant k is related to $\bar{\nu}$ by

$$k = 4\pi^2 \mu c^2 \bar{\nu}^2 \tag{8.29}$$

and k tells us the shape of the bond potential near its minimum. Combined with the rotational measurement of r_e, we can thus pin down a significant part of the bond potential by measuring *vibrational transitions*. Table 8.3 lists vibrational frequencies $\bar{\nu}$, force constants k, and bond energies D_0 for selected small molecules, while Figure 8.8 illustrates the energy levels for a vibrating, rotating diatomic molecule. Mainly due to confinement by the bond potential, the vibrational levels are generally ~10 times further apart than the rotational except for high j. The vibrational levels by themselves are nondegenerate, but a molecule in a particular (v, j) state will have the $2j + 1$ degeneracy from the rotational component of its motion.

---

* The dependence of the energy on the quantum number correlates with the shape of the potential, being more strongly increasing for potentials that are more confining as the total energy rises. The box potential has maximum confinement, giving $\varepsilon_n \propto n^2$, while the harmonic oscillator potential widens with increasing n, giving $\varepsilon_n \propto n$ as in Equation 8.25, and the Coulomb potential widens still more, giving $\varepsilon_n \propto 1/n^2$.

TABLE 8.3
Vibrational frequencies, force constants, and bond energies

Molecule	$\bar{\nu}$ (cm^{-1})[a]	k (N m^{-1})[b]	D_0 (eV)[c]
N_2	2360	2297	9.76
CO	2170	1902	11.09
O_2	1580	1176	5.12
HF	4139	966	5.87
D_2	3118	576	4.56
H_2	4395	573	4.48
HCl	2990	516	4.49
F_2	892	445	1.60
Cl_2	565	329	2.48
Li_2	351	255	1.05
Br_2	323	245	1.97
I_2	215	173	1.54
NaCl	366	109	4.27
LiH	1406	103	2.43
H_2O	3756a, 3652s, 1595b	776$_{OH}$, 69b	4.77
H_2S	2684a, 2611s, 1290b	414$_{SH}$, 45b	3.51
HCN	3312$_{CH}$, 2089$_{CN}$, 712b	580$_{CH}$, 1790$_{CN}$, 20b	4.29, 9.24
CO_2	2349a, 1388s, 667b	1550$_{CO}$, 57b	8.33
CS_2	1523a, 657s, 397b	750$_{CS}$, 23b	6.00
N_2O	2224a, 1285s, 589b	1460$_{NN}$, 1370$_{NO}$, 49b	(5.76)
NO_2	1621a, 1320s, 648b	913$_{NO}$, 152b	4.85
SO_2	1361a, 1151s, 519b	997$_{SO}$, 81b	5.55

[a] For triatomics, a = antisymmetric stretch, s = symmetric stretch, b = bend.
[b] For triatomics, the k's are no longer related to the $\bar{\nu}$'s by Equation 8.29.
[c] For triatomics, the bond energies are approximate, computed from heats of formation (see Chapter 10).

As we will see in the next subsection, N-atom polyatomic molecules that are in low vibrational states can be viewed as a collection of $3N - 5$ or $3N - 6$ harmonic oscillators called the **normal modes of vibration,** with generally distinct frequencies for each mode except in symmetrical molecules (e.g., CO_2; see Figure 8.2).

Vibrational Spectroscopy

Table 8.3 gives directly the wavelengths for $\Delta v = \pm 1$ transitions. A value of $\bar{\nu} \approx 1000$ cm^{-1} implies wavelengths of 10^{-5} m or 10^4 nm, placing these transitions in the *infrared* region of the spectrum. The discovery of the infrared region of the solar spectrum was made in 1800 by Herschel, but he refused to believe the "heat rays"

Figure 8.8
Energy levels of a rotating, vibrating diatomic molecule. To a good approximation, the total internal energy is the sum of rotational and vibrational energies, $\varepsilon = \varepsilon_v + \varepsilon_j$. As shown, in a realistic molecule the vibrational spacing decreases with increasing v, while the rotational levels are more closely spaced at higher v. Both of these effects are a result of the weaker restoring force at large r for realistic bond potentials (see Figure 8.6). This provides less constraint on the vibrational amplitude, allowing the energy to drop in accord with the uncertainty principle, and also allows the average bond length to expand slightly in the higher levels, giving a higher moment of inertia and consequently smaller rotational constant B. The arrows illustrate allowed transitions in a polar molecule; these yield the doublet structure of the HCl vibrational absorption spectrum of Figure 8.9.

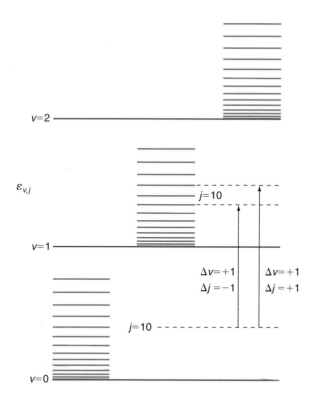

were in any way related to visible light. It was not until 1881 that Abney and Festing found that a variety of organic liquids showed absorption at specific infrared wavelengths. Twenty-five years later Lord Rayleigh speculated that such absorption is due to molecular vibration, and the technique was then extended to smaller molecules and to gaseous and solid samples. The methods of infrared spectroscopy were greatly advanced by Pringsheim, of blackbody radiation renown (see Chapter 2).

The requirement of an oscillating dipole for absorption or emission of radiation is met here by a *change* in the dipole moment that accompanies the vibrational motion. This means that, as in rotation, for diatomic molecules only those with *permanent* dipole moments can absorb or emit infrared light. However, larger molecules that have $\mu = 0$ can still absorb and emit provided the particular vibrational degree of freedom results in a distortion of the molecule's symmetry. For example, the antisymmetric stretching and bending vibrations of CO_2, which has no dipole, create temporary μ's in the course of vibration, and hence can undergo transitions.

The **vibrational selection rule** in the harmonic oscillator model is $\Delta v = \pm 1$; in real molecules, this rule is relaxed somewhat, but the so-called *overtone* transitions with $\Delta v = 2$ or 3 are substantially weaker than $\Delta v = \pm 1$. When $\Delta v = +1$ we have absorption, the most common form of **infrared spectroscopy**.* Figure 8.9 diagrams an infrared spectrometer and displays some typical vibrational spectra. The "fine structure" on the HCl and CO_2 spectra arises as a result of rotational transitions that

---

* Molecules are sometimes formed in vibrationally excited states as a result of chemical reactions, and then emit infrared photons. The study of these chemiluminescent emissions, particularly by J. Polanyi, has taught us much about molecular motions in the course of reaction. See Chapter 15 for further discussion.

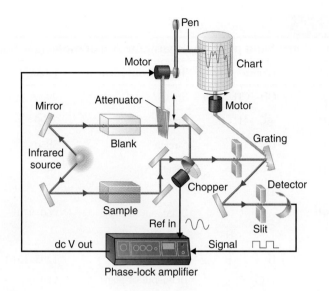

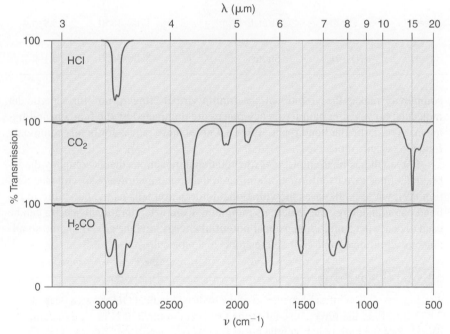

Figure 8.9
Schematic of a dual-beam infrared spectrometer, and three infrared absorption spectra. As in the rotational spectra, these are actually line spectra, but here the resolving power is too low to yield individual rotation–vibration lines, showing instead vibrational bands. (See Figure 8.8 for an explanation of the doublet structure in HCl.) In CO_2 only the antisymmetric stretching (4.3 μm) and bending (15 μm) fundamental vibrations appear, because these motions produce a temporary dipole moment in this otherwise symmetrical molecule, as shown in Figure 8.2. (The other two weaker bands are overtones/combinations.) The more cluttered formaldehyde (H_2CO) spectrum reflects its six vibrational degrees of freedom. Can you assign some of these peaks to characteristic bond vibrations from Table 8.4?

accompany the vibrational ones. The quantitative spectrum obtained depends on the physical state of the molecular sample; spectra of liquids and solids tend to be less structured with broader absorption bands (groups of lines) than those of gases. The spectra of an increasing number of molecules have been measured in the gas phase, but for the most part, especially for organic compounds, infrared spectra are taken with solid samples embedded in a KBr salt pellet.

The vibrational spectrum even of a moderately complex molecule can be astonishingly dense. Not only does each of the allowed normal modes contribute a vibrational band, but the modes also cooperate to produce **combination bands** and undergo **overtone transitions** in which one or more modes change by $\Delta v = 2$. This

TABLE 8.4

Vibrational band positions characteristic of molecular bond types

Bond	Group	Type of vibration	$\bar{\nu}$ (cm^{-1})	λ (μm)
C—H	CH$_2$, CH$_3$	Stretch	2900–3000	3.4–3.3
C—H	≡C—H	Stretch	3300	3.0
C—H	Benzene	Stretch	3030	3.3
C—H	—CH$_2$—	Bend	1465	6.8
C—H	H\C=C/H (cis)	Bend, out-of-plane	960–970	10.4–10.3
C—H	H\C=C/H (gem)	Bend, in-plane	1295–1310	7.7–7.6
		Bend, out-of-plane	690	14.5
C=O	>C=O	Stretch	1700–1850	5.9–5.4
O—H	—O—H	Stretch	3200–3650	3.1–2.7

complexity makes the vibrational spectrum a virtual "fingerprint" for a particular molecule; however, certain frequency ranges are dominated by the fundamental frequencies of particular bond types. Table 8.4 gives these ranges for bonds commonly found in organic molecules.

For small gas-phase molecules, the rotational transitions that accompany the vibrational lines give us information similar to that in the microwave spectra (see Section 8.2), while in all cases the vibrational information tells us the curvature in the bond potential energy curve near its minimum. Complete vibrational spectra can be used to construct a multidimensional **potential energy surface** governing the simultaneous vibrational motion of all the atoms in a molecule.

EXAMPLE 8.2

The low-resolution infrared spectrum of the hydroxyl radical OH shows a peak at $\lambda = 2.675$ μm. Find the force constant of the OH bond in units of N m^{-1}, dyne/cm, and eV/Å^2. Plot the harmonic potential curve for OH (eV versus Å), given $r_e = 0.970$ Å, up to the first excited vibrational level.

Solution:
From $\Delta\varepsilon_{\text{vib}} = h\nu = hc\bar{\nu}$ and Equation 8.29, we have $k = 4\pi^2\mu c^2\bar{\nu}^2 = 4\pi^2\mu c^2/\lambda^2$. The reduced mass μ is

$$\mu = \frac{m_H m_O}{m_H + m_O}$$

$$= \frac{(1.008)(15.999)}{1.008 + 15.999} \text{amu} \left(\frac{1 \text{ g}}{6.022 \times 10^{23} \text{amu}}\right)\left(\frac{1 \text{kg}}{10^3 \text{g}}\right)$$

$$= 1.575 \times 10^{-27} \text{ kg}$$

Then

$$k = \frac{(4\pi^2)(1.575 \times 10^{-27} \text{ kg})(2.998 \times 10^8 \text{ m s}^{-1})^2}{(2.675 \times 10^{-6} \text{ m})^2}$$

$$= 780.8 \text{ N m}^{-1} = 7.808 \times 10^5 \text{ dyne/cm}$$

$$= 780.8 \text{ J m}^{-2} \left(\frac{1 \text{ eV}}{1.602 \times 10^{-19} \text{ J}}\right)\left(\frac{1 \text{ m}^2}{10^{20} \text{ Å}^2}\right) = 48.73 \text{ eV/Å}^2$$

We can then tabulate the values $V(\text{eV}) = \frac{1}{2}(48.73)[r(\text{Å}) - 0.970]^2$:

$r - r_e$ (Å)	V (eV)
0.0	$0(-D_e)$
±0.01	0.0024
±0.05	0.0609
±0.10	0.2437
±0.15	0.5482
±0.20	0.9746

The first excited state occurs at $\varepsilon = h\nu(1 + \frac{1}{2}) = 1.5 hc/\lambda = 1.114 \times 10^{-19}$ J = 0.695 eV. The vibrational energies are only a small fraction of the bond energy $D_0 = 4.392$ eV. As for the rotational example (Example 8.1), we can also develop an "engineering formula" by precalculating the constants and conversion factors to find

$$k \text{ (eVÅ}^{-2}) = 3.667 \times 10^{-6} \mu \text{ (amu)} [\bar{\nu} \text{ (cm}^{-1})]^2$$

As in the pure rotational case, the Raman effect can be used to study vibrational modes of molecules for which the dipole moment does not change, especially for homonuclear diatomic molecules.

The assignments of spectra to various vibrational DFs, and the uniqueness of the information on potential curves and surfaces obtained, are often checked by **isotopic substitution.** For example, the bond potential energy curves for HCl and DCl are (almost) identical, since the electronic structures are the same; however, the reduced masses differ by a factor of 2, making the vibrational and rotational spectra dramatically different. Light-atom substitution causes the greatest change in reduced mass; thus it is typically H, C, N, and O atoms whose masses are varied; ^{2}H, ^{13}C, ^{15}N, and ^{18}O are the most popular minor isotopes.

8.4 Electronic Transitions in Molecules

The chief molecules of air, N_2 and O_2, cannot absorb rotational or vibrational quanta due to their lack of a dipole moment, but are quite able to absorb photons of shorter wavelength and undergo electronic transitions. Air is transparent, so this absorption must occur in the ultraviolet (UV) region of the spectrum. This is to be expected, and is found for many small molecules composed of nonmetals, since the spectra of their atoms also lie in the UV range.

A pioneer in the measurement and analysis of UV electronic spectra in the 1920s and 1930s was Gerhard Herzberg, whose books on the subject are standards in the field. Robert Mulliken provided orbital assignments of the transitions of entire classes of diatomic molecules using the molecular orbital theory he developed. These electronic spectra, as you'll soon appreciate, would have been uninterpretable without the aid of quantum mechanics, and the experiments and theory tended to develop in parallel, more so in this case than in pure rotational and vibrational spectroscopy, or in atomic spectroscopy (see Chapter 2).

The molecular orbital (MO) model presented in the preceding chapter does provide a convenient framework for understanding **molecular electronic transitions.** Electrons in a particular MO configuration are either excited (electronic absorption spectroscopy) or de-excited (emission) to other vacant MOs in complete analogy with atomic transitions. The process of course can have an effect on the chemical bonding in the molecule, since it may change the relative number of bonding and antibonding electrons. There are selection rules for these transitions as well; we will not go into them here, stating only that the rules for electronic transitions are not as stringent as those for rotation or vibration and a transition between a given pair of MO configurations nearly always has an "allowed" component, just as for AOs.

Molecules with single bonds, such as H_2 or HBr, may undergo dissociation upon absorption of a UV photon (**photodissociation**) if the transition occurs from the σ (bonding) to the σ^* (antibonding) MO. This would yield an equal number of bonding and antibonding electrons, and hence a repulsive potential energy between the atoms. Figure 8.10 illustrates the $\sigma \longrightarrow \sigma^*$ transition via energy levels and potential energy curves. In the nonbonded excited state, there are no quantized vibrational states since the boundary condition on the vibrational wave function at large distance

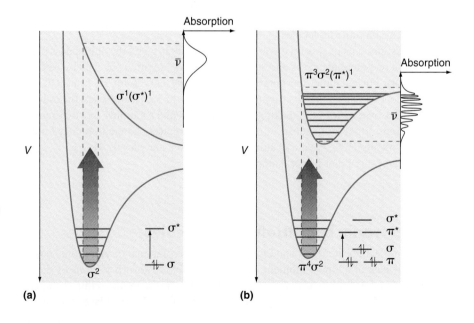

Figure 8.10
Two types of electronic transitions, shown for a diatomic molecule. In (a) a $\sigma \longrightarrow \sigma^*$ transition results in a nonbonding electron configuration, corresponding to a repulsive dissociative potential energy curve, and a continuous absorption spectrum (shown turned on its side). In (b) the absorption, $\pi \longrightarrow \pi^*$, leaves the bond weakened but intact, resulting in a dense line spectrum arising from the rotational–vibrational states of the excited electronic state. The low-resolution spectrum shown for this case displays, as in Figure 8.9, only the vibrational bands. In case (b) there would typically be several electronic absorption bands, some of them overlapping. An example of type (a) behavior is dissociative absorption of HCl(g), while case (b) typifies the ultraviolet spectrum of N_2 or CO.

has been removed. This so-called *bound-free* transition then yields a smooth, *continuous* absorption spectrum.

On the other hand, multiple-bonded molecules such as N_2 or O_2 usually undergo transitions to *bound* excited electronic states when they absorb UV light. This occurs because exciting one electron to an antibonding MO still leaves a net excess of bonding electrons; the bond is weakened but not broken by the excitation. Figure 8.10 also shows this situation. This bound–bound transition ($\sigma \longrightarrow \sigma^*$, $\sigma \longrightarrow \pi^*$, $\pi \longrightarrow \sigma^*$, or $\pi \longrightarrow \pi^*$) gives rise to a complicated line spectrum. The large number of lines results from vibrational and rotational transitions that accompany the electronic one ("rovibronic" transitions). As Figure 8.10 shows, weakening of the bond causes the bond potential energy curve of the newly excited electronic state to shift its equilibrium distance r'_e to larger values. The vertical jump from the ground state then results in the population of a variety of excited vibrational states. The thoroughly confused excited molecule will quickly (usually in microseconds to nanoseconds) reradiate the absorbed photon energy at other wavelengths, to vibrationally excited states of the ground electronic state, or often to other lower excited states. The identity between the emission and absorption spectra we saw for atoms is thus lost for molecules; the emission spectrum from an excited state accessible from the ground state is almost always shifted toward the red.

A rotationally resolved electronic band (either absorption or emission, most often the latter) can be analyzed to yield characteristics of the potential-energy curves of both of the paticipating electronic states, in a manner similar to that described for vibration and rotation, in the case of a diatomic molecule. For triatomics and larger, however, the spectrum is even denser and analysis is not straightforward; such spectra are still the subject of current research. In some cases, the orbital configurations of the states can nonetheless be identified.

Photoelectron Spectroscopy

If the photon wavelength is short enough, an electron can be removed completely from a molecule to become free. This **photoionization** excitation process is analogous to the photoelectric effect (see Chapter 2). The kinetic energy K of the photoelectron is given by a relation similar to Einstein's photoelectric equation (Equation 2.5),

$$K = h\nu - \text{IE}(n) \qquad (8.30)$$

where $\text{IE}(n)$ is the ionization energy out of a particular MO. As an example, Figure 8.11 shows a photoelectron energy spectrum for N_2 ionized by the $1s2p \longrightarrow 1s^2$ 58.43-nm line of helium, along with an apparatus schematic. Superimposed on the three MO energies (σ^*_{2s}, π_{2p}, and σ_{2p}) accessible by the He photon are vibrational progressions that arise when the bond order is altered by the ionization. The progression is most extensive for removal of a π_{2p} electron, the most strongly bonding orbital in N_2. Experiments of this sort probe the MO energies directly, and are often used to assess the quality of quantum-chemical calculations. In addition, the universality of the ionization process allows the technique to be employed as an analytic tool, called *electron spectroscopy for chemical analysis* (ESCA).

Figure 8.11
A schematic of a photoelectron spectrometer, and a photoelectron spectrum for N_2. The experiment must be performed in a high vacuum that is also free of magnetic fields. The orbital labels refer to the MOs from which electrons have been ejected, and the v' quantum numbers label the vibrational progressions in each state of the ion formed.

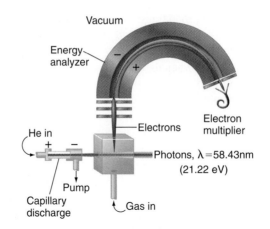

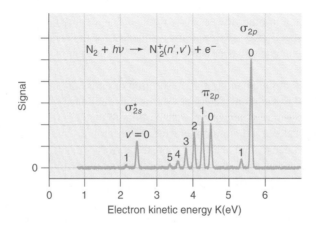

8.5 Looking Back and Peeking Ahead

This chapter concludes our survey of the theory of atoms and molecules that underlies chemistry. In our analysis of structure and reactivity we have emphasized a "one-molecule-at-a-time" approach for getting at chemical essentials. Although we have come far with this program, there are many fascinating everyday chemical properties and events that require us to consider the effects of the stupendously large number of molecules in a laboratory- or industrial-sized chemical system. Why does a gas exert pressure? Why is pressure often an important factor in gas-phase reactions? What do we really mean by temperature? Why does water evaporate? Why do certain reactions occur readily despite an energy deficit? How and to what extent can we extract useful work from a chemical reaction? Does chemistry have anything to say about living systems?

A good part of the answers to such questions involves simply scaling up from molecules to moles. What you have learned to this point is indeed a useful guide to the material to come. But to set foot in the real world—the ocean depths, the swirling atmosphere, the incredible variety of animate and inanimate matter, the planets, sun, stars, and beyond—will require you to broaden your thinking if you are to maintain a chemical perspective on it all. In the coming chapters we will attempt to build a bridge that links the proclivities of individual molecules to the behavior of *matter,* molecules in bulk. Historically, such a bridge was built and crossed, of course, from the opposite shore. We hope you will feel the wispy presence of those who first accomplished the feat, as you pass them on your way to the tangible world of chemistry.

SUMMARY

At ordinary temperatures molecules are always in motion. This motion may be analyzed in terms of **degrees of freedom** (DFs), with one DF assigned to each coordinate-momentum pair (e.g., x, p_x) needed to describe the location and path of an atomic assembly in 3-D space. Atoms have only 3 DFs, ascribed to **translation** through space, whereas molecules of N atoms have $3N$ DFs. Three of these are translational, 2 (if the molecule is linear) or 3 (if not linear) of them are **rotational,** and $3N - 5$ (if linear) or $3N - 6$ (if not linear) of them are **vibrational** DFs.

Rotational motion occurs when the atoms in a molecule tumble around their **center-of-mass** point. The rotational motion of linear molecules is analyzed in detail. It can be described as **kinetic energy of rotation** $\varepsilon_{rot} = K_{rot} = \frac{1}{2}I\omega^2$. Here I is the **moment of inertia,** given for a diatomic molecule AB as $I = \mu r_e^2$, with the **reduced mass** $\mu = m_A m_B/(m_A + m_B)$ and r_e the equilibrium bond length; ω is the angular velocity $d\theta/dt$. In the absence of external forces, K_{rot} is constant, and the molecule rotates in a plane of constant orientation in space. Planar rotation is a consequence of the conservation of angular momentum $J = I\omega$. In terms of J, $\varepsilon_{rot} = J^2/(2I)$. A consideration of the de Broglie wavelength involved in thermally excited rotation shows that quantum mechanics must govern such motion. In the quantum theory of rotation, both J and ε_{rot} are quantized, according to $J^2 = j(j + 1)h^2/4\pi^2$ and $\varepsilon_{rot} = j(j + 1)h^2/(8\pi^2 I)$, where $j = 0, 1, 2, \ldots$. The **rotational constant** B (cm^{-1}) is defined by $\varepsilon_{rot}/(hc) = Bj(j + 1)$; B values for a variety of linear molecules are given in Table 8.2. A molecule typically rotates once every 10^{-12} s (1 ps).

Dipolar molecules can absorb light in the **far-infrared** to **microwave** region of the spectrum ($\lambda \approx 0.001 - 1$ cm); if they are linear, they undergo $\Delta j = +1$ transitions. The resulting line positions are (approximately) equally spaced in inverse centimeter units, being given by $\bar{\nu} = \Delta\varepsilon_{rot}/hc = 2Bj$. The rotational constant B, and therefore the bond length, can then be obtained from the line spacing. Nonpolar molecules cannot absorb directly, but they can scatter visible light while undergoing rotational transitions to yield what is called a **Raman** rotational spectrum. By means of spectral measurements of these types, nearly all the information on bond lengths and angles presented in Chapters 5, 6, and 7 has been obtained.

Vibrational motion in molecules mainly takes the form of bond stretching, where the distances between the atoms oscillate, and bending, where the bond angles oscillate. The stretching vibration of a diatomic molecule is considered in detail. The vibrational energy is the sum of kinetic energy $\frac{1}{2}\mu v^2$ and potential energy $V(r)$ as specified by the **bond potential energy curve,** where again μ is the reduced mass. In the absence of external forces, this energy is conserved. For low vibrational energies the bond potential curve is well approximated by a Hooke's law or **harmonic oscillator** potential $V(r) = \frac{1}{2}k(r - r_e)^2$, where k is the force constant and r_e the equilibrium bond length. For this potential the bond length $r(t)$ is found to oscillate sinusoidally about r_e, implying that the bond lengths quoted in earlier chapters represent average values. In terms of the force constant k, the frequency of vibration is $\nu = (1/2\pi)\sqrt{k/\mu}$. For typical values of k and μ, a molecule is found to vibrate ~50 times during one rotational period.

Because of confinement of the vibrational amplitude by the bond potential energy curve, vibrational motion is more strongly quantized than rotation. This confinement produces a significant **zero point vibrational energy,** given by $\frac{1}{2}h\nu$, with h being Planck's constant and ν the frequency of vibration. In the harmonic oscillator model, the vibrational energy levels are given by $\varepsilon_v = (v + \frac{1}{2})h\nu$, $v = 0, 1, 2, \ldots$, with equal spacing ($h\nu$) between levels, whereas for the actual bond potential, the level spacings decrease slightly as v increases. The existence of zero point vibrational energy implies that the bond dissociation energy D_0 is always a bit less than the distance D_e from the bottom of the potential well to the dissociation limit, that is, $D_0 = D_e - \frac{1}{2}h\nu$. The vibrational levels are much more widely spaced than the rotational except for highly excited rotational states. Vibrational wavenumbers $\bar{\nu} = \nu/c$ are tabulated for a number of diatomic molecules. For polyatomic molecules, each vibrational DF, called a **normal mode of vibration,** is characterized by its own harmonic frequency, and a few triatomic examples are also tabulated.

Vibrational spectroscopy is largely confined to the infrared region of the spectrum, with wavelengths between roughly 1 and 20 μm. The **vibrational selection rule** for vibrational transitions is $\Delta v = \pm 1$ in the harmonic oscillator model, with only dipolar molecules able to emit or absorb. For real molecules, $\Delta v = 2, 3$, etc., **overtone** transitions are also possible, but occur more weakly. In polyatomic molecules, many overtone and **combination** vibrational bands are seen in absorption spectroscopy at shorter wavelengths, in addition to the fundamental $\Delta v = 1$ bands. Organic compounds show frequencies characteristic of particular bonds or groups, and the infrared absorption spectrum then becomes a means for structural analysis and compound identification.

High-resolution vibrational spectra of small molecules show resolved rotational lines comprising the vibrational band. These lines yield bond lengths and angles, as in pure rotational spectra, while the central vibrational frequency tells us the curvature in the bond potential curve near its minimum. The Raman effect can also be used to explore vibrations of nonpolar molecules. **Isotopic substitution** is frequently used to confirm assignments of vibrational frequencies and thereby enable the determination of the potential-energy surface for vibrational motion of an entire molecule.

A natural description of **molecular electronic transitions** is provided by molecular orbital theory. Small molecules composed of nonmetals typically absorb in the ultraviolet region of the spectrum when an electron jumps to an excited MO. When such a transition occurs, a molecule will change its bonding characteristics, bond lengths, and bond angles, and hence rotational and vibrational transitions normally accompany electronic ones. Molecules with single bonds often undergo **photodissociation** when they absorb a UV photon (e.g., in a $\sigma \longrightarrow \sigma^*$ transition), yielding a continuous spectrum, while multiple-bonded molecules reach other bonded states as a rule, yielding a line spectrum containing hundreds or even thousands of vibrational–rotational transitions. The identity between the emission and absorption spectra found for atoms is lost for molecules, whose emission spectra nearly always lie to the red of the absorption.

At very short wavelengths molecules, like atoms, can undergo **photoionization,** in which electrons from various MOs are ejected into the ionization continuum. The spectrum of kinetic energies of the free electrons, called a photoelectron spectrum, provides direct evidence for the existence and relative energy of MOs, and often yields structural information on the photoion produced when vibrational progressions are observed. The technique is also useful as an analytic tool, called electron spectroscopy for chemical analysis (ESCA).

EXERCISES

Degrees of freedom

1. Give the number of translational, rotational and vibrational degrees of freedom for each of the following atoms and molecules:

   He F_2 NO SO_2 H—C≡C—H

2. Same as Exercise 1, for the following atoms, molecules, or groups thereof:

   2Ne Ne_2^+ N≡N—O CH_2=C=O C_3H_8
   (two free neon atoms)

3. Same as Exercise 1, for the following atoms, molecules, or groups thereof:

   2OH H_2O_2 $Cu(NH_3)_4^{2+}$ $(H_2O)_n$ $H^+(H_2O)_n$
   (two free hydroxyl radicals)

Rotational motion

4. Using Equation 8.1 and the diagram that precedes it, show that Equations 8.2 follow.

5. By substituting Equations 8.2 into Equation 8.3, second line, show that Equation 8.4 follows.

6. Use Equation 8.9 to calculate the average rotational periods τ_{rot} at 300 K for $H_2(g)$, $HI(g)$, and $I_2(g)$, where the bond lengths are 0.741, 1.60, and 2.66 Å, respectively. If you viewed a movie of the rotational motion, with the picture centered on the molecular center of mass, how would HI rotation differ from H_2 and I_2 rotation?

7. Using the first form of Equation 8.9, calculate the average rotational period τ_{rot} for $CO_2(g)$ at 300 K. Each C=O bond length is 1.16 Å, and the center of mass is located squarely on the C nucleus. What would τ_{rot} be at 3000 K? At 30 K?

Quantized rotation

8. Calculate the de Broglie wavelength in Å for rotating H_2, HI, and I_2 at 300 K from Equation 8.10. Also find $\lambda/(2\pi r_e)$ for each, using data from Exercise 6. In which of the three molecules is the Newtonian picture of end-over-end tumbling of the atoms closest to reality?

9. In the Bohr model of the hydrogen atom described in Chapter 2, the electron moves in circular paths

about the nucleus in a way exactly analogous to the rigid-rotor model for molecular rotation. By noting that all kinetic energy is then present as rotational energy, K_{rot}, and that $K_{rot} = -E_n$, find the Bohr formula for a rigid rotor, and compare it to Equation 8.13. (*Hint:* The radius of the nth orbit r_n must be used to construct the moment of inertia in Equation 8.7.)

10. Fill in the algebra leading from Equation 8.13 to Equation 8.14.

11. Show that the Bohr transition frequency $\nu = \varepsilon_{j,j-1}/h$ becomes identical to the classical frequency of rotation $\nu_{rot} = 1/\tau_{rot}$ in the limiting case that the rotational energies are much larger than the spacing between them. (This is equivalent to allowing j to grow without bound.)

12. By approximating $j(j+1) \approx (j+\frac{1}{2})^2$, derive a formula for the rotational quantum number $\bar{j}$ corresponding to rotational energy $\varepsilon_{rot} = k_B T$. Calculate $\bar{j}$ for the molecules of Exercises 6 and 8. How do your results illustrate Bohr's correspondence principle? (The formula for $\bar{j}$ of course permits noninteger values of $\bar{j}$; you may round or truncate to the nearest integer.)

13. Use the data of Table 8.2 to compute bond lengths in Å for CsI and LiH, after verifying the "engineering" formula of Example 8.1.

14. Using data from Exercise 6, predict the positions in cm^{-1} of the three longest wavelength lines in the far-infrared spectrum of HI.

15. Which of the following molecules will show a far-infrared or microwave spectrum?

    ClF Cl$_2$ NO$_2$ CS$_2$ C$_2$H$_4$ CH$_3$Cl

Vibrational motion

16. Substitute Equation 8.16 into Equation 8.15 and show that Equation 8.17 results.

17. Use the Maclaurin series for e^x [$e^x = 1 + x + x^2/2 + x^3/6 + \ldots$] to show that the Morse bond potential energy function $V(r) = D_e[1 - e^{-\beta(r-r_e)}]^2 - D_e$ reduces to the parabolic form of Equation 8.20 when r is very near r_e. Identify the force constant k in terms of D_e and β.

18. Use data from Table 8.3 to calculate vibrational periods for H$_2$ and I$_2$, and, using the results of Exercise 6, find the ratios τ_{rot}/τ_{vib} at 300 K. (Note that finding τ_{vib} does not require the temperature.)

Quantized vibration

19. Use the uncertainty principle to estimate the harmonic oscillator zero point vibrational energy, taking $\Delta r = r_+ - r_-$ from Equation 8.23, finding K_{vib} from Δp_r, and using the condition $\langle K_{vib} \rangle = \langle V_{vib} \rangle = \frac{1}{2}\varepsilon_{vib}$, which holds for the harmonic oscillator, where the brackets denote average values.

20. Use data in Table 8.3 to compute the classical amplitudes of vibration (the distance in Å from r_e to the outer turning point) of H$_2$ and I$_2$ in their vibrational ground states. By what fractions do the bond lengths change, given $r_e(H_2) = 0.74$ Å and $r_e(I_2) = 2.66$ Å?

21. Find $k_B T$ for $T = 300$ K in wavenumber (cm^{-1}) units. By comparing $k_B T$ with the entries in Table 8.3, explain why most small molecules are found in their vibrational ground states at room temperature.

22. Based on the data in Tables 8.2 and 8.3, calculate and plot the bond potential energies on the same Å scale for O$_2$ and NaCl up to an energy of 2000 cm^{-1} above the potential minimum, and indicate any bound vibrational energy levels present by horizontal lines as in Figure 8.6.

23. Which of the molecules of Exercise 15 will show an infrared spectrum?

24. In a real molecule, all the overtone transitions ($\Delta v > 1$) are weakly allowed. Use the harmonic oscillator model to estimate the wavelength of laser light (nm) needed to excite the $v = 5$ state of a C—H stretching vibration in benzene (see Table 8.4). In what region of the spectrum is this? Would you expect your calculation to yield an overestimate or underestimate of the actual wavelength required?

25. Molecules containing different isotopes of the same elements are expected to have identical bond potential curves, since the electronic wave function and energy depend only on the charges of the nuclei and not their masses. (This holds only if the nuclei could be held stationary, which is nearly true as far as the electrons' motion is concerned.) Using the data of Tables 1.3 and 8.3 (which refer to the more abundant isotopes) predict the vibrational frequencies $\bar{\nu}$ in cm^{-1} and the bond dissociation energies D_0 in eV of ^{2}H^{35}Cl, ^{1}H^{37}Cl, and ^{14}N^{15}N.

Electronic spectroscopy

26. The halogens Cl$_2$, Br$_2$, and I$_2$ are all colored. Use a MO energy-level diagram to deduce the MOs involved in the visible absorption.

27. Shining 266-nm light on HBr causes dissociation into H and Br atoms. What MOs are involved in this transition? Given the bond energy of HBr, $D_0 = 3.751$ eV, what is the kinetic energy of the atoms after dissociation?

28. In the N_2 molecule, which transition, $\sigma_{2p} \longrightarrow \pi^*$ or $\pi \longrightarrow \pi^*$, will produce a greater change in the bond length? How is this difference reflected in the absorption bands for these two transitions?

29. As given in Appendix B, the particle-in-a-box (PiB) model is a good semiquantitative guide to electronic spectra of long-chain molecules, the longer the better. In molecules with alternating double bonds, the lowest $\pi \longrightarrow \pi^*$ transition wavelength can be predicted from the length of the molecule. Use the PiB model to estimate $\lambda_{\pi \longrightarrow \pi^*}$ for octatetraene $CH_2=CH-CH=CH-CH=CH-CH=CH_2$, for which the length of the molecular "box" is 9.8 Å. Assume a HOMO $\longrightarrow$ LUMO transition in order to determine the PiB quantum numbers. The experimental value is 286 nm.

30. Use the $v' = 0$ peaks in the HeI (58.43-nm) photoelectron spectrum of N_2 presented in Figure 8.11 to estimate molecular orbital energies (eV), and construct a roughly quantitative MO energy-level diagram for N_2. What is the ionization energy (eV) of N_2?

31. The HeI (58.43-nm) photoelectron spectrum of NO shows four bands at electron kinetic energies of 2.89, 4.65, 5.10, and 11.67 eV. Assign these to molecular orbitals based on the level ordering appropriate for O_2, construct a quantitative MO energy-level diagram for NO, and give the ionization energy of NO.

Properties of Gases and the Kinetic Molecular Theory

CHAPTER 9

The town of Magdeburg viewed from the mayor's balcony, 17th century engraving. Von Guericke's water barometer is the four-story-tall spire on the right. On the left is a replica of the top of the barometer, with the floating wooden figure.

Men occasionally stumble over the truth, but most of them pick themselves up and hurry off as if nothing had happened.

Winston Churchill

CHAPTER OUTLINE

9.1 Pressure of a Gas: Barometric Principles

9.2 The Ideal Gas Law

9.3 Gas Mixtures: Dalton's Law of Partial Pressures

9.4 The Kinetic Molecular Theory

9.5 The Behavior of Real Gases

Among the three states of matter—gas, liquid, and solid—the gaseous state was the last to be understood in a chemical sense; but, once their chemical nature was established, gases through their simplicity proved to possess the power to lead us toward a new understanding of the world. Gases, liquids, and solids may be simply (but not rigorously) distinguished by the properties of *size* (or volume) and *shape:* gases

Joseph Louis Gay-Lussac (1778–1850). After his precise confirmation and extension of Charles' law in 1802, Gay-Lussac's comprehensive experimental work and survey of gas reactions established the law of combining volumes in 1808. This law was the key that allowed Amedeo Avogadro to unlock the diatomic nature of elemental gases and to provide the last principle needed to specify the ideal gas equation of state.

have neither in definite proportion, solids have both, and liquids have definite size but not shape. Gases are elusive—hold your open hand in front of you, and try to grab some air. But with your two hands you *can* establish that air is not empty space; crossing your hands and pressing your palms together produces a hiss as something tangible escapes, and the slight pop as you separate them comes as that substance reoccupies the space it had lost.

The word *gas* ultimately derives from the ancient Greek word *chaos,* meaning both "air" and "space." But chaos also has a modern meaning—a state of disorder and confusion—that describes perfectly the modern picture of gases: *molecular chaos.* The 17th-century Belgian chemist van Helmont first used the word *gas* to describe air like vapors of other substances, although the term *airs* for gases of different composition remained in wide use well into the 19th century.

The most readily observable properties of gases turn out to have nothing to do with their chemical properties. We begin with a look at our own earthly atmosphere, Shakespeare's "most wonderful canopy," and the ubiquitous pressure it exerts.

9.1 Pressure of a Gas: Barometric Principles

In 1643 the Italian scientist Evangelista Torricelli, a student of Galileo, discovered that when a filled tube of mercury is inverted into a bowl of mercury, the mercury in the tube does not spill out, but remains suspended. If the tube is longer than about 76 cm, the mercury falls to that level but no further, leaving an empty space above, as shown in Figure 9.1. This device is now known as a **barometer,** and has survived, with improvements, in this form to the present day; you will find a mercury barometer in nearly every college chemistry laboratory building. Otto von Guericke (who is also credited with the invention of the first vacuum pump), as mayor of the German town of Magdeburg, first erected a *water* barometer in 1646, finding that a tower of more than 10 m (34 ft) was required! A little wooden figure floated within the glass-topped brass tower, rising to greet fair weather ahead and falling with foul.

Both Torricelli and von Guericke realized that the force necessary to suspend the barometric fluids arises from the pressure of the atmosphere. Pressure is defined as force exerted per unit area:

$$P = F/A \tag{9.1}$$

The force necessary to suspend a disk of thickness dz and area A of a barometric fluid of density ρ (see Figure 9.2) at a height z above the level of the reservoir is

$$dF = g\,dm = g\rho A\,dz \tag{9.2}$$

where Newton's second law has been applied for the small mass dm of the disk, whose volume is $A\,dz$, and where g is the acceleration due to gravity. Now all we have to do is *add up* all the disks from reservoir level $z = 0$ to the height of the column $z = h$; that is, we *integrate* Equation 9.2:

$$F = \int dF = \int_{z=0}^{z=h} g\rho A\,dz = g\rho A z\big|_0^h = g\rho A h \tag{9.3}$$

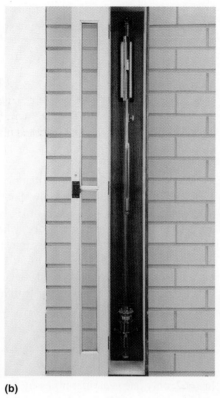

Figure 9.1
(a) Torricelli's barometer. The tube must extend at least 760 mm above the mercury reservoir in order to allow measurement of the pressure on earth near sea level. Changes in weather cause the Hg level to change by up to 30 mm; inland locales have average pressures ranging from 730 to 750 mm down to 500 mm in mountainous regions. The volume above the Hg level in the second and third tubes is, aside from a small amount of Hg vapor, a perfect vacuum. (b) A more modern mercury barometer. An adjusting screw allows the reservoir level to be restored to a fixed mark for measurements accurate to 0.1 mm Hg. Other modern barometers use slight deflection of a metal diaphragm to measure pressure differences.

or, using the definition of Equation 9.1,

$$P = \rho g h \tag{9.4}$$

The last step in Equation 9.3 can be taken, in which the product $g\rho A$ is pulled outside the integral sign, for barometer tubes of uniform cross-sectional area A. The result (Equation 9.4) is independent of A, implying that the fluid will be suspended at the same height no matter how large or small the bore of the tube for a given fluid and external pressure. The difference between water and mercury lies in their different *densities* ρ; for a given P, Equation 9.4 implies that $\rho_{Hg} h_{Hg} = \rho_{H_2O} h_{H_2O}$, or that the ratio of column heights is the inverse ratio of fluid densities. Mercury (Hg) is the densest liquid known, $\rho = 13.595$ g cm^{-3} at 0°C; whereas for H$_2$O(l), $\rho = 1.000$ g cm^{-3} at 0°C.

From Equation 9.4, the height of the Hg column is directly proportional to the pressure. For this reason the atmospheric pressure is still given as mm Hg,* although the units of P, N m^{-2}, are certainly not length units. Dry air at sea level and 0°C supports a 760-mm column of Hg, which corresponds to one **standard atmosphere** (1 atm). The density of Hg (and other fluids) varies slightly with temperature, so the unit **Torr** has been defined so that **760 Torr = 1 atm,** independent of the temperature.

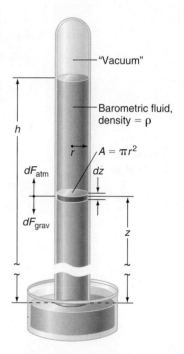

Figure 9.2
The force of gravity dF_{grav} on the disk of barometric fluid, of density ρ and thickness dz, is exactly balanced by the force dF_{atm} arising from the pressure of the atmosphere, causing suspension of the disk at height z above the fluid reservoir. This force equilibrium allows the column height h to be calculated for a uniform cross-sectional area A as given in the text.

---

* In U.S. weather reports the "barometric pressure" is usually reported in "inches," short for inches of mercury. What is a standard atmosphere in "inches"? (1 in. = 2.540 cm)

Like mm Hg, Torr and atm are not proper mechanical units. In the various unit systems, using Equation 9.4 for Hg, we have

$$1 \text{ atm} = 101{,}325 \text{ N m}^{-2} \text{ (Pa)} \qquad (9.5)$$
$$= 1{,}013{,}250 \text{ dyne cm}^{-2}$$
$$= 14.70 \text{ lb in.}^{-2}$$

where Pa is an abbreviation for Pascal, the SI unit of pressure. (Blaise Pascal's contribution to the understanding of our atmosphere was apparently to build a *wine* barometer!)

It was not recognized at first that the space above the mercury within the tube is a vacuum (nothing!). We now know that this vacuum is actually not perfect, but contains a small pressure (10^{-3} Torr) of Hg vapor. This is nicely demonstrated by touching a Tesla coil to the top of the tube, which generates the Hg spectrum.

The tremendous force of the atmosphere may be appreciated from the values given in Equation 9.5. The body of an adult human must withstand a total force of more than 15 tons! Normally rigid containers, such as metal gasoline cans or plastic milk bottles, collapse when the air within them is pumped out.

9.2 The Ideal Gas Law

In 1662, only one year after his essay *The Sceptical Chymist* had turned the world of alchemy on its ear, Robert Boyle published a treatise titled *On the Spring of the Air and Its Effects* that paved the way for a modern theory of gases. Boyle devised a *J-tube* that held a trapped gas volume in its short end, as illustrated in Figure 9.3. By adding Hg to the long end and measuring the volume of gas as well as the difference in heights of the Hg in the short and long ends, Boyle obtained the data given in Table 9.1 and plotted in Figure 9.3. The data show that, to a very good approximation,

$$PV = \text{constant} \qquad \text{or} \qquad P \propto \frac{1}{V} \qquad (9.6)$$

Either form of the *PV* relation is known as **Boyle's law**. For Equation 9.6 to hold, no air (or other gas) could be allowed to escape, and the temperature had to be held constant. Boyle also noted the dramatic effect of a candle flame on the trapped volume, but the notion of temperature was vague then, and there were as yet no reliable thermometers.

More than a century passed before Jacques Charles, in work he never published, performed a quantitative study of the dependence of the volume of an air sample on temperature at constant pressure in about 1787. Joseph Gay-Lussac (who later discovered the law of combining volumes for reacting gases) performed further, more careful studies and published his results in 1802, giving full credit to Charles. The results showed that the volume rose linearly with temperature t (°C) or that

$$V \propto t + t_0 \qquad \text{or} \qquad \frac{V}{t + t_0} = \text{constant} \qquad (9.7)$$

By fitting his data to a straight line, Gay-Lussac obtained the $V = 0$ intercept, corresponding to $t_0 = 267$°C, by extrapolation; the modern value is $t_0 = 273.15$°C.

9.2 The Ideal Gas Law

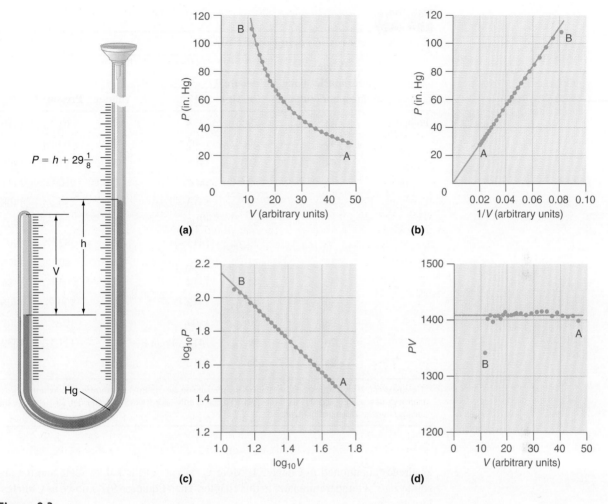

Figure 9.3
Robert Boyle's J-tube and plots of the data he obtained from it. The measured numbers are given in Table 9.1. Plotting the data in different ways can sometimes reveal a simple mathematical relationship between the variables. (**a**) Direct plot of the raw data. Generally, the data are replotted in various standard ways to obtain straight lines when a monotonic relationship like that of plot (a) is found. (**b**) Since plot (a) seems to show an inverse proportion, the abscissa variable V is inverted in plot (b), yielding a straight line with zero intercept, and implying $P = m(1/V)$, with m being the slope of the line. (**c**) A log$N$-log plot can be used to find a power-law relationship $P = aV^b$; the slope of this plot gives the power b. Here $b = -1$ and $a = m$. (**d**) Since plots (b) and (c) imply PV = constant, plotting the product PV versus either P or V on an expanded ordinate scale can show more precisely how well the inverse law is obeyed. The precision with which the measurements were made (to within 1/16 in. or 1.5 mm Hg) was critical in enabling Boyle to derive his law.

The best thermometers then available—and still in use today—exploited the nearly linear expansion of Hg(l) with temperature. What Gay-Lussac's work suggested was that the expansion of air or other gases was more perfectly linear with t than Hg; a more accurate thermometer would therefore use a gas as the expanding fluid. Years later William Thomson, later to be named Lord Kelvin, defined a new, *absolute* (kelvin, K) temperature scale based on gas expansion, a scale with the same size of degree as the Celsius scale, but with its zero set at $t = -t_0$, that is,

$$T(\text{K}) = t(°\text{C}) + 273.15 \tag{9.8}$$

TABLE 9.1
Boyle's data on the spring of the air

	Volume (index marks along uniform bore tubing)	Pressure* (in. Hg)	PV		Volume	Pressure*	PV
Point A	48	29 2/16	1398		23	61 5/16	1410
	46	30 9/16	1406		22	64 1/16	1409
	44	31 15/16	1405		21	67 1/16	1408
	42	33 8/16	1407		20	70 11/16	1414
	40	35 5/16	1413		19	74 2/16	1408
	38	37	1406		18	77 14/16	1402
	36	39 5/16	1415		17	82 12/16	1407
	34	41 10/16	1415		16	87 14/16	1406
	32	44 3/16	1414		15	93 1/16	1396
	30	47 1/16	1412		14	100 7/16	1406
	28	50 5/16	1409		13	107 13/16	1402
	26	54 5/16	1412	Point B	12	111 9/16	1339
	24	58 13/16	1412				

* The first point, 29 2/16, corresponds to atmospheric pressure on the day Boyle performed the experiment; mercury was gradually added to the J-tube, giving a height difference h which was added to 29 2/16 to give the total pressure, as illustrated in Figure 9.3.

as already mentioned in Chapter 1. More is said about the kelvin scale and the absolute zero of temperature later in this chapter. Then Equation 9.7, known as **Charles' law,** becomes

$$\frac{V}{T} = \text{constant} \quad \text{or} \quad V \propto T \tag{9.9}$$

Gay-Lussac also showed that Boyle's law continues to hold for different temperatures, and Charles' for different pressures, so that Equations 9.6 and 9.9 can be combined to give

$$\frac{PV}{T} = \text{constant} \tag{9.10}$$

where the constant depends only on the amount of gas in the sample.

EXAMPLE 9.1

The chemistry of the stratosphere has been investigated by launching helium balloons carrying instrument payloads (mass and optical spectrometers). What volume of helium at 1 atm pressure and 7°C should be blown into a 60,000-L balloon so that it will fully inflate in the stratosphere, where the temperature is −43°C and the pressure 10 Torr?

Solution:
Because both the pressure and temperature are changing, it's most convenient to use Equation 9.10, which implies

$$\frac{P_g V_g}{T_g} = \frac{P_s V_s}{T_s} \longrightarrow V_g = V_s \frac{P_s}{P_g} \frac{T_g}{T_s}$$

where g = ground and s = stratosphere. Any consistent units can be used in the ratios, as long as the temperature is absolute:

$$V_g = (60{,}000 \text{ L})\left(\frac{10 \text{ Torr}}{760 \text{ Torr}}\right)\left(\frac{280 \text{ K}}{230 \text{ K}}\right)$$
$$= 960 \text{ L}$$

where we convert atmospheres to Torrs and degrees Celsius to kelvin.

We could have used Equations 9.6 and 9.9 in succession to solve this problem, even though the intermediate result for the volume, accounting only for the adjustment in pressure, might never actually occur. We could apply Equation 9.9 first, followed by Equation 9.6 as well. This flexibility results from the fact that P, V, and T are properties that depend only on the *state* of the gas at the time of the measurement, and *not* on the *history* of the gas, that is, the *path* chosen to go from one state to another. These considerations will take on greater importance in Chapter 10.

The remaining factor in Equation 9.10, the constant on the right, was elucidated by the chemistry of reacting gases. Gay-Lussac found that in the (sometimes explosive) combination of hydrogen and chlorine to make hydrogen chloride, *one* volume of hydrogen reacts with *one* volume of chlorine to make *two* volumes of hydrogen chloride. This suggested to Amedeo Avogadro that (1) **equal volumes of any gas contain equal numbers of gas molecules** (at given temperature and pressure) and also that (2) gaseous hydrogen and chlorine are present as *diatomic* molecules H_2 and Cl_2. The latter would be necessary to account for the two volumes of HCl; if the reaction occurred between atoms, H + Cl $\longrightarrow$ HCl, only one volume of HCl would have resulted. Figure 9.4 illustrates Avogadro's analysis. Postulate 1 follows from Dalton's law of definite proportions as interpreted in terms of a definite (1:1) atomic ratio for the chloride of hydrogen. It is known as **Avogadro's law,** and can be stated as

$$\frac{V}{n} = \text{constant} \quad \text{or} \quad V \propto n \qquad (9.11)$$

for constant P and T, where n is the number of *moles* of gas, and the proportionality constant is the same for all gases. We can use moles n instead of numbers of molecules N because n and N are proportional; then $n = N/N_A$, where N_A is Avogadro's number, and the proportionality between V and N therefore implies $V \propto n$ as well. Equation 9.11 can now be combined with Equation 9.10 to yield

$$\frac{PV}{nT} = \text{constant} \equiv R \qquad (9.12)$$

Figure 9.4
Avogadro's reasoning in combining Dalton's law of definite proportions, implying a definite atomic ratio in hydrogen chloride, with Gay-Lussac's law of combining volumes, for the reaction of hydrogen and chlorine. Only diagram **(b)** is consistent with Gay-Lussac's measurement of two volumes of hydrogen chloride produced. The sizes of the molecules relative to their containers are greatly exaggerated.

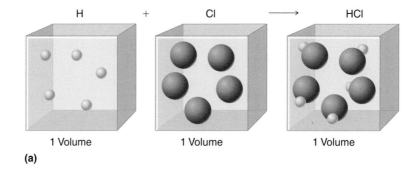
(a)

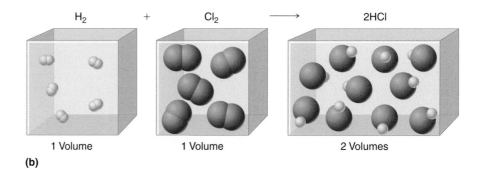
(b)

or

$$PV = nRT \qquad (9.13)$$

as it is usually stated. Equation 9.13 is now called the **ideal gas law,** and the constant of proportionality R is the **ideal gas constant.** (The "ideal" appellation is discussed later in this chapter.) Many years later, Josiah Willard Gibbs would show that the quantities P, V, T, and n exhaust the number of variables needed to describe a pure gas (see Chapter 14). Equation 9.13 is thus a "complete" equation, and is called an **equation of state,** because it describes a relationship among variables that can be established from measurements on the sample without knowledge of its history.

The gas constant R can be obtained by measuring P, V, T, and n for a gas sample and using Equation 9.12, or by other thermodynamic measurements of the sort described in Chapter 10. Its modern value is, in various units,

$$\begin{aligned} R &= 8.31447 \text{ J K}^{-1} \text{ mol}^{-1} \\ &= 0.0820574 \text{ L atm K}^{-1} \text{ mol}^{-1} \\ &= 62.3637 \text{ Torr L K}^{-1} \text{ mol}^{-1} \\ &= 1.98721 \text{ cal K}^{-1} \text{ mol}^{-1} \end{aligned} \qquad (9.14)$$

Note that the PV product has units of energy, as indicated by the first and last R values given.

It is convenient to have a reference set of conditions, T and P, for the purpose of gas stoichiometry; at constant T and P, Avogadro's law holds, $V \propto n$. A state of

standard temperature and pressure (STP) is defined by $T = 273.15$ K ($t = 0°C$) and $P = 1$ atm. At STP the ideal gas molar volume is

$$\overline{V}_{STP} \equiv \left(\frac{V}{n}\right)_{STP} = \frac{(0.0820574 \text{ L atm K}^{-1} \text{ mol}^{-1})(273.15 \text{ K})}{1 \text{ atm}} \quad (9.15)$$

$$= 22.414 \text{ L mol}^{-1}$$

as we introduced in Chapter 1. Boyle's and Charles' laws can be used to convert the actual state of a given volume of gas to STP, and the molar volume $\overline{V}_{STP}$ of Equation 9.15 then allows conversion to moles of gas n. Note that n can also be obtained from solving the ideal gas law for n under the actual conditions. The following example illustrates these two equivalent approaches.

EXAMPLE 9.2

A small sample of liquid CCl_4 is vaporized into an evacuated 0.500-L glass bulb to yield a pressure of 35.3 Torr at 18°C. Find the moles of $CCl_4(g)$ present in the bulb, and calculate the concentration $[CCl_4]$ in mol L^{-1} and in molecules cm^{-3}.

Solution:
First, the STP method:

$$n_{CCl_4} = (0.500 \text{ L})\underbrace{\left(\frac{35.3 \text{ Torr}}{760 \text{ Torr}}\right)}_{\text{Boyle}}\underbrace{\left(\frac{273 \text{ K}}{291 \text{ K}}\right)}_{\text{Charles}}\left(\frac{1 \text{ mol}}{22.4 \text{ L}}\right)$$

$$= 9.73 \times 10^{-4} \text{ mol}$$

Then the gas-law method:

$$n_{CCl_4} = \frac{PV}{RT} = \frac{(35.3 \text{ Torr})(0.500 \text{ L})}{(62.36 \text{ Torr L K}^{-1} \text{ mol}^{-1})(291 \text{ K})}$$

$$= 9.73 \times 10^{-4} \text{ mol}$$

which agrees with the STP method as it should. Then

$$[CCl_4] = 9.73 \times 10^{-4} \text{ mol}/0.500 \text{ L} = 1.95 \times 10^{-3} \text{ mol L}^{-1}$$
$$= (1.95 \times 10^{-3} \text{ mol L}^{-1})(10^{-3} \text{ L cm}^{-3})(6.02 \times 10^{23} \text{ mol}^{-1})$$
$$= 1.17 \times 10^{18} \text{ molecules cm}^{-3}$$

Any chemist who deals with gases uses Equation 9.13 on a regular basis for both static and flowing gas systems; the ideal gas law also has several important applications, among them thermometry and determination of molecular mass of gases and vapors. In a rigid container the pressure of a gas (usually He is chosen) is directly proportional to temperature (from Equation 9.13) and can be measured more precisely than the temperature itself by electromechanical pressure gauges. After a suitable calibration (say, in an ice bath at 0.00°C) the *gas thermometer* becomes a precision temperature-measuring device. For the purpose of molecular mass determination, Equation 9.13 can be rewritten as

$$M = \rho RT/P \quad (9.16)$$

where M is the molecular mass (grams per mole) and ρ is the gas density (usually in grams per liter rather than grams per cubic centimeter). The density of a gas or the vapor of a volatile compound can be determined by weighing a known volume of gas; measuring the pressure and temperature of the sample then allows M to be obtained.

9.3 Gas Mixtures: Dalton's Law of Partial Pressures

Boyle's and Charles' experiments were performed on air samples, which, as Priestley, Lavoisier, and Dalton later showed, actually consist of a *mixture* of gases. Table 9.2 gives the fractional composition by moles (the **mole fraction**) of dry air at sea level. The mole fraction X_i of a particular component i in a mixture (be it gas or not) is defined by

$$X_i = \frac{n_i}{n_1 + n_2 + n_3 + \cdots} \equiv \frac{n_i}{\sum_j n_j} = \frac{n_i}{n} \qquad (9.17)$$

where the capital sigma (Σ) stands for a *summation* over the index beneath it of all terms included to the right of it. The summation here runs over all the components of the mixture, and therefore yields the total number of moles n. Thus X_i must lie between zero and one, by inspection of Equation 9.17. Gay-Lussac's work made it clear that gas mixtures such as air obey the ideal gas law just as well as pure gases. We may now begin to draw on our knowledge of microscopic chemistry in an attempt to understand why this should be so.

TABLE 9.2

Composition of dry air near sea level*

Component		Mole fraction
Nitrogen	N_2	0.78084
Oxygen	O_2	0.20942
Argon	Ar	0.00934
Carbon dioxide	CO_2	0.00037†
Neon	Ne	0.0000182
Helium	He	0.0000052
Methane	CH_4	0.0000017
Krypton	Kr	0.00000114
Hydrogen	H_2	0.0000005
Dinitrogen oxide	N_2O	0.0000003
Xenon	Xe	0.000000087

* The 1976 standard atmosphere, adjusted to account for recent CO_2 measurements. Small but variable amounts of O_3, SO_2, NO_2, NH_3, CO, and I_2 may also be present. Moist air may contain a fraction of up to 0.05 of water vapor on a rainy day in summer. The total mass of the atmosphere is 5.1×10^{15} metric tons.
† CO_2 level has been rising steadily, and may be higher by the time you read this.

Let's consider the major constituent of air, nitrogen. As obtained from the density of liquid nitrogen (0.808 g cm^{-3}) and Avogadro's number (see Chapter 1), the mean diameter of a N_2 molecule is ~3 Å. At STP we know that 6×10^{23} molecules of N_2 occupy 22.4 L = 22,400 cm^3, or that one molecule (on the average) has 22,400 cm^3/6×10^{23} = 3.7×10^{-20} cm^3 = 37,000 Å^3 all to itself, a cube 33 Å on a side. As shown in Figure 9.5, gaseous N_2 is therefore mostly empty space. This typical result—the 37,000-Å^3 cube is the same for all gases at STP—implies that gas molecules are on the average very distant from their neighbors. This accounts nicely for the observation that all gases are **miscible** (mix uniformly) in all proportions (unlike many liquids), and why our own breathable mixture, the earth's atmosphere, behaves like any other gas, pure or not.

The isolation of one gas molecule from another, and consequently one gaseous chemical component from another, means that *each component behaves as though the gas container is otherwise empty,* and that we can therefore write gas laws for each component. In a container of volume V and temperature T we have

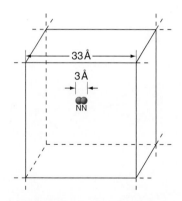

Figure 9.5
The volume per molecule of $N_2(g)$ (or other gas) at STP is about a 1000 times larger than its own molecular volume.

$$P_1 = \frac{n_1 RT}{V}, \quad P_2 = \frac{n_2 RT}{V}, \quad P_3 = \frac{n_3 RT}{V}, \quad \cdots \tag{9.18}$$

where the P_i are called **partial pressures.** The total pressure of the mixture, containing n moles of gas total, is

$$P = \frac{nRT}{V} = (n_1 + n_2 + n_3 + \cdots)\frac{RT}{V} = P_1 + P_2 + P_3 + \cdots \tag{9.19}$$

by using Equations 9.18 or, more compactly,

$$P = \left(\sum_j n_j\right)\frac{RT}{V} = \sum_j P_j \tag{9.20}$$

Equation 9.19 (or 9.20) is called **Dalton's law of partial pressures.** A useful way of stating Dalton's law follows from dividing Equation 9.18 for component i by the gas law for the entire mixture (Equation 9.19), yielding $P_i/P = n_i/n$ for any component, or

$$P_i = X_i P \tag{9.21}$$

where the mole fraction X_i is defined in Equation 9.17. For example, the partial pressure of O_2 in air can be obtained from the data in Table 9.2 as $P_{O_2} = (0.209)(760 \text{ Torr}) = 159$ Torr. Another useful corollary is that, at constant T and V, $P_i \propto n_i$, enabling a ready connection between partial pressure and reaction stoichiometry in commonly used rigid reaction vessels. But perhaps the most useful tool is the *idea* behind Dalton's law: *treat each gas component as though the others weren't there.* The following example may help to clarify these principles.

EXAMPLE 9.3

Hydrogen sulfide gas undergoes oxidation according to

$$2H_2S(g) + 3O_2(g) \longrightarrow 2SO_2(g) + 2H_2O(l)$$

In a study of this reaction, a stoichiometric mixture of H_2S and O_2 is prepared in a 0.500-L glass bulb at 298 K and 875 Torr total pressure. What are the partial pressures P_{H_2S} and P_{O_2}? Assuming complete reaction, how many moles of SO_2 can be formed under these conditions? If the temperature of the bulb is kept at 298 K, what will P_{SO_2} be?

Solution:
The reaction stoichiometry implies that, for every 5 mol of gas present, 2 will be H_2S and 3 will be O_2. This yields mole fractions $X_{H_2S} = 2/5 = 0.400$ and $X_{O_2} = 3/5 = 0.600$ and partial pressures

$$P_{H_2S} = X_{H_2S} P = (0.400)(875 \text{ Torr}) = 350 \text{ Torr}$$

$$P_{O_2} = X_{O_2} P = (0.600)(875 \text{ Torr}) = 525 \text{ Torr}$$

Note that this result does not require that you compute the actual number of moles of gas present, though you could do so using the ideal gas law of Equation 9.13. To find mol SO_2, you can use the partial pressure of the limiting reagent. In this case we have exact stoichiometry, so either reagent will do. Stoichiometry dictates that 2 mol H_2S produce 2 mol SO_2, and mol H_2S can be found from the ideal gas law by simply ignoring the O_2:

$$n_{H_2S} = \frac{P_{H_2S} V}{RT} = \frac{\left(\frac{350}{760} \text{ atm}\right)(0.500 \text{ L})}{(0.0821 \text{ L atm K}^{-1} \text{ mol}^{-1})(298 \text{ K})}$$

$$= 9.41 \times 10^{-3} \text{ mol}$$

and this is also mol SO_2. You could now apply the gas law again to find P_{SO_2} but it is easier to note that $P_i \propto n_i$ for any component i, whether reagent or product, and that therefore P_{SO_2} after reaction is identical with P_{H_2S} prior to reaction since the mole ratio is 1:1. Thus, $P_{SO_2} = 350$ Torr.

Given the vacuous nature of the gaseous state, it is easy to see why the gases of a mixture exert pressure independently. But what is the origin of this pressure? Does the ideal gas law itself have a molecular interpretation? The answers to these questions lie in an analysis of the *translational motion* of molecules, the three translational degrees of freedom. This in turn leads to a quantitative connection between temperature and molecular energy that until now you have been asked to accept as a postulate.

9.4 The Kinetic Molecular Theory

The ideas about molecular motion and degrees of freedom we discussed in Chapter 8 were first fully developed in 1859 by James Clerk Maxwell (who is even more famous for his laws of electromagnetism; see Chapter 2). Maxwell, who always liked to have mechanical models to explain observed behavior, was a ready believer in Dalton's atomic theory, and sought to explain gas behavior in terms of the mechanical energy of individual gas molecules. At that time the atomic theory was by no means universally accepted, and the work of Maxwell and that most ardent proponent of a general molecular theory of matter, Ludwig Boltzmann, was to some extent ignored. Nonetheless, the new molecular picture sounded the death knell for an older theory of heat energy, the caloric theory, which we will discuss briefly in Chapter 10.

In analyzing gas behavior, Maxwell's main postulates were our early claim (see Chapter 1) that, at any temperature above absolute zero, *molecules are in constant motion,* and further that, *collectively, this motion is completely random or chaotic.* We begin by noting that the translational kinetic energy of a molecule, $\frac{1}{2}mv^2$, arises from its mass m and velocity v, where

$$v^2 = v_x^2 + v_y^2 + v_z^2 \tag{9.22}$$

with (v_x, v_y, v_z) the Cartesian velocity components. The velocity vector $\mathbf{v}$, whose square magnitude is given by Equation 9.22, may be depicted in **velocity space** as in Figure 9.6, where the axes represent velocities rather than positions, and the tip of $\mathbf{v}$ coincides with the center of mass of the molecule. Any individual molecule at a particular time can be assigned a well-defined velocity vector (both magnitude and

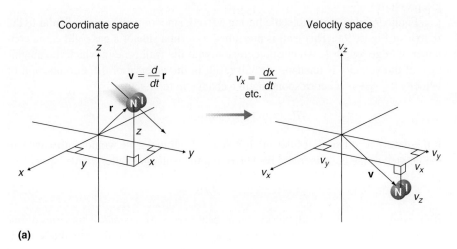

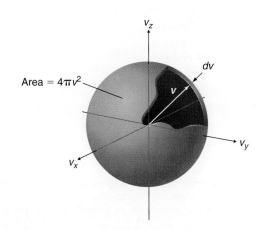

Figure 9.6
(a) From coordinate space to velocity space. A freely translating N_2 molecule is continuously changing its coordinates, but in velocity space it is located at a fixed position, equal to the vector distance it travels in 1 s. Collisions with a wall or with other gas molecules will alter the direction and sometimes the magnitude of $\mathbf{v}$, and thus change its location in velocity space as well as coordinate space. (b) The probability of finding a molecule with velocity of a given magnitude is proportional to the velocity-space volume of the spherical shell $4\pi v^2\, dv$.

Figure 9.7
Velocity-vector relations for a molecule colliding elastically (no change in the magnitude of the velocity, with plane mirror reflection) with a fixed, smooth wall perpendicular to the x axis. The net change in v_x is $-2v_x$, which may be combined with the molecular mass and the transit time in the container to yield the average force of impact as discussed in the text.

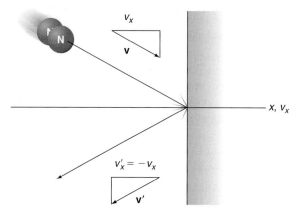

direction),* but, when viewed over a longer period, or when many molecules are observed, we need to consider the **average** or **mean velocity,** whose magnitude is denoted by $\bar{v}$. With these conditions in mind, we will now trace the molecular origin of gas pressure.

Figure 9.7 depicts a molecule hitting a fixed, smooth wall† perpendicular to the v_x axis of Figure 9.6. This wall is presumed to form a side of a gas-tight cubic container of edge length L. An elastic collision with the wall (one in which the magnitude of the velocity is unchanged) will result in the reversal of the x component of velocity, $v_x \longrightarrow -v_x$, corresponding to a change in momentum

$$\Delta P_x = m(-v_x) - mv_x = -2mv_x \qquad (9.23)$$

If there were no other molecules in the container, our molecule would scurry away to the far side of the container and back, striking the wall again after a time lapse

$$\Delta t = 2L/v_x \qquad (9.24)$$

If the wall were a drumhead, you'd hear a "beat" every Δt seconds. This series of impulses constitutes an average force on the wall given by Newton's second law,

$$F = \left|\frac{\Delta p_x}{\Delta t}\right| = \frac{2mv_x}{2L/v_x} = \frac{2}{L}\left(\tfrac{1}{2}mv_x^2\right) \qquad (9.25)$$

where in the last step we have factored out the x component of the translational kinetic energy. The pressure exerted by our molecule is then

$$P = \frac{F}{A} = \frac{2}{L \cdot L^2}\left(\tfrac{1}{2}mv_x^2\right) = \frac{2}{V}\left(\tfrac{1}{2}mv_x^2\right) \qquad (9.26)$$

where V is the volume of the cubic container.

---

* This is true as long as we don't ask for its precise position at the same time; such a measurement is limited by the uncertainty principle (see Chapter 3).
† Walls, of course, are also made of atoms or molecules and are generally not very smooth! The smoothest surface yet found by chemists is a certain face of a crystal of metallic silver.

Now suppose there are N molecules. Then, just as for Dalton's law, we find

$$P = \sum_{i=1}^{N} P_i = \frac{2}{V} \sum_{i=1}^{N} \tfrac{1}{2} m_i v_{x,i}^2 \qquad (9.27)$$

For simplicity we assume the gas is pure,* so that all of the m_i are identical and can be pulled out from the summation. We now define the x component **mean square velocity** as follows:

$$\overline{v_x^2} \equiv \frac{1}{N} \sum_{i=1}^{N} v_{x,i}^2 \qquad (9.28)$$

When Equation 9.28 is substituted into Equation 9.27, and both sides multiplied by V, we get

$$PV = 2N \left(\tfrac{1}{2} m \overline{v_x^2}\right) \qquad (9.29)$$

Equation 9.29 is beginning to look like the ideal gas law, but we still have to extend it to three dimensions. This extension is easy with the aid of the chaos postulate: collectively random motion implies no preferred direction in space, so that

$$\overline{v_x^2} = \overline{v_y^2} = \overline{v_z^2} = \tfrac{1}{3} \overline{v^2} \qquad (9.30)$$

where the last equality follows from the definition of Equation 9.22, and the mean square velocity $\overline{v^2}$ is defined as in Equation 9.28. Finally, by substituting Equation 9.30 into Equation 9.29, we obtain

$$PV = \tfrac{2}{3} N \left(\tfrac{1}{2} m \overline{v^2}\right) \qquad (9.31)$$

When this is compared to the ideal gas law, $PV = nRT$, we arrive at the *kinetic theory interpretation of temperature,*

$$\tfrac{2}{3} N \left(\tfrac{1}{2} m \overline{v^2}\right) = nRT \qquad (9.32)$$

or

$$E = \tfrac{3}{2} nRT \qquad (9.33)$$

where E is the **total kinetic energy** of the entire gas sample, given by

$$E = N \left(\tfrac{1}{2} m \overline{v^2}\right) = \sum_{i=1}^{N} \tfrac{1}{2} m v_i^2 \qquad (9.34)$$

---

* Even the purest gases commercially available are only 99.9999% pure, implying more than 10^{17} molecules of impurities per mole of gas. And, of course, gases will usually contain a mixture of isotopes with different m_i. However, neither of these complications prevents us from considering a single isotope of a pure gas.

The **molar kinetic energy** $\bar{E}$ is given by

$$\bar{E} = \frac{E}{n} = \tfrac{3}{2} RT \tag{9.35}$$

On a microscopic scale, dividing $\bar{E}$ by Avogadro's number, we get the **average energy per molecule,**

$$\bar{\varepsilon} = \tfrac{3}{2} k_B T \tag{9.36}$$

where $k_B = R/N_A$ is **Boltzmann's constant.** Any given molecule can, of course, have more or less energy than $\bar{\varepsilon}$. Thus we have arrived, thanks to the simple, universal behavior of gases, at a justification of the notion that, as postulated in Equation 1.15, the absolute temperature is a measure of the amount of molecular motion. This idea has revolutionized our understanding of energy and the world.

The kinetic theory analysis also provides a mechanically consistent interpretation of the existence of an absolute zero of temperature: according to Equations 9.33 and 9.34, at $T = 0$ K, all molecular motion ceases. Because kinetic energy is an inherently positive quantity, this implies that *temperatures colder than 0 K are impossible, since molecules cannot have less than no kinetic energy.* However, taking into consideration the quantum-mechanical character of molecular motion, as discussed in Chapter 8, requires that we allow the molecules to have the energy of their lowest state, the so-called **zero point energy.** This residual energy is "trapped" and cannot be removed by further cooling. Thus at $T = 0$ the molecules paradoxically are not completely still, but sit there shivering in the cold. (Of course, long before 0 K is approached, all molecular substances have condensed and frozen into solid crystals; see the last section of this chapter and also Chapter 14.)

Under STP conditions, the amount of energy given by Equations 9.35 and 9.36 is 3406 J mol^{-1} (0.814 kcal mol^{-1}) or 5.66×10^{-21} J (0.0353 eV molecule^{-1}). On a molecular scale this is "small potatoes," but for one mole it is quite a lot of energy, enough to lift a 1-ton weight 30 cm off the ground, or equivalent to the kinetic energy of 30 fastballs each thrown at 90 mph! It is no wonder that gases have found great use in mechanical devices as "working substances." Chapters 10 and 11 will delve into the usefulness of this energy.

Predictions of the Kinetic Theory of Gases

The **kinetic molecular theory** interprets the universality of the ideal gas law as due to the fact that at a given temperature all gases have identical kinetic energy per mole. This in turn means that the average velocities of lighter (lower molecular mass) gases are greater than those of heavier gases, in order to compensate for their lightness. From Equations 9.34 and 9.36, $\tfrac{1}{2} m\overline{v^2} = \tfrac{3}{2} k_B T$, or

$$v_{\text{rms}} = \sqrt{\overline{v^2}} = \sqrt{\frac{3k_B T}{m}} \tag{9.37}$$

where v_{rms} is the **root mean square (rms) velocity.** Some startling numbers come out of Equation 9.37. For N_2 at 273 K, $v_{\text{rms}} = 493$ m s^{-1} (1100 mph)! H_2 moves at 1840 m s^{-1} (more than 4000 mph), while I_2 vapor lumbers along at a mere 164 m s^{-1} (370 mph).

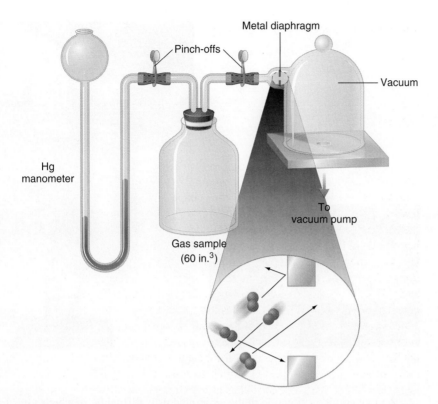

Figure 9.8
Schematic illustration of Graham's 1846 experiment. The manometer on the left is measuring the difference between the gas sample pressure and atmospheric pressure. The time required to bring the manometer reading close to 760 mm Hg was measured for O_2, H_2, N_2, and CO_2. The hole in the metal diaphragm was adjusted to allow a sample of air to escape in 1000 s. Graham devised several other apparatuses to study both effusion and diffusion, including one similar to that of Figure 9.9. The inset shows the kinetic theory picture of effusion.

Evidence for the validity of Equation 9.37 had to come from experiments that probe the velocity rather than the energy of a gas, that is, from measurement of *rates* rather than static properties. The earliest such evidence was provided by Thomas Graham, who showed in 1846 that *the rate of effusion of a gas is inversely proportional to the square root of its density*. The process defined as **effusion** is one in which gases leak from a reservoir through a small hole into a region of much lower pressure (a vacuum). Graham's experiment is illustrated in Figure 9.8. Now at constant temperature and pressure, gas density ρ is proportional to molar mass $M = N_A m$; hence, Equation 9.13 can be converted into

$$\rho = \frac{MP}{RT} = \frac{mP}{k_B T} \qquad (9.38)$$

As discussed later, the effusion rate of a gas is proportional to its average velocity; so when ratios of rates are taken for two gases A and B at the same T and P, we get **Graham's law:**

$$\frac{\text{Rate A}}{\text{Rate B}} = \frac{v_{\text{rms, A}}}{v_{\text{rms, B}}} = \sqrt{\frac{m_B}{m_A}} = \sqrt{\frac{\rho_B}{\rho_A}} \qquad (9.39)$$

where the second equality follows from Equation 9.37 and the third from Equation 9.38. This means that a mixture of gases in a container with a leak will lose the lighter gases more rapidly, a finding widely exploited in the separation of isotopes, as well as in the detection of leaks in vacuum chambers.

Figure 9.9
Graphic illustration of Graham's law of diffusion. Hydrogen gas, collected by upward displacement in an inverted beaker, diffuses through the wall of the porous ceramic cup faster than air can diffuse out (**a**). This produces a pressure rise in the otherwise airtight liquid reservoir, forcing liquid out the spout (**b**).

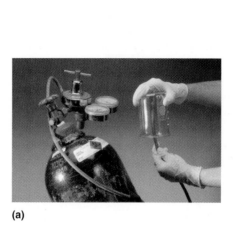

(a)   (b)

A more common situation in the air around us is **diffusion,** in which a gas from a source mixes with the air (or other gas) and spreads out with time. Diffusion rates are lower than effusion rates, but also obey Graham's law (Equation 9.39), at least approximately. Diffusion can also occur through porous solid materials, such as the walls of balloons or porous ceramics, as illustrated in Figure 9.9. Figure 9.10 shows a simple demonstration of Equation 9.39 in the reaction $NH_3(g) + HCl(g) \longrightarrow NH_4Cl(s)$ carried out in a long tube. And, of course, our sense of smell relies on diffusion to convey odoriferous molecules into our nasal passages.

Figure 9.10
Deposit of solid NH_4Cl adduct serves as a time marker for the relative diffusion rates of NH_3 and HCl through the air-filled tube. See Exercise 45 at the end of the chapter to work out where the deposit should appear according to Graham's law. Can you deduce which side the NH_3 came from?

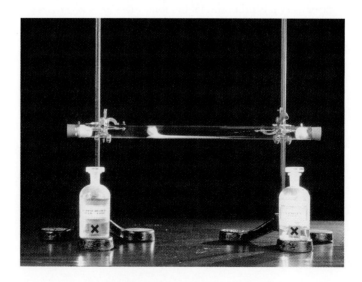

The Maxwell–Boltzmann Velocity Distribution

At several junctures we have observed that, in speaking of the kinetic properties of a large number of atoms or molecules, it is the *average value* of ε the energy or v the velocity that characterizes the group. *An average value is always derived from a distribution of values.* For example, the average score on a tough college chemistry exam might be 61%, but the scores may range from 30% to 93%, that is, there is a *distribution* of scores. To show this distribution in a small class, a histogram is constructed, in which the scores s for a given range Δs ($\Delta s = 10\%$ for example) are grouped together, as illustrated in Figure 9.11. The *probability* that your score lies in a particular range from s to $s + \Delta s$ is then $\Delta N(s)/N$, where $\Delta N(s)$ is the number of scores in the class of N students that fell in this range. (This probability becomes a certainty for you, of course, once you get your exam back, however unlikely your score. Likewise, any particular molecule will have a definite velocity and energy that may defy the odds.)

We then define the **distribution function** $f(s)$—in our example, a percent score distribution function—so that $f(s)\Delta s = \Delta N(s)/N$, where $f(s)$ is the *probability per interval* Δs of finding score s, and $f(s)\Delta s$ is the *probability* of finding score s

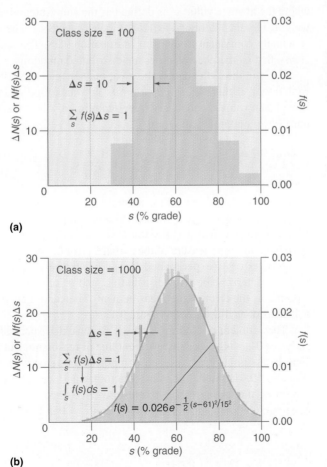

Figure 9.11
Grade distribution histograms for classes of $N = 100$ and 1000 students. As N increases, the grade bins Δs can be made smaller, and for a typical class the result approaches a Gaussian error distribution as shown in (b). Fluctuations in the histograms, particularly visible in (b), are due to random fluctuations in the populations; these fluctuations are proportional to the square root of the number of students in each bin. As N approaches, say, N_A, the bin size $\Delta s \longrightarrow ds$, the histogram becomes smooth, and the distribution function $f(s)$ becomes a well-defined, smooth function.

in this range. We have already seen an important example of a distribution function, the radial distribution function (RDF) for atomic orbitals, discussed in Chapters 3 and 4.

A typical gas sample contains a huge number of molecules (3×10^{19} molecules in only 1 cm^3 at STP), so that when discussing its properties, in particular the velocities of the gas molecules, the histogram range Δv can be made very small, say, dv. We can then define the **velocity distribution function** $f(v)$ so that $f(v)\, dv = dN(v)/N$, where $f(v)\, dv$ is the *probability* that a molecule will be found with velocity between v and $v + dv$. The passage from a coarse sample interval to a fine one is illustrated in Figure 9.11; for molecular velocities, the interval dv can be made so fine for a large gas sample that for any practical purpose $f(v)$ becomes a continuous function. Note that the units of $f(v)$ are those of v^{-1} (s m^{-1}), in order to make $f(v)\, dv$ a pure number, a probability between zero and one.

How can $f(v)$ be determined? Well, we do have two constraints to apply to $f(v)$. First, it must yield the correct $\overline{v^2}$ in accord with our analysis of the pressure of a gas (see Equation 9.37),

$$\overline{v^2} \equiv \int_0^\infty v^2 f(v)\, dv = \frac{3k_B T}{m} \tag{9.40}$$

where the average value v^2 is defined as the sum (integral) of all possible v^2 values, weighted by the likelihood $f(v)\, dv$ of their occurrence. The integral in Equation 9.40 is actually equivalent to the sum in Equation 9.28. The second constraint is that of *chaos*, that is, that the velocities are distributed randomly.

Early in the 19th century the great mathematician Karl Friedrich Gauss demonstrated that, if there are *random* errors in a measured quantity s, distributed about some mean value $\bar{s}$, and a very large sampling of s is made, the distribution function must *always* approach the form

$$f(s) = \frac{1}{\sigma\sqrt{2\pi}}\, e^{-(s-\bar{s})^2/(2\sigma^2)} \tag{9.41}$$

where e again is the base of the natural logarithms, $e = 2.718.\ldots$ The $f(s)$ of Equation 9.41 is known as the *normal distribution function*, and the exponential of the negative square is called a **Gaussian** function. The quantity σ is called the standard deviation, and is a measure of the average error or spread in possible s values. When applied to errors in experimental measurement (see Chapter 1, Section 1.3), Equation 9.41 characterizes the error distribution, and σ, or some multiple of σ, is generally reported as the error in the measurement. The curve generated by Equation 9.41 is bell shaped, as illustrated in Figure 9.11.

The problem of finding $f(v)$ was first tackled successfully by Maxwell in the early 1860s, using a slightly different line of reasoning. But his approach was equivalent to assuming that for each component of velocity, say, v_x, Gauss's error law (Equation 9.41) must hold, since the molecules are not aware of their coordinates. Moreover, the average value $\bar{v}_x$ must be zero, since the molecules may equally well be moving left or right. Taking $s = v_x$ and $\bar{s} = 0$ in Equation 9.41, we get

$$f(v_x) = \frac{1}{\sigma\sqrt{2\pi}}\, e^{-v_x^2/(2\sigma^2)} \tag{9.42}$$

Then, since the three Cartesian directions are independent, the total distribution is just the *product* of Equation 9.42 and two more identical expressions for y and z,* giving

$$f(v_x, v_y, v_z) = \frac{1}{(\sigma\sqrt{2\pi})^3} e^{-(v_x^2+v_y^2+v_z^2)/(2\sigma^2)} \qquad (9.43)$$

where we have combined the three exponents, and assumed that the standard deviation σ is independent of direction.

We would now like to express Equation 9.43 in a form that gives the probability of finding the magnitude of the velocity between v and $v + dv$, irrespective of direction. The exponent simplifies nicely, since $v^2 = v_x^2 + v_y^2 + v_z^2$. The problem is very similar to that of constructing a radial distribution function for an orbital starting with the square of the orbital wave function (see Chapter 3), except that here we are in "velocity space," where points in space are given by (v_x,v_y,v_z) instead of (x,y,z). The result we seek is simply obtained by multiplying Equation 9.43 by the volume of a spherical shell in velocity space $4\pi v^2\, dv$ (see Figure 9.6), giving

$$f(v)\, dv = 4\pi \left(\frac{1}{(\sigma\sqrt{2\pi})^3}\right) v^2\, e^{-v^2/(2\sigma^2)}\, dv \qquad (9.44)$$

where $f(v)$ is the desired distribution function.

We still need to find σ in Equation 9.44. This is where the first of our two constraints, Equation 9.40, enters. When we apply this condition [by evaluating the integral $\int v^2 f(v)\, dv$], we find

$$\sigma^2 = \frac{k_B T}{m} \qquad (9.45)$$

This finally yields, when inserted into Equation 9.44,

$$f(v)\, dv = 4\pi \left(\frac{m}{2\pi k_B T}\right)^{3/2} v^2\, e^{-\frac{1}{2}mv^2/k_B T}\, dv \qquad (9.46)$$

This formula for the velocity distribution function, first derived by Maxwell in 1862, was later obtained by Ludwig Boltzmann from more general considerations (see Chapter 10), and is known as the **Maxwell–Boltzmann velocity distribution.** Figure 9.12 graphs the distribution for N_2 molecules at three temperatures. Equation 9.46 is *normalized,* so that $\int f(v)\, dv = 1$, and from it all possible average values of velocity (or kinetic energy) can be derived. Some 60 years after its first derivation, Otto Stern (famous for his magnetic deflection of electron spin; see Chapter 4) verified Equation 9.46 with a measurement of gravitational deflection of a molecular beam.

Relation 9.45 lends added insight into the meaning of absolute temperature. Not only does temperature characterize the average energy of a molecule, but it also

---

* It is a general rule that the probabilities of independent events are multiplicative; for example, the probability of a tossed coin coming up "heads" twice in a row is just $\frac{1}{2} \times \frac{1}{2} = \frac{1}{4}$, where $\frac{1}{2}$ is the chance of finding heads on any one toss.

Figure 9.12
Maxwell–Boltzmann velocity distributions (Equation 9.46), for $N_2(g)$ at three temperatures. The areas under the curves are equal. Accompanying the shift of the peak velocity to higher values at higher temperatures are a broadening of the distribution and a reduction in the number of molecules with velocities in any particular range.

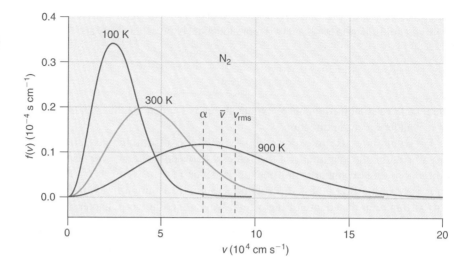

defines the *spread in energy;* as T rises, both the average velocity and the spread in velocities increases, as Figure 9.12 illustrates.

Equation 9.46 allows us to evaluate two other characteristic velocities that are important in understanding dynamic events in gases. The **most probable velocity**, usually denoted α, is obtained by finding the maximum in $f(v)$ (setting $df/dv = 0$), and is given by

$$\alpha = \sqrt{\frac{2k_B T}{m}} \tag{9.47}$$

The **mean** or **average velocity** $\bar{v}$ is obtained from the integral

$$\bar{v} = \int_0^\infty v f(v)\, dv = \sqrt{\frac{8k_B T}{\pi m}} \tag{9.48}$$

As indicated in Figure 9.12, these two velocities differ slightly, with $\alpha < \bar{v} < v_{rms}$, where the root mean square velocity, v_{rms}, is defined in Equation 9.37.

Gas Dynamics

The phenomena of gas effusion and diffusion described earlier in this section are readily explained by the kinetic theory. But more importantly, all chemical reactions, in particular those involving gases, occur as a result of collisions of one reagent molecule with another, a phenomenon that also finds a facile description in terms of moving molecules. And the energy liberated by a reaction results in "hot" molecules, which often most efficiently dissipate their excess energy in collisions with the walls of their container. All of these happenings can be analyzed in a relatively straightforward but nevertheless quantitative way using kinetic theory ideas. In what follows, we will give heuristic arguments to justify the form of the rigorously stated formulas.

For all of the phenomena mentioned, we are interested in *collision rates*, denoted by the letter Z, both individual and collective. You are already equipped to handle the concepts involved, based on the kinetic theory analysis of the ideal gas law. The essence is contained in Equation 9.24, which gives the (average) time Δt between collisions of an individual molecule X with a wall. The wall collision rate for this molecule (collisions s^{-1}) is just $1/\Delta t$; for N molecules the **wall collision rate** is then $Z_{\text{wall}} \approx N/\Delta t \approx Nv/L$, where, as before, L is a characteristic length dimension of the gas container. Writing $1/L \approx A/V$, where A is the area of the struck wall and V is the volume of the container (an exact relation for a cubic container), the result $Z_{\text{wall}} \approx Nv/L$ can be expressed in terms of the **number density** $[X] = N/V$ (molecules per unit volume) as

$$Z_{\text{wall}} = \tfrac{1}{4} [X] \bar{v} A \tag{9.49}$$

where $\bar{v}$ is the average velocity given in Equation 9.48 and the factor ¼ results from a more rigorous derivation. Equation 9.49 also accurately describes effusion of a gas through a hole of area A in a container, since molecules heading for a collision with a wall do not know about the hole, and must "collide" with the hole at the same rate per unit area. Ratios of effusion rates from Equation 9.49 yield Graham's law directly.

To get the **molecular collision rate** Z_{mol}, we can in essence replace the wall area A by the area one molecule presents to another. For molecules of mean diameter d, the mutual area is πd^2; the precise formula for the rate is then

$$Z_{\text{mol}} = \sqrt{2}\,[X]\bar{v}\pi d^2 \tag{9.50}$$

This is the average number of collisions one X molecule undergoes in gas X per unit time. This result can be used to find the average distance traveled between collisions, called the **mean free path** λ, as the average velocity $\bar{v}$ multiplied by the mean time between collisions $\Delta t_{\text{col}} = 1/Z_{\text{mol}}$:

$$\lambda = \frac{\bar{v}}{Z_{\text{mol}}} = \frac{1}{\sqrt{2}\,[X]\pi d^2} \tag{9.51}$$

The collision rates predicted by Equations 9.49 and 9.50 are, as you might expect, quite large.* The number density $[X] = N/V$ can be obtained from the ideal gas law; at STP, $N/V = 2.7 \times 10^{19}$ molecules cm^{-3}. The average velocity $\bar{v}$ for N$_2$ at 273 K is 4.5×10^4 cm s^{-1} from Equation 9.48; note that $\bar{v}$ is slightly smaller than v_{rms} for N$_2$ quoted earlier. Together these yield from Equation 9.49 $Z_{\text{wall}}/A = 3.0 \times 10^{23}$ collisions cm^{-2} s^{-1}, or roughly an Avogadro's number of collisions every second on only 1 cm^2! As we have learned from the Avogadro's number analysis of Chapter 1, a typical diameter for a small gas molecule like N$_2$ or H$_2$O(g) is $d \approx 3$ Å. The quantity πd^2 appearing in Equation 9.50 is known as a **collision cross section,** with the typical value of 30 Å^2 (30 $\times$ 10^{-16} cm^2). This gives $Z_{\text{mol}} = 5.2 \times 10^9$ collisions s^{-1}, or roughly a collision every few hundred picoseconds.

---

* To conform with usual practice in gas-phase collision chemistry, we use cgs units. Conversion to mks or SI is straightforward, especially since charges are not involved.

We can compare Z_{mol} to typical rates of rotation and vibration (see Chapter 8): if a molecular "day" is ~ 1 ps, as defined by a rotational period, then the mean time between collisions, $1/Z_{mol}$, amounts to months to years on a molecule's timescale! Finally, the mean free path for N_2 under these conditions is, from Equation 9.51, $\lambda = 8.7 \times 10^{-6}$ cm or roughly 1000 Å. If a molecule were the size of a car, it would travel about two city blocks before having an "accident." However, all of these gas-dynamic quantities can change dramatically under different conditions, as the following example shows.

EXAMPLE 9.4

A certain vacuum chamber for chemical research is 1.00 m in length. What maximum pressure in Torrs can exist inside the chamber (at 298 K) that will allow molecules injected into it to travel "wall to wall" without undergoing a collision? What is the wall-to-wall transit time of a N_2 molecule in this chamber in the absence of collisions? Assume a molecular diameter of 3.00 Å.

Solution:

The question asks that the mean free path λ (Equation 9.51) be at least 1 m. The number density $[X] = N/V$ that satisfies this requirement is

$$[X] = \frac{1}{\sqrt{2}\,\lambda\pi d^2}$$

$$= \frac{1}{\sqrt{2}\,(100.\text{ cm})(\pi)(3.00 \times 10^{-8}\text{ cm})^2}$$

$$= 2.50 \times 10^{12} \text{ molecules cm}^{-3}$$

The pressure corresponding to this is obtained from the ideal gas law:

$$P = \frac{n}{V}RT = \frac{N}{V}\frac{RT}{N_A}$$

$$= \frac{(2.50 \times 10^{12} \text{ cm}^{-3})(0.08206 \text{ L atm K}^{-1} \text{ mol}^{-1})(298 \text{ K})}{(10^{-3} \text{ L cm}^{-3})(6.022 \times 10^{23} \text{ mol}^{-1})}$$

$$= 1.02 \times 10^{-7} \text{ atm} = 7.72 \times 10^{-5} \text{ Torr}$$

At this pressure an appreciable fraction of molecules will still undergo a collision or two on average. Chemists therefore generally aim for operating pressures at least an order of magnitude (factor of 10) lower, which is readily achieved with modern high-vacuum pumps. To find the N_2 transit time, we use the average velocity of Equation 9.48, although to get a rough estimate, any of the formulas of Equations 9.37, 9.47, or 9.48 could be used. The transit time is $\Delta t = L/\bar{v}$. To use Equation 9.48 in SI units, we need the mass m (in kilograms) of an individual N_2 molecule $m = M/(1000\,N_A)$ or

$$m(N_2) = \left(\frac{28.0 \text{ g}}{1 \text{ mol}}\right)\left(\frac{1 \text{ kg}}{1000 \text{ g}}\right)\left(\frac{1 \text{ mol}}{6.022 \times 10^{23} \text{ molecules}}\right) = 4.65 \times 10^{-26} \text{ kg}$$

Equation 9.48 then yields

$$\bar{v} = \sqrt{\frac{8k_BT}{\pi m}}$$
$$= \sqrt{\frac{(8)(1.38 \times 10^{-23} \text{ J K}^{-1})(298 \text{ K})}{(\pi)(4.65 \times 10^{-26} \text{ kg})}}$$
$$= 475 \text{ m/s } (4.75 \times 10^4 \text{ cm/s})$$

The transit time $\Delta t = (1.00 \text{ m})/(475 \text{ m/s}) = 2.11 \times 10^{-3}$ s = 2.11 ms. The time between "drumbeats" for an individual molecule (Equation 9.24) is then twice this number, or 4.2 ms. All of these gas-dynamic quantities are average values, and in general the transit times would be distributed in a way consistent with the Maxwell–Boltzmann velocity distribution of Equation 9.46.

The ideas of the kinetic molecular theory are nowadays applied not only to the translational motion of gases but to every area of chemistry. The molecular collision rates given in this section determine the rates of gas-phase reactions, while similar but more complex encounters, only now beginning to be understood, govern solution chemistry. Thermal fluctuations in the conformations of large biological molecules may also dominate the rates of metabolic processes in living systems. Atomic *motion* goes hand in hand with the atomic theory itself.

9.5 The Behavior of Real Gases

Table 9.3 presents the experimental molar volumes $\bar{V} = V/n$ at STP for some common gases. If these gases all obeyed the "ideal" gas law, they would all have identical $\bar{V} = RT/P = 22.414$ L mol^{-1} (see Equation 9.15). The small deviations you can see in the table, about 2% at most, are an indication that gases do not obey $PV = nRT$ exactly, and that the chemical identity of the gas actually makes a difference, albeit slight. This **nonideal behavior** becomes more pronounced at *higher* pressures and/or *lower* temperatures. The effect of temperature is clear if we observe that all the gases of Table 9.3 can be *liquefied* at sufficiently low T, the ultimate in nonideal gas behavior. As illustrated in Figure 9.13, Charles' law plots of V versus T for real gases show a sudden drop at the condensation point (normal boiling point for the molecule as a liquid) characteristic of the gas. Charles' law predicts that $V = 0$ when $T = 0$ K, but, as you know, the molecules themselves occupy a volume in space, and although it is small, it is certainly not zero.

These features of the behavior of substances we call gases at STP invite a modification of the ideal gas law that includes the chemical identity of the gas—a *chemical* gas law. One of the earliest, and still the most widely used, suggestions along these lines was made by Johannes van der Waals in 1873.

Van der Waals assumed that the general form of the ideal gas law would still hold, but with modifications to the pressure and volume. That is, he wrote

$$P_{\text{ideal}} V_{\text{ideal}} = nRT \quad (9.52)$$

where P_{ideal} and V_{ideal} result from correcting the *actual* measured values of P and V for intermolecular interactions. Van der Waals reasoned that the *actual* pressure P

TABLE 9.3
Properties of real gases

Gas	STP molar volume (L mol^{-1})	Boiling point at 1 atm t_b (°C)	Van der Waals constants a (atm L^2 mol^{-2})	b (L mol^{-1})
H_2	22.419	−252.9	0.2421	0.02651
He	22.427	−268.9	0.0342	0.02376
Ne	22.414	−246.0	0.2056	0.01672
Ar	22.369	−185.7	1.337	0.03201
Kr	22.328	−152.3	2.295	0.03957
Xe	22.241	−107.1	4.138	0.05156
N_2	22.373	−195.8	1.349	0.03860
O_2	22.368	−183.0	1.364	0.03186
F_2	22.376	−188.1	1.149	0.02876
Cl_2	22.085	−34.6	6.501	0.05628
HF(g)	...	19.6	9.411	0.07371
HCl	22.251	−84.9	3.652	0.04061
HBr	22.196	−67.0	4.436	0.04436
HI	22.114	−35.6	5.747	0.04895
$H_2O(g)$	...	100.0	5.463	0.03048
H_2S	22.190	−60.7	4.478	0.04330
H_2Se	22.140	−41.5	5.451	0.04789
NH_3	22.176	−33.4	4.167	0.03711
PH_3	22.207	−87.7	4.634	0.05157
CH_4	22.331	−164.	2.273	0.04306
SiH_4	22.240	−111.8	4.325	0.05791
CO	22.368	−191.5	1.454	0.03952
NO	22.362	−151.8	1.439	0.02887
CO_2	22.252	−78.5$_{sublimes}$	3.610	0.04286
N_2O	22.249	−88.5	3.803	0.04435
$NO_2(g)$	...	21.2	5.294	0.04435
O_3	22.274	−111.9	3.522	0.04871
SO_2	22.066	−10.	6.776	0.05679
C_2H_6	22.171	−88.6	5.507	0.06513
C_2H_4	22.226	−103.7	4.552	0.05821
C_2H_2	22.212	−84.0	4.475	0.05241
C_3H_8	21.936	−42.1	9.267	0.09048

Source: B. E. Poling, J. M. Prausnitz, and J. P. O'Connell, *Properties of Gases and Liquids,* 5th ed. (New York: McGraw-Hill, 2000).

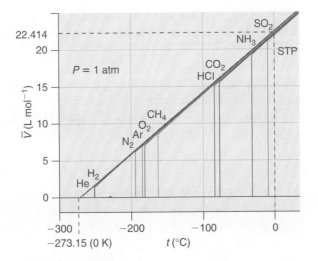

Figure 9.13
Charles' law plots for real gases. The abrupt drops occur at the boiling (condensation) point, unique for each gas (see Table 9.3), and appreciable deviations from the ideal molar volume begin to occur as that temperature is approached. Liquid and solid molar volumes are too tiny to be seen on this scale.

would be less than P_{ideal}, since a molecule heading for the wall, to do its bit toward sustaining the pressure, would be slowed down by the pull of its neighbors, that is, by **intermolecular attraction**. This is illustrated in Figure 9.14; it's akin to being late for class when you meet and talk to a friend on the way. For a single molecule this attraction is proportional to the number of molecular "friends" nearby, that is, to the molar concentration n/V; when scaled up by the total number of molecules present, it becomes proportional to $(n/V)^2$. Then the ideal pressure can be obtained from

$$P_{\text{ideal}} = P + a\left(\frac{n}{V}\right)^2 \tag{9.53}$$

where P is the *actual* pressure as read on a pressure gauge or manometer, and a is a constant, characteristic of a particular gas, that determines the magnitude of the attractive correction.

The volume correction is simpler, since each molecule occupies a small volume that excludes other molecules; that is, other molecules that approach too closely

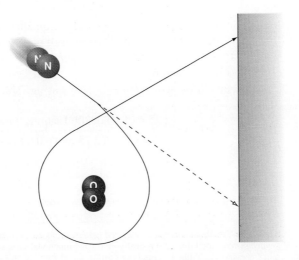

Figure 9.14
An innocent nitrogen molecule heading for the wall (dotted arrow) meets an attractive oxygen molecule on the way and, hence, is delayed in reaching the wall, resulting in a decrease in the pressure it exerts. The oxygen molecule must also swing around during the encounter (not shown).

experience strong **intermolecular repulsion.** The *actual* volume is therefore *greater* than the ideal, due to this **excluded volume,** and we write

$$V_{\text{ideal}} = V - nb \qquad (9.54)$$

where V is the *actual* volume of the gas container, and b is the excluded molar volume, again characteristic of the gas considered.

If we now substitute Equations 9.53 and 9.54 into Equation 9.52, we get

$$\left[P + a\left(\frac{n}{V}\right)^2\right][V - nb] = nRT \qquad (9.55)$$

the so-called **van der Waals equation of state.** Table 9.3 lists the van der Waals constants a and b for common gases. Note that Equation 9.55 can also be written in terms of the molar volume $\overline{V}$ as follows:

$$\left[P + \frac{a}{\overline{V}^2}\right][\overline{V} - b] = RT \qquad (9.56)$$

Of course, Equation 9.55 is not as easy to use as the ideal gas law, being equivalent to a cubic equation in V, but it is the simplest such equation that accounts for intermolecular effects. It does *not* yield perfect agreement with experimental gas data, but other equations of state with more parameters do agree within error.* The following example illustrates the use of the van der Waals equation.

EXAMPLE 9.5

Suppose 2.25 mol of CO_2 gas are compressed into a 2.00-L cylinder at 298 K. What pressure will be read on a gauge attached to the cylinder? What is the percent deviation from the ideal pressure?

Solution:
Solving Equation 9.55 for P, we get

$$P = \frac{nRT}{V - nb} - a\left(\frac{n}{V}\right)^2$$

Before putting in numbers, we can make a qualitative comparison with the ideal $P = nRT/V$. The first term on the right by itself will yield a pressure *higher* than ideal (since $b > 0$) while the second term will *decrease* the net pressure. At fixed V, the first term grows with T, and hence $P > P_{\text{ideal}}$ at high T; there is a temperature at which $P = P_{\text{ideal}}$, and below which $P < P_{\text{ideal}}$. Now using Table 9.3 for the a and b constants,

$$P = \frac{(2.25 \text{ mol})(0.08206 \text{ L atm K}^{-1} \text{ mol}^{-1})(298 \text{ K})}{2.00 \text{ L} - (2.25 \text{ mol})(0.04286 \text{ L mol}^{-1})}$$

$$- (3.610 \text{ atm L}^2 \text{ mol}^{-2})\left(\frac{2.25 \text{ mol}}{2.00 \text{ L}}\right)^2$$

$$= 28.90 - 4.57 = 24.3 \text{ atm}$$

---

* The van der Waals equation and other real gas equations are not like Schrödinger's equation; they are *semiempirical,* combining a physically motivated functional form with input from experiment. The a and b constants of Table 9.3 come from fitting actual *PVT* data for each gas.

while $P_{ideal} = 27.5$ atm. The percent deviation is

$$\frac{P - P_{ideal}}{P_{ideal}} \times 100\% = -12\%$$

For pressures greater than 65 atm, CO_2 actually condenses to a liquid. Note that the attractive correction to P is more important; this is true of all gases at STP except He, Ne, and H_2.

All gases show substantial deviation from ideal behavior at pressures of 100 atm or more; this is to be expected, since then the molar volume of the gas is starting to approach that of a liquid, and both the a and b terms in the van der Waals equation supply substantial corrections. This phenomenon will be touched on briefly in Chapter 14, and is extensively discussed in physical chemistry texts.

If you examine Table 9.3 carefully, you will see that the b values correlate with the size of the molecules (with a few exceptions, such as HF), and that the magnitudes of the a's closely parallel the boiling points. Both a and t_b measure the strength of the attractive forces, and also increase with molecular size. We leave a discussion of the nature of the attractive forces for Chapter 14. The molar volumes are almost all less than the ideal 22.414 mol L^{-1}, indicating that intermolecular attraction, which tends to shrink $\overline{V}$, is generally dominant over molecular size effects at STP.

This concludes our brief survey of the properties of gases. But don't make the mistake of tucking the information given here safely away. ("Well, that takes care of gases—just liquids and solids to go. . . .") In addition to being an important manifestation of the molecular motion that underlies all of chemistry, the gas phase will provide us with a conceptual and practical springboard to the role of energy in chemistry and to the study of rates of chemical reactions, as well as an understanding of condensed phases based on the real gas behavior just discussed. There are also remarkable parallels between gaseous molecules in a vacuum and solute molecules in a dilute solution. But even beyond all this, an understanding of gases will eventually lead us to a profound and useful way of describing *chemical affinity*—whether, and how far, a given chemical reaction will actually proceed. So, propelled by a puff of gas, we journey onward to examine our next topic, chemical energetics.

SUMMARY

A **barometer,** a tube with one end closed, filled with a liquid called the barometric fluid, and inverted into a reservoir of the same liquid, relies on the pressure P of atmospheric air to support a column of fluid of density ρ to a height h according to $P = \rho g h$, where g is the acceleration of gravity. Mercury (Hg) has been and remains the usual barometric fluid because of its high density and consequent small column height. At 0°C and sea level, the dry atmosphere supports a 760 mm Hg column; the pressure P necessary to do this is 101,325 N m^{-2} (Pascals, Pa), and is called a **standard atmosphere.** The pressure is often reported in units of atmospheres (atm) or **Torr (760 Torr = 1 atm).** The space above the mercury in the barometer tube is a vacuum, with only a tiny pressure of Hg vapor present.

In experiments with a J-shaped tube containing a trapped sample of air, Boyle established the inverse proportion between pressure and volume now known as **Boyle's law,** $PV =$ constant, which holds when the temperature and the size of the gas sample are held constant. Later Charles and Gay-Lussac showed, in what became known as **Charles' law,** that $V \propto T$ or $V/T =$ constant, with T being the absolute (kelvin scale) temperature $T(K) = t(°C) + 273.15$, as long as P and sample are constant. Avogadro postulated $V \propto N$, with N being the number of molecules in the sample, which we now write as $V \propto n$ or **V/n = constant,** where n is the number of moles, $n = N/N_A$. These three laws can be combined into the **ideal gas law** PV/nT = constant or $PV = nRT$, where R is the **ideal gas constant,** $R = 0.08206$ L atm K^{-1} mol^{-1}.

Atmospheric air is actually a mixture of gases. To a very good approximation, molecules in a sample of gas act independently owing to their small size relative to the volume available to them. This leads naturally to **Dalton's law of partial pressures,** in which the total pressure P of a gas mixture is the sum of **partial pressures** P_i of each component, $P = \Sigma_i P_i$, where $P_i = X_i P$ and X_i is the mole fraction of component i, $X_i = n_i/\Sigma_j n_j$. Solving gas mixture problems is simplified by exploiting the implication of Dalton's law that each component behaves as though the others were not there.

The **kinetic molecular theory** provides a natural explanation for these properties of gases, based on the notion of constant, chaotic molecular motion. Consideration of the force molecules exert on the walls of their container leads to Equation 9.31, $PV = \frac{2}{3}N(\frac{1}{2}m\overline{v^2})$, where $\frac{1}{2}m\overline{v^2}$ is the *average* kinetic energy of a molecule. Comparison with the ideal gas law yields the connection, heretofore assumed, between energy and absolute temperature, $E = \frac{3}{2}nRT$ or $\overline{\varepsilon} = \frac{3}{2}k_B T$, where E is the total translational energy of the gas and $\overline{\varepsilon}$ the average energy of a molecule as above. This brings us at once to a formula for the **root mean square (rms) velocity** $v_{rms} = \sqrt{3k_B T/m}$ of an individual molecule, typically hundreds of meters per second, and to predictions for the dynamic properties of gases, including effusion, diffusion, and collision rates.

In **effusion,** gas molecules move through a small orifice from a high pressure to a vacuum, whereas in **diffusion** the gas mixes with other gases already present. Diffusion is slower than effusion due to collisions, but both follow **Graham's law,** which states that the rate of effusion/diffusion is inversely proportional to the square root of the gas density, and hence to the molecular mass of a gas. Thus light molecules diffuse more rapidly than heavy ones. The rates are proportional to the average molecular speed, and Graham's law follows from $v_{rms} \propto m^{-1/2}$.

The assumption of molecular chaos leads to the **Maxwell–Boltzmann velocity distribution** law, which gives the relative number or fraction of molecules with speeds in a certain range. The peak of the distribution corresponds to the **most probable velocity** $\alpha = \sqrt{2k_B T/m}$, while the somewhat larger **mean** or **average velocity,** $\overline{v} = \sqrt{8k_B T/\pi m}$, and the larger still rms velocity, can also be obtained from it. The distribution broadens with increasing T, implying that the concept of temperature involves not only the *average* energy of a molecule but also the *range* of energies a molecule may possess.

The kinetic theory also offers predictions for **collision rates** Z of molecules with the walls of their container ($Z_{wall} = \frac{1}{4}[X]\overline{v}A$) and with other molecules ($Z_{mol} = \sqrt{2}[X]\overline{v}\pi d^2$), along with the closely related **mean free path** $\lambda = 1/(\sqrt{2}[X]\pi d^2)$. Here [X] is the concentration of gas molecules (molecules cm^{-3}), $\overline{v}$ is the average velocity, A is an area of a wall or opening, and d is a molecular diameter.

Owing to the weak but nonzero forces between gas molecules, real gases show without exception **nonideal behavior,** that is, they deviate slightly from the ideal gas law and the predictions of kinetic molecular theory. These deviations are accounted for in an approximate way by the **van der Waals equation of state** $[P + a(n/V)^2][V - nb] = nRT$, where $a(n/V)^2$ is a small pressure correction that arises from intermolecular attraction, while $-nb$ corrects the volume for the small excluded volume occupied by the molecules themselves.

EXERCISES

The atmosphere and its pressure

1. Calculate the minimum height of a barometric tube using water as the barometric fluid (the von Guericke barometer). Assume a density of 1.00 g cm^{-3}.

2. In Boyle's experiment (Table 9.1 and Figure 9.3), the initial pressure reading of 29 $\frac{2}{16}$ inches Hg (point A) corresponded to atmospheric pressure on the day of the experiment.
   a. Re-express this pressure in mm Hg (Torr), atmospheres (atm), and N m^{-2} (Pa).
   b. If the J-tube had a bore diameter of 0.375 in., what mass of Hg in grams in excess of the initial reading was supported by the air in the tube at point B?

3. The density of air at 1 atm pressure and 0°C is 1.29×10^{-3} g cm^{-3}. Imagine a column of air of uniform density and cross section (an "air barometer"). What is the height of this column? (This is called the *scale height* of the atmosphere. The atmosphere's density, however, decreases with increasing height, giving a much greater atmospheric expanse; see Exercise 7.)

4. On the surface of Mars, the pressure was measured by the Martian lander *Viking I* in 1976 as only 7.4 Torr (al-

most entirely CO_2). Let's suppose there are Martians and that they have had their own Pascal to invent an alcohol barometer (density 0.79 g cm^{-3}). It seems that they carry pocket alcohol barometers around with them, although Martians are much shorter than Earthlings. How can this be? (Mars' gravity is 38% as strong as Earth's.)

5. A generalization of the barometer is the manometer, in which a U-tube half-filled with fluid is connected to two sources of gas. Here fluids of lower density are more useful, because differences in the height of the fluid in the two arms of the tube are then very sensitive to small pressure differences. A commonly used manometer fluid is silicone oil, with a density of 1.09 g cm^{-3}. To what pressure difference, in Torrs and atmospheres, does a difference in height of 15.0 cm correspond?

6. A vacuum chamber is sealed with a 25.0-cm-diam round plate (called a vacuum flange). Calculate the force of the atmosphere in newtons and pounds on the flange when the chamber is evacuated.

7. If you assume the atmosphere is at constant temperature, then our barometric analysis implies that the pressure of the atmosphere must decrease with increasing height z above the earth's surface. Use the ideal gas law to find the density ρ of air as a function of P and T. Then substitute ρ into Equation 9.2, written in the form

$$dP = -g\rho\, dz$$

Now divide both sides by P and integrate from $z = 0$ to h to yield the **barometric formula**

$$P = P_0 e^{-mgh/k_B T}$$

where P_0 is the pressure at $z = 0$, usually taken to be sea level. [Recall that $m/k_B = M/R$, where M is the (average) molar mass (of air).] Use this formula to estimate the pressure of the atmosphere at $h = 8$ km, where commercial jet airliners fly, and thereby explain why jet passenger compartments must be pressurized.

Ideal gases

8. A cylinder of helium is used to fill balloons at the county fair. If the cylinder is 20 cm in diameter and 1 m high, and contains 172 atm of helium, how many 4.0-L balloons can be filled to a pressure of 1.5 atm? (The internal pressure of a balloon when inflated is generally greater than 1.0 atm due to the stretching of its walls.)

9. Pressure gauges for tires usually read in "pounds per square inch gauge" or psig, referring to the pressure in excess of 1 atm; that is, a tire inflated to 35 psig actually contains 50 lb in.$^{-2}$ of air. Suppose a compressor delivering 120 psig is used to inflate a tire to 160 L at 32 psig. What volume of air was delivered from within the compression chamber? What volume would this air occupy at 1.0 atm? (Assume constant temperature.)

10. A person walking briskly in $-20°C$ weather inhales 2.7 L of air in a single breath. What volume will this air occupy in the lungs if allowed to reach body temperature, 37°C? (Assume the pressure is constant.)

11. A 500-mL thermos bottle is filled with 350 mL of hot tea at 90°C, tightly capped, and forgotten. After a few days, the tea has returned to room temperature, 22°C, and the thermos is impossible to open. Why is this? (*Hint:* Calculate the air pressure in the thermos at 22°C, then find the net force, in newtons and pounds on the thermos cap, assuming an area of 30 cm^2 and an external pressure of 1.0 atm.)

12. A 74.3-mL sample of dry hydrogen gas is collected from a metal–acid reaction at 742 Torr and 23°C. What volume would the H_2 occupy at STP? Based on your calculation of the STP volume, find the moles of H_2 formed.

Ideal gas equation of state

13. In Boyle's experiment (Table 9.1 and Figure 9.3), suppose the J-tube had a bore diameter of 0.375 in., and the trapped volume readings of Table 9.1 were in units of 0.125 in. along the tube. Assuming a temperature of 15°C, find the number of moles and the mass in grams of air trapped in the tube.

14. One of the greatest explosion hazards in chemical laboratory work is a runaway gas-forming reaction in a closed container. Suppose 10.0 g $NaHCO_3$ (baking soda) is mixed with a stoichiometric amount of HCl solution at 21°C in a tightly stoppered 200-mL flask. Neglecting the volume of the condensed phases, what pressure (in atmospheres) of CO_2 will develop when the reaction goes to completion?

15. What volume will 1 mol of an ideal gas occupy at 25°C and 1.0 atm? (This is sometimes called **normal temperature and pressure, NTP**.)

16. Ultra-high-vacuum chambers suitable for studying the chemistry of surfaces must be evacuated to better than 10^{-10} Torr. At this pressure, how many molecules are present in a 100-L chamber at 25°C?

17. A flash bulb of volume 2.6 cm^3 contains O_2 at 2.3 atm and 26°C. What mass of Mg wool can be flashed (converted to MgO) by the O_2 present?

18. Each firing of a 0.050-L cylinder in an automobile engine consumes about 0.025 mL of gasoline (octane, C_8H_{18}, density 0.74 g/mL). Assuming the combustion of 1 mol gasoline yields 17 mol of gas ($8CO_2$ + $9H_2O$), what is the temperature of the hot gases if the pressure inside the cylinder reaches 10.0 atm after each firing?

19. In the Dumas method for determination of molecular mass of volatile liquids, a compound of unknown molecular mass is vaporized completely into a flask with a pinhole over its mouth, driving out all the air. The liquid is then recondensed and weighed. If the vapor from 2.428 g liquid was found to occupy a 400.-mL flask at 100°C and 753 Torr, what is the molecular mass?

20. Use the ideal gas law to develop an "engineering formula" that relates the number density ρ_N, molecules cm^{-3}, to the pressure P in Torrs at 298 K. What is the number density at 1 atm? At 10^{-14} Torr, the pressure in interplanetary space?

Dalton's law of partial pressures

21. The human lung requires about 100 Torr of oxygen for comfortable breathing. What minimum total atmospheric pressure is needed, assuming the same composition as dry air at sea level (Table 9.2).

22. A total of 34.7 mL of $O_2(g)$ is collected over water at 24°C and 749 Torr from the decomposition of potassium chlorate, $2KClO_3(s) \longrightarrow 2KCl(s) + 3O_2(g)$. The vapor pressure of water at 24°C is 22.4 Torr and can be assumed present in the gas collection vessel, an inverted graduated cylinder. What is the pressure of dry O_2? What mass in grams of $KClO_3$ has decomposed?

23. A 5.0-L bulb containing He at 210 Torr is connected by a stopcock (valve) to a 3.0-L bulb of Ar at 320 Torr. After the stopcock is opened and the gases allowed to mix thoroughly, what is the total pressure, the partial pressures of He and Ar, and their mole fractions? (Assume the temperature is constant, and neglect the volume of the stopcock.)

24. In Exercise 23, suppose the helium is replaced with H_2 and the argon with Cl_2 at the same pressures. After mixing, what are the partial pressures of H_2 and Cl_2? Now the mixture is sparked, and the reaction goes to completion, forming HCl(g). What is the total pressure and the partial pressures of all species present? (Assume the temperature has returned to its initial value prior to reaction.)

25. What partial pressure of O_2 at 50°C and 0.050 L is required to react stoichiometrically with the vapor from 0.025 mL gasoline (octane, C_8H_{18}, with a density of 0.74 g/mL) to yield $CO_2(g)$ and $H_2O(g)$? Assuming that the O_2 is at its normal partial pressure in air, what total pressure (air + fuel vapor) is present in the 0.050-L volume?

26. Many states now require emissions tests as a part of state automobile inspection. Allowable limits on unburnt hydrocarbons (HC) are 220 parts per million (ppm) and on CO emission 1.20%. Assuming a total pressure of 1.00 atm, what are the maximum partial pressures in Torrs of HC and CO?

Gas stoichiometry

27. One of the reasons Jacques Charles never published his results on gas volumes was his preoccupation with balloon flight. In 1783 he participated in the first flight using a hydrogen-filled balloon, in which the hydrogen was made "by the action of vitriol [sulfuric acid] on iron filings" to yield iron(II) sulfate and H_2 gas. What mass of iron filings would it take to fill a round balloon, 5.0 m in diameter, with H_2 at 12°C and 0.960 atm?

28. After cracking, petroleum is refined by distillation in which fractions of pure hydrocarbons are recovered. These are then characterized by chemical analysis. A certain fraction was found to have a vapor density of 2.749 g/L at 100°C and 1.00 atm. When 5.178 g was burned in pure oxygen, the $CO_2(g)$ produced filled a 5.07-L bulb to a pressure of 1354 Torr at 25°C. Find the molecular formula of the hydrocarbon.

29. The industrially valuable metal chromium is extracted by treating chromite ore, $FeCr_2O_4(s)$, with $Cl_2(g)$ to yield chromyl chloride $CrO_2Cl_2(l)$ according to

$$2FeCr_2O_4(s) + 7Cl_2(g) \longrightarrow 4CrO_2Cl_2(l) + 2FeCl_3(s)$$

The CrO_2Cl_2 is then distilled off and reduced to Cr(s) with magnesium metal. What volume of $Cl_2(g)$

at 1.20 atm and 50°C is required per kilogram of Cr produced?

30. Rocket (and other engine) fuels provide thrust through the production of hot gases that can exert a great force against the rocket engine shell. For example, the combustion of hydrazine, $N_2H_4(l) + O_2(g) \longrightarrow N_2(g) + 2H_2O(g)$, produces the gaseous products at about 2000 K. What mass of hydrazine is required to produce 4450 N (1000 lb) of thrust against a 1-m-diam hemispherical rocket shell? (*Hint:* Calculate the pressure for a complete sphere; the surface area of a sphere is $4\pi r^2$, where r is the radius, and its volume is $\frac{4}{3}\pi r^3$.)

The kinetic molecular theory

31. Equal numbers of atoms of He and Xe will exert the same pressure at constant volume and temperature, despite the great mass disparity. Explain this fact using kinetic theory analysis.

32. Suppose a single Ar atom of velocity 4.13×10^4 cm s^{-1} is in a cubical box 28.2 cm on a side. What pressure, in pascals, Torrs, and atmospheres does it exert on the walls of its container? What pressure is exerted by 6.022×10^{23} Ar atoms of this velocity?

33. Calculate the average velocities $\bar{v}$ (cm s^{-1}) of the following gases: (a) H atoms at 2800 K, (b) I_2 vapor at 114°C, and (c) octane (C_8H_{18}) vapor at 200°C.

34. When a gas molecule escapes the earth's gravity, it is like an atom or molecule being ionized. Use the gravitational potential energy $V = -GMm/r$ to find a formula for the escape velocity v_{esc}, thereby showing that v_{esc} is independent of molecular mass. (*Hint:* The threshold to escape occurs when the kinetic energy exactly cancels the potential energy. The result implies that v_{esc} is the same for a He atom and an interplanetary space payload!) Compute v_{esc} for the radius r equal to that of the earth, 6370 km, and find the temperature at which v_{rms} for He equals v_{esc}.

35. Using the escape velocity found in Exercise 33, find the ratio of numbers of molecules with velocities in the immediate vicinity of v_{esc} and of α, the peak velocity, for molecular oxygen at STP. How does your result reflect the inability of O_2 to escape into space?

36. From the graphs of Figure 9.12, estimate the full velocity width at half-maximum probability at the three temperatures, and plot these widths Δv against $\sqrt{T}$. What aspect of the meaning of absolute temperature does your plot reflect? ($\Delta v = 1.153\alpha$)

37. In chemical kinetics, energy barriers E_a are often found that slow reactions down. Then it is the energy distribution rather than the velocity distribution that is significant. Using $f(E)\,dE = f(v)\,dv$ with $f(v)$ given by Equation 9.46 and $E = \frac{1}{2}mv^2$, give the explicit form of $f(E)$, and sketch this distribution.

38. Sound needs a medium, such as air, in order to travel, and the speed of sound is limited by the speed with which air molecules move and collide. The speed of sound in dry air at 0°C is 1127 ft/s. Compare this value with the average speed of air (using an average mass of 29 amu) at 0°C.

39. Calculate the translational energy of a mole of CO_2 gas at 37°C (body temperature) in kilojoules and kilocalories.

40. For H_2 gas molecules at 300 K, calculate the collision rates Z_{wall}/A, Z_{mol}, and the mean free path λ at (a) 1.00 atm and (b) 10^{-7} atm. Assume a molecular diameter of 2.7 Å. How would you expect these results to change for heavier molecules, assuming only a small change in diameter?

41. In Boyle's experiment (Table 9.1 and Figure 9.3), under the conditions of Exercise 13, for points A and B estimate the following:

    a. The collision rates Z (collisions s^{-1}) of the trapped air molecules with the surface of the mercury.

    b. The mean free paths (Å) in the trapped air. For these calculations assume a molecular diameter of 3.3 Å.

42. In the method developed by Knudsen for measuring **vapor pressures,** a known mass of a volatile element or compound is placed in a container with a single hole of known area inside a vacuum chamber, and the mass loss with time is measured. In one experiment, potassium metal K was heated to 320°C in a steel can (a "Knudsen cell") with a 1.00-mm-diam hole in one end. After exactly 5 min the mass had decreased by 1.790 g. Calculate the vapor pressure (Torr) of K at 320°C.

43. An unknown gas is found to contain only sulfur and oxygen. A fixed volume of this gas effuses through a diaphragm (as in Figure 9.8) in 51.8 s, while the same volume of nitrogen effuses in 34.3 s. Assuming constant temperature, find the molar mass and molecular formula of the gas.

44. In the experiment depicted in Figure 9.9, how many times as fast does H_2 diffuse into the porous cup as air diffuses out? Determine the molecular mass of air from the composition given in Table 9.2.

45. For the diffusion experiment of Figure 9.10, predict the position of the NH_4Cl product deposit, assuming a 1.00-m tube length. Where would the deposit have appeared if HI were substituted for HCl? If $CH_3CH_2NH_2$ (ethyl amine) were substituted for NH_3?

46. A popular modern experimental technique is the so-called "pump-probe" experiment in which two pulsed lasers are employed to investigate some molecular process. For example an ozone molecule can be photodissociated ("pumped"), $O_3 + h\nu \longrightarrow O_2 + O$, and the O_2 molecule can then be examined ("probed") by a second laser, $O_2 + h\nu' \longrightarrow O_2^* \longrightarrow$ light emission. In this type of study, the time delay between firing the two lasers must be shorter than the time it takes O_2 to collide with another molecule. Estimate the maximum permissible time delay for $P = 1$ Torr of O_2, $T = 300$ K, and a collision diameter of 3.0 Å.

Real gases and the van der Waals equation of state

47. Gases and gas mixtures are sold commercially in cylinders filled to more than 100 atm. Under such conditions the pressure read on a gauge attached to the cylinder will not follow the ideal gas equation. Pure $N_2(g)$, 200. mol, is compressed into a 30.0-L steel cylinder at 18°C. Calculate the pressure reading using (a) the ideal gas equation and (b) the van der Waals equation. Explain the difference you find.

48. Repeat the calculation of Exercise 47 for helium gas. You will find a larger difference in the two pressures; explain why.

49. Make plots on the same set of axes of P versus V at 0°C (an isotherm) for 1 mol of $NH_3(g)$ using the ideal gas and van der Waals equations for $1.0 \text{ L} \leq V \leq 50.0 \text{ L}$. Which region of your plot shows the greatest deviation from ideal behavior?

50. Phosphine gas (PH_3) is used to make certain types of semiconductors. How many moles of $PH_3(g)$ are present in a 0.500-L cylinder at 15.0 atm and 27°C? (Use the van der Waals equation in the following way. First solve Equation 9.56 for the molar volume $\overline{V}$ using an iterative method, in which you start with the ideal volume $\overline{V}_{ideal}$ in the pressure correction term. Estimate $\overline{V}$ by solving for the linearly occurring $\overline{V}$. Then improve on this by putting your estimate into the pressure correction term and solving again. In equations,

$$\overline{V}_0 = \frac{RT}{P}$$

$$\overline{V}_1 = \frac{RT}{P + a/\overline{V}_0^2} + b$$

$$\overline{V}_2 = \frac{RT}{P + a/\overline{V}_1^2} + b$$

etc.

Then find $n = V/\overline{V}$.)

51. In Boyle's experiment (Table 9.1 and Figure 9.3), under the conditions of Exercise 13, point B deviates substantially from what is expected based on the other data and ideal gas behavior. Using the experimental conditions and results of Exercise 13, find the molar volumes using both the ideal gas and van der Waals equations of state. Can the observed deviation be interpreted as a deviation from ideal behavior?

52. Choose an example from Table 9.3 and compare the value of $\overline{V}_{STP}$ given there with the result from the van der Waals equation. Give % errors for the ideal gas and van der Waals approximations.

53. If the van der Waals b parameter represents an excluded volume, then you should be able to estimate a molecular diameter from its value. Do this for H_2O, H_2S, and H_2Se, and give a plausible explanation for the trend you observe. How does the H_2O value compare with that derived from liquid density (see Chapter 1)?

Energy Changes in Chemical Reactions

CHAPTER 10

Paddle-wheel experiment devised by James Joule in 1845 to measure the mechanical equivalent of heat. In modern units, Joule found 1 calorie = 4.17 Joules (modern value 4.184). See Section 10.1 for a description.

CHAPTER OUTLINE

10.1 Conceptual Developments

10.2 Heat and Work in a Chemical System

10.3 The First Law of Thermodynamics: The Energy of the Universe Is Constant

10.4 Measuring the Heat of a Chemical Reaction: Calorimetry

10.5 Systematics of Heats of Reaction

10.6 Origins of the Heat of Reaction: Bond Energies

10.7 Heat Capacities: A Molecular Interpretation

Heat and Cold are Nature's two hands by which She chiefly worketh.

Francis Bacon (1627)

William Thomson (Lord Kelvin) (1824–1907). As a student at Cambridge University, Thomson won prizes in both mathematics and sculling. He is most noted for his absolute temperature scale, but he was instrumental in bringing the laws of thermodynamics to their modern form. Later, as Lord Kelvin, he often extolled the achievements of 19th century science, famously declaring in 1896 that no new discoveries remained to be made.

We have spent the first part of this book learning about atoms and molecules not only for the fascination of glimpsing their tiny world, but because they are the constituents of ours. Now we will find that atoms and molecules provide a means to explain and understand the ordinary experiences of our existence, beginning here with the sensations of hotness and coldness, two of the four qualities of matter deemed central by Aristotle. (The other two, wetness and dryness, will come later!) A molecular analysis will enable us to clarify the concepts of heat and temperature, the quantitative measures of hot and cold, in terms of molecular energy. Then we can recognize the manifestations of molecular energy in its various observable forms, including that unique form involved in chemical reactions. This energy—the word derives from the Greek ενεργον, "work from within"—is a very practical thing; we use it to heat our homes, propel our cars, run our clocks and calculators, and so on. These uses all involve a *flow* of energy from a source to its surroundings; accounting for the amount and direction of that flow of energy is an important part of evaluating its utility. The concepts of heat, temperature, and energy flow taken together form the science of **thermodynamics,** again of Greek origin (probably $\theta\epsilon\rho\mu\epsilon + \delta\eta\nu\alpha\mu\iota\varsigma$, literally "heat force").

10.1 Conceptual Developments

Though the ideas of hot and cold are as old as humankind, their quantitative study began only in the 18th century with the invention of the thermometer and quantitative temperature scales by Fahrenheit and Celsius, as mentioned in Chapter 1. Even then, what it is that *makes* hot or cold was obscure. Lavoisier, the clearest thinker of his day and the father of chemical science, ascribed hotness to the presence of something he called *caloric,* "the imponderable matter of heat," and considered caloric to be one of the chemical elements. Dalton took up this idea, describing caloric as an elastic, weightless fluid present between the atoms of all matter. Noting how heating a gas made it expand (Charles' law), Dalton ascribed this to the absorption of caloric by the gas, much as a sponge absorbs water and expands.

The idea of heat as something intertwined with, but separate from, atoms was forever shattered by the brilliant experimentation of Benjamin Thompson, also known as Count Rumford (a name he bestowed on himself). At one stage in Thompson's peripatetic career, he was in charge of cannon boring for the Bavarian army. The great amount of heat produced by the drilling of the cannon barrels surprised him. Now according to the caloric theory, caloric was being removed from the metal of the barrel and the turnings to produce this heat. Thompson decided to test this idea by substituting a blunt drill for the usual one, and contriving to perform the boring operation with bit and barrel carefully insulated and completely immersed in water. He found that the water could be brought to a boil, and boiled indefinitely as long as the boring operation was continued, while very few metal turnings were produced. Noting the exhaustion of the workhorses used to turn the boring machine, he correctly deduced that it was their *work,* manifested in their motion, that produced the otherwise inexhaustible supply of heat and that, therefore, *heat has no independent existence apart from hot objects.* In his own insightful and prescient words:

> It is hardly necessary to add that anything which any insulated body or system of bodies can continually be furnished without limitation cannot possibly be a material substance, and it appears to me to be extremely difficult if not quite impossible to form any distinct ideas of anything being excited and communicated in the manner the *Heat* was excited and communicated in these experiments except it be *Motion.*

These words were written in 1798, but it would be more than 40 years before this connection between work, motion, and heat so elegantly demonstrated by Thompson would be assimilated into the fabric of science as what is known as the **law of conservation of energy:**

> Heat and work are both forms of energy. In any process, energy can be converted from one form to another, but it is never created or destroyed.

Now you may wonder about naming this law "conservation of energy." Way back in Chapter 1, we stated a law by that name, involving the kinetic and potential energy of an isolated object. We found there, however, that if friction were present (as it always is except in a vacuum) the object could not be regarded as isolated, and the energy of the object might be dissipated into its surroundings. The law just stated, as we shall see, *encompasses* the earlier version from Chapter 1; in the example of the dropped book, where the book comes to rest despite its ample supply of energy, the mechanical energy of the book, originally acquired as a result of work, is simply dissipated as heat.

The law in this form was first stated by Rudolf Clausius in 1850, based not on Thompson's experiments, but on newer investigations carried out independently by Julius Mayer and James Joule, and an analysis by Hermann von Helmholtz. Mayer and Joule, natives of Germany and England, respectively, separately measured the *mechanical equivalent of heat,* that is, the mechanical energy equivalent to that needed to raise the temperature of 1 gram of water by 1 degree Celsius, which is called a **calorie,** as introduced in Chapter 1. Joule's experiment, which came within 0.6% of the modern value of the equivalent, consisted of a paddlewheel immersed in an insulated container of water and connected by pulleys to a falling weight. As the weight fell, the paddle stirred the water and heated it ever so slightly. The potential energy lost in the fall, *mgh,* had been gained by the water in the form of heat. The modern value for the equivalent, as given in Chapter 1, is

$$1 \text{ calorie} = 4.184 \text{ kg m}^2 \text{ s}^{-2} = 4.184 \text{ J} \tag{10.1}$$

You can now appreciate the naming of the SI unit of mechanical energy after Joule. This sort of measurement is so difficult that the calorie has been redefined to be *exactly* 4.184 J.

Despite these advances, in 1850 we still had no idea what makes an object feel hot. Thompson again had it right when he connected heat with *motion,* and it was Maxwell and Boltzmann, in their work on the behavior of gases just discussed in the last chapter, who finally put the *atoms* in motion and gave us our modern molecular conception of heat and temperature that will form the basis for our look at energy in chemistry.

10.2 Heat and Work in a Chemical System

Figure 10.1 illustrates a chemical reaction occurring in a flask equipped with a balloon and a temperature probe. The reaction is one of a class of redox displacement reactions discussed in Chapter 5, in which a metal is attacked by strong acid to form a salt and hydrogen gas:

$$\text{Zn}(s) + \text{H}_2\text{SO}_4(aq) \longrightarrow \text{ZnSO}_4(aq) + \text{H}_2(g) \tag{10.2}$$

The balloon is inflating, while the probe registers a rise in temperature; these represent the communication of work and heat, respectively, that is, *energy*, from the reacting chemicals to their surroundings. The reagents and products and the water solvent of Reaction 10.2 are defined to constitute a **system**, a chemical system such as we are interested in, while everything else—the flask, the stopper, the probe, the walls of the balloon, the benchtop, the air in the room, and so on—makes up the **surroundings**. More generally, when looking at energy flow, you can define the system in any convenient way, for example including the flask, but in chemistry the chemicals themselves are our focus. Once the system is defined, the surroundings by default become everything that has been left out of the system. We can thus fairly say that

$$\text{System} + \text{Surroundings} = \text{The Universe} \tag{10.3}$$

This seems a rather pretentious statement to make; after all, no one has made measurements on the entire universe! But, as we shall see, local experience with small systems tells us that the surroundings do indeed receive (or supply) every bit of the energy flow from a process in the system, and that, even if we lose track of the energy after a while, it is still sure to be there, somewhere in the universe. Further confidence in this view is gained by thinking of the *molecules* that make up the system and surroundings as the energy carriers.

It is useful to classify systems as to their mode of communication with the surroundings. Very common in general chemistry laboratories are **open systems**, where heat, work, and matter can all be transferred between system and surroundings—the matter sometimes by accident! In performing measurements of energy changes, however, it is more common to use a **closed system**, in which only heat and/or work is exchanged, and all the chemicals remain trapped; this is the sort of system Lavoisier often used (see Chapter 1). In the extreme, a system may be completely **isolated**, with no communication of any kind with the surroundings. You might think this sort of ex-

Figure 10.1
The reaction Zn(s) + H$_2$SO$_4$(aq) ⟶ ZnSO$_4$(aq) + H$_2$(g) is run in an Erlenmeyer flask, stoppered and equipped with a sidearm balloon to collect the gas evolved. The inflation of the balloon involves work done by the reacting chemicals against the atmosphere, while the above-ambient temperature monitored by the thermocouple temperature sensor indicates the evolution of heat by the reaction as well. See the text for techniques for calculating the heat and work evolved. The reaction mixture is the system, and everything else, depicted or not, is the surroundings. As long as the balloon does not burst, the system is closed, meaning that no matter is exchanged between system and surroundings.

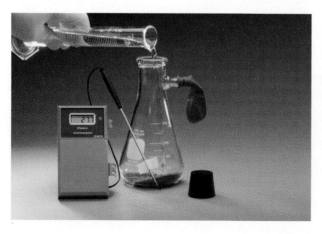

periment would be completely useless, but in fact, as we will examine in Chapter 11, such systems are among the most important of all. In view of the relationship of Equation 10.3, *an isolated system is equivalent to the universe.*

The chemical system depicted in Figure 10.1 is an example of a closed system; if the balloon were to burst, however, the system would become open, with the hydrogen gas escaping to the surrounding atmosphere. A manufacturing plant, you, and the earth are examples of open systems; both the factory and you constantly exchange materials, as well as heat and work, with the environment, while the earth loses helium and an occasional space probe to interstellar space, in addition to carrying on its love affair with the sun's radiant energy. Open systems are bad news at the grocery store, however; no one wants a leaky carton of milk or torn box of cereal. Closed or isolated systems are generally a human contrivance, a message to the universe in a corked bottle.

To understand energy flow we have to be able to *measure* it; recall Lord Kelvin's saying (Chapter 1) about the quantitative nature of scientific understanding. We've already seen the definition of work in Chapter 1: work w is force F acting through a distance x to produce a visible change, and it is given by

$$w = \int F\, dx \tag{10.4}$$

Let's try to apply this formula to the inflating balloon of Figure 10.1. The balloon surface has area A (if spherical, $A = 4\pi r^2$, where r is the radius of the balloon). As the hydrogen gas fills it, it expands a little at a time, with the wall moving outward by a distance dx. As you may know from experience, blowing up a balloon can be a lot of work; some of this is expended stretching the walls of the balloon, but most of it results from pushing back the atmosphere. As we saw in Chapter 9, the atmosphere exerts an external force $F_{ext} = P_{ext}A$ over the surface of the balloon, where P_{ext} is the pressure of the atmosphere (standard value 1 atm = 760 Torr = 101,325 N m^{-2}). This force must be supplied by the system in order to inflate the balloon against the force of the atmosphere. Equation 10.4 then becomes

$$w = -\int P_{ext} A\, dx = -\int P_{ext}\, dV \tag{10.5}$$

where dV is the small increase in volume corresponding to the displacement dx, and the minus sign is inserted to make work done *by* the system on the surroundings negative, because it represents loss of energy by the system. To get an actual amount of work, the initial and final volumes V_1 and V_2 must be specified, yielding the definite integral

$$w = -\int_{V_1}^{V_2} P_{ext}\, dV \tag{10.6}$$

In the case considered, the external pressure, as supplied by the atmosphere, is constant (unless a storm blows in during the experiment!) and can be taken outside the integral sign, giving

$$w = -P_{ext}(V_2 - V_1) \equiv -P_{ext}\Delta V \tag{10.7}$$

where the symbol Δ is used to indicate the change in a quantity. If the balloon in Figure 10.1 is inflated from 0 to 5.00 L of volume, the work is

$$w = -(1.00\text{ atm})(5.00\text{ L} - 0.00\text{ L}) = -5.00\text{ L atm} \quad \text{or}$$
$$w = -(1.013 \times 10^5\text{ N m}^{-2})(5.00 \times 10^{-3}\text{ m}^3) = -506\text{ J}$$

As a shortcut you can use the conversion 1 L atm = 101.3 J, and convert to joules after the calculation. This may be further converted to calories (cal) using Equation 10.1:

$$w = -(506 \text{ J})(1 \text{ cal}/4.184 \text{ J}) = -121 \text{ cal}$$

Again, a quick conversion factor is 1 L atm = 24.22 cal. The minus sign indicates that the work is done *by* the system.

An energy of 506 J is quite a lot; considering that a hard-thrown football might have 10 J of kinetic energy, blowing up a 5-L balloon is like throwing 50 footballs! But its thermal equivalent seems slight; 121 g of water, little more than a half cup, can be heated by 1°C with this amount of energy.

The contents of the flask in Figure 10.1 determine the work that could possibly be done; this depends on the number of moles of hydrogen gas that can be formed by the available reagents. At constant P and T, the ideal gas law yields

$$w = -P\Delta V = -\Delta n_{gas} RT \qquad (10.8)$$

where Δn_{gas} is the net change in the number of moles of gas when going from reagents to products. For Reaction 10.2, Δn_{gas} = +1 mol/(mol reaction). Example 10.1 shows how to incorporate the reaction stoichiometry.

EXAMPLE 10.1

For Reaction 10.2 (depicted in Figure 10.1), 300. mL of 1.00 M H_2SO_4 was added to 14.66 g Zn metal. Calculate the work, assuming the reaction went to completion, and the temperature of the expanding gas was 25°C.

Solution:

First you must find the limiting reagent; we will assume you still remember how to do that. Zinc turns out to be limiting. The number of moles of hydrogen gas produced is then equal to the moles of Zn consumed:

$$\Delta n_{gas} = n_{H_2} = (14.66 \text{ g Zn})\left(\frac{1 \text{ mol Zn}}{65.39 \text{ g Zn}}\right)\left(\frac{1 \text{ mol H}_2}{1 \text{ mol Zn}}\right)$$
$$= 0.2242 \text{ mol}$$

and the work is then

$$w = -(0.2242 \text{ mol})(0.08206 \text{ L atm K}^{-1} \text{ mol}^{-1})(298 \text{ K})$$
$$= -5.483 \text{ L atm} = -555.5 \text{ J} = -132.8 \text{ cal}$$

Note that $\Delta n_{gas} > 0$ implies $w < 0$; that is, work is done *by* the system on the surroundings. Note also that the calculation does not presuppose the presence of a container (the balloon) to catch the gas. An open reaction flask still does work, the evidence being the bubbles of gas evolving from the reaction mixture, but it is not useful work. A reaction such as this was used on a much larger scale by Jacques Charles to fill the early hydrogen balloons for manned flight (see Exercise 9.27).

The temperature probe shown in Figure 10.1 is registering higher than ambient temperature; the flask feels warm to the touch if above 37°C, body temperature. This warmth indicates a release of what we call *heat* by the reaction system; the heat released (or absorbed), q, is determined by the change in temperature according to

$$q = \int_{T_1}^{T_2} C \, dT \tag{10.9}$$

Although the probe is outside the flask, the rise in temperature reflects the temperature change within the system of reacting chemicals. The proportionality constant C depends on the composition of the system and is called the **heat capacity** of the system; C has units of J K^{-1} or cal K^{-1}. Unlike the case for work, w (Equation 10.6), a minus sign is not needed here; Equation 10.9 will yield a positive q when T rises, since this represents a gain of heat energy *by* the system. For nearly all situations we will encounter here, the heat capacity of the system has a constant value, and, like the external pressure, can be taken outside the integral sign to give

$$q = C(T_2 - T_1) \equiv C\Delta T \tag{10.10}$$

You should note the symmetry between the pairs of Equations 10.6, 10.7, and 10.9, 10.10, and that the temperatures refer to those of the system.

For the chemical system shown in Figure 10.1, the heat initially deposited in the system by the reaction will be given off to the surroundings, and the flask and its contents will return to ambient temperature (T_1) given enough time. The surroundings, being merely the rest of the universe, may be regarded as having an *infinite* heat capacity and, thus, suffer no change in temperature even though they have been given heat by the system. To handle this situation, it is useful to break the process down into two idealized stages. In stage I, the heat generated by the reaction is completely trapped within the system, causing it to "heat up" from T_1 to T_2; during this stage, the system is considered to be *isolated*. In stage II, this heat energy passes through the boundary between system and surroundings, and the system returns to its prereaction temperature T_1. (In practice, the complete trapping of heat is impossible, but it can be very nearly attained with a suitably designed boundary, as discussed later.) Overall, then, the system will lose heat if it initially shows a temperature rise, and this lost heat is given by $q = -C\Delta T$. Reactions that give off heat in this way are called **exothermic,** $q < 0$; reactions that cause the system to cool off, and therefore must eventually receive heat from the surroundings, are called **endothermic,** $q > 0$.

This discussion of the time it takes for things to happen brings us to a central aspect of thermodynamic analysis: *thermodynamics is timeless!* An assessment of heat and work, energy flow, has to wait until things stop happening. At that point, a steady state known as **equilibrium** has been established; before that time we cannot assign definite limits to our work and heat integrals, Equations 10.6 and 10.9. It would be like trying to balance your checking account while others are out writing checks and making deposits on it. The next section restates these ideas more formally.

To calculate the heat q from Equation 10.10, the system heat capacity C must be known. For reactions in dilute aqueous solution (concentrations 1 M or smaller), C is very closely approximated by the heat capacity of the water present. Because water formed the basis for finding the mechanical equivalent of heat outlined earlier, we know that 1 cal or 4.184 J will cause the temperature of 1 gram of water to increase by 1°C = 1 K; that is, the **specific heat** (capacity), denoted by a small c, of water is $c = 1$ cal g^{-1} K^{-1} or 4.184 J g^{-1} K^{-1}. For an aqueous system we can then readily find C from

$$C \approx mc \qquad (10.11)$$

where m is the mass of water in the system; for dilute solutions the density is close to that of pure water, 1.00 g/mL, and the mass in grams is therefore numerically equal to the volume of the system in milliliters. In general, specific heats c for most homogeneous materials are well known, and enable the heat capacity of any sample to be calculated. The following example illustrates calculation of the heat evolved in our system.

EXAMPLE 10.2

For the reaction system described in Example 10.1, the temperature was found to increase from 22°C to a maximum of 47°C, and then it began to fall again as the reaction subsided. Calculate the heat and the heat per mole of Zn.

Solution:
Using the idealized two-stage analysis and Equation 10.11, we have $q = -mc\Delta T$; for 300. mL of solution, $m \approx 300.$ g. Then

$$q = -(300.\text{ g})(1\text{ cal g}^{-1}\text{ K}^{-1})(47 - 22)\text{K} = -7.5 \times 10^3 \text{ cal}$$
$$= -7.5 \text{ kcal} = -31 \text{ kJ}$$

Note that, although the temperatures are given in Celsius, only their difference is required, and thus no conversion to the kelvin scale is necessary, since 1°C = 1 K. From the first example, 0.2242 mol Zn reacted, giving

$$q/n = -7.5 \text{ kcal}/0.2242 \text{ mol} = -33 \text{ kcal mol}^{-1}$$

You should realize that the quoted temperature rise was not as high as it could have been, since the temperature probe shown in Figure 10.1 is detecting heat coming through the flask wall while the reaction is still in progress; the escaping heat is not available to raise the temperature of the system itself. We will examine more accurate heat measurements in Section 10.5.

Notice how even the relatively modest heating of the flask that occurred involved a much greater energy than the work that was done, a factor of 60 greater in this case. This is typical of chemical reactions run under conditions like those illustrated in Figure 10.1; from a practical point of view, most of the energy of the reaction was wasted as heat. This need not be the case, however; one of the uses of thermodynamics is to enable us to find ways to reduce this waste of energy. To understand how to do this, we need to examine more carefully the ideas introduced so far.

10.3 The First Law of Thermodynamics: The Energy of the Universe Is Constant

The reaction between zinc and sulfuric acid produces both heat and work that is passed to the surroundings. This energy, as we have already hinted at in Chapter 5 and will examine further in this chapter, arises from a drop in the total energy of the molecules that comprise the system due to the rearrangement of electrons in the reaction. In fact, the production of hydrogen gas, a fuel, indicates that there is yet more energy to be garnered. The first law of thermodynamics as applied in chemical systems says that the evolution of heat and work by the system must be accompanied by a fall in the **internal energy** E of the system in order to conserve energy; that is,

$$\Delta E = q + w \qquad (10.12)$$

where as usual $\Delta E \equiv E_2 - E_1$. Equation 10.12 covers both possible directions of net energy flow and either sign for ΔE. The sum $q + w$ also represents the negative of the energy received (or given) by the surroundings, so we can also state

$$\Delta E_{sys} = -\Delta E_{surr} \quad \text{or} \quad \Delta E_{univ} = \Delta E_{sys} + \Delta E_{surr} = 0 \qquad (10.13)$$

and therefore, with Clausius, state in words,

The energy of the universe is constant.

State Variables

We will not spend too much time worrying about the universe, though, because we are more interested in statements like Equation 10.12 that concern the system we want to study. Let's return to our example, the Zn + H_2SO_4 reaction, and try to analyze it in thermodynamic terms. After the reaction is over and the flask has cooled down, what is left is a flask containing a zinc sulfate solution. A clever detective might be able to figure out that a reaction producing work and heat had just been run, from scraps of evidence here and there in the surroundings—such as bits of balloon scattered about—but no *measurement* he could make on the $ZnSO_4$ solution would reveal that it was the outcome of a reaction. The solution could have been equally well prepared by dissolving $ZnSO_4(s)$ from a bottle of reagent salt. Even if the detective had stumbled in while the flask was still warm, how could he tell that someone hadn't just heated a flask of salt solution? Likewise there would be no means to tell whether work had occurred; the trace of extra hydrogen in the room could have been planted there by a trickster with a lecture bottle of hydrogen.

These observations imply that heat and work in a process cannot be determined by merely measuring properties of the system before and after the process has occurred; one must observe the process itself, that is, follow the *path* from initial to final state. *The quantities of heat and work evolved or consumed by the system are path dependent.* But how then can energy be conserved in the process? Note that Equation 10.12 does not relate q or w individually to the system energy change, but only their sum $q + w$. In our example, if more work is done by the system, then less heat will be evolved, and vice versa. As we shall see in the next section, if our reaction were run in a rigid, sealed container, no work could occur and the container would have gotten slightly warmer.

On the other hand, the energies E_1 and E_2, and hence the energy change ΔE, must be knowable, at least in principle, from measurements made on the system before and after the process. This is complicated in a chemical system by the presence of different compounds before and after, but the energies of those compounds depend only on their nature, that is, on how strongly they are bonded together, on how the electrons are arranged in each. A sample of metallic zinc and a sulfuric acid solution have an intrinsically different internal energy from a zinc sulfate solution and gaseous hydrogen. The actual values of these energies depend on the total amounts present, as well as on how we define the zero of energy, but once these are chosen, E_1 and E_2 are completely defined by the composition of the system, along with its temperature and pressure, all knowable without recourse to the system's *history*, that is, how it got to its present state. This implies that ΔE is *independent of path*, being defined only by the initial and final **states** of the system. This path independence ensures that energy will be conserved regardless of the details of any particular process.

Those properties of a system that we can measure without knowing its previous experiences are called **state variables** (in some texts, state functions). Some examples are total mass, pressure P, volume V, number of moles n of each type of compound, and temperature T. These in turn define the energy E, as just discussed, and therefore E is a state variable as well. However, q and w require us to know a path for a process in which a system evolves from one state to another; that is, they require more than just state measurements and are, therefore, *not* state variables.

It is a simple corollary that as the system changes its state, *all* state variables will evolve in a way that is independent of path. If we choose to change the state a little at a time, say by reacting the zinc with the acid a little at a time, the energy will change by a small amount dE regardless of the bits of heat and work evolved. When we finally reach the end of the reaction, the change in energy ΔE will be the sum of all the little bits of energy dE, or

$$\Delta E = \int_{state1}^{state2} dE \tag{10.14}$$

This is the same result you would get by putting in the initial and final state values E_1 and E_2 for the limits on the integral, that is, just by knowing the start and stop of the whole process, without any information on the intermediate stages. Therefore, dE (or a small change in any state variable) is an *exact differential,* and Equation 10.14 is just another way of stating the first law.

Yet another way of considering changes in state results from analyzing a process that occurs in a **cycle,** that is, one in which the system ends in the same state in which it began. You might think that such a process would be singularly uninteresting—it seems as if nothing has happened! But in fact cyclic processes are of great practical importance; all engines, such as those inside cars, run in cycles. Much useful work can be done along the way, and, in chemistry, many fascinating transformations can occur, depending on how the cyclic path is chosen. Regarding the energy E, we can say that all the changes in E around the cycle, regardless of its nature, must be zero, or

$$\oint dE = 0 \tag{10.15}$$

where the little circle on the integral sign indicates a closed path over which the series of energy changes dE takes place. This is actually equivalent to Equation 10.14

State Changes for an Ideal Gas

We have been particularly interested in formulating the first law in a way that allows you to see how it applies in a chemical reaction, and this application will be thoroughly fleshed out as we go. But the particular example of a system consisting of an ideal gas undergoing changes in state allows us to make definite predictions of heat, work, and energy changes based on ideal gas behavior, and to relate these changes to the behavior of the gas molecules that comprise the system using kinetic molecular theory. This predictive power has already been illustrated by the example of the hydrogen gas expanding from our Zn/H_2SO_4 reaction system.

As we saw in Chapter 9, the weak intermolecular forces in a gas produce slight deviations from ideal gas behavior. In an ideal gas the potential energy arising from these forces is negligible compared to the kinetic, and the energy of an ideal gas is therefore entirely kinetic as far as translational motion is concerned. Polyatomic gases can also store energy in internal degrees of freedom, a feature that was touched on in Chapter 8 and will be considered further later in this chapter; for now we need only account for the translation of the molecule as a whole, because it is this motion that gives rise to pressure, again as established in the previous chapter. So let us first consider an ideal *monatomic* gas, as exemplified by the noble gases, where the total energy is given by

$$E = \sum_{i=1}^{N} \varepsilon_i = \tfrac{3}{2} N k_B T = \tfrac{3}{2} n R T \tag{10.16}$$

as we learned in Chapter 9 by comparing the prediction of kinetic molecular theory with the ideal gas law (Equations 9.31 to 9.33). As you can see, the energy of an ideal gas depends only on the absolute temperature, and *not* on pressure or volume; even for real gases, the P or V dependence is very weak in general. As we also discussed in Chapter 9, Equation 10.16 implies a natural choice for the zero of energy, namely, when the kinetic energy of the gas atoms becomes zero, at absolute zero temperature, 0 K.

The analysis of changes of state for an ideal gas is greatly simplified by the fact that $E = E(T)$ only. In an **isothermal** change, the temperature is held constant—by keeping the gas sample in good thermal contact with surroundings of constant temperature—during the process. This means $\Delta T = 0$, and therefore $\Delta E = 0$ from Equation 10.16. Then, according to Equation 10.12,

$$q + w = 0 \quad \text{or} \quad q = -w \quad \text{for } T = \text{constant} \tag{10.17}$$

Notice that for an isothermal process it is impossible to compute q directly, because there is no temperature change to put into Equation 10.10. On the other hand, w is easy to compute from Equation 10.7, for example, assuming a change in volume under the influence of a constant external pressure. If the gas is expanding, it is doing work on the surroundings, and its energy and temperature will drop unless heat is added to it; the right amount to maintain the starting temperature of the gas is given by Equation 10.17. This circumstance leads to a fundamental distinction between work w and heat q: although the transfer of work between system and surroundings always leads to a measurable change in the surroundings, the transfer of heat does not always do so. In the reaction system of Section 10.2, the cooling flask does not measurably change the temperature of the world. In this case we *infer* the transfer of heat from measurement of the work at constant T.

If the volume is held constant, called an **isochoric** process, the gas is now unable to inflict any visible change on the surroundings, and the only change it can undergo is a change in T. If gas trapped inside a glass bulb is heated with a Bunsen burner, its temperature must increase, in part because it cannot throw off the added energy by working on the surroundings. In this case we can say that $w = 0$ and therefore

$$\Delta E = q_V \qquad \text{for } V = \text{constant} \tag{10.18}$$

where the subscript has been added to q to indicate the constant volume. By writing Equation 10.16 for two different temperatures and subtracting, you can readily find that

$$\Delta E = \tfrac{3}{2}nR\Delta T = q_V \tag{10.19}$$

By comparing Equation 10.10, $q = C\Delta T$, with Equation 10.19, we can identify the **heat capacity at constant volume** C_V for a monatomic gas as

$$C_V = \tfrac{3}{2}nR \tag{10.20}$$

A more common occurrence in the laboratory is a system that has neither constant T nor constant V, but is subjected to constant pressure P while it undergoes a change. For an ideal monatomic gas, such an **isobaric** process allows both heat and work to be exchanged with the surroundings. In this case

$$\Delta E = q_P - P\Delta V \tag{10.21}$$

where Equation 10.7 has been invoked for the work term, and the subscript P on q indicates constant pressure. By transposing the work term, and collecting initial and final state variables, Equation 10.21 can be rewritten as

$$(E_2 + PV_2) - (E_1 + PV_1) = q_P$$

Because the quantities in parentheses are all state variables, we can now define a new state variable, the **enthalpy** H (from the Greek ενθαλπο, meaning "heat within"), by

$$H = E + PV \tag{10.22}$$

and use it to write Equation 10.21 as

$$\Delta H = q_P \tag{10.23}$$

This definition of H and the first law in the form of Equation 10.23 transcend the case of the monatomic ideal gas we are considering and will be applied next to the commonly encountered chemical reaction run at constant (ambient) pressure.

10.3 The First Law of Thermodynamics: The Energy of the Universe Is Constant

By combining the definition of Equation 10.22 with the kinetic molecular result of Equation 10.16, we readily find that

$$H = \tfrac{3}{2}nRT + PV = \tfrac{3}{2}nRT + nRT = \tfrac{5}{2}nRT \qquad (10.24)$$

where the ideal gas law was used to replace PV by nRT. We can then deduce that the **heat capacity at constant pressure** C_P for a monatomic gas is

$$C_P = \tfrac{5}{2}nR \qquad (10.25)$$

by comparing ΔH from Equation 10.24 with $q = C\Delta T$. An interesting result follows from comparing C_P and C_V:

$$C_P - C_V = nR \qquad (10.26)$$

that is, C_P is always larger than C_V. Relation 10.26 turns out to hold for all ideal gases, monatomic or polyatomic, and is consistent with the observation that adding a given quantity of energy to a gas at constant pressure will produce a smaller temperature rise than if added at constant volume, because the gas is able to expend some of the energy as work at constant P. This result does not extend to liquid or solid phases (such as water or ice), because changes in pressure produce only tiny changes in the volume of a condensed phase, and the work term becomes negligible. Thus, *for a condensed phase* $C_P \approx C_V$. Condensed-phase heat capacities are discussed further in Section 10.7.

A very useful result flows from Equation 10.18, $\Delta E = q_V$, and the fact that E is a function only of T for an ideal gas. If the temperature change ΔT of a system is known, then, *regardless of changes in P or V*, the change in energy is

$$\Delta E = C_V \Delta T \qquad (10.27)$$

When a gas expands against a nonzero pressure without the addition of heat (a so-called **adiabatic** expansion), it will cool owing to the energy being expended as work. In such a process (to be considered in more detail in Chapter 11), both V and T are changing, yet Equation 10.27 still holds for an ideal gas, because E is a state variable. The analogous relation $\Delta H = C_P \Delta T$ also holds even if the pressure is not constant, as the example illustrates.

EXAMPLE 10.3

Oftentimes synthesis of new chemical compounds must be carried out in an argon atmosphere to avoid air oxidation of reagents and/or products. Suppose Ar in a storage cylinder at 25.0 atm and 22°C is expanded isothermally into a final volume of 3.50 L and pressure of 1.00 atm at a constant external pressure of 1.00 atm inside a reactor. The reaction mixture and the Ar are then heated at constant volume to 320°C, where the reaction then proceeds isothermally. Calculate q, w, ΔE, and ΔH for the Ar, assuming ideal gas behavior, for the expansion and heating stages separately, and for the overall process.

Solution:
For the isothermal expansion, we apply Equation 10.17, with $w = -P_{ext}\Delta V = -P_{ext}V_2(1 - P_2/P_1)$ by Boyle's law, yielding $w = -340.$ J, $q = +340.$ J, $\Delta E = \Delta H = 0$.

For the heating stage, we can apply Equations 10.18 to 10.20, obtaining $w = 0$ and $\Delta E = q_V = C_V \Delta T = \frac{3}{2} nR\Delta T$. Although the pressure of the Ar is not constant in the second stage, we nonetheless have $\Delta H = C_P \Delta T = \frac{5}{2} nR\Delta T$, H being a state variable. With n calculated from the information for the first stage ($n = 0.145$ mol Ar), we get $\Delta E = q_V = +537$ J and $\Delta H = +896$ J. Tabulating the results,

Energy	Stage I	Stage II	Total
w	-340 J	0 J	-340 J
q	$+340$	$+537$	$+877$
ΔE	0	$+537$	$+537$
ΔH	0	$+896$	$+896$

Note that ΔE and ΔH are determined entirely by the second stage, because only then does T change, and they are independent of path; path details must be known to calculate q and w. You can see that $\Delta E = q + w$ holds for the entire process as well as each stage. In this case, ΔH is an artificially "inflated" energy, because it includes the work that would have been done were the second stage carried out isobarically, that is, if the system were allowed to inflate.

10.4 Measuring the Heat of a Chemical Reaction: Calorimetry

As has been hinted at in Chapter 5, and exemplified by the $Zn + H_2SO_4$ reaction, many reactions liberate energy, mostly in the form of heat, as products are formed. To make a precise measurement of the heat energy, the usual practice has been to trap the heat in a small part of the surroundings immediately adjoining the system, and measure changes in its properties. A device that accomplishes this task is called a **calorimeter** (accent on the third syllable, although really a *calorie meter*). If you can remember the very beginning of Chapter 1, you will recall that Antoine Lavoisier was a pioneer (and a master) of the confinement of chemical reactions. So you should not be surprised to learn that Lavoisier, along with Pierre Laplace (better known for his mathematical exploits), built the first precision calorimeter in 1780 (see Figure 10.2). This was an *ice* calorimeter, which measured the heat of a reaction in terms of the quantity of ice its heat would melt. For example, for the combustion reaction

$$C(s) + O_2(g) \longrightarrow CO_2(g)$$

Lavoisier and Laplace found that "one ounce of carbon in burning melts six pounds and two ounces of ice." This result, when converted to modern units, and using the known heat absorbed in melting a given amount of ice, is within 1% of the currently accepted value for the **heat of combustion** of carbon in graphite form. The melting of the ice signified a *release* of heat by the reaction, that is, an exothermic reaction ($q < 0$). Our $Zn + H_2SO_4$ example is also an exothermic reaction, as are the vast majority of known reactions. Endothermic reactions ($q > 0$) though rare, are also found; cold packs for soothing injuries and reducing swelling run on endothermic reactions. Endothermic processes are also of theoretical significance for understanding the driving forces behind chemical reactions, as we will see in the next chapter.

As introduced in Chapter 5, combustion reactions are among the most common and important reactions in daily life. In addition, the vast majority of known chemical

Figure 10.2
The ice calorimeter of Lavoisier and Laplace (dimension scale is in inches). The nested chamber *b* is first packed with ice, then a combustible sample in a dish is placed in the central screen basket *f* and ignited. The liquid water formed as a result of melting the ice by the heat of the reaction is drained into the pan P beneath by opening valve *y*. The outer jacket *a* provides thermal insulation. Knowing the tare weight of the pan and weighing it with the collected water allows the mass of ice melted to be obtained. This simple device allowed the heats of highly exothermic reactions to be measured to better than 1% accuracy (see Exercise 20 at the end of the chapter).

10.4 Measuring the Heat of a Chemical Reaction: Calorimetry

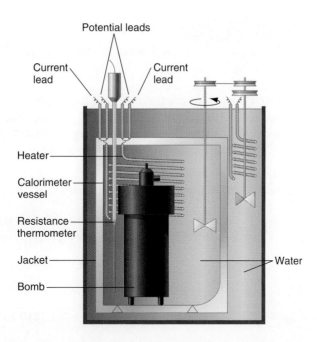

Figure 10.3
Bomb calorimeter used at the National Institute of Standards and Technology (formerly the National Bureau of Standards).

compounds are organic, containing carbon, hydrogen, and other elements, and all of them are combustible. Because of this, great effort and ingenuity has gone into the development of accurate methods for measuring heats of combustion. In 1881 Berthelot invented the **bomb calorimeter** (a good kind of bomb that never explodes). Figure 10.3 illustrates a modern version used by the National Institute of Standards and Technology (formerly the National Bureau of Standards) for combustion measurements. Because of the many components (stirrer, thermometer, inner jacket, etc.), accurate estimation of the heat capacity of the system is difficult; therefore a *calibration* procedure is used. Known electrical energy from a heating element within the inner jacket is applied to produce a measured temperature rise ΔT_c; if a voltage V is applied to a heater of known resistance R for a time t, $q_{elec} = V^2 t/R$ joules of heat energy are deposited. The heat capacity of the system is then obtained as $C = q_{elec}/\Delta T$. Then, with the combustible substance stored in a crucible inside the bomb, which is prefilled with oxygen, a spark or heated wire is used to ignite the reaction. The temperature change ΔT is measured, and the reaction heat computed from $q = -C\Delta T$. Precisions of 0.02% are regularly obtained from this procedure. Heats of reaction are reported as ΔH per mole of fuel; the calorimeter being sealed, the measured q_V is ΔE, which is converted to ΔH according to

$$\Delta H = \Delta E + \Delta n_{gas} RT \tag{10.28}$$

a result that is obtained by combining Equations 10.8, 10.21, and 10.23.

Reaction heats are also converted to so-called normal temperature (25°C or 298.15 K) and standard pressure (partial pressure of each component 1 atm), called **NTP**. To denote NTP conditions, we write the heat of reaction as $\Delta H°_{298}$, where the superscript ° implies both standard pressure and per mole of reaction, and the subscript gives the absolute temperature. The subscript is often omitted for brevity if NTP conditions are implied.

EXAMPLE 10.4

A bomb calorimeter like that illustrated in Figure 10.3 is calibrated by applying a voltage of 24.00 V to the immersed resistor, whose resistance is 5.000 Ω, for 5 min (300. s). This produces a temperature rise of 3.072°C. The bomb contains 2.000 mL benzene, $C_6H_6(l)$, density 0.8765 g/mL, with excess oxygen. After the calorimeter has returned to ambient temperature (22°C), a small heater wire ignites the combustion reaction $C_6H_6(l) + 7\frac{1}{2} O_2(g) \longrightarrow 6CO_2(g) + 3H_2O(l)$, yielding a temperature rise of 6.329°C. Calculate $\Delta H°$ for the combustion of benzene.

Solution:

The resistance heating allows the calorimeter's heat capacity to be determined from

$$C = \frac{q_{elec}}{\Delta T} = \frac{V^2 t}{R \Delta T} = \frac{(24.00 \text{ V})^2 (300.0 \text{ s})}{(5.000 \text{ }\Omega)(3.072°C)}$$

$$= 1.125 \times 10^4 \text{ J/°C} = 2689 \text{ cal/°C}$$

The heat of the reaction is then computed from

$$q_V = -C \Delta T = -(2689 \text{ cal/°C})(6.329°C) = -1.7019 \times 10^4 \text{ cal } (-7.121 \times 10^4 \text{ J})$$

where the V subscript indicates constant volume. The calorimeter thus measures ΔE, given on a per mole basis by

$$\Delta E = \frac{q_V}{n_{C_6H_6}} = \frac{-17.019 \text{ kcal}}{2.000 \text{ mL}} \frac{1 \text{ mL}}{0.8765 \text{ g}} \frac{78.113 \text{ g}}{1 \text{ mol}}$$

$$= -758.4 \text{ kcal/mol } (-3173 \text{ kJ mol}^{-1})$$

Finally, ΔE is converted to $\Delta H°$ using Equation 10.28, where R is the gas constant (not the resistance) and T is the ambient temperature 295 K. This yields

$$\Delta H° = -758.4 \text{ kcal/mol} + (-1.5 \text{ mol gas/mol})$$
$$\times (1.9872 \text{ cal K}^{-1} \text{ mol}^{-1})(1 \text{ kcal}/1000 \text{ cal})(295 \text{ K})$$
$$= -758.4 - 0.9 = -759.3 \text{ kcal/mol } (-3177 \text{ kJ mol}^{-1})$$

The work correction to get ΔH from ΔE is small, as is usual; the correction therefore need not be calculated very precisely in most cases. A final step needed in precise work is a correction from the experimental ambient temperature to 298.15 K. Later we will discuss such corrections, but they are normally very small (negligible in the present problem) if the ambient temperature is within a few degrees of normal. In combustion reactions it is customary to report the heat evolved per mole of fuel; this is why we wrote the reaction with a fractional coefficient for O_2. The reaction written with integer coefficients (multiplied by 2) would have a $\Delta H°$ twice that obtained, since 2 mol of benzene would be burned.

A simpler calorimeter for measuring heats of reactions has become popular in both research and instructional laboratories, owing to the ready availability of styrofoam cups. With the so-called **coffee cup calorimeter,** the accurate measurement of heats of solution reactions has become routine, and fledgling chemists regularly match literature values for solution reaction heats! Styrofoam is such an effective thermal insulator that the heat of reaction is nearly fully trapped within the reaction solution itself, making calibration almost unnecessary. If the solution is reasonably dilute ($\leqslant 1$ M reagent concentrations) the heat capacity is nearly that of the water solvent. In calories per degree Celsius, it is numerically equal to the mass of water in grams. Because the reaction is necessarily run under constant pressure, the measured heat q_P is ΔH directly.

10.5 Systematics of Heats of Reaction

Since the heat involved in a chemical reaction has so many practical uses, not the least of which is as a source of warmth, it is of great importance to know the magnitudes of reaction heats and to be able to predict them for new types of chemical reactions. At first you might think that there is no recourse but to measure the heat of each reaction of interest using the methods discussed in the last section. But in fact, as G. Hess first showed experimentally in 1840, heats of reaction are not independent of each other if they involve some of the same elements or compounds. From our examination of the energy change of a chemical process in light of the first law in Section 10.3, we concluded that the energy of an element or compound is an intrinsic property that depends only on the nature of its atoms and the bonds between them. So, for example, the heats of the following reactions have been measured:

Reaction	ΔH°_{298}	
	kcal/mol	kJ mol^{-1}
(a) $2C(s) + O_2(g) \longrightarrow 2CO(g)$	-52.8	$-221.$
(b) $2CO(g) + O_2(g) \longrightarrow 2CO_2(g)$	-135.2722	-565.979
(c) $C(s) + O_2(g) \longrightarrow CO_2(g)$	-94.0518	-393.513

The heat of reaction (a) is difficult to measure, because some CO_2 is always formed even when the amount of oxygen is limiting, while the full oxidation of C and CO to CO_2 is easier to achieve. These reactions have a special relationship: the sum of reactions (a) and (b) is twice that of reaction (c). In practical terms, reaction (a) is a partial oxidation of carbon, and reaction (b) finishes the job, with the overall process represented by reaction (c). As you can readily verify, the stoichiometric relationship among reactions (a), (b), and (c) [(a) + (b) = 2(c)] also holds within experimental uncertainty for the heats of reaction: $\Delta H_a + \Delta H_b = 2\Delta H_c$. This relationship between stoichiometry and ΔH is found experimentally to hold for every case tried, and is known as **Hess's law of constant heat summation.**

Although Hess's law was discovered prior to a full statement of the first law, nowadays it is regarded as following directly from the first law applied to the enthalpy H, a state variable. Just as the enthalpy change in a *physical* change of state for a gas is independent of path, so the enthalpy change in a *chemical* change of state—a chemical reaction—is independent of how it is carried out, including the use of intermediate stages in the reaction. Whether we choose to oxidize carbon a step at a time or in one fell swoop, the net energy obtained from the combustion will be identical in each case. This result can also be viewed as arising from the intrinsic and invariant chemical energy of the reaction participants $C(s)$, $O_2(g)$, $CO(g)$, and $CO_2(g)$, the electronic energies of the atoms and the bonds they make in forming macroscopic samples. The energy *differences* as measured by ΔH simply reflect the invariant character of these intrinsic molecular energies, in the same way that measuring the spectrum of a given atom or molecule reflects an invariant energy level structure. We have already seen hints of this in the analysis of metal + nonmetal reactions in Chapter 5. For this reason it is convenient to make an **enthalpy diagram** showing the enthalpy levels of the various sets of reagents and products. Such a diagram is shown in Figure 10.4 for the carbon–oxygen reaction system.

Four features of this diagram are worth pointing out explicitly. First, the energy of a macroscopic sample depends on the size of the sample, that is, on how many atoms or molecules the sample contains. Just as two CO molecules possess twice as much bond energy as one, so 2 *mol* of CO(*g*) possess twice the amount of energy of 1 mol. Enthalpy diagrams are normally given on a per-mole basis, with any stoichiometric coefficients, including fractions, indicating numbers of moles. This is why Hess's law for ΔH parallels the stoichiometric adjustments made in adding the reactions; adding reactions (a) and (b) yields 2 mol of reaction (c), and therefore twice the enthalpy of the products of reaction (c). Second, adding the same element or compound to two sets of compounds with different enthalpy levels shifts the absolute enthalpies of the two sets by the same amount—the enthalpy of the added compound—but does not affect the difference between them. This applies to the extra mole of O_2 added to the two higher enthalpy levels, whose difference is the heat of reaction (a). Thirdly, the diagram conveys more information than the three ΔH's used to construct it. For example, it shows at once that the *reverse* of an exothermic reaction such as reaction (a) is an endothermic reaction with a heat of the same magnitude, that is, $\Delta H_{\text{reverse}} = -\Delta H_{\text{forward}}$ or

$$\Delta H_{\text{forward}} + \Delta H_{\text{reverse}} = 0 \tag{10.29}$$

Moreover, if we take the three-reaction system and consider proceeding forward through the intermediate CO stage, reactions (a) and (b), and then back to the el-

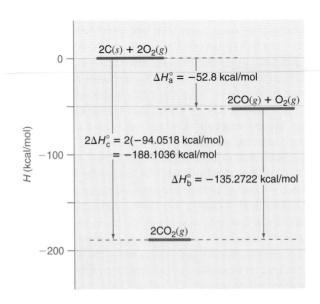

Figure 10.4
Hess's law applied to the combustion of carbon. The heat for the complete oxidation, $2\Delta H_c$, is equal to the sum of the heats for the two stages, $\Delta H_a + \Delta H_b$; therefore the measurement of any two of these heats determines the third. Considered to be a thermodynamic cycle beginning and ending with the original reagents, Hess's law is written $\Sigma \Delta H = 2\Delta H_c - \Delta H_b - \Delta H_a = 0$. See the text for other relationships implied by this diagram.

ements by way of the *reverse* of reaction (c), we will have traversed a **thermodynamic cycle,** for which

$$\Delta H_a + \Delta H_b + 2\Delta H_{c,\text{reverse}} = \Delta H_a + \Delta H_b - 2\Delta H_c = 0$$

where the forward–reverse relationship of Equation 10.29 has been used to replace $\Delta H_{c,\text{reverse}}$. Actually both of these relationships represent cycles, the first with two stages and the second with three, and the fact that the heats sum to zero is a requirement of the first law as given by Equation 10.15. Any process that terminates in the same state where it began can have no net change in energy or enthalpy. Finally, just as in spectroscopy, ΔH for a reaction measures only enthalpy differences, not the absolute enthalpy. We are therefore free to set the zero of enthalpy at any convenient point. (Recall that the zero of energy of an atom was set at the ionization limit, that for a diatomic molecule at the dissociation limit.) Two standard choices for the enthalpy zero are discussed, one in this section and another in the next.

EXAMPLE 10.5

Suppose a greenhorn chemist walks into an industrial laboratory and brashly proposes to synthesize diamonds C(dia) from methane and oxygen by carrying out the reaction

$$CH_4(g) + O_2(g) \longrightarrow C(\text{dia}) + 2H_2O(l)$$

She claims she will be able to determine the heat of this reaction by measuring the heat of combustion of CH_4 and of C(dia), so she asks for a calorimeter, and methane and diamonds to burn. (This smacks of the alchemist who, given gold to start with, "miraculously" produces it.) Will she be able to find ΔH?

Solution:
The combustion reactions (and their heats) are

(a) $CH_4(g) + 2O_2(g) \longrightarrow CO_2(g) + 2H_2O(l)$ $\Delta H = -212.7$ kcal/mol
(b) $C(\text{dia}) + O_2(g) \longrightarrow CO_2(g)$ $\Delta H = -94.4$

Inspecting the desired reaction, we see that 1 mol of methane appears on the left, and one of diamond on the right, which suggests that we should try subtracting reaction (b) from reaction (a). This works, with the O_2 and CO_2 canceling in the difference. So we conclude that the enthalpy change of the desired reaction is given by $\Delta H = \Delta H_a - \Delta H_b$ from Hess's law, and the brash young chemist will be able to get it by burning methane and diamonds. Using the experimental heats given, $\Delta H = -212.7 - (-94.4) = -118.3$ kcal/mol (-495.0 kJ mol^{-1}). However, this does not establish that the hypothetical diamond synthesis is *feasible;* as you know, when you light a Bunsen burner, you don't find diamond crystals raining down from the flame onto your lab bench! (However, scientists are now learning how to grow diamond films on surfaces in a vacuum.) The important point is that Hess's law can be used to predict reaction heats for reactions that would be difficult to study directly.

Hess's law has several practical consequences. As the preceding example showed, it can be used predictively, and it can also be used to improve the precision and accuracy of $\Delta H°$ for reactions, such as the partial oxidation of carbon, that may elude accurate measurement. In this case the precise values of $\Delta H°$ for the combustion of C(s) and CO(g) can be used to yield $\Delta H° = -52.8314$ kcal/mol for the partial oxidation reaction. This is a simple example of the great use to which a thermodynamic cycle can be put. It implies also that we really don't have to measure the heat of all known reactions, but can obtain the heats from other $\Delta H°$'s using a suitable cycle. A general way of applying this corollary is described next.

Standard Heats of Formation

A reaction used to supply pure $CO_2(g)$ for use in making Na_2CO_3 by the **Solvay process** is the decomposition of limestone with heat,

$$CaCO_3(s) \xrightarrow{\Delta} CaO(s) + CO_2(g), \qquad \Delta H° = +42.62 \text{ kcal } (+178.3 \text{ kJ})$$

This reaction is endothermic, as shown by the positive $\Delta H°$, and is one of those reactions that is difficult to study using calorimetry (although there are other ways of obtaining its $\Delta H°$, as described in the next chapter). So let's devise a cycle involving this reaction of a type that could be generalized to other reactions. From a chemical point of view, all of the compounds that participate in a chemical reaction are merely different ways of assembling the constituent elements. For example we can at least imagine making calcium carbonate (limestone), calcium oxide (lime), and carbon dioxide from the elements:

(a) $\quad Ca(s) + C(s) + \tfrac{3}{2}O_2(g) \longrightarrow CaCO_3(s)$
(b) $\quad Ca(s) + \tfrac{1}{2}O_2(g) \longrightarrow CaO(s)$
(c) $\quad C(s) + O_2(g) \longrightarrow CO_2(g)$

These reactions, in which *compounds are formed from the elements in their most stable forms at NTP,* are called **formation reactions.** They are always written to yield 1 mol of the compound, and they may involve fractional coefficients, as for $CaCO_3$ and CaO. These types of reactions of course must have $\Delta H°$'s, and these are called **standard heats of formation,** denoted $\Delta H°_{f, 298}$ or just $\Delta H°_f$ if the temperature is understood to be near normal. Reactions (b) and (c) are simple oxidations, whose heats can be measured directly. Notice that if you add reactions (b) and (c) and subtract reaction (a), you obtain the limestone decomposition reaction; all of the elements appearing in the formation reactions cancel. Hess's law then leads to the conclusion that $\Delta H° = \Delta H°_b + \Delta H°_c - \Delta H°_a = \Delta H°_f(CaO(s)) + \Delta H°_f(CO_2(g)) - \Delta H°_f(CaCO_3(s))$. A few more examples such as this provide convincing evidence that *any reaction can be assembled by an appropriate sum of formation reactions,* and that therefore the heat of any reaction is just some linear combination of heats of formation. The general formula that applies is

$$\Delta H° = \sum_P \nu_P \Delta H°_f(P) - \sum_R \nu_R \Delta H°_f(R) \qquad (10.30)$$

where P stands for products and R for reagents, and the ν's are stoichiometric coefficients. Our limestone example is easily seen to be a special case of Equation 10.30.

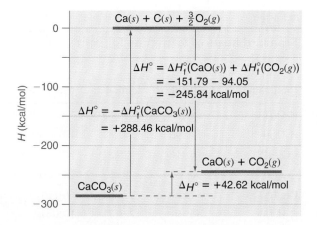

Figure 10.5
The thermodynamic cycle implied by the use of heats of formation to calculate heats of reaction, applied to the decomposition of limestone, $CaCO_3(s) \longrightarrow CaO(s) + CO_2(g)$. Note that the intermediate state at the top, where the uncombined elements are present in their standard states, does not actually occur in the reaction as normally carried out.

Figure 10.5 is an enthalpy-level diagram for the limestone decomposition that illustrates the cycle corresponding to the formation reaction analysis. The uppermost enthalpy level, with the constituent elements in their standard states, need not (and usually does not) occur in the reaction during its normal course. Adopting the formation reactions as a standard way of tabulating thermochemical information amounts to choosing the elements in their standard states as a reference enthalpy state, a zero on the enthalpy scale.

The tremendous savings in adopting a standard like this can be seen by comparing the size of the table generated if we knew the ΔH_f°'s of all known *compounds* (more than 20 million so far!) with that generated from the ΔH°'s of all known *reactions* of those compounds. The ΔH_f° table would occupy about 100 volumes, filling a small bookcase; while a table of reaction heats, assuming only binary reactions leading to a unique set of products, would occupy more than *4 million bookcases* (or hundreds of high-capacity hard-disk drives). In practice, for only a small fraction of known compounds do we have measured heats of formation; in the standard reference, *The Handbook of Chemistry and Physics* (71st edition), most of the known values are contained in a small 59-page section. As emphasized earlier in the example of $CO(g)$, direct measurement of ΔH_f° for a compound is not always possible or straightforward; chemists have often used cycles involving other reactions, usually combustions, to isolate the desired heat of formation. Table 10.1 lists the heats of formation of some common compounds; a more complete listing can be found in Appendix C.

This compact table enables the calculation of a large number of possible heats of reaction; for example, the reaction $2Na(s) + Cl_2(g) \longrightarrow 2NaCl(s)$, used to introduce chemical reactions in Chapter 5, is a formation reaction, and the heat of this reaction is therefore just twice the heat of formation of $NaCl(s)$, or -196.54 kcal/mol (-822.3 kJ/mol), since the reagents are elements in their standard states. Likewise, the heats of formation of metallic oxides, such as Fe_2O_3 or Al_2O_3, represent the reaction heats for the familiar processes of rusting and corrosion. The different values for $H_2O(l)$ and $H_2O(g)$ underline the importance of specifying the state of aggregation; modern gas furnaces utilize the heat of condensation of the water vapor formed

TABLE 10.1

Standard heats of formation for some common compounds

Compound	$\Delta H^\circ_{f, 298}$ kcal/mol	$\Delta H^\circ_{f, 298}$ kJ/mol	Compound	$\Delta H^\circ_{f, 298}$ kcal/mol	$\Delta H^\circ_{f, 298}$ kJ/mol
$H_2O(l)$	−68.315	−285.83	$SiO_2(s)$	−217.72	−910.94
$H_2O(g)$	−57.796	−241.82	$NH_3(g)$	−11.02	−46.11
$CO(g)$	−26.416	−110.52	$NO(g)$	21.57	90.25
$CO_2(g)$	−94.052	−393.51	$NO_2(g)$	7.93	33.18
$CH_4(g)$	−17.88	−74.81	$O_3(g)$	34.1	142.7
$C_2H_2(g)$	54.19	226.73	$H_2S(g)$	−4.93	−20.63
$C_2H_4(g)$	12.50	52.30	$SO_2(g)$	−70.94	−296.81
$C_2H_6(g)$	−20.24	−84.68	$HCl(g)$	−22.062	−92.31
$CH_3OH(l)$	−57.04	−238.66	$NaCl(s)$	−98.268	−411.15
$C_2H_5OH(l)$	−66.37	−277.69	$NH_4Cl(s)$	−75.15	−314.4
$C_6H_6(l)$	11.718	49.028	$NaHCO_3(s)$	−227.25	−950.81
$C_6H_6(g)$	19.821	82.93	$Na_2CO_3(s)$	−270.24	−1130.68
$C_6H_{12}O_6(s)$	−301.	−1260.	$MgO(s)$	−143.81	−601.70
$i\text{-}C_8H_{18}(l)$	−49.77	−208.2	$CaO(s)$	−151.79	−635.09
			$CaCO_3(s)$	−288.46	−1206.92
			$Fe_2O_3(s)$	−197.0	−824.2
			$Al_2O_3(s)$	−400.5	−1675.7

as a combustion product to improve efficiency. The following example illustrates the ease of use of ΔH°_f's in more complex reactions.

EXAMPLE 10.6

An historically famous example of an exothermic reaction is the rail-welding "thermite" reaction $2Al(s) + Fe_2O_3(s) \longrightarrow Al_2O_3(s) + 2Fe(s)$, introduced in Chapter 5 as an example of redox displacement. Use Table 10.1 to find ΔH° for this reaction.

Solution:
Applying Equation 10.30, we obtain

$$\Delta H^\circ = \Delta H^\circ_f(Al_2O_3(s)) + 2\Delta H^\circ_f(Fe(s)) - [\Delta H^\circ_f(Fe_2O_3(s)) + 2\Delta H^\circ_f(Al(s))]$$
$$= -400.5 + 2(0.0) - [-197.0 + 2(0.0)]$$
$$= -203.5 \text{ kcal/mol} \quad (-851.4 \text{ kJ/mol})$$

Note that the elements contribute nothing to the calculation. As a rule, $\Delta H°$ for redox *addition* reactions is just $\Delta H_f°$ for the adduct, while for redox *displacement* it is the difference between $\Delta H_f°$'s for the product and reagent compounds, weighted by stoichiometric coefficients. Later we will see that the heat of the thermite reaction is enough to produce molten iron.

Most of the reactions we examined in Chapter 5, as well as our Zn/H_2SO_4 thermodynamic example, involve an aqueous phase, sometimes exclusively. In Chapter 5 you learned how to write ionic equations, omitting spectator ions, that reveal the essence of the reaction. Both neutral and ionic species in aqueous solution are *hydrated;* that is, they are weakly bonded to a surrounding shell of water molecules. This gives rise to contributions to the heats of aqueous processes from the energies of these solute–solvent bonds. (The central role of bond energies in causing energy changes in reactions is discussed in the next section.) For the purpose of defining heats of aqueous-phase reactions, all that matters is the *difference* in hydration energies of the products and reagents. This permits us to define a standard reference state for an aqueous ion in addition to the elements in their standard states. This definition actually originated in electrochemistry (see Chapter 13), and is stated by setting

$$\Delta H_{f,\,298}°(\text{H}^+(aq)) = 0 \qquad (10.31)$$

Thus the heat of formation of any ion in solution is defined *relative* to that for aqueous protons $\text{H}^+(aq)$. Here the superscript ° refers to a standard *concentration*, taken to be 1.00 M, rather than a standard partial pressure. Writing Reaction 10.2 as

$$\text{Zn}(s) + 2\text{H}^+(aq) \longrightarrow \text{Zn}^{2+}(aq) + \text{H}_2(g) \qquad (10.2')$$

where the SO_4^{2-} ion, a spectator in this reaction, has been omitted, and applying Equation 10.30, we find that

$$\Delta H° = \Delta H_f°(\text{Zn}^{2+}(aq))$$

since Zn and H_2 are elements in their standard states and the definition of Equation 10.31 has been adopted. Thus, for those metals readily attacked by acid, the heats of formation of their positive ions can be obtained directly by measuring the heats of the metal–acid reactions. Table 10.2 lists $\Delta H_f°$'s for common aqueous solutes.

Heats of formation from Tables 10.1 and 10.2 can be combined to analyze reactions involving both aqueous ions and covalent compounds. For example, for the Arrhenius acid–base reaction $\text{H}^+(aq) + \text{OH}^-(aq) \longrightarrow \text{H}_2\text{O}(l)$,

$$\Delta H° = \Delta H_f°(\text{H}_2\text{O}(l)) - \Delta H_f°(\text{OH}^-(aq)) = -13.34 \text{ kcal/mol} \,(-55.8 \text{ kJ mol}^{-1})$$

[In fact, accurate measurements of the heat of this reaction can be made with a coffee cup calorimeter, and used to deduce the heat of formation of $\text{OH}^-(aq)$.] Appendix C presents a more complete collection of $\Delta H_f°$'s for both pure compounds and aqueous ions.

TABLE 10.2

Standard heats of formation for common molecules and ions in aqueous solution

Species	$\Delta H°_{f, 298}$ kcal/mol	$\Delta H°_{f, 298}$ kJ/mol	Species	$\Delta H°_{f, 298}$ kcal/mol	$\Delta H°_{f, 298}$ kJ/mol
$Li^+(aq)$	−66.55	−278.45	$I_2(aq)$	5.4	22.6
$Na^+(aq)$	−57.39	−240.12	$OH^-(aq)$	−54.97	−229.99
$K^+(aq)$	−60.32	−252.38	$Cl^-(aq)$	−39.95	−167.15
$Mg^{2+}(aq)$	−111.58	−466.85	$Br^-(aq)$	−29.05	−121.55
$Ca^{2+}(aq)$	−129.74	−542.83	$I^-(aq)$	−13.19	−55.19
$Ba^{2+}(aq)$	−128.50	−537.64	$HCO_3^-(aq)$	−165.39	−691.99
$Mn^{2+}(aq)$	−52.76	−220.7	$CO_3^{2-}(aq)$	−161.84	−677.14
$Fe^{2+}(aq)$	−21.3	−89.1	$CH_3COOH(aq)$	−116.16	−486.0
$Fe^{3+}(aq)$	−11.6	−48.5	$CH_3COO^-(aq)$	−116.16	−486.0
$Co^{2+}(aq)$	−13.9	−58.	$NO_2^-(aq)$	−25.0	−104.6
$Ni^{2+}(aq)$	−12.9	−54.	$NO_3^-(aq)$	−49.56	−207.4
$Cu^{2+}(aq)$	15.48	64.77	$PO_4^{3-}(aq)$	−306.9	−1284.1
$Zn^{2+}(aq)$	−36.78	−153.89	$SO_3^{2-}(aq)$	−151.9	−635.5
$Al^{3+}(aq)$	−127.	−531.	$SO_4^{2-}(aq)$	−217.32	−909.27
$NH_4^+(aq)$	−31.67	−132.51	$ClO^-(aq)$	−25.6	−107.1
$CO_2(aq)$	−98.90	−413.80	$ClO_3^-(aq)$	−24.78	−103.7
$NH_3(aq)$	−19.19	−80.29	$Cr_2O_7^{2-}(aq)$	−356.2	−1490.3
$Cl_2(aq)$	−5.6	−23.4	$MnO_4^-(aq)$	−129.4	−541.4
$Br_2(aq)$	−0.62	−2.59			

10.6 Origins of the Heat of Reaction: Bond Energies

You may have been struck by the widely varying magnitudes of the heats of formation of the species listed in Tables 10.1 and 10.2; for example, whereas water vapor, $H_2O(g)$, is formed from its elements with a large release of heat, nitric oxide, $NO(g)$, is formed endothermically from $N_2(g)$ and $O_2(g)$. The formation of NO in this way actually occurs inside internal combustion engines, with the energy supplied by the high-voltage spark from a spark plug as well as the heat of fuel combustion. For gas-phase reactions like these, the difference in energy between reagents and products (which we call the heat of reaction) results from differences in the **bond energies** of the reagents and products, as has been pointed out in a qualitative way in Chapter 5. *An exothermic reaction therefore reflects product bonds that are stronger than those of the reagents,* a characteristic of the vast majority of known reactions, *while an endothermic one indicates that the bonding weakens in going from reagents to products.* How can we analyze this conclusion in a quantitative way? Let's seek to construct a cycle analogous to the one we used for heats of formation.

Although molecules are too clever to sacrifice excess energy when partaking in a reaction, we can nonetheless imagine first breaking all of the bonds in the reagent molecules, and then making all of the bonds in the products. The first step exacts a tremendous energy cost, much larger than simply decomposing the reagents into the elements in their standard states; but most of this expenditure is returned when the product bonds are formed. This scenario requires the formation of *free gas-phase atoms,* species almost never observed among the chemically active elements under normal conditions. In forming NO(g) we imagine

$$N_2(g) + O_2(g) \longrightarrow 2N(g) + 2O(g) \longrightarrow 2NO(g)$$

The energy consumed in the first step is clearly the sum of the energies D_0 of the N≡N and O=O bonds, and is referred to as the **atomization energy.** The second step releases energy equal to twice the D_0 of an N=O bond, and the heat of reaction is the difference. This relationship is illustrated by means of potential energy curves in Figure 10.6. In general, the heat of reaction is given by

$$\Delta H° = \sum_R n_R D_0(R) - \sum_P n_P D_0(P) \tag{10.32}$$

where R represents reagents and P products as in Equation 10.30, and the n's are the number of bonds of each type. To turn Equation 10.32 into a practical relationship, we need to know the D_0's. Often this information can be obtained from molecular spectroscopy (see Chapter 8), but it is more usually the case that the D_0's are inferred from known heats of reaction. In addition, Equation 10.32 strictly applies only at $T = 0$ K; at 298 K, the gas-phase molecules have thermal energy stored in their various degrees of freedom, as we discuss further in the next section. Generally, tables of bond energies (more accurately, bond *enthalpies*) for use in Equation 10.32 are corrected for this small additional energy. For our purpose here, this is a fine point and will be ignored, especially in view of the difficulties encountered in considering bond energies for polyatomic molecules.

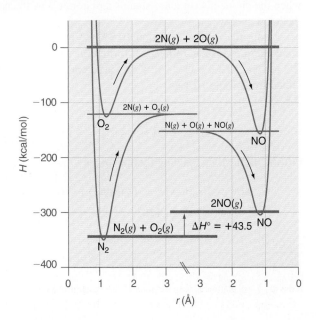

Figure 10.6
Interpretation of the heat of the endothermic reaction $N_2(g) + O_2(g) \longrightarrow 2NO(g)$ in terms of bond energies, as illustrated using bond potential energy curves. The up arrows denote the energy expended as the reagent bonds are broken, the down arrows the energy released as product bonds are formed. The bond energies are measured from the zero point vibrational energy levels rather than the minima in the potential energy curves (see Chapter 8).

In the Lewis–Pauling valence bond model for chemical bonding of Chapter 6, you were encouraged to view a polyatomic molecule "one bond at a time," as the Lewis electron-dot structure would suggest. Thus H_2O is considered to have two identical O–H bonds and, by extension, two identical bond *energies* D_0(O–H). Though the molecular orbital model of Chapter 7 suggests that such an extension should be regarded with caution, especially when "resonance" (delocalization of bonding electrons) may be present, the concept of bond energy in a polyatomic molecule has been found to be useful even though approximate. Given this assumption, bond energies such as D_0(O–H) can be readily found from the heat of formation of water and the (spectroscopically determined) H–H and O=O bond energies:

$$D_0(\text{O–H}) = \tfrac{1}{4}[2D_0(\text{H–H}) + D_0(\text{O=O}) - 2\Delta H_f^\circ(\text{H}_2\text{O}(g))]$$
$$= \tfrac{1}{4}[2(104.1) + 118.3 - 2(-57.8)] = 110.5 \text{ kcal/mol } (462 \text{ kJ/mol})$$

a result readily obtained by solving Equation 10.32, as applied to $2H_2 + O_2 \longrightarrow 2H_2O$, for D_0(O–H).

In the NO and H_2O examples, the atomization energy consisted simply of the bond energies of the gas-phase elements. When a molecule's constituent elements have liquid or solid standard states, however, the atomization step must take such elements out of their condensed phases into the gas. For example, for carbon compounds such as CO_2 and CH_4, the estimation of D_0(C–H) and D_0(C=O) from ΔH_f°'s requires the heat of atomization of carbon, $C(s,\text{graphite}) \longrightarrow C(g)$. The value of ΔH° for this process has been in dispute for many years, but an accurate value appears to be 171.3 kcal/mol (716.7 kJ/mol). To cover cases like this, the first sum on the right-hand side of Equation 10.32 is replaced by $\Sigma_R n_R \Delta H_{\text{atomiz}}^\circ(R)$ when applied to the estimation of bond energies, where now n_R is the number of atoms of each type.*

The results of such calculations yield bond energies that are approximately transferable from one molecule to another; for example, the C–H bond energy in CH_4 is about the same as that in C_2H_6 or C_3H_8. Table 10.3 presents atomization energies and the energies of both single and multiple bonds involving common elements. Note that when the element's standard state is a diatomic gas, the atomization enthalpy is half the bond energy.

Using the rule of Equation 10.32, the heat of reaction can be calculated from bond energies instead of heats of formation. For example, the heat of combustion of CO to CO_2, $2CO + O_2 \longrightarrow 2CO_2$, used earlier to introduce Hess's law, can be obtained from Table 10.1 as

$$\Delta H^\circ = 2\Delta H_f^\circ(\text{CO}_2) - 2\Delta H_f^\circ(\text{CO})$$
$$= 2(-94.052) - 2(-26.416) = -135.272 \text{ kcal/mol } (-565.98 \text{ kJ/mol})$$

or from Table 10.3, heeding footnote *a*, as

$$\Delta H^\circ = 2D_0(\text{C}\equiv\text{O}) + D_0(\text{O=O}) - 4D_0(\text{C=O})$$
$$= 2(256) + 118 - 4(192) = -138 \text{ kcal/mol } (-577 \text{ kJ/mol})$$

---

* For elements that are diatomic gases, the atomization reaction is $\tfrac{1}{2}A_2 \longrightarrow A$, and $\Delta H_{\text{atomiz}}^\circ(A) = \tfrac{1}{2}D_0(A\text{–}A)$, giving back Equation 10.32. You can therefore mix atomization and bond energies in the same calculation as long as you are careful about the coefficients.

TABLE 10.3
Atomization and bond energies in kilocalories per mole

	$\Delta H°_{atomiz}$	–H	–C	=C	≡C	–N	=N	≡N	–O	=O	–S	–Cl
H	52.1	104	99			93			110		81	103
C	171.3	99	83	147	194	70	147	213	84	174[a]	62	78
N	113.0	93	70	147	213	38	100	226	48	150[b]		48
O	59.2	110	84	174[a]	256	48	150[b]		33	118		49
S	66.6	81	62	114							55	60
P	75.2	76	63			50			84		55	79
F	18.9	135	105			65			44		68	44
Cl	29.1	103	78			48			49		60	58
Br	26.7	87	66			58					51	52
I	25.5	71	57						48			50

[a]For the double bonds in CO_2, a better value to use is 192 kcal/mol (see text).
[b]This is the bond energy of NO, which has a bond order of 2½.

Figure 10.7 compares the two cycles used to obtain these results. The heat of formation calculation gives the more precise result and is always to be preferred when accurate data are available. In this case the bond energy result could have been made of comparable precision, but most bond energies are averages over several identical bonds within a molecule and over several molecules containing that type of bond,

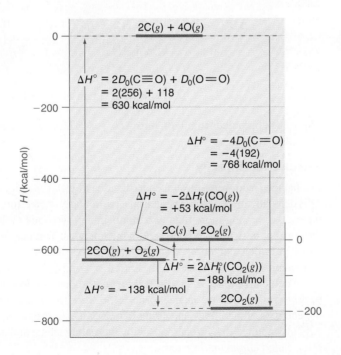

Figure 10.7
Comparison of heat-of-formation and bond-energy cycles for the oxidation of $CO(g)$ to $CO_2(g)$. Note the much greater energies involved in the bond-energy scheme. The $\Delta H°$ given for this reaction, -138 kcal/mol, is that obtained from bond energies, and it differs slightly from the more accurate value from heats of formation (see text).

making the resulting estimate less precise. Bond energies attain practical usefulness when the species involved in a reaction are not themselves stable molecules, as the following example illustrates.

EXAMPLE 10.7

A well-known *intermediate* molecule in the combustion and cracking of hydrocarbons is the methyl radical ·CH$_3$, where the dot indicates an unpaired electron. Use data from Table 10.3 to estimate the heat of formation of ·CH$_3$, and calculate the heat released in its combustion.

Solution:
Equation 10.32 can be applied to find both of the required ΔH's or, better, the result for ΔH_f° can be used in applying Equation 10.30 to the combustion. The formation reaction is

$$C(s) + \tfrac{3}{2}H_2(g) \longrightarrow \cdot CH_3(g)$$

giving, from Equation 10.32, assuming ·CH$_3$ contains three C—H bonds,

$$\Delta H_f^\circ = 171.3 + (\tfrac{3}{2})104 - 3(99) = +30 \text{ kcal/mol}$$
$$\text{(literature value 35 kcal)}$$

The balanced combustion reaction (yielding water vapor) is

$$4CH_3(g) + 7O_2(g) \longrightarrow 4CO_2(g) + 6H_2O(g)$$

Though using heats of formation is the best way to find ΔH_{comb}° per mole of ·CH$_3$, we will illustrate the use of bond energies in the calculation, heeding footnote *a* of Table 10.3:

$$\Delta H_{comb}^\circ = \tfrac{1}{4}[12D_0(\text{C--H}) + 7D_0(\text{O=O}) - 8D_0(\text{C=O})$$
$$- 12D_0(\text{O--H})]$$
$$= \tfrac{1}{4}[12(99) + 7(118) - 8(192) - 12(110)]$$
$$= -210 \text{ kcal/mol} \quad (-880 \text{ kJ/mol})$$

The value computed from literature ΔH_f°'s is -215.6 kcal/mol.

Resonance Energy

In Chapter 7 you were introduced to the idea of electron delocalization in molecules. This delocalization, in which electrons, particularly π electrons, were found to spread themselves among more than two atoms in a polyatomic molecule, tends to muddy the concept of two-atom bonds and bond energies. You may recall that molecules in which this delocalization occurs usually have two or more equivalent or nearly equivalent Lewis structures with shifted π bonds, the so-called *resonance forms*. Since delocalization is always a *stabilizing* (i.e., energy-lowering) influence, we expect molecules with resonance forms to be *more stable* than one would predict from the energies of the bonds appearing in one of these structures. This idea is most successfully applied to hydrocarbons with alternating single and double bonds; these are often called *conjugated* systems.

The most celebrated case of resonance is benzene (C_6H_6), as we saw in Chapter 7. The six-membered ring contains three C—C bonds, three C=C bonds, and six C—H bonds. The example just given illustrates how to predict ΔH_f° from bond energies. For the formation reaction $6C(s) + 3H_2(g) \longrightarrow C_6H_6(g)$, we find $\Delta H_f^\circ = +56$ kcal/mol from bond energies, while the experimental number from Table 10.1 is +19.32 kcal/mol. The discrepancy of 36 kcal/mol (150 kJ/mol) is too large to be ascribed to errors in the bond energies themselves, and is referred to as **resonance energy.** Benzene is more stable by this amount than the energies of its bonds predict it to be, owing to the delocalized π bonding around the ring. In a later course you may learn that, within a simplified MO model for the π bonding called the Hückel model, this stabilization is equivalent to the energy of an extra π bond. This extra stability makes benzene far less reactive toward electrophiles than would be expected for molecules containing π electrons. The reactions of benzene are discussed more extensively in Chapter 18.

Stabilization by delocalization can also occur in molecules with contiguous π bonds, such as CO_2, where there appear to be no resonance forms. Thus the bond energy for C=O tabulated in Table 10.3 does not apply accurately to the two double bonds in CO_2. The table entry is derived instead from *carbonyl* compounds such as $H_2C=O$, where delocalization does not occur.

10.7 Heat Capacities: A Molecular Interpretation

In Section 10.4, Example 10.4, we saw a typical method for measuring the heat capacity of a system, through the transfer of a known amount of electrical energy. Such methods can be made quite general, enabling the measurement of C for a wide variety of substances. Table 10.4 lists the specific heats c and **molar heat capacities** $\overline{C}_P \equiv C_P/n$ for some common elements, compounds, and materials; Appendix C gives a more complete list.

While the specific heats c tend to be highly variable (compare Ar and H_2 for example), the *molar* heat capacities $\overline{C}_P$ for pure elements or compounds are more similar in magnitude and vary in a systematic way. For gaseous species, $\overline{C}_P$ generally increases with the complexity of the molecule. For H_2O, and for other volatile liquids as well, the heat capacity of the liquid is generally higher than that of the vapor. For metals, $\overline{C}_P$ varies only slightly about an average value of 6 cal/mol K. These observations suggest a strong connection between $\overline{C}_P$ and the *number of degrees of freedom* possessed by an atom or molecule.

We have already seen in Section 10.3 that the kinetic molecular theory presented in Chapter 9 leads to a prediction for an ideal monatomic gas of

$$\overline{C}_P = \tfrac{5}{2}R = (2.5)(1.9872 \text{ cal/mol K}) = 4.968 \text{ cal/mol K } (20.8 \text{ J/mol K})$$

in excellent agreement with the values given in Table 10.4 for He and Ar. The diatomic gases H_2, N_2, etc., have $\overline{C}_P$'s higher by about R, except for Cl_2. The kinetic theory predicts that for each new kinetic energy term, one for each degree of freedom and therefore three per atom added, the heat capacity should increase by $\tfrac{1}{2}R$ and, in addition, that a harmonic oscillator potential will contribute another $\tfrac{1}{2}R$. The extra degrees of freedom of a polyatomic molecule provide new ways to store energy, thus increasing $\overline{C}_P$. The kinetic theory prediction of an equal amount of energy, $\tfrac{1}{2}RT$, in each degree of

TABLE 10.4
Heat capacities of common elements, compounds, and materials at 298 K

Substance	c (cal/g °C)	$\overline{C}_P$ (cal/mol K)	Substance	c (cal/g °C)	$\overline{C}_P$ (cal/mol K)
He(g)	1.241	4.968	C(graphite)	0.170	2.038
Ar(g)	0.124	4.968	C(diamond)	0.122	1.462
$H_2(g)$	3.417	6.889	Al(s)	0.216	5.82
$N_2(g)$	0.249	6.961	Si(s)	0.171	4.78
		(8.293 @ 1500 K)	S(s)	0.169	5.41
$O_2(g)$	0.219	7.016	Ca(s)	0.151	6.05
HCl(g)	0.191	6.96	Fe(s)	0.107	6.00
$Cl_2(g)$	0.114	8.104	Ni(s)	0.106	6.23
CO(g)	0.248	6.959	Cu(s)	0.092	5.84
$CO_2(g)$	0.202	8.87	Ag(s)	0.056	6.059
$O_3(g)$	0.195	9.37	Pt(s)	0.032	6.18
$SO_2(g)$	0.149	9.53	Au(s)	0.031	6.075
$NH_3(g)$	0.493	8.38	Hg(l)	0.033	6.69
$CH_4(g)$	0.527	8.439	Pb(s)	0.031	6.32
$H_2O(l)$	1.00	18.0	U(s)	0.028	6.61
$H_2O(g)$(110°C)	0.48	8.6	SiO_2(quartz)	0.18	10.62
$H_2O(s)$(−5°C)	0.50	9.0	NaCl(s)	0.21	12.07
$CH_3OH(l)$	0.61	19.5	CaO(s)	0.18	10.23
$C_2H_5OH(l)$	0.58	26.6	$CaCO_3(s)$	0.20	19.57
			$Al_2O_3(s)$	0.19	18.89
			Air	0.24	6.93
			Steel	0.11	
			Marble	0.21	
			Glass	0.20	
			Wood	0.4	
			Protein	0.4	
			Human body (avg)	0.83	

freedom is called the **equipartition theorem**. For an N-atom molecule, appealing to the degree-of-freedom analysis presented in Chapter 8, the prediction is

$$\begin{array}{cccc} & \text{trans} & \text{rot} & \text{vib} \\ \overline{C}_P = & \tfrac{5}{2}R & + R & + (3N-5)R = (3N-\tfrac{3}{2})R \text{ (linear molecule)} \\ \overline{C}_P = & \tfrac{5}{2}R & + \tfrac{3}{2}R & + (3N-6)R = (3N-2)R \text{ (bent molecule)} \end{array} \quad (10.33)$$

This analysis leads to the prediction $\overline{C}_P = 4.5\,R = 8.942$ cal/mol K for *any* diatomic molecule, a value larger than any of those given for diatomics in Table 10.4.

The problem with the kinetic analysis is that it ignores *quantization of energy* in the internal degrees of freedom; that is, each kinetic energy term contributes $\frac{1}{2}R$ only when the motion is classical. As we saw in Chapter 8, energy quantization is most dramatic in molecular vibration. But to understand how this affects the thermal energy of a molecule, we need to know how the energy is spread among the available levels. We need something akin to the Maxwell–Boltzmann velocity distribution of a gas that also applies to the internal DFs. As hinted in Chapter 9, that more general energy distribution was brought into being by the rebel genius Ludwig Boltzmann in the 1880s, *before* anything was known about the quantum of energy.

Molecular Energy Storage: The Boltzmann Distribution

What Boltzmann showed was that, if one demanded that a large collection of molecules arrange themselves in *the most probable way* among the energy levels available to them, the resulting distribution of molecules would always take on the same mathematical form. A brief sketch of Boltzmann's derivation is given in Appendix D. For a particular degree of freedom with quantum numbers $n = 0, 1, 2, \ldots$, and corresponding energy levels $\varepsilon_0, \varepsilon_1, \varepsilon_2, \ldots$, the result can be stated as

$$\frac{N_n}{N_0} = \frac{g_n}{g_0} e^{-(\varepsilon_n - \varepsilon_0)/k_B T} \tag{10.34}$$

where the N's are *populations* in the various levels and the g's are the *degeneracies* of those levels. This formula is especially simple for the harmonic oscillator model of molecular vibration given in Chapter 8. In that model, the quantum number is denoted v, the energy levels are given by $\varepsilon_v = h\nu(v + \frac{1}{2})$, and the degeneracies g_v are unity for all v. Equation 10.34 then becomes

$$N_v = N_0\, e^{-vh\nu/k_B T} \tag{10.35}$$

For a fixed T, the populations decay exponentially with increasing vibrational excitation.

To get some idea of the actual populations predicted by Equation 10.35, let's use our old friend nitrogen (N_2) as an example. From Table 8.3 of Chapter 8, the vibrational frequency in wavenumber units is $\overline{\nu} = 2360$ cm^{-1}, giving a vibrational quantum $h\nu = hc\overline{\nu}$ of 4.69×10^{-20} J or 0.293 eV. At 298 K the average thermal energy is $k_B T = 4.11 \times 10^{-21}$ J or 0.0257 eV, more than a factor of 10 smaller than the vibrational quantum. We then can obtain from Equation 10.35 the ratio $N_1/N_0 = \exp[-(1)(0.293)/(0.0257)] = 1.1 \times 10^{-5}$. This minuscule ratio means that, in a sample of 100,000 molecules at 298 K, *one molecule* might be found in the first excited vibrational state, the rest being in the ground state and *possessing no thermal energy at all*. Therefore the ability of a nitrogen molecule to accept *vibrational* energy is almost nil at 298 K. Although this prohibition depends on the frequency—high-frequency "stiff" bonds are less able to store energy than low-frequency bonds—at 298 K the great majority of small molecules can hold only a small fraction of the full RT of energy in each vibrational mode predicted classically. This is why the heat capacities of the diatomic gases listed in Table 10.4 are all lower by about R than those calculated from Equation 10.33. A good estimate of $\overline{C}_P$ for small molecules is often obtained by simply omitting the vibrational contributions given in Equation 10.33.

Note that it is the *ratio* of the energy quantum to $k_B T$ that determines the population in the excited states; the smaller this ratio, the higher the population of excited molecules. Thus changes in T as well as in $\overline{\nu}$ can change the populations. The entry in Table 10.4 for N_2 at 1500 K reflects the increase in $k_B T$ and the consequently appreciable contribution to $\overline{C}_P$ from the greater proportion of vibrationally excited molecules present at 1500 K. Figure 10.8 illustrates the Boltzmann distribution for the vibration of Cl_2, which shows some vibrational contribution to $\overline{C}_P$ even at 298 K due to its relatively weak bond and low frequency. The rotational degrees of freedom are also quantized, as discussed in Chapter 8, but ordinarily their quanta are considerably smaller than $k_B T$ at 298 K, and therefore show near-classical behavior.

In contrast to diatomic molecules, atomic solids such as those listed in Table 10.4 have frequencies of vibration within the solid lattice that range to values well below $k_B T$ at 298 K; hence, the quantization of their vibrational energy has little effect on energy storage. An atom in a solid can be modeled as a three-dimensional harmonic oscillator (HO) of very low frequency, with each DF therefore holding RT of energy (recalling that the HO potential energy must be included). The molar heat capacity of an atomic solid is therefore

$$\overline{C}_P \approx \overline{C}_V \approx 3R = 5.962 \text{ cal/mol K}$$

or approximately 6 cal/mol K, which is in good to excellent agreement with the data. This approximate constancy of the molar heat capacity at 6 cal/mol K for metals, the **law of Dulong and Petit,** was discovered empirically in 1819 and had been used in those days to correct erroneous atomic weights. Our model correctly predicts that the heat capacity of an atomic solid should remain roughly constant up to the melting point. It also suggests that on cooling a solid to very low temperature (so that $hc\overline{\nu}_{max} \gg k_B T$, where $\overline{\nu}_{max}$ is the highest lattice frequency), the heat capacity should

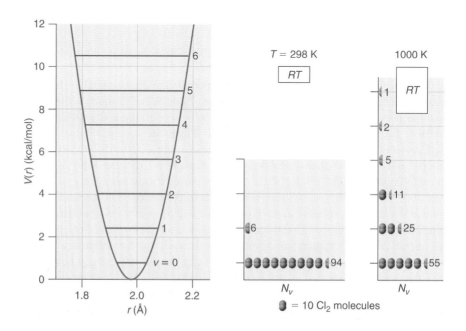

Figure 10.8
Population of the vibrational states of Cl_2 at two temperatures according to Boltzmann's formula (Equation 10.35), valid in the harmonic oscillator approximation (see Chapter 8). On the left is the harmonic oscillator bond potential energy curve for Cl_2, showing the lowest seven vibrational energy levels, and on the right the populations in those levels for a sample containing 100 Cl_2 molecules. At 298 K, the thermal energy RT is much smaller than the vibrational spacing, and most of the sample is confined to the ground state (a typical situation at room temperature), whereas at 1000 K the molecules possess a wide range of vibrational energy owing to the greater thermal energy available.

begin to fall, approaching zero as $T \longrightarrow 0$ K. This prediction was borne out early in this century with the advent of *cryogenics,* the study of material properties at very low temperatures. The measurements of $\overline{C}_P$ at low T were among those that pointed to the need for a "quantum revolution" (see Chapter 2). A quantum theory for the heat capacity of a metal was first proposed by the redoubtable Albert Einstein in 1907.

Ionic solids such as NaCl also seem to follow this simple model, with each ion in the salt crystal behaving like an atom in a metal as to its energy storage. The molar or formula heat capacity for a salt is roughly $3R$ times the number of ionic particles in its formula. Molecular liquids or solids, however, are more complex, with both internal *intramolecular* and *intermolecular* forces contributing to the total energy and hence to the heat capacity. (Intermolecular forces will be considered in Chapter 14.) It is the forces between molecules that make the heat capacity of a molecular condensed phase such as $H_2O(l)$ greater than that of its vapor.

Temperature Dependence of the Enthalpy

A knowledge of heat capacities allows you not only to compute the expected rise in temperature when a given amount of heat energy is added to or taken from a system of known composition, but also to find the change in enthalpy when the temperature of a system changes. If the heat capacity is a constant, independent of temperature, then the enthalpy change is $\Delta H = n\overline{C}_P \Delta T$, as discussed in Section 10.3. This implies that for a chemical reaction, the enthalpy level of each reagent and product will change by $\nu \overline{C}_P \Delta T$ when T changes, where $\overline{C}_P$ is the molar heat capacity of the compound in question and ν is its stoichiometric coefficient. We then get a formula for calculating $\Delta H°$ for a reaction at a temperature T_2 from that measured at temperature T_1:

$$\Delta H°_{T_2} = \Delta H°_{T_1} + \left[\sum_P \nu_P \overline{C}_P(P) - \sum_R \nu_R \overline{C}_P(R) \right](T_2 - T_1) \quad (10.36)$$

where the sums are analogous to those in Equation 10.30. The quantity in square brackets, consistent with the Δ notation, is compactly referred to as ΔC_P. Equation 10.36 can be used to correct a ΔH measured at ambient temperature to one at 298 K, as mentioned in Example 10.4, but more commonly it is used to estimate ΔH at some desired temperature $T = T_2$ from that calculated from tables ($T_1 = 298$ K). Obtaining bond energies useful at 298 K from spectroscopically determined dissociation energies, which necessarily refer to 0 K, forms another important application; this correction was mentioned in the previous section.

For example, from Table 10.1 we can calculate $\Delta H°_{298}$ for the vaporization of water, $H_2O(l) \longrightarrow H_2O(g)$, + 10.52 kcal/mol (44.0 kJ/mol). Noting from Table 10.4 that the molar heat capacity of water vapor is smaller than that of liquid water, we can anticipate from Equation 10.36 that $\Delta H°$ will be smaller at the normal boiling point of water, 100°C or 373 K. Putting this in numbers, we find

$$\Delta H°_{373} = 10.52 + (8.6 - 18.0)(373 - 298)/1000 = 9.81 \text{ kcal/mol}$$
$$(41.1 \text{ kJ/mol})$$

The directly measured experimental value is 9.72 kcal/mol at 100°C. The calculated result is approximate due to the neglect of the temperature dependence of C_P, but is

Figure 10.9
Illustration of the temperature dependence of ΔH for the vaporization of water. The small increments in H result from heating the liquid and vapor to a higher temperature; to the extent that these increments are unequal do the heats of vaporization at the two temperatures differ. The absolute enthalpy scale is determined by heats of formation.

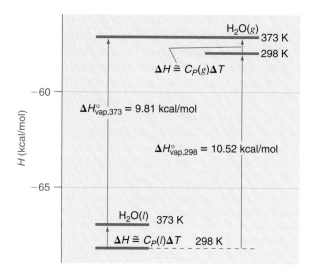

nonetheless quite accurate if the temperature range is not too large. Figure 10.9 illustrates the ΔH relationship with an enthalpy diagram.

The small change in ΔH, only 0.7 kcal for a 75°C temperature change, is typical. In many chemical reactions ΔH°_{298} is considerably larger, and the fractional change is so small that it can often be neglected, allowing ΔH to be treated as a temperature-independent quantity. We will be making some use of this approximation in the next chapter. More generally, since the number of atoms, and therefore degrees of freedom, is conserved in a chemical reaction, we can always expect ΔC_P to be small for reactions occurring in the gas phase. An appreciable ΔC_P arises mainly when the number of bonds, and hence vibrational degrees of freedom, differs between reagents and products.

As you may now be able to appreciate, the conservation of energy (the first law of thermodynamics), is a powerful tool for chemistry, a guidepost to the relation between everyday observations of hot and cold and the atomic model as well as a practical and predictive method of reaction analysis. Calling it the first law, instead of just the law, of course, implies that there's more to come—to be precise, two more laws. The second law, which will be the subject of the next chapter, is by most reckoning more profound—and certainly more mysterious. To gain some grasp of it, you will need a molecular view of its origins, much as we have begun to invoke in these last two sections.

SUMMARY

The concepts of **energy** and **temperature** central to the science of **thermodynamics** are readily interpreted by the idea that molecules are always in motion. Thermodynamics is intended to describe the flow of energy between a **system,** a defined part of the universe we wish to study, and its **surroundings,** the rest of the universe. This flow can be in the form of work w, which results in an observable change in the configuration of the surroundings, and/or of heat q, which does not. That heat and work are both forms of energy was proved experimentally first by Benjamin Thompson and then by James Joule, who measured *the mechanical equivalent of heat.* In modern mechanical units, one **calorie,** the heat necessary to raise the temperature of 1 gram of water by

1 degree Celsius, is equivalent to 4.184 joules of mechanical energy. Today both heat and work are understood to arise from *the mechanical energy of large numbers of molecules.* This idea provides a natural underpinning for the first great generalization of thermodynamics, the first law of thermodynamics, the **law of conservation of energy.** On a molecular level there is no friction, only energetic molecules exchanging their mechanical energy with other molecules. Since this mechanical energy must be conserved for the molecules making up the universe, the energy of the universe must be constant. Thus the principles of mechanics applied to molecules lead directly to principles applying to the larger world around us.

Thermodynamics provides a framework for analyzing change in terms of the initial and final **states** of a system. These states are characterized by certain measurable properties of the system called **state variables,** such as temperature T, pressure P, volume V, and chemical composition. Different states of a system correspond to different values for one or more state variables, and the passage from one state to another can be characterized by changes, denoted by a capital Greek delta Δ, in these variables, for example $\Delta T = T_2 - T_1$. Changes in state variables are independent of the path chosen to bring the system from one state to another. Work and heat, on the other hand, are path-dependent quantities and are not needed to define the state of a system.

For example, a reaction that produces a gas, such as the reaction of a metal with acid, will evolve heat and work into the surroundings in amounts that depend on how the reaction is carried out. The general formula for work is $w = -\int P_{ext} dV$; if the external pressure P_{ext} is constant, an **isobaric** process, then $w = -P_{ext}\Delta V$; whereas if the volume is held constant instead, an **isochoric** process, $w = 0$. Similarly, the general formula for heat is $q = \int C\, dT$, where C is the **heat capacity** of the system, for the case that the system reaches a new, stable temperature. If the temperature is held constant, an **isothermal** process, q becomes the negative of the work done; and q is zero in an **adiabatic** process even though the temperature may change. The first law encompasses these various types of processes through the relation $\Delta E = q + w$, where E is the system's internal energy, a state variable whose change depends only on the initial and final states of the system, and not on the path chosen to go between them. In these formulas for w, q, and ΔE, the convention is adopted that energy gains in the system are positive.

The work involved in a chemical reaction at constant pressure depends on the change in numbers of moles of gas involved (Δn_{gas}) according to $w = -\Delta n_{gas} RT$, and can therefore be readily calculated from the reaction stoichiometry. If the reaction is occurring in dilute aqueous solution, the heat capacity is simply estimated as $C = mc$, where m is the mass of water within the system in grams, and c is the **specific heat** of water, 1 cal g^{-1} K^{-1} or 4.184 J g^{-1} K^{-1}. This enables an easy estimation of the heat evolved or taken up by an aqueous chemical reaction, by trapping the heat temporarily in the solution and measuring the temperature rise or fall. Under this circumstance the heat is $q = -C\Delta T$, the minus sign owing to the eventual loss of the trapped heat to the surroundings. It is typical in chemical systems that $|q| \gg |w|$.

Considering changes in state of an ideal gas allows the prediction of q, w, and ΔE from the gas laws, and their interpretation in terms of the kinetic molecular theory of Chapter 9. In a monatomic gas, such as He or Ar, all of the energy is present in the translational degrees of freedom, giving $E = \frac{3}{2}nRT$. This is independent of pressure and volume, and sets the zero of energy at $T = 0$ K. It follows that in an isothermal change, for which $\Delta T = 0$, $\Delta E = 0$ and consequently $q = -w$; and in an isochoric change, $w = 0$ and $q \equiv q_V = \Delta E$. The heat capacity at constant volume is therefore $C_V = \frac{3}{2}nR$. In the more common case of an isobaric change, $w = -P\Delta V$, and therefore $q \equiv q_P = \Delta E + P\Delta V = \frac{3}{2}nR\Delta T + nR\Delta T = \frac{5}{2}nR\Delta T$, and the heat capacity at constant pressure is $C_P = \frac{5}{2}nR$. This form for q_P prompts the definition of the **enthalpy** H as $H = E + PV$; since E, P, and V are state variables, H is also, and all changes in state are accompanied by a ΔH. The definition yields $\Delta H = q_P = C_P\Delta T = \frac{5}{2}nR\Delta T$, and a comparison of C_P and C_V shows that $C_P - C_V = nR$, a relationship that holds for polyatomic gases as well, as long as they behave ideally. These formulas for ΔE and ΔH hold regardless of changes in pressure or volume, that is, regardless of the path chosen to get from T_1 to T_2. Complex processes can be broken down into simpler stages; for example, a change from (V_1, T_1) to (V_2, T_2) can be analyzed in terms of an isothermal change in volume and an isochoric change in temperature, in either order, without requiring that the intermediate state actually exist.

The measurement of the heat q of a chemical reaction is carried out by trapping the heat either in the system itself or in its immediate surroundings, using a device called a **calorimeter.** The earliest accurate calorimetric measurements were carried out by Lavoisier and Laplace in 1780 using an ice calorimeter; more modern experiments employ a **bomb calorimeter,** a heavy-walled steel can, containing the reagents, immersed in an insulated water bath. The bomb type has been especially useful for obtaining **heats of combustion.** In instructional labs, the favorite for solution reactions is the **coffee cup calorimeter,** a Styrofoam cup that traps heat in the reaction solution itself.

The vast majority of known reactions liberate heat ($q < 0$, **exothermic**), but many examples of **endothermic** ($q > 0$) reactions are also known. To systematize the tabulation of experimental reaction heats, the results are reported as enthalpy changes at 1.00 atm pressure and 298.15 K (25°C) (**NTP**) per mole of reaction, $\Delta H°_{298}$, in either kilocalories per mole or kilojoules per mole.

Because the energy of a given element or compound at a given pressure and temperature is an intrinsic property of that substance, the heats of reaction for reactions involving one or more of the same elements or compounds are related to each other, as first shown by Hess in 1840. **Hess's law of constant heat summation** states that the heat of any reaction that can be obtained by adding other reactions is given by the same sum of the heats of those other reactions. That is, if reaction (a) + reaction (b) = reaction (c), then $\Delta H_a + \Delta H_b = \Delta H_c$; stoichiometric multipliers that adjust the amounts of reactions to add also apply to the ΔH relation. Hess's law can be illustrated on an **enthalpy diagram,** such as Figure 10.4, where the relative enthalpy levels of the reagents and products are indicated. Analysis of the diagram shows that the heat of a reaction carried out backward (products $\longrightarrow$ reagents) is the negative of the forward heat ($\Delta H_{\text{reverse}} = -\Delta H_{\text{forward}}$) and that the reactions and their sum can be regarded as forming a **thermodynamic cycle,** for which the sum of heats must be zero: $\Delta H_a + \Delta H_b - \Delta H_c = 0$. The zero of enthalpy is arbitrary.

Hess's law implies that we do not have to measure the heat of every possible reaction as long as we know the heats of some component reactions. The standard reactions that chemists have chosen are the **formation reactions,** those in which a compound is formed from its constituent elements in their most stable states at 1 atm and 298 K. These reactions are written with stoichiometric coefficients adjusted to yield 1 mol of compound; this often results in fractional coefficients normally eschewed in chemistry; for example, $Mg(s) + \frac{1}{2} O_2(g) \longrightarrow$ $MgO(s)$. The heats of these reactions are called **standard heats of formation,** denoted $\Delta H^{\circ}_{f, 298}$, or just ΔH°_f if the temperature is near normal. The heat of any reaction then can be obtained from the heats of formation of its participating compounds according to $\Delta H^{\circ} = \Sigma_P \nu_P \Delta H^{\circ}_f(P) - \Sigma_R \nu_R \Delta H^{\circ}_f(R)$, where P stands for products, R for reagents, and the ν's are stoichiometric coefficients. It is not always possible to measure these heats of formation directly; in such cases they can often be derived from a cycle involving readily measured or known ΔH's. Table 10.1 allows the prediction of the heats for a vast number of possible reactions. Reactions in aqueous solution, which involve enormous solvation energies, are put on a similar footing by defining the heat of formation of a solute ion relative to that for $H^+(aq)$, that is, by taking $\Delta H^{\circ}_{f, 298}(H^+(aq)) \equiv 0$, where the superscript ° denotes a standard concentration, 1 M. Table 10.2 lists ΔH°_f's for common solutes, while Appendix C provides a more complete listing for both molecules and solutes. Values for both free molecules and solutes can be combined to describe multiphase reactions.

The origin of the heat of a reaction can be traced to the difference in **bond energies** (D_0's) between the product and reagent molecules. An exothermic reaction reflects product bonds that are stronger than those of the reagents; that is, the energy is released as a result of the formation, rather than the breakage, of chemical bonds. This idea is put on a quantitative basis by constructing a cycle in which the reagent molecules are first *atomized,* converted into a collection of gas-phase atoms, a process that consumes enormous energy; and the atoms are then reassembled as product molecules, whereby most of that energy is recovered. In a gas-phase reaction, the first step consists of breaking all of the reagent bonds and the second of forming all of the product bonds, giving the heat of reaction as $\Delta H^{\circ} = \Sigma_R n_R D_0(R) - \Sigma_P n_P D_0(P)$, where again R represents reagents and P products, and the n's are the number of *bonds* of each type. While the heats of an even larger number of reactions can be predicted from the bond energies listed in Table 10.3, the values obtained are somewhat less accurate than those derived from heats of formation. This lack of accuracy stems from the approximate nature of bond energies for polyatomic molecules; for example, the bond energy of the C—H bond, $D_0(C—H)$, is taken as an *average* of the four bonds in $CH_4(g)$, and differs slightly from that in any individual bond in CH_4 or in other molecules possessing that bond. In molecules such as benzene C_6H_6, a further complication occurs because of *delocalized π bonding,* which makes the molecule more stable than one would predict based on the bond energies corresponding to a single Lewis structure resonance form. The difference between the predicted and experimental heats of formation of molecules such as benzene is called the **resonance energy.**

The essential role played by molecules in the first law becomes even more evident when heat capacities are examined. While the specific heats (cal/g°C) of various pure substances vary greatly, the molar heat capacities (cal/mol K) are more similar in magnitude, and they vary in a systematic way with molecular complexity, as documented in Table 10.4. These features are readily explained by the capacity for energy storage of the *degrees of freedom* possessed by an atom or molecule, that is, the *mechanical energy* present in the various modes of molecular motion. If this motion obeyed Newton's laws, the heat capacities of all gas-phase molecules would be predictable from the formulas of Equation 10.33. These formulas, however, almost always predict too large a $\overline{C}_P$ because they neglect the *quantization* of the internal motions of a molecule, particularly its *vibrations.*

To understand the effects of quantization on a molecule's capacity to store energy, the populations in the various energy levels for a sample of molecules at a given temperature are needed. This was provided in a universal form by Boltzmann; for vibrations the relative populations are $N_v = N_0 \exp(-vh\nu/k_B T)$, where v is the vibrational quantum number $v = 0, 1, 2, \ldots$, the N's are populations, h is Planck's constant, ν is the vibrational frequency, and k_B is Boltzmann's constant. Because for typical vibrational frequencies and temperatures the Boltzmann distribution predicts that only a very small fraction of molecules will be vibrationally excited, molecules are generally unable to store the full RT of energy in vibrational degrees of freedom that is predicted by Newton's laws. On the other hand, atomic and ionic solids, which possess much smaller lattice vibrational frequencies, behave like classical 3-D harmonic oscillators at room temperature. For an atomic solid the molar heat capacity in this model is $3R \approx 6$ cal/mol K, while for a salt it is $3R$ times the number of ions in the formula. Both of these results agree well with experimental values. The near constancy of $\overline{C}_P$ at 6 cal/mol K for metallic solids is known as **the law of Dulong and Petit.** Molecular liquids and solids may store energy in both intramolecular and intermolecular degrees of freedom; no simple model can be applied in such cases.

If the heat capacities of the participants in a chemical reaction are known, the temperature dependence of ΔH can be predicted using Equation 10.36. In most cases this dependence is weak, and ΔH can often be taken as constant, independent of T.

EXERCISES

Work and heat

1. A 6.00 L volume of air at a pressure of 1.00 atm is trapped in a cylinder behind a sealed but movable piston. A force is suddenly applied to the piston, causing the volume in the cylinder to decrease to 5.45 L. Assuming ideal gas behavior and constant temperature, calculate the work in joules.

2. A mole of water (18.0 g) evaporates from a lake at 15°C and 0.97 atm. Calculate the work in joules.

3. In an accident in which the valve on a 50.0-L gas cylinder was sheared off, Ar at 170. atm was suddenly released to the atmosphere at a barometric pressure of 745 Torr. Calculate the work in joules—part of which propelled the cylinder through two brick walls! Assume constant temperature and ideal gas behavior (not highly accurate at 170 atm).

4. When you take your car to the local service station to inflate your tires, the tire pressures are read (at least at this writing) in "pounds per square inch gauge" (psig), which is the pressure in excess of 1.00 atm = 14.7 lb in.$^{-2}$; that is, 20 psig is actually 35 lb in.$^{-2}$. A compressor in the service station supplies air at 60 psig. Suppose a tire is inflated from zero to 32 psig, increasing the volume from 60 to 160 L, at 12°C and 0.96 atm. Calculate (a) the work in joules of compressing the air that will end up in the tire from 0.96 atm to 60 psig assuming the external pressure is constant at 60 psig, and (b) the work of inflating the tire, assuming the external pressure is atmospheric. (The second calculation ignores the work of stretching the tire walls.) Assume constant temperature and ideal gas behavior throughout.

5. A baseball batter hits a high pop-up. How high would the ball, of mass m, have to rise so that when it falls into the catcher's mitt it would have kinetic energy equivalent to the heat necessary to raise the temperature of an equal mass m of water by 1°C?

6. A 500. g mass of water is heated on an electric burner from 10° to 95°C to make tea for two. (a) How much heat energy in calories and joules has been added to the water? (b) The electrical energy needed to operate the burner is given by $E = iVt$, where i is the current in amperes, V is the voltage, and t is the time. Assuming no heat loss to the surroundings, how long should it have taken to heat the water if the burner draws 10.0 A at 120 V? (c) If this amount of energy were used instead to lift a 1-ton (10^3 kg) mass, to what height could it be raised?

7. Adding a certain amount of heat to 100. g water raises its temperature by 3.79°C. Adding the same amount of heat to 100. g octane, $C_8H_{18}(l)$, raises its temperature by 9.63°C. Find the specific heat of octane.

8. A house contains 2.4×10^6 L of air at 17.2°C and 1.00 atm. Assuming no heat or air is lost, how much heat in calories and joules must a furnace supply to raise the temperature of the air to 18.8°C? Use data from Table 10.4.

9. How many servings of tea could be made, under the conditions of Exercise 6, from the energy released in the accident of Exercise 3?

The first law: Physical changes

10. Find the final temperature of a 100.0-g sample of water prepared by adding (a) 50.0 g hot water at 63.4°C to 50.0 g cold water at 12.1°C; (b) 75.0 g hot to 25.0 g cold; (c) 25.0 g hot to 75.0 g cold. Assume the container is completely insulating (a property approached by a Styrofoam coffee cup).

11. A 150. g mass of boiling water (100°C) is added to a 200.-g ceramic cup of heat capacity 0.19 cal g^{-1} °C^{-1} at 22°C. If the entire cup were to reach the same temperature as the water, what would the final temperature be? (A similar conclusion would apply to pouring hot coffee or tea, although ceramic cups tend to be partially insulating.)

12. Repeat the calculation of Exercise 10, part (a), when a glass beaker weighing 150. g is used as the container. Assume the entire beaker is initially at cold-water temperature and attains the same final temperature as the water sample. Use data from Table 10.4, and neglect further heat leakage to the surroundings.

13. A 20.0-g chunk of an unknown metal at 200°C is added to 75.0 g water at 22.0°C in a Styrofoam cup. The final temperature is 26.4°C. Neglecting the heat absorbed by the cup, estimate the specific heat of the metal. Using the law of Dulong and Petit, which states that the molar heat capacity of a metal is approximately 6 cal mol^{-1} K^{-1}, estimate the atomic mass of the unknown metal.

14. All of the hisses you've heard in your life come from a gas expanding against the atmosphere. Suppose 5.00 L of hydrogen gas at 1.30 atm and 24.0°C in a balloon escapes into the atmosphere, $P = 0.98$ atm. Calculate q, w, and ΔE in joules and calories if the expansion is carried out (a) isothermally or (b) adiabatically. [Hint for part (b): You must first find the final temperature using the first law.] Assume ideal gas behavior, and use the heat capacity data of Table 10.4.

15. In a typical hand pump for inflating tires, balls, etc., the air displaced on each stroke is about 1 L. Even though such pumps are usually made of metal tubing and conduct heat to the surroundings well, more than about 20 strokes make the pump noticeably warm near its base. Assuming this is not due to internal friction, use the first law to explain the heating. If the pump walls were a perfect thermal insulator and the air were trapped inside the pump, what would be the change in temperature of the gas due to a single stroke? Assume that before the stroke the air in the pump is at 15°C and 1.00 atm, and that during the stroke the gas volume decreases from 1.2 to 0.2 L and the external pressure is constant at 6.00 atm. (A sequence like this actually occurs inside a diesel engine cylinder.) Assume ideal gas behavior, and use data from Table 10.4.

16. The molar heat of vaporization of water at constant pressure is 44.4 kJ/mol at 15°C. What is the heat q that accompanies the work w of Exercise 2? What is the internal energy change ΔE? The enthalpy change ΔH?

17. Use the first law and the definition of enthalpy to show that for 1 mol of an ideal gas heated from absolute temperature T_1 to T_2, $\Delta H \geq \Delta E$. Explain your result in words in terms of the work done. Under what condition(s) does the equals sign hold? What condition would hold for a cooling process?

18. Three argon atoms in a container have kinetic energies of 2.7×10^{-21} J, 4.1×10^{-21} J, and 5.4×10^{-21} J. What is the total energy E of this three-atom sample? What is the average energy $\bar{\varepsilon}$ of an atom in the sample? Give an estimate of the "temperature" of this sample. What would be the total energy of a 1-mol sample with this average energy?

19. In Example 9.1, a helium sample being used to carry a stratospheric payload undergoes a change in state in ascending to the stratosphere. Analyze the overall process under the conditions given there in terms of first law quantities.

The first law: Chemical changes

20. In the ice calorimeter experiment on the combustion of carbon, carried out by Lavoisier and Laplace as described in Section 10.4, the combustion of 1 oz of carbon was found to melt 6 lb, 2 oz of ice. Analyze this finding to obtain the heat of combustion $\Delta H°$ of carbon in calories per gram and kcalories per mole. Use the modern heat of fusion of water ice, 79.72 cal/g (1 lb = 16 oz = 453.6 g). Compare your result with the value given in Section 10.5.

21. Using a bomb calorimeter like that described in Section 10.4, a calibration is carried out with an immersed 100.-Ω resistor and a dc voltage of 115 V, yielding a temperature rise of 3.833°C for 120. s of joule heating. Then 1.505 g acetaldehyde, $CH_3CHO(l)$, is ignited in excess oxygen, causing a temperature rise of 9.617°C. Calculate the heat of combustion $\Delta H°$ of acetaldehyde in calories per gram and kcalories per mole.

22. A 0.401 g sample of calcium metal is added to 100.0 mL of 0.2523 M HCl in a coffee cup calorimeter. Gas bubbles are observed, and the temperature of the system rises from 23.1° to 36.0°C. Neglecting the heat absorbed by the calorimeter, and assuming the

heat capacity of the system is that of the same volume of pure water, find the heat of the reaction $\Delta H°$ per mole of Ca. What error is incurred by neglecting the heating of the coffee cup, if the actual calorimeter heat capacity is 2.0 cal/°C? What is the ratio of work to heat for this reaction? (Assume normal temperature and pressure.)

Hess's law and heats of formation

23. Modern natural gas furnaces produce more heat by condensing the water vapor that is formed as a combustion product. Use the heat of the burning of methane to produce water vapor, $\Delta H° = -191.61$ kcal/mol CH_4, and the heat of vaporization of water, $\Delta H°_{vap} = 10.52$ kcal/mol H_2O, to find the heat of combustion of methane to produce water liquid. Construct an enthalpy level diagram that reflects your calculation. By what percentage is the efficiency of the furnace increased?

24. Given the heats of the following two reactions,

	$\Delta H°$ (kcal/mol)
$N_2(g) + 2O_2(g) \longrightarrow 2NO_2(g)$	$+15.86$
$2NO(g) + O_2(g) \longrightarrow 2NO_2(g)$	-27.28

find the heat of the reaction $N_2(g) + O_2(g) \longrightarrow 2NO(g)$. Draw an enthalpy level diagram corresponding to your calculation. What is the heat of the reaction $\frac{1}{2}N_2(g) + \frac{1}{2}O_2(g) \longrightarrow NO(g)$? Of the reaction $2NO(g) \longrightarrow N_2(g) + O_2(g)$?

25. What relationship if any exists among the heats of the three acid–base reactions:

(1) $HCl(aq) + NaOH(aq) \longrightarrow NaCl(aq) + H_2O(l)$
(2) $HCl(aq) + NH_3(aq) \longrightarrow NH_4Cl(aq)$
(3) $NH_4Cl(aq) + NaOH(aq) \longrightarrow$
$NaCl(aq) + NH_3(aq) + H_2O(l)$

Construct an appropriate enthalpy level diagram, assuming all reactions are exothermic, with magnitude of heat released decreasing in the order given.

26. An approach often taken for obtaining standard heats of formation is to burn every reagent or product in the formation reaction and use Hess's law. Use the standard heats of combustion of $C(s)$, -94.052 kcal/mol, $H_2(g)$, -68.315 kcal/mol, and $CH_4(g)$, -212.8 kcal/mol, to find the standard heat of formation of $CH_4(g)$. Compare your result with the entry in Table 10.1.

27. Use standard heats of formation from Table 10.1 to calculate the heat of the sodium–chlorine reaction discussed in Chapter 5, $2Na(s) + Cl_2(g) \longrightarrow 2NaCl(s)$. (You should find this calculation to be extremely simple!) Illustrate the calculation with an enthalpy diagram.

28. Methanol ($CH_3OH(l)$) obtained from biomass conversion is proposed to be blended with gasoline ($C_8H_{18}(l)$) to extend our fossil-fuel reserves. Use data from Table 10.1 to find and compare the enthalpy released per gram of fuel for methanol and octane combustion. Will the use of "gasohol" increase or decrease automobile gas mileage?

29. Man-made atmospheric pollutants appear to be destroying the stratospheric ozone layer—which protects us by absorbing potentially fatal ultraviolet radiation of 200- to 300-nm wavelength—by catalyzing the reaction $2O_3(g) \longrightarrow 3O_2(g)$. What is $\Delta H°$ for this reaction? Per mole of ozone lost? Will this reaction heat or cool the stratosphere?

30. Although ozone may be helpful to us in the stratosphere, its presence down here in the troposphere is downright distressing. In addition to its own toxicity, it rapidly converts $NO(g)$ to $NO_2(g)$, an acid-forming gas, through the reaction $O_3(g) + NO(g) \longrightarrow NO_2(g) + O_2(g)$. Find the heat of this tropospheric reaction.

31. A final step in the Solvay process for the industrial production of sodium carbonate ($Na_2CO_3(s)$) is the decomposition by heating of sodium bicarbonate ($NaHCO_3(s)$):

$2NaHCO_3(s) \longrightarrow Na_2CO_3(s) + CO_2(g) + H_2O(g)$

Calculate $\Delta H°$ for this reaction, one we will consider further in the next chapter.

32. Write the three acid–base reactions of Exercise 25 in ionic form, and use the data of Tables 10.1 and 10.2 to calculate their heats. (Note that the heats of formation of spectator ions cancel out of the calculation.)

33. Find the heat of reaction for the redox displacement reaction

$Zn(s) + CuSO_4(aq) \longrightarrow ZnSO_4(aq) + Cu(s)$

introduced as an example in Chapter 5 using the data in Table 10.2.

34. Combine data from Tables 10.1 and 10.2 to find the heat of the water autoionization reaction $H_2O(l) \longrightarrow H^+(aq) + OH^-(aq)$. As we shall see in Chapter 12, the endothermicity of this reaction is a dominant factor in the small extent of autoionization in pure water. Of what common reaction is this reaction the reverse?

35. In Chapter 5 you learned to balance complicated aqueous reactions such as

    $KMnO_4 + 5FeCl_2 + 8HCl \longrightarrow MnCl_2 + 5FeCl_3 + KCl + 4H_2O$

    Now you can find the heats of such reactions! Try it for this one, remembering to eliminate spectator ions first.

36. We shall show in Chapter 12 that the heat of a reaction determines how the extent of reaction depends on temperature: if the reaction is exothermic, its extent will be decreased by an increase in T; the reverse is true for endothermic reactions. Find the heat of the limestone precipitation reaction $Ca^{2+}(aq) + CO_3^{2-}(aq) \longrightarrow CaCO_3(s)$, and decide whether limestone deposits in caves would form more readily at high or low temperature.

Bond energies

37. In Chapter 5 we discussed the sodium–chlorine reaction mechanism in terms of the gas-phase reaction $Na(g) + Cl_2(g) \longrightarrow NaCl(g) + Cl(g)$. Taking the bond energy of NaCl(g) to be 98 kcal/mol as given in Chapter 5, use bond energies to find the heat of this reaction. Illustrate your calculation by sketching the bond potential energy curves on the same set of axes. Estimate the bond lengths r_e for your sketch using the neutral and ionic radii from Chapter 4.

38. Confirm the value for the bond energy of a C—H bond given in Table 10.3 by using other data in that table along with the heat of formation of $CH_4(g)$ from Table 10.1. Draw an enthalpy diagram that illustrates your calculation.

39. The NH_2 radical is an important intermediate in the oxidation of ammonia. Use data from Table 10.3 to estimate the heat of formation ΔH_f° of NH_2 (literature value 44.3 ± 1.1 kcal/mol). Construct the corresponding enthalpy diagram.

40. Two important *elementary steps* in the hydrogen–oxygen explosion are $H(g) + O_2(g) \longrightarrow OH(g) + O(g)$ and $O(g) + H_2(g) \longrightarrow OH(g) + H(g)$. Use bond energies to find the heats of these reactions, and decide which of them is exothermic and will therefore contribute to accelerating the reaction.

41. A central reaction in the Ostwald synthesis of nitric acid from ammonia is

    $4NH_3(g) + 5O_2(g) \longrightarrow 4NO(g) + 6H_2O(g)$

    which occurs over a platinum catalyst. Use both heats of formation and bond energies to calculate the heat of this reaction, and compare your results. Which is more accurate?

42. The simplest case of an alternating single and double bond system is 1,3-butadiene, $CH_2\!=\!CH\!-\!CH\!=\!CH_2$. Use bond energies to predict the standard heat of formation of this molecule, and compare it to the experimental value, $\Delta H_f^\circ = 26.3$ kcal/mol. Considering that the bond energy estimate should be accurate within 1 kcal/mol for the Lewis structure given, is significant resonance stabilization present?

Heat capacities and molecular energy storage

43. Use the formulas of Equation 10.33 to predict the molar heat capacities of $HCl(g)$ and $CO_2(g)$, and compare them to the values given in Table 10.4. If they differ appreciably, explain why.

44. In some cases the formulas of Equation 10.33 work quite well if the vibrational terms are simply neglected. Show that this is true for $N_2(g)$, $O_2(g)$, and $CH_4(g)$ by comparison with the values in Table 10.4. Explain briefly.

45. Use the law of Dulong and Petit to predict the specific heat of metallic chromium $Cr(s)$. (Literature value: 0.1073 cal/g °C.) For a given increase in temperature, which metal can absorb more heat energy per unit mass, $Al(s)$ or $Cr(s)$? Give an atomic explanation for your answer.

46. Based on the extension of the law of Dulong and Petit to salts outlined in Section 10.7, estimate the molar heat capacity and the specific heat of $CaCl_2$. (Literature value: 0.1563 cal/g °C.)

47. Before natural gas, largely $CH_4(g)$, can be burned, it must be heated to 600°C, its ignition temperature.

(This is why gas furnaces and ranges require a pilot flame or heated filament.) How much heat in calories is required to bring 0.350 L $CH_4(g)$ initially at STP to 600°C at constant pressure? At constant volume? Assume the heat capacity is temperature independent.

48. During the spring and fall, diurnal temperature variations often approach 20°C. How much heat is deposited by the sun in heating the air over a city of 250 km^2 by 20°C? Assume an 8-km-thick air layer (see Exercise 9.3) at atmospheric pressure and an initial temperature of 5°C.

49. To make dry ice, $CO_2(g)$ must be cooled from 25° to $-78°C$. Calculate the amount of heat that must be removed through refrigeration at constant volume and at constant pressure in producing 1.00 kg of dry ice. Which process requires less refrigeration?

50. For a sample of hydrogen molecules at 25°C, what is the ratio of the numbers of molecules in the first excited vibrational state ($v = 1$) to that in the ground state? At what temperature would this ratio be 0.10? Use data from Table 8.3.

51. The vibrational energy distribution for $Cl_2(g)$ at 298 K is illustrated in Figure 10.8. Tabulate the Boltzmann factors $e^{-vh\nu/kT}$ for $v = 0$, 1, and 2, and add them. If these are the only states with appreciable population, this sum is an estimate of the so-called **partition function** q of Appendix D, and the fractional populations of each state are given by $f(v) = [e^{-vh\nu/kT}]/q$. The energy stored in Cl_2 vibration (in excess of its zero point energy) is given as $E_{vib} = N\Sigma_v (\varepsilon_v - \varepsilon_0)f(v)$. Compute E_{vib} at 298 K for 1 mol of Cl_2 and compare it to the classical energy RT. How does this help to interpret the heat capacity value given in Table 10.4? How could the calculation be made more accurate?

Temperature dependence of ΔH

52. What is the standard enthalpy change (298 K) in the Haber–Bosch process for the synthesis of ammonia, $3H_2(g) + N_2(g) \longrightarrow 2NH_3(g)$? Industrially, this reaction is generally run at 800 K. Use Equation 10.36 and data from Table 10.4 to develop a working formula for ΔH_T° for this reaction, and calculate ΔH_{800}°.

53. As a consequence of the temperature dependence of ΔH, heats of formation given in Table 10.1 are less accurate the greater the departure from 298 K. Use Equation 10.36 and Table 10.4 to find the heat of formation of $CH_4(g)$ at 230 K, stratospheric temperature.

CHAPTER 11

Spontaneity of Chemical Reactions

This sequence is naturally viewed top-to-bottom; even thinking about its reverse provides comic relief.

CHAPTER OUTLINE

11.1 Berthelot's Hypothesis

11.2 Carnot's Ideal Engine and the Entropy S

11.3 The Second Law of Thermodynamics: The Entropy of the Universe Tends to a Maximum

11.4 Entropy and Probability: Boltzmann's Postulate $S = k \ln W$

11.5 The Third Law of Thermodynamics: The Entropy of a Perfect Crystal at $T = 0$ K Is Zero

11.6 Focusing on the System: The Gibbs Free Energy $G = H - TS$

11.7 Use and Misuse of the Second Law

Whence is it that nature does nothing in vain; and whence arises all that order and beauty which we see in the world?

Sir Isaac Newton (1704)

Things naturally happen in a certain way. If you hold a glass of water in front of you and let go, it will fall and hit the floor. The water will spill, and the glass will probably break. You can make these statements without actually trying it, because you know *from experience* what will happen. The glass of water will not fall up, or remain suspended. Once the water spills, it will not clean itself up by climbing back into the glass. And if the glass should break, the shards will never reassemble themselves into a seamless glass again. Even to think of such possibilities is laughable; these things just do not happen. We can say that events have a *natural direction* in which they occur. This is no less true of the events that form the subject of this book, chemical reactions. Hydrogen and oxygen readily combine to make water, but life would be difficult if, without warning, a glass of water could now and then choose to decompose into its elements. This possibility is just as absurd as the reassembling glass. The determination and prediction of the natural direction of chemical events, based on this commonsense way of thinking and on *experience,* will be the subject of this chapter.

Well, you say, then where's the mystery you promised us at the end of the last chapter? Consider this. As you stand empty handed with the spilled water and broken glass at your feet, the first law of thermodynamics allows the possibility that the glass will reassemble, the water will climb in, and the full glass of water will hop back up into your hand. All the energy involved in the process—the falling glass, the spilling water, and the shattering glass—is still present, and energy conservation would not be violated were these strange things to happen. On the molecular level, when hydrogen and oxygen combine to make water, the energy that is released in this exothermic reaction *is still present* after the reaction; that is, there is still enough energy to cause water to revert to its constituents. Moreover, the laws of mechanics, either Newtonian or quantum, are *time reversible,* allowing molecular events to precisely retrace their steps backward in time. Perhaps the absurd occurrences, the ones we never observe, are not as laughable as they seem.

Josiah Willard Gibbs (1839–1903) was professor of mathematical physics at Yale University from 1871 until his death. His major work, *On the Equilibrium of Heterogeneous Substances,* published in 1876 in an obscure journal, was ignored during his lifetime, but proved to be the basis for understanding spontaneity and equilibrium in chemical systems.

11.1 Berthelot's Hypothesis

It is clear that we need another law of thermodynamics; the first law just doesn't contain the information needed to distinguish between the possible and the impossible. We can start by observing that, with respect to chemistry, the vast majority of chemical reactions that readily occur are exothermic. This led the French chemist M. Berthelot to propose in 1867 that the *only* spontaneously occurring chemical processes are exothermic, that is, that $\Delta H < 0$ is required for a reaction to happen "naturally." If this proposal were correct, this inequality would be a new law and would qualify as a second law of thermodynamics. Berthelot's hypothesis was quickly put to the test, and exceptions were found. For example, the dissolution of ammonium chloride in water,

$$NH_4Cl(s) \longrightarrow NH_4^+(aq) + Cl^-(aq), \qquad \Delta H° = +3.53 \text{ kcal/mol}$$

occurs readily despite its endothermicity, which cools the resulting solution. Even more dramatic, as illustrated in Figure 11.1, is the solid–solid acid–base reaction $2NH_4Cl(s) + Ba(OH)_2(s) \longrightarrow 2NH_3 + 2H_2O + BaCl_2$, which is so endothermic

Figure 11.1

An endothermic but spontaneous acid–base reaction. In (a) the unmixed reagents are all allowed to come to thermal equilibrium with their surroundings, that is, they reach the same temperature. In (b) the solid reagents are mixed by swirling the flask, producing a drop in temperature while the reaction mixture becomes liquefied by the water produced in the reaction. If the cold flask is placed on a wetted wooden square, it will cause the water to freeze and make the flask stick to the wood.

(a)

(b)

that it can cause water to freeze. Relatives of this reaction are used in cold packs for soothing injuries. But perhaps the most telling counterexample is the evaporation of water, $H_2O(l) \longrightarrow H_2O(g)$, $\Delta H° = +10.52$ kcal/mol. If evaporation did not occur, the lack of "weather" on earth would make it uninhabitable. If you have ever been swimming outdoors, you have experienced this endothermic but spontaneous process in the chill you receive on emerging from the water. It can also be easily demonstrated, as shown in Figure 11.2, by simply removing a metal-sheathed temperature probe from a container of water and letting it dry in the air while noting the drop in temperature.

From the point of view of the motions of individual atoms and molecules, Berthelot's hypothesis is quite sensible. An exothermic reaction is an energetically downhill process, making chemical bonds in the products that are stronger than those of the reagents. This downhill run can be pictured as taking place along a path on a *potential energy surface* for the reaction (to be discussed in more detail in Chapter 15),

Figure 11.2

An illustration of possibly the most important endothermic-but-spontaneous process, the evaporation of water. In (a) the thermocouple temperature probe is at equilibrium with a beaker of water. When removed (b), the probe cools as water evaporates from it.

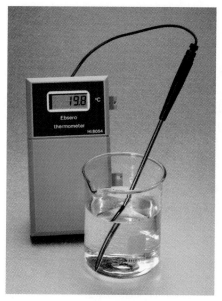
(a)

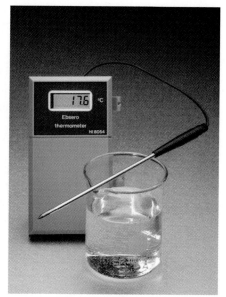
(b)

making the spontaneity of an exothermic reaction nothing more than the mechanical response of the reagents to an attractive force drawing them toward products. We can conclude that the sign of ΔH for a process must be an important factor in determining whether it will happen, but there must be something more. We need to be able to account for the few but important endothermic but spontaneous reactions. But more generally, we need to be able to rationalize our own experience with what happens and what doesn't.

11.2 Carnot's Ideal Engine and the Entropy S

To discover what's wrong with Berthelot's idea, that is, to find the missing term in the criterion for spontaneity, we must carefully define what we mean by a spontaneous process. "It happens" is correct, but not quantitative.

Spontaneity and Reversibility

We begin our quest for a quantitative assessment of spontaneity by examining the concept of work w in more detail. In the P–V work for an ideal gas considered in Chapter 10, the external pressure P_{ext} was generally constant during the change of state, giving $w = -P_{ext}\Delta V$. For an ideal gas expansion, P_{ext} would necessarily be smaller than the internal pressure P of the gas at any point in the expansion. If this were not the case, the gas would be compressed rather than expanding. Let's therefore construct a concrete, simple example that embodies this constraint: one mole of ideal gas at standard temperature (273 K) begins with $P_1 = 2.00$ atm and $V_1 = 11.2$ L (using the ideal gas equation of state), and expands isothermally to a final state characterized by STP, $P_2 = 1.00$ atm and $V_2 = 22.4$ L. If $P_{ext} = 1.00$ atm throughout the expansion, then the gas will expand freely—spontaneously—to its final state and stop there. If the surroundings are also at 273 K, then the gas is in **equilibrium** with its surroundings in state 2. The expansion process and the work done can be illustrated on a P-V diagram, as given in Figure 11.3 (long arrow in the illustration and lower curve in the P-V diagram). The work done *by* the gas, $P_{ext}\Delta V$, is just the area under the P_{ext} versus V curve.

Two observations concerning the expansion just described can be made without explicit calculation of the work, both stemming from the fact that, at all points during the expansion prior to reaching state 2, the system pressure P is appreciably greater than P_{ext}: $P > P_{ext}$. (At any point in the expansion, where the volume is V, $P = nRT/V$ from the ideal gas law; as long as $V < V_2$, then $P > P_{ext} = P_2$.)

1. More work could have been extracted from the gas if P_{ext} were brought closer to P during the expansion; in the limit where an expansion is still maintained, we would have $P_{ext} = P - dP$, where dP is an infinitesimally small difference in pressure. This limiting case yields **maximum work** by the system for the given initial and final states, that is, the maximum possible area under the P versus V curve.

2. For the expansion as given, no small change dP in the external pressure will stop the system's inexorable inflation toward the final state; that is, the expansion with $P_{ext} = 1$ atm is **irreversible**. *We will therefore associate spontaneity of a process with irreversibility.*

Figure 11.3
Irreversible and nearly reversible isothermal expansions illustrated by gas in a cylinder pushing a piston against the atmosphere. For the irreversible path, the disk weights, which compress the gas to a pressure greater than atmospheric, are removed all at once, whereas for the nearly reversible path the weights are removed one at a time. The latter path takes longer, but produces more work. A closer approximation to a reversible expansion would be achieved by replacing the weights with a tall beaker of sand, and removing the sand one grain at a time. The expansion could be reversed by replacing a grain of sand. The $P-V$ diagram at the bottom of the figure illustrates these two paths. The work done by the system (the gas) is the area under the P versus V curve in each case. The darker blue area gives the extra work gained by running the process reversibly. The stair-step curve shows the approximation to the reversible limit obtained by removing the weights one at a time. To construct the $P-V$ diagram, 1 mol of gas at 273 K is assumed to behave ideally. The cylinder would have to be in good thermal contact with an ice bath to maintain constant temperature.

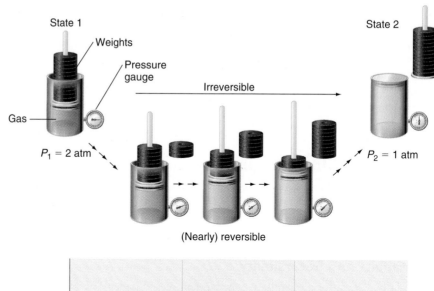

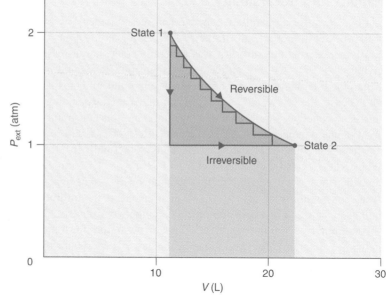

Moreover, there exists a continuum of possible paths (one of which is illustrated in Figure 11.3) between the constant pressure path and the limiting path where P_{ext} is infinitesimally less than P; that is, there are degrees of spontaneity. In the limiting case, a small increase dP in P_{ext} *can* halt the expansion, and it becomes **reversible**. If we did halt the expansion in this way, the system would be at equilibrium, with no tendency either to expand or compress. *We will therefore associate equilibrium (lack of spontaneity) with reversibility.*

The reversible expansion is a limiting case that can never be attained in practice, but it will form an indispensable standard against which real, and therefore necessarily irreversible, expansions are gauged. In this case, P_{ext} differs negligibly from $P = nRT/V$, and the reversible work can be found from

$$-w_{rev} = \int_{V_1}^{V_2} P\, dV = \int_{V_1}^{V_2} \frac{nRT}{V}\, dV = nRT \int_{V_1}^{V_2} \frac{dV}{V} = nRT \ln\frac{V_2}{V_1} \quad (11.1)$$

where the nRT factor, being constant for an isothermal expansion of a fixed sample of gas, can be taken out from under the integral sign, and the remaining integral is the standard logarithmic integral, $\int dx/x = \ln x$. We have also used the identity $\ln x_2 - \ln x_1 = \ln(x_2/x_1)$ to write the result more compactly. The result corresponds to the area under the reversible P versus V curve, also shown in Figure 11.3; the curve is just a segment of a Boyle's law plot (see Figure 9.3). For the irreversible expansion as originally formulated, we have

$$-w_{irrev} = P_2(V_2 - V_1) = nRT\left(1 - \frac{V_1}{V_2}\right) \quad (11.2)$$

where the ideal gas law has been used to replace P_2. We can now compute the work in each case in proper units using $R = 1.987$ cal K^{-1} mol^{-1} (8.314 J K^{-1} mol^{-1}) to get $-w_{rev} = 376$ cal (1570 J) and $-w_{irrev} = 271$ cal (1130 J). For this case, in which the gas doubles its volume, the ratio is $w_{rev}/w_{irrev} = 2\ln 2 = 1.386$; the reversible work is 39% greater, consistent with its being the maximum work the gas can do.

By our definition, a reversible process can just as easily be run backward; for a reversible compression, all that happens is that everything changes sign, and work is done *on* the gas instead of *by* it. However, a simple change of sign does *not* suffice for an irreversible (real) compression. If we wanted to recompress our expanded gas at constant P_{ext}, the minimum pressure needed would be 2.00 atm, and twice as much work would be done *on* the gas as had been done *by* it. By an argument similar to that given in observation 1 earlier in this section we can conclude that a reversible compression requires the **minimum work** to be done by us on the gas. These two results—that the reversible expansion yields maximum work and the reversible compression requires minimum work—are features we would expect an ideal machine or engine to possess.

The Carnot Cycle

The analysis of the previous section was first carried out by a French engineer named Sadi Carnot, in a brief monograph he published in 1824 entitled "On the Motive Power of Heat." Being an engineer, Carnot wanted a solution to the problem of how to construct the most efficient **engine,** a machine for doing useful work. In that era engines were becoming increasingly important as the industrial revolution took shape. Carnot recognized that a general feature of all engines known at that time was *the conversion of heat* (from combustion of a fuel) *into mechanical work in a repeating cycle.* The work was usually the result of the expansion of a gas (often steam) in a cylinder against a piston. A cycle was accomplished by storing some of the work done as rocking of a beam or rotation of a flywheel, which caused the piston to return to its starting position, compressing the gas as it did so. A new expansion cycle was then ready to begin. Effective operation of the engine required cooling of the gas later in the cycle. Figure 11.4 illustrates an early engine.

Carnot wished to discover whether there was any intrinsic limit to the efficiency of such an engine. So he assumed a frictionless construction, with all steps carried out reversibly using an ideal gas as the working substance, to eliminate losses due to imperfections in the machine or its method of operation. He had no laws of thermodynamics to guide him; only the "caloric" theory of heat was available then (though Count Rumford had already disproved it). Although Carnot professed a belief in the caloric theory, he ignored its conservation when it was logical to do so. Here we analyze the reversible engine using the first law because the argument is then much simpler.

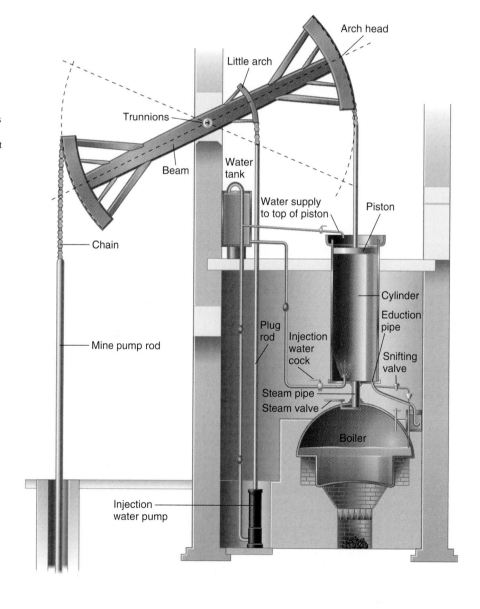

Figure 11.4
Possibly the earliest practical steam engine, the first "atmospheric" engine, built by Thomas Newcomen in 1712. The "working element" is the cylinder at the right, shown fully expanded by injecting steam from the boiler. The cylinder was made of cast brass 21 inches in diameter and 7 feet 10 inches high. The piston was sealed by a leather flap with water on top to hold it in place. The injection water cock was Newcomen's new idea for causing rapid condensation of the steam and speeding up the engine. This cock and the steam valve were attached to the plug rod, driven by the rocking overhead beam, in such a way as to cause the cock and valve to open at appropriate times in the cycle. Condensed steam flowed out the eduction pipe and was eventually returned to the boiler, while air dissolved in the boiler water was removed by the snifting valve. (The dissolved air would accumulate in the cylinder and cause the engine to run down.) This engine was used to pump water from mines in Staffordshire, England. Because a patent on the use of steam to do work had already been issued, Newcomen apparently realized little financial gain from his 25 years of development and construction. A working steam engine can still be viewed at a sawmill in Greenfield Village near Fort Dearborn, Michigan.

Carnot constructed an idealized thermodynamic cycle, now called the **Carnot cycle,** consisting of four reversible stages, which are illustrated in Figure 11.5:

I. *Isothermal expansion.* In this stage, heat q_h is added to the gas from a hot reservoir at temperature T_h. From the first law for an ideal gas, $\Delta E_I = 0$ and therefore $q_h = -w_I$. Since q_h is positive, the gas does work and expands, and Equation 11.1 applies. The states between which the expansion occurs are labeled 1 and 2, so $-w_I = nRT_h \ln(V_2/V_1)$.

II. *Adiabatic expansion.* Here the gas continues to expand, but without any further heat being added, that is, $q = 0$. Since the gas is doing work without compensating heat, the energy and temperature of the gas must fall: $\Delta E_{II} = C_V \Delta T_{II} = w_{II}$, where all quantities are negative. If the final temperature is T_c (c for cold), then $\Delta T_{II} = T_c - T_h$. The expansion occurs between states 2 and 3.

III. *Isothermal compression.* The gas has now done all the work it can, and must be brought back to its original state. This begins with the present stage, in which the heat q_c generated by the compression is deposited in a cold reservoir at temperature T_c. As in stage I, $\Delta E_{III} = 0$ and $q_c = -w_{III}$, but now q_c is negative. The compression proceeds to state 4, giving $-w_{III} = nRT_c \ln(V_4/V_3)$.

IV. *Adiabatic compression.* This final stage brings the gas back to its starting point, by compressing it without allowing heat transfer. As in stage II, $q = 0$ and $\Delta E_{IV} = C_V \Delta T_{IV} = w_{IV}$, but now all quantities are positive, and $\Delta T_{IV} = T_h - T_c = -\Delta T_{II}$.

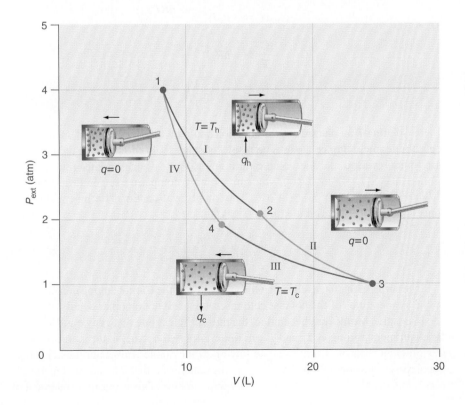

Figure 11.5
A $P-V$ diagram for the Carnot cycle. The working substance is an ideal monatomic gas, and all stages, I through IV, are assumed reversible. Stages I and III are isothermal, following Boyle's law, $PV = \text{const}$; whereas stages II and IV are adiabatic, following $PV^\gamma = \text{const}$, where $\gamma = C_P/C_V = 5/3$. Heat q_h is taken up from a hot reservoir at temperature T_h during stage I and q_c is discharged to a cold reservoir at temperature T_c during stage III. The cylinders illustrate the expansion and compression stages. The net work done by the engine is the area enclosed by the cycle. (In real cycles the isothermal and adiabatic stages are not cleanly separated.) If you know that 1 mol of gas is involved, can you use the diagram to estimate the temperatures T_h and T_c and the efficiency? (Answer: yes; 400 K, 300 K, and 0.25, respectively.)

This cycle is evidently the most efficient one that can be constructed between temperatures T_h and T_c, because all stages have been carried out reversibly, ensuring maximum work in expansion and minimum work in compression.

The **efficiency** ε of the cycle is the ratio of the net work done by the engine to the heat input, or

$$\varepsilon = \frac{-(w_I + w_{II} + w_{III} + w_{IV})}{q_h} \tag{11.3}$$

The net work, given by the sum in the numerator of Equation 11.3, is just the area enclosed by the reversible cyclic path of Figure 11.5. The work terms from adiabatic stages II and IV exactly cancel as you can deduce from their earlier descriptions; furthermore $-w = q$ for the isothermal stages I and III. Therefore,

$$\varepsilon = \frac{q_h + q_c}{q_h} = \frac{nRT_h \ln(V_2/V_1) + nRT_c \ln(V_4/V_3)}{nRT_h \ln(V_2/V_1)} \tag{11.4}$$

where we have substituted the reversible formulas given earlier for stages I and III. Bear in mind that q_c is negative. To simplify this relation further, we need a relationship among the volumes. The temperature of the cold reservoir, T_c, determines the volume change in the adiabatic expansion II, and in turn in the adiabatic compression as well. To exploit this, we use the first law in *differential form*: $dE = dq + dw$. For an adiabatic process, $dq = 0$, and therefore in stage II,

$$dE = dw \tag{11.5}$$

$$C_V\, dT = -P\, dV = -\frac{nRT}{V} dV$$

$$C_V \int_{T_h}^{T_c} \frac{dT}{T} = -nR \int_{V_2}^{V_3} \frac{dV}{V}$$

$$C_V \ln \frac{T_c}{T_h} = -nR \ln \frac{V_3}{V_2}$$

where isolating T on one side of the equation and V on the other enabled us to use the logarithmic integral once again. The result for stage IV is similar, with the temperature limits reversed and the volume ratio being V_1/V_4. Using another rule of logarithms, $\ln(x_2/x_1) = -\ln(x_1/x_2)$, you can show by comparing stages II and IV that $V_3/V_2 = V_4/V_1$ or that $V_2/V_1 = V_3/V_4$. This simplifies Equation 11.4 by enabling the cancellation of the logarithmic factors, leaving

$$\varepsilon = \frac{q_h + q_c}{q_h} = \frac{T_h - T_c}{T_h} \tag{11.6}$$

The surprise in this result is that the efficiency of the Carnot engine depends only on the temperatures of the hot and cold reservoirs, and not on other details of the reversible paths. Moreover, this engine, the most efficient one that can be made, is *not* 100% efficient as long as T_h and T_c are finite. Any *real* engine, in turn, must be *even less* efficient than the Carnot engine, owing to inevitable irreversibilities and frictional losses in its operation. This leads to the conclusion that *it is impossible to convert heat to work with unit efficiency in a cyclic process.* This is our first statement of the second law of thermodynamics. Couched as it is in terms of heat and work, which in general are not state variables, this statement is not yet directly useful to us in our quest for a better criterion for spontaneity in a chemical reaction. But Equation 11.6 contains the seed that will grow into the criterion we seek.

A New State Variable: The Entropy S

Just as energy changes in an ideal gas not undergoing chemical change were found useful in understanding the implications of the first law, so the Carnot cycle, where again the system (the ideal gas) remains chemically unchanged, will help us to a clearer perception of what the second law implies. The relation between heat and temperature of Equation 11.6 can be rearranged to read

$$\frac{q_h}{T_h} + \frac{q_c}{T_c} = 0 \tag{11.7}$$

If we wanted to, we could break up the cycle into a series of little steps with heat dq; hence, Equation 11.7 implies that the quantity dq/T, integrated over a reversible cycle, is zero, or the *there exists a quantity* S, *whose change is calculated by* $\int dq_{rev}/T$, *that is conserved in a cycle*. Clausius in 1850 called this new state variable S, **entropy**, from the Greek ευτροπος, meaning "inner tendency." We will later consider how to set the zero of S (through the third law of thermodynamics); for now entropy will be defined by its change:

$$\Delta S = \int \frac{dq_{rev}}{T} \tag{11.8}$$

Because S is a state variable, *the path chosen for the integral must always be the reversible path, regardless of the actual path followed in the process under consideration.*

For a real, irreversible heat engine, the efficiency is less than $(T_h - T_c)/T_h$, and therefore

$$\frac{q_h}{T_h} + \frac{q_c}{T_c} < 0 \tag{11.9}$$

Note that ΔS is nonetheless still zero for the cycle, irrespective of the actual path. We can therefore state in general for an arbitrary cycle that

$$\Delta S \geq \int \frac{dq}{T} \tag{11.10}$$

since ΔS is zero and the integral in negative, or zero in the reversible limit. Since real, irreversible engines are the only kind that actually work, this inequality is a requirement that all engines must satisfy; that is, it is a criterion for the spontaneous operation of an engine.

Let's see how Equation 11.8 works for individual stages of a cycle. For an isothermal expansion, the $1/T$ factor is constant and can be taken outside the integral, giving

$$\Delta S = \frac{1}{T}\int dq_{rev} = \frac{q_{rev}}{T} = \frac{-w_{rev}}{T} = nR \ln \frac{V_2}{V_1} \tag{11.11}$$

Since $-w_{rev}$ is the maximum work that can be done by the system, q_{rev} is the maximum heat that can be added, and any spontaneous expansion will therefore have $q < q_{rev}$. Once again ΔS is independent of how the expansion actually occurs, and thus the inequality of Equation 11.10 holds for the isothermal stage as well. Note that here both ΔS and $\int dq/T$ are positive, but ΔS is greater. For a reversible adiabatic expansion, $q = 0$, and therefore $\Delta S = 0$. Reversible adiabatic processes are also **isentropic**. In the case of isothermal compressions, Equation 11.11 also applies, but because the final volume will be less than the initial and the ln goes negative, ΔS will be negative. We still have Equation 11.10 nevertheless, because now q is more negative than q_{rev}. It thus appears, and turns out to be the case in fact, that Equation 11.10 is true in general for any change of state, and it is a restatement in terms of entropy of Carnot's original conclusion. Equation 11.10 is a mathematical form of the second law of thermodynamics.

From the defining equation for ΔS (Equation 11.8), you can see that S has units of calories per kelvin or joules per kelvin, the same as those of heat capacity. Like E and H, S is an extensive quantity, increasing with the size of the system, as is made clear by the formula for an isothermal expansion, Equation 11.11. If 1 mol of ideal gas expands isothermally to twice its volume, Equation 11.11 yields

$$\Delta S = nR \ln 2 = (1 \text{ mol})(1.987 \text{ cal/mol K}) \ln 2 = 1.377 \text{ cal/K } (5.763 \text{ J/K})$$

This value holds even if the expansion occurs irreversibly against a constant $P_{ext} = P_2$, which yields, using Equation 11.2,

$$q/T = -w_{irrev}/T = nR(1 - \tfrac{1}{2}) = 0.994 \text{ cal/K } (4.157 \text{ J/K})$$

The inequality of Equation 11.10 is thus satisfied.

When an ideal gas is heated in a rigid container (V fixed), it cannot work on the surroundings, and therefore the internal energy and the temperature of the gas must rise. If the heating is done slowly enough that the temperature of the entire gas sample rises by little steps dT, the heating is reversible, and $dq_{rev} = C_V dT$. This yields an entropy change

$$\Delta S = \int_{T_1}^{T_2} \frac{C_V \, dT}{T} = C_V \ln \frac{T_2}{T_1} \qquad (11.12)$$

The integral was evaluated assuming a constant C_V. If the heating were carried out at constant P instead, Equation 11.12 would apply with C_V replaced by C_P. Whether the actual heating process is reversible or not, this relation yields the correct ΔS for given initial and final states. If both the volume and temperature change in a process, then, by visualizing an isothermal stage followed by an isochoric one, Equations 11.11 and 11.12 can be applied in succession, since ΔS must be independent of path.

Equation 11.10 refers, as do the other relationships developed in this section, to properties of the system only. To obtain the most general statement of the second law, one that is comparable in scope to the universal conservation of energy, we also need to consider the surroundings. This will require us to analyze, for the first time, the properties of an **isolated** system.

11.3 The Second Law of Thermodynamics: The Entropy of the Universe Tends to a Maximum

Figure 11.6 illustrates a critical experiment carried out by James Joule in 1843; in his own words:

> I provided another copper receiver (E) which had a capacity of 134 cubic inches. ... I had a piece D attached, in the center of which there was a bore ⅛ of an inch in diameter, which could be closed perfectly by means of a proper stopcock. ... Having filled the receiver R with about 22 atmospheres of dry air and having exhausted the receiver E by means of an air pump, I screwed them together and put them into a tin can containing 16½ lb. of water. The water was first thoroughly stirred, and its temperature taken by the same delicate thermometer which was made use of in the former experiments on mechanical equivalent of heat. The stopcock was then opened by means of a proper key, and the air allowed to pass from the full into the empty receiver until equilibrium was established between the two. Lastly the water was again stirred and its temperature carefully noted.

After many trials, Joule found that there was no measurable temperature change; the conclusion he drew was that the energy of the air is independent of its volume, as long as it does no work.* The rigid containers, and the evacuation of the receiver, ensured that no work was done by the gas, since it was expanding against a vacuum, $P_{ext} = 0$. So we have $q = 0$, $w = 0$, and $\Delta E = 0$, and, according to the first law, nothing has happened in the system, which behaved as though it were isolated.

Of course, something *did* happen: we know *from experience* that the gas spontaneoulsy expanded into the evacuated receiver until the pressures on the two sides

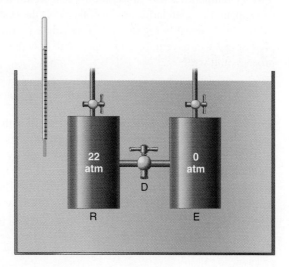

Figure 11.6
A schematic of the apparatus used by Joule to study gas expansion. See the text for Joule's own description of the experiments.

---

* In a later collaboration between Joule and William Thomson (Lord Kelvin) it was found that the gas actually cools slightly on expansion, owing to the work done in overcoming the weak intermolecular attractions in a real gas. Joule could not have seen this effect in the experiment described, because the heat capacity of his calorimeter was too large. The slight cooling, now known as the Joule–Thomson effect, is the major method used to liquefy air and other gases.

were equal. And even though this happened totally irreversibly, an entropy change, based on a reversible version of the process, an isothermal expansion, can be calculated from Equation 11.11. Note that ΔS is always *positive* for an isothermal expansion, since $V_2 > V_1$. As we have already observed in Chapter 10, changes in an isolated system are equivalent to changes in the universe, since the surroundings undergo no change. The entropy has therefore increased, and in this case we may say that it is not just the entropy of the system, but the entropy of the *universe* that has increased. Further, you know that the gas, like spilled water, would never return to its original container of its own accord in a million years; that unlikely event would correspond to an entropy *decrease*. This is just one example, but many more like it provide convincing proof that *the only spontaneous processes in an isolated system are those for which the entropy increases*. And since an isolated system represents the universe when it changes, we are led to write

$$\Delta S_{universe} \geq 0 \qquad (11.13)$$

for any spontaneous process, with the equality occurring for a system at equilibrium, that is, in the reversible limit. This is the all-encompassing statement we have been seeking, the most general way of formulating the **second law of thermodynamics.** Now we can state, as Clausius first did in 1850, the two laws of thermodynamics:

I. The energy of the universe is constant.

II. The entropy of the universe tends to a maximum.

Like a truly reversible process, you may (justly) regard a truly isolated system as a figment of the imagination. On a more practical level, no one would think of running a chemical reaction in isolation—that would indeed be useless. (In theory, one can expand the boundaries of what one calls "the system" until all the changes in entropy can be sensibly regarded as occurring within it, and Equation 11.13 must apply.) So we will have to be concerned with how the surroundings are affected by the process, and write

$$\Delta S_{universe} = \Delta S_{sys} + \Delta S_{surr} \geq 0 \qquad (11.13')$$

This is a nontrivial extension, because it implies that one can have a spontaneous process occurring in a system even though ΔS_{sys} may be negative, as long as ΔS_{surr} is positive and greater in magnitude, so as to satisfy Equation 11.13. This seemingly minor detail is actually of major importance; as we shall see, you and I would not be here if it were not for the possibility that $\Delta S_{sys} < 0$! But before we consider these real-world complications, it is helpful to explore the meaning of S itself. This will be done (you guessed it!) with a little help from our friends, the molecules.

11.4 Entropy and Probability: Boltzmann's Postulate $S = k \ln W$

Let's consider a small-scale molecular version of Joule's experiment, as illustrated in Figure 11.7. Suppose $N = 10$ molecules begin in the left-hand compartment, a *configuration* we will designate by $L = 10, R = 0$ or briefly $10L, 0R$, and the valve

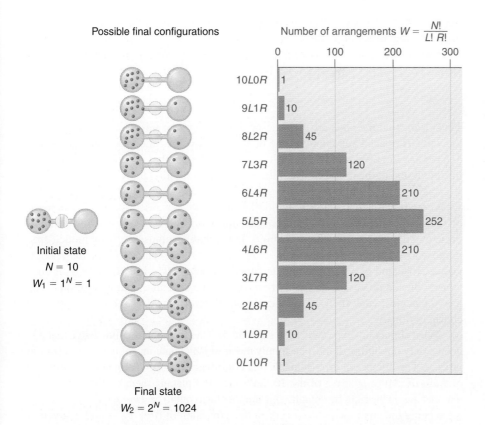

Figure 11.7
A 10-molecule simulation of the Joule experiment of Figure 11.6. The significant features are that (1) the final number of arrangements W_2 greatly exceeds the initial number W_1, and (2) the most likely final configuration has equal numbers of molecules on each side, as illustrated by the histogram on the right. The entropy change for an N-molecule simulation is $\Delta S = k_B \ln(W_2/W_1) = Nk_B \ln 2$.

is opened. At any finite temperature, the molecules are always in motion, as established by the kinetic molecular theory (Chapter 9). Since the molecular motion is random, *by chance* a particular molecule will find the passageway to the right-hand compartment and rattle through to the other side. This can be continued until we have five molecules on each side, a $5L,5R$ configuration, the expected final state of the system. But, *by chance,* a molecule, having arrived on the right, could equally well rattle through to the left again; if all the molecules on the right *by chance* did this, while those on the left remained where they were, we would end where we began. So the expected final state is just one of many possibilities ($4L,6R$; $6L,4R$; etc.), including the initial state $10L,0R$ prior to expansion. Since everything is happening by chance, the outcome must be a matter of the relative *probabilities* of these various ways of distributing the molecules between the two containers. The probability of a given configuration is proportional to the number of possible *arrangements* (ways of choosing which molecule goes where) that yield that configuration.

Let's begin by asking for the number of arrangements of the initial state $10L,0R$ *relative* to all possible arrangements available to the system after the valve is opened. For compartments of equal volume, we can take the initial state 1 to be a single possible arrangement for each molecule, namely on the left, and the final state 2 to be two possible arrangements, either left or right. (This is only true in a relative sense; actually there are *twice as many* arrangements per molecule for the expanded state.) In an ideal gas the molecules act independently, making the overall number of possibilities a *product* of the individual ones. If we denote by W the total number of arrangements possible, then

$$W_1 = 1\cdot1\cdot1\cdot1\cdot1\cdot1\cdot1\cdot1\cdot1\cdot1 = 1^{10} = 1$$

$$W_2 = 2\cdot2\cdot2\cdot2\cdot2\cdot2\cdot2\cdot2\cdot2\cdot2 = 2^{10} = 1024$$

This means that the odds are 1024 : 1 against finding the initial configuration once the valve is opened and enough time passes to allow the random molecular motion to have its effect. With only a modest increase in the number of molecules involved, the likelihood of the initial configuration diminishes rapidly; for $N = 100$ molecules, $W_2 = 2^{100} \approx 10^{30}$ while W_1 is still 1. If N is Avogadro's number $N_A = 6.02 \times 10^{23}$, W_2 is certainly much too large to fathom! This is why, in a gaseous system composed of such large numbers of molecules, we simply *never* observe it to undergo compression spontaneously—the probability is laughably minuscule. This model, first proposed by Ludwig Boltzmann in 1872, provides an explanation for the spontaneous expansion of a gas into new territory and why gases fully occupy their containers. Such a process is driven by two factors: the constant, random motion of the gas molecules, and their consequent natural tendency to explore all possible configurations.

What about our "expected" final state, $5L,5R$? We first must acknowledge that it is not the only possibility. It must be simply the *most likely* outcome of the randomization process. The number of arrangements that contribute to $5L,5R$ is equal to the number of ways of choosing any 5 of the 10 molecules and placing them on the left; the remainder are on the right by default. This number turns out to be a *binomial coefficient*—these being the coefficients that occur in the expansion of $(a + b)^N$—and is given in general by $N!/(L!R!)$, where $X! \equiv X(X - 1)(X - 2)\cdots 1$ is called X-factorial. So $W(5L,5R) = 10!/(5!5!) = 252$, while $W(10L,0R) = 10!/(10!0!) = W_1 = 1$, that is, the configuration $5L,5R$ is 252 times more likely than $10L,0R$, comprising about one-quarter of the total number of arrangements. Nearly equalized configurations, such as $6L,4R$, also have large W's [$10!/(6!4!) = 210$], but biased ones such as $8L,2R$ are less likely [$10!/(8!2!) = 45$]. Therefore the equalized configuration, and ones near it, are by far the most probable way to find the system after expansion. As discussed earlier, if we increase N, the more nearly equalized configurations are increasingly likely; for $N = 100$ the configuration $50L,50R$ is favored by more than a factor of 7 over $60L,40R$. Hence a macroscopic gas sample after expansion will show equal pressures in the two compartments within a tolerance much too small to be measurable.

This analysis is clearly connected with the second law and entropy, and it remained for Boltzmann to make the connection. Because entropies are additive while probabilities are multiplicative, it was natural to try a logarithmic relationship,

$$S = k \ln W \tag{11.14}$$

where k is a proportionality constant. An entropy change is then

$$\Delta S = S_2 - S_1 = k \ln \frac{W_2}{W_1} \tag{11.15}$$

Now, in Joule's experiment the final volume of the gas was twice the initial; if it had been, say, three times the initial volume (the receiver twice as large as the original gas container), then a gas molecule would be twice as likely to be on the right as on the left. (We could imagine partitioning this volume into two halves, now giving each molecule three possible arrangements instead of two, with two of these arrangements corresponding to being on the right.) Thus the number of arrangements per molecule is proportional to the volume available to it, and W for N molecules is therefore proportional to V^N, by analogy with our earlier reasoning. Equation 11.15 then becomes

$$\Delta S = k \ln \left(\frac{V_2}{V_1}\right)^N = Nk \ln \frac{V_2}{V_1} \qquad (11.16)$$

where we have made use of another property of logarithms, $\ln x^n = n \ln x$. Comparing Equation 11.16 with the formula for an isothermal expansion (Equation 11.11), Equation 11.16 becomes identical to it provided $k \equiv k_B = nR/N = R/N_A = 1.38 \times 10^{-23}$ J K^{-1}, where k_B is **Boltzmann's constant,** as introduced in Chapter 1 and used in the kinetic molecular theory of Chapter 9. Thus, Equation 11.14, which is engraved on Boltzmann's tombstone (in the form $S = k \log W$) at the University of Vienna, leads not only to the correct form for the entropy change, but provides a molecular explanation for the irreversible processes of nature. *The universal entropy must always increase because molecules, due to their random motion, are always tending toward states with the largest number of possible arrangements.*

From the small-scale example, you should appreciate that the macroscopic concept of entropy, drawn entirely from our experience with systems of "ordinary" size, in Boltzmann's (and our modern) view, exists only because of the incredibly numerous molecules that comprise the system. If Joule's experiment were done on a one-molecule sample, the initial and final states would be equally likely, and the single molecule would, on a very short timescale, periodically return to its initial state! This is what distinguishes entropy from the internal energy E or the enthalpy H; E and H are *mechanical* quantities that, as we saw in Chapter 10, can be understood by scaling up from the properties of individual molecules. Entropy S, on the other hand, is a purely *probabilistic* quantity that only has meaning for large numbers of molecules. In chemical reactions, the heat of reaction ΔH arises from the action of mechanical forces (breaking and making bonds); whereas, as we shall see, the entropy change reflects only how the number of ways of arranging the products compares to that of the reagents, which has nothing directly to do with the chemical forces.

11.5 The Third Law of Thermodynamics: The Entropy of a Perfect Crystal at $T = 0$ K Is Zero

Boltzmann's postulate (Equation 11.14) also provides a convenient and molecularly realistic way of defining the zero of entropy. All substances condense into liquids and, except for helium, finally into crystalline solids as they are cooled. As absolute zero, $T = 0$ K, is approached, the molecules of the crystal gradually lose their thermal energy and can no longer rattle about randomly. Provided the crystal has formed without defects (no missing atoms or dislocated rows or planes of atoms that get frozen into place), there is now only one possible way the atoms or molecules can occupy their spaces in the crystal lattice. In that case we can take the actual number of arrangements, $W = 1$. Since $\ln 1 = 0$, the entropy $S = 0$. In addition, because all random motion has ceased, the crystalline system has no tendency to change, and $\Delta S = 0$ for any possible process. Nothing can happen at all at 0 K! We can thus state the **third law of thermodynamics:**

$$S_0 = 0 \text{ for a perfect crystal} \qquad (11.17)$$

Because S is well defined only for a macroscopic crystal, it is also impossible by any practical means to cool such a crystal down to absolute zero, because this would mean that *every atom* in the crystal would have to be in its ground state. The lowest temperatures that have been achieved in the laboratory are in the nK range.

Absolute Entropies

With the aid of the third law it becomes possible to measure and tabulate **absolute entropies** for all substances at 298 K and 1 atm. The constant-pressure version of Equation 11.12 is

$$\Delta S = \int_{T_1}^{T_2} \frac{C_P \, dT}{T} \qquad (11.18)$$

If we take $T_1 = 0$ K and have measured C_P from near-zero to some higher temperature $T_2 = T$, Equation 11.18 can be used to obtain the absolute entropy S_T°. Generally, C_P must be kept under the integral sign because its temperature dependence cannot be neglected over a wide range, particularly near 0 K. If the substance is not a solid at 298 K, however, phase changes from solid to liquid (fusion) at $T = T_f$ and perhaps from liquid to gas (boiling) at $T = T_b$ must occur. This complicates the determination of S_{298}° in two ways. First, the heat capacities of the different phases will differ, requiring separate integrals for each. Second, there are enthalpy changes associated with the transitions themselves, given by $\Delta S = \Delta H/T$ for each transition. Because each new term represents heat added to the system, all terms, and therefore S_{298}°, are necessarily positive. Table 11.1 presents standard absolute entropies for the compounds whose heats of formation were given in Table 10.1; a more complete listing may be found in Appendix C. This table also includes entropies for the elements comprising those compounds, because these are no longer defined to be zero when using the third law.

This table contains many more lessons on the nature of entropy; let's consider a few of them. First, according to Boltzmann's interpretation, the fact that S° is always positive implies that more than one arrangement of the molecules of the system is possible—in fact the magnitudes of S° in the table imply that *many, many* arrangements exist. For example, consider $C(s,gr)$, which has the smallest S° in the table, 1.372 cal/mol K or 5.74 J/mol K. If we assume a 1-mol (12.0-g) sample of graphite, we then have from Equation 11.14,

$$\ln W = \frac{S^\circ}{k_B} = \frac{5.74 \text{ J/K}}{1.38 \times 10^{-23} \text{ J/K}} = 4.16 \times 10^{23}$$

$$W = e^{4.16 \times 10^{23}} = 6 \times 10^{10^{23}}$$

This number of arrangements, 6 followed by 10^{23} zeroes, dwarfs Avogadro's number; if the number W were printed on a tape as 600000000 . . . , and the tape were un-

TABLE 11.1

Standard absolute entropies $S°_{298}$ of some common elements and their compounds

Element or compound	$S°_{298}$ cal/mol K	$S°_{298}$ J/mol K	Element or compound	$S°_{298}$ cal/mol K	$S°_{298}$ J/mol K
$H_2(g)$	31.208	130.57	$NO(g)$	50.347	210.65
$O_{2(g)}$	49.003	205.03	$NO_2(g)$	57.35	240.0
$H_2O(l)$	16.71	69.91	$O_3(g)$	57.08	238.8
$H_2O(g)$	45.104	188.72	$S(s)$	7.60	31.8
$C(s,gr)$	1.372	5.74	$H_2S(g)$	49.16	205.7
$CO(g)$	47.219	197.56	$SO_2(g)$	59.30	248.1
$CO_2(g)$	51.06	213.6	$Cl_2(g)$	53.288	222.96
$CH_4(g)$	44.492	186.15	$HCl(g)$	44.646	186.80
$C_2H_2(g)$	48.00	200.8	$Na(s)$	12.24	51.2
$C_2H_4(g)$	52.45	219.5	$NaCl(s)$	17.24	72.1
$C_2H_6(g)$	54.85	229.5	$NH_4Cl(s)$	22.6	94.6
$CH_3OH(l)$	30.3	127.	$NaHCO_3(s)$	24.3	102.
$C_2H_5OH(l)$	38.4	161.	$Na_2CO_3(s)$	32.26	135.0
$C_6H_6(l)$	41.30	172.8	$Ca(s)$	9.90	41.4
$C_6H_6(g)$	64.34	269.2	$CaO(s)$	9.50	39.7
$C_6H_{12}O_6(g)$	69.0	289.	$CaCO_3(s)$	22.2	92.9
$i\text{-}C_8H_{18}(l)$	101.1	423.	$Fe(s)$	6.52	27.3
$Si(s)$	4.50	18.8	$Fe_2O_3(s)$	20.89	87.4
$SiO_2(s,quartz)$	10.00	41.8	$Al(s)$	6.77	28.3
$N_2(g)$	45.77	191.5	$Al_2O_3(s)$	12.17	50.9
$NH_3(g)$	45.97	192.3			

rolled, it would stretch 20,000 *light-years* (1 light-year ≈ 10^{16} m) into space! This stupendous number of arrangements arises just from the vibrational agitation of each carbon atom in the graphite lattice that exists at 298 K, due to its mechanical energy $\varepsilon \approx k_B T = 4 \times 10^{-21}$ J. Because the C atoms cannot move to new locations in the lattice, the arrangements that contribute to W arise instead from the ways of distributing the total energy of the sample among its atoms. Finding W this way turns out to be equivalent to counting spatial arrangements for gases, as we did in analyzing the Joule experiment, but is more general. This method, invented by Boltzmann and outlined in Appendix D, allows the restrictions imposed by quantum mechanics on the allowed energies of the atoms, as well as the atoms' inherent indistinguishability (factors Boltzmann wasn't aware of), to be accounted for. It also leads to the distribution of molecular population among the energy levels known as the **Boltzmann distribution,** which we invoked in Chapter 10 in discussing heat capacities. If we know what these allowed energies are, W and hence S can be calculated with extremely high accuracy, often better than $S°$ can be measured in the laboratory. Because this is especially true of gases, the gas-phase values of $S°$ in Table 11.1 are actually *calculated* from Boltzmann's formula, Equation 11.14.

If you inspect the entropies of the solid compounds in Table 11.1, say, the Na(s) entry and the four entries that follow it, you can see that $S°$ grows with the complexity of the units making up the solid. This is attributed to the increased number of degrees of freedom (three additional for each atom in the formula) and the consequent increase in the number of ways the atoms of the compound can be arranged. Further, comparison of solid, liquid, and gas phases of molecular compounds of similar complexity shows that $S°$ (solid) < $S°$ (liquid) < $S°$ (gas). In a solid, the molecules are fixed at their lattice sites, and can only jiggle around there, while in the liquid state the molecules are free to move to new locations, although they must remain in proximity to their neighbors. Liquids are more disordered and disorganized than solids because of this translational freedom, and they have higher entropy. In gases the molecules enjoy a great increase in this freedom, because they are no longer required to be near each other; this is the ultimate in disarray (absent breaking the molecule up into its constituent atoms), and the entropy is still higher. We thus associate high entropy with a high level of disorder; further, the second law can be interpreted as requiring an ever-greater degree of disorder in the universe. This interpretation is both useful and disturbing, and we consider both attributes in what lies ahead. But first we must consider defining standard entropies for aqueous solutes, in order to bring the entropy idea into the working laboratory.

When we depart from the world of pure compounds into the much more real world of mixtures—of solutes in solutions—it is not so easy to apply the third law, because of the interaction between solute and solvent. Besides the tremendous and hard-to-measure energetic effect of this interaction, which forced us to adopt a separate convention for ΔH_f's of solutes, there is also a profound effect on the *arrangement* of the solvent molecules in the neighborhood of the solute, particularly for solute ions. For example, water molecules will tend to form a shell around a cation such as Ca^{2+} with their oxygen ends pointing toward the ion, a much different and more constrained arrangement than pure water would adopt. This configurational change contributes to the entropy in a way that is impossible to isolate by experiment, so again we are forced to adopt a convention for aqueous solutions. We set

$$S°_{298}(H^+(aq)) = 0$$

and define all other entropies by measuring ΔS in reactions involving the species of interest with $H^+(aq)$. As for $\Delta H_f°$, the superscript zero here indicates a standard concentration, 1 M. Measuring ΔS for aqueous reactions cannot be done in the same way as that used to construct Table 11.1, however; we will learn how ΔS is obtained in the next chapter. For now, we need the results of such measurements, and some selected values are presented in Table 11.2; as usual, a more complete listing can be found in Appendix C.

Some of the entries in this table are negative, but this merely reflects a state of lower entropy than that of $H^+(aq)$. Note that the multiply charged ions tend to have the lowest $S°$'s; this arises because such ions induce a greater degree of organization in the water solvent shell, in some case attracting a second, outer shell.

TABLE 11.2
Standard entropies for common molecules and ions in aqueous solution

Species	$S°_{298}$ cal/mol K	$S°_{298}$ J/mol K	Species	$S°_{298}$ cal/mol K	$S°_{298}$ J/mol K
$Li^+(aq)$	3.2	13.4	$OH^-(aq)$	−2.57	−10.75
$Na^+(aq)$	14.1	59.0	$Cl^-(aq)$	13.5	56.5
$K(aq)$	24.5	102.5	$Br^-(aq)$	19.7	82.4
$Mg^{2+}(aq)$	−33.0	−138.	$I^-(aq)$	26.6	111.
$Ca^{2+}(aq)$	−12.7	−53.1	$HCO_3^-(aq)$	21.8	9.2
$Ba^{2+}(aq)$	2.3	10.	$CO_3^{2-}(aq)$	−13.6	−57.
$Mn^{2+}(aq)$	−17.6	−74.	$CH_3COOH(aq)$	42.7	179.
$Fe^{2+}(aq)$	−32.9	−138.	$CH_3COO^-(aq)$	20.7	86.6
$Fe^{3+}(aq)$	−75.5	−316.	$NO_2^-(aq)$	29.4	123.
$Cu^{2+}(aq)$	−23.8	−99.6	$NO_3^-(aq)$	35.0	146.
$Zn^{2+}(aq)$	−26.8	−112.1	$PO_4^{3-}(aq)$	−52.	−218.
$Al^{3+}(aq)$	−76.9	−322.	$SO_3^{2-}(aq)$	−7.	−29.
$NH_4^+(aq)$	27.1	113.	$SO_4^{2-}(aq)$	4.8	20.
$CO_2(aq)$	28.1	118.	$ClO^-(aq)$	10	42.
$NH_3(aq)$	26.6	111.	$ClO_3^-(aq)$	38.8	162.
$Br_2(aq)$	31.2	131.	$Cr_2O_7^{2-}(aq)$	62.6	262.
$I_2(aq)$	32.8	137.	$MnO_4^-(aq)$	45.7	191.

Entropy Changes in Chemical Reactions

It is normally impossible to predict, based on general principles, the heat of a reaction; you need a table of bond energies or heats of formation, along with Hess's law, even to say whether a given reaction is exothermic or endothermic, that is, to state the sign of ΔH. But thanks to Boltzmann's view of entropy, it *is* possible in many cases to deduce the sign of ΔS without further information such as that given in Table 11.1. You should note that here we are thinking about reactions occurring in a well-defined system, allowing the possibility implied by Equation 11.13' that the sign of ΔS_{sys} can be either positive or negative. The general principle behind the prediction of the sign of ΔS is simple: ΔS *will always be positive when the system proceeds to a less ordered state,* that is, for order ⟶ disorder.

Examples are the best way to see how this principle applies. For gas-producing reactions such as carbonate–acid

$$Na_2CO_3(s) + 2HCl(aq) \longrightarrow 2NaCl(aq) + CO_2(g) + H_2O(l)$$

we expect $\Delta S > 0$ due to the much greater entropy of the gaseous state. In general, $\Delta n_{gas} > 0$ yields $\Delta S > 0$. On the other hand, our favorite reaction $2H_2(g) + O_2(g) \longrightarrow 2H_2O(l)$ must have $\Delta S < 0$, because it makes 3 mol of gas disappear, $\Delta n_{gas} = −3$.

Water synthesis is also an example of the formation of larger molecules from smaller ones, a net creation of chemical bonds, the same sort of reactions that allow collections of extremely large and complex molecules like you and me to form, for which we expect $\Delta S < 0$ in general. For reactions in aqueous solution, reorganization of the solvent shell around each species makes it harder to predict the sign of ΔS. The reaction $H^+(aq) + OH^-(aq) \longrightarrow H_2O(l)$ reaction, for example, shows $\Delta S > 0$ despite the appearance of forming a larger species from smaller ones; the reagent ions in this case engender dramatic solvent organization that is lost on neutralization. Precipitation reactions, such as $Ca^{2+}(aq) + CO_3^{2-}(aq) \longrightarrow CaCO_3(s)$, on the other hand, uniformly show $\Delta S < 0$, the organization of the crystalline solid always being greater than that of free ions in solution.

Because S is a state variable, quantitative calculations of $\Delta S°$ for a reaction can be made from a table of $S°$'s using a rule analogous to that for obtaining $\Delta H°$ from $\Delta H_f°$'s,

$$\Delta S° = \sum_P \nu_P S°(P) - \sum_R \nu_R S°(R) \tag{11.19}$$

where as before P stands for products, R for reagents, and the ν's are stoichiometric coefficients. For example, $\Delta S°$ for the water synthesis reaction is given by

$$\Delta S° = 2S°(H_2O(l)) - [2S°(H_2(g)) + S°(O_2(g))]$$
$$= 2(16.71 \text{ cal/mol K}) - [2(31.21 \text{ cal/mol K}) + 49.00 \text{ cal/mol K}]$$
$$= -78.00 \text{ cal/mol K}$$

using the $S°$ values from Table 11.1. The negative value was anticipated from our qualitative analysis. You can also mix and match values from Tables 11.1 and 11.2, as the following example illustrates.

EXAMPLE 11.1

Make a qualitative assessment of the expected entropy change for the $Na(s) + H_2O(l)$ reaction, and check it by using the data of Tables 11.1 and 11.2 to calculate $\Delta S°$.

Solution:
The balanced chemical equation (Equation 5.9) is

$$2Na(s) + 2H_2O(l) \longrightarrow 2NaOH(aq) + H_2(g)$$

Based on our earlier thinking, this reaction is expected to have $\Delta S° > 0$ because H_2 gas is produced where none existed before, that is, $\Delta n_{gas} = +1$. The solvated ions $Na^+(aq)$ and $OH^-(aq)$ are products, however, and the organization of the solvent will lower the entropy of the products. In addition, the data of Table 11.1 show that $S°(H_2(g))$ is substantially smaller than that of the other gases given there. (This property is a result of the low mass of H_2, which reduces the number of energy levels available to the molecules and, hence, the number of arrangements.) The initial prediction is therefore less certain. Calculating $\Delta S°$ using Equation 11.19 and the data in Tables 11.1 and 11.2, we get

$$\Delta S° = 2S°(Na^+(aq)) + 2S°(OH^-(aq)) + S°(H_2(g)) - [2S°(Na(s)) + 2S°(H_2O(l))]$$
$$= 2(14.1) + 2(-2.57) + 31.21 - [2(12.24) + 2(16.71)]$$
$$= -3.6 \text{ cal/mol K}$$

where we have separated NaOH into its ions. Contrary to our prediction, we find $\Delta S < 0$, owing to the peculiarities just noted.

You now have seen how entropy changes can be evaluated and also some typical magnitudes of ΔS for the processes of interest to us here. Just as in the case of energy changes, for ideal gases undergoing changes of state the ΔS's are comparatively small and predictable; whereas ΔS for chemical reactions, while somewhat more predictable than ΔH, can also take on a wide range of values. You might think it peculiar that, despite the second law's requirement that the overall entropy always be on the rise, three of the five chemical examples we've just considered had *negative* entropy changes even though they were very spontaneous and facile reactions. Josiah Willard Gibbs, holder of the first Ph.D. in engineering awarded in the United States and professor of mathematical physics at Yale, thought so, too, in the early 1870s. This led him to an amalgamation of the first and second laws that has shaped the way chemists think about and analyze chemical reactions ever since.

11.6 Focusing on the System: The Gibbs Free Energy $G = H - TS$

We began our quest for a complete law of spontaneity with the observation of Berthelot that nearly all reactions that happen naturally are exothermic. This is a reflection of the properties of the individual reacting molecules, which in most cases are busy making new chemical bonds that are stronger than the old ones. Exothermic reactions also often involve making *more* bonds as well, that is, making big molecules from small ones. In either case $\Delta H < 0$ means that *a more ordered state* is being formed. This is just the opposite of the natural tendency implied by entropy, whose constant increase implies a state of greater disorder. Thus *within a system* a tension exists between chemical forces, which promote cohesion and organization, and entropic tendencies, which drive the system toward a more random arrangement of its parts. A starkly simple example will illustrate this. Consider the dissociation of molecular hydrogen into its atoms,

$$H_2(g) \longrightarrow 2\,H(g)$$

The experimental fact is that, at NTP, a 1-mol sample of hydrogen gas *will not contain a single pair of free hydrogen atoms*. It is all H_2, despite the fact that the entropy of the system (and therefore the universe if the gas were isolated) would be greatly increased were all the bonds to break. Dissociation does not occur because it is opposed by the force of covalent bonding—in this case, by the bond energy of H_2. Yet we also know—from solar blackbody radiation and Fraunhofer's dark lines—that on the 5000 K surface of the sun, hydrogen is present exclusively in atomic form. So at high temperature, dissociation of chemical bonds becomes the order of the day. Why is this? If you look back to Boltzmann's analysis of the Joule experiment, you will find that what drives the quest for more probable arrangements is the random thermal motion of the molecules in the system. As T increases, this molecular thrashing-about becomes more violent, until finally the chains of bondage are broken and atomization sets in. That, in a nutshell, is what Gibbs was the first to see clearly, and all that remains for us is to turn this line of reasoning into a quantitative statement.

Equation 11.10 states the second law in terms of system quantities only. If we assume constant temperature, then the $1/T$ factor can be taken outside the integral, and we get

$$\Delta S \geq \frac{q}{T} \qquad (11.20)$$

If we also assume the pressure to be constant, then $q = q_P = \Delta H$, which is a consequence of the first law. If we now substitute ΔH for q, multiply through by T, transpose the ΔH to the left, and then multiply by -1—which reverses the sense of the inequality—we find

$$\Delta H - T\Delta S \leq 0 \qquad \text{at constant } T \text{ and } P \qquad (11.21)$$

where H, T, and S are all state variables. The inequality must refer to the change in a new state variable, the **Gibbs free energy** G, defined by

$$G = H - TS \qquad (11.22)$$

This makes Equation 11.21 read simply

$$\Delta G \leq 0 \qquad \text{at constant } T \text{ and } P \qquad (11.23)$$

Equation 11.23 is a combined statement of the first and second laws of thermodynamics regarding the spontaneity of any process, couched in terms of system properties only, and it is the general criterion we have been seeking. Gibbs's original derivation was much more elegant and rigorous, involving the use of differentials such as dG, but the result is still Equation 11.23.

Gibbs's new inequality corrects Berthelot's hypothesis $\Delta H \leq 0$ for the influence of our new state variable S; since

$$\Delta G = \Delta H - T\Delta S \qquad (11.24)$$

at constant T, the tension just described is rendered quantitative. Equations 11.23 and 11.24 together show that a chemical reaction is encouraged to occur both when it is exothermic, $\Delta H < 0$, and when it creates greater randomness, $\Delta S > 0$, because both of these conditions tend to make $\Delta G < 0$. Furthermore, the influence of the entropy term grows linearly with temperature. When the signs of ΔH and ΔS are the same—in our example, both positive—a competition exists, in which ΔH will dominate at low temperature and ΔS at high. Gibbs's analysis thus provides a natural scheme for accounting for the observations on hydrogen dissociation—its occurrence in the sun (and other stars) but not on earth.

The $\Delta G = 0$ condition was included in Equation 11.23 to allow reversible or equilibrium processes to be described as well. Although a reversible process as considered in, say, the Carnot cycle is an idealized limit not attainable in practice, in chemistry equilibrium generally refers to the stasis that ensues when a chemical reaction has run its course. Then $\Delta G = 0$ becomes a means of predicting what the system will be like after reaction, a very powerful analytical tool that we will spend the next three chapters wielding. For now we consider the new control Gibbs's criterion allows us to exercise over reaction spontaneity itself.

Temperature Control of Spontaneity

First, let's ask why Gibbs called G (for him it was ψ; the G and the Gibbs came more recently) the *free* energy. Consider a typical chemical reaction, such as water synthesis, which is exothermic but with $\Delta S < 0$; that is, it makes more bonds that result in more order. The value of ΔH is negative but $-T\Delta S$ is positive, making ΔG less negative than ΔH according to Equation 11.24. The reaction liberates heat ΔH, but some of this energy is used up in creating the increased order, leaving only ΔG *free* for use in performing practical tasks, that is, for doing useful work. For $2H_2(g) + O_2(g) \longrightarrow 2H_2O(l)$, $\Delta H°$ is just twice the heat of formation of water from Table 10.1, -136.63 kcal (-571.7 kJ), while $\Delta S°$ was computed earlier as -78.00 cal/K (-326.4 J/K), yielding

$$\Delta G° = -136.63 \text{ kcal} - (298 \text{ K})(-78.00 \text{ cal/K})(10^{-3} \text{ kcal/cal})$$
$$= -113.39 \text{ kcal} (-474.4 \text{ kJ})$$

In the simpler case of an isothermal expansion of an ideal gas, $\Delta H = 0$ and $\Delta G = -T\Delta S = -q_{rev} = w_{rev}$; that is, ΔG *represents the maximum work available* in the process, a result that turns out to be true in general. Thus if hydrogen were to be used as a fuel to propel an automobile for example, the maximum useful energy obtainable is ΔG rather than ΔH; the closest we have come to this ideal limit is through the use of *fuel cells,* to be touched on in Chapter 13.

System temperature is a variable we can control, and its appearance as a factor in Equation 11.24 implies the potential for controlling the spontaneity of a process. In fact you have made use of, or have benefited from, this ability countless times in your life. The simple act of boiling water on the stove, useful in preparing beverages and cooking foods, is a matter of taking a process, $H_2O(l) \longrightarrow H_2O(g)$, that is not spontaneous under standard conditions at room temperature, and increasing T until it becomes so. Boiling is endothermic but has $\Delta S° > 0$, with $\Delta G° > 0$ at room temperature, and increasing T to 373 K (100°C) brings about $\Delta G° < 0$ by increasing the entropic term. Another everyday application lies in cooking and baking; an ingredient useful for creating light, soft baked goods is baking soda, sodium bicarbonate $NaHCO_3(s)$. Baking soda works by decomposing into $CO_2(g)$ at baking temperature, filling the baking dough or batter with tiny gas bubbles that lower its density. The chemical reaction is

$$2NaHCO_3(s) \longrightarrow Na_2CO_3(s) + CO_2(g) + H_2O(g)$$

From the $\Delta H°_f$'s of Table 10.1, you should have found $\Delta H° = +32.41$ kcal ($+135.6$ kJ) if you were assigned Exercise 10.31. You can guess that $\Delta S° > 0$ from $\Delta n_{gas} = +2$; a quantitative calculation using Table 11.1 yields $\Delta S° = +79.8$ cal/K ($+334$ J/K). Further, based just on everyday experience you can also guess that $\Delta G° > 0$ for this reaction at normal temperatures; otherwise baking soda could not be sold in loosely sealed boxes or stored for long periods. The calculation using Equation 11.24 at 298 K yields $\Delta G° = +8.63$ kcal ($+36.1$ kJ). So, like boiling, the reaction is endothermic, entropy producing, and normally nonspontaneous, but becomes spontaneous at sufficiently high temperature.

To make a quantitative estimate of the temperature necessary to "turn the reaction around," we will assume that $\Delta H°$ and $\Delta S°$ are independent of T, so that we can use their 298 K values. Our analysis at the end of Chapter 10 showed that $\Delta H°$

Figure 11.8
Temperature dependence of $\Delta G°$ for the decomposition of baking soda. The slope of the line is $-\Delta S°$ for the reaction. For $T > T_{ta} = 406$ K, the reaction becomes spontaneous under standard conditions [1 atm partial pressure of both $CO_2(g)$ and $H_2O(g)$].

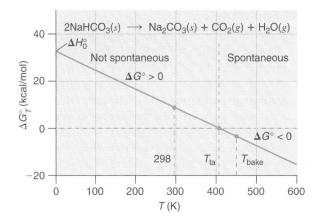

depends only weakly on T; a similar situation holds for $\Delta S°$. Therefore, as long as a relatively small range of T is considered, we can estimate $\Delta G°_T$ as a linear function of T using the standard $\Delta H°$ and $\Delta S°$. The turnaround temperature, T_{ta}, occurs when $\Delta G°$ becomes zero; as illustrated in Figure 11.8, for any $T > T_{ta}$, $\Delta G° < 0$, and the decomposition is spontaneous. Using Equation 11.24 with $\Delta G° = 0$ yields

$$T_{ta} \approx \frac{\Delta H°}{\Delta S°} \qquad (11.25)$$

For the baking soda reaction, putting in numbers yields $T_{ta} = 406$ K (133°C or 271°F). Normal baking temperatures (generally 350°F or 177°C) have to be higher than this to ensure an appreciable degree of spontaneity, as well as to accelerate other reactions that occur in baking. Note that Equation 11.25 only makes sense when the signs of $\Delta H°$ and $\Delta S°$ are the same; otherwise, $\Delta G°$ never changes sign. For molecular synthesis, where both $\Delta H°$ and $\Delta S°$ are negative, this relation predicts that beyond T_{ta}, decomposition rather than building up will occur, and the production of complex structures will be impossible at sufficiently high T. Just as for ΔH, the ΔG for a reverse reaction—in this case the decomposition of a large molecule back to small ones—is the negative of ΔG for the forward reaction; thus if you find $\Delta G° > 0$ for a reaction, then its reverse is spontaneous.

It is important to observe, before we leave this topic, that the fact that a reaction is found to be spontaneous does *not* mean it must occur quickly. H_2/O_2 mixtures can remain unreacted for years in the absence of a spark or flame despite the large, negative $\Delta G°$ found earlier. The *rate* at which products are formed cannot be determined from thermodynamic analysis. Rates of reactions will be the subject of Chapter 15.

11.7 Use and Misuse of the Second Law

Gibbs's analysis has provided a marvelously comprehensive answer to Newton's query at the start of this chapter. There is no doubt that "the order and beauty that we see in the world" of nature is due to the formation of large, complex molecular structures whose interactions give rise to the wonderful patterns of color and form in the living things we see around us. Such structures are clearly able to form despite the higher degree of order that accompanies their formation, due to the great strength of chemical bonding relative to the random thermal motions, the molecular tumbling and jiggling induced by heat. This situation, though, is all a matter of temperature; for that reason there is little hope of finding life as we know it on the surface of Venus, where a runaway greenhouse effect has produced sizzling temperatures. On the other hand, *some* tumbling and jiggling are absolutely essential to life. I could not have written these words, nor could you read them, were it not for billions of chemical reactions happening within our cells, all of which rely on molecular motion for their occurrence. In view of this requirement, there is also only a faint chance that life now exists on Mars, where even the warmest days are cold by our standards, and the nights are unbearably bitter. The earth is a very special place, with temperatures that are well adjusted for large molecules not only to exist, but also to come into being through chemical reactions, and to react further. To a great extent, this beneficent temperature range is dictated by the existence of water in liquid form, which apparently is not to be found on either the Venetian or Martian surfaces.

Entropy and Time

So powerful is this logical structure of thermodynamics, that it has been tempting—often overwhelmingly so—to use it to interpret or analyze various phenomena of our experience. The most pervasive is the passing of time, indeed the very nature of time itself. What do the notions of *past, present* ("now"), and *future* really mean, and how did they arise? Humans mark time by the passage of events, happenings, and the second law tells us that nothing "happens" unless it is accompanied by an increase in S_{univ}. Thus, as Sir Arthur Eddington observed, entropy is time's arrow. The disturbing aspect of this view is that it suggests that the universe is "running down"; some scientists have predicted that eventually the universe will suffer a "heat death," in which all matter will be at the same (high) temperature, at equilibrium, and all spontaneity will cease. Others, such as the physicist Stephen Hawking (holder of the same professorial chair at Cambridge University once held by Newton), have suggested that the second law goes hand in hand with the apparent expansion of the universe (a part of the so-called Big Bang theory). Should the universe ever stop expanding and begin to contract (toward an eventual "Big Crunch"), the sign of the second law inequality would be reversed, and time would run backwards! A broken teacup would reassemble itself and hop back up on the table; the newspaper would contain tomorrow's stock market prices; people would arise from their graves, grow young, and die by disappearing into the womb; and everyone would think this was perfectly normal. Chemists, who recognize the molecular nature of things and the true meaning of temperature, naturally scoff at this last scenario. Molecules in a teacup do not care whether the universe is expanding, contracting, or neither.

Entropy and Evolution

It has also been argued that the second law stands opposed to, in fact invalidates, the theory of evolution, whose basic tenet is that the present complex life forms on earth evolved from simpler forms in the distant past. This view also stems from the "running down" implication of the constant increase in S_{univ}, which is taken to mean that everything is becoming more disordered, and that life as we know it is gradually degenerating from a former, more perfect state. The problem with this reasoning is that it appears to confuse the earth, a tiny speck floating in an oceanic universe, with the universe itself. The earth is a system, with infinite surroundings, and it is Gibbs's combined first- and-second-law picture that then applies. The temperature of the earth is such that chemical forces, which form the focus of this entire book, dominate the randomizing tendencies of molecular motion—which is all that underlies the concept of entropy—and therefore the coming into being of order out of disorder is not at all mysterious. It's all due to Coulomb's law! Hence, from a molecular chemistry viewpoint, the evolution from the simple to the complex is to be expected, rather than its opposite, even if one were to ignore or discredit the fossil record.

Entropy and Society

If human beings could be regarded as molecules, Gibbs's analysis would also apply to societal behavior. The present-day scope and variety of statistical samplings of people's characteristics, habits, and opinions is truly mind boggling and could provide the basis for such analyses. In general, the enthalpy term corresponds to the rule of law, according to which a society bands together under a set of agreed behaviors for the mutual benefit of its members, whereas entropy is represented by the members' free choice of behaviors, including those that violate the law. Unlike molecules, however, people's actions are determined not at random, but by a complex set of internal tendencies and external influences that vary from one individual to another. While a sample of gas molecules will show very predictable behavior in its bulk properties, the same cannot be said for a sample of people. Economic predictions, for example, are notoriously difficult to make owing to the quixotic, often lemming-like responses of human beings. Further, a molecule with energy on the low end of the Boltzmann distribution is quite likely to gain significant energy through encounters with other molecules in rather short order, but the same is clearly untrue for a person at the low end of the income distribution. Molecular society is much less stratified and segregated than human society. These factors make it highly unlikely that any apparent decline in "moral values" (the supralegal enthalpic "glue" holding society together) can be thought of as a natural occurrence associated with an inevitable increase in "societal entropy." History is rife with cyclical falls and rises that make such a view untenable.

Where It All Leads

In chemistry, predicting whether a given reaction will occur is a powerful ability, but we have not yet made a quantitative estimate of how far that reaction will go toward complete conversion into products. That is, we have not yet exploited the *degree* of spontaneity inherent in the magnitude of ΔG. We have seen "weak" reactions in Chapter 5; for example, we found that some acids are strong and others are weak. The synthesis of ammonia from hydrogen and nitrogen, which is of global agricultural importance as the initial step in fertilizer production, frustratingly refuses to go to completion, and has to be run in cycles. These are cases of low degree of spontaneity, resulting in low conversion of reagents to products. However, even after a reaction is over, we are not always helpless to alter the state of the system. We have seen the effect of changing T on the direction and degree of spontaneity of a reaction as reflected in its ΔG. What we will find is that by changing conditions for a completed reaction—a system that has reached equilibrium and has $\Delta G = 0$—we can create a nonzero ΔG for a further change in the proportion of products in the system. These challenges will be taken up, and the results applied, in the next three chapters.

SUMMARY

Natural events, including chemical reactions, have a natural direction of occurrence. Being energetically downhill, that is, exothermic, is clearly a factor in making a reaction spontaneous, as postulated by Berthelot; but examples of endothermic yet naturally occurring processes force the conclusion that another factor is also important. To discover the nature of this factor, the concept of spontaneity is explored more quantitatively, in the context of P-V work w in an ideal gas. The isothermal expansion of a gas will occur spontaneously and will be **irreversible** provided the internal pressure of the gas P exceeds the external pressure P_{ext}. If P_{ext} is gradually increased toward P, the process becomes less and less spontaneous, until when P_{ext} is infinitesimally less than P the expansion becomes **reversible**. Reversibility means that an infinitesimal increase in P_{ext} can cause the expansion to become a compression instead; this is a characteristic of a system at **equilibrium**. A graph of P_{ext} versus V, called a P-V diagram, displays these differences, and also shows that the work done by the gas, which is the area under the P_{ext} versus V curve, is maximized in the reversible limit. This work is given by $-w_{rev} = nRT \ln(V_2/V_1)$. Work done on a gas in a reversible compression is the minimum required and is given by the same formula.

The concept of reversibility as the ideal limit of a spontaneous process was put to use by Carnot in constructing a reversible cycle that now bears his name. Conceived as the ideal engine, the **Carnot cycle** yields maximum net work for a given amount of heat q_h added from a hot reservoir. By analyzing the heat and work involved in each of four stages, Carnot showed that the engine's **efficiency** is given by $\varepsilon = (q_h + q_c)/q_h = (T_h - T_c)/T_h$, where q_c is heat given off at a cold reservoir, and T_h and T_c are the temperatures of these reservoirs. Since the operating temperatures must both be finite, *it is therefore impossible to convert heat to work with 100% efficiency in a cycle.* This is a statement of a new law not contained in the first law, and is one way of stating the **second law of thermodynamics**.

Carnot's expression relating heat to temperature can be rearranged to yield $(q_h/T_h) + (q_c/T_c) = 0$. If the heat is envisioned as being transferred reversibly in tiny amounts dq_{rev}, this implies $\oint dq_{rev}/T = 0$, or that there exists a state variable S, whose change is given by $\Delta S = \int dq_{rev}/T$, that is conserved in the cycle. State variable S is called the **entropy**, and its change ΔS is always calculated along the reversible path regardless of the actual process considered. For a real engine, or any real, irreversible process, it is found that $\Delta S \geq \int dq/T$, where the equals sign includes the reversible limit. This is a second, mathematical way of stating the second law. For an isothermal expansion of an ideal gas, $\Delta S = nR \ln(V_2/V_1)$, whereas for an adiabatic ($q = 0$) expansion, $\Delta S = 0$. For an isochoric ($\Delta V = 0$) change in temperature, $\Delta S = C_V \ln(T_2/T_1)$, whereas for an isobaric ($\Delta P = 0$) change in T, C_V is replaced by C_P in this formula. If a process produces changes in both T and V (or T and P), it may be broken down into isochoric (or isobaric) and isothermal stages, and the formulas already given for these stages employed, since ΔS must be independent of path.

The most general way of stating the second law is suggested by an analysis of an experiment conducted by Joule, in which air expands into an evacuated container. Joule established that (very nearly) $q = 0$, $w = 0$, and $\Delta E = 0$ for this process; that is, this expansion is not describable at all by the first law. From the formula for an isothermal expansion, it is found that $\Delta S > 0$. Because in this case the system is isolated, this is a *universal* entropy increase, and leads us to state the second law of thermodynamics in the universal form, $\Delta S_{univ} \geq 0$. Then the two laws of thermodynamics may be given as follows:

I. The energy of the universe is constant.

II. The entropy of the universe tends to a maximum.

For systems that are not isolated, the second law requires that $\Delta S_{sys} + \Delta S_{surr} \geq 0$, and therefore it is possible for a process to have $\Delta S_{sys} < 0$ and still be spontaneous.

A molecular analysis of the second law as applied to Joule's experiment is enabled by Boltzmann's postulate $S = k \ln W$, where W is the number of arrangements open to a system of N molecules and $k = k_B$ is Boltzmann's constant. This analysis shows that the requirement of increasing entropy in an isolated system (and hence in the universe) corresponds to the system of molecules seeking out states with the largest possible number of configurations, and finding the *most probable* configuration, in which the number of possible arrangements of the molecules W is maximized. This tendency to assume the most probable configuration is driven by the random motion of the individual molecules in the system at finite absolute temperature T. As the number of molecules in the system is increased, the most probable configuration becomes overwhelmingly so, leading to the apparent irreversibility we observe in naturally occurring events. Boltzmann's interpretation of S, which is now universally accepted, means that S, unlike E or H, is a purely probabilistic quantity that relies for its valid definition on the large number of molecules that comprises any macroscopic system.

Boltzmann's relation $S = k \ln W$ also leads to a way of defining **absolute entropies.** As any substance is cooled down, it eventually becomes (except for He) a crystalline solid, and the ability of the molecules to change their configuration is gradually lost, until at $T = 0$ K all random motion ceases, and there is only one possible arrangement, $W = 1$. Since $\ln 1 = 0$, we arrive at what is now known as the **third law of thermodynamics,** $S_0 = 0$ *for a perfect crystal.* Achieving 0 K involves bringing *every molecule* in the crystal to its lowest energy state, and is therefore impossible for macroscopic samples; in addition, no spontaneous changes in the system can occur if the molecules cannot move, and therefore $\Delta S_0 = 0$ for any process whatsoever. This leads to a whimsical, but nonetheless correct, way of stating the three laws of thermodynamics:

I. You can't win, you can only break even.
II. You can only break even at absolute zero.
III. You can't reach absolute zero.

Absolute entropies are formulated by assuming $S_0 = 0$, and invoking the formula given earlier for ΔS in temperature changes to calcutate $S_{298}°$, from experimental heat capacities or from the extension of Boltzmann's analysis given in Appendix D. In general, we find that gases possess the largest $S°$'s, followed by liquids and solids, and that $S°$ increases with molecular complexity, both observations consistent with Boltzmann's probabilistic interpretation. When applied to mixtures and solutions, where solute–solvent interactions may produce large changes in the arrangement of the solvent shell around a solute species, absolute entropies are presently impossible to assign, and the standard $S_{298}°(H^+(aq)) = 0$ is adopted, analogous to the convention used for $\Delta H_f°$ in solution. For aqueous solutes, the superscript ° indicates a standard concentration, 1 M.

With some thought it is possible to make qualitative predictions of the *sign* of $\Delta S°$ for chemical reactions using the general properties of S just noted. In addition, because S is a state variable, the rule $\Delta S° = \Sigma_P v_P S°(P) - \Sigma_R v_R S°(R)$ can be used to obtain quantitative entropy changes for chemical reactions from absolute or standard entropies.

Because many reactions that are found to be spontaneous have $\Delta S° < 0$, it is not sufficient to consider only the entropy change in the system to determine spontaneity. In the late 19th century Gibbs introduced the concept of **free energy** G, defined by $G = H - TS$, and showed that *the general criterion for spontaneity is* $\Delta G = \Delta H - T\Delta S \leq 0$ *at constant T and P.* The equals sign is included to account for systems at equilibrium. Gibbs's rule contains both the first and second laws of thermodynamics, and has proven to be the correct way to analyze spontaneous physical or chemical change. The free energy concept involves a tension between cohesive bonding forces (ΔH), which tend to organize the atoms in the system into molecules, and probabilistic tendencies ($-T\Delta S$), which oppose this organization.

In a typical chemical reaction that is exothermic but endoentropic, the energy represented by ΔH cannot be wholly converted to useful purposes; some of it is tied up in maintaining a greater degree of order in the system, and it is only the free energy ΔG that is useful. In general, ΔG represents the maximum work available from the system.

The appearance of temperature in the criterion for spontaneity, along with the relatively weak dependence of ΔH and ΔS on T, implies that temperature changes can control both whether a given process will be spontaneous and the degree of spontaneity. This is illustrated with a decomposition reaction that is endothermic but exoentropic. In general, provided ΔH and ΔS have the same sign, there exists a turnaround temperature $T_{ta} \approx \Delta H°/\Delta S°$ that will reverse the natural direction a reaction takes. At sufficiently high T, the entropic term begins to dominate, and the molecules of chemistry must cease to exist. Although evaluation of ΔG predicts the natural *direction* of a reaction, it cannot predict the *rate* at which products will form.

Gibbs's analysis provides a basis for comprehending the existence of the order and beauty we find in nature, including ourselves. The temperature of the earth is low enough that chemical forces dominate chaotic tendencies; this is the only reason chemistry is a subject of study. Nonetheless, chaotic molecular motion is the mechanism of chemical change, and life would not be possible without it.

EXERCISES

Reversible and irreversible work

1. Consider 2 mol of an ideal gas at 298 K expanding isothermally to three times its original volume. Calculate the reversible work in cal and J. Show that the ratio of this work to the irreversible work resulting from an isothermal expansion at 298 K against a constant external pressure equal to the final system pressure is $(3/2)\ln 3$, a number greater than unity. Accompany your calculations with a P-V diagram showing the two expansions.

2. Suppose the irreversible expansion described in Section 11.2 were run in two stages: first the mole of ideal gas at 273 K is expanded isothermally against $P_{ext} = 1.50$ atm until $P = P_{ext}$; then P_{ext} is reduced to 1.00 atm and the expansion completed. Calculate the work for this two-stage expansion in cal and J, accompanying your calculation with a P-V diagram. How does this amount of work compare to the one-stage irreversible expansion and the reversible expansion? How would $-w$ change if more such stages were added, but the initial and final states were kept the same?

3. Devise a two-step isothermal compression analogous to that for the expansion of Exercise 2, which takes a mole of ideal gas at 273 K from an initial state $P_1 = 1.00$ atm to $P_2 = 2.00$ atm. Calculate the work in cal and J and give a P-V diagram. Compare the work to that of the single-stage irreversible compression and the reversible compression, and guess how w would change if more stages were added.

4. Equations 11.5 relate the state variables for a reversible adiabatic expansion. Suppose as 1 mol of argon, initially at a pressure $P_1 = 5.00$ atm, expands adiabatically, its temperature drops from 450 to 300 K.
   a. Find the initial and final volumes V_1 and V_2 in L, the final pressure P_2 in atm, and calculate the work in cal and J.
   b. Show using your numerical results that $P_1 V_1^\gamma = P_2 V_2^\gamma$, where $\gamma = 1 + (R/C_V) = C_P/C_V$ is the heat capacity ratio. Use this result, assuming that it holds at each point in the expansion, to sketch a P-V diagram for the process.

5. Suppose 2.00 mol of $N_2(g)$ expand adiabatically from an initial temperature of 300 K and volume of 24.6 L to a final volume of 36.9 L against a constant external pressure of 1.00 atm.
   a. How much work (cal and J) results?
   b. What are the final temperature and pressure of the gas? (Use data from Table 10.4, assumed temperature independent.)
   c. Would the same final state have been reached if the adiabatic expansion occurred reversibly between the same two volumes? Why or why not?

The Carnot cycle

6. Carry through the operations indicated in Equation 11.5 for stage IV of the Carnot cycle. Show, by comparing your result to that given in Equation 11.5 for stage II, that $V_2/V_1 = V_3/V_4$ (thereby simplifying the efficiency expression Equation 11.4).

7. A typical automobile engine has an operating temperature of 145°C. If its surroundings are at 15°C, what is the maximum efficiency the engine can achieve? How could the theoretical efficiency be improved? Why do you think this is not done in practice?

8. Show algebraically that Equation 11.6 leads to Equation 11.7. Further show that the inequality for a real engine, $(q_h + q_c)/q_h < (T_h - T_c)/T_h$, leads to Equation 11.9.

9. Imagine a dismal engine, in which the isothermal stages occur with a constant external pressure equal to the final system pressure, while the adiabatic stages are replaced by a simple cooling or heating of the gas at constant volume between T_h and T_c. Take $n = 1$ mol, $T_h = 600$ K, $T_c = 300$ K, $P_1 = 10.0$ atm, $P_2 = P_4 = 5.0$ atm. Draw a P-V diagram for this ludicrous engine, and show by evaluating the efficiency that it will not be able to perform any useful work at all. Do you think such an engine is any more feasible than Carnot's engine? Explain your response briefly.

Entropy: Physical changes

10. Calculate the entropy change ΔS for the isothermal expansion of Exercise 1 in J/K and cal/K, for both the reversible and irreversible process. Establish that $\Delta S \geq q/T$ for the two expansions.

11. Use Boyle's law to show that the entropy change for an isothermal change in volume of an ideal gas can be

written as $\Delta S = nR \ln(P_1/P_2)$. Use this formula to calculate the entropy change, in cal/K and J/K, when 0.250 mol of $N_2(g)$ is compressed from 1.00 to 6.00 atm.

12. One mole of helium gas initially at 1.00 atm is heated, raising its temperature from 25° to 275°C. Calculate ΔS in J/K and cal/K assuming ideal gas behavior and (a) constant V or (b) constant P. Use data from Table 10.4, assumed T-independent. Note which process produces a larger ΔS, and explain briefly. Show that ΔS for the isobaric heating can also be obtained by considering a two-stage process, isochoric heating followed by isothermal expansion.

13. 5.00 mol of steam, $H_2O(g)$, cool from 350°C to 100°C at constant pressure without condensing. Calculate ΔS in cal/K and J/K based on ideal gas behavior. Use data from Table 10.4, assuming temperature independence.

14. The steam of Exercise 13 is allowed to condense to $H_2O(l)$ at 100°C and constant pressure. For the process $H_2O(g) \longrightarrow H_2O(l)$, $\Delta H°$ can be obtained from Table 10.1. Calculate the entropy change for the process in cal/K and J/K. (A more precise value for $\Delta H°$, which accounts for the nonnormal temperature, can be estimated as discussed in Section 10.7.)

15. For the Joule experiment as described by him in Section 11.3, find the entropy change in cal/K and J/K, assuming ideal gas behavior, equal volumes for the receivers R and E, and an initial temperature of 20°C.

16. Suppose the Joule experiment of Section 11.3 is run with two different gases, say He and Ar, in the left-hand and right-hand containers, of volumes V_{He} and V_{Ar}, at the same pressure. When the stopcock is opened, the gases will mix and, after a brief fluctuation, the pressure of the system will return to the same value as before. Assuming the gases act independently, find a formula for ΔS. Show using Boyle's law and Dalton's law of partial pressures that your formula is equivalent to $\Delta S = -nR(X_{He} \ln X_{He} + X_{Ar} \ln X_{Ar})$, where n is the total moles of gas and the X's are mole fractions. (This is known as the **entropy of mixing**, and is readily generalized to an arbitrary number of components.) What is the nature of the brief fluctuation above?

17. The common laboratory procedure of diluting a concentrated solution is quite analogous to an isothermal expansion of an ideal gas, with the volumes of the solution before and after dilution replacing the gas volumes. If 10.0 mL of a 1.00 M HCl solution is to be diluted to 0.0500 M, how much water must be added? What is the entropy change, in cal/K and J/K, that accompanies the dilution? Show that it can be written as $\Delta S = nR \ln(c_1/c_2)$, where the c's are molar concentrations.

18. A gas never spontaneously undergoes compression, two gases never spontaneously unmix, and you can make a dilute solution from a concentrated one, but not vice versa. What are the entropy changes for the reverse processes of those described in Exercises 15, 16, and 17, and how can they be interpreted in terms of their failure ever to occur? (You may use the Answer Key for any of these exercises you did not attempt.)

Entropy and probability

19. Consider a four-molecule model of the Joule experiment, constructed as in Section 11.4. Find W_1 and W_2, sketch the initial configuration and all possible final configurations as in Figure 11.7, and give the number of arrangements contributing to each configuration. Plot these numbers of arrangements against the number of molecules on the left in each configuration. Calculate the entropy change for the four-molecule expansion process in J/K and cal/K. What is the significance of the sign of ΔS?

20. Joule's experiment can be generalized into an experiment with many compartments of equal volume. For three compartments, labeled *left, center,* and *right,* and $N = 6$ molecules, choose an "interesting" configurations at random, for example, $3L,1C,2R$. Then, using numbered molecules or colored molecules, sketch out all possible arrangements contributing to it. Confirm that for the configuration you chose, the number of arrangements is given by the factorial formula $N!/(L!C!R!)$. What is the total number of arrangements possible? Calculate the entropy change in J/K and cal/K for the initial state $6L,0C,0R$ and a final state in which the compartments are all connected. Which configuration in the final state is most likely?

21. Boltzmann's method for finding the most probable distribution of molecules among energy levels is outlined in Appendix D. This method is more easily analyzed when the energy levels available are equally spaced (as in a harmonic oscillator) and there are

small numbers of molecules and quanta. Consider a system of three such energy levels, seven molecules, and a total of four quanta of energy (one quantum being the spacing between adjacent levels) available to the system. First, construct all possible *configurations* (level populations) of the seven molecules among the three levels. Then, cast out all those that have either fewer or more than four quanta of energy, assuming molecules in the ground state have no quanta. Finally, using a labeling scheme, construct all possible *arrangements* for each of the remaining configurations. (The number of arrangements for a given configuration is given by the formula of the previous exercise.) Which configuration is most probable? For this configuration, plot the population against the energy of the level.

Absolute entropy and chemical change

22. Without looking in tables, choose the substance from each pair with higher absolute entropy S°_{298}:
    a. $SO_2(g)$, $SO_3(g)$
    b. $Cl_2O(g)$, $Cl_2(g)$
    c. $CH_3OH(l)$, $CH_3OH(g)$
    d. $Na(l)$, $Na(s)$

23. For the following chemical reactions, surmise the sign of ΔS° without calculating it:
    a. $2Na(s) + Cl_2(g) \longrightarrow 2NaCl(s)$
    b. $Mg(s) + 2HCl(aq) \longrightarrow MgCl_2(aq) + H_2(g)$
    c. $HCl(g) + NH_3(g) \longrightarrow NH_4Cl(s)$
    d. $CaO(s) + SO_2(g) \longrightarrow CaSO_3(s)$
    e. $H_2(g) + Br_2(g) \longrightarrow 2HBr(g)$

24. For the reactions of Exercise 23, use Tables 11.1 and 11.2, or Appendix C, to calculate ΔS°. Compare your results with the predictions made in Exercise 23. For 23(b), you must write the reaction in ionic form. Reactions 23(a) through (e) are spontaneous as written. What can you conclude about the sign and magnitude of ΔS°_{surr} in those cases where $\Delta S^\circ < 0$?

Free energy changes and temperature control

25. Carry out explicitly the algebraic steps described following Equation 11.20 that lead to Equation 11.21. Since the free energy G is a state variable, what must be true about the free energy change for any observable process? For a cyclic process? Assuming that a free energy of formation ΔG°_f can be defined for any compound, what would the formula look like for calculating ΔG° for a reaction in terms of the ΔG°_f's?

26. Calculate the free energy change ΔG for the ideal gas expansion of Exercise 1. In light of the role of ΔG as a criterion for spontaneity, what is the significance of its sign? How is it related to work in this case?

27. Use the result of Exercise 11 to show that $\Delta G = nRT \ln(P_2/P_1)$ for an isothermal change of state in an ideal gas, and calculate ΔG in cal and J for the compression of $N_2(g)$ described there, assuming normal temperature. What is the significance of your calculation in terms of the capacity of the gas to do work?

28. Use the result of Exercise 17 to show that for an isothermal dilution process $\Delta G = nRT \ln(c_2/c_1)$, where c_1 is the molar concentration before dilution and c_2 after. Calculate ΔG in cal and J for the dilution of HCl described there, assuming normal temperature, and point out the significance of its sign.

29. Calculate the standard free energy change ΔG°_{298} in kcal/mol and kJ/mol for the combustion of graphite, $C(s) + O_2(g) \longrightarrow CO_2(g)$, from ΔH° and ΔS° using data in Table 10.1 and 11.1. State whether this reaction is spontaneous under standard conditions. [Since this is a formation reaction, the quantity you have calculated is called the **standard free energy of formation** $\Delta G^\circ_{f, 298}$ of $CO_2(g)$.]

30. An important industrial source of lime, $CaO(s)$, as well as a key step in the Solvay process discussed in Chapter 10, is the thermal decomposition of limestone $CaCO_3(s)$, which yields $CaO(s)$ and $CO_2(g)$. What would you predict to be the sign of ΔS° for this reaction? Use Table 10.1 to calculate ΔH° and Table 11.1 to find ΔS° for the decomposition. Further, use these values to find ΔG°_{298}. Is the reaction spontaneous at 298 K? Does there exist a temperature T_{ta} at which its natural direction will reverse? If so, estimate T_{ta}.

31. Based on the values of ΔH° and ΔS° for the water synthesis reaction given in Section 11.6, use a cycle to find the corresponding values for the production of water vapor, and decide whether this friendly all-gas explosion would cease to occur at some other temperature. If you decide in the affirmative, estimate the temperature T_{ta} above which this would occur. Above this temperature, what might happen to a sample of $H_2O(g)$?

32. For the hydrogen dissociation reaction $H_2(g) \longrightarrow 2H(g)$, find $\Delta H°$ and $\Delta S°$ from Appendix C, and use them to find $\Delta G°_{298}$. Interpret the sign of $\Delta G°_{298}$ in terms of the spontaneity of the dissociation at NTP. Estimate a temperature above which this reaction may readily occur. How do these results relate to the observations concerning this reaction in Section 11.6?

33. Combustion reactions involving carbonaceous fuels all tend to be spontaneous at room temperature, and are often also unable to be controlled by temperature changes. Find $\Delta H°$, $\Delta S°$, and from them $\Delta G°_{298}$ for the combustion of the common fuel propane $C_3H_8(g)$ to $CO_2(g)$ and $H_2O(g)$. Show that the reaction is spontaneous at 298 K and standard pressure and that it remains so at all temperatures.

Free Energy and Chemical Equilibrium

CHAPTER 12

Limestone cave with stalactites (top) and stalagmites (bottom). These form when, due to the evaporation of dripping water, the equilibrium solubility of $CaCO_3$ is exceeded. See Section 12.4 for a discussion of precipitation equilibria.

A little learning is a dangerous thing;
Drink deep, or taste not the Pierian spring:
There shallow draughts intoxicate the brain,
And drinking largely sobers us again.

Alexander Pope (1711)

We naturally think of events as discrete, with a beginning and an end. Everyone waits with common anticipation for a stage play to start, and everyone knows when it's over and it's time to leave the theater. Much the same can be said of our experience with chemical events, though in some cases it can be hard to tell from direct observation whether a chemical reaction is over or has not yet even begun. Moreover, the actors in chemical dramas are incredibly tiny, ceaselessly dancing molecules, and it is difficult to believe that *their* play would ever truly come to an end.

It is curious but telling that achieving **equilibrium** in a chemical system, getting to the end of a reaction, was first formulated in a general way by the Swedish chemists Guldberg and Waage in 1862 as a *dynamic balance* between the forward reaction and its reverse. Both forward and reverse reactions occur with certain *rates*, v_{forw} and v_{rev}, that depend on the amounts of reagent and product present. When in the course of a reaction the products have built up to a critical level and the reagents have been correspondingly depleted, the reverse rate precisely balances the forward rate, $v_{forw} = v_{rev}$. At that point, any measurements we make on a system will reveal a static situation,

CHAPTER OUTLINE

12.1 The Attainment of Equilibrium

12.2 LeChâtelier's Principle: Controlling the Position of Equilibrium

12.3 Equilibria in Aqueous Acid–Base Chemistry

12.4 Equilibria in Multiphase Systems

12.5 Deviations from Ideal Behavior

JACOBUS HENRICUS VAN 'T HOFF
1852–1911

Jacobus H. van't Hoff (1852–1911) was a professor at Amsterdam and later Berlin. His research was wide ranging in both gas-phase and solution chemistry. He proposed an equation relating the equilibrium constant for a reaction to the temperature that was later verified by a free-energy analysis. He was awarded the first Nobel Prize in chemistry in 1901.

unchanging amounts of the chemicals involved—a system at equilibrium—even though, as you might have anticipated, the molecules themselves never cease to react. Chemical equilibrium is thus macroscopically static but microscopically dynamic. Even in the simple gas expansion experiment of Joule discussed in the previous chapter, equilibrium on the molecular scale is a dynamic balance, with molecules moving to the right- and left-hand compartments at equal rates. We will return to this intriguing connection between action and stasis in Chapter 15.

12.1 The Attainment of Equilibrium

In the last chapter we learned a thermodynamic criterion for determining whether an event will occur, namely, $\Delta G < 0$ at constant T and P, where G is the Gibbs free energy. What this means is that, before a spontaneous reaction begins, the free energy of the reagents G_R is higher than that of the products G_P, and the process must release free energy as it occurs until $G_R = G_P$. At that point the reaction must cease on a macroscopic level, since then $\Delta G = G_P - G_R = 0$; this is the **equilibrium condition.** Thus far we have learned to calculate *standard* free energy changes, $\Delta G°$, where the degree symbol requires that all reagents and products be present at 1 atm partial pressure for gases, or 1 M concentration for solutes in solution. If we set up a reaction system in its standard state, then it is plain that if $\Delta G° < 0$, the reaction will proceed (at an unspecified rate, taking anywhere from seconds to centuries), using up the reagents and forming products. At the end of the reaction, then, the reagents and products are no longer in their standard states. Furthermore, you would seldom if ever *start* a reaction in its standard state; rather, you would naturally begin with pure reagents and no products at all! If we are to understand the end, the equilibrium that follows the reaction, as well as spontaneity in the general case, we must learn how to deal with free energies of *non*standard states.

Dependence of Free Energy on Pressure

Let's first consider a specific example in order to establish a method for finding G under nonstandard conditions; later we will state our results in more general terms. An historically much studied example of gas-phase chemical equilibrium is the dimerization of nitrogen dioxide,

$$2NO_2(g) \rightleftharpoons N_2O_4(g) \qquad (12.1)$$

NO_2 is an orangy-brown, extremely toxic gas, responsible for the brownish haze of urban smog, while N_2O_4 is colorless.* Absorption spectroscopy can therefore be readily used to monitor the partial pressure of NO_2 and to determine whether, and to what extent, Reaction 12.1 has occurred. We will be using the double arrow $\rightleftharpoons$ to join reagents and products for systems at equilibrium, an acknowledgment of the

---

* The color of NO_2 indicates that it absorbs strongly in the visible region of the spectrum; solar radiation excites NO_2, enabling it to react rapidly with small organic molecules also present in smog to make peroxyacyl nitrate, a severe irritant. Both molecules are powerful oxidizing agents, especially in the presence of moisture; the dimer is currently used as an oxidizer in the fuel systems aboard various space probes.

12.1 The Attainment of Equilibrium

(a)

$$2NO_2 \rightleftharpoons N_2O_4$$

(b)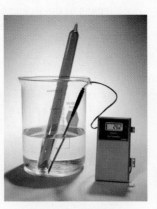

Dry-ice isopropanol   Ice water   Normal temperature   Boiling water

Figure 12.1
The NO_2 dimerization equilibrium under various conditions. In (a), increased pressure results in a greater fraction of dimer, and the intensity of the color in the pressurized syringe, due to undimerized NO_2, is not as great as it would be had no shift in equilibrium occurred. In (b), dimerization is decreased by an increase in the temperature, as indicated by the increased coloration; in the $-75°C$ bath not only is dimerization nearly complete, but the dimer has condensed. The pressure is also rising from left to right because the gas tube is rigid and sealed, but from (a) this would lead to a decrease in color, the opposite of what is seen.

dynamic balance between the reaction and its reverse that then exists. Figure 12.1 illustrates this equilibrium under various conditions. From your knowledge of chemical structure and thermodynamics already acquired, you should be able to make qualitative predictions concerning this reaction. Structurally, NO_2 is a free radical, with an unpaired electron localized on N; dimer formation corresponds to the combination of two unpaired electrons, making a new N—N bond. You might therefore predict that the dimerization is exothermic, $\Delta H° < 0$; in fact, $\Delta H° = -13.67$ kcal/mol (-57.2 kJ/mol), indicating a rather weak single bond. Further, you should have noted that $\Delta n_{gas} = -1$, leading you to expect $\Delta S° < 0$ as well; the standard entropies yield $\Delta S° = -42.00$ cal/K mol (-175.7 J/K mol). Since the signs of $\Delta H°$ and $\Delta S°$ are the same, this is a classic case of the competition between bonding and chaos, with a turn-around temperature of only 325 K (52°C). Then $\Delta G°_{298} = -1.15$ kcal/mol (-4.8 kJ/mol); the negative sign means that, when the partial pressures of both NO_2 and N_2O_4 are 1.00 atm, the reaction will proceed forward, that is, to the right, toward more dimer. However, the relatively small *magnitude* of $\Delta G°$ implies that there is only a small rightward driving force, and that *the reaction may not go to completion*.

However far the reaction *does* go beginning from standard conditions, it is certain that P_{NO_2} will be less than 1 atm when equilibrium is reached. What is the free energy of NO_2 under these conditions? First, let us define the **standard free energy of formation** $\Delta G°_{f,298}$ of $NO_2(g)$ in a way that is analogous to defining its standard enthalpy of formation, as the free energy change for its formation reaction (see Exercises 11.25 and 11.29). As for $\Delta H_{f,298}^\circ$, $\Delta G_{f,298}^\circ$ of an element in its natural standard state is defined to be zero. These standard free energies of formation can be computed from ΔH°s and ΔS°s and tabulated; Appendix C lists these values for reference. The ΔG_{298}° for a reaction can then be readily evaluated from the Hess's law formula

$$\Delta G^\circ = \sum_P \nu_P \Delta G_f^\circ(P) - \sum_R \nu_R \Delta G_f^\circ(R) \tag{12.2}$$

where as before the ν's are stoichiometric coefficients, P stands for products, and R for reagents.

Thanks to our analysis of engines at the start of the last chapter, we know what happens to the free energy when the pressure of an ideal gas changes at constant T; ΔG is just the reversible work involved in the isothermal change (see Exercise 11.27), given by

$$\Delta G = nRT \ln \frac{P_2}{P_1} \tag{12.3}$$

This is the standard reversible work formula of Equation 11.1 with Boyle's law used to replace the volumes with pressures. Now we can correct the standard free energy of formation for the nonstandard pressure by using Equation 12.3; the free energy of formation of NO_2 at some arbitrary partial pressure P_{NO_2} is given by

$$\Delta G_f = \Delta G_f^\circ + RT \ln \frac{P_{NO_2}}{P_{NO_2}^\circ} \tag{12.4}$$

where $P_{NO_2}^\circ$ is the standard partial pressure, 1 atm. As long as we use atmospheres as our unit of pressure, the P° denominator can be omitted for convenience; you should bear in mind, though, that the argument of a natural log must be a unitless number (a ratio of pressures in this case). Equation 12.4 in effect gives the free energy of NO_2 at arbitrary pressure, referred to its constituent elements in their standard states. It is of course valid for any gas that behaves (nearly) ideally, and it is the basic relation that will allow us to track the free energy of reactions under actual conditions in the lab and in nature. It is worthwhile to recall that the logarithmic pressure dependence of G arises directly from that of S, that is, from the relative number of ways of arranging the gas molecules at various pressures.

Now with the aid of Equation 12.4, which applies to N_2O_4 as well as to NO_2, we can use Equation 12.2, written without the degree symbols, to find ΔG for the reaction *under conditions of arbitrary partial pressures:*

$$\begin{aligned}\Delta G &= \Delta G_f(N_2O_4) - 2\Delta G_f(NO_2) \tag{12.5}\\ &= \Delta G_f^\circ(N_2O_4) + RT \ln P_{N_2O_4} - 2\Delta G_f^\circ(NO_2) - 2RT \ln P_{NO_2} \\ &= [\Delta G_f^\circ(N_2O_4) - 2\Delta G_f^\circ(NO_2)] + [RT \ln P_{N_2O_4} - RT \ln P_{NO_2}^2] \\ \Delta G &= \Delta G^\circ + RT \ln \frac{P_{N_2O_4}}{P_{NO_2}^2}\end{aligned}$$

In the second line of Equation 12.5, Equation 12.4 is substituted; in the third the standard free energy terms are collected in the left square brackets and the work terms in the right, with the stoichiometric coefficients (in this case only that of NO_2) being converted to exponents within the logarithm; and finally in the fourth line the contents of the left bracket are identified as the standard free energy change $\Delta G°$ and the ln's are combined. The partial-pressure quotient that forms the argument of the ln is called the **reaction quotient** Q. Before we explore the new wonders presented by the last line of Equation 12.5, let's restate our result in more general terms. For a general ideal gas-phase reaction

$$a\text{A} + b\text{B} + c\text{C} + \ldots \rightleftharpoons d\text{D} + e\text{E} + f\text{F} + \ldots$$

where A, B, C, ... are the reagents with stoichiometric coefficients $a, b, c, \ldots$, and likewise for the products D, E, F, ... and $d, e, f, \ldots$, the free energy change per mole reaction for arbitrary partial pressures $P_A, P_B, \ldots$ is given by

$$\Delta G = \Delta G° + RT \ln Q, \qquad \text{where } Q = \frac{P_D^d P_E^e P_F^f \cdots}{P_A^a P_B^b P_C^c \cdots} \qquad (12.6)$$

You should be able to fathom the origin of the result Equation 12.6 by reference to the procedure followed in Equations 12.5. Note that when the system is in its standard state, each partial pressure is 1 atm, causing the logarithmic term to vanish. One aspect of the usefulness of Equation 12.6, of which Equation 12.5 is a particular case, is illustrated by the following example.

EXAMPLE 12.1

Find the free-energy change ΔG_{298} of the dimerization reaction Equation 12.1 under the initial conditions of (a) $P_{NO_2} = 1.00$ atm, $P_{N_2O_4} = 0.100$ atm; (b) $P_{NO_2} = 0.100$ atm, $P_{N_2O_4} = 1.00$ atm; and (c) $P_{NO_2} = 1.00$ atm, $P_{N_2O_4} = 0.000$ atm. From your values, decide in which direction the reaction will proceed.

Solution:
We can apply Equation 12.5 directly, using $\Delta G°$ as given earlier, to these three cases:

a. $\Delta G = -1150$ cal/mol + (1.987 cal/K mol)
    $(298 \text{ K}) \ln[0.100/(1.00)^2]$
    $= -1150 + 592 \ln 0.100 = -1150 - 1363 = -2510$ cal/mol
    Since $\Delta G < 0$, the reaction will proceed forward, to the right.

b. $\Delta G = -1150 + 592 \ln[1.00/(0.100)^2] = -1150 + 2727$
    $= +1580$ cal/mol
    $\Delta G > 0$, so the reaction will run backward, to the left.

c. $\Delta G = -1150 + 592 \ln[0.000/(1.00)^2] \longrightarrow -\infty$

$\Delta G < 0$ but it is singular, since $\ln 0 = -\infty$; reaction should proceed to the right. To get the actual work available, these ΔG's would be multiplied by the actual number of moles of reaction that occur. In case (b) the reaction goes in reverse, the opposite of the direction expected on the basis of $\Delta G°$, and a clear indication of the importance of the work correction. In case (c), where no products are present, Equation 12.5 (or 12.6) fails and cannot be used, as discussed later, although the predicted direction of reaction is correct.

It is somewhat surprising that Equation 12.6 can be used to predict free-energy changes for all situations except the most common one, that is, when no products are present at the start of the reaction, for which it yields $\Delta G = -\infty$. This unrealistic behavior arises because in arriving at Equation 12.5 we have assumed that finite amounts of both NO_2 and N_2O_4 are present; if there is no product, its contribution to the total free energy of the system is zero, and should not be included. However, as $P_{N_2O_4}$ becomes very small, ΔG, considered as a tendency to react, does tend to grow in magnitude without bound. The singularity occurs in the logarithmic term, which is ultimately traceable to the entropy and Boltzmann's relation $S = k_B \ln W$. As the partial pressure of products approaches zero, it must eventually become smaller than that exerted by a single molecule of product; if there are no molecules to arrange, the entropy of the products cannot be defined. Thermodynamics, however—as based on the statements of the laws and the associated mathematical relationships that follow from them—makes no reference to the existence of molecules, and S must be defined to ensure the universal validity of the second law. Moreover, there is an implicit assumption in the use of calculus, particularly the differential, in deriving thermodynamic relationships, including Equation 12.6; namely, that a chemical substance is infinitely divisible—a view held by Aristotle, but definitely out of favor today! If you go back to the beginning of the last chapter, where reversible work was computed from $-w = \int P\, dV$, the differential dV is permitted mathematically to be smaller than the volume occupied by a single molecule. We will see problems arising from this inconsistency between thermodynamics and the atomic theory crop up again in treating systems at equilibrium.

The Equilibrium Constant $K = \exp[-\Delta G°/RT]$

As a reaction proceeds, the free energy that guarantees its spontaneity is expended, and eventually the system must reach equilibrium. As illustrated in Figure 12.2 for the $2NO_2 \rightleftharpoons N_2O_4$ reaction, in general the free energy of the reagents G_R falls as they are depleted, while that of the products G_P rises as their partial pressures build up, until they meet. When equilibrium has been attained, we have

$$\Delta G = G_P - G_R = 0 = \Delta G° + RT \ln Q_{eq} \qquad (12.7)$$

from applying Equation 12.6, where Q_{eq} is the reaction quotient at equilibrium. Denoting Q_{eq} by K, the **equilibrium constant,** Equation 12.7 results in

$$\Delta G° = -RT \ln K \quad \text{or} \quad K = e^{-\Delta G°/RT} \qquad (12.8)$$

At a fixed temperature, K is truly a constant, since $\Delta G°$ depends only on T, the nature of the system, and the definition of the standard state. Equilibrium constant K determines the ratio of product to reagent partial pressures at equilibrium; whatever

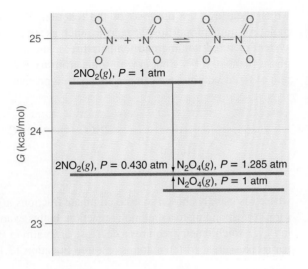

Figure 12.2
Illustration of the changes in free energy that accompany the attainment of equilibrium in NO_2 dimerization, starting from standard conditions and holding the temperature constant at 298 K. In general, the free energy of the reagents falls while that of the products rises until they become equal. The fall in G for NO_2 is larger than the rise for N_2O_4 both because the equilibrium position lies to the right under standard conditions and because the reaction quotient of Equation 12.5 is more sensitive to P_{NO_2}.

the starting conditions, this ratio must eventually prevail. Note that $\Delta G° < 0$ corresponds to $K > 1$; that is, there will be a preponderance of products at equilibrium, whereas for $\Delta G° > 0$, $K < 1$, and the reagents will rule. This definition of K also allows us to rewrite Equation 12.6 as

$$\Delta G = RT \ln(Q/K) \qquad (12.9)$$

Equation 12.9 allows you to decide on the spontaneity of a process based directly on comparing Q under your own set of initial conditions to K. If $Q < K$, that is, you have less reaction product than required for equilibrium, then $\Delta G < 0$ and the reaction will proceed spontaneously in the forward direction. For $Q > K$, that is, you have more reaction product than at equilibrium, $\Delta G > 0$, and the reaction will run in reverse. Equation 12.9 also provides an alternative way of computing the magnitude of ΔG, as in Example 12.1 earlier.

The Position of Equilibrium in Gas-Phase Reactions

For our $2NO_2 \rightleftharpoons N_2O_4$ example, K may be computed from Equation 12.8: $K = 6.97$ at 298 K. Note that K is a unitless number, because each pressure appearing in the reaction quotient is implicitly divided by $P° = 1$ atm; but for problem-solving purposes, it is often convenient to assign units to K. In this example

$$K = \left(\frac{P_{N_2O_4}}{P_{NO_2}^2}\right)_{eq} = 6.97 \text{ atm}^{-1} \qquad (12.10)$$

If we know K, relationships like Equation 12.10 can allow us to determine the composition of the system at equilibrium starting from arbitrary initial partial pressures, an extremely useful and practical piece of information, in this reaction and many others. To accomplish this, we must make use of the stoichiometry of the reaction, which relates the depletion of reagents to the accumulation of products. These are related quantitatively by defining an *extent of reaction variable x*, which measures the progress of the reaction. Reaction variable x has units of moles, or units proportional to moles, such as atmospheres of partial pressure in a system of fixed volume and temperature, a situation common in laboratory gas-phase chemistry. If we begin with some arbitrary $P_{NO_2} = P_0$ and $P_{N_2O_4} = 0$ in a system of fixed volume, we can set up a partial pressure table:

$$\begin{array}{lccc} & 2NO_2 \rightleftharpoons & N_2O_4 & \text{Total pressure} \\ \text{Initial} & P_0 & 0 & P_0 \\ \text{Equilibrium} & P_0 - 2x & x & P_0 - x \end{array} \quad (12.11)$$

where x is usually chosen so as to eliminate fractions in the equilibrium partial pressure, corresponding to a unit molar reaction. In this example, for an increase of x atm in $P_{N_2O_4}$, stoichiometry requires a decrease of $2x$ atm in P_{NO_2}. If we now substitute the partial pressures written in terms of x into Equation 12.10, we find

$$K = \frac{x}{(P_0 - 2x)^2} \quad (12.12)$$

which can be written in the quadratic form

$$x^2 - \left(P_0 + \frac{1}{4K}\right)x + \frac{P_0^2}{4} = 0 \quad \text{or} \quad (12.13)$$

$$x^2 - 1.0359x + 0.2500 = 0$$

where numerical values have been inserted for $2NO_2 \rightleftharpoons N_2O_4$, with $P_0 = 1.00$ atm. This is in the standard form $ax^2 + bx + c = 0$, and the quadratic formula

$$x = \frac{-b \pm \sqrt{b^2 - 4ac}}{2a}$$

can then be applied to solve it for x. The quadratic formula yields two roots, *only one of which is chemically acceptable;* in addition, *you should carry as many digits as your calculator will hold,* because this formula can be "ill conditioned," sometimes giving nonsensical answers for truncated quadratic coefficients. (These are things you never had to worry about in high-school algebra!) Here the two roots are $x = 0.383$ and 0.653 atm; only the first is realistic, since $x = P_{N_2O_4}$ cannot exceed $P_0/2 = 0.5$ atm. Thus at equilibrium $P_{N_2O_4} = 0.383$ atm, $P_{NO_2} = P_0 - 2x = 0.234$ atm, and the total pressure is $P = 0.383 + 0.234 = 0.617$ atm.

Note that $P_{N_2O_4}$ exceeds P_{NO_2} at equilibrium, and the reaction is 76.6% complete, as required by $K > 1$. Note also that the pressure is lower at equilibrium than initially (0.617 versus 1.00 atm) because the reaction running forward reduces moles of gas present. As we will see in the next section, if $\Delta G°$ were lower (more negative and of greater magnitude), then K would be $\gg 1$, and the reaction would be essentially complete, $x \longrightarrow 0.5$.

In many cases a chemist is presented with a system already at equilibrium. In our example, the NO_2 in the equilibrium mixture is readily detected by absorption spectroscopy, which allows its partial pressure to be determined. In a two-component system like this one, the partial pressure of the remaining component is then obtained by solving the equilibrium expression, for example, Equation 12.10, directly: $P_{N_2O_4} = K P_{NO_2}^2$. It is even simpler experimentally to monitor the total pressure P of the system; this provides a diagnostic for the attainment of equilibrium as well as a means to determine composition. When P is known, if $P_{N_2O_4} = x$, then $P_{NO_2} = P - x$ from Dalton's law of partial pressures; the equilibrium condition can then be set up as in Equation 12.12, and solved for x, as illustrated in the following example.

EXAMPLE 12.2

Suppose the $2NO_2 \rightleftharpoons N_2O_4$ system is brought to equilibrium at a total pressure of 1.00 atm at 25°C. What are the partial pressures of NO_2 and N_2O_4?

Solution:
With $P_{NO_2} = P - x$ and $P_{N_2O_4} = x$, the equilibrium expression analogous to Equation 12.12 is $K = x/(P-x)^2$. Here x may vary from 0 to 1 atm. The resulting quadratic is $x^2 - (2P + 1/K)x + P^2 = 0$, or $x^2 - 2.1435x + 1.0000 = 0$. The roots are $x = 0.686$, 1.457; only 0.686 is acceptable, yielding $P_{N_2O_4} = 0.686$ atm and $P_{NO_2} = 0.314$ atm. You can always check your result by substituting back into the equilibrium quotient: $0.686/(0.314)^2 = 6.96$, the small discrepancy due to rounding error. Again the position of equilibrium lies to the right, as expected.

Reactions with more complex stoichiometry, such as the synthesis of ammonia $3H_2 + N_2 \rightleftharpoons 2NH_3$, will often give rise to extent-of-reaction equations of third degree or higher. While such systems cannot be handled with simple algebra, they are readily analyzed by numerical methods on a computer or a calculator with an equation solver. In cases like this, it is nonetheless often possible to obtain approximate equilibrium concentrations with less labor by exploiting limiting cases such as $K \gg 1$ or $K \ll 1$, and to improve them by successive approximations. We will illustrate this approach later in this chapter.

Achieving equilibrium, the exhaustion of the free-energy supply of a reaction, also means that the free energy itself must have reached a minimum value; otherwise there would exist a reaction direction that would further reduce G, and further reaction would occur spontaneously. The equilibrium position, therefore, corresponds to the extent of reaction x that makes $G(x)$ a minimum, or $dG/dx = 0$, as well as where $\Delta G = 0$. In fact it can be shown (it was first shown by Gibbs) that $\Delta G \propto dG/dx$, that is, ΔG is proportional to the slope of the curve $G(x)$ versus x. This is illustrated for the $2NO_2 \rightleftharpoons N_2O_4$ reaction in Figure 12.3. Roughly parabolic near its minimum, G acts as a restoring potential for the reaction, much as a bond potential energy curve keeps the atoms in a chemical bond close to their equilibrium position at the bottom of its well. Thus Gibbs dubbed the free energy of each component of a chemical system the **chemical potential** μ of that component, identical to the free energy of formation given by Equation 12.4. The details of this formalism, which consist of the analysis of the "total differential" dG, are reserved for a more advanced course. It is found that the minimum in G is due to the entropy of mixing—dilution of each gas by the other.

Figure 12.3
Dependence of G and ΔG on the extent of reaction variable x, starting on the left from standard conditions, and holding the temperature constant at 298 K. Where ΔG passes through zero, G passes through its minimum value; G thus acts as a chemical potential that determines the position of equilibrium and prevents the system from departing from it. The ΔG curve becomes singular (approaching $+\infty$) as $x \longrightarrow 0.5$.

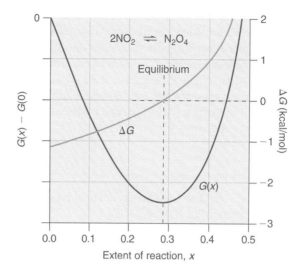

12.2 LeChâtelier's Principle: Controlling the Position of Equilibrium

Recall from our early discussion of reversible work in the previous chapter that an equilibrium state is a reversible state; that is, small changes in external conditions (in the case of work, changes in P_{ext}) can make a process go forward or backward, depending on the sense of the change. For a chemical system at equilibrium, we have at our disposal several ways to change the conditions under which equilibrium exists: changes in total pressure P or volume V, changes in temperature T, or changes in the partial pressures of the chemical components. When any of these variables is altered, the system may be temporarily thrown into a state of disequilibrium; there will now be a nonzero ΔG driving the system to a new equilibrium state, which it will achieve in due course. What will this new state be like? In which direction will the perturbation on the system push the equilibrium position? A guide to the answers to these questions is again reversible work: An *increase* in P_{ext} on a gas at equilibrium will cause compression, a decrease in the volume of the system, and an *increasing opposing pressure*. Vice versa, a decrease in P_{ext} will result in further expansion, an increase in volume, and a decrease in opposing pressure. In either case, *an equilibrium system responds to the stress of an external change by opposing it.* This statement turns out to be general, and was first enunciated by Henri LeChâtelier in 1888:

> Any change in one of the variables that determines the state of a system in equilibrium causes a shift in the position of equilibrium in a direction that tends to counteract the change in the variable under consideration.

The inevitability of **LeChâtelier's principle** can be seen by imagining that the opposite were true. Then a decrease in P_{ext} would result in an increased internal pressure, and you could compress a gas to hundreds of atmospheres by exerting less and less force on it, an obvious, almost comical violation of the conservation of energy. This principle seems to apply across a wide spectrum of human affairs as well, from individual behavior under stress to political prodding of government bureaucracy.

First let's consider changes in pressure or volume. Under isothermal conditions these two variables can be analyzed together, owing to Boyle's law. Using the

$2NO_2 \rightleftharpoons N_2O_4$ system as an example, a decrease in V (or an increase in P) will induce a response, because in this reaction the number of moles of gas changes, $\Delta n_{gas} = -1$, implying that the system can alter its own pressure through a shift in equilibrium position. According to LeChâtelier's principle, "squeezing" the system will cause it to make fewer moles of gas, thereby relieving the stress; it can do this by shifting its equilibrium to the right, toward more dimer. You can verify this qualitative conclusion by repeating Example 12.2 for $P = 10.0$ atm (approximately equivalent to reducing the volume by a factor of 10); the result is $P_{N_2O_4} = 8.87$ atm, 89% dimer as opposed to 69% at $P = 1.00$ atm. Conversely, lowering P (or increasing V) decreases the extent of reaction; try inventing your own conditions to test this statement. These changes, however, will only have an effect if the system is capable of responding; the equilibrium $2HI(g) \rightleftharpoons H_2(g) + I_2(g)$ cannot be altered by changes in external pressure or volume, because $\Delta n_{gas} = 0$, and the system has no means to change its pressure or volume by shifting equilibrium position.

A much more powerful and general method for shifting an equilibrium is through changes in T. We have already explored the effect of temperature on the spontaneity of a reaction at standard conditions in the last chapter. Varying T was found to vary the *magnitude* as well as the sign of $\Delta G°$, and we now know how to relate $\Delta G°$, through K, to the position of equilibrium. When T changes, the equilibrium constant is no longer constant, since it depends exponentially on $\Delta G°$; moreover, even a modest change in $\Delta G°$ might change K by orders of magnitude. We can therefore expect dramatic effects on equilibrium systems of changing T. Before we explore this question in a quantitative way, however, let's ask what M. LeChâtelier has to say. Increasing T increases the amount of thermal energy or enthalpy within a system; most chemical reactions, as we have seen in Chapter 10, have their own means of supplying energy, because they are exothermic. An exothermic reaction running in reverse becomes endothermic. Since a chemical system at equilibrium is reversible, it is free to run in either direction, evolving or absorbing heat energy as the situation demands. Thus an exothermic reaction at equilibrium can oppose the stress of an increase in temperature by running backward, thereby consuming the heat energy that has been added. The equilibrium position will then shift to the left, toward reagents. The converse would be true for a normally endothermic reaction. Our $2NO_2 \rightleftharpoons N_2O_4$ example is typical of the exothermic case; as illustrated in Figure 12.1, heating the system drives it toward the left, toward more NO_2, while cooling it results in more dimerization.

For a quantitative analysis of the effect of changing T, we need to examine how the equilibrium constant changes. Then we can carry out an extent-of-reaction analysis to find the new equilibrium state. Let's combine Equation 11.24 under standard conditions, $\Delta G° = \Delta H° - T\Delta S°$, with Equation 12.8, $\Delta G° = -RT \ln K$:

$$-RT \ln K = \Delta H° - T\Delta S° \quad \text{or} \quad \ln K = -\frac{\Delta H°}{RT} + \frac{\Delta S°}{R} \quad (12.14)$$

If we now assume, as we did in Chapter 11, that $\Delta H°$ and $\Delta S°$ are independent of T, we can write Equation 12.14 for two different temperatures, T_1 and T_2, and subtract the T_1 equation from the T_2 equation. When this is done, you can see that the $\Delta S°/R$ terms will cancel, since they do not depend on T. This algebra yields

$$\ln \frac{K_2}{K_1} = -\frac{\Delta H°}{R}\left(\frac{1}{T_2} - \frac{1}{T_1}\right) \quad (12.15)$$

Another way of arriving at this result is by noting from Equation 12.14 that a plot of $\ln K$ versus $1/T$ will be a straight line with slope $-\Delta H°/R$, or that

$$\frac{d \ln K}{d(1/T)} = -\frac{\Delta H°}{R} \tag{12.16}$$

When this equation is integrated between the limits T_1 and T_2, Equation 12.15 results. Thus Equations 12.15 and 12.16 are different ways of stating the same relationship; either of them is sometimes called the **van't Hoff equation.** The van't Hoff equation is itself actually equivalent to Equation 11.24, reexpressed in terms of K, and is a quantitative reflection of LeChâtelier's principle in the way it relates K to $\Delta H°$.

We can apply the van't Hoff Equation (Equation 12.15) to analyze quantitatively the $2NO_2 \rightleftharpoons N_2O_4$ experiments depicted in Figure 12.1, by calculating K at the various temperatures as illustrated by the following example.

EXAMPLE 12.3

Use LeChâtelier's principle to predict the sense of the shift in equilibrium for the $2NO_2 \rightleftharpoons N_2O_4$ system when it is cooled to $-75°C$, and use the van't Hoff equation to confirm your prediction with a quantitative calculation of K at $-75°C$, given $\Delta H° = -13.67$ kcal/mol (-57.2 kJ/mol) and $K = 6.97$ at $25°C$.

Solution:
Cooling an exothermic reaction causes it to produce more heat by running forward; hence, more dimerization will occur, as observed. Equation 12.15 yields

$$\ln(K_2/K_1) = +[13{,}670 \text{ cal/mol}/1.9872 \text{ cal/(K mol)}](1/198K - 1/298K)$$
$$= 11.66$$
$$K_2 = (1.16 \times 10^5)(6.97) = 8.06 \times 10^5$$

This large value of K implies nearly complete reaction. If you were asked to calculate the extent of reaction in this case, the setup used in Equations 12.11 to 12.13 would fail numerically. When K is large, it is better to assume complete conversion to products as a starting point, and to take x to be the extent of the reverse reaction; that is, $P_{N_2O_4} = P_0 - x$ and $P_{NO_2} = 2x$, where P_0 is now the partial pressure of N_2O_4 expected if reaction were complete. If $P_0 = 1.00$ atm, $x = 0.00056$ atm at $-75°C$, or nearly complete dimerization. In this case, you would also be justified in neglecting x with respect to P_0 in $P_0 - x$, thereby eliminating the linear term in the quadratic equation and simplifying the algebra.

The van't Hoff equation is also often used in the other direction, since equilibrium concentrations, or just total pressures, can often be measured, and K deduced from them. Then a plot of $\ln K_T$ versus $1/T$ yields a straight line with slope $-\Delta H°/R$ and intercept $\Delta S°/R$, thereby allowing the heat and entropy of reaction to be measured without calorimetry. Figure 12.4 illustrates this for the dissociation of fluorine, $F_2(g) \rightleftharpoons 2F(g)$. This technique is also an indispensable component in the extraction of standard entropies of molecules and ions in aqueous solution, as discussed and tabulated in the previous chapter. We will now apply the equilibrium ideas de-

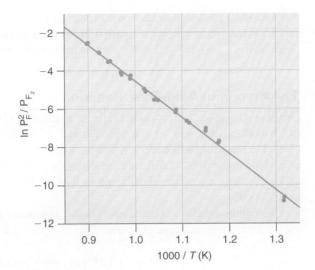

Figure 12.4
Van't Hoff plot for the dissociation of fluorine. The partial pressures for the evaluation of K were inferred from total pressure measurements. The slope of the best straight line through the points yielded $\Delta H° = 37.7 \pm 0.4$ kcal/mol.

veloped so far to reactions in solution, in the process exposing perhaps their most important application and bringing them within your grasp in introductory laboratory experiments.

12.3 Equilibria in Aqueous Acid–Base Chemistry

As we have noted in Chapters 10 and 11, taking thermodynamics into the aqueous phase involves the energetic and entropic effects of the interactions between the solute and the solvent, water, as well as a new standard state, 1.00 M concentration of all solutes present. For reasons such as limited solubility that you may already be aware of and that will certainly become apparent as we consider the varieties of aqueous reactions, this definition of standard state is much less practical than that of 1 atm partial pressures in a gas mixture. We nonetheless need a reference state for the assembly and use of solution-thermodynamic tables, and of necessity we also need a way of representing the enthalpy and entropy of real, nonstandard aqueous reaction systems.

Departing from a standard state in solution can be represented by dilution, for example, when a 1 M $CuSO_4$ solution is diluted to 0.1 M. The blue color characteristic of $Cu^{2+}(aq)$ diminishes as water is added; the act of pouring the water causes enough turbulent mixing to distribute the Cu^{2+} ions uniformly throughout the diluted solution. (This uniformity would come about even without stirring given sufficient time for random diffusion of the ions.) In terms of the number of arrangements available to the ions, dilution is nearly identical to gas expansion, with the gas volumes replaced by solution volumes. Thus the entropy change for dilution can be written in the same way as for gas expansion (Equation 11.11), $\Delta S = nR\ln(V_2/V_1)$. Now the molarity c is n/V; and because n is constant during dilution, $V_2/V_1 = c_1/c_2$. Further, when solutions are reasonably dilute (~1 M or less), the heat of dilution is usually negligibly small, and $\Delta G \approx -T\Delta S$. Thus we can write the free-energy change for dilution as

$$\Delta G = nRT\ln\frac{c_2}{c_1} \qquad (12.17)$$

and the results presented in Section 12.1, particularly Equations 12.6, 12.8, 12.9, and 12.15, all follow, as long as the partial pressures are replaced by molar concentrations. Now we are ready to launch onto the uncharted waters of aqueous equilibrium.

The Arrhenius Acid–Base Reaction

Our plan is to dissect the two major types of reaction, acid–base and redox (discussed in Chapter 5), with our new thermodynamic tools. In this chapter we focus on acid–base reactions, while redox reactions, under the guise of electrochemistry, will be covered in the next. At first we concentrate on purely aqueous-phase acid–base reactions, leaving mixed-phase reactions such as gas forming and precipitation for the next section.

Within the confines of reactions in water, the most fundamental type is without question the Arrhenius acid–base reaction:

$$H^+(aq) + OH^-(aq) \rightleftharpoons H_2O(l) \qquad (12.18)$$

The previous two chapters have equipped us to find the standard free-energy change for this reaction by combining the enthalpy change obtained using Tables 10.1 and 10.2 (see Exercise 10.34), $\Delta H° = -13.34$ kcal/mol (-55.85 kJ/mol), and the entropy change from Tables 11.1 and 11.2, $\Delta S° = +19.28$ cal/K mol ($+80.67$ J/K mol), to yield $\Delta G° = -19.09$ kcal/mol (-79.87 kJ/mol). You might regard the values for both $\Delta H°$ and $\Delta S°$ as mildly surprising. Reaction 12.18 appears to be forming a new O—H bond, for which an energy release of 100 kcal/mol or more might be expected, making the value found seem anomalously low. The forming of the new bond might also be construed as increasing the order in the system, suggesting a negative $\Delta S°$. Both of these apparent anomalies point to the importance of ion solvation; as discussed briefly in Chapter 5, $H^+(aq)$ is shorthand for what we wrote there as H_3O^+, hydronium ion, and is most likely a complex web with the formula $H_9O_4^+$, as illustrated in Figure 12.5. Likewise, OH^- is better represented as $H_7O_4^-$, and the Arrhenius reaction, which turns these two ionic aggregates into a neutral molecule, now takes on a different complexion. The term $\Delta H°$ now represents a difference between the solvating bond energies in the complexes and the product bond, which is necessarily considerably smaller than the product bond itself; while the loss of organization upon breakup of the solvation clusters makes the positive $\Delta S°$ seem plausible.

The $\Delta G°_{298}$ leads to $K = 1.0 \times 10^{14}$, from which we can extract the concentrations of $H^+(aq)$ and $OH^-(aq)$ at equilibrium; the large value of K implies minuscule amounts of reagents remaining for exactly stoichiometric initial concentrations. Exact stoichiometry is most closely achieved in a titration, to be discussed in detail later in this section. In setting up the reaction quotient Q, we note that the product of the reaction, $H_2O(l)$, is also the solvent, whose concentration is large (the molarity of H_2O in itself is 56 M) and practically unaffected by the small amount of reaction product formed in "dilute" solution ($\gtrsim 1$ M reagent concentrations). Thus H_2O changes its free energy negligibly as a result of the reaction and need not be included in the reaction quotient, since it begins and ends the reaction in its standard state at 1 atm external pressure. Water effectively behaves as a *pure condensed phase* in dilute aqueous reactions, a condition that implies that its free energy per mole is nearly invariant to the reaction process, as we will see more generally in the next section. For a reaction that goes nearly to completion, the equilibrium concentrations are best

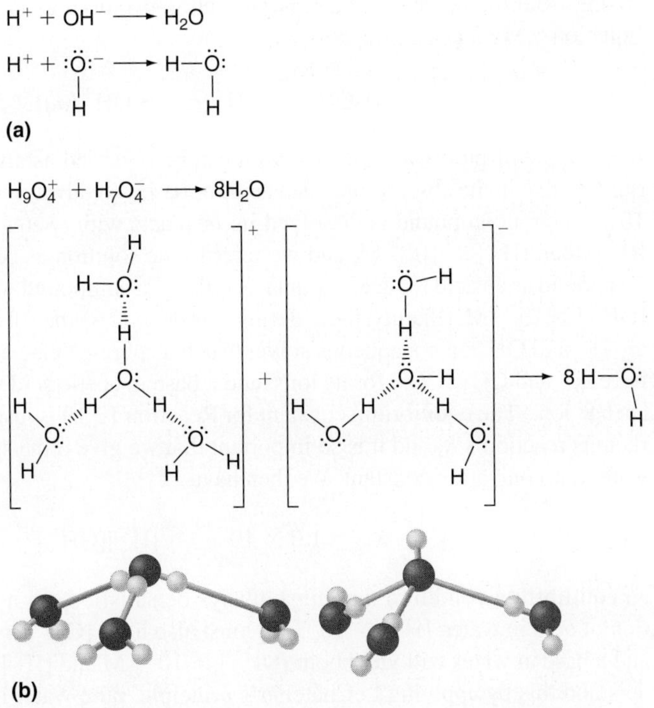

Figure 12.5
Simple (a) and hydrated (b) pictures of the Arrhenius acid–base reaction. The complex and organized structure of the hydrated ions helps to explain why the Arrhenius reaction has a positive entropy change, as well as a release of enthalpy that is relatively small compared to an OH bond energy.

expressed using the amount of reagent remaining as the extent of reaction variable x (now a concentration rather than a partial pressure, but still proportional to the number of moles); that is, we treat the reaction as its weakly occurring reverse. So for exact stoichiometry we write

$$H^+ + OH^- \rightleftharpoons H_2O$$
$$x \; x$$

noting that nothing need be written beneath the H_2O product, since it is the solvent. Then we have

$$K = Q_{eq} = \frac{1}{[H^+][OH^-]}$$

$$1.0 \times 10^{14} = \frac{1}{x^2}$$

$$x = [H^+] = [OH^-] = 1.0 \times 10^{-7} \text{ M}$$

Thus at the equivalence point of the Arrhenius acid–base reaction, the concentrations of reagents H^+ and OH^- remaining at equilibrium are equal and very small, as you might have anticipated from the large K. In practice, the conditions represented by this equilibrium calculation are very hard to achieve; if you started with 1 M acid and base, you would have to pinpoint stoichiometric equivalence to one part in 10 million!

Now, aside from the salt produced in the Arrhenius reaction, all that is left in the solution is the solvent water, and the small residual H^+ and OH^- can be viewed as arising from the reverse reaction, water autoionization, as introduced in Chapter 5, Equation 5.31, written here as

$$H_2O(l) \rightleftharpoons H^+(aq) + OH^-(aq) \qquad (12.19)$$

In a sample of pure water this reaction can be regarded as always being at equilibrium, so that in the absence of added acids or bases, we always have $[H^+] = [OH^-] = 10^{-7}$ M. If a compound is dissolved in, or reacts with, water to produce additional $[H^+]$, then $[H^+] > 10^{-7}$ M, and we refer to the solution as being acidic, and to the compound as an acid (or acid former). Similarly, a compound we call a base will yield $[OH^-] > 10^{-7}$ M. (Plainly, these definitions depend on the solvent water, whose ions are H^+ and OH^-; a nonaqueous solvent such as pure—"glacial"—acetic acid would have H^+ and CH_3COO^- for its ions, and a base in acetic acid would produce excess acetate ion.) The equilibrium constant for Reaction 12.19 is just the inverse of the Arrhenius reaction's K, and it is so important that we give it the special symbol K_w, the water autoionization constant. We then have

$$K_w = 1.0 \times 10^{-14} = [H^+][OH^-] \qquad (12.20)$$

an equilibrium condition that must always be satisfied. As a consequence, a solution of acid in water, $[H^+] > 10^{-7}$ M, must also have $[OH^-] = K_w/[H^+] < 10^{-7}$ M, and a base in water will yield both $[OH^-] > 10^{-7}$ M and $[H^+] < 10^{-7}$. You can understand this by applying LeChâtelier's principle: pure water is a system in equilibrium with its ions, and adding *either* acid, H^+, or base, OH^-, will cause the equilibrium to shift to the left, toward neutral water, that is, water ionization is inhibited in acidic or basic solutions.

To avoid having to deal with the minuscule concentrations often involved in acid–base equilibria, chemists have adopted a logarithmic scale known as the **pH** scale (standing for the negative of the **p**ower of the **H**ydrogen ion concentration), defined by

$$\text{pH} = -\log_{10}[H^+] \qquad (12.21)$$

Thus the pH of a pure water sample (neutral) is $-\log_{10}(1.0 \times 10^{-7}) = -(-7.0) = 7.0$, an acidic solution will have pH < 7, and a basic solution pH > 7. If we generalize this log scale so that $p(\text{Anything}) = -\log_{10}[(\text{Anything})]$, we can take $-\log_{10}$ of both sides of the equilibrium condition (Equation 12.20) to get

$$\text{pH} + \text{pOH} = pK_w = 14.00 \qquad (12.22)$$

The value of K_w (or pK_w) given is, of course, temperature dependent and must obey the van't Hoff equation (Equation 12.15); but because $\Delta H_w° = +13.34$ kcal/mol is reasonably small, the temperature dependence is weak over small ranges of T (for example, neutral pH = 6.81 at body temperature 37°C or 310 K).

pH, or $[H^+]$, is readily measured with an electrical device known as a *pH meter*, consisting of an electrode probe that is inserted into the solution to be measured, connected to a voltmeter calibrated to read out the pH directly, as illustrated in Figure 12.6. For now the pH meter will be treated as a "black box" (a convenient gizmo whose inner workings we don't understand), but in the next chapter we will briefly describe the essence of its operation. You may already be able to understand

12.3 Equilibria in Aqueous Acid–Base Chemistry

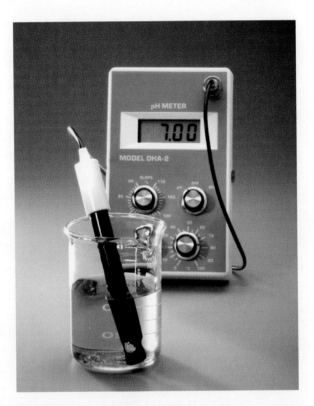

Figure 12.6
The pH meter. The probe, which contains a pair of sensitive electrodes encased in a protective plastic sheath, is directly immersed in the solution whose pH is to be measured. One of the electrodes contains a standard acid solution, and the difference in free energy between the standard and the unknown solution generates a voltage that is read by the meter. Here pure water at 298 K is being measured. A small amount of bromthymol blue acid-base indicator has been added. Bromthymol blue is blue in base, yellow in acid, and blue-green in neutral solution. The theory of operation of the pH meter will be given in the next chapter.

the principle, namely, that the pH meter is comparing the free energy of $H^+(aq)$ in the unknown solution to that of a reference solution of known $[H^+]$ already present within the probe, exploiting Equation 12.17.

We call dissolved compounds such as $HCl(aq)$ and $NaOH(aq)$ "strong" acids and bases because their ionization equilibrium lies far to the right; for example,

$$HCl(aq) \rightleftharpoons H^+(aq) + Cl^-(aq), \qquad K = K_a \approx 10^7 \qquad (12.23)$$

(The K_a here is quite uncertain, because the small amount of neutral HCl remaining at equilibrium is difficult to measure.) Therefore, a 1 M HCl solution will have pH = 0, and a 1 M NaOH solution pH = 14. The following example will give you some practice in calculating pH once you know $[H^+]$ or $[OH^-]$.

EXAMPLE 12.4

Calculate the pH of the following solutions: (a) 0.100 M HCl; (b) 0.500 M HNO_3 (also a strong acid); (c) 0.00200 M NaOH; (d) 0.00210 M NaOH; and (e) a mixture of 30.0 mL 0.25 M KOH and 20.0 mL 0.25 M HBr (again a strong acid).

Solution:
With today's calculators, finding pH from $[H^+]$ using the definition of Equation 12.21 is as straightforward as entering the numerical value of the concentration and pressing the "log" key, followed by the "+/−" key (the sign, not the operation). Another way is to enter $[H^+]$, then "1/x," and then "log." (Why does this

give the same result?) All the examples are strong acids and bases, so you can assume 100% ionization in each case.

a. $pH = -\log_{10}(1.00 \times 10^{-1}) = -(-1.00) = 1.00$

b. HNO_3, like HCl, is **monoprotic,** that is, it yields one proton per ionizing molecule, so

$$[H^+] = [HNO_3]_0 \text{ and } pH = -\log_{10}(5.00 \times 10^{-1}) = 0.30$$

c. Here you could either use Equation 12.20 to find $[H^+]$ from $[OH^-]$ followed by Equation 12.21, or use the generalized form of Equation 12.21 to find pOH, followed by Equation 12.22 to get pH. (The latter is generally preferred.)

$$pOH = -\log_{10}(2.00 \times 10^{-3}) = 2.70; \qquad pH = 14.00 - pOH = 11.30$$

d. This tests the sensitivity of the log scale.

$$pOH = 2.68; \qquad pH = 14.00 - pOH = 11.32$$

e. When these solutions are mixed, an Arrhenius acid–base reaction will occur nearly to completion, according to the analysis presented earlier. Here, however, the base is in excess; remembering that excess strong base will inhibit water ionization, the pH will be entirely determined by the concentration of excess base. The final $[OH^-]$ is then

$$[OH^-] = [(0.03 \text{ L})(0.25 \text{ M}) - (0.020 \text{ L})(0.25 \text{ M})]/0.050 \text{ L}$$
$$= 0.050 \text{ M}$$
$$pOH = 1.30; pH = 12.70$$

Some of these concentrations may seem small, but at the level of individual ions the numbers are plentiful; for example a 1-mL sample of pure water contains $(10^{-7})(0.001)(6 \times 10^{23}) = 6 \times 10^{13}$ (60 trillion) ion pairs, and even 1 mL of 1 M NaOH still has six million H^+ ions floating in it. Thus in aqueous acid–base chemistry there are large numbers of species of all types at equilibrium, and the statistical underpinning of the equilibrium ideas remains intact. This allows us to examine with some confidence the behavior of acids and bases that do not produce 100% ions at equilibrium.

Ionization of Weak Acids and Bases

While HCl, HBr, and HI are all strong acids, HF is a **weak acid,** meaning that the equilibrium

$$HF(aq) \rightleftharpoons H^+(aq) + F^-(aq), \qquad K \equiv K_a = 7.0 \times 10^{-4}$$

lies far to the left, favoring undissociated HF. The reasons for this are again surprising. While HF has the strongest single bond known, $D_0(H\text{—}F) = 135$ kcal/mol, the ionizing reaction is nonetheless *exothermic,* $\Delta H° = -3.0$ kcal/mol, which favors its occurrence. Thus it is a decrease in entropy, $\Delta S° = -24.5$ cal/K mol, which owes to the increased solvent organization engendered by the ions, that makes this ionization weak. The resulting ΔG°_{298} is +4.3 kcal/mol, leading to the K_a value given earlier.

As usual, we will attempt to determine the equilibrium composition and therefore the pH of an HF solution by use of an extent of reaction variable x and the constraints imposed by stoichiometry. As in the strong acid case, we will neglect at first the H^+ supplied by water. Let's denote the initial concentration of HF by $c \equiv [HF]_0$, and make a table as in Equation 12.11:

$$\begin{array}{ccc} HF & \rightleftharpoons & H^+ + F^- \\ c - x & & x \quad\quad x \end{array} \tag{12.24}$$

This leads to the equilibrium expression

$$K_a = \frac{[H^+][F^-]}{[HF]} = \frac{x \cdot x}{c - x} \tag{12.25}$$

Note that if we start with un-ionized HF, the free energy change is not defined, but once equilibrium is reached and there are large numbers of all species present, the thermodynamic analysis is on firm ground. As with our NO_2 example, Equation 12.25 can be put in the form of a quadratic equation. But before we solve it for $x\,(=[H^+])$ using the quadratic formula, we can anticipate that x will be small; instead of $x = c$ as we would would have for a strong acid, we expect that $x \ll c$, leading to a substantially lower $[H^+]$ and higher pH than if HF were strong. This characteristic is illustrated for acetic acid, another weak acid, in Figure 12.7. This small extent of ionization leads directly to an easily solved *approximation* to Equation 12.25, obtained by neglecting x in the denominator, that is, by setting $c - x \approx c$. Then

$$K_a \approx \frac{x^2}{c}, \quad x^2 \approx cK_a, \quad x = [H^+] \approx \sqrt{cK_a} \tag{12.26}$$

$$pH \approx \tfrac{1}{2}(pc + pK_a)$$

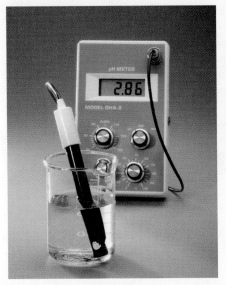
(a) CH$_3$COOH

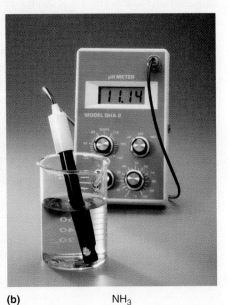
(b) NH$_3$

Figure 12.7
0.10 M solutions of a weak acid and a weak base. The pH values indicate that only a small fraction of the acetic acid (CH$_3$COOH) and ammonia (NH$_3$) molecules have produced ions in solution. Bromthymol blue has been added to each solution.

where the necessity of a positive concentration forces us to choose the positive square root. Note that Equation 12.26 must always yield too large an [H^+], or too small a pH, compared to the exact solution. (Can you surmise why?) Thus we can estimate the pH of a 0.100 M HF solution as pH ≈ ½(1.00 + 3.15) = 2.08, corresponding to [H^+] ≈ 8.4 × 10^{-3}. Thus x = [H^+] is only 8% of the original concentration c, and we expect our approximate pH to be only slightly smaller than the exact answer.

You can readily obtain the exact result for weak acid ionization by putting Equation 12.25 in quadratic form and applying the quadratic root formula, which yields

$$x = [H^+] = \frac{-K_a + \sqrt{K_a^2 + 4cK_a}}{2} \tag{12.27}$$

where the choice of the "plus" sign in the quadratic formula is again dictated by the positivity of [H^+]. For 0.100 M HF the result is [H^+] = 8.02 × 10^{-3} M, smaller, as expected, than the approximate result from Equation 12.26 by 5%; the pH is 2.096. Let's examine briefly how the quadratic formula of Equation 12.27 reduces to Equation 12.26. You may have noticed, if you ran the numbers through your calculator to check our quadratic answer, that within the square root, $4cK_a \gg K_a^2$, causing the square root to reduce approximately to $2\sqrt{cK_a}$. Further, $2\sqrt{cK_a} \gg K_a$, although this inequality is not as strong as the first one, leading finally to Equation 12.26. This might lead you to suspect that Equation 12.26 could be readily improved on, without having to use the quadratic formula, by not neglecting K_a with respect to $2\sqrt{cK_a}$. This leads to an improved approximation:

$$[H^+] \approx \sqrt{cK_a} - \frac{K_a}{2} \tag{12.28}$$

Equations 12.26 and 12.28 can be derived more rigorously by using the binomial theorem to expand the square root in Equation 12.27. We observe that this improved approximation reduces [H^+] as it should. Furthermore, for 0.100 M HF, Equation 12.28 yields [H^+] = 8.02 × 10^{-3} M, identical to three significant figures with the quadratic solution.

Another method for getting a more exact answer is the method of *successive approximations*, also called the *iterative method*, in which a first estimate is improved on by iteration using the exact equation. In this case you would use Equation 12.26 to make the estimate x_0, then obtain an improved value x_1 from $x_1 = \sqrt{(c - x_0)K_a}$, and continue if necessary with $x_{i+1} = \sqrt{(c - x_i)K_a}$, where i = 1, 2, . . . , watching for *convergence*, that is, agreement between successive values. In this case the successive values converge rapidly, x_0 = 8.37, x_1 = 8.01, x_2 = 8.02, x_3 = 8.02 × 10^{-3} M; the successive approximations method is most often used in computer programs designed to handle reaction quotients of arbitrary complexity.

EXAMPLE 12.5

Find the pH resulting from the ionization of acetic acid (CH_3COOH) in a solution originally 0.0200 M in the acid; K_a = 1.76 × 10^{-5}.

Solution:
Let's compare the four ways of computing the pH given earlier. The (exact) quadratic formula of Equation 12.27 yields

$$[H^+] = 0.5\{-1.76 \times 10^{-5} + [(1.76 \times 10^{-5})^2 + 4(0.0200)(1.76 \times 10^{-5})]^{1/2}\}$$

$$= -8.8 \times 10^{-6} + 5.934 \times 10^{-4} = 5.846 \times 10^{-4}$$

$$pH = 3.233$$

while the easy-to-use formula of Equation 12.26 gives

$$[H^+] \approx [(0.0200)(1.76 \times 10^{-5})]^{1/2} = 5.933 \times 10^{-4}; \; pH = 3.227$$

an error of less than 2% in $[H^+]$. This is generally considered to be acceptable accuracy. It is clear that the modification of Equation 12.28 will be essentially exact, since the correction $-K_a/2$ is the first term in the quadratic formula, and Equation 12.26 agrees with the square root second term. The iterative method yields the successive values $[H^+] = 5.933, 5.844, 5.846 \times 10^{-4}$, again in agreement with the quadratic formula.

The reason for all the acrobatics in solving a simple quadratic equation is to give you a set of tools to use in tackling more complex equilibrium expressions. In nearly all cases the simplest formula (Equation 12.26) is adequate for weak acid ionization. It will fail, however, in well-defined limits—for example, when the initial concentration c is small enough to be comparable to or less than K_a in magnitude. You can see this by examining how the quadratic formula of Equation 12.27 reduces to Equation 12.26. Of greater importance is the failure that occurs when any of the Formulas 12.26 to 12.28 predicts a pH close to or greater than 7. In our HF example, this would happen if the initial concentration of HF were $\sim 10^{-6}$ M or less. This implies that the H^+ produced by water itself cannot be neglected, and gives us our first look at *competing equilibria*.

In a situation where water is competitive with a weak acid such as HF in producing H^+, we must consider the equilibrium conditions of Equations 12.20 and 12.25 simultaneously. There are now four unknown concentrations—$[H^+]$, [HF], $[F^-]$, and $[OH^-]$—to be determined, and without further information, these unknowns cannot be found. We cannot construct a table like that of Equation 12.24, because now $[H^+]$ must be greater than, rather than equal to, $[F^-]$, because of the $[H^+]$ from H_2O. Fortunately, there are more general relations that must be obeyed, namely, **charge balance** and **mole balance** (in some texts, mass balance or material balance) arising from the conservation laws for charge and mass. Charge balance requires that the sum of all positive charges in the solution must equal that of all negative charges, whereas mole balance requires that the initial amount of acid (or other centrally involved species) must appear at equilibrium in either ionized or un-ionized form (in general as a reacted or unreacted species). These two conditions supply the additional constraints needed to allow the four unknowns to be determined. Let's write the two equilibrium conditions and the charge balance and mole balance relations together:

(a) $\quad K_a = \dfrac{[H^+][F^-]}{[HF]}$ (12.29)

(b) $\quad K_w = [H^+][OH^-]$

(c) $\quad [H^+] = [F^-] + [OH^-] \quad$ (charge balance)

(d) $\quad [HF]_0 = [HF] + [F^-] \quad$ (mole balance)

Solving these four equations simultaneously for the four unknown concentrations is not easy, because conditions (a) and (b) are *nonlinear*. Nonetheless the algebra can still be reduced to solving a single equation in one variable through the method of elimination of variables: conditions (b), (c), and (d) can be used to express, [HF], [F$^-$], and [OH$^-$] in terms of [H$^+$], and the results substituted in condition (a). If we let $x = [H^+]$ and $c = [HF]_0$ as before, we obtain

$$K_a = \dfrac{x\left(x - \dfrac{K_w}{x}\right)}{c - x + \dfrac{K_w}{x}}$$ (12.30)

This result is a more exact form of Equation 12.25; when multiplied out to remove all x-containing denominators, it yields a cubic equation. Cubic equations are not easily solved analytically, and we don't want you to devote time and worry to the exact solution of this one. Rather let's examine how it reduces to Equation 12.25. This happens, as you can see by inspection, when the correction terms K_w/x become negligible compared to x. From condition (b), of Equation 12.29, $K_w/x = [OH^-]$, which in turn is equal to the [H$^+$] arising from water autoionization. This means that Equation 12.30 reduces to Equation 12.25 when $[H^+]_{H_2O} \ll [H^+]_{HF}$, as you may have anticipated. The lesson to be taken home is the method of setting up the conditions: use all equilibria plus charge and mole balance.

A large number of compounds exhibit weak acidity, so we haven't been devoting all this effort to the problem in vain. A number of the simple inorganic acids besides HF are weak, most of them the so-called **oxyacids,** for example, nitrous acid (HNO$_2$). Other examples, such as hydrosulfuric acid (H$_2$S) and phosphoric acid (H$_3$PO$_4$), can donate more than one proton in solution and are called **polyprotic** (in some texts polybasic); that case will be discussed later in this chapter. Many examples of **monoprotic acids** are organic compounds containing the *carboxyl group* –COOH, with the generic structure

$$R-\underset{O}{\overset{\overset{O}{\|}}{C}}-O-H$$

where R represents a substituent group, usually an *alkyl group* C_nH_{2n+1}, where $n=0$, 1, 2, ... The hydrogen atom on the right is slightly acidic; that is, in water a few of the O—H bonds will cleave heterolytically to yield H$^+$(*aq*) and a *carboxylate anion* RCOO$^-$(*aq*). Carboxylate has two resonance forms, implying delocalized π bonding in the O—C—O linkage. The simplest of these acids is that for R = H, written HCOOH and called *formic acid*. (The modern systematic name is methanoic acid; a brief introduction to organic nomenclature is given in Chapter 18.) Formic acid is the famous distillate obtained from red ants. By far the most common is *acetic acid* (ethanoic acid, CH_3COOH), well known as the ingredient in vinegar responsible for its pungent odor; we have already met its anion, the *acetate ion* (CH_3COO^-), in Chapter 5 and in the example just given. Next in the series is *propionic acid* (propanoic acid, C_2H_5COOH), whose anion, in the form of calcium propionate, is frequently used in packaged foods as a spoilage-retarder.

Next in line are the rancid-smelling butyric acid (butanoic acid, C_3H_8COOH) and every possible value of n in C_nH_{2n+1}; those for $n = 5\text{–}18$ are known as *fatty acids*, found in natural animal and vegetable fat, and are often used for facial and bath soap and skin care products. Fatty acids may also contain carbon–carbon double bonds (in so-called unsaturated fats). The carboxyl group may also replace an H on benzene, yielding *benzoic acid* (C_6H_5COOH), and a whole family of related compounds, including aspirin. Other hydrogens may be substituted by halogens F, Cl, etc., or by *amine groups*—NH_2; when an NH_2 group is bonded to the carbon neighboring the carboxylic carbon, the compound is called an *α-amino acid*. The simplest of the 20 amino acids that are used to make proteins in living organisms is *glycine* (H_2NCH_2COOH), which is just acetic acid with H replaced by NH_2. It might be apparent to you that carboxylic acids, too, can be polyprotic, if another carboxyl group is strung onto the R substituent.

Table 12.1 lists ionization equilibrium constants K_a for selected monoprotic acids, strong or weak. Some trends are evident in the K_a values that lend insight into the relation between acidity and structure:

1. In the hydrohalic acids, acid strength increases as the group is descended (although the solvent leveling effect, discussed in Chapter 5, makes HCl, HBr, and HI all appear equally strong in water), due in this case to the decreasing H—X bond strength, which dominates over the slightly less favorable hydration energy of the larger anions.

2. In the oxyhalic acids (HXO_n), for a given halogen, such as Cl, the acidity increases with the number of nonhydroxyl oxygens bonded to the halogen, increasing by a factor of 10^5 or so for each O added. This effect owes to the electron-withdrawing character of the nonhydroxyl O's, which makes the X atom increasingly starved for electron density, which it can acquire by losing H as H^+. The argument is similar to that given in Chapter 5 as to whether a given X—O—H linkage will be acidic or basic. Here, the added O's are like additional, powerful members of a tug-of-war team pulling electron density away from H and making it easier to ionize. Inspection of the table shows that this effect is also present in HNO_n.

3. The discussion of the oxyhalic acids in trend 2 also aids in understanding why the carboxylic acids R—COOH are much more acidic than alcohols R—CH_2OH. Further, when R is an alkyl group rather than H, the acidity is reduced; it is well known in organic chemistry that alkyl groups donate rather than withdraw electron density, thereby making the hydroxyl H less positive. On the other hand, when Cl is substituted for H as in the chloroacetic acids, this is reversed and electrons are again withdrawn, making the acid stronger.

TABLE 12.1

Acid ionization constants K_a for selected monoprotic acids in aqueous solution

Acid name	Formula	K_a	pK_a
Hydrofluoric	HF	7.0×10^{-4}	3.15
Hydrochloric	HCl	$\sim 1. \times 10^{7}$	$-7.$
Hydrobromic	HBr	$\sim 1. \times 10^{9}$	$-9.$
Hydroiodic	HI	$\sim 1. \times 10^{11}$	$-11.$
Hypochlorous	HClO	3.0×10^{-8}	7.52
Chlorous	$HClO_2$	1.1×10^{-2}	1.96
Chloric	$HClO_3$	$\sim 1. \times 10^{3}$	$-3.$
Perchloric	$HClO_4$	$\sim 1. \times 10^{7}$	$-7.$
Hypobromous	HBrO	2.1×10^{-9}	8.68
Bromic	$HBrO_3$	$> 1.$	$< 0.$
Hypoiodous	HIO	$\sim 1. \times 10^{-11}$	11.
Iodic	HIO_3	1.6×10^{-1}	0.80
Periodic	H_5IO_6	5.1×10^{-4}	3.29
Nitrous	HNO_2	4.6×10^{-4}	3.34
Nitric	HNO_3	$2. \times 10^{1}$	-1.3
Hypophosphorous	H_3PO_2	$1. \times 10^{-2}$	2.0
Hydrocyanic	HCN	4.9×10^{-10}	9.31
Formic	HCOOH	1.77×10^{-4}	3.75
Acetic	CH_3COOH	1.76×10^{-5}	4.75
Propionic	CH_3CH_2COOH	1.34×10^{-5}	4.87
Butyric	$CH_3(CH_2)_2COOH$	1.54×10^{-5}	4.81
Chloroacetic	$CH_2ClCOOH$	1.40×10^{-3}	2.85
Dichloroacetic	$CHCl_2COOH$	3.32×10^{-2}	1.48
Trichloroacetic	CCl_3COOH	$2. \times 10^{-1}$	0.70
Benzoic	C_6H_5COOH	6.46×10^{-5}	4.19
Glycine	H_2NCH_2COOH	1.67×10^{-10}	9.78

As for weak acids, a number of compounds when dissolved in water produce a yield of OH^- that is small compared to their initial concentration, and therefore are classified as **weak bases**. However, nearly all of these compounds act as bases not by donating OH^- to the solution, but by splitting or **hydrolyzing** water to produce OH^-. (Here we are excluding sparingly soluble metal hydroxides; they will be treated in the last section.) The prototypical example is ammonia NH_3, whose reaction with water was discussed in Chapter 5 in Equations 5.29 and 5.30. Let's promptly put our equilibrium tools to use:

12.3 Equilibria in Aqueous Acid–Base Chemistry

$$NH_3 + H_2O \rightleftharpoons NH_4^+ + OH^-$$
$$c - x \qquad\qquad x \qquad x$$

$$K \equiv K_b = \frac{[NH_4^+][OH^-]}{[NH_3]} = \frac{x^2}{c - x} \quad (12.31)$$

Here we again let c be the initial concentration $[NH_3]_0$ and x the extent of reaction; K_b is the **weak base ionization constant,** with a value of 1.79×10^{-5} for ammonia. Note that H_2O as the solvent does not enter the reaction quotient. By comparing with Equation 12.25 for HF ionization, you can see that, algebraically speaking, the problem of finding the equilibrium concentrations is identical to the weak acid case. The only difference is that here $x = [OH^-]$ and the pOH may be found directly. Another great relief comes from realizing that the approximations of Equations 12.26 to 12.28 and the competing equilibrium relation (Equation 12.30) can be applied as well with this simple change in the definition of x and the replacement of K_a by K_b. (The charge balance and mole balance relations of Equation 12.29 look different, reflecting the new chemistry, but the form of Equation 12.30 is not affected.) The following example illustrates the weak base problem.

EXAMPLE 12.6

A basic solution of pH = 10.500 is needed for pH calibration. What concentration of ammonia NH_3, $K_b = 1.79 \times 10^{-5}$, will yield this pH?

Solution:
For the weak base equilibrium exemplified in Equation 12.31, the extent of reaction variable $x = [OH^-]$. So we find pOH = 14.000 − pH = 3.500, and $[OH^-]$ = 3.16×10^{-4} M. Solving the equilibrium expression in Equation 12.31 for $c = [NH_3]_0$, we find

$$c = x + x^2/K_b = 3.16 \times 10^{-4} + (3.16 \times 10^{-4})^2/1.79 \times 10^{-5} = 5.90 \times 10^{-3}\,M$$

The approximation of Equation 12.26 becomes pOH = $\frac{1}{2}$(pc + pK_b) for weak bases, yielding

$$pc = 2pOH - pK_b = 2(3.500) - 4.747 = 2.253; \qquad c = 5.58 \times 10^{-3}\,M$$

which is too low by 5%. Can you find the pH that results from using the approximate c? (*Answer:* 10.512) The improved approximation of Equation 12.28 is $[OH^-] = \sqrt{cK_b} - K_b/2$ for a weak base; this yields the same c to three figures as in Equation 12.31 but in this problem is more cumbersome to use. The pH required is far enough from 7 that ionization of water does not contribute significantly; pH = 10.5 means $[H^+] = 3 \times 10^{-11}$ M, and this is also the $[OH^-]$ coming from H_2O. Solutions of exact and stable pH are better prepared by using a buffer system, to be described later in this section.

Most examples of weak bases are again organic compounds called *amines*, in which one or more of the hydrogens on NH_3 are substituted by alkyl groups, for example, *methylamine* (CH_3NH_2). Beginning with *butylamine* ($C_4H_9NH_2$), these smell like rotting fish. Again a benzene ring (phenyl group) can also be substituted, yielding *aniline* ($C_6H_5NH_2$). A very important class of bases involve N atoms substituted for carbon in ring compounds; for example, *pyridine* (C_5H_5N) is a benzene-like ring where N has replaced CH. Four such bases—known as adenine, guanine, cytosine,

TABLE 12.2

Weak base ionization constants K_b for selected monobasic nitrogen bases in aqueous solution

Base name	Formula	K_b	pK_b
Ammonia	NH_3	1.79×10^{-5}	4.75
Methylamine	CH_3NH_2	4.4×10^{-4}	3.36
Ethylamine	$CH_3CH_2NH_2$	4.7×10^{-4}	3.33
Propylamine	$CH_3(CH_2)_2NH_2$	3.8×10^{-4}	3.42
Butylamine	$CH_3(CH_2)_3NH_2$	4.1×10^{-4}	3.39
Dimethylamine	$(CH_3)_2NH$	5.1×10^{-4}	3.29
Trimethylamine	$(CH_3)_3N$	0.6×10^{-4}	4.22
Glycine	$HOOCCH_2NH_2$	2.2×10^{-12}	11.66
Aniline	$C_6H_5NH_2$	4.2×10^{-10}	9.38
Pyridine	C_5H_5N	2.3×10^{-9}	8.64

and thymine—are essential components of the double-stranded molecule deoxyribonucleic acid or DNA, our genetic material. DNA contains "molecular blueprints" for building proteins from *amino acids,* which you just met earlier. The juxtaposition of an amine group ($-NH_2$), which is basic, with a carboxyl group ($-COOH$), which is acidic, gives amino acids some unusual properties that we will discuss in the next subsection. And, as you might surmise, there are amine bases with two or more amine groups, the simplest being *ethylene diamine* ($H_2NCH_2CH_2NH_2$), that can accept two or more protons; these will be touched on briefly later in this chapter.

Table 12.2 lists ionization constants for a selection of nitrogen weak bases. As for acids, trends in these values arise from some general structural properties. It is most useful here to invoke the Lewis base idea, the availability of the lone pair on nitrogen for donation being the essential ingredient in the production of OH^- in water. Alkyl amines are stronger bases than ammonia for the same reason that alkyl carboxylic acids are weaker than formic acid: the alkyl group donates electron density to N, making the lone pair more attractive to acids. Further, as is also the case for the acids, this effect is almost independent of the size of the alkyl group—although adding more alkyl groups, as in di- or trimethylamine, does not increase basicity appreciably, and even reduces it in the trimethyl case. The much weaker character of the ring bases aniline and pyridine has to do with the rehybridization of N in these compounds, a subject you may learn about in an organic chemistry course.

Brønsted–Lowry Acid–Base Reactions

All of these weak acids and bases, true to their names, can undergo acid–base reactions; for example, the weak acid acetic acid reacts readily with the strong base NaOH:

$$CH_3COOH + NaOH \rightleftharpoons CH_3COO^-Na^+ + H_2O$$

This reaction can be regarded as a sum of the weak acid ionization and standard Arrhenius reactions (omitting the Na^+ spectator ion),

$$CH_3COOH \longrightarrow CH_3COO^- + H^+ \qquad K_a = 1.76 \times 10^{-5} \qquad (12.32)$$

$$\underline{H^+ + OH^- \longrightarrow H_2O \qquad \qquad K = 1/K_w = 1.0 \times 10^{14}}$$

$$CH_3COOH + OH^- \longrightarrow CH_3COO^- + H_2O \qquad K = K_a/K_w = 1.8 \times 10^9$$

The equilibrium constant for the overall reaction was obtained by multiplying those of the component reactions. This follows from the additivity of the $\Delta G°$'s combined with their logarithmic relation to the corresponding K's (Equation 12.8); you can also verify the resulting K by multiplying the two equilibrium conditions governing the component reactions. While the K derived for this reaction is large, and the equilibrium lies far to the right, the reaction is not as spontaneous as the Arrhenius acid–base reaction itself due to the weakness of acetic acid. The overall reaction of Equation 12.32 is a proton transfer between CH_3COOH and OH^-, and therefore is of the Brønsted–Lowry (BL) type.

The important BL concept that we want to put into an equilibrium context is that of conjugate acids and bases, discussed at some length in Chapter 5. In Reaction 12.32 the acetate ion CH_3COO^- is the conjugate base to the acid CH_3COOH, and similarly H_2O is the conjugate acid to the base OH^-. When the equilibrium in a BL reaction lies far to the right, the conjugate base is always weaker than the original base (CH_3COO^- is weaker than OH^-) and the conjugate acid is weaker than the original acid (H_2O is weaker than CH_3COOH). The salient point is that CH_3COO^-, the anion of a weak acid, is indeed a base, although it must be a weak base, since it is not as strong as OH^-. A solution of sodium acetate CH_3COONa should therefore show a pH greater than 7; that is, it should produce a slight excess of OH^- in water by abstracting H^+ from it. The reactions that occur when sodium acetate is dissolved in water are first 100% dissociation into aqueous ions, $CH_3COONa(s) \longrightarrow CH_3COO^-(aq) + Na^+(aq)$, followed by

$$CH_3COO^- + H_2O \rightleftharpoons CH_3COOH + OH^- \qquad (12.33)$$

This is again a BL reaction, and just the reverse of the highly spontaneous Reaction 12.32. The equilibrium here must therefore lie far to the left, as you might expect for a weak base, with an equilibrium constant K_b that is just the reciprocal of that for Reaction 12.32, $K_b = K_w/K_a$. The reciprocal relation between the K's follows from the fact that $\Delta G°$ for the reverse reaction is just the negative of the forward $\Delta G°$, again combined with $\Delta G° = -RT \ln K$ (Equation 12.8). You could also convince yourself of this by noting that the reaction quotient Q for the reverse reaction is the reciprocal of that for the forward.

The BL conjugate relationship holds for any weak acid HA, described by K_a, and its anion A^-, described by K_b, allowing us to state in general that

$$K_a K_b = K_w \qquad (12.34)$$

Since K_w is a constant, this is an inverse proportionality between K_a and K_b, implying that the weaker an acid is, the stronger the base its anion will be, and conversely. In the case of a strong acid like HCl, its anion Cl^- is essentially ineffective as a base;

NaCl solutions are pH neutral. Since acetic acid is not too weak, $K_a = 1.76 \times 10^{-5}$, the acetate ion is a very weak base, $K_b = 1.0 \times 10^{-14}/(1.76 \times 10^{-5}) = 5.7 \times 10^{-10}$. HCN, on the other hand, is very weak, making CN^- (as in NaCN) a more effective base. Given a table of K_a's such as Table 12.1, Equation 12.34 allows K_b for the anions to be computed readily. The BL conjugate base concept also allows us to expand the list of compounds that act as weak bases to include the salts of all weak acids listed in Table 12.1.

Because reactions of anions of weak acids are identical (except for the charges) to those of neutral weak bases, for example, Equation 12.31, equilibrium analysis follows along the same lines. That is, identifying x with OH^- and using K_b in place of K_a, Equations 12.26 to 12.28 and Equation 12.30 can all be applied in their domains of validity.

In analogy to Reaction 12.32, weak bases typified by NH_3 react with strong acids such as HCl—this being the archetypical BL reaction, Equation 5.37:

$$NH_3 + HCl \rightleftharpoons NH_4^+ + Cl^-$$

Again, if we omit the spectator ion Cl^-, the resulting reaction

$$NH_3 + H^+ \rightleftharpoons NH_4^+ \tag{12.35}$$

is just the sum of the weak base ionization reaction, Reaction 12.31, and the Arrhenius acid–base reaction as discussed earlier. The equilibrium constant for this reaction is then $K = K_b/K_w = 1.8 \times 10^9$, again making it an essentially complete reaction. NH_4^+ is now the conjugate acid, and will react weakly in water, again by the reverse of the acid–base reaction of Reaction 12.35, with equilibrium constant $K_a = K_w/K_b = 5.6 \times 10^{-10}$. We thus find that once again Equation 12.34 holds, where now K_a refers to the protonated cation of the weak base. Finding the pH of a solution of NH_4^+, obtained by dissolving $NH_4Cl(s)$ in water, for example, is again a matter of applying the weak acid algebra of Equations 12.26 to 12.30, this time without any change in the meaning of x or K_a. And finally, Equation 12.34 once again implies that weaker bases when protonated make stronger acids, and conversely.

Weak acids may of course also react with weak bases, for example,

$$CH_3COOH + NH_3 \rightleftharpoons CH_3COO^- + NH_4^+ \tag{12.36}$$

By writing the weak acid ionization, weak base ionization, and Arrhenius acid–base reaction under each other, you should convince yourself that Reaction 12.36 is the sum of those three reactions. Then the equilibrium constant is given by

$$K = \frac{K_a K_b}{K_w} \tag{12.37}$$

This is going to yield smaller values of K than the weak acid + strong base or the weak base + strong acid cases because both K_a and K_b are small; in this case we find $K = 3.2 \times 10^4$, still a quite spontaneous reaction, in which the products predominate at equilibrium. Because of the presence of four species with variable free energy, finding the position of equilibrium here is a problem distinct from those treated so far, but, as you should readily be able to work out, it still yields a quadratic equation for the extent of the reaction variable.

Reaction 12.36 leads us to look at the unique compounds known as amino acids; using the simplest of these, glycine (H_2NCH_2COOH), as an example, we have both a weak acid –COOH and a weak base –NH_2 group present in the same molecule. The molecule can then undergo an *internal* proton transfer

$$H_2NCH_2COOH \rightleftharpoons {}^+H_3NCH_2COO^- \tag{12.38}$$

a reaction we anticipate should be spontaneous with $K \approx 10^4$ by analogy with Reaction 12.36. The charged form of the acid is called a *zwitterion* (double ion) and is the predominant form at near-neutral pH. This helps to account for the anomalously low K_a value for glycine given in Table 12.1; since the ionizable proton is now bonded to the amine nitrogen rather than the carboxyl oxygen, ionization corresponds to that of NH_4^+, which as we've just seen has a K_a of very similar magnitude. Likewise, the very small K_b in Table 12.2 arises from protonation of the carboxylate anion of the zwitterion, rather than the amine group, which is already protonated.

Common Ions and Buffers

Thus far we have considered ionization of pure compounds and stoichiometric reactions as they come to equilibrium. What happens if we disturb that equilibrium by adding a new solute to the system? As you can guess, LeChâtelier's principle is going to serve as a powerful qualitative guide as to what we should expect; in addition it can often help us in attacking the quantitative determination of concentrations when the new equilibrium point is reached.

Let's illustrate the general ideas involved with a simple example. Suppose you have 50 mL of 0.2 M HF at equilibrium, a problem you may know how to approach by now. What happens when 50 mL of 0.2 M HCl is added? To attack this problem you first need some chemical knowledge, such as that contained in Table 12.1. HF is a weak acid, producing a small amount of H^+ as a product of its ionization. HCl is a strong acid, 100% ionized to H^+ and Cl^- in water. The H^+ from HCl affects the HF equilibrium because H^+ is a product of HF ionization; it increases the free energy of H^+ in the system by increasing its concentration and thereby upsets the free energy balance between HF and $H^+ + F^-$. LeChâtelier's principle tells us that the HF equilibrium will react to this change by opposing it, that is, by trying to use up the added H^+. The result is that the HCl inhibits HF ionization; the added H^+ is referred to as a **common ion,** that is, an ion that is the same as one involved in the equilibrium that has been disturbed. The extent-of-reaction table becomes

	HF	$\rightleftharpoons$	H^+	+	F^-	
Initial	c_1		0		0	
Equil, no HCl	$c_1 - x$		x		x	
Equil, HCl added	$c_1 - x$		$c_2 + x$		x	(12.39)

where we have set $c_1 = [HF]_0$ and $c_2 = [HCl]_0 = [H^+]_0$. Note that the *value* of x will change when HCl is added. The usual reaction quotient then leads to the relation

$$K_a = \frac{(c_2 + x)x}{c_1 - x} \tag{12.40}$$

Equation 12.40 again yields a quadratic equation that may be solved, but in practice it almost never is. LeChâtelier's principle allows us to say that the extent of reaction x, already small in pure HF, will be yet smaller with HCl added, and may therefore be neglected with respect to both c_1 and c_2. This has separate implications for $[H^+]$ and $[F^-]$: $[H^+]$ is determined essentially entirely by the amount of added HCl, and Equation 12.40 yields $[F^-] \approx K_a c_1/c_2$. In our example, we must use values of c_1 and c_2 that reflect the diluting effect of mixing the solutions (recall how this reduces the free energy). The final solution volume is 100 mL and therefore the initial concentrations of HF and HCl are each halved. Then $[H^+] \approx [HCl]_0 = 0.100$ M and pH = 1.00, while $[F^-] \approx (7.0 \times 10^{-4})(0.100)/(0.100) = 7.0 \times 10^{-4}$. Referring back to our earlier treatment of pure HF ionization, we found $[H^+] = [F^-] = 8.0 \times 10^{-3}$. Thus adding HCl decreases $[F^-]$, which equals $[H^+]_{HF}$, by more than an order of magnitude, in agreement with the prediction of LeChâtelier's principle. Further, our neglect of x with respect to c_1 and c_2 is good within an error less than 1%. This type of common ion problem has an important application: it enables us to "tune" $[F^-]$ over a range of values by changing the pH, a useful tool in carrying out controlled precipitations, to be discussed in Section 12.4.

From this example you may be able to perceive that adding common product ions is a general way to inhibit an already weak ionization reaction; we could also have added F^- ions [in the form of, say, KF(s) or KF(aq)] to HF, thereby again reducing the extent of reaction in a similar way. In that case, the pH of the solution is actually increased, while $[F^-] \approx [KF]_0$. The resulting solution has the remarkable property of resisting changes in pH when strong acids or bases are added in limited amounts, and is called a **buffer**. The analysis of a buffer solution, containing large concentrations of a weak acid and its salt, proceeds analogously to our earlier example. Now using a generalized notation HA for the weak acid, we have

$$\begin{array}{cccc} \text{HA} & \rightleftharpoons & H^+ + & A^- \\ [HA]_0 - x & & x & [A^-]_0 + x \end{array} \qquad K_a = \frac{x([A^-]_0 + x)}{[HA]_0 - x} \approx \frac{x[A^-]_0}{[HA]_0} \qquad (12.41)$$

By taking $-\log_{10}$ of both sides of the approximate equilibrium expression on the right, we obtain a formula yielding the pH directly:

$$\text{pH} = pK_a + \log_{10}\frac{[A^-]_0}{[HA]_0} \qquad (12.42)$$

This relationship is known as the **Henderson–Hasselbalch equation;** note that it corresponds exactly to the equilibrium condition in the limit where reaction can be neglected, since it uses the original concentrations of acid and salt. When moles of acid and salt are equal, Equation 12.42 yields pH = pK_a, the center point of the pH range where this analysis applies. This is also the point of greatest resistance to changes in pH, as the following example illustrates.

EXAMPLE 12.7

Find the pH of acetic acid (HA)/sodium acetate (NaA) buffer solutions made from

a. 50. mL 0.20 M HA and 50. mL 0.20 M NaA
b. 25. mL 0.20 M HA and 75. mL 0.20 M NaA.

Calculate the change in pH when 80 mg (0.002 mol) NaOH(s) is added to each of these solutions, and compare the results with adding the same amount of NaOH to an unbuffered solution of the same pH and volume as solution (a). Neglect the change in volume on adding NaOH(s).

Solution:
In applying the Henderson–Hasselbalch equation (Equation 12.42), we can take advantage of the fact that the concentrations appear only in a ratio, causing the (identical) final solution volumes to cancel; we therefore only need to know moles of salt and acid. This is an advantage as you will see when we need to obtain pH changes upon adding acid or base.

a. $n_{HA} = n_{NaA} = 0.010$ mol, and pH = pK_a = 4.75

b. $n_{HA} = 0.005$ mol, $n_{NaA} = 0.015$ mol; pH = $4.75 + \log_{10} 13.01 = 5.23$.

Note that solution (b), which contains more salt, has a higher pH as expected since the salt is basic. The added strong base will react stoichiometrically with HA (see Equation 12.32), thereby reducing the amount n_{HA} and increasing n_{NaA}.

a. $n_{HA} = 0.010 - 0.002 = 0.008$; $n_{NaA} = 0.010 + 0.002 = 0.012$ mol

$$\text{pH} = 4.75 + \log_{10}(0.012/0.008) = 4.93, \text{ an increase of } 0.18$$

b. $n_{HA} = 0.005 - 0.002 = 0.003$; $n_{NaA} = 0.015 + 0.002 = 0.017$ mol

$$\text{pH} = 4.75 + \log_{10}(0.017/0.003) = 5.50, \text{ an increase of } 0.27$$

So we see that solution (a), with equal amounts of acid and salt, is a more effective buffer than (b), since it yields a smaller change in pH. If the solution is unbuffered, the acid present to yield pH = 4.75 is negligible compared to the added base, and the pH is determined entirely by the NaOH: pOH = $-\log_{10}(0.002 \text{ mol}/0.100 \text{ L}) = 1.70$ and pH = 12.30, a swing of more than 7.5 pH units! An experiment similar to this example is illustrated in Figure 12.8.

The buffering action shown in the example and figure can be understood qualitatively by appealing to LeChâtelier's principle once again. Adding OH^- uses up H^+ (to make water), and the system then tries to replace the H^+ that was lost by ionizing more HA, thereby maintaining the pH. A similar situation would occur if strong acid were added, increasing moles of H^+: the reaction would run in reverse, A^- using up the added H^+ and reforming HA. A buffer has a finite capacity to absorb these acid and base doses; in the example with solution (b) more than 0.005 mol base added would have destroyed the buffer by reacting with all the HA, and any excess base would then cause a large jump in pH as in the unbuffered case. More robust buffers are made by using higher concentrations of acid and salt, or by using larger volumes. The buffer concept will also help you to understand titration of a weak acid by a strong base to be discussed later; and the common ion effect, a more general concept, is also useful in understanding solubility and precipitation of salts, treated in Section 12.4.

Weak bases can also be mixed with their salts to make buffers; we will leave it to you in the end-of-chapter exercises (Exercises 40 and 41) to see how the

Figure 12.8
Illustration of buffering action by an acetic acid/sodium acetate buffer. (**a**) A few drops of concentrated NaOH solution is added to the buffer solution, resulting in only a minor change in pH. (**b**) The same number of drops of base is now added to a dilute acid solution of nearly the same pH as the buffer, yielding a substantial change in pH. Universal indicator, which is red in acid and blue in base, has been added to each solution.

(a)

(b)

Henderson–Hasselbalch equation works in that case. By choosing either a weak acid or a weak base to build a buffer, effective buffer solutions can readily be made to cover the pH range of 3 to 11; outside this range the acid or base is too strong to allow effective buffering action by its salt. Many aqueous reactions are pH sensitive and benefit from the pH control a buffer affords; examples range from electroplating of metals to metabolic reactions in living cells.

Ionization of Polyprotic Acids

Many common acids are polyprotic, that is, they can supply more than one proton per ionizing molecule; the mineral acids sulfuric acid (H_2SO_4) and phosphoric acid (H_3PO_4) are examples of **oxyacids** with this property. In oxyacids the acidic protons are always part of hydroxyl –OH groups, each capable of heterolytic cleavage when the central atom is more electronegative than H. Successive protons are never lost with equal ease, however; as each proton departs, it leaves behind an increasingly negatively charged ion, and further ionization is opposed by an increasing Coulomb attraction. As a consequence, the first proton is always the most readily lost, and each one thereafter is increasingly difficult to lose. (This situation should remind you of the increasing ionization energies in an isoelectronic sequence of atomic ions of increasing positive charge, or of each successive electron removed from a single atom; see Chapter 4.)

For example, when CO_2 dissolves in water, *carbonic acid* (H_2CO_3) is formed, which can be written as $OC(OH)_2$ to indicate the two hydroxyl groups. Both H's are acidic, and carbonic acid ionizes in a stepwise fashion:

Step 1: $H_2CO_3 \rightleftharpoons H^+ + HCO_3^-$ $\quad K_{a1} = \dfrac{[H^+][HCO_3^-]}{[H_2CO_3]} = 4.3 \times 10^{-7}$ (12.43)

Step 2: $HCO_3^- \rightleftharpoons H^+ + CO_3^{2-}$ $\quad K_{a2} = \dfrac{[H^+][CO_3^{2-}]}{[HCO_3^-]} = 4.8 \times 10^{-11}$

This reasoning is reflected in the inequality $K_{a2} \ll K_{a1}$, the K's being the stepwise equilibrium constants. Each step has the form of ionization of a monoprotic weak acid, and the conjugate base HCO_3^- of the first step becomes the acid of the second step. At equilibrium all species involved in the two steps will have some finite concentrations, although in a solution of pure H_2CO_3 we expect that $[CO_3^{2-}]$ will be much smaller than $[HCO_3^-]$ because of the small magnitude of K_{a2}. It is useful to think of the problem of finding the equilibrium concentrations by analogy with the problem of accounting for water ionization as discussed earlier in this section. Here we will be content to neglect the contribution of water to the pH, but we still have two simultaneous equilibria to satisfy, and, as in Equations 12.29, there will also be charge balance and mole balance conditions. The result is again a cubic equation in $x \equiv [H^+]$ that we would like to avoid solving if we can! So once again we call on M. LeChâtelier for help.

Both ionization steps in Equation 12.43 produce protons, but the first step is more efficient by a factor of 10^4. Thus the pH of the solution is determined essentially completely by step 1, that is, by the first monoprotic ionization step. Incredibly, this means you can use the monoprotic weak acid ionization formulas, Equations 12.26 to 12.28, to find the pH of the solution, substituting K_{a1} for K_a, and further we have $[HCO_3^-] \approx [H^+]$ and $[H_2CO_3] \approx [H_2CO_3]_0 - [H^+]$. The only remaining unknown is $[CO_3^{2-}]$, and it is readily obtained from the step 2 equilibrium condition:

$$[CO_3^{2-}] = K_{a2}\dfrac{[HCO_3^-]}{[H^+]} \approx K_{a2} \quad (12.44)$$

This is another improbable result: $[CO_3^{2-}]$ is independent of the initial concentration $[H_2CO_3]_0$, and is equal to K_{a2}, no calculation required! These results hold for any weak acid donating two or more protons, as long as $K_{a2} \ll K_{a1}$.

Carbonate, and most of the other anions of polyprotic acids, are ordinarily found as salts, usually with the cations Na^+, K^+, or Ca^{2+}. A diprotic acid has two possible salts, with Na^+ for example replacing one proton, as in $NaHCO_3$, or both, as in Na_2CO_3, two of nature's and the chemical industry's most common compounds. When salts like these are dissolved in water, they ionize completely to yield the anions HCO_3^- and CO_3^{2-}. Because these anions are part of a weak acid equilibrium, as acid and/or conjugate base, solutions containing them will not be pH neutral. The completely deprotonated anion CO_3^{2-} is the easiest to analyze, since it acts only as a base, producing OH^- in solution according to

$$CO_3^{2-} + H_2O \rightleftharpoons HCO_3^- + OH^- \quad (12.45)$$

This reaction has the same form, except for the extra negative charge, as that for the anion of a weak monoprotic acid, as in Equation 12.33; its equilibrium constant is therefore given by $K_{b2} = K_w/K_{a2}$. In this case $K_{b2} = 2.1 \times 10^{-4}$, and an appreciable

[OH$^-$] is expected. HCO$_3^-$ can react further to make more OH$^-$, but that reaction will be negligible because of its much smaller $K = K_{b1} = K_w/K_{a1} = 2.3 \times 10^{-8}$. Thus the pOH and pH are determined nearly completely by the equilibrium of Reaction 12.45, and the calculation is exactly as we have discussed for the salt of a monoprotic acid.

The hydrogen-containing salts are not as easy to analyze, because they can either lose their remaining protons or gain the ones they're missing. HCO$_3^-$ for example can react with water to form H$_2$CO$_3$ or ionize to form CO$_3^{2-}$:

$$\text{HCO}_3^- + \text{H}_2\text{O} \rightleftharpoons \text{H}_2\text{CO}_3 + \text{OH}^- \qquad K_{b1} = K_w/K_{a1} = 2.3 \times 10^{-8} \qquad (12.46)$$

$$\text{HCO}_3^- \rightleftharpoons \text{H}^+ + \text{CO}_3^{2-} \qquad K_{a2} = 4.8 \times 10^{-11}$$

From the relative sizes of the K's, you would conclude that a NaHCO$_3$ solution will be basic. This turns out to be true, but the exact treatment of the equilibria is complex. We can develop an approximate analysis by noting that HCO$_3^-$, as both acid and base, can react with *itself* according to

$$2\,\text{HCO}_3^- \rightleftharpoons \text{H}_2\text{CO}_3 + \text{CO}_3^{2-} \qquad K = \frac{K_a K_b}{K_w} = \frac{K_{a2}}{K_w}\left(\frac{K_w}{K_{a1}}\right) = \frac{K_{a2}}{K_{a1}} \qquad (12.47)$$

$$= 1.1 \times 10^{-4}$$

where we have drawn on our earlier analysis of the reaction of a weak acid with a weak base for the K formula (see Equation 12.37), using the appropriate K_a and K_b from Equation 12.46. You can verify this by taking the reactions of Equation 12.46 along with water ionization and adding them in a way that yields Equation 12.47. The resulting K for the self-reaction is considerably larger than those of Equation 12.46, implying that the equilibrium of Equation 12.47 will dominate. An extent-of-reaction analysis shows that this in turn leads to $[\text{CO}_3^{2-}] \approx [\text{H}_2\text{CO}_3]$ at equilibrium. Further, because $K \ll 1$, $[\text{HCO}_3^-] \approx [\text{HCO}_3^-]_0$. These two approximations, when combined with the equilibrium conditions governing Reaction 12.47 and the second reaction of Equation 12.46 (using simple algebra that you can carry through yourself), lead to the results

$$[\text{CO}_3^{2-}] \approx [\text{H}_2\text{CO}_3] \approx [\text{HCO}_3^-]_0 \sqrt{\frac{K_{a2}}{K_{a1}}} \qquad (12.48)$$

$$[\text{H}^+] \approx \sqrt{K_{a1} K_{a2}} \qquad \text{or}$$

$$\text{pH} \approx \tfrac{1}{2}(pK_{a1} + pK_{a2})$$

The result for the pH is surprising, being independent of the initial concentration of NaHCO$_3$; putting in numbers yields pH $= 8.34$, basic as we guessed earlier. As for the ionization of the acid itself, these equilibrium formulas for solutions of pure salts hold for any polyprotic weak acid system with $1 \gg K_{a1} \gg K_{a2}$. Now we are ready to look at mixtures and the common ion effect.

If you add a strong acid to a weak polyprotic acid, the same common ion effect will occur as discussed in the previous subsection. Ionization of the weak acid will be inhibited, and the pH will be determined by the amount of strong acid added. To find out what happens to the doubly charged anion in this case, you can add the two ionization steps of Equation 12.43 to get the total ionization reaction, using the generalized notation H_2A,

$$H_2A \rightleftharpoons 2H^+ + A^{2-} \qquad K = K_{a1}K_{a2} = \frac{[H^+]^2[A^{2-}]}{[H_2A]} \qquad (12.49)$$

Because $[H^+]$ is already fixed, and $[H_2A] \approx [H_2A]_0$, $[A^{2-}]$ can be computed directly from the overall equilibrium condition of Equation 12.49; while $[HA^-]$ can be computed similarly from the first-step equilibrium. Because again we can "tune" the concentrations of the anions to any value by varying the amount of acid added, this has an important application in controlled precipitation of salts of the anion.

EXAMPLE 12.8

Hydroselenic acid (H_2Se), an important intermediate in the production of ultrapure Se for photocopying machines, is a weak diprotic acid with ionization constants $K_{a1} = 1.29 \times 10^{-4}$ and $K_{a2} = 1.0 \times 10^{-11}$. Find the pH and the concentration of Se^{2-} for the following solutions:

a. 0.100 M H_2Se
b. 0.100 M Na_2Se
c. 0.100 M NaHSe
d. 0.100 M in both H_2Se and HCl.

Solutions:
The two ionization steps are $H_2Se \rightleftharpoons H^+ + HSe^-$ and $HSe^- \rightleftharpoons H^+ + Se^{2-}$.

a. Since $K_{a2} \ll K_{a1}$, the ionization is dominated by the first step, so $[H^+] \approx [cK_{a1}]^{1/2} = [(0.100)(1.29 \times 10^{-4})]^{1/2} = 3.59 \times 10^{-3}$ M; pH = 2.44. Equation 12.44 fixes $[Se^{2-}] = K_{a2} = 1.0 \times 10^{-11}$ M; since this is also $[H^+]$ from step 2, the assumption of step 1 dominance is warranted.

b. Here Na_2Se is assumed to be 100% ionized into $2Na^+ + Se^{2-}$, and Se^{2-} acts as a base as in Equation 12.45. First-step domination yields

$$[OH^-] \approx [cK_w/K_{a2}]^{1/2} = [(0.100)(10^{-14})/(10^{-11})]^{1/2} = 0.0100 \text{ M}$$

the exact answer is 0.0095 M;

$$\text{pOH} = 2.02 \text{ and pH} = 11.98.$$

$$[Se^{2-}] = [Se^{2-}]_0 - [OH^-] = 0.0905 \text{ M}.$$

c. Again assuming NaHSe produces 100% HSe^-, and using Equations 12.48, we have

$$[H^+] \approx [K_{a1}K_{a2}]^{1/2} = [(1.29 \times 10^{-4})(1.0 \times 10^{-11})]^{1/2} = 3.59 \times 10^{-8} \text{ M}$$

$$\text{pH} = 7.44$$

$$[Se^{2-}] \approx [HSe^-]_0[K_{a2}/K_{a1}]^{1/2} = (0.100)[(1.0 \times 10^{-11})/(1.29 \times 10^{-4})]^{1/2}$$

$$= 2.8 \times 10^{-5} \text{ M}$$

d. The H^+ from HCl (100% ionized) inhibits H_2Se ionization, so pH = 1.00 and, using Equation 12.49,

$$[Se^{2-}] \approx ([H_2Se]_0/[H^+]^2)K_{a1}K_{a2} = [(0.100)/(0.100)^2](1.29 \times 10^{-4})(1.0 \times 10^{-11})$$
$$= 1.3 \times 10^{-14} \text{ M}$$

three orders of magnitude smaller than part (a).

A polyprotic acid is far more flexible for making up buffer solutions, since now you can choose a pK_a from any of the ionization steps. For example, carbonic acid buffers can be made for use near pH = pK_{a1} = 6.37, or near pH = pK_{a2} = 10.32. The former, which is a major component of the buffering system in human blood, relies on large concentrations of H_2CO_3 and HCO_3^-, while the latter would use HCO_3^- and CO_3^{2-}. In either case the Henderson–Hasselbalch equation (Equation 12.42) would apply with the appropriate K_a and choice of acid and conjugate base. (Note that the conjugate base always resides in the numerator of the log term.)

Table 12.3 collects stepwise equilibrium constants for some representative diprotic and triprotic acids. They all show first-step dominance, as we reasoned at the

TABLE 12.3
Stepwise equilibrium constants for selected polyprotic acids

Acid name	Formula	pK_{a1}	pK_{a2}	pK_{a3}
Hydrosulfuric	H_2S	7.04	13.89	
Hydroselenic	H_2Se	3.89	11.	
Sulfurous	H_2SO_3	1.81	6.99	
Sulfuric	H_2SO_4	−2.	1.92	
Selenous	H_2SeO_3	2.46	7.31	
Selenic	H_2SeO_4	−2.	1.92	
Phosphorous	H_3PO_3	2.00	6.59	
Phosphoric	H_3PO_4	2.12	7.21	12.67
Arsenic	H_3AsO_4	2.25	6.77	11.60
Carbonic	H_2CO_3	6.37	10.32	
Silicic	H_4SiO_4	9.66	11.7	12.
Chromic	H_2CrO_4	0.74	6.49	
Oxalic	HOOCCOOH	1.23	4.19	
Malonic	$HOOCCH_2COOH$	2.83	5.69	
Succinic	$HOOC(CH_2)_2COOH$	4.16	5.61	
Glutaric	$HOOC(CH_2)_3COOH$	4.31	5.41	
Adipic	$HOOC(CH_2)_4COOH$	4.43	5.41	
Phthalic	$C_6H_4(COOH)_2$	2.89	5.51	
Citric	$C_3H_4(OH)(COOH)_3$	3.14	4.77	6.39
Ascorbic (vitamin C)	$C_6H_6O_4(OH)_2$	4.10	11.79	

beginning of this subsection. As for the oxyhalo acids, the group VI oxyacids of S and Se as well as the group V oxyacids of P and As show the acidifying effect of additional nonhydroxyl oxygens. (In H_3PO_3 one of the H's is bound directly to P, so that both H_3PO_3 and H_3PO_4 have one nonhydroxyl O.) The *dicarboxylic acids* oxalic through adipic are an interesting test of our reasoning regarding the stepwise nature of the ionization of diprotic acids. In this sequence, the carbon-chain "spacer" is growing in length, increasing the spatial separation of the two carboxyl groups. As this happens, K_{a1} decreases and then stabilizes at a value close to those for monocarboxylic acids (see Table 12.1); this trend is a result of the decreasing influence of the second carboxyl group, which is an electron-withdrawing group, on the ionization of the first. Then K_{a1} begins to approach K_{a2}, which is relatively unperturbed by the spacer length; the similar values for long chains result from the large separation of the negative carboxylate formed in the first step from the second ionizable carboxyl. The carboxylate therefore exerts less Coulombic restraint on the second ionization. This contrasts with the case of the inorganic acids, where the protons all must come from the same center, and unfortunately causes failure of many of the approximations we have developed in this subsection. Nature is not always kind to its lovers.

Just as two or more carboxyl groups may be found in the same molecule, so may two or more amine groups, giving rise to polybasic compounds. These are far less common than the acids; in addition the equilibrium treatment of the acid–base behavior of these compounds and their salts mirrors that for the polyprotic acids, just as we found in the monoprotic case; we therefore will not dwell on such a treatment here. Table 12.4 lists four of the more common dibasic nitrogen bases; as in the acids, the first ionization step is more favorable than the second. Hydrazine is more noted for its use as a rocket fuel, ethylene diamine as a *bidentate ligand* in compounds with transition metal ions, hexamethylene diamine as a reactant in the manufacture of nylon (along with adipic acid), and adenine as one of the purine bases that are part of DNA.

Acid–Base Titrations

In a **titration,** solutions are mixed in a precise manner to produce as nearly as possible exact stoichiometry in a reaction, and acid–base titrations are by far the most common. The equipment is simple: a **buret,** a long tube with volume graduations and a stopcock at one end, which generally holds a base solution of known concentration (called the **titrant**), and a beaker or flask containing an acid to be analyzed (called the **analyte**), also in solution. A few drops of **indicator** solution are usually added to the flask to make the stoichiometric equivalence point (called the **endpoint**) visible through a color

TABLE 12.4

Stepwise equilibrium constants for selected dibasic nitrogen bases

Base name	Formula	pK_{b1}	pK_{b2}
Hydrazine	H_2NNH_2	6.05	14.
Ethylene diamine	$H_2N(CH_2)_2NH_2$	3.39	6.44
1,6-diaminohexane	$H_2N(CH_2)_6NH_2$	2.14	3.24
Adenine	$C_5N_5H_5$	4.17	9.88

change. You are expected to have learned the basics of doing a titration in your laboratory sessions, most likely during the first term. Here you will learn to understand how titrations work by analyzing how the pH changes while a titration is in progress; it is also likely that you will carry out titration experiments in which you measure the pH.

Prior to beginning a titration, it is typical for the analyte sample to be either dissolved in water, if it is a solid, or for the solution containing it to be diluted. In the usual case of an acid analyte, this means that the pH of the solution is determined by both the amount of acid present and the total solution volume, although the moles of acid present is naturally independent of how much extra water is added. Then, as base is added from the buret, the total volume of the solution is constantly increasing, and must be accounted for in calculating the pH as the titration progresses. As long as you bear these minor complications in mind, there are no new calculations to master here, only a few new concepts. The most important of these is known as the **pH titration curve,** simply a plot of pH versus volume of added titrant.

First consider the titration of a strong acid, such as HCl, by a strong base, such as NaOH. We know the Arrhenius acid–base reaction is strongly spontaneous, with equilibrium lying far to the right; this means that any OH^- added will react stoichiometrically with the H^+ present (see Example 12.4). If the amount of base added is not yet *equivalent* to the amount of acid in the analyte, the acid will be in excess and the base will be the limiting reagent. The pH is then determined by the amount of acid still present, that is, not yet titrated. Exactly at the endpoint (a condition very difficult to attain, because of the finite drop size, ~0.05 mL, a buret can deliver), we have a solution containing only NaCl, giving pH 7.00. For volumes beyond the endpoint (generally not important in a practical titration; being here means you "overshot" the endpoint) the base titrant is now in excess, and it now determines the pH. The implication is that there is a large swing in pH very near the endpoint, as Example 12.9 and Figure 12.9 will now be used to illustrate.

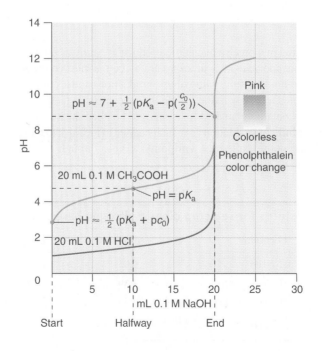

Figure 12.9
pH titration curves for a strong acid (HCl) and a weak acid (CH_3COOH) titrated by a strong base (NaOH). The pH formulas (from Equations 12.26 and 12.42) all refer to the weak acid at the three important points discussed in the text. The K_a of the weak acid may be obtained directly from the pH at the halfway point. The factor of 2 in the $p(c_0/2)$ term in the endpoint formula is a dilution factor that applies when the volumes of acid and base are equal; in general, this factor would be the ratio of the endpoint volume to the initial volume of the analyte solution. Beyond the weak-acid endpoint the strong- and weak-acid curves are indistinguishable.

EXAMPLE 12.9

Titrations are often used to determine more precisely the concentrations of solutions that have been made up to an approximate molarity. Suppose 50.00 mL of a solution of approximately 0.1 M HCl (obtained by **pipet**) is titrated with 0.2000 M NaOH solution, and 25.35 mL NaOH is required to reach the endpoint, as shown by the indicator phenolphthalein turning from colorless to pink. Find the actual concentration of the HCl solution, and calculate the pH for the following volumes of NaOH added: (a) 0.00 mL (before starting); (b) 10.00 mL; (c) 20.00 mL; (d) 25.00 mL; (e) 25.30 mL (one drop away from the endpoint); (f) 25.40 mL (one drop beyond the endpoint); (g) 30.00 mL; and (h) 40.00 mL. Plot your results as pH versus volume added.

Solution:
Since the reaction is NaOH + HCl $\longrightarrow$ NaCl + H_2O, with equal stoichiometric coefficients for acid and base, the endpoint corresponds to equal moles. Working in millimoles (mmol), we have

$$\text{mmol NaOH} = (0.2000 \text{ M})(25.35 \text{ mL}) = 5.070 \text{ mmol} = \text{mmol HCl}$$

therefore,

$$\text{M HCl} = (5.070 \text{ mmol HCl})/(50.00 \text{ mL}) = 0.1014 \text{ M}$$

Now the pH can be calculated for various volumes of added base. Recall that prior to the end point the base limits the reaction, and after the acid does. The HCl was not diluted beforehand, so the starting volume is 50.00 mL (neglecting the few drops of indicator).

a. $\text{pH} = -\log_{10}(0.1014) = 0.99$.

b. Because of the changing solution volume, it is always safest to work in moles rather than concentrations; this is normal in limiting reagent problems. From above, initial mmol HCl = 5.070; so

$$\text{mmol HCl left} = 5.070 \text{ mmol} - \text{mmol NaOH added}$$
$$= 5.070 \text{ mmol} - (0.2000 \text{ M})(10.00 \text{ mL})$$
$$= 5.070 - 2.000 = 3.070 \text{ mmol}$$
$$\text{M HCl} = (3.070 \text{ mmol})/(50.00 \text{ mL} + 10.00 \text{ mL}) = 0.05117 \text{ M}; \text{pH} = 1.29$$

Note that even though the titration is 40% complete, the pH, being a logarithmic quantity, has risen only a little.

c. Proceeding as in (b),

$$\text{mmol HCl left} = 5.070 - (0.2000 \text{ M})(20.00 \text{ mL}) = 1.070 \text{ mmol}$$
$$\text{M HCl} = (1.070 \text{ mmol})/(50.00 \text{ mL} + 20.00 \text{ mL}) = 0.01529 \text{ M}; \text{pH} = 1.82$$

d. Again as in (b)

$$\text{mmol HCl left} = 5.070 - (0.2000 \text{ M})(25.00 \text{ mL}) = 0.070 \text{ mmol}$$
$$\text{M HCl} = (0.070 \text{ mmol})/(50.00 \text{ mL} + 25.00 \text{ mL}) = 9.3 \times 10^{-4} \text{ M}; \text{pH} = 3.03$$

The pH is finally showing significant change at 98.6% completion.

e. Yet again as in (b)

$$\text{mmol HCl left} = 5.070 - (0.2000 \text{ M})(25.30 \text{ mL}) = 0.010 \text{ mmol}$$
$$\text{M HCl} = (0.010 \text{ mmol})/(50.00 \text{ mL} + 25.30 \text{ mL}) = 1.3 \times 10^{-4} \text{ M}$$

Note the loss of significant figures due to subtraction. Here we have to be careful when calculating the pH, for very near the endpoint H^+ from H_2O might not be negligible. Here it is, however, so pH = 3.88.

f. Now OH^- will be in excess, so

$$\text{mmol NaOH} = (0.2000 \text{ M})(25.40 \text{ mL}) - 5.070 = 0.010 \text{ mmol}$$
$$\text{M NaOH} = (0.010 \text{ mmol})/(50.00 \text{ mL} + 25.40 \text{ mL}) = 1.3 \times 10^{-4} \text{ M}$$
$$\text{pOH} = 3.88; \text{pH} = 14.00 - 3.88 = 10.12$$

The pH has increased by more than six units in only two drops! This implies that the slope of the titration curve is extremely large near the endpoint.

g. As in (f),

$$\text{mmol NaOH} = (0.2000 \text{ M})(30.00 \text{ mL}) - 5.070 = 0.930 \text{ mmol}$$
$$\text{M NaOH} = (0.930 \text{ mol})/(50.00 \text{ mL} + 30.00 \text{ mL}) = 0.01163 \text{ M}$$
$$\text{pOH} = 1.93; \text{pH} = 12.07$$

h. Again as in (f)

$$\text{mmol NaOH} = (0.2000 \text{ M})(40.00 \text{ mL}) - 5.070 = 2.930 \text{ mmol}$$
$$\text{M NaOH} = (2.930 \text{ mol})/(50.00 \text{ mL} + 40.00 \text{ mL}) = 0.03256 \text{ M}$$
$$\text{pOH} = 1.49; \text{pH} = 12.51$$

The pH changes are now more gradual, as they were at the start. The results are similar to those shown in Figure 12.9.

The sudden change in pH near the endpoint is a consequence of the large equilibrium constant for the reaction, and also makes the endpoint easy to locate with the proper indicator. In this case a single drop of base can cause the solution to go from colorless to a deep pink.

Far more common is the titration of a weak acid by a strong base; as discussed under BL reactions, the equilibrium constant is now K_a/K_w, and can vary over a wide range. The weaker the acid the smaller the acid–base equilibrium constant, and the more slowly the pH changes near the endpoint of the titration. There are three characteristic points in a weak-acid titration that you should become acquainted with: the start, the halfway point, and the endpoint. Once you know the pH at these points, sketching a titration curve is easy to do.

1. *The start.* Here we have a pure solution of a weak acid; knowing moles and volume, you can generally use Equation 12.26 to find the pH.
2. *The halfway point.* When exactly half of the acid has reacted, we now have a solution with equal concentrations of the acid and its salt—a 1:1 buffer solution with pH = pK_a. Note that the volume of the solution need not be known to get the pH at this point; moreover, the Henderson–Hasselbalch equation (Equa-

tion 12.42) applies in the vicinity of the halfway point, and requires only a mole ratio. Because a buffer resists changes in pH, the pH titration curve will be relatively flat (small slope) near this point.

3. *The endpoint.* Here the solution has become that of the pure salt of the weak acid. From our analysis of this conjugate base, you know that the pH will be greater than 7; that is, *the endpoint does not correspond to a pH neutral solution.* The pH is readily calculated knowing moles of base added and total volume, using our analysis of weak base ionization.

For the same concentrations of analyte and titrant, the pH is thus higher at every stage prior to the endpoint than in the titration of a strong acid; beyond the endpoint, the strong-acid and weak-acid curves become indistinguishable. Figure 12.9 compares titration curves for weak and strong acids.

Just as understanding the titration of a monoprotic weak acid draws on the three main results of our acid–base equilibrium analysis—the weak acid itself, its salt, and its buffer—so does the titration of a polyprotic acid draw on the more extensive set of results that pertain to it. And just as a weak diprotic acid with $K_{a2} \ll K_{a1}$ loses its "easiest" proton almost exclusively, so in a titration of it with a strong base the easiest proton will be the first to react, the second proton reacting only after the first has been exhausted. To deduce how the pH will change during the titration, you need consider only where you are in terms of acid and salt concentrations, assuming as before that added base has reacted completely. The result is that, for a diprotic acid, there are five points of interest along the titration curve instead of three: the start, the first midpoint (where half of the easier protons have reacted), the first endpoint (all easier protons reacted), the second midpoint (where half of the harder protons have also reacted), and the second endpoint (all protons reacted). As you may have deduced, the two midpoints correspond to the two possible 1:1 buffers, while the pH at the first endpoint is governed by the self-reaction, as in Reaction 12.47, and given by Equation 12.48 as the average of the pH's at the midpoints. The pH titration curve for the diprotic case resembles two monoprotic acid titration curves joined end to beginning, as illustrated in Figure 12.10.

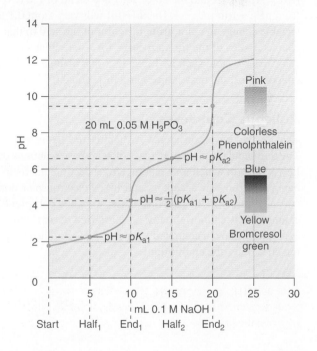

Figure 12.10
pH titration curve for a weak diprotic acid (H_3PO_3) titrated by a strong base (NaOH). The pH formulas at the two halfway points and the first endpoint are Equations 12.42 and 12.48, respectively; the starting and ending pH formulas would be similar to those of Figure 12.10. Note that both endpoints can be detected by using the appropriate indicators.

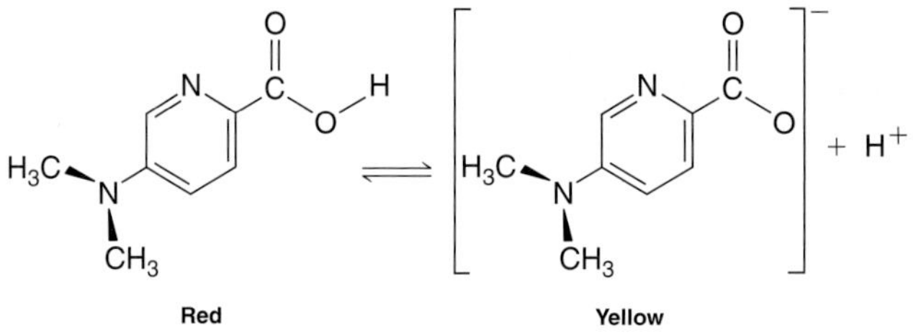

Figure 12.11
The structure of the acid–base indicator methyl red in its un-ionized and ionized forms, which have different absorption spectra in the visible and, therefore, different colors. The loss of a proton frees up a $p\pi$ orbital on the carboxyl oxygen, which then participates in the delocalized π bonding, decreasing the HOMO–LUMO gap and shifting the absorption maximum toward the red.

Red ⇌ Yellow + H⁺

As a practical matter, one of the two reactants in an acid–base titration ought to be strong; otherwise the pH swing at the endpoint will be too small to detect accurately. This is particularly true if an indicator is used to detect the equivalence point. An acid–base indicator is generally a large organic molecule containing "conjugated" double bonds, that is, alternating single and double bonds as in benzene (see Chapter 7), that also happens to be a weak acid itself. If we write the indicator as HIn (In standing for the rest of the large molecule, not the element indium), then in solution the equilibrium

$$\text{HIn} \rightleftharpoons \text{H}^+ + \text{In}^- \qquad K_{\text{In}} = \frac{[\text{H}^+][\text{In}^-]}{[\text{HIn}]} \qquad (12.50)$$

will be established. Figure 12.11 illustrates the structure of the common indicator methyl red (HIn) and its anion (In⁻). Formation of the anion frees up p electrons on a hydroxyl oxygen, which then can participate in the delocalized π bonding that is the result of conjugation. This makes the π electron "box" bigger, and lowers the π HOMO ⟶ LUMO transition frequency, thereby giving the anion a different color. Now the indicator is present in very low concentration, so that the pH is determined by the titration reaction. According to the equilibrium condition of Reaction 12.50 this in turn fixes the [In⁻]/[HIn] ratio, and therefore the color of the titration solution. When pH = pK_{In}, the indicator and its anion are present in equal concentrations, [In⁻]/[HIn] = 1, and the color seen is a blend of the two colors. As the pH changes from $pK_{\text{In}}-1$ to $pK_{\text{In}}+1$, [In⁻]/[HIn] will vary from 0.1 to 10, giving a visually detectable change in color, from the color of the acid to that of the anion. Knowing the characteristics of the titration curve for a particular system, you can choose an indicator from a table such as Table 12.5 that changes color in the pH range where the endpoint lies; the best choice has a pK_{In} equal to the pH estimated at the endpoint.

Aqueous Complexes

Thus far we have considered Arrhenius and Brønsted–Lowry acid–base equilibria. The third, and most general, class of acid–base reactions discussed in Chapter 5 is the Lewis type, in which lone pair electrons become the focus of the reaction, with acids being electron pair acceptors and bases electron pair donors. Among reactions occurring entirely in the aqueous phase, so-called **complexation** reactions are the chief example that cannot already be classified as Arrhenius or Brønsted–Lowry. For example, when concentrated ammonia NH_3 is added to an aqueous solution containing Ni^{2+} ions, the equilibrium

$$Ni^{2+} + 6NH_3 \rightleftharpoons Ni(NH_3)_6^{2+} \qquad K_f = \frac{[Ni(NH_3)_6^{2+}]}{[Ni^{2+}][NH_3]^6} = 1 \times 10^9 \qquad (12.51)$$

TABLE 12.5

A selection of acid–base indicators

Name	pK_{In}		Color change with increasing pH
Crystal violet	0.9		Yellow ⟶ blue
Thymol blue	2.0	8.6	Orange ⟶ yellow ⟶ blue
Erythrosin B	2.9		Orange ⟶ red
Bromphenol blue	3.8		Yellow ⟶ blue
Methyl orange	3.9		Orange ⟶ yellow
Bromcresol green	4.6		Yellow ⟶ blue
Methyl red	5.4		Red ⟶ yellow
Eriochrome black T	5.8		Red ⟶ blue
Bromcresol purple	6.0		Yellow ⟶ purple
Alizarin	6.4	11.8	Yellow ⟶ red ⟶ violet
Bromthymol blue	6.8		Yellow ⟶ blue
Phenol red	7.6		Yellow ⟶ red
Phenolphthalein	9.1		Colorless ⟶ pink
Thymolphthalein	10.0		Colorless ⟶ blue
Alizarin yellow	11.0		Yellow ⟶ orange

is established. The reaction is readily observed by the change in color from a bluish-green characteristic of aqueous Ni^{2+} to a blue-violet characteristic of the nickel (II)-ammonia **transition metal complex,** whose proper name is hexaamminenickel (II). The nomenclature, structure, and properties of these complexes will be discussed at length in Chapter 17; for now we note that in Reaction 12.51 Ni^{2+} is the Lewis acid and NH_3, known as a **ligand** in this type of reaction, is the Lewis base. The complex **formation constant,** is in this case large enough to ensure the ready occurrence of this reaction, provided $[NH_3]$ is sufficiently large. The position of equilibrium is hypersensitive to $[NH_3]$ owing to the sixth power of $[NH_3]$ occurring in the reaction quotient; note that if $[Ni^{2+}] = 0.1$ M and $[NH_3] = 0.01$ M at equilibrium, very little complex will be present, whereas if $[NH_3] = 1$ M, Reaction 12.51 is essentially complete. In the latter situation, the concentration of "free" (hydrated) Ni^{2+} can be found directly from the equilibrium condition of Reaction 12.51.

Reactions such as Reaction 12.51 are more than mere chemical curiosities; your body creates complexes in order to store and use the metallic minerals it needs to function. The most familiar example is hemoglobin, a complex between a large organic molecule with nitrogen-base sites (the heme protein) and Fe ions, which gives your blood its red color. Oxygen (O_2) can bind as a ligand to hemoglobin, providing a transport mechanism for oxygen throughout your body. Biochemical applications of complexation will be discussed further in Chapter 18.

Just as a polyprotic acid does not give up all its protons at once, neither do all six ligands in Reaction 12.51 bind simultaneously; rather they add in six separate steps, each with its own equilibrium constant! The use of the overall equilibrium in Reaction 12.51 is therefore restricted to the case for which the ligand NH_3 is present in high concentration; the example with 0.01 M NH_3 mentioned previously would be a messy one, with nonnegligible concentrations of free Ni^{2+} and all possible complexes $Ni(NH_3)_n$, where n = 1–6, present at once. A simpler case is a two-ligand complex such as $Ag(NH_3)_2^+$, diamminesilver(I), which forms in two steps:

$$Ag^+ + NH_3 \rightleftharpoons AgNH_3^+ \qquad K_{f1} = \frac{[AgNH_3^+]}{[Ag^+][NH_3]} = 2.1 \times 10^3 \qquad (12.52)$$

$$AgNH_3^+ + NH_3 \rightleftharpoons Ag(NH_3)_2^+ \qquad K_{f2} = \frac{[Ag(NH_3)_2^+]}{[AgNH_3^+][NH_3]} = 8.2 \times 10^3$$

This pair of stepwise reactions may remind you of the diprotic acid reactions of Reaction 12.43; in fact when charge balance and mole balance conditions are enforced, a cubic equation in the chosen variable, usually taken as the free Ag^+ concentration, results. Here, however, we cannot make the sorts of approximations that were appropriate for the acid, because the K_f's for each step are comparable. Nonetheless, when these two steps are added, an overall complexation reaction analogous to Reaction 12.51 is obtained, with equilibrium constant $K_f = K_{f1}K_{f2} = 1.7 \times 10^7$, and the overall equilibrium relation can be used to find $[Ag^+]$ when $[NH_3]$ is high. We will reserve a table of K_f's and a discussion of their magnitudes for Chapter 17, where these quantities will be interwoven with the striking color changes and other intriguing properties displayed by these complexes and their reactions. As you might recall from Chapter 5, Ag^+ commonly forms precipitates; an important application of complexation in aqueous ion chemistry is to prevent precipitation (or dissolve insoluble metal salts) by storing metal ions in stable complexes.

12.4 Equilibria in Multiphase Systems

In our discussion of equilibrium in the Arrhenius acid–base reaction at the beginning of the last section, we argued that the $H_2O(l)$ product can be omitted from the reaction quotient, because the free energy of the liquid solvent is changed negligibly by the occurrence of the reaction in dilute solution. Here let's consider the simpler equilibrium, between water and its vapor

$$H_2O(l) \rightleftharpoons H_2O(g) \qquad (12.53)$$

This is just a *physical* change—albeit arguably the most important physical change that occurs on the surface of the earth—but it can be analyzed in terms of free energy and equilibrium in the same way as any chemical change. As we noted in Chapter 1, liquid water, like most other liquids and solids, is close to *incompressible;* that is, its volume changes very little as the external pressure on it is changed. This implies that changes in pressure do negligible work on a condensed phase, and that *negligible work is required to bring a pure condensed phase from standard pressure to the actual pressure.* Since the partial-pressure or concentration factors in the reaction quotient arise directly from this type of work, we can conclude that *pure condensed phases never appear in reaction quotients.* So the equilibrium condition for water evaporation is simply $K = P_{H_2O}$, where P_{H_2O} is called the *equilibrium vapor pressure* of water. So-called phase equilibria such as that of Equation 12.53 are so important that we will devote the better part of a future chapter to their understanding. The principle that comes from them is what we need here.

Some of the earliest reactions we considered in Chapter 5 involved more than one phase. Let's apply the principle of condensed phases to the writing of reaction quotients for some of these multiphase reactions. Reaction 5.1, $2Na(s) + Cl_2(g) \rightleftharpoons 2NaCl(s)$, has a reaction quotient $Q = 1/P_{Cl_2}$, since $Na(s)$ and $NaCl(s)$ represent pure condensed phases. For Reaction 5.9, $2Na(s) + 2H_2O(l) \rightleftharpoons 2NaOH(aq) + H_2(g)$, $Q = [NaOH]^2 P_{H_2}$, since again $Na(s)$ is a pure condensed phase and $H_2O(l)$ is the solvent, which is nearly a pure condensed phase. In this example, note that molarity and partial pressure occur together. Although this appears to be mixing apples and oranges, no inconsistency arises because, as you may recall, division of each quantity by its standard-state value is implied, making each factor a unitless number. Reaction quotients are not useful for these reactions under near-normal conditions because their equilibria lie so far to the right; the baking soda decomposition $2NaHCO_3(s) \rightleftharpoons Na_2CO_3(s) + CO_2(g) + H_2O(g)$, considered at the end of the previous chapter, represents a more "balanced" reaction, with a reaction quotient leading to the equilibrium condition

$$K = P_{CO_2} P_{H_2O} = 4.7 \times 10^{-7} \quad (12.54)$$

where the value of K has been computed from $\Delta G°_{298} = 8.63$ kcal/mol, as found in Chapter 11.

The form of the equilibrium condition, in which some of the reaction participants are not included, has important implications not only for the calculation of the position of equilibrium but also for the application of LeChâtelier's principle in deciding how to control that position. *When a reagent or product does not appear in Q, changing its amount cannot affect the position of equilibrium.* This means that adding more baking soda to your sample will *not* shift the equilibrium to the right, as you might have thought; adding more $NaHCO_3(s)$ does not change its free energy *per mole*. This situation changes if we dissolve $NaHCO_3$ in water; now it has a concentration that will drop as decomposition occurs, and you *can* affect the position of equilibrium by increasing its concentration, say, by adding more solid *to the solution*. Remember that the solid is dispersed randomly when it dissolves, thereby acquiring a variable entropy, and entropy changes drive the equilibrium position; the reaction quotient is just a handy way of representing the entropic balance. You may recognize the solution reaction that corresponds to the decomposition above as Reaction 12.47; it becomes a homogeneous reaction except at higher temperature, where the solubility of CO_2 in water is exceeded and $CO_2(g)$ is evolved.

Solubility and Precipitation

Aside from gas-forming reactions, the most important type of two-phase reaction involving an aqueous medium is **precipitation,** a reaction type in the Lewis category that we also met in Chapter 5. Precipitation is nearly always involved in isolating the product of a chemical synthesis, where it is also known as *crystallization,* and is an important reaction in formation of natural deposits as well as in water purification technology. In Chapter 5 deciding on the spontaneity of a precipitation reaction was a matter of consulting the solubility rules; with our new, powerful equilibrium methods, we can now make quantitative statements concerning the extent of reaction, or whether a precipitate will form at all. It is always possible (though sometimes difficult) to make a solubility rule fail by reducing the reagent concentrations, and therefore their free energy, below a certain level; we will be able to calculate what that level is. This sort of predictive power is used routinely in both laboratory chemistry and industrial processes.

Our example in Chapter 5 was the reaction of $Pb^{2+}(aq)$ with $Cl^-(aq)$ to make the sparingly soluble salt $PbCl_2(s)$. Since the product is a pure solid, it will not appear in the reaction quotient, and we can write

$$Pb^{2+}(aq) + 2Cl^-(aq) \rightleftharpoons PbCl_2(s) \quad K = \frac{1}{[Pb^{2+}][Cl^-]^2} = 8.55 \times 10^4 \quad (12.55)$$

The large value of K (though not as large as many) ensures that the reaction will go nearly to completion; if stoichiometric amounts of Pb^{2+} and Cl^- solutions have been used, after equilibrium has been reached the mixture will contain small postreaction $[Pb^{2+}]$ and $[Cl^-]$. The equilibrium is then most readily analyzed by assuming that a small amount of $PbCl_2(s)$ has dissolved, that is, by considering the reverse of Reaction 12.55. (This should remind you of our analysis of the Arrhenius acid–base reaction in Equations 12.18 to 12.20.) The reverse reaction's equilibrium constant is

$$K_{sp} = [Pb^{2+}][Cl^-]^2 = 1/K = 1.17 \times 10^{-5}$$

where K_{sp} is known as the **solubility product.** Note that, like water ionization, the reaction quotient for the reverse reaction has a unit denominator; that is, it is just an ionic concentration product or *ion product* Q_{ion}.

Two useful practical applications flow from the precipitation equilibrium condition. First, whether a precipitate will form at all can be predicted. From Equation 12.9, we can derive the condition $Q_{ion} > K_{sp}$ as a criterion for precipitation. For example, if equal volumes of 0.100 M $Pb(NO_3)_2$ and 0.200 M HCl are mixed, $Q_{ion} = [Pb^{2+}]_0[Cl^-]_0^2 = (0.0500)(0.100)^2 = 5.00 \times 10^{-4} > K_{sp}$, and a precipitate will form; however if the concentrations are reduced to 0.0100 M Pb^{2+} and 0.0200 M Cl^-, Q_{ion} will be less than K_{sp}, and no reaction will occur. If no precipitate is present, the equilibrium condition cannot hold, leading us to add a corollary to our condensed-phase principle: the condensed phase in question *must be present* for equilibrium to be established. This is a sensible constraint; in the present example, if there is no $PbCl_2(s)$, it cannot supply the ions necessary to fulfill the solubility product. Also note that exact stoichiometry is not required; one can use large $[Cl^-]$ (>1 M) to precipitate very small amounts of lead, and most of the Cl^-, as the excess reagent, will remain in solution at equilibrium.

Secondly, the amount of a solid salt that can be dissolved in a given volume of water at equilibrium can be both predicted and controlled. When dissolving a single pure salt in pure water, the resulting concentration is called the **solubility** of the salt. For this example, the extent-of-reaction analysis leads to

$$\text{PbCl}_2(s) \rightleftharpoons \text{Pb}^{2+}(aq) + 2\text{Cl}^-(aq)$$
$$\phantom{\text{PbCl}_2(s) \rightleftharpoons} x \qquad 2x$$

$$K_{sp} = (x)(2x)^2 \qquad (12.56)$$
$$x = (K_{sp}/4)^{1/3} = 0.0143 \text{ M}$$

Since $x = [\text{Pb}^{2+}] = [\text{PbCl}_2]$, the solubility is 0.0143 M, or 0.40 g/100 mL. When equilibrium has been reached, the aqueous phase is said to be **saturated.** If the solution is made 1.00 M in HCl, the solubility drops because of the common ion effect; if c is the added $[\text{Cl}^-]$, then at equilibrium $[\text{Cl}^-] = 2x + c \approx c$ if c is large, and $x \approx K_{sp}/c^2 = 1.17 \times 10^{-5}$ M in this example, more than a thousand times smaller than in pure water. Once again we can "tune" the equilibrium to almost any desired solubility by controlling a common ion. A calculation similar to this latter one would be done if the precipitation reaction were run nonstoichiometrically, with one reagent in excess, since this produces the same sort of concentration pattern as in common ion addition.

If the anion of a sparingly soluble salt is a good base, the solubility becomes pH dependent, more salt dissolving at lower pH. Metal ions from group IIA and the transition series, for example, generally have insoluble hydroxides, but these are readily taken up by acidic solutions. Other anions are the ones you might expect by combining the solubility rules and weak acid ionization constants; the most important are sulfides and carbonates, both of which are used in metal ion separation and identification as discussed briefly in the next section.

Table 12.6 presents a compilation of solubility products; this table is a quantitative reflection of the solubility rules of Chapter 5, slightly augmented. Space does not permit the discussion of the relative magnitudes of K_{sp}, but it can be said that their widely varying values, particularly among the sulfides, allow *differential precipitation* of sulfides, a backbone of the classic scheme for the separation and identification of metal ions in solution.

Qualitative Metal Ion Analysis

About 20 of the most common metal ions in solution can be separated into five groups on the basis of differing solubilities in a process called **qualitative metal ion analysis.** In general, when two or more equilibria occur with differing degrees of completion, the one that is most spontaneous, in this case with the smallest K_{sp}, will occur preferentially. The scheme employs three anions, Cl^-, S^{2-}, and CO_3^{2-}, as precipitating agents, along with pH control of S^{2-} to achieve a fourth stage of separation. The alkali and ammonium ions do not precipitate at all, forming the solutes of the final aqueous phase. Spot tests such as colored precipitates, complexes, or flame tests are then used for individual identification. The scheme is best appreciated by referring to the diagram in Figure 12.12 in light of the data in Table 12.6.

The few insoluble chlorides (group I) are first separated from the rest, followed by the highly insoluble sulfides (group II), in each case using centrifugation to separate the precipitate from the remaining solution. Then addition of base increases $[\text{S}^{2-}]$, bringing down the more soluble sulfides and a few hydroxides (group III), followed by carbonate precipitation of the remaining alkaline earths (group IV). Group V comprises the alkali and ammonium ions remaining in solution.

TABLE 12.6

Solubility product constants K_{sp} at 25°C for sparingly soluble salts

Formula	K_{sp}	Formula	K_{sp}	Formula	K_{sp}
Fluorides		**Hydroxides** (*continued*)		**Phosphates** (*continued*)	
MgF_2	7.4×10^{-11}	$Cu(OH)_2$	5.5×10^{-20}	$Ni_3(PO_4)_2$	4.7×10^{-32}
CaF_2	1.5×10^{-10}	$AgOH$	1.8×10^{-8}	$Cu_3(PO_4)_2$	1.4×10^{-37}
SrF_2	4.3×10^{-9}	$Zn(OH)_2$	7.7×10^{-17}	$Cd_3(PO_4)_2$	2.5×10^{-33}
BaF_2	1.8×10^{-7}	$Cd(OH)_2$	5.3×10^{-15}	**Carbonates**	
Hg_2F_2	3.1×10^{-6}	$Hg(OH)_2$	3.1×10^{-26}	$MgCO_3$	7.0×10^{-6}
PbF_2	7.1×10^{-7}	$Sn(OH)_2$	5.5×10^{-27}	$CaCO_3$	5.0×10^{-9}
Chlorides		$Pb(OH)_2$	1.4×10^{-20}	$SrCO_3$	5.4×10^{-10}
$CuCl$	1.7×10^{-7}	**Sulfides**		$BaCO_3$	2.6×10^{-9}
$AgCl$	1.8×10^{-10}	MnS	3.8×10^{-15}	$MnCO_3$	2.2×10^{-11}
Hg_2Cl_2	1.5×10^{-18}	FeS	1.3×10^{-20}	$FeCO_3$	3.1×10^{-11}
$PbCl_2$	1.2×10^{-5}	CoS	8.2×10^{-22}	$NiCO_3$	1.4×10^{-7}
Bromides		NiS	9.2×10^{-23}	$CuCO_3$	1.8×10^{-10}
$CuBr$	6.3×10^{-9}	CuS	1.1×10^{-37}	$ZnCO_3$	1.2×10^{-10}
$AgBr$	5.4×10^{-13}	Ag_2S	1.1×10^{-49}	$CdCO_3$	6.2×10^{-12}
Hg_2Br_2	6.4×10^{-23}	ZnS	2.5×10^{-26}	Hg_2CO_3	3.7×10^{-17}
$PbBr_2$	6.6×10^{-6}	CdS	1.2×10^{-30}	$PbCO_3$	1.5×10^{-13}
Iodides		HgS	5.4×10^{-54}	**Oxalates**	
CuI	1.3×10^{-12}	SnS	2.7×10^{-29}	$MgC_2O_4 \cdot 2H_2O$	4.8×10^{-6}
AgI	8.5×10^{-17}	PbS	7.5×10^{-30}	$CaC_2O_4 \cdot H_2O$	2.3×10^{-9}
Hg_2I_2	5.3×10^{-29}	**Sulfates**		$MnC_2O_4 \cdot 2H_2O$	1.7×10^{-7}
PbI_2	8.5×10^{-9}	$CaSO_4$	7.1×10^{-5}	CuC_2O_4	4.4×10^{-10}
Hydroxides		$SrSO_4$	3.4×10^{-7}	$Ag_2C_2O_4$	5.4×10^{-12}
$Mg(OH)_2$	5.6×10^{-12}	$BaSO_4$	1.1×10^{-10}	$ZnC_2O_4 \cdot 2H_2O$	1.4×10^{-9}
$Al(OH)_3$	2.5×10^{-32}	Ag_2SO_4	1.2×10^{-5}	$CdC_2O_4 \cdot 3H_2O$	1.4×10^{-8}
$Ca(OH)_2$	4.7×10^{-6}	$PbSO_4$	1.8×10^{-8}	$Hg_2C_2O_4$	1.8×10^{-13}
$Cr(OH)_3$	4.7×10^{-38}	**Phosphates**		PbC_2O_4	8.5×10^{-10}
$Mn(OH)_2$	2.1×10^{-13}	$Mg_3(PO_4)_2$	9.9×10^{-25}	**Chromates**	
$Fe(OH)_2$	4.9×10^{-17}	$AlPO_4$	9.8×10^{-21}	$BaCrO_4$	1.2×10^{-10}
$Fe(OH)_3$	2.6×10^{-39}	$Ca_3(PO_4)_2$	2.1×10^{-33}	Ag_2CrO_4	1.1×10^{-12}
$Co(OH)_2$	1.1×10^{-15}	$FePO_4 \cdot 2H_2O$	9.9×10^{-29}	$PbCrO_4$	2.8×10^{-13}
$Ni(OH)_2$	5.6×10^{-16}	$Co_3(PO_4)_2$	2.1×10^{-35}		

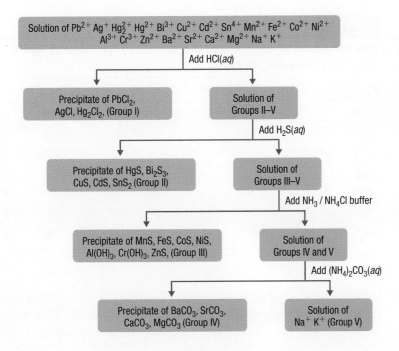

Figure 12.12
Broad outline of the classical qualitative metal ion analysis separation scheme, showing the five ion groups, I–V, based on solubilities. Further separation of ions within a group is done by further selective precipitation or complexation combined with spot tests.

12.5 Deviations from Ideal Behavior

The relationships we have used to find changes in free energy from logarithmic changes in pressure or concentration, especially the basic Equations 12.3 and 12.17, assume ideal behavior. That is, we have neglected any enthalpy changes associated with, for example, allowing gas molecules or solute ions to move further apart within the system. Real gases, as we have examined in Chapter 9, actually attract each other weakly at large distances (corresponding to low pressures) and repel strongly when closer (at higher pressures). Thus you might expect the free-energy change in the isothermal expansion of a real gas to be less negative than the ideal at moderate pressure (less work done) because the molecules have to overcome their mutual attraction in order to expand; that is, the expansion is very slightly endothermic. The situation is similar for ions in aqueous solution. Because of the Coulombic attraction between ions—highly attenuated by water's large dielectric constant—a positive ion, for example, tends to be surrounded by negative ions, all solvated and at some distance. When the ion concentration decreases, this average distance grows, and because the ions attract, again the process is slightly endothermic.

These intermolecular or interionic energies cause deviations from our predictions concerning free energy and the position of equilibrium. Consider the classic case of the synthesis of ammonia from a stoichiometric mixture of hydrogen and nitrogen. Ideally we expect the equilibrium reaction quotient to be constant, independent of pressure, but, as Table 12.7 shows, reality is different, due to intermolecular forces. The conditions are extreme, but well within the bounds of industrial synthesis. G. N. Lewis in 1923 proposed a reformulation of the free-energy expressions that would include deviations from ideality while retaining the convenient logarithmic form of the dependence on partial pressure. In this revamping, the free energy of

TABLE 12.7

Experimental equilibrium in the ammonia synthesis at 450°C with 3:1 ratio of H_2 to N_2

P_t, atm	%NH_3 at equilibrium	K_P	K_γ	K
10	2.04	4.34×10^{-5}	0.990	4.29×10^{-5}
30	5.80	4.57×10^{-5}	0.951	4.34×10^{-5}
50	9.17	4.76×10^{-5}	0.893	4.23×10^{-5}
100	16.36	5.26×10^{-5}	0.774	4.04×10^{-5}
300	35.5	7.81×10^{-5}	0.473	3.70×10^{-5}
600	53.6	1.67×10^{-4}	0.247	4.12×10^{-5}
1000	69.4	5.42×10^{-4}	0.188	1.02×10^{-4}

formation (equivalent to Gibbs' chemical potential) for a component compound in a system that is analogous to the ideal expression of Equation 12.4 is written as

$$\Delta G_f = \Delta G_f^\circ + RT \ln(\gamma P) \tag{12.57}$$

where γ is called the **activity coefficient,** and the product γP is called the **activity** of that component. (Lewis dubbed the product γP the *fugacity* of the gas, a term you may meet again in a more advanced course.) An analogous expression is postulated for ions in solution, with P replaced by the concentration c. As we will see, if the predominant interparticle interactions are attractive, γ will be less than unity, and this is ordinarily found to be the case. Using the ammonia synthesis as an example, Equation 12.57 leads to the equilibrium condition

$$K = \frac{(\gamma_{NH_3} P_{NH_3})^2}{(\gamma_{N_2} P_{N_2})(\gamma_{H_2} P_{H_2})^3} = \frac{P_{NH_3}^2}{P_{N_2} P_{H_2}^3} \times \frac{\gamma_{NH_3}^2}{\gamma_{N_2} \gamma_{H_2}^3} = K_P K_\gamma \tag{12.58}$$

In Table 12.7, appropriate γ values for N_2, H_2, and NH_3 have been inserted into the K_γ, producing much-improved constancy in the equilibrium reaction quotient. So that you don't think of γ as simply a fudge factor, we will now briefly describe two ways of obtaining γ using simple models.

For gases, we can use the van der Waals equation of state, Equation 9.55, to calculate the work in expanding a real (van der Waals) gas isothermally. With a few simplifying approximations, this work can be written as the ideal gas work plus a correction term, yielding the free-energy relation

$$\Delta G_f = \Delta G_f^\circ + RT \ln P - P\left(\frac{a}{RT} - b\right) \tag{12.59}$$

Since $\ln(\gamma P) = \ln P + \ln \gamma$, we can identify the activity coefficient as

$$\ln \gamma = -\frac{P}{RT}\left(\frac{a}{RT} - b\right) \tag{12.60}$$

Notice that γ is pressure and temperature dependent, as you might expect, and further, if $a/(RT) > b$, then $\ln\gamma < 0$ and $\gamma < 1$. For NH_3 at 1 atm and 298 K, the van der Waals constants of Table 9.3, $a = 4.17$ atm L^2 mol^{-2} and $b = 0.0371$ L mol^{-1}, yield $\gamma = 0.995$. For $P = 10$ atm, $\gamma = 0.947$; 100 atm, 0.580.

For ions in solution, Debye and Hückel adopted Lewis's activity coefficient model for the free energy. After a lengthy derivation in which the Boltzmann distribution and Poisson's equation for obtaining electric fields were employed, they arrived at a relatively simple form for the free energy of an ion in solution given by

$$\Delta G = \Delta G_f^\circ + RT \ln c - N_A \frac{e^2 z^2}{2Dr_d} \tag{12.61}$$

where the correction term on the right has the form of a Coulomb's law potential energy (written in cgs-esu, with N_A scaling up the microscopic Coulombic interaction to a per mole basis). Here e is the electronic charge, z is the charge on the ion, D is the dielectric constant of the solvent (for water, $D = 78.5$), and r_d is called the *Debye length*, and is given by

$$r_d^2 = \left(\frac{1000\, DRT}{8\pi N_A^2 e^2}\right)\left(\frac{1}{2}\sum_i c_i z_i^2\right)^{-1} \tag{12.62}$$

The quantity within the rightmost parentheses is called the *ionic strength I* of the solution, and the sum extends over all ions present at concentration c_i with charge z_i. The factor of 2 in the denominator of the Coulombic term in Equation 12.61 prevents double counting when the free energies of all the ions present are included. Equations 12.61 and 12.62 are readily interpreted as the result of the ion in question surrounding itself with a shell of counter ions of charge equal and opposite to its own at an average distance r_d. Note that, according to Equation 12.62, as the concentration of ions in the solution—the ionic strength—grows, r_d decreases, the Coulombic term becomes more important, and the free energy deviates further from the ideal. The activity coefficient can be identified in the same manner as above:

$$\ln\gamma = -\frac{N_A e^2 z^2}{2RTDr_d} \tag{12.63}$$

As an example, let's consider Na^+ ions in 0.100 M NaCl solution. The ionic strength $I = c = 0.100$ M, Equation 12.62 gives $r_d = 3.042$ Å $I^{-1/2} = 9.62$ Å, and Equation 12.63 then yields $\ln\gamma = -3.570$ Å $(z^2/r_d) = -0.371$ and $\gamma = 0.690$. As expected, an identical result is obtained for Cl^-, since only the magnitude of the charge on the ion is significant in the Debye–Hückel model. The model is only applicable to dilute solutions; its range of validity can be extended to higher concentrations by replacing r_d in Equations 12.61 and 12.63 by $r_d + d$, where d is a cutoff radius, corresponding to an ion "touching" distance. Note that, due both to the much stronger forces between ions in solution, and to their closer proximity, solution activities deviate from ideality more strongly for "normal" conditions than do their gas-phase counterparts. The interaction between a pair of ions in solution is schematically illustrated in Figure 12.13.

It's important to realize that these corrections do not affect the concepts we have developed in this chapter concerning free energy and the tendency toward equilibrium. We have employed ideal gases and solutions as illustrations of these ideas

Figure 12.13

Illustration of ion interaction in solution. The little teddy-bear heads are water molecules, which tend to align their dipoles along the electric field lines generated by the Coulomb force between the ions, much as iron filings line up in the lines of force between two poles of a magnet. This alignment reduces the attraction between ions by a factor of 80, the so-called *dielectric constant* of water liquid. Nonideal corrections to the free energy result from this residual attraction in the Debye–Hückel model, as discussed in the text.

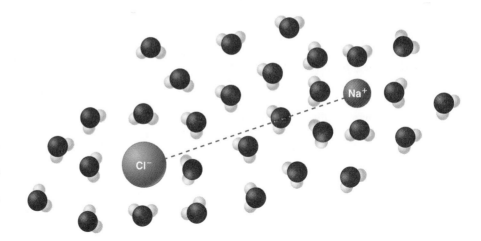

because they both give a relatively transparent picture of the roles played by enthalpy and entropy in the attainment of equilibrium, and are reasonably accurate, failing only for strongly interacting species at high concentration. Such "strong" conditions are often employed in the laboratory, however, making it important for you to know that corrections for nonideal behavior often need to be made, and *can* be made.

We will nonetheless continue to use the ideal approximation in our treatment of the other great class of chemical reactions, oxidation–reduction, in the next chapter. Those reactions, we will find, are possessed of much greater chemical potential in general; and this potential can be made to take on its electrical meaning—a *voltage*—with far-reaching practical consequences.

SUMMARY

In a chemical system **equilibrium** corresponds to concentrations of reagents and products that are constant in time, usually at the end of a spontaneous reaction. Equilibrium can be viewed as a dynamic balance between the reaction in question and its reverse, as indicated by using a double arrow $\rightleftharpoons$ in the chemical equation. In thermodynamics equilibrium is reached as a result of expending free energy G until no more is available. If a reaction is spontaneous as written, with $\Delta G < 0$, it will proceed until $\Delta G = 0$; that is, until the free energy of the reagents G_R equals that of the products G_P. If the system begins in its standard state, 1 atm partial pressures or 1 M concentrations of all reaction participants, it must end in some other, nonstandard state as reagents are converted to products; moreover, there is no need, nor is it even possible in many cases, to begin in a standard state. This problem is attacked by first defining a **standard free energy of formation** ΔG_f°, the free energy change per mole of a formation reaction for a particular compound. Then the ideal gas isothermal work formula of Chapter 11 is invoked, leading to a formula for the free energy of formation for arbitrary pressure, $\Delta G_f = \Delta G_f^\circ + RT \ln P$, where P is expressed in atmosphere units, and the standard pressure is $P^\circ = 1$ atm. The logarithmic term is a correction for the work necessary to bring the compound from standard pressure to the actual pressure being considered, and arises from the entropic contribution to G. For a gas-phase reaction, this leads to a general expression for the free-energy change for arbitrary partial pressures of each participant of the form $\Delta G = \Delta G^\circ + RT \ln Q$, where ΔG° is the standard free-energy change per mole for the reaction, computed either from $\Delta H^\circ - T\Delta S^\circ$ or from the standard free energies of formation of the participants using a Hess's law formula; and Q is the **reaction quotient**. For a gas reaction $aA + bB + \ldots \rightleftharpoons cC + dD + \ldots$, Q has the form $Q = (P_C^c P_D^d \cdots)/(P_A^a P_B^b \cdots)$, where the P's are partial pressures. These relations allow the spontaneity of a reaction to be assessed at arbitrary pressure according to the general criterion derived in Chapter 11, $\Delta G < 0$. If $\Delta G < 0$, the reaction will proceed spontaneously as written, toward the right, making products, whereas if $\Delta G > 0$, the reaction will run backward, toward the left, reforming reagents. If $\Delta G = 0$, the system is at equilibrium, a state that is analyzed in this chapter.

If $\Delta G = 0$, then $\Delta G° = -RT \ln Q_{eq} \equiv -RT \ln K$, where K is the **equilibrium constant,** the reaction quotient under equilibrium conditions after reaction is over. The general free-energy change can be expressed in terms of K by $\Delta G = RT \ln(Q/K)$. If $\Delta G° < 0$, $K > 1$, and the product partial pressures will be greater than those of the reagents at equilibrium, whereas if $\Delta G° > 0$, $K < 1$, and the reagents will predominate. Using an extent-of-reaction analysis with known reaction stoichiometry, finding the equilibrium partial pressures of all species is reduced to an algebraic problem in which an extent-of-reaction variable x is found in terms of the initial partial pressures and K. Simple stoichiometry often leads to a quadratic equation for x, with only one of the two roots chemically acceptable. The equilibrium state corresponds to a minimum in $G(x)$. The analysis breaks down when absolutely no product molecules are present at the start of a process, or when no reagent molecules remain at equilibrium.

LeChâtelier's principle states that a system at equilibrium will respond to the stress of an external change by opposing it or by counteracting it. An increase in pressure, or decrease in volume, will cause the equilibrium position of a gas reaction to shift in the direction that makes fewer moles of gas; for example, if $\Delta n_{gas} > 0$, the shift will be to the left, while if $\Delta n_{gas} = 0$, the equilibrium will not shift. Increasing temperature will cause an exothermic reaction to run backward, the direction that removes the heat that has been added. This may be quantitatively characterized by the **van't Hoff equation,** in differential form $d(\ln K)/d(1/T) = -\Delta H°/R$. The van't Hoff equation points to a technique for obtaining heats and entropies of reaction without calorimetry, through measuring the temperature dependence of the equilibrium constant.

These results can be extended to solution chemistry by replacing partial pressures with concentrations. This allows us to treat the spontaneity and equilibrium of aqueous acid–base reactions, the Arrhenius reaction $H^+ + OH^- \rightleftharpoons H_2O$ being the most common. The low magnitude of $\Delta H°$ and the positivity of $\Delta S°$ are analyzed in terms of the hydrated structures of the H^+ and OH^- ions in water. They lead to an equilibrium constant $K = 1.0 \times 10^{14}$ at 298 K, implying that the equilibrium ion concentrations in pure H_2O are small, $[H^+] = [OH^-] = 1.0 \times 10^{-7}$ M. These concentrations are determined by noting that water, as the solvent, undergoes a negligible change in free energy when the reaction occurs, and thus is omitted from the reaction quotient. The equilibrium constant for the reverse reaction, ionization of water, is $K_w = 1/K = 1.0 \times 10^{-14}$, the so-called water ionization constant, and it is related to the ion concentrations by $K_w = [H^+][OH^-]$. Acid solutions are defined by $[H^+] > 10^{-7}$ M and bases by $[H^+] < 10^{-7}$ M. These are usually expressed on a logarithmic scale by $pH = -\log_{10}[H^+]$, and the ionization equilibrium can be written $pH + pOH = pK_w = 14.00$. An electrochemical device known as a pH meter allows the ready measurement of pH.

Strong acids and bases have large equilibrium constants that govern their ionization in water, making them essentially 100% ionized. The pH of the solution is then pc (or $14.00 - pc$), the negative logarithm of c, the initial concentration of the acid (or base). Despite the small concentrations involved, for example, 10^{-14} M H^+ in 1 M NaOH, the numbers of individual ions present in normal samples is still sufficiently large as to preserve the validity of thermodynamic analyses. Weak acids remain mostly un-ionized at equilibrium, as characterized by acid ionization constants $K_a \ll 1$. Equilibrium analysis leads to a quadratic equation for the extent-of-reaction variable, with the approximate but generally sufficiently accurate solution $[H^+] \approx \sqrt{cK_a}$ or $pH \approx \frac{1}{2}(pc + pK_a)$. Improved solutions are available by use of the quadratic formula (positive root), by the improved approximation $pH \approx \sqrt{cK_a} - K_a/2$, or by iteration. In cases where H^+ from the ionization of water itself cannot be neglected, a more general treatment is needed; this invokes **charge balance** and **mole balance** in addition to the equilibrium conditions, yielding a cubic equation in the extent-of-reaction variable. This is shown to reduce to the standard quadratic equation when $[H^+] \gg K_w/[H^+]$, that is, when $[H^+] \gtrsim 10^{-6}$ M. A large number of compounds exhibit weak acidity, many of them of the form RCOOH, where R is an alkyl or substituted alkyl group and COOH is the *carboxyl group*. In the oxyhalic acids HXO_n, acid strength (K_a) increases as the number of nonhydroxyl oxygens increases, by about a factor of 10^5 for each O.

Weak bases often contain the nitrogen atom with a lone pair of electrons, for example, RNH_2, which produces OH^- by *hydrolyzing* water, $RNH_2 + H_2O \rightleftharpoons RNH_3^+ + OH^-$. The algebra behind finding equilibrium concentrations is identical to the weak-acid case, provided $[H^+]$ is replaced by $[OH^-]$ and K_a by K_b, the weak base ionization constant.

Acid–base reactions involving weak acids or bases, or both, are of the Brønsted–Lowry type, and reaction produces conjugate acids and bases. The equilibrium in these reactions still lies to the right, but not as strongly as the Arrhenius acid–base reaction, with $K = K_a/K_w$ for a weak acid + strong base reaction, and $K = K_b/K_w$ for weak base + strong acid. The anion A^- of a weak acid is a weak base; a solution of the salt NaA is therefore basic. As in the case of the nitrogen bases, A^- produces OH^- by hydrolyzing water, $A^- + H_2O \rightleftharpoons HA + OH^-$, and finding the resulting pH is a problem identical in form to weak base ionization, taking $K_b = K_w/K_a$. The inverse relationship between the K_a of the acid and the K_b of its anion implies that the weaker the acid, the stronger its conjugate base. A precisely complementary analysis holds for a weak base B and its conjugate acid BH^+. Weak acid + weak base reactions have equilibrium constants $K = (K_aK_b)/K_w$.

LeChâtelier's principle may also be applied to acid–base equilibria, to analyze the shifts in equilibrium that occur when a **common ion** is added to the solution. The effect of adding H^+ or

A^- to a solution of weak acid HA is treated in detail; the latter produces a **buffer** solution, containing both HA and its salt NaA, which has the desirable property of being resistant to changes in pH. The pH of a weak-acid buffer solution is approximately given by the **Henderson–Hasselbalch equation** $pH = pK_a + \log_{10}([A^-]_0/[HA]_0)$. Weak-base buffers can be created and analyzed in a similar way. Effective buffer solutions can be made to cover the range $pH \approx 3–11$; most effective buffering action occurs when $pH \approx pK_a$.

Polyprotic acids contain more than one ionizable proton; when they ionize, general Coulombic considerations suggest that the first proton is most readily lost, while each succeeding H^+ is removed with increasing difficulty. Diprotic acids H_2A are analyzed in detail. Separate ionization equilibria govern each ionization step, with equilibrium constants K_{a1} and K_{a2}, respectively. When $1 \gg K_{a1} \gg K_{a2}$, the pH is dominated by the first step, and can be found in the same way as in the monoprotic case. Further, $[HA^-] \approx [H^+]$ and $[A^{2-}] \approx K_{a2}$. A diprotic acid has two possible salts, for example, NaHA and Na_2A, both of which will influence the pH of their solutions. A^{2-} will hydrolyze water to produce OH^- in the same way as the anion of a monoprotic acid, allowing the pH to be determined as before. HA^- can act as both acid or base, and the self-reaction $2HA^- \rightleftharpoons H_2A + A^{2-}$ dominates the equilibrium, with the result that $pH \approx \frac{1}{2}(pK_{a1} + pK_{a2})$. Adding H^+ to H_2A inhibits H_2A ionization, making the overall equilibrium condition $K = K_{a1}K_{a2} = ([H^+]^2[A^{2-}])/[H_2A]$ useful in determining $[A^{2-}]$, by replacing $[H^+]$ with that resulting from the strong acid added, and $[H_2A]$ with its initial value. Buffers can be made for pH's near pK_{a1} or pK_{a2}. Equilibrium data for a number of polyprotic acids are presented and their structural connections discussed briefly. Polybasic bases can be treated in a complementary way, but no explicit analysis is given here.

Acid–base **titrations** are readily analyzed by equilibrium methods. For a strong monoprotic acid titrated by a strong base, the pH is readily calculated from the amount of acid remaining prior to the endpoint, and from the excess base and the water equilibrium condition beyond the endpoint; the pH changes rapidly near the endpoint, making it easy to detect with an appropriate **indicator**. These properties are readily examined on a **pH titration curve**, a plot of pH versus volume of added base. If the acid is weak, the pH is higher at every point prior to the endpoint than the strong acid case, with the midpoint corresponding to a 1:1 buffer solution. The pH change at the endpoint is then not as steep, requiring a careful choice of indicator; phenolphthalein is usually used. For a diprotic weak acid H_2A with $K_{a1} \gg K_{a2}$, the titration curve shows two distinct regions, each resembling the monoprotic weak acid; the midpoint of each region is a 1:1 buffer solution of HA^-:H_2A (first step) or A^{2-}:HA^- (second step), and the first break, corresponding to complete reaction of the first proton, is a solution of HA^-. For both monoprotic and diprotic cases, the pH is readily estimated at these characteristic points from the preceding analysis. Indicators are themselves weak acids of organic molecules large enough to have visible absorption; as the pH increases, the indicator is converted to its ionized form when $pH \approx pK_{In}$, shifting the absorption maximum and changing the observed color.

Lewis acid–base reactions in solution are mainly typified in **complexation** reactions of the general form $M + nL \rightleftharpoons ML_n$, where M is a transition metal ion, the Lewis acid; L is a **ligand** molecule, such as NH_3, the Lewis base; and ML_n is a **transition metal complex**, the Lewis adduct. Although complex formation takes place in n separate steps, the overall equilibrium condition $K_f = [ML_n]/([M][L]^n)$ is useful for finding the uncomplexed [M] when [L] is large. The overall complex **formation constant**, K_f, is the product of the n stepwise equilibrium constants. Further discussion is given in Chapter 17.

In multiphase equilibria, pure condensed phases and solvents in dilute solutions do not appear in the equilibrium quotient, because negligible work is done in bringing these species from standard pressure to the actual pressure due to their relative incompressibility. This simplifies many complicated multiphase equilibrium problems, but makes it impossible to shift the equilibrium position by changing the amounts of pure condensed phases present. The most common type of two-phase reaction is **precipitation**. Precipitation reactions, in which an insoluble salt is formed by the combination of cations and anions in solution, are readily analyzed in terms of the **solubility product** K_{sp}. For a sparingly soluble salt M_xL_y, $K_{sp} = [M]^x[L]^y$; this condition can be used to assess both whether and how much precipitate will form, and the solubility of the salt in water. The addition of common ions, either M or L, will reduce the solubility; LeChâtelier's principle allows a simplified calculation of solubility in the presence of a common ion. A table of solubility products is given, and is seen to reflect the solubility rules given in Chapter 5. Knowledge of these solubilities has allowed a separation scheme to be developed for 20 common metal ions, thereby enabling **qualitative metal ion analysis**.

In dense gases or concentrated solutions, interactions between the particles involved in equilibrium causes the free energy to deviate from a logarithmic dependence on pressure or concentration. This deviation is accounted for by introducing an **activity coefficient** γ, and writing the free energy of formation as $\Delta G_f = \Delta G_f^\circ + RT \ln(\gamma P)$. For gases, the van der Waals equation can be used to obtain γ, while in solution the Debye–Hückel Coulombic model of interacting ions is used. Under normal conditions, $\gamma < 1$, and the **activity** γP or γc is less than expected based on stoichiometry.

EXERCISES

Gas-phase spontaneity and equilibrium

1. For the equilibrium governing the conversion of ozone to normal oxygen, $2O_3(g) \rightleftharpoons 3O_2(g)$, carry through the algebra of Equation 12.5 to find a general formula for ΔG for the reaction for constant temperature and arbitrary partial pressures of O_3 and O_2. Verify that your expression conforms to the general formula of Equation 12.6. Using ΔG_f° from Appendix C, find ΔG_{298} when $P_{O_3} = 0.0010$ atm and $P_{O_2} = 0.20$ atm. Will this reaction be spontaneous in the direction written under these conditions?

2. Give the general form of the free-energy change ΔG at constant temperature for the synthesis of ammonia, $3H_2(g) + N_2(g) \rightleftharpoons 2NH_3(g)$. Use Appendix C and $T = 298$ K to write your formula in numerical form. What is ΔG under standard conditions? When $P_{H_2} = P_{N_2} = 1.00$ atm and $P_{NH_3} = 0.00$ atm? When $P_{H_2} = 3.00$ atm, $P_{N_2} = 1.00$ atm, and $P_{NH_3} = 0.10$ atm? At equilibrium?

3. A step in the synthesis of sulfuric acid is the oxidation of sulfur dioxide, $2SO_2(g) + O_2(g) \rightleftharpoons 2SO_3(g)$. Write the reaction quotient Q for this reaction in terms of the partial pressures of reagents and product, and state a formula for ΔG at constant temperature. Using Appendix C and $T = 298$ K, find ΔG° and K for this reaction. From Equation 12.9, predict whether the reaction will be spontaneous as written when $P_{SO_2} = 0.50$ atm, $P_{O_2} = 0.10$ atm, and $P_{SO_3} = 0.25$ atm.

4. It has been remarked that it is fortunate for us that the potential atmospheric reaction $N_2(g) + O_2(g) \rightleftharpoons 2NO(g)$ is *not* spontaneous, thereby allowing us oxygen to breathe. Find ΔG_{298}° and K for this reaction, and assess its spontaneity under atmospheric conditions (see Table 9.2 of Chapter 9), assuming no NO is present in the atmosphere. What minimum partial pressure of NO is required to assure the veracity of the above statement? Would our oxygen supply be seriously depleted by the extent of reaction implied by your value of P_{NO}?

5. At equilibrium in the ammonia synthesis $3H_2(g) + N_2(g) \rightleftharpoons 2NH_3(g)$ at 450°C, the following partial pressures were found: $P_{H_2} = 21.20$ atm, $P_{N_2} = 7.07$ atm, and $P_{NH_3} = 1.74$ atm. Find the equilibrium constant K and ΔG° at this temperature.

6. At 1375 K H_2S dissociates according to $2H_2S(g) \rightleftharpoons 2H_2(g) + S_2(g)$. When 0.224 mol H_2S is placed in a 24.0-L vessel and heated to 1375 K, the total pressure is found to be 1.173 atm. Calculate K and ΔG° for this reaction at 1375 K.

7. When pure halogen gases are mixed, *interhalogen* compounds are formed; for example $Cl_2(g) + Br_2(g) \rightleftharpoons 2BrCl(g)$. For initial partial pressures of the pure halogens of 200. Torr each, set up an extent-of-reaction analysis, and find the partial pressures of all species when equilibrium has been reached at 298 K.

8. At 900 K the reaction $C_2H_6(g) \rightleftharpoons C_2H_4(g) + H_2(g)$ has $\Delta G^\circ = 5.35$ kcal/mol. Calculate the partial pressure of H_2 at equilibrium if pure C_2H_6 is passed over a dehydrogenation catalyst at this temperature and 1.00 atm total pressure.

LeChâtelier's principle and the van't Hoff equation

9. Using data from Appendix C, predict the effects of increases in temperature and pressure on the position of equilibrium in the following reactions:
   a. $COCl_2(g) \rightleftharpoons CO(g) + Cl_2(g)$
   b. $N_2O(g) + 4H_2(g) \rightleftharpoons 2NH_3(g) + H_2O(g)$
   c. $ClO_2(g) + O_3(g) \rightleftharpoons ClO(g) + 2O_2(g)$
   d. $CH_4(g) + Cl_2(g) \rightleftharpoons CH_3Cl(g) + HCl(g)$
   e. $C_3H_8(g) + 3H_2O(g) \rightleftharpoons 3CO(g) + 7H_2(g)$ (propane reforming)

10. For our $2NO_2 \rightleftharpoons N_2O_4$ text example, use the van't Hoff equation (Equation 12.15) to predict K at 100°C, decide whether your value confirms what is expected from LeChâtelier's principle, and find the equilibrium partial pressures at that temperature and a total pressure of 1.30 atm.

11. The equilibrium constant for the dissociation of chlorine, $Cl_2(g) \rightleftharpoons 2Cl(g)$, is found to be $K = 2.48 \times 10^{-5}$ at $T = 1200$ K and $K = 0.0129$ at $T = 1600$ K.
   a. What would you predict from LeChâtelier's principle for the pressure and temperature dependence of the dissociation?
   b. From an extent-of-reaction analysis for a given total pressure P at equilibrium, derive a formula for P_{Cl}/P as a function of K and P, and show that it conforms to the result of part (a).

c. Find P_{Cl}, P_{Cl_2}, and the fractional dissociation of Cl_2 at 1600 K for $P = 1.00$ atm and for $P = 10.00$ atm.

d. Estimate $\Delta H°$ for the dissociation from the data at the two temperatures. Compare it to the value obtained from Appendix C and discuss any difference.

12. Assuming the rapid attainment of equilibrium, which of the following conditions would be most favorable for the production of ethanol from the reaction of ethylene with steam,
$$CH_2=CH_2(g) + H_2O(g) \rightleftharpoons CH_3CH_2OH(g)$$
   a. $P = 5$ atm, $T = 600$ K
   b. $P = 50$ atm, $T = 400$ K
   c. $P = 50$ atm, $T = 600$ K
   d. $P = 5$ atm, $T = 400$ K

13. Use LeChâtelier's principle to predict what combination of nonstandard temperature and pressure will optimize the yield of ammonia from the Haber–Bosch synthesis $3H_2(g) + N_2(g) \rightleftharpoons 2NH_3(g)$. Do the actual conditions used industrially, several hundred atmospheres and 700 K, correspond to your prediction? Use the data of Appendix C to predict the equilibrium constant for this reaction at 700 K.

14. The synthesis of $SO_3(g)$ from SO_2 and O_2 is carried out industrially at $T = 950$ K. Use data from Appendix C to estimate the equilibrium constant for this reaction at 950 K. Will the yield of SO_3 at this temperature, assuming equilibrium is achieved, be higher or lower than at 298 K?

15. *Thermal cracking,* employed by the petrochemical industry to reduce large hydrocarbon molecules to smaller ones, is typified by the reaction $CH_3CH_2CH_2CH_3(g) \rightleftharpoons CH_3CH_3(g) + CH_2=CH_2(g)$, which occurs at 500°C. Estimate the equilibrium constant for this reaction at 500°C, and find the cracking fraction at 1.00 atm total pressure. Would the yield be higher or lower if the reaction could be run at 25°C? Would an increase in total pressure increase or decrease the yield at equilibrium?

16. In the derivation of the van't Hoff equation (Equation 12.15), we assumed that $\Delta H°$ is independent of temperature. In Chapter 10, Section 10.7, we learned how to estimate the temperature dependence of $\Delta H°$ from Equation 10.36. Using this relationship and starting from the differential form of the van't Hoff equation (Equation 12.16), derive an improved version of Equation 12.15 by integration, assuming that the heat-capacity difference ΔC_P is temperature independent. Use your result, along with heat capacities from Appendix C, to obtain an improved estimate of K at $-75°C$ in Example 12.3.

Acids, bases, and pH

17. A sample of HCl(*aq*) is diluted from 1.00 M to 0.025 M at 298 K. What is the change in free energy (kcal/mol or kJ/mol) that accompanies this dilution?

18. Find the pH of the following solutions, and classify them as acidic, basic, or neutral:
    a. 1.00 M HBr; 2.50×10^{-3} M HCl; 1.46×10^{-11} M HNO_3
    b. 0.0500 M NaOH; 7.66×10^{-4} M $Ca(OH)_2$; 9.2×10^{-21} M CsOH
    c. 20.0 mL 0.200 M NaOH added to 10.0 mL 0.500 M HI
    d. 25.0 mL 0.0150 M H_2SO_4 added to 25.0 mL 0.050 M $Sr(OH)_2$
    e. 4.602 g BaO dissolved and diluted to 500.0 mL

19. Find the concentrations $[H^+]$ and $[OH^-]$ in the following aqueous solutions:
    a. Household bleach, pH = 10.68
    b. Human blood, pH = 7.40
    c. Cider vinegar, pH = 2.45
    d. Rainwater, pH = 5.7
    e. Orange juice, pH = 3.5
    f. Household ammonia, pOH = 2.55

20. Adding 0.25 mL of a strong monoprotic acid solution to 500. mL of water produced a pH of 2.52. What was the concentration of the strong acid?

21. Use the van't Hoff equation to find pK_w and the pH corresponding to a neutral aqueous solution at the following temperatures: 0°C, 37°C, 100°C.

Weak acids and bases

22. Using data from Table 12.1, find $\Delta G°$ for the ionization of nitrous acid HNO_2. Use Appendix C to find $\Delta H°$ for this ionization, and decide whether K_a will be strongly temperature dependent and in which direction.

23. Weak acids have ionization constants in the range $K_a \approx 10^{-4}$ to 10^{-10} at 25°C. To what range of $\Delta G°$, (kcal/mol or kJ/mol) does this range of K_a correspond?

24. The pH of a formic acid (HCOOH) solution is measured to be 2.10. What was the molarity of HCOOH before ionization? At equilibrium?

25. The pH of a 0.150 M solution of an unknown monoprotic acid is found to be 2.09. What is K_a for this acid?

26. Find the pH of a 0.0500 M $HClO_2$ solution, comparing the approximations of Equations 12.26 and 12.28 to the quadratic solution of Equation 12.27.

27. Draw Lewis structures of appropriate geometry, including lone pair electrons and any resonance forms, for the generic carboxylic acid (RCOOH) and its anion ($RCOO^-$), and give a MO–VB interpretation of the σ and π bonding. Use data from Appendix C for acetic acid ($CH_3COOH(aq)$) and acetate ion ($CH_3COO^-(aq)$) to characterize the enthalpic and entropic components of the ionization reaction, and discuss them in terms of your bonding analysis and solvent effects.

28. In the interpretation of acidity of an O—H bond, first given in Section 5.4 and again here in Section 12.3, it is primarily $\chi(X)$ in the X—O—H linkage that affects acid behavior, larger $\chi(X)$ more readily allowing loss of H^+. An important secondary effect is exerted by other atoms or groups that may also be bonded to X, as discussed here for the oxyacids. Use this line of reasoning to arrange the haloacetic acids $CH_2ClCOOH$, $CH_2BrCOOH$, and CH_2ICOOH in order of increasing acidity (increasing K_a value).

29. For the ionization of ammonia (NH_3), Equation 12.31, write the equilibrium, charge balance, and mole balance conditions that are analogous to those for HF given in Equation 12.29. Show that these conditions lead to a single equation identical in form to Equation 12.30, where $x = [OH^-]$ and K_a is replaced by K_b. What condition must be met to allow your equation to reduce to the simpler Equation 12.31?

30. Draw geometrically realistic structures for the weak base methylamine (CH_3NH_2) and its protonated cation methylammonium. Find the pH of a 0.0250 M solution of methylamine.

31. Use Tables 12.1 and 12.2 to find equilibrium constants and $\Delta G°$'s for the following aqueous reactions, identifying the acid, base, conjugate acid, and conjugate base:

    a. $CH_3NH_2 + HCl \rightleftharpoons CH_3NH_3^+ + Cl^-$
    b. $HNO_2 + KOH \rightleftharpoons KNO_2 + H_2O$
    c. $NH_3 + HCOOH \rightleftharpoons NH_4^+ + HCOO^-$
    d. $2NH_4Cl + Ba(OH)_2 \rightleftharpoons 2NH_3 + BaCl_2 + 2H_2O$
    e. $CH_3COONa + HIO_3 \rightleftharpoons CH_3COOH + NaIO_3$

32. Find the pH of the following solutions:

    a. 0.100 M KCN
    b. 0.0200 M NH_4Cl
    c. 0.0015 M $(CH_3CH_2COO)_2Ca$
    d. 0.74 M NaOCl

33. Write down the equilibrium, charge balance, and mole balance conditions for the ionization of the salt NaA of a weak acid HA analogous to Equations 12.31.

34. Write down the equilibrium, charge balance, and mole balance conditions for the ionization of the salt BHCl of a weak base B analogous to Equations 12.31.

Common ions and buffers

35. Calculate the pH and the CH_3COO^- concentration in a solution made by mixing 100. mL 0.25 M CH_3COOH and 100. mL 0.050 M HBr.

36. What should happen to the concentration of NH_4^+ if aqueous NaOH is added to an NH_3 solution? Compare $[NH_4^+]$ in a 0.125 M NH_3 solution to that for a solution 0.125 M in NH_3 and 0.125 M in NaOH.

37. Use the more exact form of the buffer equilibrium condition given in Equation 12.41 to derive a quadratic formula solution for $x = [H^+]$. Check the accuracy of the Henderson–Hasselbalch equation (Equation 12.42) by using your result to compute $[H^+]$ and pH for the two buffer solutions of Example 12.7.

38. Neglecting the change in volume, what mass of KNO_2 crystals should be added to 500. mL 0.200 M HNO_2 to produce a buffer solution of pH = 3.34? What is the effect on the pH of adding 0.02 mol KOH to this solution? Compare it with adding the same amount of KOH to an unbuffered solution of the same pH.

39. A buffer solution can be made by partially neutralizing a weak acid solution with strong base. Suppose 20.0 mL 0.200 M NaOH is added to 25.0 mL 0.500 M $CH_2ClCOOH$. (a) Write the neutralization reaction and find its equilibrium constant. (b) Calculate the pH of the resulting solution. (c) What further added volume of NaOH would be required to make a buffer of greatest effectiveness?

40. For a weak-base buffer composed of known concentrations of base B and its salt BHCl, derive a formula

for the pH analogous to the Henderson–Hasselbalch equation (Equation 12.42). (*Hint:* Set the problem up in a way analogous to Equation 12.41.) Show that your equation is identical in form to Equation 12.42, with K_a given by the ionization constant of the conjugate weak acid BH^+.

41. What is the pH of a solution 0.25 M in NH_3 and 0.25 M in NH_4Cl? 0.10 M in NH_3 and 0.50 M in NH_4Cl? (Use the result of Exercise 40.) Based on LeChâtelier's principle, what would you predict for the effect on the pH of the added NH_4Cl? Compare your pH values to those that would be obtained for the same solutions without NH_4Cl.

42. a. A buffer solution of pH 4.00 is often required for pH meter calibration. Given ample quantities of 0.200 M HCl, 0.200 M NaOH, and 0.200 M solutions of a weak monoprotic acid or base of your choice, decide which solutions and what volumes to mix in order to prepare 500. mL pH 4.00 buffer. Try to choose your weak acid or base to yield a buffer of maximum effectiveness.

    b. Repeat part (a) for a buffer of pH 10.00.

Polyprotic acids

43. Find the pH and the equilibrium concentrations of $H_2PO_3^-$ and HPO_3^{2-} in a 0.200 M H_3PO_3 solution.

44. Using the approximations leading to Equations 12.48, carry through the algebra for a solution of the monosodium salt NaHA of a diprotic acid H_2A to verify those relations and express them in generalized form. Indicate clearly the assumptions made in your derivation.

45. A saturated H_2S solution has a concentration of 0.10 M prior to ionization. (a) Find the pH and the concentrations of HS^- and S^{2-} at equilibrium. (b) If the pH is adjusted to 3.00 by adding a few drops of concentrated HCl, find $[HS^-]$ and $[S^{2-}]$ when equilibrium is reached.

46. Estimate the pH of a 0.150 M sodium oxalate $Na_2(COO)_2$ solution.

47. As Table 12.3 indicates, the second proton of H_2SO_4 does not ionize completely, making the pH of a 0.050 M solution of H_2SO_4 greater than 1.00. Find the pH of this solution, as well as the concentrations of HSO_4^- and SO_4^{2-}, by assuming that the first proton does ionize completely, that is, that the solution starts from $H^+ + HSO_4^-$.

48. Using the approximations leading to Equation 12.44 for a diprotic acid, analyze the ionization of a triprotic acid H_3A, such as H_3PO_4 or H_3AsO_4. Begin by writing the three ionization steps and their associated equilibrium expressions. Assume that $1 \gg K_{a1} \gg K_{a2} \gg K_{a3}$, and that the first two steps can be treated in the same way as in the diprotic case. Restate the approximate relationships for the first two steps in terms of the new neutral and anion formulas H_3A, H_2A^-, and HA^{2-}. Show that the equilibrium concentration of A^{3-} is approximately given by $[A^{3-}] \approx K_{a2}K_{a3}/[[H_3A]_0 K_{a1}]^{1/2}$.

49. By making approximations analogous to those used for a diprotic acid with $K_{a2} \ll K_{a1} \ll 1$, find the pH and the concentration of $N_2H_6^{2+}$ in a 0.250 M N_2H_4 (hydrazine) solution.

Acid–base titrations

50. 10.00 mL of a solution of formic acid HCOOH required 31.23 mL of 0.1034 M NaOH to titrate it. (a) Find the (un-ionized) concentration of the formic acid solution. (b) Find the pH at the three characteristic points (the start, the midpoint, and the endpoint), and sketch a titration curve on suitable labeled axes. (c) Select a suitable indicator for this titration from Table 12.5.

51. A student was given an unknown monoprotic weak acid to analyze by titration with standard NaOH solution. Unfortunately the concentration of NaOH was also unknown, the label on the bottle having been obliterated by dribbles of base. Titration of 20.00 mL of the acid solution required 27.86 mL NaOH. When 13.93 mL of NaOH was added to another 20.00 mL of acid, the pH was found to be 4.19. Find the K_a of the acid, and sketch a titration curve.

52. The titration of a weak base by a strong acid has a pH titration curve that is "upside down" (a reflection through pH = 7) compared to that of the weak acid plus strong base. Sketch a titration curve for the analysis and standardization of a 0.1 M NH_3 solution by standardized 0.1000 M HCl, and select an appropriate indicator for the titration from Table 12.5.

53. 10.00 mL of a selenous acid (H_2SeO_3) solution was titrated with 0.2000 M NaOH, yielding endpoints at 15.71 and 31.38 mL. Find the concentration of the H_2SeO_3, and sketch a titration curve, indicating numerical values of the pH at the five characteristic points.

54. Solutions of NaOH to be used for acid analysis are usually standardized against potassium hydrogen phthalate, HOOCC$_6$H$_4$COOK(s), abbreviated KHP, the monopotassium salt of phthalic acid. In a typical procedure, 0.7835 g KHP was dissolved in 50. mL water, and required 36.85 mL NaOH solution to titrate it. (a) What is the molarity of the NaOH? (b) Calculate the pH at the three characteristic points, and sketch the titration curve. (Recall that phthalic acid is diprotic.) (c) Select an appropriate indicator for the titration from Table 12.5.

55. HCl solutions can be standardized by titrating sodium carbonate Na$_2$CO$_3$. 16.07 mL of an HCl solution was required to titrate 0.1714 g Na$_2$CO$_3$ dissolved in 50. mL water, using methyl orange indicator. (a) Recalling that CO$_3^{2-}$ acts as a dibasic weak base, write the overall titration reaction in ionic form and find its equilibrium constant. (b) Compute the molarity of the HCl. (c) Calculate the pH at the five characteristic points, and sketch the titration curve. (d) Select an indicator from Table 12.5 appropriate for titrating to the HCO$_3^-$ endpoint.

Aqueous complexes

56. The Ni(NH$_3$)$_6^{2+}$ complex is used as an example in Section 12.3. Concentrated ammonia NH$_3$ is added to a 0.10 M NiCl$_2$ solution until it is 1.0 M in NH$_3$. Neglecting the diluting effect of the added volume, what is the concentration of free (hydrated) Ni^{2+} at equilibrium? What would it be if [NH$_3$] were increased to 2.0 M?

57. Cu^{2+} ion forms a complex Cu(NH$_3$)$_4^{2+}$ when ammonia is added to a 0.025 M CuSO$_4$ solution. The formation constant for this complex is $K_f = 1 \times 10^{12}$. What minimum concentration of ammonia at equilibrium is needed to keep [Cu^{2+}] below 10^{-15} M?

Multiphase equilibria

58. Write reaction quotients for the following chemical reactions:
    a. 2K(s) + 2H$_2$O(l) $\rightleftharpoons$ 2KOH(aq) + H$_2$(g)
    b. NH$_3$(l) $\rightleftharpoons$ NH$_3$(g)
    c. NH$_3$(aq) $\rightleftharpoons$ NH$_3$(g)
    d. 4Al(s) + 3O$_2$(g) $\rightleftharpoons$ 2Al$_2$O$_3$(s)
    e. 2Al(s) + $\frac{3}{2}$O$_2$(g) $\rightleftharpoons$ Al$_2$O$_3$(s)
    f. 2KClO$_3$(s) $\rightleftharpoons$ 2KCl(s) + 3O$_2$(g)
    g. SO$_2$(g) + H$_2$O(l) $\rightleftharpoons$ H$_2$SO$_3$(aq)

59. Each of the following reactions is important in the industrial manufacture of chemicals. Write reaction quotients for each.
    a. 4HF(g) + SiO$_2$(s) $\rightleftharpoons$ SiF$_4$(g) + 2H$_2$O(g)
       (step in making H$_2$SiF$_6$ for fluoridation)
    b. 2NH$_3$(g) + H$_3$PO$_4$(g) $\rightleftharpoons$ (NH$_4$)$_2$HPO$_4$(s)
       (phosphate fertilizer)
    c. Fe$_2$O$_3$(s) + 3CO(g) $\rightleftharpoons$ 2Fe(s) + 3CO$_2$(g)
       (smelting of iron ore)
    d. C(s) + H$_2$O(g) $\rightleftharpoons$ CO(g) + H$_2$(g)
       (production of "synthesis gas")
    e. H$_2$S(g) + 2NaOH(aq) $\rightleftharpoons$ Na$_2$S(aq) + 2H$_2$O(l)
       ("scrubbing" of H$_2$S from cracking of high-sulfur oil)
    f. CO$_2$(g) + Na$_2$CO$_3$(aq) + H$_2$O(l) $\rightleftharpoons$ 2NaHCO$_3$(aq)
       (removal of CO$_2$ from CO$_2$/H$_2$ mixtures)

60. Lime CaO(s), whose chief use is now for removal of slag in steelmaking, is produced by heating limestone:

    $$CaCO_3(s) \underset{\Delta}{\rightleftharpoons} CaO(s) + CO_2(g)$$

    Lime kilns are usually operated at 1200°C. Use data from Appendix C, or from Tables 10.1 and 11.1, to estimate the partial pressure of CO$_2$(g) were equilibrium to be reached in this reaction at 1200°C. Why are high temperatures required? Why do you think lime kilns are always vented to the atmosphere?

61. When a chemist purchases HI(g) [or HI(aq)] for laboratory use, she must use it quickly, or the decomposition 2HI(g) $\rightleftharpoons$ H$_2$(g) + I$_2$(s) will tend to equilibrium. Assuming she starts with a flask containing 5.00 atm of pure HI, set up an extent-of-reaction analysis, and find the equilibrium partial pressures of all species and the total pressure at equilibrium at 298 K.

62. The equilibrium CO$_2$(g) $\rightleftharpoons$ CO$_2$(aq), or equivalently, CO$_2$(g) + H$_2$O(l) $\rightleftharpoons$ H$_2$CO$_3$(aq), is significant in applications ranging from pH control in the blood to carbonation of beverages. (a) Use LeChâtelier's principle to predict the effect of the following changes on this equilibrium: raise the temperature; raise P_{CO_2}; add NaOH to the aqueous phase; add Na$_2$CO$_3$ to the aqueous phase. (b) Use data from Appendix C to predict the concentration [CO$_2$] or [H$_2$CO$_3$] at equilibrium at 298 K when P_{CO_2} is held at 1.00 atm.

63. Acid rain is produced in large part when gaseous SO$_2$ produced by burning high-sulfur coal and oil dissolves

in rain. Suppose the air contains 0.050 Torr SO_2 during a rainstorm. Find the equilibrium concentration of $SO_2(aq)$, or $H_2SO_3(aq)$, at 298 K in the rainwater, and estimate its pH.

64. By taking the equilibrium $H_2S(aq) \rightleftharpoons H_2S(g)$ into account, find the maximum concentration of H_2S in water that can be obtained at 298 K and 1.00 atm external pressure. (*Hint:* H_2S will spontaneously pass into the gas phase when $P_{H_2S} = P_{ext}$.)

65. The decomposition of baking soda has been discussed in the text of both Chapters 11 and 12. Decide whether baking soda will decompose spontaneously in a cupboard at 298 K if the partial pressure of $CO_2(g)$ is 0.38 Torr and that of water vapor is 3.0 Torr. Would it help to breathe on your baking soda to preserve it? If your conclusion was that it is stable, at what water vapor pressure would it begin to decompose?

Solubility and precipitation

66. How many grams of silver acetate (CH_3COOAg) will dissolve in 1.0 L water? For silver acetate K_{sp} is 1.94×10^{-3}.

67. Limewater is a saturated solution of $Ca(OH)_2$. What is the molarity of Ca^{2+} and the pH of limewater?

68. The maximum water hardness allowed by the Environmental Protection Agency corresponds to a calcium ion (Ca^{2+}) concentration of 1.0×10^{-4} M. What mass in grams of Na_2CO_3 must be added to 500. mL maximum-hardness water to produce a $CaCO_3$ precipitate? Neglect the change in volume due to the addition of carbonate.

69. Show that a $Cu(OH)_2$ precipitate will form when 50.0 mL 0.200 M NaOH is added to 50.0 mL 0.0500 M $CuCl_2$, and find $[Cu^{2+}]$ at equilibrium. Can this precipitation reaction be considered quantitative for purposes of stoichiometric analysis?

70. Hydroxide precipitation is naturally pH sensitive, tending not to occur at low pH, where $[OH^-]$ is small. What is the maximum allowable pH that will prevent 0.100 M Fe^{3+} from precipitating as $Fe(OH)_3$?

71. A common spot test for carbonate rocks is the dissolution with fizzing that occurs when acid is poured on them. The mineral *witherite* is $BaCO_3$. At what pH will witherite be soluble to the extent of 0.10 M Ba^{2+}? (*Hint:* First find the required $[CO_3^{2-}]$ from the solubility product; then, assuming the pH is low enough that H_2CO_3 is the major carbonate species in solution, use the overall acid ionization equilibrium.)

72. Separation of group I chloride precipitates in qualitative analysis exploits the fact that $PbCl_2$ is more soluble in hot water than in cold. What does this imply about the sign of $\Delta H°$ for the $PbCl_2(s) \rightleftharpoons Pb^{2+} + 2Cl^-$ reaction? Use data from Table 12.6 and Appendix C to estimate K_{sp} for $PbCl_2$ at 95°C.

73. Sulfide precipitation, used in qualitative analysis, can be controlled by adding acid to an H_2S solution. What range of pH will prevent the precipitation of CoS but allow that of NiS in a saturated (0.10 M) H_2S solution if each metal ion is present at 0.050 M concentration prior to adding H_2S?

74. Precipitation can be used as a titration method for determining extremely small ion concentrations. In a procedure for determining the formation constant K_f of $Ag(NH_3)_2^+$, a solution of the complex was prepared by mixing 5.00 mL 0.100 M $AgNO_3$, 20.00 mL 0.500 M NH_3, and 25.00 mL water. The free Ag^+ was then titrated with 0.0500 M NaCl until a slight opalescence signaled AgCl precipitate formation. If the titration required 10.31 mL NaCl, find the formation constant of the complex. (Assume $K_f \gg 1$, so that the complexation reaction is complete, and $[Ag(NH_3)_2^+] \approx [Ag^+]_0$ before titration. Remember to include the diluting effect of the NaCl.)

Nonideal behavior

75. Compute the activity coefficient γ and the product γP for $H_2(g)$ at 298 K and (a) 1.00 atm; (b) 10.0 atm; (c) 1000. atm. Assume van der Waals gas behavior. What can you conclude about the pressure dependence of nonideality in H_2? About the dominant forces between H_2 molecules?

76. Use the result of the previous exercise to calculate the work done (J and cal) in compressing 1 mol of $H_2(g)$ from 1.00 atm to 10.0 atm isothermally; compare it to the ideal gas case. Briefly discuss any difference you find.

77. What is the activity coefficient of steam ($H_2O(g)$) at 1.00 atm and 100°C? 200°C? Assume van der Waals gas behavior. What can you conclude about the temperature dependence of nonideality in steam? about the dominant forces between H_2O molecules?

78. Calculate and compare the Debye lengths r_d and the activity coefficients γ for 0.0100 M solutions of KBr and $CuSO_4$. Discuss briefly any differences in terms of the interionic forces.

Electrochemistry

Bubbles of $H_2(g)$ forming at the cathode of a photoelectrolysis cell. See Section 13.5 for a discussion of electrolysis.

Thunder is good, thunder is impressive, but it is lightning that does the work.

Mark Twain (1908)

Oxidation–reduction reactions, wherein electrons are transferred from a donor, the reducing agent, to an acceptor, the oxidizing agent, are often much more spectacular than their sedate cousins, the acid–base reactions. Flames and explosions, luminescence and violent action, are the nearly exclusive province of redox reactions. These properties reflect a substantially larger release of energy, especially free energy, as the reaction proceeds; thermodynamically, we think of this enormous free energy drop as the impetus behind the spectacular events. If any chemical reactions can be useful to us, in the sense of giving a lot of "bang for the buck," they must be redox, a fact humankind has exploited since before the dawn of recorded history. People have warmed themselves, cooked their food, and created explosives for both entertainment and destruction with the aid of redox chemistry. Widespread use of a phenomenon so readily attained as fire was bound to predate an understanding of the underlying chemistry; a grasp of redox principles only came with the harnessing of electricity. By 1767, Joseph Priestley could presciently observe:

CHAPTER OUTLINE

13.1 Voltage from the Free Energy of a Redox Reaction

13.2 Electrochemical Cells: Principles and Practice

13.3 Electrode Potentials

13.4 Concentration Cells

13.5 Electrolysis: Redox Chemistry in Reverse

13.6 The Electrochemistry of Corrosion

Alessandro Volta (1745–1827) was a professor of physics at the University of Pavia, Italy. His long-standing interest in static electricity led him to investigate Galvani's "animal electricity" beginning in 1794. He built the first "electric pile" (battery) in 1800, establishing that animal matter was not required for electrical action. He was lionized by Napoleon after demonstrating his pile in Paris in 1801. The unit of electric potential, the volt, V (1 V = 1 J/C) was named in his honor in 1881.

... for chymistry and electricity are both conversant about the latent and less obvious properties of bodies; and yet their relation to each other has been but little considered, and their operations hardly ever combined. . . .

Learning how to turn the electron flow that accompanies a redox reaction to our own use is today still the subject of lively research efforts on a variety of fronts, from the development of more powerful batteries to solar energy conversion and storage. It is in exploiting this connection that we can finally hope to capture and use all of the free energy that redox reactions have to offer. Hence arises the title of this chapter, in which we intend to still the chemical "thunder" and try to get the "lightning" to do the work.

13.1 Voltage from the Free Energy of a Redox Reaction

In the year 1800, precisely a century before Planck's revolutionary hypothesis, Alessandro Volta began a revolution of his own, with a discovery that has changed the face of chemistry and civilization perhaps even more than has quantum theory. Analyzing some puzzling observations of Luigi Galvani in 1786, in which frogs' legs hung by copper hooks twitched when they brushed against an iron railing, Volta realized that it was the dissimilar metals, copper and iron, that caused an electric current to flow through the leg and make it twitch. This prompted him to build a stack of metal plates, alternating silver and zinc, separated by blotting paper soaked in brine, that became known as a Voltaic pile (see Figure 13.1). When the opposite ends of the pile were connected by a wire, Volta found that a current flowed. Figure 13.1 shows the ends connected by strips of copper to cups of brine, and the other end to a metal wire. When Volta dipped his fingers in one of the cups and touched the bare metal strip on the other end, he experienced a continuous mild shock; when he put the metal strip in his mouth, he noted a bitter taste. The latter was the first successful **electrolysis** of water (an experiment later demonstrated in more proper fashion by Carlisle and Nicholson in England by immersing two wires from the ends of a pile in a salt solution). Electrolysis causes chemical reactions to run backward, a topic to be discussed in Section 13.5.

Volta had constructed and used the first **battery,** composed of a series of what we refer to in this chapter as **electrochemical cells.** One cell would consist of a single pair of plates of dissimilar metals separated by a cardboard spacer. Such an arrangement is also referred to as a *voltaic cell* or *galvanic cell,* in honor of its discoverers. The key innovation in the construction of the cell was the use of the brine-soaked spacers (replacing Galvani's frog legs); these, as we can now deduce with our modern knowledge of the nature of salt solutions, allowed ions, but not electrons, to flow from one metal to the other. Volta's battery was not particularly stable; large currents could be drawn only for a brief time, and as the spacers dried out, the current decreased unsteadily to zero. Before we discuss improvements to Volta's battery, let's analyze the redox reaction that made it function.

The brine caused ions of silver and zinc (always present in the form of thin oxide layers, Ag_2O and ZnO, on the metal plates) to dissolve. The redox displacement reaction

$$2Ag^+(aq) + Zn(s) \longrightarrow 2Ag(s) + Zn^{2+}(aq) \tag{13.1}$$

Figure 13.1
Volta's piles, as presented to the Royal Society in 1800. Each pile consists of repeating trios of silver, zinc, and brine-soaked blotting paper disks. The voltage of the piles increases as more stacks are added; adjacent stacks are in reverse order (in series), joined by copper strips, which also connect the ends of the piles to cups of brine. Using the standard voltage of the single Zn/Ag cell given in the text, can you estimate the voltage of the four-stack pile? (*Answer:* 44 V; you must count only the number of cardboard spacers.)

could then occur slowly through the diffusion of silver ions through the cardboard spacer to the surface of the Zn plate. (Volta believed that direct contact between the Ag and Zn plates within the pile was required, but this was later shown not to be the case. His pile could have been constructed by connecting the plates within the pile with a wire or plate of copper or other metal.) When the Zn and Ag plates at the opposite ends of the pile were connected by an external conductor, however, the reaction could occur rapidly, since the Ag^+ ions could now obtain electrons from the silver plate itself, flowing freely from the Zn plate through the conductor to the Ag. (This rapidly exhausted the small supply of Ag^+, deadening the battery until further oxidation and dissolution of the Ag could occur.) In Chapter 5 you learned how to use ion-electron half-reactions in the analysis of redox reactions; here these half-reactions take on a genuine physical significance. Ag^+ is reduced in Reaction 13.1 by acquiring electrons indirectly supplied by Zn, but these ions don't "know" where the electrons they receive are coming from; likewise, Zn is being oxidized by the loss of electrons to Ag, but is not in direct contact with Ag. We thus have a physical separation of the half-reactions

$$Zn(s) \longrightarrow Zn^{2+}(aq) + 2e^-$$
$$2 \times (Ag^+(aq) + e^- \longrightarrow Ag(s))$$

where the factor of 2 for Ag^+ reduction brings the electrons into balance.

We have written Reaction 13.1 in its spontaneous direction with foreknowledge; though Volta apparently noted no chemical transformation when using his pile, selective corrosion of the Zn plates was found in later studies. From Appendix C we find the standard free energy change at 298 K,

$$\Delta G° = -2\Delta G_f°(Ag^+(aq)) + \Delta G_f°(Zn^{2+}(aq))$$
$$= -2(18.433) + (-35.14) = -72.01 \text{ kcal } (-301.3 \text{ kJ})$$

recalling that elements in their standard states have zero free energies of formation. This is quite a large magnitude for $\Delta G°$, dwarfing the -19 kcal of the Arrhenius acid–base reaction, and approaching the energy of a chemical bond. The equilibrium constant for Reaction 13.1 is $K = 6.1 \times 10^{52}$, a stupendously large number (170,000 times larger than the *square* of Avogadro's number). The equilibrium condition

$$K = \frac{[Zn^{2+}]}{[Ag^+]^2} \tag{13.2}$$

then implies that, for any finite starting concentrations, the reaction will truly go to completion. In fact, even assuming a 1-L volume of 1.0 M Ag^+ salt solution, with a piece of zinc metal large enough to make Ag^+ the limiting reagent, the large K implies that *not a single Ag^+ ion will remain at equilibrium!* Equation 13.2 yields $[Ag^+]_{eq} = \{[Zn^{2+}]_{eq}/K\}^{1/2} = \{0.5 / 6.1 \times 10^{52}\}^{1/2} = 2.9 \times 10^{-27}$ mol/L $= 0.0017$ Ag^+ ions/L. Because Ag^+ ions are indivisible, this answer is clearly nonsense in a 1-L solution, and represents a breakdown of thermodynamic relationships due to the atomic nature of matter akin to that we encountered in Chapter 12, Section 12.1. Recall that we obtained infinite free-energy changes when absolutely no product was present at the start of a reaction, which was traceable to the vitiation of the entropy concept in the absence of large numbers of molecules of all participants in a reaction. This stricture on the use of entropy holds no less at equilibrium than at any other extent of reaction. All we can conclude is that, in a system of ordinary size, this reaction is atomically complete, and the precise value of the colossal K obtained does not carry its usual meaning.

Although we may not be able to give a thermodynamic description of Volta's cell reaction after equilibrium is reached, that state represents a "dead" cell, which is no longer capable of doing useful work and is therefore less interesting. A freshly prepared cell *is* capable of doing work, in this case the work of "pushing" electrons through the external connecting wire, and under the right conditions we *can* describe it. These driven electrons in turn can be made to run electrical devices, making the cell a useful energy source. In electrical circuit analysis the impetus for electron flow is an electrical **potential difference** $\mathscr{E}$, measured in volts (V), the unit named after Volta. The **volt** is defined by the work w in joules (J) done in moving a charge Q in coulombs (C) across a potential difference $\mathscr{E}$, so that 1 J of work is expended in driving 1 C with a 1-V potential or, more generally, $w(J) = Q(C) \times \mathscr{E}(V)$. Knowing that ΔG represents the maximum work available from a chemical reaction, we can state that, for a redox (electron-driving) reaction,

$$\Delta G = w_{max} = Q\mathscr{E} \tag{13.3}$$

This simple relationship bears the message that P-V work is not the only kind of work a chemical reaction can do; we now have electrical work as well, and its efficient extraction depends on how well we construct the cell and the external circuit. Recall from Chapter 10 that typically P-V work represents a small fraction of the energy evolved by most reactions, and the prospect of turning to electrical work to extract more of the free energy becomes even more inviting. In effect, we can now think of changing the "thunder" of the H_2–O_2 explosion into the "lightning" of a smoothly running electrical device driven by the free energy released by this powerful reaction. This potential was realized in the *Apollo* lunar missions of 1969–1972, when H_2–O_2 cells were used to provide both energy for spacecraft appliances and pure drinking water for the astronauts. These **fuel cells,** to be described in the next section, are now being used in power plants and automobiles.

Equation 13.3 can be cast into a more useful form by incorporating the stoichiometry of the reaction into the charge Q. If N electrons are transferred in the reaction, then $Q = -Ne = -nN_Ae$, where e is the magnitude of the electronic charge, n is the number of moles of electrons, and N_A is Avogadro's number. The quantity N_Ae is the charge on 1 mol of elementary charges, and is called the **Faraday constant** F, in honor of Michael Faraday, whose electrochemical legacy is described in Section 13.5. The Faraday has the value

$$F = (6.02214 \times 10^{23}\ \text{mol}^{-1})(1.602176 \times 10^{-19}\ \text{C}) = 96{,}485\ \text{C/mol}$$
$$= 96.485\ \text{kJ mol}^{-1}\ \text{V}^{-1} = 23.0605\ \text{kcal mol}^{-1}\ \text{V}^{-1}$$

where the second line results from $1\ \text{C} = 1\ \text{J V}^{-1}$. This transforms Equation 13.3 into

$$\Delta G = -nF\mathscr{E} \tag{13.4}$$

For a single pair of plates in Volta's silver-zinc pile, we can now compute the standard voltage $\mathscr{E}°$ from Equation 13.4, knowing from the half-reactions that 2 mol of electrons are transferred per unit reaction, $n = 2$, giving $\mathscr{E}° = +1.561$ V. Because of the negative sign in Equation 13.4, the reaction being analyzed will have $\mathscr{E}° > 0$ when written in its spontaneous direction; but $\mathscr{E}°$ for the reverse reaction will still be the negative of that for the forward. Since ΔG increases with moles of reaction, and hence moles of electrons n, the cell voltage $\mathscr{E} \propto \Delta G/n$ is an *intensive* property, independent of the number of electrons transferred. This is a characteristic of practical importance as we will point out later. Volta's pile consisted of a number of cells connected in series, which yields an amplification of the voltage by the number of pairs of plates in the stack.

Drawing current through the external wire allows the reaction to proceed in its spontaneous direction, thereby lowering its free energy by reducing the amounts of reagents and increasing those of the products. As we saw in Chapter 12, this progress is described by Equation 12.6, $\Delta G = \Delta G° + RT \ln Q$. You can readily show that substituting Equation 13.4 into this relation yields

$$\mathscr{E} = \mathscr{E}° - \frac{RT}{nF} \ln Q \tag{13.5}$$

known as the **Nernst equation,** after Walther Nernst, who first established it experimentally in the late 1880s by measuring cell voltages and their dependence on concentration. For practical work, the Nernst equation is usually written

$$\mathscr{E} = \mathscr{E}° - \frac{0.05916}{n} \log_{10} Q \tag{13.6}$$

where we have used the relationship $\ln x = \ln(10^{\log_{10} x}) = (\log_{10} x) \ln 10 = 2.3026 \log_{10} x$, and assumed $T = 298.15$ K. For Reaction 13.1, let's suppose that the ions are present at the (nonstandard) concentrations $[Ag^+] = [Zn^{2+}] = 0.100$ M. Equation 13.6 then yields

$$\mathscr{E} = \mathscr{E}° - \frac{0.05916}{n} \log_{10} \frac{[Zn^{2+}]}{[Ag^+]^2}$$

$$= 1.561 - \frac{0.05916}{2} \log_{10} \frac{0.100}{(0.100)^2}$$

$$= 1.531 \text{ V}$$

Note that, due to the small coefficient of $\log_{10} Q$, many orders of magnitude deviation from standard conditions are needed to alter the cell voltage $\mathscr{E}$ significantly, unless $\mathscr{E}°$ happens to be small. It is normally difficult to reverse the direction of electron flow merely by changing concentrations. For cells with usefully large voltages, as in this case, $\mathscr{E}$ varies little with use and remains nearly equal to $\mathscr{E}°$ until the cell is almost spent.

When electrons have been pushed to the extent that the free-energy change allows, the cell is "dead," a condition we call *equilibrium*. The criterion for equilibrium, $\Delta G = 0$, is equivalent to $\mathscr{E} = 0$, and the Nernst equation (Equation 13.6) then leads to the equilibrium relation

$$\log_{10} K = \frac{n\mathscr{E}°}{0.05916} \tag{13.7}$$

where as before $K = Q_{eq}$. Thus measurement of the cell voltage under standard conditions allows us to obtain directly the equilibrium constant for the cell reaction, that is, the voltage of a "live" cell can tell us what it will be like when it is "dead." The equilibrium calculation for Volta's silver-zinc cell given earlier could equally well have been carried out using Equation 13.7 and a measured standard cell voltage. Redox equilibrium can be analyzed in the same way you have seen in Chapter 12, using an extent-of-reaction analysis, but almost always with the simplifying assumption that a reaction with $\mathscr{E}°$ on the order of a volt is essentially complete. As you can readily find from Equation 13.7, one electron transferred with $\mathscr{E}° = 1$ V yields $K \sim 10^{17}$. You must generally have in mind as well the principles used in dealing with reactions involving pure condensed phases (see Chapter 12, Section 12.4), particularly the corollary that all pure condensed phases must be present at equilibrium.

Armed with a thermodynamic analysis of the redox chemistry that underlies a cell's function, we are now ready to take a more informed look at how the electrochemical cell has evolved since Volta's day. Included will be examples that illustrate the application of these principles to the "classic" cells to be described.

13.2 Electrochemical Cells: Principles and Practice

The main improvement in cell design since Volta's pile has been based on the recognition of the importance of the **electrolyte,** the conducting solution necessary to make a complete circuit and at the same time provide reagent ions for the electron transfer.

The cell reaction is not a reaction between two metals, but between a positive metal ion, a **cation,** and another metal. In Volta's invention the necessary Ag^+ ions were present by accident. The next major advance substituted copper, a less precious metal, for silver, and was developed into a useful, if somewhat inconvenient, source of energy by John Daniell in 1836.

The Daniell Cell

Drawing on the storehouse of electrochemical lore gathered by Sir Humphry Davy, Faraday, and a host of others since Volta's work, Daniell provided his cell with a plentiful supply of Cu^{2+} to enable the displacement reaction

$$Cu^{2+}(aq) + Zn(s) \longrightarrow Cu(s) + Zn^{2+}(aq) \tag{13.8}$$

As illustrated in Chapter 5, if you dip a strip of zinc into a copper sulfate solution, the strip immediately blackens from the deposit of finely divided copper; on the other hand, a strip of clean copper dipped into $ZnSO_4(aq)$ shows no change. These simple experiments unambiguously establish the spontaneous direction of the reaction, and at the same time dictate the essential ingredients of a cell based on it.

From Appendix C you can readily find the standard free-energy change for this reaction, $\Delta G° = -50.80$ kcal (-212.5 kJ); from Equation 13.4 with $n = 2$, this yields $\mathcal{E}° = +1.10$ V. This is the voltage that will be generated by a cell, constructed as illustrated in Figure 13.2(a), with **half-cells** containing the separated $Zn \mid Zn^{2+}$ and $Cu \mid Cu^{2+}$ half-reaction *couples,* where each half-cell comprises an **electrode** of the pure metal and an electrolyte solution of standard 1.0 M concentration of the matching metal salt. A **salt bridge,** a glass U-tube filled with an inert electrolyte such as KNO_3 and plugged on each end with a porous material (replacing Volta's moist spacer), joins the two half-cells electrically, but keeps Cu^{2+} ions from directly contacting the Zn electrode. A porous barrier is also often used, particularly in practical cells. The structure of the cell can be indicated schematically by the line notation

$$Zn \mid Zn^{2+} (1.0 \text{ M}) \parallel Cu^{2+} (1.0 \text{ M}) \mid Cu \tag{13.9}$$

where the half-cell in which oxidation occurs is by convention placed on the left, and is called the **anode** compartment; while the right-hand half-cell supports reduction and is called the **cathode** compartment. The central double line indicates the salt bridge or other conductive barrier between the half-cells. Within the cell, the Zn anode serves as a sink for electrons and the Cu cathode as a source. (The words *electrode, electrolyte, electrolysis, anode, cathode, ion, cation,* and *anion* were all first introduced by Michael Faraday.)

As indicated in Figure 13.2(a), electrons in the voltmeter circuit travel from the Zn to the Cu electrode. As current I (amperes, A) $= dQ/dt$ (C s^{-1}) flows through the circuit, the half-reactions $Zn \longrightarrow Zn^{2+} + 2e^-$, in which Zn leaves its $4s$ electrons on the anode to become a solvated ion, and $Cu^{2+} + 2e^- \longrightarrow Cu$, in which Cu adds to its valence shell and deposits itself on the cathode, occur continuously. Both Zn^{2+} and Cu^{2+} are moving toward the cathode, and that is why they are called cations. If sulfate salts are used for the electrolyte solutions, the sulfate ions SO_4^{2-} all move toward the anode, and hence are called **anions.** Within the salt bridge, the inert ions do the same dance with, for example, K^+ moving toward the cathode, and NO_3^- toward the anode.

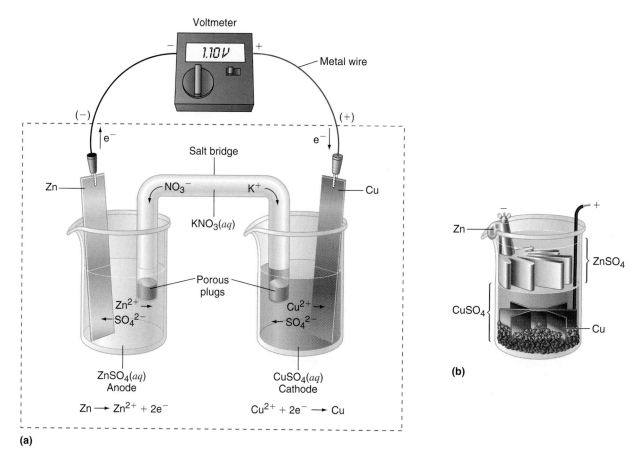

Figure 13.2
Two versions of the Daniell cell. (a) A laboratory-scale version illustrating the mechanism of action of the cell. All cations, including those in the salt-bridge electrolyte, move toward the Cu cathode and away from the Zn anode, while all anions move toward the Zn anode and away from the Cu cathode. The energy of ion motion is provided by the free-energy change of the cell reaction, Equation 13.8. The various parts of the cell are labeled as given in the text; the dashed box encloses the complete cell, which could be packaged as a single-cell battery with + and − terminals. (b) The high-current gravity cell developed by Daniell for laboratory and commercial use. Making the $ZnSO_4$ electrolyte less concentrated than $CuSO_4$ reduces its density, and it "floats." The cell is not portable, and must be constructed at the location of use.

The Nernst equation, (Equation 13.6) like the free-energy equation (Equation 12.6), only applies in a static situation, when no current is flowing. This means that voltage measurements will be accurate only if negligible current is drawn in the process, a requirement now generally met by modern, sensitive voltmeters. Two things happen when a current I (in amperes) is drawn. First, the internal resistance R_{int} of the cell—due mainly to the salt bridge, but also to finite electrode area, finite concentrations, limits on ion motion through a solvent, etc.—causes the output voltage to drop according to $\mathscr{E}_I = \mathscr{E} - IR_{int}$. Second, the ion motion described earlier causes the cell to *polarize*, that is, the + and − charges separate, thereby increasing R_{int}. The net result is that $\mathscr{E}$ continues to drop even as a constant current is drawn, by an amount far greater than that predicted by the Nernst equation from depletion of reagent ions.

Daniell's "solution" to this problem was to build a cell with minimal internal resistance, illustrated in Figure 13.2(b), in which the half-cell solutions are separated only by gravity. The Daniell cell can deliver large currents without significant voltage drop, and in the 19th century found wide use in telegraph offices and in powering home appliances such as doorbells, becoming the first successful battery. Perhaps it is most famous for being the cell—actually thousands of cells connected in series—that supplied the voltage for Millikan's oil drop experiment (see Chapter 1). Even today, if a truly steady voltage, without "ac ripple," is needed, one turns to a cell rather than a commercial "dc power supply" that runs on AC current from a wall outlet. The Daniell cell is definitely not portable, however, and can be disturbed even by undue vibration.

EXAMPLE 13.1

Daniell achieved gravitational isolation of the Cu and Zn half-cells of Figure 13.2 by simply making the Zn half-cell more dilute and therefore less dense. Given the standard cell voltage $\mathscr{E}° = 1.10$ V, what voltage will the Daniell cell provide if $[Zn^{2+}] = 0.0100$ M and $Cu^{2+} = 1.00$ M?

Solution:
To predict the voltage with a nonstandard Zn half-cell, we use the Nernst equation (Equation 13.6). In the overall cell reaction (Reaction 13.8), 2 mol of electrons are transferred per mole of reaction, $n = 2$, and the reaction quotient is simply $Q = [Zn^{2+}]/[Cu^{2+}]$. The nonstandard cell voltage is therefore

$$\mathscr{E} = 1.10 - \frac{0.0592}{2} \log_{10} \frac{0.0100}{1.00}$$
$$= 1.10 + 0.06 = 1.16 \text{ V}$$

Note that using a reduced product concentration increases the voltage—by 0.03 V for each factor of 10 reduction in this case—and therefore the spontaneity of the reaction, just as it would lower the free-energy change ΔG.

The Daniell cell will generate a voltage even if the cathode is not made of copper; copper will plate out on any conductive electrode with which it does not react. Likewise, it is not necessary for the anode compartment to contain a zinc salt; any electrolyte inert toward the zinc anode, such as NaCl, can be used. After only a short period of use, allowing the buildup of $[Zn^{2+}]$, the cell voltage will be close to $\mathscr{E}° = 1.10$ V.

When used as a current source, a cell has a lifetime dictated by the amounts of reagents it contains. It is advantageous to make the anode the limiting reagent; in that way the cell voltage, which drops due to loss of reagent ion concentration but not of anode mass, remains high until the cell "dies." For example, if a 10.0-g (0.153-mol) zinc plate is used as the anode, it will supply 0.306 mol $e^- = 29,500$ C of charge. If a current of 0.10 A is drawn continuously from the cell, the cell will survive $(29,500 \text{ C})/(0.10 \text{ C/s}) = 2.95 \times 10^5$ s $= 82$ hr. The Daniell cell could be kept "alive" by periodically replacing the anode, and by adding crystals of $CuSO_4$ to the cathode electrolyte.

Leclanché's Dry Cell

The modern portability of batteries that we now take for granted was pioneered by Georges Leclanché in the 1860s. His so-called *dry cell* is still sold today, one of the longest lived electrical inventions on record. The modern dry cell, illustrated in Figure 13.3, utilizes a zinc anode jacket, a porous fiber divider, a damp electrolyte paste of NH_4Cl and MnO_2, and an inert carbon cathode down the center, all encased

Figure 13.3
An inside view of the Leclanché dry cell. You are encouraged to take a dead flashlight battery and saw it open with a hacksaw to see for yourself what it is like inside.

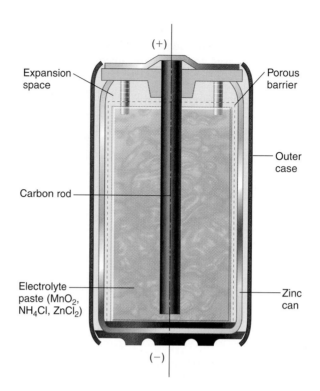

in a paper, plastic, or metal jacket to make a portable, easily handled chemical power source. The half-cell reactions are

Anode: $\quad Zn \longrightarrow Zn^{2+} + 2e^-$ (13.10)

Cathode: $\quad 2MnO_2 + 2NH_4^+ + 2e^- \longrightarrow Mn_2O_3 + 2NH_3 + H_2O$

These generate a (slightly nonstandard) cell voltage of 1.5 V. The carbon rod serves only as a source of electrons for the MnO_2, and is not consumed as the cell is used. In this case MnO_2 is the limiting reagent, to avoid breaching the Zn jacket and causing leakage. The cell reaction produces NH_3 vapor and a concentrated ammonia solution, which can corrode through the jacket, "ooze" into the device the battery is meant to power, and damage it; dry cells should be removed from flashlights, etc., when being stored for extended periods. The use of a paste electrolyte limits mobility; after extended use the battery voltage will drop due to depletion of MnO_2 near the C cathode, and recover after some time is allowed for diffusion of fresh MnO_2 to the center. An interesting aspect of the dry cell is that the cathode is inert; the cathode reaction involves not a metallic element, but two different oxides of manganese.

The voltage and lifetime of the dry cell have been greatly improved by using a high-porosity Zn anode and KOH instead of NH_4Cl in the electrolyte to make an *alkaline dry cell*. The half-reactions in the alkaline cell become

Anode: $\quad Zn + 2OH^- \longrightarrow ZnO + H_2O + 2e^-$ (13.11)

Cathode: $\quad 2MnO_2 + H_2O + 2e^- \longrightarrow Mn_2O_3 + 2OH^-$

also yielding 1.5 V. Owing to the more conductive and less corrosive alkaline paste, these cells can sustain higher currents, and last five to eight times longer than the acid dry cell.

EXAMPLE 13.2

A dry cell is to be used as a flashlight battery, in which a current of 100. mA is to be drawn continuously. What is the lifetime of a dry cell per gram of zinc under these conditions?

Solution:
Whether the cell is alkaline or not, 2 mol e^- are yielded by 1 mol Zn, allowing us to find the charge Q supplied by 1 g of Zn. When Q is divided by the current I, we obtain the time t required to deliver that charge, the lifetime of a 1-g Zn dry cell. Thus,

$$t = \frac{Q}{I} = 1 \text{ g Zn} \left(\frac{1 \text{ mol Zn}}{65.4 \text{ g Zn}}\right)\left(\frac{2 \text{ mol } e^-}{1 \text{ mol Zn}}\right)\left(\frac{96{,}500 \text{ C}}{\text{mol } e^-}\right)\left(\frac{1 \text{ s}}{0.100 \text{ C}}\right)$$
$$= 2.95 \times 10^4 \text{ s} = 8.2 \text{ hr}$$

Normally two cells in series are used in flashlights, to double the voltage, and the calculation applies to each cell. Here we have specified the current; this result does not apply when two cells are used instead of one on a load of given resistance. The 30-Ω bulb of a flashlight draws twice as much current when the voltage is doubled, using up the anode twice as fast. In addition, the cell will polarize with steady use, making the actual continuous operating time somewhat shorter.

The fact that dry cell batteries may be purchased in a variety of sizes, all supplying the same voltage, is a reflection of the intensive character of the cell voltage $\mathscr{E}$ discussed earlier. These cells differ only in their current capacity—how much current can be supplied for how long—with the larger cells supplying more. Nine-volt batteries used to power solid-state electronics employ Volta's pile principle, containing a stack of six dry cells connected in series. The advent of the alkaline battery has engendered the development of a variety of metal/metal oxide cells, many in miniature sizes ("button" batteries), which you may meet in the end-of-chapter exercises. The nickel oxide–cadmium ("nicad") battery is the first rechargeable "dry" cell; it is actually wet, but it is hermetically sealed to prevent leakage of the alkaline electrolyte. Rechargeable cells are also called "secondary" batteries or "storage" batteries. Recharging a cell involves applying an external voltage opposite in sign and large enough to force the cell reaction to run in reverse; it is akin to electrolysis, to be examined in Section 13.5.

Planté's Lead-Acid Storage Battery

Another electrical invention of unusual longevity is the lead-acid storage battery, developed by Gaston Planté in 1859; you will find one in nearly every automobile made today (a practice begun in 1915 or so). Figure 13.4 illustrates the usual automotive lead-acid battery, consisting of six cells in series driven by the half-reactions

Anode: $\quad Pb(s) + SO_4^{2-}(aq) \longrightarrow PbSO_4(s) + 2e-$ (13.12)

Cathode: $\quad PbO_2 + SO_4^{2-}(aq) + 4H^+(aq) + 2e^- \longrightarrow PbSO_4(s) + 2H_2O(l)$

Figure 13.4
Detail of a single cell of the six used in the lead-acid automotive battery. The large electrode areas and the lack of any physical barrier between the electrodes allow large currents to be drawn.

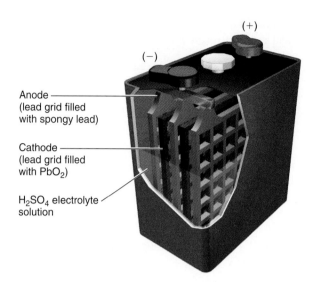

The unusual feature of the lead-acid cell is that the lead-containing compounds remain as deposits on their respective electrodes; the electrolyte is simply a sulfuric acid solution (4.8 M in a fresh battery), and no salt bridge or porous barrier is needed. The lead-acid battery thus has very low R_{int}, and is capable of delivering very high currents, up to 1000 A in the latest models. By adding the half-reactions of Equation 13.12 you can readily write the overall cell reaction and find using Appendix C that its standard voltage is $\mathscr{E}° = 2.04$ V. Thus six cells in series will yield 12 V, the commercial voltage. As long as H_2SO_4 is the limiting reagent, the battery is rechargeable, until too much $PbSO_4$ has flaked away from the electrodes to allow the cell reaction to run in reverse.

EXAMPLE 13.3

Find the overall cell reaction for the lead-acid battery, and calculate the sulfate ion concentration at equilibrium, assuming an initial concentration of H_2SO_4 of 4.8 M.

Solution:
The anode and cathode reactions of Equation 13.12 can be added directly, since the electrons are already in balance. The result is

$$Pb(s) + PbO_2(s) + 4H^+(aq) + 2SO_4^{2-}(aq) \longrightarrow 2PbSO_4(s) + 2H_2O(l)$$
$$4x2x$$

Since two electrons are transferred per unit reaction, the equilibrium condition (Equation 13.7) becomes

$$\log_{10} K = \frac{2\mathscr{E}°}{0.05916} = 69.0$$
$$K = 1.0 \times 10^{69}$$

With the equilibrium lying very far to the right, the extent-of-reaction analysis assumes small amounts of H^+ and SO_4^{2-} as indicated. Equilibrium requires

$K = 1/([H^+]^4[SO_4^{2-}]^2) = 1/(1024x^6)$, giving $x = 1.0 \times 10^{-12}$ M or $[SO_4^{2-}]_{eq} = 2.0 \times 10^{-12}$ M. Note that in this case the equilibrium concentrations are chemically realistic despite the large K value; this is so because of the larger stoichiometric coefficients, which yield a reaction quotient of high degree, six in this problem. (The nth root of any number approaches 1 as n increases.) Thus the electrolyte of a dead lead-acid battery contains very little lead or acid, being essentially pure water and nonconducting. The small $[H^+]$ implied by the value of x means that the cell reaction must be rewritten for basic conditions when the cell is nearly run down, but as a battery it will have ceased to function well before this point. The electrolyte can be restored by recharging; see Section 13.5.

Because of environmental concerns stemming both from the toxicity of lead and excessive burning of fossil fuels, a lightweight replacement for the lead-acid battery that would be capable of both starting and running an automobile has been avidly sought. In addition, Pb is the heaviest commonly occurring element, and batteries made from lighter elements might have higher *energy density,* that is, energy yield per unit mass, a critical load factor in vehicular propulsion. A promising class of batteries is the alkali-sulfur cells (Li-S or Na-S), which require molten metal and sulfur for their operation; a set of such batteries weighing 250 kg has been shown to be capable of driving an electric car for 150 km at 100 km/hr without recharging. The cells must be operated at elevated temperature (250°C), and are currently too expensive for widespread commercial use. Another fascinating possibility is an aluminum-air battery, in which a plate of aluminum is catalytically oxidized by the O_2 in flowing air using an alkaline electrolyte. Instead of recharging, the Al plate must be periodically replaced, $Al(OH)_3$ product removed, and water added. This inspires a futuristic scenario: as you drive along, dropping empty aluminum soda cans into a compacter leading to the battery, you stop now and then to take on water from a spigoted tower reminiscent of that used by steam locomotives of yore, while a trap beneath your car opens to receive a glob of insoluble $Al(OH)_3$. Aluminum takes the place of a fuel (hydrogen or hydrocarbon) in this sort of battery, leading to our next example.

Fuel Cells

Combustion of fuels, fire, is among the oldest chemical transformations known to us. In controlled combustion, such as that within a combustion engine, the fuel is added gradually and burnt a little at a time, allowing the power developed by the engine to be regulated. Likewise, in a fuel cell, a combustion reaction occurs with the gradual addition of fuel to the anode compartment and oxidizer to the cathode compartment, that is, the cell is *fueled* from a storage tank, in addition to burning a fuel to obtain electrical energy. Within a few years after Volta's pile, several researchers investigated the feasibility of using the simplest combustion reaction, the hydrogen–oxygen reaction, to generate electricity. But it is only since the 1960s that this simplest of fuel cells has been turned into a practical device, finding a role in the U.S. space program as we have already mentioned; and thus far it is the only fuel cell that has been made to function reliably.

Given our experience with improved cell design as outlined earlier, it is not surprising that the H_2–O_2 (or H_2–air) cell is alkaline, using concentrated KOH solution

Figure 13.5
Schematic diagram of an alkaline hydrogen–oxygen fuel cell. The gases must be supplied continuously from reservoirs (not shown), and the water product must also be removed continuously. In a multicell battery, the gases and electrolyte flow from one cell to the next. Diffusion of the gases into the porous barrier/electrodes limits the current that may be drawn. Great improvements in barrier materials have been made recently.

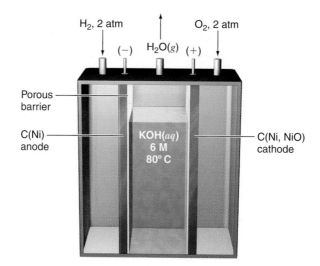

or molten KOH as the electrolyte. A schematic diagram of the cell is shown in Figure 13.5. The electrodes here are of critical importance, because the H_2–O_2 reaction does not occur without a spark or a catalyst (see Chapter 15). Inert porous graphite electrodes impregnated with nickel (anode) or a mixture of nickel and nickel oxide (cathode) have proven to be successful in catalyzing the reaction at ordinary temperatures. The half-reactions in alkaline medium are

Anode: $\quad\quad\quad\quad H_2(g) + 2OH^-(aq) \longrightarrow 2H_2O(l) + 2e^- \quad\quad\quad$ (13.13)

Cathode: $\quad\quad\quad\quad O_2(g) + 2H_2O(l) + 4e^- \longrightarrow 4OH^-(aq)$

Multiplying the anode reaction by 2 and adding, the familiar $2H_2(g) + O_2(g) \longrightarrow 2H_2O(l)$ is obtained, whose standard free-energy change at 298 K has already been found in Chapter 11 to be $\Delta G° = -113.39$ kcal (-474.4 kJ). Using Equation 13.4 with $n = 4$ yields $\mathscr{E}° = +1.229$ V, the voltage supplied by a standard H_2–O_2 fuel cell. Note that standard conditions here imply $P_{H_2} = P_{O_2} = 1$ atm, and that running the cell as illustrated, or with 1 atm of air instead of pure O_2, would produce a different $\mathscr{E}$.

The only barrier to converting other fuels to electrical energy is a practical one: thus far suitable electrodes, electrolyte, and operating conditions that can be readily employed have not been found. Methane (CH_4) can be used provided it is first converted to H_2 by *steam reforming* the fuel to CO_2 and H_2:

$$CH_4(g) + 2H_2O(g) \longrightarrow CO_2(g) + 4H_2(g) \quad\quad\quad (13.14)$$

This reaction requires both a catalyst and free-energy input; the latter reduces the overall efficiency below that for a true methane–air cell. Nonetheless a bank of cells designed to run on the reforming output [with CO_2 trapped out of the gas stream; see Exercise 12.59(f)], has been available as an emergency backup in a power plant near Tokyo. Such large-scale fuel cells are likely to become a major source of electrical power in the future. Nonetheless, in combustion reactions, as in many other redox systems, there is certainly a need for further imaginative ideas and creative design work aimed at tapping a larger fraction of their tremendous chemical potential.

13.3 Electrode Potentials

The physical isolation of half-reactions in an electrochemical cell suggests that it might be useful to think of a half-reaction as possessing its own free-energy change. The overall cell reaction's free energy would then be the appropriately weighted and signed sum of the half-reaction free energies. For this purpose we conventionally write each half-reaction as a reduction; using the Volta cell reaction of Equation 13.1 as an example, the $Zn^{2+} + 2e^- \longrightarrow Zn$ ΔG would be subtracted from twice that for $Ag^+ + e^- \longrightarrow Ag$ to give the cell ΔG:

$$\Delta G_{cell} = 2\Delta G_{Ag^+|Ag} - \Delta G_{Zn^{2+}|Zn}$$

When Equation 13.4 is used to replace the ΔG's with $\mathscr{E}$'s, noting that $n = 1$ for the Ag half-reaction but $n = 2$ for the others, we get

$$-2F\mathscr{E}_{cell} = 2(-F\mathscr{E}_{Ag^+|Ag}) - (-2F\mathscr{E}_{Zn^{2+}|Zn}) \quad \text{or}$$

$$\mathscr{E}_{cell} = \mathscr{E}_{Ag^+|Ag} - \mathscr{E}_{Zn^{2+}|Zn}$$

Note how the stoichiometric coefficients have disappeared from the final result for the cell voltage, while the anode voltage, where oxidation occurs, must be subtracted. This outcome turns out to be general:

$$\mathscr{E}_{cell} = \mathscr{E}_{cathode} - \mathscr{E}_{anode} \tag{13.15}$$

for any cell, where the cathode and anode potentials are reduction potentials by convention.* If you have written the cell reaction in its spontaneous direction, then $\mathscr{E}_{cell}$ from Equation 13.15 will be positive; if you get a negative $\mathscr{E}_{cell}$, then the reaction will occur in reverse, and the identities of the cathode and anode must be interchanged.

Equation 13.15 implies that, by tabulating half-cell **standard reduction potentials,** we can find the potential for a cell constructed from any pair of half-cells; this advantage is similar to that gained by tabulating enthalpies and free energies of formation. In constructing such a table we need a reference half-cell against which other half-cells can be measured; by convention the $H^+|H_2$ couple is used.

---

* Suppose the reduction half-reactions are

Cathode: $\qquad A^{x+} + xe^- \longrightarrow A$
Anode: $\qquad B^{y+} + ye^- \longrightarrow B$

A common multiple (perhaps not the least) that will balance the electrons is xy, so multiplying the first reaction by y, the second by $-x$, and adding yields a balanced redox reaction

$$yA^{x+} + xB \longrightarrow yA + xB^{y+}$$

and the corresponding free-energy relation

$$\Delta G = y\Delta G_{cathode} - x\Delta G_{anode}$$

Noting that the number of electrons transferred per reaction is xy, Equation 13.4 then leads to

$$-xyF\mathscr{E}_{cell} = y(-xF\mathscr{E}_{cathode}) - x(-yF\mathscr{E}_{anode})$$

Dividing both sides by $-xyF$ then gives Equation 13.15.

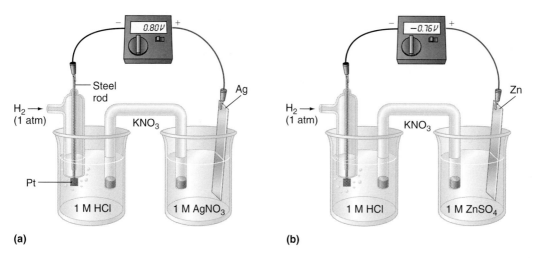

(a) **(b)**

Figure 13.6
In a standard hydrogen electrode, Hydrogen gas is slowly bubbled through at 1 atm, and adsorbs on the inert Pt electrode. Standard cells, incorporating the hydrogen electrode coupled to **(a)** a silver half cell and **(b)** a zinc half cell, are shown. For metallic partner half-cells, a negative cell voltage when the hydrogen electrode is wired as the anode indicates that the metal is susceptible to attack by acid.

The Standard Hydrogen Electrode

Figure 13.6 illustrates cells incorporating a hydrogen half-cell. The problem of having a gas as an electrode is resolved by using a noble metal, Pt, which is itself inert but which strongly adsorbs hydrogen on its surface, and continuously bubbling hydrogen gas at 1 atm over its surface. When this electrode is immersed in 1 M H^+ solution, we have a **standard hydrogen electrode** (SHE), whose half-reaction is the reduction of H^+ to H_2, and whose half-cell potential is defined to be zero at 25°C:

$$2H^+(aq) + 2e^- \longrightarrow H_2(g), \qquad \mathscr{E}° \equiv 0.00 \text{ V} \qquad (13.16)$$

It is this definition that is the origin of our setting $\Delta H_f°(H^+(aq)) = \Delta G_f°(H^+(aq)) = S°(H^+(aq)) = 0.0$ as a reference for solutes in aqueous solution at 25°C. Because only the voltage of a full cell, representing a complete chemical reaction, is measurable in practice, and is the *difference* of half-cell potentials, we are free to adopt this standard, which is in essence the same arbitrary setting of an energy zero as in the formation convention for enthalpy and free energy. This will become clearer when we examine cells that contain the SHE.

Figure 13.6 also illustrates two cells in which the two standard half-cells of the Volta pile are coupled with the SHE. When the voltmeter is connected with the same polarity to the electrodes of each cell, the $(Pt)H_2|H^+||Ag^+|Ag$ cell registers +0.80 V, while the $(Pt)H_2|H^+||Zn^{2+}|Zn$ cell reads −0.76 V. These voltages tell us that the reaction

$$2Ag^+(aq) + H_2(g) \longrightarrow 2Ag(s) + 2H^+(aq) \qquad (13.17)$$

is spontaneous under standard conditions as written, but that the reaction

$$Zn^{2+}(aq) + H_2(g) \longrightarrow Zn(s) + 2H^+(aq) \qquad (13.18)$$

is not. Not only must the cell notation for Reaction 13.18 be written in reverse order, $Zn|Zn^{2+}||H^+|H_2(Pt)$, but you should recognize that the reverse of Reaction 13.18 is quite spontaneous; that is, zinc reacts readily with acid, the reaction we used to introduce the first law in Chapter 10. Silver, on the other hand, is inert to attack by $H^+(aq)$. You may also know that copper is not attacked by H^+ either; a $(Pt)H_2|H^+||Cu^{2+}|Cu$ cell will therefore show a positive $\mathscr{E}°$. Moreover, if you subtract the voltage of the $Zn|H_2$ cell from that of the $Ag|H_2$ cell, you obtain $\mathscr{E}°$ for Volta's cell, +1.56 V, a result expected because Volta's cell reaction is also the difference of Reactions 13.17 and 13.18. Figure 13.7 illustrates these relationships on a "voltage ladder"; they translate via Equation 13.4 into an experimental confirmation of Hess's law applied to free energies. The ladder also shows that the placement of a voltage zero is arbitrary, and does not change the prediction of cell voltage.

Table 13.1 is an abbreviated collection of standard half-cell reduction potentials relative to SHE, referring to electrolyte solutions containing 1 M concentrations of all ions in a neutral or acidic aqueous medium. If the electrons involved in the reduction are taken to have zero energy, that is, they are ignored, these potentials are directly related through Equation 13.4 to the standard free energies of formation given in Appendix C. In particular, for reduction of a metal ion to its elemental form, the half-cell $\mathscr{E}°$ corresponds to the negative of the standard free energy of formation of the aqueous ion. To construct a full cell that will have a standard voltage $\mathscr{E}°$, however, it is not necessary to use standard half-cells, but only to ensure that the reaction quotient Q is unity. For example, the Daniell cell will yield $\mathscr{E} = \mathscr{E}° = 1.10$ V as long as $[Zn^{2+}] = [Cu^{2+}]$. In practice, the SHE is a somewhat hazardous electrode to use, Pt being a known catalyst for the H_2–O_2 explosion. Fortunately, it is not necessary to measure the voltage of all cells containing the SHE; once the voltage of a more convenient cell is well known relative to SHE, that cell can be used as a secondary standard. The highly stable **calomel (Hg_2Cl_2) electrode,** whose reduction potential is known to be +0.2444 V with respect to SHE at 25°C, and whose temperature and electrolyte concentration dependence are also well established, has played this role for many decades. This electrode has become a standard part of the pH meter, and its composition will be given in the next section.

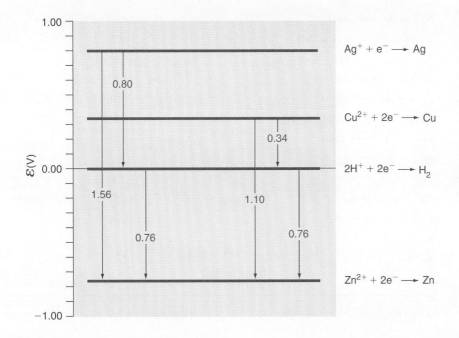

Figure 13.7
A "voltage ladder" illustrating the relationship between half-cell voltages. Like the free-energy change, under standard conditions the voltage between two half-cells, the distance between the two rungs, is immutable, and independent of where zero volts is set. Setting the voltage of the standard hydrogen electrode to zero implies that the other half-cell voltages are actually those generated by connecting each in turn to the hydrogen electrode to make a full cell.

TABLE 13.1

Standard reduction potentials in aqueous solution relative to SHE at 298 K

Half-reaction (couple)	$\mathscr{E}°(V)$
$F_2 + 2e^- \longrightarrow 2F^-$	2.890
$PbO_2 + 4H^+ + SO_4^{2-} + 2e^- \longrightarrow PbSO_4 + 2H_2O$	1.69
$MnO_4^- + 8H^+ + 5e^- \longrightarrow Mn^{2+} + 4H_2O$	1.49
$Cl_2 + 2e^- \longrightarrow 2Cl^-$	1.360
$Cr_2O_7^{2-} + 14H^+ + 6e^- \longrightarrow 2Cr^{3+} + 7H_2O$	1.33
$O_2 + 4H^+ + 4e^- \longrightarrow 2H_2O$	1.229
$Br_2 + 2e^- \longrightarrow 2Br^-$	1.087
$NO_3^- + 4H^+ + 3e^- \longrightarrow NO + 2H_2O$	0.96
$Ag^+ + e^- \longrightarrow Ag$	0.80
$Fe^{3+} + e^- \longrightarrow Fe^{2+}$	0.77
$I_2 + 2e^- \longrightarrow 2I^-$	0.535
$O_2 + 2H_2O + 4e^- \longrightarrow 4OH^-$	0.401
$Cu^{2+} + 2e^- \longrightarrow Cu$	0.34
$Hg_2Cl_2 + 2e^- \longrightarrow 2Hg + 2Cl^-$	0.27
$AgCl + e^- \longrightarrow Ag + Cl^-$	0.22
$SO_4^{2-} + 4H^+ + 2e^- \longrightarrow H_2SO_3 + H_2O$	0.20
$2H^+ + 2e^- \longrightarrow H_2$	0.000
$Pb^{2+} + 2e^- \longrightarrow Pb$	-0.13
$Sn^{2+} + 2e^- \longrightarrow Sn$	-0.14
$Ni^{2+} + 2e^- \longrightarrow Ni$	-0.23
$PbSO_4 + 2e^- \longrightarrow Pb + SO_4^{2-}$	-0.35
$Fe^{2+} + 2e^- \longrightarrow Fe$	-0.41
$S + 2e^- \longrightarrow S^{2-}$	-0.51
$Zn^{2+} + 2e^- \longrightarrow Zn$	-0.76
$2H_2O + 2e^- \longrightarrow 2OH^- + H_2$	-0.828
$Mn^{2+} + 2e^- \longrightarrow Mn$	-1.03
$Al^{3+} + 3e^- \longrightarrow Al$	-1.68
$Mg^{2+} + 2e^- \longrightarrow Mg$	-2.36
$Na^+ + e^- \longrightarrow Na$	-2.71
$Ca^{2+} + 2e^- \longrightarrow Ca$	-2.76
$Ba^{2+} + 2e^- \longrightarrow Ba$	-2.90
$K^+ + e^- \longrightarrow K$	-2.92
$Li^+ + e^- \longrightarrow Li$	-3.05

EXAMPLE 13.4

A $Cl_2|Cl^-$ half-cell can be built by saturating a NaCl solution with $Cl_2(g)$ and inserting an inert electrode such as $Pt(s)$ or $C(gr)$. What voltage will be generated by a standard $Cl_2|Cl^-$ half-cell combined with a SHE? With a standard $Al^{3+}|Al$ half-cell? Give the balanced cell reactions and the line notation for these cells.

Solution:
For $Cl_2|Cl^-$ combined with $H^+|H_2$, we find and record the half reactions and their reduction potentials from Table 13.1, and subtract the one with the lower voltage from that with the higher, in accord with Equation 13.15:

	$\mathscr{E}°(V)$	
$Cl_2 + 2e^- \longrightarrow 2Cl^-$	1.360	(cathode)
$2H^+ + 2e^- \longrightarrow H_2$	0.000	(anode)
$H_2 + Cl_2 \longrightarrow 2H^+ + 2Cl^-$	1.360 V	$= \mathscr{E}°_{cell}$

The line notation for this cell, since $H^+|H_2$ is the anode, is $(Pt)H_2|H^+||Cl^-|Cl_2(C)$; in this cell both electrodes are inert. Note that for any cell constructed with a hydrogen electrode, the magnitude of $\mathscr{E}°_{cell}$ is given by that of the reduction potential for the other half-reaction, since $\mathscr{E}° = 0.000$ V for SHE. For $Cl_2|Cl^-$ with $Al^{3+}|Al$, we reuse the $Cl_2|Cl^-$ $\mathscr{E}°$, and apply the factors of the least common multiple needed to balance the electrons, but, faithful to Equation 13.15, *not* when subtracting the $\mathscr{E}°$'s:

	$\mathscr{E}°(V)$	
$Cl_2 + 2e^- \longrightarrow 2Cl^-$	1.360	(cathode)
$Al^{3+} + 3e^- \longrightarrow Al$	-1.68	(anode)
$2Al + 3Cl_2 \longrightarrow 2Al^{3+} + 6Cl^-$	3.04 V	$= \mathscr{E}°_{cell}$

For this cell the line notation is $Al|Al^{3+}||Cl^-|Cl_2(C)$.

Table 13.1 is arranged from highest (most positive) to lowest (most negative) reduction potential, that is, from the most easily reduced to the most difficult. When analyzing an electrochemical cell, the couple you find to lie higher in the table will always become the cathode. For the metallic couples, you may recognize that, bottom to top, the order is exactly that given by the **activity series** of Chapter 5; in fact, the series arises directly from the relative reduction potentials of the elemental ions in it. This electrochemical series is not identical with that you might derive from gas-phase ionization energies, because the metals as solids vary in their binding energy within the crystal lattice, and the solvation energies of the ions differ due to differences in charge and ionic radius. But these vicissitudes in atomic environment cause only minor alterations in the more global left-to-right trends in reactivity, which are dominated by the intrinsic properties of the atoms themselves.

Many of the examples we considered in the previous section employed a basic electrolyte, beginning with the alkaline dry cell. A metallic half-cell in basic medium will often involve an insoluble metal hydroxide, in line with the solubility rules of

Chapter 5 and the solubility products of Chapter 12. For example, the anode of the aluminum–air battery mentioned earlier is described by the half-reaction, in reductive form

$$Al(OH)_3(s) + 3e^- \longrightarrow Al(s) + 3OH^-(aq)$$

This half-reaction is related to the reduction of $Al^{3+}(aq)$ through the solubility equilibrium of $Al(OH)_3$, since the reactions

$$Al^{3+}(aq) + 3e^- \longrightarrow Al(s) \qquad \Delta G° = -3F\mathscr{E}°$$
$$= 116 \text{ kcal (485 kJ)}$$
$$Al(OH)_3(s) \rightleftharpoons Al^{3+}(aq) + 3OH^-(aq) \qquad \Delta G° = -RT \ln K_{sp}$$
$$= 43 \text{ kcal (180 kJ)}$$

and their free energies can be added to give the above alkaline half-reaction, its free energy, +159 kcal (+665 kJ), and a half-cell voltage $\mathscr{E}° = -2.30$ V. We have used values from Tables 13.1 and 12.6; alternatively, the $\Delta G_f°$'s of Appendix C would suffice. In deriving new half-cell potentials, it is good practice to convert all other quantities to ΔG's, perform the appropriate reaction sum, and then convert the desired ΔG back to $\mathscr{E}$ with Equation 13.4.

13.4 Concentration Cells

The idea that free energy is concentration dependent was introduced in Chapter 12 by considering the drop in free energy that accompanies the dilution of a Cu^{2+} solution with water. If instead of water, a more dilute Cu^{2+} solution were added, the solutions would mix and eventually become a single, uniform solution with some intermediate concentration, again (because this is also spontaneous) accompanied by a fall in free energy. Where there is a free-energy drop, a voltage ought to be generated. Thus, if we set up a cell with identical half-cell couples, but with different electrolyte concentrations, we expect a nonzero cell voltage, as illustrated in Figure 13.8. The cell reaction for this **concentration cell** is

$$Cu^{2+}(aq) + Cu(s) \longrightarrow Cu(s) + Cu^{2+}(aq)$$

a nonreaction, but because the more concentrated half-cell tends to be diluted, the Cu^{2+} in that cell will tend to plate on the copper electrode, thereby lowering $[Cu^{2+}]$, while in the more dilute half-cell $Cu(s) \longrightarrow Cu^{2+} + 2e^-$ will occur, thereby raising $[Cu^{2+}]$. If the half-cells were shorted by a wire, this would continue until the two half-cells were of equal $[Cu^{2+}]$. The more concentrated half-cell thus becomes the cathode and the less concentrated the anode, described by $Cu|Cu^{2+}(c_1)||Cu^{2+}(c_2)|Cu$ with $c_1 < c_2$. The standard potential for a concentration cell is necessarily zero, so the entire voltage of this type of cell is due to the logarithmic term in the Nernst equation:

$$\mathscr{E} = 0.00 - \frac{0.05916}{2} \log_{10} \frac{[Cu^{2+}]_{anode}}{[Cu^{2+}]_{cathode}}$$

Each factor of 10 difference in the concentrations yields 0.03 V of cell potential. Though such small voltages do not allow such a cell to be used as a battery, concen-

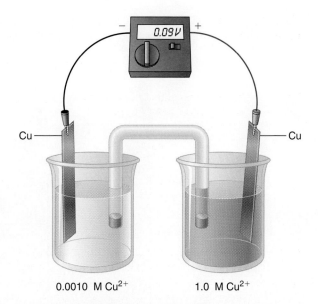

Figure 13.8
Example of a concentration cell. The only difference between the two half-cells is the electrolyte concentration. The entropy of mixing that would result if the electrolytes were combined into a single solution of uniform concentration is responsible for the small voltage produced.

tration cells have several analytical applications. The most important of these is the electrochemical measurement of pH.

The pH Meter

By analogy with the copper concentration cell of the previous section, we could assemble a cell from a *pair* of hydrogen electrodes in which one of the electrodes is a SHE, and the electrolyte solution of the other is of unknown concentration; in line notation this would be $(Pt)H_2(1 \text{ atm})|\ [H^+]||H^+(1\text{ M})|H_2(1\text{ atm})(Pt)$, with $[H^+]$ being the unknown concentration. The unknown half-cell will be the anode for any $[H^+]$ less than 1 M. The voltage of this cell is simply

$$\mathscr{E} = -\frac{0.05916}{2}\log_{10}\frac{[H^+]^2}{1^2}$$
$$= -0.05916 \log_{10}[H^+]$$
$$= +0.05916 \text{ pH}$$

A change in the pH of one unit (a factor of 10 in $[H^+]$) thus yields a 0.06-V increase in voltage; a simple change in the scale or digital readout of a sensitive voltmeter allows the pH to be read directly. This cell is thus a *pH meter*, requiring that the anode and a salt bridge to the cathode be inserted into the unknown solution.

It is not safe or practical to employ hydrogen electrodes for the routine measurement of pH; the invention of the **glass electrode** in the 1930s led to a compact, portable pH cell that uses no gases. A thin membrane made from a specially formulated soda glass was found to be permeable to H^+ but not to larger ions; placed as a barrier between two acid solutions of different concentration, the glass membrane allowed a measurable voltage to be developed that was sensitive only to $[H^+]$ and not to the concentrations of other ions such as Cl^-. In the modern pH meter cell illustrated in Figure 13.9, this membrane is incorporated into the cathode, a $Ag|AgCl(s)$

Figure 13.9
The pH meter electrodes. In practice, these are usually combined into a single assembly, somewhat inaccurately referred to as a pH electrode. The known HCl solution within the glass electrode and the unknown solution in the beaker form a concentration cell whose voltage is proportional to the pH of the unknown solution.

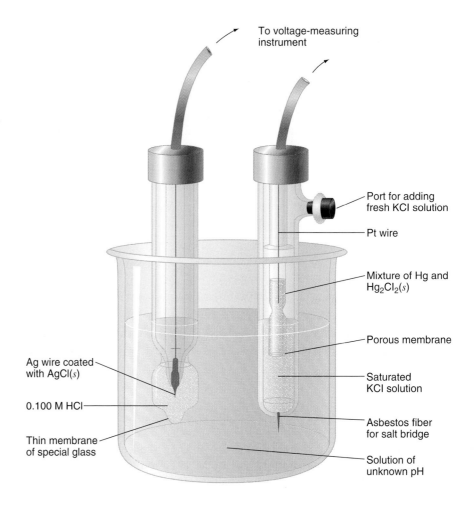

reference electrode with 1 M HCl electrolyte, while the anode is a calomel electrode (Pt)Hg|Hg$_2$Cl$_2$(s), with saturated (4.6 M) KCl electrolyte, instead of a hydrogen electrode. A capillary arm of the calomel electrode forms part of the salt bridge. The glass membrane and the unknown solution effectively also become part of the salt bridge, and the cell voltage contains a contribution from the voltage drop (free-energy difference) across the membrane. Various authors differ on the cell notation for this cell, but a sensible rendering seems to be

Ag|AgCl(s), HCl(0.1 M)|glass membrane |[H$^+$]|||KCl(*satd*), Hg$_2$Cl$_2$(s) | Hg(Pt)

The Ag|AgCl and Hg|Hg$_2$Cl$_2$ electrodes have well-known reduction potentials, $+0.2223$ and $+0.2444$ V, respectively, that nearly cancel. The voltmeter that reads out the pH is calibrated against a standard (buffer) solution of known pH; this effectively adjusts the meter to compensate for the electrode voltages, internal resistance, nonnormal temperature, etc.

Membranes that serve as sensors for specific ions now abound; these **ion-selective electrodes** or **pIon electrodes** operate on the same principle as the pH electrode, through the exclusive transmission of particular ions across the membrane. One can now detect, for example, Na^+, NH_4^+, F^-, Ca^{2+}, NO_3^-, and other simple ions selectively down to 10^{-6} M concentrations with commercially available electrodes. Gas-permeable and biocatalytic membranes also allow electrochemical detection of neutral molecules ranging from O_2 and CO_2 to proteins (chains of amino acids).

In a concentration cell, even extremely small concentrations produce readily measurable voltages; for example, the pH meter can successfully measure pH approaching 14.0, or $[H^+] = 10^{-14}$ M. This great sensitivity makes possible the electrochemical measurement of solubility products that would be far too small to obtain by, say, titrating the ions left in solution, as the following example illustrates.

EXAMPLE 13.5

A copper concentration cell is built in which one half-cell is standard and the other contains 1.0×10^{-3} M Cu^{2+}. Enough NaOH is added to the dilute cell to bring it to 0.100 M $[OH^-]$, producing a $Cu(OH)_2$ precipitate. A cell voltage of 0.511 V is measured. What is the solubility product of $Cu(OH)_2$?

Solution:
The precipitation reaction leaves only a small $[Cu^{2+}]$ in the electrolyte, which is responsible for the voltage reading. We can therefore determine this concentration from the Nernst equation:

$$\mathscr{E} = -(0.05916/2) \log_{10}([Cu^{2+}]/1) = +0.02958 \, pCu$$
$$pCu = 0.511/0.02958 = 17.27; \, [Cu^{2+}] = 5.3 \times 10^{-18} \, M$$

The solubility product is then

$$K_{sp} = [Cu^{2+}][OH^-]^2 = (5.3 \times 10^{-18})(0.100)^2 = 5.3 \times 10^{-20}$$

As usual, the more dilute half-cell is the anode. Note the relationship between this half-cell and the alkaline Al half-cell of the previous section; what we have made is an alkaline copper half-cell whose reduction potential is 0.511 V *lower* than that of a standard copper half-cell. A similar lowering occurs in an alkaline zinc half-cell, making an "alkaline Daniell cell" a distinct possibility.

Redox titrations can also be followed by monitoring the potential of the ionic solution containing the unknown analyte; Pt and calomel electrodes are conventionally used in this method, known as *potentiometric titration*. For example, when a solution of Fe^{2+} is titrated by an oxidizing agent to give Fe^{3+}, the Pt electrode forms a half-cell with the analyte solution; according to the Nernst equation (Equation 13.6), the voltage will change as the ratio $[Fe^{3+}]/[Fe^{2+}]$ increases.

Now we have considered electrochemical cells driven by a spontaneous redox reaction ($\mathscr{E}° > 0$) and by only a concentration difference ($\mathscr{E}° = 0$); a cell with $\mathscr{E}° < 0$ is simply mislabeled as to the identity of the anode and cathode. Further, any cell with $\mathscr{E} = 0$ represents a system at equilibrium, a "dead" cell. But what happens when, instead of a passive external circuit, an external voltage source is connected to the cell electrodes? In that event, we can do work on the cell. As we now examine, we can bring "dead" cells back to life ("recharge" them), reverse the spontaneous direction of a live cell, and take any stable (equilibrium) electrolyte solution and create a free-energy imbalance in a process known as *electrolysis*.

13.5 Electrolysis: Redox Chemistry in Reverse

One of the earliest uses of electrochemical cells was in the decomposition of salt solutions. When Volta submitted his paper on the zinc-silver pile to the Royal Society in England, it immediately caused a stir, and within a few months the electrolysis of water had been demonstrated. Sir Humphry Davy, perhaps the original daredevil of science, immediately built a 250-cell battery, nearly electrocuting himself, and set about to show that the compounds we now call alkali hydroxides actually contained "a new principle," which turned out to be the alkali metals. He found that these metals could only be recovered by electrolyzing the slightly moist solid compound. Within a span of 8 years, Davy had discovered and isolated all of the alkali metals and most of the alkaline earths through electrolysis of the corresponding moist solid or molten salt electrolytes. This is still the way we obtain these elements. More importantly for the field of chemistry, these developments led Davy, J. Berzelius, and others to our earliest notions concerning the electrical nature of chemical interactions.

In each of these historic experiments, a normally stable compound, say H_2O or NaOH, was being decomposed into elements or other species by the action of an externally applied voltage. Water electrolysis, as illustrated in Figure 13.10, produces $H_2(g)$ and $O_2(g)$ in a 2:1 volume ratio in the reaction

$$2H_2O(l) \longrightarrow 2H_2(g) + O_2(g)$$

just the reverse of our well-known explosive, highly spontaneous water synthesis reaction. Electrolysis stands as one of the most dramatic illustrations of our control over

Figure 13.10
The electrolysis of water. In addition to the decomposition of water into H_2 and O_2, acid is produced at the anode and base at the cathode, as can be established by adding an indicator such as bromthymol blue to the electrolyte. The salt used in the electrolyte must be more stable against electrolysis than water itself.

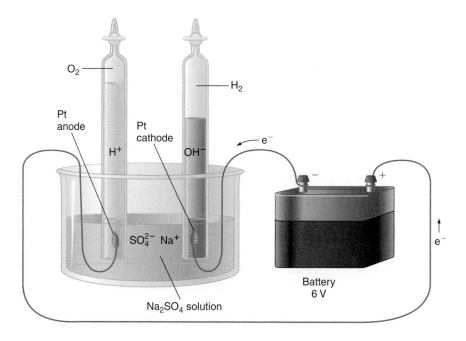

a chemical reaction, far outshining the mere changing of the system temperature in controlling reaction direction. The principle behind electrolysis is simple: the external voltage, or its equivalent free-energy change, is being pitted against the free energy of the normally spontaneously formed end product of a reaction relative to the reagents that produced it. If the external voltage exceeds in magnitude by some small amount (called the **overvoltage**) the voltage of the reaction that spontaneously formed the system to which it is applied, electrolysis will occur, forcing the system-formation reaction to run in reverse.

A cell set up with an external voltage source in the circuit is called an **electrolytic cell.** In some books you will also find it referred to as an electrochemical cell of the electrolytic type. The cell can be operated as a single compartment containing the electrolyte to be electrolyzed with the two electrodes immersed in it. The voltage applied can have either polarity, the choice determining which electrode will be the anode and which the cathode. If useful gases are formed as a result, they can be collected, as Figure 13.10 illustrates. Special electrolysis product collection naturally requires that one know the identity of anode and cathode, that is, that the external voltage be applied with the correct polarity, and frequently some type of porous divider, analogous to the salt bridge, must be present to prevent diffusive mixing of the pure electrolysis products in prolonged operation of the cell.

Let's illustrate the principles of electrolysis with a simple salt electrolyte. Figure 13.11 shows an electrolytic cell in operation with a NaI solution as electrolyte. A divider has been inserted into the initially uniform solution, and a voltage is being applied, plus-terminal to the left-hand electrode. The sense of the voltage causes electrons to flow away from this electrode and toward the other; thus electrons will be taken away from species at this electrode, and it therefore must be the anode, where oxidation takes place. The species to be oxidized is usually neutral or an anion, because only anions can be attracted to the anode. The possible anode reactions are

$$\begin{array}{ll} & \mathscr{E}°(V) \\ 2I^- \longrightarrow I_2 + 2e^- & -0.54 \\ 2H_2O \longrightarrow O_2 + 4H^+ + 4e^- & -1.23 \end{array}$$

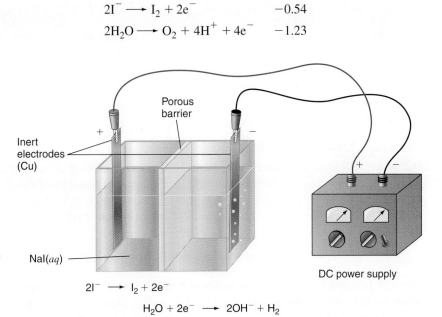

Figure 13.11
The electrolysis of aqueous NaI. Iodine formed at the anode causes that compartment to turn orangy-brown, while phenolphthalein added to the initially clear solution turns pink as water is reduced at the cathode, producing OH^- and bubbles of $H_2(g)$.

where the half-reactions and their voltages have been obtained from Table 13.1. Note that they are written as oxidations, necessitating a change in sign for the voltages. Barring a large overvoltage effect, the reaction with the least free-energy hill to climb ($\mathscr{E}° < 0 \Rightarrow \Delta G° > 0$) will be the one that predominates; thus I_2 rather than O_2 is formed at the anode.

We normally expect that any anode half-reaction with voltage lying above -1.23 V will occur on electrolysis of an aqueous electrolyte, but that any with voltage below this limit, such as $2F^- \longrightarrow F_2 + 2e^-$, $\mathscr{E}° = -2.87$ V, will not, and water itself will be oxidized. Anions that do not appear as reduction products in Table 13.1 include NO_3^- and SO_4^{2-}; these are virtually impossible to oxidize, because the N and S atoms already are present in their highest possible oxidation states. This oxidation resistance makes these anions useful if water electrolysis is desired. A significant exception to this rule is Cl^-, which will be oxidized instead of water if it is present at sufficiently large concentration, due to overvoltage for the water oxidation; we will examine an industrially important electrolysis cell that takes advantage of this anomaly later.

At the cathode the possible reduction reactions that can occur in our electrolysis cell are

$$\begin{array}{ll} & \mathscr{E}°(V) \\ Na^+ + e^- \longrightarrow Na(s) & -2.71 \\ 2H_2O + 2e^- \longrightarrow 2OH^- + H_2 & -0.828 \end{array}$$

Here it is clear that reduction of water must occur; we already know that sodium metal reacts violently with water (see Chapter 5), making any $Na(s)$ formed immediately susceptible to oxidation by water. As shown in Figure 13.11, phenolphthalein indicator added to the cathode partition turns pink due to OH^- formation, and hydrogen gas bubbles are evolved, confirming this conclusion. The overall cell reaction is then the sum of the I^- oxidation and water reduction,

$$2I^- + 2H_2O \longrightarrow I_2 + 2OH^- + H_2, \qquad \mathscr{E}° = -1.37 \text{ V}$$

where the electrolysis voltage is the sum of the anode and cathode voltages, and $\mathscr{E}°$ is negative, indicating the nonspontaneity of the cell reaction and representing the absolute minimum applied voltage (of opposite polarity) needed to cause it to occur. In this cell a thermodynamically absolutely stable electrolyte solution is forced to decompose.

Faraday's Laws of Electrolysis

Our NaI electrolysis cell could be the basis for a practical method of making elemental iodine (I_2). This would involve applying a voltage sufficient to initiate electrolysis, then running the cell for a certain time t (s) during which an average current I (A = C/s) passes through the cell. The total charge passing through the cell would then be

$$Q = It = nF \qquad (13.19)$$

where as before n is the number of moles of electrons and F is Faraday's constant. Since it takes 2 mol of electrons to produce 1 mol of I_2, Equation 13.19 determines the mass of I_2 that will be formed. Faraday first announced this quantitative relationship in 1833 as two **laws of electrolysis,** which we state in modern terms as follows:

I. The amount of chemical action is directly proportional to the amount of charge passed through the cell.

II. The amount of charge necessary to deposit 1 mol of substance is 96,485n Coulombs, where n is the number of moles of electrons that 1 mol of substance requires for its deposit.

The first law, as Faraday established experimentally, marked the first understanding that the *extent* of electrolysis arises from charge rather than voltage (then called electric intensity), quantities that his mentor Davy had confused. Faraday originally expressed the second law in terms of *equivalents,* mass per unit charge, rather than moles. (The electron and its charge had not yet been discovered nor measured in Faraday's time.) The use of Faraday's laws, as embodied in Equation 13.19, is illustrated by the following example.

EXAMPLE 13.6

Suppose a current of 2.0 A is drawn through the NaI cell for a total of 300. s (5 min). What mass of I_2 will be deposited at the anode?

Solution:
Applying Equation 13.19,

$$g_{I_2} = (2.0 \text{ A})(300. \text{ s}) \left(\frac{1 \text{ mol e}^-}{96,500 \text{ C}} \right) \left(\frac{1 \text{ mol I}_2}{2 \text{ mol e}^-} \right) \left(\frac{254 \text{ g I}_2}{1 \text{ mol I}_2} \right)$$
$$= 0.79 \text{ g}$$

Faraday's second law yields the moles of electrons, and his first then prescribes the conversion to grams of deposit.

Faraday's laws also determine the upper limit to the lifetime of a battery, for given current drawn and masses of battery chemicals, as briefly described in Section 13.2 for the Daniell cell.

Applications of Electrolysis

In addition to playing an important role in the discovery of the electrical nature of chemical bonding, electrolysis has so many significant modern applications as to rival or even overshadow the electrochemical cell of Volta. In industry the production of high-energy and high-purity chemicals and the plating of metals on other surfaces for protection and decoration are some of many ways electrolysis is used. In chemical research, electrolysis plays important roles in areas ranging from highly precise chemical analysis to the search for a means of using solar radiation to perform photoelectrolysis of water. Closer to home, recharging a storage battery is nothing more (or less) than electrolysis of a spent batch of battery chemicals, causing them to regain their original high-energy form.

Figure 13.12
The so-called Downs cell for the electrolytic production of elemental sodium and chlorine from a molten $NaCl/CaCl_2$ mixture. $Na(l)$ is less dense than the molten salt, and floats to the top where it is piped away. The current necessary for large-scale electrolysis also serves to keep the salt molten through *Joule heating*.

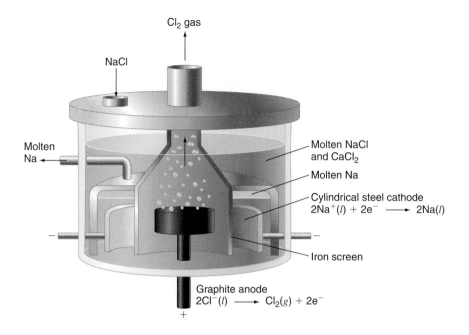

In a *Downs cell,* shown in Figure 13.12, molten sodium chloride $NaCl(l)$, containing a small amount of $CaCl_2$ to lower its melting point to 600°C, is electrolyzed to make liquid sodium metal $Na(l)$ and chlorine $Cl_2(g)$,

$$2NaCl(l) \longrightarrow 2Na(l) + Cl_2(g)$$

a reaction which is effectively the reverse of the spectacular chemiluminescent reaction of Na and Cl_2 introduced as our first redox reaction in Chapter 5. The important property that is exploited here is that when an ionic salt is melted, the ions are set free from the ionic lattice and are able to migrate to the electrodes, and we have a *molten electrolyte.* If instead a concentrated NaCl solution (*brine*) is electrolyzed, H_2 and Cl_2 (because of the overvoltage for water oxidation) gases will be produced along with NaOH (*caustic soda*) solution; for this purpose a *diaphragm cell* very similar to the one illustrated in Figure 13.11 for NaI is used. The H_2 and Cl_2 are then combined photochemically (see Chapter 15) to make ultra-pure $HCl(g)$. This cell is the chief source of NaOH for the chemical industry, where it is employed mainly in pulp and paper processing and in the manufacture of organic chemicals.

The development of an electrolytic method for extracting aluminum metal from bauxite ore independently by Charles Hall and Paul Heroult in 1886 revolutionized the metals industry. A mixture of $Al_2O_3(s)$ and *cryolite* (Na_3AlF_6) proved to have a sufficiently low melting point to create a suitable molten electrolyte, whose electrolysis yields molten Al, which in this case is denser than the salt mixture and sinks to the bottom, as Figure 13.13 shows. Hall, a 21-year-old college student whose chemistry professor had challenged his class that "anyone who can manufacture aluminum cheaply will make a fortune," performed his first successful experiments with a frying pan and electrochemical cells constructed from canning jars. Aluminum is valuable as a structural component because of its light weight and resistance to corrosion; it is most heavily used in the housing and aircraft industries.

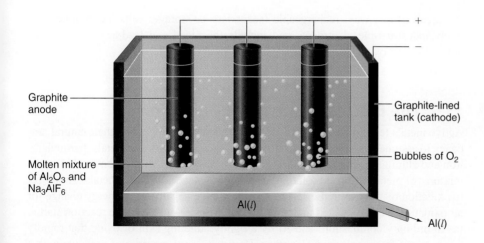

Figure 13.13
Simplified schematic of an electrolytic cell for the production of aluminum metal by the Hall–Heroult process. Oxygen gas is also produced; a "crust-breaker" (not shown) is attached to the top of the cell to poke holes in the nonmolten electrolyte crust, permitting the escape of O_2 and preventing the buildup of explosive gas pressures.

Electrolysis is also used extensively in metal refining and the *electroplating* of metals on other metals. If an impure metal slab, such as a copper ingot produced by conventional chemical reduction, is made the anode in a cell, more active impurities, such as Fe and Zn, will be oxidized along with the copper, while the noble metal impurities Ag, Au, and Pt will remain in metallic form. By using a $CuSO_4$ electrolyte solution and controlling the applied voltage, pure copper can be made to plate out on the cathode, while Fe^{2+} and Zn^{2+} remain in solution and the noble metals fall off the anode to make an extremely valuable "anode sludge." The most common types of electroplating, nickel, silver, and chromium ("chrome"), operate on much the same principle. The metal object to be plated must first be cleaned and polished chemically; active metals such as Al must first be protected with an electrolytic "strike" bath, which deposits a thin coat of a protective metal such as Ni or Cu, prior to the final plating operation. Active metals that readily air-oxidize can also be *anodized,* used as anodes in an alkaline electrolytic cell, causing a thick layer of tightly adhering oxide to form that protects against further corrosion. The former U.S. Steel Building in Pittsburgh, Pennsylvania, is constructed entirely of glass and anodized steel.

Electroplating or *electrodeposition* can also be an efficient and accurate method of chemical analysis. For example, the percent copper in an alloy such as brass can be determined by dissolving a sample in nitric acid and plating the copper onto a platinum cathode; the voltage of the cell must be controlled to prevent the plating of the more active Zn and Sn components. Weighing the Pt electrode before and after plating directly determines the mass of copper in the sample. For species such as halogens that do not adhere to an electrode, *coulometric analysis,* in which the current and time of electrolysis are monitored, can be used to obtain concentrations and amounts in a dissolved sample, as well as to prepare standard halogen solutions. The potential applied to the cell can be varied in a definite way (for example, a ramp or a sawtooth waveform) while the current is monitored in a set of related techniques collectively known as *voltammetry;* this is useful in analyzing multicomponent samples or, if the ramp speed rivals the rates of oxidation or reduction, discovering mechanistic information and revealing the presence of reaction intermediates.

The electrolysis cell itself could be a spent electrochemical cell, a dead battery, in which case an applied voltage of the proper polarity will *recharge* the cell, restoring its free energy by running the cell reaction in reverse. Not all cells can be recharged in this way; rechargeable (secondary or storage) cells incorporate relatively indestructible electrodes, and include the nicad and lead-acid batteries, as noted previously.

13.6 The Electrochemistry of Corrosion

Active metals tend to be anodes in electrochemical cells, reflecting their natural tendency to assume combined forms, to oxidize. Many of these same metals, particularly iron and aluminum, are used in pure form or more often in alloys (nonstoichiometric mixtures of metals in a solid solution) as structural materials. These metals are subject to oxidation by oxygen in the air but, despite a large negative free-energy change for air oxidation, Fe is protected in dry air by the difficulty of getting the reaction started (as in ordinary combustion) and Al by a tightly adhering skin of oxide that rapidly forms over a freshly cleaned Al surface. However, in the presence of moisture with a few ions dissolved in it, both metals will rapidly oxidize in the process commonly known as **corrosion.** Corrosion is a major problem in the industrialized world, causing the destruction of tens of billions of dollars worth of structures and equipment each year in the United States alone. How does it happen?

As illustrated in Figure 13.14, suppose we have a painted iron object, such as an automobile fender, with a few holes in the paint leading to the metallic surface. The presence of water and ions provides both a medium for dissolved Fe^{2+} and a "salt bridge" leading to a site for the reduction of O_2, while the iron metal itself completes the electrical circuit. The half-reactions and their sum are

$$Fe(s) \longrightarrow Fe^{2+}(aq) + 2e^- \qquad (13.20)$$

$$\underline{O_2(g) + 2H_2O(l) + 4e^- \longrightarrow 4OH^-(aq)}$$

$$2Fe(s) + O_2(g) + 2H_2O(l) \longrightarrow 2Fe^{2+}(aq) + 4OH^-(aq)$$

The standard voltage of the overall reaction is $\mathscr{E}° = +0.81$ V, making it highly spontaneous; under these conditions a piece of iron will continue to corrode until no metal is left. Aqueous Fe^{2+} will be rapidly oxidized to Fe^{3+} by O_2 diffusing into the electrolyte; Fe^{3+} then combines with the OH^- to make the familiar orange precipitate $Fe(OH)_3(s)$, or $Fe_2O_3 \cdot 3H_2O(s)$, which we know as rust. Reduction of O_2 tends to occur at the larger holes, where air has easier access, while erosion or pitting occurs at the smaller holes as the iron oxidizes and the ions diffuse away, toward the cathode. The rust naturally concentrates at the cathode sites as these ions oxidize and react with O_2 and OH^- present there.

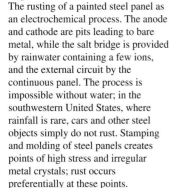

Figure 13.14
The rusting of a painted steel panel as an electrochemical process. The anode and cathode are pits leading to bare metal, while the salt bridge is provided by rainwater containing a few ions, and the external circuit by the continuous panel. The process is impossible without water; in the southwestern United States, where rainfall is rare, cars and other steel objects simply do not rust. Stamping and molding of steel panels creates points of high stress and irregular metal crystals; rust occurs preferentially at these points.

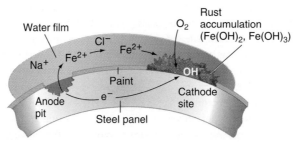

A number of environmental factors accelerate the rusting process. In most urban areas road salt (usually a mixture of NaCl and $CaCl_2$) is used to melt ice on roads; when the resulting saltwater wets automobile panels, it makes a very effective salt bridge. Acid rain, particularly prevalent in the northeastern United States, acts in a similar way; in addition, iron and steel are readily attacked by acid directly, since the $Fe^{2+}|Fe$ couple lies below the $H^+|H_2$ couple in reduction potential. Even a low $[H^+]$ in rain will remove the OH^- product from Reaction 13.20, making it still more spontaneous. Heat combined with moisture also increases the rate of corrosion; for example, automobile mufflers rust out quickly if cars are driven only short distances, since the muffler never becomes hot enough to keep the H_2O combustion product in the vapor phase, and thus allows it to accumulate as hot liquid.

Rust prevention is a major industry in its own right. The most common means of protecting steel from rust is electrolytic plating with zinc, a process called **galvanizing.** Because zinc lies below iron in reduction potential, it will corrode preferentially, acting as a *sacrificial* anode. To protect underground iron pipe, a block of active metal, usually Mg, is buried with it, connected by an insulated wire. Again, the more active metal oxidizes first, making iron the cathode in this "cell," in which moist soil forms the salt bridge; the reaction at the iron surface then becomes the reduction of water itself. Industrial laboratories are continually developing paints and coatings that give improved rust protection through nonporosity, blending in of reducing agents, and resistance to chipping.

In all of redox chemistry, as in acid–base chemistry, the state of aggregation or the *phase* of the reaction participants must be explicitly given, because these phases change not only the thermodynamics (and as we shall see in Chapter 15, the kinetics) of a reaction, but also its very appearance and perhaps even its viability. The thermodynamics changes because the intrinsic energies and entropies of different phases of a substance are different. This leads us to the next chapter, in which we consider the energetics involved in changes of phase. This exploration will lead us to the very basis for the existence of condensed phases: *intermolecular forces*.

SUMMARY

The thermodynamics of redox reactions is exemplified by the properties of **electrochemical cells,** in which two physically separated redox half-reactions are coupled together electrically in a closed electric circuit. This allows us to measure the voltage generated by the reaction and the electron flow that accompanies its occurrence. This voltage, or *redox potential* $\mathscr{E}$ (in volts, V) is directly proportional to the free-energy change ΔG, since $Q\mathscr{E}$ is the work done in pushing electrons through the circuit, where Q is the charge (in coulombs, C) that has passed. Therefore, $\Delta G = -nF\mathscr{E}$, where n is the number of moles of electrons that have passed, and F is the **Faraday constant,** 96,485 C/mol. From this result we find that $1\text{ V} = 96.485\text{ kJ mol}^{-1} = 23.0605\text{ kcal mol}^{-1}$. This identification also enables the general free-energy relationship for a redox reaction to be written in terms of electric potentials as $\mathscr{E} = \mathscr{E}° - (RT/nF)\ln Q$, where $\mathscr{E}°$ is the cell voltage under standard conditions, and Q is the reaction quotient. The criterion for sponaneity then becomes $\mathscr{E} > 0$. In working form at 298 K we obtain the **Nernst equation,** $\mathscr{E} = \mathscr{E}° - (0.0592/n)\log_{10}$, which leads directly to the equilibrium $\mathscr{E} = 0$ condition, $\log_{10}K = n\mathscr{E}°/0.0592$.

A working electrochemical cell consists of two **half-cells** called the **anode** and **cathode** compartments, joined by a barrier, the **salt bridge,** which allows ions to pass through it but prevents large-scale mixing of the compartment contents. Each compartment contains an **electrode,** a solid conductor made either of a reagent or product metal or an inert support, and an **electrolyte,** a solution, paste, or melt containing mobile ions. Connecting the electrodes with a conducting load allows current I (in amperes, A) to flow from the anode to the cathode as the redox reaction releases its free energy. When current is drawn, positive ions or cations move toward the cathode, where they may be reduced, while negative ions or **anions** move toward the anode, where they may be oxidized. Cathode reactions can also involve reduction of neutrals or anions that leave the cathode, while anode reactions can also be oxidation of neutrals or cations that leave the anode. In either case we have oxidation (loss of electrons) at the anode and reduction (gain of electrons) at the cathode. Inert materials may be substituted for the anode electrolyte and/or the cathode electrode. The cell may be succinctly described in line notation as *anode electrode | anode electrolyte (conc) || cathode electrolyte (conc) | cathode electrode,* where the identities of the anode and cathode are determined by the spontaneous direction of the reaction. The Nernst equation describes the dependence of cell voltage on concentration, while the amounts of reagents used to build the cell determine how much charge can be drawn from it.

Practical electrochemical cells—**batteries**—often employ paste or molten electrolytes, with ion-conducting membranes instead of salt bridges, and the most successful ones are *alkaline,* with the electrolyte paste containing OH^-. Cells described in some detail include the dry cell and the lead-acid storage battery, which is rechargeable. Though not yet household items, **fuel cells,** which run on the combustion of hydrogen or small hydrocarbons, are increasingly employed in the electric power and automotive industries.

Using the **standard hydrogen electrode** (SHE) as a reference, half-cell **standard reduction potentials** can be tabulated, as measured directly with voltmeters or inferred from measured free energies. The SHE is assigned a standard reduction potential of $\mathscr{E}° = 0.00$ V, and all others are then referred to it. A selection of these potentials in aqueous medium is given in Table 13.1; they range from $+2.89$ V for the reduction of F_2 to -3.05 V for the reduction of $Li^+(aq)$, and they form the quantitative basis for the *activity series*. The potential of any cell or redox reaction is then $\mathscr{E}_{cell} = \mathscr{E}_{cathode} - \mathscr{E}_{anode}$, with the inclusion of stoichiometric factors rendered unnecessary by the intensive character of the half-cell voltages. Cell voltages for reactions involving precipitates or complexes can be obtained by combining those from Table 13.1 with equilibrium constants or standard free energies of precipitation or complex formation. In general this involves converting all quantities into free energies and using Hess's law. This conversion must also be done when new half-cell voltages are to be found by combining known half-cell voltages.

Even for two half-cells containing the same ingredients, the Nernst equation predicts that a voltage will result if the concentrations or partial pressures differ; cells for which this is the only source of voltage are called **concentration cells.** The voltage generated by a solution of arbitrary pH in electrical contact through a glass membrane with a standard 1 M HCl solution is the basis for the pH meter. Concentration cells are also useful for measuring solubility products, wherein ions in one of the half-cells are selectively removed by adding a precipitating agent.

In **electrolysis,** the external circuit of a cell is changed from a passive load to a voltage source. If the applied voltage is large enough to overcome some natural redox voltage characteristic of the cell, *reverse chemistry will occur;* that is, reactions can be forced to run in normally nonspontaneous directions. Stable electrolyte solutions or melts can be decomposed, and spent electrochemical cells can be recharged, that is, restored to their original activity. The voltage is often somewhat in excess of the minimum needed to overcome the natural free-energy barrier; the excess is called **overvoltage.** Electrolysis is the major method by which highly active elements such as Na and Cl_2 are isolated, and it is also used to purify less active metals and to electroplate one metal onto another.

Faraday's **laws of electrolysis** state that (I) the amount of chemical change is directly proportional to the amount of charge passed through the cell, and (II) the amount of charge needed to deposit 1 mol of substance is $96,485n$ C, where n is the number of moles of electrons transferred per mole of substance deposited. These laws also apply to the functioning of electrochemical cells; they determine battery lifetime.

Corrosion is the oxidation of metals by the action of atmospheric oxygen and moisture. It often occurs by the formation of natural electrochemical cells. The corrosion or rusting of iron, and its electrochemical prevention, is analyzed in some detail.

EXERCISES

Electrochemical cells: stoichiometry and thermodynamics

1. For Volta's cell reaction (Equation 13.1), determine the minimum mass of Zn metal required to make 500. mL 0.100 M Ag^+ the limiting reagent. If Ag^+ is the limiting reagent under these conditions, use the equilibrium condition of Equation 13.2 to find the equilibrium concentration of Ag^+ in mol/L and ions/L. How many Ag^+ ions will be present in the system, assuming no volume change? What will be the concentration of Zn^{2+} at equilibrium?

2. Find the work done (J and cal) in pushing electrons through a 100-W light bulb with a potential difference of 120 V for 1 hr. Proceed by finding the current I from the relationship power $P = I\mathscr{E}$, then find the charge $Q = It$ in coulombs to be used in Equation 13.3. By analyzing your calculation, suggest an easier way to do it.

3. How many electrons passed through the light bulb of Exercise 2? How many moles of electrons? What mass of Ag metal would be deposited in each cell of a 120-V Ag/Zn voltaic pile used to supply the current?

4. Combine Equations 12.6 and 13.4 to derive Equation 13.5, and verify the numerical coefficient in Equation 13.6. Use Equation 13.6 to compute the cell voltage for a Ag/Zn cell with $[Ag^+] = 0.50$ M and $[Zn^{2+}] = 0.020$ M.

5. To what does Equation 13.6 reduce at equilibrium? Use your result to confirm the value of the equilibrium constant for the reduction of Ag^+ by Zn given just above Equation 13.2.

6. Will any reaction at all take place when, as described below Reaction 13.8, a strip of Cu is dipped in 200. mL of a 1.0 M $ZnSO_4$ solution? (*Hint:* Calculate $[Cu^{2+}]$ at equilibrium using Equation 13.8 and an appropriate reaction quotient.)

7. Sketch a $Zn\,|\,Zn^{2+}\,\|\,Ag^+\,|\,Ag$ cell using $Zn(NO_3)_2$ and $AgNO_3$ electrolyte solutions, and a KNO_3 solution in the salt bridge. On your sketch, label the anode and cathode compartments, indicate the direction of flow of ions in the anode, cathode, and salt-bridge electrolytes, and of electrons in the external circuit. If your cell were to be used as a battery, what signs would you use to label the electrodes?

8. In the cell of Exercise 7, which electrode could be replaced by an inert material? Which electrolyte could be replaced by an inert electrolyte? Which of these substitutions, if any, would affect the cell voltage? In which direction?

9. Most electrochemical cells and procedures are designed to operate at the normal temperature, 25°C, and nearly all the work in this chapter assumes conditions that are near 25°C. By analyzing the origins of the Nernst equation, decide which of its terms are temperature dependent. Using data from Appendix C, evaluate the standard voltage of Volta's cell (Exercise 7) at 50°C, and state the Nernst equation in working form that governs his cell at this temperature.

Practical electrochemical cells and their characteristics

10. After use, the copper electrode of a large Daniell cell is found to have increased in mass by 41.76 g. What charge was drawn from the cell? If the cell had been operated for 5.0 hr at constant current, what current was drawn? What maximum work could have been done by the cell, assuming a standard cell voltage?

11. The half-reactions of the Leclanché dry cell are given in Equation 13.10. What is the overall cell reaction? Assuming the cell potential of 1.50 V is close to standard, use Table 13.1 to find the standard half-cell reduction potential for the cathode reaction. Give the line notation for the dry cell.

12. What is the overall cell reaction for the alkaline dry cell, whose half-reactions are given in Equation 13.11? Use data from Appendix C to find standard half-cell reduction potentials and the standard full-cell potential for the alkaline dry cell. (*Hint:* In the half-reactions the electrons can be treated as if they had no free energy, since their free energy cancels in the overall cell reaction.) Give the line notation for this cell. Discuss briefly why this cell might be expected to provide a more constant potential during its lifetime than the Leclanché cell.

13. A typical 1.5-V size AA alkaline dry cell weighs 23.0 g. Assuming 70.0% of this mass consists of the cell reagents in stoichiometric amounts (the "active ingredients"), how much charge can this cell deliver? How much work can it do? If a steady current of 0.050 A is drawn from it, how long will it last?

14. In many portable devices, cells are connected in series to provide a higher potential. Compare drawing a given charge Q from such a battery with drawing the same charge from a single cell containing the same mass of reagents as the multicell battery. Which yields more work? Which consumes the larger amount of reagents?

15. Among the many miniature cells ("button" batteries) that have proliferated in recent years is the zinc-mercuric oxide cell, in which Zn is oxidized to ZnO and HgO is reduced to Hg in basic medium. Write the half-reactions and the overall reaction for the Zn/HgO cell, and guess the line notation for the cell. Sketch a possible picture of the "guts" of the button battery, label the anode and cathode, the + and − terminals, and indicate the direction of electron flow in an external circuit. Using data from Appendix C, find the standard cell potential for this cell.

16. Another common "button" battery is the alkaline zinc-silver oxide cell. Guess the cell reaction for this cell, give its line notation, sketch a picture of the cell's innards, and use data from Appendix C to find its standard potential.

17. Miniature batteries using ZnO and Li as reagents in alkaline medium (often simply called lithium batteries) have recently become popular. Assuming the products are Zn and Li_2O, what are the half-reactions and the overall cell reaction for this cell? Which electrode is the anode and which the cathode, and which signs $(+, -)$ would label them on the battery case? Use data from Appendix C to find the standard cell potential.

18. In the lead-acid battery (Equation 13.12 and Figure 13.4) the H_2SO_4 electrolyte is gradually consumed. When you buy a new car battery, fresh 4.8 M acid is added. What is the voltage $\mathscr{E}$ of a freshly charged six-cell lead-acid battery? Assuming the volume of electrolyte remains constant, what will the voltage be if 50% of the acid is consumed? If 99% is consumed? Use your results to construct a rough plot of $\mathscr{E}$ versus $[H_2SO_4]$. (See Example 13.3.)

19. On a very cold day, a lead-acid battery may have trouble starting your car. Possible causes include increased engine friction and lower mobility of electrolyte ions in the battery. Evaluate the single-cell standard voltage of the lead-acid storage battery at 25°C and −15°C using data from Appendix C, and decide whether voltage changes are part of the difficulty. (Presumably a lower voltage would present a starting problem.)

20. Lithium is favored for use in vehicular batteries under development because of its low atomic weight, which gives cells that use it a high energy density (energy per unit mass). The lithium-sulfur battery operates at high temperature with molten Li and S, graphite electrodes, and a porous β-alumina (β-Al_2O_3) separator. The β-alumina allows Li^+ but not S^{2-} ions (which tend to form polysulfides S_n^{2-}) to migrate between compartments. Give half-reactions and the overall reaction for this cell, and use data from Table 13.1 to calculate its $\mathscr{E}°$. Compare this result to that obtained for the cell reaction at 25°C involving solid-state reagents and product from data in Appendix C. How will the cell voltage change with temperature?

21. Molten sodium-sulfur batteries preceded lithium-sulfur because of Na's lower melting point (98° versus 186°C) and lower cost. State half-reactions and overall reaction for a Na/S cell, and compute its $\mathscr{E}°$ from data of Table 13.1. Also find $\mathscr{E}°$ for this cell assuming solid reagents and products using data from Appendix C.

Electrode potentials

22. For the aluminum-chlorine cell of Example 13.4, use standard free energies for the half-reactions and Equation 13.4 to establish the validity of Equation 13.15 for the cell voltage. Compute these half-cell free energies from data in Appendix C, and show that both the half-cell potentials and the resulting standard cell voltage agree with those obtained from Table 13.1.

23. For each of the following aqueous redox reactions (given in ionic form) deduce the half-reactions, compute the standard potential from data in Table 13.1, and deduce whether or not the reaction will proceed spontaneously to the right under standard conditions:

   a. $Pb^{2+} + Ni \longrightarrow Ni^{2+} + Pb$
   b. $2Ag^+ + Fe \longrightarrow 2Ag + Fe^{2+}$
   c. $Mg + 2H^+ \longrightarrow Mg^{2+} + H_2$
   d. $Sn^{2+} + 2Br^- \longrightarrow Sn + Br_2$
   e. $Cl_2 + 2I^- \longrightarrow 2Cl^- + I_2$

24. Repeat Exercise 23 for the following reactions:

   a. $MnO_4^- + 5Fe^{2+} + 8H^+ \longrightarrow Mn^{2+} + 5Fe^{3+} + 4H_2O$
   b. $2Al^{3+} + 2Cr^{3+} + 7H_2O \longrightarrow 2Al + Cr_2O_7^{2-} + 14H^+$
   c. $Hg_2Cl_2 + 2Ag \longrightarrow 2AgCl + 2Hg$
   d. $3Cu^{2+} + 2NO + 4H_2O \longrightarrow 3Cu + 2NO_3^- + 8H^+$
   e. $2H_2 + O_2 \longrightarrow 2H_2O$

25. Use Table 13.1 to deduce which of the following metals will be attacked by acid: Ag, Fe, Mn, Cu, Ni.

26. In introductory electrochemical laboratory exercises, a standard $Zn|Zn^{2+}$ electrode is often used as a secondary reference, and a series of half-cells are built and their voltages measured when coupled to the Zn electrode by a salt bridge. A certain standard half-cell yields a voltage of $+1.53$ V with standard Zn when Zn is wired as the anode. What is its standard reduction potential with respect to the standard hydrogen electrode (SHE)? Sketch a voltage ladder to illustrate the relationship between the two voltages. Why don't you need to know the number of electrons involved in the unknown half-reaction in order to find its $\mathscr{E}°$?

27. Most modern electrochemical cells operate in an alkaline medium. For nickel electrodes, the alkaline half-reaction is usually $Ni(OH)_2 + 2e^- \longrightarrow Ni + 2\,OH^-$. Show that this reaction is the sum of the standard reduction of Ni^{2+} and the solubility reaction of $Ni(OH)_2$. Use this fact together with data from Table 13.1 and 12.6 to deduce the standard reduction potential for this reaction.

28. Verify the standard half-cell reduction potential for the anode of the lead-acid storage battery given in Table 13.1 by combining the standard reduction potential of Pb^{2+} with the solubility product K_{sp} of $PbSO_4$ from Table 12.6. (*Hint:* First find the relationship between the anode half-reaction and the reactions suggested.)

29. The half-reaction by which water is reduced to OH^- and H_2, as given in Table 13.1, is actually the sum of the hydrogen electrode reduction and twice the water autoionization reaction, written as $H_2O \longrightarrow H^+ + OH^-$. What does this imply about the standard free energy of water reduction? The standard reduction potential? Does the voltage tabulated agree with your deduction?

30. Find the standard half-cell reduction potential for the half-reaction $Fe^{3+} + 3e^- \longrightarrow Fe$ by combining those for the steps $Fe^{3+} + e^- \longrightarrow Fe^{2+}$ and $Fe^{2+} + 2e^- \longrightarrow Fe$ as given in Table 13.1. Why can't you just add the half-cell potentials for the stepwise reduction?

31. The free-energy change in the half-reaction for the reduction of oxygen to water by H^+ and e^- given in Table 13.1 is the same as that for the combustion of hydrogen. Show why this is so.

32. What familiar reaction is needed to establish the relationship between the reduction of oxygen in acid,
$$O_2 + 4H^+ + 4e^- \longrightarrow 2H_2O,$$
and its reduction in neutral or basic medium,
$$O_2 + 2H_2O + 4e^- \longrightarrow 4OH^-?$$
How must their voltages be related? Verify the voltage in base given in Table 13.1 from that in acid.

33. It is often useful to consider nonstandard half-cell potentials. These may be computed from the Nernst equation by setting up a half-reaction quotient, done in the usual way except that the electrons are omitted. Consider the cathode of the H_2/O_2 fuel cell, Equation 13.13: write the half-reaction quotient and substitute it into the Nernst equation. Calculate the half-cell potential when

   a. $[OH^-] = 1.00$ M, $P_{O_2} = 159$ Torr
   b. $[OH^-] = 0.050$ M, $P_{O_2} = 760$ Torr

34. Use the balanced combustion reaction of CH_4 to water and carbon dioxide to deduce the anode half-reaction in a methane-oxygen fuel cell, assuming the cathode reaction is as given in Equation 13.13 for the H_2–O_2 cell. Use data from Table 13.1 and Appendix C to find the reduction potential for this half-cell (the standard anode potential), and the overall cell potential. Combine this cell potential with the H_2–O_2 potential to deduce the standard free-energy change of the steam reforming reaction, Equation 13.14.

35. Fuel cells based on methanol CH_3OH have recently been developed. Assuming the cell is alkaline, and the cathode reaction (reduction of oxygen) is the same as in the H_2–O_2 cell, find the anode half-reaction by using the balanced combustion equation [to $CO_2(g)$ and $H_2O(l)$] for CH_3OH. From Appendix C and Table 13.1, find the standard half-cell reduction potential for this half-reaction, and the cell $\mathscr{E}°$.

The standard hydrogen electrode and concentration cells

36. Consider coupling a standard Br_2/Br^- half-cell with the standard hydrogen electrode SHE. Write down the anode and cathode half-reactions, find the overall cell reaction, and compute the cell voltage from data in Table 13.1.

37. Consider coupling a standard Mg/Mg^{2+} half-cell with the standard hydrogen electrode SHE. Write down the anode and cathode half-reactions, find the overall cell reaction, and compute the cell voltage from data in Table 13.1.

38. Using the results of the previous two exercises, what are the half-reactions and the standard cell potential for a cell built on the reaction $Mg(s) + Br_2(aq) \longrightarrow MgBr_2(aq)$?

39. What is the pH in a standard hydrogen electrode? Consider the cell $(Pt)H_2|H^+\|Cl^-|Cl_2(C)$, $\mathscr{E}° = 1.360$ V, from Example 13.4. Suppose all concentrations and partial pressures are held at standard values except $[H^+]$. Use the working form of the Nernst equation, Equation 13.6, to write an expression for the cell voltage that directly shows its dependence on pH. Use your relationship to show that $\mathscr{E} = \mathscr{E}°$ when the hydrogen electrode is standard, and to find $\mathscr{E}$ for this cell at (a) $[H^+] = 0.10$ M; (b) 1.0×10^{-7} M; and (c) 1.0×10^{-14} M.

40. Use general relationships to show that the voltage generated by a concentration cell is a direct measure of the entropy change that would result were the

(ideal) electrolytes from the two cell compartments to be mixed. Use the Cu^{2+} concentration cell discussed in the text as a concrete example. Discuss your finding briefly in terms of the statistical (Boltzmann) interpretation of entropy given in Chapter 11.

41. A pH meter is basically a concentration cell in which one electrode is a SHE and the other is formed by using the unknown solution as electrolyte. Presuming that the pH of the unknown solution is greater than zero, identify the anode and cathode and give the line notation for this cell. Set up the Nernst equation in a form that directly contains the pH of the unknown solution. To what cell voltages do the pH values 4.00, 7.00, and 10.00 correspond? (These pH's are commonly used for meter calibration.)

42. Concentration cells are useful for measuring solubility products, as illustrated in Example 13.5. To measure K_{sp} for $PbSO_4$, a Pb^{2+} concentration cell was set up, with one compartment containing 0.100 M Pb^{2+} and the other containing 0.00100 M Pb^{2+}. Enough SO_4^{2-} was then added to the anode compartment to bring the final $[SO_4^{2-}]$ at equilibrium to 0.100 M. The cell voltage was measured to be 0.170 V. Find K_{sp} for $PbSO_4$.

43. Cell voltages are useful in monitoring redox titrations just as pH meters are for acid–base. A common analytical procedure is the determination of iron by oxidizing Fe^{2+} to Fe^{3+} with permanganate solution. One half-cell consists of a Pt electrode inserted into the analyte solution, and the other of a saturated calomel electrode (SCE), whose half-cell reduction potential is known and constant. Set up the Nernst equation for the Pt|Fe^{2+}, Fe^{3+} half-cell, and predict the voltage of this half-cell near the start of the potentiometric titration, where $[Fe^{2+}] \approx [Fe^{2+}]_0 = 0.100$ M and $Fe^{3+} = 1.0 \times 10^{-4}$ M, 1/4 through, 1/2 through, 3/4 through, and just prior to the endpoint. To simplify the calculations, neglect the volume of titrant added. Plot your results as $\mathscr{E}$ versus extent of reaction, and compare the shape of the curve to that for pH versus extent of reaction in an acid–base titration. At what point is the half-cell voltage at its standard value? (The influence of any excess oxidizing agent may be neglected prior to the endpoint, since the redox equilibrium lies very far to the right.)

Electrolysis

44. Find the minimum applied voltage needed to carry out the electrolysis of water to hydrogen and oxygen (see Figure 13.10). Presuming your voltage supply terminals are marked + and −, at which terminal will hydrogen gas appear? What reactions are occurring at the respective electrodes, assuming a neutral salt solution is used? Why is no reaction observed if no electrolyte is present? What would happen if you used an alternating current (ac) voltage supply?

45. If a concentrated NaCl solution is electrolyzed, $H_2(g)$ and $Cl_2(g)$ (instead of O_2) are produced. What is the overall electrolysis equation and standard voltage for this process? What happens at the anode of the cell? At the cathode? How many C of charge must be passed through the cell to produce 200. L of Cl_2 at STP?

46. Suppose an external voltage is applied to an electrolysis cell consisting of two copper electrodes, a $CuSO_4$ electrolyte solution, and no divider. Using Table 13.1 for necessary information, describe what will happen at the electrode to which positive voltage is applied. Is this the anode or cathode? Do the same for the negative electrode. What is the overall cell process? Why is water not directly involved in the process?

47. The electrolytic cell of the previous exercise is an example of a *coulometer*, since passing a known charge through it will result in the deposit of a known mass of Cu at the cathode. What mass of Cu will be plated out at the cathode if a current of 0.145 A is passed through the cell for 45 min?

48. Faraday and Berzelius accurately determined the *equivalent weights* of many metals (the ratio of atomic mass to atomic charge in their salts) by the use of the *silver coulometer*. An electrolysis cell containing an unknown metal ion would be connected in series with a cell containing two Ag electrodes and a $AgNO_3$ electrolyte solution; the mass of unknown metal plated on the cathode combined with the mass of Ag plated within the Ag|$AgNO_3$ cell allowed the equivalent weight of the metal to be found. Find the equivalent weight of an unknown metal M if 3.108 g of the metal plates out while 3.236 g Ag is deposited. If this metal is known to make dihalide salts (MCl_2), what is its atomic weight?

49. A typical NaCl(*aq*) electrolysis plant can produce up to 200 metric tons (200×10^3 kg) of NaOH—industrially known as *caustic soda*—per day. Assuming operation around the clock, what current in amperes is required?

50. A dam on the Columbia River hydroelectrically delivers 2×10^8 A continuously. How many kg of Al metal

can be produced in the Hall–Heroult process in 1 hr if all this current is used for that purpose?

Corrosion chemistry

51. Equation 13.20 produces $Fe^{2+}(aq)$ from iron, air, and water. Suggest a similar set of half-reactions and overall reaction for producing $Fe(OH)_3(s)$ by further air oxidation of Fe^{2+}. Find the standard potential $\mathscr{E}°$ for this reaction by any convenient method.

52. The air oxidation of aluminum is much slower than that of iron, but occurs by direct conversion of $Al(s)$ to $Al(OH)_3$ without an intermediate oxidation state. Give half-reactions and overall reaction for this conversion in the presence of moisture and find its standard potential.

CHAPTER 14

States of Matter and Intermolecular Forces

A snowflake is a uniquely beautiful form of ice, $H_2O(s)$. See Figure 14.31 for the molecular structure of ice.

The propensity of many people to imagine water as naturally and essentially liquid is a prejudice contracted from the habit of seeing it much oftener in this state than in the solid state.

Joseph Black (posthumous, 1803)

CHAPTER OUTLINE

14.1 It's All a Matter of Temperature

14.2 Heating Curves

14.3 The Free Energy of Phase Changes

14.4 Phase Transitions, Molecular Structure, and Intermolecular Forces

14.5 Structure and Bonding in Solids

14.6 Liquids and Solutions, Featuring Water

Not a day passes that you don't experience the three states of matter: gas, liquid, and solid. You breathe the gases of air, drink liquid water and other water-based beverages, and eat with solid metal sticks—knives and forks—from solid ceramic dishes. You naturally come to associate these states with the materials they are made of; water is one of the few examples where you commonly experience the other states or **phases,** solid ice and gaseous steam. Water is indeed special, but only because its three states can exist in a range of pressure and temperature readily accessible on the earth's surface. It turns out that all simple substances, and their mixtures, display at least these three phases under the right conditions. The fact that metals can be melted or *fused* into liquid form was known to the earliest metal workers dating to prehistoric times, but the fact that gases could also be condensed into liquids and even solids had to await the pioneering work of Faraday and later Dewar in the 19th century. Today we know that every gas, even the most gaslike of all, helium, can be condensed, while even the most refractory metal can be melted and then vaporized to gaseous metal atoms and molecules. These transformations, or **phase transitions,** all involve the transfer of heat—a fact discovered by Joseph Black—and thus fall under the aegis of thermodynamics, that redoubtable edifice we have been examining for the last four chapters. The vaporization of water, a liquid $\longrightarrow$ gas transition that happens everywhere on earth, has been used as a cardinal thermodynamic example. Under other conditions, however, what is considered common might be quite different, because . . .

Joseph Black (1728–1799) was a British chemist and a practicing physician. In addition to his own discovery of specific heat and latent heat, he was a master of the world's chemical literature, and the preeminent teacher of chemistry in his time. One of his students translated Lavoisier's *Elements of Chemistry* into English. Black himself wrote very little; most of his work was published by his students posthumously from compilations of his lecture and laboratory notes.

14.1 It's All a Matter of Temperature

Let's consider three common but distinctive substances: nitrogen, water, and copper. You would never consider copper as a liquid, and yet, copper metal can be melted, and even boiled, if the temperature is high enough. Likewise you would not normally think of nitrogen as a liquid, yet nitrogen gas can be condensed, and even frozen, if the temperature is low enough. Figure 14.1 compares the temperature ranges over which each of these three substances displays each of its three phases, at standard pressure.

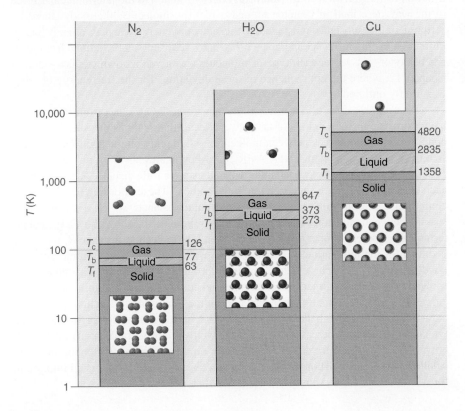

Figure 14.1
Ranges of temperature in which solid, liquid, and gas phases exist for three typical chemical substances, under standard pressure $P = 1.00$ atm, where T_f is the normal freezing point, T_b the normal melting point, and T_c the critical point, above which the gas cannot be liquefied regardless of P. (T_c for Cu is estimated.) The insets show schematically the arrangement of molecules in the solid and gas phases; the liquid phase is similar to the solid, except there is no long-range order. Note the similarity of the three diagrams on this logarithmic temperature scale, particularly the relatively narrow range of T for the liquid phases.

As solids, copper, water ice, and "nitrogen ice" each consist of a regular, repeating arrangement of its constituent atoms or molecules we call a **crystal.** In such crystals, the molecular units are considered to be "touching," that is, the molecules are packed like oranges in a crate; as we saw in Chapter 1, the typical diameter of a small molecule like water is about 3 Å, and this is roughly the distance between molecules in solids. In the liquid phase, the molecules are no longer bound to remain in such geometrically ordered arrays, but nonetheless remain in proximity to one another. This loosening up of the intermolecular arrangement makes a liquid pourable, while the continued closeness is reflected in similar densities for the solid and liquid phases. In the gas phase, which we have considered in some detail in Chapter 9, the molecules are on the average much farther apart, roughly 10 times farther at standard temperature and pressure. This greater intermolecular distance is reflected in much lower density (by about a factor of 1000) and by the ease with which a gas can be compressed, compared to which, liquids and solids are essentially incompressible.

We found in Chapter 9 that a single universal law, the ideal gas law, provides a good description of any gas. The reason this law works so well is that the forces between molecules, the **intermolecular forces,** can be neglected as contributors to the total thermodynamic energy of the gas. This in turn is traceable to the large intermolecular distances prevailing in a gas under ordinary conditions (see Section 9.3 of Chapter 9 for a calculation). In liquids and solids, however, the molecules are in intimate contact—they are "feeling" each other—and are exerting significant intermolecular forces. No universal rule is going to be able to describe all such substances, since molecules of different substances have different shapes and sizes and, consequently, as we shall see, exert different kinds and strengths of forces on each other. The energy of a substance in a condensed phase now will contain a substantial contribution from **intermolecular potential energy** (related to the intermolecular force in the way described in Chapter 1), and a realistic molecular thermodynamic description of that substance must include this potential energy. We have already had an intimation of this influence in Chapter 9, where we discussed deviations from ideal gas behavior—in which each type of molecule was given its own gas law—and in Chapter 12, where the consequence of these deviations for the free energy of a gas was briefly outlined.

What is the nature of these intermolecular forces? Our clues from real gas behavior provide the beginnings of an answer. Recall that real gas pressure had to account for attractive forces between molecules, while real volume had to include the size of the molecules themselves. As Richard Feynman puts it (in his *Lectures on Physics*), molecules "attract each other when they are a little distance apart, but repel each other when they are squeezed into one another." These opposing forces reach a compromise at some intermediate intermolecular distance, which determines the spacing between molecules in condensed phases. Figure 14.2 shows how a typical intermolecular potential energy $V(R)$ depends on intermolecular distance R; the curve is drawn for the interaction of two argon atoms. The graph should remind you of a bond potential energy curve (see Chapters 5 through 8); that for the Cl_2 molecule is shown for comparison. Intermolecular potential curves are indeed similar to bonding curves, except that normally the distance R_m at the minimum of the curve (where the attraction and repulsion balance) is typically 3 Å or more, considerably larger than a typical bond length, whereas the depth of the well ε is considerably smaller, typically only 1 to 10 kcal/mol for substances that are liquids under normal conditions, and smaller still for gases. Though weak, the attractive part of the curve is all important, for without it condensed phases could not exist at all!

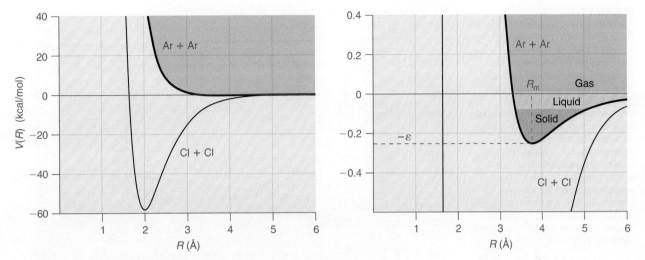

Figure 14.2
Intermolecular potential energy curve for the interaction of two Ar atoms (heavy line), compared to the bond potential energy curve for Cl_2, shown on two different energy scales. The R_m notation marks the position of the minimum in the Ar + Ar curve (3.8 Å), while ε is its well depth (0.25 kcal/mol). The ranges of energy and distance occupied by Ar atoms in the three phases of Ar are indicated schematically by the boundaries above the Ar_2 curve. Argon melts at 83.8 K and boils at 87.3 K; at these temperatures the average mechanical energy is $RT \approx 0.17$ kcal/mol, comparable to the attractive potential energy between Ar atoms.

So why are nitrogen, water, and copper so different? The answer lies in the strength of the attractive forces, that is, in the depth of the intermolecular potential well. Atoms and molecules can only "stick together" to make a condensed phase if their average *thermal energy* $k_B T$ is much less than the intermolecular "bond" energy ε. When T is increased, the average energy increases in proportion until the intermolecular forces can no longer hold the molecules in place. At that point the solid melts or fuses, or the liquid vaporizes or boils. We can say at once from inspecting Figure 14.1, or just knowing the melting and boiling points, that the attractive forces between copper atoms are considerably stronger (ε is larger) than those between water molecules, which are in turn several times stronger than those between nitrogen molecules. It truly is all a matter of temperature, then, that causes us to regard these three common substances so differently. But how do these all-important attractive (and repulsive) forces arise? That remains to be seen. First we must as usual be more quantitative concerning the heat, work, and energy involved in changes of phase.

14.2 Heating Curves

You have already learned in Chapter 10 what happens when you heat a substance at constant pressure; its temperature will rise by an amount proportional to the amount of heat q added, with a proportionality constant C_P called the heat capacity at constant pressure. Equation 10.10, now without an exothermic chemical process producing heat from within, gives us

$$q_P = \Delta H = C_P(T - T_0) \qquad (14.1)$$

where T_0 is some initial temperature (°C or K), and we have neglected any dependence of C_P on temperature. Recall also that ΔH automatically includes any P-V work the substance does while it is being heated. Equation 14.1 implies that a plot of q versus T, a **heating curve,** will be a straight line of slope C_P. We also found in Chapter 10 (Section 10.7) that in general the heat capacity of a substance is a unique property of that substance, so that the slope of the line depends on what it is you're heating. For a given substance, C_P also depends on what phase it happens to be in; beginning with water as a familiar but important example, you can find in Table 10.4 the molar heat

Figure 14.3
The heating curve for 1 mol (18 g) water, q_P versus T. The slopes of the curve in the three phase regions are the heat capacities for those phases, and the heights of the jumps at the freezing point T_f and boiling point T_b are the latent heats or enthalpy changes accompanying the melting and boiling phase transitions. The figure is drawn for heating ice beginning at $-50°C$, while the absolute enthalpy ($H°$) scale is set by the standard heat of formation of $H_2O(l)$ at 25°C. Because T_b is pressure dependent (see the text), the right-hand side of the figure can change, with the boiling jump occurring at lower T for $P < 1$ atm and higher for $P > 1$ atm.

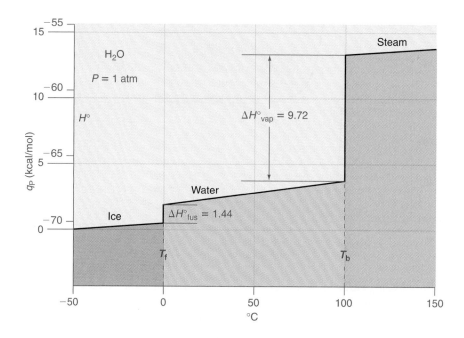

capacities for ice, water, and steam: 9.0, 18.0, and 8.6 cal/mol K, respectively. The different C_P's imply that the slope of the q versus T curve will change as we pass into a new phase with increasing T.

In fact, something more profound happens when ice melts or water boils, as first noted (and measured quantitatively) by Joseph Black, a Scottish chemist, in about 1760. As illustrated in Figure 14.3, the heating curve for water shows discontinuous jumps* at the melting point T_f and the boiling point T_b. These jumps correspond to adding heat without changing the temperature, in order to effect the change in phase; Black referred to the quantity of heat consumed during the change in phase as *latent heat*. We now recognize these heats as enthalpy changes that accompany changes in phase, $\Delta H°_{fus}$ for melting a solid and $\Delta H°_{vap}$ for boiling a liquid. As you can see from the heights of the jumps in Figure 14.3, these two ΔH's are quite different for water—as they are for all other substances—with $\Delta H°_{vap} > \Delta H°_{fus}$. This difference can be understood by referring to the intermolecular potential energy diagram of Figure 14.2: to melt a solid, you need just enough energy that a few molecules can crawl out of one well into a second one nearby, while to boil a liquid you must completely "break" all the intermolecular "bonds," which requires much greater energy. As discussed in some detail later in this chapter, $\Delta H°_{vap}$ is a good measure of the depth of the potential well in Figure 14.2; that is, the bonds that hold water molecules in place in the liquid have a strength of about 10 kcal/mol (0.4 eV/molecule).

---

* You may have learned about discontinuous functions like $q(T)$ in differential calculus; the heating curve provides a real-life application of the theorems surrounding such functions. For example, as you approach the melting point from either side, neither the function $q(T)$ nor its derivative $dq/dT = C_P$ will have the same values. This means that $q(T)$ is neither continuous nor differentiable at a "transition temperature."

EXAMPLE 14.1

Acetone, CH_3COCH_3, is a common ingredient in paint strippers and nail-polish removers, as well as a widely used laboratory solvent. Acetone boils at 56.0°C under atmospheric pressure; its heat of vaporization is $\Delta H°_{vap} = 6.96$ kcal/mol, while $\overline{C}_P(l) = 30.2$ cal/K mol and $\overline{C}_P(g) = 17.9$ cal/K mol. Calculate the amount of heat (joules or calories) needed to raise the temperature of 74.0 mL acetone from 25.0° to 70.0°C at a constant external pressure of 1.00 atm, and sketch a heating curve q versus T for the process. (The density of liquid acetone is 0.784 g/mL.)

Solution:
The number of moles of acetone can be obtained from the liquid volume V and density ρ (see Chapter 1) as $n = \rho V/M = (74.0$ mL$)(0.784$ g/mL$)/(58.0$ g/mol$) = 1.00$ mol. Since the temperature interval includes the boiling point T_b, the calculation must be carried out in these steps:

a. First liquid acetone is heated from 25.0°C to T_b, 56.0°C; using Equation 14.1,

$$q_a = n\overline{C}_P(l)(T_b - T_1)$$
$$= (1.00 \text{ mol})(30.2 \text{ cal/K mol})(56.0 - 25.0)\text{K} = 936 \text{ cal}$$

b. The acetone then boils at T_b, absorbing heat:

$$q_b = n\Delta H°_{vap} = (1.00 \text{ mol})(6960 \text{ cal/mol}) = 6960 \text{ cal}$$

c. Now the acetone is 100% vapor or gas; to achieve 70°C requires

$$q_c = n\overline{C}_P(g)(T_2 - T_b)$$
$$= (1.00 \text{ mol})(17.9 \text{ cal/K mol})(70.0 - 56.0)\text{K} = 251 \text{ cal}$$

The total heat required is therefore $q = q_a + q_b + q_c = 8150$ cal; 85% of the heat goes into boiling the acetone. If heat is added at a constant rate, a substantial pause in temperature will be observed when T reaches 56.0°C, as you have surely experienced with water at 100°C. The heating curve will be similar in appearance to that portion surrounding the boiling point of water in Figure 14.3; as in that figure, the slope of the liquid portion of the curve is steeper due to the larger heat capacity of the liquid. The volume will also increase greatly upon vaporization; using the ideal gas law for the vapor, $\Delta V = 27{,}000$ mL $(g) - 74$ mL $(l) = 26{,}900$ mL $\approx V_g$, and the volume ratio is $V_g/V_l = 365$.

We also have to attend to our other important thermodynamic factor, the entropy S. Since ΔS is the sum of all the bits of heat added as the temperature increases divided by the temperature at which each is added, we have $\Delta S = \int dq/T > 0$ for a reversible addition of heat. Thus we know that S, like q or H, will also be increasing with T. Since, from Equation 14.1, $dq = C_P dT$ at constant P, we arrive at a result you've already seen in Chapter 11 (cf. Equations. 11.12 and 11.18), namely

$$\Delta S = \int_{T_0}^{T} \frac{C_P \, dT}{T} = C_P \ln \frac{T}{T_0} \tag{14.2}$$

where, as in Equation 14.1 (and in Equation 11.12), we have neglected the dependence of C_P on T to carry out the integration by taking C_P out from under the $\int$ and recognizing our old friend the logarithmic integral. The increase in S with T predicted by Equation 14.2 is slower than linear, but it does increase; note that only absolute T can be used here. For each new phase you again have to substitute a different value of C_P, and Equation 14.2 in integrated form will only hold *between* phase changes. Right at a transition temperature (T_f or T_b), the heat is added without change in T—T is constant during addition of each bit dq—making $\Delta S = q/T$, where q is the total heat needed to change the phase. Thus for the two common phase changes,

$$\Delta S°_{fus} = \frac{\Delta H°_{fus}}{T_f} \quad \text{and} \quad \Delta S°_{vap} = \frac{\Delta H°_{vap}}{T_b} \quad (14.3)$$

where we have invoked standard pressure (1 atm). Figure 14.4 illustrates Equations 14.2 and 14.3 for water. It is useful for thinking about the increase in S with T to recall that S is a measure of the level of disorder in the system. The quantity of molecular motion increases with T, resulting in more chaotic behavior within a given phase and a gradual rise in S; whereas at a phase transition there is a qualitative jump in the kinds of random motion that are possible, yielding a quantitative leap in S. In melting a solid, you "free" the molecules from their lattice positions, while in boiling a liquid you free them from each other completely. Thus $\Delta S°_{vap}$ is by far the larger change. This process of accumulating entropy as the temperature rises corresponds exactly to the experimental method for obtaining the absolute entropies of Chapter 11.

Figure 14.4
Behavior of the entropy S with temperature T as 1 mol of water is heated through its freezing and boiling points at atmospheric pressure. Away from the jumps at T_f and T_b, the behavior of $S(T)$ is given approximately by Equation 14.2, while the heights of the jumps are given by Equations 14.3. The entropy scale is absolute, based on the third law.

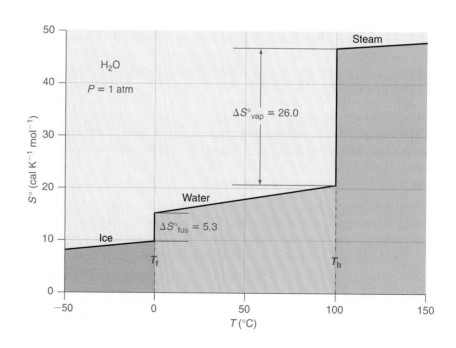

EXAMPLE 14.2

Use the data given in Example 14.1 for acetone, CH_3COCH_3, to find ΔS (J/K or cal/K) for the heating process specified there.

Solution:
Again, since the temperature interval includes the boiling point T_b, the calculation must be carried out in three steps:

a. First liquid acetone is heated from 25.0°C to T_b, 56.0°C; using Equation 14.2

$$\Delta S_a = n\overline{C}_P(l) \ln(T_b/T_1)$$
$$= (1.00 \text{ mol})(30.2 \text{ cal/K mol}) \ln(329 \text{ K}/298 \text{ K}) = 2.99 \text{ cal/K}$$

Note that kelvin temperature units must be used here!

b. The acetone then boils at T_b; from Equation 14.3 the entropy increases by

$$\Delta S_b = n\Delta H°_{vap}/T_b = (1.00 \text{ mol})(6960 \text{ cal/mol})/(329 \text{ K})$$
$$= 21.2 \text{ cal/K}$$

Again, kelvin units are needed.

c. Heating the acetone vapor from T_b to 70°C produces

$$\Delta S_c = n\overline{C}_P(g) \ln(T_2/T_b)$$
$$= (1.00 \text{ mol})(17.9 \text{ cal/K mol}) \ln(343 \text{ K}/329 \text{ K})$$
$$= 0.75 \text{ cal/K}$$

The total entropy generated is therefore $\Delta S = \Delta S_a + \Delta S_b + \Delta S_c = 24.9$ cal/K; 85% of ΔS arises from boiling the acetone.

The diagrams of Figures 14.1, 14.3, and 14.4 carry a "fact of life" in the way they are drawn. You never find ice forming in a room temperature glass of water by itself; if you add ice it will quickly melt. Likewise a room temperature glass of water never boils by itself; any steam would quickly condense at room temperature. That is, at a given pressure, the phases of matter have their own well-defined, nonoverlapping range of temperatures. This discussion of what is possible and impossible should remind you of the way we were in Chapter 11; we need a way to analyze these articles of common experience. Once again, Gibbs to the rescue!

14.3 The Free Energy of Phase Changes

A change of phase for a pure substance can be regarded as a special case of a chemical reaction in which the product is the same as the reagent except for its state of aggregation. The free energy as defined in Chapter 11 therefore applies to phase changes as well, as was briefly mentioned there; in fact, Gibbs first used his new ideas to describe such changes, in his two-part paper *On the Equilibrium of Heterogeneous Substances* published in 1876–1877. You are already familiar with the principles needed to analyze them: when $\Delta G < 0$ for a particular change of phase, then it will happen spontaneously; if $\Delta G > 0$, then the reverse phase change will occur, and if $\Delta G = 0$, the two phases are in equilibrium, that is, they coexist. We add the superscript ° if the phase change occurs at 1 atm pressure.

Thus the fact that ice melts rather than forming in a room temperature glass of water at 1 atm implies that $\Delta G° > 0$ for $H_2O(l) \longrightarrow H_2O(s)$ at 298 K. Likewise it must be that $\Delta G° > 0$ for $H_2O(l) \longrightarrow H_2O(g)$ at 298 K, where $P_{H_2O} = 1$ atm. (Note that the latter does not mean that water cannot *evaporate*, only that it cannot *boil* at 1 atm.) We know that these changes do take place spontaneously at or below 0°C and at or above 100°C; the temperatures T_f and T_b, according to the analysis of the temperature dependence of $\Delta G°$ of Section 11.6, must be *turnaround temperatures*— temperatures at which the spontaneity of freezing and boiling reverses directions, respectively. So if you place your glass of water in a freezer at $-20°C$, it will freeze, or if in an oven at 200°C, it will boil. Right at a transition temperature, $\Delta G°$ is zero, and the two phases can coexist; since $\Delta G° = \Delta H° - T\Delta S°$, then $\Delta S° = \Delta H°/T_{trans}$, a relation we had already deduced earlier from the definition of ΔS and the experimental fact of latent heat.

EXAMPLE 14.3

Using standard enthalpies and entropies from Appendix C, estimate the boiling point of water by assuming it is a turnaround temperature for the process $H_2O(l) \longrightarrow H_2O(g)$.

Solution:
From Appendix C we obtain $\Delta H°_{298} = 10.52$ kcal/mol and $\Delta S°_{298} = 28.39$ cal/K mol for the vaporization of water. If, as in Chapter 11, we assume that $\Delta H°$ and $\Delta S°$ are nearly independent of temperature, we can use the condition $\Delta G° = \Delta H° - T\Delta S° = 0$ at $T = T_{ta}$ to find $T_{ta} = T_b \approx \Delta H°/\Delta S° = (10520$ cal$)/(28.39$ cal/K$) = 370.5$ K $= 97.4°C$. The result is not exactly 100°C because of the slight T dependence of $\Delta H°$ and $\Delta S°$.

Interpreting T_f and T_b as turnaround temperatures explains why freezing and boiling occur at these sharply defined temperatures, because only at some unique temperature will the free energies of the initial and final phase be exactly equal, for example, $G°(l) = G°(g)$, as implied by $\Delta G° = 0$. If we combine the information for water on H from Figure 14.3 and S from Figure 14.4, we can form $G = H - TS$ for each phase and graph its dependence on T at standard pressure, as is done in Figure 14.5. The slope of G versus T for each phase is $-S$ for that phase, and since S is always positive, all the G curves must decline with increasing T. For solids, S is very small, and $G(T)$ is nearly flat. The liquid phase is somewhat more entropic, and shows a more rapid falloff, while the enormous entropy of the gas phase causes $G(T)$ to decrease sharply. Further, because $G = H$ at 0 K, the three free-energy curves originate at successively higher energy for solid, liquid, and gas. Thus, as T increases, first the gradually falling $G(l)$ curve cuts through the nearly flat $G(s)$, and in turn the steeply decreasing $G(g)$ curve cuts through $G(l)$. These intersection points, where the free energies of the pairs of phases (s, l) and (l, g) are equal, define the transition temperatures.

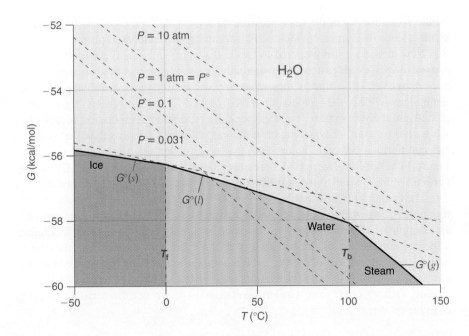

Figure 14.5
The molar Gibbs free energy G of the three phases of water as a function of temperature. The solid curve represents the standard free energy $G°$ of the phase with lowest free energy at any T, that is, the most stable phase at that T. The dashed curves show the behavior of G for the phases that are not the most stable, or, for the vapor (gas) phase, under nonstandard pressure, as indicated next to each curve. Freezing and boiling are the result of an intersection of G curves for the two phases involved in the transition. Nonnormal boiling points result from the intersection of $G(g)$ at nonstandard pressure with $G(l)$; for $P = 0.031$ atm the intersection of $G(g)$ with $G(l)$, and hence the boiling point, occurs at 25°C.

What happens if we depart from standard pressure? You may have heard that it's tough to hard-boil an egg in the Rocky Mountains; this is due to the lower-than-standard pressure at those high elevations, which *causes water to boil at a lower temperature*. On the other hand, in a pressure-cooker, the water and steam are sealed in, creating a higher pressure and causing the water to boil at a *higher* temperature. These observations point to a general fact: the boiling point of water is not an immutable constant, but depends on the external pressure. We associate boiling with the spontaneous formation of bubbles of water vapor that burst at the surface of the liquid and release their contents. The bubbles could not form unless the pressure of water vapor inside them were equal to, or a little greater than, the opposing pressure of the surroundings—usually the air—for the same reason that a gas cannot expand against the atmosphere unless its pressure exceeds atmospheric. That is, boiling requires $P_{H_2O} \geq P_{ext}$, with the equals sign implying equilibrium; water boils at 100°C only at standard pressure $P_{H_2O} = P_{ext} = 1$ atm. The variable boiling point corresponds to a varying point of intersection between $G(l)$ and $G(g)$; as illustrated in Figure 14.5, this comes about because $G(g)$ depends on pressure, a fact we have already exploited in Chapter 12 in analyzing gas equilibrium.

Making use of our analysis of the pressure dependence of G for a gas from Chapter 12, specifically Equation 12.3, we can say that if $G°$ is the vapor free energy at standard pressure $P°$, then G at other pressures is given by

$$G(g) = G°(g) + RT \ln \frac{P}{P°} \qquad (14.4)$$

If atmosphere units are used, $P°$ can be omitted. Note that P here is the pressure of the vapor itself rather than the external pressure; when boiling occurs, $P = P_{ext}$. Equation 14.4 allows us to find the $G(g)$ versus T curve at other pressures, as

Figure 14.5 shows. For the same pressures shown in the figure, $G(l)$ and $G(s)$ barely change, since they represent the nearly incompressible condensed phases.

Now at last we are prepared to tackle that most important and most subtle of phase changes, **evaporation.** After you bathe or shower, your bathtub, towel, etc., are all wet. Yet you know that by tomorrow they will be dry again. Surely the water did not boil away, since the temperature never reached T_b. But nonetheless the moisture passed into the vapor phase, a process we call evaporation. Evaporation occurs when the *partial pressure* of water vapor in the gas phase, P_{H_2O}, is less than that required for equilibrium between liquid and gas at a given temperature. Referring to Figure 14.5, we can always find a pressure that will make $G(g)$ intersect $G(l)$ at any chosen T; that P is called the **equilibrium vapor pressure** of the liquid at that T.

The evaporation of water $H_2O(l) \longrightarrow H_2O(g)$ is such an important and instructive process that we have already had occasion to use it twice, first in Chapter 11 as an example of an endothermic yet spontaneous process, and then again in Chapter 12 to develop the rule that says we may omit any pure condensed phase from the reaction quotient. As we found there, the reaction quotient for evaporation is simply $Q = P_{vap}$, where P_{vap} is the partial pressure of the vapor of interest, and by subtracting $G(l) \approx G°(l)$ from $G(g)$ of Equation 14.4, we get the usual free-energy change:

$$\Delta G_{vap} = \Delta G°_{vap} + RT \ln P_{vap} \qquad (14.5)$$

This equation holds for the evaporation of any liquid, by substituting the appropriate $\Delta G°$ and vapor pressure values. For water, the standard free energies of formation for $H_2O(l)$ and $H_2O(g)$ from Appendix C yield $\Delta G°_{vap,298} = +2.053$ kcal/mol. The positive value means that the evaporation of water is nonspontaneous *under standard conditions*, that is, $P_{H_2O} = 1$ atm. At equilibrium, $\Delta G_{vap} = G(g) - G(l) = 0$, and, as in Chapter 12, Equations 12.7 and 12.8, we get

$$\Delta G° = -RT \ln P_{eq} \qquad \text{or} \qquad P_{eq} = P° \, e^{-\Delta G°/RT} \qquad (14.6)$$

The role of equilibrium constant is played by the equilibrium vapor pressure P_{eq} itself; at 298 K for water vapor, we find $P_{eq} = 0.0313$ atm, or 24 Torr. This number means that if a sample of liquid water is allowed to come to equilibrium with its vapor at 298 K (which could be achieved by tightly sealing a bottle or jar partially filled with water at 298 K and waiting a short time), it will exert a pressure of 24 Torr. Note that the external pressure does not enter the calculation; it does not matter that 1 atm of air is also trapped in the bottle. The only effect of the air is to prevent boiling; evaporation will continue, and at equilibrium, the rate of evaporation will exactly match the rate of condensation. As for chemical equilibrium in general, phase equilibrium implies not a cessation of molecular happenings, but a balancing of forward and reverse rates.

The fact that puddles and wet towels will evaporate to dryness must imply, then, that the partial pressure of water vapor in the air is usually *less* than its equilibrium value. If we write the free-energy pressure dependence as we did in Equation 12.9, we obtain

$$\Delta G_{vap} = RT \ln \frac{P_{vap}}{P_{eq}} \qquad (14.7)$$

from which it is plain that $P_{vap} < P_{eq}$ will make $\Delta G_{vap} < 0$, implying spontaneous evaporation. Meteorologists use a *relative humidity* (RH) index, defined as RH = $P_{H_2O}/P_{eq} \times 100\%$, to report the amount of water vapor in the air. From the RH, you can compute ΔG_{vap} and assess the degree of spontaneity of evaporation. Typically RH approaches 100% (i.e., liquid–vapor equilibrium is reached) only during a rainstorm or in dense fog—or in the shower!

We already know that in general equilibrium constants K show a strong temperature dependence, and we therefore expect that P_{eq}, our K for evaporation, will as well. Recall from Chapter 12, Equation 12.14, that if we regard ΔH°_{vap} and ΔS°_{vap} as constant, independent of T, we can obtain

$$\ln \frac{P_2}{P_1} = -\frac{\Delta H^\circ_{vap}}{R}\left(\frac{1}{T_2} - \frac{1}{T_1}\right) \quad \text{or} \quad \frac{d \ln P}{d(1/T)} = -\frac{\Delta H^\circ_{vap}}{R} \quad (14.8)$$

Here P_1 and P_2 are the equilibrium vapor pressures at temperatures T_1 and T_2, respectively. These relations are special cases of the more general van't Hoff equation (Equations 12.15 and 12.16); for vapor pressures they are equivalent forms of the **Clausius–Clapeyron equation.** Like the van't Hoff equations, Equations 14.8 lead to a downward-sloping straight line when $\ln P_{eq}$ is plotted against $1/T$, with slope $-\Delta H^\circ_{vap}/R$. It then becomes possible to measure ΔH° *without* doing thermochemistry, if we can measure $P_{eq}(T)$, just as we found for chemical equilibrium in general in Chapter 12.

For phase transitions it is customary simply to plot P_{eq} versus T; from Equation 14.6 we obtain

$$P = P^\circ\, e^{-\Delta H^\circ_{vap}/RT} e^{\Delta S^\circ_{vap}/R} \quad (14.9)$$

This yields a curve that is increasing exponentially with T when T is low enough that $\Delta H^\circ_{vap} \gg RT$. Figure 14.6 is a plot of P_{eq} versus T for water and three structurally similar common compounds: isobutane, $(CH_3)_2CHCH_3$, a component of lighter fluid; acetone, $(CH_3)_2C{=}O$, found in paint thinners and strippers, and isopropanol, $(CH_3)_2CHOH$, the major ingredient in rubbing alcohol. These curves describe the locus of points of equilibrium between liquid and vapor; you can also think of them as

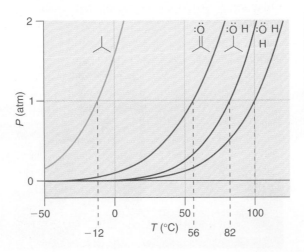

Figure 14.6

Equilibrium vapor pressure versus temperature for water and three structurally similar organic compounds from the left, isobutane, acetone, and isopropyl alcohol. The stick figures beside each curve are semistructural depictions, which represent carbon atoms by a vertex or terminus, and omit hydrogens bonded to carbon.

mapping out all possible boiling points for each liquid. The **normal boiling point** T_b of the liquid is the point on the curve where $P_{eq} = 1$ atm, as we reasoned earlier. For isobutane, $T_b < 25°C$, and hence it exists as a gas under normal conditions; however, isobutane can be stored in a lighter case as a liquid, as long as the pressure inside the case exceeds its vapor pressure at 25°C. Moreover, water and the other liquids can be made to *boil* at any temperature above their freezing points by adjusting the external pressure to equal their equilibrium vapor pressure at that temperature, as we discussed at the outset, and as Figure 14.7 illustrates for water. A liquid, however, will *evaporate* at any external pressure, up to its equilibrium vapor pressure at the prevailing temperature. In Section 14.4 we discuss why the curves of Figure 14.6 lie where they do in terms of the intermolecular attractions in these compounds.

EXAMPLE 14.4

Estimate the minimum pressure at which isobutane will maintain a liquid phase at 25°C, given its normal boiling point, $-12°C$, and heat of vaporization, $\Delta H^°_{vap} = 5.35$ kcal/mol.

Solution:

The minimum P needed must be the equilibrium vapor pressure P_{eq} of isobutane at 25°C; at any $P < P_{eq}$, the liquid phase will boil (if $P = P_{ext}$) or evaporate (if $P = P_{vap}$) to completion. The information given allows us to use the Clausius–Clapeyron equation (Equation 14.8) in its integrated form, taking (T_1, P_1) to be the normal boiling point (1.00 atm, $-12°C$):

$$\ln \frac{P_2}{1.00 \text{ atm}} = -\frac{5350 \text{ cal/mol}}{1.99 \text{ cal/K mol}} \left(\frac{1}{298 \text{ K}} - \frac{1}{261 \text{ K}} \right)$$
$$= +1.28$$
$$P_2 = 3.59 \text{ atm}$$

This is the pressure that would exist inside a sealed container (such as a lighter) containing pure isobutane at equilibrium at 25°C. If the container were opened to the atmosphere, the liquid would boil away completely. As we shall see later, the liquid phase can be maintained only if $T < T_c$, where T_c is the *critical temperature*. For isobutane, $T_c = 135°C$, making our calculation valid here.

Figure 14.7
Boiling water under different conditions of temperature and pressure. On the left, water is boiled in the normal way by heating an open container of liquid under atmospheric pressure. On the right, a cold-water aspirator is used to remove the air above the liquid, thereby reducing the pressure below the equilibrium vapor pressure of the liquid at ambient temperature and allowing normal evaporation to become boiling.

There is an old saying you may have heard: "Warm air holds more moisture." Now you can understand why it is true. The temperature of the air is the temperature of the water vapor in it, and the higher T is, the higher P_{vap} can be before condensation will occur, as dictated by the curve for water in Figure 14.6. When warm air containing nearly the maximum amount (RH $\rightarrow$ 100%) of water vapor is cooled (for example, during the night), fog, condensed water droplets, will form if the temperature has dropped below that where $P_{vap} \leq P_{eq}$ (the so-called *dew-point* temperature, where RH = 100% and condensation becomes spontaneous). Humidity, fog, dew, and rain are all determined by that same P_{eq} versus T curve. What about snow, sleet, and hail? For those, we need to probe further, into the other major phase transformation, melting or fusion.

The Phase Diagram: Water

Below 0°C, the lowest free energy, most stable form of water is ice, as we saw in Figure 14.5; at 1 atm external pressure, 0°C (273.15 K) is the freezing, or fusion, point of water, the **normal freezing point** T_f. At this point the enthalpy jumps by $\Delta H°_{fus}$ as ice melts (or falls by $-\Delta H°_{fus}$ as water freezes) as shown in Figure 14.3; but the free energies are equal, $G°(l) = G°(s)$, or $\Delta G°_{fus} = 0$, as shown in Figure 14.5. We expect the freezing point to be essentially independent of the pressure, since neither liquid nor solid phases are very compressible, that is, $G(l)$ and $G(s)$ are nearly pressure independent. In this limit, the boundary between liquid and solid, the line on a P versus T graph that describes their equilibrium or coexistence, is a nearly vertical line that extends from the point where it strikes the equilibrium vapor pressure curve upward, as illustrated in Figure 14.8. The line cannot pass below the vapor pressure curve, since below that curve liquid water would spontaneously vaporize. What about solid water? Can it also pass into the vapor phase?

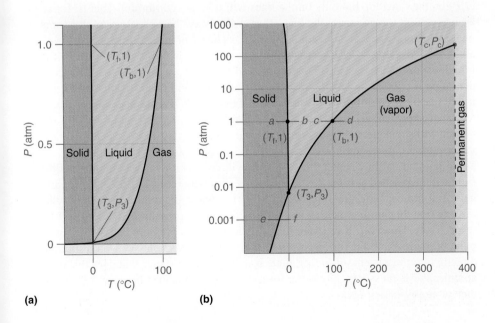

Figure 14.8
Two views of the phase diagram for water. The phases are separated by lines along which two adjacent phases are at equilibrium. Labeled points are the normal freezing point (T_f, 1), the normal boiling point (T_b, 1), the triple point (T_3, P_3), and the critical point (T_c, P_c). (a) A linear pressure scale is used to show the phase diagram in the freezing/boiling domain. (b) A logarithmic pressure scale allows behavior at very low and very high P to be included. Beyond the critical temperature T_c no phase transitions are possible, and H$_2$O becomes a permanent gas. The lines *a–b* and *c–d* represent the normal (isobaric) melting and boiling transitions, while the line *e–f* shows sublimation of ice at low T and P, conditions found at the outer edge of the earth's atmosphere known as the *mesosphere*.

Ice cubes stored for a long time in the freezer will appear shrunken; your author once tried to store a 4-cm-diameter hailstone in the freezer, and it disappeared in a few days, leaving only a smudge of dirt behind. In very cold climes, banks of snow gradually grow smaller even when the outdoor temperature remains below T_f. These observations are evidence that the process $H_2O(s) \longrightarrow H_2O(g)$, called **sublimation,** is occurring spontaneously. From a thermodynamic viewpoint, any process that can be written down may possibly occur; all we have to do is to compute the free-energy change for the process. In fact, the analysis we have carried through for the liquid $\longrightarrow$ vapor transition applies equally well to solid $\longrightarrow$ vapor. In Equations 14.8 and 14.9, we need only replace ΔH°_{vap} with ΔH°_{sub}. Hess's law says that the energy required for sublimation, $s \longrightarrow g$, should be the sum of those required for fusion, $s \longrightarrow l$, and vaporization, $l \longrightarrow g$, or

$$\Delta H^\circ_{sub} \approx \Delta H^\circ_{fus} + \Delta H^\circ_{vap} \tag{14.10}$$

This would be an exact relation except for the slight difference in T for the various processes. Thus ΔH°_{sub} is always greater than ΔH°_{vap}—it takes more energy to pull a molecule out of a solid lattice than to free it from a crawling liquid, as Figure 14.2 is intended to convey.

Right at the point where the solid–liquid and liquid–gas curves of Figure 14.8 meet, all three phases must be in equilibrium; we call this the **triple point,** with coordinates T_3, P_3. At this point, the vapor pressure exerted by the solid must be the same as that by the liquid, since $G(s) = G(l)$. At lower temperature the solid exerts a lower vapor pressure than the liquid would at the same T, because the greater heat needed to sublime the solid makes the exponential factor in Equation 14.9 smaller. So we finally arrive at a diagram, called the **phase diagram,** shown in Figure 14.8 for water, in which all possible phase equilibria are indicated by curves relating the pressure and temperature at which at least two free energies become equal.

A system consisting of one pure substance can in principle exist at any point (T,P) on the diagram, but only on the lines will it be at equilibrium. That is, for an arbitrary point not on one of the lines, there will always be a negative ΔG for some phase transition to occur, until equilibrium is reached. For instance, if we have steam at 25°C and 1 atm $(T,P) = (298, 1)$, it will spontaneously move toward the liquid–vapor line, that is, it will condense until $P_{vap} = P_{eq}$, a foggy event. Since condensation is exothermic, the temperature of the steam will rise as it condenses, and it will reach the equilibrium line at some T above 298 K. Far more common, if water vapor is present at, say, 0.016 atm (12 Torr) and 298 K, any liquid water in the system will spontaneously evaporate, to dryness if necessary, in striving to bring the vapor pressure to its equilibrium value of 24 Torr. Evaporation is endothermic, so, as we pointed out in Chapter 12, the water will cool as it evaporates, and equilibrium will be reached at some $T < 298$ K. If we start with ice at $T < 0°C$ (273 K) and heat it at atmospheric pressure, we traverse the phase diagram horizontally, witnessing first the familiar melting and then boiling as we pass through 0° and 100°C, respectively. However, if we were to do the heating at a pressure below the triple-point pressure $P_3 = 4.6$ Torr, *we would not see the ice melt!* The ice would sublime directly into the gas phase! In fact, there is a common substance that does this trick at atmospheric pressure: CO_2 in the form of *dry ice,* so named because it does not melt at atmospheric pressure.

There is one other important point on the phase diagram that we normally don't witness for water: the so-called **critical point** (T_c, P_c). This point is the terminus of the liquid–vapor phase line; at temperatures greater than T_c, there are no longer separate liquid and vapor phases, and the substance becomes a **permanent gas.** Therefore T_c is the highest possible boiling point a substance can show. As illustrated in Figure 14.8, the critical point occurs for water at $T_c = 374°C$ (647 K) and $P_c = 218$ atm, conditions not easily reached except with specialized equipment. ("Critical water" has unique solvent properties that only recently have begun to be explored by chemists.) Because the liquid phase ceases to exist for $T > T_c$, the heat associated with the $l \longrightarrow g$ transition must become zero at $T = T_c$, that is, $\Delta H°_{vap,T_c} = 0$. This is consistent with what we know about the general behavior of $\Delta H°_{vap,T}$; applying Equation 10.36 to evaporation (as we already did in Chapter 10), we find

$$\Delta H°_{vap,T} \approx \Delta H°_{vap,T_b} + [\overline{C}_P(g) - \overline{C}_P(l)](T - T_b) \qquad (14.11)$$

Because $\overline{C}_P(g)$ is nearly always less than $\overline{C}_P(l)$, due to loss of heat storage by the intermolecular potential energy when liquid vaporizes, $\Delta H°_{vap}$ is generally a *decreasing* function of temperature. The decrease is more rapid than linear, however, when T is near T_c, and Equation 14.11 breaks down.

Phase Behavior in General

Every stable substance has its own phase diagram, with a characteristic set of points $(T_f, 1)$, $(T_b, 1)$, (T_3, P_3), and (T_c, P_c); there are many more compounds that decompose when they are heated, prior to reaching T_c, or often even T_b. If you take a look in the *Handbook of Chemistry and Physics* at pages showing physical constants of organic or inorganic compounds, you will find relatively few boiling points. Still other compounds spontaneously burn when heated in air instead of undergoing phase changes; you never find logs melting in the fireplace. These are examples of chemically favorable free energies taking precedence over the physical free energy of a change in phase; if the turnaround temperature for a decomposition occurs prior to a phase-change temperature, the former will occur preferentially. Nonetheless there exists a variety of important compounds that exhibit behavior qualitatively similar to that of water; Table 9.3 of Chapter 9 introduced you to some properties of compounds that are gases at NTP. Further phase data on some of these, as well as substances that are liquids and solids at NTP, are collected in Table 14.1.

An example where behavior uncommon in water is readily observed is carbon dioxide (CO_2), whose phase diagram is displayed in Figure 14.9. The triple-point pressure P_3 occurs above 1 atm, implying that CO_2 does not display a liquid phase at 1 atm, subliming directly to the gas at $-78°C$. Moreover, its critical point occurs at $T_c = 31°C$, making behavior near this point easier to witness. A sealed tube of pure CO_2 (no air) will be a liquid at 25°C, since it is free to come to equilibrium with its vapor at a much higher pressure. As T_c is approached, the sample will follow the liquid $\longrightarrow$ gas equilibrium line until T_c is reached, during which the meniscus separating liquid and vapor phases gradually rises to the top of the tube and disappears altogether at T_c. Above T_c, CO_2 is a permanent gas. As given in Table 14.1, other common gases have very low T_c's and are therefore permanent at NTP. Michael Faraday was the first to liquefy Cl_2 by pressurizing it, but was unsuccessful with other gases;

TABLE 14.1
Phase properties of selected elements and compounds[a]

	$\overline{C}_P(s)$ (cal/K mol)	$\rho(s)$ (g/cm³)	T_f (K)	$\Delta H°_{fus}$ (kcal/mol)	$\overline{C}_P(l)$ (cal/K mol)	$\rho(l)$ (g/cm³)	T_b (K)	$\Delta H°_{vap}$ (kcal/mol)	$\overline{C}_P(g)$ (cal/K mol)	T_3 (K)	P_3 (atm)	T_c (K)	P_c (atm)
He	⋯	0.198	⋯	⋯	3.9	0.147	4.2	0.02	5.0	2.2[b]	0.050[b]	5.2	2.24
Ne	6.5	1.43	24.6	0.08	8.7	1.25	27.1	0.43	5.0	24.6	0.43	44.4	27.2
Ar	8.4	1.62	83.8	0.27	10.7	1.40	87.3	1.61	5.0	83.8	0.68	150.9	48.3
Kr	8.6	2.79	115.8	0.38	11.1	2.44	119.9	2.17	5.0	115.8	0.72	209.4	54.3
Xe	8.5	3.39	161.4	0.52	11.0	2.97	165.1	3.02	5.0	161.4	0.81	289.7	57.6
H_2	1.4	0.081	13.8	0.03	4.6	0.070	20.3	0.22	6.9	13.8	0.069	33.0	12.8
F_2	17.3	1.3	53.5	0.12	13.7	1.108	85.0	1.62	7.5	53.5	0.0022	144.1	51.0
Cl_2	13.4	1.9	172.2	1.53	15.6	1.367	238.6	4.86	8.1	172.	0.014	416.9	78.9
Br_2	11.2	3.4	265.9	2.52	18.1	3.119	331.9	7.07	8.6	265.6	0.057	588.	102.0
I_2	13.0	4.93	386.7	3.77	19.4	4.0	457.5	10.03	8.8	386.7	0.12	819.	⋯
HF	⋯	⋯	189.8	0.26	14.6	0.987	293.	6.19	7.0	189.	0.004	461.	63.9
HCl	12.1	⋯	159.0	0.33	14.4	1.187	188.	3.86	7.0	159.3	0.14	324.7	82.0
HBr	11.4	⋯	184.7	0.48	15.5	2.77	206.8	4.44	7.0	185.	0.29	363.2	84.4
HI	11.4	⋯	222.4	0.77	15.3	2.85	237.6	4.98	7.0	219.3	0.42	424.	82.0
O_2	11.0	1.36	54.4	0.11	13.0	1.14	90.2	1.63	7.0	54.4	0.0015	154.6	49.8
O_3	⋯	⋯	80.3	⋯	⋯	1.614	161.8	3.37	9.4	80.	6×10^{-6}	261.1	55.0
$S_{(rhomb)}$	5.4	2.07	388.4	0.41	7.3	1.82	717.8	4.42	5.7	393.	0.0006	1314.	204.2
SO_2	⋯	⋯	197.6	0.54	20.8	1.435	263.1	5.96	9.5	197.7	0.017	430.8	77.8
SO_3	⋯	⋯	290.0	1.90	65.0	1.88	317.9	9.72	12.1	290.0	0.21	⋯	⋯
H_2O	9.0	0.917	273.15	1.436	18.0	1.000	373.15	9.716	8.6	273.16	0.0060	647.1	217.7
H_2S	⋯	⋯	187.6	5.22	16.2	0.96	213.6	4.67	8.2	185.3	0.19	373.2	88.2
H_2Se	⋯	⋯	207.4	⋯	⋯	2.12	231.9	4.71	⋯	⋯	⋯	411.	88.0
H_2Te	⋯	⋯	224.	1.67	13.8	2.57	271.	4.59	⋯	⋯	⋯	⋯	⋯
N_2	11.0	1.027	63.2	0.17	⋯	0.804	77.4	1.33	7.0	63.2	0.125	126.2	33.5
$P_{(white)}$	5.7	2.69	317.3	0.16	⋯	1.745	550.	2.96	16.1	⋯	⋯	994.	⋯
N_2O	⋯	⋯	182.3	1.56	18.5	1.226	184.7	3.95	9.2	182.3	0.87	309.6	71.6
NO	8.7	⋯	109.5	0.55	15.0	1.269	121.4	3.31	8.9	109.0	0.20	180.	64.0

Compound															
NO$_2$	...	...	...	261.9	3.50	...	16.4	1.49	294.3	9.11	9.2	261.9	0.18	431.	99.7
NH$_3$	11.7	...	...	195.4	1.35	...	18.0	0.682	239.8	5.70	8.4	195.4	0.060	405.5	112.0
PH$_3$	...	...	...	140.	2.00	...	...	0.746	185.4	3.49	8.9	138.7	0.034	324.5	64.6
AsH$_3$	...	...	...	157.	...	...	...	1.689	210.6	3.99	9.1	...	...	373.1	...
SbH$_3$	...	...	...	185.	...	...	...	2.26	256.1	5.09	9.8	...	...	...	...
C(*gr*)	2.0	2.03	4098.^{c}	25.0^c	...	...	...	...	...	...	5.0	4247.	99.7	...	...
C(*dia*)	1.5	3.17	...	...	...	...	...	...	...	...	...	4150.	120	...	...
Si	4.8	2.33	1687.	12.00	...	2.53	3538.	...	...	...	5.3	...	...	...	...
Ge	5.6	5.32	1211.4	8.83	...	...	3106.	...	...	79.8	7.3	...	...	...	...
Pb	6.3	11.3	600.6	1.14	...	10.8	2022.	...	...	42.9	5.0	...	...	...	...
CO	12.3	0.929	68.1	0.22	14.4	0.793	81.6	1.52	...	...	7.0	68.1	0.15	132.9	34.5
CO$_2$	13.0	1.56	194.7^c	5.9^c	18.8	1.101	...	...	...	3.99	8.9	216.6	5.15	304.1	72.8
CS$_2$	...	1.554	161.6	1.05	18.3	1.256	319.	...	...	6.39	10.9	...	...	552.	78.0
SiO$_2$	10.6	2.65	1883.	2.03	...	...	2503.	...	...	...	...	...	...	...	...
CH$_4$	10.2	0.52	90.7	0.29	12.8	0.44	111.7	...	...	1.96	8.4	90.7	0.11	190.5	45.4
SiH$_4$	...	...	88.	...	...	0.68	161.	...	...	2.89	10.2	88.	0.00007	269.7	48.4
GeH$_4$	...	...	108.	...	...	1.523	183.	...	...	3.36	10.8	...	...	312.2	48.9
SnH$_4$	...	...	123.	...	...	...	...	...	221.3	4.55	11.7	...	...	...	...
C$_2$H$_2$	...	...	189.6^c	5.22^c	...	...	...	...	...	4.08	10.5	192.8	1.27	308.3	60.6
C$_2$H$_4$	...	...	104.	...	...	...	169.4	...	...	4.17	10.4	104.	0.0011	282.3	49.8
C$_2$H$_6$	...	...	90.3	0.68	17.2	0.55	184.5	...	...	3.51	12.6	90.3	0.00001	305.4	48.2
C$_3$H$_8$	...	...	85.5	0.84	23.4	0.59	231.0	...	...	4.55	17.4	85.5	9×10^{-9}	369.8	42.0
n-C$_4$H$_{10}$	...	...	134.9	1.11	34.2	0.60	272.6	...	...	5.70	23.3	134.9	9×10^{-6}	425.1	37.4
i-C$_4$H$_{10}$	...	...	134.8	1.11	34.1	0.55	261.2	...	...	5.35	23.1	134.8	0.00004	407.9	35.8
n-C$_5$H$_{12}$	...	...	143.4	2.01	40.0	0.621	309.2	...	...	6.16	26.6	143.4	2×10^{-6}	469.7	33.2
n-C$_6$H$_{14}$	...	...	177.8	3.13	46.7	0.655	341.9	...	...	6.90	31.6	...	...	507.7	29.7
n-C$_7$H$_{16}$	...	...	182.5	3.38	53.7	0.680	371.6	...	...	7.59	36.5	...	...	540.3	27.2
n-C$_8$H$_{18}$	...	...	216.3	4.94	60.9	0.699	398.8	...	...	8.22	41.5	...	...	568.9	24.6
i-C$_8$H$_{18}$	...	...	165.8	2.16	57.1	0.688	372.4	...	...	7.36	35.2	...	...	544.0	25.4
C$_6$H$_6$	...	...	278.7	2.38	32.5	0.874	353.2	...	...	7.34	19.7	278.5	...	407.9	35.8
C$_6$H$_5$CH$_3$	...	...	178.2	1.64	37.6	0.862	383.8	...	...	7.93	24.9	...	...	591.8	40.5

(*continued*)

TABLE 14.1
Phase properties of selected elements and compounds[a] (continued)

	$\overline{C}_P(s)$ (cal/K mol)	$\rho(s)$ (g/cm^3)	T_f (K)	$\Delta H°_{fus}$ (kcal/mol)	$\overline{C}_P(l)$ (cal/K mol)	$\rho(l)$ (g/cm^3)	T_b (K)	$\Delta H°_{vap}$ (kcal/mol)	$\overline{C}_P(g)$ (cal/K mol)	T_3 (K)	P_3 (atm)	T_c (K)	P_c (atm)
CH$_3$F	...	...	131.4	...	...	0.843	194.8	3.99	9.0	...	...	317.8	58.0
CH$_3$Cl	...	...	176.1	1.54	...	0.916	249.0	5.11	9.8	...	...	416.2	65.9
CH$_3$Br	...	...	179.6	1.43	...	1.676	276.7	5.72	10.1	...	...	...	...
CH$_3$I	...	...	206.8	...	30.1	2.279	315.6	6.53	10.5	...	...	528.	...
CH$_3$OH	...	...	175.5	0.76	19.4	0.787	337.7	8.42	10.5	175.5	7. × 10^{-6}	512.6	79.9
CH$_3$NH$_2$	...	...	179.7	1.47	24.4	0.699	266.9	6.12	12.0	...	...	430.7	75.2
CH$_2$Cl$_2$	...	...	178.0	1.43	24.2	1.317	313.	6.71	12.2	...	...	510.	60.2
C$_2$H$_5$OH	...	...	159.0	1.20	26.8	0.785	351.4	9.22	15.6	...	...	513.9	60.5
i-C$_3$H$_7$OH	...	...	183.6	1.28	37.4	0.781	355.4	9.52	21.3	...	...	508.3	47.0
(CH$_3$)$_2$O	...	...	131.6	1.18	...	...	248.3	5.14	15.4	...	...	400.0	53.0
(C$_2$H$_5$)$_2$O	...	...	156.8	1.74	42.0	0.708	307.6	6.34	28.6	...	...	466.7	35.9
CH$_3$CO$_2$H	...	...	289.7	2.76	29.5	1.044	391.0	5.66	15.9	289.7	0.013	592.7	57.1
CH$_3$CN	...	...	229.3	1.95	21.8	0.777	354.8	7.11	12.5	...	...	545.5	47.9
(CH$_3$)$_2$CO	...	...	178.3	1.36	30.2	0.784	329.2	6.96	17.9	...	...	508.1	46.4
(CH$_3$)$_2$SO$_{(DMSO)}$			291.7	3.43	36.6	1.096	462.	10.31	...	...	...	...	...
C$_3$H$_5$(OH)$_{3(glycerol)}$			291.3	4.37	52.3	1.257	563.	14.6	...	...	...	...	...
(CH$_3$)$_2$NCHO$_{(DMF)}$			212.7	3.86	36.0	0.945	426.	9.19	...	...	...	649.6	...
C$_4$H$_8$O$_{(THF)}$	...	...	164.8	2.04	29.6	0.880	338.	7.12	18.2	...	...	540.1	51.2
B	2.7	2.34	2348.	12.00	...	...	4273.	114.7	5.0				
Al	5.8	2.70	933.5	2.66	...	2.40	2792.	70.3	5.1				
B$_2$O$_3$	15.0	2.46	318.	5.75	15.0	...	2130.	100.0	16.0				
Al$_2$O$_3$	18.9	3.97	2327.	26.6	...	...	...	...	...				
B$_2$H$_6$	...	...	107.6	...	...	...	185.6	3.41	13.6			289.8	40.0
Zn	6.1	7.14	692.7	1.75	...	6.58	1180.	29.	5.0				
Hg	6.7	14.19	234.3	0.55	6.7	13.534	629.9	14.13	5.0	234.3	4. × 10^{-9}	1765.	1490.
Cu	5.8	8.96	1357.8	3.17	6.5	7.99	2835.	75.	5.0				
Ag	6.1	10.5	1234.9	2.70	8.1	9.32	2435.	63.	5.0				

Au	6.1	19.3	1337.3	3.00	6.5	17.3	3129.	77.4	5.0	...	...	...
Ni	6.1	8.90	1728.3	4.18	7.9	7.81	3186.	95.	5.6	...	...	...
Fe	6.0	7.87	1811.	3.30	...	7.13	3134.	90.	6.1	...	...	...
Mn	6.3	7.3	1519.	3.09	...	5.4	2334.	58.	5.0	...	...	...
Cr	5.6	7.15	2180.	3.14	...	6.46	2944.	...	5.0	...	...	...
W	5.8	19.3	3695.	12.50	...	17.6	5828.	180.	5.1	...	...	...
V	6.0	6.0	2183.	5.14	...	5.5	3680.	114.	6.2	...	...	...
Ti	6.0	4.5	1941.	3.38	...	4.1	4560.	104.	5.9	...	...	...
Sc	6.1	2.99	1814.	3.37	...	...	2830.	81.	5.3	...	...	...
Mg	6.0	1.74	923.	2.03	...	1.58	1363.	31.5	5.0	...	...	...
Ca	6.2	1.54	1115.	2.04	...	1.37	1757.	37.7	5.0	...	...	...
CaO	10.0	3.34	2887.	14.10	...	...	3120.					
$CaCl_2$	17.4	2.22	1045.	6.82	...	...	2208.6					
Li	5.9	0.534	453.6	0.72	7.6	0.51	1615.	32.2	5.0	...	3223.	680.
Na	6.7	0.97	370.9	0.62	7.6	0.93	1156.	21.4	5.0	370.7	2083.	253.
K	7.1	0.89	336.5	0.55	7.6	0.83	1032.	18.4	5.0	336.1	2220.	162.
Rb	7.4	1.53	312.5	0.55	7.6	1.39	961.	16.5	5.0	312.	2083.	144.
Cs	7.7	1.93	301.6	0.50	...	1.86	944.	15.8	5.0	301.8	2051.	116.
LiCl	11.5	2.07	883.	4.76	...	1.49	1656.	35.9				
NaCl	12.0	2.17	1073.8	5.28	...	1.55	1738.	42.6				
KCl	12.3	1.99	1044.	6.34	...	1.52	1770.	40.2				
NaF	11.2	2.17	1269.	7.97	...	1.94	1977.	54.8				
NaBr	12.3	3.20	1020.	6.24	...	2.33	1660.	38.1				
NaI	12.4	3.67	933.	5.64	...	2.74	1577.	38.1				

[a] $\overline{C}_P$ and ρ values for condensed phases refer to 298 K, unless that condensed phase is not stable at 298 K; in that case they refer to T_f. $\Delta H°$ applies at the corresponding transition temperature. $\overline{C}_P$, ρ, and $\Delta H°_{vap}$ are all T dependent, more so at very low T.
[b] Superfluid/liquid/vapor triple point.
[c] Sublimation; $\Delta H°_{vap}$ in this case refers to T_3.

he came to realize that generally liquefaction could occur only if the gas were cooled well below ambient T. James Dewar was the first to liquefy the other gases—except helium—in the 1880s, by cooling them first below their T_c. Helium was not liquefied until 1908, by Heike Kamerlingh-Onnes; its T_c is only 5.2 K! Helium is also the only substance that does not freeze at $P = 1$ atm; instead, at 2.2 K it undergoes a transition to a new state of He(l) called a *superfluid*.

A variable we have so far ignored in these phase diagrams is the molar volume $\overline{V}$. For a permanent gas, the ideal gas law can be used to obtain $\overline{V} = RT/P$; for constant T this yields the hyperbolic isotherms $P = RT/\overline{V}$ we have already seen in Chapters 9 and 11. As the critical temperature T_c is approached, however, the isotherms begin to take on a peculiar shape, as illustrated for CO_2 in Figure 14.9, in anticipation of the large change in $\overline{V}$ that must occur when condensation to a liquid becomes possible. The ideal gas equation can no longer be used, and must be replaced by a more realistic equation of state, such as the van der Waals equation introduced in Chapter 9. When the P-T phase diagram and P-V isotherms of CO_2 are combined, we obtain a two-dimensional surface $P(T,V)$ like that also shown in Figure 14.9. Assuming the equation of state of the substance has been found, $\overline{V}$ is determined once P and T are specified.

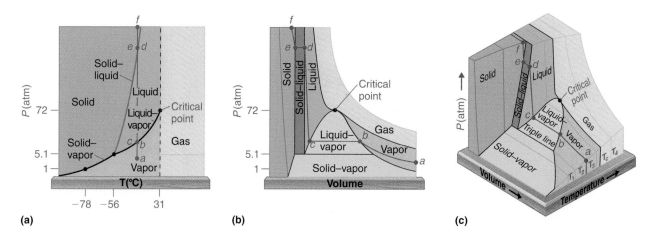

Figure 14.9

Phase diagrams for carbon dioxide (CO_2). Pressure and temperature scales are nonuniform, schematic only. **(a)** The normal P versus T phase diagram. CO_2 is one of a small number of substances whose triple point lies above 1 atm; therefore CO_2 does not display normal melting and boiling points, but only a single sublimation point at $-78°C$. In addition, its critical point occurs at the readily accessible temperature of 31°C. The line af depicts what happens when a CO_2 sample is compressed at constant T, where T lies between T_3 and T_c: first condensation (point bc) and then freezing (point de). Freezing occurs because solid CO_2 (dry ice) is denser than liquid CO_2, making the liquid–solid equilibrium line slope to the right; freezing does not occur for H_2O under these conditions. **(b)** P versus V phase diagram for CO_2; each phase equilibrium point on the P-T diagram becomes a horizontal line on P versus V, due to the change in V (ΔV) that accompanies a phase change, as illustrated by the path $abcdef$ of (a). The phase transitions bc and de are shown by the heavier lines. The phase boundaries then "fold over" to produce equilibrium areas instead of lines; that for liquid–vapor has a maximum at the critical point. Away from the phase equilibrium curves, constant T defines an isotherm, which becomes that for an ideal gas (a graph of Boyle's law) as T increases beyond T_c. **(c)** If the features of (a) and (b) are combined, we obtain a phase surface $P(V,T)$ that summarizes all aspects of phase behavior for a pure substance; the line $abcdef$ is a constant-T cut through this surface.

The change in molar volume that accompanies a phase change in fact plays an important role in defining the shape of the phase equilibrium lines, as first shown by Clausius in 1848. By expressing the laws of thermodynamics in *differential* form,

(I) $$dE = dq - P\,dV \qquad (14.12)$$
(II) $$dS = dq/T$$

where *P-V* work and a reversible path are assumed, Clausius was able to show that, along a phase boundary

$$\frac{dP}{dT} = \frac{\Delta H}{T\Delta V} \qquad (14.13)$$

where ΔH and ΔV are the enthalpy and volume changes, respectively, for the phases involved in the equilibrium. This relation is sometimes called the *Clausius equation*. When one of the phases is vapor, $\Delta V \approx V_{gas}$, and Equation 14.8 is recovered under ideal gas conditions. Equation 14.13 applies directly to the solid–liquid (fusion) equilibrium; except for very low-temperature transitions, ΔV is nearly independent of T and P, being determined by the difference in density between solid and liquid. Equation 14.13 can then be integrated readily to obtain the form of the solid–liquid equilibrium line,

$$P = P° + \frac{\Delta H_{fus}}{T_f \Delta V_{fus}}(T - T_f) \qquad (14.14)$$

This is the equation of a straight line of very steep slope, since ΔV_{fus} is tiny. If the solid form is more dense than the liquid (the usual case), ΔV_{fus} will be positive, and the phase line will lean slightly to the right. For water, however, the liquid is more dense, making the line lean to the left, meaning that increased pressure will cause ice to melt at a lower temperature.

The form or *topology* of the phase diagram is quite general, as first shown by Gibbs in the two-part paper cited earlier. In that work Gibbs formulated an equilibrium relation called the **phase rule,** given by

$$f = c - p + 2 \qquad (14.15)$$

where c is the number of components (distinct chemical substances) present, p is the number of phases present, and f is the number of thermodynamic degrees of freedom, that is, the number of state variables that can be freely assigned to the system. We have been considering a single substance, $c = 1$. If there are two phases 1 and 2 in equilibrium, $p = 2$, then $f = 1 - 2 + 2 = 1$, and only one thermodynamic variable may be assigned without constraint. If we choose to specify T, then the pressure P is already determined; that is, there exists a curve $P(T)$ along which the equilibrium condition $G_1 = G_2$ holds. If there are three phases in equilibrium, $p = 3$, then $f = 0$ (that is, no variable can be freely assigned) and the equilibrium must be confined to a single point (T_3, P_3) which we call the triple point. Since any pair of phases chosen from the three is also at equilibrium, the three possible phase lines must meet at the triple point. The phase rule can be applied to mixtures, solutions, and multiphase chemical systems, as well as to pure substances that undergo solid–solid phase changes.

14 States of Matter and Intermolecular Forces

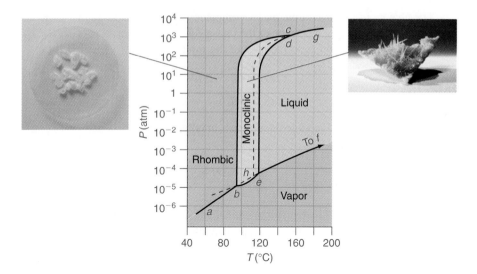

Figure 14.10
Phase diagram for elemental sulfur (S). Note the logarithmic pressure scale. The two crystalline allotropes of S give rise to an extra phase boundary line in the diagram, the leftmost vertical line, along which the two solid phases coexist. The dashed lines represent metastable states, which are encountered because of the sluggishness with which the solid forms interconvert. See the text for a description of sulfur's phase behavior.

Many solids show more than one crystalline form, particularly when cooled to very low T or subjected to very high P. An example of crystalline allotropy that occurs at ordinary temperature and pressure is solid sulfur, which can exist as either *rhombic* or *monoclinic* crystals. The phase diagram must then include transitions between these solid phases, as shown in Figure 14.10. The rhombic form of S is more stable at NTP, but when a rhombic sulfur crystal is heated to 95°C (line AB in Figure 14.10), it will spontaneously (but slowly) rearrange its atoms into the monoclinic form. This less stable allotrope then melts at 120°C (line BE) upon further heating. Sulfur actually exhibits *three* separate triple points, one involving the two solid forms and the vapor (point B), another at high pressure with the two solids and the liquid (C or D), and the usual (monoclinic) solid–liquid–vapor point (E). If the rhombic form is heated quickly through the solid–solid transition (the usual situation), it does not have time to recrystallize and simply melts at 114°C (point H). This latter melting is a nonequilibrium or *metastable* process, and corresponds to a *supercooled state* of liquid S. Likewise, if molten sulfur above 120°C is quickly cooled, it will freeze at 120°C into the monoclinic form, pass by the monoclinic ⟶ rhombic transition at 95°C without enough time to recrystallize, and end up as supercooled monoclinic sulfur at NTP. The two types of crystals are readily distinguished visually. The failure of sulfur to show reversible behavior when heated up and cooled back is called *hysteresis*.

Water and most other liquids can also show *supercooling* (liquid continuting to exist below the freezing point) as well as *superheating* (liquid above the boiling point). These both arise from the combination of relatively rapid cooling or heating and the lack of nucleation sites needed to initiate the phase change in a particular system, for example, water in a container with very smooth walls. Superheating can be particularly dangerous in the laboratory, because a superheated liquid, like someone with a pent-up sneeze, can explode in a sudden paroxysm of boiling. Substances consisting of larger molecules can more readily be supercooled into an amorphous (noncrystalline) nonequilibrium state called a "glass," by analogy with ordinary window glass, which has no definite crystal structure.

Solid–solid transitions that have attracted recent interest involve changing electrical and magnetic characteristics. Transitions from a normal to a *superconducting* state, for which there is no ΔH, involve a sudden drop in electrical resistance from "normal" to near-zero values at a "critical" temperature T_c. Kamerlingh-Onnes, who first liquefied helium, also discovered the first superconductor, solid mercury, for which T_c is 4.0 K. Prior to 1986, only certain metals and a number of alloys had been shown to exhibit superconductivity, with the highest T_c being about 23 K. Then binary and ternary metal oxides, usually containing Cu, were found to have much higher T_c's. The most famous of these is the so-called "1-2-3" superconductor $YBa_2Cu_3O_{7-x}$, for which $T_c = 90$ K, above the boiling point of nitrogen. This means that a cable or a printed-circuit board made with 1-2-3 and cooled by liquid nitrogen will conduct current with no resistance. The possible commercial applications are far-flung, from faster computers to magnetically levitated trains using superconducting electromagnets. Stimulated by these possibilities, the search for ever higher T_c goes on; the current record is 138 K. The research methods used, however, border on alchemy, since we as yet do not have a working theory for oxide superconductors. In addition, the oxides are ceramic materials, prone to fracture and not easily fabricated into wires. (In the most recent designs, the superconducting material is packed into hollow metal wires.)

14.4 Phase Transitions, Molecular Structure, and Intermolecular Forces

Though Gibbs' phase rule may fix the general form of the phase diagram for a pure substance, the actual magnitudes of the properties that specify it depend intimately on the structure of the atoms or molecules of which the substance consists. As discussed briefly earlier, the melting and boiling points of a compound reflect the strength of the intermolecular attraction; we further observe that the heat of vaporization ΔH°_{vap} is a direct measure of the depth of the well in the intermolecular potential curve, Figure 14.2. You might therefore expect a positive correlation between T_b and ΔH°_{vap}; if you examine those two columns in Table 14.1 (which were deliberately placed side by side) the correlation is strikingly evident, especially for homologous substances such as the sequence of hydrocarbons CH_4, C_2H_6, C_3H_8, ... In fact, for a large number of substances that are liquids in a range not too far from normal temperature, the *ratio* of transition heat to transition temperature is approximately constant,

$$\frac{\Delta H^\circ_{vap}}{T_b} \approx 22 \text{ cal/K mol} \qquad (14.16)$$

as first noted by F. Trouton in 1884 and now known as **Trouton's rule.** The constant, 22 cal/K mol, is nothing other than the entropy change ΔS°_{vap}, and what Trouton's rule is really saying is that, *for any transition from liquid to vapor, about the same increase in molecular chaos is involved.* In a liquid, molecules already have translational freedom, and when they vaporize, this freedom increases by about the same amount per molecule, and therefore per mole. A general statement such as Equation 14.16 will always have its exceptions, and these invariably prove to be instructive. Water shows a somewhat larger ΔS°_{vap} (26.0 cal/K mol), for reasons we are soon to discuss. Just as the variability in atomic structure of the elements gives rise to the variety of reactions and compounds we observe in chemistry, so the underlying molecular structure is responsible for the range of phase properties collected in Table 14.1.

Attractive Intermolecular Forces

To begin our examination of the cohesive forces that produce condensed phases, we will focus on nonbonded atoms, such as the noble gases, or covalently bonded small molecules, excluding metals and ionic salts. The latter are readily identified for the most part by melting and boiling points well above room temperature, often greater than 1000 K, where bonds of "chemical" strength are being broken. The forces holding the molecules of low-boiling liquids together are substantially weaker—as reflected in their small $\Delta H°_{vap}$'s—allowing us to regard the molecular constituents in such condensed phases as independent entities.

Maxwell (1860s) realized that there ought to be repulsive forces acting between any pair of molecules, as a result of his investigation of the gas property called *viscosity,* the resistance to gas flow. Later van der Waals (1873) formulated his real gas equation of state based on the recognition that molecules both attract and repel each other; the attraction was absolutely essential to allow a gas to condense into a liquid. But until the advent of quantum mechanics, the origins of neither the attractive nor the repulsive forces could be fully understood. What was required was a basic knowledge of the structure of atoms and molecules, their electronic wave functions, electrons represented by the standing waves you know as orbitals. Then most of the intermolecular forces, particularly the attractive ones, could be analyzed in terms of the *electrostatic* (coulombic) interactions between the charge clouds (squares of orbitals) comprising each molecule. These forces had been known from laboratory-scale experiments and analysis for more than a century, beginning with Coulomb's own experiments, as introduced in Chapter 1. They are all "long range" in nature, and their potential energies all have an inverse-power dependence on the intermolecular distance R, that is,

$$V(R) \propto \frac{1}{R^s} \tag{14.17}$$

where s can take on positive integer values beginning with $s = 1$, the coulomb interaction itself.

In more advanced texts, you can learn how all long-range forces arise from the electron and nucleus distributions of two interacting molecules A and B; here we will simply tabulate the most important ones in order of decreasing range (larger s), accompanied by a simplified illustration in Figure 14.11. All of the formulas are stated in cgs-esu.

1. *Charge–charge interaction* ($s = 1$). If both species carry a charge, their interaction is mainly coulombic, $V(R) = e^2 z_A z_B / R$, where e is the magnitude of the electronic charge, and z_A, z_B are the charges on the two ions in units of e. We have already explored this potential in describing the interactions within an atom (Chapters 1 through 4), in ionic molecules (Chapter 5), and in the interaction of solute ions in solution (Chapter 12). Here of course, each ion may bear a multiple charge, and may be either atomic or molecular. As usual, if the charges are unlike, attraction results, and if like, repulsion.

2. *Charge–dipole interaction* ($s = 2$). When an ion A of charge z_A interacts with a neutral molecule B possessing a dipole moment μ_B, such as Na^+ interacting with H_2O, the magnitude of the interaction potential is given by $V(R) = e z_A \mu_B / R^2$. This interaction depends on the angle between the dipole vector and the line joining A and B. Assuming the ion is a cation ($z = +1, +2, \ldots$), if the negative end of the dipole is closer to the ion, attraction will result, and if the positive end is closer, repulsion. The reverse would hold if A were an anion.

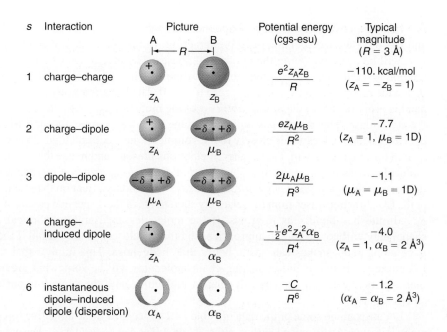

Figure 14.11
Classification and depiction of long-range attractive intermolecular forces, where s is the inverse power of the intermolecular separation R in the intermolecular potential energy. The pictures are schematic representations of the charge distributions on the two interacting molecules. The first three interactions are *electrostatic*, the fourth is *inductive*, and the last is *dispersive*, as indicated. The potential energy formulas are valid for the orientations of the dipolar molecules shown, where z is the charge in units of $|e|$, μ is the dipole moment, and α is the polarizability (see the text).

3. *Dipole–dipole interaction* ($s = 3$). When two dipolar molecules interact, the magnitude of the potential is $V(R) = 2\mu_A\mu_B/R^3$. This interaction depends in a complicated way on the orientation angles of the dipoles with respect to the line joining A and B, but is attractive when the negative end of one dipole is nearer the positive end of the other, and repulsive if both negative or both positive ends are nearer. The dipole–dipole interaction is identical in form to that between two magnetic dipoles, for example, two bar magnets.

4. *Charge–induced dipole interaction* ($s = 4$). If an ion of charge z_A interacts with a neutral molecule B that has no dipole moment, the dominant interaction is $V(R) = -\frac{1}{2}e^2 z_A^2 \alpha_B/R^4$, where α_B is the *polarizability* of the molecule B. The polarizability α of a molecule is a measure of the ease with which its electron cloud is distorted; α has units of volume, length cubed. This is a qualitatively different type of interaction from the first three, because here the ion A *induces* the formation of a dipole in B, by distorting or *polarizing* B's electron cloud. This interaction, unlike the first three, is always attractive.

5. *Dispersion (instantaneous dipole–induced dipole) interaction* ($s = 6$). This potential has the form $V(R) = -C/R^6$ and is always present, even if there are other terms such as those listed earlier. First elucidated by F. London (who with Heitler also gave us the first quantum-mechanical model of the chemical bond), this attraction arises from short time *fluctuations* in the charge clouds of a pair of molecules, which create small temporary or instantaneous dipole moments. An instantaneous dipole induces a dipole in the neighboring molecule, that is, it causes the neighbor's charge cloud to distort as well. London gave an approximate formula $C = \frac{3}{2}\alpha_A\alpha_B[\mathrm{IE}_A\mathrm{IE}_B/(\mathrm{IE}_A + \mathrm{IE}_B)]$, that is, the product of the polarizabilities multiplied by the "reduced" ionization energy (IE). This interaction is also always attractive and is responsible for the condensed phases of the noble gases, which cannot interact in the ways listed earlier.

In a pure covalently bonded compound, only the $s = 3$ (dipole–dipole or DD) and $s = 6$ (London dispersion or LD) interactions can occur, since few if any free ions will be present even in the condensed phases. The magnitude of the DD attraction is fixed by the size of the molecular dipole (if it exists at all), and can take on a wide range of values, including zero if $\mu = 0$. The LD attraction on the other hand is proportional to the *polarizability* α of the molecules, as defined earlier for the $s = 4$ case. The α term has units of volume, usually expressed as angstroms cubed (Å^3), and corresponds closely to the volume of the electron cloud. This volume in turn increases with the number of electrons present, and hence the size and molecular mass of the molecule. Thus *London dispersion attraction increases as the mass of the interacting molecules increases.* A secondary effect on the strength of the LD attraction arises from the relatively short range of the dispersion potential—R^{-6} diminishes rapidly as R grows. If the molecules are roughly spherical in shape, they will not be able to approach each other as closely as they could if they were elongated—snakes can nestle better than hedgehogs. This implies that the LD attraction between linear or long-chain molecules will be somewhat greater than that between more spherically shaped molecules with the same number of electrons.

Let us examine some of the data in Table 14.1 to see how these attractive forces are manifested. As discussed earlier, either T_b or ΔH°_{vap} can be used as a measure of the strength of the attraction. Figure 14.12 plots these quantities as a function of pe-

Figure 14.12
Dependence of **(a)** boiling point and **(b)** heat of vaporization on the period n of the central atom in the hydrides of Groups IV through VII and the noble gases. The general increase in T_b and ΔH°_{vap} as n increases is ascribed to increasing London dispersion (LD) attraction, while the anomalously large values for NH$_3$, H$_2$O, and HF are evidence for the existence of hydrogen bonding (HB) in these substances.

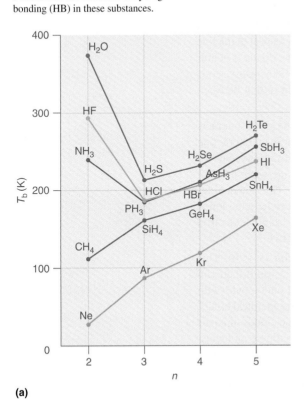

(a)

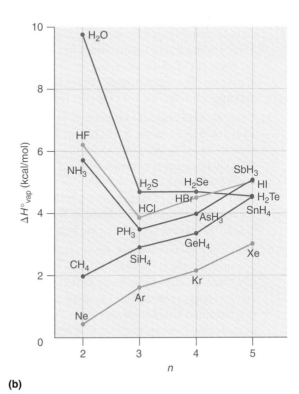

(b)

riod n for the noble gases and the hydrides of Groups IV through VII. The noble gases and Group IV hydrides are both nonpolar, and their behavior is consistent with only LD forces acting. The steady increase in both T_b and $\Delta H°_{vap}$ with period is attributed to the larger number of electrons and concomitant greater polarizability for the later periods. The diameter and hence volume of the Group IV hydride of a given period is necessarily larger than its noble gas counterpart, due both to the larger central atom and to the spreading out of the electron density by the chemical bonds; thus the curves for Group IV lie above those for the noble gases. The growth of the LD contribution to the attraction for the other groups is also evident for $n > 2$, until in the fifth period it is clear that LD is the dominant force. For the polar molecules of Groups V through VII, DD forces must contribute in addition, and must account for the nonmonotonic behavior in those families.

Another important example of the way LD forces work is the series of hydrocarbons with the formula C_nH_{2n+2}, where $n = 1, 2, 3, \ldots$ (These are called *alkanes,* saturated hydrocarbons, or less commonly *paraffins.*) In the *normal (n-)* hydrocarbons, the C atoms are connected in a single chain, with the rest of the carbon valences saturated with C—H bonds. Figure 14.13 plots their melting and boiling points against n, here the number of carbons in the chain; both curves have positive slopes on the whole, again reflective of the increase in LD attraction for higher n. As the figure illustrates, as n goes from 1 to 20, the alkanes pass from gases to liquids to solids under normal conditions. Beginning with $n = 4$, butane, *structural isomers* become possible, in which the carbon chain is branched. This makes a less snakelike molecule, as alluded to earlier, thereby reducing the boiling point below that for the straight-chain isomer, as you can see from the butane and octane (C_8H_{18}) examples included in Table 14.1. (More about structural isomerism can be found in Chapter 18.)

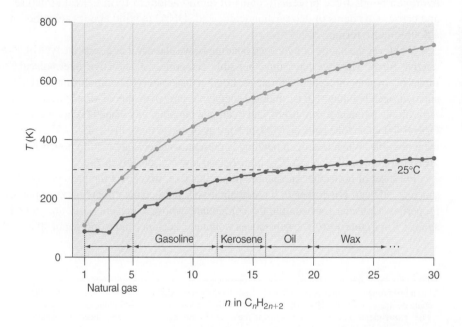

Figure 14.13
Variation of melting point and boiling point with the number of carbon atoms in the straight-chain hydrocarbons. As in Figure 14.12, the general increase in these temperatures is attributed to increasing London dispersion (LD) attraction as the size of the molecule grows.

Hydrogen Bonding

Returning to Figure 14.12, the most striking features of the plots are the "spikes" that occur for the smallest of the polar hydrides of Groups V, VI, and VII, especially that for water. The dipoles are largest for the smallest members due to the greater electronegativity difference and therefore more polar bonds, thus maximizing the product $\mu_A\mu_B$ that enters the numerator of the DD potential; in addition the molecular radius is smallest due to the minimal size of the central atom, thus minimizing the distance R to which a given pair of molecules can approach, and again maximizing the R^{-3} DD term. NH_3, H_2O, and HF are thus expected to have especially enhanced DD forces. Spectroscopic measurements (such as the infrared frequencies) on these molecules in "dimers" (bound pairs of molecules), and X-ray diffraction experiments (see the next section) on the liquid and solid phases, show that the hydrogen atom is intimately involved in the intermolecular "bonding." A given pair of molecules usually has (on the average) a collinear X—H----Y geometry, where the covalent X—H bond is about 1 Å in length, and the H----Y distance is about 1.9 Å. This latter distance is unusually small for an intermolecular separation, and chemists and biologists refer to the H----Y interaction as a **hydrogen bond.** This type of intermolecular bond is generally thought to be largely or wholly electrostatic (DD plus smaller correction terms), and it is controversial at present whether there is any component of covalent electron delocalization across this "bond." It certainly does resemble a classic Lewis acid–base interaction A----:B (see the discussion of the HF----HF dimer in Chapter 7).

From the ΔH°_{vap} values for NH_3, H_2O, and HF, we can deduce that the strength of a hydrogen bond is about 5 kcal/mol; water has a higher ΔH°_{vap} because it has two hydrogens and two lone pairs for making hydrogen bonds, while NH_3 has only one lone pair and HF only one H. *The hydrogen bond is the strongest type of attractive intermolecular force;* if it were much stronger, we wouldn't be able to distinguish it from a normal chemical bond.* As we will describe in Chapter 18, hydrogen bonds force proteins (chains of amino acids) to form helical structures and cause two chains of deoxyribonucleic acid (DNA) to wind about each other in a double-helix form.

Whatever the nature of hydrogen bonding is, we classify it as a separate type of attractive force. Thus we have three possible categories of attractive intermolecular forces in covalent molecular substances: London dispersion (LD), dipole–dipole (DD), and hydrogen bonding (HB). London dispersion is universal, dipole–dipole exists only if the dipole moment μ is nonzero, and hydrogen bonding only when H is bonded to N, O, or F and when lone pairs on N, O, or F are available to make the X–H----:Y arrangement possible. When hydrogen bonding is present, its effects are readily noticeable in the form of unusually large ΔH°_{vap}, high T_b, and low P_{vap} at a given T.

For a set of molecular substances of similar size, we expect the presence or absence of DD and HB forces to have qualitatively predictable effects on their phase properties. Consider for example the compounds whose vapor pressures are plotted against T in Figure 14.6. The substances isobutane, acetone, and isopropanol all have

---

* For a brief time in the late 1960s a few chemists were convinced that something called "polywater" could be made, where the H's formed stronger "bridge" bonds between H_2O units; but this idea faded when experiments purportedly synthesizing the strange water could not be reproduced and accurate quantum-chemical calculations failed to show anything beyond the normal hydrogen bond.

similar numbers of electrons and bonding geometry, and hence are expected to have about the same LD attraction; thus the differences among them you see in the figure are due to DD and HB forces.

1. Isobutane, $(CH_3)_2CHCH_3$, is bonded only with nonpolar C—C ($\Delta\chi = 0$) and C—H ($\Delta\chi = 0.3$) bonds; hence isobutane has no measurable dipole moment, and LD is the dominant attraction between isobutane molecules. In comparison with polar or hydrogen-bonded substances, we expect isobutane to have a lower $\Delta H°_{vap}$ and T_b, and a higher P_{vap} at 25°C.

2. Acetone $(CH_3)_2C=O$, with its polar C=O bond, has a substantial dipole moment μ (2.88 D), and hence will have DD forces *in addition to* LD forces. Acetone does not enjoy hydrogen bonding, because, although there exist O lone pairs and H atoms, the H's are all bonded to C and are nonpolar. We nonetheless expect acetone to have a higher $\Delta H°_{vap}$ and T_b, and a lower P_{vap} at 25°C than isobutane due to DD attraction; Figure 14.6 confirms the latter two comparisons, and Table 14.1 the first.

3. Isopropanol, $(CH_3)_2CHOH$, not only has a dipole moment (1.66 D) due to its polar C—O and O—H bonds, but it also satisfies the requirements for the presence of hydrogen bonding, an O—H bond and lone pairs on O, leading to O—H----:O attraction. We thus expect isopropanol to show all three types of attraction (LD, DD, and HB), and to display higher $\Delta H°_{vap}$, higher T_b, and lower P_{vap} at 25°C than either isobutane or acetone. Again these expectations are confirmed by Figure 14.6 and Table 14.1.

Water itself has yet higher $\Delta H°_{vap}$, T_b, and lower P_{vap} than any of these molecules, despite its fewer electrons and therefore smaller LD attraction. This testifies to the strength of the HB interaction, and the fact that a given water molecule, with its two H's and two lone pairs, can engage in four such hydrogen bonds. The bulky methyl CH_3 groups on isopropanol prevent a given molecule from making more than one hydrogen bond; this effect is called *steric hindrance*. By identifying the types of intermolecular forces present, you can often predict the ordering of $\Delta H°_{vap}$, T_b, and/or P_{vap} for a given set of molecular substances, as illustrated in the following example.

EXAMPLE 14.5

Arrange the compounds methyl fluoride (CH_3F), methane (CH_4), methanol (CH_3OH), and ethane (CH_3CH_3) in order of increasing normal boiling point T_b.

Solution:
Both methane and ethane are nonpolarly bonded, and therefore have only LD forces; however, ethane has more electrons than methane, and therefore has stronger LD attraction and a higher T_b. Methyl fluoride has a dipole moment due to its polar C—F bond and therefore DD attraction, but does not have HB despite lone pairs on F because C—H is nonpolar. It has the same number of electrons as ethane, balancing LD, and hence it must have a higher T_b than ethane due to DD. Methanol has LD comparable to methyl fluoride and ethane, also has DD due its polar C—O and O—H bonds, and shows HB due to O—H----:O; it thus has the highest T_b. We thus deduce the order of T_b's $CH_4 < C_2H_6 < CH_3F < CH_3OH$; Table 14.1 confirms this with T_b's of $-161°$, $-89°$, $-78°$, and $+65°C$ in the order given.

The existence of a hydrogen-bonded network ought to endow liquid H$_2$O with increased order compared to, say, a noble gas liquid. Thus we expect a somewhat larger entropy change on vaporization, as noted earlier, since in the gas the molecules are so far apart (and freely rotating) that all of the intermolecular attractions become largely ineffective.

Repulsive Intermolecular Forces and the Intermolecular Potential

A difficulty with the inverse-power potential, Equation 14.17, is that it dives down without limit as the intermolecular distance R goes to zero; recall the discussion of the coulomb potential in Chapter 1. But long before this can happen, the molecules "bump into" each other, roughly speaking. More accurately, the electron clouds of a pair of molecules begin to overlap when R decreases to a certain value, typically 3 Å or so for small molecules. If the molecules have filled orbitals, the electrons of each molecule are forbidden to enter the orbital space of the other by the Pauli principle, causing the molecules to repel each other. The potential energy arising from this "Pauli repulsion" or "overlap repulsion" is well represented by an exponential function of R, Ae^{-BR}, where A and B are positive constants that depend on the spatial extent and "stiffness" of the electron clouds of the particular molecules. The exponential form is sometimes referred to as the *Born–Mayer potential*. The overall intermolecular potential for noble gases or nonpolar molecules is then approximately

$$V(R) = Ae^{-BR} - \frac{C}{R^6} \tag{14.18}$$

where it is assumed the molecules are roughly spherical. Figure 14.14 illustrates how the repulsive and attractive terms behave separately, and how the well in the potential is a result of a compromise between them. Also shown are the intermolecular potentials for the noble gases, derived from data on nonideal gas behavior. For polar molecules, the DD term would have to be added, including its angle dependence; with small correction terms not described here, hydrogen bonding can also be emulated. For more "shapely" molecules such as ethane (CH$_3$CH$_3$), each methyl group (CH$_3$), or even each atom in the molecule, can be assigned its own potential function like that of Equation 14.18. These potential functions are now widely used in *molecular dynamics simulations,* in which the molecules are allowed to move according to Newton's laws subject to the pair potentials in a large-scale computer simulation of the actual condensed phase. Intermolecular potentials have been developed by W. Jorgensen and his students that allow the accurate prediction of properties such as the density ρ and ΔH°_{vap} from such calculations.

The attractive and repulsive intermolecular forces, represented by Equation 14.18 and its elaborations, are fittingly called **van der Waals forces;** the attractive terms correspond mainly to the van der Waals "*a*" constant of Chapter 9, while the repulsive term determines mainly the van der Waals "*b*." For many purposes, in particular for describing deviations from ideal gas behavior, the interaction even between molecules with complex shapes can be represented by the spherically symmetric potential function of Equation 14.18. The intermolecular distance R at which the attractive and repulsive terms compromise, and the potential energy curve reaches

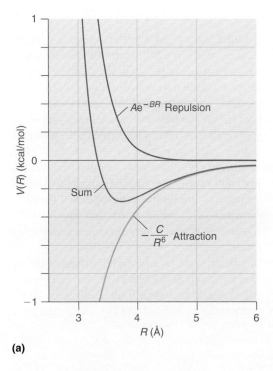
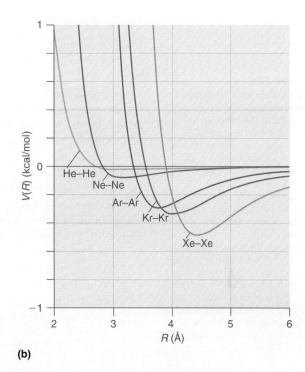

Figure 14.14
(a) Repulsive and attractive contributions to the intermolecular potential for neutral, nonpolar molecules. Note that the repulsion persists well beyond the position of the minimum in the overall potential. (b) Comparison of the intermolecular potentials for the noble gases. As in the two previous figures, the growth in the well depth with period of the noble gas is ascribed to increasing London dispersion attraction, while the increase in the position of the minimum is a reflection of the increasing atomic radius of the noble gas atom.

a minimum, is denoted R_m, while the depth of the well thus formed is denoted ε. (The counterparts of these for bond potentials are r_e and D_e.) Good estimates of R_m can be obtained by adding the so-called **van der Waals radii** r_{vdW} of the two interacting molecules, that is, $R_m \approx r_{vdW}(A) + r_{vdW}(B)$. In Chapter 1 we used Avogadro's number N_A to estimate the diameter of a molecule; the van der Waals radius is half of this diameter. No convenient sum rule like this can be applied to ε, however, owing to the variety of possible attractive interactions.

As pointed out in Section 14.1, the location and depth of the intermolecular potential well determine the distance between molecules in condensed phases and the stability of those phases. The atoms and molecules that form solids at NTP, however, are necessarily bound in most cases by forces stronger than nonbonded van der Waals types. This most important state of matter, literally the foundation on which humankind stands, is examined next.

14.5 Structure and Bonding in Solids

A cursory perusal of Table 14.1 shows that the elements and simple compounds that are solids at NTP largely consist of metals and salts, with a few additions to the group coming from elements just to the right of the metal–nonmetal divide. There are also an almost limitless number of "macromolecules," both natural and synthetic that are normally solids, not listed in the table; these are nearly always van der Waals solids, however, just as the lower melting nonmetals and small covalent molecules are. In this section we first examine the stronger forces that hold metals, metalloids, and salts in place, followed by a brief look at molecular solids.

Metal Crystals

Metal atoms are normally found aggregated into a **crystal lattice,** a regular geometric array of points each occupied by a single atom, in which each atom has anywhere from 6 to 12 "nearest neighbors." The number of neighbors in the smallest shell of lattice points surrounding a particular atom is called its **coordination number.** The two most common crystal systems are **cubic** and **hexagonal,** among the seven crystal systems that are known. (Two others are the rhombic and monoclinic systems we met earlier for sulfur.) The geometric designations refer to the shape of the **unit cell,** the smallest unit that has the complete packing pattern of the whole crystal. Figure 14.15 illustrates the three predominant crystal lattice types: two from the cubic system, **body-centered cubic (bcc)** and **face-centered cubic (fcc)** (also called *cubic closest-packed, ccp*); and one hexagonal, **hexagonal closest-packed (hcp).** In bcc each atom has a coordination number of 8, while in fcc and hcp it is 12. Figure 14.16 is a periodic chart of the representative and transition metals, indicating the most stable type of crystal each forms.

These crystal structures are usually determined by a powerful technique known as **X-ray diffraction,** suggested by von Laue and developed by W. and L. Bragg in 1912. Recall from Chapter 2 (Figure 2.3) that X-rays are a form of electromagnetic radiation of very short wavelength, on the order of an ångström (10^{-8} cm). The main

Figure 14.15
Illustration of the unit cells for the three most common types of crystal structures formed by the elemental solids. Note that the unit cell for the hexagonal case is actually one-third of an hexagonal solid as shown by the dashed lines, and that the base of the cell is a 60° rhombus rather than a square. The space-filling models illustrate the nearest neighbors, which can be used to relate the atomic radius r to the edge length of the unit cell a as in the next line. As shown by the dash-dot lines, in bcc the nearest neighbors lie along an interior diagonal, in fcc along a face diagonal, and in hcp along an edge. The number of atoms per unit cell is determined in the cubic cases by counting one-eighth for each vertex atom, one-half for each face atom, and one for an interior atom.

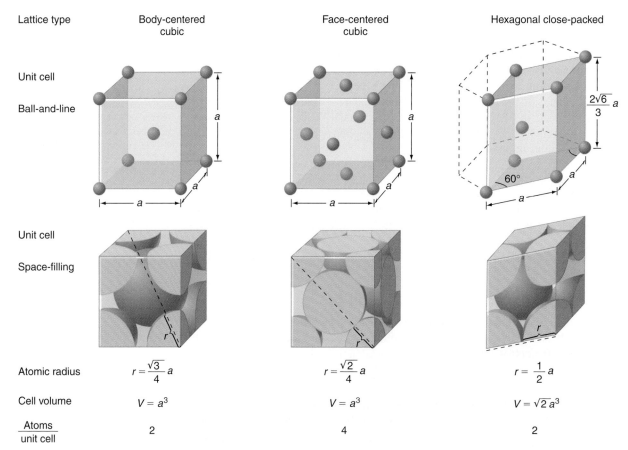

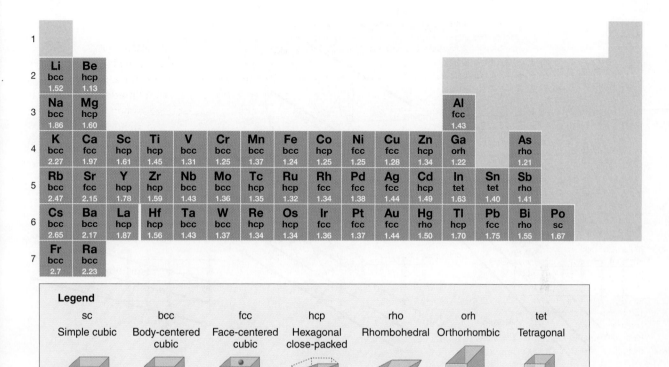

Figure 14.16
The most stable crystal types formed by the representative and transition metals at normal T and P. Also given are atomic radii determined from X-ray analysis of the crystals, as illustrated in the next figure.

source of X-rays in the laboratory is cathode-ray-excited photon emission from metal atoms with vacancies in their 1s shells (originally called *K-shells*). You can readily calculate from the Bohr formula, Equation 2.17 of Chapter 2, that when $Z \gtrsim 20$, the wavelength of the $n = 2 \longrightarrow n = 1$ transition becomes of the order of a few angstroms. As illustrated in Figure 14.17, when a beam of X-rays of known wavelength λ is reflected from atoms in adjacent planes, interference will occur because of the different path lengths traveled by the light waves. The situation is quite analogous to Young's two-slit experiment, illustrated in Figure 2.2, except that both the wavelength of the light and the path-length differences are much smaller than for visible light passing through macroscopic slits. As in the two-slit experiment, bright interference fringes will be seen at angles with respect to the direction of the incident X-ray beam where constructive interference has occurred, due to a path-length difference that is an integral number of X-ray wavelengths, $n\lambda$. As Figure 14.17 shows, the path-length difference is $2d \sin\theta$ for atomic planes a distance d apart scattering X-rays incident on the metal crystal at angle θ with respect to the atomic planes. The constructive interference condition is then

$$n\lambda = 2d \sin\theta, \qquad n = 1, 2, \ldots \qquad (14.19)$$

Equation 14.19 is known as the **Bragg law** of diffraction, and n is called the *order* (not to be confused with the principal quantum number). Because the angle of incidence is equal to the angle of reflection, a bright fringe will appear at an angle 2θ. For

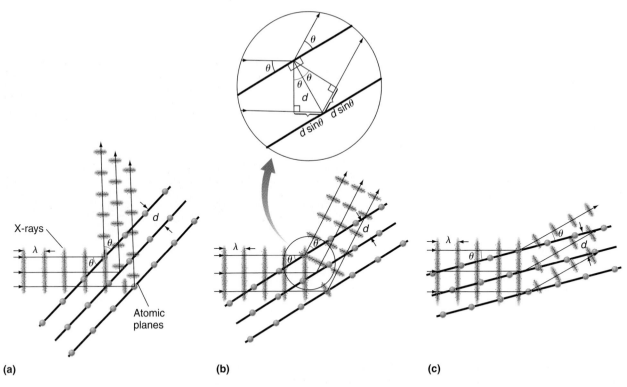

Figure 14.17
Reflection of X-rays by planes of atoms in a crystal, for three different angles θ between the incident X-ray beam and the crystal planes. The X-rays have wavelength λ, and the spacing between adjacent atomic planes is d. Portions of the X-ray beam reflect from different planes, breaking the beam up into partial waves that recombine as they leave the crystal. Usually the partial waves are out of phase and interfere destructively, as in (a) and (c), but if the angle θ is just right, they will emerge in phase, as in (b), and interfere constructively. The condition for constructive interference is that the path-length difference for reflection from two adjacent planes must be an integral number of wavelengths, $n\lambda$; the inset shows from geometry that this difference is $2d \sin\theta$, and hence $n\lambda = 2d \sin\theta$. The constructive interference occurs on a cone of half-angle 2θ, producing a bright ring on a photographic film surrounding the X-ray beam, as illustrated in the next figure.

a powdered metal sample, the individual tiny crystals have arbitrary orientation with respect to the X-ray beam; only those whose planes make the proper angle with the incident X-rays as given by the Bragg law will produce a bright fringe. Furthermore, since all azimuthal angles (around the direction of the incident rays) occur with equal likelihood, the reflections occur on a cone of half-angle 2θ, and the fringe takes the form of a *ring* around the beam direction. Figure 14.18 shows a typical setup for an X-ray powder diffraction experiment, and photographs of segments of interference rings for copper and iron. In Figure 2.10 you have already seen complete rings generated by X-ray diffraction from Al metal. More than one ring is seen due to the different possible planes within each crystal, which in turn depends on the symmetry of the crystal. Thus the number and angles of the diffraction rings determine both the size and the type of unit cell; from these data the atomic radius can be determined precisely. The example illustrates how the Bragg law is used to find interatomic plane spacings.

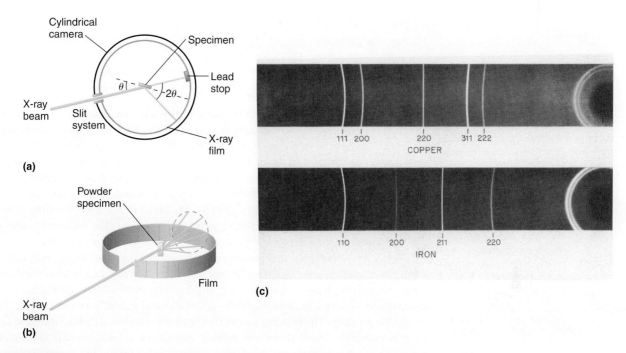

Figure 14.18
(a), (b) Setup for an X-ray diffraction experiment. (c) Diffractograms from two common metals. As indicated, if constructive interference occurs for an angle of incidence θ with respect to the crystal plane (see the previous figure), the X-ray film will be exposed at an angle 2θ. If the sample is polycrystalline, as in a powder, all azimuthal angles are equally likely to occur, and a bright interference ring results. In practice, the sample is slowly rotated during the experiment to produce uniform rings. The negatives of two diffractograms are shown in (c) resulting from X-rays of Co of wavelength $\lambda = 1.793$ Å (Co Kα_2) incident on powdered Cu and Fe. The rings are assigned to various atomic planes within the crystal using **Miller indices** $hkl = 111, 200$, etc. The Miller indices are the reciprocals of the lattice coordinates xyz within the unit cell through which the plane cuts; they can be used to obtain the cell edge length a from the interplane distance d determined by the angle 2θ of the diffraction ring and the Bragg formula $\lambda = 2d \sin\theta$, as in Example 14.6. The pattern of rings uniquely determines the crystal type, the pattern for Cu being characteristic of fcc and that for Fe of bcc.

EXAMPLE 14.6

The scattering angle 2θ for the smallest diffraction ring for Co X-rays of $\lambda = 1.793$ Å from Fe powder (the leftmost ring segment shown in Figure 14.18) is 52.62°. Calculate the spacing between Fe atomic planes responsible for this ring.

Solution:
Assuming a first-order diffraction ring $n = 1$ (higher orders must appear at wider angles or else not at all), Equation 14.19 yields

$$d = n\lambda/(2 \sin\theta) = (1)(1.793 \text{ Å})/(2 \sin 26.31°) = 2.023 \text{ Å}$$

This ring is assigned to the "110" set of planes in Fe(s), where 110 refers to the *Miller indices hkl*, the reciprocals of the three points of the unit cell through which the plane passes, as noted in the caption to Figure 14.18. It can be shown that the interplanar distance between any hkl planes is related to the unit cell edge length

a by $d = a/[h^2 + k^2 + l^2]^{1/2}$. This yields the Fe($s$) cell edge length $a = 2.861$ Å. Further, the pattern of rings can be uniquely fit only to a bcc unit cell; using the bcc formula of Figure 14.15 for the nearest neighbor interatomic distance yields $r = a\sqrt{3}/2 = 2.478$ Å, and an Fe atomic radius of 1.239 Å.

In more recent years, **X-ray crystallography** has been used to determine the structures of proteins, DNA, and other macromolecules; in these experiments a *single crystal* of the sample is prepared, and the X-rays produce interference spots rather than rings. The structure of the molecular subunits as well as the geometry of the unit cell must be found using mathematical iteration beginning with an "educated guess" of the structure based on bonding theories (Chapters 6 and 7) and experience with simpler molecules. These structures form the backbone of our understanding of biochemical structure and reactivity; an example is given later. The basic principles of the X-ray method, however, are the same as in the simple atomic crystals.

X-ray analysis shows that metals generally adopt coordination numbers of 8 or 12, as mentioned at the beginning of this section. When this fact is combined with their generally high melting points, a strong bonding interaction is indicated that is qualitatively different than the highly directional σ and π bonds we have considered in small molecules. The bonding theory that has been developed to describe these features is called the **band theory of solids,** where *band* refers to nearly continuous ranges or bands of allowed energy levels. As introduced briefly in Chapter 5, the valence electrons in metals become part of an "electron sea"; now that you have seen more sophisticated models of bonding, you can appreciate that the "sea" corresponds to *delocalized bonding* taken to an extreme. The bonds in metals extend over the entire metal crystal in all three dimensions, making the crystal one giant molecule! This amazing idea, that an electron wave could extend over an entire piece of metal, is enormously successful at explaining metallic properties such as high bond strengths and high electrical conductivity, as we now briefly examine.

To simplify things a little, let's consider a one-dimensional model using the simplest metal, lithium. We know from our analysis of Li_2 in Chapter 7 that when two Li atoms are brought together, their $2s$ atomic orbitals form a pair of energy levels and molecular orbitals, one bonding σ_{2s}, with energy below that of the separated atoms, and the other antibonding σ_{2s}^*, with energy above the asymptote. What happens when a third Li atom is brought up along the bond axis of the Li_2? The lesson we learned in Chapter 7 is that electrons naturally delocalize if given the chance, so we expect that three new MO's will form that extend over the entire line of three atoms. The lowest energy of these will have no nodes between the atoms, the next one node, and the highest two nodes, all of which can be built up from linear combinations of the $2s$ orbitals, as illustrated in Figure 14.19. The quantum rules that are being followed are the very simple ones of the particle-in-a-one-dimensional-box model of Appendix B. In the ground state of Li_3, only the lower orbitals are occupied, an important feature as we shall see. If this process is continued by adding more Li atoms, say, N of them, we will get an increasing number of energy levels and orbitals located in a band bounded by the totally nodeless orbital of lowest energy and the $(N - 1)$-node orbital of highest energy. If this line of atoms is 1 cm long, $N \approx 10^8$, the levels become so dense as to be nearly continuous, as Figure 14.19 illustrates. The particle-in-a-box model tells us that if the box is widened, the energy-level spacings collapse. Because electrons occupy the MOs pairwise, the N valence electrons fill the $N/2$ lowest or-

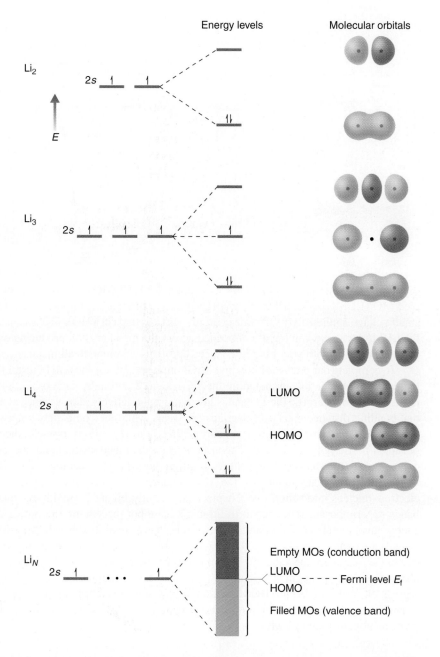

Figure 14.19
Molecular orbital development of the band theory of metals, shown for lithium.

bitals, leaving the other $N/2$ empty; we thus have a *half-filled band* of energies. The model is readily generalized to three dimensions (just as the particle-in-a-box model is) with its main features intact.

This model makes two principal predictions, both in good accord with experience. First, all of the occupied orbitals are more or less strongly bonding, leading to substantial stability of the solid. In fact, we would expect the heat of vaporization of Li to be comparable to the single-bond strength of Li_2; from Table 14.1 we find $\Delta H°_{vap} = 32.2$ kcal/mol, while from Chapter 7 $D(Li-Li) = 26.3$ kcal/mol. Second,

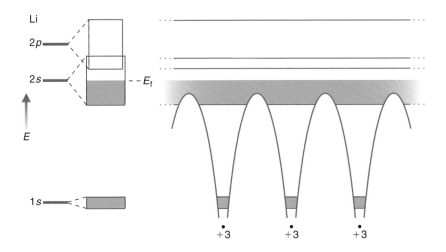

Figure 14.20
Energy bands in a lithium crystal; on the right these are superimposed on a 2-D coulomb potential that the electrons experience from the +3 nuclear charges on the Li nuclei. The shaded bands are occupied with electrons. Note that the 2s and 2p bands overlap in energy. This same picture applies qualitatively to Be, except that the 2s band is completely full. Beryllium is a conductor nonetheless because of the 2s–2p band overlap.

the energy spacing between the highest occupied MO (HOMO) and lowest unoccupied MO (LUMO) is vanishingly small, and only a tiny nudge, provided, say, by a small voltage applied across the line of atoms, suffices to push a HOMO electron into the LUMO, where it can travel unimpeded across the metal crystal, producing electrical conduction with very little resistance, a property common to all metals.

To complete our picture of bonding in a lithium crystal, we show in Figure 14.20 the energies obtained from including the 1s, 2s, and 2p orbitals. Each orbital gives rise to a band of energies: the 1s band is low lying and relatively narrow, since at normal bonding distance the 1s–1s overlaps are weak, while the completely empty 2p band (which contains three times as many orbitals), *overlaps in energy* with the unoccupied portion of the 2s band. This occurs when the metal–metal bond energy is comparable to the valence orbital energy splitting (here due to screening as described in Chapter 4). Also shown is the one-dimensional coulomb picture of the electron–nucleus interaction (see Chapter 7, especially Figure 7.4) with the energy bands superimposed. If we move to lithium's neighbor, beryllium, the orbital and energy-level pictures are similar, but because Be has a filled 2s subshell, the s-band orbitals are completely occupied. Beryllium is still an electrical conductor, however, due to the overlap of the valence s- and p bands. In the transition metals, d bands are also present, as you might anticipate. In all cases there is an uppermost occupied energy level, analogous to the HOMO in small covalent molecules, that is called the **Fermi level;** electrical conduction always occurs in the lowest unoccupied energy level, analogous to the LUMO.

Nonmetallic Network Solids

In a few special cases near the metal–nonmetal divide, an interesting change in the nature of the bonding occurs. In Group IV, C, Si, and Ge all form solids that can be characterized as **three-dimensional networks,** in which all the atoms have their normal valence, making four bonds with neighboring atoms. The most well-known case is the diamond allotrope of carbon, in which each C is tetrahedrally bonded to four nearest neighbor C's. As Figure 14.21 illustrates, a diamond crystal can be built up by replacing the hydrogens in methane by other carbons, and the bonding modeled as

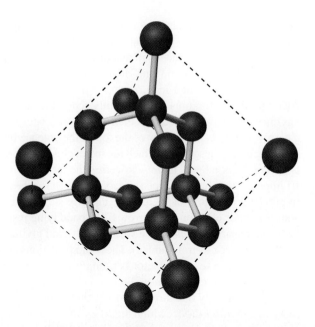

Figure 14.21
The structure of diamond. Each carbon is tetrahedrally bonded to four others, making the geometry similar to methane. A cubic unit cell can be built up from four tetrahedral bonding arrangements about the central C's; these C's themselves are arranged tetrahedrally in space. Each central C shares three of the C's bonded to it with another central C. These shared atoms are located on the six faces of the unit cell. The remaining atoms bonded to these central C's form four of the eight vertices, again tetrahedrally arranged. The four C's at the remaining vertices are not directly bonded to the others. The unit cell therefore contains eight C atoms. The other group IV elements form similar structures save for Pb.

an sp^3–sp^3 overlap. As in metals, the entire diamond crystal is a single giant molecule. When energy bands are constructed for a diamond structure, however, a sizable energy gap or **band gap** is found to occur between the highest filled bonding orbitals, called the **valence band,** and the lowest unfilled antibonding orbitals, the **conduction band,** making diamond an electrical insulator.

As we descend down Group IV, the metallic character of the elements is increasing, and the band gaps grow smaller, as shown in Figure 14.22. Beginning with silicon, we find electrical conductivity that lies between a perfect insulator, like diamond, and a metal; Si, Ge, and gray Sn, all with diamond-like lattices, are called **semiconductors.** In Si for example, the band gap is 1.1 eV, still a considerable energy jump for an electron, but one that can be bridged by a "normal" voltage. Silicon

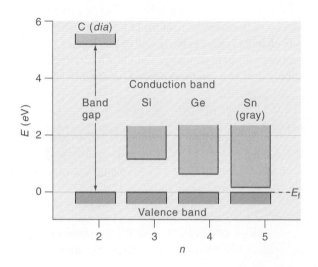

Figure 14.22
Energy gap between the valence and conduction bands in the diamond-like Group IV elements. The lower coordination number (4) compared to metals prevents 3-D delocalization of the electrons and gives rise to narrower, nonoverlapping ns and np bands. The gap shrinks with increasing n due to the larger orbitals and lower ionization energies of the heavier IV's, which decreases the localization of the covalent bonds.

in particular has found an enormous usefulness in all types of electrical devices; we are living in a "Silicon Age" that will stretch well into the 21st century.

It has been found that small amounts of carefully chosen impurities implanted in a Si crystal can be used to "tune" its semiconductor properties. If, say, a few parts per million (ppm, 10^{-6} mol impurity/mol Si) of a group III element such as boron are added to Si, the B atoms will randomly replace Si atoms in the lattice. However, B has one fewer valence electrons than Si, which creates vacancies or "holes" that form a narrow empty band just above the top of the valence band. Electrons are more easily excited into these holes, which are much closer in energy to the filled orbitals than is the conduction band, thereby lowering the resistance dramatically. Si with a group III impurity is called a *p*-type semiconductor (*p* for the *p*ositive holes created by the impurity). Each atom of a group V impurity such as arsenic As, on the other hand, adds a valence electron to a narrow filled band lying just below the conduction band, again decreasing the resistance substantially, and giving an *n*-type semiconductor (*n* for the *n*egative electrons added). These modifications are illustrated in Figure 14.23.

If an *n*-type semiconductor is brought in contact with a *p*-type, a special electrical connection called a *p-n junction* is created, which preferentially allows current to flow in only one direction, from the *n* side to the *p* side. Such a junction can serve as a *rectifier* diode, which modifies an ac voltage by eliminating the negative-going part of the waveform. These junctions have also found use in *light-emitting diodes* (LEDs), which emit light when electrons jump across the junction (from *n* to *p*) into a lower energy level, and as *solar cells,* which absorb visible light by reversing the action of the LED, converting the photons into electrical energy. Using ultra-clean high-purity manufacturing methods, solar cells with more than 10% conversion efficiency have been produced and are now commonly supplied in calculators and other light-powered devices. Some scientists have suggested that the sun could readily provide us with unlimited energy, if huge solar-cell panels were put into earth orbit, beaming their energy down at longer wavelengths to ground-based receivers.

The *transistor* is in essence a sandwich with *n*-type bread and a thin layer of *p*-type filling, an *n-p-n* triode junction. By applying suitable positive bias voltages to the *p* layer, called the *base,* and one of the *n* electrodes, called the *collector,* the transistor can act as a *current amplifier.* This property results because the *p* layer acts as a variable barrier to electron flow, through which the electrons must "tunnel"; a small change in the occupancy of the holes in the *p* layer (due to small base voltage or current changes) changes the height of this barrier, on which the tunneling collector current depends exponentially. The other *n* electrode, called the *emitter,* stands ready to supply the extra electrons needed for amplification when the barrier is reduced.

Figure 14.23
Band structures for *p*- and *n*-type semiconductors with a silicon substrate. The holes from the *p* dopant form a new conduction band much closer to the Si valence band, while the electrons from the *n* dopant form a new valence band much closer to the Si conduction band. In each case, the electrical resistivity is decreased by a factor of 10^5 or more. When these are joined to make an *n-p* junction, electrons naturally flow downhill from the new valence band in the *n* layer into the new conduction band in the *p* layer. The energy mismatch between these new bands is the basis for rectifiers, light-emitting diodes, and solar cells, as described in the text.

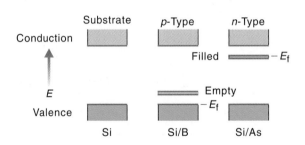

Allotropy in Elemental Solids

As inspection of Table 14.1 shows, there is a dramatic change in the strength of binding in elemental solids as the metal–nonmetal divide is crossed. Consider, for example, the difference between C(*s*) and N$_2$(*s*). The left-to-right decline in stability of the solid phase can be correlated with a decrease in coordination number, a reluctance to share electrons in many directions that naturally accompanies an increase in atomic valence-shell occupancy and in electronegativity χ. Elements near the metal–nonmetal divide tend to display at least two crystal structures, one metallic and one nonmetallic (network or molecular). Figure 14.24 is a broken periodic table that displays the resulting **allotropy.**

The most well-known allotropy is that of carbon, which comes in the energetically similar but geometrically distinct forms called *graphite* and *diamond*. We have just illustrated the diamond structure in Figure 14.21, in which C is sp^3 hybridized; Figure 14.25 shows graphite, which consists of stacked planes of fused benzene rings of sp^2 C atoms. The planes are held together only by van der Waals (LD) forces, and can readily slide past one another. This property is what makes pencils so easy to write with; a pencil with diamond "lead" would be beautiful but useless. While diamond is an insulator, graphite is nearly as good a conductor as a metal, due to the delocalized π bonding in the fused-ring plane. In 1985 a third allotrope of carbon was discovered by R. Smalley and H. Kroto, consisting of round shells or "balls" of 60 atoms each. As shown in Figure 14.25, these atoms are bonded in fused five-membered and six-membered rings in a "truncated icosahedron," a pattern identical to that of a soccer ball. The new allotrope was named *buckminsterfullerene,* after the

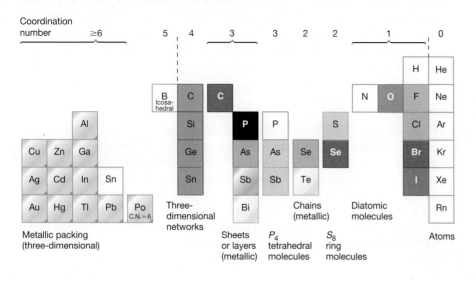

Figure 14.24
A broken portion of the periodic table showing allotropy of the elements near the metal–nonmetal divide. The crystals of lower coordination number tend to be nonmetallic in their properties. The colors shown from P to the right are the colors of the crystalline allotropes.

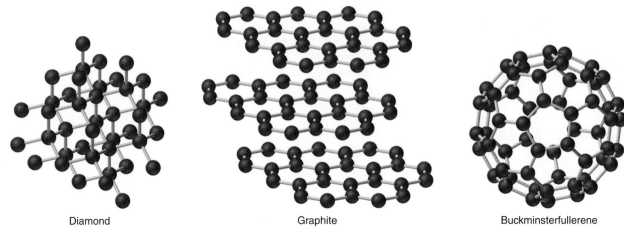

Figure 14.25
Three allotropic forms of carbon; diamond was shown earlier in Figure 14.21. The graphite form is the most stable, the buckyball (fullerene) form the least.

philosopher-architect R. Buckminster Fuller, inventor of the geodesic dome, the structure of which resembles that of C_{60}. Since then C_{60} has become known as a "buckyball," and a growing number of larger closed shells of C atoms, some round and some oval, have been made. The chemistry of C_{60} and its relatives continues to occupy a number of research laboratories, in which atoms have been encapsulated within a buckyball, and "buckybowls," buckyballs split in half, have been formed, along with "nanotubes," long pipes of C atoms bonded as in graphite, sometimes capped on one end with a buckybowl to make a "nano-test tube" or "nanorope." The number of allotropes of C, once thought forever two, now seems to be increasing without end, approaching the infinite planes of graphite in the limit of extremely large bonded structures.

In hindsight, these new forms of carbon might have been expected, since carbon's neighbor boron has three major allotropes and several more minor ones. The major ones are all formed from B_{12} shells, icosahedra consisting of two stacked five-membered rings with atoms capping the top and bottom. Each atom in B_{12} has a coordination number of 5 rather than the 3 of graphite and buckminsterfullerene; this makes B balls smaller. A minor B allotrope actually has a discernible B_{60} shell as part of its structure. Two B icosahedral structures are illustrated in Figure 14.26, along with some other structures typical of the allotropic metalloids and central nonmetals.

Ionic Solids

As discussed in Chapter 5 and 6, the bonding between ions is purely coulombic, without a contribution from electron sharing, although such bonds are of comparable strength to covalent and polar covalent bonds. When an ionic compound, a salt, exists as a solid crystal, the coulomb potential continues to be responsible for the crystalline properties, although the problem of evaluating these forces is complicated by the array of ionic neighbors surrounding a particular ion. Nonetheless, we expect high melting and boiling points, as the examples near the bottom of Table 14.1 indicate. Most binary 1:1, 1:2, and 2:1 salts are found to exist in cubic lattices;

14.5 Structure and Bonding in Solids

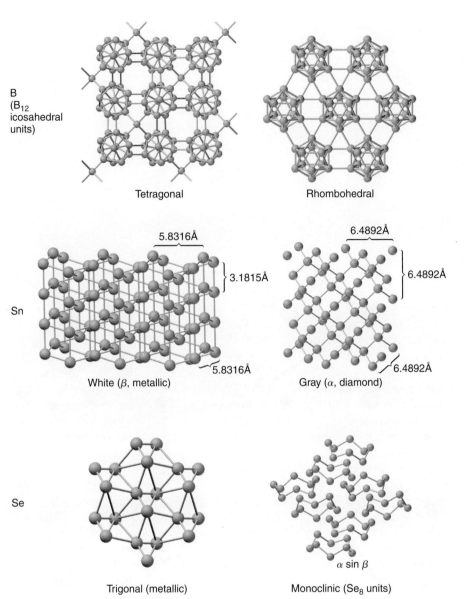

Figure 14.26
Crystal structures of allotropic elements.

Figure 14.27 illustrates some of the common lattice types. The type of crystal formed depends on the relative ionic radii, and for less ionic compounds such as ZnS, on the formation of polar covalent bonds. Just as in the bonding between two atoms in the framework of Chapters 5 and 6, a continuum exists from the purely electrostatic bonding in ionic solids, through polar-covalent bonding, to pure covalent networks such as diamond or silicon. Note that in any case, individual molecules cannot be identified, and the entire crystal can be considered one giant molecule, as in the metallic solids.

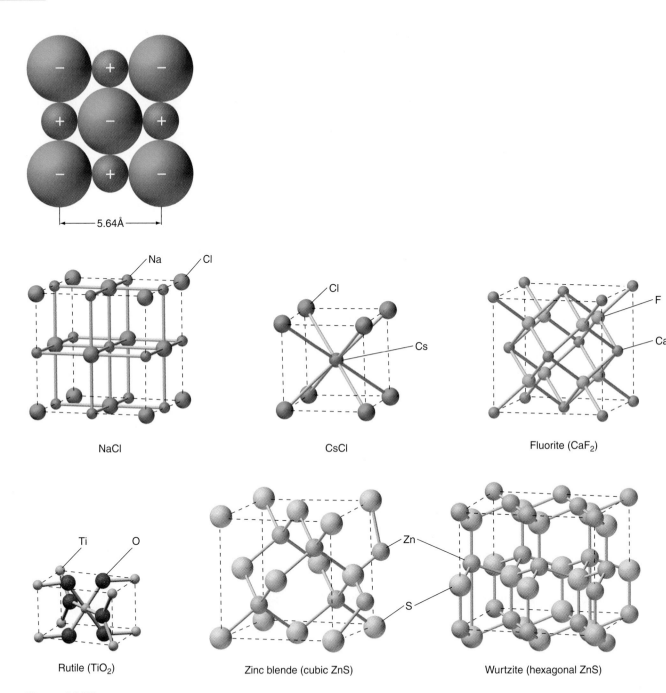

Figure 14.27
Common crystal structures for metal–nonmetal ionic and polar-covalent compounds. Upper left is a space-filling model for NaCl.

The potential energy V_i of an individual ion in a purely ionic crystal such as NaCl(s) can be written down in terms of the coulombic terms and exponential Born–Mayer repulsions as

$$V_i = A \sum_{j \neq i} e^{-Br_{ij}} + \sum_{j \neq i} \frac{z_i z_j e^2}{r_{ij}} \qquad (14.20)$$

where r_{ij} is the distance between the ion selected and its neighbors. The coulomb sum extends over the entire crystal, while the sum of repulsions need only include the equidistant nearest neighbors because of its short range, and thus reduces to a single term. It is of interest that the coulomb sum includes repulsion from ions j of the same charge as ion i, which typically constitute the next-to-nearest neighbors, and that the ion feels the overlap repulsion from all of its nearest neighbors. The coulomb sum can actually be carried out exactly for cubic lattices and the result expressed in terms of nearest neighbor distance. The total potential energy of the crystal, called the **lattice energy** E_L, can be found by summing Equation 14.20 over all ions, taking care to count the coulombic and repulsive terms between ions i and j only once.

To see how this model is tested, let's use NaCl as an example. The lattice energy E_L is actually the energy change for the reaction $Na^+(g) + Cl^-(g) \longrightarrow NaCl(s)$, and can be obtained from measurable energy changes by use of a suitable energy cycle, the **Born–Haber cycle**. As illustrated in Figure 14.28, the crystallization of the gas-phase ions can also be accomplished by neutralizing them, converting them to elements in their natural states, and running the $Na(s) + \frac{1}{2}Cl_2(g) \longrightarrow NaCl(s)$ reaction of Chapter 5, all of whose energy changes are well known. For NaCl, E_L derived from the cycle is -188.0 kcal/mol, compared to -190.0 kcal/mol calculated using Equation 14.20. This agreement is good evidence for the validity of the ionic model.

In addition to their high boiling and melting points, as found in Table 14.1, two other properties are also characteristic of ionic crystals: they are brittle and electrically insulating. The brittleness of salts contrasts with the malleability of most metals. As illustrated in Figure 14.29, when two layers of ions in a salt are displaced, like charges may be brought in proximity, and the coulomb repulsion may cause the crystal to fracture. In a metal, however, the electron sea shields the positive metal ion cores from each other, and the metal crystal will simply deform rather than break. In the band model the localized charges of the ionic crystal correspond to a

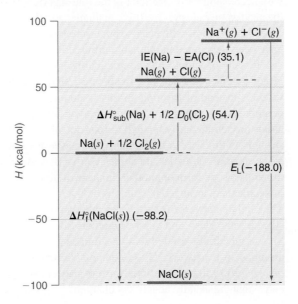

Figure 14.28
The Born–Haber cycle for ionic crystals, illustrated for NaCl.

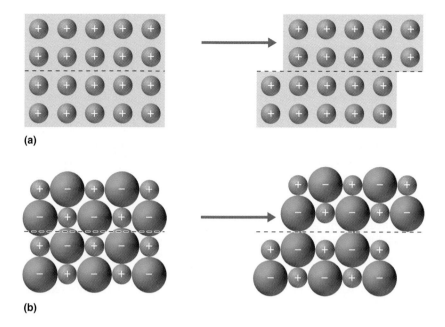

Figure 14.29
Effect of deformation by sliding of crystal planes for (a) a metal and (b) an ionic salt. The juxtaposition of like charges in the salt results in increased repulsion between planes and crystal fracture.

filled narrow valence-like band, built from the filled anion atomic orbitals, well separated in energy from an empty conduction band, arising from the empty cation atomic orbitals.

The size of the band gap for alkali halides, I–VII salts, is similar to that in diamond, making them electrical insulators. For II–VI and III–V compounds, however, the bonding acquires some covalent character due to the decreasing electronegativity difference $\Delta \chi$. This broadens the bands, bringing the top of the valence band closer to the bottom of the conduction band, and gives these compounds, such as CdS or GaAs, semiconducting properties. Figure 14.30 illustrates the trend in band structure with decreasing ionic character in the fourth period. II–VI compounds have been synthesized as tiny clusters ~ 100 Å in diameter called *quantum dots,* which behave like super-heavy atoms with a dense but discrete set of energy levels, while III–V crystals have been developed into tunable infrared lasers.

In salts of simple polyatomic anions such as CO_3^{2-} or SO_4^{2-}, the entire anionic group occupies a lattice site. In silicates, SiO_3^{2-} compounds, however, the "extra" Si valence, instead of forming an Si=O double bond as in CO_3^{2-}, tends to bond with neighboring oxygens to form *polysilicate* anions such as $Si_3O_9^{6-}$, in which the Si atoms are tetrahedrally bonded to four neighboring O atoms. The silicates can bond strongly in this way to form chains, as in asbestos fibers, or sheets, as in the mineral mica; most of the rocks and stones you've picked up in your life have been silicate minerals of one form or another. In these minerals the metal cations (and often water molecules) are arranged in columns or layers alternating with the polysilicate anions. The sand that covers the earth's beaches and ocean floors is a grainy mixture of polysilicate salts, glass, and quartz. Quartz is SiO_2 in a perfect crystalline form, in which Si is tetrahedrally bonded to 4 O's, and each O to 2 Si's, with polar-covalent bonds arranged in a network. Glass contains similar structures on a small scale, but is *amorphous* (without a definite form or pattern) on a larger scale, with many "loose ends" that are usually negatively charged.

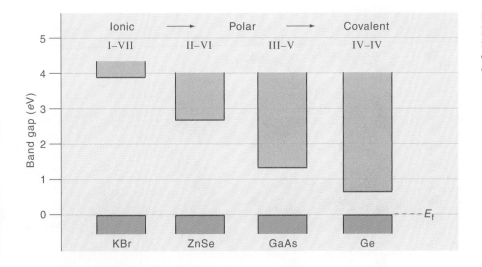

Figure 14.30
Behavior of the band gap in isoelectronic fourth-period 1:1 binary compounds as the group difference decreases.

Molecular Solids

With the exception of large molecules of high molecular mass, most molecular substances are either liquids or gases at NTP; these species are characterized, as we have seen, by strong covalent and polar-covalent intramolecular bonds but relatively weak intermolecular attractions of the hydrogen-bonding, dipole–dipole, and London dispersion types. We include in this group the noble gases, the only examples that remain atomic but nonbonded in the condensed phases. As indicated in Table 14.1, all of these except He will crystallize at standard pressure as the temperature is reduced; He forms an hcp crystal only above $P = 25$ atm near $T = 0$ K. Molecular crystals fall into one of the seven crystal classes, with each molecular unit occupying a lattice site. The new feature not encountered in our survey of metallic and ionic solids is the *orientation* of each molecule, which generally differs from site to site but shows an overall repeating pattern. The orientational pattern is determined by maximizing the attractive intermolecular forces, that is, by arranging each molecule to be most favorably disposed toward its neighbors. An important example is water ice, whose cubic crystal structure is illustrated in Figure 14.31. Overall $H_2O(s)$ has a diamond-like lattice, but each water molecule is oriented so that its two O—H bonds point at O atoms of neighboring molecules so as to make hydrogen bonds. Thus each molecule makes four such bonds with its neighbors, two with its H's and two with its lone pairs, that are tetrahedrally arranged about O, identical in form to C(*dia*). As another example, N_2 freezes at 63 K to form an fcc crystal, but with the N_2 molecules oriented in T-shaped pairs; a facial plane of crystalline N_2 is shown in Figure 14.1. In this case the orientation is determined mainly by electrostatic interactions involving the *quadrupole moment* of N_2, since N_2 has no dipole.

Because the molecules in a molecular solid retain their identity, and in most cases their basic structure, the spectrum of a molecular solid is quite similar to its gas-phase counterpart, except for a slight broadening and often a loss of the rotational degrees of freedom. Thus any band structure that exists in such solids is narrow, and the solid is invariably an electrical insulator. This preservation of molecular character has made low-temperature solids a widely used medium to study energy transfer between molecules and the structures of normally unstable molecules trapped in solid–noble-gas matrices.

Figure 14.31
The crystal structure of water ice.

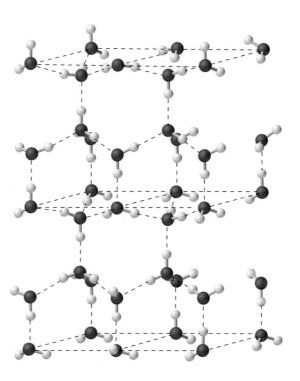

As mentioned earlier in this section, X-ray crystallography of molecular crystals has been used to elucidate both the overall crystal class and the structure of an ever-increasing number of macromolecules, with a particular focus on those of biological importance. Figure 14.32 is a structure of hydrated Vitamin B_{12} determined by Dorothy Hodgkin and her research group using the X-ray method. Here is her 1959 description of their findings:

> In outline, the crystal structure is simple. The roughly sperical molecules are arranged in approximately hexagonally close packed layers, the two layers in each crystal unit being slid a little distance from the formal positions required by close-packing, with the spaces between the layers occupied by water molecules. But in detail the atomic arrangement is very intricate, a consequence of the exact molecular geometry. In wet B12, where the crystals are suspended in their mother liquor, the acetamide and propionamide groups all make direct intermolecular hydrogen-bonded contacts—the B12 molecules are in contact with one another on every side. Between these, the water molecules, approximately 22 per B12 molecule, make additional hydrogen bonds with the active groups, particularly with the amide and phosphate groups and with one another. Along particular lines in the crystal structure, water is continuous from one side of the crystal to another and is probably trickling through it during X-ray photography in narrow streams, one molecule deep. . . .

In addition to this sort of ground-breaking work, many organic and inorganic chemists who synthesize new molecules regularly form them into single crystals and make use of X-ray and other methods to determine their structures. In this way molecular crystals have become a central research tool at the frontiers of chemistry.

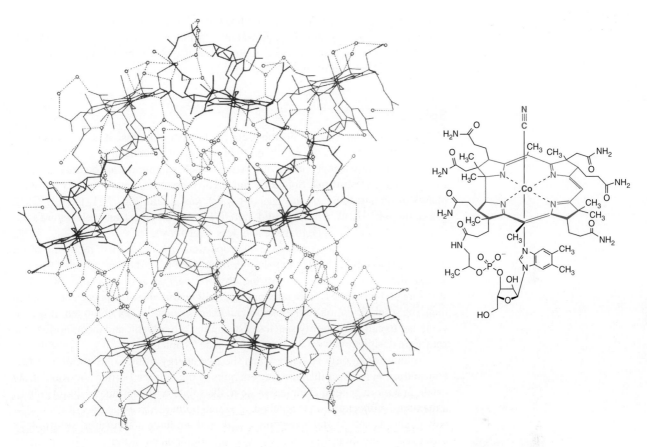

Figure 14.32
The structure of wet vitamin B_{12} in the form of a molecular crystal, determined from X-ray crystallographic analysis. The solid lines are the C and N atom skeleton, the open dots are water molecules, and the dotted lines are hydrogen bonds. A set of nine B_{12} molecules is depicted; these molecules are arranged in a nearly hexagonal close-packed structure like that shown in Figure 14.15. The inset shows the bonded structure of an individual dry B_{12} unit, which involves a cobalt ion bonded to four N's, in a so-called porphyrin ring complex.

14.6 Liquids and Solutions, Featuring Water

Once a few atoms or molecules in a solid acquire enough thermal energy to escape their lattice sites, the resulting changes in intermolecular forces rapidly lead to breakup of the crystal, that is, to melting. Displacing a single molecule from its lattice position can disrupt the symmetrical pattern of hundreds of nearby molecules; a rough equivalent is pulling a single playing card out of the middle of a house of cards. This involves a relatively small change in the potential energy of the condensed phase, and thus, as Figure 14.5 showed for water, the free energy of a liquid is typically only slightly lower than that of its parent solid above T_f. It seems realistic therefore to consider a liquid as possessing roughly a crystalline structure over a short range, but without the long-range order characteristic of crystalline solids. This notion is supported by X-ray diffraction experiments on liquids, which typically show only one sharp diffraction ring and diffuse scattering otherwise. The sharp ring corresponds to relatively fixed nearest neighbor distances, which are found to be close to those in the parent solid, while the diffuse scattering occurs from a disordered range of distances for molecules farther apart. For example, in water ice the intermolecular distance is 2.76 Å, while in liquid water it ranges from 2.88 Å at 1.5°C to 3.00 Å at 83°C. Water is more dense than ice due to the increase in the average number of nearest neighbors from 4 in ice to 4.5 to 5 in the liquid, and not because of any decrease in intermolecular distance. H. Frank proposed a "flickering cluster" model of liquid water, in which groups of five or so H_2O molecules form a temporary ice-like aggregate that quickly breaks up and reforms nearby with a new set of molecules.

Solutions

Mixtures of compounds are all-important in chemistry, because it is only when two or more compounds are physically brought in contact that chemical transformations are possible. Those special kinds of mixtures where mixing occurs on the molecular level are called **solutions.** Solutions can involve combinations of any of the three phases of matter; there are gas–gas (e.g., air), gas–liquid [e.g., $NH_3(g)$ in $H_2O(l)$], gas–solid [e.g., $H_2(g)$ in $Pd(s)$], solid–liquid (e.g., sugar water), and solid–solid solutions (e.g., Ni and Cr in Fe to make stainless steel alloy).

While all gases mix perfectly in all proportions, the intimacy of the liquid phase prevents this from happening in solution. Instead we have the general rule "like dissolves like," which is drawn from the observation that only molecules of similar structure and properties, especially polarity, will readily dissolve in each other. Even then, the solubility of one compound in another is generally not unlimited. It is conventional to call the majority species of a solution the **solvent,** and the minorities the **solutes,** a designation we have been using since Chapter 1.

Solutions can only be made when the free energy of the solution, of a certain concentration, is lower than the free energies of the unmixed solute and solvent. As usual, in a more general chemical system, the solute will continue to dissolve until a maximum concentration is reached, at which point the free energies of the undissolved and dissolved states become equal, and we have a **saturated solution,** as discussed in Section 12.4, for slightly soluble salts in water. As illustrated in Figure 14.33, the enthalpy and entropy changes that determine this limit can be analyzed in terms of three hypothetical stages: first cavities are made in the solvent, then the solute particles are separated from each other, and finally the solute particles are inserted into the solvent cavities. The first two stages are always endothermic, and the dissolving process is exothermic overall only if the insertion stage releases more energy than the first two consume. The dissolving process would at first blush always have a favorable, positive entropy change at every stage, were it not for the fact that the third step can result in organizing the solvent molecules in solvation shells around the solute particles, an especially significant effect for ionic solutes in water.

The upshot is that these competing enthalpy and entropy factors produce variations in solution thermodynamics that are hard to predict by inspecting the molecules of solute, solvent, and their solution. Consider, for example, dissolving the three common ionic compounds NH_4Cl, $CaCl_2$, and $Ca(OH)_2$ in water at 298 K:

	$\Delta H°$ (kcal/mol)	$\Delta S°$ (cal/K mol)	$\Delta G°$ (kcal/mol)
$NH_4Cl(s) \rightleftharpoons NH_4^+(aq) + Cl^-(aq)$	+3.5	+18.	−1.8
$CaCl_2(s) \rightleftharpoons Ca^{2+}(aq) + 2\,Cl^-(aq)$	−19.4	−11.	−16.2
$Ca(OH)_2(s) \rightleftharpoons Ca^{2+}(aq) + 2\,OH^-(aq)$	−4.0	−38.	+7.2

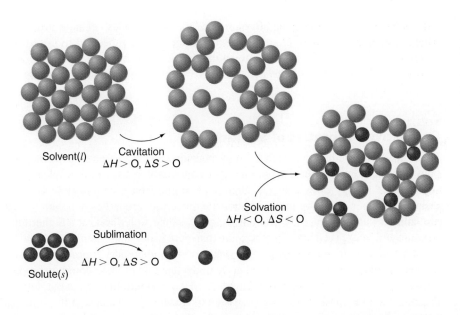

Figure 14.33
Schematic of the energetics of solution formation. This 2-D illustration uses spherical solvent and solute particles; when polar molecules are involved, molecular orientation becomes an important factor as well.

Both NH_4Cl and $CaCl_2$ dissolve readily in water, as indicated by the negative $\Delta G°$'s, but for different reasons. In NH_4Cl it is the positive $\Delta S°$ that dominates—dispersing the ions in water produces a less ordered state than the crystal and pure water display—while for $CaCl_2$ it is the energy regained from ion-solvent attraction, as reflected in the negative $\Delta H°$, that overcomes an unfavorable $\Delta S°$ due to the accompanying solvent organization. On the other hand, $Ca(OH)_2$ is on Chapter 12's slightly soluble-salt list, Table 12.6; although the solution process is slightly exothermic, the opposition of entropy prevents concentrated solutions of the hydroxide from forming. The more negative $\Delta S°$ compared to $CaCl_2$ results from the greater solvent organization engendered by $OH^-(aq)$ due to the strong hydrogen bonds that are formed (see Figure 12.6).

In many familiar mixtures known collectively as **colloids,** the dispersion of one substance in another is uniform to the unaided eye, but the "solute" particles are actually aggregates of molecules bound together by one or more types of intermolecular forces. These molecular aggregates can range in diameter from 100 to 10,000 Å, encompassing the range of wavelengths of visible light. This makes such particles efficient scatterers of light, and causes the colloid to appear cloudy, and to scatter a beam of light passed through it off to the side, the *Tyndall effect.* Everyday examples of colloids are milk, paint, mayonnaise and other creamy dressings, and aerosol sprays, where air is the "solvent." Most colloids are thermodynamically unstable; for example, if you shake a bottle of oil and water, you can form a temporary colloidal suspension, but soon the system separates once more into two distinct phases. In seemingly stable colloids such as milk, the solute particles are charged and maintain the suspension due to coulombic repulsion; this repulsion creates a free-energy barrier to sedimentation (settling out) of the milk solids and fats, resulting in a *metastable equilibrium.* In 1857 Michael Faraday prepared a suspension of gold particles that to this day shows no sign of sedimentation.

Colloids can also be formed from single "macromolecules" of high molecular weight suspended in a solute. Large molecules are often synthesized by reactively bonding a large number of identical or similar molecules into a single molecule with repeating units called a *polymer;* we reserve discussion of such molecules for Chapter 18.

Colligative Properties of Solutions

Up to this point we have neglected any changes in solvent free energy that accompany equilibria in solution. The fact is, when you dissolve a solute in a solvent, the solvent free energy *does* change, albeit very slightly on a chemical scale. This change, however, can be of major importance for certain properties of solutions that are familiar parts of everyday life. For example, salted water boils at a higher temperature and freezes at a lower temperature than pure water; and when placed against a semipermeable membrane with pure water on the other side, it produces a phenomenon called osmotic pressure. To make understanding of these common properties accessible, we will adopt a very simple model of a solution: a solute in a solvent is analogous to a small amount of one gas mixed in another, except that the volumes occupied by each component are not variable; that is, the solute, solvent, and solution are incompressible. This model results in predictions that are in fair to good agreement with experimental data, and yields laws whose forms are independent of the solute and solvent being considered, just as the ideal gas law and Dalton's law of partial pressures apply to all gases and gas mixtures. For this reason these laws are called the **colligative properties,** *colligative* being a synonym for collective, or universal.

For clarity and simplicity of notation in what follows, let's denote the solvent by A and the solute by B. In a dilute solution the solvent A is in great excess, so that the mole fraction $X_B \ll X_A$, where $X_A + X_B = 1$. In our model, when B dissolves in A, the volume of the solution V_{AB} is slightly larger than that of pure A V_A, by the extra volume occupied by B itself. If A and B are composed of particles of similar sizes (true for many common aqueous solutions), then $V_A/V_{AB} \approx n_A/(n_A + n_B) = X_A$. We learned in Chapters 11 and 12 that when the volume of a system increases, $\Delta V = V_{AB} - V_A > 0$, it does work on the surroundings, and its free energy per mole must fall by $-RT \ln(V_{AB}/V_A)$ or $+RT \ln(V_A/V_{AB}) \approx +RT \ln X_A$. The free energy of liquid A as solvent $G_A(l)$ is therefore less than its free energy $G_A^\circ(l)$ as a pure compound according to

$$G_A(l) = G_A^\circ(l) + RT \ln X_A \qquad (14.21)$$

Recall from Chapter 11 that the logarithmic term really represents an increase in entropy. To maintain equilibrium with its vapor $A(g)$, we must have $G_A(l) = G_A(g)$, that is, the vapor pressure P_A of A must adjust itself to maintain equilibrium with the solution phase. (For convenience we assume the vapor pressure of solute B is negligibly small.) Using Equation 14.4 for $G_A(g)$ (with $P^\circ \equiv 1$ atm) and applying the equilibrium condition, we get

$$G_A^\circ(l) + RT \ln X_A = G_A^\circ(g) + RT \ln P_A \quad \text{or} \qquad (14.22)$$
$$\Delta G_{vap}^\circ = -RT \ln(P_A/X_A)$$

But from Equation 14.6 we know that $\Delta G°_{vap} = -RT \ln P°_A$, where $P°_A$ is the equilibrium vapor pressure of pure A. Thus the logarithmic terms $-RT \ln(P_A/X_A)$ and $-RT \ln P°_A$, and therefore their arguments P_A/X_A and $P°_A$, must be equal, giving

$$P_A = X_A P°_A \tag{14.23}$$

That is, the vapor pressure of A as solvent is lower than that of pure A(*l*) by a factor of its mole fraction. This relationship was discovered experimentally by F. Raoult in 1886, and is known as **Raoult's law**. It holds very well when solvent and solute molecules have similar size and structure.

The consequence of Raoult's law of vapor-pressure lowering is that an aqueous solution at any temperature T will have a lower H_2O vapor pressure than pure water at that T; at $T = 100°C$, therefore, the solution will have a vapor pressure less than 1 atm and will not boil. This is illustrated by the phase diagrams of Figure 14.34. We expect the solution to reach boiling at some higher temperature given by $T_b + \Delta T_b$, where T_b is the boiling point of the pure solvent and ΔT_b is called the **boiling point elevation**. If it is assumed as before that $\Delta H°_{vap}$ and $\Delta S°_{vap}$ are T independent, ΔT_b can readily be found. Equation 14.22 can be rewritten as

$$\ln \frac{P_A}{X_A} = -\frac{\Delta G°_{vap}}{RT} \tag{14.24}$$

$$= -\frac{\Delta H°_{vap}}{RT} + \frac{\Delta S°_{vap}}{R}$$

$$= -\frac{\Delta H°_{vap}}{R}\left(\frac{1}{T} - \frac{1}{T_b}\right)$$

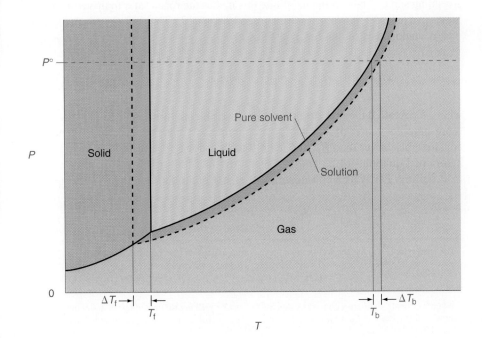

Figure 14.34
Phase diagram for a pure solvent compared to that for a dilute solution of a nonvolatile solute in that solvent. The downward displacement of the liquid–vapor curve for the solution reflects Raoult's law, Equation 14.23, and results in higher boiling and lower freezing temperatures for the solution.

where in the last line we have inserted $\Delta S^\circ_{vap} = \Delta H^\circ_{vap}/T_b$ from Equation 14.3. Equation 14.24 is a generalization of the Clausius–Clapeyron equation (Equation 14.8) for solutions; thus the liquid–vapor phase boundary for the solution will have the same shape as that for the pure liquid, while at every T Raoult's law (Equation 14.23) holds. At the solution boiling point $P_A = 1$ atm. To find ΔT_b the following approximations are customarily made: $(1/T - 1/T_b) = -\Delta T_b/(TT_b) \approx -\Delta T_b/T_b^2$, and $-\ln X_A = -\ln(1 - X_B) \approx X_B$.* The first is good when as expected $\Delta T_b \ll T_b$ (in kelvin units), and the second is good when $X_B \ll 1$, as we assumed at the outset. Finally, the mole fraction X_B is replaced by the **molality** m_B, moles of solute per kilogram of solvent, where $m_B \approx 1000 X_B/M_A$, and M_A is the molecular mass of A. Making these approximations and substitutions in Equation 14.24, we obtain

$$\Delta T_b = T - T_b = K_b m_B \qquad \text{where} \qquad K_b = \frac{RT_b^2 M_A}{1000 \Delta H^\circ_{vap}} \qquad (14.25)$$

The constant K_b is called the **boiling point elevation constant,** and is readily calculated from the expression given. The experimental values for K_b are generally within a few percent of those predicted.

When a solution freezes, normally the solvent alone will crystallize, leaving behind the solute, which remains dissolved. This implies that at the *solution* freezing point the solution must be at equilibrium with the crystallized pure solvent, and hence the meeting of the solution vapor pressure curve with the pure solid vapor pressure curve marks the solution's freezing point. As Figure 14.34 shows, these curves meet below the triple point T_3, and thus generally below the normal freezing point T_f, and we have **freezing point depression** ΔT_f in addition to boiling point elevation. If you live in a climate where ice and snow form in winter, you have probably seen the phenomenon of freezing point depression exploited to melt ice and snow on roads by spreading salt crystals on them. (While this makes the roads safer to drive on, it promotes corrosion of the vehicles driven; see Chapter 13.) The analysis proceeds completely analogously to that for ΔT_b, and yields

$$\Delta T_f = T_f - T = K_f m_B \qquad \text{where} \qquad K_f = \frac{RT_f^2 M_A}{1000 \Delta H^\circ_{fus}} \qquad (14.26)$$

where K_f is called the **freezing point depression constant.** Table 14.2 gives some representative values of K_b and K_f for common solvents. Note that K_f is generally considerably larger than K_b, due mainly to the appearance of ΔH° in the denominators of Equations 14.25 and 14.26 and the fact that melting requires less heat than boiling. Partly as a consequence, freezing point depression has been applied

---

* The term $-\ln(1 - x)$ has the well-known power series expansion (see your calculus text):

$$-\ln(1 - x) = x + \tfrac{1}{2}x^2 + \tfrac{1}{3}x^3 + \tfrac{1}{4}x^4 + \cdots$$

When x is small, this series converges rapidly and can be well approximated by retaining only its first term. In this case, $x = X_B$ is typically 0.01 or less, and the first term represents the series sum to better than 1%.

TABLE 14.2
Freezing point depression and boiling point elevation constants

Solvent	T_f (°C)	K_f (°C kg/mol)[a]	T_b (°C)	K_b (°C kg/mol)[a]
Water (H_2O)	0.0	1.86	100.0	0.512
Carbon tetrachloride (CCl_4)	−22.9	32.	76.7	5.03
Chloroform ($CHCl_3$)	−63.5	4.7	61.7	3.63
Carbon disulfide (CS_2)	−111.5	3.83	46.2	2.34
Acetic acid (CH_3COOH)	16.6	3.9	117.9	3.07
Ethanol (C_2H_5OH)	−114.2	1.9	78.3	1.22
Benzene (C_6H_6)	5.5	4.9	80.1	2.53
Phenol (C_6H_5OH)	43.	7.4	181.7	3.56
Camphor ($C_{10}H_{16}O$)	178.4	39.7	208.0	5.95

[a]More fully, °C kg solvent/mol solute.

far more widely than boiling point elevation. In addition to its use in thawing icy roads and in automotive antifreeze, freezing point depression has been employed in finding molecular masses of unknown compounds. A well-established technique called the *Rast method* uses the organic compound *camphor* ($C_{10}H_{16}O$) as a solvent. Camphor melts at 178.4°C and has a very high K_f of 39.7°C kg solvent/mol solute; thus a very small amount of unknown compound dissolved in camphor will produce a readily measured freezing point depression and a fairly accurate molecular mass. In modern organic chemistry, this technique has been supplanted by mass spectrometry.

EXAMPLE 14.7

What is the freezing point of an aqueous solution made by dissolving 1.00 g acetamide (CH_3CONH_2) in 100. mL of water?

Solution:
The molality m_B of the solute, acetamide, is

$$m_B = (1.00 \text{ g acetamide})/[(59.1 \text{ g acetamide/mol})(0.100 \text{ kg } H_2O)]$$
$$= 0.169 \text{ mol/kg}$$

We then obtain from Equation 14.26 and Table 14.2

$$\Delta T_f = K_f m_B = (1.86°\text{C kg/mol})(0.169 \text{ mol/kg}) = 0.314°\text{C}$$

The freezing point of the solution is therefore

$$T = T_f - \Delta T_f = 0.000 - 0.314 = -0.314°\text{C}$$

When NaCl dissolves in water, it ionizes, producing 2 mol of particles for every mole of salt. In the 1880s van't Hoff first observed that the freezing points of aqueous salt solutions were anomalously low, with ΔT_f approaching a multiple of that predicted from Equation 14.26 as the solutions were diluted. This multiple was generally found to equal the number of ions in the simplest formula of the salt, and is now called the *van't Hoff factor i*. It was a significant strand in the web of evidence that led Arrhenius to propose his theory of ions in solution. In a dilute ionic solution we therefore replace Equation 14.26 with

$$\Delta T_f = i K_f m \tag{14.27}$$

and similarly for Equation 14.25. For NaCl, $i = 2$, while for $CaCl_2$, $i = 3$. ($CaCl_2$ is a more effective road salt than NaCl because of its large exothermic heat of solution, not because of its large van't Hoff factor. On a per gram basis, $CaCl_2$ actually yields fewer ions than NaCl.) Nonionizing solutes such as alcohols or glycols—the "antifreeze" solutes—have $i=1$.

The phenomenon of **osmotic pressure** has the same origin in solvent free-energy change as freezing point depression and boiling point elevation. As illustrated in Figure 14.35, osmotic pressure is exerted when a solution of concentration c is brought in contact with the pure solvent by means of a **semipermeable membrane,** which allows only solvent molecules to pass through, and not solute. The solvent molecules are found to pass from the pure solvent into the solution, thereby diluting it, and simultaneously raising the liquid level of the solution above that of the solvent. The osmotic pressure is proportional to the height to which the solution level is raised when equilibrium is reached.

According to Equation 14.21, the free-energy difference per mole between the solution and the pure solvent is $\Delta G = RT \ln X_A$. Recall that ΔG measures the work available, in this case that used to raise the level of the solution against gravity. On a per mole basis, this work is $w = M_A gh = \Pi \overline{V}_A$, where h is the height of the solution above the solvent level, $\overline{V}_A$ is the molar volume of the solvent, and Π is defined to be

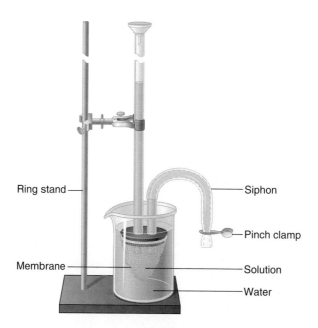

Figure 14.35
A simple apparatus for the measurement of osmotic pressure.

the osmotic pressure. Thus we must have $\Pi \overline{V}_A = RT \ln X_A$. Employing the same approximations as those used earlier for boiling point elevation, and here using (by convention) molarity rather than molality as the unit of concentration, we finally obtain

$$\Pi = cRT \tag{14.28}$$

where c is the solute concentration in moles per liter. The first accurate measurements of Π were reported by W. Pfeffer in the 1880s using sugar solutions in an apparatus similar to that shown in Figure 14.35; van't Hoff noted that Pfeffer's results were approximately reproduced by Equation 14.28, which is identical in form to the ideal gas law $P = (n/V)RT$. For example, with a 0.100 M aqueous sucrose solution, Pfeffer measured $\Pi = 2.59$ atm, while Equation 14.28 predicts 2.45 atm. The logarithmic form given earlier, before approximations are made, is somewhat more accurate. Substantial osmotic pressures are generated even by solutions of modest concentration; if the solution vessel of Figure 14.35 were sealed, the membrane would be likely to burst.

Osmosis is essential for the survival of living things. A plant absorbs most of its water by osmosis; in animals osmosis is used to regulate the flow of water between body fluids and cells. *Reverse osmosis* is a widely used water-purification method; if pressure in excess of Π is applied to the solution in Figure 14.35, pure water will flow out. The factors involved in osmosis and its reverse are analogous to those in electrochemical concentration cells and electrolysis. A familiar example of the effects of osmosis is a pickle; pickles are made by soaking cucumbers in brine, and osmosis of water (the pure solvent) present within the cuke through its skin (the semipermeable membrane) into the brine (the solution) causes it to shrink to pickle size.

Examination of the colligative properties of solutions completes our analysis of the role of thermodynamics, and free energy in particular, in chemical phenomena. The same basic relationships appear again and again in a different guise, a tribute to the universality of Gibbs's formulation, and the ideas of thermodynamics will continue to permeate our discussion of the remaining topics. As powerful as it is, however, thermodynamics is a science of states at equilibrium, and cannot be used in general to describe systems during the turmoil of irreversible change. In the next chapter we shall turn to chemical change in the making, in the lively subject known as chemical kinetics.

SUMMARY

The three states of matter, solid, liquid, and gas, can often be realized for a single substance by varying its temperature. As the temperature increases at atmospheric pressure, solids melt into liquids at their normal freezing point T_f, and liquids then boil into gases at their normal boiling point T_b. These physical transformations are called phase changes or **phase transitions.** Water is a common example of a substance that exhibits this behavior, but many chemically stable substances do likewise. The temperature at which these phase transitions occur are determined by the strength of the attractive interatomic or intermolecular forces; the stronger the attractive forces, the higher the melting (or fusion) and boiling points. Intermolecular interactions are conveniently represented by an **intermolecular potential energy** curve, which bears some resemblance to the bond potential energy curves encountered earlier, but, for substances that are liquids or gases at 25°C and 1 atm (NTP), it has a much smaller well depth (<10 kcal/mol) and larger intermolecular distance (>3 Å) where the potential minimum occurs.

When a substance is heated at constant pressure, it absorbs energy $q_P = C_P(T - T_0)$, where C_P is the heat capacity of the sample at constant pressure. If q_P is plotted against T, a straight line of slope C_P results. The value of C_P depends on which state the substance is in; in addition, when a phase transition occurs, heat is added without a change in T until the change in phase is complete. Thus a plot of q_P versus T, a **heating curve,** will show three linear regions of different slopes separated by discontinuous upward jumps at the melting or fusion and boiling points. The heights of these jumps are the heats required to fuse and boil the substance, called the heat of fusion, $\Delta H°_{fus}$, and the heat of vaporization, $\Delta H°_{vap}$, respectively. Typically $\Delta H°_{vap} \gg \Delta H°_{fus}$, since to vaporize a substance the intermolecular attractions must be overcome completely. This also makes $\Delta H°_{vap}$ a good measure of the depth of the intermolecular potential. Adding heat also causes the entropy S to increase; a plot of $S°$ versus T is similar in appearance to a heating curve, with jumps $\Delta S° = \Delta H°/T$ at T_f and T_b.

Combining the behavior of H and S allows the free energy $G = H - TS$ to be analyzed as phase transitions are encountered. At a transition temperature for standard P the two phases involved (1 and 2) are in equilibrium and $\Delta G° = 0$, or $G_1° = G_2°$; away from these T's, the free energies for each possible phase are varying with T as dictated by the competition between H and S. Below T_f, the solid phase has the lowest G, and is therefore the most stable phase; between T_f and T_b, G of the liquid is lowest; above T_b, the gas (or vapor) phase has lowest G. Phase transitions therefore correspond to intersections of these $G(T)$ curves, and the transition temperatures are turnaround temperatures $T_{ta} = \Delta H°/\Delta S°$, where the spontaneous direction of phase transformation is reversed. For example, below $T_f = 0°C$ water spontaneously freezes into ice, but above $0°C$ ice spontaneously melts into liquid water.

Due to the pressure dependence of S and therefore G, the boiling point of water or other substance is pressure dependent, with boiling occurring at lower T if the external P is lower. For an ideal vapor, $G(T,P) = G°(T) + RT \ln P_{vap}$, and it is the intersection of this G curve with $G°(l)$ that dictates the boiling point. Thus for the $l \longrightarrow g$ transition $\Delta G = G(g) - G(l) = \Delta G° + RT \ln P_{vap}$. In a sealed container containing the liquid and some air space, equilibrium will quickly be established between liquid and vapor, giving $\Delta G = 0$ and $\ln P_{eq} = -\Delta G°/RT$, where P_{eq} is called the **equilibrium vapor pressure**. If the external pressure $P_{ext} < P_{eq}$, active boiling will occur until these pressures are equal, whereas if $P_{ext} > P_{eq}$ equilibrium will be attained by **evaporation**, a reversible molecular-level process. As in other chemical equilibria, $l \rightleftharpoons g$ equilibrium is a dynamic balance between the rates of evaporation and condensation, the reverse process. Normally in the earth's atmosphere, the partial pressure of water vapor P_{vap} is less than its equilibrium value P_{eq}; *relative humidity* (RH) is defined as $P_{vap}/P_{eq} \times 100\%$, and when RH $< 100\%$, evaporation is spontaneous, $\Delta G = RT \ln(P_{vap}/P_{eq}) < 0$, regardless of the total pressure. The temperature dependence of P_{eq} is expressed by an analogue of the van't Hoff equation known as the **Clausius–Clapeyron equation,** $\ln(P_2/P_1) = -(\Delta H°_{vap}/R)(1/T_2 - 1/T_1)$, where P_i is the equilibrium vapor pressure at T_i. A plot of P versus T is exponentially rising, while a plot of $\ln P$ versus $1/T$ is linear with a slope of $-\Delta H°_{vap}/R$. The **normal boiling point** T_b can be located on the P versus T curve by finding the point where $P = 1$ atm, and the *dew point,* the temperature where condensation becomes spontaneous, by knowing the RH at the prevailing T.

Because condensed phases are nearly incompressible, the melting or fusion transition occurs at a nearly fixed temperature regardless of P, described graphically as a nearly vertical line on a P versus T plot separating solid and liquid phases. A third possibility is direct conversion of solid to vapor, called **sublimation,** whose P versus T equilibrium curve is also given by the Clausius–Clapeyron equation, but with $\Delta H°_{sub}$, the heat of sublimation, replacing $\Delta H°_{vap}$, where $\Delta H°_{sub} \approx \Delta H°_{fus} + \Delta H°_{vap}$. This gives the P versus T sublimation curve a slightly steeper slope than that for evaporation. The equilibrium curves for $l \rightleftharpoons g$ and $s \rightleftharpoons g$, $s \rightleftharpoons l$ meet at a single point called the **triple point** (T_3, P_3), where all three phases are simultaneously in equilibrium—for H_2O, reaching this point corresponds to making ice water boil. A P versus T plot that includes these three equilibrium curves for a given pure substance is called a **phase diagram** for the substance. Experimentally it is found that above a certain temperature called the **critical temperature** T_c, pure substances can no longer be liquefied, and become **permanent gases.** The $l \rightleftharpoons g$ equilibrium line therefore terminates at this point (T_c, P_c), while $\Delta H°_{vap} \longrightarrow 0$ there. The topology of the phase diagram is dictated by Gibbs's **phase rule,** $f = c + p - 2$, where f is the number of thermodynamic degrees of freedom, c is the number of chemically distinct components, and p is the number of phases present. Many substances also display more than one solid form and require additional lines on the phase diagram to represent solid $\rightleftharpoons$ solid phase equilibria.

Intermolecular forces are at the root of phase behavior. Both $\Delta H°_{vap}$ and T_b correlate with the strength of the binding forces in condensed phases; **Trouton's rule** states that $\Delta H°_{vap}/T_b \approx 22$ cal/K mol, and implies that $\Delta S°_{vap}$ is approximately constant. Attractive intermolecular forces all display inverse-power dependence on the intermolecular distance R—potential $V(R) \propto R^{-s}$—and include charge–charge ($s = 1$), charge–dipole ($s = 2$), dipole–dipole ($s = 3$), charge–induced dipole ($s = 4$), and instantaneous dipole–induced dipole or London dispersion ($s = 6$). In a pure covalently bonded compound, only the dipole–dipole (DD) and London dispersion (LD) attractions can operate, DD only for polar molecules, LD always. DD depends on the dipole moment μ and can be attractive or repulsive for different dipole orientations, while LD depends on the polarizability α and is always attractive. The polarizability of a molecule is proportional to the volume of its electron cloud; therefore, LD increases with the number of electrons and the molecular mass of the species. When the hydrides of a given group are compared, it is found that DD is more important for the lighter central atoms while LD dominates for the heavier ones.

When the bonding in a hydride is sufficiently polar and the central atom also has one or more lone pairs, a special type of DD attraction occurs called **hydrogen bonding** (HB). In this interaction, a polar hydrogen atom is situated in a nearly collinear arrangement between its bonding partner and the lone pair atom, as X—H----:Y, producing the strongest intermolecular attraction known, with a "bond" energy of about 5 kcal/mol. HB is most evident when X and Y are chosen to be some combination of N, O, and F, and is of central importance in determining the conformations of large molecules where it is present, especially of proteins and the nucleic acids. Because of its unique strength and geometric requirements, HB is classified separately; we therefore can have LD, DD, and/or HB attractive forces between covalent molecules.

When a pair of molecules is squeezed together, repulsion due to overlap of filled orbitals on the two molecules overwhelms the attraction, effectively accounting for the finite size of molecules. This so-called *Born–Mayer* repulsion falls off exponentially with distance, $V = Ae^{-BR}$, and the complete intermolecular potential for a pair of nonpolar molecules is then $V = Ae^{-BR} - CR^{-6}$. This function displays a minimum where the repulsive and attractive forces balance, at roughly the average distance between nearest neighbors in the liquid or solid state. The overall attractive and repulsive forces are called **van der Waals forces,** and correlate with the "a" and "b" constants in the van der Waals equation of state given in Chapter 9.

Simple substances that are solids at NTP include all but one of the elemental metals and metalloids, and their salts with nonmetals. These species generally form crystals, regular geometric arrays described as **crystal lattices** with atoms occupying the lattice sites. Metals normally form one of two types of **cubic** crystals, **body-centered cubic (bcc)** or **face-centered cubic (fcc),** or a **hexagonal** crystal, **hexagonal close-packed (hcp).** The geometric designations refer to the shape of the **unit cell,** the smallest figure which, when repeated in all three dimensions, generates the entire crystal lattice. In these three lattice types, the atoms are extensively bonded, to 8 nearest neighbors for bcc, and 12 for fcc and hcp; the number of nearest neighbors is called the **coordination number.**

Most of our information about the geometry of crystals comes from **X-ray diffraction** experiments, in which a beam of X-rays of selected wavelength is reflected by a solid sample. Rays reflected from different planes in a crystal will destructively interfere unless the difference in path length they traverse is an integral number n of X-ray wavelengths λ. This leads to the **Bragg law** of X-ray diffraction, $n\lambda = 2d\sin\theta$, where d is the distance between adjacent reflecting planes and θ is the angle of incidence of the X-rays relative to the plane. Reflection from a powdered sample produces a bright ring at angle 2θ with respect to the beam direction for a given set of planes. The Bragg law (with $n = 1$) can then be used to determine d, while the number and spacing of the rings, which correspond to the possible atomic planes, determine the crystal lattice type.

The bonding in metals is described by the **band theory of solids,** a generalization of molecular orbital theory, in which the valence electrons occupy delocalized molecular orbitals that extend over the entire crystal lattice, and the orbital energies form a continuous band of energies. Each metal crystal can be thought of as a giant molecule. The highest band of occupied orbitals forms the **valence band,** while the lowest unoccupied orbitals comprise the **conduction band;** electrons excited into the latter are responsible for conducting electricity. The highest energy in the valence band is called the **Fermi level.** In a metal there is no energy gap between these two bands, and electrical resistance is low.

The metalloids and central nonmetals, particularly the elements of Group IV, can form **nonmetallic network solids,** in which the atoms have a lower coordination number and bond directionality as in their small molecular compounds. Group IV examples are C(*dia*) and Si(*s*). This results in a **3-D network** that is usually modeled with sp^3-hybridized atoms. An energy gap or **band gap** exists between the valence and conduction bands in these crystals that ranges from 5.2 eV in C(*dia*) to only 0.06 eV in gray Sn(*s*). In the middle of this range, for Si and Ge, electrical properties lie between a perfect insulator and a metallic conductor; these elements are **semiconductors.** By adding small amounts of Group III or V impurities, the resistance of Group IV semiconductors can be controlled; Group III causes positive "holes" to form a small empty band just above the valence band, while Group V forms a small filled band just below the conduction band. The former is called a *p*-type semiconductor, the latter an *n*-type, and junctions between them are used to make rectifiers, transistors, light-emitting diodes, and solar cells.

A substantial number of elements near the metal–nonmetal divide display **allotropy,** more than one stable crystal type, often with one metalloid and one network-type structure. The most stable form of carbon is not diamond but graphite, in which C is three- rather than four-coordinate, and forms infinite sheets of fused benzene-like rings. Due to the delocalized π bonding, C(*gr*) is a good conductor in contrast to C(*dia*). A third allotropic carbon family has recently been discovered in which C forms shells containing 60 C atoms (C_{60}), a form named *buckminsterfullerene;* many analogues of C_{60} now have been synthesized, and the number of allotropes of C is increasing with no end in sight.

Ionic compounds have uniformly high melting and boiling points due to the coulomb force. They tend to form cubic lattices in which each ion has a shell of nearest neighbors of opposite charge, and the **lattice energy** can be evaluated by summing the Born–Mayer repulsions between nearest neighbors combined with the coulomb interactions extending across the entire crystal. The results agree well with Hess's law values based on the

Born–Haber cycle, confirming the validity of the ionic model. As the metal and nonmetal groups draw together, the bonding passes from ionic to polar covalent; correspondingly, the band gap, which is large for ionic salts, decreases to values in the semiconductor range for III–V compounds such as GaAs.

Polar-covalent or pure-covalent molecules form solid crystals in which the identity of the molecules is preserved, due to the relative weakness of the intermolecular forces. For polar molecules, the orientation of each molecule in the lattice is important to the stability of the crystal; in water ice, for example, each H_2O unit is oriented so as to make four hydrogen bonds with its neighbors in a diamond-like structure. Crystals of larger molecules have become tools for studying their 3-D structures through the use of X-ray crystallography; this is the main source of information on the structure of biologically important compounds.

Liquids are a relatively poorly understood phase from a structural viewpoint. Their free energies are generally only marginally lower than those of their solid phases; X-ray scattering experiments suggest they have short-range order similar to that in the corresponding solid, but no long-range order.

The dissolving of a solute in a solvent can be pictured as a three-step process in which the solvent undergoes cavitation, the solute is dispersed, and then solvated in the solvent cavities. The balance among the enthalpies and entropies of these steps is delicate, making both solubility and its main driving force highly variable even for ionic solutes dissolving in water. Some properties of solutions, such as **freezing point depression, boiling point elevation,** and **osmotic pressure,** are found to be universal or **colligative,** being relatively independent of the nature of the solvent and solute. These properties can be interpreted in terms of a gas-like model, in which the free energy of the solvent A is lowered by $RT \ln X_A$ relative to pure $A(l)$, where X_A is the mole fraction of A in the solution. This leads to **Raoult's law** $P_A = X_A P_A^\circ$; that is, the vapor pressure of the solvent is lowered. Lower vapor pressure implies that a higher temperature will be required to reach $P_A = 1$ atm, that is, there will be boiling point elevation. Making a series of approximations appropriate for a dilute solution leads to $\Delta T_b = K_b m_B$, where ΔT_b is the increase in the boiling point; m_B is the **molality** of solute B, mol B/kg A and K_b, the boiling point elevation constant, is characteristic of the solvent. Similar reasoning leads also to freezing point depression $\Delta T_f = K_f m_B$. A solution will generate osmotic pressure when brought in contact with pure solvent due to this same free energy lowering, provided the solutions are separated by a **semipermeable membrane** that allows only solvent and not solute to pass through it. The solvent will tend to diffuse into the solution compartment, a process called **osmosis,** thereby raising its liquid level or, if the volume is constant, increasing the pressure. This pressure is called the osmotic pressure Π, and is given approximately by $\Pi = cRT$, where c is the molarity of the solute; this form is exactly analogous to the ideal gas law $PV = nRT$.

EXERCISES

Heating curves and phase changes

1. Freezer units in refrigerators typically achieve a temperature of $-15°C$. (a) How much heat (cal or J) must be removed from 1.00 L of water at 25°C and 1 atm to produce ice cubes at $-15°C$ and 1 atm? (b) Find the entropy and free-energy changes accompanying this cooling process. Use data from Table 14.1, and assume the heat capacities are temperature independent.

2. Ammonia (NH_3) is used as a refrigerant fluid in gas refrigerators. Use information from Table 14.1 to construct a heating curve (q_P versus T) for NH_3 between $-100°$ and 25°C at 1 atm, assuming the heat capacities are temperature independent. Use your curve to estimate the heat involved (cal or J) when 1 mol NH_3 passes from $-50°$ to 0°C at 1 atm.

3. Dry ice, $CO_2(s)$, is frequently used as a refrigerant in shipping frozen foods and other perishables. Use information from Table 14.1 to construct a heating curve for CO_2 between $-78.5°$ and $-15°C$ at 1 atm, including sublimation of the solid. How does your curve differ from a "normal" heating curve like that for water in Figure 14.3? How much heat (cal or J) can 1.00 kg of $CO_2(s)$ at $-78.5°C$ and 1 atm absorb and still remain at or below $-15°C$?

4. Welding requires the melting of metal. How much heat (cal or J) is required to melt a 20.0-g iron rod initially at 25°C? Can this heat be supplied by producing this amount of iron from the thermite reaction (see Example 10.6)?

5. Liquid sodium, $Na(l)$, may be used as a coolant in future nuclear power plants. (a) How much heat is required to melt 100. kg Na at 1 atm, starting from 25°C? (b) What are the entropy and free-energy changes accompanying the heating process? Use data from Table 14.1, and assume the heat capacities are temperature independent.

Free energy and transition temperatures

6. Boron trichloride (BCl_3) is a gas at room temperature, but is readily condensed to a liquid with moderate cooling. Estimate the condensation temperature at 1 atm for BCl_3 using data from Appendix C.

7. Ethanol (CH_3CH_2OH), the toxin in alcoholic beverages and mouthwashes, is a liquid at 25°C and 1 atm in pure ("absolute") form and is often used as a nonaqueous solvent. Use data from Appendix C to estimate the normal boiling point of pure ethanol, and compare with the experimental result in Table 14.1. Briefly explain any discrepancy you find.

8. Dichloromethane (CH_2Cl_2, commonly methylene chloride) is often used as an extraction solvent in biochemical analyses. Use data from Appendix C to estimate the normal boiling point of CH_2Cl_2, and compare with the experimental T_b of Table 14.1. If the temperatures differ, explain briefly.

Vapor pressure, evaporation, and boiling

9. Estimate the boiling point of water in the Rocky Mountains, where atmospheric pressure is 510 Torr, using data from Table 14.1, in two ways. (a) Assume that T_b (rockies) is a turnaround temperature, and find the appropriate ΔS from $\Delta S°$ and the pressure dependence of S. You may assume that ΔH is pressure independent, that is, that $\Delta H \approx \Delta H°$. (b) Use the integrated form of the Clausius–Clapeyron equation (Equation 14.8) with the normal boiling temperature as a reference point.

10. Although propane (C_3H_8) is a gas at room temperature and atmospheric pressure, it is usually stored as a liquid in pressurized tanks for use as a fuel. What minimum pressure must a propane tank withstand in order to store liquid propane on a hot day in summer when the temperature reaches 40°C?

11. On a hot day in summer, the relative humidity might be 70% at 90°F (32°C). (a) What is the partial pressure (atm and Torr) of water vapor in the atmosphere under these conditions? (b) Is the liquid water in a swimming pool at equilibrium with its vapor under these conditions? If not, what is the free-energy change per mole of water for evaporation? (c) Find the dew point (°C and °F).

12. On a cold day in winter, the relative humidity might be 40% at 0°F (-18°C). (a) What is the partial pressure (atm and Torr) of water vapor in the atmosphere under these conditions? (*Hint:* You will need $\Delta H°_{sub}$, which you can obtain from Equation 14.10 using data in Table 14.1. Use the triple point as a reference state.) (b) Find the **ice point**, the temperature at which the water vapor will freeze. (c) Find the relative humidity in a home heated to 65°F (18°C), assuming the only source of water vapor is that from outdoors.

13. At a height of 50 km, the pressure of the atmosphere is 0.75 Torr. What will happen to an ice cube if it is heated at this altitude? Find the temperature at which sublimation becomes spontaneous.

14. In the training of jet pilots, a glass of water is secured to a ledge within sight of the pilot. When in flight, the water begins to boil, signaling loss of cabin pressure. If the cockpit temperature is 10°C, what is the air pressure in the cockpit?

Phase behavior and phase diagrams

15. Estimate $\Delta H°_{sub}$ for $CO_2(s)$ from the triple point (T_3, P_3) and T_{sub} given in Table 14.1, and compare with the value given there. (Recall that CO_2 sublimes rather than melting at 1 atm.)

16. At pressures above $P_3 = 5.15$ atm, solid CO_2 will melt instead of subliming. Estimate $\Delta H°_{fus}$ for CO_2 from the enthalpy data given in Table 14.1. What limits the accuracy of your answer?

17. Like $\Delta H°_{vap}$ and T_b, the critical temperature T_c is a measure of the strength of the intermolecular attraction. Compute the ratio T_c/T_b for the noble gases Ne, Ar, Kr, and Xe, and the hydrogen halides HF, HBr, HCl, and HI. Noting that T_c/T_b is nearly constant, use the average value from these eight substances to predict T_c for H_2O, PH_3, and SiH_4 from their T_b's in Table 14.1. Give a percent error where experimental values of T_c are available for comparison.

18. The integrated form of the Clausius–Clapeyron equation (Equation 14.8) is approximate because it neglects the T dependence of $\Delta H°$. By substituting Equation 14.11 into the differential form of the Clausius–Clapeyron equation and integrating both sides, derive an improved approximation. Use it to estimate the vapor pressure of water at 25°C using the boiling point as reference and data from Table 14.1. Compare your result with that obtained from the normal form of the Clausius–Clapeyron equation and with the experimental value of 23.8 Torr.

19. Use information from Table 14.1 to sketch phase diagrams for the following pure substances:
    a. HCl
    b. Hg
    c. NH_3
    d. Na
    e. O_2

20. As discussed in Section 14.3, the solid–liquid equilibrium line on the phase diagram leans slightly to the left for H_2O, as given by Equation 14.14. Use this equation to estimate the pressure needed to reduce the freezing point of water by 0.1°C.

21. Using Equations 14.12 and the definitions of H and G, derive the Clausius equation, Equation 14.13, from the equilibrium condition in differential form, $dG_1 = dG_2$. (Hint: Eliminate dq from Equations 14.12; then find expressions for the differentials dH and dG. You should end up with $dG = V\,dP - S\,dT$, which may be applied to each of the two phases at equilibrium.)

Phase behavior and intermolecular forces

22. Normally Trouton's rule, Equation 14.16, is applied to compounds that are liquids or gases at NTP, but it works nearly as well for solids. Check Trouton's rule for the elements Pb and Ge, and use it to predict ΔH°_{vap} for Si and SiO_2; necessary data can be found in Table 14.1.

23. Combine Trouton's rule and the Clausius–Clapeyron equation to derive the approximate expression $\ln P = 11(1 - T_b/T)$, where P is the vapor pressure of a liquid in atm and T_b is its boiling point in K. Use it to estimate the vapor pressure of acetone (CH_3COCH_3) at 25°C from data in Table 14.1; compare this with P calculated from the Clausius–Clapeyron equation with boiling as the reference point. Which of these do you expect to be more accurate? (The experimental value is 231 Torr.)

24. Compare ΔS°_{vap} for CH_3CH_3, CH_3F, and CH_3OH as computed from data in Table 14.1. Which of these compounds shows the most ordered liquid state? The least ordered?

25. How strongly are ions solvated in a solution? Calculated the attractive ion-dipole potential energy (kcal/mol or kJ/mol) between Na^+ and H_2O using $\mu(H_2O) = 1.85$ D, and the van der Waals radii $r(Na^+) = 0.85$ Å and $r(H_2O) = 1.50$ Å. Repeat the calculation for $Cl^- + H_2O$, where $r(Cl^-) = 1.50$ Å. (Hint: The potential-energy formulas in Section 14.4 are given in cgs-esu units; use the definitions 1 esu $\equiv$ 1 [erg·cm]$^{1/2}$ and 1 D $\equiv 10^{-18}$ esu·cm, and the conversions 1 erg/molecule $= 1.439 \times 10^{13}$ kcal/mol $= 6.022 \times 10^{13}$ kJ/mol.) Which ion is more strongly solvated? Why? (Note that what you have calculated is not the intermolecular "bond energy" ε, because it does not include the Born–Mayer repulsion.)

26. Compute the attractive dipole–dipole (DD) potential energy (kcal/mol or kJ/mol) between two water molecules, using data and hints from Exercise 25. Is this attractive energy sufficient to account for the magnitude of ΔH°_{vap} for water? What forces have been neglected in your calculations?

27. Ammonia dissolves readily in water, but most of the NH_3 remains in molecular form. Using $\mu(NH_3) = 1.49$ D, $r(NH_3) = 1.73$ Å, and other data from Exercise 25, calculate the attractive dipole–dipole potential energy (kcal/mol or kJ/mol) between ammonia and water. Compare it to the result of Exercise 26.

28. In addition to the ion–dipole interaction between Na^+ and H_2O of Exercise 25, Na^+ can *induce* a further dipole in water. Use $\alpha(H_2O) = 1.45$ Å^3 and data from Exercise 25 to compute the ion–induced dipole attractive potential energy (kcal/mol or kJ/mol), and compare it to the result of Exercise 25.

29. An approximate formula for coefficient C in the London dispersion (LD) energy $V = -C/R^6$ is

$$C = \tfrac{3}{2}\alpha_A\alpha_B \frac{IE_A\, IE_B}{IE_A + IE_B}$$

where the α's are polarizabilities and the IE's are ionization energies. Here is a short table of data for computing the dispersion energy:

Molecule	α (Å^3)	IE (eV)	r_{vdW} (Å)	μ (D)
Ar	1.64	15.76	1.88	0
H_2O	1.45	12.6	1.50	1.85
NH_3	2.10	10.2	1.73	1.49
CH_4	2.59	12.6	1.96	0
HF	2.46	15.77	1.61	1.82
HCl	2.77	12.74	1.85	1.08
HBr	3.61	11.62	1.95	0.82
HI	5.35	10.39	2.10	0.44

Use data from this table to compute C (eV Å^6) and the London dispersion (LD) energy (eV and kcal/mol or kJ/mol) at the van der Waals minimum for the following cases:

a. Ar----Ar
b. H_2O----H_2O
c. NH_3----NH_3
d. HF----HF
e. CH_4----CH_4
f. H_2O----NH_3
g. H_2O----HF

30. For the interaction of two Xe atoms, the van der Waals well depth ε is 0.48 kcal/mol. Use this value along with $r_{vdW}(Xe) = 2.20$ Å and the dispersion constant $C = 287$ eV Å^6 to determine the A (eV and kcal/mol) and B (Å^{-1}) constants in the Born–Mayer repulsion Ae^{-BR}. [Hint: Using $R_m = 2r_{vdW}(Xe)$, the minimum in the intermolecular potential $V(R)$, Equation 14.18, occurs at $(R_m, -\varepsilon)$, where $dV(R)/dR = 0$. This yields two equations that can be solved for A and B.] By what percentage does the repulsion cancel the attraction at the potential minimum?

31. For the attraction between two water molecules, add the result of Exercise 29(b) (LD) to that of Exercise 26 (DD) to obtain the total attractive energy (kcal/mol or kJ/mol) at the van der Waals distance. The strength of a hydrogen bond (HB) was given in Section 14.4 as 5 kcal/mol (21 kJ/mol); how does the DD + LD energy compare to this? What significant energy has been omitted, and what would its effect be? (See Exercise 30.)

32. If the dipole moments of two water molecules happen to be opposed, for example, H$_2$O--OH$_2$, $\mapsto\ \mapsto_2$, the DD potential energy becomes positive, $V = +2\mu^2/R^3$. Is there any residual attraction between the two molecules in this orientation? You can use the results of Exercises 26 and 29(b).

33. In a small table, compare the total attractive energy (DD + LD) between two H$_2$O molecules to that of H$_2$O----NH$_3$, H$_2$O----CH$_4$, and CH$_4$----CH$_4$. Use data and results of Exercises 26, 27, and 29. How does this help us to explain why water is a liquid while methane is a gas at NTP? In light of the discussion of solutions in Section 14.6, how does this help us to understand the rule *like dissolves like?* (In this case H$_2$O dissolves NH$_3$ more readily than CH$_4$.)

34. Use data and results of Exercise 29 to make a small table showing distances, DD, LD, and total attractive energies for the hydrogen halides. Compare HF and HI in terms of the percentage of the total attraction from LD. Use your results to interpret the vaporization data of Figure 14.12, in particular, the anomalously high T_b and ΔH°_{vap} of HF.

Metallic solids and X-ray diffraction

35. X-rays of wavelength $\lambda = 1.476$ Å (from a W anode) are scattered from a polycrystalline nickel metal sample, producing a diffraction ring at an angle (2θ) of 42.2° with respect to the X-ray beam direction. Assuming first-order diffraction, find the spacing d between atomic planes in the nickel lattice that caused this ring. Noting that angles 2θ greater than 180° cannot occur, do you expect a second-order ring to appear from this set of planes? At what angle (2θ) should it occur?

36. The spacing between two atomic planes in a silver crystal is 2.351 Å. At what angle(s) 2θ will X-rays of wavelength $\lambda = 2.070$ Å (from a Cr anode) be scattered? Note that $0° < \theta < 90°$.

37. The diffractogram for copper metal of Figure 14.18 shows a ring (labeled by $hkl = 111$) at an angle (2θ) of 50.98° for an X-ray wavelength of 1.793 Å. What is the spacing d between atomic planes in Cu that gave rise to this ring? Assuming that the pattern of rings shown for Cu is characteristic of a fcc lattice, use the formula $a = d[h^2 + k^2 + l^2]^{1/2}$ to find the edge length of the unit cell a, and the information in Figure 14.15 to find the atomic radius r of Cu. Do any of the other diffraction rings arise from the 111 planes?

38. Using information given in Figure 14.15, determine the *packing fraction f* for the bcc and fcc lattices, where f is the ratio of occupied volume to total volume within the unit cell. Which lattice type shows more efficient packing?

39. From X-ray diffraction we know that silver crystallizes in a face-centered cubic lattice with a unit cell edge length (*lattice parameter*) $a = 4.0862$ Å. The density of silver is 10.501 g/cm^3. Use these data and information given in Figure 14.15 to derive a value for Avogadro's number N_A.

Band theory of solids

40. Extend Figure 14.19 to the case of five Li atoms on a line; qualitatively sketch the energy levels and their associated orbitals, along with occupancies. Generalize your results as to the number of nodes n in the middle and highest energy orbitals as a function of the number of atoms N.

41. Use the $2p\sigma$ orbitals of Li to sketch the p-band analogue to the four-atom s-band energies and orbitals of Figure 14.19. A useful principle to use in making your orbital sketches is that the number of adjacent pairs of atoms forming a bonding (rather than antibonding) overlap is $N - 1$ in the lowest level, decreases by one for each step up the energy ladder, and ends with no pairs bonded in the highest level. How do the nodal patterns compare with those for $2s$ bonding? What

will happen to the lowest energy $2p\sigma$ and highest energy $2s$ levels as the number of atoms is increased?

42. An approximate analysis of the N-atom linear molecule as shown in Figure 14.19 leads to a formula for the energy levels,

$$E_k = \alpha + 2\beta \cos\left(\frac{k\pi}{N+1}\right), \quad k = 1, 2, 3, \ldots, N$$

where α is the energy of the atomic orbital, β is the exchange energy discussed in Chapter 6, and k is the quantum number. (a) Compare the $N = 2$ and $N \longrightarrow \infty$ cases as to the spacing between the lowest and highest energy levels. (b) Give a formula for the spacing between adjacent energy levels k and $k + 1$, where $k = (N + 1)/2$, that is, in the middle of the band. What happens to the spacing as N grows? [Hint: Use the trigonometric identity $\cos(x + y) = \cos x \cos y - \sin x \sin y$ to simplify the difference formula. For large N the approximation $\sin x \approx x$, valid for small x, will be useful.]

43. Construct a diagram analogous to that for Li and Be in Figure 14.20 that describes the energy bands in Na(s) and Mg(s) crystals. Given that Mg is a good electrical conductor, what must be the relationship between the $3s$ and $3p$ bands?

44. For a portion of a carbon-atom plane of graphite, as shown in Figure 14.25, give the hybridization of C and a sketch of the lowest π molecular orbital. (Consult Chapter 7 for the example of benzene.) Since this is only the lowest π orbital in the π band, it accommodates only two electrons. Where are the other π electrons?

Ionic solids

45. Note that, as shown in Figure 14.27, the NaCl crystal lattice can be viewed as a face-centered cubic lattice of Cl^- ions, as in Figure 14.15, with the gaps or holes along the edges of the unit cell, as well as in the very center, filled by the smaller Na^+ ions. X-ray diffraction on the crystal yields a cell edge length of $a = 5.64$ Å. If the Cl^- ions along a face diagonal "touch," and the Na^+ and Cl^- ions along an edge also "touch," (see the space-filling model in Figure 14.27) what are the effective ionic radii r_+ and r_-?

46. As Figure 14.27 suggests, the CsCl crystal lattice can be viewed as a body-centered cubic lattice, in which the Cl^- ions occupy the vertices of the unit cell, and the Cs^+ ion the body center. X-ray diffraction on the crystal yields a cell edge length of $a = 4.20$ Å. If the Cl^- ions along an edge "touch," and the Cs^+ and Cl^- ions along the interior body diagonal also "touch," as in the bcc space-filling model in Figure 14.15, what are the effective ionic radii r_+ and r_-?

47. Construct a Born–Haber cycle for NaI analogous to that given in Figure 14.28 for NaCl, using data from Figures 4.9 and 4.11 and Appendix C. What is the lattice energy E_L predicted by the cycle? The theoretical lattice energy based on the ionic model is -163 kcal/mol.

Solutions and colligative properties

48. Using data from Appendix C, find $\Delta H°$, $\Delta S°$, and $\Delta G°$ for dissolving (a) $MgCl_2$ and (b) $Mg(OH)_2$ in water to form ionic solutions. What is the dominant factor affecting the solubility in each case? Discuss briefly in terms of solute–solvent interactions.

49. Using the approximations described beneath Equation 14.24 to simplify it, show that Equation 14.25 is the result.

50. Use Equation 14.25 and the data in Table 14.1 to calculate the boiling point elevation constant K_b for water. Compare with the value given in Table 14.2. Predict K_b for acetone, CH_3COCH_3.

51. As Figure 14.34 shows, the freezing point of a solution is determined by the point of intersection of the depressed liquid–vapor equilibrium line with the solid–vapor line. At this point the free energies of the solution and the pure crystallized solvent are equal, that is, $G_A°(l) + RT \ln X_A = G_A°(s)$. Use this equation to show that

$$\ln X_A = -(\Delta H_{fus}°/R)(1/T - 1/T_f)$$

a relationship analogous to Equation 14.24. (From this equation flows Equation 14.26, using the same set of approximations used to get Equation 14.25.)

52. The various ways of expressing concentration of a solute B in a solvent A—molarity $c_B = n_B/V$, molality $m_B = 1000\, n_B/g_A$, and mole fraction $X_B = n_B/(n_B + n_A)$—are all related, provided the densities ρ and ρ_A and molecular masses M_A and M_B are known. Here V is the volume in L, g is the mass in g, the factor of 1000 converts between kg and g, and the lack of a subscript indicates the solution itself. The mass of solvent A is $g_A = 1000 \rho_A V_A = n_A M_A$, and the mass of solution is $g = g_A + g_B = g_A + n_B M_B = 1000 \rho V$.

From these relations and the definitions, show that

$$c_B = 1000\rho n_B/(g_A + n_B M_B)$$

By eliminating g_A between the definition of m_B and this result for c_B, show that

$$m_B = c_B/(\rho - c_B M_B/1000)$$

Also show that $m_B \approx 1000 X_B/M_A$ for $X_B \ll 1$, as assumed in the derivation of the boiling point elevation formula Equation 14.25.

53. In candy-making, large amounts of sucrose, $C_{12}H_{22}O_{11}(s)$, are dissolved in water to make syrup. Estimate the boiling point of a syrup made by dissolving 100. g sucrose in 200. mL water.

54. An antifreeze solution consists of 25% by volume ethylene glycol $C_2H_6O_2$ in water. The density of ethylene glycol is 1.110 g/mL and it has negligible vapor pressure. Estimate the freezing point of the antifreeze.

55. Suppose you want to boil an egg in the Rocky Mountains, where atmospheric pressure is 510 Torr. As shown in Exercise 9, water boils well below 100°C there. Estimate the mass of salt (NaCl) to be dissolved in 1.00 L of water that will raise its boiling point back to 100°C. (*Hint:* You will have to use the local boiling point as a reference point, and recalculate K_b from Equation 14.25.)

56. Compare the aqueous freezing point depressions produced by equal masses of NaCl(s) (rock salt) and $CaCl_2(s)$. Per unit mass, which will be more effective as a road salt?

57. 50.0 mg of an unknown organic compound is mixed with 1.00 g camphor. The melting point of the mixture is found to be 171.3°C. Estimate the molecular mass of the unknown compound. (See Table 14.2 for necessary data.)

58. What will be the osmotic pressure exerted by a 0.200 M solution of sucrose in water at 37°C? If the density of the solution is 1.09 g/cm^3, and the apparatus of Figure 14.35 is used, how high a column of solution would be supported by this pressure? (Recall from Chapter 9 that the pressure necessary to support a column of barometric fluid at a height h is $\Pi = \rho g h$.)

59. An osmotic-pressure apparatus like that shown in Figure 14.35 can be used to determine the molecular weights of macromolecules such as polymers. In one determination, 7.00 g of polyvinyl chloride (PVC, [—CH_2—$CHCl$—]$_n$) was dissolved in 1.00 L of a cyclohexanone ($C_6H_{10}O$) solvent, and the solution ($\rho = 0.980$ g/cm^3) rose in the capillary by 5.10 cm above solvent level at 295 K. What is the osmotic pressure Π in Torr and atm? Find the molecular mass of the polymer, and estimate the number of monomer units in the polymer (the value of n above).

60. The molarity of solute ions in seawater is approximately 2 mol ions/L. In the desalination of seawater by reverse osmosis, estimate the minimum external pressure that must be applied to the salt solution to force pure water through a semipermeable membrane into a tank of pure water. Assume a temperature of 15°C.

CHAPTER 15

Rates and Mechanisms of Chemical Reactions

The hydrogen-filled airship *Hindenburg* seconds after ignition of the H_2-O_2 reaction by a static spark as it was about to land in New Jersey after a trans-Atlantic flight from Frankfurt, Germany on May 6, 1937. In only 34 seconds the airship—ten times the size of modern "blimps"—and its 212,000,000 L of H_2 were completely consumed. Most of the passengers and crew survived, nonetheless, by dropping to the sandy landing area below. See Section 15.5 for a discussion of flames and explosions.

Chemical phenomena must be treated as if they were problems in mechanics.

Lothar Meyer (1868)

"Oh, ye'll tak' the high road an' I'll tak' the low road An' I'll be in Scotland afore ye . . . "

Refrain from *Loch Lomond*, old Scottish folk song

CHAPTER OUTLINE

15.1 On the Way to Equilibrium

15.2 Rate Laws and Rate Constants

15.3 Reaction Mechanisms

15.4 Molecular Theories of Elementary Reactions

15.5 Chemical Catalysis and Complex Reactions

Now we come to the play itself. Imagine viewing only the beginning and the end of Shakespeare's *Hamlet*. At the start we are introduced to a cast of curious characters. Then, in the end, many of them are strewn over the stage, dead. What sequence of events led to this? How did the characters interact? And who were the most important ones? Were there any characters who arrived in the middle, unbeknownst to us? How did these interlopers influence the outcome?

After some playful electron pushing early on in Chapter 5, we have mainly been describing only the beginning and the end of chemical dramas in Chapters 10 through 14. This is where thermodynamics reigns supreme. As in a play, however, much of the modern interest in chemical transformations lies in how they actually happen, what the molecules actually do. It is through that understanding that we can achieve the ultimate predictive power and control over the outcome of a chemical event. Intimately coupled to this molecular-level detail is the macroscopic *rate* at which the reaction converts its reagents into products. If the molecules are having a hard time getting enough energy or getting into the right orientation for reaction, the reaction is observed to be slow. Thus the study of reaction rates, known as **chemical kinetics,** and its offspring, chemical reaction dynamics, has played a paramount role in unraveling the mystery surrounding the chemical act itself. As we will see, the nature of the reagents as well as the appearance of unstable interlopers known as *reaction intermediates* play essential roles in the chemical outcome. Foreign substances known as *catalysts,* not part of the balanced chemical equation, can dramatically increase the rate. Aside from its practical value, chemical kinetics has the ultimate goal of allowing us to look upon what we have learned, and see how the nature of the players leads to their eventual fates, that is, to correlate the structural features of the reagent molecules with the new molecules they yield and their rates of production.

Svante Arrhenius (1859–1927) eventually became a professor of physics at the Royal Institute of Technology in Stockholm, Sweden. Most famous for his controversial theory of free ions in solution, for which he was nearly failed in his doctoral examination, Arrhenius proposed an exponential form for the temperature dependence of the rate constant of an elementary step in a chemical reaction that has since permeated every area of chemical dynamics.

15.1 On the Way to Equilibrium

You were introduced to the idea of the rate of molecular processes in Chapter 9, where predictions of the kinetic molecular theory as to the rates of effusion and molecular collisions were developed. Here we would like to introduce the notion of reaction rate in a more macroscopic, phenomenological way that relates to your knowledge of equilibrium, that is, to ask how the system gets to equilibrium and how long it takes. The new variable is the *time t,* a strictly nonthermodynamic quantity. As time passes, a reaction proceeds toward equilibrium, as characterized by the approach of the extent-of-reaction variable x to its equilibrium value. *During* the reaction, x depends on the time elapsed, that is, $x = x(t)$.

As in Chapter 12, we will start with an example from gas-phase chemistry. Here we will follow a long-standing convention in gas-phase chemical kinetics and express the amounts of reagent and product in terms of concentrations c (usually moles per liter or molecules per cubic centimeter) rather than partial pressures P. At constant volume, either c or P is proportional to the number of moles present, and we have $P = cRT$ for ideal gases. Consider, for example, the thermally induced gas-phase decomposition of chloroethane into ethylene and hydrogen chloride in a rigid container at 530°C. In terms of concentration the extent-of-reaction analysis is

$$\underset{c_0 - x}{C_2H_5Cl(g)} \longrightarrow \underset{x}{C_2H_4(g)} + \underset{x}{HCl(g)} \tag{15.1}$$

where $c_0 \equiv [C_2H_5Cl]_0$ is the initial concentration in moles per liter, and the concentrations of possible reaction intermediates are neglected. This reaction is typical of decompositions, endothermic ($\Delta H° = +17.24$ kcal) but favored by entropy ($\Delta S° = +31.16$ cal/K), and spontaneous at sufficiently high temperature; at 530°C, $\Delta G° = -7.78$ kcal/mol and $K = 131$. The K value assures us that very little starting material will remain at equilibrium; what interests us now is how and how fast the system achieves it. Since $\Delta n_{gas} = +1$ mol, the total pressure in the reaction vessel will rise as equilibrium is approached, and by monitoring the pressure with time, we can learn about the time dependence of the extent of reaction, $x(t)$, by using the ideal gas law and Dalton's law of partial pressures. If the reaction were complete, the pressure would double after sufficient time had passed for the attainment of equilibrium, corresponding to $x \approx c_0$. From $x(t)$, we can derive the concentrations of both the reagents and the products from the extent-of-reaction analysis of Equation 15.1; a plot of these concentrations against t, shown in Figure 15.1 for this reaction, is called the **time profile** of the reaction.

As illustrated in Figure 15.1, the reaction is "over" after several minutes. We define the **reaction rate** by the time derivative dx/dt, the slope of the $x(t)$ curve. As equilibrium is approached, the slope of $(x)(t)$ gets smaller and smaller, finally going to zero; the condition $dx/dt = 0$ at long times provides us with a new definition of the equilibrium state. As we mentioned at the outset of Chapter 12, however, on a microscopic scale the reaction is never truly over; as the products build up, they begin to undergo the reverse reaction, and $dx/dt = 0$ actually reflects a balance between forward and reverse rates, an observation we will analyze in Section 15.3. The concentrations that are eventually reached at long times are just those we would calculate using the equilibrium constant at the chosen reaction temperature.

As the reaction proceeds, we can relate the decay and growth of concentrations in the system to the reaction rate based on the extent-of-reaction analysis of Equation 15.1, neglecting reaction intermediates. By taking the time derivative of each of the concentrations, we find

$$\text{Rate} = \frac{dx}{dt} = -\frac{d[C_2H_5Cl]}{dt} = +\frac{d[C_2H_4]}{dt} = +\frac{d[HCl]}{dt} \quad (15.2)$$

Because of the stoichiometric relation given by the balanced chemical equation, the rates of reagent decay and product buildup are related; decay rates are always negative, buildup rates positive. If the stoichiometric coefficients ν are not unit, they multiply x (as we have seen particularly for precipitation reactions in Chapter 12), and therefore enter these rate relations according to

Figure 15.1
Time profile for the decomposition of chloroethane $C_2H_5Cl \longrightarrow C_2H_4 +$ HCl at 530°C. The rates of change of the concentrations are the slopes of the time profiles, as illustrated. Equilibrium is reached at long times, where the slopes go to zero.

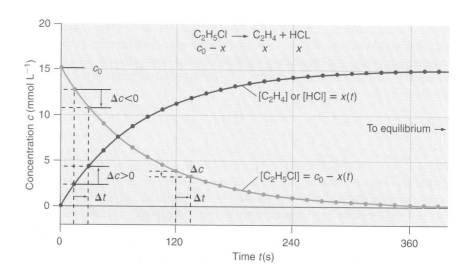

$$\frac{dx}{dt} = -\frac{1}{\nu_R}\frac{d[R]}{dt} = +\frac{1}{\nu_P}\frac{d[P]}{dt} \qquad (15.3)$$

where R stands for reagents, P for products.

EXAMPLE 15.1

The decomposition of N_2O_5

$$2N_2O_5 \longrightarrow 4NO_2 + O_2$$

has been studied both in the gas phase and in CCl_4 solution. For an initial concentration $[N_2O_5]_0 \equiv c$, set up an extent-of-reaction analysis and state the rate of the reaction dx/dt in terms of the rate of change of $[N_2O_5]$, $[NO_2]$, and $[O_2]$. Could the gas-phase rate of this reaction be studied by monitoring the pressure in a rigid container?

Solution:
We normally define the extent-of-reaction variable x to represent a mole of reaction; that way the stoichiometric coefficients can be used directly as multipliers of x. We write

	$2N_2O_5$	$\longrightarrow$	$4NO_2$	$+$	O_2
Initially,	c		0		0
At time t,	$c - 2x$		$4x$		x

noting that $0 < x < c/2$. By differentiating each concentration with respect to time, or by using Equations 15.3, we find

$$\text{Rate} = \frac{dx}{dt} = -\frac{1}{2}\frac{d[N_2O_5]}{dt} = +\frac{1}{4}\frac{d[NO_2]}{dt} = +\frac{d[O_2]}{dt}$$

At time t, the total molar concentration is $c - 2x + 4x + x = c + 3x$. Since the total pressure is then $P = (c + 3x)RT$, the pressure will increase with time, and can be used to monitor the rate of reaction. Reaction intermediates are known but may be neglected.

15.2 Rate Laws and Rate Constants

It is found experimentally that the reagent and product concentrations for many reactions show a smooth time profile like that in Figure 15.1. This suggests that the reaction rate follows a mathematical law of some form. Let's focus on the decay of $[C_2H_5Cl]$ in Reaction 15.1. Table 15.1 presents numerical time-profile data corresponding to the points shown in Figure 15.1, which might be obtained from pressure measurements at regular time intervals. We can get estimates of the rate of decay from these discrete data by finite difference, that is, $d[C_2H_5Cl]/dt \approx \Delta[C_2H_5Cl]/\Delta t$, where the Δ's represent differences between discrete data points; this is done at six points along the time profile in the table. Note that the magnitude of the rate shrinks with time, as we observed earlier. The concentration itself is also decaying with time, and when the decay rate is divided by the average concentration in the selected time

TABLE 15.1

Time profile data for $C_2H_5Cl \longrightarrow C_2H_4 + HCl$ decomposition at 530°C[a]

t (s)	$[C_2H_5Cl]$ $(c_0 - x)$ (mmol L^{-1})	$[HCl]$ (x) (mmol L^{-1})	$\dfrac{\Delta[C_2H_5Cl]}{\Delta t}$ (mmol L^{-1} s^{-1})	$\dfrac{-1}{[C_2H_5Cl]} \dfrac{\Delta[C_2H_5Cl]}{\Delta t}$ (s^{-1})
0	15.17	0.00		
15	12.79	2.38		
			−0.133	0.0113
30	10.79	4.39		
45	9.10	6.08		
			−0.095	0.0114
60	7.67	7.50		
75	6.47	8.71		
			−0.068	0.0114
90	5.45	9.72		
105	4.60	10.58		
			−0.048	0.0113
120	3.88	11.30		
135	3.27	11.90		
150	2.76	12.42		
165	2.32	12.85		
			−0.024	0.0112
180	1.96	13.21		
240	0.99	14.18		
300	0.50	14.67		
			−0.0042	0.0111
360	0.25	14.92		
420	0.13	15.04		
∞	0.00	15.17		

[a] Values in the fourth and fifth columns of this table are obtained by *finite difference*, as indicated by the Δ's, from the times t and concentrations $[C_2H_5Cl]$ straddling them. For example, the rate entry in column 4 between $t = 15$ and $t = 30$ is obtained by computing $\Delta[C_2H_5Cl] = 10.79 − 12.79 = −2.00$ mmol/L, $\Delta t = 30 − 15 = 15$ s, and taking their ratio. This approximates the *slope* of the time profile of $[C_2H_5Cl]$ in Figure 15.1. The entry in column 5 is obtained by finding the *average* concentration in the time interval, $[C_2H_5Cl] = (12.79 + 10.79)/2 = 11.79$, and dividing the negative of the rate of column 4 by this average. The near-constancy of the values in column 5 demonstrates *first-order* kinetics for this reaction; the average of these values yields the *rate constant k*.

interval, we find a result that is nearly constant, independent of time, given in the rightmost column. That is, for this reaction,

$$\frac{1}{[C_2H_5Cl]} \frac{\Delta[C_2H_5Cl]}{\Delta t} \approx -k \quad \text{or} \quad \frac{\Delta[C_2H_5Cl]}{\Delta t} \approx -k[C_2H_5Cl] \quad (15.4)$$

where k is a positive constant. The second equality states that the rate of decay of reagent is proportional to its concentration; as the concentration declines, so does the rate. This makes good sense: fewer molecules yield fewer opportunities for decay in any given Δt. The second of Equations 15.4 is an example of a **rate law,** an equation showing how the rate depends on the concentrations in the system. In this case, the rate is proportional to the *first power* of the concentration, and we call it a **first-order** rate law. The constant k is called the first-order **rate constant.** To make the overall units of the rate relations Equation 15.4 consistent, k must have units of $(\text{time})^{-1}$, where the time units used are usually seconds (s), but could be minutes, hours, days, or years depending on how fast things happen.

Corresponding to the rate law of Equation 15.4 is a unique time profile for the decay of $[C_2H_5Cl]$. We derived the rate law by differentiating the original time-profile data; therefore, to get the mathematical form of the time profile, we must *integrate* the rate law. To simplify the notation, let $c(t)$ denote $[C_2H_5Cl]$ at time t. The first-order rate law then reads

$$\frac{dc}{dt} = -kc \tag{15.5}$$

Mathematically this is known as a first-order differential equation for the unknown function $c(t)$. This equation can be simply solved by *separation of variables,* isolating the variables c and t on opposite sides. Multiplying Equation 15.5 through by dt/c gives

$$\frac{dc}{c} = -k\,dt \tag{15.6}$$

We now can integrate both sides, using the chemical limits of integration supplied by our problem. The concentration goes from its initial value c_0 to some value $c(t)$, while the time goes from the start of the reaction, taken as $t = 0$, to some time t; therefore,

$$\int_{c_0}^{c} \frac{dc'}{c'} = -k \int_0^t dt' \tag{15.7}$$

$$\ln c' \Big|_{c_0}^{c} = -kt' \Big|_0^t$$

$$\ln \frac{c}{c_0} = -kt \quad \text{or} \quad c = c_0 e^{-kt}$$

The integral over c on the left is our familiar logarithmic integral; primes are used on the integration variables to avoid confusing them with the limits. The bottom line is the result, the integrated form of the rate law or just the **integrated rate law.** It gives the reagent time profile $c(t)$. The extent-of-reaction variable for our reaction is then given by $x(t) = c_0 - c(t) = c_0(1 - e^{-kt})$.

The integrated rate law in its logarithmic form shows that if $\ln c$ is plotted against t (a semilog plot), the result is a straight line with slope $-k$; this is illustrated in Figure 15.2. In general this is a better way to determine k than that illustrated in Table 15.1, since it does not involve finite-difference approximations. The errors in the concentration data can also be accounted for by using the method of least-squares to obtain the slope.

First-order rate results are often reported in terms of the **half-reaction time** $t_{1/2}$, the time needed for half of the reagent molecules to react. At time $t = t_{1/2}$, $c = c_0/2$, and Equation 15.7 then gives us $\ln(\tfrac{1}{2}) = -kt_{1/2}$, or

$$t_{1/2} = \frac{\ln 2}{k} = \frac{0.693}{k} \tag{15.8}$$

Figure 15.2
A plot of $\ln[C_2H_5Cl]$ versus t, based on data of Table 15.1. According to the integrated form of the first-order rate law (Equation 15.7), this plot should be linear with intercept $\ln[C_2H_5Cl]_0$ and slope $-k$, where k is the first-order rate constant. The slope of the best straight line drawn through the points yields $k = 0.0112 \text{ s}^{-1}$.

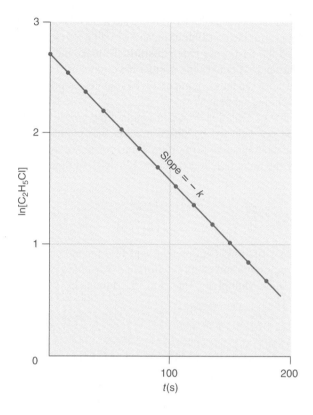

For a first-order decay, $t_{1/2}$ is independent of the initial concentration c_0, and is also sometimes called the *half-life* of the reaction. This independence is another way of testing whether first-order behavior is present; if you examine the data of Table 15.1, you will find that, no matter where you start in time, the time required to lose half the reagent concentration then present is approximately the same.

Not all decomposition reactions are necessarily first order; for example, the reaction

$$2NO_2(g) \longrightarrow 2NO(g) + O_2(g) \tag{15.9}$$

has been found to follow the rate law

$$\frac{d[NO_2]}{dt} = -k[NO_2]^2 \tag{15.10}$$

Since the rate depends on the second power of the concentration, this is called a **second-order** reaction. Using the same separation-of-variables method as for the first-order case, we arrive at the integrated form

$$\frac{1}{c} = \frac{1}{c_0} + kt \tag{15.11}$$

where again c is used to denote the concentration of the disappearing reagent. In this case a plot of $1/c$ versus t will yield a straight line of slope $+k$. Unless a wide range of times is sampled, however, the plot appears nearly linear even if the reaction is actually first order, and therefore may not be a good criterion for determining reaction order. To find the half-reaction time $t_{1/2}$, we substitute $c = c_0/2$ in Equation 15.11 and find $t_{1/2} = 1/(kc_0)$. Unlike the first-order case, here $t_{1/2}$ depends on the initial concentration; larger c_0 yields a shorter half-life, a good diagnostic for a reaction order higher than first.

A few gas reactions, principally those which require a *catalytic surface* to proceed at an appreciable rate, are **zero order.** For example, the decomposition of ammonia on hot tungsten, $2NH_3 \longrightarrow N_2 + 3H_2$, proceeds with

$$-\frac{d[NH_3]}{dt} = k \qquad (15.12)$$

that is, the reaction occurs at a *constant rate*. Integration of a zero-order rate law is particularly easy, and leads to

$$c = c_0 - kt \qquad (15.13)$$

where again we have used the generalized notation c for $[NH_3]$. A characteristic of a zero-order reaction is a well-defined end-of-reaction time, when the right-hand side of Equation 15.13 becomes zero, at $t_{end} = c_0/k$. This implies a half-life $t_{1/2} = c_0/(2k)$, and the reaction takes longer if the initial concentration is higher, the opposite of a second-order reaction. The physical process of evaporation is a common example of a zero-order reaction; in addition, more complex reactions are often zero order in one or more of the participating reagents. We will examine these more complicated cases shortly.

Table 15.2 collects results for various reaction orders, and Figure 15.3 compares the reagent-decay time profiles for orders 0, 1, and 2. The story of reaction order and rate is akin to the fable of the tortoise and the hare; although the higher order reaction, like the hare, begins at a greater rate, the lower order reaction, the tortoise, reaches the "finish-line," equilibrium, sooner. Here we have assumed that equilibrium lies far to the right (as may be predicted from the $\Delta G°$ for the reaction), so that we can ignore the reverse reaction for all times except very near the end of

TABLE 15.2

Rate laws and integrated forms for various reaction orders

Order	Rate law	Integrated form (time profile)	Half-reaction time $t_{1/2}$	Linear plot
0	$\dfrac{dc}{dt} = -k$	$c = c_0 - kt$	$\dfrac{c_0}{2k}$	c versus t
1	$\dfrac{dc}{dt} = -kc$	$c = c_0 e^{-kt}$	$\dfrac{\ln 2}{k}$	$\ln c$ versus t
2	$\dfrac{dc}{dt} = -kc^2$	$\dfrac{1}{c} = \dfrac{1}{c_0} + kt$	$\dfrac{1}{kc_0}$	$\dfrac{1}{c}$ versus t
$n \geq 2$	$\dfrac{dc}{dt} = -kc^n$	$\dfrac{1}{c^{n-1}} = \dfrac{1}{c_0^{n-1}} + (n-1)kt$	$\dfrac{2^{n-1} - 1}{(n-1)kc_0^{n-1}}$	$\dfrac{1}{c^{n-1}}$ versus t

Figure 15.3
Time profiles for different orders n in the rate law $dc/dt = -kc^n$. The rate constants k have been chosen to yield the same half-reaction time $t_{1/2}$. See Table 15.2 for the algebraic forms of these curves.

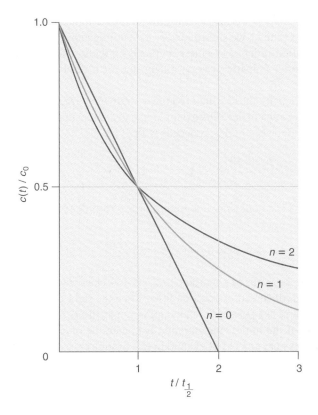

the reaction. The important consequences of the rate balance that must obtain at equilibrium will be examined in Section 15.3.

More commonly, chemical reactions involved at least two and sometimes more distinct reagents. One of the earliest reactions to be studied in a systematic way was the aqueous oxidation of iodide ion by hydrogen peroxide,

$$H_2O_2 + 2I^- + 2H^+ \longrightarrow I_2 + 2H_2O \tag{15.14}$$

Depending on initial conditions, this reaction can take several minutes to reach completion, though the equilibrium lies extremely far to the right. The extent of reaction is readily monitored by adding starch indicator and observing the waxing blue color of the I_2–starch complex. In general, we expect the rate law for the reaction to involve all of the reagents, and the postulate is generally made that

$$\frac{d[I_2]}{dt} = k[H_2O_2]^p[I^-]^q[H^+]^r \tag{15.15}$$

where the exponents p, q, and r are called the orders with respect to each reagent, and the sum $p + q + r$ is called the *overall order* of the reaction. In Reaction 15.14, experiments like those we are about to describe show that $p = 1$, $q = 1$, and $r = 1$, making the reaction first order with respect to $[H_2O_2]$, first order with respect to $[I^-]$, first order with respect to $[H^+]$, and third order overall. This example shows that the reaction orders need not reflect the stoichiometric coefficients (as the previous examples might have led you to conclude); further, the orders are not necessarily integers. Occasionally, a species not involved in the balanced chemical equation can appear in the rate law as well; this is typical for *homogeneous catalysts*.

Determining the Rate Law: The Method of Initial Rates

In more complicated reactions such as Reactions 15.14, the time profiles can become quite difficult to derive, particularly if nonstoichiometric initial concentrations are used, or the reverse reaction begins to interfere at later times. In these situations the rate law can often be obtained by measuring the **initial rate** for various initial concentrations of the reagents. In general this means doing many experimental runs under different initial conditions, rather than a few runs in which the concentrations are followed out to longer times. Often considerable experimentation is required to determine initial concentrations that yield readily measurables rates—neither too fast nor too slow, nor otherwise difficult to measure. In typical laboratory experiments you may do as part of an introductory chemistry course, this sort of work has been done ahead of time, allowing you to focus on the reaction at hand. In addition, many reactions, particularly decompositions, are not "clean," giving a variety of products, and following them for long times may give results that are impossible to interpret.

As an example of the method of initial rates, let's consider the aqueous inverse disproportionation reaction

$$BrO_3^- + 5Br^- + 6H^+ \longrightarrow 3Br_2 + 3H_2O \qquad (15.16)$$

Table 15.3 shows the results of a series of measurements for various initial concentrations; spectrophotometry on the orange Br_2 product was used to obtain the rate as $dx/dt \approx \frac{1}{3}\Delta[Br_2]/\Delta t$ through the measurement of the time Δt required to attain a given absorbance. The strategy normally adopted is, when possible, to vary only one of the concentrations from a reference run, leaving the others fixed, and observe its effect on the rate. In the present case, HBr solution is used, and the H^+ and Br^- are

TABLE 15.3

Initial rate data for the reaction
$BrO_3^- + 5Br^- + 6H^+ \longrightarrow 3Br_2 + 3H_2O$

Run[a]	$[BrO_3^-]_0$ (M)	$[Br^-]_0$ (M)	$[H^+]_0$ (M)	Δt[b] (s)	Rate = $\Delta x/\Delta t$[c] (mol L^{-1} s^{-1})
I	0.020	0.10	0.10	48.2	4.1×10^{-5}
II	0.040	0.10	0.10	24.2	8.3×10^{-5}
III	0.020	0.20	0.10	23.6	8.5×10^{-5}
IV	0.020	0.20	0.20	6.1	32.8×10^{-5}

[a] The four runs were prepared by mixing solutions of $KBrO_3$, KBr (for Run III), and HBr.
[b] The reaction mixture waxes orange as $[Br_2]$ builds up; Δt is the time required to yield $[Br_2]$ = 0.0060 M as measured by a calibrated spectrophotometer.
[c] Since $[Br_2] = 3x$, $\Delta x/\Delta t = \frac{1}{3}\Delta[Br_2]/\Delta t = \frac{1}{3}(0.0060 \text{ M})/\Delta t$.

therefore added together. If we take the rate law to be analogous to that given in Equation 15.15, namely,

$$\text{Rate} = \frac{1}{3}\frac{d[\text{Br}_2]}{dt} = k[\text{BrO}_3^-]^p[\text{Br}^-]^q[\text{H}^+]^r \tag{15.17}$$

then when the ratio of two rates is taken, all of the concentrations that have been held constant will cancel. For example, comparing runs I and II from Table 15.3, in which only $[\text{BrO}_3^-]_0$ was changed, we can write

$$\frac{\text{Rate II}}{\text{Rate I}} = \frac{\Delta t\,\text{I}}{\Delta t\,\text{II}} = \left(\frac{[\text{BrO}_3^-]_\text{II}}{[\text{BrO}_3^-]_\text{I}}\right)^p$$

$$\frac{8.3 \times 10^{-5}\ \text{mol L}^{-1}\text{s}^{-1}}{4.1 \times 10^{-5}\ \text{mol L}^{-1}\text{s}^{-1}} = \left(\frac{0.040\ \text{mol L}^{-1}}{0.020\ \text{mol L}^{-1}}\right)^p$$

$$2.0 = (2.0)^p; \qquad p = 1$$

It is clear that p has to be 1 here, because doubling $[\text{BrO}_3^-]_0$ doubles the rate; in general, you would have to solve $y = x^p$ for p. This is readily done using either natural logarithms (ln) or Briggsian logarithms ($\log_{10}$), with the result $p = \log(y)/\log(x)$. (In a more complete study, the initial rate would be measured for a series of $c_0 \equiv [\text{BrO}_3^-]_0$; then the slope of a plot of log(rate) versus $\log c_0$ would be the order p.) Runs I and II therefore show that the reaction is *first order* with respect to BrO_3^-. Varying the other concentrations also affects the rate as Table 15.3 indicates: runs I and III show that doubling $[\text{Br}^-]_0$ also doubles the rate, while runs III and IV show that doubling $[\text{H}^+]$ quadruples the rate. These results yield $q = 1$ and $r = 2$ in the rate law of Equation 15.17 through an analysis similar to that given earlier for BrO_3^-. This allows us to state the resulting rate law as

$$\text{Rate} = k[\text{BrO}_3^-][\text{Br}^-][\text{H}^+]^2 \tag{15.18}$$

making Reaction 15.16 first order with respect to both BrO_3^- and Br^-, second order with respect to H^+, and fourth order overall.

The rate data in Table 15.3 represent an experimentally favorable outcome, with small errors in the measured rates. More often, you have to cope with much larger errors, and the rate law exponents are correspondingly less certain and usually nonintegral. What is normally done in such cases is to round the exponent to the nearest integer if the errors permit; occasionally half-integers occur.

Once the rate law is established, the data of Table 15.3 also allow us to find the rate constant k. Solving Equation 15.18 for k, and using the data for run I, we find $k = 2.05\ \text{L}^3\ \text{mol}^{-3}\ \text{s}^{-1}$; the other runs produce very nearly the same result. With noisier data, a k would be calculated from each run and the results averaged.

EXAMPLE 15.2

A classic example in the kinetics of organic reactions is the substitution reaction

$$(\text{CH}_3)_3\text{C}\text{—Cl} + \text{OH}^- \longrightarrow (\text{CH}_3)_3\text{C}\text{—OH} + \text{Cl}^-$$

which converts *t*-butyl chloride, abbreviated *t*BuCl (properly 2-methyl-2-chloropropane) to the corresponding alcohol *t*BuOH. The following times were measured at 23°C for complete reaction of a small amount of OH^-, observed by the blue-to-yellow color change of added bromthymol blue acid–base indicator:

Run	$[tBuCl]_0$ (M)	$[OH^-]_0$ (M)	t (s)	$-\Delta[OH^-]/\Delta t$ (M s^{-1})
1	0.020	0.0040	28	1.43×10^{-4}
2	0.010	0.0040	53	0.75×10^{-4}
3	0.020	0.0020	13	1.54×10^{-4}

Use these data to estimate the form of the rate law for this reaction, and calculate the rate constant.

Solution:
We assume the rate law is of the form

$$\text{Rate} = -d[OH^-]/dt = k[tBuCl]^m[OH^-]^n$$

and use the data to estimate m and n. The fifth column in the data table is an estimate of the rate, where $\Delta[OH^-] = [OH^-]_0$ and $\Delta t = t$. Comparing runs 1 and 2, we see that halving $[tBuCl]$ roughly halves the rate, while runs 1 and 3 show that halving $[OH^-]$ has almost no effect on the rate. A more precise analysis using logs (see earlier discussion) yields $m = 0.92 \approx 1$ and $n = 0.11 \approx 0$. The reaction is therefore first order in $tBuCl$, zeroth order in OH^-, and first order overall, with a rate law of

$$\text{Rate} = k[tBuCl]$$

Assuming this rate law, each of the runs can be used to find the first-order rate constant k, from $k \approx \text{rate}/[tBuCl]_0$, and the results averaged. We obtain from runs 1, 2, and 3 the values $k = 7.15 \times 10^{-3} \text{ s}^{-1}$, $7.5 \times 10^{-3} \text{ s}^{-1}$, and $7.7 \times 10^{-3} \text{ s}^{-1}$, giving an average $k = 7.45 \pm 0.30 \times 10^{-3} \text{ s}^{-1}$. The surprising aspect is the lack of dependence on OH^-. The mechanistic implications for this reaction and its relatives are discussed in Section 15.3.

Determining the Rate Law: The Isolation Method

Provided the reaction under study can be followed for longer times without interference from competing reactions or from the reverse reaction, another convenient and more accurate method for determining rate laws and rate constants is the **isolation method**. In this technique, one of the reagent concentrations is purposely kept very much smaller than the others, so that in the course of the reaction only the small concentration changes appreciably, with the others remaining nearly constant. The analysis of the time profile then proceeds as in the single-reagent cases we considered at the outset, and an "effective" rate constant is obtained that includes the rate law factors from the large concentrations. Measurements with different initial large concentrations then fill out the rate law, as we will now illustrate.

The isolation method is frequently used when one reagent is harder to prepare in large concentrations, or is readily detected in small quantity by sensitive methods. For example, the atmospherically important gas reaction

$$H + O_3 \longrightarrow OH + O_2 \qquad (15.19)$$

was studied under conditions where [H] ≪ [O$_3$], and the H atom concentration was followed by detecting fluorescence from atoms excited by UV light. A reliable method for reactions like this one that involve a free radical is the *discharge-flow* technique, illustrated in Figure 15.4. Separate runs were made for different initial O$_3$ concentrations. Some time profiles are presented in Figure 15.5. Since [O$_3$]$_0$ is more than 100 times greater than [H] in each case, the change in [O$_3$] is always less than 1% and it becomes a constant factor for a given run. If we assume that the rate law is of the form $-d[\text{H}]/dt = k[\text{H}]^p[\text{O}_3]^q$, and $[\text{O}_3] \approx [\text{O}_3]_0$, then

$$-\frac{d[\text{H}]}{dt} \approx (k[\text{O}_3]_0^q)[\text{H}]^p \equiv k_{\text{eff}}[\text{H}]^p \qquad (15.20)$$

where k_{eff} is an "effective" rate constant, valid for the chosen [O$_3$]$_0$. As Figure 15.5 shows, the plot of $\log_{10}[\text{H}]$ (and therefore $\ln[\text{H}] = 2.303 \log_{10}[\text{H}]$) versus t is linear, and the reaction is therefore first order ($p = 1$) in [H]. The effective *pseudo-first-order* rate constant k_{eff} changes as [O$_3$] is changed, and when k_{eff} is plotted against [O$_3$], another straight line results with zero intercept. This indicates that $k_{\text{eff}} \approx k[\text{O}_3]$, and the reaction is first order in [O$_3$], that is, $q = 1$. The slope of this second line yields the numerical value of the true *second-order* rate constant k.

The flow tube method for gas reactions allows the study of second-order reactions on a millisecond (10^{-3} s) timescale, much faster than you could measure with an ordinary stopwatch. More recently, the use of pulsed lasers and fast spectroscopic detection has enabled time profiles to be measured in the nanosecond to femtosecond (10^{-9} to 10^{-15} s) regime for selected processes such as dissociation and recombination. These new techniques, which approach the timescale of molecular motion itself, will be discussed further in Section 15.4.

The rate law is of value in allowing us to predict the rate of product formation knowing the concentrations of the participating reagents. In the upper atmosphere, experiments flown on rockets, airplanes, and balloons have allowed us to measure the concentrations of various atmospheric molecules and radicals, and these along with rate constants measured in the laboratory are used in elaborate kinetic models that predict the future fate of molecules such as O$_3$. Of equal importance, the rate law car-

Figure 15.4
A discharge-flow apparatus for the measurement of fast gas-phase reactions. Reactive species such as free radicals or ions are produced in the discharge region, and other reactants are introduced downstream by a movable injector. With a detector monitoring either loss of reagent or buildup of product at a fixed position further downstream, different injector positions yield different distances x over which reaction occurs. Independent measurement of the *flow velocity* v_f then allows the reaction time $t = x/v_f$ and consequently the time profile to be obtained. See Figure 15.5 for illustrative data.

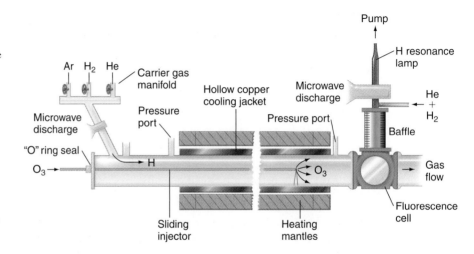

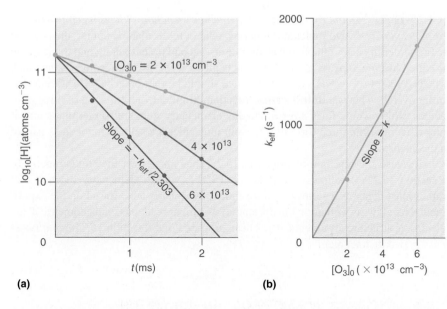

Figure 15.5
Kinetic data for the H + $O_3 \longrightarrow$ OH + O_2 reaction, taken with the apparatus shown in Figure 15.4. **(a)** Logarithmic time profiles of [H] for various $[O_3]_0$, chosen to be in greater than 100-fold excess. The straight lines indicate first order with respect to [H], and the slopes yield k_{eff} of Equation 15.20. Note that the slope is steeper (greater rate of loss of [H]) at higher $[O_3]_0$. **(b)** Variation of k_{eff} with $[O_3]_0$. The linear variation indicates first order with respect to $[O_3]$, and the slope gives the overall bimolecular rate constant, $k = 2.8 \pm 0.2 \times 10^{-11}$ cm^3 molecule^{-1} s^{-1} (1.7×10^{10} L mol^{-1} s^{-1}).

ries in its form information about the *molecular mechanism* of the reaction, an often controversial but deeply insightful aspect of chemical kinetics to be examined next.

15.3 Reaction Mechanisms

Among the kinetic examples we have considered in the previous section, only the H + O_3 reaction is thought to proceed directly by a molecular-level encounter between an atom of hydrogen and an ozone molecule, that is, by a single **elementary reaction**. The other reactions, even the first-order decomposition of C_2H_5Cl, are thought to proceed by a *series* of molecular steps we shall refer to as a **reaction mechanism**. To define what is meant by this, let's use another well-known example, whose rate law was found in Example 15.2.

Elementary Reactions

The aqueous conversion of *t*-butyl chloride to *t*-butyl alcohol by reaction with OH^-,

$$(CH_3)_3C\text{—}Cl + OH^- \longrightarrow (CH_3)_3C\text{—}OH + Cl^- \tag{15.21}$$

can be monitored by following the declining pH of the reaction solution as base is consumed. In Example 15.2, we found the rate law for this reaction to be

$$-\frac{d[(CH_3)_3CCl]}{dt} = k[(CH_3)_3CCl] \tag{15.22}$$

that is, the reaction is zeroth order in OH^-. If this reaction occurred in a single elementary step, we would expect the rate to depend on $[OH^-]$ as well as $[(CH_3)_3CCl]$, since the rate at which the two reagent molecules encounter each other should depend on how many of each are present. That is, a true *molecular* encounter between $(CH_3)_3CCl$ and OH^- should show a rate that is first order in *both* $[(CH_3)_3CCl]$ and

[OH$^-$]. In general, *the rate law for an elementary reaction reflects its stoichoimetry or molecularity.* If two molecules react by a direct molecular encounter, the rate law is first order in each of the two concentrations, second order overall, and the reaction is called **bimolecular.** If a molecule reacts all by itself, as in an isomerization such as CH$_3$—N≡C ⟶ CH$_3$—C≡N, the rate should be first order in CH$_3$NC and first order overall. This is a **unimolecular** reaction. Simultaneous collision of three molecules, which would form a **termolecular** elementary step, are so rare that elementary reactions with three (or more) reagent molecules are seldom invoked as part of a mechanism.

Because the rate law of Equation 15.22 does *not* reflect the stoichiometry of Reaction 15.21, it follows that Reaction 15.21 is *not* an elementary reaction, but rather must occur as a series of two or more elementary steps. Based on the experimental rate law, as well as on kinetic studies of similar reactions, and geometrical and electronic structure arguments, it has been proposed that Reaction 15.21 *actually occurs* in two distinct elementary steps:

1. $(CH_3)_3C-Cl \xrightarrow{k_1} (CH_3)_3C^+ + Cl^-$

2. $(CH_3)_3C^+ + OH^- \xrightarrow{k_2} (CH_3)_3C-OH$

This is a simple example of a reaction mechanism. It postulates the existence of a **reaction intermediate,** the *t*-butyl cation $(CH_3)_3C^+$, that is created in the course of the reaction but is too reactive to remain and be counted among the reaction products. The constants k_1 and k_2 are rate constants for the elementary steps, which by definition have rate laws that reflect their stoichiometry, that is, the first step is *unimolecular* and the second step is *bimolecular.* If the two elementary reactions are added, the intermediate drops out, and the overall reaction of Equation 15.21 is recovered; this is a universal property of properly written reaction mechanisms.

Rate-Limiting Step

If we further assume that step 1 is much *slower* than step 2 (that is, that k_1 is very small), step 1 will *limit* the rate of conversion of *t*-butyl chloride to the alcohol, and the overall (observed) rate will reflect the rate of the slow step. The **rate-limiting** effect of the slow step is similar to the effect of a constriction or **bottleneck** in a pipe on the rate of fluid flow through it. As Figure 15.6 illustrates, no matter how big the pipe is either upstream or downstream of the constriction, the constriction itself fixes the fluid flow rate. If the overall reaction rate reflects the rate of the slow step, then it must also behave according to the *rate law* of the slow step, that is,

$$-\frac{d[(CH_3)_3CCl]}{dt} = k_1[(CH_3)_3CCl] \qquad (15.23)$$

But this is identical in form to the observed rate law of Equation 15.22 as long as we identify the observed rate constant k as k_1, the unimolecular rate constant of the first step. If this mechanism is correct, then what we have learned is not about the overall reaction, but only about the rate at which *t*-butyl chloride ionizes. The rate measure-

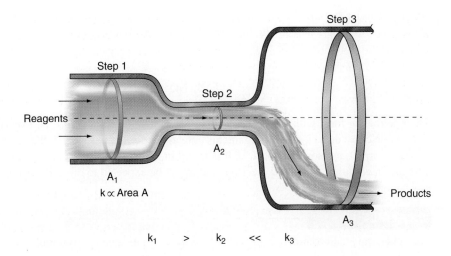

Figure 15.6
Water-pipe analogy for the rate-limiting step in a mechanism. Here step 2 is rate limiting, regardless of the much larger areas (rate constants) of the other steps.

ments contain no information on the second step or any subsequent occurrence, as long as it is much faster than step 1.

The fact that the preceding mechanism yields a rate law of the same form as observed experimentally does *not* prove the correctness of the mechanism. Any number of fast steps could follow step 1 without changing the prediction. Another sort of slow unimolecular process, say, involving the migration of the Cl to a different carbon atom, might be occurring as well. It is only through the study of several similar reactions that chemists have been led to postulate the ionization step. In fact, the reaction of the corresponding straight-chain butyl chloride is found to be still slower than Reaction 15.21, and furthermore it is a second-order reaction, first order in butyl chloride and first order in OH^-, and probably occurs in a single step. Reaction 15.21 is referred to as a S_N1 reaction (substitution, nucleophilic, first order), while its straight-chain analogue is S_N2.

The rate-limiting step need not be the first one in the mechanism. Consider the atmospheric reaction and rate law

$$2NO + O_2 \longrightarrow 2NO_2 \qquad \frac{1}{2}\frac{d[NO_2]}{dt} = k[NO]^2[O_2] \qquad (15.24)$$

Given the overriding influence of the slow step, we might expect it to be termolecular, involving the simultaneous encounter of two NO molecules with an O_2. However, as stated earlier, termolecular collisions are so rare that we look first for an alternative mechanism involving only unimolecular or bimolecular steps. One possibility is

1. $2NO \rightleftharpoons N_2O_2$ Fast equilibrium, constant = K_1
2. $N_2O_2 + O_2 \xrightarrow{k_2} 2NO_2$ Slow

In the first step, called a *dimerization* reaction, an equilibrium is assumed to be rapidly established, while the second step slowly consumes the dimer N_2O_2, a reaction intermediate. The rate law for the second step is

$$\frac{1}{2}\frac{d[NO_2]}{dt} = k_2[N_2O_2][O_2] \qquad (15.25)$$

It is bimolecular, but involves the reaction intermediate N_2O_2. We now use the equilibrium reaction quotient from the first step, $K_1 = [N_2O_2]/[NO]^2$, to find $[N_2O_2]$, and substitute into Equation 15.25 to find

$$\frac{1}{2}\frac{d[NO_2]}{dt} = k_2 K_1 [NO]^2 [O_2] \tag{15.26}$$

which agrees in form with the observed rate law as long as we identify the experimental rate constant as $k = k_2 K_1$. Again this agreement does not prove that the mechanism is correct, but some support for it is found in the fact that the NO molecule is a *free radical* with an unpaired electron, and might be expected to dimerize readily, as we already know NO_2 does.

EXAMPLE 15.3

The acid-catalyzed bromination of acetone (propanone) in aqueous solution,

$$CH_3COCH_3 + Br_2 \xrightarrow{H^+} CH_3COCH_2Br + H^+ + Br^-$$

is found to obey the following rate law: rate = $k[CH_3COCH_3][H^+]$. The reaction is zero order in the Br_2 reagent, but first order in the H^+ catalyst, which is not consumed in the reaction. Is this reaction "elementary" in the sense we have introduced here? Why or why not? If not, suggest a possible rate-limiting step and a two-step mechanism. Verify for your mechanism that the two steps add up to the overall reaction, with any catalyst or reaction intermediate that appears in the mechanism dropping out in the sum.

Solution:
The reaction cannot be elementary, because if it were its rate law would be first order in *both* CH_3COCH_3 and Br_2. By the bottleneck principle, the slow, rate-limiting step must involve CH_3COCH_3 and H^+, $CH_3COCH_3 + H^+ \longrightarrow$ intermediate(s); since both are present at the start, it is reasonable to make this the first step in the mechanism. Because Br_2 is not involved, the product(s) of this step cannot be the final reaction products—it (they) must be reaction intermediate(s). It seems reasonable chemically that H^+ would protonate a lone pair on acetone's O atom, making a single reaction intermediate $CH_3COCHCH_3^+$. A possible mechanism is therefore

1. $CH_3COCH_3 + H^+ \xrightarrow{k_1} CH_3COHCH_3^+$ Slow bimolecular

2. $CH_3COHCH_3^+ + Br_2 \xrightarrow{k_2} CH_3COCH_2Br + 2H^+ + Br^-$ Fast bimolecular

The rate of the reaction will be the rate of the slow step, $k_1[CH_3COCH_3][H^+]$, and the measured rate constant k is k_1. When these two steps are added, the overall reaction is recovered, and the extra proton and the protonated acetone intermediate drop out. Nothing is learned concerning the nature or rate of the second step, only that it must be faster than the first. (You would be right to think that the second step, although bimolecular, looks too complicated to be elementary. Organic chemists favor three more steps after the first slow one: isomerization, bromination, and elimination!)

Scattered throughout this book, and particularly in Chapters 5 and 13, we have met reactions with large and not easily predicted stoichiometric coefficients. If the constraint is observed that the steps in a mechanism must add up to the overall reaction in such a case, and we admit only unimolecular and bimolecular steps, then the mechanism is forced to contain many steps. Generally what is done in such cases is to be explicit until the slow step is reached (as suggested by the rate law observed), and then to "lump" the remaining steps into one equation for the remainder of the reaction.

The Steady-State Approximation

While it is plausible that the slow step in a mechanism determines the rate law, we certainly have not proven this to be the case. In general the steps in a mechanism each contribute a rate law equation to a *system* of equations; these equations have to be solved together to obtain the time profiles for all the reagents, intermediates, and products. The problem should remind you of those based on simultaneous equilibria from Chapter 12, except that in rate problems we are required to solve a set of *differential* equations. This is much more difficult, and in all but a few special cases the rate equations must be solved by numerical methods on a computer. However, in cases where the reaction intermediates never reach concentrations comparable to those of the stable molecules—which covers the great majority of multistep reactions—we can fruitfully apply an approximation to the rates of change of the intermediate concentrations known as the **steady-state approximation.**

To see what this approximation consists of and how it simplifies the rate equations, let's use an example similar in form to the reaction of *t*-butyl chloride with base (Equation 15.21). We write the reaction generically as

$$A + B \longrightarrow C + D$$

and the two-step mechanism as

1. $A \xrightarrow{k_1} E + D$
2. $E + B \xrightarrow{k_2} C$

Here E is the intermediate, which is produced at a certain rate $k_1[A]$ in step 1, and lost at a rate $k_2[E][B]$ in step 2. Because E occurs in both steps, its concentration *couples* the rate equations. The set of rate equations that must be solved is as follows:

$$-\frac{d[A]}{dt} = \frac{d[D]}{dt} = k_1[A] \qquad (15.27)$$

$$\frac{d[E]}{dt} = k_1[A] - k_2[E][B]$$

$$-\frac{d[B]}{dt} = \frac{d[C]}{dt} = k_2[E][B]$$

where the first term in $d[E]/dt$ occurs with a plus sign because it forms E, and the second occurs with a minus sign because it consumes E. If we now postulate that E is always present at very low concentration, its rate of change will likewise always be small, and we can set

$$\frac{d[E]}{dt} \approx 0 \longrightarrow [E] \approx \frac{k_1[A]}{k_2[B]} \qquad (15.28)$$

where the expression for [E] arises from Equation 15.27 with zero rate. Note that what we are really saying is that, while $d[E]/dt$ may not be exactly zero, it is certainly much less than either of the terms that contribute to it. This is the steady-state approximation (SSA), where the term *steady-state* applies to the intermediate(s). Figure 15.7 illustrates the time profiles that are appropriate for the SSA in our model reaction.

With the aid of the SSA, we can now formulate a single rate law that represents the two-step mechanism. Substituting the result from the SSA of Equation 15.28 for [E] into $d[C]/dt$, we find

$$\frac{d[C]}{dt} = k_2 \left(\frac{k_1[A]}{k_2[B]} \right)[B] = k_1[A] \qquad (15.29)$$

where k_2 and [B] cancel, leaving us with the first-order rate law that we would have written based on the slow-step rule. Note that a necessary condition for the validity of the SSA is that [E] never builds up to an appreciable extent, and this can only happen if step 2 uses up E as quickly as it is formed by step 1, that is, if step 2 is much faster than step 1. The SSA thus provides us with a general justification of the rate-determining character of the slow step. If step 1 were faster than step 2, [E] would build up to a large value, making the reverse reaction of step 1 important, and we would arrive at the pre-equilibrium case considered in the previous subsection. If the two rates are comparable, however, there will still exist a significant range of times during which $d[E]/dt$ is small, since [E] must always pass through a maximum in the course of the reaction. Later we shall see a famous example of this last situation, in which the SSA is still applied.

Figure 15.7
Time profiles appropriate for using the steady-state approximation for the generic A + B ⟶ C + D reaction with intermediate E. Except near $t = 0$, both [E] and $d[E]/dt$ remain small throughout.

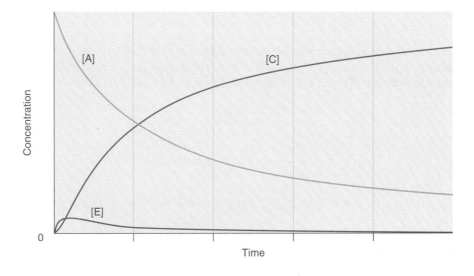

EXAMPLE 15.4

In Example 15.3, a two-step mechanism for the acid-catalyzed bromination of acetone was proposed; in abbreviated notation

1. $AH + H^+ \xrightarrow{k_1} AH_2^+$

2. $AH_2^+ + Br_2 \xrightarrow{k_2} ABr + 2H^+ + Br^-$

Use the steady-state approximation to show that this mechanism leads to rate = $d[ABr]/dt = k_1[AH][H^+]$, in agreement with observation.

Solution:
Since ABr is produced only in step 2, $d[ABr]/dt = k_2[AH_2^+][Br_2]$. This rate law cannot be compared with the experimental result because it contains the intermediate $[AH_2^+]$. The SSA allows us to express $[AH_2^+]$ in terms of concentrations of reagents (and perhaps products). The mechanism leads to

$$d[AH_2^+]/dt = k_1[AH][H^+] - k_2[AH_2^+][Br_2]$$

and this is approximxately zero in the SSA. Transposing the k_1 term and solving, we find

$$[AH_2^+] \approx (k_1[AH][H^+])/(k_2[Br_2])$$

When this is substituted into $d[ABr]/dt$, $k_2[Br_2]$ cancels leaving

$$d[ABr]/dt = k_1[AH][H^+]$$

which agrees with experiment (and the bottleneck rule) provided we identify the bimolecular rate constant k_1 with the experimental second-order rate constant k. Note that if we were to include deprotonation of AH, the reverse of step 1, in the mechanism, the expression for $d[AH_2^+]/dt$ would contain an additional loss term $-k_{-1}[AH_2^+]$. Carrying through the algebra as before, the resulting rate law would not be of the "power law" form, and would consequently disagree with experiment. An example of this type is the Lindemann mecahnism discussed in Section 15.4.

The Principle of Detailed Balance

At several junctures we have stated that equilibrium in a chemical system is a microscopically dynamic state, in which ceaseless molecular action occurs, but with no macroscopic effect. This owes to a balancing of forward and reverse rates that does not allow a *net* change in either reagent or product concentrations, and corresponds to the far right end of the time profile for the reaction. As an example, let's consider the common case of the ionization of an acid in water,

$$HA \underset{k_{-1}}{\overset{k_1}{\rightleftharpoons}} H^+ + A^- \qquad (15.30)$$

We now invoke the explicit meaning of the double arrow ($\rightleftharpoons$) as implying rate laws and rate constants for the forward and reverse reactions. The reagent HA is lost when ionization occurs, but reformed when H^+ and A^- recombine. Let's assume that both forward and reverse reactions are elementary, that is, they occur molecularly just as it appears from the chemical equation. Then, as in the case of the reaction intermediate of the previous section, the effect of both steps is expressed by writing both gain and loss terms on the right-hand side to represent the rate $d[HA]/dt$:

$$\frac{d[HA]}{dt} = k_1[HA] - k_{-1}[H^+][A^-] \tag{15.31}$$

At equilibrium, [HA] is no longer varying with time, and therefore $d[HA]/dt = 0$. As Equation 15.31 explicitly indicates, the forward and reverse rates must exactly balance to fulfill this condition, but neither is zero by itself. This is the underpinning for the idea that equilibrium does not mean a cessation of reaction, but only an adjustment, a "tuning" of the forward and reverse rates to make them equal. If we set the right-hand side of Equation 15.31 to zero and rearrange the factors, we find

$$\frac{k_1}{k_{-1}} = \frac{[H^+][A^-]}{[HA]} \tag{15.32}$$

You should recognize the right-hand side of this equation as the *reaction quotient Q* for the acid ionization reaction. Since Equation 15.32 is only true at equilibrium, the reaction quotient Q must equal the equilibrium constant K, and therefore

$$K = \frac{k_1}{k_{-1}} \tag{15.33}$$

This relationship between the rate constants and the equilibrium constant is known as the **principle of detailed balance.** The principle generalizes for a reaction represented by a multistep mechanism, where each elementary step has a forward and reverse rate law, to read

$$K = \frac{k_1 k_2 k_3 \cdots}{k_{-1} k_{-2} k_{-3} \cdots} \tag{15.34}$$

where the rate constant ratio for each step is just the equilibrium constant for that step. Recall that the steps in a mechanism must add to give the overall reaction, and in that event the equilibrium constants for each step are multiplied to give the overall K.

Let's pause to consider the implications of this principle. A large K (a strong acid in our example) means that $k_1 \gg k_{-1}$, that is, for given concentrations, the forward reaction must be much *faster* than the reverse. Then the only means to balance the rates is through adjusting the concentrations appearing in the rate law (Equation 15.31). If k_1 is large, then $[HA]_{eq}$ must be very small, and $[H^+]_{eq}$ and $[A^-]_{eq}$ must be large to compensate for the small k_{-1}. This is precisely the requirement of the law of mass action relating K to the equilibrium reaction quotient. Thus what we have is not only a dynamical *interpretation* of equilibrium, but a *justification* of its properties. But why should one rate constant be larger than another? What is it about the reaction that causes the disparity? Is there any limit to how big a k can be? And why should a thermodynamic quantity like K be involved? The next subsection will give us insight into these questions, and the generalized molecular models to be presented in Section 15.4 are the chemist's attempt to answer them.

The Temperature Dependence of Reaction Rates

The rates of chemical reactions have long been known to be very sensitive to temperature. In their pioneering 1866 study of the oxidation of oxalic acid by permanganate (see Exercise 1), Harcourt and Esson went to great lengths to keep the temperature of their solutions absolutely constant. They observed that even the heat liberated by a small amount of exothermic reaction would cause a slight increase in T and proportionally a much larger change in the rate, usually greatly increasing it. This natural heating of the system by exothermic reactions remains a major source of error in measured rates to this day. If we presume that the rate law remains the same over a chosen range of temperature, the T dependence must reside in the rate constant k, that is $k = k(T)$. In 1889, our old ionic friend Svante Arrhenius, then a pioneering young physical chemist, suggested that the rate constant should depend on T according to

$$k = A\, e^{-E_a/RT} \qquad (15.35)$$

where A is called the **frequency factor** or simply the preexponential factor, and E_a is called the **activation energy.** Arrhenius based this hypothesis not only on the well-known sensitivity of k to T, but on the principle of detailed balance, as we shall see shortly.

Already in Arrhenius's day there were enough kinetic data including temperature dependence to put his postulate (Equation 15.35) to the test; this form predicts that a plot of $\ln k$ versus $1/T$ will be a straight line with intercept $\ln A$ and slope $-E_a/R$. When data were plotted in this way, now called an *Arrhenius plot,* straight lines were indeed obtained, as shown in Figure 15.8, allowing ready determination of A and E_a. The Arrhenius plot should remind you of the van't Hoff plot of $\ln K$ versus $1/T$ of Chapter 12; the two are isomorphic. The difference lies in the experimental fact that for nearly all known cases, the slope of the Arrhenius plot is negative, implying $E_a > 0$. In cases where the reaction has several elementary steps, the Arrhenius plot typically yields the activation energy of the rate-limiting step. A table of Arrhenius parameters for elementary reactions will be given in the next section.

Figure 15.8
Arrhenius plots in the form $\log_{10} k$ versus $1,000/T$ for three unimolecular and three bimolecular reactions. The Arrhenius equation (Equation 15.35) predicts that these plots should be linear with slopes $-E_a/[(1000)(2.303)R]$ and y-intercepts $\log_{10} A$. Experimental values of E_a and $\log_{10} A$ for these and other reactions can be found in Table 15.4. When measurements extend over larger temperature ranges, slight curvature is sometimes found, usually attributed to a slight temperature dependence of A (as predicted by reaction rate theory; see Section 15.4).

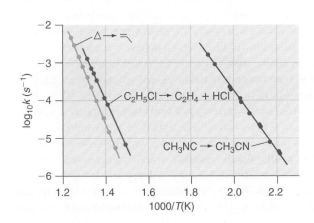
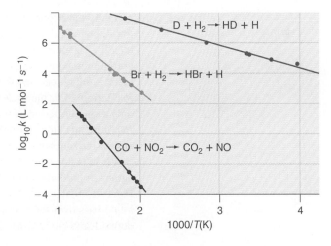

EXAMPLE 15.5

In Example 15.2, kinetic data were presented for the solution reaction

$$(CH_3)_3CCl + OH^- \longrightarrow (CH_3)_3COH + Cl^-$$

A further experiment was performed using the concentrations of Run 1 but with a reaction mixture heated to 37°C; the measured reaction time was $\Delta t = 7$ s. Assuming the rate law is the same as that given in Example 15.2, find the rate constant k at 37°C. Further assuming that $k(T)$ is given by the Arrhenius form (Equation 15.35), combine these data to estimate the activation energy E_a and the preexponential factor A.

Solution:
The rate law was found to be rate = $k[(CH_3)_3CCl]$; using the finite-difference method outlined in Example 15.2, $k(37°C) = 29 \times 10^{-3}$ s^{-1}. To find E_a, we use Equation 15.35, modified by taking the ln of both sides to give $\ln k = \ln A - E_a/RT$. Writing this for each of the two temperatures $T_1 = 23°C$ (296 K) and $T_2 = 37°C$ (310 K) and subtracting, we find

$$\ln(k_2/k_1) = -(E_a/R)(1/T_2 - 1/T_1)$$

This is identical to the integrated form of the van't Hoff equation, Equation 12.15, and may be solved readily for E_a. Using the average value $k(23°C) = 7.45 \times 10^{-3}$ s^{-1} from Example 15.2 and absolute temperatures, we find $E_a = 18$ kcal/mol. Then $A = k \exp(E_a/RT) = 8.7 \times 10^{10}$ s^{-1}, using the 23°C data. The terms E_a and A can be interpreted as a barrier height and an attempt rate for the bottleneck step, the ionization of tBuCl, as discussed in the text. A more thorough study of $k(T)$ would involve rate measurements over a range of temperatures, analyzed by an Arrhenius plot of $\ln k$ versus $1/T$. For solution reactions, the range of T is naturally limited to lying between the freezing and boiling points of the solvent.

An important part of the postulate of Equation 15.35, which we now call the *Arrhenius form* of k, was the picture Arrhenius had in mind. In his view, k is the product of an *attempt rate* or *collision rate*, and the *fraction of molecules with energy E_a or greater*. The former is represented by the frequency factor A, while the latter is given by the exponential $\exp(-E_a/RT)$, a "Boltzmann factor" that, as we have seen in Chapter 10, determines the distribution of energy among the molecules in the system. If E_a is more than a few kilocalories per mole, the Boltzmann factor is very small at room temperature, and the reaction will be observed to have a modest rate. In this picture, which became the so-called collision theory of reaction rates to be examined in Section 15.4, the molecules are constantly colliding and attempting to react, but only a small fraction of them has sufficient energy, given by E_a, to do so. This energy requirement can be thought of, in our favorite way of analyzing things, as a *potential energy barrier* separating reagents from products that must be surmounted if a reaction is to occur. This barrier is incorporated into our old thermodynamic enthalpy diagram to produce an **energy profile** for the reaction; some examples are shown in Figure 15.9. In these diagrams, we imagine the energy of the system to vary smoothly as we pass over the activation hill on the way to products, as a function of a loosely defined "reaction coordinate." In the simplest case, this coordinate would first follow

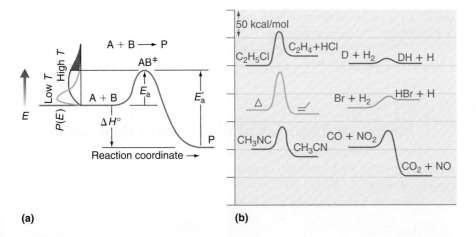

Figure 15.9
(a) Energy profile for a "typical" exothermic elementary reaction A + B ⟶ P. AB‡ is the "transition state" and E'_a is the activation energy for the reverse reaction. Also shown are Boltzmann energy distributions at two temperatures, illustrating the great increase in the fraction of molecules with $E > E_a$ as T rises. (b) Energy profiles for the reactions whose Arrhenius plots were shown in Figure 15.8.

the breaking of a reagent bond, and then the making of a product bond. We will elaborate further on this in the next section.

The energy profile can be used for the forward reaction or its reverse equally well. Just as the original enthalpy diagram tells us that the heats of the forward and reverse reactions (reverse is denoted here by a prime, ′) are the negative of each other, $\Delta H° = -\Delta H°'$, so the energy profile relates the activation energies for the forward and reverse reactions by

$$E'_a = E_a - \Delta H° \tag{15.36}$$

(Here we are ignoring the small difference between E and H.) Thus, for an exothermic forward reaction, $\Delta H° < 0$, the reverse reaction has a higher activation energy, $E'_a > E_a$, and is thus often *slower* than the forward reaction by many orders of magnitude. On the other hand, an endothermic reaction shows $E'_a < E_a$, and the reverse reaction is generally *faster* than the forward. If we assume that the forward and reverse rate constants k and k' can each be written in the Arrhenius form of Equation 15.35, their ratio is given by

$$\frac{k}{k'} = \frac{A \exp(-E_a/RT)}{A' \exp(-E'_a/RT)} \tag{15.37}$$

$$= \frac{A}{A'} \exp(-\Delta H°/RT)$$

where Equation 15.36 has been used to simplify the exponential. This rate constant ratio has the same temperature dependence as the equilibrium constant, as it must if the principle of detailed balance, Equation 15.33, is to be satisfied. Arrhenius's postulate is therefore consistent with this principle. Furthermore, since $K = \exp(-\Delta G°/RT) = \exp(\Delta S°/R - \Delta H°/RT)$, Equation 15.37 implies that the ratio of the frequency factors is

$$\frac{A}{A'} = \exp(-\Delta S°/R) \tag{15.38}$$

The interesting implication of this relationship is that the ratio of attempt rates depends on the relative entropies of the reagents and products. In the often-encountered case of $\Delta S° < 0$, the reagents attempt to react with a lower frequency than the products attempt to fall apart to reagents. Equation 15.38 also implies that the A factors must have a slight temperature dependence, since $\Delta S°$ does; however, the overriding T-effect resides in the Boltzmann factor.

15.4 Molecular Theories of Elementary Reactions

Armed with Arrhenius's findings, we are now ready to address the questions raised earlier concerning the magnitudes of reaction rates, by examining two widely used models for describing the rates of elementary processes.

Collision Theory

By far the simplest way of interpreting Equation 15.35 is to treat E_a as a potential-energy barrier as suggested earlier, and to develop a molecular model for the preexponential factor A. In gas-phase kinetics, this model has come to be known as the **collision theory** of elementary reactions. In Chapter 9, we briefly justified a gas-kinetic formula, Equation 9.50, for the collision rate Z_{mol} of a particular molecule in a gas of certain density. Now let us take the molecule to be a reagent A, and the gas to be another reagent B that can react with A. The average velocity $\bar{v}$ appearing in Equation 9.50 must be replaced by the *average relative velocity* $\bar{v}_{rel}$ between A and B, given by

$$\bar{v}_{rel} = \sqrt{\frac{8k_B T}{\pi \mu}} \tag{15.39}$$

where μ is the collisional *reduced mass*, $\mu = m_A m_B/(m_A + m_B)$, a quantity we have met before in Chapter 8. The total collision rate per unit volume between A and B is then given by

$$Z_{AB} = \bar{v}_{rel}\, \pi d_{AB}^2\, [A][B] \tag{15.40}$$

where the factor $\sqrt{2}$ appearing in Equation 9.50 is canceled by the reduced mass and the distinguishability of A and B, and d_{AB} is a collision diameter, given by $d_{AB} = (d_A + d_B)/2$, with d_A and d_B the molecular diameters of A and B. These diameters may be obtained from a variety of sources, for example, from nonideal gas behavior (Chapter 9) or from condensed-phase densities (Chapters 1 and 14). The quantity πd_{AB}^2 is called a *collision cross section*, since it measures the cross-sectional area that the molecules present to each other. We take the rate of potentially reactive collisions to be given by Z_{AB}, and multiply it by the fraction of molecules with enough energy to react, $\exp(-E_a/RT)$, yielding the collision theory expression for the rate of an elementary bimolecular reaction,

$$\text{Rate} = \bar{v}_{rel}\, \pi d_{AB}^2\, e^{-E_a/RT}[A][B] \tag{15.41}$$

Since the rate law for a bimolecular reaction is rate = $k[A][B]$, we may identify the bimolecular rate constant k and preexponential factor A_{ct} as

15.4 Molecular Theories of Elementary Reactions

$$k = \bar{v}_{rel} \pi d_{AB}^2 \, e^{-E_a/RT} \tag{15.42}$$

$$A_{ct} \equiv \bar{v}_{rel} \pi d_{AB}^2 = \sqrt{\frac{8k_B T}{\pi \mu}} \pi d_{AB}^2$$

where in the second line we have introduced the formula for $\bar{v}_{rel}$ from Equation 15.39 in order to show the explicit dependence of k and A_{ct} on the chemical variables. We note that the preexponential factor A_{ct} is predicted to increase with temperature as $T^{1/2}$, a mild temperature dependence that is easily masked by the exponential. The rate constant also increases with decreasing reduced mass, just as the rate of effusion does according to Graham's law (Section 9.4); elementary reactions where H· or H_2 is one of the reagents are potentially the fastest. Factor A_{ct} also represents our sought-after upper limit to the magnitude of a rate constant, corresponding to zero activation energy, or *reaction on every collision*. Depending on the reaction system, this upper limit can vary over a sizable range, but is typically $\sim 10^{-10}$ cm^3 molecule^{-1} s^{-1} or $\sim 10^{11}$ L mol^{-1} s^{-1}. Few reactions are observed to be this fast; those that are generally lead to reaction flames or, in the extreme case, explosions.

The collision theory cannot predict the activation energy E_a, but other than that everything in Equation 15.42 is known from other sources. Table 15.4 compares observed A factors, derived from measuring bimolecular reaction rates over a range of T and fitting to Arrhenius form (Equation 15.35) (see Figure 15.8), with A_{ct} (at 298 K) predicted from the collision theory expression (Equation 15.42) for a number of reactions. As you can see, agreement (within a factor of 3 or so) is achieved in only a few cases; the observed A's are generally smaller than the collision theory predictions. The disagreement is usually expressed as the ratio $p = A/A_{ct}$, the last column in Table 15.4, and is known as the *steric factor*. If you accept the collision theory premise that every reaction must occur by way of a collision, then the results imply that even if a collision occurs with sufficient energy to climb the activation hill, it may not result in reaction. This failure to react is attributed to an orientational or *steric* effect, in which the molecules encounter each other with sufficient energy, but are "facing the wrong way." For example, in the reaction O + ClO $\longrightarrow$ O_2 + Cl, if the O collides with the Cl end of ClO, O $\longrightarrow$ ClO, it is likely to be less successful in abstracting an O atom than it would be if it attacked the O end, O $\longrightarrow$ OCl. The collision theory gives both of these encounter types the same weight, thereby overestimating the attempt rate. It is thus readily understood why $p < 1$ as a rule. The drawback is that the p factor is not easily calculated, relegating the collision theory to a descriptive and conceptual rather than predictive role in chemical kinetic theory.

It might also seem that collision theory omits a large number of reactions, since it is inherently bimolecular with a second-order rate law, while many reactions are known to be first order. Until the early 1920s first-order decompositions and isomerizations were thought to be unimolecular. It was then that F. Lindemann proposed that even these reactions are mediated by collisions. Consider the generic first-order reaction A $\longrightarrow$ P. The **Lindemann mechanism** for this reaction consists of two steps:

1. $A + A \underset{k_{-1}}{\overset{k_1}{\rightleftharpoons}} A^* + A \tag{15.43}$

2. $A^* \xrightarrow{k_2} P$

TABLE 15.4

Arrhenius parameters for selected bimolecular gas-phase reactions[a]

Reaction	E_a	$\log_{10}A$	$\log_{10}A_{ct}$	Steric factor p
Na + Cl$_2$ ⟶ NaCl + Cl	0	11.6	11.58	1.0
K + Br$_2$ ⟶ KBr + Br	0	12.0	11.47	3.4
K + HBr ⟶ KBr + H	0.2	11.6	11.46	1.3
Br + H$_2$ ⟶ HBr + H	19.7	11.43	11.60	0.7
Cl + H$_2$ ⟶ HCl + H	4.3	10.92	11.59	0.22
F + H$_2$ ⟶ HF + H	1.13	11.06	11.55	0.32
O + H$_2$ ⟶ OH + H	11.3	10.4	11.53	0.07
D + H$_2$ ⟶ HD + H	7.61	10.64	11.60	0.11
O + ClO ⟶ O$_2$ + Cl	0.1	10.35	11.06	0.19
Cl + O$_3$ ⟶ ClO + O$_2$	0.5	10.21	11.22	0.10
O + O$_3$ ⟶ O$_2$ + O$_2$	4.09	9.68	11.27	0.026
O + NO$_2$ ⟶ O$_2$ + NO	0.24	9.59	11.24	0.022
O + CO$_2$ ⟶ O$_2$ + CO	54.2	10.28	11.28	0.10
OH + H$_2$ ⟶ H$_2$O + H	4.17	9.67	11.63	0.011
OH + CO ⟶ CO$_2$ + H	1.08	8.62	11.33	0.0019
CO + O$_2$ ⟶ CO$_2$ + O	51.0	9.54	11.34	0.004
NO + O$_3$ ⟶ NO$_2$ + O$_2$	2.72	9.04	11.35	0.005
CO + NO$_2$ ⟶ CO$_2$ + NO	27.8	8.75	11.35	0.0025
C$_2$H$_4$ + H$_2$ ⟶ C$_2$H$_6$	43.0	6.09	11.81	0.0000019

[a] Activation energy E_a in kcal/mol and A in L mol^{-1} s^{-1}. A_{ct} prediction of collision theory (Equation 15.42 at T = 298 K) using molecular diameters d derived from viscosity measurements. Rate constants at a given T may be generated from $\log_{10}k = 1000E_a/(2.303RT) + \log_{10}A$, generally reproducing actual experimental values within 20%.

In the first step, known as *collisional activation,* two A's collide with enough energy to cause one of the molecules to become internally (vibrationally) excited, denoted by the asterisk. The reverse reaction, deactivation, must also be included, because it is generally faster than activation. The energy deposited in the intermediate A* is assumed to be sufficient to allow A* to decompose to products P in step 2, a truly unimolecular process. To find the rate law predicted by this mechanism, we invoke the steady-state approximation for the intermediate A*:

$$\frac{d[A^*]}{dt} = k_1[A]^2 - k_{-1}[A^*][A] - k_2[A^*] \approx 0 \quad (15.44)$$

Note that here there are two loss terms (those with minus signs), since we have included the possibility that A* may be deactivated by collision before it has a chance to decompose to products. When the rate $d[A^*]/dt$ is set to zero as indicated, we can solve the resulting equation for [A*]:

15.4 Molecular Theories of Elementary Reactions

$$[A^*] = \frac{k_1[A]^2}{k_{-1}[A] + k_2} \tag{15.45}$$

Products P are formed only in the second step; the rate of product formation is then

$$\frac{d[P]}{dt} = k_2[A^*] = \frac{k_1 k_2 [A]^2}{k_{-1}[A] + k_2} \tag{15.46}$$

where we have substituted for $[A^*]$ Equation 15.45 from the SSA. This nonstandard rate law was first given by Lindemann in 1922.

Though the Lindemann rate law is not of the normal form, two limiting cases can be recognized in which the order of the reaction can be defined. If the rate of deactivation of $[A^*]$ greatly exceeds the decomposition rate, that is, $k_{-1}[A] \gg k_2$, the denominator of the rate expression of Equation 15.45 reduces to its first term, one factor of [A] cancels and we have

$$\frac{d[P]}{dt} = \frac{k_1 k_2}{k_{-1}}[A] \equiv k_\infty [A] \tag{15.47}$$

Because the deactivation rate increases with [A], it is always possible to choose a sufficiently high concentration or pressure of A that deactivation dominates. Decomposition then becomes the rate-limiting step in the high-pressure limit, and the reaction is first order and apparently unimolecular even though collisions remain necessary to reaction. You can also obtain Equation 15.47 by assuming that the collisional activation step has reached equilibrium, as discussed earlier. Because these "high" pressures typically lie in the range of a few torr or greater, first-order behavior is what is normally seen. The notation k_∞ indicates the limiting value of the first-order rate constant at infinite pressure. This limiting value is in turn a reflection of the time it takes for A^* to "channel" its energy into the bonds that must be broken to achieve reaction.

On the other hand, if the pressure is decreased, the opposite limit may be achieved, $k_2 \gg k_{-1}[A]$, and the second term in the denominator of Equation 15.45 is now the important one. The reaction then becomes *second order*

$$\frac{d[P]}{dt} = k_1[A]^2 \tag{15.48}$$

as expected, since now collisional activation is the rate-limiting step. You could also arrive at this limit by using the bottleneck principle, as discussed earlier. These two cases can be combined by defining an "apparent" first-order rate constant $k = k_1 k_2/(k_{-1} + k_2[A])$ that becomes pressure dependent at low pressures (low [A]). The relative rate of reaction $(d[P]/dt)/[A]$ is therefore predicted to decrease at low [A], since decomposition must "wait" for collisional activation to supply activated molecules A^*. This decrease in the effective first-order rate constant was a critical test of Lindemann's model. Figure 15.10 displays data for the isomerization of cyclopropane to propylene:

$$\underset{H_2C-CH_2}{\overset{CH_2}{\diagup\!\!\diagdown}} \longrightarrow CH_3-CH=CH_2 \tag{15.49}$$

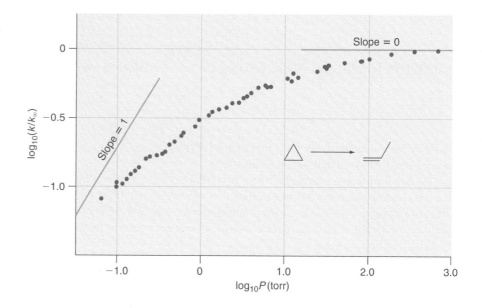

Figure 15.10
Dependence of the apparent first-order rate constant k on pressure for the isomerization of cyclopropane. At low pressure the slope of the plot of log k/k_∞ versus log pressure approaches unity, reflecting second-order kinetics, while at high pressure it goes to zero, and the reaction becomes first-order. *Source:* Pritchard et al., Proc. Roy. Soc. (1953).

over a wide range of pressures, in which the "falloff" in the rate constant at low pressure is evident. Similar curves have been found for a variety of apparent first-order reactions, a resounding vindication of Lindemann's basic idea. Given that we can use collision theory to predict the rate of activation, the shape of the falloff region contains new information on the unimolecular decomposition step, spawning a new theory for this step developed over a span of 20 years by Rice, Ramsperger, Kassel, and Marcus now known as RRKM theory. This theory more accurately described how A* distributes its energy among its degrees of freedom, leading to an improved account of the gradual falloff in k at low pressure; it is closely related to the theory we are about to present, as we delve more deeply into the "mechanism" of an elementary step itself.

Transition State Theory

As we have seen, the unpredictable steric factor p makes the collision theory of reaction rates less useful than one might hope. In 1935 H. Eyring formulated an alternative theory for bimolecular reactions, based on the known connections between equilibrium and rate, between thermodynamics and kinetics. Examining the energy profile of a typical activated, exothermic reaction (see Figure 15.9), Eyring recognized that one could identify the transitory species that exists as the system passes over the activation barrier as a separate thermodynamic state of the system, called the **activated complex** or the **transition state**. For a bimolecular reaction A + B ⟶ P (products), the transition state is denoted AB‡. Because this state lies atop a barrier crest, it is in *metastable equilibrium,* and we can write a "mechanism" for an elementary bimolecular reaction:

$$A + B \rightleftharpoons AB^\ddagger \xrightarrow{k_c} P$$

If we assume that the transition state is in equilibrium with reagents, that is, that decomposition of the activated complex into products is rate limiting, then we can write $K^{\ddagger} = [AB^{\ddagger}]/[A][B]$, where $K^{\ddagger}$ is the equilibrium constant for complex formation.

Now the rate of product formation is the rate at which activated complexes decay into products, an inherently unimolecular process, given by $d[P]/dt = k_c[AB^{\ddagger}] = k_c K^{\ddagger}[A][B]$. The following line of reasoning allows us to deduce k_c, the rate constant for decomposition of the complex. The complex flies apart when a critical bond is broken. A breaking bond has a force constant that is going to zero, and therefore its energy quantization is relaxed and it can take on the classical energy $\varepsilon = k_B T$. The frequency corresponding to this energy is $\nu = \varepsilon/h = k_B T/h$, by Planck's hypothesis. In the classical limit, this frequency becomes the classical frequency of motion along the direction of the dissolving bond, and Eyring therefore identified ν as the unimolecular rate constant k_c, giving us the following simple expression for the bimolecular rate constant:

$$k = \frac{k_B T}{h} K^{\ddagger} \tag{15.50}$$

The equilibrium constant $K^{\ddagger}$ is related to a free-energy change for complex formation by $-RT \ln K^{\ddagger} = \Delta G^{\circ\ddagger} = \Delta H^{\circ\ddagger} - T\Delta S^{\circ\ddagger}$ in the way we have seen for normal thermodynamic states in Chapter 12, except that here for gas reactions we are using concentrations instead of partial pressures, and $K^{\ddagger}$ has units of liters per mole instead of inverse atmospheres. With the definition of standard state being 1 mol/L concentrations, Equation 15.50 can then be written

$$k = \frac{k_B T}{h} e^{\Delta S^{\circ\ddagger}/R} e^{-\Delta H^{\circ\ddagger}/RT} \tag{15.51}$$

From this equation we can identify the Arrhenius activation energy, defined as the slope of a $\ln k$ versus $1/T$ plot, as

$$E_a \equiv \Delta E^{\circ\ddagger} = \Delta H^{\circ\ddagger} - (\Delta n^{\ddagger} - 1)RT \tag{15.52}$$

where $\Delta n^{\ddagger}$ is the change in number of moles of gas in going from reagents to the transition state. For bimolecular gas-phase reactions as in our example, $\Delta n^{\ddagger} = -1$, while for unimolecular reactions it is zero, as it is for all condensed-phase reactions. At normal temperatures the difference between $\Delta H^{\circ\ddagger}$ and E_a is a few kilocalories per mole at most. With this relationship, the preexponential factor A can also be identified from Equation 15.51 as

$$A = \frac{k_B T}{h} e^{\Delta S^{\circ\ddagger}/R} e^{(1-\Delta n^{\ddagger})} \tag{15.53}$$

For gas reactions, in order to make the **entropy of activation** $\Delta S^{\circ\ddagger}$ correspond to a standard state of 1 atm (as we have used in thermodynamics of gases), the right-hand sides of Equations 15.51 and 15.53 should be multiplied by $(RT/P^{\circ})^{-\Delta n^{\ddagger}}$. For a bimolecular gas reaction at 298 K, this yields a factor of 24.47 L mol^{-1}, the molar volume at NTP, or 4.063×10^{-20} cm^3 molecule^{-1}. These numbers change if $T \neq 298$ K. No modification is necessary for solution reactions, save to recall that when using these equations the units of $K^{\ddagger}$ are implicitly present. A gas-phase example will help to clarify the units and numerics.

EXAMPLE 15.6

Although you could use the results of Appendix D along with a guessed transition state to compute the entropy of activation $\Delta S^{\circ\ddagger}$ for use in Equations 15.51 and 15.53 (see the following discussion), this is complicated, and we will leave such exercises for a more advanced course. But you *can* obtain $\Delta S^{\circ\ddagger}$ from measured A's using Equation 15.53, an exercise that conveys useful information about the nature of the activated complex. Find $\Delta S^{\circ\ddagger}$ at 298 K for the O + H$_2$ $\longrightarrow$ OH + H reaction using Equation 15.53 and the data in Table 15.4. Compare your result with ΔS° for the gas reaction O + H$_2$ $\longrightarrow$ H$_2$O computed from data in Appendix C.

Solution:
Solving Equation 15.53 for $\Delta S^{\circ\ddagger}$ with $\Delta n^\ddagger = -1$ and $A = 2.5 \times 10^{10}$ L mol^{-1} s^{-1} from Table 15.4,

$$\Delta S^{\circ\ddagger} = R\{-2 + \ln[A(h/k_B T)(P^\circ/RT)]\}$$
$$= (1.99 \text{ cal mol}^{-1} \text{ K}^{-1})\{-2 + \ln[2.5 \times 10^{10} \text{ L mol}^{-1} \text{ s}^{-1}$$
$$\times (6.63 \times 10^{-34} \text{ J s})/((1.38 \times 10^{-23} \text{ J K}^{-1})(298\text{K}))$$
$$\times (1 \text{ atm})/((0.8206 \text{ L atm K}^{-1} \text{ mol}^{-1})(298\text{K}))]\}$$
$$= (1.99)\{-2 + \ln[1.65 \times 10^{-4}]\}$$
$$= -21.3 \text{ cal K}^{-1} \text{ mol}^{-1}$$

The loss of entropy on forming the OH$_2^\ddagger$ complex (noting that $\Delta n^\ddagger = -1$) is the usual result. ΔS° for O + H$_2$ $\longrightarrow$ H$_2$O from Appendix C is -24.75 cal K^{-1} mol^{-1}. Comparing we see that the OH$_2^\ddagger$ complex is not as "well organized" as H$_2$O, consistent with the less symmetric O—H—H geometry one might expect to promote breakup into OH + H. But the parity of these two ΔS's does imply a well-bonded "tight" transition state.

At first blush, it would appear that transition state theory is even less useful than collision theory at predicting reaction rates, since now there are two unknowns, E_a or $\Delta H^{\circ\ddagger}$ and $\Delta S^{\circ\ddagger}$. Something has been gained, nonetheless, because by casting transition state theory in thermodynamic terms, Eyring was able to use the tools of *statistical thermodynamics* (see Appendix D) to predict the entropy of activation $\Delta S^{\circ\ddagger}$. There is still guesswork involved, in that one has to postulate a particular geometry and a set of vibrational frequencies for the activated complex AB‡ in order to find its partition function, but the predictions of this theory are vastly superior to those of the collision theory. If the reacting molecules are assumed to be structureless, transition state theory reassuringly gives a result identical to the collision theory rate constant of Equation 15.42.

For reactions between more complex molecules, there is a correspondingly greater reduction in entropy S, $\Delta S^{\circ\ddagger} < 0$, in assembling a collision complex with the proper geometry. According to Equation 15.53, this makes the preexponential factor shrink rapidly, providing a natural way of accounting for the steric factor, even given the uncertainty in our knowledge of the properties of AB‡. For example, for the reaction NO + O$_3$ $\longrightarrow$ NO$_2$ + O$_2$ (of stratospheric significance), transition state theory predicts $\log_{10}A = 8.64$, compared to the collision theory value of 11.35 and experimental value of 9.04 (see Table 15.4). This prediction is based on a structure for the activated complex in which the N atom bonds to a terminal O on O_3 to make a five-atom chain

$$O-N-O-O-O$$

As the NO + O_3 example shows, chemically intuitive notions concerning bonding and vibration are readily employed in the estimation of the structure of $AB^\ddagger$, especially if the stuctures of the reagent and product molecules are well known. Often there is some latitude, in that one can make the complex "look" more like the reagents or more like the products; using both to calculate $\Delta S^{o\ddagger}$ and comparing with experiment can convey molecular-level detail concerning the location of the activation barrier along the reaction coordinate. In recent work, molecular orbital computer codes have been developed that are capable of locating least-energy paths through the transition-state region and the activated complex structures that occur along those paths. It is often the case that exothermic reactions have "early" transition states that resemble the reagents, while endothermic reactions need to become product-like to reach the top of the barrier. While these two choices of transition state can produce significantly different predictions for $\Delta S^{o\ddagger}$, they exert an even stronger influence on the actual molecular motions that a reacting system undergoes, a recently revealed facet of chemical kinetics that we'll examine next.

Molecular Reaction Dynamics

In Chapters 8 and 9 we learned of the ceaseless molecular motion that occurs within any sample of matter at a finite temperature, while in Chapter 14 the forces between molecules were brought to light. Combining these two fundamental properties allows us to imagine myriad collisions between pairs of *moving molecules subject to intermolecular force*. That is, molecular encounters of the type that have drawn our attention in this chapter must be more than collisions between spheres of given diameters, which bounce from each other if they approach closely enough. Molecular reality is both more complex and more graceful than this "billiard ball" picture suggests. You can appreciate this if you ponder how an activated complex with a specific structure can be formed from freely flying reagents. Even in the simplest example of two atoms colliding, the long-range attractive dispersion forces first cause the atoms to pull toward each other until, if the encounter is sufficiently close, the gradually rising Pauli repulsion gently turns the atoms aside. They exit the same way they came, again passing through the attractive part of the interaction at larger distances. A typical **trajectory** is shown in Figure 15.11.

For the molecular encounters that produce chemical transformations, the presence of dipole moments or acid–base interactions will cause molecular orientations to change as a result of reagent approach, perhaps "locking on" to the geometry that will lead eventually to permanent changes in the bonding, the hallmark of a chemical process. The bond breaking and making that occurs can be pictured, in this same trajectory framework, as the motion of an atom or group in transit between its donor and its acceptor. We have seen pictures of this sort in Chapter 7 when discussing the frontier molecular orbital model; here we are going to ask you to think more deeply about the actual *molecular dynamics*, molecular motions subject to forces, that underlie the birth of new chemical substances.

Figure 15.11
Collision trajectories for two atoms under the influence of van der Waals forces. The numbers indicate the passage of time; the total duration of the collision, from $t = 1$ to $t = 5$, is of the order of a picosecond or less. With the center of mass of the atoms held stationary, the atoms approach each other and recede after collision along straight lines (Newton's first law) from opposite directions. For identical atoms, the trajectories exactly mirror each other, the force on one atom always accompanied by an equal and opposite force on the other (Newton's third law). The shapes of the trajectories are determined by Newton's second law. When the atoms are still rather far apart ($t = 2$), the attractive London dispersion (LD) force begins to accelerate them toward each other. The overlap (Pauli) repulsive force then takes over until the two atoms "touch" at $t = 3$, and they recede with gracefully symmetric motion through the attractive region ($t = 4$) before resuming undisturbed linear paths. Each trajectory is symmetric about the "apse line." Molecular trajectories are more complex due to orientational effects and rotation/vibration within each molecule, but are roughly of the same form.

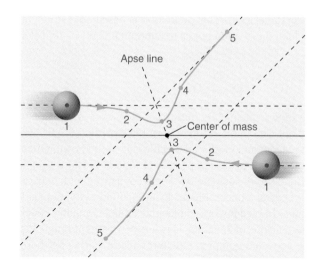

We must begin this perusal with a caveat. In Chapter 8 we noted that classical or Newtonian mechanics, though a venerable theory, does not suffice to describe atomic and molecular motion. Motions along a path, the *classical trajectories,* are to some extent a figment of an overactive chemical imagination, and ought to be replaced by quantum-mechanical wave functions that satisfy the Schrödinger equation. Nonetheless, because nuclei are so much heavier than electrons, the behavior of an atom in a force field is much more nearly classical than that of an electron. This means that wave functions that describe nuclear motion, in particular the time-dependent ones (called *wavepackets*), are well localized in space around the classical path that Newton's laws would give. This is particularly true for *translational* degrees of freedom, for which energy quantization is not a significant factor. The "least classical" atoms are the lightest ones, hydrogen atoms, and chemists have discovered that they are able to "tunnel" through the activation energy barrier, thereby increasing the reaction rate. Tunneling (passing from one side of a barrier to the other without going over it) cannot be described by Newton's laws, since the tunneling particle (a "tunnicle") would have negative kinetic energy (or imaginary velocity) while passing through the barrier.

If we neglect such quantum effects, we can approach reaction dynamics in the mechanical spirit urged on us by Lothar Meyer. Let's first consider the simplest type of chemical reaction from a dynamical point of view, an *atom-molecule reaction* of the type A + BC $\longrightarrow$ AB + C. This type of reaction is often an elementary step in gas reactions between simple molecules. For example, important steps in the explosive reaction $H_2 + F_2 \longrightarrow$ 2HF are F + $H_2 \longrightarrow$ HF + H and H + $F_2 \longrightarrow$ HF + F. In this type of reaction, one chemical bond is broken and one is made, and the molecular mechanics of the reaction can be imagined to occur in three stages,

1. A $\cdots\cdots$ B—C
2. A $\cdots$ B $\cdots$ C
3. A—B $\cdots\cdots$ C

that is, A approaches BC, the new AB bond begins to form while the old BC bond begins to break, and finally the liberated atom C departs from the newly born AB. These

15.4 Molecular Theories of Elementary Reactions

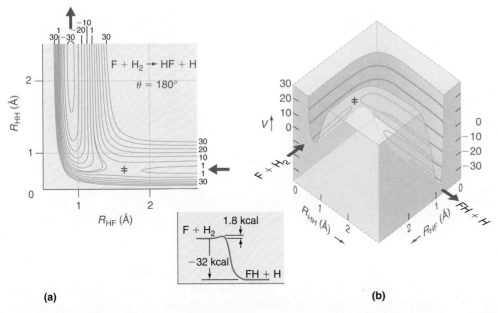

Figure 15.12
Depictions of the potential energy surface for the reaction F + H$_2$ ⟶ HF + H based on a recent highly accurate molecular orbital calculation (Stark & Werner, *J. Chem. Phys.*, 1996). **(a)** Potential energy contour map for collinear F⋯H⋯H geometry ($\theta = 180°$ is the bond angle). Contours are drawn every 5 kcal/mol for +30 to −30 kcal, with one extra contour added for $V = 1$ kcal in order to delineate the activation barrier, whose location is indicated by ‡. Reagents enter from the lower right, where R_{HF} is large and R_{HH} is at equilibrium for H$_2$ at the lowest point, and products emerge at the upper left, where R_{HH} is large and R_{HF} is now at equilibrium. The difference in energy between these extremes is 31.8 kcal/mol, which is the exothermicity of the reaction; the barrier height is 1.8 kcal, in good agreement with E_a from Table 15.4. **(b)** This surface is shown as a groove cut into a transparent block; for clarity, only half of the contours are drawn, with the positive-energy ones on the large-R edge of the groove omitted. It is more easily seen here than in (a) that the extremes represent bond potential energy curves for the two diatomics. The inset shows the corresponding energy profile.

stages are smoothly jointed by continous motion of all three atoms subject to Newton's laws of motion. At all times the center-of-mass point of the three-atom system is held fixed, and for simplicity we assume that the reaction occurs in a collinear geometry. The potential energy becomes a function of both bond distances, $V = V(R_{AB}, R_{BC})$, and can be depicted as a *surface,* rather than simply a curve, as illustrated in Figure 15.12 for the F + H$_2$ reaction. The surface becomes a bond potential curve for BC in the "entrance valley," and the same for AB in the "exit valley," while in between it displays a barrier (a point of maximum potential energy). The heat of reaction is the difference in potential energy level between the two valley floors. Motions of the atoms can then be represented by a mass point moving on this surface, a sort of molecular bobsled run, and the classical laws of motion will tell us if and how a point in the entrance valley (stage 1 above) of this surface arrives at the "transition state" atop the barrier where the bonds are switching (stage 2), and by and by emerges from the exit valley (stage 3), completing a successful reaction. Figure 15.13 illustrates both a successful "reactive" trajectory and an unsuccessful one. Trajectories like these, which are not badly "snarled," typically take less than a picosecond to complete.

This microscopic view of the reaction mechanism engenders two interesting new questions concerning the energetics of the reaction. First, what is the best way to overcome the activation barrier? In interpreting out earlier energy profiles (Figure 15.9), we spoke of increasing the temperature and thereby endowing an increasing fraction of the reagents with enough energy to pass over the activation barrier. When the reagents are heated, however, both the translational and vibrational degrees of freedom receive additional energy, and it may be that one of these forms of energy is more effective than the other in enabling the system to climb the hill, as hinted at in Figure 15.13. In an extensive series of computer-generated trajectory studies in the late 1960s, J. Polanyi and his students established that this is indeed the case, and which form of energy, translational or vibrational, will be more effective in promoting reaction is determined by which valley the barrier crest occupies. In our F + H$_2$

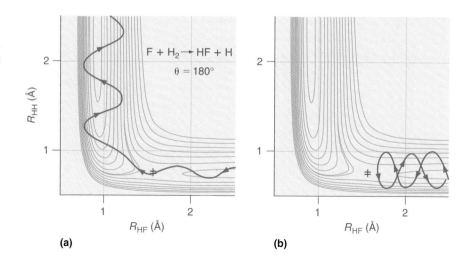

Figure 15.13
Classical trajectories on the F + H$_2$ potential energy surface. **(a)** A reactive trajectory that begins at the lower right as F and H$_2$, successfully negotiates the activation barrier, rounds the bend where both interatomic distances are changing simultaneously, and emerges at the top as HF and H. The total duration of the trajectory is less than a picosecond. Motion along the valley floors represents translational energy, while oscillatory motion between the walls of the valleys indicates vibrational energy. This implies that the newly formed HF on the upper left is vibrationally excited. **(b)** An unreactive trajectory that is unable to surmount the barrier and must therefore double back on itself to re-form reagents. Although the reagents have enough energy to surmount the barrier, too great a fraction of it is present as H$_2$ vibration.

example (as in many exothermic atom–molecule reactions), the barrier crest lies in the entrance valley, and in this "early barrier" case increasing the translational energy of F relative to H$_2$ is far more effective than vibrational excitation of H$_2$ in getting the system around the corner. On the other hand, when the crest lies in the exit valley, a "late barrier" (as in many endothermic reactions), vibrational energy becomes the preferred means of finding that valley. This property is known as **selective energy consumption.** These correlations are readily understood by considering the way the different forms of energy affect the motion of the trajectory point along the surface; one of the exercises at the end of this chapter will walk you through the reasoning.

The second question regards the disposal of the heat of reaction in the products. Ordinarily we think of an exothermic reaction as heating up the system, that is, increasing its temperature by endowing both translational and vibrational degrees of freedom with increased average energy. But it may be that the reaction preferentially deposits the energy released into vibration or translation, and what is observed in the system is the result of *collisional relaxation* of the enhanced degree of freedom until equilibrium (a well-defined final temperature) is reached. Polani's studies of exothermic systems have shown us that, again, this is indeed the case, and which product degree of freedom gets the lion's share of the energy is likewise determined by the location of the crest of the activation barrier. Here early barriers preferentially yield vibrationally excited products, while late barriers (a much rarer occurrence for exothermic reactions) put the bulk of the energy into translation. This property is called **specific energy disposal.** As in selective energy consumption, these correlations are related to the behavior of the trajectories as they round the bend in the potential energy surface (an end-of-chapter exercise may clarify this for you). Specific energy disposal is surprising, since it is not predicted by the prince of all kinetic theories, transition state theory (TST). TST assumes a statistical distribution of energy among all the degrees of freedom in the intermediate complex, and this is expected to be maintained in the products. The trajectory studies as well as modern experiments have made us realize, however, that this last assumption disregards the possibility that specific forces between the departing fragments—for example, the steep downhill slide that characterizes highly exothermic reactions—may channel the available energy into specific degrees of freedom, giving a decidedly *nonstatistical* energy distribution.

15.4 Molecular Theories of Elementary Reactions

These results have been the wellspring for modern studies, both experimental and theoretical, that exploit this selectivity and specificity. F. Crim and his students have now succeeded in carrying out "bond-selective chemistry," using a laser to excite bond vibrations that predictably enhance certain reactions over others. G. Pimentel and his students have turned the specific disposal of energy as HF vibration in the F + H$_2$ reaction into a "chemical laser," in which the highly populated excited vibrational levels of HF cause stimulated emission at the infrared wavelengths of vibrational transitions.

Experimental methods for studying the molecular dynamics of elementary reactions have progressed hand in hand with these theoretical descriptions. Many of these methods make use of *molecular beams*, directed rays of gas-phase reagent molecules that intersect in a vacuum chamber, with product molecules or emitted light from the reactive encounters monitored by sensitive detectors. The detection of chemiluminescent light under these "single-collision" conditions was pioneered by Polanyi, while D. Herschbach, Y. Lee, and R. Zare blazed the trail of detecting the product molecules themselves using a mass spectrometer, or a laser to induce product fluorescent emission. The directionality provided by the use of molecular beams has made possible the measurement of *angular distributions* of product molecules, very much like those which led Rutherford and his students to the nuclear model of the atom (see Chapter 1). As of old, these modern angular experiments reflect the forces acting during the fleeting existence of the collision complex, and also bring in rotation and the angular momentum of the colliding partners. The most recent development, of the mid-1980s, has been the introduction of ultrafast pulsed lasers (pulsewidth 10–100 fs, as short as a vibrational period—see Chapter 8) by A. Zewail and coworkers, who have now "caught" reactive complexes in the act of dissociating into products. Fast lasers have also begun to shed new light (forgive the pun!) on reactions occuring in solution, to be examined next.

Reactions in Solution

Although the presence of a solvent such as water makes possible many reactions that would not occur in the gas phase, a reaction between solutes in a solution is slowed by the need to plow through the ever-present solvent shell around the reagents. Instead of colliding as a result of free translational motion, the reagent molecules come together by *diffusing* through solvent. Picture two lovers on a crowded train station platform—they would readily find each other were they alone, but here they must struggle through a milling horde of other passengers. This is illustrated on a molecular level in Figure 15.14. The coming together of A and B is clearly going to take much longer in a crowd of solvent molecules. Equation 15.40 contains the essential factors governing the collision rate: relative velocity of approach, target area or cross section, and the product of the concentrations. The factor that must be altered is the relative velocity, and to understand how to do this we must delve into the phenomenon of diffusion on the molecular level.

In Chapter 9 we introduced the ideas of effusion and diffusion in gases; the difference between these is that in diffusion the progress of the mobile species is inhibited by collisions with the ambient gas. For example, when bleach is sprayed into the air in a room, within a few seconds the odor of Cl$_2$ gas reaches the nostrils of anyone in the room. The Cl$_2$ is said to **diffuse** through the air. As is true for the A molecule

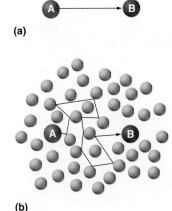

Figure 15.14
Molecule A meets molecule B in the gas phase (**a**) and in solution (**b**), where the unlabeled particles are solvent molecules. In solution, A makes slower progress toward B, but once A and B meet, the solvent will keep them together longer—the *solvent cage effect*.

in Figure 15.14, on a microscopic level the Cl_2 molecules collide with molecules of air at random locations in space, and as a result of those collisions are deflected in random directions. A rigorous treatment of diffusion in 3-D is very difficult; we would need to solve a partial differential equation similar in form to the Schrödinger equation! Instead, as we did for gas dynamics in Chapter 9, we will use a simple intuitive model that gets us very close to the same result.

On the average, the distance the diffusing molecule travels between collisions is the *mean free path* λ, defined in Chapter 9. The haphazard progress of the diffusing molecules may be modeled by a *random walk* process. With an eye to applying the random walk idea to diffusion of reagents through a liquid solvent, we will consider a simple one-dimensional random walk model. In one dimension, the steps of a *molecular random walk* can only occur to the left or to the right, corresponding to colliding at random with a solvent molecule on either side of the starting point, $x = 0$. We will take each step in the random walk to be of length λ, the mean free path. We would like to know the time Δt it will take for the molecule to diffuse a distance Δx from the origin. Because of the random nature of the process, there of course will be a distribution of distances traveled after some time lapse Δt; in fact, because the probability of being hit from the left or right is the same, the most likely outcome is no progress at all, and the molecule remains at the origin.

Using a computer program incorporating a random number generator, we can simulate a set of random walks, as shown in Figure 15.15. The displacement from the origin can be represented by the number of steps in excess in a particular direction (right or left) $\Delta n = \Delta x/\lambda$, and the time elapsed by the total number of steps $N = \bar{v}\Delta t/\lambda$, where $\bar{v}$ is the average velocity of the diffusing molecule between collisions. As Figure 15.15 indicates, the "trajectories" fan out from the origin with passing time, forming a distribution of displacements that is similar to that for the

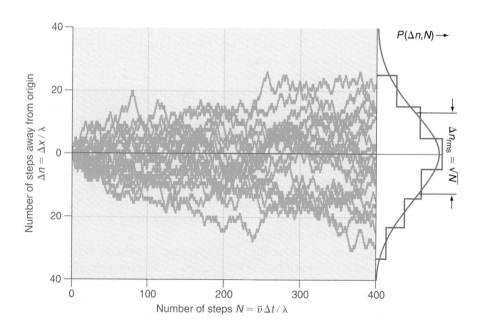

Figure 15.15
Computer-generated set of 20 one-dimensional random walks, beginning at the left; λ is the mean free path. The root mean square displacement $\Delta x_{rms} = \lambda \Delta n_{rms}$ increases as $\lambda\sqrt{N}$; the distribution of displacements $P(\Delta n, N)$ is given on the right as a histogram. For a large number of walks, this distribution becomes the Gaussian error distribution function of Equation 15.54 as illustrated.

distribution of errors in a measurement. For a large number of walks, this distribution takes the form of a Gaussian function, as we have already discussed in Chapter 9 for the one-dimensional distribution of velocities in a gas. The random walk distribution is given by either of the equivalent forms

$$P(\Delta n, N) = \sqrt{\frac{2}{\pi N}}\, e^{-(\Delta n)^2/2N} \quad \text{or} \quad (15.54)$$

$$P(\Delta x, \Delta t) = \sqrt{\frac{2}{\pi \lambda \bar{v} \Delta t}}\, e^{-(\Delta x)^2/(2\lambda \bar{v} \Delta t)}$$

The premultiplying factors normalize these distributions to probability densities. Because the distribution $P(\Delta x, \Delta t)$ is symmetrical about the origin, on the average $\Delta x = 0$; however, the *root mean square displacement* Δx_{rms} (which is like a standard deviation or standard error) is nonzero, being given by

$$\Delta x_{\text{rms}} = \sqrt{\overline{(\Delta x)^2}} = \sqrt{\lambda \bar{v} \Delta t} \equiv \sqrt{2D\Delta t} \quad (15.55)$$

Thus, the distance traveled from the origin goes as the square root of the time lapse Δt, rather than linearly with Δt. The quantity D defined by $D = \frac{1}{2}\lambda \bar{v}$ is called the **diffusion coefficient;** D is susceptible to experimental measurement, for example, by injecting a colored solute at a certain point in a solution and monitoring the time it takes to reach a nearby point spectrophotometrically. For a small solute species in a solvent such as water, λ is of the order of the solvent diameter $\sim 10^{-8}$ cm, while $\bar{v}$ is $\sim 10^4$ cm/s, making $D \approx 5 \times 10^{-5}$ cm^2/s. Table 15.5 lists values of D for some common molecules and ions in water. Using D in our formulation enables a more general treatment of the diffusion problem that includes attractive interaction between solute and solvent as well as "many-body" effects wherein a solute is interacting with several solvent molecules at once as it diffuses; both of these effects tend to make diffusion more difficult, reducing D. Equation 15.55 was first derived by Albert Einstein in 1905 in his analysis of *Brownian motion*.

TABLE 15.5

Diffusion coefficients D (10^{-5} cm^2 s^{-1}) for common solutes in water

Cations	D	Anions	D	Neutrals	D
H^+	9.31	OH^-	5.30	NH_3	1.77
Li^+	1.03	F^-	1.46	CO_2	1.71
Na^+	1.33	Cl^-	2.03	CH_3COOH	0.99
K^+	1.96	Br^-	2.08	CH_3OH	1.58
Mg^{2+}	0.71	I^-	2.05	Cl_2	1.3
Ca^{2+}	0.79	CH_3COO^-	1.09	Br_2	0.9
Ba^{2+}	0.85	CO_3^{2-}	0.96	I_2	1.38
Cu^{2+}	0.71	SO_4^{2-}	1.06	$C_6H_{12}O_6$	0.67

This random walk analysis of diffusion enables us to define a **diffusion velocity** $\bar{v}_D$ by

$$\bar{v}_D \equiv \frac{\Delta x_{\text{rms}}}{\Delta t} = \Delta x_{\text{rms}} \left(\frac{\lambda \bar{v}}{(\Delta x_{\text{rms}})^2} \right) \quad (15.56)$$

$$= \bar{v} \left(\frac{\lambda}{\Delta x_{\text{rms}}} \right)$$

$$= \frac{2D}{\Delta x_{\text{rms}}}$$

where in the first line we have substituted Δt from Equation 15.55. Here you can see that the diffusion velocity can be much smaller than the average velocity $\bar{v}$ if the molecule must traverse distances large compared to λ, while we can also write $\bar{v}_D$ in terms of the diffusion coefficient.

When we use the diffusion velocity $\bar{v}_D$ in place of $\bar{v}$ in the collision theory expression for the rate constant with zero activation energy, we find $k_D = (2D/\Delta x_{\text{rms}})\pi d_{AB}^2$, where the diffusion coefficient D now corresponds to the *mutual* diffusion of A and B toward each other, that is, $D = D_A + D_B$. Now the rate is determined by how quickly A and B can diffuse over a distance d_{AB}; we therefore replace Δx_{rms} by d_{AB}, which cancels one factor of d_{AB}. In addition, if diffusion from all possible directions toward the surface of a sphere of radius d_{AB} is accounted for, we get an additional factor of 2, to yield the simple expression.

$$k_D = 4\pi d_{AB} D \quad (15.57)$$

Using $d_{AB} \approx 10^{-8}$ cm and $D = D_A + D_B \approx 10^{-4}$ cm^2/s, a typical magnitude for k_D is $\sim 10^{-11}$ cm^3/s or 10^{10} L mol^{-1} s^{-1}. For the Arrhenius acid–base reaction H$^+$(aq) + OH$^-$(aq) $\longrightarrow$ H$_2$O(l), using diffusion coefficients for H$^+$ and OH$^-$ from Table 15.5 and $d_{AB} \approx d(H_2O) = 3.0$ Å, we find $k_D = 3.3 \times 10^{10}$ L mol^{-1} s^{-1}, a factor of 4 *smaller* than the observed value of k. The Coulombic attraction between the ions, not accounted for in our simple treatment, would undoubtedly increase d_{AB}. On the whole this implies that the acid–base reaction occurs with little or no activation energy, consistent with a reaction between oppositely charged ions. Equation 15.57 thus gives us an approximate upper limit for the rate constant analogous to that for gas-phase reactions from the collision theory, and reactions such as H$^+$ + OH$^-$ that occur at this limiting rate are said to be **diffusion-controlled reactions.** Table 15.6 collects some bimolecular rate constants and Arrhenius parameters for aqueous reactions; you can see that a number of the simpler ones are at or near the diffusion-controlled limit.

The diffusion-controlled limit can be readily combined with the transition-state theory given earlier by recognizing that, when the decomposition of AB‡ is faster than its formation, the rate of the reaction is limited by the diffusion rate, and the rate constant $k = k_D$, given by Equation 15.57. On the other hand, under the assumption of rate-limiting decomposition of AB‡ as supposed in our transition-state derivation, then $k = (k_B T/h)K^{\ddagger}$ as in Equation 15.50. Intermediate cases can be treated by invoking the steady-state approximation, as exemplified in the Lindemann mechanism, Equations 15.43 through 15.46. For solution-phase reactions of more "normal" rate, particularly those involving organic compounds and complex transition states, diffusion is so

TABLE 15.6
Rate constants at 298 K and Arrhenius parameters for aqueous bimolecular reactions[a]

Reaction	$k(298\ \text{K})$ (L mol^{-1} s^{-1})	A (L mol^{-1} s^{-1})	E_a (kcal mol^{-1})
$H^+ + OH^- \longrightarrow H_2O$	1.35×10^{11}		
$H^+ + NH_3 \longrightarrow NH_4^+$	4.3×10^{10}		
$H^+ + HS^- \longrightarrow H_2S$	7.5×10^{10}		
$OH^- + CO_2 \longrightarrow HCO_3^-$	3.3×10^3	1.5×10^{10}	9.1
$OH^- + HCO_3^- \longrightarrow CO_3^{2-} + H_2O$	$6. \times 10^9$		
$OH^- + NH_4^+ \longrightarrow NH_3 + H_2O$	3.4×10^{10}		
$OH^- + C_2H_5Br \longrightarrow C_2H_5OH + Br^-$	8.7×10^{-5}	4.3×10^{11}	21.4
$CH_3COOC_2H_5 + OH^- \longrightarrow CH_3COO^- + C_2H_5OH$	0.10	1.7×10^7	11.2
hemoglobin $\cdot\ 3O_2 + O_2 \longrightarrow$ hemoglobin $\cdot\ 4O_2$	$2. \times 10^7$		

[a] Where A and E_a are not given, either they have not been measured or $E_a = 0$ and $A = k$.

much faster that it does not influence the observed rate, and TST is nearly universally used. In all events we assume the reagents are perfectly mixed prior to reaction, a condition difficult to achieve for solution reactions. You may have observed, when doing acid–base titrations, that near the endpoint the reaction occurs rapidly around the entry point of the base titrant, as shown by the trail of indicator color change around it. The rates of the ultrafast reactions of Table 15.6 have been measured by the *temperature-jump method,* a technique invented by M. Eigen and coworkers in the 1950s that subjects a perfectly mixed equilibrium system to a quick change in T, thereby changing both forward and reverse rate constants, but by different amounts.

Modern studies intended to probe the details of solution reactions have focused on unimolecular reactions, such as isomerization or dissociation, that can be readily initiated by a fast laser; the dissociation and subsequent recombination of I_2 in CCl_4 solution has been extensively examined using sub-picosecond laser pulses. Combined with theoretical simulations (massive trajectory studies in which the motions of the reagent along with dozens of solvent molecules are followed with classical mechanics), these experiments have revealed how the shape and dynamics of the solvent "cage" around the reactants influence the outcome.

15.5 Chemical Catalysis and Complex Reactions

In all of the preceding theoretical considerations, we have allowed the reaction to proceed unaided, granting the great influence of the solvent in solution reactions. The fascinating question of whether and how we may introduce "chaperones" in order to effect a more rapid union of the reagents is addressed in the final section of this chapter. Much as Shakespeare's playmaker henchmen Rosenkrantz and Guildenstern both thickened and hastened the plot of *Hamlet,* a cleverly chosen chemical interloper can both complicate and enhance reaction.

Catalysis: Homogeneous and Heterogeneous

The acceleration of the rate of a chemical reaction by the addition of small amounts of a compound that is not itself consumed by the reaction has become increasingly common in modern chemical technology. Such a "magical" compound is known as a **catalyst**: to the ancient Arabian alchemist this was the avidly sought "xerion, aliksir, noble stone, magisterium, that heals the sick, and turns base metals into gold, *without itself undergoing the least change.*" The word *catalyst* was coined by Berzelius in the early 19th century. A catalyst operates by providing a new reaction path with an activation energy lower than that of the unaided reagents. Typically it works by bonding to one (or more) of the reagents, converting it to a form more readily attacked by the other; it then is released in a subsequent product-forming step. **Homogeneous** catalysts are present in the same phase as the reagents, while **heterogeneous** ones intrude from a different phase. An ubiquitous example of a heterogeneous catalyst is the automotive catalytic converter (Figure 15.16), which catalyzes (among other reactions) the oxidation of CO, resulting from incomplete combustion of gasoline, into CO_2, $2CO + O_2 \longrightarrow 2CO_2$. This catalyzed conversion was first studied by Irving Langmuir in 1922. The finely divided solid platinum Pt catalyst does not change the position of equilibrium, which greatly favors the CO_2 product, but simply accelerates the rate of product formation by "activating" CO and O_2 *adsorbed* on (bonded to) the metal surface. Normally CO will not "burn" unless heated to 600°C, but in the presence of the Pt catalyst the reaction occurs rapidly at temperatures below 200°C. In view of our analysis of energy profiles, this implies that Pt has greatly lowered the activation energy of the rate-limiting step—in this case, the mechanism changes completely. (The kinetics of surface-catalyzed reactions will be discussed later.)

Although catalysts come in many forms and structures, the heterogeneous ones are typically transition metals and their oxides or halides, while homogeneous ones are often free radicals in gas-phase reactions or acids (in the most general Lewis sense) in solution reactions. A homogeneous gas-phase example that has become societally significant in modern times is the catalytic conversion of stratospheric ozone O_3 to ordinary oxygen O_2 by atomic chlorine radicals. Ozone partially dissociates to $O_2 + O$ in the presence of solar ultraviolet light, and the rate-limiting step in the $2O_3 \longrightarrow 3O_2$ conversion is the recombination of "odd oxygen"

$$O_3 + O \longrightarrow 2O_2$$

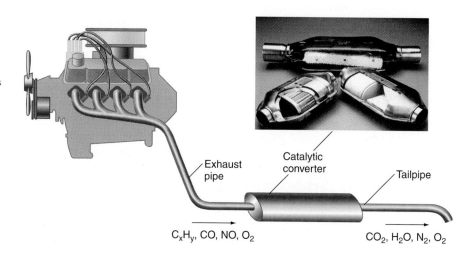

Figure 15.16
Catalytic converter for automobile exhaust gas conversion. If you look under a car made after 1990, you will find it in the exhaust train between the engine and the muffler; the inset shows a cutaway view. The catalyst within, usually cemented to a solid support, is expected to accelerate a number of conversions as indicated, and may contain Pt and Pd metals, along with the oxides V_2O_5, Cr_2O_3, and CuO, in finely divided form. Pt catalyzes the conversion of CO to CO_2 as discussed in the text.

which has an activation energy of $E_a = 4.1$ kcal/mol. The Arrhenius factor $\exp(-E_a/RT)$ is less than 0.0001 at stratospheric temperatures ($\sim$220 K), and the unaided reaction therefore occurs very slowly. This slow recombination allows an appreciable concentration of O_3 to build up. In the presence of Cl atoms, however, the two-step sequence

$$\begin{aligned} \text{Cl} + \text{O}_3 &\longrightarrow \text{ClO} + \text{O}_2 \\ \underline{\text{ClO} + \text{O} \longrightarrow \text{Cl} + \text{O}_2} \\ \text{O}_3 + \text{O} &\longrightarrow 2\text{O}_2 \end{aligned}$$

can occur; each step has negligibly small E_a, making the sequence more than 1,000 times faster than the uncatalyzed reaction. (See Table 15.4 for rate constant data.) Note that the two reactions add up to the erstwhile slow reaction, while Cl does not appear in the products. Cl consumed in the first step is regenerated in the second, and can then attack another O_3. This is known as a *catalytic cycle;* it enables a small concentration of Cl to produce a large ozone depletion. The energy profiles for the uncatalyzed and catalyzed reaction paths are compared in Figure 15.17.

Some atmospheric Cl is produced naturally, but it now appears that most of the Cl atoms in the stratosphere arise from solar photodissociation of *freons,* a commercial name for chlorofluorocarbons such as CF_2Cl_2, which have been widely used as refrigerant gases and spray propellants. The same sunlight that dissociates ozone causes the reaction $CF_2Cl_2 \longrightarrow CF_2Cl + Cl$, and the Cl atoms then become the catalysts for ozone loss. Free Cl is gradually converted to the stable "reservoir molecules" HCl and $ClNO_3$, but over the North and South Poles icy clouds that form during the dark polar winters heterogeneously catalyze the reaction $HCl + ClNO_3 \longrightarrow Cl_2 + HNO_3$, and the advent of sunlight then dissociates the Cl_2, causing dramatic "ozone holes" in the polar summers (see Figure 15.17). Should these holes spread, the result might be catastrophic for us, because the ozone layer protects the delicate biochemicals we are made of from ultraviolet dissociation. Though freon production has been phased out around the world, the holes are expected to worsen before they begin to shrink, owing to the chemical inertness and large molecular mass of the freons already present in the troposphere. The first property ensures their long-term survival, while the second prolongs their upward diffusion into the stratosphere.

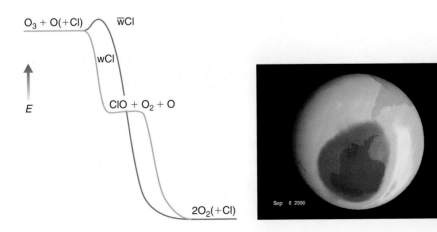

Figure 15.17
Chlorine-atom-catalyzed stratospheric ozone loss. On the left are energy profiles for the uncatalyzed and catalyzed recombination of odd oxygen. The little activation bump on the uncatalyzed pathway ($\overline{w}$Cl) is the critical feature. On the right is a Southern Hemisphere map of the "ozone hole" over Antarctica, which has been worsening since this map was made in 2000; darker areas indicate greater ozone depletion.

Enzyme Catalysis

A factor-of-1,000 enhancement in rate appears quite impressive until we examine the processes going on literally right under our noses! As you sit reading this, the very act of taking information and storing it in your brain—not to mention breathing, maintaining your body temperature, and digesting your lunch—is made possible by a series of chemical reactions accelerated by incredibly efficient biochemical catalysts known as **enzymes**. Enzymes belong to a class of molecules called *proteins,* long chains of amino acids covalently bonded together. You have already met a simple amino acid, glycine, in Chapter 12, and more details of the molecular structures of amino acids and proteins will be presented in Chapter 18. Even a relatively inefficient enzyme "merely" accererrates certain reactions by a factor of a million, 10^6, and acceleration factors $> 10^{17}$ are known! To cite a simple example, the enzyme known as *urease* catalyzes the breakdown of urea, $CO(NH_2)_2$, a by-product of digesting your lunch, according to the reaction

$$CO(NH_2)_2(aq) + H^+(aq) + 2H_2O(l) \xrightarrow{\text{urease}} 2NH_4^+(aq) + HCO_3^-(aq)$$

Temperature variation studies show that the activation energy for this reaction is 33 kcal/mol without enzyme, but only 9 kcal/mol with urease present; the ratio of the $\exp(-E_a/RT)$ factors is 2×10^{17} at body temperature, 37°C.

How does an enzyme accomplish this amazing speed-up? Enzymes are the present-day result of a long evolutionary process in which these molecules were selectively designed to catalyze specific reactions or types of reaction. After being themselves produced in a series of amino-acid linking reactions to be discussed in more detail in Chapter 18, the protein chain *folds* into a compact shape that contains a cavity of the approximate size and shape of the molecules whose reaction is to be catalyzed. The cavity typically includes hydrogen bonding groups as well as hydrophobic (water-avoiding) regions that hold the reagents in a particular spatial arrangement, weaken appropriate bonds, and allow the products to escape before back-reaction can occur. A possible sequence of events is illustrated in Figure 15.18.

The kinetic mechanism governing enzyme action was first formulated by Michaelis and Menten in 1913. The enzyme E first reversibly forms a complex with the key reagent, here denoted as the **substrate** S, followed by reaction to form a specific set of products P as directed by the type of enzyme:

$$E + S \underset{k_{-1}}{\overset{k_1}{\rightleftharpoons}} ES$$

$$ES \xrightarrow{k_2} E + P$$

where ES is the enzyme–substrate complex. The net reaction is just S $\longrightarrow$ P. Applying the steady-state approximation to the concentration of complex [ES], and invoking the mole balance condition $[E]_0 = [E] + [ES]$, we have

$$\frac{d[ES]}{dt} = k_1[E][S] - k_{-1}[ES] - k_2[ES]$$

$$= k_1[E]_0[S] - k_1[ES][S] - k_{-1}[ES] - k_2[ES] = 0$$

$$[ES] = \frac{k_1[E]_0[S]}{k_{-1} + k_2 + k_1[S]}$$

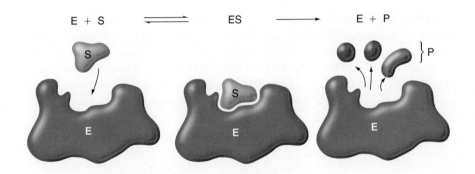

Figure 15.18
Schematic illustration of enzyme catalysis. In this case the decomposition of the substrate S into three fragments, as in the urease-catalyzed reaction, is accelerated by first forming the enzyme–substrate complex ES.

where we have applied mole balance and the SSA in the second line and solved for [ES] in the third line. The rate law for product formation is then

$$\frac{d[P]}{dt} = k_2[\text{ES}] = \frac{k_1 k_2 [E]_0 [S]}{k_{-1} + k_2 + k_1 [S]} \tag{15.58}$$

By defining the *Michaelis constant*

$$K_m \equiv \frac{k_{-1} + k_2}{k_1} \tag{15.59}$$

the rate can be written in terms of only two rate parameters, k_2 and K_m:

$$\frac{d[P]}{dt} = \frac{k_2 [E]_0 [S]}{K_m + [S]} \tag{15.60}$$

Note the similarity between the Michaelis–Menten rate law of Equation 15.58 and the Lindemann model of Equation 15.46. Thinking of this mechanism as applied to the digestion of food S, we can identify two limits, feast and famine. In the famine limit, S is in short supply, and the rate becomes first order in [S], while when food is plentiful, the rate approaches a constant value, independent of [S]. Over what range of [S] the behavior switches between these limits is determined by the magnitude of the Michaelis constant K_m.

Surface Catalysis

Because an enzyme is such a large molecule, enzyme catalysis is similar to heterogeneous catalysis by a solid surface. Metallic and metal oxide surfaces also have specific preferred binding sites, often at atomic-scale surface irregularities such as steps, pits, or protrusions. While the catalytic activity of solids may extend to several reaction types, surfaces may often be chemically tailored to catalyze specific reactions. Like the docking of the substrate to the enzyme, reagent molecules must be *adsorbed* onto the surface of the catalyst prior to reaction. Reaction normally occurs entirely on the surface, followed by *desorption* of the products.

The kinetics of surface-catalyzed reactions were first analyzed by Irving Langmuir in 1916. Langmuir assumed that the surface offered a fixed number of effective binding sites, and that at a particular gas-phase or dissolved concentration of an adsorbable species A, there would be a certain fraction of these sites θ occupied by adsorbed A molecules. Maximum coverage, $\theta = 1$, would imply a monolayer of A on the surface. The rate of desorption of A from the surface is proportional to θ, or equal to $k_d\theta$. The rate of adsorption depends on both the fraction of *unoccupied* sites $1 - \theta$ and the concentration [A], $k_a[A](1 - \theta)$. If the ensuing reaction is slow enough, equilibrium is reached, making these rates equal, $k_d \theta = k_a[A](1 - \theta)$. Solving for θ, Langmuir obtained

$$\theta = \frac{k_a[A]}{k_d + k_a[A]} = \frac{K_{ad}[A]}{1 + K_{ad}[A]} \tag{15.61}$$

where $K_{ad} = k_a/k_d$ is the equilibrium constant for adsorption, and is known as the *adsorption (sticking) coefficient*. A plot of θ versus [A] or P_A at constant T is now known as a *Langmuir adsorption isotherm*. At low [A], the coverage increases linearly with concentration, while at high [A], $\theta \longrightarrow 1$; what is "low [A]" and "high [A]" is determined by the sticking coefficient K_{ad}.

Let's suppose we are dealing with a surface-catalyzed decomposition, $A \longrightarrow P$. Two often-encountered cases are first-order reactions, such as the decomposition $2HI \longrightarrow H_2 + I_2$ on Pt, and zeroth-order reactions, such as $2NH_3 \longrightarrow N_2 + 3H_2$ on W. In Langmuir's model, these two cases correspond respectively to *weakly adsorbed* (small K_{ad}) and *strongly adsorbed* (large K_{ad}) reagents. In either case the reaction rate is assumed to be proportional to the surface coverage, $d[P]/dt = k\theta$. When K_{ad} is small, Equation 15.61 yields $\theta \approx K_{ad}[A]$, and $d[P]/dt = kK_{ad}[A]$, first-order kinetics; while when K_{ad} is large, $\theta \longrightarrow 1$, and the reaction is zeroth order. Figure 15.19 illustrates these two limits. The latter generally results in greater catalytic action, since the heat released by the adsorption process is greater for larger K_{ad}, and serves to reduce the apparent activation energy for the reaction. For example, W reduces E_a for NH_3 decomposition from 90 to 39 kcal/mol. The number of possibilities grows for bimolecular reactions: both reagents may be adsorbed prior to reaction, called a *Langmuir mechanism*, while an adsorbed species may react directly with an impacting molecule from the gas phase, known as an *Eley-Rideal mechanism*. We now know that strongly adsorbed species often undergo extensive rearrangment, including dissociation, that can facilitate reaction. For example, both H_2 and O_2 dissociate and bond atomically to a Pt surface.

By its very nature, a catalyst promotes *kinetic control* of the outcome of a reaction, that is, the products need not be those that are thermodynamically most favored. In the *Fischer-Tropsch process*, small organic molecules are produced by the reaction

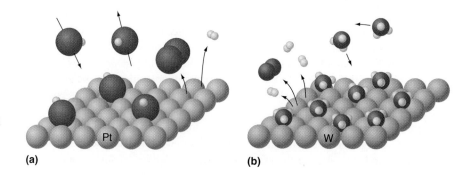

Figure 15.19
Two limiting cases in surface-catalyzed decomposition. (a) *Weak adsorption limit* (small K_{ad}) in Pt-catalyzed HI decomposition, in which the decomposition step is fast compared to adsorption, and few adsorption sites are occupied. (b) *Strong adsorption limit* in W-catalyzed NH_3 decomposition; here the decomposition step is slow, and many sites are filled.

of CO and H_2; choosing the proper catalyst can result in a particular product to the exclusion of others. Some typical reactions and their catalysts are

$$CO + 3H_2 \xrightarrow{Ni} CH_4 + H_2O \qquad \Delta G° = -34 \text{ kcal/mol CO}$$

$$CO + 2H_2 \xrightarrow{Cu+ZnO} CH_3OH \qquad \Delta G° = -6 \text{ kcal/mol CO}$$

$$3CO + 7H_2 \xrightarrow{Co} C_3H_8 + 3H_2O \qquad \Delta G° = -24 \text{ kcal/mol CO}$$

Although all three reactions are spontaneous under standard conditions, the Ni-catalyzed production of methane is thermodynamically most likely. Yet the mixed Cu/ZnO catalyst yields methanol nearly exclusively, while Co promotes formation of the less stable higher hydrocarbons. Kinetic control is an essential part of modern organic synthesis of complex molecules, and the search for selective catalysts continues unabated.

Chain Reactions, Explosions, and Flames

If finely divided Pt is inserted into a mixture of H_2 and O_2, an explosion will result. In this case the Pt is simply an *initiator* rather than a catalyst, and exerts little influence over the reation outcome. The same result is produced by introducing a spark or a flame; for other reactions, light can serve as an initator. After ignition, explosive reactions and combustion flames are *self-sustaining*, requiring no further external input. Such reactions are characterized by an initial step with very high activation energy, but fast, strongly exothermic subsequent steps. The energy released by the fast steps may heat the system to temperatures where the activation barrier of the initial step is readily overcome; in addition one of the fast steps may produce the same species that the first step did, bypassing it and increasing the rate of conversion. The second possibility is called a **chain reaction.** Chain reactions are characterized by complex, nonstandard rate laws and mechanisms, so, after a brief description of a difficult example, we will discuss one of the few well-characterized cases.

In Chapter 13 we described the electrolytic production of NaOH from aqueous NaCl by electrolysis. Electrolysis products include H_2 and Cl_2, some of which is piped to a *photochemical reactor,* where they are explosively combined to make high-purity gaseous HCl. You might think the reaction $H_2 + Cl_2 \longrightarrow 2HCl$ would be an elementary bimolecular process proceeding in a single bond-switching step from a squarish transition state, but in fact, the direct bimolecular pathway has a very high activation energy of ~50 kcal/mol. In the absence of light or other stimulus, a mixture of H_2 and Cl_2 is essentially unreactive, and can be stored indefinitely, despite a very negative $\Delta G°$ of -46 kcal/mol. The reaction is extremely light sensitive, however, and can be detonated even by room lighting. Kinetic measurements on this reaction have been notoriously difficult; in addition to its senstivity to light, its rate depends on the wall composition and area of its container, and reaction is poisoned by trace amounts of oxygen.

These problems are muted in the related $H_2 + Br_2 \longrightarrow 2HBr$ reaction, however, and the thermal kinetics of this reaction were reported as early as 1907 by Bodenstein and Lind over the temperature range of 205° to 302°C. By using the isolation method, they showed that the rate law is

$$-\frac{d[H_2]}{dt} = \frac{k[H_2][Br_2]^{1/2}}{1 + m([HBr]/[Br_2])} \qquad (15.62)$$

where k and m are constants. Unusual features of this rate law include the fractional power of [Br_2], the inhibition by [HBr] product, and reduction of the inhibition by [Br_2]. This interesting result lay in the literature for 12 years, when what is now the accepted mechanism was postulated:

$$1.\ Br_2 \underset{k_{-1}}{\overset{k_1}{\rightleftharpoons}} 2\ Br$$

$$2.\ Br + H_2 \underset{k_{-2}}{\overset{k_2}{\rightleftharpoons}} HBr + H$$

$$3.\ H + Br_2 \overset{k_3}{\longrightarrow} HBr + Br$$

The mechanism starts with the dissociation of Br_2 into Br atoms, which under normal circumstances would limit the rate, having an activation energy of 45 kcal/mol, the bond energy of Br_2. The Br atom, however, then attacks H_2, releasing the highly reactive atomic H radical, which then reacts rapidly with Br_2 to liberate another Br atom. Like the catalytic Cl atom of the ozone depletion mechanism, this new Br atom bypasses the initial step, reacting with H_2 in the second step. Reactions 2 and 3 form a closed chain that terminates only when reaction -2 removes the H atom or reaction -1 removes the Br. Reaction 1 is called **initiation,** reactions 2 and 3 **chain propagation,** and reactions -1 and -2 **termination.** Reactions 2 and 3 are so fast that a single pair of Br atoms from reaction 1 can cycle through the chain thousands of times before termination. This sort of "runaway" chain reaction is an essential part of explosions. Note that the reverse of reaction 3 is not included, since 3 is exothermic by 41 kcal/mol and the reverse rate is negligibly small.

In the preceding mechanism, both the H and Br radicals are intermediates, and after including all production and loss terms for each, we can apply the steady-state approximation twice:

$$\frac{d[H]}{dt} = k_2[Br][H_2] - k_3[H][Br_2] - k_{-2}[H][HBr] = 0$$

$$\frac{d[Br]}{dt} = 2k_1[Br_2] - k_2[Br][H_2] + k_3[H][Br_2] + k_{-2}[H][HBr] - 2k_{-1}[Br]^2 = 0$$

The large number of terms for [Br] results from its involvement in every step of the mechanism. Applying these two conditions by way of some algebraic legerdemain, it can be shown that the resulting rate law is

$$-\frac{d[H_2]}{dt} = \frac{k_2 K_1^{1/2}[H_2][Br_2]^{1/2}}{1 + (k_{-2}/k_3)[HBr]/[Br_2]} \tag{15.63}$$

where $K_1 = k_1/k_{-1}$ is the equilibrium constant for step 1. This expression is identical in form to the experimental rate law given earlier, provided $k = k_2 K_1^{1/2}$ and $m = k_{-2}/k_3$. This reaction, like that of $H_2 + Cl_2$, can also be initiated by light, which acts by *photodissociating* the Br_2 into atoms, stimulating step 1. It is now known that the light sensitivity of $H_2 + Cl_2$ is also due to photodissociation of Cl_2, and that the gas-phase component of that reaction follows a similar mechanism.

15.5 Chemical Catalysis and Complex Reactions

Especially violent explosions can result if the chain propagation steps produce more than one free radical. This is called *chain branching,* and the most well-known example is the $2H_2 + O_2 \longrightarrow 2H_2O$ reaction, our favorite from Chapter 1. Here is the pair of chain propagation steps that typify this behavior:

$$H + O_2 \longrightarrow OH + O$$
$$O + H_2 \longrightarrow OH + H$$

The H and O products propagate the chain loop formed by these two reactions, while the hydroxyl radicals OH can go off and start chains of their own. If each OH reacted with H_2 as follows,

$$OH + H_2 \longrightarrow H_2O + H$$

only 50 cycles of the branching chain, which can occur in a split second, would be required to make nearly a mole of water! This branched chain mechanism thus accounts for the explosive character of H_2/O_2 mixtures, once a spark or flame is introduced to produce H and O and start the chain rolling. Here we must admit, however, that the mechanism of this simplest and most venerable of all combustion reactions, the "calxing" of hydrogen, is not well understood. Like the $H_2 + Cl_2$ reaction, the $2H_2 + O_2$ explosion is very sensitive to the walls of its container (before it obliterates them). Moreover, the reaction is so fast that considerable self-heating occurs, which causes the rate constants to increase over the brief duration of the reaction; under these circumstances a rate law is difficult if not impossible to devise.

The self-heating phenomenon is a nearly universal feature of combustion, the oxidation of a cabonaceous fuel, and is readily controlled to create a nonexplosive reaction *flame.* You have doubtlessly seen many fires in your life, and Bunsen burner (natural gas) flames are legion in general chemistry laboratories. While a rate law for a reaction flame cannot be derived from kinetics experiments due to the variable "rate constants," we can relate the structure of a "premixed" flame, shown in Figure 15.20, to the possible reactions occuring within it. The flame can be ignited either by preheating or by a spark, which produces O atoms from O_2 and free radicals from the fuel; after ignition, the flame temperature may reach anywhere from 1700°C to more than 3000°C depending on the overall ΔH and stoichiometry. In the *inner cone* or *primary combustion zone* a hydrocarbon begins to lose its H atoms, giving rise to a characteristic blue flame arising from emission by electronically excited C_2, CH, and other radicals produced in a sequence of *abstraction reactions* by O atoms and OH radicals. In methane flames these are typified by

$$CH_3 + O \longrightarrow CH_2 + OH$$
$$CH_2 + OH \longrightarrow CH + H_2O$$

These reactions are invariably exothermic due to the strong O—H bond, and supply the heat necessary to sustain the flame. The similarity of these reactions to those in the combustion of H_2 should alert you to the possibility of explosion, a well-known if dreaded event with natural gas. Some CO and CO_2 are already formed in this region. In the *interconal region,* in addition to some "finished" combustion products, there are a large number of atomic and small molecular radicals, as well

 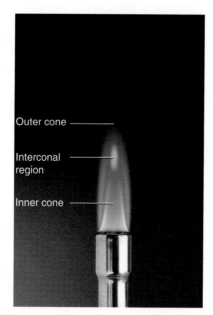 

Figure 15.20
Structure of a flame. On the left is a well-aerated (premixed) methane flame, which has very little diffusion component; in the center the aeration is reduced, and the three regions as labeled and described in the text appear; on the right no air is premixed with the fuel, and combustion takes place entirely by diffusive mixing.

as a few ions. The bare or unsaturated carbon "soot" begins to oxidize, releasing more heat as carbon–oxygen bonds begin to form. Finally, in the *outer cone*, the remaining radicals and bare and partially oxidized carbon begin to react with O_2 diffusing in from the surrounding air, being converted to CO, CO_2, and H_2O, and liberating still more heat. The maximum flame temperature generally occurs in the interconal region, where cooling by the external atmosphere begins to balance heating by the exothermic reactions, and is generally less than half the adiabatic flame temperature (see Chapter 10). A well-aerated Bunsen burner flame (mostly methane combustion) is dominated by the primary and interconal regions, with very little "mopping up" needed above, while candle flames or various types of fires, in which much larger hydrocarbons and hydroxycarbons are burning, have extensive yellow-orange outer-cone diffusion flames. The bright warm glow of a candle arises from carbon soot, formed by hydrogen abstraction, being heated to incandescence as it oxidizes in the surrounding air.

Oscillating Reactions

If you watch a candle for a while, you will notice that occasionally the flame begins to flicker or oscillate, perhaps perturbed by a draft in the room, and then becomes steady again. Oscillations in a flame from a gaseous fuel can be the prelude to an explosion. Many of the things that happen in our world run in cycles, from dripping faucets to the *vibrato* of an operatic singer to a beating heart to day and night to economic well-being. Some of these are natural consequences of mechanical forces, but often oscillatory behavior is associated with instability, a special realm where forces periodically get out of balance. When a faucet drips, a drop of water builds up to such a mass that the cohesive forces holding the drop to the mouth of the faucet are overcome by gravity. After the drop falls, a new one is born and begins to grow, held to the faucet against gravity, until once again the balance of forces is tipped. If you open the valve, the water then begins to flow smoothly; the flow rate is now too fast for a single drop to form.

Despite an abundance of analogues in nature, the first report of an oscillatory chemical reaction, discovered by the Russian chemist Belousov in 1950, was met not only with skepticism but with outright rejection. The editor of the Russian journal to which Belousov submitted his report returned it at once, with a terse note about violating the second law of thermodynamics, despite Belousov's having included a chemical recipe for testing his claim! Belousov's old, yellowed manuscript was discovered many years later by a student named Zhabotinsky, who verified the older chemist's finding, and finally gained acceptance for it. Now known as the Belousov–Zhabotinsky or BZ reaction, it involves a complex aqueous solution of BrO_3^-, Br^-, malonic acid ($HOOCCH_2COOH$), and cerium ions (Ce^{4+}), whose ultimate reaction product is bromomalonic acid ($HOOCCHBrCOOH$). The mechanism has been elucidated only recently, and involves 18 separate steps and 21 different chemical species!

About 20 of these oscillating reactions are now known, most of them involving bromate or iodate, with apparently similar, very complicated mechanisms. A former honors chemistry student at the University of Pittsburgh has shown that a variety of organic carbonyl compounds can substitute for malonic acid, so there are undoubtedly many more than 20 chemical oscillators. Simple mathematical kinetic models have been developed that show oscillatory behavior, but these usually have no chemical mechanistic counterpart. The main feature that appears to make these reactions oscillatory is the production of a species later in the mechanism that was a reagent in an earlier step, a situation very similar to that present in a chain reaction in the gas phase. For example, in the BZ reaction, a series of elementary steps leads to *inverse disproportionation*,

$$1. \quad BrO_3^- + 5Br^- + 6H^+ \longrightarrow 3Br_2 + 3H_2O$$

and the Br_2 product then slowly attacks malonic acid,

$$2. \quad Br_2 + HOOCCH_2COOH \longrightarrow HOOCCHBrCOOH + H^+ + Br^-$$

regenerating Br^- for reaction 1. Since reaction 2 is slow, reaction 1 accumulates a large excess of Br_2 before reaction 2 begins to speed up. Reaction 2 then acts like a drop leaving the faucet, stimulating reaction 1, whose rate depends on $[Br^-]$. The net result is that $[Br_2]$ (an intermediate, not a product) oscillates in time. This oscillation is shown in especially dramatic fashion using a redox indicator, as illustrated in Figure 15.21.

The doubts of the Russian editor about the possibility of a reaction like this centered on the notion that the free energy G must always fall in the course of reaction, or else the process would not happen spontaneously. This appears to imply time profiles like those we have been dealing with in this chapter, smoothly falling reagent concentrations and smoothly rising products. In fact, the products do build up in time monotonically; it is only the intermediate that oscillates. In our dripping faucet analogue, the pool of water in the sink continues to rise; it is only the drop that oscillates in size.

Figure 15.21
Oscillatory kinetics in the Belousov–Zhabotinsky bromate/malonic acid redox reaction. The color changes due to temporal oscillations in the [Br$_2$] intermediate are amplified with ferroin redox indicator.

This chapter concludes our examination of the foundations of chemical thought on which all of modern chemistry rests. But "resting" is not at all descriptive of modern chemistry or chemists. They (including your author) are always exploring new realms of structure and reactivity, and in the 20th century they have blurred the boundaries that once separated them from physicists and biologists. What remains for us to do is to glimpse briefly three of the most far-reaching fields to which chemists have made and are making seminal contributions: the nature of the nucleus and nuclear transformations, the behavior of transition metals, and the structure and reactions of organic and biological compounds. The next three chapters can serve you either as final looks at what chemistry has wrought to the present day, or as springboards to further exploration of the real world with the intellectual and practical tool called chemistry. We hope you will go on, not only in further courses of study, but throughout your life!

SUMMARY

Chemical kinetics is the study of reactions in progress. Extent-of-reaction analysis, introduced earlier as a method of equilibrium analysis, is adapted to the study of reaction rates by allowing the extent-of-reaction variable to become a function of time, $x = x(t)$. As a reaction proceeds, the reagent concentrations decay in time while the products build up, until equilibrium is reached; typically this makes $x(t)$ a monotonically increasing function of time. The **reaction rate** is defined as dx/dt, and is related to the rates of change of the concentrations of reagents R and products P by $dx/dt = -(1/\nu_R)d[R]/dt = +(1/\nu_P)d[P]/dt$. By measuring [R] or [P] as a function of time, we can deduce the reaction rate, and derive **time profiles** of all the concentrations $c(t)$ from $x(t)$ and the stoichiometry of the reaction.

It is usually found that $x(t)$ is a well-behaved, smooth function of time, suggesting that it follows a mathematical law. Analysis of time profiles by finite difference often reveals a dependence of the reaction rate on some power of the reagent concentrations; for a single reagent (decomposition or isomerization reaction), $dx/dt = -dc/dt = kc^n$. This is called a **rate law,** the proportionality constant k is called the **rate constant,** and the exponent n is the **order** of the reaction, and is usually an integer. When this differential equation is integrated, it yields theoretical time profiles that can be compared against experimental ones and used to determine the order n and the rate constant k; integrated forms are given in Table 15.2. Zeroth-, first-, and second-order reactions are discussed, and their time profiles compared. A **first-order** reaction yields a linear plot of $\ln c$ versus t, and a **half-reaction time** $t_{1/2} = \ln 2/k$, whereas for a second-order reaction $1/c$ versus t is linear, and $t_{1/2} = 1/(kc_0)$.

For more complex reactions, the rate law may contain a product of concentrations raised to various powers, rate $= k[A]^a[B]^b[C]^c \cdots$. Then a is the order with respect to [A], b is the order with respect to [B], c is the order with respect to [C], . . . , and the overall order of the reaction is $a + b + c + \ldots$ The orders $a, b, c, \ldots$ do not bear any necessary relation to the stoichiometric coefficients, unless the reaction is elementary. Rate laws for complex reactions are determined by systematically changing the concentration of each reagent with respect to some reference, and either measuring the **initial rate,** wherein a point near the start of the time profile is measured, or using the **isolation method,** wherein all reagents are in great excess except one, and the time profile of the limiting reagent is measured. Occasionally a reagent not present in the overall balanced chemical equation affects the rate; its concentration must then be varied with the others, and the rate law modified to include it.

The form of the rate law found from experiment is a reflection of the **reaction mechanism,** a series of **elementary reaction** steps by which the reaction proceeds on a molecular level. An elementary step is itself a chemical reaction whose balanced equation reflects the number and kind of molecules required to make it happen in a single molecular event. The rate law for an elementary step follows from its stoichiometry: for A $\longrightarrow$ products, the rate law is first order in A and overall, and the step is called **unimolecular;** for A + B $\longrightarrow$ products, the rate law is first order in both A and B, second order overall, and is called **bimolecular;** etc. A multistep mechanism always entails one or more temporary, unstable species known as **reaction intermediates.**

It may be that the reaction whose rate law has been found is actually elementary; it is necessary but not sufficient that the orders equal the stoichiometric coefficients. In the usual case the orders differ from the coefficients; then two methods can be used to deduce mechanistic information from the rate law. The first is the **rate-limiting** step or **bottleneck** principle, by which the rate law is determined by that of the rate limiting or slowest step in the mechanism. In low-order rate laws, the rate-limiting step is usually the first one, while in higher order ones it often is preceded by a fast step that comes to equilibrium. In each case, the overall rate law is identical in form to that of the slow step, with suitable substitution from equilibrium constants and quotients to replace reaction intermediates with starting materials.

The second method, which may be used to confirm or modify the prediction of the bottleneck principle, is the **steady-state**

approximation, which uncouples the rate equations associated with a mechanism by requiring that $d[\text{intermediate}]/dt = 0$. The SSA allows the inclusion of reverse elementary reactions without requiring that they reach equilibrium, and can explain rate laws of nonstandard form as well as those considered above. Because the rate law senses mainly the slow step, there are an infinite number of mechanisms consistent with a given rate law; a mechanism remains a hypothesis until confirmed by studies of similar reactions or detection of postulated intermediates.

Both elementary reactions and multistep reactions obey the **principle of detailed balance,** wherein the equilibrium state results from a balancing of forward and reverse reaction rates. This principle leads to the relation $K = k/k'$ for an elementary reaction, where K is the equilibrium constant, k the rate constant for the forward reaction, and k' the rate constant for its reverse. For a multistep reaction with steps 1, 2, 3, ..., this generalizes to $K = (k_1 k_2 k_3 \cdots)/(k_{-1} k_{-2} k_{-3} \cdots)$.

The observed strong dependence of rate on temperature prompted Arrhenius to propose $k = A \exp(-E_a/RT)$ or $\ln k = \ln A - E_a/RT$, where E_a is the **activation energy** and A is a constant, the preexponential factor; a plot of $\ln k$ (or $\log_{10} k$) versus $1/T$ is found to be linear or near linear for all elementary reactions studied to date. Table 15.4 lists Arrhenius parameters for a number of bimolecular reactions. We interpret E_a as the height of a potential energy barrier that must be surmounted for reaction to occur, and A as an attempt rate. Combining E_a with the heat of reaction allows us to draw an **energy profile** for a reaction, a heat-of-reaction diagram with a hump of height E_a with respect to the energy of the reagents smoothly connecting the reagent and product levels (see Figure 15.9).

Two theories of elementary reaction rates are considered. In both, the activation energy is treated as an experimental quantity and a theory for the A factor is developed. In the **collision theory** (CT, Equation 15.42), A is the kinetic-molecular collision rate A_{ct}, leading to an upper limit ($E_a = 0$) for rate constants of $\sim 10^{11}$ L mol^{-1} s^{-1}. As displayed in Table 15.4, most reactions have A factors smaller than A_{ct}; the so-called *steric factor* $p = A/A_{ct}$ varies from ~ 1 to 10^{-5} or smaller. Smaller p's correlate with more complex reagents. Nonetheless, collision theory explains the pressure dependence of "unimolecular" reactions through a preliminary collision that energizes a molecule prior to decomposition, in the **Lindemann mechanism.** The nonstandard Lindemann rate law (Equation 15.46) is derived using the SSA.

Transition state theory (TST, Equations 15.50 through 15.53) formulates the rate constant by assuming equilibrium between the reagents A + B and the **activated complex** or **transition state** AB‡, defined to be a metastable state atop the activation barrier. The result is that $A \propto \exp(\Delta S^{o\ddagger}/R)$, where $\Delta S^{o\ddagger}$ is the **entropy of activation.** For assumed structureless reagents, TST reduces to CT, while for intricate transition states, $\Delta S^{o\ddagger} \ll 0$, thereby reducing A. In many cases the p factor is nearly completely accounted for by TST, implying greatly improved agreement with experiment. To implement TST requires a structure for AB‡, which has stimulated both theoretical and experimental work in an attempt to characterize the activated complex directly.

In molecular reaction dynamics we attempt to picture the actual motions or **trajectories** of individual atoms and groups during an elementary reaction. To do this for the simple bond-switching reaction A + BC $\longrightarrow$ AB + C requires a potential energy surface $V(R_{AB}, R_{BC}, \theta)$, where the R's are interatomic distances and θ is the bond angle. The collinear F + H$_2$ $\longrightarrow$ HF + H reaction is used as an example; the surface, derived from an accurate molecular orbital calculation, resembles a grooved bobsled run, and reactive trajectories are ones that successfully negotiate the bend in the groove, where the activation barrier lies and where both bond lengths are changing at once. Reactions with "early" barriers (just before the bend) are promoted by translational energy and not vibrational, while the reverse is true for reactions with "late" barriers; this is called **selective energy consumption.** Exothermic reactions with early barriers also funnel the reaction energy preferentially into product vibration, while late barriers put it into translation, examples of **specific energy disposal.** These nonequilibrium tendencies (beyond prediction even by TST) have been exploited in learning to control reactivity by selective excitation and in creating chemical lasers. Molecular beam experiments have provided the soil in which these ideas have taken root and grown.

Reactions in solution are complicated by diffusion through solvent, but a development comparable to that of CT (Equation 15.57) yields an upper limit of $\sim 10^{10}$ L mol^{-1} s^{-1} for rate constants of **diffusion-controlled reactions.** For solution reactions between more complex species, TST still provides a good description, and TST concepts concerning the activated complex continue to be widely used.

Catalysts are chemical species that accelerate selected reactions without being consumed. A catalyst does not affect ΔG for a reaction, but operates by providing a path with lower E_a. Catalysts may be **homogeneous**—in the same phase as the reagents—or **heterogeneous**—from a different phase. Examples discussed include biological catalysts called **enzymes** and metallic and oxide surfaces. The former operate according to the Michaelis–Menten mechanism (Equation 15.60), derived again using the SSA, while the later may be described by Langmuir's adsorption theory (Equation 15.61).

Certain reactions have feedback built into their mechanisms. In the gas phase these are called **chain reactions,** characterized by **initiation, chain propagations,** and **termination** steps, complex rate laws, and explosive tendencies. The reactions between H_2 and Cl_2 or Br_2 are simple chain reactions, in which the propagation steps produce a radical species (Cl· or Br·) that closes the chain loop, while $2H_2 + O_2$ is a branched chain reaction in which pairs of radicals are produced within the primary chain loop that can start new chain loops. These reactions are difficult to characterize because of their high rates, exothermicity (causing heating and changes in rate constants), and sensitivity to walls, and the Cl_2 and O_2 reactions are still not fully understood. The situation is even less well defined for combustion of carbon compounds; combustion flame chemistry is briefly discussed. In solution, feedback can give rise to oscillations, and recently discovered temporal oscillations in hybrid redox/substitution reactions are described.

EXERCISES

Rates and time profiles

1. The pioneering kineticists Harcourt and Esson began their investigations in 1865 with the oxidation of oxalic acid by permanganate in acidic solution. They did not have the benefit of Arrhenius's ionic theory, but we will write the reaction nonetheless in ionic form:

   $$2MnO_4^- + 5H_2C_2O_4 + 6H^+ \longrightarrow 2Mn^{2+} + 10CO_2 + 8H_2O$$

   a. Assuming a stoichiometric mixture of permanganate and oxalic acid in an excess of acid, formulate an extent-of-reaction analysis for this reaction, and express the rate of the reaction dx/dt in terms of the rates of change of $[MnO_4^-]$, $[H_2C_2O_4]$, and $[Mn^{2+}]$.

   b. Roughly sketch the time profiles of $[MnO_4^-]$, $[H_2C_2O_4]$, and $[Mn^{2+}]$, indicating the definition of the rate on the MnO_4^- profile, and the equilibrium region, assuming it greatly favors products. (Harcourt and Essen eventually abandoned the study of this reaction due to complicated kinetics and self-heating effects. They discovered that this reaction is *autocatalytic;* that is, the Mn^{2+} product accelerates the reaction, an anti-LeChâtelier result.)

2. Table 15.3 lists kinetic data for the aqueous reaction

   $$BrO_3^- + 5Br^- + 6H^+ \longrightarrow 3Br_2 + 3H_2O$$

   a. Under the conditions of run I from Table 15.3, set up an extent-of-reaction analysis, give numerical values with units for the rates of change of $[BrO_3^-]$, $[Br^-]$, and $[H^+]$, and estimate these concentrations after the reaction time has elapsed.

   b. On the same set of labeled axes, graph your results as the beginnings of time profiles for each of these species.

3. For the gaseous decomposition of ozone, $2O_3 \longrightarrow 3O_2$:

   a. Set up an extent-of-reaction analysis for a given initial concentration $[O_3]_0$, and state the reaction rate dx/dt in terms of the rates of change of $[O_3]$ and $[O_2]$.

   b. Roughly sketch the expected time profiles, assuming the reaction equilibrium favors products overwhelmingly.

   c. Assuming the reaction is run in a rigid container at constant temperature and ideal gas behavior, state the pressure at time t in terms of $[O_3]_0$ and $x(t)$.

Rate laws: Single reagent

4. The decomposition of N_2O_5, $2N_2O_5 \longrightarrow 4NO_2 + O_2$, was a workhorse system in the early days of chemical kinetics. At 55°C the following data were obtained from pressure measurements:

time (min)	0	5	10	15	20	25
$[N_2O_5]$ (mmol L^{-1})	14.7	9.3	5.9	3.7	2.4	1.5

   Find the order of the reaction, state the rate law in both differential and integrated form, find the rate constant k with appropriate units, and give the half-reaction time $t_{1/2}$. You may use finite-difference or graphical methods.

5. The vapor of di-*t*-butyl peroxide decomposes into acetone and ethane at 147°C according to

   $$(CH_3)_3COOC(CH_3)_3 \longrightarrow 2\,CH_3COCH_3 + C_2H_6$$
   or
   $$D \longrightarrow 2A + E$$

in abbreviated notation. Using the time-profile data:

time (s)	0	360	720	1080	(∞)
[D] (mmol L^{-1})	6.62	5.90	5.27	4.71	0.01

a. Determine the order of the reaction, state the rate law, and calculate the rate constant k and the half-reaction time $t_{1/2}$.

b. Sketch the expected time profiles of [D], [A], and [E] on the same set of labeled axes, and indicate $t_{1/2}$ and the equilibrium region.

c. Predict the time required for 90% decomposition.

6. Hydrogen iodide gas decomposes slowly at high temperature, $2HI(g) \longrightarrow H_2(g) + I_2(g)$. At 427°C the time required to decompose 10% of the HI, $t_{0.10}$, was measured for different initial HI concentrations using the visible absorption of I_2 vapor:

[HI]$_0$ (mmol L^{-1})	6.9	11.5	16.0
$t_{0.10}$ (min)	231	139	100

a. Determine the order of the reaction, state the rate law, and calculate the rate constant. Remember the factor of 2 connecting $d[HI]/dt$ and the rate.

b. For the third [HI]$_0$, sketch the time profiles for [HI], [H$_2$], and [I$_2$] on the same set of axes, indicating the half-reaction time $t_{1/2}$, assuming no back reaction.

Rate laws: Two or more reagents

7. The rate law for the reduction of nitric oxide by hydrogen, $2NO + 2H_2 \longrightarrow N_2 + 2H_2O$, is found to be rate = $k[NO]^2[H_2]$. Give the orders with respect to [NO] and [H$_2$] and the overall order, and the units of k. By what factor does the rate change on doubling [NO]? Halving [H$_2$]? Tripling both [NO] and [H$_2$]?

8. In the oxidation of bisulfite ion by bichromate,

$$2HCrO_4^- + 3HSO_3^- + 5H^+ \longrightarrow 2Cr^{3+} + 3SO_4^{2-} + 5H_2O$$

doubling the initial concentration $[HCrO_4^-]_0$ doubles the initial rate, halving $[HSO_3^-]_0$ reduces the rate by a factor of 4, and decreasing the initial pH by one unit increases the rate by a factor of 10. What is the rate law for this reaction? Give the orders with respect to each reagent, the overall order, and the units of the rate constant. Why aren't the orders equal to the stoichiometric coefficients?

9. When ozone decomposition $2O_3 \longrightarrow 3O_2$ is studied in O_3/O_2 mixtures and the initial rates are measured, it is found that doubling [O$_3$]$_0$ quadruples the rate, while doubling [O$_2$]$_0$ halves the rate. What is the rate law for this reaction? Give the orders with respect to [O$_3$] and [O$_2$], the overall order, and the units of k.

10. In the chlorination of chloroform, $CHCl_3 + Cl_2 \longrightarrow CCl_4 + HCl$, it is found that quadrupling $[CHCl_3]_0$ quadruples the initial rate, but quadrupling $[Cl_2]_0$ only doubles the rate. What is the rate law? Give the reaction orders with respect to [CHCl$_3$] and [Cl$_2$], the overall order, and the units of k.

11. When our first example reaction, the endothermic decomposition $C_2H_5Cl \rightleftharpoons C_2H_4 + HCl$, is run at lower temperatures, it is slower, but due to the decreased K (by LeChâtelier), the reverse reaction is more important at equlibrium. Assuming that the rate law for the forward reaction is $dc/dt = -kc$ ($c \equiv [C_2H_5Cl]$) and that for the reverse reaction is $dc/dt = k'[C_2H_4][HCl]$, find an expression for the net rate dx/dt in terms of x and the initial concentration c_0 by adding the rates for the forward and reverse reaction. How would the time profiles of Figure 15.1 be modified?

12. In the H + O$_3$ example of Section 15.2, the run with $[O_3]_0 = 4 \times 10^{13}$ cm^{-3} yielded the following [H] time profile:

time (ms)	0.0	0.5	1.0	1.5	2.0
[H] (10^{11} cm^{-3})	1.50	0.86	0.49	0.28	0.16

Use these data to show that the reaction is first order with respect to [H], and find the effective first-order rate constant k_{eff} (s^{-1}). Then, assuming the reaction is also first order with respect to [O$_3$], calculate the second-order rate constant k (cm^3 molecule^{-1} s^{-1}).

13. A study of the acid-catalyzed bromination of acetone

$$CH_3COCH_3 + Br_2 \xrightarrow{H^+} CH_3COCH_2Br + H^+ + Br^-$$

yielded the following initial rate data (AH ≡ CH$_3$COCH$_3$):

Run	[AH]$_0$ (M)	[H$^+$]$_0$ (M)	[Br$_2$]$_0$ (M)	t (s)
I	4.0	1.0	0.0024	47
II	2.0	1.0	0.0024	91
III	4.0	2.0	0.0024	23
IV	4.0	1.0	0.0048	97

where t is the time for complete removal of Br_2. Use these data to estimate the rate law and calculate the rate constant. (Note that this reaction is autocatalytic. To avoid complications, $[H^+]_0$ is kept much larger than that produced by the reaction.)

14. An interesting and important method in solution kinetics employs so-called *clock reactions,* in which a fast but noninterfering reaction is used to scavenge a product of a slower reaction to be studied. One of the earliest known applications of this method was in the study of the oxidation of iodide by peroxydisulfate ion

$$2I^- + S_2O_8^{2-} \longrightarrow I_2 + 2SO_4^{2-}$$

a comparatively slow reaction, whose rate is clocked by the fast reduction of I_2 by thiosulfate ion

$$I_2 + 2S_2O_3^{2-} \longrightarrow 2I^- + S_4O_6^{2-}$$

Using a small amount of thiosulfate, and starch indicator to detect the presence of I_2 by formation of a blue complex, the solution will remain clear until the thiosulfate has completely reacted, then "flash" blue as the excess I_2 complexes with starch. (Without the clocking reaction, the blue color would gradually deepen as the slow oxidation progressed.) Clock reactions are most useful for measuring initial rates. A student obtained the following initial rate data for this reaction:

Run	$[I^-]_0$ (M)	$[S_2O_8^{2-}]_0$ (M)	$[S_2O_3^{2-}]_0$ (M)	t to blue flash (s)
I	0.080	0.040	0.0010	38
II	0.060	0.040	0.0010	52
III	0.080	0.025	0.0010	61
IV	0.020	0.010	0.0005	???

Use these data to find the rate law and calculate the rate constant for the reaction. Make a numerical prediction of the flash time for run IV.

Reaction mechanisms: Rate-limiting step

15. In our first example of Section 15.3, the substitution reaction of *t*-butyl chloride with hydroxide, a two-step mechanism was proposed. Consider instead the three-step mechanism:

1. $(CH_3)_3C-Cl \xrightarrow{k_1} (CH_3)_3C^+ + Cl^-$ Slow

2. $(CH_3)_3C^+ + H_2O \xrightarrow{k_2} (CH_3)_3C-OH + H^+$ Fast

3. $H^+ + OH^- \xrightarrow{k_3} H_2O$ Very fast

a. Identify each elementary step as unimolecular or bimolecular, and give its rate law. Verify that the sum of the three elementary reactions is the overall reaction of Equation 15.21.

b. What is the prediction of the bottleneck (slow step) principle for the overall rate law for the reaction? Does it agree with observation? What conclusion can you draw concerning the uniqueness of a proposed mechanism?

16. For the reduction of NO by H_2, $2NO + 2H_2 \longrightarrow N_2 + 2H_2O$, whose rate law was given in Exercise 7 as rate = $k[NO]^2[H_2]$, the following mechanism has been proposed:

1. $NO + NO \rightleftharpoons N_2O_2$ Fast equilibrium, K_1
2. $N_2O_2 + H_2 \xrightarrow{k_2} N_2O + H_2O$ Slow
3. $N_2O + H_2 \xrightarrow{k_3} N_2 + H_2O$ Fast

Verify that the steps add up to the overall reaction. Use the bottleneck principle to show that this mechanism is consistent with the observed rate law, and identify the observed rate constant k in terms of constants in the mechanism.

17. You were asked to deduce the rate law for ozone decomposition ($2O_3 \longrightarrow 3O_2$) in Exercise 9. You were correct if you found rate = $k[O_3]^2[O_2]^{-1}$, an unusual form that shows *inhibition* of the reaction by O_2. This puzzling result could not be explained until atoms and radicals were recognized as possible intermediates. The accepted mechanism for this reaction is:

1. $O_3 \underset{k_{-1}}{\overset{k_1}{\rightleftharpoons}} O_2 + O$ Fast equilibrium, K_1

2. $O_3 + O \xrightarrow{k_2} O_2 + O_2$ Slow

Identify the elementary steps as unimolecular or bimolecular (including the reverse of step 1) and verify that they add up to the overall reaction. Use the bottleneck principle to show that this mechanism is consistent with the observed rate law, and identify the rate constant k in terms of constants in the mechanism. Argue qualitatively why increasing $[O_2]$ inhibits decomposition. (Here the mechanism has been verified by directly observing the intermediate O atoms spectroscopically.)

18. A few reactions exhibit fractional orders, for example, the chlorination of chloroform, $CHCl_3 + Cl_2 \longrightarrow CCl_4 + HCl$ (Exercise 10), with rate $= k[CHCl_3][Cl_2]^{1/2}$. The one-half order with respect to $[Cl]_2$ suggests that the rate-limiting step involves a Cl *atom*. Propose a mechanism for this reaction in which the slow step is the abstraction of an H atom from $CHCl_3$ by Cl. Supply the Cl atoms with an initial fast equilibrium step, and add a fast "mopping up" final step that converts the product of the abstraction to an overall reaction product. Show that your mechanism adds up correctly, and use the bottleneck principle to show that it is consistent with the observed rate law.

19. Unlike its Cl_2 and Br_2 cousins, the $H_2 + I_2 \longrightarrow 2HI$ gas reaction has the simple second-order rate law rate $= k[H_2][I_2]$, and for 70 years was used as a textbook example of a simple bimolecular process. However, in 1967 J. Sullivan showed that atomic iodine was involved in the mechanism (as we knew Cl and Br were in their reactions) by shining ultraviolet light on the reaction mixture to dissociate I_2, and thereby greatly increasing the rate. The mechanism of this reaction is now believed to be

1. $I_2 \underset{k_{-1}}{\overset{k_1}{\rightleftharpoons}} 2I$ Fast equilibrium, K_1
2. $I + H_2 \underset{k_{-2}}{\overset{k_2}{\rightleftharpoons}} H_2I$ Fast equilibrium, K_2
3. $H_2I + I \overset{k_3}{\longrightarrow} HI + HI$ Slow

Verify that this mechanism adds up correctly, and use the bottleneck principle to show that it accounts for the observed second-order law. (This sort of result illustrates forcefully that a rate law agreeing with stoichiometry is necessary but not sufficient to establish the elementary nature of a reaction.)

Reaction mechanisms: The steady-state approximation (SSA)

20. In Example 15.4 a two-step mechanism for the bromination of acetone was analyzed using the steady-state approximation. Add the reverse of step 1, the deprotonation of AH_2^+, to the mechanism, and reanalyze using the SSA. Is your result consistent with the rate law given? If not, which term in the SSA expression must be neglected to bring the results into agreement? What would your result become if step 1 reached equilibrium?

21. Analyze the ozone decomposition mechanism of Exercise 17 using the steady-state approximation for the intermediate concentration [O]. Remember to include the reverse of step 1. Is your result consistent with the observed rate law? If not, what inequality must be satisfied to bring the SSA and observed laws into agreement? Is this inequality consistent with the assumption of equilibrium in step 1?

22. Esters (E), compounds made from carboxylic acids (C) and alcohols (A), are the good smells in perfumes; they must be dissolved in alcohol rather than water (W) to prevent hydrolysis back to their not-so-fragrant components, $E + W \longrightarrow C + A$, with a rate law of rate $= k[E]$. This hydrolysis is catalyzed by acid according to the following mechanism:

1. $E + H^+ \underset{k_{-1}}{\overset{k_1}{\rightleftharpoons}} EH^+$
2. $EH^+ + W \overset{k_2}{\longrightarrow} C + A + H^+$

Noting that water (W) does not enter the rate law in aqueous solution, use the SSA to derive a rate law for the acid-catalyzed hydrolysis, and show that it differs from that of the uncatalyzed reaction given. Identify the factors that would comprise the observed rate constant in the catalyzed reaction.

23. In reactions with more than one intermediate, applying the SSA becomes increasingly formidable (though still easier than solving the coupled rate equations). Write the steady-state conditions for all intermediates appearing in the $H_2 + I_2$ mechanism of Exercise 19. Deriving a rate law in this case is complicated by the appearance of both [I] and $[I]^2$ in the SSA conditions, but you can still deduce the condition(s) under which they yield the second-order law observed.

24. Carry through the algebra to establish the SSA result for the rate law of the $H_2 + Br_2$ reaction discussed in Equation 15.63 of Section 15.5.

Temperature dependence, $k = A \exp(-E_a/RT)$, and the principle of detailed balance

25. For our first example, the decomposition of chloroethane, the first-order rate constant is $k = 1.12 \times 10^{-2} \text{ s}^{-1}$ at 530°C, but falls to $4.20 \times 10^{-5} \text{ s}^{-1}$ at 430°C.
    a. Use these data to estimate the activation energy E_a (kcal/mol) and the frequency factor A (s^{-1}) for this reaction.

b. Predict the rate constant at 25°C, and comment on why there are no experimental data for this reaction at room temperature.

c. Combine E_a with the heat of reaction given in the text or computed from data in Appendix C to construct an energy profile for this reaction. What is the activation energy E_a' for the addition of HCl to C_2H_4?

26. From the results of Exercise 25 along with the standard free-energy change $\Delta G°$ and equilibrium constant at 530°C, evaluated using data given in the text or from Appendix C, use the principle of detailed balance to estimate the rate constant k' and preexponential factor A' for the reverse reaction $C_2H_4 + HCl \longrightarrow C_2H_5Cl$ at 530°C. Note that you must convert K from a standard state of 1 atm to that of 1 mol L^{-1} by multiplying it by $(RT)^{-\Delta n_{gas}}$ before applying detailed balance.

27. An old rule of thumb in chemical kinetics states that the rate of reaction doubles for every 10°C rise in temperature. What is the activation energy E_a (kcal/mol) of the rule-of-thumb reaction if the two temperatures are 25° and 35°C?

28. Consider the following temperature dependence of the rate constant for $2HI(g) \longrightarrow H_2(g) + I_2(g)$:

T (K)	629	647	666	683	700	716
k (10^{-5} L mol^{-1} s^{-1})		3.0	8.6	22	51	116	250

a. Using graphical or numerical methods, find E_a (kcal/mol) and A (L mol^{-1} s^{-1}) for this reaction, and predict the rate constant at 25°C.

b. Combine E_a with the heat of reaction found from data in Appendix C to construct an energy profile. What is the activation energy of the reverse reaction?

29. Table 15.4 includes Arrhenius parameters for two reactions that are the reverse of each other. Are the activation energies E_a and E_a' consistent with the heat of reaction obtained from Appendix C? Use the principle of detailed balance to find the equilibrium constant and free-energy change at 298 K implied by these data. Check them against the same computed from data in Appendix C.

Collision theory (CT)

30. Confirm the estimate of a typical upper limit for a collision theory bimolecular rate constant k ($\equiv A_{ct}$)

given in Section 15.4. Take $m_A = m_B \approx 30$ amu, $T = 298$ K, and $d_A = d_B \approx 3$ Å. Compute values for the relative velocity $\bar{v}_{rel}$ (cm s^{-1}) and cross section πd_{AB}^2 (Å^2 and cm^2), as well as k (cm^3 molecule^{-1} s^{-1} and L mol^{-1} s^{-1}). Also develop an "engineering formula" of the form $k = A_{ct} = \text{const}(T/\mu)^{1/2}d_{AB}^2$ by finding values of the constant that will yield k in either of the two conventional units when T is given in K, μ in amu, and d_{AB} in Å.

31. In an earlier exercise you may have found the Arrhenius A from experimental data for $2HI \longrightarrow H_2 + I_2$. Calculate the collision theory prediction A_{ct} for this reaction at 680 K using radius data from Exercise 14.29. For identical molecules reacting you have to divide A_{ct} by 2 to correct for double-counting the collisions. Find the steric factor $p = A/A_{ct}$, and discuss briefly.

32. Look up the $O + O_3 \longrightarrow O_2 + O_2$ reaction in Table 15.4, and find its steric factor p. Rationalize p by sketching potentially successful and unsuccessful regions of attack of O on O_3. The factor in A_{ct} that must be reduced to agree with experiment is the "reaction cross section" Q, given in collision theory by $p\pi d_{AB}^2$. Use A_{ct} given in the table to find Q (cm^2 and Å^2) using $Q \approx pA_{ct}/\bar{v}_{rel}$ at 298 K.

Transition state theory (TST)

33. The "worst case" for collision theory in Table 15.4 is the hydrogenation of ethylene (an industrially important reaction): $C_2H_4 + H_2 \longrightarrow C_2H_6$. Use transition state theory to find the entropy of activation $\Delta S^{°\ddagger}$ at 298 K for this reaction from the data. As preparation for doing this, develop "engineering formulas" of the form $A = \text{const}\cdot T^2 \exp(\Delta S^{°\ddagger}/R)$ and $A_{298} = \text{const}' \exp(\Delta S^{°\ddagger}/R)$, by finding values for the constants that will give A in either cm^3 molecule^{-1} s^{-1} or L mol^{-1} s^{-1} units (see Example 15.6). Compare your result with $\Delta S°$ for the same reaction computed from Appendix C; discuss briefly in terms of the expected structure of the activated complex.

34. At the frontier of our current understanding of chemical kinetics at the molecular level is the *chain branching* reaction $OH + H_2 \longrightarrow H_2O + H$, a central step in hydrogen combustion. Use transition state theory to find the entropy of activation $\Delta S^{°\ddagger}$ for this reaction at 298 K from the data in Table 15.4. If you have already developed "engineering formulas" for this calculation in the previous exercise, please use them! Compare your result to the entropy change for the reaction itself

computed from data in Appendix C, and draw any qualitative conclusions you can concerning the nature of the activated complex.

35. The fact that CT and TST yield the same result when the reagents are assumed structureless has been used to give a rough interpretation of the steric factor p using statistical ideas (see Appendix D). Because the entropy S relates logarithmically to the partition functions q, $S = R \ln q$, for the various degrees of freedom (see Chapter 8), and because in TST A relates exponentially to the entropy, the partition functions end up as factors in the TST expression for A, and the steric factor $p = A/A_{ct}$ comes out to be a ratio of products of vibrational q_{vib}'s for the activated complex divided by similar products of rotational q_{rot}'s for the reagents; that is, $p = (q_{vib}/q_{rot})^n$. Now roughly $q_{vib} \approx 2$ while $q_{rot} \approx 10$, so this ratio will be 0.2, 0.04, 0.008, ..., depending on the number of factors n, that is, on how many rotational degrees of freedom are converted into vibrations in the activated complex. In general $n = n_{vib, AB} - n_{vib, A} - n_{vib, B} - 1$, where the -1 is the vibration that becomes the dissociation coordinate for $AB^\ddagger$.

    a. Make a rough estimate of p for atom + diatomic molecule reactions for both linear and bent transition states. Compare with Table 15.4, and decide whether any of the systems there can be assigned as linear or bent transition states.

    b. Do the same for the $OH + H_2$ (linear and bent TS), $NO + O_3$ (bent only), and $C_2H_4 + H_2$ (bent only) reactions, and note any trends you find.

Molecular reaction dynamics

36. In the text we discussed the properties of *selective energy consumption* and *specific energy disposal* in elementary bimolecular reactions. In this and the following exercise you will be asked to apply the "Polanyi rules" relating these properties to the location of the activation barrier, "early" or "late" on the potential energy surface. For reference, we show here these two types of potential energy surface, each representing a thermoneutral reaction having a barrier of 7 kcal/mol, for a generic bond-switching reaction $A + BC \longrightarrow AB + C$. The early barrier surface is similar to that for $F + H_2$ of Figures 15.12 and 15.13 in the entrance valley (lower right). As illustrated by the trajectories, translational energy is more effective than vibrational in promoting reaction in the early barrier case, while the reverse is true for a late barrier. The general dynamic principle underlying these properties is that the mass point sliding on the suface has to be moving *toward* the barrier rather than perpendicular to it in order to surmount it. For which surface, early, late, neither, or both, are the following combinations of translational energy E_t and vibrational energy E_v (kcal/mol) likely to give a reaction? For choices not represented in the illustration, sketch your own surface and a likely trajectory.

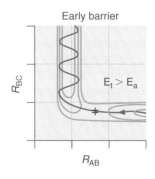

Early barrier

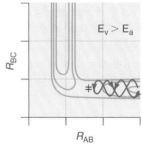

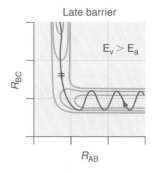

Late barrier

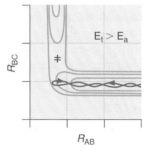

a. $E_t = 8$, $E_v = 2$ c. $E_t = 2$, $E_v = 2$
b. $E_t = 2$, $E_v = 8$ d. $E_t = 8$, $E_v = 8$

37. In *specific energy disposal* an "early" or "late" barrier in the potential energy surface channels the energy preferentially into product vibration E_v' or translation E_t', respectively. These rules may be readily understood by applying the principle of **microscopic reversibility,** the molecular-level analogue of the principle of detailed balance, which states that every trajectory can be run backward, and will retrace its path back to its origin. The early barrier, which selectively *consumed* translational energy, becomes a late barrier for the reverse reaction, and therefore specifically *produces* translational energy. For an assumed total energy of 10 kcal/mol and a barrier height of 7 kcal/mol on a thermoneutral surface as in Exercise 36, deduce the location of the barrier in the surface that would give rise to the following combinations of E_t' and E_v' (kcal/mol). Roughly sketch the surfaces and likely trajectories in each case.

    a. $E_t' = 8$, $E_v' = 2$
    b. $E_t' = 2$, $E_v' = 8$
    c. $E_t' = 5$, $E_v' = 5$

Reactions in solution

38. Assume hydrogen ions H^+ are injected into the center of a 20-mm-diameter test tube of water. Roughly how long will it take these ions to reach the walls of the test tube? Use data from Table 15.5.

39. Use the rate constant given in Table 15.6 for the Arrhenius reaction $H^+ + OH^- \longrightarrow H_2O$ to find the half-reaction time $t_{1/2}$ for a perfectly mixed solution initially containing 0.10 M of both H^+ and OH^-.

40. Use diffusion coefficients from Table 15.5 and the molecular diameter of NH_3 from Exercise 14.29 to compute the diffusion-limited rate constant of the reaction $H^+ + NH_3 \longrightarrow NH_4^+$. Compare it to the rate constant given in Table 15.6. Would you have expected the agreement here to be better or worse than that for $H^+ + OH^- \longrightarrow H_2O$ discussed in the text?

41. The diffusion limit can be used to estimate rate constants of aqueous reactions using the principle of detailed balance for which the reverse reaction is at this limit. For the weak acid ionization $CH_3COOH \rightleftharpoons CH_3COO^- + H^+$, combine the equilibrium constant $K_a = 1.8 \times 10^{-5}$ with the diffusion-limited rate constant k' for the reverse reaction using $d_{AB} \approx 5$ Å to estimate k for the ionization, with appropriate units. In a solution initially 0.10 M in CH_3COOH, what is $t_{1/2}$ for ionization?

42. Precipitation reactions often seem to be instantaneous to the unaided eye. For the "hard-water" reaction $Ca^{2+} + SO_4^{2-} \longrightarrow CaSO_4(s)$, compute the diffusion-limited rate constant using D's from Table 15.5 and $d_{AB} \approx 5$ Å. Use this rate constant to compute the half-reaction time $t_{1/2}$ for the precipitation, assuming a perfectly mixed solution with $[Ca^{2+}]_0 = [SO_4^{2-}]_0 = 0.010$ M.

43. Combine the diffusion limit with transition state theory to describe reactions in solution, assuming a mechanism

    1. $A + B \underset{k_D'}{\overset{k_D}{\rightleftharpoons}} (A \cdots B)$

    2. $A \cdots B \overset{k_{TST}}{\longrightarrow} P$

    where $k_D' \approx k_D$. Use the steady-state approximation on the "caged complex" $(A \cdots B)$ to derive an expression for the rate $d[P]/dt$. Show that this expression reduces to the diffusion rate when the rate of step 2 is faster than diffusion, and to the usual TST rate when step 2 is rate limiting.

Catalysis

44. Construct a catalytic cycle similar to that for Cl-catalyzed destruction of O_3 where NO replaces Cl. Using data from Table 15.4, sketch an energy profile like that of Figure 15.17 for NO catalysis. Do you expect NO to be as effective a catalyst as Cl? Why or why not?

45. Consider the feast and famine limits for the Michaelis–Menten rate law for enzyme catalysis, Equation 15.60. Show that in the small-[S] limit (famine), the rate reduces to the "first-step bottleneck" rate, while for large [S] (feast), the preequilibrium "second-step bottleneck" rate is obtained. (These limits, rather than the general law, were given by Michaelis and Menten in 1913.)

46. The decomposition $2N_2O \longrightarrow 2N_2 + O_2$ on a gold surface is found to be first order in $[N_2O]$. Interpret this in terms of adsorption versus reaction on Au(s).

C H A P T E R

16

The Nucleus

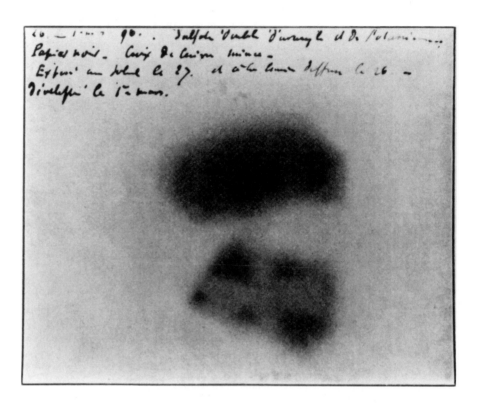

A telltale exposure. During his 1896 investigation of solar-induced fluorescence, Henri Becquerel inadvertently stored his photographic film, wrapped in black paper, in a closed drawer together with some uranium salt crystals. After a few days the film, shown here, was exposed by radiation from the uranium.

We have here a proof that there is in the atom a fundamental quantity, which increases by regular steps as we pass from one element to the next. This quantity can only be the charge on the central positive nucleus. . . .

Henry Moseley (1914)

CHAPTER OUTLINE

16.1 A Chemical Microcosm: Nuclear Structure and Bonding

16.2 Nuclear Stability and Radioactivity

16.3 Kinetics of Nuclear Decay

16.4 Nuclear Reactions

16.5 Applying Our Nuclear Knowledge

In Chapter 1 we learned that the chemical identity of an element is determined by the number of positive charges, the **atomic number** Z, contained in its **nucleus.** An element may gain, lose, or share electrons in the course of chemical reactions, but its Z remains invariant—or does it? We also learned of the existence of **isotopes,** atoms with the same Z but different **mass numbers** A; for an element X we wrote $^{A}_{Z}X$. Rather than asking why an element should have more than one isotope, you might ask why the elements have so few isotopes. The ones we don't find in Nature are absent for a reason: their nuclei are *unstable.* If an unstable isotope of an atom resides in a molecule, its Z will change as it decays, thereby altering the composition and bonding of the molecule itself. In a chemical reaction of an unstable isotope, it may transmute in the course of reaction, and what is present at the end may not be what the chemistry suggested. These may sound like rare occurrences, but their implications in our present "nuclear era" are profound. This chapter attempts to provide you with a glimpse into the world of the nucleus, a world so small that it is dwarfed by the size of an atom. Amazingly, we will learn the most about these tiny balls of charge using the chemist's most basic tool: the balance. We will analyze what we find by summoning the basic structural, thermodynamic, and kinetic principles you have been exposed to in our chemical forays.

The alchemists, those protochemists who held sway in the preweighing, prenuclear ages, had always before them the ultimate goal of *transmutation* of base metals into gold. After the nucleus was discovered by Rutherford, and the centrality of its Z to the chemical identity of an element by his student Moseley, we understood that what the alchemists really wanted was to change Z from whatever they started with to 79, the atomic number of gold. They never succeeded because *inducing* a change in Z turns out to be impossible by chemical means, no matter how cunning. In modern times we have learned not only how to transmute nuclei, but also how to synthesize new ones; this new alchemy has become the province of a few nuclear chemists and a much larger corps of nuclear physicists. The keys to this new kingdom lay in recognizing nuclear instability.

Marie Sklodowska Curie (1867–1934) was a professor of physics at the Sorbonne in Paris, France. As a graduate student at the Sorbonne studying with her husband Prof. Pierre Curie, she discovered the radioactive elements radium and polonium by mastering their chemistry. She was awarded her PhD degree and shared the third Nobel Prize in physics in the same year, 1903, and later won the chemistry Nobel in 1911. Her daughter Irène also shared the Nobel Prize in physics in 1935, one year after her death from the long-term effects of radiation.

16.1 A Chemical Microcosm: Nuclear Structure and Bonding

A model that has maintained its usefulness throughout our 20th-century exploration of nuclear science is that atomic nuclei (sometimes called **nuclides**) are composed of positively charged **protons** (p) and uncharged **neutrons** (n) (collectively **nucleons**), with the number of protons, the atomic number Z, determining the total positive charge—and hence elemental identity—of the atom, and the sum of the numbers of protons and neutrons determining the mass number A of a particular isotope of the element. The number of neutrons in a nucleus is therefore $N = A - Z$. In this model protons and neutrons are the building blocks of the nucleus just as atoms are the building blocks of molecules.

One facet of the mystery that Rutherford's nuclear atom generated was the stability of the nucleus, that is, how the building blocks are able to hang together. In an atom of gold, how can 79 protons be packed into a tiny sphere less than 0.0001 Å in diameter and remain stable despite the enormous Coulomb repulsion? The answer slowly dawned in the 1930s that there must be a new force in nature, now known simply as the **strong force,** that acts over a very short range and binds nucleons together in a cohesive whole. The problem of the myriad chemical interactions that the chemist's 100+ elements can undergo is greatly reduced in scope in the nucleus: only three "bonding" interactions need be considered: p–p, n–p, and n–n; and our best evidence indicates that the strong force is identical for all three.

Rutherford's α-particle scattering measurements, along with more recent experiments scattering protons, neutrons, and electrons from nuclei, have led to the following simple relationship between nuclear radius r_{nuc} and mass number A:

$$r_{\text{nuc}} = 1.33 \times 10^{-13} \text{ cm} \times A^{1/3} \qquad (16.1)$$

This relationship reflects the fact that nuclei, like atoms, are round; only a few exceptions are known. The cube-root dependence on A reflects a linear relationship between A and nuclear volume, and implies a closely packed arrangement of nucleons, as schematically illustrated in Figure 16.1. This radius relation gives us an estimate of the radius of a nucleon itself, $r_p = r_n \approx 1.33 \times 10^{-13}$ cm or 1.33 fm (femtometers or **fermis**), by setting $A = 1$. Although the radius formula is not as accurate for small nuclei, if we consider the **deuteron** ^2_1d, the nucleus of "heavy hydrogen" ^2_1H, or usually just D, Equation 16.1 leads to $r_{nuc} = 1.7$ fm. This suggests a slight overlap of the two nucleon radii, a situation similar to that in covalent bonding of atoms; scattering experiments imply a nucleon–nucleon "bond length" of 1.5 fm, roughly 1 fm shorter than "touching" distance.

Assuming the strong force is what holds the deuteron together, we can readily estimate the **binding energy** (BE) (analogous to the bond energy in a diatomic molecule) of the deuteron from the **mass defect** Δm introduced in Chapter 1. Binding nucleons (or small nuclei) together to make a larger nucleus is called a **fusion reaction** (again analogous to assembling a molecule from isolated gas-phase atoms or smaller molecules). Nuclear binding energies are so large that fusion reactions show a loss in mass, according to Einstein's special theory of relativity, which leads to

$$\Delta E = \Delta m c^2 \qquad (16.2)$$
$$= 931.5 \text{ MeV/amu} \times \Delta m$$

In our usual convention, Δ means difference, products minus reagents, and the second line results from converting c^2 into "engineering" units (MeV $\equiv 10^6$ eV). Equation 16.2 tells us that 1 amu of matter can yield an energy a billion times that which we normally encounter in chemistry, and that even 0.001 amu, a normal margin of error on a mole scale in a chemistry lab, carries a million times more energy. The fusion reaction for the deuteron ^2_1d is

$$^1_1p + ^1_0n \longrightarrow ^2_1d \qquad (16.3)$$

or, as it is usually written,

$$^1_1p + ^1_0n + ^0_{-1}e \longrightarrow ^2_1H \qquad (16.4)$$

where ^2_1H represents the entire isotopic atom, including its electron. We can use the mass change in Equation 16.4 to find the nuclear BE provided the binding energy of the electron (the ionization energy, IE) can be neglected, that is, IE $\ll$ BE. The bind-

Figure 16.1
Schematic illustration of a few nuclei. The relative sizes of the nuclei follow Equation 16.1, but the nucleons are in constant rapid motion, circulating throughout the nuclear volume in very short times. These "bunch of grapes" ("bog") pictures have about the same validity as Lewis structures.

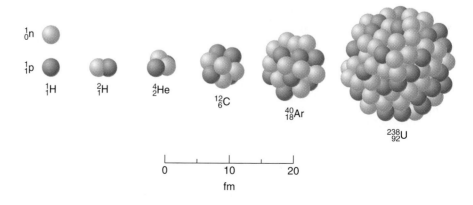

ing energy is the negative of the energy change for the fusion reaction, BE = $-\Delta E_{\text{fus}}$, and only those nuclei for which $\Delta m_{\text{fus}} < 0$ have any chance of being stable. Using Equation 16.4 and the masses in amu given in Table 1.3, we find

$$\Delta m_{\text{fus}} = -(1.007276 + 1.008665 + 0.0005486) + 2.014102$$
$$= -0.002388 \text{ amu}$$

The negative sign indicates that the deuteron is stable with respect to its constituent proton and neutron, and we know from its lack of radioactivity (see the next section) that it is absolutely stable. Using Equation 16.2, we find a binding energy of BE = 2.224 MeV; this is roughly a million times greater than a chemical bond energy, and it is our best measure of the strength of the nucleon–nucleon bond. The nuclear bond energy is also more than 100,000 times greater than the ionization energy of deuterium, justifying our neglect of the IE.

Combining the nucleon–nucleon bond length and bond energy in $^2_1$d allows us to sketch a rough potential energy curve for the interaction between two nucleons, provided we allow for zero point energy (ZPE). Unlike the small ZPE correction in a chemical bond, here ZPE is a major factor owing to the very short confinement length. The ZPE may be estimated from the particle-in-a-box model of Appendix B, suitably corrected for the finite wall at large $r_{\text{p-n}}$, to be about 50 MeV, more than 20 times larger than the BE. All this leads to the potential energy estimates given in Figure 16.2 for the three bonds; the p–p curve differs from the others mainly in the presence of a small barrier at larger r due to Coulomb repulsion. This **Coulomb barrier,** which exists between any two collections of nucleons carrying positive charges, is important in nuclear collisions and transformations, and we will refer to it several times in the following

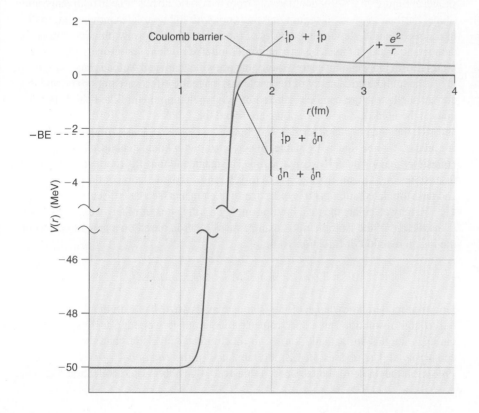

Figure 16.2

Nucleon–nucleon potential energy curves. The p–n binding energy is shown. Due to the high ZPE (note the broken energy scale) and extensive tunneling into the outer wall of the potential, the nucleons are more likely to be found at the outer turning point rather than in the middle of the well.

sections. We draw the curves without any repulsive core, although high-energy p–n scattering experiments appear to require repulsion at very short range. These basic interactions, taken to act pairwise among neighboring nucleons, account well for the binding energies of larger nuclei, although they are clearly an oversimplification of the actual situation, since n–n and p–p combinations have never been observed.

In our treatment of chemical bonding, we have given a detailed account of how bonds form in terms of electron exchange or delocalization. It occurred to H. Yukawa in the 1930s that a similar exchange of particles could be mediating the force between nucleons. Yukawa dubbed this exchanged particle a **meson,** the name indicating a particle with a mass in between those of nucleons (called **baryons** or massive ones) and electrons (called **leptons** or light ones). This model has been verified with the discovery of these mesons, now called pi-mesons or simply **pions.** In bonding together, a proton and a neutron exchange a negatively charged pion, denoted π^-, which implies that p and n are constantly interchanging their identities in the deuteron. More recent high-energy electron–proton scattering experiments suggest that nucleons, like atoms, have an interior structure, now believed to consist of a trio of very massive particles called **quarks,** which exert such strong forces on each other that most of their mass is converted into binding energy. As yet, a "bare quark" has not been isolated. The interested reader is referred to any modern physics text that reviews the quark theory, known as *quantum chromodynamics* after the "color charges" that mediate the forces between quarks.

In larger nuclei, the nucleons bond to several neighbors, somewhat like the bonding in a liquefied metal. Both protons and neutrons have **spin** of ½, however, as discussed in more detail in Section 16.5, and must behave like the electrons in an atom do, pairing up their spins within the nucleus. The actual details differ, since the neutron constituents do not repel each other as do electrons, and the nucleon potentials shown in Figure 16.2 differ qualitatively from the Coulomb potential in both range and shape. Further, there is no central massive core with light particles moving around it, like a queen and her court; the density of the nucleus is uniform within its volume, a democratic society of nucleons. In the **shell model of nuclear structure,** neutrons and protons form independent energy shells that obey the Pauli and Aufbau principles; owing to the high kinetic energy of the nucleons, magnetic effects are important, and the correct shell structure can be predicted only by including them. Filled shells (analogous to the noble gas configurations) are found for Z or $N = 2, 8, 20, 28, 50, 82, 126,$ and 184; the nuclei with either Z or N or both equal to one of these "**magic numbers**" are predicted to be unusually stable, in agreement with known enhanced binding energies for "magic" nuclei, except for 184. The magic number 184, corresponding to N in superheavy elements in the $Z = 114$ to 120 range, is a prediction of the shell model, and has stimulated attempts to synthesize such elements (see later discussion).

We may calculate the total binding energy for the nucleus of any isotope $^A_Z X$ from a knowledge of the isotopic mass using a mass relation based on a generalization of the fusion reaction of Equation 16.4,

$$BE = [Z(m_p + m_e) + (A - Z)m_n - M(A,Z)]c^2 \quad (16.5)$$

Since our modern knowledge of the isotope masses $M(A,Z)$ is extremely precise owing to high-resolution mass spectrometry (our ultimate weighing technique), these binding energies are actually known to much greater precision than their chemical counterparts! The masses $M(A,Z)$ are tabulated in the *Handbook of Chemistry and Physics* in the Table of the Isotopes; Table 16.1 is an abbreviated collection that will

TABLE 16.1

Masses of selected particles and isotopes in atomic mass units

Isotope	Mass	Isotope	Mass	Isotope	Mass	Isotope	Mass
$^{0}_{0}\nu, ^{0}_{0}\bar{\nu}$	$<6. \times 10^{-10}$	$^{28}_{14}$Si	27.976927	$^{99}_{43}$Tc*	98.906254	$^{224}_{88}$Ra*	224.020186
$^{0}_{-1}$e, $^{0}_{+1}\bar{\text{e}}$	0.00054858	$^{30}_{14}$Si	29.973770	$^{99}_{44}$Ru	98.905939	$^{226}_{88}$Ra*	226.025402
$^{1}_{1}$p	1.00727647	$^{30}_{15}$P*	30.975362	$^{116}_{46}$Pd*	115.914	$^{228}_{88}$Ra*	228.031064
$^{1}_{0}$n*	1.00866490	$^{31}_{14}$Si*	29.978307	$^{125}_{52}$Te	124.904433	$^{227}_{89}$Ac*	227.027750
$^{1}_{1}$H	1.00782505	$^{31}_{15}$P	30.973762	$^{125}_{53}$I*	124.904620	$^{228}_{89}$Ac*	228.031015
$^{2}_{1}$H	2.014102	$^{32}_{15}$P*	31.973907	$^{131}_{53}$I*	130.906114	$^{227}_{90}$Th*	227.027703
$^{3}_{1}$H*	3.016049	$^{32}_{16}$S	31.972070	$^{131}_{54}$Xe	130.905072	$^{228}_{90}$Th*	228.028715
$^{3}_{2}$He	3.016029	$^{33}_{16}$S	32.971456	$^{143}_{54}$Xe*	142.933	$^{230}_{90}$Th*	230.033127
$^{4}_{2}$He	4.002602	$^{35}_{16}$S*	34.969032	$^{139}_{56}$Ba*	138.908826	$^{231}_{90}$Th*	231.036298
$^{6}_{3}$Li	6.015121	$^{35}_{17}$Cl	34.968852	$^{145}_{60}$Nd	144.912570	$^{232}_{90}$Th*	232.038054
$^{7}_{3}$Li	7.016003	$^{37}_{17}$Cl	36.965903	$^{145}_{61}$Pm*	144.912743	$^{233}_{90}$Th*	233.041577
$^{8}_{4}$Be*	8.005305	$^{36}_{18}$Ar	35.967545	$^{193}_{77}$Ir	192.962917	$^{234}_{90}$Th*	234.043593
$^{9}_{4}$Be	9.012182	$^{40}_{18}$Ar	39.962384	$^{197}_{78}$Pt*	196.967315	$^{231}_{91}$Pa*	231.035880
$^{9}_{5}$B*	9.013328	$^{39}_{19}$K	38.963707	$^{197}_{79}$Au	196.966543	$^{233}_{91}$Pa*	233.040242
$^{10}_{5}$B	10.002937	$^{40}_{19}$K*	39.963999	$^{201}_{80}$Hg	200.970277	$^{234}_{91}$Pa*	234.043303
$^{11}_{5}$B	11.009305	$^{40}_{20}$Ca	39.962591	$^{201}_{81}$Tl*	200.970794	$^{233}_{92}$U*	233.039268
$^{9}_{6}$C*	9.031039	$^{48}_{20}$Ca	47.952533	$^{204}_{82}$Pb	203.973020	$^{234}_{92}$U*	234.040946
$^{10}_{6}$C*	10.016856	$^{55}_{25}$Mn	54.938047	$^{205}_{82}$Pb*	204.974458	$^{235}_{92}$U*	235.043924
$^{11}_{6}$C*	11.011433	$^{55}_{26}$Fe*	54.938296	$^{206}_{82}$Pb	205.974440	$^{236}_{92}$U*	236.045562
$^{12}_{6}$C	12.000000	$^{56}_{26}$Fe	55.934939	$^{207}_{82}$Pb	206.975872	$^{238}_{92}$U*	238.050784
$^{13}_{6}$C	13.003355	$^{59}_{26}$Fe*	58.934877	$^{208}_{82}$Pb	207.976627	$^{239}_{92}$U*	239.054289
$^{14}_{6}$C*	14.003241	$^{56}_{27}$Co*	55.939841	$^{210}_{82}$Pb*	209.984163	$^{240}_{92}$U*	240.056587
$^{13}_{7}$N*	13.005738	$^{59}_{27}$Co	58.933198	$^{214}_{82}$Pb*	213.999798	$^{237}_{93}$Np*	237.048167
$^{14}_{7}$N	14.003074	$^{60}_{27}$Co*	59.933819	$^{209}_{83}$Bi	208.980374	$^{239}_{93}$Np*	239.052933
$^{15}_{7}$N	15.000108	$^{56}_{28}$Ni*	55.943124	$^{210}_{83}$Bi*	209.984095	$^{238}_{94}$Pu*	238.049554
$^{15}_{8}$O*	15.003065	$^{60}_{28}$Ni	59.930788	$^{211}_{83}$Bi*	210.987255	$^{239}_{94}$Pu*	239.052157
$^{16}_{8}$O	15.994915	$^{78}_{36}$Kr*	77.920396	$^{214}_{83}$Bi*	213.998691	$^{240}_{94}$Pu*	240.053848
$^{17}_{8}$O	16.999131	$^{82}_{36}$Kr	81.913482	$^{209}_{84}$Po*	208.982404	$^{241}_{94}$Pu*	241.056845
$^{18}_{8}$O	17.999160	$^{94}_{36}$Kr*	93.938	$^{210}_{84}$Po*	209.982848	$^{243}_{94}$Pu*	243.061998
$^{19}_{8}$O*	19.003577	$^{82}_{37}$Rb*	81.918195	$^{211}_{84}$Po*	210.986627	$^{244}_{94}$Pu*	244.064199
$^{17}_{9}$F*	17.002095	$^{87}_{37}$Rb*	86.909187	$^{212}_{84}$Po*	211.988842	$^{241}_{95}$Am*	241.056823
$^{18}_{9}$F*	18.000937	$^{82}_{38}$Sr*	81.918414	$^{214}_{84}$Po*	213.995176	$^{242}_{95}$Am*	242.059541
$^{19}_{9}$F	18.998403	$^{87}_{38}$Sr	86.908884	$^{216}_{84}$Po*	216.001889	$^{243}_{95}$Am*	243.061375
$^{20}_{10}$Ne	19.992435	$^{90}_{38}$Sr*	89.907738	$^{218}_{84}$Po*	218.008966	$^{242}_{96}$Cm*	242.058830
$^{23}_{11}$Na	22.989767	$^{90}_{39}$Y*	89.907152	$^{210}_{85}$At*	209.987469	$^{247}_{96}$Cm*	247.070347
$^{24}_{11}$Na*	23.990961	$^{90}_{40}$Zr	89.904703	$^{222}_{86}$Rn*	222.017570	$^{247}_{97}$Bk*	247.070300
$^{24}_{12}$Mg	23.985042	$^{98}_{42}$Mo	97.905406	$^{223}_{87}$Fr*	223.019733	$^{245}_{98}$Cf*	245.068037
$^{27}_{13}$Al	26.981539	$^{99}_{42}$Mo*	98.907711	$^{223}_{88}$Ra*	223.018501	$^{251}_{98}$Cf*	251.079580
		$^{98}_{43}$Tc*	97.907215				

* Radioactive.

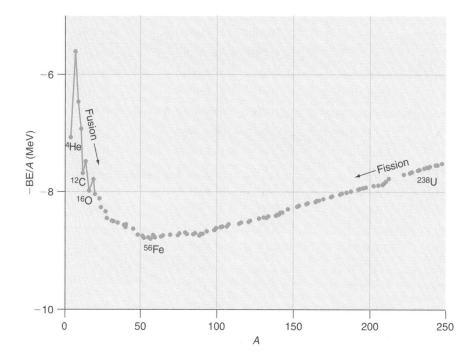

Figure 16.3
Plot of binding energy per nucleon, $-BE/A$ versus A, for the most abundant isotopes of the elements. The most stable nucleus is ^{56}Fe. Directions of exothermic fusion and fission reactions are indicated by the arrows.

be useful here. In Figure 16.3 we plot the quantity $-BE/A$, the negative of the binding energy per nucleon, for the most abundant isotopes versus A, as computed from Equation 16.5. We use a negative sign to emphasize the relative stability of the nuclei, the ones near the minimum, at $^{56}_{26}$Fe, being the most stable at -8.8 MeV/nucleon. This implies on the average 8 bonds to nearest neighbors, using the deuteron bond as our yardstick. The shape of the BE/A versus A curve is readily understood as a competition between increasing numbers of nearest-neighbor interactions and increasing Coulomb repulsion between the more distant protons as A increases. The enhanced neighborliness of the nucleons is due in part to an increase in the number of "interior" nucleons as A increases, which causes a rapid drop in $-BE/A$. The increasing Coulomb repulsion between the protons in the nucleus with increasing Z (and A) eventually outweighs the bonding effects, causing $-BE/A$ to bottom out at $A = 56$ and rise slowly beyond that point. This Coulomb repulsion ultimately destabilizes the nuclei of all elements beyond $Z = 83$; its consequences are described in the next section.

16.2 Nuclear Stability and Radioactivity

Unlike the case for the atom itself, in which charge neutrality dictates the "normal" form of an element, the stability of a collection of nucleons is always a relative issue. The reason for this lies in the possible *decay*, "uninuclear decomposition," of one nucleus into another, accompanied by the emission of subatomic particles. The criterion used to judge whether this can happen spontaneously is simply the sign of the mass change for the decay process. In nuclear decay at normal temperatures, entropy plays no role, and $\Delta G = \Delta H = \Delta E$, so that $\Delta G < 0$, our chemical criterion for spontaneity, becomes $\Delta m < 0$ for nuclei.

16.2 Nuclear Stability and Radioactivity

The discovery of unstable isotopes was an accident. In 1896 Henri Becquerel noted that a sample of uranium salt would expose a photographic plate even in a closed drawer. Becquerel incorrectly believed that the radiation that caused the exposure consisted of X-rays, which had been discovered the previous year by W. Röntgen. Marie Sklodowska, an emigré from Poland and the top science student at the Sorbonne in Paris, decided to make this mysterious finding the subject of her doctoral thesis, on the advice of her good friend and husband-to-be, Pierre Curie, a young professor there. The Curies coined the term **radioactive** to describe this phenomenon, and in her thesis Marie showed that the radiation carries a charge, unlike X-rays, and is an intrinsic property of elemental uranium, not dependent on its chemical state. Noting that uranium ore, pitchblende, is more radioactive than purified uranium, the couple then proceeded to isolate two new intensely radioactive elements, which they named polonium (Po) and radium (Ra), from immense quantities of ore. Earnest Rutherford, who had been under the tutelage of J. J. Thomson at the Cavendish Laboratory in Cambridge, England, and was then at McGill University in Canada, and Frederick Soddy, McGill's chemical demonstrator, showed that these new elements were intermediates in a series of radioactive decays beginning with U and ending with a stable isotope of Pb. Soddy's skill at chemical separations was essential to this groundbreaking effort; occasionally decay products of different origin could not be separated chemically, leading Soddy to suggest the existence of isotopes, later confirmed by mass spectrometry. Rutherford also identified three kinds of high-energy radiation, which he named α-, β-, and γ-rays, emitted during the series of decays. By 1908 the Curies' and Rutherford's investigations, which preceded the discovery of the nucleus itself, had provided nearly all the essential facts that underlie our present description of the decay process.

Figure 16.4 plots the number of protons Z versus the number of neutrons $N = A - Z$ for the stable and mildly unstable nuclei of the known elements. It is known

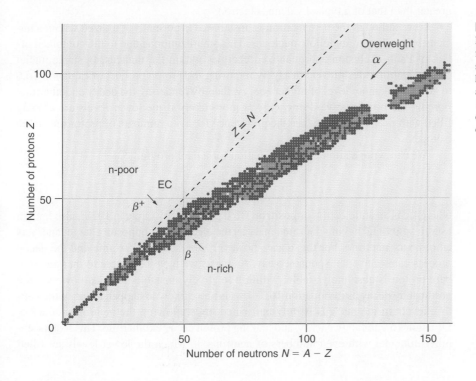

Figure 16.4

Nuclear stability diagram, Z versus N, for the known elements. Absolutely stable nuclei are in red, nuclei with half-lives greater than 1 day in green, and nuclei with half-lives between 1 minute and 1 day in blue. Dominant decay modes for n-poor, n-rich, and overweight nuclei are indicated.

as a **stability diagram.** The stable nuclei, the red points in the figure, follow a tight band that for small Z has a slope of unity, that is, equal numbers of neutrons and protons, $N = Z$, up to roughly $Z = 20$, and deviates toward smaller slope, $N > Z$, thereafter. Closer examination reveals a stair-step appearance in the pattern of stable isotopes, with a definite preference for *even numbers* of protons, neutrons, or both, and especially deep steps near the magic numbers. This **stability band** breaks off abruptly beyond bismuth, $Z = 83$. The unstable nuclei that have been observed (both in nature and as a result of our nuclear experiments) cluster tightly around the stability band, and even those beyond Bi follow an extrapolation of the curve established by the stable isotopes. In time, sometimes a very long time, these nuclei decay into more stable forms, eventually stopping at an absolutely stable nucleus, or one with an extremely long lifetime. We discuss the rate of nuclear decay in the next section.

For $Z < 84$, nuclear decay is dominated by the so-called *weak interaction,* yet another force of nature—perhaps the only other besides gravity, Coulomb, and strong—postulated to account for the decay phenomenon. It is this interaction that makes a bare neutron unstable; neutrons are known to decay into protons, electrons, and antineutrinos ($_{0}^{0}\bar{\nu}$) according to

$$_{0}^{1}n \longrightarrow\; _{1}^{1}p + _{-1}^{0}e + _{0}^{0}\bar{\nu} \tag{16.6}$$

where the antineutrino (with the overbar signifying its antiparticle status) is a nearly massless, chargeless particle whose presence is required to conserve momentum. The mass change for this decay is $\Delta m = -0.00084$ amu, yielding a spontaneous process with $\Delta E = -0.78$ MeV. Due to momentum conservation, most of this energy is released in the form of translational energy of the electron, still known as a **β particle** from the early work of Rutherford. Note that this energy is about 100,000 times greater than that of a typical valence electron.

Now, although the bare neutron is unstable, it is normally stabilized within a nucleus by binding to its fellow nucleons. This stabilization is due to the combined effect of Pauli exclusion and quantized energies within the nucleus. In stable nuclei these quantum rules prevent decay by requiring the proton product of Equation 16.6 to go into an empty level of higher energy, thereby rendering the decay endothermic. Forcing the proton upward in energy is more likely if the lower levels are already filled with proton pairs—hence the preference for even numbers of protons in stable isotopes.

We can write a similar decay equation for a bare proton,

$$_{1}^{1}p \longrightarrow\; _{0}^{1}n + _{+1}^{0}e + _{0}^{0}\nu \tag{16.7}$$

where $_{+1}^{0}e$ is an antielectron or **positron** (first predicted by Dirac in 1928), also known as a **β⁺ particle,** having the same mass as the electron but opposite charge; and $_{0}^{0}\nu$ is an ordinary neutrino. For this decay, however, Δm is $+0.00194$ amu and the decay process cannot happen spontaneously. As far as we have been able to measure, the bare proton is absolutely stable; within a relatively unstable nucleus, however, the proton may decay as in Equation 16.7. This is more likely to happen if one of the neutron levels in the nucleus is only half-filled, thus allowing the neutron product of Equation 16.7 to fall in energy, and making proton decay exothermic. This fosters stability in nuclei with even numbers of neutrons, in which the lower levels are filled with neutron pairs. The dominance of even numbers is striking: of the 279 known stable isotopes, only 4 have both odd Z and odd N: $^{2}_{1}$H, $^{6}_{3}$Li, $^{10}_{5}$B, and $^{14}_{7}$N.

Even numbers don't help, though, if the ratio of neutrons to protons is not optimum, as Figure 16.4 indicates. Too many neutrons leave vacant proton levels lying below those of the highest energy neutrons, encouraging n $\longrightarrow$ p decay whether Z is even or odd. Just the reverse occurs with too few neutrons: now there are vacant neutron levels below the highest energy proton levels, promoting p $\longrightarrow$ n decay. The increasing number of neutrons required at high Z results from the increased p–p repulsion, which increases the spacing and raises the energy of the proton levels; more neutrons than protons are thus required to match up the energies of the highest occupied levels.

Nuclei lying below the stability band of Figure 16.4 are neutron rich, whereas those lying above it are neutron poor. In either case, the weak interaction causes these nuclei to decay by emitting high-energy particles: they are radioactive. For example ^{14}C is a neutron-rich unstable isotope of carbon, occurring below the stability band, that undergoes radioactive decay according to

$$^{14}_{6}\text{C} \longrightarrow {}^{14}_{7}\text{N} + {}^{0}_{-1}\text{e} + {}^{0}_{0}\bar{\nu} \tag{16.8}$$

Note that the sums of the presubscripts, the charges on each species, and the presuperscripts, the mass numbers or **baryon numbers,** are separately conserved in the balanced nuclear decay reaction. From this point on, we will generally omit the neutrinos from these equations, because they do not affect the conservation of charge or baryon number, nor do they contribute to Δm. The ^{14}N **daughter nucleus** lies within the stability band, and does not undergo further decay. By comparing this reaction, known as **β-decay,** with the decay of the neutron (Equation 16.6), you can see that ^{14}C has achieved stability by converting one of its excess neutrons into a proton, which remains bound within the nucleus. In this way the nucleus avoids making the energetic sacrifice that would accompany the direct ejection of a neutron—which could happen only by breaking several nucleon–nucleon bonds— and more efficiently adjusts its n/p ratio, moving diagonally up and to the left toward the stability band. For nuclei lying more than one decay step away from the stability band, stepwise decay continues until a stable nucleus is reached. Extremely unstable nuclei, not shown in Figure 16.4, may directly disgorge one or more neutrons.

The general equation for β-decay of a neutron-rich isotope is

$$\beta\text{–decay} \qquad {}^{A}_{Z}\text{X} \longrightarrow {}^{A}_{Z+1}\text{Y} + {}^{0}_{-1}\text{e} \tag{16.9}$$

The nucleus decays to an element lying to its right on the periodic table. Calculating the mass change Δm for β-decay is easy once you realize that a $+1$ positive ion of the new element has been formed. The mass of the missing electron is compensated by that of the emitted electron, so that

$$\beta\text{-decay} \qquad \Delta m = M(A, Z+1) - M(A, Z) \tag{16.10}$$

where as before, the M's are isotope masses, including the electrons. For the decay of ^{14}C,

$$\Delta m = 14.003074 - 14.003241 = -0.000167 \text{ amu}$$

negative as expected for a spontaneous decay. The Δm also specifies the maximum kinetic energy the β particle can possess, here 0.156 MeV.

On the other hand, an isotope with too few neutrons, lying above the stability band, will convert one of its protons into a neutron. There are two ways it can do this: by emitting a positron, β^+**-decay,** or by capturing one of its own electrons, **electron capture** (EC).* For example, ^{82}Rb has a neutron-poor nucleus that undergoes β^+-decay,

$$^{82}_{37}\text{Rb} \longrightarrow {}^{82}_{36}\text{Kr} + {}^{0}_{+1}\text{e} \qquad (16.11)$$

while ^{82}Rb itself is produced by EC in ^{82}Sr,

$$^{82}_{38}\text{Sr} + {}^{0}_{-1}\text{e} \longrightarrow {}^{82}_{37}\text{Rb} \qquad (16.12)$$

In each case the net result is the conversion of a proton into a neutron in the parent; β^+-decay is in essence the (ordinarily nonspontaneous) proton decay (Equation 16.7). The ultimate decay product, ^{82}Kr, is stable and does not decay further. In both cases a negative ion is formed; thus the mass of the products of β^+-decay is larger than $M(A,Z-1)$ by *two* electron masses, since the masses of the electron and positron are equal. Therefore,

$$\beta^+\text{-decay} \qquad \Delta m = M(A,Z-1) - M(A,Z) + 2m_e \qquad (16.13)$$

For EC on the other hand, the mass of the excess electron on the right is balanced by that consumed, and

$$\text{EC} \qquad \Delta m = M(A,Z-1) - M(A,Z) \qquad (16.14)$$

For the β^+-decay of Equation 16.11, for example, using Equation 16.13,

$$\Delta m = 81.913482 - 81.918195 + 2(0.0005486) = -0.003616 \text{ amu}$$

again negative as required for a spontaneous decay; the reaction liberates β^+ particles with up to 3.37 MeV of kinetic energy. While some β^+ radiation can be directly detected, what normally happens is **annihilation,** the collision of the positron with a nearby orbital electron in the sample, which sends out a pair of γ-ray photons of energy 511 keV in opposite directions, and destroys both particles in the process. The alternative EC process can occur even when β^+ emission is not energetically possible, since Δm of Equation 16.14 is less by $2m_e$ than that from Equation 16.13. It is usually a $1s$ electron that is "swallowed" by the nucleus, since the $1s$ orbital has the highest amplitude there. Despite the lack of emitted particles, EC is readily detected, since the vacancy produced in the $1s$ gives rise to X-ray emission by the product isotope when a higher lying electron falls into the hole. Some neutron-poor nuclei decay by both routes, with a certain branching fraction between them. In either case, the nucleus moves diagonally down and to the right, efficiently approaching the stability band. Again, as for the neutron-rich case, extremely unbalanced nuclei can emit protons directly.

What happens for $Z > 83$ is qualitatively different. None of these "overweight" nuclei is absolutely stable, even those located on the extrapolation of the stability band; they must lose weight, and they do so by undergoing **fission,** that is, splitting

---

* You can think of EC as the reverse of ordinary β-decay; a possible alternative to β-decay would have been "PC," the reverse of β^+-decay, but there are no positrons in the atom.

up into smaller nuclei. Fortunately for us on earth, where these superheavy elements exist in some quantity, fission into two nuclei of comparable size is relatively rare, owing to the enormous Coulomb barrier separating the nuclear binding region from possible fission fragments. Instead, a typical decay route is **α-decay,** that is, ejection of a helium nucleus, the smallest "magic" nucleus of high stability. α-Decay minimizes the height of the Coulomb barrier, as dictated by the product of the charges; this product is $2(Z-2)$ for α-decay, whereas it would be $(Z/2)^2$ for two equal fragments, a factor of ~ 10 larger for $Z > 83$. For example, the major uranium isotope ^{238}U decays by α-emission,

$$^{238}_{92}\text{U} \longrightarrow {}^{234}_{90}\text{Th} + {}^{4}_{2}\text{He} \qquad (16.15)$$

producing thorium and an α particle (helium nucleus). Conveniently, the two extra electrons left when U decays are accounted for by the mass of the ^{4}He isotope, giving a mass change of

$$\alpha\text{-decay} \qquad \Delta m = M(A-4, Z-2) + M(4,2) - M(A,Z) \qquad (16.16)$$

For ^{238}U decay, this yields -0.004589 amu, with an α particle of maximum energy 4.27 MeV. It is the α-decay process for ^{226}Ra that Rutherford exploited to make a beam of α particles for his nuclear scattering experiments described in Chapter 1. Thorium-234 is also unstable, but undergoes β-decay instead of further α-decay, since loss of $2p + 2n$ has made a neutron-rich isotope lying slightly below the extrapolated stability band. Note that loss of an α moves the nucleus diagonally to the left and down on the stability diagram, perpendicular to the directions traveled in β- and $β^+$-decay or EC. Unlike the β-decays and EC, the strong force mediates α-decay, which can occur only with breaking of nucleon bonds.

EXAMPLE 16.1

Give plausible equations for the decay of the radioisotopes ^{17}F, ^{35}S, and ^{212}Po, and calculate Δm for each.

Solution:
The decay mode of a radioisotope can usually be deduced from its location on the stability diagram, or more readily, by comparing its A to the "average" A value reflected in the atomic mass on the periodic table, and using the "overweight" rule $Z > 83$ to decide whether α emission is possible. The isotope ^{17}F has two fewer neutrons than are reflected in its atomic mass of 19.0, that is, it is *neutron poor*, so we deduce that it will undergo either $β^+$-decay or EC. Either decay process will reduce Z by one and leave A the same, producing ^{17}O, which happens to be stable. The two possible decays are

$$^{17}_{9}\text{F} \longrightarrow {}^{17}_{8}\text{O} + {}^{0}_{1}\text{e} \qquad (β^+)$$
$$^{17}_{9}\text{F} + {}^{0}_{-1}\text{e} \longrightarrow {}^{17}_{8}\text{O} \qquad (\text{EC})$$

Checking with the Table of the Isotopes, we find that $β^+$-decay is what is observed. On the other hand, ^{35}S has three more neutrons than the atomic mass indicates, that is, it is *neutron rich*, so we expect β-decay to ^{35}Cl, again a stable isotope,

$$^{35}_{16}\text{S} \longrightarrow {}^{35}_{17}\text{Cl} + {}^{0}_{-1}\text{e} \qquad (β)$$

Po (polonium) has $Z = 84$, so it is *overweight* and all of its isotopes are unstable; and except for a few with a great excess of neutrons, these will decay by α emission, producing an isotope of $_{82}$Pb. Here

$$^{212}_{84}\text{Po} \longrightarrow \, ^{208}_{82}\text{Pb} + \, ^{4}_{2}\text{He} \qquad (\alpha)$$

To calculate the mass changes you need to look up these isotopes in Table 16.1, and use Equations 16.10, 16.13, 16.14, and 16.16, or, better, to understand the charge- and mass-counting behind these formulas. For ^{212}Po decay,

$$\Delta m = M(208,82) + M(4,2) - M(212,84)$$
$$= 207.976627 + 4.002602 - 211.988842 = -0.009613 \text{ amu}$$

The other mass changes are ^{35}S, -0.000180; ^{17}F, -0.002964 (EC), -0.001867 (β^+). In each case, Δm is negative as required for spontaneity, and it sets an upper bound on the amount of energy ($931.5\Delta m$ MeV) the α or β can carry away as kinetic energy. (In the cases chosen for this example, no γ-rays are emitted; see the following discussion.)

Often nuclear decay leaves the daughter nucleus in an excited state. Much like chemiluminescence from excited products of a highly exothermic chemical reaction, the excited product nucleus may radiate its excess energy, emitting **γ-rays** and thereby eventually reaching its ground state. γ-rays are photons whose wavelengths (λ) are much shorter than visible light, owing to the large spacings between nuclear energy levels; a typical spacing of 1 MeV yields $\lambda = 0.0124$ Å. Excited decay products are more common for larger nuclei, where the larger nuclear radius allows nuclear energy levels to draw closer together, and one or more excited levels to lie within the range of ΔE for the decay in question. For example, in 23% of the α-decays of ^{238}U, Equation 16.15, the α particle is accompanied by a γ-ray of energy 0.0496 MeV, wavelength $\lambda = 0.25$ Å, emitted by the lowest excited state of the ^{234}Th daughter. The energy released by the decay ΔE, still given by Equation 16.16, is then shared by the α and γ emissions; in this case a second group of α's of lower energy is observed. In the β-decay of neutron-rich ^{60}Co,

$$^{60}_{27}\text{Co} \longrightarrow \, ^{60}_{28}\text{Ni} + \, ^{0}_{-1}\text{e} + 2\gamma \qquad (16.17)$$

^{60}Ni is initially produced in its second excited state, and the bulk of the energy is released in the form of *two* γ-rays as ^{60}Ni reaches its stable ground state. Figure 16.5 illustrates the energetics of these decays. Measuring the wavelengths and energies of the γ-rays is part of *nuclear spectroscopy*, and the energy-level information it yields has aided in constructing the shell theory of nuclear structure mentioned in the previous section.

If all unstable nuclei are constantly striving to move into the stability band, you might ask why there are any naturally occurring unstable isotopes, such as ^{14}C or ^{238}U. After all, the earth is approximately 4.5 billion years old, plenty of time, one might think, for nuclei to roll downhill. It turns out that some nuclei (such as ^{238}U) decay extremely *slowly,* and are still trying to become stable. In other cases, the constant bombardment of the earth by *cosmic rays,* light and matter of incredibly high energy, is a steady initiator of nuclear reactions that create unstable isotopes (such as ^{14}C). Further, in the 20th century, we have learned how to do alchemy, that is, nuclear transmutation,

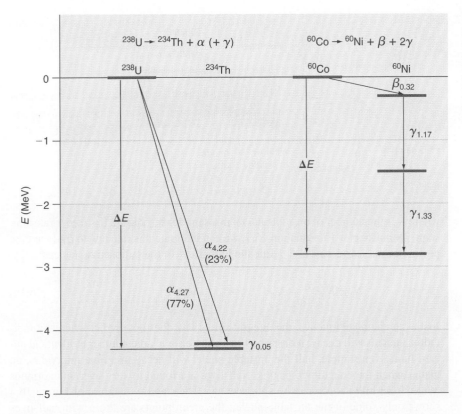

Figure 16.5
Nuclear energy-level diagrams for decay of ^{238}U and ^{60}Co. In each case the energy release, ΔE, is partitioned between the particles and the photons. The subscripts on the particle labels are their energies in MeV. Two decay routes occur for ^{238}U, while only one is observed for ^{60}Co.

by means of high-energy particle experiments and thermonuclear reactions, creating an entire menagerie of radioisotopes that occupy the fringes of the stability diagram. In any of these examples, simply knowing *whether* a given decay is spontaneous does not settle the question of *how long* a given unstable nucleus will exist before decaying.

16.3 Kinetics of Nuclear Decay

A decade after his brilliant investigations proving the existence of the nucleus in Rutherford's lab, H. Geiger invented a detector for the high-energy particles from decay now known as a **Geiger counter,** a gas-filled tube with a charge collector, as diagrammed in Figure 16.6. Each penetration of the tube "window" by an energetic particle produces a shower of ion-electron pairs from the gas fill; a positively biased wire in the center of the tube attracts the electrons and collects them, producing a current pulse for each particle that enters. With a suitable amplifier and counter, these pulses can be counted for timed intervals, thereby establishing the *rate,* decays s^{-1}, at which a given sample radiates.

At ordinary temperatures, nuclear decay is strictly a first-order "uninuclear" process, not depending on collisions for its activation. Whatever the decay products, we can write the first-order rate law

$$\frac{dN}{dt} = -kN \qquad (16.18)$$

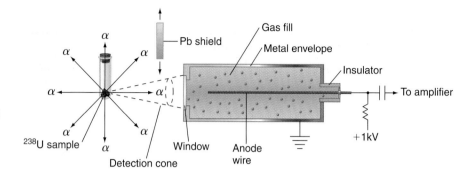

Figure 16.6
Geiger counter and a radioactive sample. Electrons, produced when the gas fill is ionized by high-energy particles entering the Geiger tube, are collected on the anode wire, creating a current pulse that is counted. The counter can detect only those particles emitted into the detection cone's "solid angle." The tube "window" is usually made of very thin aluminum plate, which allows high-energy particles to enter with minimal loss of energy.

where N is the number of unstable nuclei in a sample, and k is the first-order rate constant. As in Chapter 15, Equation 15.7, we can integrate this rate law to find the time profile for nuclear decay, and define the **half-life** $t_{1/2}$ of the radioactive nucleus:

$$N = N_0 e^{-kt} = N_0 e^{-\ln 2(t/t_{1/2})} \quad \text{where} \quad t_{1/2} = \frac{\ln 2}{k} \quad (16.19)$$

Just as in chemical kinetics, the integrated rate law does not predict when any individual nucleus will decay; it describes the behavior of a large number of identical nuclei. In a given sample, half the unstable nuclei will decay in a time interval $t_{1/2}$, no matter when the countdown begins. Half-lives are found to vary over an enormous range, from billions of years to split seconds. The stability diagram of Figure 16.4 shows only some of the unstable nuclei; the green points are those with half-lives longer than 1 day, while the blue points are those which live between 1 min and 1 day. A half-life of 1 min is judged to be a limit below which an isotope ceases to be "chemically useful." In particular the gap in the diagram beyond bismuth is occupied by isotopes with half-lives of a few seconds or less. Table 16.2 gives a sampling of half-lives and decay modes and energies for radioisotopes.

In nuclear decay both the rate law (Equation 16.18) and its integrated form (Equation 16.19) can be useful. For a very long-lived nucleus such as ^{238}U, normal human timescales do not lead to measurable changes in the population of unstable nuclei. For such nuclei, the decay rate or **activity** $-dN/dt$ is essentially constant and can be measured with a Geiger counter, provided a correction is made for its detection of only a small fraction of all decays from a radioactive sample—those whose particles pass through its window. Let us predict the activity of a 1.00-g (0.0042-mol) sample of ^{238}U from its half-life $t_{1/2}$ as given in Table 16.2:

$$-\frac{dN}{dt} = kN = \frac{N \ln 2}{t_{1/2}}$$

$$= \frac{(0.0042 \text{mol})(6.02 \times 10^{23} \text{mol}^{-1})(0.693)}{(4.46 \times 10^9 \text{ yr})(3.16 \times 10^7 \text{ s yr}^{-1})}$$

$$= 1.2 \times 10^4 \text{ s}^{-1}$$

Despite the minuscule decay rate constant (long half-life) and negligible fractional loss of sample activity, the decays occur at a readily detectable rate. The decay rate of a 1-g sample of a radioisotope is called its **specific activity.** Nuclear radiation is

TABLE 16.2
Decay properties of selected radioisotopes

Isotope	Half-life $t_{1/2}$	Decay mode	$-\Delta E$ (MeV)
$^{1}_{0}n$	14.8 min	β	0.78
$^{3}_{1}H$	12.32 yr	β	0.02
$^{11}_{6}C$	20.3 min	β^{+}	0.96
$^{14}_{6}C$	5715. yr	β	0.16
$^{15}_{8}O$	122. s	β^{+}	2.75
$^{19}_{8}O$	26.9 s	β,γ	4.82
$^{32}_{15}P$	14.28 days	β	1.71
$^{40}_{19}K$	1.28×10^{9} yr	$\beta(89\%)$	1.31
		EC(11%)	1.50
$^{60}_{27}Co$	5.272 yr	β,γ	2.82
$^{82}_{37}Rb$	1.273 min	$\beta^{+}(96\%),\gamma$	4.36
		EC(4%)	
$^{87}_{37}Rb$	4.9×10^{10} yr	β	0.27
$^{82}_{38}Sr$	25.6 days	EC	0.21
$^{90}_{38}Sr$	29. yr	β	0.55
$^{125}_{53}I$	59.9 days	EC,γ	0.18
$^{131}_{53}I$	8.04 days	β,γ	0.97
$^{210}_{84}Po$	138. days	α,γ	5.41
$^{222}_{86}Rn$	3.82 days	α,γ	5.59
$^{226}_{88}Ra$	1599. yr	α,γ	4.87
$^{234}_{90}Th$	24.1 days	β,γ	0.27
$^{235}_{92}U$	7.04×10^{8} yr	α,γ	4.68
$^{238}_{92}U$	4.46×10^{9} yr	α,γ	4.27
$^{239}_{94}Pu$	2.41×10^{4} yr	α,γ	5.24
$^{242}_{96}Cm$	163. days	α,γ	6.23

isotropic, being emitted with equal likelihood in any direction; if a detector window subtends a typical fraction 10^{-2} of the surface area of a hypothetical sphere surrounding the sample, roughly 100 counts s^{-1} will be measured. Isotropic radiation obeys an **inverse-square intensity law;** the intensity (particles cm^{-2} s^{-1}) measured or felt by a detector or other object drops off as the inverse square of the distance from the source. If the Geiger counter is moved away from the U sample, the counting rate rapidly drops to *background level,* a few counts per minute, due mainly to cosmic rays. As the half-life shortens, the specific activity increases; for example, a 1-g (0.0044-mol) sample of radium ^{226}Ra, with a half-life of 1600 yr, emits 3.7×10^{10} α's per second, a specific activity defined to be 1 Curie (Ci). Still shorter lived isotopes such as ^{82}Sr undergo measurable loss of sample activity over a modest time period, with a very high specific activity, and the integrated form of the decay law is used. Measurements on long-lived isotopes are ideal examples of the *initial rate* method of Chapter 15, while for short-lived species we can follow the time profile by periodically measuring the (high) counting rate for brief intervals Δt.

EXAMPLE 16.2

Tritium, $^3_1$H or just T, is widely used as a radioactive tracer in blood analysis. Use data from Table 16.2 to determine the concentration of T in a 10.0-mL water sample that will show an activity of 50.0 mCi. If a minimum activity of 30 mCi is required for a certain analysis, what is the useful shelf-life of this solution?

Solution:
The activity $-dN/dt = kN = N \ln 2 / t_{1/2}$, so the number of radionuclides required is

$$N = (t_{1/2}/\ln 2)(-dN/dt)$$
$$= [(12.32 \text{ yr})(3.16 \times 10^7 \text{s yr}^{-1})/(0.693)] (50.0 \times 10^{-3} \text{Ci})(3.7 \times 10^{10} \text{ s}^{-1} \text{ Ci}^{-1})$$
$$= 1.04 \times 10^{18}$$

and the concentration is

$$[\text{T}] = (1.04 \times 10^{18})/[(6.022 \times 10^{23} \text{ mol}^{-1})(0.0100 \text{ L})] = 1.73 \times 10^{-4} \text{ M}$$

The shelf-life is the time required for the activity to decay to 30 mCi. This isn't quite down to half, so we expect t to be a bit less than 12.32 yr. Since the activity is proportional to N, this will occur when $N/N_0 = 30.0/50.0 = 0.600$. From the integrated rate law,

$$t = -(t_{1/2}/\ln 2) \ln(N/N_0)$$
$$= -(12.32 \text{ yr}/0.693) \ln(0.600) = 9.08 \text{ yr (9 yr 1 mo)}$$

The β-active solution would need to be stored in a well-shielded area away from personnel; see Section 16.5 for the health effects of radiation.

Though half-lives of radioisotopes vary greatly in magnitude, some patterns emerge. Lifetimes are generally shorter the further the nucleus lies from the stability band, while for α- and β^+-decays for fixed Z there is an inverse correlation between $t_{1/2}$ and $-\Delta E$. This latter correlation was first explained by G. Gamov in the late 1920s, after quantum mechanics was invented, by means of the concept of **tunneling**. Recall from our discussion of orbitals in Chapter 3, that there is a finite probability that an electron can be found in regions where Newton's laws do not allow it to be, namely, outside the *outer turning point*. In the case of a stable atom, the electrons can tunnel to large distances, but are energetically forbidden to escape the atom because they are in bound states. In an unstable nucleus, however, the possible fragments, say an α particle plus a daughter nucleus, have enough energy to escape, but not enough to pass over the Coulomb barrier; we say the nucleus is in a **quasibound state**. For overweight nuclei, this confers some degree of stability; in fact, in Newtonian mechanics, these nuclei should live forever! However, the α particle can tunnel into the barrier; that is, its wave function has amplitude inside the barrier, in a classically forbidden region, where its kinetic energy is negative. Because the barrier has a far side where there is a second allowed region, with positive kinetic energy, this wave function "tail" allows the α to be found "outside" the barrier with some small probability. Then, given enough time, the odds are that the α will escape, and decay will have occurred. This scenario is illustrated for decay of uranium in Figure 16.7; Gamov's model produces semiquantitative agreement with experiment for both the long half-lives and the difference in $t_{1/2}$ between ^{235}U and ^{238}U listed in Table 16.2.

Decay of the overweight elements continues until the stability band is reached, by way of a multistep pathway first elucidated by Soddy and Rutherford and illustrated in Figure 16.8. For uranium, the first step is rate limiting, the rest occurring virtually instantaneously on the billion-year timescale of U α-decay. Thus pitchblende ore, from which uranium can be extracted, contains mainly U and stable Pb, the end product of the decay sequence, with other elements such as the longest lived Ra and Po, present in minute quantities—just like reaction intermediates in chemical mechanisms. This is why the Curies had to process tons of pitchblende to extract a few grams of Ra and Po. Now we can also say why U is the heaviest naturally occurring element: heavier nuclei may have been among the primordial matter forming our planet, but they are too short lived, and have decayed within the lifetime of our solar system—most likely into the methuselaic U and Th. Improved detection methods have allowed us to see decay in lighter isotopes formerly thought stable, as well.

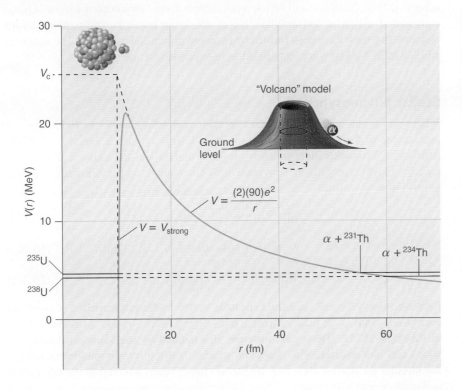

Figure 16.7
Gamov's "volcano" potential energy model for the α-decay of overweight nuclei, drawn for the decay of uranium. The crest of the Coulomb barrier occurs where the decay products Th + α just "touch." Because ^{235}U has slightly higher energy than ^{238}U, it has less of the Coulomb barrier to tunnel through, giving it a shorter lifetime. The inset shows that the α-particle "lava," instead of flowing over the lip of the volcano, simply appears on the outside after a suitable time, and then rolls downhill on the Coulomb repulsion, giving it a kinetic energy equal to the energy of the decaying nucleus.

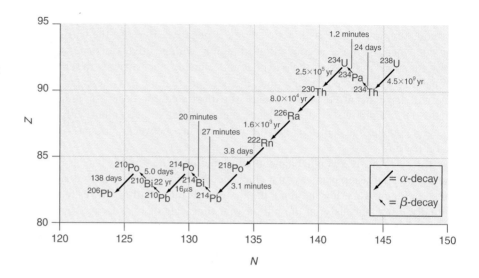

Figure 16.8
The major decay pathway of ^{238}U as first deduced by Rutherford and Soddy, with modern names for the intermediate elements. It was this work that provided the first evidence for the existence of isotopes. Decay sequences similar to this were later found for ^{235}U and ^{232}Th. Because α-decay always decreases A by 4, and β-decay not at all, all the isotopes in this sequence have $A = 4n + 2$, where n is an integer.

16.4 Nuclear Reactions

In considering the energetics and kinetics of nuclear stability and decay, it becomes clear why the alchemists never succeeded in transmuting one element into another. Starting from "base" materials, such as sulfur, mercury, or lead, this can only be done through nuclear fusion or fission processes. In either case, the energy required for surmounting the Coulomb barrier is millions of times greater than the alchemists could achieve in their crucibles, and fusion is not subject to the "philosopher's stone" chemical catalyst they tried to formulate. Nonetheless, our planet is well populated with elements that must have been formed by fusion of primordial hydrogen. The only crucibles within our ken where this could have happened are the stars.

Stellar Nucleosynthesis

Where do the sun and its fellow stars get their enormous energy? In 1939, after we had learned of the mechanism and stupendous energy of nuclear binding, H. Bethe and C. von Weiszäcker independently proposed that the sun and stars are giant fusion reactors, in which protons are converted into helium nuclei by the following sequence of reactions:

$$^{1}_{1}p + ^{1}_{1}p \longrightarrow ^{2}_{1}d + ^{0}_{+1}e + \nu \quad (16.20a)$$

$$^{2}_{1}d + ^{1}_{1}p \longrightarrow ^{3}_{2}He + \gamma \quad (16.20b)$$

$$^{3}_{2}He + ^{3}_{2}He \longrightarrow ^{4}_{2}He + 2\,^{1}_{1}p \quad (16.20c)$$

Each step is exothermic, and the net result, called **hydrogen burning,** is to covert four protons into a ^{4}He nucleus, with a total energy release of 23.7 MeV, half of this in Reaction 16.20c. These reactions are all inhibited by the Coulomb barrier V_c, which acts like an activation energy; they can only occur by surmounting or tunneling through the barrier. Reaction 16.20a has never been observed in the laboratory, and must therefore be extremely slow. The density of protons in stars is so great, however, that even an intrinsically slow reaction between them will occur at a large rate.

Tunneling is greatly enhanced by energy "resonance" with a quasibound state inside the barrier; this accelerates Reaction 16.20b and the γ-ray is emitted as the quasibound state decays to the ground state of ^{3}He. Coulomb barrier V_c is (nearly) four times greater for Reaction 16.20c, slowing it down. Recent advances in our understanding of neutrinos have resolved an apparent discrepancy in solar neutrino flux yield, confirming Reaction 16.20a as the root source of solar energy. We cannot actually observe these reactions even in our own sun, since they occur deep within its interior, where the densities and temperatures are high enough ($\rho > 10^4$ g cm^{-3} and $T > 10^7$ K) to allow the hydrogen burning reactions of Equation 16.20 to proceed at a rate consistent with the known radiative energy output of the sun. These extreme conditions were created through the heat generated by inward gravitational acceleration of the enormous mass of hydrogen in the protosolar epoch.

Hydrogen burning satisfactorily accounts for the sun's radiant energy, but it is all that our sun is capable of for the foreseeable future, and it is only the beginning of the story of how the elements of which we and our earth are made were formed. In more massive stars, called *red giants,* the density and temperature are higher ($T \approx 10^8$ K), enabling **helium burning** to commence after hydrogen burning has produced sufficient helium. Minor amounts of Li, Be, and B are made in a nuclear reaction sequence beginning with the fusion of ^{3}He and ^{4}He; these nuclei are relatively less stable and consequently less abundant. The main route to the heavier elements of C and higher appears to be a two-step reaction sequence

$$^4_2\text{He} + {}^4_2\text{He} \rightleftharpoons {}^8_4\text{Be} \tag{16.21a}$$

$$^8_4\text{Be} + {}^4_2\text{He} \longrightarrow {}^{12}_6\text{C} + \gamma \tag{16.21b}$$

that is, a net fusion of three ^{4}He nuclei into ^{12}C. Reaction 16.21a is our first example of an endothermic nuclear reaction ($\Delta E = +0.09$ MeV), and is made possible by the large thermal energy available. If we assume this reaction reaches equilibrium, then according to LeChâtelier's principle, increasing T will drive the reaction to the right. Reaction 16.21b is exothermic, but has a very high Coulomb barrier V_c that would ordinarily make it too slow to take advantage of the small amount of ^{8}Be available. Remarkably, however, as was the case for Reaction 16.20b, the binding energies of ^{8}Be and ^{4}He are again nearly "resonant" with the second excited state of the ^{12}C nucleus, allowing Reaction 16.21b to proceed by tunneling through the Coulomb barrier at modest kinetic energy, thereby greatly increasing its rate. The excited ^{12}C then radiates one or two γ-ray photons to reach its stable ground state.

Once the ^{12}C nuclei begin to accumulate, a series of fusion reactions involving protons, initiated by fusing ^{12}C with a proton, rapidly and exothermically produces a mixture of C, N, O, and F isotopes. The higher temperature produced by this energy release enables ^{12}C to fuse directly with ^{4}He to make ^{16}O, a "doubly-magic" nucleus, and at still higher temperatures, $T > 10^8$ K, ^{12}C and ^{16}O will themselves begin to "burn" to make Mg, S, and other products lighter than these by one or more α's or neutrons. From these fusion products, further reactions with p, n, and α create a host of heavier elements. It is in these massive fusion reactions that free neutrons begin to appear as products; these are essential for producing the odd-A isotopes of the elements beyond F, and for creating and stabilizing the nuclei of the heavier elements that fall below the $Z = N$ line of Figure 16.4, through *neutron capture* followed by β-decay. Once nuclei near the bottom of the binding energy curve of Figure 16.3 are formed, however, further fusion reactions become endothermic in addition to having ever-larger V_c, and the stellar furnace begins to run down. This leaves at least two important questions not addressed by nucleosynthesis in stars: where did the elements with $Z > 26$ come from, and how did any of the heavier elements with $Z > 2$ end up here on earth? Even the largest stars may not generate interior temperatures high enough to form trans-iron nuclei. Further, the accepted mechanisms for formation of sub-iron nuclei appear to doom these elements to remain forever trapped by the prodigious gravity of the stellar interior. The answers to both of these questions are now thought to lie in the most cataclysmic event in the universe, the explosion of an entire star known as a **supernova.**

Depending on the mass of the stellar crucible, nucleosynthesis will halt at Fe and Ni, or at an earlier point, and the star will "burn out" like a spent light bulb. With its source of heat removed, the star will begin to *collapse* or implode to a very small size under the relentless force of gravity. As collapse proceeds, the interior density and temperature begin to increase again, as they did when the star was first formed; if these become high enough ($T > 10^9$K), the star may explode, producing a bright flash—a nova or supernova—while spewing nuclear matter into space in whole form and, in the case of a supernova, also as disintegrated fragments from its innermost recesses. Nucleons regenerated in the disintegration include energetic neutrons, which engage even the largest of the stable nuclei in neutron capture reactions followed by β-decay, that appear to account for the trans-iron nuclei. Only the most massive stars produce supernovae, which serve both to liberate the heavier elements they have made during their lifetime and to enhance the production of still heavier nuclei.* In ending their bright existence, these giant stars become the primordial birthgivers. Our solar system is likely to have formed from a gas cloud ejected from such a supernova, with our own sun a "second-generation" star. The scarcity of uncombined hydrogen and helium on our own planet is due to its feeble gravity, which has allowed these gases to escape into space, leaving behind mainly the "ashes" from their fusion. We ourselves are the ashes of burnt stars.

In our solar system, this confluence of stellar events has produced the abundances of the elements shown in Figure 16.9. The preference for even numbers of protons and the stabilizing influence of magic numbers is readily apparent, although we are far from having achieved "nuclear equilibrium," which would make ^{56}Fe the most abundant nucleus.

Nuclear Reactions in the Laboratory

The beginnings of "nuclear chemistry," the systematic study of nuclear reactions in the laboratory, may be traced to the first artificial transmutation, carried out almost single-handedly—due to the first world war—by Rutherford at Manchester, England, in 1919:

$$^{14}_{7}\text{N} + ^{4}_{2}\text{He} \longrightarrow ^{17}_{8}\text{O} + ^{1}_{1}\text{H} \tag{16.22}$$

Rutherford discovered this reaction by detecting the high-energy protons formed when he bombarded nitrogen gas with high-energy α's from decay of "radium C"—which we now know to be a mixture of ^{214}Bi and ^{214}Po. Figure 16.10 shows the

---

* The fate of the stellar remnant is mass dependent as well, and includes simple burnout and collapse into a *white dwarf,* or post-nova collapse into a *neutron star* or even a *black hole* after the nova event carries away the excess mass and energy. Our sun is probably destined for dwarfdom.

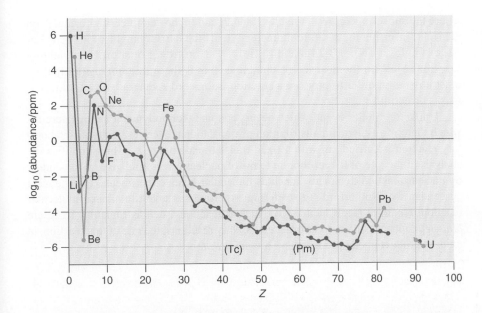

Figure 16.9
Semilogarithmic plot of relative atomic abundances of the elements (parts per million) versus Z for our solar system. Even-Z elements are in green, odd-Z in blue. The points represent a sum of isotopes. The elements Tc and Pm, as well as those between Bi and Th, lack sufficiently stable isotopes.

experimental arrangement. Not only is this reaction endothermic, but, as in stellar nucleosynthesis, there is a large Coulomb barrier V_c to overcome. The height of this barrier can be estimated by combining the nuclear radius formula (Equation 16.1) with Coulomb's law to yield

$$V_c = \frac{Z_1 Z_2 e^2}{r_1 + r_2} = \frac{1.44 Z_1 Z_2}{r_1 + r_2} \qquad (16.23)$$

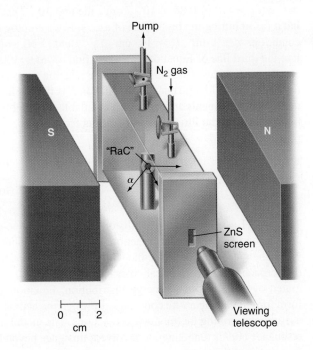

Figure 16.10
Apparatus used by Ernest Rutherford in discovering the first nuclear reaction, $^{14}\text{N} + \alpha \longrightarrow {}^{17}\text{O} + p$. The RaC α source was much closer to the ZnS screen than shown here. The reaction was initially observed as an *increase* in the number of scintillations on the ZnS screen when the chamber was filled with N_2; later experiments showed this to be due to high-energy protons. The magnet was used to deflect away the β-rays; the brass chamber actually fit snugly between the poles, which are shown pulled back for clarity.

where the subscripts represent nuclei 1 and 2, and in the "engineering formula" on the right, inserting the r's in femtometers yields the barrier height V_c in MeV. For our example, $r(^{14}N) = 3.2$ fm and $r(^4He) = 2.1$ fm from Equation 16.1, yielding $V_c = 3.8$ MeV. In the absence of tunneling, this is the minimum energy the α's must possess in order to produce a reaction, subject to a generally small reduced-mass correction.

Unlike the stellar event, which occurs at some thermodynamic temperature for which there is a Boltzmann distribution of translational energies, the typical laboratory reaction is carried out with a fixed translational energy E_t. This simplifies the analysis of the rate of the reaction in terms of nuclear forces, and gives us a window on nuclear reaction dynamics, much as molecular beam experiments, the chemical descendants of Rutherford's α-scattering experiments, have illuminated molecular reaction dynamics (see Chapter 15). Nuclear reactions such as Equation 16.22 are *binuclear*, intrinsically second-order elementary reactions; the rate of this reaction, in units of product nuclei produced per second dN/dt, is

$$\frac{dN}{dt} = \tau k[\alpha][^{14}N] \tag{16.24}$$

$$= \tau v_{rel}\sigma[\alpha][^{14}N]$$

$$\approx \tau \sigma I_\alpha [^{14}N]$$

where τ is the reaction volume, defined by the dimensions of the α beam and the path length through the nitrogen sample, k is the binuclear rate constant, v_{rel} is the relative velocity, approximately the velocity of the α beam v_α, σ is the **nuclear reaction cross section,** analogous to that introduced for chemical reactions in Chapter 15, and the brackets [ ] represent concentrations in particles per unit volume. The product $v_\alpha[\alpha]$ is the *flux density* I_α (particles cm^{-2} s^{-1}) of the α beam. A quick "collision theory" estimate of σ for Reaction 16.22 can be obtained from the sum of the nuclear radii as $\sigma \approx \pi(r_1 + r_2)^2 = 0.88 \times 10^{-24}$ cm^2; the unit 10^{-24} cm^2 is amusingly known as a **barn** (as in hitting the broad side of a. . .). If we employ Rutherford's experimental conditions $\tau \approx 1$ cm^3, $I_\alpha \approx 10^8$ cm^{-2} s^{-1}, and $[^{14}N] \approx 10^{18}$ cm^{-3} and insert these into the third line of Equation 16.24, we find $dN/dt \approx 100$ s^{-1}.

A more sophisticated analysis of the reaction cross section, taking into account that some of the kinetic energy of collision E is tied up in the centrifugal (rotational) motion of the reagents and hence is unavailable for climbing the potential energy barrier V_c, leads to the **line-of-centers model** for σ,

$$\sigma = \begin{cases} \pi(r_1 + r_2)^2\left(1 - \dfrac{V_c}{E}\right), & \text{for } E > V_c \\ 0, & \text{for } E \leq V_c \end{cases} \tag{16.25}$$

This model assumes that all collisions with radial (line-of-centers) kinetic energy greater than V_c will produce reaction, and is also frequently used in describing activated chemical reactions. The model builds in the necessity for surmounting V_c, and σ approaches the geometrical cross section $\pi(r_1 + r_2)^2$ at high energy. When integrated over a Boltzmann energy distribution, Equation 16.25 yields an Arrhenius form for the binuclear rate constant k, in which the activation energy $E_a \approx V_c$; this is useful in describing thermonuclear reactions. The model must be modified if energy resonance and/or tunneling are involved; this goes beyond what you have been prepared to do here. The general consequence of tunneling is that the reaction "turns on" at energies well below V_c.

The discovery of the neutron in 1932 by J. Chadwick, after an arduous decade of searching, was a direct outgrowth of Rutherford's series of α-particle experiments; the reaction observed was

$$^{9}_{4}\text{Be} + {}^{4}_{2}\text{He} \longrightarrow {}^{1}_{0}\text{n} + {}^{12}_{6}\text{C} \tag{16.26}$$

The neutron was evident as an extremely penetrating ray, more so than any α-, β-, γ-, or p-ray yet seen; as a neutral massive particle it could easily pass through a centimeter-thick lead plate, and then knock protons out of paraffin wax. This event at the Cavendish laboratory in England—now headed by Rutherford himself—catalyzed the study of nuclear physics, much as Thomson's discovery of the electron in the same laboratory 35 years earlier had done for atomic physics. Irène Joliot-Curie, Marie's elder daughter, and her husband Frédéric showed radioactive products as well as neutrons could be produced from nuclear reactions; for example,

$$^{27}_{13}\text{Al} + {}^{4}_{2}\text{He} \longrightarrow {}^{1}_{0}\text{n} + {}^{30}_{15}\text{P} \tag{16.27}$$

where ^{30}P is neutron poor and decays by emitting the then newly discovered positron. Chadwick then produced neutrons by direct photodissociation of deuterons with γ-rays, a year after H. Urey's discovery of "heavy hydrogen" or **deuterium** (D) in 1933. Neutrons made ideal nuclear reagents, their reactions with any nucleus being very fast due to the lack of a Coulomb barrier, and E. Fermi and others soon began to study the neutron reactions of a large fraction of the elements. The result would typically be neutron capture by a stable isotope to form a neutron-rich nucleus, which would then undergo β-decay, creating an isotope of the next element in the periodic table, in just the way stars are now believed to have done it. Along the way, unstable isotopes of two previously unknown elements with $Z < 92$, technetium ($_{43}$Tc) and promethium ($_{61}$Pm), were created; these elements turn out to have no isotopes stable enough to have survived from the birth of the solar system until now.

Neutron capture by uranium produced some strange results, however: O. Hahn and F. Strassman found an isotope of Ba among the products. When their colleague Lise Meitner was informed, she correctly speculated that **nuclear fission** had occurred, a term she coined to describe the splitting of the U nucleus into two massive pieces. It was later found that only the minor ^{235}U isotope (0.72% of natural uranium) had fissioned, while the major isotope ^{238}U had undergone a "normal" neutron capture followed by two β-decays to make the first **transuranic element,** plutonium (^{239}Pu).

Nuclear fission of sufficiently heavy nuclei is inevitably a spontaneous process, because it takes the system down in energy from the right-hand limb of the binding-energy curve of Figure 16.3. However, the same Coulomb barrier that inhibits fusion to make post-iron nuclei also stands in the way of fission. What made fission possible for ^{235}U was the highly unstable ^{236}U isotope produced by neutron-capture, allowing the system to pass readily over or through V_c for all but the highest barrier fission products, those of equal Z near Pd. Because the heavy nuclei contain an excess of neutrons already, free neutrons are produced along with unstable isotopes of the fragments; among approximately 30 significant fission channels is

$$^{235}_{92}\text{U} + {}^{1}_{0}\text{n} \longrightarrow {}^{94}_{36}\text{Kr} + {}^{139}_{56}\text{Ba} + 3{}^{1}_{0}\text{n} \tag{16.28}$$

Averaged over the range of fission products, about 2.5 neutrons are produced per fission event. The extra neutrons can cause further fission of ^{235}U, creating a **nuclear chain reaction** similar to the branched chain mechanism underlying chemical explosions, as illustrated in Figure 16.11. In ordinary uranium, the population of ^{235}U is too sparse to sustain the chain reaction, but modest enrichment to about 3% yields

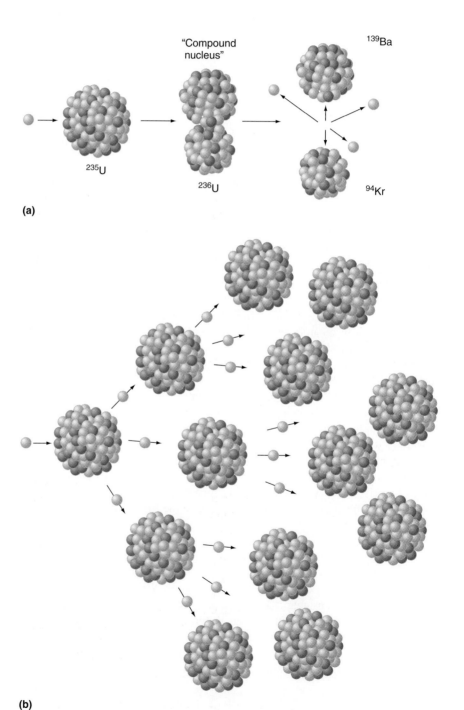

Figure 16.11
Schematic illustration of neutron-induced nuclear fission. (**a**) An individual fission event. A low-energy neutron gets stuck in the "bog" of nucleons in the fissile nucleus, which then distorts. The "compound nucleus," analogous to the transition state in chemical reactions, then splits into fission fragments and neutrons; only one possible set of fragments among many is shown. The neutron emission, although shown here as occurring simultaneously with fission, is actually delayed, occurring from the highly radioactive fragments. (**b**) A chain reaction results if at least one of the neutrons from a fission induces fission in a neighboring nucleus. In the ideal case shown here, where all three neutrons cause fission, only 50 "generations" of neutrons would be required to fission more than a mole of ^{235}U. The fission fragments from each event have been omitted for clarity.

a sustainable chain loop. The rapid release of energy per reaction a million times larger than that in a chemical explosion is bound to have significant consequences, to be discussed in the last section of this chapter.

Synthesizing still heavier transuranic elements requires carrying out supernova-like neutron capture and endothermic fusion reactions; these were pioneered by E. McMillan and G. Seaborg at Berkeley in the late 1940s using a combination of intense neutron sources and heavy-ion accelerators—*cyclotrons*—which produced α particles with energies 10 times those of Rutherford's radioactive sources. For example, the element californium ($_{98}$Cf) was first made through a succession of neutron pickups and a final α fusion reaction,

$$^{239}_{94}\text{Pu} \xrightarrow{n,\gamma} {}^{240}_{94}\text{Pu} \xrightarrow{n,\gamma} {}^{241}_{94}\text{Pu} \xrightarrow{\beta} {}^{241}_{95}\text{Am} \xrightarrow{n,\gamma} {}^{242}_{95}\text{Am} \xrightarrow{\beta} {}^{242}_{96}\text{Cm} \xrightarrow{\alpha,n} {}^{245}_{98}\text{Cf} \quad (16.29)$$

where the new transuranic elements americium and curium were made along the way, and the abbreviated notation indicates particles gained and lost in each reaction. Improvements in accelerator technology, in particular the use of superconducting magnets, have allowed the use of heavier nuclear projectiles, such as ^{12}C or ^{16}O, and have now carried us past $Z = 110$, although in many cases only a few atoms of the new elements have been made, and all of them are much shorter lived than U. A 2004 report from Russia by a joint Russian–U.S. team of scientists indicates that four atoms of element 115 may have been formed from the fusion of the rare isotopes ^{48}Ca and ^{243}Am. Thus we may be arriving at the "island of stability" near $Z = 120$ predicted by the shell model. These elements all have perfectly well-defined, stable electronic structures. As remarked in Chapter 1, Coulomb's law allows atoms of any Z to exist, although it may now be added that substantial relativistic (magnetic) effects make orbital descriptions of these elements less accurate. Mass numbers in parentheses on the periodic table indicate the most stable isotope for elements that do not occur naturally. In addition to synthesizing these new elements, we have learned to emulate many of the fusion reactions believed to occur in the stars, thus helping us to understand the origins of the stuff we are made of, our "nuclear roots."

16.5 Applying our Nuclear Knowledge

Almost from the very beginning of the "nuclear age" a century ago, we became aware that the high-energy radiations from nuclear processes would be both a blessing and a curse. Radiation therapy began with using radium to cure skin cancer in 1906, while the Curies, who more than anyone else came into contact with massive amounts of radioactive material, reported from firsthand knowledge the deleterious health effects of large doses of radiation. And although at the start no one believed that nuclear energy could be released at a rate great enough to be a useful source of energy, it was recognized that the energy per nuclear reaction was millions of times that of any known chemical reaction. Rutherford was among the earliest to realize that the lifetime of radioisotopes would be useful in gauging the age of the earth. In the same era, refinements in atomic spectroscopy revealed a "hyperfine" structure in atomic lines for certain elements, later shown to be due to what we call "nuclear spin" magnetism. Studying the fundamentals and applications of these facets of the nucleus has attracted brilliant scientists and engineers around the world, who continue to develop remarkably beneficial as well as apocalyptic uses for nuclear properties. The benefits of nuclear knowledge have in particular reached back to the chemical science from which our first bits of knowledge came.

Isotopic Chemistry

Except for hydrogen, different isotopes of the same element have nearly the same chemistry, since the electronic wave function of a molecule depends only on the charges of its nuclei, and not on small differences in their masses. This might lead you to believe that isotopic chemistry is not a very important or interesting area; nothing could be further from the truth. **Isotopic substitution** is universally used in molecular spectroscopy as a means of assigning quantum numbers to spectra, thereby establishing the true energy level and geometrical structure of a molecule—invaluable knowledge to the chemist. As we saw for diatomic molecules in Chapter 8, vibrational frequencies and rotational constants both change in a well-defined way when, say, D is substituted for H in a molecule. In general, only the correct quantum-number assignment will reproduce both the normal and isotopic spectra. In this application stable minor isotopes are used.

Isotopic labeling is also of key importance in the investigation of reaction mechanisms, particularly in reactions with the same type of atom in each of two reagents. Here either stable or unstable isotopes can be employed; the latter are called **radioactive tracers.** In the end, the reaction products are separated and analyzed, either by mass spectrometry or by counting scintillations. For example, in the *esterification* of carboxylic acids (RCOOH) and alcohols (HOR'), labeling the alcohol with the stable isotope ^{18}O leads to the appearance of the label in the ester, and not in the water solvent,

$$\underset{\text{RCOH}}{\overset{\text{O}}{\|}} + \text{H}^{18}\text{OR}' \longrightarrow \underset{\text{RC}^{18}\text{OR}'}{\overset{\text{O}}{\|}} + \text{H}_2\text{O} \qquad (16.30)$$

The presence of ^{18}O in the ester is shown by its mass spectrum, and its location in the C—O—R chain by a frequency shift in its vibrational spectrum. If the ^{18}O label is instead put on the carboxylic hydroxyl, it ends up in the water. This establishes that the reaction occurs by nucleophilic attack of the alcoholic oxygen on the carboxylic carbon, enabling us to "push electrons" in the right way, and opening the way to understanding and prediction of a host of reactions involving carbonyl carbon as a nucleophilic attack site. Symmetric electron transfer reactions, such as $Fe^{3+} + Fe^{2+} \longrightarrow Fe^{2+} + Fe^{3+}$, can be studied only by isotopic labeling; in this case the iron(II) reagent was labeled with radioactive ^{55}Fe, and the rate of appearance of the activity in the chemically separated Fe^{3+} was measured. Radio methods are extremely sensitive, allowing low radioactive mole fractions and concentrations to be employed; for more usual concentrations, the reaction is too fast to measure by separation. If the isotopic label has a nuclear spin, nuclear magnetic resonance, described later in this section, can also be used to analyze the products.

Isotopic substitution can affect the chemistry itself. By far the largest isotope effects on the rates and thermodynamics of chemical processes occur for hydrogen isotopes; the normal effect is in the direction of a reduced reaction rate for deuterated compounds, owing to lower zero point energies, slower molecular motion, and reduced tunneling rate. This is why "heavy water," D_2O, is a deadly poison: it slows down the metabolic acid–base chemistry essential for sustaining life.

Using recoil atoms formed as products of nuclear reactions has allowed the study of simple chemical processes in a new, high-kinetic-energy regime. Prototypical of such **hot atom chemistry** studies are reactions of tritium, ^3_1H, formed by neutron bombardment of ^3He or ^6Li. The new atom rapidly loses its energy through collisions with reagent gas, but some reaction will occur while the relative translational energy is in the range of 1 to 50 eV (roughly 25 to 1000 kcal/mol), exceeding the activation barrier for most chemical reactions. Atom abstraction and dissociation, rather than substitution, tend to dominate at high energies; for example, $^3H + CH_4$ yields mainly $^3H\text{—}H + CH_3$ rather than $^1H + CH_3\text{—}^3H$.

Rutherford's original α-particle scattering experiment (see Chapter 1) remains a useful tool for the noninvasive elemental analysis of solid samples. In the **Rutherford backscattering** method, the translational energies of the scattered α's are analyzed. Heavier elements impart more recoil to the α's, giving them higher energy, and the spectrum of energies can be interpreted as a mass spectrum. The first soft lunar lander, *Surveyor 5,* was equipped with a Rutherford backscattering analyzer using a 100-mCi ^{242}Cm 5-MeV α source, and through it we first learned that the moon is not made of green cheese after all, but of rocks similar to those found on earth.

Nuclear Medicine

Nowhere has the two-edged character of the nuclear sword been more evident than in health and medicine. Our society has been taught to fear the very word "nuclear," and even "radiation" is often misused in that little distinction is made between nuclear radiation and more ordinary electromagnetic energy of modest energy and longer wavelength. The fears are of "radiation sickness" from extremely high exposures, but more pervasively, of the unprovable notion that *any* amount of exposure to high-energy rays is harmful to your health. On the other hand, the use of radioisotopes in beneficial therapies and "tomographies" (3-D internal images), as well as in biomedical and biochemical research, is widespread and extremely useful.

First, let us outline what is known about nuclear radiation's health effects. The unit of exposure to radiation is called the **rad** (**r**adiation **a**bsorbed **d**ose), which corresponds to taking in 100 ergs of radiant energy per gram of body weight (100 erg g^{-1} or 0.01 J kg^{-1}). Thus the source of the radiation, which determines its energy, is as important as the rate of decay; the rad depends on the product of the two. What appears to be important in organismic response to radiation is the amount of ionization that is caused; α radiation is more powerful in this regard. The **rem** (**r**oentgen **e**quivalent in **m**an) is based on the roentgen, defined as that amount of radiation that will produce 1 esu of charged particles in 1 cm^3 of air at STP. The rem is adjusted slightly to account for observed levels of biological damage. For β and γ radiation, 1 rad = 1 rem very closely, while for α-rays, 1 rad $\approx$ 10 rem. Because of the inverse-square intensity law, and the stopping power of air for all the rays, especially α, dangerous exposures usually result from direct skin contact with or ingestion of radioactive material. Just being "near" a source of radiation is generally not nearly so risky. Rutherford and his students regularly breathed radon gas emanating from their (minuscule, milligram quantity) radium samples over a period of years. Despite inhaling the strong α-emitter, not one of them was known to have suffered ill effects, probably because, as a noble gas, Rn cannot be trapped in body tissue.

All of us are constantly exposed to high-energy radiation of natural origin. We are continuously bathed in **cosmic rays** from interstellar space: protons, electrons, and γ-rays having average energies of 200 MeV. Uranium and its more active decay intermediates are widely distributed in rocks and clays from which stone, brick, and mortar are made; Rn decay product can accumulate in stagnant air in dwellings made from these materials. In addition, the human body contains roughly 5×10^{-4} mol of the naturally occurring radioisotope ^{40}K, which produces 5000 β-decays s^{-1}, and makes each of us a walking source of nuclear energy. Combined, these sources give rise to an annual human dose of roughly 100 millirem, which has been with us since the dawn of civilization; modern developments, such as X-rays and other medically prescribed radiation, and airplane flights, which increase cosmic ray exposure, may double this figure. Radiation sickness and possible death will result from a *one-time exposure* of several hundred rem, more than 1000 times what we normally receive *in a year*. However, the threshold for biological damage, such as genetic mutation, is not known, if it exists at all. If there is no threshold, then we are a naturally radiation-damaged lot.

Radioisotopes are widely used in both medical diagnosis and therapy, especially in cancer and tumor treatment. We mention three well-known applications among a myriad. Neutron-rich ^{131}I is commonly used in treating thyroid gland problems; iodine naturally accumulates in the gland, providing both a diagnostic of thyroid activity using a β-ray detector, and, in larger amounts, a localized radiation treatment for thyroid cancer. Positron emission tomography (PET) is a useful way to obtain three-dimensional images of internal organs and tumors; among the isotopes currently used for this method are neutron-poor ^{11}C, ^{13}N, ^{15}O, and ^{82}Rb. To use these short-lived isotopes, larger hospitals may have their own small accelerators for generating them on site, in addition to arrays of annihilation radiation detectors and computer-controlled electronics for "picture taking." This is also true for the most common cancer therapy isotope, neutron-rich ^{60}Co, whose β- and γ-radiation is administered externally. Except for radiotherapy, dosage for these various treatments is typically in the millirem range, adding a small percentage to a person's natural dose unless carried out numerous times in a year.

Nuclear Magnetic Resonance (NMR)

Protons and neutrons, like electrons, have an intrinsic property called "spin," that is, they can be imagined to be rotating around their own axes like tiny planets. The nuclear shell model introduced in Section 16.1 is centered around this property. **Nuclear spin** was first suggested in 1924 by W. Pauli, famous for his principles of atomic structure (see Chapter 4), when new atomic spectroscopy experiments under very high resolution showed "hyperfine structure." This was a further splitting of atomic lines into "multiplets" with much finer spacings than the "fine structure" attributed to the electron's spin–orbit magnetic interaction discussed in Chapter 4. In the simplest case of the hydrogen atom, the proton spin angular momentum, denoted by a capital I, is equal to ½ (in units of $h/2\pi$), and it may only point "up" or "down," with projection quantum number $m_I = +½$ or $-½$. Nucleon spin, like electron spin, produces a **magnetic moment;** that is, it makes each nucleon a tiny magnet with north and south poles; for the proton, this means it can only point its north pole "up" or "down" with respect to the magnetic field set up by the electron's spin. These two ways of "coupling" the nuclear and electron spins have very slightly different energies (by a small fraction of an inverse centimeter), and cause even the $1s$ ground state of hydrogen to "split" into two very closely spaced levels that contribute to the hyperfine multiplet structure observed. Later in the 1930s, when the Raman rotational spectra of homonuclear molecules like H_2 and N_2 were measured, the rotational line intensities were found to alternate in a way that could be explained by spatial degeneracy of nuclear spins.

The spin of the neutron is also ½, and, surprisingly, it also has a magnetic moment. Unlike the electron and the proton, which are spinning charges and therefore tiny electromagnets, the neutron has no net charge. Moreover, while the proton's north pole points *along* the direction of its spin angular momentum, as expected for a rotating *positive* charge, the neutron's north pole is *opposed* to its spin, like the electron's. These facts imply that there exists a charge distribution *within* the neutron, a diffuse shell of negative charge surrounding a more compact positive shell, all spinning together while exactly canceling each other's charge. This is consistent with the exchange of a negatively charged pion between n and p; the π^- can be thought to reside in the "outer" region of the bare neutron. While Dirac's relativistic theory correctly predicts the magnetic moment of the electron, it fails for both the proton and neutron, whose moments are unequal in magnitude and larger than predicted. They are nonetheless much smaller than the electron's moment, due to the larger mass of the nucleons and hence lower rotational velocity for a given angular momentum. These peculiarities may one day be understood within the quark model, but for now we must simply accept them, because they enable us to interpret the experiments. In the spirit of J. J. Thomson, we will take the nuclear spin model to be "a matter of policy rather than a creed."

The spin and magnetic moment of a nucleus arise from the combined spin and angular momentum of rotation of an unpaired nucleon about the nuclear center of mass; this is determined by the details of the nuclear shell model, and will not be discussed here. As a rule, a nucleus will have a nonzero spin if either Z or N is odd. As mentioned in Section 16.1, magnetic effects are much more important *within* the nucleus than they are within the atom. In the deuteron, our simplest nuclear molecule, the spins of p and n are *unpaired*, giving a total nuclear spin $I = 1$, since this allows their magnetic poles to align in the attractive north-to-south way. Table 16.3 lists the spins and magnetic moments of particles and of the nuclei of some common isotopes.

When placed in an *external* magnetic field, any of the nuclei with nonzero spin will align their spins in some definite way with the field; for spin-½ nuclei such as ^1_1H, the possible alignments are "up" or "down," just as for the electron in the Stern–Gerlach experiment. Here, however, a more powerful magnet is needed, because the nuclear magnetic moments are so small. If "spin up" corresponds to the north pole's being along the direction of the field, that will be higher in energy, and spin down will be lower, creating an *energy splitting* between spin-up and spin-down, just as in the spin–orbit effect (Chapter 4). For a proton in a 10,000-Gauss (1-Tesla, T) external field, this splitting is only 0.0014 cm^{-1} or 42.6 MHz, in the radio-frequency (RF) range. If, however, we now apply a RF field (a "radio signal"), by means of a transmitter coil surrounding the sample, that has an energy $h\nu$ equal to the energy splitting, we can drive transitions from the spin-down to the spin-up state, according to the Bohr postulate. This absorbs energy from the RF coil, and the resulting decrease in RF intensity is picked up inductively by a separate receiver coil at right angles to the RF coil; after amplification we have a signal

TABLE 16.3

Nuclear spins I and magnetic moments μ for selected particles and isotopes

Species	I	$\mu/\mu_N{}^a$	Species	I	$\mu/\mu_N{}^a$
$_{-1}^{0}\text{e}$	1/2	-1836	$_{7}^{14}\text{N}$	1	0.404
$_{1}^{1}\text{p}$	1/2	2.793	$_{8}^{16}\text{O}$	0	0
$_{0}^{1}\text{n}$	1/2	-1.913	$_{8}^{17}\text{O}$	5/2	-1.894
$_{1}^{2}\text{H}$	1	0.854	$_{9}^{19}\text{F}$	1/2	2.629
$_{1}^{3}\text{H}$	1/2	2.979	$_{11}^{23}\text{Na}$	3/2	2.218
$_{2}^{3}\text{He}$	1/2	-2.128	$_{15}^{31}\text{P}$	1/2	1.132
$_{2}^{4}\text{He}$	0	0	$_{16}^{32}\text{S}$	0	0
$_{6}^{12}\text{C}$	0	0	$_{16}^{33}\text{S}$	3/2	0.643
$_{6}^{13}\text{C}$	1/2	0.702	$_{19}^{39}\text{K}$	3/2	0.391

[a] In units of the **nuclear magneton** $\mu_N = eh/(4\pi m_p c)$ in cgs-esu; $\mu_N = 5.05 \times 10^{-24}$ erg G^{-1} (5.05×10^{-27} J T^{-1}). In a magnetic field of strength B, the energy splitting between adjacent magnetic levels is $(\mu/I)B$ or $(\mu/I\mu_N)\mu_N B$. For a field $B = 1$ T (10 kG), this yields a splitting of 5.05×10^{-27} J $\times \mu/I\mu_N$; for a proton the corresponding transition energy is 2.82×10^{-26} J, 1.42×10^{-3} cm^{-1}, 42.6 MHz frequency. This last is also the "frequency of precession" of the spin vector about the field direction, known as the **Larmor frequency.**

that magnetic nuclei are present in the sample. The experimental setup is illustrated in Figure 16.12. This is called **nuclear magnetic resonance** (NMR), or simply magnetic resonance, omitting the fearful "nuclear" adjective. The effect was discovered by F. Bloch and E. Purcell in 1945. A high-sensitivity refinement of this technique, called Fourier transform NMR, uses a *pulsed* RF field; nearly all commercial NMR spectrometers are now of the Fourier transform type. More expensive NMR instruments have more powerful magnets, making the splittings larger, and thereby easier to detect and resolve; they are rated by their proton Larmor frequency, as defined in the footnote to Table 16.3.

Since each nucleus in Table 16.3 has a different magnetic moment, a given magnetic field will cause each type of nucleus to resonate at a different frequency, and we can obtain an assignable spectrum of the magnetic nuclei present in the sample without disturbing it by sweeping the radio frequency (like tuning a radio to different stations). Figure 16.13 is a NMR spectrum of a glass test tube containing water and some copper/aluminum alloy shot. Even more diagnostic, if the nuclei—most commonly H atoms—are present within a molecule such as ethanol (CH_3CH_2OH), they will each see a slightly smaller "local" field than that applied, owing to small "eddy currents" in the electron density set in motion by the field. These eddy currents in turn produce tiny fields that oppose the applied field. The magnitude of the eddy currents depends on how large the electron density is near each nucleus, a property that varies with the local chemical environment. Chemically distinct protons thus will experience slightly different fields, producing what are called **chemical shifts** in their resonance frequencies. As illustrated in Figure 16.14, ethanol has three peaks in its proton NMR spectrum, corresponding to its three "kinds" of hydrogen. Under higher

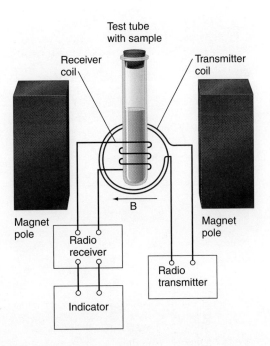

Figure 16.12
Experimental setup for nuclear magnetic resonance measurements. Either the magnetic field or the radio frequency may be varied to generate a spectrum. Modern NMR spectrometers employ electromagnets wound with superconducting wire that generate fields of 10 T (10^5 G) or more, and pulsed RF fields. After a RF pulse, the nuclear spins in the sample relax, and the receiver coil signal is the Fourier transform in time of the spectrum. The spectrum is recovered by a computer-generated inverse Fourier transform.

resolution, two of these peaks split into multiplets; we leave the explanation of *that* for you to learn when you study NMR spectroscopy of organic compounds.

The human body is 80% water, and hence is loaded with protons. If you are placed in a powerful magnetic field, all of these protons will split their energies, and you could be made to yield a huge NMR signal. If the field varies in a known way, having a slightly different value at each point in your body, your protons will "give away" their location by the calculable shift in their resonance frequencies, and you can be "mapped." This is called **magnetic resonance imaging** (MRI), and is made possible by the speed and memory of modern computers combined with powerful but readily tuned magnets (accomplished with a big magnet and an array of smaller electromagnets that are independently adjustable). This imaging method is noninvasive, and you don't have to ingest a radioactive substance or absorb high-energy radiation.

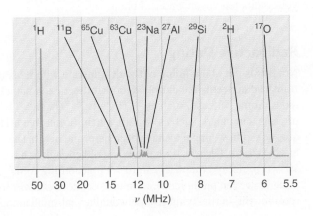

Figure 16.13
NMR spectrum ($B = 1$ T) of a glass test tube containing water and Cu/Al alloy shot. The glass contains both Na and B in addition to Si and O, and the small amount of D in water is readily picked up.

Figure 16.14
NMR spectra of ethanol, CH₃CH₂OH, where δ represents the chemical shift in parts-per-million (ppm), referred to tetramethyl silane Si(CH₃)₄ at frequency ν_0. The definition of δ makes it independent of magnetic field strength, and allows spectra from different instruments to be compared, as shown. **(a)** "Low resolution" spectrum—still very high resolution compared to the spectrum in Figure 16.13—in which the frequencies of the transitions for the three kinds of protons are resolved to about 5 parts in 10^7. The areas under the peaks are in the ratio 1:2:3, that is, they count the number of protons of each type. **(b)** High-resolution spectrum, frequencies resolved to about 5 parts in 10^8, shows splittings within each proton peak, due to magnetic couplings to the spins of protons bound to neighboring atoms.

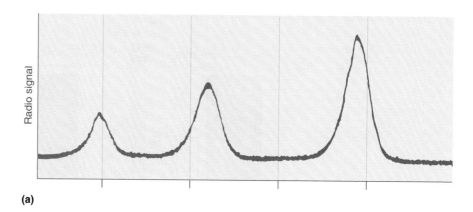

(a)

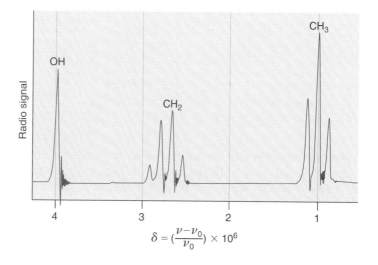

$$\delta = \left(\frac{\nu - \nu_0}{\nu_0}\right) \times 10^6$$

(b)

Recall that atoms and molecules are held together by electrical rather than magnetic forces; your body cannot "feel" even an extremely powerful magnet. Further, the photons in a RF field have minuscule energies, and the fields used are very weak, only a little stronger than the signals that freely come into your room and are picked up by your radio, television, and cell phone.

Radioactive Dating

No, this is not about going out with someone who lives near a nuclear power plant. It instead concerns the curious fact that there exist naturally occurring unstable isotopes. Some of these, such as ^{238}U, ^{235}U, ^{232}Th, and ^{40}K, have half-lives of billions of years, but others, such as ^{14}C, are much shorter lived. The long-lived ones raise the question of how long they have been here, and in turn shed light on how long the earth has been here. Clearly, the earth is not old enough to have allowed ^{238}U, $t_{\frac{1}{2}} = 4.46 \times 10^9$ yr, to decay completely into ^{206}Pb according to the scheme depicted in Figure 16.8. However, pitchblende, uranium ore, does contain some lead, and analysis of a rock you might find lying around would reveal small amounts of both ^{238}U and ^{206}Pb.

In particular, it is found that Pb and U are finely mixed in natural U-containing rocks, a good indication that the Pb came from the U, and was not simply there the whole time. We can use the proportion of the two isotopes, found by mass spectrometry, and the independently measurable $t_{1/2}$ for ^{238}U decay, to deduce how long the U has been decaying within that rock, and hence the age of the rock, in the following way.

Let's couch the problem in general terms. Suppose a parent radioisotope P decays to a stable daughter D, P $\longrightarrow$ D, with a certain half-life, or first-order decay constant k. Then at time t we can find the amount of D, because each P that decays yields a D:

$$N_P = N_{P0}\, e^{-kt} \tag{16.31}$$

$$N_D = N_{D0} + (N_{P0} - N_P) = N_{D0} + N_P(e^{kt} - 1)$$

where N_{D0} is the initial number of D's, those not resulting from decay. From the second equation we find

$$\frac{N_D - N_{D0}}{N_P} = e^{kt} - 1 \quad \text{or} \quad t = \frac{1}{k}\ln\left(1 + \frac{N_D - N_{D0}}{N_P}\right)$$

If a rock shows 50% decay, $(N_D - N_{D0})/N_P = 1$, then $t = (\ln 2)/k = t_{1/2}$; smaller fractional decay corresponds to a younger rock. The initial daughter abundance is determined by comparing isotopic abundances in a reference daughter sample not originating from decay, in this case a sample of lead ore, galena, with the sample under study. An isotope in natural lead, ^{204}Pb is not formed by decay, and allows an estimate of initial ^{206}Pb, $N_{D0} = N_{204}(N_{206}/N_{204})_{\text{galena}}$. In general the correction is found to be small, as shown in the case of a uranium sample by negligible amounts of ^{204}Pb, and we can set $N_{D0} = 0$.

The oldest known rocks on earth dated by this method are 3.7 billion years old, setting a lower bound for the age of the "cooled" earth; moon rocks brought back by the Apollo astronauts are 4.5 billion years, generally thought to be close to the earth's age, $4.57 \pm 0.02 \times 10^9$ yr, found by dating meteorites. To reduce systematic error, ^{238}U, ^{235}U, and ^{232}Th are determined for the same sample, since these produce distinct isotopes of Pb. Other isotopes and their daughters that can be used for this purpose are ^{40}K/^{40}Ar and ^{87}Rb/^{87}Sr; ^{40}K EC decay is thought to be responsible for the entire amount of ^{40}Ar in our atmosphere, since ^{40}Ar is not likely to have been formed by stellar nucleosynthesis. Uranium is exceptional in having two methuselaic isotopes, with ^{235}U being shorter lived; extrapolating back to the earth's beginning, we can deduce that the ratio of abundances ^{238}U/^{235}U, now 138, was once closer to 1. For an interesting consequence of this, see the next section.

The isotopes with half-lives close to the age of the earth are not useful for dating relatively young objects. However, cosmic rays entering the earth's atmosphere continuously produce radioactive ^{14}C, with $t_{1/2} = 5715$ yr, by reaction of secondary slow neutrons with ^{14}N, a reaction first studied by Fermi in the early 1930s. Cosmic rays have been roughly constant in intensity for at least 10^6 yr, and an equilibrium has long been established between ^{14}C production and decay. Based on the steady-state abundance of ^{14}C, W. Libby in the early 1950s developed **radiocarbon dating** of more recent (<40,000 years old) remains of once-living beings. Living organisms tend to acquire the same isotopic abundances as their environs while alive. When an organism dies, it stops exchanging C with its surroundings, and the ^{14}C decays. Fossil fuels, the remains of plants and animals dead for 10^6 yr or longer, show no ^{14}C activity.

Comparison tests on wooden objects and other artifacts of historically verifiable ages show excellent agreement, and the method is an essential tool of archaeologists and paleontologists. Recently radiocarbon dating was used to show that the Shroud of Turin, believed by some to have been the burial cloth of Christ, was made from cloth and pigments of the mid-14th century, coincidentally the time of its first public display. Future investigators, however, will have difficulty using ^{14}C to date 20th-century denizens, owing to our extensive burning of fossil fuels, rapidly adding "dead" carbon to the atmosphere, and our atmospheric testing of nuclear weapons between 1945 and 1963, which has produced large amounts of ^{14}C from the same neutron reaction that cosmic rays engender.

EXAMPLE 16.3

The *haniwa* are ancient Japanese wooden statues that served as temple guardians. Analysis of a statue sample in 1983 revealed a ^{14}C specific activity of 14.1 decays m^{-1} g^{-1}, while the ^{14}C standard decay rate (determined from modern but prenuclear samples) is 15.3 decays m^{-1} g^{-1}. Estimate the age of the *haniwa*.

Solution:
Assuming that the ^{14}C activity would have been equal to the standard value when the statues were carved, and using the proportionality between decay rate $-dN/dt$ and ^{14}C population N, this becomes a backward extrapolation of the decay, allowing us to find t_0, the time of their creation:

$$t - t_0 = -(t_{1/2}/\ln 2) \ln(N/N_0) = -(t_{1/2}/\ln 2) \ln[(dN/dt)/(dN/dt)_0]$$
$$= -(5715/0.693) \ln(14.1/15.3) = 670 \text{ yr}$$

If $t = 1983$ A.D., then $t_0 = 1983 - 670 = 1310$ A.D. Normally, decay counts are accumulated over many hours to provide better statistics, that is, lower random error. Recently high-sensitivity mass spectrometry has extended the range of this method to samples more than 40,000 years old.

Nuclear Energy Sources

It is ironic—and more than a little frightening—that we first used our newfound ability to release nuclear energy rapidly in the fission process to make weapons of mass destruction. To some extent that was an accident of our fractious history in the 20th century, both the liveliest and the deadliest century we have ever known. Before we close out this chapter with an examination of "the bomb," let us relate a little of the glorious hope of a limitless energy supply that our tapping of the nuclear energy reservoir has inspired.

At the end of the second world war in 1945, whose end was hastened by the use of nuclear weapons, scientists around the world united in a clarion call to turn nuclear energy to peaceful purposes. The leader in this effort was Niels Bohr, who recognized that only in what he called an "open world" could the nuclear nightmare be turned into a dream for the future. A series of "Atoms for Peace" conferences was held, and out of these grew the idea that nuclear energy might serve as the next great source of energy for society, in the face of finite and dwindling fossil fuel reserves. Britain and the United States were the first to bring commercial nuclear power reactors online in the late 1950s, using energy released by the controlled fission chain reaction of ^{235}U in place of combustion to drive a steam turbine. The principles behind controlled fission combine the naturally "delayed" neutron emission from fission products with carefully chosen fuel density to create a situation where only one of the neutrons from each fissioning ^{235}U produces another fission. A reactor fulfilling the one-neutron requirement is said to be **critical,** and the chain reaction will not "run away" and produce an explosion. It is impossible for a modern nuclear reactor to undergo a nuclear explosion like a bomb, but, as described below, insufficient attention to safety in design and operation can produce lesser disasters.

An early controlled reactor, constructed at Los Alamos, New Mexico, in 1944, was simply a stainless-steel sphere 30 cm in diameter containing a 1.2 M uranyl sulfate UO_2SO_4 aqueous solution, isotopically enriched to 15% ^{235}U. In the reactor, the solvent water served as a neutron **moderator,** slowing down the neutrons. This serves to inhibit their reaction with ^{238}U, which prefers fast neutrons for pickup due to a resonance, as well as enhance it with ^{235}U, which fissions more efficiently with slow neutrons. Modern fission reactors, like the one illustrated in Figure 16.15, use one of three available fissile materials, ^{235}U, ^{233}U ("bred" from ^{232}Th by reacting it with neutrons), or ^{239}Pu (bred from ^{238}U with neutrons), in the form of oxide pellets. The pellets are packed into long hollow rods of stainless steel or zirconium alloy; in operation these **fuel rods** are immersed in an array in the moderator, usually H_2O or D_2O. **Control rods,** made of neutron-absorbing boron or cadmium, can be inserted into the fuel rod assembly to control the neutron flux. The majority of these reactors, known as PWRs, use pressurized water as both moderator and coolant (heat energy carrier). A typical commercial reactor generates about 1000 MW (10^9 J s^{-1})

Figure 16.15
Nuclear fission pressurized water power reactor (PWR) (United States). The right-hand panel is a schematic of PWR reactor design, while the photograph shows the containment building (the small domed building on the right) and the water cooling tower. The size scale may be gauged by noting the lamp posts and traffic signals in the foreground.

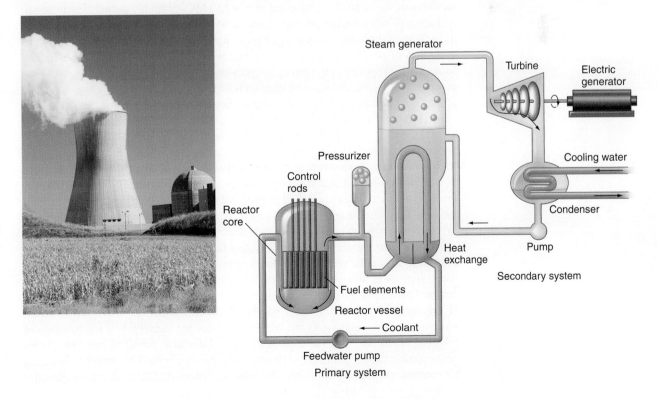

of electrical power continuously, consuming in the process about 3 kg of fuel per day—an amount you could hold in your hand—and an equal mass of radioactive fission products. By contrast, about 10,000 *tons* (10^7 kg) of coal would have to be burned per day to yield this much energy, producing a still-larger mass of air pollutants and greenhouse gases. As of the turn of the century 435 fission reactors were in operation worldwide.

Safe operation of a reactor involves mainly backup cooling systems in the event of a failure of primary coolant flow or pressure; failing that, the reactor must be housed in an impenetrable containment shell that, in addition to being able to withstand a "meltdown" of the reactor core and the extreme pressures that might be reached, must be proof against natural disasters, especially earthquakes. The moderator/coolant is recirculated to avoid releasing dissolved traces of radioactivity into the environment. In a 1979 loss-of-coolant emergency at a PWR plant at Three Mile Island near Harrisburg, Pennsylvania, the emergency cooling system functioned properly, but was prematurely interrupted manually, causing a partial meltdown of the reactor core. The containment building was not breached, but a pressure relief valve in the emergency system opened automatically, venting a small amount of radioactive material, mainly volatile ^{131}I, into the atmosphere. Radiogenic effects on neighboring populations were negligible, with exposures estimated at 100 millirem in the worst cases, less than the average annual dose. A similar loss of coolant at a graphite-moderated reactor in Chernobyl, Ukraine, in 1986, due in this instance to deliberate disabling of safety interlocks during an "experiment," led to a disaster, since the reactor had not been housed in a containment building. Overheating caused steam to form in the reactor core; this reacted with the fuel and moderator rods, producing hydrogen gas, which then exploded—a chemical event, not a nuclear one. Nonetheless, the force of the explosion dispersed nearly 180 MCi of highly radioactive material, including primary fission products, over a wide area, producing high levels of radioactivity in the water and soil in a heavily agricultural region, and exposing tens of thousands of people living along prevailing wind patterns to >25 rem of radiation. More than 200 people, all plant employees or members of cleanup crews, were exposed to >100 rem, and showed symptoms of radiation sickness. The explosion also ignited the graphite moderator, creating a reactor fire that was very difficult to extinguish. Firemen, sent in to put out the fire by air and on the ground when robotic devices failed, died of radiation poisoning shortly thereafter, as did a filmmaker documenting the aftermath; 29 deaths were recorded as a direct result of the mishap. Radioactive fallout, formerly associated only with nuclear bombs, spread over much of eastern Europe and Scandinavia in the month following the explosion. The reactor itself has since been entombed in steel and concrete, a containment building built too late. The difference in outcome of the Three Mile Island and Chernobyl events evidently owes much to differences in reactor engineering.

Outside the former Soviet Union safety has been a more important factor in reactor design (see the containment building in Figure 16.15); disposal of radioactive waste appears to be a more critical safety issue. Encasing waste in solid glass cylinders and providing for deep burial in geologically stable sites seems the only choice at present; finding suitable sites has been difficult, however, owing to stringent longevity requirements set by the long half-lives of the transuranic waste products of neutron capture by ^{238}U. These isotopes are still useful as power sources, but chemical reprocessing of spent fuel to recover them has not been done extensively owing to international security concerns. The most hazardous fission products, prominently

^{90}Sr and ^{137}Cs, have half-lives of 30 yr or less, and require much shorter storage times for decay to "safe" levels of radioactivity.

Nature has apparently beaten us to the building of a reactor by over a billion years! In 1972 French scientists discovered that some of the uranium ore from a large mine at Oklo in Gabon, West Africa, was depleted in ^{235}U by up to 70%; analysis of these ore samples revealed the now-stable remains of the primary fission products of ^{235}U. Five separate "natural reactors" seem to have been in operation simultaneously, in the period from 1.8 to 2.0×10^9 yr ago, for a duration of about a million years. This was feasible then because the abundance of ^{235}U was higher, near the "critical" 3%, due to its shorter lifetime relative to ^{238}U; the "reactors" were under water, which served as a moderator as in modern man-made facilities. Equally amazing, there was no sign that the fission products, except the highly soluble and short-lived Sr and Cs, had migrated away from their point of origin in the intervening 2 billion years, an observation that may have implications for nuclear waste storage.

Research continues at U.S. national laboratories and other labs around the world to improve the efficiency and safety of fission reactors and to reduce the amount of waste they generate. At the same time, turning nuclear fusion—the energy source of the stars—into a practical source of energy has also been actively pursued. The combination of greater fuel availability, lower production of radioactivity, and higher energy release per unit mass of fuel, makes fusion an appealing alternative to fission. The main impediment to rendering fusion practical is very simply identified: it is the Coulomb barrier V_c. A thermal fusion reactor must operate with densities and temperatures characteristic of stellar interiors, in order to overcome V_c. Attention has naturally focused on highly energy-releasing reactions for which V_c is as small as possible, such as D,D fusion:

$$^2_1H + ^2_1H \longrightarrow ^3_2He + ^1_0n \atop \longrightarrow ^3_1H + ^1_1H \qquad (16.32)$$

Since no container can withstand the required temperatures, the hot deuterium plasma must be trapped away from its walls. The best results so far have been achieved using a toroidal magnetic field as a plasma trap in a doughnut-shaped reactor called a *tokamak*. The plasma is created and heated by high-energy ion beams. Another major effort to attain practical fusion has employed *laser implosion,* shining ultrahigh-power pulsed lasers on pellets of fusion fuel. After 40 years and great expenditure of ingenuity, effort, and funds, however, a practical fusion reactor is still a dream. Unfortunately, we have succeeded in building a fusion bomb, although it has never yet been used as a weapon.

The Genie Is Out of the Bottle

In his widely acclaimed 1980 television series *Cosmos,* the late Carl Sagan viewed us and our planet Earth from an extraterrestrial perspective, ours being one civilization among many in our Milky Way galaxy. Like other civilizations, we had discovered the secrets of the nucleus; the question was whether we were wise enough to channel its enormous power to build our society rather than destroy it. The evidence then available suggested we would not survive our own inventiveness. With the disintegration of one of the "superpowers" threatening the others and the world with nuclear weapons, this danger has abated, although it has by no means disappeared.

Unfortunately it is considerably easier to create explosions with nuclear fission or fusion than it is to release the energy slowly. This happenstance has a chemical analogue: explosively combining H_2 and O_2 preceeded practical fuel cells by two centuries.

The discovery of nuclear fission described earlier happened to coincide with the rise of Nazism and Fascism in the mid- to late 1930s. When war in Europe seemed inevitable, Albert Einstein, who had already emigrated to the United States from Nazi Germany, wrote to President Franklin Roosevelt warning that fission could produce a bomb of enormous power, and that the Nazis would be actively engaged in building one. As a result, the code-named Manhattan Project was begun, and the first sustained ^{235}U chain reaction was created by a team at the University of Chicago headed by Enrico Fermi late in 1942. Huge plants were then built to produce U enriched in ^{235}U and to breed ^{239}Pu from ^{238}U by neutron bombardment. Meanwhile, World War II had engulfed Eurasia, Northern Africa, and the Pacific. On July 16, 1945, a ^{235}U fission bomb, built by a team headed by J. Robert Oppenheimer, was exploded over the desert sands at Alamogordo, New Mexico; three weeks later two bombs were dropped, one U and one Pu, on the cities of Hiroshima and Nagasaki, Japan, demolishing their centers and killing nearly 140,000 people, but bringing the last great war quickly to a close.

The war ended, unfortunately, with the replacement of one set of bellicose adversaries by another, and a nuclear arms race ensued, resulting in the development of a fusion bomb, the so-called hydrogen bomb, with a destructive power a thousand times that of the U bomb, by 1952. The energy necessary to ignite the fusion reaction was provided by employing a small fission bomb as a kind of match. Mainly as a result of spying, the major adversaries kept pace with each other, and bomb-building spread to several nations. Atmospheric testing of nuclear weapons continued until the International Nuclear Test Ban Treaty was signed in 1963, limiting nuclear tests to the underground. Linus Pauling's efforts toward banning nuclear weapons testing, which abetted the treaty, earned him the Nobel Peace Prize in 1962. China and France refused to sign, however, and have carried out atmospheric tests sporadically since then. In 1996 many nations signed a Comprehensive Test Ban Treaty banning all nuclear weapons testing; nonsignatory nations, however, have continued underground tests, most recently India and Pakistan in 1998.

Was Sagan right? Fortunately, the jury is still out. In a persistent but still fragile world, you, dear reader, will have a chance to influence the verdict.

SUMMARY

The **nucleus** of an atom is a tiny ball a few femtometers in diameter containing Z protons and N neutrons; Z is its **atomic number** and elemental identity, and $Z + N = A$, A its **mass number**. Protons and neutrons are collectively designated **nucleons**. **Isotopes** of a given element $^{A}_{Z}X$ have the same Z but different A, and hence different N. The mystery of how a nucleus with $Z > 1$ can exist is resolved by postulating the **strong force**, which acts with equal strength between any two nucleons, and overwhelms the Coulomb repulsion. The range of the strong force is very short, as reflected in the small radius of the nucleus, $r_{nuc} = 1.33$ fm $A^{1/3}$, and its strength is so large that nucleon binding (bonding) results in a measurable loss of mass $\Delta m = \Delta E/c^2 = \Delta E/(931.5$ MeV/amu$)$. Here $-\Delta E$ is the **binding energy** (BE) of the nucleus, analogous to the sum of the bond energies of a molecule. The nuclear binding energy of a particular isotope specified by Z and A (or N) can be calculated by subtracting the mass of the isotope—including its electrons—from the sum of its constituent particle masses: BE $= -\Delta E = [Z(m_p + m_e) + (A - Z)m_n - M(A,Z)]c^2$. Assembling a nucleus from nucleons or smaller nuclei is called **fusion;** fusion of smaller nuclei each containing protons is inhibited by Coulomb repulsion at large distance. The Coulomb repulsion is overcome at "touching" distance by attraction due to the strong force, forming a **Coulomb barrier.** The radius of a nucleus is about 10^{-5} that of an atom, while its binding energy is about 10^6 times that of a molecule.

The most stable nucleus is ^{56}Fe; smaller nuclei are less stable because each nucleon has fewer nearest neighbors to bond with, while larger are less stable due to residual interproton repulsion. Protons and neutrons each have a **spin** of ½, and must obey the Pauli principle when they form a nucleus; the consequence is a **nuclear shell structure,** with "**magic numbers**" Z and N corresponding to noble-gas-like closed nuclear shells. These numbers are 2, 8, 20, 28, 50, 82, 126, and 184, and nuclei with N or Z equal to one of these have especially large binding energies. Nuclei with N and/or Z even are also more stable.

A given element has only one or a small number of stable isotopes; the rest, called **radioisotopes** or radionuclides, are **radioactive,** and decay by radiating subatomic particles, a process that continues until a stable nucleus is reached. Decay is spontaneous only when $\Delta m < 0$ for the decay process. The nuclei of stable isotopes show $N/Z \approx 1$ for light elements; N/Z gradually increases beyond $Z = 20$, reaching 1.6 for ^{238}U. The stable isotopes lie along a narrow **stability band** in a plot of Z versus N, a **stability diagram.** If N is too large—a neutron-rich nucleus—decay will produce a β **particle,** a high-energy electron, thereby converting a neutron into a proton, whereas if N is too small—a neutron-poor nucleus—a β^+ **particle,** a high-energy positron, will be emitted, or alternatively, an orbital electron will be captured, EC, either path converting a proton to a neutron. For $Z > 83$, there are no known stable isotopes, with most of the heavy elements decaying by emitting an α **particle,** a high-energy He nucleus. Each decay of a parent radionuclide produces a **daughter nucleus,** which itself may or may not be stable against further decay. The daughter nucleus may be formed in an excited state, a more likely event for heavier daughters; if so, it will emit one or more γ**-rays,** photons of very short wavelength. In the decay process, as in all other nuclear transformations, the **baryon number** (mass number) and charge are conserved.

The mere fact that a nucleus can spontaneously decay does not tell us the rate of decay. Nuclear decay is intrinsically a first-order, uninuclear process, whose rate, or **activity,** is given by $-dN/dt = kN$, or $N = N_0 e^{-kt}$ in integrated form, where N is the number of radioactive nuclei, and k is the first-order rate constant. The **specific activity** is defined to be the activity per gram of radioisotope. Radioisotope decay rates are usually tabulated as **half-lives** $t_{1/2} = \ln 2/k$, and are readily measured with a Geiger counter. It is found empirically that $t_{1/2}$ is shorter the further the isotope lies from the stability band, and that for α-decay, $t_{1/2}$ is inversely related to ΔE for the decay of isotopes of a given element.

The Gamov "volcano" model posits that α-decay takes place by **tunneling** through the Coulomb barrier, and quantitatively accounts for the correlation between $t_{1/2}$ and ΔE. Decay of the overweight elements takes place by a multistep process, and the elements with Z between 83 and 92 exist only as decay intermediates in the U $\longrightarrow$ Pb decay pathway. If the U lifetimes were shorter, none of the overweight elements would occur naturally.

In the process of producing the energy by which they shine, the sun and its fellow stars are constantly fusing protons into heavier nuclei. This takes place in stages, because fusing heavier nuclei requires higher temperature to overcome the increasing Coulomb barriers. The early stages are called **hydrogen burning,** wherein four protons fuse into ^{4}He, and **helium burning,** 3 ^{4}He $\longrightarrow$ ^{12}C. The second-row elements are then formed in cycles that begin with p + ^{12}C. In giant stars fusion proceeds further, reaching the most stable nucleus ^{56}Fe. Thereafter fusion becomes endothermic, and the larger stars run down, collapse, and explode in **supernovae,** catapulting the elements they have made in their lifetime, along with trans-iron elements formed in the violence of the explosion, into interstellar space. Our solar system and we ourselves are debris from such an explosion, the ashes of a burnt star.

The first nuclear transmutation induced in the laboratory was the nuclear reaction $\alpha + {}^{14}\text{N} \longrightarrow {}^{17}\text{O} + \text{p}$. Nuclear reactions are analogous to bimolecular elementary chemical reactions, obeying second-order rate laws with rate constants based on collision rates. **Cross sections** for nuclear reactions are measured in **barns** (b), 10^{-24} cm^2, about 10^8 times smaller than chemical reaction cross sections. Nuclear cross sections σ may be estimated from the **line-of-centers model,** $\sigma = \pi(r_1 + r_2)^2(1 - V_c/E)$, $E > V_c$, where the r_i are nuclear radii, V_c is the height of the **Coulomb barrier** $Z_1 Z_2 e^2/(r_1 + r_2)$, and E is the energy of collision. As in nuclear decay, tunneling plays a role in determining the rate as well. Studies of such reactions led to the discovery of the neutron and **nuclear fission,** the opposite of fusion, in which heavy nuclei split into two large fragment nuclei plus two or three neutrons. Fission is often induced by adding a neutron to an already unstable nuclide, to enable it to overcome the enormous Coulomb barrier to form two large-Z fragments; the consequence can be a **nuclear chain reaction,** in which product neutrons induce further fissions. The most common fissile nuclei are ^{233}U, ^{235}U, and ^{239}U; only ^{235}U occurs naturally, and there is evidence that "natural fission chain reactors" may have existed on earth in the distant past. Using high-energy particle accelerators, we earthlings have produced **transuranic elements,** and we may now have reached element 115, on the edge of a plateau of stability predicted by the nuclear shell model.

Our nuclear knowledge has spawned a number of applications. In chemistry these include **isotopic substitution** and **labeling** in spectroscopic and kinetic studies, hot atom kinetics, and

Rutherford backscattering for noninvasive elemental analysis. Medical connections include both the effects of exposure to radiation and the use of radioisotopes in medical diagnosis and therapy. The energy of absorbed radiation is measured by the **rad**; 1 rad = 0.01 J/kg of body tissue. A measure more meaningful for health effects is the **rem**, approximately that amount of radiation that will produce 1 esu of ionization in 1 cm^3 of air at STP. For β- and γ-rays, 1 rad ≈ 1 rem, while for α-rays, 1 rad ≈ 10 rem, that is, the same energy dose causes greater tissue damage with α exposure. Harmful exposures are generally caused by ingestion of radioactive material. Humans receive more than 100 mrem yr^{-1} naturally from cosmic rays, ^{40}K within the body, and radiation from naturally occurring heavy isotopes, especially radon gas in brick or stone dwellings. Other activities, such as medical radiation and air flights, may double this exposure. Several hundred rem in a one-time exposure can cause radiation sickness and death. Among many medical uses of radioisotopes and nuclear radiation are tumor therapy and radiotomography, such as PET scanning.

Nuclei with odd Z or odd N have nonzero **magnetic moments** μ due to unpaired nucleon spin and rotational angular momenta. When placed in a strong magnetic field, such nuclei will generate a set of magnetic sublevels whose energy splittings are in the radio-frequency range. By employing radio transmitters and receivers, transitions between these levels can be stimulated and detected, giving rise to the analytical technique called **nuclear magnetic resonance** (NMR). **Magnetic resonance imaging** (MRI) of proton-containing beings employs a magnetic field that varies in a known way as a function of spatial location.

Naturally occurring radioisotopes can be used to date rocks, and in the case of cosmic-ray-generated ^{14}C, the age of once-living beings, by both the remnant radioactivity of the sample and the accumulated stable decay product. For parent ⟶ daughter decay, the ratio $N_D/N_P = e^{kt} - 1$, or $t = k^{-1} \ln(1 + N_D/N_P)$, where t is the time elapsed since the object was first formed. Up to five separate, independent dates can be derived for the same sample from very long-lived isotopes. The age of the earth and its solar system is found to be 4.5 billion years.

The enormous and rapid energy release of the fission chain reaction has been used to supply electrical power and to make nuclear weapons. Radioactive waste from nuclear fission power plants is difficult to dispose of. Fusion energy release has been demonstrated in a weapon, but thus far has not been harnessed in a controlled way at useful power levels.

EXERCISES

Nuclear structure and bonding

1. What is the density of nuclear material (amu fm^{-3}, g cm^{-3} and tons cm^{-3}) implied by Equation 16.1?

2. Taking the deuteron $^2_1$d as the simplest nuclear "molecule," assume the n–p potential curve of Figure 16.2 dictates its vibrational motion.

    a. Estimate its vibrational kinetic ZPE in MeV using the particle-in-a-box PiB ground-state energy of Appendix B, with a box length of $a = 1.5$ fm and an appropriate reduced mass. How does it compare with that shown in Figure 16.2? Why is it different? Based on PiB, do you expect bound vibrationally excited states of d to exist?

    b. Estimate the vibrational period τ_{vib} in zeptoseconds (zs = 10^{-21} s), as the time to vibrate to and fro once between the origin and 1.5 fm, using the known energetics of d shown in Figure 16.2.

3. Use the rigid-rotor model of Chapter 8 to describe the rotational energy levels of the deuteron, taking $r_e = 1.5$ fm and an appropriate reduced mass. Find the energy spacing between the $j=0$ and $j = 1$ levels in MeV. Do you expect d to possess bound rotationally excited states?

4. Compute the Coulomb potential energy (Coulomb barrier) in MeV between two protons at a distance $r = 1.5$ fm. Find a convenient computational formula for the Coulomb potential between two collections of nucleons containing Z_1 and Z_2 protons, respectively, of the form $V(r) = aZ_1Z_2/r$, where r is the distance between the collections in fm units and V is in MeV.

5. Find the binding energy per nucleon BE/A for ^{3}He. Based on BE/A for $^2_1$H (the deuteron), does your number suggest a linear or triangular arrangement of nucleons? Use spin and magnetic moment data from Table 16.3 to suggest that the two protons in ^{3}He have their spins paired.

6. Find the binding energy per nucleon BE/A for ^{4}He. If the nucleons are (on the average) arranged at the corners of a tetrahedron, how many bonds will there be? Is this number consistent with that expected based on the bond energy in $^2_1$d? If not, to what can the discrepancy be attributed?

7. Find the binding energy per nucleon BE/A for ^{12}C. If all nucleons could be arranged to lie within bonding distance of all others, how many bonds would there be? Use your result to compute an average nucleon–nucleon bond energy in ^{12}C, and compare it to the p–n bond energy in ^{2}H.

8. Compute mass changes Δm for the processes (a) $H^+ + e^- \longrightarrow H(1s)$ and (b) $H + H \longrightarrow H_2$, and show why we have been able to ignore them in chemistry.

9. In arguing for the existence of the *meson*, Yukawa used the energy–time uncertainty principle $\Delta E \Delta t \geq h/2\pi$ and the known finite range a of the strong force in the following way: Δt can be thought of as the duration of the force, during which a particle of energy $\Delta E = mc^2$ can be created (a "virtual" particle). This particle communicates the force from one nucleon to another at the speed of light c, determining the range a by $a = c\Delta t = hc/(2\pi\Delta E) = h/(2\pi mc)$. For $a = 1.5$ fm, what mass m (amu) is implied? (For the Coulomb force, the virtual particles are massless photons. What range a is implied?)

Nuclear stability and radioactivity

10. Write general nuclear equations like Equation 16.9 for β^+-decay, electron capture (EC), and α-decay of an isotope $^A_Z X$.

11. Predict possible decay modes for the following radioisotopes, and write balanced nuclear equations for them: ^{3}H, ^{11}C, ^{125}I, ^{228}Th. Resolve any ambiguity by checking against tables. Which of these radionuclides would be useful for PET scanning?

12. Write a balanced particle equation for the annihilation of an electron and a positron to produce two γ-rays. If the γ-rays each have the same wavelength λ, calculate the energies and wavelengths of the γ-rays produced. If the positron involved in the annihilation is produced by decay of ^{13}N, what assumption has been made in your calculation? (*Hint:* Write a nuclear equation and compute Δm for ^{13}N decay.)

13. Find Δm for the electron capture (EC) decay of ^{82}Sr, Equation 16.12. Would β^+-decay or α-decay be feasible for ^{82}Sr? (*Hint:* Write balanced decay equations for these modes, and find their Δm's.)

14. One of the two sub-uranic elements that is wholly man-made, the radioisotope ^{99}Tc (technetium-99) undergoes β-decay. Write a balanced equation for this decay, and show by computing its Δm that the decay is spontaneous. Technetium-99 is added in trace amounts to aluminum siding to retard corrosion. Why does this work?

15. The radioisotope ^{40}K is unique in dividing its decay between β emission, appropriate for a neutron-rich nucleus, and EC, normal for a neutron-poor nucleus. Write nuclear equations describing these two decays, and verify that they are both spontaneous by computing Δm for each. Is β^+-decay possible for ^{40}K?

16. If you examine the stability diagram of Figure 16.4, you will find many more unstable isotopes lying above the stability band than below it. Explain.

17. In discovering the first artificial transmutation, Rutherford used α particles from the decay of ^{214}Po. Compute the energy of the α's, assuming that they carry the entire energy change ΔE for the decay. (Actually, the energy ΔE becomes the *relative* kinetic energy between the α and the ^{210}Pb daughter, with the α's "laboratory" kinetic energy being $K_\alpha = [(1 + x)/(1 + x^2)]\Delta E$, $x = m_\alpha/m_{210}$.)

18. Is α-decay energetically possible for the only stable gold isotope, ^{197}Au? If so, speculate on why it is not observed.

19. In the β-decay of ^{60}Co, Equation 16.17 and Figure 16.5, two γ-rays with energies of 1.17 and 1.33 MeV are emitted. What are the wavelengths of these rays? If the total energy release ΔE is 2.82 MeV, what is the maximum energy the β particle may carry?

20. Try to fulfill the alchemists' dream. Starting with the unstable but long-lived lead isotope ^{205}Pb, devise a decay scheme using only α- and β-decay that will produce the only stable isotope of gold, ^{197}Au. State nuclear decay equations, and find Δm's for them to assess the feasibility of this scheme.

Kinetics of nuclear decay

21. Use Equations 16.18 and 16.19 to show that $-dN/dt = -(dN/dt)_0 e^{-kt}$, where $-dN/dt$ is the activity at time t, and $-(dN/dt)_0$ is the initial activity.

22. a. Use data in Table 16.2 to find the specific activity (decays $s^{-1} g^{-1}$) for ^{82}Sr.
    b. What mass of ^{82}Sr will yield an activity $-dN/dt$ of 100 μCi?
    c. ^{82}Sr is a precursor for use in PET scanning, which for a time was supplied by the former Soviet Union. Suppose a 100-μCi sample is shipped and arrives 5.0 days later. What is its activity on arrival?

23. Suppose a patient about to undergo a PET scan drinks a beverage containing 70. amol (attomole, 10^{-18} mole) ^{82}Rb. Using data from Table 16.2, find the initial activity in β^+-decays s^{-1}. How long will it take for the activity to decrease to 5000 decays s^{-1}, the level the patient ordinarily receives from decay of bodily ^{40}K?

24. Natural carbon shows a specific activity of 15.3 decays min^{-1} g^{-1}. Assuming this is due to the β-decay of ^{14}C (Equation 16.8), find the mole fraction of ^{14}C in natural carbon.

25. Find the radiation dose in rem yr^{-1} resulting from the human body's natural ^{40}K content using data from Table 16.2. Assume only β-decay yielding β's of the maximum energy possible, a body mass of 50. kg, 0.35% by mass of K in the body, a natural abundance of ^{40}K of 0.012% and 1 rem = 1 rad.

26. In minerals from the Pre-Cambrian Era, the ratio of ^{206}Pb to ^{238}U is 1.0 g Pb per 3.5 g U. Estimate the age of these minerals.

27. The longest lived isotope of technetium, ^{98}Tc, has a half-life of 4.2 million years against β-decay. No trace of Tc has ever been found in the earth's crust. This implies, given the sensitivity of present analytical instrumentation, that less than 10^{-12} mol fraction ^{98}Tc remains from its original proportion in the solar system based on the abundances of neighboring odd-Z elements (see Figure 16.9). What limit does this place on the age of the solar system?

28. The present abundance of the minor isotope ^{235}U in natural uranium, a mixture of ^{235}U and ^{238}U, is 0.72%. Because the half-life of ^{235}U is shorter, it has been decaying faster than ^{238}U, and its abundance has therefore been higher in the past. Using data from Table 16.2, estimate how long ago the ^{235}U abundance was 3.0%, the amount necessary to sustain a fission chain reaction.

29. Find Δm's for the α-decay of the 232, 230, and 228 isotopes of thorium (Th), and use the Gamov "volcano" model to arrange them in order of increasing half-life, shortest first. Look up the actual half-lives in a table to verify your order. Compute the height of the Coulomb barrier from Equations 16.1 and 16.23, and sketch a potential energy diagram like that of Figure 16.7 to illustrate the model.

30. The qualitative idea behind the Gamov model for α-decay can be extended to β^+-decay. Compute the Δm's for β^+-decay of ^{9}C, ^{10}C, and ^{11}C, and use them to arrange these isotopes in order of increasing half-life, shortest first.

31. Despite its short half-life, the polonium isotope ^{210}Po could be isolated by the Curies because of its continual production in the decay chain of ^{238}U.

   a. Assuming the decay chain leading from ^{238}U to ^{210}Po involves only α and β emission, how many of each particle are given off in the decay sequence?

   b. Using data from Table 16.2, determine what fraction of a sample of pure ^{210}Po isolated by the Curies in 1898 had turned to lead by the year 1903, when they shared the Nobel Prize in physics with Becquerel.

   c. If the sample of Po of part (b) weighed 200. mg, and was stored in a sealed glass ampoule of volume 100. mL, what pressure in atm of He would be present in the ampoule in 1903, assuming none of the α's escaped? (Assume ideal gas behavior and 298 K.)

32. The fissile isotope ^{233}U may be "bred" from ^{232}Th by neutron capture to make neutron-rich ^{233}Th, which then undergoes two β-decays. The second β-decay is rate limiting (the "bottleneck"), with a half-life of 27.0 days. Assume a sample of ^{232}Th has just been irradiated with neutrons. How long will it take for 99% of the ^{233}Th formed to breed ^{233}U?

33. The first "soft" lunar lander, *Surveyor 5*, carried a 100-mCi ^{242}Cm α-particle source for Rutherford backscattering analysis of lunar rocks and soil. What mass of ^{242}Cm was used in the source?

34. What mass of ^{131}I would have to be ingested to produce a radiation exposure of 100 mrad (100 mrem) in a person weighing 70.kg? Assume all of the ^{131}I decays within the body.

35. The first nuclear explosion at Alamogordo on July 16, 1945, produced several hundred grams of the radioisotope ^{90}Sr. What fraction of this isotope remained as of July 16, 1995?

Nuclear reactions

36. a. Add the hydrogen-burning Equations 16.20a, 16.20b, and 16.20c with multipliers adjusted to give an overall equation for hydrogen burning that shows four p on the left and a single ^{4}He on the right.

b. Compute Δm and ΔE for the overall equation of part (a). If this energy is released as kinetic energy of the products, which products will carry most of the energy?

c. Our sun radiates 3.9×10^{23} W (J s^{-1}) of power. Use the result of part (b) to estimate how many protons must be "burned" each second to supply this radiant energy, and how many neutrinos s^{-1} are emitted by the sun. (Note that neutrinos, having little or no mass and very weak interactions with other matter, will largely escape the sun without depositing their energy as heat.)

37. Add Equations 16.21a and 16.21b to find the overall equation for helium burning, and find its Δm. What is the minimum wavelength of γ-rays produced by this reaction?

38. Use Equations 16.1 and 16.23 to compute the height of the Coulomb barrier V_c in MeV for Reaction 16.21b. What is the Boltzmann factor $\exp[-V_c/RT]$ for this reaction at $T = 10^8$ K?

39. In the so-called CN cycle of stellar nucleosynthesis, ^{12}C reacts with a proton to make ^{13}N, which then β^+-decays to stable ^{13}C; ^{13}C in turn reacts with a second proton to make stable ^{14}N.

    a. Write equations for each of these reactions, and for the overall production of ^{14}N, and find their ΔE's in MeV.

    b. Give steady-state conditions for the intermediate nuclei in terms of forward rate constants k_1, k_2, and k_3, for the three nuclear reactions and the concentrations of species in the overall reaction.

40. An important source of energy in the OF cycle of stellar nucleosynthesis is the reaction between ^{19}F and p to yield ^{16}O and α. Find Δm and ΔE for this reaction, and use the shell model to explain the large magnitude of ΔE.

41. In the total fusion of ^{12}C, $2\,^{12}$C $\longrightarrow$ ^{24}Mg, which occurs in giant stars and has also been observed at the Argonne National Laboratory heavy-ion accelerator, γ-rays are produced. What is the minimum wavelength these rays may possess?

42. Analyze the energetics of the first laboratory transmutation, Equation 16.22.

    a. Find its Δm and ΔE, and combine with the result of Exercise 17 for the energy of the bombarding α-particles and the Coulomb barrier height given in the text to sketch a simplified energy profile (MeV scale) for this nuclear reaction. On your profile indicate the initial kinetic energy and the final energy of the products, recalling that the total energy must be constant. Does your analysis indicate what Rutherford's experiment implied, that the protons produced are energetic and therefore highly penetrating and readily detected?

    b. Embellish the energy profile of part (a) by considering the "intermediate nucleus" formed by the addition of α to ^{14}N, as well as the Coulomb barrier between the reaction products. Calculate ΔE for forming the intermediate nucleus and the height of the product Coulomb barrier, and add these to the diagram of part (a).

43. Use the line-of-centers model (Equation 16.25) the α-particle energy of Exercise 17, and the Coulomb barrier height given in the text to estimate the collision cross section σ in "barns" (b = 10^{-24}cm^2) for the nuclear reaction of Equation 16.22. By what fraction is the estimate of the reaction rate given in the text reduced in this model?

44. a. Find ΔE and the Coulomb barrier height in MeV for the first neutron-producing reaction, Reaction 16.26.

    b. Assuming a 7.8-MeV α-particle reagent and a stationary ^{9}Be target, calculate the reaction cross section in barns, and estimate the final kinetic energy of the products.

45. a. Repeat the calculations of Exercise 44 for Reaction 16.27.

    b. Calculate ΔE for forming the intermediate nucleus by fusing α with ^{27}Al. (Note that in Reactions 16.26 and 16.27 there is no Coulomb barrier for the products.)

46. A binary alloy of ^{9}Be and ^{210}Po is an intense source of neutrons.

    a. Give nuclear equations that describe how this source functions.

    b. Predict the maximum energy of neutrons emitted.

47. Write a nuclear equation for the γ-ray photodissociation of $^2_1$d. What is the maximum wavelength (Å) of γ-rays that will cause photodissociation?

48. The isotope ^{99}Tc can be formed by neutron bombardment of ^{98}Mo followed by β-decay. Write balanced nuclear equations for this sequence of reactions, and compute their Δm's.

49. Write a balanced nuclear equation for the fission of ^{235}U by a neutron to yield two identical isotopes of Pd plus four free neutrons. Compute ΔE and the Coulomb barrier in MeV for this fission, along with those for the fission reaction, Reaction 16.28. Why is fission to Pd less likely to occur?

50. A recent report indicates that a single atom of element 114 may have been formed by bombarding a ^{244}Pu target with high-energy ^{48}Ca ions.

    a. Write a balanced nuclear equation for this fusion reaction that yields the $A = 289$ isotope and excess free neutrons. Is this a "magic" nucleus?

    b. Estimate the Coulomb barrier for the reaction of part (a) and, assuming the reaction is thermoneutral, the mass of the new isotope.

51. Find and compare ΔE's per unit mass of fuel (MeV/amu and J/kg or cal/g) for the fission reaction, Reaction 16.28, and the first of the fusion reactions, Reaction 16.32.

Nuclear magnetic resonance

52. A modern NMR spectrometer is equipped with a 12-T (120,000-G) superconducting magnet. Given that the proton resonates at 42.6 MHz in a 1-T B-field, what will be its resonance frequency if the nuclear spin energy-level splitting is proportional to B?

53. The energy splitting between nuclear spin energy-levels is proportional to μ/I, where μ is the magnetic moment of the nucleus and I its spin. For a proton in a 1-T B-field, $\mu/I = 2.793/(\frac{1}{2}) = 5.586$ nuclear magnetons from Table 16.3, yielding a transition frequency of 42.6 MHz. What are the corresponding frequencies for ^{13}C and ^{19}F?

54. Photons can cause upward transitions between two energy levels 0 and 1 only if the population of the upper level N_1 is smaller than that of the lower level N_0; the closer the populations, the fewer transitions. In a system initially at thermal equilibrium, the population ratio is governed by the Boltzmann factor, $N_1/N_0 = (g_1/g_0) \exp(-h\nu/k_B T)$, where the g_i's are degeneracies and ν is the transition frequency. What is this ratio at $T = 298$ K for equal degeneracies and $\nu = 42.6$ MHz? 500 MHz? At which frequency will more transitions be possible for a given photon intensity? (If the populations are inverted, the incoming radiation will stimulate downward transitions instead, and increase its own intensity. That is called a laser.)

55. In pulsed NMR, the RF pulse duration Δt governs the range of energies ΔE available to the system through the energy–time uncertainty principle $\Delta E \Delta t \geq h/2\pi$. Give an upper limit to the pulse duration Δt that will give an energy span ΔE sufficiently large to cover the range of proton frequencies found in organic molecules. This range is specified by the range of chemical shifts $\delta = 10^6(\nu - \nu_0)/\nu_0$ observed, $\Delta\delta = 14$ parts-per-million (ppm) frequency shift. Use the Bohr postulate, and take $\nu_0 = 500$ MHz, as in a modern NMR spectrometer.

CHAPTER 17

The Transition Metals

Copper ions Cu^{2+} forming a complex with ammonia NH_3. An initial light-blue hydroxide precipitate gives way to the dark indigo complex. See Sections 17.3 and 17.4 for a discussion.

*If... we are to assume species which are formed in such a manner that **six** water molecules, **six** ammonia molecules, or **six** monovalent groups are arranged around the metal atom, then we may ask with what spatial configuration we can represent the entire molecular **complex**. If we think of the metal atom as the center of the whole system, then we can most simply place the six molecules which are bound to it at the corners of an **octahedron**.... Thus many phenomena which previously evaded every explanation are suddenly clarified.*

—Alfred Werner, 1893 (emphasis added)

CHAPTER OUTLINE

17.1 The d-Block Metals: Energies, Charge States, and Ionic Radii

17.2 Chemistry of the Early Transition Metals: Oxyions

17.3 Chemistry of the Late Transition Metals: Coordination Complexes

17.4 The Spectrochemical Series and Bonding in Complexes

17.5 Chemical Kinetics of Complexes

17.6 The Lanthanides and Actinides

17 The Transition Metals

Alfred Werner (1866–1919) was a professor of chemistry at the University of Zürich, Switzerland. Using purely chemical techniques he was able to isolate and characterize isomers of transition metal compounds; these led him to formulate the idea of the coordination complex that has since become an essential part of our picture of transition–metal ion solutions and crystals, with unforeseen biological implications. He summarized his work in a monograph, *New Ideas in Inorganic Chemistry* in 1911, and two years later was awarded the Nobel Prize. Section 17.3 describes some of his work.

In Chapters 4, 5, 13, and 14 we briefly touched on the atomic properties and chemistry of the **transition metals**, those elements lying between the s- and p-block elements in the 10-column block beginning in period $n = 4$ on the conventional periodic table, reprised in Figure 17.1. Recall first that these elements are classified as metals owing to the "accidental" interleaving of the $n = 4, 5, 6, 7$ s and p subshells by the $n = 3, 4, 5, 6$ d-subshell energy levels. This causes them to occur after the alkaline earths instead of after the noble gases, with typical electron configurations $ns^2(n-1)d^x$.

The metallic properties result from the loosely bound valence ns electrons, possessed by all the transition elements, with the consequence that the first ionization energies of the neutral elements increase very slowly across a period. This accident of nature suggests that these elements do not differ greatly in their chemistry, but while this is true on a superficial level, there are remarkable variations across each period and between periods that demand a closer look. Just comparing two 3d transition metals, iron (Fe) and copper (Cu), produces many examples of contrasting behavior. Skyscrapers cannot be built of copper, but iron with a little carbon mixed in works wonders. Metallic iron is silvery-gray and tarnishes quickly, whereas copper is orange and can retain a high luster. The redox activity series of Chapters 5 and 13 gives definite rankings to the transition metals, with copper considerably less active than iron; this explains rusty versus shiny, and why copper and not iron is useful for coinage.

Nearly all transition metals have known catalytic activity, but seldom for the same reactions; iron is able to catalyze the synthesis of ammonia from hydrogen and nitrogen, but copper is not. Only iron, and not copper, is used by warm-blooded animals to carry oxygen. Aqueous solutions of transition metal salts are nearly all colored, but they are all different; iron salts are light blue-green or yellow-orange, while those of copper are deep blue, and these colors can be mysteriously altered by adding certain reagents to each. We attempt to show in this chapter that underlying all of these differences are systematic trends and correlations among properties that are just as firmly based in elemental identity as those for the representative elements. As in the world of the nucleus explored in the last chapter, we are in a position to analyze such relationships using the structural, thermodynamic, and kinetic tools now at our disposal.

Figure 17.1
Periodic table emphasizing the transition metals.

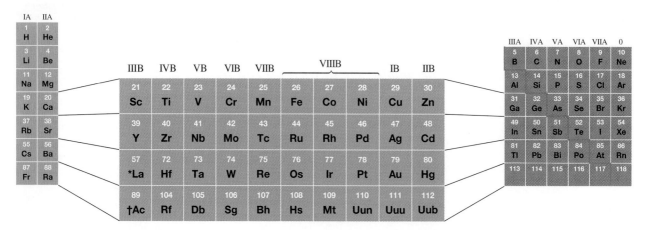

17.1 The *d*-Block Metals: Energies, Charge States, and Ionic Radii

Many of the contrasting properties of iron and copper just mentioned reflect the properties of their compounds with nonmetals; in such compounds the transition metals are present as *cations,* having given up their *s* electrons and perhaps one or more *d* electrons. (Recall from Chapter 4 that when a transition metal forms a cation, it loses its *s* electrons first.) Even in their pure metallic states, the *s* electrons are delocalized into the conduction band, and the metallic lattice consists of *d*-configuration cations of the metals. The differences in chemical properties of the transition metals thus lie to a great extent in the remaining *d* electrons of their cation states. On an atomic level, the *d* electrons are generally not disturbed until the third ionization step, $M^{2+} \longrightarrow M^{3+} + e^-$, $\Delta E = IE_3$; Figure 17.2 graphs the ionization energies IE_1, IE_2, and IE_3 versus Z for the 3*d* transition series. The striking lack of variation in IE_1 is what we found in Chapter 4; equally striking, the variation in IE_3 is strong, and we can learn much about the systematics of transition metal chemistry from it. Recalling that higher IE reflects a more stable orbital, we can see plainly in IE_3 the effect of increasing Z at constant *n* and *l* on orbital stability, with the enhanced stability of half-filled d^5 and filled d^{10} subshells superimposed. The IE_3 graph in fact resembles those for IE_1 of the *p*-block elements of Chapter 4, Figure 4.9, which show peaks at p^3 and p^6 configurations. IE_2 also shows some variation, with peaks reflecting d^5 and d^{10} M^+ ground states formed upon the loss of a single *s* electron.

For the purposes of this introductory chapter, we will focus on the more abundant 3*d* metals; properties of the lower periods will be similar within a column, with some exceptions in the 5*d* metals owing to the lanthanide contraction (see Chapter 4), which tends to make them less active and more refractory. Examples drawn from the lower periods will help you to see the similarities as we proceed.

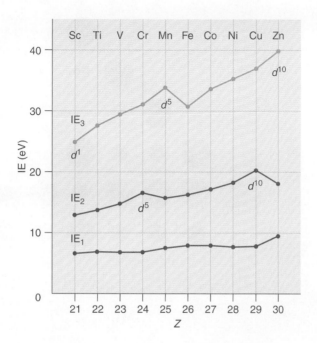

Figure 17.2
The first three ionization energies IE_{1-3} for the 3*d* transition metals versus Z. The *d*-electron configurations refer to the species being ionized, that is, for IE_3, M^{2+} is being ionized. IE graphs for 4*d* are similar, with IE_3 being lower by an average of 3 eV. IE_3's are unknown for most of the 5*d* and all the 6*d* elements.

Transition metal d-orbital stability is reflected most directly in the dominant **charge states** the ions of these elements display, given in Figure 17.3. The first $3d$ metal Sc loses its d as well as its s electrons when it ionizes, due to its low IE_3, and shows only a $3+$ charge state, Sc^{3+}. As IE_3 increases in the Ti, V, Cr sequence, the $2+$ state becomes observable, but $3+$ is far more common. Mn, with its half-filled d subshell in the $2+$ state, has the highest IE_3 of the "early" transition metals; correspondingly, the dominant charge state of Mn is $2+$, with $3+$ now less common. All of the electrons left in the d subshell are unpaired in the ions of the early metals, in accord with Hund's rule (again, see Chapter 4). As indicated in Figure 17.3, the early metals, in polar-covalent combination with O, Cl, or F, commonly show higher **oxidation states** all the way up to and including complete loss of the d shell. These states do not represent ionic charges on free metal ions, however, since oxidation numbers are really just a bookkeeping device, and they overstate the true charge on the metal when the bonding is polar-covalent. The historically assigned transition metal chemical group labels IB, IIB, IIIB, etc., reflect common charge or oxidation states, except for the catchall Fe/Ni/Co Group, Group VIIIB.

In the "late" transition metals, the filling of the latter half of the d subshell forces electrons to pair. In Fe, the d^6 configuration of the $2+$ state shows one electron pair, lowering IE_3 relative to Mn due to increased electron–electron repulsion; this makes Fe^{3+} more common than Fe^{2+}, as in Cr. With Co, the increase in IE_3 across the period resumes, and the dominant charge state becomes $2+$ rather than $3+$, similar to Mn. Ni^{2+} is found almost exclusively, while for Cu both IE_2 and IE_3 are high enough that Cu^+ ($3d^{10}$) occurs, although Cu^{2+} is far more common. We note that the $4d$ group IB element Ag shows only the $1+$ state, also d^{10}. The Group IB metals Cu, Ag, and Au are also relatively inert, which is useful for coinage, and their ions can act as oxidizing agents. Finally, Zn shows exclusively $2+$ ions, leaving a filled d^{10} subshell with the highest IE_3 in the period. These late metals are re-

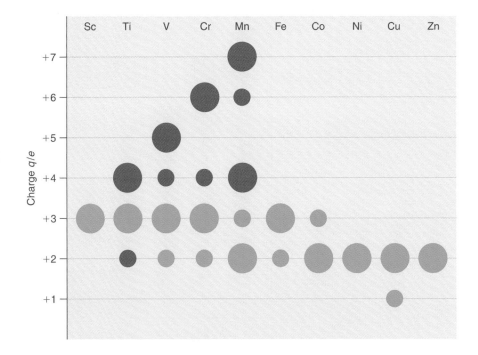

Figure 17.3
Charge states of the $3d$ transition metals. Green symbols indicate charged ions M^{n+}; blue symbols indicate oxidation numbers in polar-covalent molecules and molecular ions. Larger symbols indicate more commonly occurring states. Chemists have also created compounds with other oxidation states, even negative ones, but these are rare and less stable. The lower IE_3's of the $4d$ metals allow higher charge states in general, including M^{4+} ions.

sistant to further oxidation, owing to the generally higher stability of their d subshells, although the Group VIIIB compound OsO_4 is known, in which Os has a +8 oxidation number.

Figure 17.4 plots ionic radii for the 2+ and 3+ charge states of the $3d$ metals versus Z. When we examined radii of the representative atoms and ions in Chapter 4, Figures 4.12 and 4.13, we found that the orbital blending that occurs in covalent and polar-covalent chemical bonding blurs the effects of half-filled p subshells in a period, and the radii vary smoothly across it with increasing Z. The behavior shown in Figure 17.4 is less regular, showing anomalously large radii for d^5 and d^{10} configurations. The $3d$ transition metal ions do use their $3d$ orbitals to engage in some covalent bonding in addition to the obvious ionic interactions in their salts, as we will examine in more detail in a later section. Due to the multiple charges and "inner" character of the d orbitals, however, covalency is relatively less important in transition metal bonding, particularly if the ion's electron configuration is especially stable. As we observed in Chapter 4, covalent bonding yields smaller radii due to the requisite orbital overlap, a partial merging of the bonding atoms. We may thus conclude that the *enhanced radii of the d^5 and d^{10} ions reflect a reduced contribution from covalency* in their interactions relative to the others. These ions (Mn^{2+}, Zn^{2+}, and Fe^{3+}) derive less energetic benefit from covalent "disturbance" of their more stable configurations.

The relations among and trends in the properties of the $3d$ transition metal *ions* are thus analogous to those among the representative *neutrals*. Now we are ready to examine the chemistry of these metals, with an eye to correlating their chemical behavior with these ionic properties as we did in Chapter 5 for the representative elements.

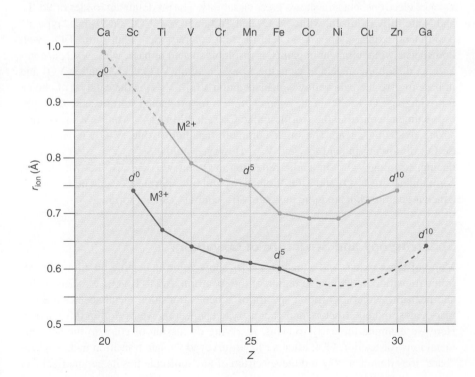

Figure 17.4
Dependence of the ionic radii for M^{2+} and M^{3+} ions on Z for the $3d$ metals. The Ca^{2+} and Ga^{3+} radii are added to give a more complete picture. As in Chapter 4, these radii are averages over a range found in typical compounds.

17.2 Chemistry of the Early Transition Metals: Oxyions

By far the most common compounds and aqueous solute species of the early 3d transition metals, Sc through Mn, involve **oxyions,** metals combined with oxygen to form a polyatomic molecular ion. In Chapter 5 we encountered two important examples, the bright orange dichromate ion $Cr_2O_7^{2-}$ and the deep purple permanganate ion MnO_4^-. In these oxyions, the oxidation numbers are $+6$ for Cr and $+7$ for Mn, corresponding to "loss" of the entire $4s$ and $3d$ shells as indicated in the charge-state diagram of Figure 17.3. Another common Cr oxyion is bright yellow CrO_4^{2-}, again a $+6$ case, which exists in pH-neutral solutions, and is interconvertible with dichromate by adding acid:

$$2CrO_4^{2-} + 2H^+ \rightleftharpoons Cr_2O_7^{2-} + H_2O \qquad (17.1)$$

Adding concentrated H_2SO_4 to chromate salt produces black chromium(VI) oxide CrO_3, commonly called "chromic acid," since it produces H_2CrO_4 when added to water. A similar treatment of permanganate salt with H_2SO_4 produces the oxide Mn_2O_7, an explosive green oil that slowly loses oxygen on standing. Both $Cr_2O_7^{2-}$ and MnO_4^- are powerful oxidizing agents in solution; as given in Chapter 5, dichromate is readily reduced to green Cr^{3+}, while permanganate becomes insoluble brown-black MnO_2 in neutral or basic solution and pale pink Mn^{2+} in acid. The aqueous "bare" metal ions actually form well-defined **complexes** with water, as discussed in the next two sections. Other oxides, green Cr_2O_3 and deep red Mn_2O_3, are common, CrO and MnO less so, while the aqueous ion Cr^{2+} is much less prevalent than Mn^{2+}, as discussed in the previous section.

The earlier transition metals have not achieved the same status as laboratory reagents that Mn and Cr have. Their chemistry is limited by their smaller number of valence electrons, but still shows great variability. The predominant oxides of Sc, Ti, and V are fully oxidized, white Sc_2O_3 and TiO_2, and red V_2O_5. Sc_2O_3 is a "typical" basic metal oxide, dissolving in acid to yield colorless Sc^{3+}, but rutile TiO_2, well known as a white pigment in paints, is insoluble in acid or base, while V_2O_5 is *amphoteric,* dissolving in acid to yield the dioxyvanadium(V) (vanadyl) cation VO_2^+ and in base to give the pale yellow vanadate anion VO_4^{3-}. V_2O_5 is the catalyst of choice for oxidizing SO_2 to SO_3 in the *contact process* for producing sulfuric acid, the highest volume chemical in the world; unlike CrO_3 and Mn_2O_7 or MnO_2, V_2O_5 is not a strong oxidizing agent. The other common vanadium oxide, V_2O_3, is similar to Sc_2O_3, dissolving in acid to yield V^{3+}. However, unlike Sc^{3+}, V^{3+} is strongly reducing, forming VO^{2+} in acidic reduction. VO and VO_2 are also known but less common, as are Ti_2O_3 and TiO. Violet Ti^{3+} and blue-purple V^{2+} can be formed in aqueous solution, but are quite unstable; Sc^{2+} and Ti^{2+} do not form.

Spectroscopy and Structure of Oxyanions

Many of these oxides and aqueous species show color, indicating an allowed electronic transition in the visible region of the spectrum, $\Delta E = 1.8–3.1$ eV. The visible and near-ultraviolet absorption spectra of aqueous transition metal species are primary clues to their structure and chemistry, a connection we exploit in this chapter. Here we examine the early oxyions, reserving discussion of color in the aqueous metal ions for Section 17.4, after we have surveyed the late transition metals where "bare" ions dominate. Behind the spectrum of any molecule lies its geometrical and electronic structure, as we have seen in Chapters 6, 7, and 8. Geometrical considerations bring to mind orbital hybridization and the VSEPR (electron pair repulsion) model, while the energy levels necessary for understanding a molecular electronic transition can be formulated at least approximately in the molecular orbital model.

As examples let's focus on the isoelectronic species VO_4^{3-}, CrO_4^{2-}, and MnO_4^-. Their ultraviolet-to-visible (UV-vis) solution absorption spectra are shown in Figure 17.5. The spectral peaks are broad and relatively featureless, instead of one or a series of sharp lines as you would see from a gas-phase atom or ion. There are two reasons for this. As we saw in Chapter 8, when a *molecular* electron hops to a new energy level, it changes the forces on the nuclei, causing a range of *vibrations* within the molecule to become excited, and giving rise to a *band* of wavelengths associated with the electronic transition. Further, when the absorbing molecules are in solution, each one finds itself in a slightly different random arrangement of solvent molecules, causing each of the underlying "vibronic" lines to shift to a slightly different frequency for each absorber. It then becomes impossible to resolve these underlying components of the band, and the spectrum turns into a "blob." Occasionally such a blob can contain more than one electronic band.

The color you see in a solution is the complement of the color(s) absorbed; the deep purple of permanganate arises from absorption of a broad band of wavelengths covering the central colors green and yellow, leaving behind red at long wavelengths and blue-violet at short, which blend to make purple. Vanadate on the other hand is pale yellow owing to its absorption being confined mainly to the UV region. Taken together, these three ions absorb at progressively longer wavelengths λ as Z increases.

How can we understand this trend? First we must have an idea of the bonding geometry. Each of these ions has a Lewis structure obeying the octet rule, four oxygen atoms singly bonded to a central metal ion. The geometries of these ions follow VSEPR rules, all being tetrahedral; in fact, they bear a striking resemblance to the common nonmetal ions of the previous period: PO_4^{3-}, SO_4^{2-}, and ClO_4^-. A full molecular orbital analysis of these tetrahedral ions is difficult, but we can construct a simple model that will give us insight into the spectra and also serve as a starting point for further developments.

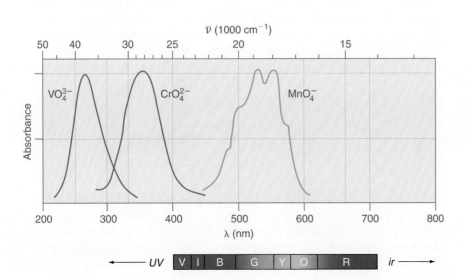

Figure 17.5
Absorption spectra of the oxyanions VO_4^{3-}, CrO_4^{2-}, and MnO_4^- in aqueous solution. The absorbance A is defined as $A = \log_{10}(I_0/I)$, where I_0 (I) is the light intensity in the absence (presence) of the absorbing species. Peak absorbances vary from one compound to another; here and in Figure 17.9 the spectra have been normalized to the same A for ease of comparison of peak wavelengths.

In molecules or ions with a central atom and a set of symmetrically arranged terminal atoms or groups, a molecular orbital energy level diagram is usually constructed by placing the central atom levels on the left, and the terminal atom levels as a degenerate group on the right. The molecular orbital energies are then placed in the center as usual. For our MO_4^{x-} ions, the $3d$ transition metals are represented by $3d$, $4s$, and $4p$ levels and their associated orbitals, and the four oxygens by four degenerate $2p\sigma$ levels of lower energy than any of the metal levels, as illustrated in Figure 17.6, where the O orbitals are preselected to have the correct bonding orientation. Bonding and antibonding combinations of the central and terminal orbitals are then formed following the LCAO-MO rules of Chapter 7.

There are two aspects to the diagrams of Figure 17.6 that you may find puzzling. First, despite the normal Aufbau filling order of the s-block and d-block elements, the $3d$ orbitals always lie lower than the $4s$ orbitals in oxidized states of the transition metals. On the other hand, the $4s$ and $4p$ orbitals, being of a higher n, have a greater spatial extent than the $3d$ orbitals, and overlap with the O atom orbitals more strongly. We thus have the peculiar situation that, while the energy match between $M(3d)$ and $O(2p)$ is better, bonding occurs largely with the $n = 4$ metal orbitals. In this simplified model we assume that these four orbitals are the ones that combine with the four

Figure 17.6
Simplified molecular orbital energy-level diagram for the tetrahedral oxyanions. The upper panel illustrates the construction of MO levels in general for this case, and the lower shows the effect of the downward shift in $3d$ energy as Z increases in the V, Cr, Mn sequence. ΔE_{CT} is the HOMO-LUMO energy spacing governing the peak absorption wavelength in these charge-transfer transitions. In these diagrams possible bonding interactions of the $3d$ orbitals have been neglected, a good approximation for the early metals.

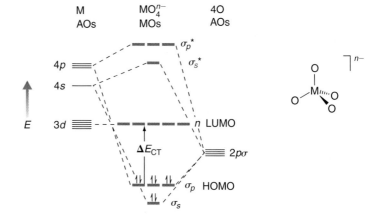

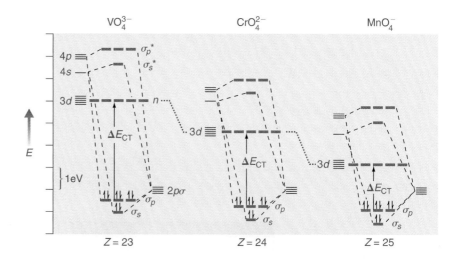

O orbitals, forming four bonding and four antibonding molecular orbitals. Although deducing the forms of these MOs—which are in general delocalized over all five atoms—can be a difficult exercise, it is easy to locate their energies qualitatively using the rules for heteronuclear bonds of Chapter 7. The bonding orbitals will lie lower in energy than the O atom levels, while the antibonding orbitals will lie above their respective metallic parents, as Figure 17.6 shows. You can usefully think of the metal atom as being sp^3 hybridized, and forming bonds by overlapping in a σ fashion with O $2p\sigma$. The diagram accounts only for electrons involved in the bonds plus any extra left on the metal atom; the terminal atom's lone pairs are omitted.

For our fully oxidized metal centers, we find that all electrons are accounted for in filling the σ_s and σ_p bonding MOs, leaving empty the nonbonding d orbitals as well as the antibonding σ_s^* and σ_p^* MOs. Recall from Chapter 7 that in polar covalent MOs the lower energy AO dominates the bonding combination and the higher energy AO the antibonding. In this case the bonding orbitals will resemble the O atom $2p$'s, and the nonbonding and antibonding ones will be metal-like. As Z increases in the V, Cr, Mn sequence, the metal orbitals drop in energy, as expected from the IE trend of Figure 17.2, reducing the spacing between the filled O-like and unfilled M-like orbitals, as shown in Figure 17.6. According to the Bohr postulate $\Delta E = hc/\lambda$, electronic transitions between the HOMO (σ_p) and LUMO ($3d$) will occur at successively longer wavelengths, in qualitative agreement with the spectra shown in Figure 17.5. Because electrons are jumping from an O-like orbital to an M-like orbital, these are known as **charge-transfer transitions,** wherein the metal atom reduces its oxidation number at the expense of the O atoms. In charge-transfer spectra the interaction between the electric field of the light and the electronic motion is maximized, making the absorption very strong. As you may have seen, even quite dilute $KMnO_4$ solutions are intensely purple.

We now have the beginnings of a picture that relates atomic properties and the trends of the transition metals themselves to those of their compounds. In the next section we introduce you to the notion of a "complex" between what up until now we have regarded as a "bare" transition metal ion in solution and two or more atoms or groups bonded to it. In these complexes additional characteristic properties, such as magnetism and relative stability, come into play, and we will find that structural and orbital analysis will enable us to connect these properties to the colorful spectra of these species.

17.3 Chemistry of the Late Transition Metals: Coordination Complexes

In the late $3d$ metals, Fe through Zn, in contrast to their early neighbors, common oxidation numbers in the binary oxides are identical to the common charge states of the atomic ions found in solution, as indicated in Figure 17.3. Higher oxidation states are rare, owing to the increasingly stable d subshell, as discussed in Section 17.1. Accordingly, the common oxides are Fe_3O_4 and Fe_2O_3, CoO and Co_3O_4, NiO, Cu_2O and CuO, and ZnO. Fe_3O_4 is a 1:1 mixture of FeO and Fe_2O_3 with its own crystal structure; it is called *magnetite,* known to the ancients as *lodestone,* nature's own permanent magnet. All of these oxides will dissolve in acid to produce the corresponding aqueous metal ions; ZnO is amphoteric, also dissolving in base to yield the zincate ion, $Zn(OH)_4^{2-}$.

As in the early transition metals, the dissolved ions usually form colored solutions: $Fe^{2+}(aq)$ is pale blue-green, Fe^{3+} yellow-orange, Co^{2+} pink, Co^{3+} (*unstable*) blue, Ni^{2+} green, Cu^{2+} blue, Cu^+ (*unstable*) colorless, and Zn^{2+} colorless. Again, the color of a solution is the complement of the color absorbed; for broadband absorption the complementary colors may be uniquely associated with the major absorption band of the ion, as presented in Table 17.1. This enables you to use your own eyes as a spectrophotometer; by "eyeball spectroscopy" you can say, for example, that blue $Cu^{2+}(aq)$ absorbs at a longer wavelength than orange $Fe^{3+}(aq)$. Further, you can note that colorless $Zn^{2+}(aq)$ and $Cu^+(aq)$ probably absorb in the ultraviolet; both happen to be d^{10} ions. The colors in the aqueous ions are, however, more difficult to explain than they are in the oxyions. The energy spacings between the $3d$ and ($4s$, $4p$) atomic ion energy levels are known to be in the UV; thus the bare ions are colorless, and water is colorless. The visible spectra force us to consider some type of bonding beyond mere random solvation between the metal ions and water as the only way to produce the more closely spaced energy levels required; that is, the spectra point to the existence of a metal ion + water **coordination complex.**

Realizing the significance of color in transition metal chemistry presupposes a knowledge of the quantized nature of light and the standing-wave motion of electrons in molecules—of quantum mechanics. As the dated quote at the start of this chapter shows, however, one scientist had deduced that coordination complexes must exist long before the coming of the quantum. Based on clever chemical experiments and an uncanny ability to keep dozens of findings simultaneously at the fore in his thoughts, the Swiss chemist Alfred Werner almost singlehandedly, and against considerable opposition, deduced most of the compositional and structural facts about transition metal compounds and chemistry we are about to examine. Let's look briefly at some of the chemical evidence, and the conclusions Werner drew from it.

TABLE 17.1

The "eyeball spectroscopy" table for broadband electronic absorption

Color you see	Color absorbed	Wavelength λ at band center (nm)	Energy level difference $\Delta E/hc$ (cm^{-1})
Colorless	Ultraviolet	<400	>25,000
Lemon yellow	Violet	410	24,400
Yellow	Indigo	430	23,200
Orange	Blue	480	20,800
Red	Blue-green	500	20,000
Purple	Green	530	18,900
Violet	Lemon yellow	560	17,900
Indigo	Yellow	580	17,300
Blue	Orange	610	16,400
Blue-green	Red	680	14,700
Green	Purple-red	720	13,900
Yellow-green	Purple	750	13,300

Stoichiometry, Isomerism, and Geometry of Complexes

Werner found he could make a variety of stoichiometric chromium compounds by adding NH_3 to $CrCl_3$ crystals: $Cr(NH_3)_3Cl_3$, $Cr(NH_3)_4Cl_3$, $Cr(NH_3)_5Cl_3$, and $Cr(NH_3)_6Cl_3$. However, he was *unable* to make $CrNH_3Cl_3$ or $Cr(NH_3)_2Cl_3$, or any compounds containing more than six NH_3's. Further, he found that the cobalt compound $Co(NH_3)_4Cl_3$ exists in two and only two forms, one green and one violet. These two pieces of evidence, combined with a host of other observations of this sort, convinced Werner that he was dealing with a coordination complex involving **six** monovalent species, here Cl^- or NH_3, bonded to the central metal ion, Cr^{3+} or Co^{3+}, and that the geometry of bonding must be **octahedral**. A perfect octahedron, from the Greek for *eight faces,* consists of eight equilateral triangles joined at their edges to form two square pyramids, which are in turn joined at their bases (see Figure 17.7). This figure has six vertices, at which the ligands are located; the metal atom resides within, at the geometrical center of the figure.

The number six, the **coordination number,** follows from the absence of $Cr(NH_3)_nCl_3$ compounds with $n = 1$ or 2, that is, containing only one or two NH_3's; in those cases there are fewer than six coordinating species. Werner postulated that, for $n > 3$, the NH_3 *displaces* Cl^- and coordinates directly to the metal ion, while the diplaced Cl^- becomes simply a counterion necessary to balance the charge. These displacements can occur up to a maximum of $n = 6$, where all the coordinating positions are occupied by NH_3's. For $n = 5$, for example, the formula $Cr(NH_3)_5Cl_3$ is rewritten as $[Cr(NH_3)_5Cl]Cl_2$, where the brackets enclose the coordination complex. The coordinating groups NH_3 and Cl^- are called **ligands,** from the Latin *ligare,* meaning "to bind."

Once the coordination number 6 was established (in Cu, Pt, and others Werner also found coordination number 4, which we will consider later), it was natural to ask, as Werner did in the quotation, how the ligands are arranged about the (assumed) central metal ion. The answer to this question lay in the existence of **isomers,** compounds with the same formula but different properties. We shall have much more to say about isomers in the next chapter on organic compounds; here we will invoke and define them as we go. Given six ligands around the Co^{3+} in $Co(NH_3)_4Cl_3$, the formula is rewritten $[Co(NH_3)_4Cl_2]Cl$, indicating four ammonia and two chloride ligands. In arranging the

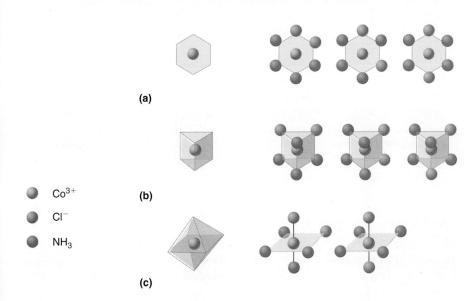

Figure 17.7
Three possible geometries for $Co(NH_3)_4Cl_2^+$ complexes: (a) hexagonal planar, (b) triangular prism, and (c) octahedral. Only the octahedral geometry yields only two geometrical isomers, in agreement with experiment.

ligands, the two chlorides can be nearest neighbors, or separated from each other by ammonia. The number of distinct spatial arrangements of the chlorides, that is, the number of **geometrical isomers,** depends on how the ligands are arrayed about the central atom. Figure 17.7 illustrates the possible isomers of $[Co(NH_3)_4Cl_2]^+$ for hexagonal planar, triangular prism, and octahedral arrangements of the ligands; as Werner first showed, only the octahedral geometry leads to two and only two geometrical isomers. In one of these, the Cl^-'s are next to each other, called *cis,* and in the other they are across, or *trans.* Armed with several examples of this sort, Werner postulated that all complexes of the general form ML_6, where M = metal and L = ligand, possess octahedral symmetry. Further, based on stoichiometric formulas for hydrated salts of the transition metals, Werner suggested that even a simple salt such as $NiCl_2$, when dissolved in water, would form a $Ni(H_2O)_6^{2+}$ complex.

Felicitously, Werner lived to see his ideas verified; the method of X-ray diffraction (see Chapter 14) was developed in the last decade of his life, and the octahedral arrangement of ligands was found to be the only one consistent with measured diffraction patterns.

Nomenclature

Before we describe further properties of these complexes, it is useful for us to have a systematic way of naming them, a way that unambiguously connects a name with its chemical formula. Consistent with their Germanic Swiss origins, all names of complexes are given as a single word. In this word the ligands are listed first in alphabetical order, preceded by a Greek prefix *di-, tri-, tetra-, penta-, hexa-* to indicate the number of ligands of each type. Then the metal is appended, followed by its oxidation state as a Roman numeral in parentheses. If the complex carries an overall negative charge, the suffix *-ate* is added to the metal name, and the metal name is sometimes truncated (if it ends in *-um*) or changed to its Latin form, for example, platinate, ferrate, or cuprate. The full name of the compound includes any counterions in the usual way we learned in Chapter 5, with positive ions always named first. The very common ligand NH_3 is given the special name **ammine** (am' · meen), water is given the Latin **aqua,** and another common ligand, CO, is called **carbonyl.** For negatively charged ligands the usual *-ide* ending is replaced by *-o;* for *-ite* and *-ate* anions, only the silent *e* is replaced by *-o.* Chlor*ide* then becomes chlor*o,* and acetat*e* becomes acetat*o.* The names of neutral ligands usually are not modified, but the larger ones are often abbreviated when used in formulas. When a ligand has *di* or *tri* in its own name, the complex name uses *bis* or *tris* followed by the ligand name in parentheses. Table 17.2 lists some commonly encountered ligands.

As in Chapter 5, we do not want to turn nomenclature into a stumbling block; you should acquire it as we go by seeing examples. The isomeric cobalt complexes just described would be named tetraammine-*cis*-dichlorocobalt(III) and tetraammine-*trans*-dichlorocobalt(III); the nickel complex $Ni(H_2O)_6^{2+}$ would be hexaaquanickel(II), while NiF_6^{4-} is hexafluoronickelate(II).

The Nature of Ligands

What all ligands have in common is at least one electron pair that can be shared with the metal; as we described in Chapter 12, the association of metal and ligand is viewed as a Lewis acid–base interaction, the ligand being the base. In Table 17.2

TABLE 17.2
Some common ligands

Name	Abbreviation	Formula/structure[a]
Iodo, bromo, chloro, fluoro		I^-, Br^-, Cl^-, F^-
Hydroxo		OH^-
Nitro, nitrito		$:NO_2^-, :ONO^-$
Acetato		CH_3COO^-
Thiocyanato, isothiocyanato		$:SCN^-, :NCS^-$
Aqua		H_2O
Ammine		NH_3
Cyano		$:CN^-$
Carbonyl		$:CO$
Nitrosyl		$:NO^+$
Oxalato	ox	$^-:OOCCOO:^-$
Acetylacetonato	acac	$CH_3COCHCOCH_3^-$
Ethylenediamine	en	$H_2NCH_2CH_2NH_2$
Pyridine	py	C_5H_5N
Bipyridyl	bipy	$(C_5H_4N)_2$
o-Phenanthroline	phen	$C_{12}H_8N_2$
Porphyrins		$C_{20}H_{12}N_4^{2-}$ et al.[b]
Ethylenediamine tetraacetato	EDTA	$(^-OOCCH_2)_2NCH_2CH_2N(CH_2COO^-)_2^{b}$

[a] Lone pairs are indicated where ambiguity exists. The organic ligands beginning with acac have many variants or *derivatives*, in which other atoms or groups substitute for H. Tridentate ligands patterned after en and bipy are omitted for brevity.

[b] For structures, see the figures.

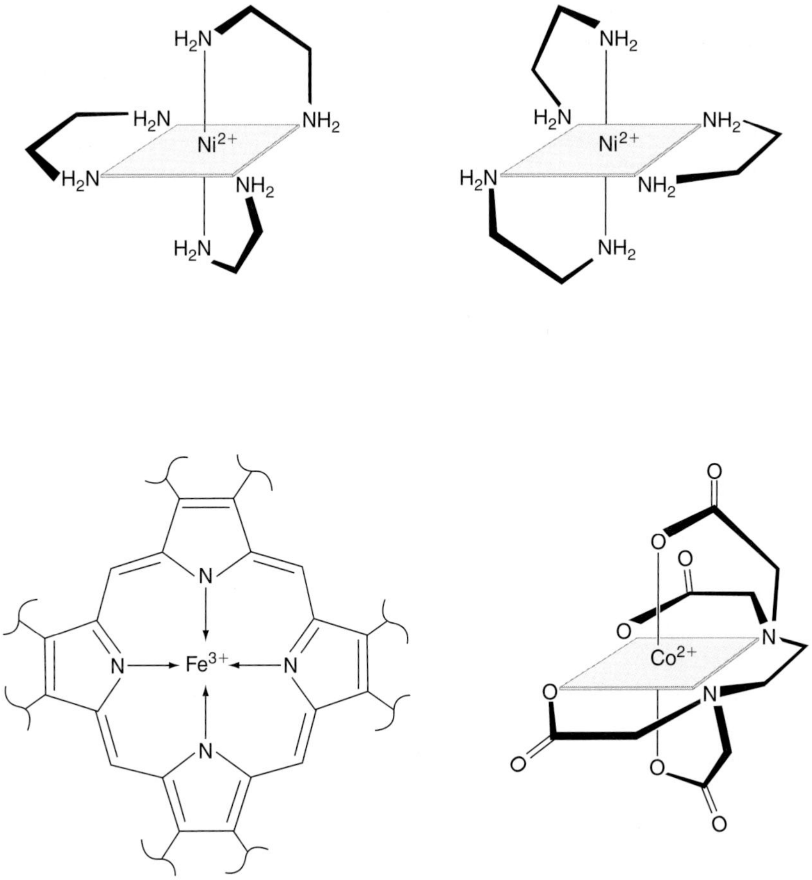

Figure 17.8
Complexes with multidentate ligands. In these structural depictions, all unlabeled vertices represent carbon atoms, and any unsaturated valence implies one or more bonds to hydrogen atoms. The bidentate complex Ni(en)$_3^{2+}$ exists as a pair of **optical isomers;** they are nonsuperimposable mirror images of each other. The tetradentate iron–porphyrin complex is planar, and the central structure of the *heme group,* the oxygen carrier in blood. In the hexadentate CoEDTA complex, all binding sites are occupied by a single ligand, called a *chelating agent.*

there are ligands that have lone pairs located on more than one atom, with carbon-based *spacer* groups between them. These are called **multidentate** ("many-toothed") ligands, and are capable of occupying or bridging two or more octahedral sites simultaneously. The simplest example of these is the *bi*dentate ligand ethylenediamine, $H_2N-CH_2CH_2-NH_2$, abbreviated en, in which each N can donate a lone pair to a metal center.* It is no coincidence that en is also a dibasic base, as listed in Table 12.4. Adding sufficient en to an aqueous Ni^{2+} solution produces the complex Ni(en)$_3^{2+}$, tris(ethylenediamine)nickel(II), in which the en's occupy pairs of adjacent octahedral sites, as shown in Figure 17.8.

There are several biological *tetra*dentate ligands called *porphyrins* that also employ N lone pairs, four of them lying in a plane, to bind metal ions; in your blood a porphyrin strongly binds Fe^{2+} (making what is known as a *heme*), another nitrogen base takes the fifth Fe coordination site, and dioxygen becomes the sixth—life-giving—Fe ligand (see Figure 17.8). The extreme case is the *hexa*dentate ligand ethylenediaminetetraacetate, abbreviated EDTA, which can bind all six sites itself, as illustrated in Figure 17.8. EDTA is sometimes called a *chelating agent* (from the

---

* The family of organic bases known as **amines** (a ·meens'), containing N with three single bonds and a lone pair, all can act as ligands, and are not to be confused with the ammine ligand NH_3.

Greek *chela*, "claw"), since it effectively encapsulates the metal ion, preventing further interaction with other ligands.

In recent times we have come to realize that the shared electron pair in the M—L association need not be a lone pair. A variety of π-bonded molecules, such as ethylene H_2C=CH_2, acetylene HC≡CH, and conjugated π ring systems such as cyclopentadienyl ($C_5H_5^-$) and even benzene (C_6H_6), have been found to bond to metals side-on, with their π clouds being the donors. Cyclopentadienyl makes especially stable "sandwich" compounds like *ferrocene* ($Fe(C_5H_5)_2$), in which the planar C_5H_5 rings are the bread and the metal ion the filling. Even H_2 has been coaxed into being a ligand, sharing its only electron pair in a side-on geometry. Metal ions will take their electron pairs where they can find them, even accepting the noble gas Xe as a ligand if there is nothing better around.

With this small sampler of possible compounds in mind, let's take a look at two other important properties of complexes that provide insight into their nature: magnetism and lability, or reactivity.

Magnetism

Given that unpaired electrons exist in the free transition metal ions, except in groups IIIB and IIB, it is not surprising that many transition metal complexes exhibit **paramagnetism,** reflecting a nonzero magnetic moment due to the unpaired electron spins. Complexes of Zn^{2+} are uniformly **diamagnetic,** without a magnetic moment, as expected for a d^{10} configuration, but this is exceptional. What is surprising is that certain complexes with partially filled d subshells are also diamagnetic, for example $Co(NH_3)_6^{3+}$ and $Fe(CN)_6^{4-}$, both of which have d^6 configurations. Others, such as $Co(NO_2)_6^{4-}$ and $Fe(CN)_6^{3-}$, show fewer unpaired electrons than the corresponding bare metal ions. These are all examples of **low-spin** complexes, while those whose magnetic moments conform to the bare metal ion are called **high spin.** The existence of low-spin complexes, like the spectral colors, requires a metal–ligand interaction that breaks the degeneracy of the d orbitals.

Lability

Werner noted in his earliest work that certain ligands were more easily displaced than others; for example, Cl^- could readily be displaced by NH_3, but to reverse the process was difficult or impossible. Further, certain charge states of complex ions, such as complexes of Cr^{2+} or Mn^{3+}, readily lose or gain electrons. Having been exposed to thermodynamics and kinetics, you know that two factors are needed to make a compound highly reactive, or **labile.** First, the contemplated reaction, such as ligand displacement, must lie downhill in free energy, that is, $\Delta G < 0$, implying that the product complex is more stable thermodynamically. Second, the activation energy E_a for the reaction must be low, to allow rapid conversion to product. To be labile, a complex must be both thermodynamically and kinetically predisposed to reaction. This page you are reading is thermodynamically *unstable* with respect to combustion, $\Delta G_{comb} < 0$, but E_a is so high that nothing appears to happen: this page is *not labile.*

Every generality in chemistry has its exceptions, but we can say that, because of the strongly attractive electrostatic and inductive forces acting between a metal ion and possible ligands (see Chapter 14), very low activation energies for ligand displacement are expected. When E_a is low, a reaction that is thermodynamically favored also becomes rapid, while its reverse is slow in accord with the principle of detailed balance, as discussed in Chapter 15. If you observe that a complex undergoes slow ligand displacement, you may therefore fairly conclude that it lies at a lower G than the product complex, whereas if ligand substitution is rapid the product complex has lower G. Complexes are not like paper.

Our standard for comparison of relative stability is the aqua complex $M(H_2O)_m^{x+}$, where m is generally 6; we will denote this complex simply as $M^{x+}(aq)$. The displacement of water by ligand L is then written as though it were a simple addition,

$$M^{x+}(aq) + nL^{y-}(aq) \rightleftharpoons ML_n^{x-ny}(aq) \tag{17.2}$$

In Reaction 17.2 it may be that not all of the water ligands have been displaced in the ML_n complex, implicit in the (aq) tag on the complex. It is also possible that all of the waters are displaced, but that $n < 6$, as in the tetrachlorocobaltate(II) complex,

$$Co^{2+}(aq) + 4Cl^- \rightleftharpoons CoCl_4^{2-} \tag{17.3}$$

where $Co^{2+}(aq)$ is actually $Co(H_2O)_6^{2+}$, and all six aqua ligands are lost in forming the tetrahedral chloro complex. For this simple introduction to complex stability, we will ignore the "mixed" complexes such as the ammine-chloro examples given above. In Chapter 12 complex formation equilibria were briefly introduced; you might want to reread the last subsection of Section 12.3. Just to remind you, the equilibrium constant for Reaction 17.2 is called the **formation constant** K_f and is logarithmically related to the standard free-energy change, $\Delta G_f^\circ = -RT \ln K_f$. Our purpose here is to use K_f values to assess relative stability/lability: the larger K_f, the more stable the complex. Further, $K_f > 1$ implies that the chosen ligand will displace H_2O spontaneously. Table 17.3 presents formation constants for some common ligands; many of the values are neither precise nor accurate, owing to the difficulty in measuring very small or very large ones, and the need to use high, nonideal concentrations of ligand. Nonetheless it is noted, for example, that the ready displacement of Cl^- by NH_3 found by Werner is reflected in generally larger K_f's for L = NH_3. Further, the bidentate ligand en tends to have larger K_f's than the monovalent ones, due in part to a more favorable entropy change, while the cyano CN^- ligand makes practically indestructible complex ions. Equilibrium constants for displacement of one ligand by another are given by the *ratio* of K_f's for each. The trend in K_f thus implies a ligand "pecking order," an important factor in our upcoming discussion of bonding and structure in complexes.

17.4 The Spectrochemical Series and Bonding in Complexes

Nickel(II) chloride as a reagent chemical is usually provided as the green hexahydrate $NiCl_2 \cdot 6H_2O$. You may now realize, as Werner first did more than a century ago, that the crystalline form of this compound contains the octahedral complex hexaaquanickel(II). If you place a few crystals in a beaker and add a few milliliters of concentrated (12 M) HCl, the crystals dissolve to give a yellow-green solution that now contains the hexachloronickelate(II) complex. Despite the minuscule K_f for this complex (Table 17.3), a sufficiently large Cl^- concentration can tip the scales—á la

TABLE 17.3

Formation constants $K_f = [ML_n]/[M][L]^n$ for selected metal ions and ligands in aqueous solution

Complex	K_f	Complex	K_f
L = Cl$^-$		**L = en (H$_2$N(CH$_2$)$_2$NH$_2$)**	
CoCl$_4^{2-}$	$8. \times 10^{-4}$	Mn(en)$_3^{2+}$	$5. \times 10^5$
CuCl$_4^{2-}$	$1. \times 10^{-5}$	Fe(en)$_3^{2+}$	$4. \times 10^9$
CrCl$_6^{3-}$	$7. \times 10^{-7}$	Co(en)$_3^{2+}$	$8. \times 10^{13}$
FeCl$_6^{3-}$	$6. \times 10^{-4}$	Ni(en)$_3^{2+}$	$4. \times 10^{18}$
NiCl$_6^{4-}$	$3. \times 10^{-8}$	Cu(en)$_2^{2+}$	$2. \times 10^{20}$
L = SCN$^-$		Zn(en)$_3^{2+}$	$1. \times 10^{13}$
FeSCN^{2+}	1.4×10^2	**L = CN$^-$**	
CuSCN$^+$	$8. \times 10^1$	Ag(CN)$_2^-$	$2. \times 10^{20}$
L = OH^{-a}		Ni(CN)$_4^{2-}$	$1. \times 10^{31}$
AgOH	$1. \times 10^3$	Zn(CN)$_4^{2-}$	$4. \times 10^{21}$
Zn(OH)$_4^{2-}$	$5. \times 10^{14}$	Cd(CN)$_4^{2-}$	$6. \times 10^{18}$
L = NH$_3^b$		Hg(CN)$_4^{2-}$	$4. \times 10^{41}$
Ag(NH$_3$)$_2^+$	1.7×10^7	Fe(CN)$_6^{4-}$	$3. \times 10^{35}$
Cu(NH$_3$)$_4^{2+}$	$1. \times 10^{13}$	Fe(CN)$_6^{3-}$	$4. \times 10^{43}$
Mn(NH$_3$)$_4^{2+}$	$2. \times 10^1$		
Zn(NH$_3$)$_4^{2+}$	$2. \times 10^9$		
Co(NH$_3$)$_6^{2+}$	$1. \times 10^5$		
Co(NH$_3$)$_6^{3+}$	$1. \times 10^{35}$		
Ni(NH$_3$)$_6^{2+}$	$1. \times 10^9$		

^{a}Most transition metal ions form *precipitates* with OH$^-$ rather than complexes.
^{b}The OH$^-$ formed when NH$_3$ is added to a solution of an aqua complex causes the early transition metals, plus iron, to form very stable hydroxide precipitates, preventing complex formation.

LeChâtelier—away from the aqua complex, since the equilibrium quotient for creating the chloro complex contains the sixth power of [Cl$^-$] in its denominator. Adding water then produces a subtle but well-defined change to a true green, as the aqua complex originally present in the crystal reforms upon dilution. If you now add NH$_3$(*aq*), the solution turns a deep blue, signaling the formation of the hexamminenickel(II) complex, NH$_3$ having displaced H$_2$O. Adding excess ethylenediamine en solution again changes the color to lavender (purple-red), and we now have tris(ethylenediamine)nickel(II). In this case adding controlled amounts of en produces stages of indigo and violet, each stage resulting from partial displacement of NH$_3$ by en. Let's summarize this magical series of transformations:

$$\text{NiCl}_6^{4-} \longrightarrow \text{Ni(H}_2\text{O)}_6^{2+} \longrightarrow \text{Ni(NH}_3\text{)}_6^{2+} \longrightarrow \text{Ni(en)}_3^{2+}$$
yellow-green   green   blue   purple-red

The spontaneity of the displacements is predictable from the formation constants of Table 17.3. By consulting Table 17.1 you can also observe that the color of each new complex corresponds to absorption at a shorter wavelength λ, or a larger energy level spacing ΔE; that is, *the more stable the complex, the larger its energy-level spacing in the visible spectrum.* The absorption spectra of the four nickel complexes are shown in Figure 17.9.

This correlation between stability and color holds almost perfectly for all of the transition metals, with ligands arranged in the same sequence. This ligand sequence is known as the **spectrochemical series,** which we arrange in order of increasing complex stability:

$$I^- < Br^- < Cl^- < F^- \sim OH^- < H_2O < SCN^- < NH_3 < en < CN^- \sim CO \quad (17.4)$$

In this sequence both K_f and ΔE increase; although the sequence applied to different transition metals leads to a different set of K_f's and colors, the trend remains the same. For example, the aqua complex of Cu^{2+}, hexaaquacopper(II), is blue, while the more stable NH_3 complex, tetraamminecopper(II), is indigo. You might tentatively conclude that better bases make better ligands, but there are clearly other factors involved; OH^- is certainly a stronger base than H_2O or NH_3, but its complexes are less stable.

In our example using Ni^{2+} complexes, the paramagnetism remains fixed across the spectrochemical series at two unpaired electrons, consistent with a d^8 metal ion. For other ions, however, particularly those with d^5 or d^6 configurations, the magnetic character may change in the series. For example, blue-green $Fe(H_2O)_6^{2+}$ is paramagnetic, a high-spin d^6 complex showing four unpaired e^-'s, as in the bare ion, while yellow $Fe(CN)_6^{4-}$ (hexacyanoferrate(II), common name ferrocyanide ion) is completely diamagnetic, a low-spin complex, as well as exceptionally stable, judging by its large K_f (Table 17.3). In general, less stable complexes tend to be high spin, and more stable low spin.

Finally, the d^4 configuration is found to be especially unstable to either oxidation or reduction, as in the Cr^{2+} and Mn^{3+} cases cited earlier, while d^3 and d^6 configura-

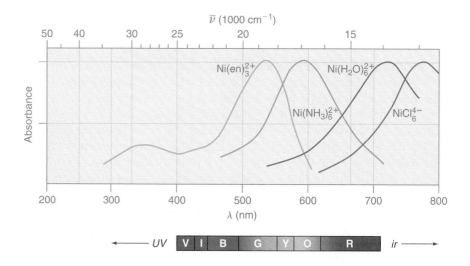

Figure 17.9
Solution absorption spectra of four nickel(II) complexes, illustrating the spectrochemical series. The UV portion of the $Ni(en)_3^{2+}$ complex spectrum is shown as an example; all the others absorb in the UV as well. A shift toward the blue indicates spontaneous ligand displacement and a more stable complex.

tions are often unusually stable. There is nothing "special" about these configurations based on the behavior of the bare ions; the latter do not correspond to a half-filled or filled d subshell.

Now assembled before us is an intricately correlated web of facts concerning this new type of molecular assembly, the coordination complex. A good model for these species should be able to explain both the trends observed and the correlations among them, in both a qualitative and semiquantitative way. Here we examine an early physical model that in a straightforward way helps us to understand the color–magnetism correlation, but is not as useful in explaining why a given ligand sits where it does in the spectrochemical series. Then we attempt to put chemistry into the model, that is, to provide an interpretation that invokes the bonding concepts developed in Chapter 7 and already used here to interpret the spectra of the oxyanions.

The Crystal Field Model

We have already noted that compounds of transition metals make colored solids as well as solutions; transition metal impurities in glass and clear crystals also impart color. H. Bethe, a physicist more noted for his later work in nuclear physics and astrophysics, saw in 1929 that the newly discovered d orbitals that transition metals were now known to possess, when placed in the electric field of a surrounding lattice of negative charges, would become split in energy. This might in turn explain both the visible absorption and the peculiar magnetic properties of transition metals in such environments; Bethe was particularly interested in the latter. Thus was born the **crystal field model** of coordination complexes.

Figure 17.10 reprises the boundary surfaces of the $3d$ orbitals from Figure 3.5. Imagine that you are an electron wave forming one of these multi-lobed orbitals, and a spherical shell of charge, equivalent to six pairs of electrons, is brought up, surrounding you. Because of Coulomb repulsion, your energy will rise by the same amount, no matter which d orbital you happen to be, and the orbitals remain degenerate. Now imagine these six electron pairs to be "concentrated" along the x, y, and z coordinate axes, as in an octahedral complex. Now it *does* matter which orbital you have chosen to be. If your lobes point *toward* the ligand electron pairs, you will suffer more Coulomb destabilization (energy increase) than if they point *between* the pairs. The octahedral electric field thus causes the degeneracy of the five d orbitals to be lifted, that is, it *splits their energies* into two groups. The d_{z^2} and $d_{x^2-y^2}$ orbital lobes point along the coordinate axes, toward the ligands; they are therefore raised in energy* relative to the d_{xy}, d_{yz}, and d_{xz} orbitals, whose lobes point exactly between the ligands. The two new energy levels created are also illustrated in Figure 17.10. The labels t_{2g} and e_g are taken from the theory of *symmetry point groups,* which you may learn more about in an advanced course; and Δ_o denotes the energy spacing between the two levels.

The splitting Δ_o in the d-orbital manifold induced by an octahedral electric field has two consequences. First, electronic transitions may now occur between these two new levels, of wavelength $\lambda = hc/\Delta_o$, as long as there are orbital vacancies.

---

* It may not be obvious to you that the d_{z^2} and $d_{x^2-y^2}$ orbitals should rise in energy by the same amount. The reason they do is that the d_{z^2} "doughnut" in the xy plane suffers the same extent of Coulomb repulsion from the four in-plane ligands as a pair of lobes would from two ligands. (Doughnut repulsion is rare among humans, though.)

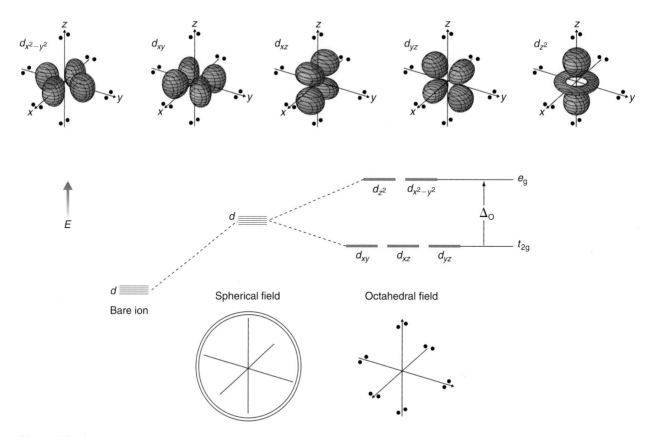

Figure 17.10
The five d orbitals in an octahedral field. The pairs of dots along the coordinate axes in the upper panel represent ligand electron pairs in an octahedral complex. The lower panel illustrates the effect of these pairs on the energies of the orbitals, and is called a crystal field energy diagram, where Δ_o is the crystal field energy splitting.

This provides a plausible interpretation of the visible absorption spectra, and also explains why complexes of Zn^{2+}, Cu^+, and Ag^+ are colorless. In these d^{10} ions, the t_{2g} and e_g levels are completely filled, and hence no transitions between them are possible. Second, electrons occupying these new levels must as usual obey the Pauli exclusion principle; if the splitting Δ_o is large enough, the electrons will also observe the Aufbau principle, and fully occupy the t_{2g} orbitals prior to the e_g. The compaction this causes may reduce the number of unpaired electrons, creating what we have called a low-spin complex. On the other hand, if Δ_o is smaller than the destabilization caused by pairing the electrons in t_{2g}, the *electron pairing energy*, the electrons will spread among all the available orbitals, both t_{2g} and e_g, in accord with Hund's rule. This produces a number of unpaired electrons identical to that in the bare metal ion, a characteristic of a high-spin complex. Furthermore, a correlation is predicted between color and spin: larger crystal field splittings should give rise to both shorter absorption wavelength *and* low spin, in agreement with the observations given earlier.

Those ligands that yield complexes with absorption in the red and high spin are called **weak field ligands;** they lie to the left in the spectrochemical series. The converse, those ligands producing low-spin blue absorbers, are **strong field ligands** and lie to the right. A particularly enlightening comparison can be made between two cobalt(III) d^6 compounds, the pale yellow-green paramagnetic complex CoF_6^{3-} and the bright orange diamagnetic $Co(NH_3)_6^{3+}$; their characteristics are given in

TABLE 17.4
A tale of two complexes

Complex	Color	Δ_o (cm^{-1})	Unpaired spins	K_f
CoF_6^{3-}	Yellow-green	13,000	4	$\ll 1$
$Co(NH_3)_6^{3+}$	Orange	22,900	0	1×10^{35}

Table 17.4. The peak absorption wavelengths and energies given there correspond closely to estimates you could make based on Table 17.1, and also to the relative locations of F$^-$ and NH$_3$ in the spectrochemical series of Equation 17.4. Magnetism and stability also correlate with position in the series. Crystal field energy diagrams for the two complexes are shown in Figure 17.11. F$^-$ clearly behaves as a weak-field ligand, producing a small level spacing that allows the electrons to spread over the two levels and maximize unpaired spin. NH$_3$, on the other hand, is a strong-field ligand in this particular case, yielding a large Δ_o and forcing the electrons to pair, filling the t_{2g} orbitals. The crystal field model thus connects color and magnetism in a natural way.

Further, the d^3 and d^6 configurations found to be especially stable are readily identified as half-filled and filled t_{2g} levels; the source of their stability is the same as that for the half-filled and filled atomic subshells. The low lability of the hexaamminecobalt(III) complex reflects its noble-gas-like filled t_{2g} orbitals. On the other hand, the *in*stability (redox lability) of d^4 reflects the "unhappy electron" problem: the fourth electron must either pair with one in t_{2g} or occupy a higher energy e_g orbital. The two ways out of this dilemma lie in either losing the fourth electron to yield a half-filled $(t_{2g})^3$, or gaining a fifth to make a half-filled d^5. As in Co(II) complexes, the d^7 the configuration is also found to be unstable toward oxidation for strong-field ligands, since the seventh electron must be promoted to the e_g level, whence it is readily lost. On standing in air, the violet $Co(NH_3)_6^{2+}$ complex slowly turns orange.

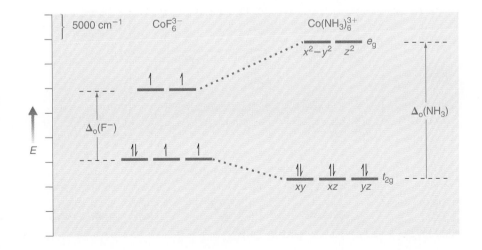

Figure 17.11
Crystal field energy diagrams for the fluoro and ammine complexes of Co(III).

Tetrahedral and Square-Planar Complexes

The crystal field model can also be readily adapted to treat the other dominant coordination number, 4, which occurs in either tetrahedral or square-planar ML_4 complexes. Transition metals sometimes show tetrahedral halo or hydroxo complexes, such as $CoCl_4^{2-}$ or $Zn(OH)_4^{2-}$, while Cu^{2+}, Pd^{2+}, and Pt^{2+} display coordination number 4 and square-planar geometry in the majority of their complexes. With only four ligand electron pairs, the overall crystal field will be weaker than the octahedral case for a given ligand, and which orbitals suffer greater repulsion will change, as illustrated in Figure 17.12.

If the ligands in a *tetrahedral* complex are arranged at diagonal corners of a cube surrounding the metal ion, you can readily see that now the orbitals that point *between* the axes are pointing *at* the ligand pairs, and are raised in energy: the lower level now contains the z^2 and x^2-y^2 orbitals, just the opposite of the octahedral field. In *square-planar* geometry, with the ligand electrons in the xy plane along the axes, only x^2-y^2 points at the ligands and is strongly destabilized; the others are perturbed to lesser extents, in-plane xy more than z^2, which leaves only its doughnut in-plane, and out-of-plane xz and yz least. Square-planar complexes have often been found to show absorption peaks in the infrared as well as visible regions; this is readily interpreted in terms of the four energy levels in the d manifold. This level pattern gives special stability to d^8 configurations, as typified by Pt(II). The square-planar complex diammine-*cis*-dichloroplatinum(II), a neutral compound without a counterion known commercially as *cisplatin*, can survive the human digestive system intact, and has been found to have antitumor activity.

The easily obtained and intuitive energy-level patterns predicted by the crystal field model are useful tools. But, to return to our Co(III) examples, what the crystal field model fails to explain is *why* F^- produces a weaker field than NH_3. Once you inform the model of the geometry and splittings, then predictions of magnetism and lability can be made—or *vice versa*—but the model cannot predict both. In fact, you might expect that F^-, carrying a negative charge, should produce a stronger electric

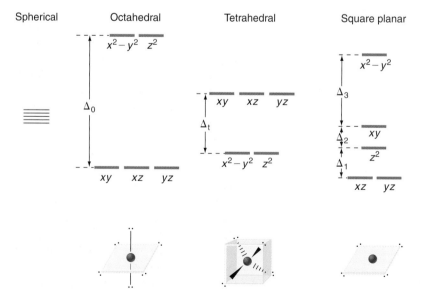

Figure 17.12
Comparison of crystal field energy diagrams for octahedral, tetrahedral, and square-planar complexes.

field around the metal ion than NH_3, which has only a dipole moment. F^- should be the ligand showing more stable, less magnetic complexes—in conflict with the facts. The difficulty is that the crystal field model does not consider the chemical nature of the ligands, that is, that their electrons are also orbitals, and that a degree of covalent bonding might be involved. A better model is needed.

The Ligand Field Model

Surely the chemists of the 1920s and 1930s wouldn't permit such a purely physical theory as the crystal field model to dominate their thoughts. In fact, they didn't; the crystal field theory was largely ignored for more than 20 years after it was proposed. We have already pointed out how the metal ion + ligand interaction can be viewed as a Lewis acid–base pairing. It shouldn't surprise you, then, that Linus Pauling soon extended his hybrid orbital valence bond model of Chapter 6 to include coordination complexes; in octahedral complexes the metal ion would employ d^2sp^3 hybrids in making the requisite six equivalent bonds.

By the late 1930s, Pauling's views had become dominant among transition metal chemists, as they had among chemists in other areas, particularly organic. The virtue of the valence bond (VB) model was its emphasis on electron sharing and covalent bonding in complexes, a facet ignored by the crystal field analysis; at the least VB could rationalize octahedral geometry in terms of orbital overlap. The new molecular orbital (MO) theory of Mulliken and Hund, not accommodating to Lewis's or Pauling's ideas, remained likewise in the background, though as early as 1935 J. Van Vleck, a physicist specializing in magnetism, pointed out that molecular orbitals might be useful for complexes. For us, the virtue of the MO approach will be its emphasis on energy levels, which are essential for understanding color and magnetism, and the chemistry of their origins. The application of the molecular orbital model to coordination complexes began in earnest in the late 1950s, particularly under the influence of C. Ballhausen, and has come to be called the **ligand field** model. As you will see, this model has the same advantages (and shortcomings) as the MO theory of Chapter 7 did for smaller molecules.

The procedure for applying the ligand field model to coordination complexes is similar to that we've already seen for the oxyanions. The metal-ion levels are stacked up on the left, the ligand levels all together on the right, and the MOs resulting from linear combinations (overlapping) of the fragment orbitals take center stage. The ligand is often a molecule or molecular ion itself, giving this new scheme a complexity beyond what you may have seen before. So let's start by assuming an atomic ligand, such as a halide ion, which presents a σ-symmetry valence orbital for bonding to the metal, as in the oxyanions. Figure 17.13 illustrates the resulting MO level diagram for an octahedral complex. To make six M-L bonds, the d orbitals must get involved, as Pauling had already shown, but only two of them have the necessary symmetry (called e_g) to overlap with the ligand orbitals. The two chosen orbitals are $d_{x^2-y^2}$ and d_{z^2}, *the very same orbitals that were destabilized in the crystal field model*. Here pointing at the ligands is favorable, that is, stabilizing, since delocalization of the electrons becomes possible through overlap. Because the $3d$ orbitals are from an inner shell, however, the overlap is weaker than it is for $4s$ and $4p$, and the MOs that result are not as strongly bonding. Nonetheless six bonding and six antibonding MOs are formed that fall into three bonding and three antibonding levels as shown. We label them as σ and σ^*, with subscripts reflecting their metal-ion parentage. As in the

Figure 17.13
Ligand field MO energy-level diagram for an octahedral complex. Electron occupancy is shown for an 18-electron weak-field complex such as CoF_6^{3-}. The identification with the crystal field levels is shown at the right.

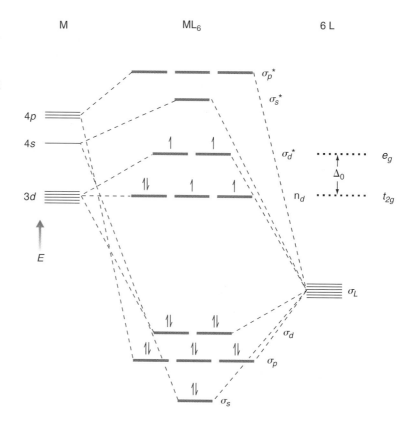

oxyanions, all the bonding levels must lie below the ligand donor levels, and all the antibonding above their respective metal-ion parents. The remaining d_{xy}, d_{xz}, and d_{yz}, orbitals become *nonbonding* (n) in this σ-only picture.

As we learned to do in Chapter 7, we now feed the available electrons into the energy levels, lowest first (Aufbau principle) and only two spin-paired electrons in the same orbital (Pauli principle). Let's suppose the ion is Co^{3+} (d^6); this gives a total of 18 electrons, 12 from 6L and 6 from M. The first 12 fit nicely in the bonding MOs, and the last 6 must occupy the nonbonding and perhaps antibonding levels. As you can see, the n and σ_d^* levels can be identified with the t_{2g} and e_g levels in the crystal field model, and the spacing between them with the crystal field splitting Δ_o. How those levels are occupied then involves the same considerations: if the splitting Δ_o is small, the electrons will spread and unpair, giving a high-spin complex, while pairing in the $n(t_{2g})$ and low spin will result if Δ_o is large. Now, as the covalent bonding becomes stronger (i.e., the complex becomes more stable), the σ levels will fall in energy *and the σ^* levels will rise in energy,* a reflection of more strongly interacting orbitals. This increases the spacing (Δ_o) between the n and σ_d^* levels; therefore, *the splitting Δ_o reflects the strength of the bonding and the stability of the complex.* Here we have the ingredient that was missing in the crystal field model. In this σ-only model, what counts is the ability to donate electrons, and we already know that, for example, NH_3 is a better donor than H_2O, which in turn is better than F^-, in accord with their positions in the spectrochemical series. In comparing CoF_6^{3-} and $Co(NH_3)_6^{3+}$, we can now rationalize the differences in their properties, beginning with an examination of the chemical nature of the ligands.

Other factors are also at work in determining the spectrochemical ordering. Negatively charged ligands repel each other, destabilizing their complexes, especially those that carry an overall negative charge. This coulombic factor makes OH^- a less effective ligand than H_2O. Further, the ligand field model can be refined to include π-type M—L interactions. These engage the otherwise nonbonding t_{2g} orbitals, which have π symmetry with respect to the ligand bond axes, and thereby influence Δ_o. As illustrated in Figure 17.14, if the ligand orbital is a filled $p\pi$, as in the halogens, it will destabilize the t_{2g} orbitals, and decrease Δ_o, while if it is an empty π^* ligand MO, as in CN^- or CO, it will stabilize them and increase Δ_o, by what is termed **back-bonding**. In back-bonding, electrons residing in the t_{2g} level delocalize into the ligand π^*, giving a new source of bond strength and stabilizing the complex as a whole. Experimental proof of the back-bonding phenomenon has been obtained from electron spectroscopy of CO bound to a Ni surface, which shows an extra peak at the expected energy of the π^* orbital. This interaction accounts in part for the great stability of cyano and carbonyl complexes. When bound to iron in human blood, these ligands completely block out O_2, making them deadly poisons.

The ligand field model also can be formulated for tetrahedral and square-planar ML_4 complexes, and carries the same additional dimension that incorporates the ligand chemical properties. As we have just seen for the octahedral case, the energy-level pattern that LF yields in the vicinity of the d level is the same as in the crystal field model. Tetrahedral complexes such as $CoCl_4^{2-}$ are a mild extension of our simple MO model for the tetrahedral oxyanions, adding a d orbital splitting due to involvement of the xy, xz, and yz orbitals in bonding. Further analysis of bonding in ML_4 complexes is reserved for the end-of-chapter exercises.

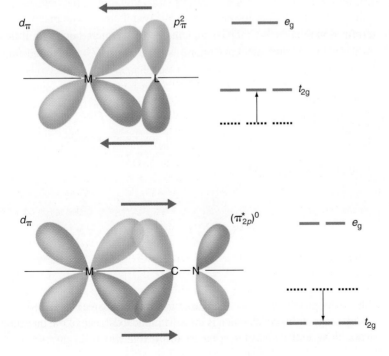

Figure 17.14
The effect of π-type interactions on the otherwise nonbonding t_{2g} orbital energies. Filled π-symmetry ligand orbitals destabilize t_{2g}, while unfilled ones, such as the π^* orbitals in CN^- or CO, stabilize it. A more complete treatment would involve a new pair of bonding and antibonding π-type MOs in each case.

17.5 Chemical Kinetics of Complexes

The colorful ligand displacement reactions of Ni^{2+} described earlier require no long waits; to the unaided eye they appear to be instantaneous, occurring as fast as the solutions can be mixed. Rapid reactions like that are a sure sign of low activation energies E_a, as we have already claimed for ligand displacement, usually called **ligand substitution** in the literature. As we will see in this subsection, rapid rates go hand in hand with relative instability. Most ligand substitution reactions are so fast that, until the development of methods for ultrafast solution kinetics in the late 1950s (see Chapter 15), the rates of only a few such reactions could be measured. These studies, dating back to the days of Werner, nearly universally employed the especially stable ammine complexes of Co^{3+} and Pt^{2+}, where the ligands are bound more tightly and can be exchanged only slowly; the kinetics of these complexes remain cardinal examples for understanding substitution rates.

Ligand Substitution

Ligand substitution (LS) occurs stepwise, as we outlined in Chapter 12, and the most informative studies isolate the displacement of a single ligand L by another X,

$$ML_n + X \longrightarrow ML_{n-1}X + L \tag{17.5}$$

For low [X] the rate law for LS is typically found to be first order with respect to both $[ML_n]$ and [X], second order overall:

$$\text{Rate} = k_{LS}[ML_n][X] \tag{17.6}$$

For octahedral aqua complexes $M(H_2O)_6^{2+}$ the rate constants k_{LS} are all large, but vary over a wide range, roughly 10^2 to 10^9 L mol^{-1} s^{-1}; the rates for the corresponding 3+ complexes are lower, 10^{-6} to 10^5 L mol^{-1} s^{-1}. The magnitude of k_{LS} is determined mainly by the identity and charge of the metal ion, with little dependence on the nature of the attacking ligand X. The largest k's are found for $Cu(H_2O)_6^{2+}$, where the rate is *diffusion controlled* (see Chapter 15); while $Cr(H_2O)_6^{3+}$ shows the lowest rate. (The aqua complex of Co^{3+} is unstable toward spontaneous reduction to Co^{2+}.) For large [X], the measured k's decline, and the reaction no longer follows a simple second-order rate law.

These observations, combined with key experiments on slow-reacting Co(III) complexes such as $[Co(NH_3)_5H_2O]^{3+}$, suggest a *dissociative interchange* (I_d) mechanism for H_2O dispalcement:

$$ML_6 + X \underset{k_{-OC}}{\overset{k_{OC}}{\rightleftharpoons}} [ML_6\text{---}X] \tag{17.7}$$

Slow $\quad [ML_6\text{---}X] \xrightarrow{k_{-L}} [ML_5\text{---}X] + L$

Fast $\quad [ML_5\text{---}X] \xrightarrow{k_a} ML_5X$

where L = H_2O and X can be a variety of ligands, usually NH_3, SCN^-, or CN^-. A key element of this mechanism is the postulated existence of the intermediate species $[ML_6\text{---}X]$, called an **outer-sphere complex,** in which X is in a secondary shell surrounding the ML_6 complex. The other positions in this outer shell are occupied by water molecules, and the structure of this shell is much less well defined in number and arrangement than ML_6 itself. Formation of the outer-sphere complex prior to loss of an aqua ligand is required if X is to displace H_2O; otherwise, with no X "waiting in the wings," a solvent H_2O will be exchanged instead, leading to no apparent reaction. Using the slow-step (bottleneck) rule, you can readily show that the rate law of Equation 17.6 can be recovered if you take

$$k_{LS} = K_{OC} k_{-L} \tag{17.8}$$

where $K_{OC} = k_{OC}/k_{-OC}$ is the equilibrium constant for forming the outer-sphere complex.

The loss of H_2O, which is the only endothermic step in the mechanism of Reaction 17.7, is a natural bottleneck. In view of Equation 17.8, this ensures that k_{LS} will be insensitive to the identity of X, since K_{OC} is determined mainly by random diffusion, and is typically close to unity. (Exceptions can occur for a 3+ or 4+ complex forming an outer-sphere complex with a negatively charged ligand, which increases K_{OC} and leads to a more complicated rate expression, such as those we examined using the steady-state approximation in Chapter 15.) The slow step will have an activation energy roughly equal to the bond strength of the M—L bond, $E_a \approx D(M—L)$; thereby *the rate varies inversely with the stability of the aqua complex,* as we argued earlier on more general grounds. This stability in turn can be traced to the complex-forming tendencies of the metal ions involved. The large k_{LS} of $Cu(H_2O)_6^{2+}$ reflects the preference of Cu^{2+} for square-planar complexes, implying that the two axial H_2O ligands are particularly easily lost. The next largest k_{LS} among common aqua complexes occur for $Mn^{2+}(d^5)$ and $Zn^{2+}(d^{10})$, whose already stable configurations discourage strong bonds to ligands. At the other extreme, the $Cr^{3+}(d^3)$ aqua complex reacts most slowly (though it appears fast in a human time frame), owing to the greater charge on the metal ion and the half-filled t_{2g} subshell. H_2O always forms weak-field octahedral complexes, which stabilize d_3 but not d^6 configurations; accordingly, we find that d^6 $Fe(H_2O)_6^{2+}$ is just as labile as d^7 $Co(H_2O)_6^{2+}$, although not at the level of Mn^{2+} or Zn^{2+}.

As expected, k_{LS} decreases with growing stability of the reagent complex, due to the increased difficulty of breaking the M—L bond. Complexes with multidentate ligands require that two or more bonds to M must be broken contemporaneously before substitution can occur, raising E_a and further shrinking k_{LS}; the $Ni(en)_3^{3+}$ complex is $\sim 10^9$ times more inert to LS than $Ni(NH_3)_6^{2+}$.

Oxidation–Reduction

Oxidation–reduction (redox) reactions among metal ions in solution are also generally very fast, making them useful in analytic titrations; a common example is the $MnO_4^- + Fe^{2+}$ reaction given in Chapter 5. You can now appreciate that this reaction really involves the complex $Fe(H_2O)_6^{2+}$, which is oxidized to $Fe(H_2O)_6^{3+}$. In redox of complexes, the question naturally arises as to whether the electrons that are transferred must travel through the shell of ligands, or whether perhaps the loss or displacement of a ligand is involved. Reactions involving species like MnO_4^-, which require multiple electron transfer steps, show complicated kinetics, as Harcourt and Essen discovered in the early days (see Exercise 1 of Chapter 15), while, as in the LS reactions, extensive kinetic studies were not feasible until recently owing to the rapidity of many of these reactions.

Here we consider simple one-electron transfer reactions, such as the isotopically labeled reaction between $Fe^{2+}(aq)$ and $Fe^{3+}(aq)$ used as an example in Chapter 16. Such reactions are also found to obey second-order kinetics, again with a wide range of rate constants that in this case are sensitive to both metal ion and ligand. For octahedral complexes we can write the general reaction as

$$ML_6^{n+} + M'L_6'^{m+} \longrightarrow ML_6^{(n+1)+} + M'L_6'^{(m-1)+} \qquad (17.9)$$

where the unprimed complex is the reducing agent, $Fe(H_2O)_6^{2+}$ in our example, and the primed complex the oxidizing agent, the Fe^{3+} aqua complex. If M and M' are different metals (a more common case), Reaction 17.9 is called a *cross reaction*. The second-order rate law,

$$\text{Rate} = k_{OR}[ML_6][M'L_6'] \qquad (17.10)$$

is found to hold over a wide range of systems and concentrations. With a few exceptions, such as the slow-reacting amminecobalt(III) complexes, the rate constants k_{OR} range from 1 to 10^7 L mol^{-1} s^{-1}. Reactions like Reaction 17.9 in which the ligand shells are not disturbed are thought to occur by an **outer-sphere mechanism.** Much like the ligand substitution mechanism, electron transfer is preceded by formation of a loosely associated outer-sphere intermediate:

$$ML_6^{n+} + M'L_6'^{m+} \underset{k_{-OC}}{\overset{k_{OC}}{\rightleftharpoons}} [ML_6^{n+} \text{---} M'L_6'^{m+}]$$

Slow $\quad [ML_6^{n+} \text{---} M'L_6'^{m+}] \xrightarrow{k_e} [ML_6^{(n+1)+} \text{---} M'L_6'^{(m-1)+}] \qquad (17.11)$

Fast $\quad [ML_6^{(n+1)+} \text{---} M'L_6'^{(m-1)+}] \xrightarrow{k_d} ML_6^{(n+1)+} + M'L_6'^{(m-1)+}$

Again, by the slow-step rule, you can easily show that the result is also analogous to LS, a second-order rate law with

$$k_{OR} = K_{OC}k_e \qquad (17.12)$$

where $K_{OC} = k_{OC}/k_{-OC}$ is the equilibrium constant for forming the precursor complex.

While evidence in support of the outer-sphere mechanism comes mainly from studies on a few key systems, its "slow" electron transfer step, with unimolecular rate constant k_e, enables the great sensitivity of the rate to both metal and ligand to be readily rationalized. Metals that are able to donate and accept t_{2g} electrons are found to react more rapidly. As illustrated in Figure 17.15, because the lobes of the t_{2g} orbitals protrude *between* the ligands, into the edges or faces of the octahedron, they more readily donate or accept electrons by overlapping neighboring orbitals from the other complex. The e_g orbitals are busy bonding with the ligands and are not as exposed to external overlap. Moreover, unlike ligand transfer, the actual transfer of the electron occurs in a very short time (~1 fs), during which the nuclei in the complex can move very little. Therefore, if the bond lengths $r(M\text{—}L)$ in the product complex are appreciably longer or shorter than in the reagent, electron transfer will be inhibited by the energy needed to make a dis-

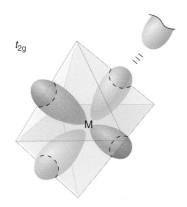

t_{2g}

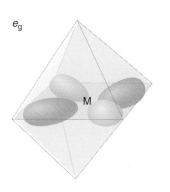

e_g

Figure 17.15
Rationale for the relative ease of electron transfer by t_{2g} orbitals. Pointing between the vertices of the octahedron where the ligands are located, these orbitals more readily overlap other donor or acceptor orbitals than do e_g ones.

torted product. This distortion energy further mitigates against transfer of e_g electrons, which are actually antibonding orbitals in the ligand-field picture; gaining or losing these electrons will change the bond order and, hence, also the bond lengths. The extent of bond-length change depends on the ligands; strong-field ligands form more robust bonds to the metal, and a change in the e_g electron population will perturb their bond lengths more strongly, further reducing the rate of redox through the e_g level.

The basic electron transfer step in the outer-sphere mechanism occurs *through space,* that is, through the direct interaction of the "hairy tails" of the donor and acceptor orbitals. Thus, as you might expect, bulky molecular ligands that hold the complexes in the outer-sphere precursor further apart also reduce the electron transfer rate. However, the large bidentate ligand *phenanthroline* (phen), three fused benzene rings with nitrogens at site-bridging positions, is found to accelerate redox; its π-bonded network apparently provides a *through-bond* pathway for electrons to flow from one metal center to the other—rather like connecting the two metal ions with a wire!

The possibility of a through-bond electron transfer pathway brings us to a second, more recently discovered class of redox reaction mechanism, the **inner-sphere mechanism.** A hallmark of this mechanism is *ligand exchange* between the reacting complexes accompanying electron transfer. Here it is best to use a particular example, to avoid drowning in a sea of primes. Consider the following reaction:

$$Co(NH_3)_5Cl^{2+} + Cr(H_2O)_6^{2+} + 5H^+ \longrightarrow Co(H_2O)_6^{2+} + Cr(H_2O)_5Cl^{2+} + 5NH_4^+$$
(17.13)

The essential result is a transfer of one electron from the Cr(II) complex (with a highly redox-labile d^4 configuration) to the Co(III). Carrying the reaction out in acid causes protonation and loss of the remaining ammine ligands from the reduced Co(II) center. The transfer of the Cl^- ligand from Co to Cr strongly suggests that, subsequent to the formation of an outer-sphere complex, the more labile Cl^- becomes a *bridging ligand,* forming the *inner-sphere* intermediate $[(NH_3)_5Co$ -- Cl^- -- $Cr(H_2O)_5]^{4+}$ in a unimolecular second step

$$[Co(NH_3)_5Cl^{2+}\text{---}Cr(H_2O)_6^{2+}] \xrightarrow{k_b} [(NH_3)_5Co\text{--}Cl^-\text{--}Cr(H_2O)_5]^{4+} + H_2O \quad (17.14)$$

that precedes the electron transfer step. This requires that both the bridging ligand and the one it displaces be relatively labile toward ligand substitution. The Cl^- ion is primarily a σ donor, and its bridging role puts the e_g orbitals on the Cr and Co centers in direct contact, facilitating electron transfer between them. The redox rate constant for the overall reaction of Reaction 17.13 is $k_{OR} = 5 \times 10^4$ L mol^{-1} s^{-1}, many orders of magnitude faster than typical rate constants for either Co(III) OR or Cr(III) LS, further suggesting a new reaction pathway involving through-bond transfer.

In biological electron transfer involving a metal center, the metal is frequently bound in a porphyrin-like ring incorporated into an enzyme, and may be 10 Å or more away from its redox partner. At this great distance, through-space electron transfer is nearly impossible, and the electron must travel through bonds instead in order to permit reaction to occur at a biologically useful rate.

17.6 The Lanthanides and Actinides

We conclude this chapter with a brief look at the bottom fringe of the periodic table, (Figure 17.16) at the elements usually left to "dangle" below the main groups in 14-element rows for periods 6 and 7. Which elements are "qualified" to be listed as lanthanides (after the first element lanthanum, La, which is not dangled with its followers!) and actinides (likewise named for but not including actinium, Ac) is a matter of some debate among chemists who study them. It is your author's opinion that lanthanum and actinium should join their dangling neighbors, for reasons given shortly, although Figure 17.16 shows the conventional arrangement. (The question would be moot were we to use the "long form" of the periodic table of Figure 4.6.) These elements are sometimes called the *inner transition metals,* and the lanthanides are still known in many circles as the *rare earths,* although they are not particularly rare compared to their near neighbors.

The lengths of the lanthanide and actinide rows reflect the 14-electron capacity of the $4f$ and $5f$ orbitals being occupied there; the f orbitals are shown in Figure 17.17. A typical electron configuration for these elements is $ns^2(n-2)f^x$, although several of them have an anomalous electron or two in the neighboring $(n-1)d$ level (see Table 4.2 of Chapter 4). Like the d-block elements, however, the chemical properties of these elements are determined mainly by their ionic forms, and these are all pure $(n-2)f^y$ configurations. Here we will briefly point out trends and familial properties of the lanthanides, and confine our discussion of actinide chemistry to the (in)famous element uranium.

The energies of the $4f$ orbitals are probed in the lanthanides by IE_3, just as those of the $3d$ orbitals were for the first transition series. They show a pattern quite analogous to $3d$ (Figure 17.2), generally increasing across the row, with exceptionally high IE_3's for europium (Eu) $(4f^7)$ and ytterbium (Yb) $(4f^{14})$. Lutetium (Lu), the last lanthanide element listed, has a lower IE_3 than Yb, well suited to begin the $5d$ series.

Figure 17.16
Periodic table again, now emphasizing the f-block elements.

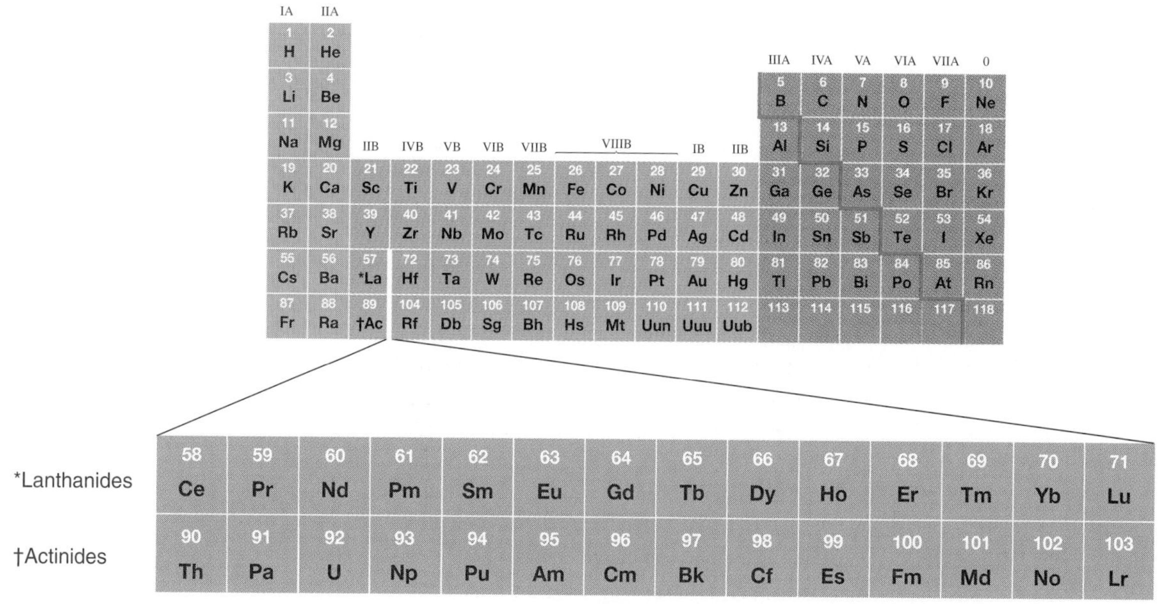

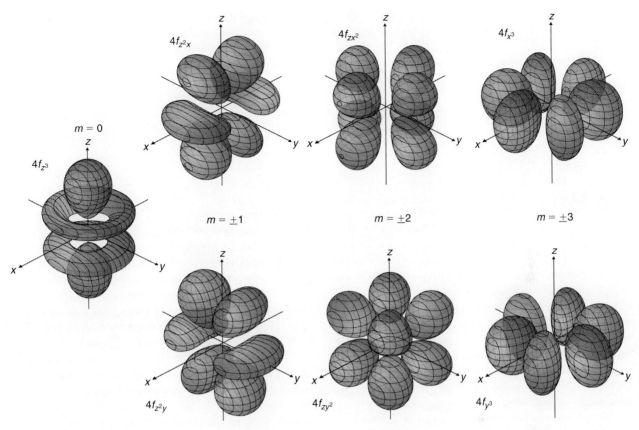

Figure 17.17
Boundary surfaces for the 4f orbitals. The orbital functions are constructed from Table 3.1, using Euler's formula to form real orbitals as in Equation 3.10.

All the lanthanides show +3 as the most stable charge state, although Ce^{4+} is kinetically stable in the absence of a reducing agent. Unlike their *d*-block counterparts, the M^{3+} ions have radii that slowly, monotonically decrease across the row. Here this smooth behavior of r_{ion} reflects the nearly total absence of covalent interaction involving the 4f subshell. These orbitals are now *two* principal quantum numbers lower than the valence shell, and as a result are small, tightly bound, and not readily overlapped by ligand orbitals. Of all the metals, save groups IA and IIA, these make the most naturally cationic species in compounds and in solution. Several of them show sharp line-like spectra as impurities in glasses and even in aqueous solution, another indication of the lack of covalent bonding to water or other species.

As pure metals, the lanthanides are very active; all of them react with water to yield $M(OH)_3(aq)$ and H_2 gas. Owing to the similar radii of their ions, they occur naturally mixed in ores such as *monazite* (MPO_4). Different formation constants K_f for their chelates with EDTA, increasing by about a factor of 10^4 from La to Yb, are used to separate them by a technique known as *ion-exchange chromatography*. The high spins possible with the *f* subshell make the metals, especially samarium (Sm), valuable as constituents of powerful permanent magnets. The chief oxides of all the lanthanides have the expected formula M_2O_3, and all dissolve in acid. The M^{3+} ions in solution form weakly bound complexes with variable numbers of ligands; since these involve very little covalent bonding, repulsive or attractive π-type interactions are absent, and a spectrochemical series is not well defined for the *f*-block metals.

By contrast, the actinides show a greater variety of charge states and oxides, none more so than uranium, whose known oxides include UO, UO_2, U_4O_9, U_2O_5, U_3O_8, and UO_3. The U_3O_8 oxide is the principal constituent of *pitchblende*, the ore made famous by the Curies. When the oxide is dissolved in nitric acid, the uranyl ion UO_2^{2+} is formed, having a linear O—U—O structure. The oxides can be converted to fluorides with suitable fluorinating agents; UF_6 is a liquid at room temperature that boils at 64°C. The combined stability and volatility of UF_6 made it possible to use gaseous diffusion through hundreds of stages to achieve enrichment of the fissionable ^{235}U isotope.

SUMMARY

Differences in the chemical behavior of the transition metals can be traced to the d-electron configurations of their 2+ and 3+ ions. The ions show periodic properties such as ionization energy and radius analogous to those of the neutral representative elements, and their charge states and bonding propensities can be understood in terms of these properties. In the 3d period, the early **transition metals** often exist as oxyions such as VO_2^+ or CrO_4^{2-}, while the later ones are more often found as "bare" metal ions in solution. The tetrahedral MO_4 oxyanions are analyzed using a simplified molecular orbital picture, and the bright colors of CrO_4^{2-} and MnO_4^- are attributed to **charge-transfer transitions** between oxygen-like MOs and empty, nonbonding metal 3d orbitals. The variety of colors seen in solutions of the late transition metals, on the other hand, is attributed to the formation of **coordination complexes** between the metal ion and surrounding water or other electron donors called **ligands.** Color is a consequence of absorption of visible light by electrons within the absorbing species, and Table 17.1 enables the association of various observed colors with the energy change ΔE of the electronic transition, so-called "eyeball spectroscopy."

The earliest evidence for the existence of 4- and 6-coordinate complexes was chemical rather than spectroscopic. In the 1890s Alfred Werner used peculiar stoichiometries and the existence of geometrical isomers of transition metal compounds to argue for the existence of octahedral ML_6 and square-planar ML_4 complexes. A nomenclature was developed for these complexes, and in the intervening century a variety of electron-donor ligands has been found to form complexes with the metals of varying stability, including multidentate ligands that can bind two or more sites around the metal simultaneously. Properties such as UV-visible absorption, magnetism, and lability were found to be correlated with the identity of the ligand, and the **spectrochemical series** (Equation 17.4) was developed, in which absorption moved to the blue, magnetism decreased, and both kinetic and thermodynamic stability increased along the series of ligands.

The first theory to be advanced to explain the phenomenon of complexation was the **crystal field model,** in which the degeneracy of the d orbitals is broken by the electric field (crystal field) generated by a regular geometric array of ligand electron pairs. In the **octahedral** case, the d_{z^2} and $d_{x^2-y^2}$ are found to be raised in energy relative to the d_{xy}, d_{xz}, and d_{yz}, forming two new energy levels denoted e_g and t_{2g}, respectively. These new levels enable electronic transitions if a vacancy exists in the e_g, immediately explaining why aqua **complexes** of all the mid to late transition metals show color except the d^{10} ions Zn^{2+} and Ag^+. The model also predicts that d^3 and d^6 ions are stable configurations, as observed. The ligands on the left of the spectrochemical series are known as **weak-field ligands,** and those on the right as strong field. **Strong-field ligands** often cause a reduction in the magnetism of the complex by creating a large crystal field splitting Δ_o, and forcing the electrons to pair up in the t_{2g} level.

The crystal field model cannot explain why neutral ligands such as NH_3 or ethylenediamine en are "stronger" than ions such as Cl^- or OH^-. For this we turn to the **ligand field** (molecular orbital) **model,** in which a more comprehensive account of the bonding, including its covalent component, can be given. As introduced in Chapter 7, the MO model involves the formation of bonding and antibonding molecular orbitals between metal and ligand orbitals of the proper symmetry. In a σ-bonding model the electron-donating ability rather than the charge of the ligand becomes paramount. The e_g energy level of the crystal field model can be identified with the σ_d^* antibonding molecular orbital, and the t_{2g} with nonbonding 3d (n_d); the splitting between these levels is now linked to the strength of the bonding in the complex, thus reproducing the observed correlation between blue-shifted absorption and complex stability. π-Symmetry ligand orbitals can perturb the n_d level; filled orbitals, such as halide $(np_\pi)^2$ destabilize it, while empty ones, such as the $(\pi_{2p}^*)^0$ MO of CN^-, stabilize it by enabling delocalization of the n_d electrons in an effect known as **back-bonding.** These additional factors complete the rationalization of the spectrochemical series.

The chemical kinetics of **ligand substitution** and **oxidation-reduction (redox) reactions** involving complexes is characterized by second-order rate laws and typically large rate constants. Proposed mechanisms involve formation of a precursor **outer-sphere complex** prior to ligand or electron transfer, with the slow step involving either loss of a ligand or through-space or through-bond electron transfer.

The lanthanides, also known as rare earths, are chemically much more similar to one another than the *d*-block metals, owing to the deep inner-shell nature of their 4*f* orbitals. They are active metals, reacting readily with water, adopting 3+ charge states, and forming basic oxides. In contrast, the actinide uranium forms a variety of oxides, the oxyion UO_2^{2+}, and a stable hexafluoride with a high vapor pressure, useful for isotope enrichment by effusion.

EXERCISES

Configurations, charge states, and radii

1. Give electron configurations and number of unpaired electrons for the following atoms or ions, using noble gas notation for core electrons:

   Fe^{2+} V^{3+} Zn Sc^{3+} Co^{2+}

2. Repeat Exercise 1 for:

   Ru^{3+} Mo^{4+} Pt^{2+} Bh^{2+} Au

3. State the oxidation number of vanadium (V) in each of the following compounds or ions:

   V_2O_3 VO_4^{3-} VO^{2+} VO_2^+ $V(NO_3)_3$

4. Among the 3*d* metals, manganese (Mn) displays the widest range of oxidation states. Give the oxidation number of Mn in each of the following:

   MnO_4^- $Mn(CO)_5$ MnO_2 MnO_4^{2-} Mn_3O_4 $MnCl_4^{2-}$

5. For the following pairs of ions, pick the one with the smaller radius:
   a. (Fe^{2+}, Fe^{3+}) (Co^{2+}, Fe^{2+}) (Pd^{4+}, Rh^{3+})
      (Sc^{3+}, Y^{3+}) (Cr^{3+}, Ti^{3+})
   b. Using atomic concepts state your reasoning for the pair of iron ions.

The early transition metals

6. Give balanced equations, classifications, and transition metal oxidation states for
   a. the formation of Mn_2O_7 from $KMnO_4$ and H_2SO_4.
   b. the reduction half-reaction converting MnO_2 to Mn^{2+} in acid solution.
   c. dissolving Sc_2O_3 in aqueous acid.
   d. dissolving V_2O_5 in aqueous acid.
   e. dissolving V_2O_5 in aqueous base.
   f. the reduction half-reaction converting V^{3+} to VO^{2+} in acid solution.

7. Rank the following oxides as potential oxidizing agents, strongest first:

   CrO_2 VO_2 MnO_2 TiO_2

   Briefly rationalize your ranking.

8. The absorbance $A = \log_{10}(I_0/I)$ of a solute in solution is proportional to both the path length L the light must travel through the solution and the concentration c of absorbing solute: $A = \varepsilon Lc$, where ε is a proportionality constant known as the *extinction coefficient*, and the relation stated is called the *Beer–Lambert law*. Justify the Beer–Lambert law by considering a thin layer of solute, thickness dL, which attenuates a light beam of intensity I by an amount $-dI = \alpha IcdL$, and integrating from 0 to L, where α is a proportionality constant. How is α related to ε? What are the units of ε in the cgs system?

9. A white solid does not absorb visible light. Rationalize the observation that Sc_2O_3 and TiO_2 are white oxides, while V_2O_5 is red. Assuming ionic bonding and a HOMO-LUMO charge-transfer transition, sketch energy-level diagrams to support your explanation.

10. The pertitanate ion TiO_4^{4-} has never been observed. Which nonmetal oxyanion does it most closely resemble? If it were synthesized, what would you predict for its color and absorption spectrum in solution? Draw a MO energy-level diagram to back up your answer.

11. Thus far chemists have been unable to make the iron oxide FeO_4. If it were to be made, what would you predict for its color and absorption spectrum? Draw a MO energy-level diagram to justify your prediction.

The late transition metals

12. State balanced equations for dissolving Fe_2O_3 in acid, ZnO in acid, and ZnO in base to form $Zn(OH)_4^{2-}$. To what class do these three reactions belong?

13. Aqueous Co(II) solution is pink (a pale red-violet) while aqueous Ni(II) is green.
    a. Which of these absorbs at the shorter wavelength?
    b. Give approximate sketches of the absorption spectra on appropriately labeled axes.
    c. Estimate the energy spacings ΔE in wavenumbers cm^{-1} between the energy levels responsible for the absorption for each solution.

14. Among the evidence used by Werner to establish the nature of octahedral coordination complexes were halide precipitation reactions with $Ag^+(aq)$. Only those halides not directly complexed to the metal ion could be precipitated. How many moles of AgCl per mole of chloro compound would be expected for the following salts?
    a. $Co(NH_3)_3Cl_3$
    b. $Co(NH_3)_4Cl_3$
    c. $Co(NH_3)_6Cl_3$

15. Although the VSEPR model predicts that four-coordinate complexes ML_4 will be tetrahedral, Werner's finding that $Pt(NH_3)_2Cl_2$ (diamminedichloroplatinum(II)) exists in two isomeric forms proved otherwise. By sketching both tetrahedral and square-planar geometries for this complex, show that the square-planar geometry leads to two geometrical isomers, but the tetrahedral to only one.

16. Give proper names to the following compounds:
    a. $Ni(CO)_5$
    b. $Co(NH_3)_6Cl_3$
    c. $[Cr(H_2O)_6](NO_3)_3$
    d. $K_3Fe(CN)_6$
    e. $K_4Fe(CN)_6$
    f. $[Ni(en)_2(NH_3)_2]SO_4$
    g. K_2CuCl_4
    h. $Zn(H_2O)_6[Zn(CN)_4]$

17. Based on the equilibrium constants given in Table 17.3, arrange the following copper(II) complexes in order of increasing stability, least stable first:
    a. $Cu(NH_3)_4^{2+}$ $Cu(H_2O)_6^{2+}$ $Cu(en)_2^{2+}$ $CuCl_4^{2-}$
    b. Choose a pair of complexes from part (a) and write a balanced ligand displacement reaction in its spontaneous direction.

Crystal field model

18. Using crystal field energy diagrams, illustrate the electron configurations of the high-spin blue-green complex $Fe(H_2O)_6^{2+}$ and the low-spin yellow complex $Fe(CN)_6^{4-}$, drawing on Table 17.1 to estimate a relative energy scale in cm^{-1}. Classify H_2O and CN^- as weak- or strong-field ligands. How does the crystal field model explain the correlation between color and spin?

19. Assuming the formation of octahedral complexes, which of the following ions will show some loss of magnetism across the spectrochemical series? Predict the number of unpaired spins in each case.
    a. Co^{3+}
    b. Cr^{2+}
    c. Fe^{2+}
    d. V^{3+}
    e. Ni^{2+}
    f. Sc^{3+}
    g. Co^{2+}
    h. Fe^{3+}
    i. Cr^{3+}
    j. Mn^{2+}

20. Predict the number of unpaired electrons for $Ni(en)_2^{2+}$ based on both tetrahedral and square-planar structures, assuming that en acts as a strong field ligand. Based on the correlation between spin and stability, which geometry do you predict will be preferred?

Ligand field molecular orbital model

21. Give ligand field energy-level diagrams for the octahedral complexes $NiCl_6^{4-}$ and $Ni(NH_3)_6^{2+}$, using information from Figure 17.9 to establish a relative energy scale in cm^{-1}. How is the enhanced stability of the ammine complex reflected in your diagrams?

22. When $NH_3(aq)$ is added to $Co^{2+}(aq)$, the color changes from pale violet to red. Interpret this observation using the ligand field model, giving energy-level diagrams with orbital occupancies and sketches of the solution absorption spectra.

23. Use ligand field energy-level diagrams to compare the aqua complexes of Ti^{3+} (violet) and Cr^{3+} (blue). Draw your diagrams side by side with ligand levels unchanged. How can the difference in color of these two complexes be explained? Which do you predict will be more stable?

24. Compare the common iron complexes $Fe(H_2O)_6^{2+}$ (blue-green) and $Fe(H_2O)_6^{3+}$ (yellow-orange) using ligand field energy-level diagrams with energy scale

estimated from Table 17.1. Leave the ligand energy levels fixed, and note that, as electrons are removed from the Fe $3d$ level, its energy falls due to decreased electron–electron repulsion. Assume that H_2O acts as a weak-field ligand in each case.

25. As illustrated in Figure 17.4, the radius of Fe^{3+} (0.60 Å) is smaller than that of Fe^{2+} (0.70 Å). This affects both the electrostatic and covalent contributions to the Fe-ligand bonding in complexes.

    a. Compute and compare the electrostatic attraction between each of these ions and a water ligand using the formula for ion-dipole attraction from Chapter 14 and the radius (1.50 Å) and dipole moment (1.85 D) of the water molecule.

    b. What effect will the smaller ionic radius of Fe^{3+} have on covalent bonding in the aqua complex? How is the effect reflected in the ligand-field diagrams of Exercise 24?

26. Although F^- is negatively charged and the conjugate base of the weak acid HF, it lies to the left of water in the spectrochemical series. Give two reasons for this anomaly.

27. For which of the following possible ligands would you expect stabilization of complexes by π backbonding to be significant?

    H_2, CO, N_2, O_2, Br_2, $H_2C{=}O$, NO, NO^+

    For a case where you expect it, give an orbital sketch of the back-bonding interaction.

Kinetics of complexes

28. Use the steady-state approximation (SSA) to analyze the I_d ligand substitution mechanism of Reactions 17.7. Under what conditions does the SSA rate reduce to the second-order rate law of Equation 17.6?

29. a. Write a mechanism analogous to Reactions 17.7 that describes the reverse of the ligand substitution reaction (Equation 17.5).

    b. Given that the slow step now involves the breaking of the M—X metal-ligand bond, of energy $D(M{-}X)$, what is the expected activation energy E_a' of the reverse reaction? How are the E_a's for forward and reverse reactions related to the heat of reaction $\Delta H°$?

    c. If the forward reaction is highly spontaneous, implying that $D(M{-}X) \gg D(M{-}L)$, what is the consequence for the overall rate of the reverse reaction?

30. The rate constant for a typical redox reaction between metal complexes is 10^4 L mol^{-1} s^{-1}. Estimate the half-reaction time for initial 0.10 M concentrations of each reagent.

31. Use the steady-state approximation (SSA) to analyze the outer-sphere electron transfer mechanism of Equations 17.11. Under what conditions does the SSA rate reduce to the second-order rate law (Equation 17.10)?

32. Sketch an orbital picture of electron transfer between Cr^{2+} and Co^{3+} occurring through the bridging ligand Cl^-, as discussed in the text.

Lanthanides and Actinides

33. Give the expected electron configuration and number of unpaired electrons expected in each of the following ions:

    La^{3+} Sm^{3+} Er^{3+} Ce^{4+} Yb^{3+} Lu^{3+} U^{4+}

34. Write a balanced chemical equation describing the reaction of

    a. dysprosium metal with water.

    b. dysprosium oxide with hydrochloric acid.

    c. pitchblende with nitric acid.

35. Give electron dot and valence bond descriptions of the linear uranyl ion UO_2^{2+}.

36. Use Graham's law (Chapter 9) to predict the relative rates of effusion of $^{235}UF_6$ and $^{238}UF_6$. How many effusion stages are required to enrich the 235 isotope from its natural 0.72% to 3.0% (as used in power plants)? To 14% (as used in bombs)?

CHAPTER 18

The Chemistry of Carbon

Dr. Stanley Miller observes an experiment he first carried out with H. C. Urey in 1953. Passing a discharge through gases believed to be present in the earth's early atmosphere in the presence of water produces amino acids and nucleotide bases. See Section 18.8 for a discussion.

CHAPTER OUTLINE

18.1 What Makes Carbon Special?

18.2 Building Blocks of Organic Chemistry: The Hydrocarbons

18.3 Derivatives of the Hydrocarbons: Functional Groups

18.4 Benzene and Its Derivatives: The Aromatics

18.5 Orbitals in Organic Reactions: The Diels–Alder Reaction

18.6 Polymers

18.7 The Molecules of Life: Organic Motifs in Biology

18.8 The Chemical Dynamics of Life

Organic chemistry . . . gives me the impression of a primeval tropical forest, full of the most remarkable things, a monstrous and boundless thicket . . . .

Friedrich Wöhler (1835)

Recently the American Chemical Society's registry of chemical compounds recorded its 20-millionth entry. Of this number only 1 million or so are classified as inorganic compounds; the other 19 million are **organic.** Organic compounds are defined to be those containing carbon (C) in combination with other elements; inorganics include *everything else*, all the other 100+ elements in any possible combination. The overwhelming dominance of organic compounds is certainly unexpected on purely statistical grounds, yet we have every reason to believe this dominance will continue and perhaps increase in the future. Even more unlikely, the organic chemist's world revolves around a remarkably small number of the lighter elements, as whimsically illustrated in Figure 18.1. Just to learn what's behind the ubiquity of carbon compounds will be one of our principal tasks in this, the last chapter of this book.

This chapter is not only an end, but a beginning, the first to provide a glimpse of the area of greatest impact of "modern" chemical science. It is one thing to claim that all the world is made of atoms and molecules, and that in understanding them we may understand our own surroundings and ourselves. It is another to be able to say exactly how the molecules make our world what it is, a knowledge that, in the broad sweep of scientific history, we have only recently acquired. Our knowledge is not—and will never be—perfect. But the remaining questions about how the world works—for which chemistry may supply the answers—have reached a remarkably deep level as we begin the new millennium. Nowhere has this chemical impact been felt more strongly than in the sister subject of biology; we now can speak boldly of the chemistry of life.

It is a delightful irony that, in the early 19th century, organic chemistry was first wrested from the then-mysterious workings of living organisms by a simple experiment. The terms *organism* and *organic* both derive from the Greek root *organon* (οργανον), meaning a working tool or instrument. By the early 1820s chemists had isolated many substances produced by organisms, but these were believed to be possessed of "vital" powers, removed from the inanimate matter in the chemist's retort. Then, in 1828, a German medicinal chemist named Friedrich Wöhler accidentally discovered that heating crystals of the salt ammonium cyanate (NH_4OCN) produced urea, a compound commonly found in urinary excretions. A modern semistructural chemical equation for this transformation is

$$NH_4^+OCN^- \xrightarrow{\Delta} H_2N-CO-NH_2 \quad (18.1)$$

Emil Harmann Fischer (1852–1919) was a professor of chemistry at the University of Berlin. His resolution of the isomers of glucose using only chemical reactions gained him great notoriety, but he also was the first to synthesize a protein from amino acids and show that it could be broken down by enzymes in the same way as natural proteins. He was awarded the second Nobel Prize in chemistry in 1902.

Wöhler dryly noted that urea had been formed "without benefit of a kidney, a bladder, or a dog." Within 50 years of Wöhler's breakthrough, thousands of organic compounds had been synthesized in the laboratory, and the discovery of synthetic organic dyes had spawned the first chemical industry in Germany. Chemical investigations had indeed developed organic chemistry into a "working tool," one that remains a cornerstone of industry and society today.

The importance of organic chemistry, both historically and in today's world, is reflected in the prominent position it holds in college chemistry curricula, generally occupying a full year of lecture and laboratory experience. "Orgo" is required not only for a student choosing chemistry as her major field of study, but for biochemistry, biology, chemical engineering, pharmacology, and premedicine as well. In this chapter only the barest introduction can be given; your author hopes of course that you will pursue this most fascinating area further, regardless of your goals in life.

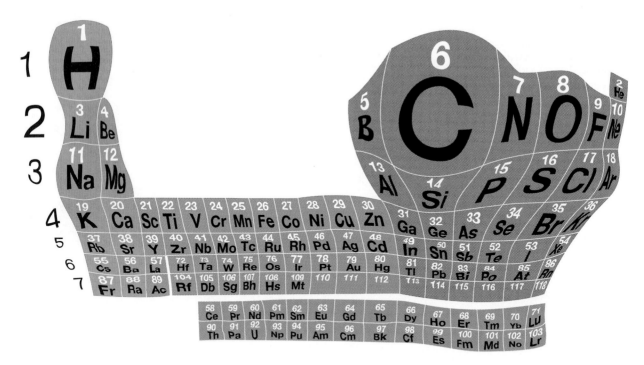

Figure 18.1
The organic chemist's periodic table.

18.1 What Makes Carbon Special?

Carbon compounds are so common that you have already seen many of them used as examples. In Chapter 5, they were the center of interest in combustion reactions. They showcased Pauling's hybridization theory and epitomized σ and π bonding in Chapter 6. In Chapter 10, the table of bond energies, Table 10.3, was constructed to highlight the bonding in organic compounds, with more entries for bonds to C than to any other element. The variety of compounds that can be made with an element depends primarily on its valence. Within the constraint of valence, the existence of any particular compound is a question of relative stability, as you have seen several times before, and stability in turn is mainly determined by the strength of the bonding. These two factors conspire to place carbon at the top compared to its nearest neighbors, boron, nitrogen, and silicon, which might be expected to vie with carbon for the title of "most prolific." Table 18.1 compares valences, bond energies, and bond lengths for B, C, N, and Si bonding to the lighter atoms.

Inspection of the table reveals that carbon's competitors fall short in a number of important bonding attributes. Perhaps the most important is the homopolar single bond. The C—C bond is by far the strongest. This allows carbon to form chains C—C—C—..., "molecular backbones" that are both strong and flexible, while leaving maximum unsaturated valence. Another important feature C alone displays is a well-balanced set of single-bond energies with other atoms, all comparable to the C—C energy, which allows a variety of heteropolar bonds to exist in one carbon-based molecule. Finally, carbon alone is able to make strong double and triple bonds both to itself and to other atoms without saturating its valence; multiple bonds can then be embedded in large molecules as foci of π-electron chromophoric and chemical activity. The shortcomings of B, N, and Si as progenitors of large and varied mo-

TABLE 18.1
Comparison of atomic and bonding properties of carbon and its nearest neighbors

	B	C	N	Si
Valence e⁻ configuration	$2s^2 2p^1$	$2s^2 2p^2$	$2s^2 2p^3$	$3s^2 3p^2$
Electronegativity χ	2.0	2.5	3.0	1.9
Normal valence	$3(sp^2)$ or $4(sp^3)$ in 3c2e bonds or with Lewis bases	4 (sp to sp^3)	$3(sp$ to $sp^3)$ $4(sp^3)$ with Lewis acids	$4(sp^3$ only)
Bond energies (kcal/mol)				
X—X	69	83	38	42
X—H	81	99	93	71
X—C		83	70	69
X—N		70	38	—
X—O		84	48	88
X=C		147	147	—
X=N		147	100	—
X=O		174	—	—
X≡C		194	213	—
X≡N		213	226	—
B—B—B	91			
B—H—B	105			
Chain length in hydrides	None (icosahedral shells)	Unlimited	3	6

lecular structures have fascinating causes, worth examining in their own right. To make our perusal brief but pertinent, we will focus on the hydrides and oxides of these elements.

Boron: Making Do with Too Few

The strength of the B—B bond given in Table 18.1 is only an estimate, because bonding among boron atoms is usually of the *three-center, two-electron* (3c2e) variety, that is, involving three boron atoms simultaneously. This peculiar behavior stems from a mismatch between orbitals and electrons; B has the requisite atomic orbitals to be sp^3 hybridized, but has only three valence electrons, not enough to supply the normal four two-center, two-electron (2c2e) bonds its hybrids would require. Boron is referred to as *electron deficient*. A promising series of boron hydrides or **boranes**, B_2H_6, B_4H_{10},

B_5H_9, B_5H_{11}, B_6H_{10}, B_6H_{12}, B_9H_{15}, $B_{10}H_{14}$, etc., has been prepared, but X-ray analysis of the structures of these compounds reveals that there are no B—B—B—... chains. Instead, triangular arrangements $_B^B{}_B$ as well as $_B^H{}_B$ are found, indicators of 3c2e bonding.

Figure 18.2 illustrates the structures of diborane, B_2H_6, and pentaborane-9, B_5H_9. With three valence electrons boron should be trivalent, but BH_3 is unstable. Instead B_2H_6 is found, having the same formula as ethane C_2H_6, but without a normal Lewis structure owing to the presence of only six pairs of electrons for binding the eight atoms together. The bridged geometry of diborane suggests the 3c2e B—H—B "banana" bond, which requires simultaneous overlap of orbitals from three atomic centers—sp^3's from each B and $1s$ from H—as the figure shows. This new type of bond is readily understood in a molecular orbital framework; combining atomic orbitals on three centers instead of the usual two to make a bonding MO is required in triatomics (see the discussion of MO bonding in H_2O in Chapter 7).

Pentaborane-9 has a beautifully symmetric structure containing five 3c2e bonds, four B—H—B and the equivalent of one B—B—B (involving the foremost B), in addition to five "normal" B—H bonds; the 14 atoms are bonded together with only 12 pairs of electrons. Figure 18.2 illustrates building a 3c2e B—B—B bond from hybrids on a triangle of three B's. The angular demands of the (roughly) sp^3-hybridized B atoms when engaged in 3c2e bonding cause the structure to turn in on itself, forming a cup or nest; B_5H_9 is sometimes called *nido*-pentaborane, from the Latin *nidus*, meaning "nest." The structures of the tetra- and higher boranes all turn out to be frag-

Figure 18.2
Ball-and-stick structures of diborane (B_2H_6) and *nido*-pentaborane-9 (B_5H_9). Unlabeled atoms are hydrogens. Shown below each structure is a schematic of the three-center, two-electron orbital overlaps involved in bonding in these molecules.

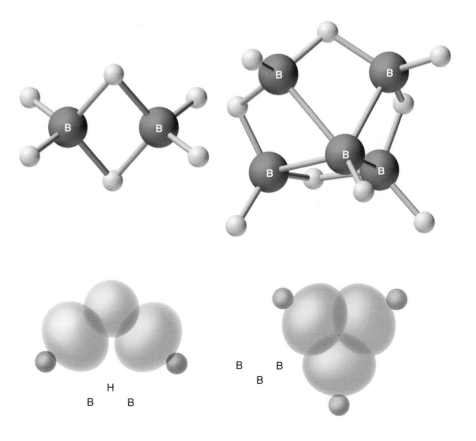

ments of an *icosahedron*, a 12-faced assembly of equilateral triangles with a B atom at each vertex that forms the basis for many of the common crystalline allotropes of B(s) (Chapter 14, Figure 14.26). These turned-in structures place a natural limit on the size of boranes.

In addition, the 3c2e bonds are rather delicate; about half of the known boranes spontaneously ignite when exposed to air, forming the much more stable $B_2O_3(s)$. This *kinetic* instability contrasts with carbon hydrides (alkanes), which are also thermodynamically unstable with respect to combustion, but are kinetically stable. In addition to requiring a "folded-in" structure, 3c2e bonding also virtually rules out multiple bonding—from the point of view of a B atom, a multiple bond is a gross waste of scarce electrons. Owing to the lack of stable chains and geometrical constraints, boron thus cannot produce the structural diversity and stability characteristic of carbon compounds.

Nitrogen: Three's a Crowd

Nitrogen does bond singly to itself, but the bond is exceptionally weak. There are a number of reasons for this weakness, but principally it may be laid to N's high electronegativity, which prevents it from sharing electrons generously with its own kind unless it is highly compensated. The N—N bond, as exemplified in hydrazine, H_2N—NH_2, is quite similar to that in F_2 and the other halogens; in the "united-atom limit" (Chapter 7) where the H atoms collapse into the N nuclei, hydrazine becomes molecular fluorine. As you might guess from this incriminating evidence, nitrogen hydrides beyond NH_3, N_2H_4, and HN_3 (hydrazoic acid, H—N=N=N, no single N—N bonds) are unknown. Clever use of carbon-containing side groups, denoted by "R," allows the synthesis of chains such as R_2N—NR—NR_2, a *triazane*, RN=N—NR_2, a *triazene*, but the analogues with R = H have never been made. We do find double bonds, however, unlike the case for boron. Any compound with even two adjacent N's is extremely unstable, decomposing under all conditions to form, among other products, molecular nitrogen N≡N. The strength of the N≡N triple bond, which is a factor of 6 greater than N—N, renders all N chains kinetically as well as thermodynamically unstable. Perhaps the most well-known example is hydrazoic acid, whose salt $NaN_3(s)$, sodium azide, is the explosive, N_2-gas-forming "active ingredient" in automobile airbags. "Sodazide" is so unstable that airbags sometimes detonate without sufficient cause, and safe disposal of old cars with their unexploded airbags has become a serious concern.

Even were N chains to be formed, their potential for diverse structures would be limited owing to N's surfeit of electrons. The lone pair every uncharged N atom must bear can only bond to a Lewis acid, and these "coordinate covalent" bonds (typified by the metal–ligand complexes we saw in the previous chapter) are also notoriously weak. This severely constrains the amount of branching possible. While in combination with carbon, isolated nitrogens can commonly be incorporated into chains—as we shall see—it is clear that N alone could never be the basis for 19 million compounds.

Silicon: More Than Just Computer Chips?

Finally we encounter an atom with the same valence as carbon. Because Si falls in the carbon group IVA family, you might expect it to form similar sorts of molecules. So it does: the silicon hydrides or **silanes** through hexasilane, Si_6H_{14} are well known, and there prove to be actual Si—Si—Si—... chains in these molecules. Limitations on chain length here owe to the weak Si—Si bond, which is only slightly stronger than that of N—N; moreover, like the boranes, silanes are prone to violent oxidation when exposed to air, the largest molecules being the most reactive. As is implied for boron, the Si—O bond strength overwhelms the Si—Si bond, resulting in kinetic as well as thermodynamic instability.

Here we may invoke intrinsic atomic similarities between B and Si, since they lie on a common downward-sloping diagonal of the periodic table. The difference lies mainly in Si's greater valence and larger $n = 3$ orbitals. The more diffuse orbitals make multiple bonding of Si to itself or to other atoms virtually unknown; π-type overlaps are simply too weak. Both B—O and Si—O bonds have considerably more ionic character than C—O, further disrupting any complex bonded structure; in neither case can the element bind to both H and O in the same molecule. As is the case with N, incorporation of carbon into the structure does allow larger molecules to form; the **siloxanes** are molecules containing long alternating chains ···Si—O—Si—O—Si—O···, with carbonaceous "R" groups saturating the Si valence. When R = CH_3, the molecules are known as **silicones,** a group of inert oils and greases (depending on molecular mass) used extensively in industrial and academic laboratories as pump oils, sealants, and lubricants. Note the avoidance of Si—Si bonds.

The very weakness of the Si—Si bond is what turns a silicon crystal into a semiconductor (see Chapter 14, Figure 14.22), by allowing the valence and conduction bands to draw closer together in energy. In that role silicon indirectly aids (and is aided by) us carbon-based life forms, in the guise of the computer that in our modern day helps us to think more complex thoughts and communicate more easily. But we cannot expect silicon to *become* the basis for a life form on its own; even after we program a computer, it can only advise us!

With this backdrop perhaps you can better appreciate the carbon chemistry we are about to describe. Carbon, that even-handed atom, is free of the handicaps that prevent its near neighbors from forming complex molecules.

18.2 Building Blocks of Organic Chemistry: The Hydrocarbons

In organic chemistry, the **hydrocarbons,** chains and branches of carbons atoms festooned with hydrogen, are the building blocks rather than stumbling blocks. The strong C—H and C—C bonds inhibit decomposition or attack by oxygen, making the hydrocarbons kinetically very stable; consequently they can be transported safely for use as fuel, and in longer chains they become everything from road pavement to beverage containers. Before we take the plunge into this world, please be reassured. It is not the intent here to present you with lists of names and structures to memorize, but rather simply to introduce you to the language and scope of carbon chemistry. The best learning stems from interest combined with gradual, repeated exposure to the vocabulary of a subject; whether or not this is your first exposure to organic chemistry, regard this chapter as a survey rather than a manual.

Many of the small hydrocarbons have already appeared in these pages, but for completeness, Table 18.2 displays some common examples of **saturated hydro-**

TABLE 18.2

Some common saturated hydrocarbons (alkanes or paraffins)

Name	Formula	Carbon skeleton
Methane[a]	CH_4	
Ethane[b]	CH_3—CH_3	—
Propane[c]	CH_3—CH_2—CH_3	∧
n-Butane[d]	CH_3—CH_2—CH_2—CH_3	∧∨

Isobutane	CH_3—CH—CH_3
	           |
	          CH_3
		Y

n-Hexane	CH_3—CH_2—CH_2—CH_2—CH_2—CH_3	∧∨∧∨
n-Octane	CH_3—CH_2—CH_2—CH_2—CH_2—CH_2—CH_2—CH_3	∧∨∧∨∧∨

Isooctane	CH_3
	                |
	 CH_3—CH—CH_2—C—CH_3
	        | |
	       CH_3 CH_3	

[a] Based on *methyl*, derived from the Greek *methy*, wine, plus *hyle*, wood, as in wood alcohol (CH_3OH).
[b] From the Greek *aithein*, to ignite, blaze; likely the origin of the *-ane* suffix.
[c] Based on *propionic acid* (CH_3CH_2COOH), from the Greek *pro*, before, plus *pion*, fat.
[d] Based on *butyric acid* ($CH_3CH_2CH_2COOH$), a component of rancid butterfat, derived from the Latin *butyrum*, fermented butterfat.

carbons. The word *saturated* means saturated with hydrogen; all of the C's bond singly to other C's, C—C, and all remaining bonds are C—H. Saturated hydrocarbons are also known as **alkanes,** and their names invariably end in *-ane;* they have also long been called **paraffins,** from the Latin for "little reactivity"—well suited to these inert compounds. Methane through butane are named by early convention, while pentane and onward combine a Greek numeric prefix for the number of C atoms with the *-ane* suffix. Their atomic constituencies can all be represented by the single formula C_nH_{2n+2}, $n = 1, 2, 3, \ldots$. All the carbon atoms in alkanes are sp^3 hybridized, with all bond angles close to the ideal 109.5°, and it is very useful and convenient to think of them in a tinker-toy context. Molecular orbitals can, of course, be constructed for these molecules, but in these cases MOs are not nearly so helpful in understanding their chemistry. The approach taken by organic chemists to questions of electronic structure and reactivity is pragmatic; Lewis structures and the valence bond model aided by curly arrows are the rule, with molecular orbital ideas inserted where they will help.

A characteristic of organic compounds containing C—C single bonds is the phenomenon of **free internal rotation** about each C—C bond axis. This allows the hydrocarbons beyond methane (CH_4) to arrange their hydrogen atoms and relative orientations of C—C axes in a variety of ways. In the simplest example, ethane (C_2H_6) written semistructurally as H_3CCH_3, the H's on one **methyl group** —CH_3 can adopt different azimuthal orientations relative to those of the other. The azimuthal angles for the two important limits, the so-called *staggered* and *eclipsed* forms, are best viewed end-on in what is called a *Newman projection,* on the right.

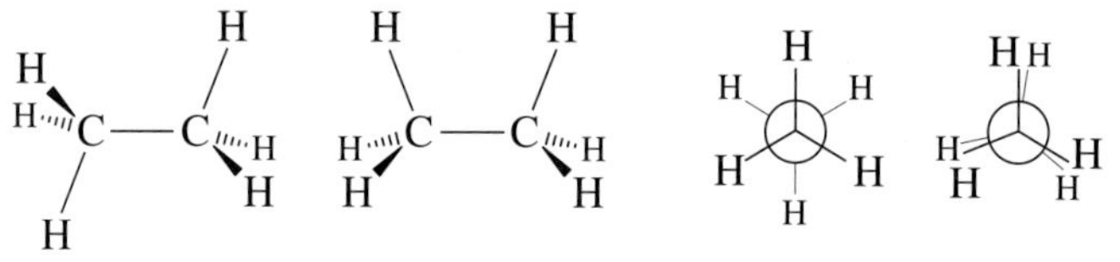

Staggered   Eclipsed   Staggered   Eclipsed

These two structures are examples of **conformational isomers** (or *conformers*) generated by the free internal rotation. Actually the rotation is not quite free, owing to nonbonded repulsions between the H's on adjacent methyls, which make the staggered conformer more stable, but only by about 3 kcal/mol. Even less constrained are nonadjacent C—C bond orientations in butane, hexane, etc. Mainly for convenience, we draw the structures of hydrocarbons as a series of zigzag line segments in a shorthand skeletal notation already used in Chapters 14 and 17. Each terminus or vertex represents a carbon atom, and hydrogen atoms are assumed to be present in sufficient numbers to sate the tetravalent C's, as shown in Table 18.2.

The lighter hydrocarbons, C_1 through C_4, are components of natural gas, while the heavier hydrocarbons, C_5 through C_{20}, are obtained from petroleum ("crude oil") by a combination of "catalytic cracking" and fractional distillation. In Chapter 14, Table 14.1 and Figure 14.13, the physical properties of the hydrocarbons were displayed. *Gasoline* or *naphtha* is a mixture of the liquid hydrocarbons, C_5 through C_{12}; *kerosene,* C_{12} through C_{18}; motor, heating, and other oils, C_{16} through C_{19}; and paraffin waxes and *asphalt,* C_{20} and above.

Structural Isomers

As Table 18.2 indicates, beginning with butane (C_4H_{10}), two or more distinct carbon skeletons with the same molecular formula can be drawn, the *straight chain* or **normal** alkanes, abbreviated *n-*, and one or more *branched chain* alkanes, which typically have both a common and a systematic name. As is typical for carbon compounds, if you can draw it (or build it with a model kit), it probably exists, as a consequence of the even strength of carbon's bonds to itself and to other atoms. This is another type of isomerism, in which the *connectivity* or bonding arrangement differs, called **structural isomerism.** Here in skeletal form are the five structural isomers of hexane (C_6H_{14}):

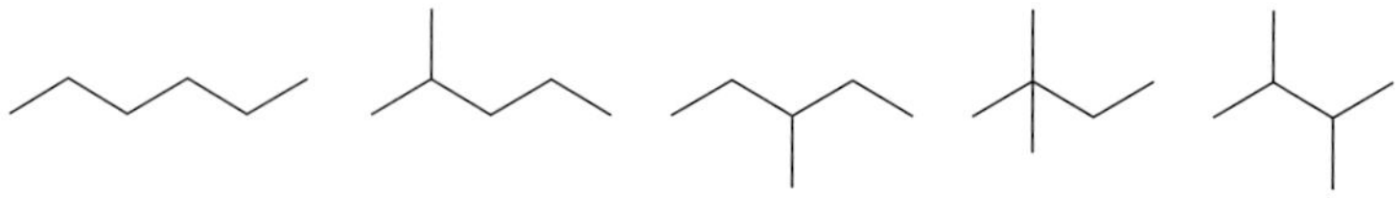

Structural isomers of the hydrocarbons were among the first types to be discovered, since each isomer has distinct physical properties, such as boiling point, and can be separated from its fellows by fractional distillation. The number of structural isomers of the alkanes grows rapidly with the number of C atoms; for octane (C_8H_{18}) there are 18 isomers, and a systematic scheme for naming them becomes essential. The branching groups may be thought of as ligands (though they are strongly cova-

TABLE 18.3

Common alkyl side chains (R groups)

Name	Abbreviation	Formula	Skeletal structure
Methyl	Me	—CH_3	
Ethyl	Et	—CH_2CH_3	
n-Propyl	n-Pr	—$CH_2CH_2CH_3$	
Isopropyl	i-Pr	—$CH(CH_3)_2$	
n-Butyl	n-Bu	—$CH_2CH_2CH_2CH_3$	
Isobutyl	i-Bu	—$CH_2CH(CH_3)_2$	
sec-Butyl	s-Bu	—$CH(CH_3)CH_2CH_3$	
tert-Butyl	t-Bu	—$C(CH_3)_3$	

lently bonded), and the smaller ones have names based on the common names of the corresponding alkanes, as presented in Table 18.3. When one wishes to be general, these **alkyl groups** are often abbreviated R—. The systematic naming of a branched chain alkane proceeds by numbering the atoms in the longest chain beginning at the end nearest the first branch, and then listing the "ligands," preceded by a number representing their position on the chain, in numerical order, followed by the name of the long chain. For example, what is commonly called *isobutane* is systematically named *methylpropane,* and what is often called *isooctane* is better termed *2,2,4-trimethylpentane.* German chemists played a large role in the early synthesis of carbon compounds, and the single giant words that name them reflect this history, as we've already seen for metal complexes. In what follows, we usually give the common name first, with the systematic name in parentheses.

For any other element, *rings* are extremely unstable, and you are normally discouraged from drawing them when learning Lewis structures. But not for C! The stablest and most well-known alkane ring compound is *cyclohexane* (C_6H_{12}), a six-membered carbon ring saturated with hydrogens. Although cyclohexane is drawn as a planar hexagon in skeletal fashion, the geometry around each C is tetrahedral, and the 109° angles force the ring to pucker, as illustrated in Figure 18.3. As shown there, cyclohexane has two main conformers, the *chair* and the *boat* forms. The chair form is more stable by 3.8 kcal/mol, since the boat form puts hydrogens bonded to the "prow" carbons close enough to generate nonbonded repulsion. Unlike the eclipsed and staggered forms of ethane, which interconvert readily at room temperature, cyclohexane, once in the chair form, must cross a 10.1 kcal/mol barrier to convert to the boat form, passing through a "half-chair" transition state with "strained" bond angles. The strain involved in this interconversion is readily appreciated when using a model kit.

Figure 18.3
"Chair" and "boat" conformations of cyclohexane (C_6H_{12}). Larger atoms are carbons.

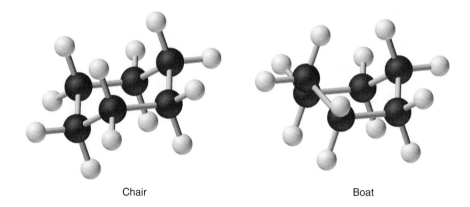

Chair   Boat

Speaking of strained bond angles, cyclohexane has the least among possible cyclic alkanes (all of which have the formula C_nH_{2n}). The smallest possible ring is cyclopropane C_3H_6, which was not synthesized until 1970, and cyclobutane through cyclooctane are well known. Not surprisingly, cyclopropane with its 60° C—C—C angles has the most strain, and this makes it considerably more reactive than the larger cycles. There are also well-known bicycles and tricycles (though these cannot be ridden), such as *decalin* ($C_{10}H_{18}$), a bicyclic compound with two cyclohexane rings *fused* along one edge. Alkyl groups may be freely substituted for H around the ring, making a multitude of new compounds to add to the already huge list of possible alkanes. The cyclohexane skeleton, with one —CH_2— replaced by —O—, is the building block for the hexose sugars, and fused-ring systems are common among biologically active molecules.

Alkane Reactions

As we averred at the outset, alkanes are relatively inert, and their reactions fall mainly into three categories: **combustion, halogenation,** and **dehydrogenation,** all of which generally require extreme conditions to achieve useful reaction rates. Combustion, written in general as

$$C_nH_{2n+2} + \tfrac{1}{2}(3n+1)O_2 \longrightarrow nCO_2 + (n+1)H_2O \tag{18.2}$$

requires heating to 600°C for ignition (provided readily by a match or sparkplug), but as you know it is highly exothermic and therefore practically irreversible. It has taken geologic eons to form the natural reserve of hydrocarbons we now wantonly burn, and no one yet knows how to turn back the CO_2. It is hard to say whether we will succumb to greenhouse heating before or after we have exhausted nature's fuel. (Please excuse; just had to get that in.) Branched chain hydrocarbons are more reactive toward oxygen than straight chain; gasoline and jet fuel contain large fractions of branched chains, such as isooctane. This higher reactivity can be traced to the slightly lower bond energies for the C—H bonds at substituted positions along the chain.

As we alluded to in Chapters 5 and 15, the mechanism of combustion is complex, but undoubtedly involves **free radicals,** species such as atomic oxygen (O), hydroxyl radicals (·OH), and *alkyl radicals* (R·). The halogenation of alkanes, for example,

$$CH_4 + Cl_2 \longrightarrow CH_3Cl + HCl \tag{18.3}$$

also proceeds by a free radical mechanism, but a much simpler one. Like combustion, halogenation requires heat to occur, but it can also be initiated with blue or ultraviolet light. The mechanism is similar to that for the $H_2 + Br_2$ reaction: Cl· radicals are generated upon breaking the weak Cl—Cl bond with heat or light, and these attack the alkane, abstracting H. The alkyl radical is then able to attack Cl_2, abstracting a Cl atom and forming the product alkyl halide. In equations,

$$Cl:Cl \rightleftharpoons Cl\cdot + \cdot Cl \tag{18.4}$$

$$Cl\cdot + H:CH_3 \longrightarrow Cl:H + \cdot CH_3$$

$$Cl:Cl + \cdot CH_3 \longrightarrow Cl\cdot + Cl:CH_3$$

Note that this is a chain reaction, like $H_2 + Br_2$, since the chlorine atomic radical formed in the third step can now take part in step 2. Among the evidence for this mechanism is the appearance of a small amount of ethane among the products, which would arise from the radical–radical recombination reaction

$$H_3C\cdot + \cdot CH_3 \longrightarrow H_3C:CH_3 \tag{18.5}$$

Note that in these mechanistic steps we have illustrated the rearrangement of the "active" electrons with curly arrows in the way introduced in Chapter 5. The C—Cl bond is one of the strongest covalent bonds Cl makes except for H—Cl (see Chapter 10, Table 10.3), and the combined bond strengths of the products relative to the reagents make the reaction exothermic and spontaneous. This is a displacement or **substitution** reaction; like combustion, it is also an oxidation–reduction reaction, with Cl going from 0 to -1 and C from -4 to -2 in oxidation number. In discussing organic reactions, redox character is usually not emphasized, since the bonds are generally all covalent. The product methyl chloride (chloromethane) is an example of an alkane *derivative;* derivatives are discussed more fully in the next section. Alkyl halides are gateways for making many other organic compounds.

Dehydrogenation, the elimination of molecular hydrogen from an alkane, again requires drastic conditions: over a Cr_2O_3 catalyst at 500°C ethane loses hydrogen to form **ethylene,**

$$H_3C-CH_3 \longrightarrow H_2C=CH_2 + H_2 \tag{18.6}$$

This reaction is endothermic but releases entropy, and the high temperature helps to increase its spontaneity. It is an example of *catalytic cracking,* a general method for breaking down alkanes of higher molecular weight. Unlike the earlier reactions, this one is thought to involve a *concerted* mechanism, with a four-membered $_{CC}^{HH}$ ring as an intermediate. The reverse of this reaction, ethylene hydrogenation, was used in Chapter 15 as an example with a highly organized (low-entropy) transition state.

Unsaturated Hydrocarbons: Alkenes and Alkynes

In Chapter 6 ethylene, $H_2C=CH_2$ (ethene), **acetylene,** $HC\equiv CH$ (ethyne), were displayed as prototypical examples of double and triple bonds. Recall that a double bond consists of one σ and one π bond, while a triple bond has a σ and two π components.

In ethylene the σ bonding is described by sp^2-hybridized carbons overlapping with each other and with H $1s$, while in acetylene sp hybrids play that role. These bonds are very strong (C=C, 147; C≡C, 194 kcal, compared to C—C, 83), and yet the exposed π density forms a reactive center for attack by electrophiles, making these compounds more versatile chemically than the paraffins. The H's can be replaced by various alkyl (R) groups, forming a family of hydrocarbons containing multiple bonds that is even larger than the alkanes; those containing a double bond are called **alkenes** (formula C_nH_{2n}, the same as alkane rings) and the triply bonded are **alkynes** (formula C_nH_{2n-2}). Isomerism now involves the location of the multiple bond as well as branching; for example, in skeletal form the four isomers of **butene** (C_4H_8) are

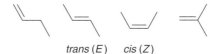

trans (E)   *cis* (Z)

In the middle two isomers the connectivity is the same, but the arrangement in space differs; these are *cis* and *trans* **geometrical isomers** of the same sort as those in transition metal complexes (Chapter 17). [*cis* is from the Latin, meaning "on this side," while *trans*, also Latin, means "across"; modern notation for these isomers replaces *cis* with Z (German *zusammen*, "together") and *trans* with E (German *entgegen*, "across from").] The difference between these isomers and the conformers of the alkanes lies in their ease of interconversion: twisting the double bond—necessary to interconvert *cis* and *trans*—causes the breakage of the π component of it, and hence requires considerably more energy, ~60 kcal/mol instead of only 3. At ordinary temperatures the *cis* and *trans* forms remain distinct, and can be separated by distillation or chromatography. In 2-butene the *trans* form is slightly more stable, because it keeps the bulky methyl groups farther apart. The repulsion between larger groups is an example of a *steric effect*. The double bond also enforces coplanarity of the four σ-bonded atoms with C=C. Alkynes do not show *cis-trans* isomerism; there can be only one atom attached to each end of C≡C, and the four atoms must be collinear.

More than one double or triple bond can exist in the same molecule, for example, H_2C=CH—CH=CH_2, 1,3-butadiene. Double bonds can form on adjacent carbons, as in C=C=C, but are more stable when they *alternate* with single bonds, as in 1,3-butadiene. The so-called alternant or conjugated hydrocarbons, which include biological molecules involved in the mechanism of vision, have a unique spectroscopy and chemistry that requires a molecular orbital description of their delocalized π bonding. Rings containing double bonds, cycloalkenes, are quite common, but owing to the strain of enforced collinearity around a triple bond, cycloalkynes are relatively rare. The special case of cyclohexatriene, better known as benzene, is covered separately in Section 18.4.

Alkenes, like alkanes, are obtained mainly from petroleum, by means of *thermal cracking* of higher hydrocarbons at 800°C; in addition to producing the lower alkanes, the high temperature also induces dehydrogenation to yield small alkenes. Production of ethylene by catalytic cracking was used as an example earlier, but uncatalyzed thermal cracking is the dominant source of alkenes for the chemical industry. The physical properties of the alkenes are very similar to those of alkanes: ethylene, propylene (propene), and butene are gases; pentene and beyond are liquids.

Addition Reactions: Carbocations and Formal Charge

A variety of electrophiles react readily with the alkene π bond by **addition** rather than substitution. For example, when an aqueous solution of Br_2 is poured into a test tube containing neat (pure) cyclohexene, $C_6H_{10}(l)$, initially the orangy aqueous solution sinks to the bottom, forming a two-phase system, since polar water and nonpolar cyclohexene are immiscible. Upon shaking, however, the orange color disappears as nonpolar Br_2 is *extracted* into the cyclohexene layer; in the absence of water it immediately reacts according to

$$\text{cyclohexene} + Br_2 \longrightarrow \text{1,2-dibromocyclohexane} \tag{18.7}$$

This reaction contrasts with the corresponding substitution reaction between bromine and cyclohexane, which requires strong heating to occur at all. Here, although 64 kcal/mol is required to break the π bond and 53 kcal to break the Br—Br bond, *each* C—Br bond pays back 66 kcal; further, the π bond is readily accessible spatially, since the molecule must be locally planar in its vicinity, allowing electrophilic attack from above or below. Instead of a four-center concerted attack of Br_2 on the π bond, the mechanism of addition is believed to involve a *bromonium ion* intermediate, in which a single Br^+ intially accepts the entire π electron pair, eventually creating a positively charged C atom (see below) that accepts the remaining Br^-. We must leave the evidence for such a seemingly improbable chain of events as one of many intrigues that a course in organic chemistry holds in store for you.

Other small molecules, such as water and the hydrogen halides, will also add across the π bond. In these additions one of the most ubiquitous intermediates in all of organic chemistry, the **carbocation** (car′ bo cat″ ion), is thought to play a central role. The Br_2 reaction just discussed appears to proceed by formation of an α-bromocarbocation, the α referring to the adjacent C. The acid-catalyzed addition of water to ethylene is initiated by the attack of a proton (actually a hydronium ion, and a good electrophile) on the π cloud to form an ethyl carbocation,

$$H_2C=CH_2 + H^+ \longrightarrow H_2C^+-CH_3 \tag{18.8}$$

followed by attack of water (a fair nucleophile) on the C^+,

$$H_2C^+-CH_3 + :OH_2 \longrightarrow CH_3-CH_2-OH + H^+ \tag{18.9}$$

forming the product ethyl alcohol (ethanol) and releasing the proton. Note that the overall reaction is just a hydration of ethylene, with no protons appearing in it. The ethyl cation is called a *primary* carbocation, since the C^+ is bonded to only one other carbon, and is less stable than a *secondary* carbocation such as isopropyl ion ($CH_3-CH^+-CH_3$); *tertiary* carbocations such as *t*-butyl ion (($CH_3)_3C^+$) are the most stable. This stability sequence owes to a small but increasing amount of π bonding that occurs between the empty C^+ p orbital and orbitals on adjacent C's as their number grows.

As the preceding discussion of the relative stability of carbocations suggests, assigning all of the positive charge of a covalent cationic molecule to a single atom in it is certainly oversimplified. It is more realistic to think of the carbocation as having a partially charged C atom, $C^{+\delta}$. The simplification is based on the notion of **formal charge.** The formal charge of an atom in a molecule is assigned by comparing the number of electrons within its valence shell in the molecule with the number it has as a free neutral atom. All bonds the atom makes are assumed to be pure covalent, so that one electron from each bond is counted as "belonging" to it, in addition to any lone pairs. On this basis, the left-hand C in H_2C^+—CH_3 is assigned a formal charge of $+1$, since it makes three bonds and has no lone pairs, while a free C atom has four valence electrons. You should note the relationship between formal charge and oxidation number (see Chapter 5); the difference is that in assigning oxidation numbers the bonds are assumed to be pure ionic rather than pure covalent. Neither is generally the case, and both formal charge and oxidation number are just simple and convenient limiting models of a more complex charge distribution.

Stronger oxidizing agents, such as $KMnO_4$, will convert the π bond into a dihydroxyl compound called a *diol*,

$$\text{C=C} \xrightarrow{KMnO_4} \underset{OH\ \ OH}{\text{C--C}} \tag{18.10}$$

or its dehydrated analogue, the *epoxide*, containing a three-membered $\underset{C\ \ \ C}{\overset{O}{\triangle}}$ ring. Double bonds can be completely "chewed up" by oxidizing agents such as O_3; in *ozonolysis*, the alkene is severed into two fragments. As we will discuss further in separate sections, double bonds are also readily *polymerized,* and are the naturally preferred sites for metabolic oxidation. The ready and flexible reactivity of the alkenes makes them, rather than the alkanes, a prime feedstock for the chemical industry.

18.3 Derivatives of the Hydrocarbons: Functional Groups

The organic compounds produced in the reactions we have just briefly surveyed, which all contain new atoms or groups that replace or add to the C and H already present, are called hydrocarbon **derivatives,** that is, they are derived from hydrocarbons. (These are at once much simpler and infinitely more complicated than those other derivatives you've learned about in calculus.) By means of further reactions of the derivatives we've just seen, many others can be made, a "boundless thicket" of complexity made possible by the democratic carbon atom. We've already used the example in chemical kinetics of the substitution reaction between *t*-butyl chloride (2-chloro-2-methyl propane) and hydroxide ion,

$$(CH_3)_3C-Cl + :OH^- \longrightarrow (CH_3)_3C-OH + :Cl^- \tag{18.11}$$

which yields an **alcohol,** generically **ROH**, here *t*-BuOH. Alcohols are perhaps the most well-known class of hydrocarbon derivatives, particularly ethyl alcohol (ethanol), CH_3CH_2OH. The acid-catalyzed addition of water to ethylene, whose mechanism was examined earlier, is the chief source of ethanol for industrial use. Table 18.4 lists names and structures of common substituent groups found in derivatives.

18.3 Derivatives of the Hydrocarbons: Functional Groups

TABLE 18.4
Common hydrocarbon derivatives

Derivative	Functional group	General formula	Examples	
Halide	—Cl, —Br	R—Cl	CH_2Cl_2, methylene chloride (dichloromethane)	$CH_3CHClCH_3$ isopropyl chloride (2-chloropropane)
Alcohol	—OH	R—OH	CH_3OH methanol	CH_3CH_2OH ethanol
Ether	—O—	R—O—R′	CH_3CH_2—O—CH_2CH_3 Diethyl ether	CH_3—O—$C(CH_3)_3$ Methyl *t*-butyl ether (MTBE)
Ketone	$-\overset{\overset{O}{\|}}{C}-$	$R-\overset{\overset{O}{\|}}{C}-R'$	$CH_3-\overset{\overset{O}{\|}}{C}-CH_3$ Acetone (propanone)	$CH_3-\overset{\overset{O}{\|}}{C}-CH_2CH_3$ Methyl ethyl ketone (MEK) (butanone)
Aldehyde	$-\overset{\overset{O}{\|}}{C}-H$	$R-\overset{\overset{O}{\|}}{C}-H$	$H-\overset{\overset{O}{\|}}{C}-H$ Formaldehyde (methanal)	$CH_3-\overset{\overset{O}{\|}}{C}-H$ Acetaldehyde (ethanal)
Carboxylic acid	$-\overset{\overset{O}{\|}}{C}-OH$	$R-\overset{\overset{O}{\|}}{C}-OH$	$CH_3-\overset{\overset{O}{\|}}{C}-OH$ Acetic acid (ethanoic acid)	$CH_3CH_2-\overset{\overset{O}{\|}}{C}-OH$ Propionic acid (propanoic acid)
Ester	$-\overset{\overset{O}{\|}}{C}-O-$	$R-\overset{\overset{O}{\|}}{C}-O-R'$	$CH_3-\overset{\overset{O}{\|}}{C}-O-CH_2CH_3$ Ethyl acetate (ethyl ethanoate)	
Amine	—NH_2	R—NH_2	CH_3—NH_2 Methyl amine (aminomethane)	$CH_3CHNH_2CH_3$ isopropyl amine (2-aminopropane)
Amide	$-\overset{\overset{O}{\|}}{C}-NH_2$	$R-\overset{\overset{O}{\|}}{C}-NH_2$	$CH_3-\overset{\overset{O}{\|}}{C}-NH_2$ Acetamide (ethanamide)	
α-Amino acid		$H_2N-\overset{\overset{R}{\|}}{CH}-\overset{\overset{O}{\|}}{C}-OH$	$H_2N-\overset{\overset{H}{\|}}{CH}-\overset{\overset{O}{\|}}{C}-OH$ Glycine	$H_2N-\overset{\overset{CH_3}{\|}}{CH}-\overset{\overset{O}{\|}}{C}-OH$ Alanine

Like their parent hydrocarbons, derivatives have also appeared as examples throughout this book. We saw **carboxylic acids (RCOOH)** and **amines (RNH$_2$)** as the most numerous members of the weak acid and weak base families in Chapter 12. Amines may be formed directly by reacting alkyl iodides with ammonia, a reaction analogous to Reaction 18.11, although a mixture of mono- and polyalkylated amines results, since an alkyl amine is a better nucleophile than ammonia itself. Alcohols (ROH), **aldehydes (RCHO),** and acids (RCOOH) are stages in the enzyme-catalyzed oxidation process known as *fermentation;* for example, ethanol is converted to acetaldehyde (ethanal) and then to acetic acid (ethanoic acid), according to

$$CH_3CH_2OH + \tfrac{1}{2}O_2 \longrightarrow CH_3CHO + H_2O \tag{18.12}$$
$$CH_3CHO + \tfrac{1}{2}O_2 \longrightarrow CH_3COOH$$

From alcohol to acid involves a change of +4 in oxidation number of the substituted carbon (from −1 to +3), a four-electron oxidation. Alcohols also undergo *dehydration* to alkenes, the reverse of water addition. The hydroxyl proton in alcohols is much less acidic than even that in water, but will undergo a gentle reaction with alkali metals to liberate hydrogen,

$$R-O\!:\!H + Na \longrightarrow R-O\!:^-Na^+ + \tfrac{1}{2}H_2 \tag{18.13}$$

The product RO$^-$ is called an **alkoxide,** and is the main precursor to the synthesis of **ethers (ROR′)** in reactions such as

$$R-Cl + Na^+\!:\!O-R' \longrightarrow R-O-R' + Na^+\,Cl^- \tag{18.14}$$

This is called *Williamson synthesis* of ethers. Many organic reactions are given names like this, after their (at times disputed) discoverers, making organic chemistry seem like a "primeval forest" full of exotic species. In fact, organic reactions run on a very few basic principles, which, when understood, clear the thicket considerably.

Substitution Reactions

Reactions 18.11 and 18.14 are examples of **nucleophilic substitution,** and you should observe how similar they are in form—just replace H by R! The mechanisms by which they occur are clarified by kinetic studies (see the discussion of Reaction 18.11 in Chapter 15), which show them to be either first- or second-order reactions, depending mainly on the R group of the alcohol. The first-order reactions are abbreviated S$_N$1 (Substitution, $_N$ucleophilic, 1st order) and the second-order, S$_N$2. We saw in Chapter 15 that Reaction 18.11 is a first-order reaction, S$_N$1, and interpreted our observation in terms of a carbocation intermediate, the *t*-butyl cation (($CH_3)_3C^+$), formed in a unimolecular rate-limiting step. Now we can understand the chemistry of this result in a little more detail. As noted earlier, tertiary carbocations like *t*-butyl are the most stable; this lowers the activation energy of the rate-limiting ionization step and makes it occur more rapidly. Further, the bulky methyl groups prevent the approach of worthy nucleophiles from the "far side" of *t*-butyl in order to displace the Cl; this spatial restriction is known as *steric hindrance.*

Consider by contrast a concrete example of Reaction (18.14), the reaction of methyl chloride with sodium *t*-butoxide:

$$CH_3—Cl + Na^{+-}:O—C(CH_3)_3 \longrightarrow CH_3—O—C(CH_3)_3 + Na^+Cl^- \quad (18.15)$$

The product is methyl *t*-butyl ether (MTBE), until recently a popular "clean air" additive to gasolines. This reaction is found to be S_N2, first order in both the chloride and the alkoxide, and the carbocation intermediate gives way to a *pentavalent carbon species*,

$$(CH_3)_3C—O:\text{------}\overset{\overset{\displaystyle H \; H}{|}}{\underset{\underset{\displaystyle H}{|}}{C}}\text{------}Cl \quad]^- \quad (18.16)$$

Two factors mitigate against a possible S_N1 route for this reaction: the methyl ion (CH_3^+) is the least stable of all carbocations, and the low-profile hydrogens cannot prevent a "backside attack" by the nucleophile (they provide little steric hindrance). Note that the *t*-butyl group is on the wrong end to stabilize the carbocation; in this reaction it is simply "along for the ride."

Another "theme group" for a potpourri of organic reactions is the **carbonyl group, C=O**, which attractively presents both a π bond for electrophiles and an electropositive C atom for nucleophiles. Carbonyl groups are found in aldehydes, ketones, acids, and amides, and the reactions of these derivatives often share common facets of carbonyl reactivity. For example, **esters (R—COO—R′)** are formed by acid-catalyzed reaction between acids and alcohols,

$$R-\overset{\overset{\displaystyle O}{\|}}{C}-OH + R'-OH \xrightarrow{H^+} R-\overset{\overset{\displaystyle O}{\|}}{C}-O-R' + H_2O \quad (18.17)$$

The mechanism of this substitution was discussed briefly in Chapter 16, where isotopic labeling was used to show that the alcoholic oxygen attacks the carbonyl carbon, a classic S_N2 case. The acid catalyst acts by protonating the carbonyl π bond on its O end, converting it into a carbocation, an even more inviting target for nucleophilic attack. **Amides (R—CO—NH₂)** can be made by reaction of ammonia with acids or esters, going by way of a similar attack on the carbonyl carbon by :NH₃. The amide linkage —CO—NH— occurs frequently in polymers and proteins, as we will see later.

The Asymmetric Carbon Atom: Enantiomers

Among the molecules used as examples in Table 18.4, one has a property not shared by any of the others. Alanine, an α-amino acid with R = CH_3, displays a new type of isomerism unlike any we've yet seen in this chapter; that is, there exist isomers of alanine that are neither conformational, structural, nor geometric. As illustrated in Figure 18.4, the alanine molecule contains a tetrahedral carbon bonded to four different groups; and as a result of this fundamental asymmetry, the mirror image of this molecule cannot be superimposed on it, and hence is a *different molecule,* called an **enantiomer** (from the Greek *enantios,* opposite). The original molecule and its

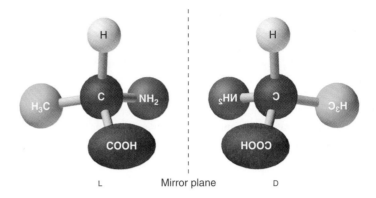

Figure 18.4
Bonding around the asymmetric carbon atom in the amino acid alanine and its mirror image. Only L-alanine occurs naturally in proteins.

nonsuperimposable mirror image form an enantiomeric pair, and the molecule is said to be **chiral.** This type of isomerism is often called *handedness,* since the relationship between an enantiomeric pair is the same as between your pair of hands. If you hold your right hand up to a mirror, its reflection looks like your left hand, yet your two hands cannot be superimposed finger for finger—either the palms are on opposite sides or the thumbs stick out on opposite sides.

Why don't any of the other molecules in Table 18.4 exhibit chirality? You can easily show by example that making any two groups bonded to a tetrahedral C identical allows the mirror image to be superimposed, eliminating the possibility of enantiomeric pairs. In all the other examples, the most variegated carbon is bonded to least two identical groups. A sure test for enantiomers is the nonexistence of a *mirror plane,* passing through the molecule, that divides it into two identical halves. All the molecules in the table except alanine have a mirror plane. A less general test is simply to locate at least one **asymmetric carbon,** bonded to four distinct atoms or groups; only alanine, among the examples given, has one. Another handy rule to remember is that C atoms engaged in multiple bonds cannot be chiral, even when all the bonded groups differ.

The discovery of this unique form of isomerism is generally credited to Louis Pasteur, a French scientist who is more well known for his theory of microbe-borne disease and his methods of sterilization. In 1848, while Pasteur was still in his mid-twenties, he was examining crystals of a substance then known as a salt of *racemic acid,* obtained by heating *tartar,* a product of fermented grape justice. Under a magnifying glass, he was amazed to find two different types of crystal, a "right-handed" and a "left-handed" form; with a fine tweezers, he meticulously separated the two crystalline forms. Now, racemic acid was known to be an isomer of *tartaric acid,* which we now know by the formula

$$\text{HO}-\underset{}{\overset{\overset{\text{O}}{\|}}{\text{C}}}-\underset{\underset{\text{H}}{|}}{\overset{\overset{\text{OH}}{|}}{\text{C}}}-\underset{\underset{\text{H}}{|}}{\overset{\overset{\text{OH}}{|}}{\text{C}}}-\underset{}{\overset{\overset{\text{O}}{\|}}{\text{C}}}-\text{OH}$$

As you can see, the two middle carbons are asymmetric, but because the molecule has two equal halves, there are only two (rather than four) enantiomers. By separating the crystals, Pasteur had accomplished the first resolution of an enantiomeric pair into its components.

Twenty years earlier, the French physicist J. Biot had discovered the phenomenon of **optical rotation,** wherein solutions of certain biologically derived substances would cause the plane of polarization of linearly polarized light to rotate *in a definite* *direction,* either to the right (clockwise or positive) or to the left (counterclockwise or negative). After Pasteur's report of *equal and opposite* optical rotation by solutions of equal concentration prepared from the two types of racemic acid crystals, Biot invited him to his own laboratory, and verified Pasteur's finding himself, the first instance of such behavior from isomers of a single substance. When equal volumes of the two solutions were mixed, the resulting solution showed no optical rotation; just so, racemic acid solutions were already known to be optically inactive. Consequently, enantiomers are also called *optical isomers,* and to this day, mixtures of equal amounts of an enantiomeric pair are called **racemic mixtures.** Using reasoning similar to that later employed by Werner in deducing the existence of octahedral complexes, J. van't Hoff, the pioneering physical chemist we have already noted in thermodynamics and kinetics, used optical activity and other isomeric evidence to posit the tetrahedral carbon atom. Like Werner's, van't Hoff's ideas were at first scorned, but their power to explain and correlate so many facts won them eventual acceptance.

Living things naturally use and produce only one of two possible enantiomers when asymmetric carbons are present. Fresh tartaric acid from fermented grapes is present as a single enantiomer, and shows positive optical rotation (to the right, or *dextro,* D-). Alanine occurs naturally only as a single enantiomer with negative optical rotation (to the left, or *levo,* L-). Though chirality clearly plays an important role in biological chemistry, the origins of this natural handedness are difficult to fathom except as a statistical accident. Enantiometric pairs have identical physicochemical properties, such as boiling and melting points and heats and free energies of formation, and there is no thermodynamic or kinetic reason why one isomer should be favored over its mate.

We shall return to this topic later in this chapter, but there are a few more terms you should encounter here. All isomers except structural, that is, conformational, geometric, and enantiomeric, are called **stereoisomers** (from the Greek *stereos,* a three-dimensional solid; this is also the root of *steric*). Stereoisomers share the property of having the same bonding arrangement or atomic connectivity, but differing in the spatial arrangement of the atoms. Conformational and geometric isomers do not of themselves generate optical activity, and are called **diastereomers** (Greek *dia,* "across"), while molecules with asymmetric C's do generate optical activity, and are called enantiomers. Thus isomers comprise structural and stereo, and stereoisomers in turn comprise diastereomers and enantiomers.

Further, it is of interest to specify the *absolute spatial configuration* of an enantiomer in the same way that it is known for diastereomers. Using rules too detailed to go into here, enantiomeric pairs are assigned a handedness, either *R* (from the Latin *rectus,* "right") or *S* (from the Latin *sinister,* "left"). These designations do not imply a particular direction of optical rotation; the determination of the absolute configuration of an isomer with a known optical activity is a problem that must be solved by chemical means for each case. Recently, a useful extension of molecular orbital theory has been developed by D. Beratan and P. Wipf that predicts the sense and magnitude of optical rotation of an enantiomer from its spatial configuration.

18.4 Benzene and Its Derivatives: The Aromatics

In Chapter 7, the structure of the six-membered ring compound **benzene** ($C_6H_6(l)$) was discussed as an example of *delocalized π bonding.* The resonance forms predicted from the Lewis structure of the molecule, which shows alternating single and double bonds, are replaced by a set of delocalized π orbitals extending around the ring. This is indicated schematically by drawing a dashed or thin solid circle within the hexagonal ring representing the delocalized π electrons; in skeletal form,

The planar hexagonal skeleton is assumed to be built in a valence bond fashion from sp^2 C hybrid orbitals bonded to each other and to $1s$ H atom orbitals, as discussed in Chapter 7. As shown in Figure 7.20, the six C $2p$ orbitals perpendicular to the plane of the ring combine to make six π MOs, three bonding and three antibonding. The bonding orbital energies all lie *below* the energy of an isolated (ethylenic) π bond, and their occupancy by the six π electrons in benzene gives it a special stability that has come to be called **aromaticity.** This term has nothing to do with benzene's smell, but is rather a mark of thermodynamic stability endowed by the delocalized π-bond network. Several other cyclic molecules and ions share this special feature; the requirements are ring planarity and 2, 6, 10, 14, . . . , $(4n+2)$ π electrons. Here we restrict ourselves to benzene and its derivatives.

The cyclic triene structure for benzene was first proposed by A. Kekulé in 1865, but experimental facts like those to be discussed next prevented its immediate acceptance. It was not until E. Hückel worked out the MO theory of conjugated π bonding in the 1930s, explaining for the first time why benzene should be especially stable, that chemists fully accepted Kekulé's postulate.

This conferred stability rationalizes benzene's relative inertness toward electrophilic reagents despite its rich π-electron cloud. For example, in Reaction 18.7, cyclohexene (C_6H_{10}) readily adds Br_2 across its double bond. When the same procedure described there is tried with C_6H_6, the Br_2 is readily extracted into the floating benzene layer, but *no reaction occurs*. You might expect all three double bonds in the Lewis structures above to react, yielding $C_6H_6Br_6$. Benzene's π cloud is different; all that C_6H_6 can be induced to do with Br_2, under harsh conditions with the aid of a catalyst, is to undergo substitution rather than addition,

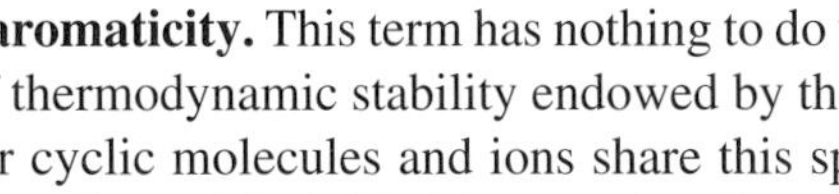

$$\text{C}_6\text{H}_6 + \text{Br}_2 \xrightarrow{\text{FeBr}_3} \text{C}_6\text{H}_5\text{-Br} + \text{HBr} \tag{18.18}$$

that is, a reaction more like that of an alkane than an alkene. The mechanism again appears to involve a bromonium ion (Br^+), and in the high-energy transition state the aromaticity is temporarily lost. As in the alkanes, once the bromobenzene derivative is made, other reactions to form a variety of derivatives become possible.

The term **aromatic** arose because many of benzene's derivatives in fact have pleasant odors. For example, the good-smelling ingredients in vanilla and oil of wintergreen are the benzene derivatives

Vanillin   and   Methyl salicylate

The monosubstituted benzenes can be categorized in the same way as alkane derivatives in Table 18.4, by identifying the **phenyl group** (C_6H_5-) as an "R." When an alkyl or alkenyl group is bonded to benzene, it becomes the "functional group"; those for the alkanes are listed in Table 18.3. Many of the benzene derivatives had common names bestowed on them long ago; for example, methylbenzene ($C_6H_5-CH_3$) is better known as toluene, hydroxybenzene (C_6H_5-OH) is called phenol, phenylamine (or aminobenzene) ($C_6H_5-NH_2$) is aniline, and ethenylbenzene (or phenylethylene) ($C_6H_5-CH=CH_2$) is styrene. Disubstituted benzenes have three structural isomers owing to the choice of three unique ring positions, *ortho, o-; meta, m-;* or *para, p-;* for the second substituent. The isomers of dimethyl benzene are

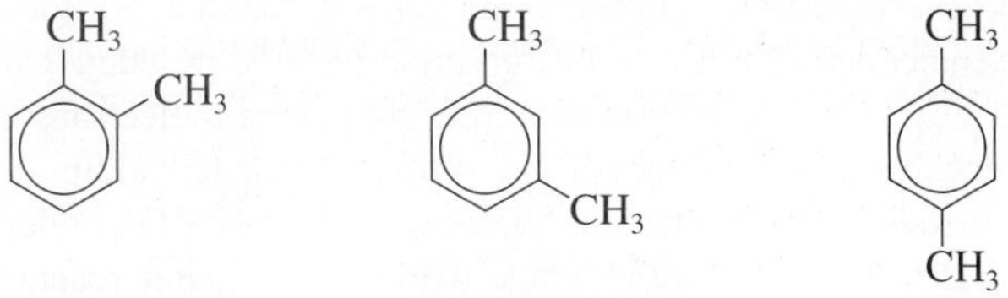

o-Dimethylbenzene
1,2-Dimethylbenzene
o-Xylene

m-Dimethylbenzene
1,3-Dimethylbenzene
m-Xylene

p-Dimethylbenzene
1,4-Dimethylbenzene
p-Xylene

The numbering system allows derivatives with three or more substituents to be named without ambiguity, while "xylene" is again a time-honored common name.

The stable benzene ring serves as a nucleus for a variety of useful compounds. Two examples are *aspirin* (acetylsalicylic acid), an analgesic, and potassium acid phthalate (KHP), familiar to fledgling chemists as the solid acid used for standardization of NaOH solutions, with structures

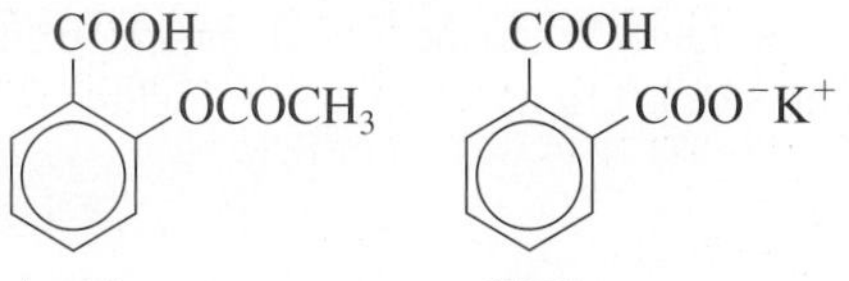

Aspirin

KHP

As we shall see, 3 of the 20 naturally occurring α-amino acids have aromatic side chains, and the cyclic bases that form part of our genetic material are derived from aromatic rings. Benzene for use in the chemical industry is obtained both directly from distillation of petroleum, and from catalytic dehydrogenation and ring closure of the gasoline fraction.

18.5 Orbitals in Organic Reactions: The Diels–Alder Reaction

Among the most powerful reactions in synthetic organic chemistry are the *ring-closing* or **cyclization** reactions, in which two noncyclic reagents combine to form a ring where none existed before. One of the earliest and still extremely useful cyclizations was discovered by O. Diels and K. Alder in 1928. The **Diels–Alder cycloaddition reaction** is written in general as a reaction between a *conjugated diene* and an alkene (here 1,3-butadiene and ethylene) to give a cycloalkene (here cyclohexene),

(18.19)

Note that two of the curly arrows point to the location of the two new bonds that close the ring. The curly arrows are ambiguous in this case; they could also be drawn in a clockwise fashion to yield the same result. The author likes to think of this ambiguity as "mechanistic resonance forms." As was true for resonance structures in an individual molecule, mechanistic resonance forms require a molecular orbital description to unify them. The Diels–Alder reaction is quite general; the diene or the double bond can already be embedded in a ring itself, giving a *bicyclic* product. This reaction is called a "4 + 2" cycloaddition, the numbers referring to the π electrons involved, and is facile relative to 2 + 2 (ene + ene) or 4 + 4 (diene + diene) additions.

Our object here is to take the frontier molecular orbital (FMO) model presented in a simple form in Chapter 7, and show how it works in an organic reaction. The FMO model is intended to supplement, and in some cases replace, the curly-arrow "electron-pushing" picture given earlier. Recall the problem we faced with curly-arrow mechanisms in Chapter 5: where do the electrons actually *go* when they are moving toward a molecule that already satisfies the octet rule? In the curly-arrow picture, this problem is swept aside by making other electrons move *in concert;* that is, there must always be more than one curly arrow. The preceding mechanism for the Diels–Alder reaction is a case in point. The FMO model, on the other hand, is *sequential* rather than concerted, always beginning with electrons in a HOMO (highest occupied MO) on one molecule seeking to occupy a LUMO (lowest unoccupied MO) on the other. In a localized-electron-pair picture, the HOMO ought always to be at the tail of the curly arrow, while the LUMO ought to be at its head. When the orbitals are delocalized over several atoms, however, this correspondence is lost as we shall see. Figure 18.5 shows the π molecu-

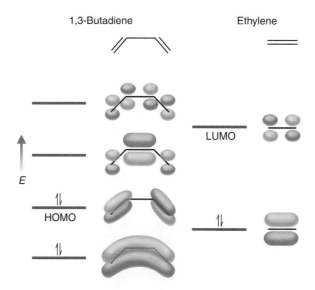

Figure 18.5
π Molecular orbitals for 1,3-butadiene and ethylene.

lar orbitals and energy levels for butadiene and ethylene, with the frontier orbitals indicated. Very important here is the *nodal rule,* which requires the number of nodes in the orbital to increase by one for each step up in energy.

By examining only the Lewis structures of butadiene and ethylene or the curly-arrow mechanism, you have no way of telling which of the reagents will supply the HOMO and which the LUMO. This information comes from comparing the π energy levels of the two reagents: *the higher energy HOMO is the one that will be employed in the reaction,* since the electrons in it are "freer" and more easily donated. In conjugated π molecules, the more π electrons, the higher the HOMO energy will be. (This situation is the even-number π analogue of the "line of lithium atoms" used to develop the band theory of solids; see Chapter 14, Figure 14.19.) Thus the diene is the natural choice to provide the HOMO, and by default, the alkene the LUMO. Figure 18.6 illustrates the interaction between these frontier orbitals. Note that the signs or *phases* of the outer lobes of the diene HOMO and ethylene LUMO match; this enables constructive overlap and delocalization of the HOMO electrons into the LUMO. A natural description of the breakage of the ethylenic double bond is provided as well, since the LUMO is a π^* orbital and its occupancy sends the bond order of the π bond to zero. Further, the overlapping of the terminal $2p$ orbitals is a natural precursor to closing the ring with two new bonds. This drains the HOMO of its electrons, and the rehybridization of the terminal carbons on the diene ($sp^2 \longrightarrow sp^3$) forces the still-occupied lowest π orbital on the diene to localize between the two inactive carbons, giving the final cyclohexene.

To make the σ bonds that close the ring, the diene and the ene must approach in different, roughly parallel planes, so that the $2p$ orbitals can overlap in a σ fashion; an approach in the same plane would result in π type overlap, which is energetically less favorable. The new six-membered ring is thus formed initially in a boat conformation. There is considerable stereochemical evidence to support this picture of the transition state, which we must again leave for a later course.

How does this picture compare with the curly arrows above? The most obvious difference is that the electrons don't go clockwise or counterclockwise, but in both directions at once, in the FMO picture. This relationship is analogous to that between the MO and VB models themselves in cases of resonance. Further, the curly arrows show electrons leaving the ene, whereas the initiating FMO step is donation of electrons to the ene. These are fundamental conceptual changes in our view of the mechanism. Further, FMOs lead to an understanding of the case of the Diels–Alder cycloaddition relative to ene + ene and diene + diene in terms of

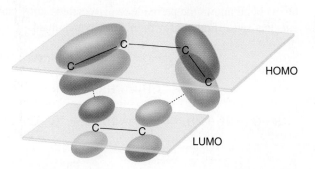

Figure 18.6
Interaction between the HOMO of butadiene and the LUMO of ethylene.

HOMO and LUMO phases; the phase-matching requirement, which is in essence an orbital symmetry requirement, is part of a system for describing concerted reactions known as the **Woodward–Hoffmann rules.** The predictions of product stereochemistry of the FMO model have been confirmed by experiment in every case. Because many biologically active compounds are particular stereoisomers, orbital analysis of this sort is particularly important in biochemistry and in the pharmaceutical industry.

18.6 Polymers

The talents of carbon know no bounds. There seems to be no upper limit to the length of a hydrocarbon chain "backbone" that will remain stable, just as there seems to be no end to the variety of allotropes of carbon the element (see Chapter 14). Even chain heteroatoms, such as N or O, and intercalated five- or six-membered rings, are tolerated as long as they remain between C's. Further, a chain of sp^3 carbon leaves two valence positions for branch groups and chains along the main line, which are stable as long as they also incorporate the ubiquitous C. What we have here is a prescription for a huge class of **macromolecules** that makes 19 million known compounds seem a small number. In the 20th century we have learned how to make enormous molecules by using reactive subunits as building blocks. This yields a long chain with a repeating pattern known as a **polymer.** Natural chemical evolution over eons has resulted in macromolecules of such complexity that they still bewilder and challenge us; that we shall save for the end.

The simplest type of synthetic polymer is that built from a single subunit, and by far the most important of these is the class built from ethylene and its derivatives,

$$n\ H_2C=CH\!-\!R \longrightarrow \cdots\!-\!CH_2\!-\!CH(R)\!-\!CH_2\!-\!CH(R)\!-\!CH_2\!-\!CH(R)\!-\!\cdots \quad (18.20)$$

$$\text{or } +\!CH_2\!-\!CH(R)\!+_n$$

When R = H, we have polyethylene itself, while R = CH_3 gives polypropylene, R = Cl polyvinylchloride (PVC), R = CN polyacrylonitrile (PAN), and R = C_6H_5 (phenyl) polystyrene. All except PAN are plastics; PAN can be made into wool-like fibers that go under the trade name *Orlon*. These are called **addition polymers;** there are no products other than the polymeric chain itself. They are made with the help of a small amount of *free-radical initiator*, usually an alkyl peroxide (R—O—O—R). The weak O—O bond allows the peroxide to dissociate readily, yielding a pair of *alkoxy radicals* that sets off rapid growth of the polymer chain:

$$R\!-\!O\!-\!O\!-\!R \longrightarrow R\!-\!O\cdot + \cdot O\!-\!R$$

$$R\!-\!O\cdot + CH_2\!=\!CHR \longrightarrow R\!-\!O\!-\!CH_2\!-\!CHR\cdot$$

$$R\!-\!O\!-\!CH_2\!-\!CHR\cdot + H_2C\!=\!CHR \longrightarrow R\!-\!O\!-\!CH_2\!-\!CHR\!-\!CH_2\!-\!CHR\cdot$$
$$\vdots$$

As many as 1500 ethylenic units polymerize in a fraction of a second. Steric hindrance by the R group generally forces the substituted carbons to alternate. Straight chain ethylene polymers are known as high-density polyethylene (HDPE), while branched chain polymers cannot pack as well, yielding a lower density. If tetrafluoroethylene is used, the polymer that results is polytetrafluoroethylene (PTFE), also known commercially as Teflon. Polymerization of butadiene and its derivatives yields *synthetic rubber* and other stretchable (elastomeric) materials.

Condensation polymers, the other major type besides addition polymers, form with the elimination of a small molecule, usually H_2O. For example, a polymer known as *nylon-6* is made from 6-aminohexanoic acid,

$$H_2N-(CH_2)_5-COOH + NH_2-(CH_2)_5-COOH \longrightarrow \qquad (18.22)$$
$$H_2N-(CH_2)_5-CO-NH-(CH_2)_5-COOH + H_2O$$

Note the amide linkage formed upon elimination of water; repeated reaction yields a *polyamide*. When the monomer is a disubstituted derivative where the functional groups can undergo an S_N2 reaction, the dimer is functionalized in the same way, and you can see how polymerization will occur naturally. Monomers like this are quite common, and a problem frequently encountered in organic synthesis is unwanted runaway polymerization. Condensation polymers are more commonly made by combining two different disubstituted monomers, each with identical functional groups that are reactive between the pair, a procedure invented by W. Carothers in 1930. The result is called a **copolymer,** and a well-known example is the polyamide *nylon-66,* of nylon stocking fame, made by the reaction of adipic acid (1,6-hexanedioic acid) and 1,6-diaminohexane,

$$n HOOC-(CH_2)_4-COOH + n H_2N-(CH_2)_6-NH_2 \longrightarrow \longrightarrow \longrightarrow \qquad (18.23)$$
$$[-CO-(CH_2)_4-CO-NH-(CH_2)_6-NH-]_n + n H_2O$$

If a diol instead of a diamine is used, the result is a *polyester;* the synthetic fabric *Dacron* is the copolymer of ethylene glycol (1,2-ethanediol) and *p*-phthalic acid (1,4-benzenedioic acid),

$$[-O-CH_2CH_2-O-CO-C_6H_4-CO-]_n \qquad (18.24)$$

Among the earliest commercially successful polymers were the *phenolic plastics*. The most famous of these, *Bakelite,* was discovered by L. Baekelund in 1907; it is a *cross-linked* copolymer made by reacting formaldehyde (methanal), H_2CO, with phenol, C_6H_5OH. Instead of reacting directly with the OH group on phenol, the formaldehyde *adds* to the *meta* and *para* positions on the benzene ring to make tri- and tetrasubstituted benzenes; these can start polymer chains in several directions at once.

In one sense, the world of polymers offers great variety, because many possible monomers and combinations of monomers can be imagined. But in another sense, it is hopelessly limited, because the result is simply the starting material repeated again and again and . . . . What if one could build a polymeric molecule with specific sequences of *different* monomer units, each with its own functionality? We now know of such molecules; they are made by all living things, and we call them *proteins*. They are one of the molecular miracles wrought by life, our last topic in this chemical odyssey.

18.7 The Molecules of Life: Organic Motifs in Biology

While not everyone may be able to agree on a definition of life—the author likes I. Asimov's criterion: *life makes an effort*—there is no doubt that the **cell,** a little chemical factory with organic walls that contains an organic soup and an organic nucleus with its own organic walls, is the unit of life. Our intent here is not to explore cell biology on the scale of the cell, but to look at those molecules that are important to its function, that make it what it is. Of the billions of choices for a molecular basis of life, even if the choices are "limited" to carbon compounds, life as we know it here on earth has selected a remarkably small set of structural motifs for the molecules that make the cell run. We can actually tell you about all the essential ones in this small section of a small chapter in a small book. Our examination of the molecules of life will be selfishly centered around our own needs; if we were plants and not animals, the plan of attack would be different, but the molecules would be the same.

To gain a little insight into the organization of this section, think about how you live. You need food and water to survive; food—organic molecular nutrients—gives you energy. This energy comes from the *controlled combustion,* biologically known as **metabolism,** of the food you eat; this is a chemical job done within the cell, and requires a set of controller molecules called **enzymes.** The energy generated must be *transduced* into a readily available form so that you can "make an effort": energy-storage molecules, known as **ATP** and **GTP,** are needed. You now have the energy to find more food, and the food ⟶ energy ⟶ more food ⟶ ... cycle ensures that you and your cells will continue to exist. To run a cycle like this requires organization, that is, information about which molecules are required when to do what and about how to make the molecules that are needed, and a mechanism for executing the instructions. The kingpin of the cellular world is known as **DNA,** an enormous molecule that is able to store the information required, regulate the energy-producing process, and pass this information along to future generations of cells and the larger beings made from them.

Carbohydrates and Fats

Most of our life energy comes from the combustion of only two types of organic compounds, carbohydrates and fats. **Carbohydrates,** as their name implies, have the formula $C_m(H_2O)_n$; the simplest and most common of these have $m=n=5$, the **pentose sugars,** and $m=n=6$, the **hexose sugars.** These simple sugars are also called **monosaccharides.** Their distinguishing structural feature is tetrahedral carbon in a chain, bonded to H and OH as side groups, —CHOH—; carbohydrates are actually polyalcohols. They can exist in either a straight chain or ring form, enabled by the presence of a carbonyl group, either terminal or on C-2. Those sugars with terminal (aldehydic) carbonyls are called **aldoses,** while the others are called **ketoses.** An *aldohexose* has the structure

$$O=\overset{1}{C}H-\overset{2}{C}H-\overset{3}{C}H-\overset{4}{C}H-\overset{5}{C}H-\overset{6}{C}H_2OH \quad\quad (18.25)$$
$$\underset{OH}{|}\underset{OH}{|}\underset{OH}{|}\underset{OH}{|}$$

Numbering of the carbons begins from the carbonyl end. Each of the four nonterminal C's is asymmetric, implying $2^4 = 16$ stereoisomers, eight pairs of enantiomers. Amazingly, only 3 of the 16 occur naturally, D-(+)-glucose, D-(+)-mannose, and D-(+)-galactose, and they are shown in Figure 18.7 in the form of **Fischer projections,** an easy method for displaying and distinguishing stereoisomers. The first pre-

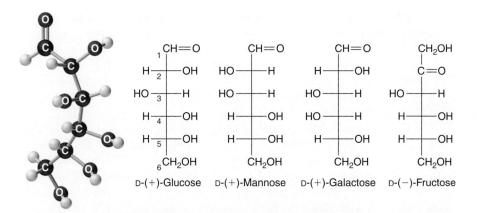

Figure 18.7
Natural sugars (monosaccharides) shown as Fischer projections. The carbon atoms are numbered by convention beginning with the end nearest the carbonyl group, as illustrated for glucose. The Fischer projection reflects the orientation of the H and OH groups when the carbon spine of the molecule is bent backward so as to pop the side groups outward, as illustrated on the left for glucose (unlabeled atoms are hydrogens).

fix may be D- or L-, as determined by whether the OH on C-5 lies to the right (D- for *dextro*) or left (L- for *levo*), and the second prefix (+) or (−) gives the direction of rotation of plane-polarized light. A clever set of experiments designed by E. Fischer in the 1880s narrowed the possible configurations of (+)-glucose to within one pair of enantiomers, that is, he eliminated the other seven pairs. Not until 1948 did X-ray diffraction experiments show that the D-enantiomer is the biological compound, as correctly guessed by Fischer 60 years earlier. Also shown in Figure 18.7 is the only naturally occurring ketohexose of eight possibilities, D-(−)-fructose. Our bodies can digest these 4 natural sugars and use their energy; the other 20 are totally indigestible, despite being connectively and energetically identical to their stereo partners.

The lone pair electrons on a hydroxyl oxygen can attack the carbonyl carbon; this allows the sugars to *cyclize,* forming either five- or six-membered rings. In aldohexoses, six-membered rings called *pyranoses* are formed from attack of the C-5 OH on C=O, and five-membered rings called *furanoses* by attack of the C-4 OH. In either case, the former carbonyl C (C-1) becomes asymmetric, creating a new pair of stereoisomers designated by prefix α- if the new hydroxyl lies to the right, and β- if to the left; both forms of glucose are shown in Figure 18.8. The interconversion between

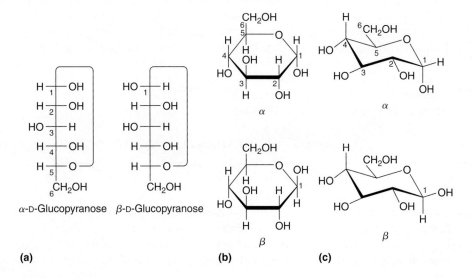

Figure 18.8
Cyclized forms of glucose shown as (a) Fischer projections, (b) Haworth projections, and (c) as more realistic chair forms. In each case note the reversed orientation of H and OH on the former aldehydic C (C-1), at the top of the Fischer projections, on the right of the Haworth and chair depictions).

the straight chain and ring forms occurs readily in aqueous solution, and quickly reaches equilibrium; at 40°C, (+)-glucose is nearly entirely in pyranose ring form, 36% α and 64% β, with only a trace of chain or furanose ring forms. The ring-adapted Fischer projection and the planar-ring **Haworth projection** are both given in Figure 18.8, along with the more realistic chair form of the six-membered ring. An aldopentose important in nucleic acid structure is D-(−)-ribose, which also exists nearly exclusively as a furanose ring.

The C_6 sugars are relatively rarely found as monosaccharides; more often they occur in dimers, **disaccharides,** or polymers, **polysaccharides.** For example **sucrose**—common table sugar—is a dimer of α-D-glucose and β-D-fructose, while **lactose**—milk sugar—is a dimer of β-D-glucose and β-D-galactose; both are illustrated in Figure 18.9. They are made in plants by enzyme-catalyzed condensation synthesis (loss of water) from the monomers, and are held together by an ether-like C—O—C linkage called a **glycoside bond.** The two most common polysaccharides are cellulose and starch; **cellulose,** the most common organic compound on the face of the earth, is a polymer of β-D-glucose; while **starch,** the major carbohydrate in plants eaten as food, is a polymer of α-D-glucose. As Figure 18.9 shows, they differ only in the chirality of their glycoside bonds, yet animals are able to break down only starch and not cellulose into energy-supplying glucose. In one way or another, nearly all of our energy comes from starch; either we eat it directly, or feed it to other animals before we eat them.

Unlike the proper isomers of sugars and starches, which are readily metabolized and a source of "quick energy," **fats,** also called **lipids,** store energy for later use. They are triesters made from glycerol (1,2,3-propanetriol) and carboxylic acids with long hydrocarbon chains known as **fatty acids.** The sodium salts of fatty acids are familiar as major ingredients in soap. A typical fatty acid is *stearic acid,* $CH_3(CH_2)_{16}COOH$, and the saturated fat *glyceryl stearate* has the structure

Figure 18.9
Common disaccharides and polysaccharides. In starch the α-glycoside bond angles cause the chain to curl into a spiral form, while in cellulose the β-glycoside bond allows the chain to remain straight.

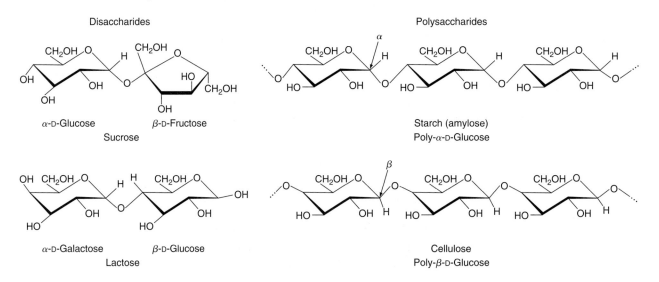

$$\begin{array}{l} \text{CH}_2-\text{O}-\overset{\overset{\displaystyle O}{\|}}{\text{C}}-(\text{CH}_2)_{16}\text{CH}_3 \\ | \quad\quad\quad O \\ \text{CH}-\text{O}-\overset{\|}{\text{C}}-(\text{CH}_2)_{16}\text{CH}_3 \\ | \quad\quad\quad O \\ \text{CH}_2-\text{O}-\overset{\|}{\text{C}}-(\text{CH}_2)_{16}\text{CH}_3 \end{array} \quad (18.26)$$

Fatty acids containing double bonds, such as *trans*-oleic acid, make *unsaturated* fats; the reactivity of the π bond makes these fats easier to metabolize. Generally body fat is "burned" only when sufficient glucose is not available.

Fats in the form of **phospholipids** are also employed by cells as cell-wall material; in a phospholipid one of the ester linkages in the structure 18.26 is taken by a phosphate ion (PO_4^{3-}), a motif we will see gain in nucleic acids. The polar ester end of the phospholipid is *hydrophilic*, interacting favorably with water, while the two remaining fatty-acid chains are *hydrophobic*. The cell wall is a phospholipid *bilayer*, with the fatty chains intermeshed within and the polar head groups exposed to the aqueous environment. The greasy middle provides a waterproof seal, and only certain proteins embedded in the cell wall provide "doors and hallways" for the passage of water and solutes back and forth.

All of this food cannot be turned into useful energy without a mechanism for burning it and storing the energy in a useful form. This leads us to the next set of molecules for life.

Amino Acids and Proteins

Instead of flasks, clamps, and heat, all a living cell has to work with is the molecules within it, to carry out the energy-producing reactions so necessary to its existence. The molecules that perform these nanoscopic miracles, that serve simultaneously as chemical lab bench, reaction flask, and catalyst, are the **enzymes** we've already met in Chapter 15. Here we consider their composition and structure in somewhat more detail. Enzymes are a special subclass of the macromolecules called **proteins.** Proteins are unbranched chains of α-**amino acids** (NH_2—CHR—COOH), the "α" indicating that the amine group is on C-2, next to the carboxyl carbon. Of the countless possibilities for R-groups, only 20 amino acids are found in natural proteins; these are given in Table 18.5. They fall into several groups: R = H or alkyl(6, including the cyclic proline); R = aromatic(4); R containing —OH, —SH, or —SCH_3(4); R containing COOH(2); R containing $CONH_2$(2); and R = aminoalkyl(2). In neutral aqueous solution the amino acids all exist in *zwitterionic* (double ion) form $^+NH_3$—CHR—COO^-, as we discussed in Chapter 12. All except glycine are chiral at the α carbon, and every one is *left-handed* L-; right-handed amino acids are simply not found in nature. They have retained their common names, which have standard three-letter and one-letter abbreviations as listed in Table 18.5.

Proteins are formed by condensation polymerization (loss of water), in the nucleophilic attack of the amine group of one amino acid on the carboxyl carbon of another, similar to the polymerization of 6-aminohexanoic acid given earlier. This

TABLE 18.5

The natural amino acids $NH_2-CH-COOH$
                               $\quad\quad\quad\quad\quad\quad\quad |$
                               $\quad\quad\quad\quad\quad\quad\quad R$

Name and abbreviations	R	Name and abbreviations	R
Glycine, Gly, G	$-H$	Serine, Ser, S	$-CH_2-OH$
Alanine, Ala, A	$-CH_3$	Threonine, Thr, T	$-CH(CH_3)-OH$
Valine, Val, V	$-CH(CH_3)_2$	Methionine, Met, M	$-CH_2CH_2-S-CH_3$
Leucine, Leu, L	$-CH_2CH(CH_3)_2$	Cysteine, Cys, C	$-CH_2-SH$
Isoleucine, Ile, I	$-CH(CH_3)-CH_2CH_3$		
Proline, Pro, P (entire acid)	(pyrrolidine)—COOH	Aspartic acid, Arp, D	$-CH_2-COOH$
		Glutamic acid, Glu, E	$-CH_2CH_2-COOH$
Phenylalanine, Phe, F	$-CH_2-$(phenyl)	Asparagine, Asn, N	$-CH_2-CO-NH_2$
		Glutamine, Gln, Q	$-CH_2CH_2-CO-NH_2$
Tryptophan, Trp, W	$-CH_2-$(indole)		
		Lysine, Lys, K	$-(CH_2)_4-NH_2$
		Arginine, Arg, R	$-(CH_2)_3NHC(NH_2)=NH$
Tyrosine, Tyr, Y	$-CH_2-$(phenyl)$-OH$		
Histidine, His, H	$-CH_2-$(imidazole)		

forms an amide linkage —CO—NH— known biochemically as a **peptide bond.** An example of a *polypeptide* is the three-acid protein

$$^+NH_3-\underset{\underset{CH_3}{|}}{CH}-\overset{\overset{O}{\|}}{C}-NH-\underset{\underset{CH(CH_3)_2}{|}}{CH}-\overset{\overset{O}{\|}}{C}-NH-\underset{\underset{CH_2CH_2CH_2CH_2NH_2}{|}}{CH}-\overset{\overset{O}{\|}}{C}-O^- \quad (18.27)$$

This molecule would be called alanylvalyllysine, with abbreviations Ala-Val-Lys or A-V-K; the sequence is always read from the amino end. The molecule is shown in its predominant zwitterionic form. Each amino acid in the chain is called a **residue;** most naturally occurring protein chains contain hundreds to thousands of residues, with molecular masses ranging from 10,000 to 1,000,000 amu and the residues arranged in a specific sequence. As the length of the chain grows, the number of possible arrangements of a given set of amino acids grows roughly as the *factorial, n!,* of the number of residues *n;* for our three-residue peptide there are 6 possibilities; for a 20-residue chain using all 20 amino acids, there are more than 2,000,000,000,000,000,000 (2×10^{18}) distinct structural isomers. Yet we know of less than 10^5 natural proteins, each considerably larger than 20 residues, with specific sequences and associated chemical properties. The chain sequence of a protein is called its **primary structure;** an example of a small protein is the 129-residue enzyme *lysozyme,* found in egg white; its primary structure is shown in Figure 18.10. Lysozyme catalyzes the destruction of bacterial cell walls, protecting the chick embryo from infection. As shown in the figure, the primary structure includes four *cross-links* made from joining cysteine residues on adjacent chain segments through oxidation of the *thiol* —S—H groups,

$$\overset{|}{C}H-CH_2-S-H + \tfrac{1}{2}O_2 + H-S-CH_2-\overset{|}{C}H \longrightarrow \qquad (18.28)$$
$$\overset{|}{C}H-CH_2-S-S-CH_2-\overset{|}{C}H + H_2O$$

The resulting **disulfide bonds** cause the chain to fold in a specific way, enabling further changes in its conformation, to be discussed shortly. The S—S bond is relatively weak ($D = 55$ kcal/mol) and easily broken; when egg white is cooked, the lysozyme loses its cross-links and along with them its catalytic activity.

The sequence of amino acids in a protein can be determined chemically by selectively derivatizing the carboxy end, cleaving the residue, and identifying the amino acid by chromatography. Automation of this procedure has made it possible to deduce the primary structures of even the larger proteins. However, the plethora of single bonds in a protein chain implies countless possibilities for different comformations, arrangements of the chain in 3-D space. These arrangements have been classified into three further categories, called simply secondary, tertiary, and quaternary structures. Much of what we know about the 3-D "shapes" of protein molecules is obtained from

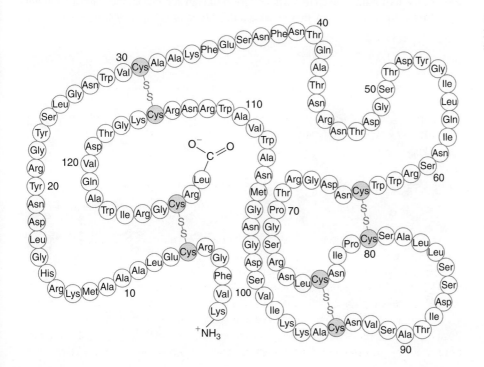

Figure 18.10
Primary structure (amino-acid sequence) of egg white lysozyme.

X-ray crystallography (see Chapter 14). Obtaining pure single crystals of a protein is difficult, and analyzing their X-ray patterns even more challenging. Nonetheless, genomic research has stimulated the rapid growth of our "library" of known protein comformations, about 20,000 at this writing.

Among the early findings from X-ray analysis of protein single crystals were two distinct types of scattering patterns, labeled simply "α" and "β." The former was first analyzed in 1948 by Linus Pauling, who deduced the existence of what is known as the **α-helix**. As illustrated in Figure 18.11, the amino acid chain winds around a straight axis in a *clockwise* or *right-handed* fashion when viewed from the amino end. In this type of **secondary structure,** the α-helix is held in place by *hydrogen bonds* between carbonyl oxygens and amino hydrogens,

$$\begin{matrix} \diagdown \\ N-H \text{\tiny{IIIIIIIIIIII}} :\ddot{O} \\ \diagup \quad\quad\quad\quad\quad \diagdown\!\!\!\!\!\diagdown \\ \quad\quad\quad\quad\quad\quad\quad C- \\ \quad\quad\quad\quad\quad\quad \diagup \end{matrix} \qquad (18.29)$$

While it isn't geometrically possible to attain the collinear N—H --- :O arrangement for hydrogen bonding within a single residue, winding the chain helically without straining the bond angles allows collinear geometry and the right H --- O distance (about 2 Å) to be attained every *fourth* residue; that is, residue 1 hydrogen-bonds to residue 4, residue 2 to residue 5, etc. As Figure 18.11 shows, the H bonds lie along the spiral axis, maintaining its linearity; it is said that Pauling, bored with reading detective novels while recovering from a cold, discovered the structure by building paper models in bed.

Once the nature of the α secondary structure was known, Pauling guessed that H bonds would be involved in the β secondary structure as well, and deduced that this would take the form of a **pleated sheet,** in which adjacent chain segments would line up with N—H's in one segment opposite O=C's in the next. The pleated sheets are created by folding the chain back and forth like an accordion; the folds are known as **β-turns.** In most proteins there is also a third type of secondary structure called a

Figure 18.11
The two main types (α-helix and β-sheet) of secondary structures of protein chains. Both are organized by repeating sequences of hydrogen bonds.

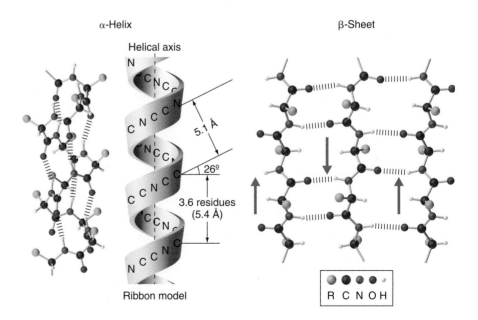

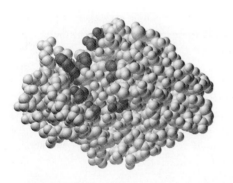
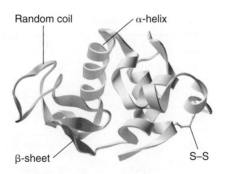

Figure 18.12
Space-filling (left) and ribbon (right) models of the secondary and tertiary structure of egg white lysozyme.

random coil; as its name implies, the residue chain is (seemingly) folded randomly, but also mainly under the influence of H bonds. In recent depictions, these three types of secondary structure are drawn with ribbons representing the chain, as shown for lysozyme in Figure 18.12.

How the three secondary structures are interspersed and arranged in space to give the overall shape of the protein is called its **tertiary structure.** Proteins that make up tendons and hair are "molecular ropes" containing three or more α-helices woven around each other; while most that serve as enzymes, such as lysozyme, are **globular,** roughly spherical in shape, with occasionally interesting terrain, such as deep pockets, where their catalytic activity is centered. The urease example of Chapter 15 is a globular case. The globular form of enzymes is encouraged by water solvent, which interacts well with hydrophilic polar groups, but excludes the "greasy" (hydrophobic) hydrocarbon groups. The first seven amino acids in Table 18.5 have greasy side chains, which are forced to turn inward, away from water, when they are residues in a protein. This in turn causes the protein chain to curl or fold in on itself.

Certain proteins, such as myoglobin, the oxygen carrier in blood, form **quaternary structures,** aggregates of separate protein molecules, held in place by nonbonded forces (H bonds and dispersion attraction). Hemoglobin is a tetrahedral assembly of four myoglobin molecules, each containing a heme group (see Chapter 17).

When in its properly folded conformation, with specific catalytic activity, an enzyme is said to be in its **native** form. If it is removed from water or subjected to stress (e.g., low pH and/or heat), however, it may unfold, or become **denatured.** In early experiments on purified proteins in aqueous solution, it was shown that a denatured protein folds into its active form *without assistance,* that is, *the sequence of amino acids in a protein completely determines its native form.* We are still learning through time-resolved kinetics experiments and molecular dynamics simulations how this works; if we ever succeed in deciphering the rules of the folding game, we will be able to design our own enzyme catalysts!

Cyclic Bases, Nucleotides, and Nucleic Acids

The energy released in the enzyme-catalyzed combustion of food cannot be allowed to dissipate as heat; rather, it must be captured and stored in some useful form. The role of an energy reservoir in living systems is played almost exclusively by a single molecule, **adenosine triphosphate (ATP),** a member of a small family of compounds known as the **nucleotides,** so named because they are found mainly in cell nuclei. To

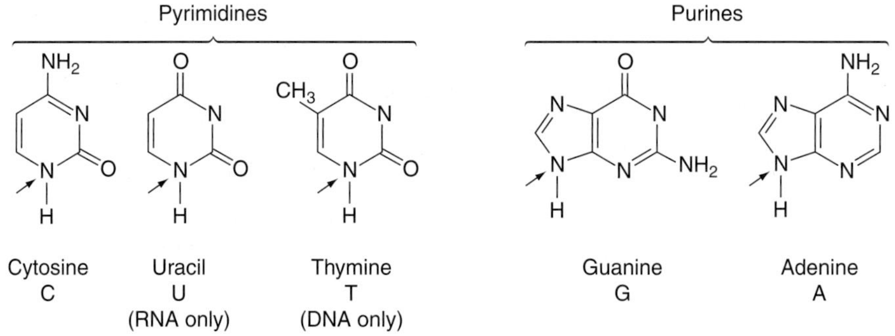

Figure 18.13
The five natural nucleotide bases. The arrows point to the nitrogens where attachment to the ribose sugar ring occurs. See Figure 18.14 for an assembled nucleotide. Uracil (U) occurs only in ribonucleic acids (RNA) and thymine (T) only in deoxyribonucleic acids (DNA).

understand how ATP is able to store energy, we need to know the structure of nucleotides. A nucleotide is a single molecule with three components: a **cyclic nitrogen base**, a **sugar** in its cyclic form, and a **polyphosphoric acid** group. The sugar ring binds to the polyphosphoric acid via an ester-like linkage C—O—P, similar to that in phospholipids; and to the cyclic base by a C—N bond, as in amines.

The nucleotides found in living systems employ only five different bases: cytosine, uracil, thymine, guanine, and adenine, frequently abbreviated by their initials C, U, T, G, and A; their structures are given in Figure 18.13. The first three are called **pyrimidines**, having structures derived from the monocyclic aromatic base pyrimidine, a benzene ring with N substituted for C at the 1 and 3 positions. The latter two are called **purines**, derived from the bicyclic aromatic base purine. Only the aldopentose sugar **ribose** appears in natural nucleotides; ribose binds at its aldehydic carbon (C-1′) to a ring nitrogen in the base. The bonded combination of base and sugar rings is called a **nucleoside**. The terminal —CH$_2$—OH (C-5′) of the ribose can combine with a monohydrogen phosphate group (H—O—PO$_3^{2-}$) with loss of water to form the phosphate ester linkage, and we now have a nucleotide. A **ribonucleotide** has a ribose ring in it, while in a **deoxyribonucleotide** the ribose ring has the OH at the carbon adjacent to the aldehydic C (C-2′) replaced by H. Up to two more phosphates can be added to the phosphate of the nucleotide, producing di- and triphosphate species. ATP is a ribonucleotide; its structure is depicted in Figure 18.14. Each of the other four bases also forms ribonucleotides of similar structure. All the nucleotides are right-handed enantiomers, due to the ribose sugar; in this case, left-handed nucleotides are not found in nature.

The most labile bonds in ATP are the P—O—P linkages between the phosphate groups, and this is where metabolic free energy is stored.* ATP is readily hydrolyzed to ADP, adenosine diphosphate,

$$\text{Adenosine} - O - \underset{\underset{O}{|}}{\overset{\overset{O^-}{|}}{P}} - O - \underset{\underset{O}{|}}{\overset{\overset{O^-}{|}}{P}} - O - \underset{\underset{O}{|}}{\overset{\overset{O^-}{|}}{P}} - O^- + H_2O \longrightarrow \tag{18.30}$$

$$\text{Adenosine} - O - \underset{\underset{O}{|}}{\overset{\overset{O^-}{|}}{P}} - O - \underset{\underset{O}{|}}{\overset{\overset{O^-}{|}}{P}} - O^- + H_2PO_4^-$$

---

* The biochemical standard state is defined to be pH = 7 rather than pH = 0. This affects $\Delta G°$ for all reactions involving H$^+$; for example, $\Delta G°$ for ATP hydrolysis (Reaction 18.30) is *positive* for pH = 0.

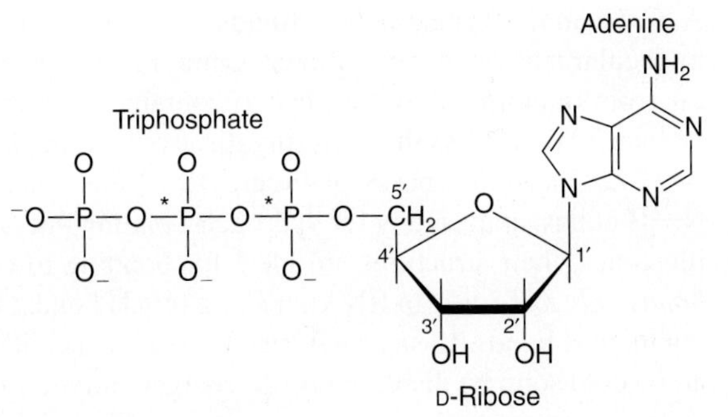

18.7 The Molecules of Life: Organic Motifs in Biology

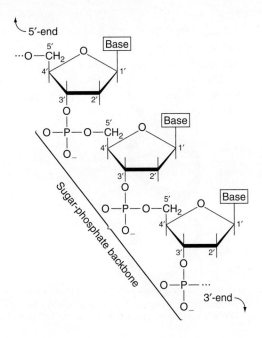

Figure 18.14
Adenosine triphosphate, a ribonucleotide. The two P—O bonds marked by an asterisk can be readily hydrolyzed with a release of 7.3 kcal/mol of free energy for each. In a deoxyribonucleotide, the OH group at C-2' is replaced by H.

releasing 7.3 kcal/mol of free energy. During metabolism, the combustion energy is used to run this reaction in reverse, ADP + $H_2PO_4^-$ ⟶ ATP + H_2O, thereby storing 7.3 kcal/mol; while when you use your muscles, ATP is there supplying the energy needed. The reaction mechanism is similar to the hydrolysis of P_4O_{10} to H_3PO_4 discussed in Chapter 5.

Whether we have a ribonucleotide or a deoxyribonucleotide, each contains a "free" hydroxyl group at C-3' on the ribose ring, and a phosphate end, making it possible to *polymerize* nucleotides through formation of a phosphate ester linkage just like the one binding the phosphate group to C-5' in a nucleotide unit. The polymer built from ribonucleotides is called a **ribonucleic acid (RNA),** while that made from the deoxy version is a **deoxyribonucleic acid (DNA).** The generic form of the polymeric chain of DNA is illustrated in Figure 18.15; it always terminates on one end with a hydroxyl at C-3' (the so-called 3' end) and on the other with a phosphate group bonded to C-5' (the 5' end). Like the proteins, nucleic acids can have an arbitrary sequence of bases as side groups, but there are only 5 choices instead of 20. Natural RNA employs only the four bases, A, C, G, and U, while DNA uses only

Figure 18.15
Generic structure of the DNA polymer; base = A, C, G, or T, as given in Figure 18.13. In an RNA polymer, the down hydrogen at C-2' is replaced by an OH group, and base = A, C, G, or U.

Figure 18.16
Selective hydrogen bonding between the pairs of nucleotide bases A≡≡≡T and G≡≡≡C. Note that the overall length of the H-bonded complex is nearly the same in the two cases.

A, C, G, and T. Typical DNA strands may contain thousands of nucleotides, giving molecular masses in the millions. Again, like the proteins, these base sequences are far from random—they turn out to contain codes that specify protein residue sequences! We discuss this amazing discovery in the next section.

The nucleotide bases of Figure 18.13 are bristling with lone pairs and polar N—H bonds, at the ready for hydrogen-bonding interactions. In fact, as Figure 18.16 illustrates, their structures are ideal for bonding to each other in selected pairs: *A bonds only to T* (or U in RNA), making two H bonds; while *C bonds only to G,* making three H bonds. If, say, an A tried to bond to a G, it could make only two H bonds, and would soon be displaced by an energetically more favorable C. Note also that a bicyclic purine always bonds to a monocyclic pyrimidine; this implies that the A-T and C-G pairs occupy the same amount of space when H bonded.

This selective and spatially uniform H bonding of the bases allows DNA to exist as a *double strand,* provided the second strand contains a sequence of bases complementary to the first. Because of the enantiomeric nucleotide sugars, the DNA double strand winds around itself to make a **right-handed double helix.** This structure was first recognized by J. D. Watson and F. C. Crick in 1954, based on the X-ray crystallographic data of R. Franklin and M. Wilkins, and is pictured in the schematic "ribbon" form and also in a space-filling model in Figure 18.17. (It is said that if Linus Pauling had seen the X-ray diffractograms, he would have been the first to deduce the structure of DNA as well! But he had already done more pioneering work than any three scientists one could name. . . .) The elucidation of the structure of nucleic acids presently represents an apex in our chemical knowledge of living things. All plants and animals, down to the tiniest one-celled creatures, contain DNA, which comprises what are known as **genes,** the dictators of life's myriad forms. To understand the connections between the life we experience in ourselves and our surroundings, and its molecular basis in DNA, RNA, enzymes,

Figure 18.17
The structure of the DNA double helix. The two polymers are held in place by the selective hydrogen bonding shown in Figure 18.16. The strands run antiparallel to each other. In the schematic structure on the left, the sugar-phosphate backbones are ribbons, while in the space-filling model on the right, the sugar and phosphate groups are colored to outline the backbone. The bases stack on each other with their rings perpendicular to the helical axis.

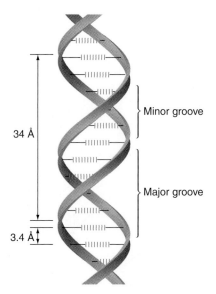

and carbohydrates, requires that we connect the structures of these molecules with the roles they play. This finally brings us to. . . .

18.8 The Chemical Dynamics of Life

Where do sugars and starches come from? How do we burn them to provide vital energy? Where is the lab manual that tells how to carry out this life-giving chemical synthesis and combustion of food? In the 20th century, these questions have turned from ones of near-theological dimension to ones of chemistry. It has been claimed that all of life boils down to about 120 essential chemical reactions. In this ultimate little section, we will explore a select few of these, to give you a flavor of our present knowledge of the chemical dynamics of life.

Photosynthesis: The Light and Dark Reactions

In a way, all of life revolves around a single highly spontaneous reaction, the combustion of glucose:

$$C_6H_{12}O_6 + 6O_2 \longrightarrow 6CO_2 + 6H_2O \qquad \Delta G° = -686 \text{ kcal/mol} \qquad (18.31)$$

Plant life performs the stupendous feat of running this reaction *backward,* executing a formidable uphill traverse in free energy, in the synthesis of glucose from atmospheric CO_2 and water. Combustion and its reverse are redox reactions, so that living systems have to be capable of redox chemistry, as well as the acid–base chemistry involved in ATP energy storage. The molecule that performs the task of flexibly being oxidized or reduced as the occasion demands is **nicotinamide adenosine dinucleotide phosphate (NADP),** a nucleotide dimer made from AMP and a new base-sugar nucleoside containing the base nicotinamide, and depicted in Figure 18.18. Its

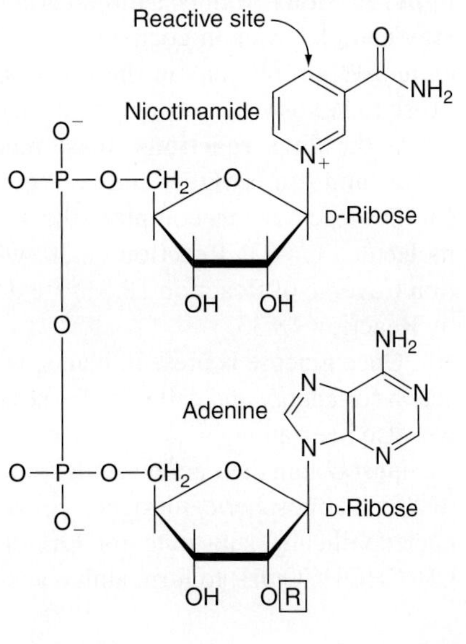

Figure 18.18
The structure of the oxidized form of nicotinamide adenine dinucleotide (NAD^+) and of $NADP^+$, the ubiquitous redox reagents of life. In NAD^+, R = H; in $NADP^+$, R = PO_3^{2-}.

reversible oxidation occurs on the nicotinamide ring, while an aldehyde (or other carbonyl group) is reduced to an alcohol:

$$R-N: CH_2 + O=CH-R' + H^+ \rightleftharpoons (18.32)$$
$$CONH_2$$

$$R-\overset{+}{N}CH + HO-CH_2-R'$$
$$CONH_2$$

or, in shorthand notation, $NADPH + RCHO + H^+ \rightleftharpoons NADP^+ + RCH_2OH$, where NADPH is the reduced form of NADP. In **photosynthesis,** light energy is absorbed by a Mg^{2+}–porphyrin complex called *chlorophyll,* creating excited electrons. These electrons are taken up by $NADP^+$, reducing it to NADPH, and reinjected when needed through Reaction 18.32 in order to reduce CO_2. Part of the solar energy is also used to convert ADP to ATP, the reverse of Reaction 18.30. All of this is accomplished, naturally, at the active site of a photosynthetic enzyme assembly called simply *photosystem I.* In this staged reduction process, NADPH and ADP act as **coenzymes.** Neither Reaction 18.32 nor the reduction of NADP by chlorophyll involve molecular oxygen; this is called an *anaerobic* reduction pathway.

A notable advance in photosynthesis took place about 3 billion years ago, when a new enzyme assembly called *photosystem II* evolved that could use light to reduce $NADP^+$ to NADPH while oxidizing water itself to O_2 and H^+:

$$NADP^+ + H_2O + 4h\nu \longrightarrow NADPH + \tfrac{1}{2}O_2 + H^+ (18.33)$$

The O_2 produced as part of this *aerobic* reduction pathway could bubble out of system, making Reaction 18.33 irreversible, and greatly increasing the availability of NADPH and thereby the overall rate of reduction of CO_2 to glucose. Both the anaerobic and aerobic pathways continue to coexist in plants; these so-called **light reactions** produce a three-carbon sugar, a *triose phosphate,* while storing energy and electrons in coenzymes ATP and NADPH—consuming a total of forty-eight 700-nm photons in the process. The building up of the carbon chain from triose to hexose then occurs stepwise in a complex reaction sequence known simply as the **dark reactions;** these reactions were elucidated by M. Calvin in the 1950s, and use NADPH and ATP generated during the light reaction phase. The energy necessary to complete the synthesis is supplied by the hydrolysis of 18 molecules of ATP, Reaction 18.30, while the 24 electrons required for the reduction (reverse of Reaction 18.31) are donated by 12 molecules of NADPH supplied by Reaction 18.33.

Once glucose is present, plants, with the aid of more enzymes, are able to make starch for energy and cellulose for structural members, stems, and leaves. They, like we, also need amino acids for the enzymes that catalyze these reactions; they, unlike we, must obtain their amino acids from soil bacteria, which use *nitrogenase* enzymes to "fix" atmospheric nitrogen, converting it into ammonia. Ammonia then can nucleophilically substitute for OH on hydroxylated acids (such as L-lactic acid, $CH_3CHOHCOOH$) to form amino acids.

Enzyme Catalysis and Carbohydrate Metabolism

The symbiosis between animals and plants is so pervasive that it is often taken for granted. They supply us with food and oxygen; we supply them with carbon dioxide and fertilizer from our decaying remains; the combination maintains the balance of O_2 and CO_2 in the atmosphere. To play our role in life, we are equipped to run Reaction 18.31 in its spontaneous direction, not nearly so remarkable as the feat plants perform, but nonetheless miraculous in the way its free energy is captured for later use.

The burning of glucose and other carbohydrates by living things is called **metabolism.** Like photosynthesis, metabolism occurs by both an anaerobic and an aerobic pathway, the aerobic being apparently a recent development associated with the buildup of oxygen in the atmosphere. In the first stage of energy retrieval, known as **glycolysis,** glucose is first anaerobically oxidized to the α-keto acid **pyruvic acid** ($CH_3COCOOH$) by direct coupling with the reduction of a close relative of $NADP^+$, **nicotinamide adenosine dinucleotide (NAD^+),** just $NADP^+$ without the phosphate ester group. The overall glycolysis reaction is

$$C_6H_{12}O_6 + 2NAD^+ \longrightarrow 2CH_3COCOOH + 2NADH + 2H^+ \quad (18.34)$$
$$\Delta G° = -35 \text{ kcal/mol}$$

The multistep process is catalyzed by 10 different enzymes; the initial step is the phosphorylation of glucose, which requires the free energy of ATP hydrolysis, but in subsequent steps ATP is formed, resulting in a net gain of 2 ATP, about 15 kcal/mol, but maintaining a strongly negative overall $\Delta G°$. Figure 18.19 illustrates the glycolytic pathway. The free energy released in forming pyruvic acid is only 5% of the available energy from glucose combustion; only this first stage of Reaction 18.31 provided free energy to living things before the atmosphere became oxidizing, and the relatively small yield inhibited development of larger, multicellular organisms.

It was the appearance of O_2 in the atmosphere that enabled tapping of the remaining 95% of the combustion energy still present in pyruvic acid. But O_2 could not be allowed to "burn" pyruvic acid; the whole process had to be arranged as an indirect electron transfer involving intermediate coenzymes, much as we have devised fuel cells to tap combustion free energy without wastefully and dangerously burning hydrogen. Indeed, O_2 is a threat to a cell's existence if allowed to wield its oxidizing power indiscriminately; all of the molecules we have have been describing react vigorously with oxygen under the right conditions! The first step in achieving this delicate maneuver is

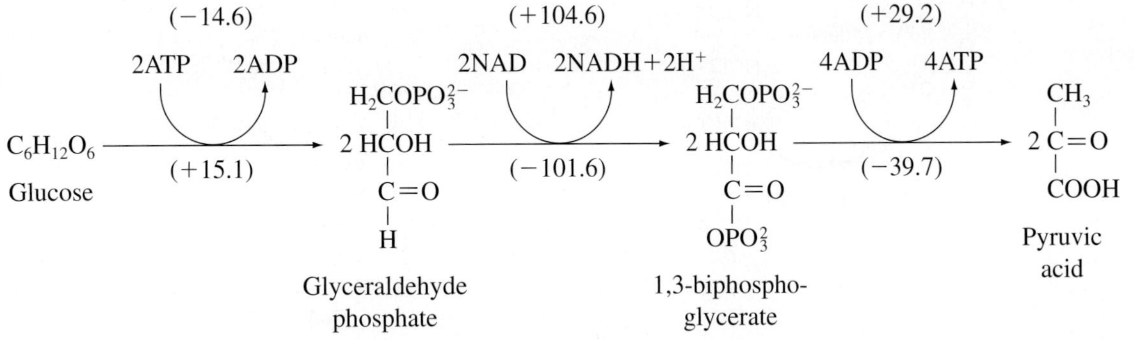

Figure 18.19
The glycolytic (anaerobic) pathway for extraction of energy from glucose, in abbreviated form. The numbers in parenthesis are free-energy changes $\Delta G°$ in kilocalories per mole.

the conversion of pyruvate ion (CH_3COCOO^-) to an acetyl–coenzyme complex called **acetyl CoA,** short for acetyl-coenzyme A complex. The overall reaction is

$$CH_3COCOO^- + CoA\text{—}SH + NAD^+ \longrightarrow \qquad (18.35)$$
$$CH_3CO\text{—}S\text{—}CoA + CO_2 + NADH \qquad \Delta G° = -8 \text{ kcal/mol}$$

In this complex reaction, the entire carboxyl group is eliminated as CO_2, the first CO_2 molecule of three, while NAD^+ again plays oxidizing agent. The reaction is enabled by a *dehydrogenase complex* consisting of three types of enzyme (60 or more individual proteins!) and five coenzymes all held in place by nonbonded forces. Coenzyme A has a familiar component, ADP, bonded through a phosphate ester linkage to an unusual diamidothiol chain containing *pantothenic acid* as a subunit.

Acetyl CoA then enters an incredibly intricate process known as the **citric acid cycle** (or Krebs cycle, after its discoverer). In this cycle of nine identifiable separate enzyme-catalyzed reactions, illustrated in Figure 18.20, a new oxidizer comes on the scene, **flavin adenine dinucleotide (FAD).** FAD has not one but two reducible ring nitrogens in the tricyclic base flavin, but otherwise is very similar to NAD^+. The overall reaction that takes place in the cycle can be written as

$$3NAD^+ + GDP^{3-} + HPO_4^{2-} + FAD + CH_3CO\text{—}S\text{—}CoA + 2H_2O \longrightarrow \qquad (18.36)$$
$$3NADH + GTP^{4-} + FADH_2 + HS\text{—}CoA + 2CO_2 + 2H^+$$
$$\Delta G° = -211 \text{ kcal/mol}$$

Figure 18.20
The citric acid cycle (also known as the Krebs cycle or the tricarboxylic acid cycle). (a) Structures of the oxidized form of FAD and CoA. (continued on 719)

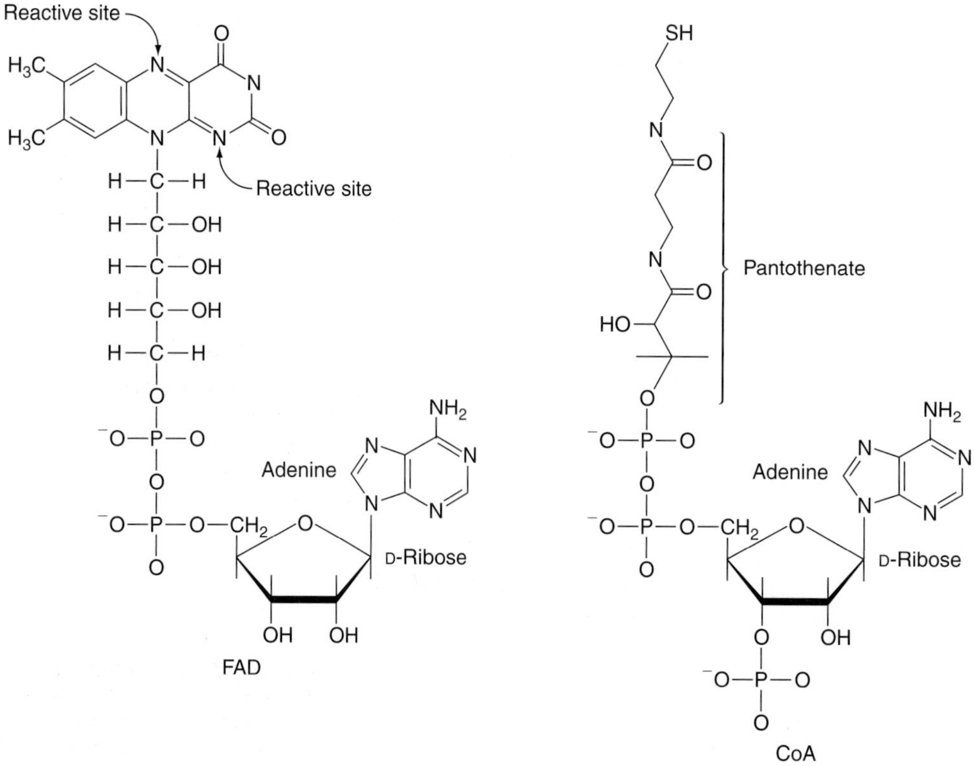

(a)

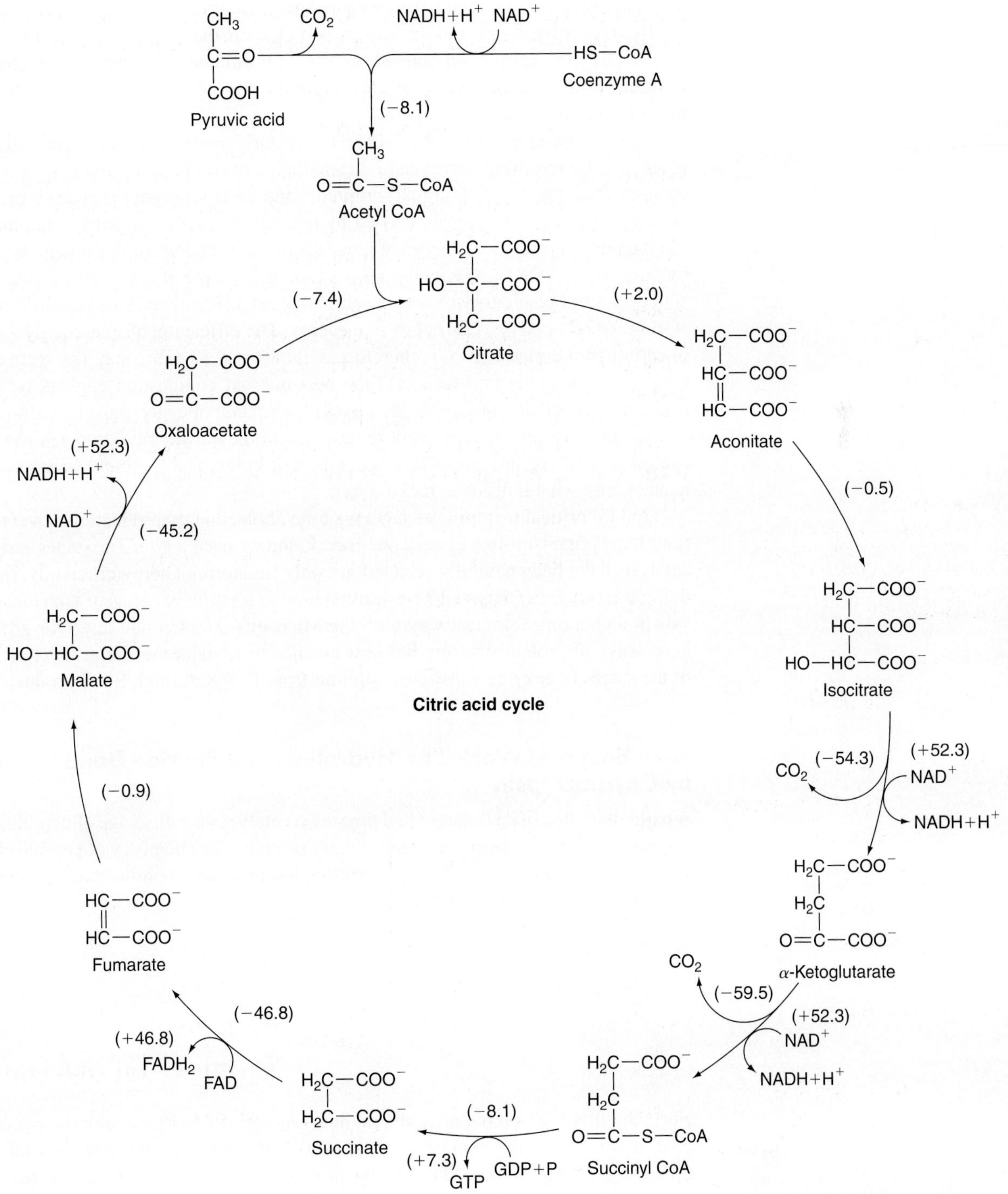

18.8 The Chemical Dynamics of Life

Figure 18.20 (continued)
(b) The nine-step cycle. Numbers in parentheses are standard free-energy changes $\Delta G°$. GTP is guanosine triphosphate.

The other two CO_2 molecules are produced here, completing the carbonic oxidation, but the excess hydrogen and the electrons pulled off the sugar are stored in NADH and $FADH_2$; O_2 has not yet entered the scene. Fortunately the cycle is net strongly downhill in free energy, while still leaving more for the final stage of carbohydrate metabolism, the **respiratory chain**.

In the respiratory chain, the $3NADH + FADH_2$ produced in the citric acid cycle (each glucose requiring two cycles) donate their excess electrons and hydrogen ions through five stages, involving *flavoproteins* and metalloporphyrin-bearing proteins called *cytochromes,* to oxygen carried by hemoglobin, reducing it to water, and releasing the rest of the free energy of glucose combustion. Part of this energy is stored by converting ADP to ATP as before. Including the three stages—glycolysis, citric acid cycle, and respiratory chain—a net total of 30 ADP ⟶ ATP conversions occur, storing $30(7.3) = 219$ kcal/mol of free energy. The efficiency of free-energy storage in carbohydrate metabolism is therefore $219/686 = 0.32$. This may not seem spectacular, but consider two facts: (1) the best internal combustion engines typically achieve an efficiency of only 0.25, and (2) each metabolic step must be sufficiently downhill in free energy to prevent reverse reactions. Furthermore, oxygen has been excluded until the last possible stage, where it enters stably bound to iron and is handed only what it needs to make water.

A third critical factor in the success of metabolic energy production is metabolic *rate;* large organisms like us need our energy, and we need it *now*. The stages and their catalysts have been naturally selected not only for storing energy efficiently, but for doing it rapidly. In Chapter 15 we learned how to treat the kinetics of enzyme catalysis in a phenomenological way with the Michaelis–Menten rate law. Our growing knowledge of protein structure has now enabled us to paint molecular-level pictures of the stages in enzyme-substrate evolution from $E + S$ through ES to products P.

How Enzymes Work: The Hydrolysis of a Peptide Bond by Chymotrypsin

A particular class of enzymes called *proteases* catalyzes the cleavage of peptides into fragments; proteases help you digest dietary protein. The chemistry of peptide cleavage is the reverse of peptide bond formation, a simple nucleophilic attack on the carbonyl carbon by a water lone pair:

$$\cdots\text{—CH}(\text{R})\text{—C}(=\text{O})\text{—NH—CH}(\text{R}')\text{—}\cdots + H_2O \longrightarrow \cdots\text{—CH}(\text{R})\text{—C}(=\text{O})\text{—OH} + H_2N\text{—CH}(\text{R}')\text{—}\cdots \quad (18.37)$$

Normally this reaction requires several hours in hot 55% sulfuric acid; the enzyme's task is to make this happen quickly at body temperature and neutral pH. *Chymotrypsin* is a globular, 245-residue protease enzyme that catalyzes this reaction, *but only when* R is aromatic, that is, the left-hand amino acid residue is phenylalanine or tryptophan. As illustrated in Figure 18.21, chymotrypsin has an **active site** comprising the R groups of three amino acids, Asp[102], His[57], and Ser[195], and an adjacent hydrophobic pocket lined with aliphatic and aromatic R's. The pocket is the right width

18.8 The Chemical Dynamics of Life 721

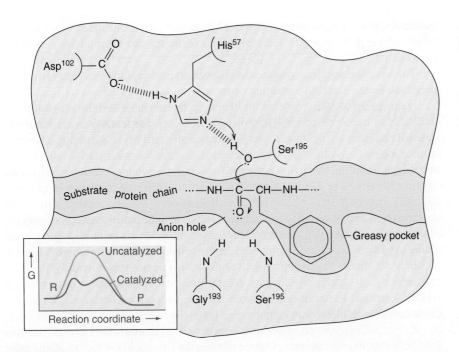

Figure 18.21
The operation of the active site of the protein-cutting enzyme chymotrypsin. The curly-arrow mechanism illustrates the first step in a two-step catalytic pathway. The inset at lower left shows the reduction in free energy of activation enabled by the stabilization of transition states by the enzyme.

and depth to accommodate a phenyl group, and allows, the substrate peptide to "drop anchor" in just the right position to permit reaction-promoting interactions with the active-site residues. It is this greasy pocket that endows the enzyme with **specificity** for the aromatic residues. The reaction has been shown to occur in two basic steps. First, the serine OH group attacks the carbonyl of the aromatic residue, severing the peptide bond and forming an enzyme ester intermediate:

$$\cdots\text{—CH}\underset{\underset{\text{R}}{|}}{\text{—}}\overset{\overset{\text{O}}{\|}}{\text{C}}\text{—NH—CH}\underset{\underset{\text{R}'}{|}}{\text{—}}\cdots + \text{HO—CH}_2\text{—chymo} \longrightarrow \qquad (18.38)$$

$$\cdots\text{—CH}\underset{\underset{\text{R}}{|}}{\text{—}}\overset{\overset{\text{O}}{\|}}{\text{C}}\text{—O—CH}_2\text{—chymo} + \text{H}_2\text{N—CH}\underset{\underset{\text{R}'}{|}}{\text{—}}\cdots$$

The amino end, not being anchored, drifts away, allowing a molecule of water to approach, which hydrolyzes the ester bond,

$$\cdots\text{—CH}\underset{\underset{\text{R}}{|}}{\text{—}}\overset{\overset{\text{O}}{\|}}{\text{C}}\text{—O—CH}_2\text{—chymo} + \text{H}_2\text{O} \longrightarrow \qquad (18.39)$$

$$\cdots\text{—CH}\underset{\underset{\text{R}}{|}}{\text{—}}\overset{\overset{\text{O}}{\|}}{\text{C}}\text{—OH} + \text{HO—CH}_2\text{—chymo}$$

The reaction over, the acid end raises anchor and departs as well. When Reactions 18.38 and 18.39 are added, the overall reaction, Reaction 18.37 results. The enzyme, although forming a covalently bonded intermediate, is regenerated by the ester hydrolysis, ready for the next cleavage. The reaction rate is enhanced by a factor of 10^6, implying a lowering of the activation energy for the reaction by more than 8 kcal/mol.

The ease and specificity with which chymotrypsin enables cleavage raises two questions. First, how is the lowering of the activation energy achieved? This is evidently due to the attractive interactions between the substrate and the active site, including both hydrogen bonding and dispersion forces, particularly in the transition states. Recall that a single hydrogen bond has a strength of ~5 kcal/mol. These favorable interactions allow low-energy transition-state structures that would be very unstable in isolation. Secondly, if the active site involves so few residues, why does chymotrypsin require 245 residues? This question is difficult to answer in any specific way, but the answer clearly lies in protein folding. It is the tertiary structure of chymotrypsin that places the three active residues and the hydrophobic pocket in the proper spatial relation to each other. Thus, most of a protein's primary sequence and secondary structure is needed simply to ensure proper folding to its native form.

In ways similar to chymotrypsin, the thousands of enzymes found within a living cell are each specifically adapted to catalyze a particular reaction on a particular type of substrate. Furthermore, enzymes are able to work cooperatively, creating a chain of reactions like that in carbohydrate metabolism that achieves a specific overall process. Although a cell of our own does not have the machinery to make its own food, it is able to *make its own enzymes* that run the tiny chemical factory. Cells must therefore contain blueprints, detailed plans that spell out exactly what is needed to keep life going. These plans represent the ultimate in organization, "negative entropy" stored in molecular form by living things against all the random tendencies of matter at a finite temperature.

Making the Right Enzymes: DNA, RNA, and the Code of Life

The discovery of the molecular basis used by a cell to store and pass on information certainly ranks as one of the more revolutionary developments in biology. It occurred in the early 1960s, while your author was still an undergraduate, completely ignorant of the revolution as it unfolded. It was an inevitable consequence of the discovery of the structure of DNA by Watson and Crick in 1954; soon thereafter Crick suggested that enzymes must be created by coded sequences of bases in nuclear DNA, which are copied by complementary base pairing onto **messenger RNA (mRNA).** mRNA then travels out into the cell cytoplasm, where it attracts a host of **transfer RNA (tRNA)** molecules, each bearing a base sequence complementary to some coded section of mRNA, along with a corresponding amino acid residue. The linear sequence of bases in DNA is thus **replicated** and **translated** into a specific sequence of amino acids, which is assembled by a coordinated series of peptide bond-forming reactions into a biologically useful enzyme. This scenario, dubbed the *central dogma of molecular genetics* by Crick, is now an established fact of life.

Since there are 20 amino acids to choose from, the DNA code must involve an *ordered sequence* of the four DNA bases. Pairs of bases are not adequate, since this gives only $4^2 = 16$ ordered combinations. Trios of bases provide too many combinations, $4^3 = 64$, but this leaves room for *punctuation,* trios that start and stop an amino acid sequence, and also future expansion, should new types of amino acids be found useful. The earliest experiments to begin cracking the code were carried out in the laboratory of M. Nirenberg, using small polynucleotides made of all uracil ribonucleotides, U_n. (Recall that RNA uses U instead of thymine T.) In 20 test tubes were a cell extract and a mixture of the 20 natural amino acids, with each test tube containing a different radioactive amino acid. The U_n polynucleotide was added to each tube, and in only one of the tubes was a radioactive polypeptide produced, that containing radiolabeled phenylalanine (Phe). The U_n polynucleotide is acting as a mRNA that codes for a single amino acid, implying that UUU is a coding sequence, a **codon,** for Phe. In a similar way, CCC was found to be a codon for proline (Pro), and AAA for lysine (Lys). Once this start was made, further progress came from the use of binary mixtures of nucleotides, and trinucleotides of known composition. H. Khorana completed the cracking of the code by devising methods for synthesizing RNA with known repeating sequences, followed by radiolabeled experiments. The results are given in Table 18.6.

This "genetic dictionary" presents a number of curious features. Every one of the 64 possible codons is used, making the code *degenerate,* that is, more than one base triplet codes for the same amino acid. In most cases, the first two bases form the code, and the third base, called the *wobble base,* is variable. Only methionine (Met) and tryptophan (Trp) have unique codons. On the other hand, all codons code for one and only one amino acid. There is only one *start* codon, AUG, but three *stop* codons, UAA, UAG, and UGA. The mRNA code is complementary to the code stored on the DNA template, with U replaced by T; for example, the mRNA codon for Thr, ACU, corresponds to the DNA triplet TGA. The genetic code is universal; the dictionary is the same for bacteria and humans alike.

Once the genetic code had been broken, it became possible to investigate the dynamics of **protein synthesis** in more detail, starting with the DNA template. The

TABLE 18.6

The standard genetic code (mRNA codons and their amino acids)

UUU	Phe	UCU	Ser	UAU	Tyr	UGU	Cys
UUC	Phe	UCC	Ser	UAC	Tyr	UGC	Cys
UUA	Leu	UCA	Ser	UAA	Stop	UGA	Stop
UUG	Leu	UCG	Ser	UAG	Stop	UGG	Trp
CUU	Leu	CCU	Pro	CAU	His	CGU	Arg
CUC	Leu	CCC	Pro	CAC	His	CGC	Arg
CUA	Leu	CCA	Pro	CAA	Gln	CGA	Arg
CUG	Leu	CCG	Pro	CAG	Gln	CGG	Arg
AUU	Ile	ACU	Thr	AAU	Asn	AGU	Ser
AUC	Ile	ACC	Thr	AAC	Asn	AGC	Ser
AUA	Ile	ACA	Thr	AAA	Lys	AGA	Arg
AUG	Met,Start	ACG	Thr	AAG	Lys	AGG	Arg
GUU	Val	GCU	Ala	GAU	Asp	GGU	Gly
GUC	Val	GCC	Ala	GAC	Asp	GGC	Gly
GUA	Val	GCA	Ala	GAA	Glu	GGA	Gly
GUG	Val	GCG	Ala	GAG	Glu	GGG	Gly

sequence of DNA bases that specifies one protein is called a *gene;* human DNA contains about 30,000 genes, while that of a one-celled bacterium has about 4000 genes. Genes in turn are assembled into units called **chromosomes;** one chromosome may contain thousands of genes. A bacterium has only 1 chromosome, while human cells have 46. The sum of all the chromosomes of an organism is called its **genome.** Most of our information about these structures and how they enable protein synthesis comes from study of the infectious bacterium *Escherichia coli* (*E. coli*).

Step 1: Making RNA. The DNA in the nucleus of *E. coli* is present as a *supercoiled* (coiled coil, like a twisted telephone cord) double helix formed into a closed loop. The supercoils occur in reverse senses around the region where the DNA base template is read, allowing the double helix to unwind and separate along a short region, one or two helical turns, as illustrated in Figure 18.22. The RNA nucleosides ATP, CTP, GTP, and UTP then bond to their complementary DNA bases T, G, C, and A, thus faithfully replicating the base sequence (in mirror-image form) that will specify the protein to be synthesized. The entire process is overseen and catalyzed by the **DNA-directed RNA polymerase** enzyme, which moves the so-called *transcription bubble* along the DNA chain as the RNA chain grows, and catalyzes the formation of phosphate ester bonds between the RNA nucleosides. Once started, RNA transcription in *E. coli* proceeds at a rate of 50 bases per second. Messenger RNA, which is to transmit the coded sequence of amino acids, is terminated by so-called *hairpin* structures; in a hairpin the base sequence at an end of the RNA strand is complementary to a nearby sequence, and can fold in on itself to make a terminal double strand that is not part of the protein coding.

Two other types of RNA are also made in this way: tRNA and **ribosomal RNA (rRNA).** All of them are processed further by **nuclease** enzymes into their so-called mature forms. tRNA also undergoes modification of its bases to produce its final tertiary structure. All three RNAs make their way through special passageways in the nuclear membrane into the cell cytoplasm, where they each play a different and essential role in protein synthesis.

Step 2: Aminoacylating tRNA. The modified bases of tRNA give it a unique "twisted" form, enabling it to be recognized by one of the **aminoacyl tRNA synthetase** enzymes. These enzymes, one for each amino acid, recognize the codon sequence in tRNA, and catalyze esterification of the appropriate amino acid at the 3'

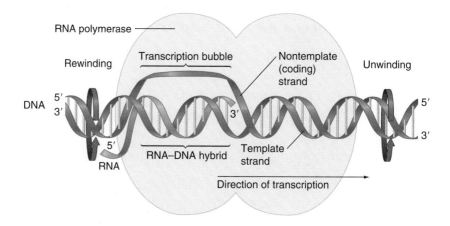

Figure 18.22
Copying the code of DNA onto a messenger RNA (mRNA), catalyzed by the enzyme RNA polymerase. (*Source:* From Lehninger, Nelson, & Cox, *Principles of Biochemistry.*)

end of the nucleic acid chain. The same synthetase is generally able to process all of the degenerate codons for a given amino acid; it also appears to recognize particular base pairs in tRNA that are not part of the codon, implying a secondary genetic code. This critical step creates an aminoacylated tRNA, for example Ala-tRNA^{Ala}, which carries the codon on a hairpin turn and its correct amino acid on the opposite end.

Step 3. Assembling Ribosomes.   Meanwhile, rRNa has been taken up by certain proteins to make hybrid globular enzymes called **ribosomes.** Ribosomes serve as catalysts for building the protein chain, and must contain rRNA in order to bind to mRNA and tRNA through the usual base-pair hydrogen bonding. In *E. coli* ribosomes consist of about 65% rRNA and 35% protein, and are about 180 Å in diameter. Ribosomes (and other large aggregates) are characterized by their *sedimentation coefficient nn*S, giving the relative rate at which they may be separated out of solution in a centrifuge. The larger the ribosome, the larger its sedimentation coefficient. *E. coli* ribosomes are 70S, and are assembled from two subunits, the 30S and 50S ribosomes, each containing one (30S) or two (50S) large molecules of rRNA and dozens of individual proteins. Ribosomes are arguably the most complex "chemical laboratories" in living things, and all life that survives by making proteins must have them.

Step 4: Initiation.   To initiate the synthesis of a desired protein, a number of molecular ingredients must be brought together. This and the following two steps are schematically illustrated in Figure 18.23. The process begins when mRNA attaches to the so-called *peptidyl site* P on the 30S ribosome at the initiation codon AUG. This attachment is catalyzed by an enzyme called **initiation factor 3 (IF-3),** which has been previously activated by **IF-1.** Once the initial attachment is complete, methionine tRNA, tRNA^M, which bears the anticodon (complementary base code) for the startup sequence, is attracted to the site with the help of another enzyme, **IF-2,** also previously activated by IF-1. IF-2 bears a molecule of **guanosine triphosphate (GTP),** which is used exclusively in protein synthesis as a source of free energy. The hydrolysis of one GTP ($GTP^{4-} + H_2O \longrightarrow GDP^{3-} + H_2PO_4^-$) allows binding of 50S ribosome on the opposite side of the mRNA chain, completing the assembly of the full 70S ribosome. The mRNA–ribosome complex is now ready to commence building the protein chain. The overall appearance, makeup, and operation of the complex is very similar to that in RNA synthesis itself.

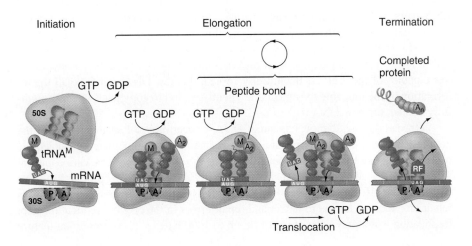

Figure 18.23
Overview of protein synthesis, steps 4 through 6 as discussed in the text.

Step 5: Elongation. tRNA^2, bearing the anticodon for the second mRNA codon—held in place at the *aminoacyl site* A, adjacent to the peptidyl site on 30S—is brought in by three **elongation factors (EFs),** using one GTP, and selective base-binding occurs. This is followed by 50S-catalyzed peptide bond formation between the carbonyl of the first amino acid M and the amine of the second A$_2$, the new one brought in by tRNA^2. Thus, the protein chain begins from the amino end. The S$_N$2 activation barrier is overcome with the aid of GTP hydrolysis and catalyzed by 50S ribosome. The breaking of the ester link between tRNA^M and M sets off GTP-enabled **translocation,** in which the mRNA is shifted by one codon so that tRNA^2 resides at the P site, making room for the next tRNA, tRNA^3 bearing A$_3$, at the A site and releasing now-spent tRNA^M. The process then repeats as needed, with two GTPs hydrolyzed per step; the total energy demand is four GTPs per step, 29 kcal/mol, when the aminoacylation of tRNA is included.

Step 6: Termination. Elongation continues until a stop codon, UAA, UAG, or UGA, enters the A site; there are no tRNAs for these. Instead, three **release factors (RFs)** are set into action, hydrolyzing the ester bond between the last tRNA, tRNA^n, at the P site and its A$_n$. This frees the completed protein, while releasing tRNA$_n$ from the P site and dissociating the 70S ribosome into its 50S and 30S pieces, ready to begin another synthesis. The protein then begins to fold into its native form, ready to fulfill its destiny as a catalyst.

Despite the enormous size of the molecules involved and the complexity of the chain-building process, adding one amino acid takes only 50 ms in *E. coli,* and an average enzyme is completed in less than 20 s. The rate is enhanced by "parallel processing"—several ribosomes can be producing amino acid chains simultaneously at different locations along the same mRNA strand. Precise recognition at each stage yields an error rate of roughly 1 incorrect amino acid in 10,000; this means that 29 out of every 30 enzymes are "perfect." Greater perfection in code translation would come only at the expense of speed; in addition, enzymes and other proteins are rapidly metabolized, bad along with good, so that enzymes that are no longer needed or nonfunctional do not "hang around."

How is all information required for promoting and managing the complex processes we have described preserved in DNA and passed along to future generations of an organism? It is clear that heredity must involve **DNA replication,** the underlying molecular basis of cell division. From all that we have seen, particularly RNA synthesis, a **DNA polymerase** enzyme must organize and catalyze the replication process. Further, the code for making this enzyme must be stored in—DNA. DNA represents the ultimate in *self-reference,* containing the instructions for perpetuating itself. This is how microbial diseases spread: infectious cells automatically make copies of themselves. If faithfully copied, the new DNA will have all the information of the old, able to run cellular chemistry, and to copy itself again; if faithfully copied, the new DNA will . . . .

In broad outline, the mechanism of DNA replication involves unwinding of the double helix (as in RNA synthesis) and the mating of complementary nucleotide bases by selective hydrogen bonding. Both strands of the original DNA are copied at once, as established by isotopic labeling studies. In *E. coli* the copying process itself errs once in 10^6 to 10^8 base pairs, and this accuracy is improved to about 1 in 10^{10} by a repair enzyme that is able to detect and repair errors in the original replication step. Replication is not only accurate but fast, up to 1000 bases s^{-1}. An important feature of DNA replication is that it appears to be spontaneous, $\Delta G° < 0$, without expending ATP or GTP free energy, during the elongation stage; this results in an increased rate, since two coupled reactions are not required.

Continuously active "sentinel" nuclease enzymes are constantly checking for and repairing damage to the DNA helix. DNA repair, like repair of any complex machine such as an automobile, is energetically expensive, ranging from complete replacement of damaged or defective nucleotides to restoration of bases with improper substituents. For example, guanine can become methylated at its six-membered-ring hydroxyl. It now can only make two hydrogen bonds instead of three, and will tend to bond to thymine rather than cytosine in the double helix. If not repaired, this will lead to a GC $\longrightarrow$ AT **mutation** when the DNA is replicated; this may yield an erroneous codon if the base affected does not wobble and, in turn, one type of protein with an incorrect residue, made by the copying mechanism we have just outlined. This protein will then probably not fold properly, and be nonfunctional. Note how much more damaging to cell function a DNA error is as compared to a protein transcription error: *every* protein made by copying the DNA template is defective, and, assuming the cell survives long enough to divide, all future generations will contain the same defect. The methylation error, fortunately, can be corrected by a **methyltransferase** enzyme, which transfers the methyl group to the sulfur atom of one of its cysteine residues. This enzyme is not strictly a catalyst, since it is deactivated by a single reparation. Destruction of an entire protein for the sake of repair exemplifies the importance of DNA maintenance to a cell—auto mechanics do not often die in the line of duty!

The Molecular Roots of Life

Several "chicken-and-egg" questions are raised by all this. How did it all start? Where did the DNA come from? Where did DNA and RNA polymerases come from, if they are themselves proteins, and are also needed to copy DNA and begin protein synthesis? How were the first ribosomes made, if they are themselves part protein and are also the machinery for producing proteins? Where did the building blocks, amino acids, nucleotide bases, etc., come from?

There is strong evidence that the early atmosphere of the earth contained no uncombined oxygen, but instead consisted mainly of hydrogen, methane, ammonia, and water vapor, molecules known to comprise the surfaces of the Jovian planets of our solar system. In a classic experiment, S. Miller and H. Urey in 1953 subjected such a mixture to electrical sparks (to simulate lightning), and found amino acids among the products in the aqueous phase after less than 2 weeks of sparking. Subsequent experiments with more sensitive analysis of the chemical residue showed, in addition to CO and CO_2 in the gas phase, 11 of the 20 essential amino acids, a variety of sugars through hexoses, dozens of carboxylic acids, including small fatty acids, four of the five nucleotide bases—those used by RNA—and many other species. Phosphate in the aqueous phase, also thought to be present in the prebiotic era, allowed ribonucleotides and ribonucleic acids to form under these conditions.

Coupled with the more recent discovery of catalysts made of pure RNA, so-called **ribozymes**, these findings have led to the speculation that life began with small bits of RNA, which could both replicate themselves, by a primitive form of complementary base pairing, and catalyze the synthesis of small polypeptides, marking the beginnings of the genetic code. These small protoenzymes might assist RNA in synthesizing other proteins, leading to the precursors of ribosomes. Darwin's principle of natural selection began to take hold: those molecular assemblies that were able to extract material and energy from their environment more quickly and efficiently grew and multiplied more rapidly, and eventually dominated. It became increasingly important to preserve the hard-won information concerning the most advantageous chemical pathways, leading to cell structure and the less-readily-metabolized DNA motif. The molecular mechanisms of life were and are continuously being refined; less efficient pathways are discarded when better ones appear, even at the expense of added complexity.

Where Are We Heading?

Being able to peer into the molecular details of the DNA ⟶ RNA ⟶ protein ⟶ metabolism life cycle also puts into human hands an enormous "chemical potential" for both good and evil. Pathology, the study of the origins of disease, is now being refined at the molecular level; the complete eradication of hereditary diseases may be possible with a map of the entire human genome, nearly complete as this is written. **Genetic engineering,** the splicing and replacing of genes, has allowed us to custom-modify the genes of bacteria to produce useful enzymes and digest harmful pollutants. It has also given a new and frightening tool to the eugenicists, those who would modify our own species according to some predetermined set of "desirable" traits. At this writing, **cloning,** inserting absolutely identical copies of the DNA of a living organism into an embryo, is an active area of investigation. The cloning of mammals has already been accomplished, and the possible cloning of a human being cannot be far down the road. Where all of this will lead is hard to say—in the words of Niels Bohr, "Predictions are always difficult, especially of the future"—but your own personal knowledge of what is involved can go a long way toward putting public decisions on a rational foundation, avoiding both fear-mongering and reckless adventurism. As never before, the future is yours to determine.

Have we answered the question "What is life?"? No, we have simply given an abbreviated *description* of the chemical processes that living things use in their daily existence, a description that even in more detailed form is in many ways still incomplete. If we thoroughly understood life, we might be able to assemble in a test tube a "little Frankenstein monster," a living creature, something not yet achieved. In doing that, we would simply be imitating what has developed through billions of years of chemical evolution. The human species, so far as we know the "highest form of life," is so complex that, even given an essentially complete genome, we are decades, perhaps a century, away from knowing the details of its chemical workings. In closing, we can say that our world, ourselves included, owes its very form and function to the structures, interactions, and dynamics—the chemistry—of the atoms and molecules that comprise it. We hope this is not the end, but just the beginning of your education in chemistry!

SUMMARY

Organic chemistry is the chemistry of carbon compounds, which are far more numerous than all others combined. Carbon's neighbors, boron, nitrogen, and silicon, are prevented by valence, electronegativity, or bonding radius from making long chains, strong homopolar single bonds, and multiple bonds that are stable in larger assemblies. The building blocks of organic molecular structure are the hydrocarbons, including **alkanes** or paraffins with single C—C bonds, **alkenes** with one or more C=C bonds, and **alkynes** featuring C≡C bonds. The hydrogen atoms in alkanes, for example, ethane (H_3C—CH_3), can have different relative spatial orientations, giving rise to conformational isomers. Beyond C_3 compounds, straight and branched chain **structural isomers** become possible, and there appears to be no limit to the length of the chain. Rings are also formed beginning with C_3; the most common are six-membered cyclohexane rings, with "boat" and "chair" (more stable) conformations. Common reactions of alkanes include **combustion,** halogenation, and **dehydrogenation,** all of which occur only under extreme conditions when uncatalyzed. Alkenes and alkynes are more reactive than alkanes, undergoing electrophilic **addition** and oxidation at the multiple bond site under mild conditions.

Hydrocarbons form a number of stable derivatives, in which a hydrogen in a C—H bond is replaced by another group. Common derivatives include halides such as R—Cl, **alcohols** (R—OH), **ethers** (R—O—R′), ketones (R—CO—R′), aldehydes (R—CO—H), **caboxylic acids** (R—COOH), **esters** (R—COO—R′), **amines** (R—NH_2), **amides** (R—CO—NH_2), and α-amino acids (H_2N—CHR—COOH), where R is an alkyl, alkenyl, or alkynyl group. Alkenes and their derivatives can show geometrical isomerism, in which groups on each of the two C atoms of C=C are adjacent (*cis*) or across (*trans*). Common reactions of derivatives include first-order (S_N1) and second-order (S_N2) nucleophilic displacement of one substituent by another; this reaction mechanism is particularly facile for halide and carbonyl C=O derivatives.

Derivatives may also show a different type of isomerism, called enantiomerism or optical isomerism, usually connected with an asymmetric or chiral C atom, one bonded to four different groups. **Enantiomers** are pairs of molecules that differ from one another only in that their mirror images are not superimposable, like left and right hands. Solutions of a pure enantiomer will rotate the plane of plane-polarized light to the left (levo or L-) or to the right (dextro or D-), while a 50:50 mixture of an enantiomeric pair, called a **racemic mixture,** shows no optical activity. Naturally occurring organic compounds are often pure enantiomers. For a molecule with n chiral C atoms and no center of symmetry, there are 2^n pairs of enantiomers. In general, isomers are classified as structural and **stereoisomers,** where stereoisomers comprise enantiomers and **diastereomers;** diastereomers include **conformational** and **geometrical isomers.**

An especially stable type of delocalized π bonding confers a property called **aromaticity** on **benzene** (C_6H_6) and its derivatives. In aromatic compounds all of the bonding π levels are completely delocalized and have energies below that of an isolated π bond; this includes planar rings with $4n + 2$ π electrons. The reactivity of aromatic compounds is low, similar to that of alkanes, and they undergo **substitution** rather than addition reactions.

Organic reaction mechanisms are usually depicted with curly arrows, but in cases of π-bond reactions this can lead to ambiguity. A frontier molecular orbital (FMO) analysis of the ring-forming **Diels–Alder cycloaddition reaction** between a diene and an ene is presented. Critical to a successful analysis is recognition of the importance of π MO phases in overlapping the diene HOMO and the ene LUMO. The FMO picture eliminates the curly-arrow ambiguity, and leads to predicted stereochemistry of the product that has been confirmed by experiment.

Macromolecules with repeating organic molecular subunits are called **polymers.** Common examples of **addition polymers** are plastics and fibers made from ethylene monomers and their derivatives, such as polyethylene and polyvinylchloride. **Condensation polymers** are made from difunctional monomers, such as amino acids, that can undergo end-to-end nucleophilic substitution with elimination of water. **Copolymers** are made by condensing pairs of difunctional monomers such as diamine and a diacid; a well-known example is nylon. Hard plastics such as Bakelite have polyfunctional monomers that form two- and three-dimensional network structures.

Living organisms mainly use a few basic types of organic compounds. **Carbohydrates** and fats are sources of metabolic energy. The most important carbohydrates are the **pentose** and **hexose sugars** ribose and glucose, and the polymers derived from glucose, **starch,** and **cellulose. Fats,** also called **lipids,** are esters made from **fatty acids,** long-chain carboxylic acids, and glycerol. In **phospholipids,** used as cell wall material, one of the acids is replaced by an inorganic phosphate ester.

Proteins, ordered chains comprised of 20 natural α-amino acids linked together by peptide bonds (—CO—NH—), serve both structural and catalytic purposes. Catalytic proteins are called **enzymes.** Enzymes exhibit four types of structure. The primary structure is the particular sequence of amino acid residues, about 350 on the average, along with any disulfide cross-links between cysteine residues. Secondary structures include the α-helix, pleated β-sheets, and random coils, all governed by hydrogen-bonding interactions. Tertiary structure is the final folded or native state of the enzyme, and quaternary structure refers to aggregates of individual proteins; both are also determined by nonbonded forces.

The cyclic nitrogen bases adenine (A), cytosine (C), guanine (G), thymine (T), and uracil (U) combine with ribose and phosphate groups to make compounds called **nucleotides,** which reside mainly in the nucleus of a cell. The ribonucleotide **adenosine triphosphate (ATP)** is the main reservoir of energy for cell chemistry. This energy is stored in the triphosphate P—O—P linkage, and released when ATP hydrolyzes, $ATP^{4-} + H_2O \longrightarrow ADP^{3-} + H_2PO_4^-$, $\Delta G° = -7.3$ kcal/mol. Phosphate ester bonds link nucleotides together to make nucleic acids; **ribonucleic** acid (**RNA**) is made from **ribonucleotides,** while **deoxyribonucleic acid (DNA)** is made from **deoxyribonucleotides,** in which the **ribose** OH group at C-2′ is replaced by H. RNA employs only the bases A, C, G, and U, while DNA contains A, C, G, and T. DNA forms a **right-handed double helix,** where one stand has a certain sequence of bases, and the other has a complementary sequence selected by hydrogen bonding. Adenine selectively bonds to T, forming three H bonds, while C bonds to G with two H bonds. The H bonds occur either as N—H --- :O=C or N—H --- :N.

Photosynthesis of glucose from CO_2 and water by plants is initiated by light absorbed by chlorophyll, creating an excited electron. In the anaerobic pathway, this electron is collected by **nicotinamide adenosine dinucleotide phosphate (NADP$^+$),** reducing it to NADPH; NADPH then reduces CO_2. In the aerobic pathway, NADPH is produced from NADP$^+$ by the light-assisted oxidation of H_2O to O_2, while some light energy is also stored in ATP. The entire process is controlled by collections of proteins called photosystems I and II. The eventual product of the **light reactions** is a triose phosphate; this is converted to glucose through a complex series of **dark reactions** that make use of the energy and electrons stored during the light phase. Altogether forty-eight 700-nm photons, the energy of 18 ATPs, and electrons from 12 NADPHs are required to make one molecule of glucose.

Metabolism of glucose by animals back into CO_2 and water occurs in three stages. The first stage, glycolysis, is anaerobic, and converts glucose to pyruvic acid with the aid of 10 enzymes and nicotinamide adenine dinucleotide (NAD), resulting in a net gain of two ATP and storage of four electrons in two NADH. After conversion of pyruvic acid to acetyl coenzyme A with evolution of one molecule of CO_2, the citric acid cycle creates two more molecules of CO_2 while storing the electrons from the oxidation in NAD and in **flavin adenine dinucleotide (FAD),** converting it to $FADH_2$. In the respiratory chain, NADH and $FADH_2$ then pass H$^+$ and e$^-$ to O_2 through a series of five redox proteins, producing H_2O in the end. A total of 30 ADP $\longrightarrow$ ATP conversions occur for each molecule of glucose metabolized, an energy efficiency of 32%.

The mechanism of enzyme catalysis is illustrated by the selective cleavage of **peptide bonds** by chymotrypsin, a 245-residue protease enzyme. After docking of the polypeptide substrate at the enzyme's active site, the amino end of the peptide bond is cleaved away in forming an ester bond with a serine **residue** at the active site. Then the carboxyl end is released by catalyzed hydrolysis of the serine ester. The enzyme achieves a lowering of the activation energy at each stage through nonbonded interactions that stabilize the two transition states.

Information concerning protein synthesis is stored in the DNA double helix in the form of three-base codons that specify particular amino acids. A section of the helix is unwound by an RNA polymerase enzyme and the sequence of codons transcribed by selective H bonding to make mRNA. **mRNA, tRNA,** and **rRNA** all assist in protein synthesis, tRNA being aminoacylated with the amino acid corresponding to its anticodon by the synthetase enzymes, and rRNA being incorporated into protein-synthetic catalysts called **ribosomes.** The mRNA carrying the code for the enzyme residue sequence becomes bound between two ribosome subunits, and one by one, tRNAs bearing the corresponding amino acids selectively hydrogen bond to mRNA, and peptide bonds are formed catalytically between adjacent amino acids with energy from hydrolysis of **guanosine triphosphate (GTP).** Special punctuation **codons** begin and end the synthesis. The preservation and transmission of information stored in DNA is the job of nuclease and **DNA polymerase** enzymes. DNA is **replicated** in a manner similar to RNA synthesis, making two new complementary strands with an error rate of only one mistake in 10^{10} bases.

Experiments simulating prebiotic conditions on earth have shown that most of the molecules used by living things are spontaneously synthesized, including amino acids, sugars, and ribonucleotide bases. It is speculated that life originated from small bits of RNA that could act as both catalysts for peptide synthesis and templates for reproducing themselves.

EXERCISES

1. Draw five structural isomers of octane (C_8H_{18}) in both semistructural form showing all hydrogens, and in skeletal form. Why can't ring compounds appear among your isomers? Choose an isomer with two or more branches and give its proper name.

2. For *n*-butane and beyond, the 109° C—C—C bond angles can be arranged in various ways in space relative to each other, even if we consider only those structures where all C atoms lie in the same plane. This is a form of conformational isomerism. For butane only the zigzag and folded conformers can be drawn. Try to draw all possible in-plane bond angle conformers for *n*-hexane in skeletal form.

3. What is the smallest alkane hydrocarbon that can show chirality? Draw its structure and give its proper name.

4. Use information given in the text to find the ratio of concentrations of boat to chair conformations of cyclohexane at 25°C.

5. Assuming that the boat ⟶ chair conversion in cyclohexane vapor has a preexponential factor of 1.0×10^{13} s^{-1} and an activation energy of 10.1 kcal/mol, estimate the half-reaction time for unimolecular conversion at 25°C. What will happen to $t_{1/2}$ as the temperature is raised? As the pressure is reduced?

6. Show that the second and third elementary steps in the mechanism for the halogenation of methane (Reaction 18.4) add up to the overall reaction (Reaction 18.3). Identify the reaction intermediates.

7. Estimate the heats of reaction from bond energies for Reaction 18.3 and the second and third steps in the mechanism of Reaction 18.4. How are these heats related? Based on your $\Delta H°$s for Reaction 18.4, which step do you expect to be rate limiting?

8. Estimate the temperature above which the dehydrogenation of ethane (Reaction 18.6) will become spontaneous at standard pressure.

9. Draw two straight chain and one branched chain structural isomer of pentene (C_5H_{10}). Which of these will also show geometrical (*cis-trans*) isomerism?

10. When HBr adds across the double bond of propene (CH_2=CH—CH_3), 2-bromopropane is formed nearly exclusively. Rationalize this observation by writing a mechanism in which propene is first protonated to form a carbocation, and assessing the relative stability of the two carbocations that could be formed. Bromide will then add to the more stable carbocation. (In general, the bromide always adds to the more substituted carbon; this is called *Markovnikov's rule*.)

11. Give a semistructural depiction of the first three steps involved in acid-catalyzed ester formation; the overall reaction is Reaction 18.17. In the first step, the carbonyl oxygen is protonated; in the second, the π bond becomes a lone pair on the carbonyl oxygen, creating a carbocation; in the third the cationic C is attacked by a lone pair on the alcoholic oxygen. Use curly arrows to indicate electron movement at each step, and assign formal charges where they are not zero. What species must "leave" the adduct formed in step 3 in order to form the final product?

12. Give an alternative set of reagents that might be used in Reaction 18.15, by switching the Cl and O$^-$ groups. Do you expect this reaction to be more or less facile than that of Reaction 18.15? Is any change in the mechanism expected? State your reasoning.

13. How many enantiomeric pairs exist with the formula C_2H_4FCl?

14. Which of the following compounds can show optical activity?

    a. chlorobromoethylene
    b. malonic acid (HOOCCH$_2$COOH)
    c. 2-butanol
    d. 1-butanol
    e. 1,2-dichloropropane

15. Unlike alkyl alcohols, the hydroxyl hydrogen of phenol (C_6H_5OH) shows measurable acidity in aqueous solution. Rationalize this observation by drawing resonance forms that stabilize the phenolate ion ($C_6H_5O^-$) by delocalizing the negative charge. In each structure assign formal charges where they are nonzero.

16. A common π-type ligand is the cyclopentadienyl anion $C_5H_5^-$. Decide whether or not $C_5H_5^-$ is aromatic.

17. There is evidence that a bromonium ion (Br$^+$) initiates the electrophilic addition of Br$_2$ to double bonds. What are the HOMO and LUMO orbitals for the attack of Br$^+$ on a π bond? Sketch an orbital picture of the interaction between the frontier orbitals.

18. Teflon is a trade name for polytetrafluoroethylene. Draw the structures of the monomer and polymer of Teflon analogous to Equation 18.20.

19. Write a reaction between ethylene glycol and *p*-phthalic acid to make one unit of the copolymer dacron (Equation 18.24). Use curly arrows to illustate the mode of coupling of the comonomers.

Blackbody Radiation: The Origin of the Quantum Theory

By nature I am peacefully inclined and reject all doubtful adventures. But a theoretical interpretation had to be found at any cost, no matter how high. . . . I was ready to sacrifice every one of my previous convictions about physical laws.

Max Planck

In this famous quote, Planck was referring to the vexing problem of **blackbody radiation,** introduced in Chapter 2. Here we will provide a brief derivation of Planck's formula for the electromagnetic energy emitted by an ideal blackbody as a function of wavelength λ and absolute temperature T, as well as the standard wave-theory formula, also known as the Rayleigh–Jeans law, which gave rise to the "ultraviolet catastrophe" in the form of a divergence to infinity at small λ. These formulas are the functions behind the graphs of Figure 2.6. The derivation will make use of the geometric series and its sum, and differential calculus.

To make the desired formulas independent of the actual size of the hot object (the blackbody), we find the *energy density* per unit wavelength $\rho_E(\lambda)$, energy per unit volume per length (erg cm^{-4} or J m^{-4}) of light energy within the body. This energy is emitted to give the observed spectrum. The hot body is assumed to form a set of standing waves of electromagnetic energy within its boundaries. The energy density $\rho_E(\lambda)$ is then given by a product of two factors: the density of standing waves of a particular wavelength $\rho(\lambda)$ and the average energy at that wavelength $\overline{\varepsilon}(\lambda)$, that is,

$$\rho_E(\lambda) = \rho(\lambda)\,\overline{\varepsilon}(\lambda) \tag{A.1}$$

For convenience we will assume the body is a cube of edge length L, but the result we will obtain is independent of the actual shape.

First let's find the wavelength density $\rho(\lambda)$. In one dimension, say along the x axis, the standing-wave condition is

$$n_x\left(\frac{\lambda}{2}\right) = L \tag{A.2}$$

that is, an integral number n_x of half-wavelengths of the light wave must "fit" within the body, where $n_x = 1, 2, 3, \ldots$. Each n_x gives a new wavelength or a new frequency of light, with number of nodes $n_x - 1$ in the standing wave. Solving for n_x, we get

$$n_x = \frac{2L}{\lambda} \tag{A.3}$$

that is, for a given λ there exists an n_x, but it is not necessarily an integer. For our problem, though, the dimensions of a typical blackbody are of the order of $L = 1-10$ cm, while the light inside it has $\lambda \approx 10^{-4}$ cm. This makes n_x a large number, and whether we consider it an integer or not makes little difference to the function $n_x(\lambda)$, that is, $n_x(\lambda)$ can be considered a continuous function. Now, n_x (actually $n_x - 1$) is not only a node count for wavelength λ, but it also gives the total number of waves with wavelength $\geq \lambda$. If we decrease λ by a little bit, $-d\lambda$, the number of wavelengths n_x will increase by a little bit, dn_x, and the ratio of these bits, divided by the length L, is the one-dimension wavelength density $\rho_x(\lambda)$:

$$\rho_x(\lambda) \equiv -\frac{1}{L}\frac{dn_x}{d\lambda} = \frac{2}{\lambda^2} \tag{A.4}$$

where the last equality is obtained by differentiating Equation A.3. Our conclusion is that the density grows without bound as $\lambda \longrightarrow 0$; that is, there are more and more possible standing waves in a given interval $\Delta\lambda$ as λ shrinks. Going to three dimensions does not alter this conclusion.

In three dimensions, the standing waves take on a vector character, and we can assign node counts for each dimension n_x, n_y, and n_z. The number of nodes in the vector wave is then $n = [n_x^2 + n_y^2 + n_z^2]^{1/2}$, where $n = 2L/\lambda$ as in Equation A.3. The total number of wavelengths in this case is given by $\frac{1}{8}$ the volume of a sphere of radius n, $(\frac{1}{8})(\frac{4}{3}\pi n^3)$; that is, this sphere will contain all possible combinations of n_x, n_y, and n_z that yield n. The factor of $\frac{1}{8}$ is necessary because all the n's are necessarily positive, restricting the counting procedure to a single octant in (n_x, n_y, n_z) space. This number, divided by the volume of the container L^3, gives the total number of waves per unit volume, and as in the one-dimensional case, its derivative with respect to λ yields the three-dimensional wavelength density $\rho(\lambda)$. We have

$$\rho(\lambda) = -\frac{1}{L^3}\frac{d}{d\lambda}\left(\frac{\pi}{6}n^3\right) \tag{A.5}$$

$$= -\frac{1}{L^3}\frac{\pi}{6}8L^3\frac{d}{d\lambda}\left(\frac{1}{\lambda^3}\right)$$

$$= \frac{4\pi}{\lambda^4}$$

where in the second line we have used Equation A.4 and taken all constants outside the derivative operation. The final result is beautifully simple, as is often the case for formulas that are in some sense "true."

Our result for $\rho(\lambda)$ does not quite yet fit the case of electromagnetic waves; note we haven't yet said what kind of waves they are. For each standing wave, the electric field of the light can point along either of the two transverse directions; for example, if the wave is propagating along x, the electric field can point along either y or z. This means that for each n there are two possible standing waves, and our formula for $\rho(\lambda)$ should be multiplied by 2, giving

$$\rho(\lambda) = \frac{8\pi}{\lambda^4} \tag{A.6}$$

As in the one-dimensional case, this density diverges as $\lambda \longrightarrow 0$, only more strongly. It is this divergence that, as we shall see later, gave rise to the "ultraviolet catastrophé."

So far we have seen no hint of any quanta lurking around. This idea arose from Planck's calculation of the average energy factor $\bar{\varepsilon}(\lambda)$ in Equation A.1. Planck was thoroughly acquainted with Ludwig Boltzmann's methods (ca. 1885) for computing the average energy of an ensemble of objects, based on a most likely distribution of the total energy among the objects we now know as the Boltzmann distribution. Boltzmann was a staunch advocate of the atomic theory; you might say he staked his entire career on the existence of the atom. This caused him to distrust the old classical theories of heat and heat energy, which never invoked atoms, and made extensive use of calculus. Boltzmann avoided the use of calculus in his work, because in his view one might be ignoring some important feature having to do with the discrete nature of matter by blithely integrating over some variable. In fact, it was just such a feature that Planck stumbled on in using Boltzmann's methods!

Let's see how Planck obtained $\bar{\varepsilon}(\lambda)$ using Boltzmann's ideas. At a particular λ, there will be a distribution of energy among the possible waves. Some will have the minimum possible energy, some will have energy ε_0 above the minimum, some will have energy $2\varepsilon_0$, some $3\varepsilon_0$, etc. Here ε_0 is an "energy grain," a discrete amount of energy of (as yet) arbitrary size. Boltzmann proved that at an absolute temperature T the relative populations of waves with these energies must stand in the ratio

$$1 : e^{-\varepsilon_0/k_B T} : e^{-2\varepsilon_0/k_B T} : e^{-3\varepsilon_0/k_B T} : \ldots \tag{A.7}$$

Here e is the base of the natural logarithms, e = 2.718..., and k_B is Boltzmann's constant. The negative exponents mean that as the energy grows, the population dies away. The quantity $\exp(-\varepsilon/k_B T)$ is known as a Boltzmann factor. Appendix D presents a derivation of this result.

Let's suppose there are N_0 waves with minimum energy. The Boltzmann factors then allow us to find the total energy at wavelength λ by knowing both the energy of a set of waves and the number of waves in that set. First, the total number of waves N is

$$N = N_0(1 + e^{-\varepsilon_0/k_B T} + e^{-2\varepsilon_0/k_B T} + e^{-3\varepsilon_0/k_B T} + \ldots) = \frac{N_0}{1 - e^{-\varepsilon_0/k_B T}} \tag{A.8}$$

where the rightmost expression is found by identifying the sum as a geometric series. Notice that this is done without the aid of integral calculus; we have explicitly made a sum using a discrete, finite energy grain ε_0. The total energy E of all the waves is obtained by multiplying the energy of each set of waves by the number of waves with that energy and summing:

$$E = N_0(\varepsilon_0 e^{-\varepsilon_0/k_B T} + 2\varepsilon_0 e^{-2\varepsilon_0/k_B T} + 3\varepsilon_0 e^{-3\varepsilon_0/k_B T} + \ldots) \tag{A.9}$$

$$= \frac{N_0 \varepsilon_0}{e^{\varepsilon_0/k_B T}(1 - e^{-\varepsilon_0/k_B T})^2}$$

Note here that the first set of waves is missing, because it contributes no energy above the minimum. The explicit sum on the right is obtained by noting that each term on the left is the negative of the first derivative of each term on the left in Equation A.8, with respect to the variable $\beta = 1/k_B T$. This allows us to find the sum in Equation A.9 by differentiating the right side of Equation A.8. If we now divide the total energy of the waves by the total number of waves (Equation A.9 divided by Equation A.8), we attain our goal, the average energy per wave,

$$\bar{\varepsilon}(\lambda) = \frac{E}{N} = \frac{\varepsilon_0}{e^{\varepsilon_0/k_B T} - 1} \tag{A.10}$$

Again, a simple, lovely result. The curious thing is that the wavelength λ does not appear on the right-hand side of Equation A.10; all we have is the energy grain ε_0. In Maxwell's wave theory of light, the energy of the light has nothing to do with its wavelength, depending only on the square of the wave amplitude.

The next step in finding the average energy per wave, following Boltzmann further, is to let the arbitrary energy grain ε_0 become vanishingly small. The exponential function in Equation A.10 can then be expanded in a Maclaurin power series and truncated after the second term: $\exp(\varepsilon_0/k_B T) \approx 1 + \varepsilon_0/k_B T$. Substituting this approximation into Equation A.10 gives simply $\bar{\varepsilon}(\lambda) = k_B T$, that is, the average energy is independent of wavelength and is always $k_B T$. (This result was well known to both Boltzmann and Planck, and is called the *classical equipartition theorem*.) If we proceed to combine this result with our standing-wave density (Equation A.6) to find the blackbody spectrum of Equation A.1, we obtain

$$\rho_E(\lambda) = \frac{8\pi k_B T}{\lambda^4} \tag{A.11}$$

Again, a beautifully simple result known as the Rayleigh–Jeans law. In this case, however, simplicity is not truth: Equation A.11 diverges as $\lambda \longrightarrow 0$, a "catastrophic" result in gross disagreement with experiment; it produces the dashed curves in Figure 2.6. However, it is found to agree with experiment at very long wavelengths in the far-infrared region.

What to do? Planck went back to his expression with the energy grain, Equation A.10. Knowing the behavior of exponentials, he saw that if ε_0 were inversely proportional to wavelength (directly proportional to frequency $\nu = c/\lambda$), the exponential in the denominator of Equation A.10 would grow without bound as $\lambda \longrightarrow 0$, and do away with the ultraviolet catastrophé by forcing $\rho_E(\lambda)$ to go to zero. So, against all reason, he tried

$$\varepsilon_0 = h\nu = \frac{hc}{\lambda} \tag{A.12}$$

where h is a proportionality constant, now known as Planck's constant, whose value was to be determined by fitting the experimental data of Lummer and Pringsheim. Equation A.12 is what we now call *Planck's quantum hypothesis*. Substituting Equation A.12 into Equation A.10, and again combining Equation A.10 with Equation A.6 in Equation A.1, we get Planck's expression for the blackbody spectrum:

$$\rho_E(\lambda) = \frac{8\pi hc}{\lambda^5} \frac{1}{e^{hc/\lambda k_B T} - 1} \tag{A.13}$$

This equation is certainly not as pretty as Equation A.11, but it has stood the test of time and many precise experiments, and is what in science we call "true." Planck was quickly able to estimate a value for h by using Wien's law, $\lambda_{max}T = 0.288$ cm K: he differentiated Equation A.13 to find an expression for the wavelength λ_{max} at maximum intensity, and found h by knowing the constant in Wien's law. He then found that, with minor adjustments, Equation A.13 with $h = 6.57 \times 10^{-27}$ erg s could represent all of the experimental data exceedingly well; the solid curves in Figure 2.6 are computed using this formula. Our modern (1998) value is $h = 6.6260688 \times 10^{-27}$ erg s (in cgs units). Planck's formula continued to work at large λ as well, where it conveniently reduced to the Rayleigh–Jeans law. Interestingly, Planck could not have found Equation A.13 without Boltzmann's atomistic distrust of integral calculus in physical problems: using integration rather than discrete sums in Equations A.8 and A.9 can lead only to Equation A.11.

Why do we call Equation A.12 a "quantum" hypothesis? If you go back to Equations A.8 and A.9, in light of Equation A.12, you see that Boltzmann's energy grain is finite and takes on a real meaning, and the sums truly are discrete sums. The energy of a particular group of waves can only be $E = nh\nu$, where $n = 1, 2, 3, \ldots$, and $\varepsilon_0 = h\nu$ is the minimum size of "energy packet" or "photon" available. A wrinkle that has been added since Planck's epochal finding is the notion of "zero point energy": even the lowest energy group of waves has an energy per wave of $\frac{1}{2}h\nu$, and in general the energy of the nth group is $E = (n + \frac{1}{2})h\nu$. Zero point energy has no effect on Planck's formula (Equation A.13), however, since it cannot be radiated away.

FURTHER READING

Planck, Max, *The Theory of Heat Radiation* (1912; Dover edition, 1959).
Moore, Walter J., *Physical Chemistry*, 3rd ed. (Prentice Hall, 1962), Chap. 12.

EXERCISES

1. Verify the right-hand-sides of Equations A.8, A.9, and A.10. For Equation A.8, use the formula for the sum of a geometric series $s = a + ar + ar^2 + ar^3 + \ldots = a/(1 - r)$. For Equation A.9 carry through the differentiation of both sides of Equation A.8 as indicated in the text.

2. For this and the following exercises, it will be convenient to express the Planck distribution law of Equation A.13 in terms of the variable $x = hc/\lambda k_B T$. Prove Wien's law, $\lambda_{max}T = $ constant, in the process finding an expression for the constant. Then using Wien's value for it, evaluate Planck's constant h. (You must set $d\rho_E/d\lambda = 0$; use the chain rule $d/d\lambda = (dx/d\lambda)d/dx$ to carry out the differentiation.)

3. Based on heat measurements by Tyndall, Josef Stefan in 1879 proposed what is now known as the Stefan–Boltzmann law, which gives the total rate of energy emission per unit area of a hot object as $R = \sigma T^4$, where $\sigma = 5.67 \times 10^{-8}$ J m^{-2} s^{-1} K^{-4}. This rate is related to the total energy per unit volume E_V by $R = \frac{1}{4}cE_V$, where

$$E_V = \int_0^\infty \rho_E(\lambda)\, d\lambda$$

Show that this law follows from Planck's distribution (Equation A.13) by carrying out the preceding integration; give an expression for σ and evaluate it using modern constants. Convert the integral to one over $x = hc/\lambda k_B T$, and use the well-known result

$$\int_0^\infty \frac{x^3 dx}{e^x - 1} = \frac{\pi^4}{15}$$

4. By use of the Maclaurin series

$$e^x = 1 + x + \frac{x^2}{2!} + \frac{x^3}{3!} + \cdots$$

   a. Show that Planck's distribution (Equation A.13) reduces to the Rayleigh–Jeans expression (Equation A.11) at large λ.
   b. By retaining a correction term from part (a) and using the binomial expansion

$$(1 + x)^p = 1 + px + \frac{p(p-1)}{2!}x^2 + \frac{p(p-1)(p-2)}{3!}x^3 + \cdots$$

   find the value of $x = hc/\lambda k_B T$ at which the Rayleigh–Jeans law agrees with the Planck law within 10%.

5. Planck made use of Wien's proposed distribution $\rho_E(\lambda) = (\alpha/\lambda^5)e^{-\beta/\lambda T}$ in deciding how to relate ε_0 and λ.

   a. Show that Planck's radiation law reduces to Wien's for small λ, and give formulas for Wien's constants α and β.
   b. By using the binomial expansion given in the previous problem, retain the first correction term from part (a), and estimate the value of $x = hc/\lambda k_B T$ at which the Wien distribution agrees with Planck within 10%. Is this value smaller than that at the maximum in the Planck curve? (See Exercise 1.)

6. Use L'Hôpital's rule to show that the Planck distribution of Equation A.13 rigorously avoids the ultraviolet catastrophé as $\lambda \longrightarrow 0$. To simplify the problem, use $x = hc/\lambda k_B T$ as the variable.

Ultraviolet catastrophe
    Came on us unaware
    Came up the Strand
    To Carlton Terrace, down the stair
    At Piccadilly Underground, where I, Max Planck,
    Am wont to play an anharmonic air
    Upon my violin, so marvellously tuned
    To modes of hydrogenic unison.

But where are the short-wave rays that fail to light
    The incandescent blackness? I have found
    Degrees of freedom powerlessly bound
    In chains of integer constraint,
    And atoms, governed by the selfsame laws
    Quiescent, in a cold ground state.

And when the spectrum slides
    Beyond the violet end,
    Touched by the last dim rays
    The silver surface spits electrons and displays
    The undivided energy of light!

O James Clerk Maxwell I can sometimes see
    A wave, like you, sometimes a particle like me
    —Can both of us be right?

John Lowell, —excerpt from "The Waste Lecture"
Physics Today, Vol. 42, No. 4, April 1989

A Particle in a Box
A (Relatively) Simple Quantum Mechanical Problem
Displaying Many of the Features of More Complicated Problems

Introduction

Many of the important aspects of quantum mechanics as applied to electrons, atoms, and molecules—features involving the energy levels and their associated wave functions—can be appreciated by appealing to simple examples. The most famous of these is the case of a particle confined to a one-dimensional box. If the box is of length a, then Heisenberg's uncertainty principle tells us at once that an uncertainty in the momentum of the particle $\Delta p \simeq h/a$ must accompany its confinement in the box, where h is Planck's constant. This in turn implies that the lowest (ground-state) energy $E \simeq (\Delta p)^2/2m \neq 0$, where m is the particle's mass; that is, the particle cannot be "at rest" in the box! This amazing result, which also holds for electrons placed in atomic and molecular "boxes," arises in a natural way in solving the Schrödinger equation for this problem, as given in this appendix.

Setting Up the Problem

To solve any mechanical problem using Schrödinger's wave mechanics, you first have to figure out how to write down explicitly the Hamiltonian for the system, $H = K + V$, where K is the kinetic energy ($K = p^2/2m$) and V the potential energy. If we assume that the particle cannot escape the walls of the box, but is doomed to bounce back and forth within the box forever (a classical particle picture of what's happening!), then the potential energy is

$$V = \begin{cases} 0 & \text{for } 0 < x < a \\ +\infty & \text{for } x < 0 \quad \text{or} \quad x > a \end{cases} \quad \text{(B.1)}$$

where x is the position of the particle. This means that within the box there is no force on the particle until it strikes a wall (at $x = 0$ or $x = a$), where the force is infinite. Figure B.1 shows a graph of V.

Next we must use the special form of the momentum p recommended by Schrödinger for use in his equation. Instead of $p = mv = m\,dx/dt$, as Sir Isaac Newton would tell us, we use

$$\hat{p} = \frac{h}{2\pi i}\frac{d}{dx}$$

that is, we replace p by a **differential operator** $\hat{p}$ (a thing that will automatically take the first derivative of any function you put on its right). Here h is again Planck's constant and $i = \sqrt{-1}$. It is this somewhat mysterious transcription of p that begins the conversion of this problem from Newton's mechanics to the wave mechanics of Schrödinger. The kinetic energy now becomes

Figure B.1
The box.

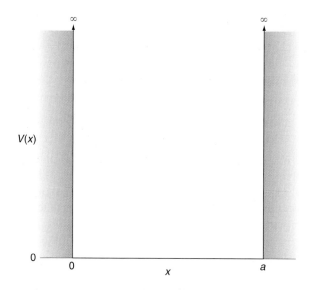

$$\hat{K} = \frac{\hat{p}^2}{2m} = \frac{1}{2m}\left(\frac{h}{2\pi i}\frac{d}{dx}\right)^2 = -\frac{h^2}{8\pi^2 m}\frac{d^2}{dx^2}$$

an operator that takes second derivatives of functions.

Now we can write down the Schrödinger equation, $\hat{H}\psi = (\hat{K} + V)\psi = E\psi$, explicitly for our problem. It is

$$-\frac{h^2}{8\pi^2 m}\frac{d^2\psi}{dx^2} + V\psi = E\psi \tag{B.2}$$

where V is defined in Equation B.1, ψ is the **wave function** for the particle (a sort of one-dimensional orbital), and E is the energy of the particle. As long as we stay away from the walls, V is zero, and the Schrödinger equation simplifies to

$$\frac{d^2\psi}{dx^2} = -\frac{8\pi^2 mE}{h^2}\psi \equiv -k^2\psi \tag{B.3}$$

where k is defined by the identity sign: $k^2 = 8\pi^2 mE/h^2$.

Solving the Schrödinger Equation

An important motivation for writing this appendix for you is to show you that there is nothing mysterious about solving the wave equation. Equation B.3 says that the solution ψ must be such that when you differentiate it twice you get a negative constant times ψ. You can verify by direct substitution into Equation B.3 that the trigonometric function

$$\psi(x) = A \sin kx + B \cos kx \tag{B.4}$$

satisfies Equation B.3 identically, with A and B (as yet) arbitrary constants ("constants of integration").

The function $\psi(x)$ of Equation B.4 cannot be calculated without specifying A, B, and k (or E). It is at this point that the physics (or chemistry) of the problem must be introduced. We expect that, if the wave function squared, $|\psi(x)|^2$, is to represent the probability density for finding the particle at position x, then at the walls of the box, $x = 0$ and $x = a$, ψ must become zero, because the particle cannot penetrate beyond these points. These requirements on ψ, $\psi(0) = 0$ and $\psi(a) = 0$, are called **boundary conditions.** The first of these,

$$\psi(0) = 0 = A \sin 0 + B \cos 0 \tag{B.5}$$

yields $B = 0$ because $\cos 0 = 1$. The second gives

$$\psi(a) = 0 = A \sin ka \tag{B.6}$$

Because B is already zero, we don't need to include it. We don't want to set $A = 0$, because this would make the entire wave function zero, a rather trivial and meaningless solution to our problem! Instead, we ask what k has to be to satisfy this condition. We know the sine function oscillates, becoming zero when the argument is $n\pi$, where n is an integer, $n = 1, 2, 3, \ldots$ (Why can't $n = 0$ be used?). We can thus satisfy the second boundary condition, Equation B.6, by requiring that

$$ka = n\pi, \quad n = 1, 2, 3, \ldots \tag{B.7}$$

As you may have guessed, n is a **quantum number,** the only one we will get in this one-dimensional (1-D) problem. Condition (B.7) is equivalent to requiring "standing waves" inside the box, that is, $n(\lambda/2) = a$, where $\lambda = h/p$ is the deBroglie wavelength. You can easily prove this equivalence by using the definition of k from Equation B.3. Using this definition in Equation B.7, we can easily find the **energy levels** E_n to be

$$E_n = \frac{n^2 h^2}{8ma^2} \tag{B.8}$$

(Note that the sign of E_n is positive, even though we are talking about a bound particle; this arises because of our choice of zero potential V at the bottom of the box.) The boundary conditions are seen to *impose* quantization of the energy; this feature is shared by all wave mechanical problems of bound particles, including electrons in atoms and molecules. The ground state (or "zero point") energy $E_1 = h^2/8ma^2$ is not zero, but a finite value, as we argued from the uncertainty principle in the introduction. Unlike electronic energy levels in atoms, these levels increase in spacing as n increases. Figure B.2 illustrates the lowest four levels. The energy level formula (Equation B.8) says that the levels get closer together for heavier particles (large m) or for larger boxes (large a).

Figure B.2
Energy levels for a particle in the box.

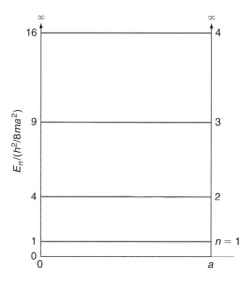

The Wave Functions

The second boundary condition (Equation B.7) not only specified the energy levels, but also completed the determination of the wave function within a constant factor. Using Equations B.4 through B.7, we get

$$\psi_n(x) = A \sin\left(\frac{n\pi x}{a}\right) \tag{B.9}$$

where A is the amplitude of the wave, and we can get its value by asking that $|\psi(x)|^2 dx$ be the *probability* of finding the particle between x and $x + dx$. Because the particle has got to be in the box, the sum of all these probabilities for different values of x must be one, that is,

$$\int |\psi_n|^2 dx = A^2 \int_0^a \sin^2\left(\frac{n\pi x}{a}\right) dx = 1 \tag{B.10}$$

This is called a **normalization condition**; evaluation of the integral leads to $A = \sqrt{2/a}$. Finally we get

$$\psi_n(x) = \sqrt{\frac{2}{a}} \sin\left(\frac{n\pi x}{a}\right) \tag{B.11}$$

These "1-D orbitals" are plotted in Figure B.3. The increase in the number of nodes (places where ψ is zero) as the energy increases is a familiar feature from orbitals in atoms. The situation is most similar to that for the radial parts of electron s orbitals. There are several other chemical applications of the particle-in-a-box model; perhaps the most significant is the analogy with molecular orbitals in

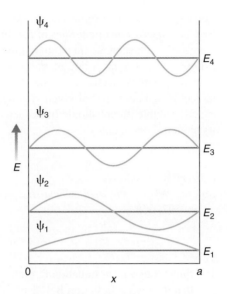

Figure B.3
Wave functions for a particle in the box.

linear and long-chain molecules. The simplest example is of course H_2, where the σ and σ^* orbitals

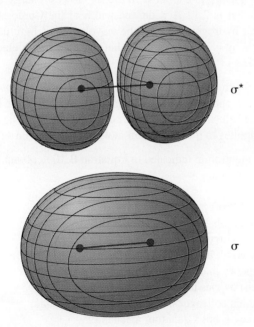

can be roughly represented by the two lowest particle-in-a-box wave functions rotated about the horizontal axis to sweep out solid shapes. Thinking of orbitals as "boxes" where electrons can be stored is not such a bad analogy after all!

The particle-in-a-box model can be extended readily to two and three dimensions, in which the wave functions grow into *products* of sine functions like that of

Equation B.11, with x replaced successively by y and z in each factor, while the energy becomes a *sum* of energies of the form of Equation B.8 in each of the added dimensions, with separate quantum numbers n_x, n_y, and n_z characterizing the x, y, and z dimensions. The most important application of the 3-D result is in treating the effects of quantum mechanics on the number of allowed translational states of a gas molecule in a container of given volume. It is indeed wondrous that such a simple, exactly soluble model should have such a wide range of applications in addition to its obvious conceptual value.

FURTHER READING
Moore, W. J., *Physical Chemistry*, 4th ed. (Prentice-Hall, 1972), Chap. 13, pp. 597–608.

EXERCISES
1. Substitute ψ from Equation B.4 into the wave equation (Equation B.3) and show that it satisfies Equation B.3 identically.

2. Show that Equation B.7 is equivalent to the standing-wave condition $n(\lambda/2) = a$, where λ is the deBroglie wavelength.

3. Calculate the spacing in eV between the two lowest energy levels for an electron in a box that is 2.12 Å wide, the width of the Coulomb potential in the ground state of hydrogen. What is the wavelength of light necessary to excite the electron? Why is it different than that required for an electron in hydrogen ($\lambda = 122$ nm)?

4. If there are N electrons in a box, the lowest $N/2$ energy levels would be filled with electron pairs. Suppose $N = 10$. What minimum width of the box is required so that a "valence" electron (in level $N/2$) can absorb visible light ($\lambda = 400$ nm)? (This model applies to the long-chain molecules in vitamin A.)

5. Carry out the integration indicated in Equation B.10 to obtain the formula given for the wave amplitude A.

Schrödinger, Erwin! Professor of physics!
   Wrote daring equations! Confounded his critics!
(Not bad, eh? Don't worry, this part of the verse
   Starts off pretty good, but it gets a lot worse.)
Win saw that the theory that Newton'd invented
   By Einstein's discov'ries had been badly dented.
What now? wailed his colleagues. Said Erwin, "Don't panic,
   No grease monkey I, but a quantum mechanic!
Consider electrons. Now these tiny articles
   Are sometimes like waves, and then sometimes like particles.
If that's not confusing, the nuclear dance
   Of electrons and suchlike is governed by chance!
No sweat, though—my theory permits us to judge
   Where some of 'em is and the rest of 'em was!"

Not everyone bought this. It threatened to wreck
    The comforting linkage of cause & effect.
E'en Einstein had doubts, and so Schrödinger tried
    To tell him what quantum mechanics implied.
Said Win to Al, "Brother, suppose we've a cat
    And into a tube we have put that cat at—
Along with a solitaire deck and some Fritos
    a bottle of Night Train, a couple mosquitos
(Or something else rhyming) and, oh, if you got 'em
    One vial prussic acid, one decaying ottom
Or atom—whatever—but when it emits,
    A trigger device blasts the vial into bits.
Which snuffs out poor kitty. The odds of this crime
    Are 50 to 50, per hour each time.
The cylinder's sealed. The hours pass away. Is
    Our pussy still breathing, or pushing up daisies?
Now you'd say the cat either lives or it don't,
    But quantum mechanics is stubborn and won't.
Statistically speaking, the cat (goes the joke)
    Is half a cat living and half a cat croaked.
To some this may seem a ridiculous split,
    But quantum mechanics must answer "Tough ----!"
We may not know much, but one thing's for sho'!
    There's things in the cosmos that we cannot know.
Shine light on electrons, you'll cause them to swerve.
    The act of observing disturbs the observed—
Which ruins your test. But then if there's no testing
    To see if a particle's moving or resting
Why try to conjecture? Pure useless endeavor!
    We know probability—certainty, never.
The effect of this notion? I very much fear
    'Twill make doubtful all things that were formerly dear.
Till soon the cat doctors will say in reports,
    "We've just flipped a coin and we've learned he's a corpse."
So said Herr Erwin. Quoth Albert, "You're nuts.
    God doesn't play dice with the universe, putz.
I'll prove it!" he said, and the Lord knows he tried—
    In vain, until finally he more or less died.
Win spoke at the funeral! "Listen, dear friends,
    Sweet Al was my buddy. I must make amends.
Though he doubted my theory, I'll say of this saint,
    Ten-to-one he's in heaven, but five bucks says he ain't!"

—Cecil Adams, *The Straight Dope*

APPENDIX C

Selected Values of Thermodynamic Properties at 298.15 K and 1 atm

Units used are kcal/mol for ΔH_f° and ΔG_f°, and cal/(K mol) for S° and $\overline{C}_P$. The table is arranged by chemical groups. To convert to SI units, in most cases all that is required is multiplication of each entry by 4.184 (J/cal or kJ/kcal). To use the SI unit of 1 bar = 100 kPa for standard pressure, instead of 1 atm, where 1 atm = 1.01325 bar, ΔG_f° and S° may need small corrections of the form

$$\Delta G_f^{\text{bar}} = \Delta G_f^\circ - 0.007799 \Delta n_{\text{gas}} \text{ kcal/mol}$$

$$S^{\text{bar}} = S^\circ + 0.02616 \text{ cal}/(\text{K mol})$$

These corrections are to be applied only when the substance is a gas (for S°) or when gases are involved in the formation reaction (for ΔG_f°); they may be neglected if the number of decimal places in the entry is too small to be affected by them, as is often the case.

The use of calories in this table is both historic and realistic: historic, because these are the units in which nearly all of the data gathered in the 20th century have been presented; and realistic, because the precision of the values is directly reflected in the number of significant figures. For example, for $\Delta H_f^\circ(H_2(aq))$, the SI entry would have to be −4.184 kJ/mol, to be exactly convertible to the experimentally determined −1.0 kcal/mol.

Species	ΔH_f°	ΔG_f°	S°	$\overline{C}_P$
Hydrogen				
$H_2(g)$	0.0	0.0	31.208	6.889
$H(g)$	52.095	48.581	27.391	4.968
$H^+(g)$	367.161			
$H^+(aq)$	0.0	0.0	0.0	
$H^-(g)$	33.39			
$H_2^+(g)$	357.23			
$H_2(aq)$	−1.0	4.2	13.8	
GROUP IA				
Lithium				
$Li(s)$	0.0	0.0	6.96	5.92
$Li(g)$	38.09	30.28	33.143	4.968
$Li^+(g)$	162.42			
$Li^+(aq)$	−66.554	−70.10	3.2	
$Li_2(g)$	47.6	37.6	47.06	8.52
$LiO(g)$	18.1	12.5	50.40	7.75
$Li_2O(s)$	−142.91	−134.13	8.98	12.93
$LiH(g)$	30.7	25.2	40.77	7.06

Selected Values of Thermodynamic Properties at 298.15 K and 1 atm

Species	ΔH_f°	ΔG_f°	S°	$\overline{C}_P$
Lithium, continued				
LiH(s)	−21.64	−16.34	4.782	6.66
LiOH(s)	−115.90	−104.92	10.23	11.87
LiF(s)	−147.22	−140.47	8.52	9.94
LiCl(s)	−97.66	−91.87	14.18	11.47
LiBr(s)	−83.942	−81.74	17.75	12.
LiI(s)	−64.63	−64.60	20.74	12.20
Li_2S	−106.9	−104.94	15.0	9.23
Li_2SO_4(s)	−343.33	−315.91	27.5	28.10
$LiNO_3$(s)	−89.0	−72.2	23.	
Li_3PO_4(s)	−500.9			
Li_2CO_3(s)	−290.54	−270.58	21.60	23.69
$LiBH_4$(s)	−45.6	−29.9	18.13	19.73
$LiAlH_4$(s)	−24.2			
Sodium				
Na(s)	0.0	0.0	12.24	6.75
Na(g)	25.65	18.354	36.712	4.968
Na^+(g)	145.55			
Na^+(aq)	−57.39	−62.593	14.1	11.1
Na_2(g)	33.97	24.85	55.02	
NaO(g)	25.	19.7	54.6	8.3
NaO_2(s)	−62.2	−52.2	27.7	17.24
Na_2O(s)	−99.00	−89.74	17.94	16.52
Na_2O_2(s)	−122.10	−107.00	22.70	21.33
NaH(g)	29.88	24.78	44.93	7.002
NaH(s)	−13.49	−8.015	9.564	8.70
NaOH(s)	−101.723	−90.709	15.405	14.23
NaF(s)	−137.105	−129.902	12.30	11.20
NaCl(s)	−98.268	−91.815	17.24	12.07
NaBr(s)	−86.296	−83.409	20.75	12.28
NaI(s)	−68.78	−68.37	23.55	12.45
$NaClO_3$(s)	−87.422	−62.697	29.5	
$NaClO_4$(s)	−91.61	−60.93	34.0	
$NaBrO_3$(s)	−79.85	−58.04	30.8	
$NaIO_3$(s)	−115.150			22.0
Na_2S(s)	−87.2	−83.6	20.0	
Na_2SO_3(s)	−263.1	−242.0	34.88	28.74
Na_2SO_4(s)	−331.52	−303.59	35.75	30.64
$Na_2SO_4 \cdot 10H_2O$(s)	−1034.24	−871.75	141.5	
$Na_2S_2O_3$(s)	−268.4	−245.7	37.	
$NaHSO_4$(s)	−269.0	−237.3	27.0	
NaN_3(s)	5.19	22.41	23.15	18.31
$NaNO_2$(s)	−85.72	−68.02	24.8	
$NaNO_3$(s)	−111.82	−87.73	27.85	22.20
$NaNH_2$(s)	−29.6	−15.3	18.38	15.31
Na_3PO_4(s)	−458.27	−427.55	41.54	36.68

Species	ΔH_f°	ΔG_f°	S°	$\overline{C}_P$
Sodium, continued				
$NaH_2PO_3(s)$	−367.3	−331.3	30.47	27.93
$Na_2HPO_3(s)$	−336.8			
$Na_2HPO_4(s)$	−417.8	−384.4	35.97	32.34
$Na_2CO_3(s)$	−270.24	−249.64	32.26	26.84
$Na_2C_2O_4(s)$	−315.0	−308.		34.
$NaHCO_3(s)$	−227.25	−203.4	24.3	20.94
$HCOONa(s)$	−159.3	−143.4	24.80	19.76
$NaOCH_3(s)$	−87.9	−70.46	26.43	16.60
$NaOC_2H_5(s)$	−98.90			
$CH_3COONa(s)$	−169.41	−145.14	29.4	19.1
$NaCN(s)$	−20.91	−18.27	27.63	16.82
$NaCNS(s)$	−40.75			
$Na_2SiO_3(s)$	−371.63	−349.19	27.21	
$Na_2SiF_6(s)$	−677.	−610.4		
$Na_2B_4O_7 \cdot 10H_2O(s)$	−1503.0	−1318.5	140.	147.
$NaBH_4(s)$	−45.08	−29.62	24.21	20.74
$NaBF_4(s)$	−440.9	−418.30	34.73	28.74
$NaAlO_2(s)$	−271.30	−256.06	16.90	17.52
$NaAl(SO_4)_2 \cdot 12H_2O$	−1434.69			
$Na_2CrO_4(s)$	−320.8	−295.17	42.21	33.97
$Na_2Cr_2O_7(s)$	−472.9			
Potassium				
$K(s)$	0.0	0.0	15.34	7.07
$K(g)$	21.33	14.49	38.295	4.968
$K^+(g)$	122.92			
$K^+(aq)$	−60.32	−67.70	24.5	5.2
$K_2(g)$	30.8	22.1	59.69	
$KO_2(s)$	−68.10	−57.23	27.9	18.53
$K_2O(s)$	−86.4	−77.14	24.38	17.79
$K_2O_2(s)$	−118.4	−102.4	26.31	22.91
$KH(s)$	−13.82	−8.132	12.	9.06
$KH(g)$	30.0	25.1	47.3	
$KOH(s)$	−101.521	−90.61	18.85	15.51
$KF(s)$	−135.58	−128.53	15.91	11.72
$KCl(s)$	−104.385	−97.79	19.74	12.26
$KBr(s)$	−94.210	−90.98	22.92	12.50
$KI(s)$	−78.370	−77.651	25.41	12.65
$KHF_2(s)$	−221.72	−205.48	24.92	18.39
$KClO_3(s)$	−95.06	−70.82	34.2	23.96
$KClO_4(s)$	−103.43	−72.46	36.1	26.86
$KBrO_3(s)$	−86.10	−64.82	35.65	28.72
$KIO_3(s)$	−119.83	−100.00	36.20	25.45
$K_2S(s)$	−90.	−86.7	27.5	17.85
$K_2SO_4(s)$	−343.64	−315.83	41.96	31.42
$KHSO_4(s)$	−277.4	−246.5	33.0	

Selected Values of Thermodynamic Properties at 298.15 K and 1 atm

Species	ΔH_f°	ΔG_f°	S°	$\overline{C}_P$
Potassium, continued				
$KNO_2(s)$	−88.39	−73.28	36.35	25.67
$KNO_3(s)$	−118.22	−94.39	31.80	23.04
$K_3PO_4(s)$	−475.2	−444.28	50.6	39.34
$K_2HPO_4(s)$	−424.42	−391.16	42.8	33.77
$KH_2PO_4(s)$	−374.84	−338.39	32.23	27.86
$K_2CO_3(s)$	−275.1	−254.2	37.17	27.35
$CH_3COOK(s)$	−172.8			
$KHCO_3(s)$	−230.2	−206.4	27.6	
$KCN(s)$	−27.0	−24.35	30.71	15.84
$KCNS(s)$	−47.84	−42.62	29.70	21.16
$K_2SiF_6(s)$	−706.5	−668.9	54.0	
$K_3Fe(CN)_6(s)$	−59.7	−31.0	101.83	
$K_4Fe(CN)_6(s)$	−142.0	−108.3	100.1	79.40
$KMnO_4(s)$	−200.1	−176.3	41.04	28.10
$K_2CrO_4(s)$	−335.5	−309.7	47.83	34.89
$K_2Cr_2O_7(s)$	−492.7	−449.8	69	52.4
Rubidium				
$Rb(s)$	0.0	0.0	18.35	7.424
$Rb(g)$	19.330	12.690	40.626	4.968
$Rb^+(g)$	117.137			
$Rb^+(aq)$	−60.03	−67.87	29.04	
$Rb_2O(s)$	−81.0	−71.72	30.	17.7
$RbH(s)$	−12.5			
$RbOH(s)$	−99.95	−87.1		
$RbF(s)$	−133.3	−126.3	19.2	12.1
$RbCl(s)$	−104.05	−97.47	22.92	12.52
$RbBr(s)$	−94.31	−91.25	26.28	12.63
$RbI(s)$	−79.78	−78.60	28.30	12.71
$Rb_2S(s)$	−86.2			
$Rb_2SO_4(s)$	−343.12	−314.76	47.19	32.04
$RbNO_3(s)$	−118.32	−94.61	35.2	24.4
$Rb_2CO_3(s)$	−271.5	−251.2	43.34	28.11
$RbHCO_3(s)$	−230.2	−206.4	29.0	
Cesium				
$Cs(s)$	0.0	0.0	20.37	7.69
$Cs(g)$	18.180	11.748	41.942	4.968
$Cs^+(g)$	109.456			
$Cs^+(aq)$	−61.73	−69.79	31.80	
$Cs_2O(s)$	−82.64	−73.65	35.10	18.16
$CsH(s)$	−12.95			
$CsOH(s)$	−99.6	−88.6	23.6	16.22
$CsF(s)$	−132.3	−125.6	22.18	12.21
$CsCl(s)$	−105.89	−99.08	24.18	12.54
$CsBr(s)$	−96.99	−93.55	27.02	12.65
$CsI(s)$	−82.84	−81.40	29.41	12.62

Species	ΔH_f°	ΔG_f°	S°	$\overline{C}_P$
Cesium, continued				
Cs_2S	−86.0			
$Cs_2SO_4(s)$	−344.89	−316.36	50.65	32.24
$CsNO_3(s)$	−120.93	−97.18	37.1	
$Cs_2CO_3(s)$	−274.2	−254.3	48.87	29.60
$CsHCO_3(s)$	−230.9			
Francium				
$Fr(s)$	0.0	0.0	22.8	
$FrF(s)$			26.	12.6
$FrCl(s)$			27.	12.8
$FrBr(s)$			31.	12.9
$FrI(s)$			33.	12.9
GROUP IIA				
Beryllium				
$Be(s)$	0.0	0.0	2.27	3.93
$Be(g)$	77.7	68.5	32.543	4.968
$Be^+(g)$	293.965			
$Be^{2+}(g)$	715.398			
$Be^{2+}(aq)$	−91.5	−90.75	−31.0	
$BeO(s)$	−145.7	−138.7	3.38	6.10
$BeH_2(s)$	−4.60			
$Be(OH)_2(s)$	−215.7	−194.8	12.4	15.7
$BeF_2(s)$	−245.4	−234.1	12.75	12.39
$BeCl_2(s)$	−117.2	−106.5	19.76	15.50
$BeBr_2(s)$	−85.0	−80.64	24.	15.79
$BeI_2(s)$	−45.1	−44.74	28.8	16.48
$BeS(s)$	−56.0	−55.7	8.85	8.14
$BeSO_4(s)$	−288.05	−261.44	18.62	20.48
$Be_3N_2(s)$	−133.4	−127.36	8.157	15.43
$Be_2C(s)$	−27.96	−27.37	3.9	10.34
$BeCO_3(s)$	−245.			
Magnesium				
$Mg(s)$	0.0	0.0	7.81	5.95
$Mg(g)$	35.30	27.04	35.502	4.968
$Mg^+(g)$	213.100			
$Mg^{2+}(g)$	561.299			
$Mg^{2+}(aq)$	−111.58	−108.7	−33.0	
$MgO(s)$	−143.81	−136.10	6.44	8.88
$MgH_2(s)$	−18.0	−8.6	7.43	8.45
$Mg(OH)_2(s)$	−220.97	−199.23	15.10	18.41
$MgF_2(s)$	−268.5	−255.8	13.68	14.72
$MgCl_2(s)$	−153.28	−141.45	21.42	17.06
$MgBr_2(s)$	−125.3	−120.4	28.0	14.03
$MgI_2(s)$	−87.0	−85.6	31.0	17.89
$MgS(s)$	−82.7	−81.7	12.03	10.19

Selected Values of Thermodynamic Properties at 298.15 K and 1 atm

Species	ΔH_f°	ΔG_f°	S°	$\overline{C}_P$
Magnesium, continued				
$MgSO_4(s)$	−307.1	−279.8	21.9	23.06
$MgSO_4 \cdot 7H_2O(s)$	−809.92	−686.4	89.	
$Mg_3N_2(s)$	−110.1	−95.7	21.	25.
$Mg(NO_3)_2(s)$	−188.97	−140.9	39.2	33.92
$Mg_3(PO_4)_2(s)$	−903.6	−845.8	45.22	51.02
$MgC_2(s)$	21.	20.3	13.	13.44
$MgCO_3(s)$	−261.9	−241.9	15.7	18.05
$MgC_2O_4 \cdot 2H_2O(s)$		−375.9		
$Mg_2Si(s)$	−18.6	−18.0	18.	17.6
$MgSiO_3(s)$	−370.22	−349.46	16.19	19.45
Calcium				
$Ca(s)$	0.0	0.0	9.90	6.05
$Ca(g)$	42.6	34.5	36.992	4.968
$Ca^+(g)$	185.05			
$Ca^{2+}(g)$	460.29			
$Ca^{2+}(aq)$	−129.74	−132.30	−12.7	
$CaO(s)$	−151.79	−144.37	9.50	10.23
$CaH_2(s)$	−42.3	−33.0	9.90	9.80
$Ca(OH)_2(s)$	−235.68	−214.74	19.9	20.91
$CaF_2(s)$	−291.5	−279.0	16.46	16.02
$CaCl_2(s)$	−190.2	−178.8	25.0	17.35
$CaCl_2 \cdot 2H_2O(s)$	−335.3			
$Ca(ClO_4)_2$	−176.09			
$CaBr_2(s)$	−163.2	−158.6	31.	18.0
$CaI_2(s)$	−127.5	−126.4	34.	18.4
$CaS(s)$	−115.3	−114.1	13.5	11.33
$CaSO_3(s)$	−277.1	−257.2	24.23	21.92
$CaSO_4(s)$	−342.6	−315.93	25.5	23.82
$CaSO_4 \cdot \frac{1}{2}H_2O(s)$	−376.85	−343.41	31.2	28.54
$CaSO_4 \cdot 2H_2O$	−483.42	−429.60	46.4	44.46
$CaSe(s)$	−88.0	−86.8	16.	11.5
$Ca(N_3)_2(s)$	3.5			
$Ca_3N_2(s)$	−103.2	−88.1	25	22.5
$Ca(NO_3)_2(s)$	−224.28	−177.63	46.1	35.70
$Ca(NO_3)_2 \cdot 4H_2O$	−509.64	−409.53	89.7	
$Ca_3P_2(s)$	−121.	−115.	29.6	27.8
$Ca_3(PO_4)_2(s)$	−984.9	−928.5	56.4	54.45
$CaHPO_4(s)$	−433.65	−401.83	26.62	26.30
$Ca(H_2PO_4)_2(s)$	−742.04			
$CaC_2(s)$	−14.3	−15.5	16.72	14.99
$CaCO_3(s, calcite)$	−288.46	−269.80	22.2	19.57
$CaC_2O_4 \cdot H_2O(s)$	−400.30	−361.85	37.4	36.52
$Ca(CN)_2(s)$	−44.1			
$CaSiO_3(s)$	−390.76	−370.39	19.58	20.38

Appendix C

Species	ΔH_f°	ΔG_f°	S°	$\overline{C}_P$
Strontium				
Sr(s)	0.0	0.0	12.5	6.3
Sr(g)	39.3	31.3	39.32	4.968
Sr^+(g)	172.11			
Sr^{2+}(g)	427.96			
Sr^{2+}(aq)	−130.45	−133.71	−7.8	
SrO(s)	−141.5	−134.3	13.0	10.76
SrO_2(s)	−151.4	−137.	14.1	16.5
SrH_2(s)	−43.1	−33.4	11.9	9.61
$Sr(OH)_2$(s)	−231.5	−210.6	23.2	17.9
SrF_2(s)	−290.7	−278.4	19.63	16.73
$SrCl_2$(s)	−198.1	−186.7	27.45	18.07
$SrBr_2$(s)	−171.5	−166.6	32.29	18.01
SrI_2(s)	−134.2	−133.3	38.03	18.63
SrS(s)	−112.9	−111.8	16.3	11.64
$SrSO_4$(s)	−347.3	−320.5	29.1	24.3
Sr_3N_2(s)	−93.5	−76.7	29.5	31.0
$Sr(NO_3)_2$(s)	−182.3	−186.		
Sr_3P_2(s)	−152.			
$Sr_3(PO_4)_2$(s)	−985.4			
SrC_2(s)	−18.			
$SrCO_3$(s)	−291.6	−272.5	23.2	19.46
SrC_2O_4(s)	−327.6			
Barium				
Ba(s)	0.0	0.0	15.0	6.71
Ba(g)	43.	35.	40.663	4.968
Ba^+(g)	164.67			
Ba^{2+}(g)	396.86			
Ba^{2+}(aq)	−128.50	−134.02	2.3	
BaO(s)	−133.4	−126.3	16.8	11.34
BaO_2(s)	−151.6	−139.2	22.25	16.1
BaH_2(s)	−45.43	−36.16	15.06	11.0
$Ba(OH)_2$(s)	−226.2	−205.4	25.6	24.3
$Ba(OH)_2 \cdot 8H_2O$(s)	−718.6			
BaF_2(s)	−288.5	−276.5	23.03	17.02
$BaCl_2$(s)	−205.2	−193.7	29.56	17.96
$BaBr_2$(s)	−181.0	−176.1	35.5	18.4
BaI_2(s)	−144.7	−143.7	39.47	18.5
BaS(s)	−110.	−109.	18.7	11.80
$BaSO_4$(s)	−352.1	−325.6	31.6	24.32
BaSe(s)	−89.			
$Ba(NO_3)_2$(s)	−237.11	−190.42	51.1	36.18
BaC_2(s)	−17.92	−18.92	21.0	15.27
$BaCO_3$(s)	−290.7	−271.9	26.8	20.40
$(CH_3COO)_2Ba$	−354.8			

Selected Values of Thermodynamic Properties at 298.15 K and 1 atm

Species	ΔH_f°	ΔG_f°	S°	$\overline{C}_P$
Barium, continued				
$BaSiO_3(s)$	−388.05	−368.13	26.2	21.51
$BaCrO_4(s)$	−345.6	−321.53	37.9	28.73
Radium				
$Ra(s)$	0.0	0.0	17.	
$Ra(g)$	38.	31.	42.15	4.97
$Ra^+(g)$	161.22			
$Ra^{2+}(g)$	396.70			
$Ra^{2+}(aq)$	−126.1	−134.2	13.	
$RaO(s)$	−125.			
$RaCl_2 \cdot 2H_2O(s)$	−350.	−311.4	51.	
$Ra(IO_3)_2$	−245.4	−207.6	65.	
$RaSO_4(s)$	−351.6	−326.4	33.	
$Ra(NO_3)_2(s)$	−237.	−190.3	53.	
GROUP IIIB				
Scandium				
$Sc(s)$	0.0	0.0	8.28	6.10
$Sc(g)$	90.3	80.32	41.75	5.28
$Sc^{3+}(aq)$	−146.8	−140.2	−61.	
$Sc_2O_3(s)$	−456.22	−434.85	18.4	22.52
$Sc(OH)_3(s)$	−325.9	−294.8	24.	
$ScCl_3(s)$	−221.1	−203.8	30.9	21.9
Yttrium				
$Y(s)$	0.0	0.0	10.62	6.34
$Y(g)$	100.7	91.1	42.87	6.18
$Y^{3+}(aq)$	−161.2	−153.9	−57.	
$Y_2O_3(s)$	−433.7	−412.7	31.8	27.57
$YCl_3(s)$	−229.0	−221.7	32.7	22.0
Lanthanum				
$La(s)$	0.0	0.0	13.6	6.48
$La(g)$	103.0	94.07	43.563	5.438
$La^{3+}(aq)$	−169.0	−163.4	−52.0	
$La_2O_3(s)$	−428.7	−407.7	30.43	26.00
$LaCl_3(s)$	−256.0			
$LaCl_3 \cdot 7H_2O$	−759.7	−648.5	110.6	103.0
GROUP IVB				
Titanium				
$Ti(s)$	0.0	0.0	7.32	5.98
$Ti(g)$	112.3	101.6	43.066	5.893
$TiO^{2+}(aq)$	−164.9			
$TiO_2(s, rutile)$	−225.8	−212.6	12.03	13.15
$TiCl_4(l)$	−192.2	−176.2	60.31	34.70

Species	ΔH_f°	ΔG_f°	S°	$\overline{C}_P$
Zirconium				
$Zr(s)$	0.0	0.0	9.32	6.06
$Zr(g)$	145.5	135.4	43.32	6.37
$ZrO_2(s)$	−263.04	−249.24	12.04	13.43
$ZrO(OH)^+(aq)$	−270.1			
$ZrCl_4(s)$	−234.35	−212.7	43.4	28.63
GROUP VB				
Vanadium				
$V(s)$	0.0	0.0	6.91	5.95
$V(g)$	122.90	108.32	43.544	6.217
$VO^{2+}(aq)$	−116.3	−106.7	−32.0	
$VO_2^+(aq)$	−155.3	−140.3		
$V_2O_5(s)$	−370.6	−339.3	31.3	30.51
$HVO_4^{2-}(aq)$	−227.0	−233.0	4.	
$VCl_3(s)$	−138.8	−122.2	31.3	22.27
Tantalum				
$Ta(s)$	0.0	0.0	9.92	6.06
$Ta(g)$	186.9	176.7	44.241	4.985
$Ta_2O_5(s)$	−489.0	−456.8	34.2	32.30
$TaO_2^+(aq)$		−201.4		
$TaCl_5(s)$	−205.3	−178.4	53.0	35.35
GROUP VIB				
Chromium				
$Cr(s)$	0.0	0.0	5.68	5.58
$Cr(g)$	94.8	84.1	41.68	4.97
$Cr^{2+}(aq)$		−34.3		
$Cr^{3+}(aq)$		−51.5	−73.5	
$Cr_2O_3(s)$	−272.4	−252.9	19.4	28.38
$CrO_3(s)$	−140.9	−122.5	17.5	16.57
$HCrO_4^-(aq)$	−209.9	−182.8	44.0	
$CrO_4^{2-}(aq)$	−210.60	−173.96	12.00	
$Cr_2O_7^{2-}(aq)$	−356.2	−311.0	62.6	
$Cr(OH)_3(s)$	−254.3	−215.3	19.2	
$CrCl_2(s)$	−94.5	−85.1	27.56	17.01
$CrCl_3(s)$	−133.0	−116.2	29.4	21.94
$CrO_2Cl_2(l)$	−138.5	−122.1	53.0	
Molybdenum				
$Mo(s)$	0.0	0.0	6.85	5.75
$Mo(g)$	157.3	146.4	43.461	4.968
$MoO_3(s)$	−178.08	−159.66	18.58	17.92
$MoO_4^{2-}(aq)$	−238.5	−199.9	6.5	
$MoS_2(s)$	−56.2	−54.0	14.96	15.19
$Mo(CO)_6(s)$	−234.9	−209.8	77.9	57.90

Selected Values of Thermodynamic Properties at 298.15 K and 1 atm

Species	ΔH_f°	ΔG_f°	S°	$\overline{C}_P$
Tungsten				
W(s)	0.0	0.0	7.80	5.80
W(g)	203.0	192.9	41.549	5.093
$WO_3(s)$	−201.45	−182.62	18.14	17.63
$WO_4^{2-}(aq)$	−257.1			
$W(CO)_6(s)$	−227.9			
GROUP VIIB				
Manganese				
Mn(s)	0.0	0.0	7.65	6.29
Mn(g)	67.1	57.0	41.49	4.97
$Mn^{2+}(aq)$	−52.76	−54.5	−17.6	
MnO(s)	−92.07	−86.74	14.27	10.86
$MnO_2(s)$	−124.29	−111.18	12.68	12.94
$MnO_4^-(aq)$	−123.9	−101.6	45.4	
$MnO_4^{2-}(aq)$	−156.	−119.7	14.	
$Mn_2O_3(s)$	−229.2	−210.6	26.4	25.73
$Mn_3O_4(s)$	−331.7	−306.7	37.2	33.38
$Mn(OH)_2(s)$	−166.2	−147.0	23.7	
$Mn(OH)_3(s)$	−212.	−181.	23.8	
$MnCl_2 \cdot 4H_2O(s)$	−403.3	−340.3	72.5	
MnS(s)	−51.2	−52.2	18.7	11.94
$MnSO_4(s)$	−254.60	−228.83	26.8	24.02
$MnCO_3(s)$	−213.7	−195.2	20.5	19.48
Rhenium				
Re(s)	0.0	0.0	8.81	6.09
Re(g)	184.0	173.2	45.131	4.968
$ReO_4^-(aq)$	−188.2	−166.	48.1	
$ReCl_3(s)$	−63.	−45.	29.6	22.08
Iron				
Fe(s)	0.0	0.0	6.52	6.00
Fe(g)	99.5	88.6	43.112	6.137
$Fe^{2+}(aq)$	−21.3	−18.85	−32.9	
$Fe^{3+}(aq)$	−11.6	−1.1	−75.5	
$Fe_2O_3(s, hematite)$	−197.0	−177.4	20.89	24.82
$Fe_3O_4(s, magnetite)$	−267.3	−242.7	35.0	34.28
$Fe(OH)_2(s)$	−136.0	−116.3	21.	
$Fe(OH)_3(s)$	−196.7	−166.5	25.5	
$FeCl_3(s)$	−95.48	−79.84	34.0	23.10
$FeCl_3 \cdot 6H_2O(s)$	−531.5			
FeS(s)	−23.9	−24.0	14.41	12.08
$FeS_2(s)$	−42.6	−39.9	12.65	14.86
$FeSO_4(s)$	−221.9	−196.2	25.7	24.04
$FeSO_4 \cdot 7H_2O(s)$	−720.50	−599.97	97.8	94.28
$FePO_4 \cdot 2H_2O(s)$	−451.3	−396.2	40.93	43.15
$FeCO_3(s)$	−177.00	−159.35	22.2	19.63

Species	$\Delta H_f°$	$\Delta G_f°$	$S°$	$\overline{C}_P$
Iron, continued				
$Fe(CO)_5(l)$	−185.0	−168.6	80.8	57.5
$Fe(CN)_6^{3-}(aq)$		178.7		
$Fe(CN)_6^{4-}(aq)$	126.7	170.4		
$Fe(SCN)^{2+}(aq)$	5.6	17.0	−31.	
$Fe_2SiO_4(s)$	−353.7	−329.6	34.7	31.76
Ruthenium				
$Ru(s)$	0.0	0.0	6.82	5.75
$Ru(g)$	153.6	142.4	44.550	5.144
$RuO_2(s)$	−72.9	−60.4	13.9	13.5
$RuCl_3(s)$	−55.0	−38.2	30.4	22.0
Osmium				
$Os(s)$	0.0	0.0	7.8	5.9
$Os(g)$	189.	178.	46.000	4.968
$OsO_4(s)$	−94.2	−72.9	34.4	19.1
$OsCl_3(s)$	−45.5			
Cobalt				
$Co(s, hexagonal)$	0.0	0.0	7.18	5.93
$Co(s, face\text{-}centered\ cubic)$	0.11	0.06	7.34	
$Co(g)$	101.5	90.9	42.879	5.502
$Co^{2+}(aq)$	−13.9	−13.0	−27.	
$Co^{3+}(aq)$	22.	32.	−73.	
$Co_3O_4(s)$	−213.	−185.	24.5	29.5
$Co(OH)_2(s)$	−129.0	−108.6	19.	36.5
$Co(OH)_3(s)$	−171.3	−142.6	20.	
$CoCl_2(s)$	−74.7	−64.5	26.09	18.76
$CoCl_2 \cdot 6H_2O(s)$	−505.6	−412.4	82.	
$CoS(s)$	−19.3	−19.8	16.	
$CoSO_4(s)$	−212.3	−187.0	28.2	24.7
$CoSO_4 \cdot 6H_2O(s)$	−641.4	−534.35	87.86	84.46
$Co(NO_3)_2(s)$	−100.5	−55.1	46.	
$Co(NH_3)_6^{2+}(aq)$		−57.7		
$Co(NH_3)_6^{3+}(aq)$	−139.8	−38.9	40.0	
$Co(NH_3)_5Cl^{2+}(aq)$	−162.1	−86.2	96.1	
$Co_3(PO_4)_2(s)$		−573.3		
Rhodium				
$Rh(s)$	0.0	0.0	7.53	5.97
$Rh(g)$	133.1	122.1	44.383	5.022
$Rh_2O_3(s)$	−85.	−66.	25.4	24.8
$RhCl_3(s)$	−71.51	−54.45	30.3	22.0
$RhCl_6^{3-}(aq)$	−202.8			
Iridium				
$Ir(s)$	0.0	0.0	8.48	6.00
$Ir(g)$	159.0	147.7	46.240	4.968

Selected Values of Thermodynamic Properties at 298.15 K and 1 atm

Species	ΔH_f°	ΔG_f°	S°	$\overline{C}_P$
Iridium, continued				
$IrO_2(s)$	−58.	−45.	14.	14.3
$IrF_6(s)$	−138.54	−110.34	59.2	
$IrCl_3(s)$	−58.7	−40.5	27.45	20.52
$IrCl_6^{3-}(aq)$	−176.			
Nickel				
$Ni(s)$	0.0	0.0	7.14	6.13
$Ni(g)$	102.7	91.9	43.519	5.583
$Ni^{2+}(aq)$	−12.9	−10.9	−30.8	
$NiO(s)$	−57.3	−50.6	9.08	10.59
$Ni(OH)_2(s)$	−126.6	−106.9	21.	
$NiF_2(s)$	−157.2	−145.9	17.59	15.30
$NiCl_2(s)$	−72.976	−61.918	23.34	17.13
$NiCl_2 \cdot 6H_2O(s)$	−502.67	−409.54	82.3	
$NiS(s)$	−19.6	−19.0	12.66	11.26
$NiSO_4(s)$	−208.63	−181.6	22.	33.
$NiSO_4 \cdot 6H_2O(s)$	−641.21	−531.78	79.94	78.36
$Ni_3(PO_4)_2(s)$		−562.4		
$NiCO_3(s)$	−166.0	−146.4	20.6	20.6
$Ni(CO)_4(l)$	−151.3	−140.6	74.9	48.9
Palladium				
$Pd(s)$	0.0	0.0	8.98	6.21
$Pd(g)$	90.4	81.2	39.90	4.968
$Pd^{2+}(aq)$	35.6	42.2	−44.	
$Pd_2H(s)$	−4.7			
$PdO(s)$	−27.6	−20.4	9.3	7.5
$PdCl_2(s)$	−47.5	−36.3	24.9	18.0
$PdCl_4^{2-}(aq)$	−131.5	−99.7	40.	
$PdCl_6^{2-}(aq)$	−143.	−102.8	65.	
$PdS(s)$	−16.9	−15.95	13.16	10.37
Platinum				
$Pt(s)$	0.0	0.0	9.95	6.18
$Pt(g)$	135.1	124.4	45.96	6.102
$Pt_3O_4(s)$	−39.			
$PtCl_2(s)$	−25.5	−22.3	52.49	18.02
$PtCl_4^{2-}(aq)$	−119.3	−86.4	37.	
$PtCl_6^{2-}(aq)$	−159.7	−115.4	52.5	
cis-$Pt(NH_3)_2Cl_2(aq)$		−54.66		
trans-$Pt(NH_3)_2Cl_2(aq)$		−53.25		
$PtS(s)$	−19.5	−18.2	13.16	10.37
GROUP IB				
Copper				
$Cu(s)$	0.0	0.0	7.923	5.840
$Cu(g)$	80.86	71.37	39.74	4.968

Species	ΔH_f°	ΔG_f°	S°	$\overline{C}_P$
Copper, continued				
$Cu^+(aq)$	17.13	11.95	9.7	
$Cu^{2+}(aq)$	15.48	15.66	−23.8	
$Cu_2(g)$	115.72	103.24	57.71	8.75
$CuO(s)$	−37.6	−31.0	10.19	10.11
$Cu_2O(s)$	−40.3	−34.9	22.26	15.21
$CuH(s)$	5.1			
$Cu(OH)_2(s)$	−105.9	−85.8	20.8	21.0
$CuCl(s)$	−32.8	−28.65	20.6	11.6
$CuCl_2 \cdot 2H_2O(s)$	−196.3	−156.8	40.	
$CuBr(s)$	−25.0	−24.1	22.97	13.08
$CuI(s)$	−16.2	−16.6	23.1	12.92
$CuS(s)$	−12.7	−12.8	15.9	11.43
$CuSO_4(s)$	−184.36	−158.2	26.	23.9
$CuSO_4 \cdot 5H_2O(s)$	−544.85	−449.344	71.8	67.
$Cu(NH_3)_4^{2+}(aq)$	−83.3	−26.60	65.4	
$Cu_3(PO_4)_2(s)$		−490.3		
$CuCO_3(s)$	−142.2	−123.8	21.	
$CuC_2O_4(s)$		−158.2		
Silver				
$Ag(s)$	0.0	0.0	10.17	6.059
$Ag(g)$	68.01	58.72	41.321	4.968
$Ag^+(aq)$	25.234	18.433	17.37	
$Ag_2O(s)$	−7.42	−2.68	29.0	15.74
$AgOH(s)$		−29.7		
$AgF(s)$	−48.9	−45.6	20.	12.4
$AgCl(s)$	−30.370	−26.244	23.0	12.14
$AgBr(s)$	−23.99	−23.16	25.6	12.52
$AgI(s)$	−14.78	−15.82	27.6	13.58
$Ag_2S(s)$	−7.79	−9.72	34.42	18.29
$Ag_2SO_4(s)$	−171.10	−147.82	47.9	31.40
$AgNO_3(s)$	−29.73	−8.00	33.68	22.24
$Ag(NH_3)_2^+(aq)$	−26.60	−4.12	58.6	
$Ag_3PO_4(s)$		−210.		
$AgCN(s)$	34.9	37.5	25.62	15.95
$Ag_2CO_3(s)$	−120.97	−104.48	40.0	26.8
$Ag_2C_2O_4(s)$	−159.1	−139.6	48.0	
$CH_3COOAg(s)$	−93.41	−74.2	33.8	
$Ag_2CrO_4(s)$	−176.2	−153.4	51.8	34.0
Gold				
$Au(s)$	0.0	0.0	11.33	6.075
$Au(g)$	87.5	78.0	43.115	4.968
$AuO_3^{3-}(aq)$		−12.4		
$Au(OH)_3(s)$	−101.5	−75.77	45.3	22.5
$AuCl(s)$	−8.7	−3.5	20.5	11.65

Selected Values of Thermodynamic Properties at 298.15 K and 1 atm

Species	ΔH_f°	ΔG_f°	S°	$\overline{C}_P$
Gold, continued				
$AuCl_3(s)$	−28.1	−11.4	35.4	22.66
$AuCl_4^-(aq)$	−77.0	−56.22	63.8	
$AuBr_4^-(aq)$	−45.8	−40.0	80.3	
$Au(CN)_2^-(aq)$	57.9	68.3	41.	
GROUP IIB				
Zinc				
$Zn(s)$	0.0	0.0	9.95	6.07
$Zn(g)$	31.245	22.748	38.450	4.968
$Zn^{2+}(aq)$	−36.78	−35.14	−26.8	
$ZnO(s)$	−83.24	−76.08	10.43	9.62
$Zn(OH)_2(s)$	−153.74	−132.68	19.5	17.3
$Zn(OH)_4^{2-}(aq)$		−205.23		
$ZnF_2(s)$	−182.7	−170.5	17.61	15.69
$ZnCl_2(s)$	−99.20	−88.296	26.64	17.05
$ZnBr_2(s)$	−78.55	−74.60	33.1	15.7
$ZnI_2(s)$	−49.72	−49.94	38.5	15.7
$ZnS(s)$	−49.23	−48.11	13.8	11.0
$ZnSO_4(s)$	−234.9	−209.0	28.6	23.7
$ZnSO_4 \cdot 7H_2O(s)$	−735.60	−612.59	92.9	91.64
$Zn(NO_3)_2(s)$	−115.6			
$Zn(NO_3)_2 \cdot 6H_2O(s)$	−551.30	−423.79	109.2	77.2
$Zn_3(PO_4)_2(s)$	−691.3			
$ZnCO_3(s)$	−194.26	−174.85	19.7	19.05
$ZnC_2O_4 \cdot 2H_2O(s)$		−321.7		
$Zn(CH_3)_2(l)$	5.6			
$Zn(C_2H_5)_2(l)$	2.5			
Cadmium				
$Cd(s)$	0.0	0.0	12.37	6.21
$Cd(g)$	26.77	18.51	40.066	4.968
$Cd^{2+}(aq)$	−18.14	−18.542	−17.5	
$CdO(s)$	−61.7	−54.6	13.1	10.38
$Cd(OH)_2(s)$	−134.2	−113.2	23.	
$CdCl_2(s)$	−93.57	−82.21	27.55	17.85
$CdBr_2(s)$	−75.57	−70.82	32.8	18.32
$CdI_2(s)$	−48.6	−48.13	38.5	19.11
$CdS(s)$	−38.7	−37.4	15.5	11.6
$CdSO_4(s)$	−223.06	−196.65	29.407	23.80
$CdSe(s)$	−34.6	−33.83	19.9	11.85
$CdTe(s)$	−22.1	−22.0	24.	11.92
$Cd(NH_3)_4^{2+}(aq)$	−107.6	−54.1	80.4	
$Cd_3(PO_4)_2(s)$		−587.1		
$CdCO_3(s)$	−179.4	−160.0	22.1	
$CdC_2O_4 \cdot 3H_2O(s)$		−360.4		

Species	ΔH_f°	ΔG_f°	S°	$\overline{C}_P$
Mercury				
$Hg(l)$	0.0	0.0	18.17	6.688
$Hg(g)$	14.655	7.613	41.79	4.968
$Hg^{2+}(aq)$	40.9	39.30	−7.7	
$Hg_2^{2+}(aq)$	41.2	36.70	20.2	
$HgO(s)$	−21.71	−13.995	16.80	10.53
$Hg(OH)_2(s)$		−70.68		
$Hg_2F_2(s)$		−104.1		
$HgCl_2(s)$	−53.6	−42.7	34.9	17.7
$Hg_2Cl_2(s)$	−63.39	−50.377	46.0	24.4
$HgBr_2(s)$	−40.8	−36.6	41.	18.
$Hg_2Br_2(s)$	−49.45	−43.278	52.	25.
$HgI_2(s)$	−25.2	−24.3	43.	18.6
$Hg_2I_2(s)$	−29.00	−26.53	55.8	25.3
$HgS(s, red)$	−13.9	−12.1	19.7	11.57
$HgS(s, black)$	−12.8	−11.4	21.1	
$Hg_2SO_4(s)$	−177.61	−149.589	47.96	31.54
$Hg_2CO_3(s)$	−132.3	−111.9	43.	
$Hg_2C_2O_4(s)$		−141.8		
$Hg(CH_3)_2(l)$	14.3	33.5	50.	
$Hg(C_2H_5)_2(l)$	7.2			
$Hg(CN)_2(s)$	63.0	74.3	27.4	
$Hg(CN)_4^{2-}(aq)$	125.8	147.8	73.	
$Hg_2CrO_4(s)$		−155.75		
GROUP IIIA				
Boron				
$B(s)$	0.0	0.0	1.40	2.65
$B(g)$	134.5	124.0	36.65	4.971
$BO_2^-(aq)$	−184.60	−162.27	−8.9	
$B_2O_3(s)$	−304.20	−285.30	12.90	15.04
$BH_4^-(aq)$	11.51	27.31	26.4	
$B_2H_6(g)$	8.5	20.7	55.45	13.60
$H_3BO_3(s)$	−261.55	−231.60	21.23	19.45
$H_3BO_3(aq)$	−256.29	−231.56	38.8	
$H_2BO_3^-(aq)$	−251.8	−217.6	7.3	
$B(OH)_4^-(aq)$	−321.23	−275.65	24.5	
$H_2B_4O_7(aq)$		−650.1		
$B_4O_7^{2-}(aq)$		−622.6		
$BF_3(g)$	−271.75	−267.77	60.71	12.06
$BF_4^-(aq)$	−376.4	−355.4	43.	
$BCl_3(l)$	−102.1	−92.6	49.3	25.5
$BCl_3(g)$	−96.50	−92.91	69.31	14.99
$BBr_3(l)$	−57.3	−57.0	54.9	30.6
$BBr_3(g)$	−49.15	−55.56	77.47	16.20
$BI_3(g)$	17.00	4.96	83.43	16.92

Selected Values of Thermodynamic Properties at 298.15 K and 1 atm

Species	ΔH_f°	ΔG_f°	S°	$\overline{C}_P$
Boron, continued				
BS(g)	81.74	69.02	51.65	7.18
BN(s)	−60.8	−54.6	3.54	4.71
B_4C_3(s)	−17.	−17.	6.48	12.62
$B(CH_3)_3$(l)	−34.2	−7.7	57.1	
$B(CH_3)_3$(g)	−29.7	−8.6	75.2	21.15
Aluminum				
Al(s)	0.0	0.0	6.77	5.82
Al(g)	78.0	68.3	39.30	5.11
Al^{3+}(aq)	−127.	−116.	−76.9	
Al_2(g)	116.14	103.57	55.7	8.7
AlO_2^-(aq)	−219.6	−196.7	−5.	
Al_2O_3(s, corundum)	−400.5	−378.2	12.17	18.89
AlH_3(s)	−2.73	11.11	7.18	9.64
$Al(OH)_3$(s)	−304.9	−271.9	17.	
AlF_3(s)	−359.5	−340.6	15.88	17.95
$AlCl_3$(s)	−168.3	−150.3	26.45	21.95
$AlCl_3 \cdot 6H_2O$(s)	−643.3	−540.4	76.	70.8
$AlBr_3$(s)	−122.2	−116.8	43.	24.0
AlI_3(s)	−75.0	−71.9	38.	23.6
Al_2S_3(s)	−173.	−170.5	27.9	27.0
$Al_2(SO_4)_3$(s)	−822.38	−740.95	57.2	62.0
$Al_2(SO_4)_3 \cdot 6H_2O$(s)	−1269.53	−1104.82	112.1	117.8
AlN(s)	−76.0	−68.6	4.82	7.20
$Al(NO_3)_3 \cdot 6H_2O$(s)	−681.28	−526.74	111.8	103.5
AlP(s)	−39.3	−37.7	11.3	10.
$AlPO_4$(s)	−414.4	−386.7	21.70	22.27
AlAs(s)	−27.8	−27.5	14.4	10.95
Al_4C_3(s)	−49.9	−46.9	21.26	27.91
$Al(CH_3)_3$(l)	−32.6	−2.4	50.05	37.19
$Al(CH_3)_3$(g)	−17.7			
Gallium				
Ga(s)	0.0	0.0	9.77	6.18
Ga(l)	1.33			
Ga(g)	66.2	57.1	40.38	6.06
Ga^{3+}(aq)	−50.6	−38.0	−79.	
Ga_2O_3(s)	−260.3	−238.6	20.31	22.00
$Ga(OH)_3$(s)	−230.5	−198.7	24.	
$GaCl_3$(s)	−125.4	−108.7	34.	17.9
Ga_2S_3(s)	−123.4	−120.9	34.	25.
GaN(s)	−26.2	−18.6	7.1	9.74
GaP(s)	−24.	−21.8	12.28	10.54
$GaPO_4$(s)		−310.1		
GaAs(s)	−17.	−16.2	15.34	11.05
GaSb(s)	−10.0	−9.3	18.18	11.60

Appendix C

Species	ΔH_f°	ΔG_f°	S°	$\overline{C}_P$
Indium				
In(s)	0.0	0.0	13.82	6.39
In(g)	58.15	49.89	41.51	4.98
$In^{3+}(aq)$	−25.	−23.4	−36.	
$In_2O_3(s)$	−221.27	−198.55	24.9	22.
InS(s)	−33.0	−31.5	16.	11.5
$In_2S_3(s)$	−102.	−98.6	39.1	28.20
InN(s)	−4.1	3.75	10.4	9.96
InP(s)	−21.2	−18.4	14.3	10.86
InAs(s)	−14.0	−12.8	18.1	11.42
InSb(s)	−7.3	−6.1	20.6	11.82
Thallium				
Tl(s)	0.0	0.0	15.45	6.29
Tl(g)	43.55	35.24	43.225	4.968
$Tl^+(aq)$	1.28	−7.74	30.0	
$Tl^{3+}(aq)$	47.0	51.3	−46.	
$Tl_2O(s)$	−40.4	−34.3	34.7	18.85
$Tl_2O_3(s)$	−94.3	−74.5	38.	25.2
TlOH(s)	−57.1	−46.8	21.	
TlCl(s)	−48.79	−44.20	26.59	12.17
$TlCl_3(s)$	−75.3	−57.7	36.4	26.0
TlBr(s)	−41.4	−40.00	28.8	12.07
TlI(s)	−29.6	−29.97	30.5	12.55
$Tl_2S(s)$	−23.2	−22.4	36.	19.19
$Tl_2CO_3(s)$	−167.3	−146.9	37.1	
GROUP IVA				
Carbon				
C(s, graphite)	0.0	0.0	1.372	2.038
C(s, diamond)	0.4533	0.6930	0.568	1.4615
C(s, buckminsterfullerene)	2.032			1.49
C(g)	171.291	160.442	37.7597	4.9805
CO(g)	−26.416	−32.780	47.219	6.959
$CO_2(g)$	−94.052	−94.254	51.06	8.87
$CO_2(aq)$	−98.90	−92.26	28.1	
CH(g)	142.4	134.	43.75	6.976
$CH_2(g)$	92.35	88.25	46.35	8.269
$CH_3(g)$	34.82	35.36	46.41	9.250
$CH_4(g)$	−17.88	−12.13	44.492	8.439
$CH_4(aq)$	−21.28	−8.22	20.0	
$C_2H_2(g)$	54.54	50.0	47.997	10.499
$C_2H_4(g)$	12.496	16.282	52.45	10.41
$C_2H_5(g)$	25.7	31.8	60.2	11.16
$C_2H_6(g)$	−20.236	−7.860	54.85	12.585
$C_3H_6(g, propylene)$	4.88	15.0	63.83	15.265
$C_3H_6(g, cyclopropane)$	12.72	24.9	56.89	21.56

Selected Values of Thermodynamic Properties at 298.15 K and 1 atm

Species	ΔH_f°	ΔG_f°	S°	$\overline{C}_P$
Carbon, continued				
$C_3H_8(g)$	−24.821	−5.614	64.51	17.44
$n\text{-}C_4H_{10}(g)$	−29.811	−3.755	74.10	23.30
$i\text{-}C_4H_{10}(g)$	−31.453	−4.295	70.41	23.13
$n\text{-}C_8H_{18}(l)$	−49.77	3.92	111.67	44.88
$i\text{-}C_8H_{18}(l)$	−53.77	3.07	101.09	45.03
$C_6H_6(l)$	11.718	29.756	41.30	32.53
$C_6H_6(g)$	19.821	30.99	64.34	19.725
$HCO(g)$	−4.12	−7.76	53.68	8.26
$HCHO(g)$	−28.	−27.	52.26	8.46
$HCOOH(l)$	−101.51	−86.38	30.82	23.67
$HCOOH(g)$	−90.48	−83.89	59.48	10.81
$HCOOH(aq)$	−101.68	−89.0	39.	
$HCOO^-(aq)$	−101.71	−83.9	22.	
$H_2CO_3(aq)$	−167.22	−148.94	44.8	
$HCO_3^-(aq)$	−165.39	−140.26	21.8	
$CO_3^{2-}(aq)$	−161.84	−126.17	−13.6	
$CH_3OH(l)$	−57.04	−39.76	30.3	19.5
$CH_3OH(g)$	−47.96	−38.72	57.29	10.47
$CH_3OH(aq)$	−58.779	−41.92	31.8	
$CH_2CO(g)$	−14.6	−14.8	59.16	12.37
$H_2C_2O_4\,[(COOH)_2](s)$	−197.7	−166.8	28.7	28.
$H_2C_2O_4(aq)$	−195.57	−166.8	10.9	
$HC_2O_4^-(aq)$	−195.6	−166.93	35.7	
$C_2O_4^{2-}(aq)$	−197.2	−161.1	12.2	
$CH_3CHO(l)$	−45.96	−30.64	38.3	
$C_2H_4O(l, \text{ethene oxide})$	−18.60	−2.83	36.77	21.02
$C_2H_4O(g)$	−12.58	−3.12	57.94	11.45
$CH_3COOH(l)$	−115.8	−93.2	38.2	29.7
$CH_3COOH(g)$	−103.31	−89.4	67.5	15.9
$CH_3COOH(aq)$	−116.16	−94.8	42.7	
$CH_3COO^-(aq)$	−116.16	−88.29	20.7	
$NH_2CH_2COOH(aq)$	−112.28	−75.25	28.54	
$^+NH_3CH_2COO^-(aq)$	−122.85	−88.59	37.84	
$HCOOCH_3(l)$	−90.60			29.
$CH_3CH_2OH(l)$	−66.37	−41.80	38.4	26.64
$CH_3CH_2OH(g)$	−56.19	−40.29	67.54	15.64
$CH_3CH_2OH(aq)$	−68.9	−43.44	35.5	
$CH_3CH_2O^-(aq)$		−24.5		
$CH_3OCH_3(g)$	−44.3	−27.3	63.72	15.76
$CH_2OHCH_2OH(l)$	−108.70	−77.25	39.9	35.8
$CF_3(g)$	−112.4	−109.2	63.36	11.91
$CF_4(g)$	−221.	−210.	62.50	14.60
$CH_3F(g)$			53.25	8.96
$CH_2F_2(g)$	−106.8	−100.2	58.94	10.25
$CHF_3(g)$	−164.5	−156.3	62.04	12.20
$COF_2(g)$	−151.7	−148.0	61.78	11.19

Species	ΔH_f°	ΔG_f°	S°	$\overline{C}_P$
Carbon, continued				
$CCl_3(g)$	14.			
$CCl_4(l)$	−32.37	−15.60	51.72	31.49
$CCl_4(g)$	−24.6	−14.49	74.03	19.91
$CH_3Cl(g)$	−19.32	−13.72	56.04	9.74
$CH_3Cl(aq)$	−24.3	−12.3	34.6	
$CH_2Cl_2(l)$	−29.03	−16.09	42.5	23.9
$CH_2Cl_2(g)$	−22.10	−15.75	64.56	12.18
$CHCl_3(l)$	−31.5	−17.1	48.5	27.8
$CHCl_3(g)$	−24.	−16.	70.86	15.73
$CH_2CHCl(g)$	8.5	12.4	63.07	12.84
$CH_3CH_2Cl(l)$	−32.63	−14.20	45.60	24.94
$CH_3CH_2Cl(g)$	−26.81	−14.45	65.94	15.01
cis-$CHClCHCl(l)$	−6.6	5.27	47.42	27.
trans-$CHClCHCl(l)$	−5.53	6.52	46.81	27.
$COCl_2(g)$	−52.3	−48.9	67.74	13.78
$CF_3Cl(g)$	−166.	−156.	68.16	15.98
$CF_2Cl_2(g)$	−114.	−105.	71.86	17.27
$CFCl_3(l)$	−72.02	−56.61	53.86	29.05
$CFCl_3(g)$	−66.	−57.	74.05	18.66
$CBr_4(s)$	4.5	11.4	50.8	34.5
$CBr_4(g)$	16.	4.873	21.79	
$CH_3Br(g)$	−8.4	−6.2	58.86	10.14
$CH_3I(l)$	−3.7	3.2	39.0	30.
$CH_3I(g)$	3.1	3.5	60.71	10.54
$CS_2(l)$	21.44	15.60	36.17	18.1
$CS_2(g)$	28.05	16.05	56.82	10.85
$COS(g)$	−33.96	−40.47	55.32	9.92
$CH_3SH(l)$	−11.08	−1.85	40.44	21.64
$CH_3SH(g)$	−5.34	−2.23	60.96	12.01
$CN(g)$	109.	102.	48.4	6.97
$(CN)_2(g)$	73.84	71.07	57.79	13.58
$HCN(l)$	26.02	29.86	26.97	16.88
$HCN(g)$	32.3	29.8	48.20	8.57
$HCN(aq)$	36.0	41.2	22.5	
$CN^-(aq)$	36.1	39.6	28.2	
$HCNO(aq)$	−36.90	−28.0	34.6	
$CNO^-(aq)$	−34.9	−23.3	25.5	
$ClCN(g)$	34.5	32.9	56.31	10.70
$BrCN(s)$	33.58			
$BrCN(g)$	44.5	39.5	59.32	11.22
$ICN(s)$	39.71	44.22	23.0	
$ICN(g)$	53.9	47.0	61.35	11.54
$HSCN(aq)$		23.31		
$SCN^-(aq)$	18.27	22.15	34.5	
$CH_3CN(l)$	7.50	18.46	35.76	21.86
$CH_3CN(g)$	15.59	19.74	58.59	12.48

Selected Values of Thermodynamic Properties at 298.15 K and 1 atm

Species	ΔH_f°	ΔG_f°	S°	$\overline{C}_P$
Carbon, continued				
$CH_3NC(g)$	35.61	39.60	59.02	12.65
$CH_3NH_2(l)$	−11.3	8.5	35.90	
$CH_3NH_2(g)$	−5.49	7.67	58.15	12.7
$CH_3NH_2(aq)$	−16.77	4.94	29.5	
$CH_3NH_3^+(aq)$	−29.86	−9.53	34.1	
$CO(NH_2)_2(s)$	−79.71	−47.19	25.00	22.26
$CS(NH_2)_2(s)$	−21.1			
$NH_2CH_2CH_2NH_2(aq)$	−13.32			
Silicon				
$Si(s)$	0.0	0.0	4.50	4.78
$Si(g)$	108.9	98.3	40.12	5.318
$SiO(g)$	−23.8	−30.2	50.55	7.15
$SiO_2(s, quartz)$	−217.72	−204.75	10.00	10.62
$SiH_4(g)$	8.2	13.6	48.88	10.24
$Si_2H_6(g)$	19.2	30.4	65.14	19.31
$SiF_4(g)$	−385.98	−375.88	67.49	17.60
$SiF_6^{2-}(aq)$	−571.0	−525.7	29.2	
$SiCl_4(l)$	−164.2	−148.16	57.3	34.73
$SiCl_4(g)$	−157.03	−147.47	79.02	21.57
$Si_3N_4(s)$	−177.7	−153.6	24.2	23.78
$SiC(s)$	−15.6	−15.0	3.97	6.42
Germanium				
$Ge(s)$	0.0	0.0	7.43	5.580
$Ge(g)$	90.0	80.3	40.103	7.345
$GeO(s)$	−62.6	−56.7	12.	
$GeO_2(s)$	−131.7	−118.8	13.21	12.45
$GeH_4(g)$	21.7	27.1	51.87	10.76
$GeCl_4(l)$	−127.1	−110.6	58.7	
$GeS(s)$	−16.5	−17.1	17.	11.42
$GeS_2(s)$	−37.5	−36.95	20.9	15.70
Tin				
$Sn(s, white)$	0.0	0.0	12.32	6.45
$Sn(s, gray)$	−0.50	0.03	10.55	6.16
$Sn(g)$	72.2	63.9	40.243	5.081
$Sn^{2+}(aq)$	−2.1	−6.5	−4.	
$Sn^{4+}(aq)$	7.3	0.6	−28.	
$SnO(s)$	−68.3	−6.14	13.5	10.59
$SnO_2(s)$	−138.8	−124.2	12.5	12.57
$SnH_4(g)$	38.9	45.0	54.39	11.70
$Sn(OH)_2(s)$	−134.1	−117.5	37.	
$Sn(OH)_4(s)$	−265.3			
$SnCl_4(l)$	−122.2	−105.2	61.8	39.5
$SnCl_6^{2-}(aq)$	−231.9			
$SnS(s)$	−24.	−23.5	18.4	11.77

Appendix C

Species	ΔH_f°	ΔG_f°	S°	$\overline{C}_P$
Lead				
$Pb(s)$	0.0	0.0	15.49	6.32
$Pb(g)$	46.6	38.7	41.889	4.968
$Pb^{2+}(aq)$	−0.4	−5.83	2.5	
$PbO(s, yellow)$	−51.466	−44.91	16.42	10.94
$PbO(s, red)$	−52.34	−45.16	15.9	10.95
$PbO_2(s)$	−66.3	−51.95	16.4	15.45
$Pb(OH)_2(s)$	−123.0	−108.1	21.	
$PbF_2(s)$	−158.7	−147.5	26.4	17.28
$PbCl_2(s)$	−85.90	−75.08	32.5	18.43
$PbCl_4(l)$	−78.7			
$PbBr_2(s)$	−66.6	−62.60	38.6	19.15
$PbI_2(s)$	−41.94	−41.50	41.79	18.49
$PbS(s)$	−24.0	−23.6	21.8	11.83
$PbSO_4(s)$	−219.87	−194.36	35.51	24.667
$PbSe(s)$	−24.6	−24.3	24.5	12.0
$PbCO_3(s)$	−167.1	−149.5	31.3	20.89
$PbC_2O_4(s)$		−179.3	34.9	25.2
$Pb(CH_3)_4(l)$	23.4			
$Pb(CH_2CH_3)_4(l)$	12.6			
$Pb(CH_2CH_3)_4(g)$	26.19			
GROUP VA				
Nitrogen				
$N_2(g)$	0.0	0.0	45.77	6.961
$N(g)$	112.979	108.886	36.613	4.968
$NO(g)$	21.57	20.69	50.347	7.133
$NO_2(g)$	7.93	12.26	57.35	8.89
$N_2O(g)$	19.61	24.90	52.52	9.19
$N_2O_3(l)$	12.02			
$N_2O_3(g)$	20.01	33.32	74.61	15.68
$N_2O_4(l)$	−4.66	23.29	50.0	34.1
$N_2O_4(g)$	2.19	23.38	72.70	18.47
$N_2O_5(s)$	−10.3	27.2	42.6	34.2
$N_2O_5(g)$	2.7	27.5	85.0	20.2
$NH_3(g)$	−11.02	−3.94	45.97	8.38
$NH_3(aq)$	−19.19	−6.35	26.6	
$NH_4^+(aq)$	−31.67	−18.97	27.1	
$NH_2OH(aq)$	−21.7	−5.60	40.	
$N_2H_4(l)$	12.10	35.67	28.97	23.63
$N_2H_4(g)$	22.80	38.07	56.97	11.85
$N_2H_4(aq)$	8.20	30.6	33.	
$N_2H_5^+(aq)$	−1.8	19.7	36.1	
$HN_3(l)$	63.1	78.2	33.6	
$HN_3(g)$	70.3	78.4	57.09	10.44
$HN_3(aq)$	62.16	76.9	34.9	

Selected Values of Thermodynamic Properties at 298.15 K and 1 atm

Species	ΔH_f°	ΔG_f°	S°	$\overline{C}_P$
Nitrogen, continued				
$N_3^-(aq)$	60.3	77.7	32.	
$HNO_2(g)$	−19.0	−11.0	60.7	10.9
$HNO_2(aq)$	−28.5	−12.1	32.4	
$NO_2^-(aq)$	−25.0	−7.7	29.4	
$HNO_3(l)$	−41.61	−19.31	37.19	26.26
$HNO_3(g)$	−32.28	−17.87	63.64	12.75
$NO_3^-(aq)$	−49.56	−26.61	35.0	
$NH_4NO_3(s)$	−87.37	−43.98	36.11	33.3
$(NH_4)_2O(l)$	−102.94	−63.84	63.94	59.08
$NF_3(g)$	−29.8	−19.9	62.29	12.7
$NH_4F(s)$	−110.89	−83.36	17.20	15.60
$NH_4Cl(s)$	−75.15	−48.51	22.6	20.1
$NH_4ClO_4(s)$	−70.69	−21.2	44.0	30.6
$NH_4Br(s)$	−64.73	−41.9	27.	23.
$NH_4I(s)$	−48.14	−26.9	28.	
$NH_4HS(s)$	−37.5	−12.1	23.3	
$NH_2SO_3H(s)$	−161.3			
$NH_2SO_3H(aq)$	−156.3			
$(NH_4)_2SO_4(s)$	−281.86	−215.56	52.65	44.81
$NOF(g)$	−15.7	−12.0	59.3	9.76
$NOCl(g)$	12.36	15.79	62.52	10.68
$NOBr(g)$	19.56	19.70	65.16	10.87
Phosphorus				
$P(s, white)$	0.0	0.0	9.82	5.698
$P(s, red)$	−4.2	−2.9	5.45	5.07
$P(g)$	75.20	66.51	38.978	4.968
$P_2(g)$	34.5	24.8	52.108	7.66
$P_4(g)$	14.08	5.85	66.89	16.05
$P_4O_{10}(s)$	−713.2	−644.8	54.70	50.60
$PH_3(g)$	1.3	3.2	50.22	8.87
$PH_3(aq)$	−2.27	6.05	28.7	
$PH_4^+(aq)$		22.0		
$H_3PO_3(aq)$	−232.2	−204.8	40.	
$H_2PO_3^-(aq)$		−202.35	19.	
$HPO_3^{2-}(aq)$	−233.8	−194.0		
$H_3PO_4(s)$	−305.7	−267.5	26.41	25.35
$H_3PO_4(aq)$	−308.2	−274.2	42.1	
$H_2PO_4^-(aq)$	−311.3	−271.3	21.3	
$HPO_4^{2-}(aq)$	−310.4	−261.5	−8.6	
$PO_4^{3-}(aq)$	−306.9	−243.5	−52.	
$PF_3(g)$	−219.6	−214.5	65.28	14.03
$PF_5(g)$	−381.1	−363.5	71.9	20.28
$PCl_3(l)$	−76.4	−65.1	51.9	
$PCl_3(g)$	−68.6	−64.0	74.49	17.17
$PCl_5(g)$	−95.35	−77.57	84.3	26.68

Species	ΔH_f°	ΔG_f°	S°	$\overline{C}_P$
Phosphorus, continued				
$POCl_3(g)$	−141.5	−130.3	77.59	20.30
$PBr_3(l)$	−44.1	−42.0	57.4	
$PBr_3(g)$	−33.3	−38.9	83.17	18.16
$PN(g)$	26.26	20.97	50.45	7.10
Arsenic				
$As(s)$	0.0	0.0	8.4	5.89
$As(g)$	72.3	62.4	41.61	4.968
$As_2(g)$	53.1	41.1	57.2	8.366
$As_2O_5(s)$	−221.05	−187.0	25.2	27.85
$AsH_3(g)$	15.88	16.47	53.22	9.10
$HAsO_2(aq)$	−109.1	−96.25	30.1	
$AsO_2^-(aq)$	−102.54	−83.66	9.7	
$H_3AsO_3(aq)$	−177.4	−152.92	46.6	
$H_2AsO_3^-(aq)$	−170.83	−140.33	26.4	
$H_3AsO_4(aq)$	−215.7	−183.1	44.0	
$H_2AsO_4^-(aq)$	−217.39	−180.01	28.	
$HAsO_4^{2-}(aq)$	−216.62	−170.79	−0.4	
$AsO_4^{3-}(aq)$	−212.27	−154.97	−38.9	
$AsF_3(g)$	−187.80	−184.22	69.07	15.68
$AsCl_3(g)$	−62.5	−59.5	78.17	18.10
$As_2S_3(s)$	−40.4	−40.3	39.1	27.8
Antimony				
$Sb(s)$	0.0	0.0	10.92	6.03
$Sb(g)$	62.7	53.1	43.06	4.97
$Sb_2O_5(s)$	−232.3	−198.2	29.9	28.10
$SbH_3(g)$	34.681	35.31	55.61	9.81
$HSbO_2(aq)$	−116.6	−97.4	11.1	
$SbO_2^-(aq)$		−81.32		
$H_3SbO_3(aq)$	−184.9	−154.1	27.8	
$H_3SbO_4(aq)$	−216.8			
$SbCl_3(s)$	−91.34	−77.37	44.0	25.8
$SbCl_3(g)$	−75.0	−72.0	80.71	18.33
$Sb_2S_3(s)$	−41.8	−41.5	43.5	28.65
Bismuth				
$Bi(s)$	0.0	0.0	13.56	6.10
$Bi(g)$	49.5	40.2	44.669	4.968
$Bi^{3+}(aq)$		19.8		
$Bi_2(g)$	52.5	41.2	65.43	8.83
$Bi_2O_3(s)$	−137.16	−118.0	36.2	27.13
$BiCl_3(s)$	−90.6	−75.3	42.3	25.
$BiCl_4^-(aq)$		−115.1		
$BiOCl(s)$	−87.7	−77.0	28.8	17.7
$Bi_2S_3(s)$	−34.2	−33.6	47.9	29.2

Selected Values of Thermodynamic Properties at 298.15 K and 1 atm

Species	ΔH_f°	ΔG_f°	S°	$\overline{C}_P$
GROUP VIA				
Oxygen				
$O_2(g)$	0.0	0.0	49.003	7.016
$O(g)$	59.553	55.389	38.647	5.237
$O_3(g)$	34.1	39.0	57.08	9.37
$O_2(aq)$	−2.8	3.9	26.5	
$O_3(aq)$	30.1	41.6	35.	
$OH(g)$	9.31	8.18	43.890	7.143
$OH^-(g)$	−33.67			
$OH^-(aq)$	−54.970	−37.594	−2.57	
$HO_2(g)$	2.5	5.4	54.7	8.34
$H_2O(l)$	−68.315	−56.687	16.71	17.995
$D_2O(l)$	−70.411	−58.195	18.15	20.16
$HDO(l)$	−69.285	−57.817	18.95	
$H_2O(g)$	−57.796	−54.634	45.104	8.025
$D_2O(g)$	−59.560	−56.055	47.404	8.191
$HDO(g)$	−58.628	−57.805	47.684	8.08
$H_2O_2(l)$	−44.88	−28.78	26.20	21.3
$H_2O_2(g)$	−32.58	−25.24	55.6	10.3
$H_2O_2(aq)$	−45.69	−32.05	34.4	
Sulfur				
$S(s, \text{rhombic})$	0.0	0.0	7.60	5.41
$S(s, \text{monoclinic})$	0.08	0.023	7.79	
$S(g)$	66.636	56.951	40.084	5.658
$S_2(g)$	30.68	18.96	54.51	7.76
$SO(g)$	1.496	−4.741	53.02	7.21
$SO_2(l)$	−76.6			
$SO_2(g)$	−70.944	−71.748	59.30	9.53
$SO_2(aq)$	−77.194	−71.871	38.7	
$SO_3(s)$	−108.63	−88.19	12.5	
$SO_3(l)$	−105.41	−88.04	22.85	
$SO_3(g)$	−94.58	−88.69	61.34	12.11
$H_2S(g)$	−4.93	−8.02	49.16	8.18
$H_2S(aq)$	−9.5	−6.66	29.	
$HS^-(aq)$	−4.22	3.01	14.6	
$S^{2-}(aq)$	7.2	21.966	6.8	
$H_2SO_3(aq)$	−145.51	−128.56	55.5	
$HSO_3^-(aq)$	−149.67	−126.15	33.4	
$SO_3^{2-}(aq)$	−151.9	−116.3	−7.	
$H_2SO_4(l)$	−194.548	−164.938	37.501	33.20
$HSO_4^-(aq)$	−212.08	−180.69	31.5	
$SO_4^{2-}(aq)$	−217.32	−177.97	4.8	
$S_2O_3^{2-}(aq)$	−155.0	−124.9	16.0	
$S_4O_6^{2-}(aq)$	−292.6	−248.7	61.5	
$S_2O_8^{2-}(aq)$	−320.0	−265.4	59.3	
$SF_4(g)$	−185.2	−174.8	69.77	17.45
$SF_6(g)$	−289.	−264.2	69.72	23.25

Appendix C

Species	$\Delta H_f°$	$\Delta G_f°$	$S°$	$\overline{C}_P$
Sulfur, continued				
$SOCl_2(g)$	−50.8	−47.4	74.01	15.9
$SO_2Cl_2(g)$	−87.0	−76.5	74.53	18.4
$(CH_3)_2SO(l)$	−48.6	−23.7	45.0	35.2
$(CH_3)_2SO(g)$	−35.96	−19.48	73.20	21.26
Selenium				
$Se(s, black)$	0.0	0.0	10.144	6.062
$Se(s, red)$	1.6			
$Se(g)$	54.27	44.71	42.21	4.978
$Se_2(g)$	34.9	23.0	60.2	8.46
$SeO(g)$	12.75	6.41	55.9	7.47
$SeO_2(s)$	−53.86			
$SeO_3(s)$	−39.9			
$H_2Se(g)$	20.5	17.0	52.9	
$H_2Se(aq)$	18.1	18.4	39.9	
$HSe^-(aq)$	24.6	23.57	42.3	
$Se^{2-}(aq)$	31.6	37.2	20.0	
$H_2SeO_3(aq)$	−121.29	−101.87	49.7	
$HSeO_3^-(aq)$	−123.5	−98.3	30.4	
$SeO_3^{2-}(aq)$	−122.39	−89.33	3.9	
$HSeO_4^-(aq)$	−143.1	−108.2	22.0	
$SeO_4^{2-}(aq)$	−145.3	−105.42	5.7	
$SeF_6(g)$	−267.	−243.	74.99	26.4
Tellurium				
$Te(s)$	0.0	0.0	11.88	6.15
$Te(g)$	47.02	37.55	43.65	4.968
$TeO(g)$	15.6	9.2	57.7	7.19
$TeO_2(s)$	−77.1	−64.6	19.0	
$H_2Te(g)$	36.9	33.1	56.0	
$H_2Te(aq)$		34.1		
$HTe^-(aq)$		37.7		
$Te^{2-}(aq)$		52.7		
$TeF_6(g)$	−315.	−292.	80.67	
GROUP VIIA				
Fluorine				
$F_2(g)$	0.0	0.0	48.44	7.48
$F(g)$	18.88	14.80	37.917	5.436
$F^-(g)$	−61.04			
$F^-(aq)$	−79.50	−66.64	−3.3	
$F_2^+(g)$	366.6			
$OF(g)$	26.	25.	48.47	7.48
$OF_2(g)$	−5.2	−1.1	59.11	10.35
$HF(g)$	−64.8	−65.3	41.508	6.963
$HF(aq)$	−76.5	−70.94	21.2	
$HF_2^-(aq)$	−155.34	−138.18	22.1	

Selected Values of Thermodynamic Properties at 298.15 K and 1 atm

Species	ΔH_f°	ΔG_f°	S°	$\overline{C}_P$
Chlorine				
$Cl_2(g)$	0.0	0.0	53.288	8.104
$Cl(g)$	29.082	25.262	39.457	5.220
$Cl^-(g)$	−55.72			
$Cl^-(aq)$	−39.952	−31.372	13.5	
$Cl_2(aq)$	−5.6	1.65	29.	
$ClO(g)$	24.34	23.45	54.14	7.52
$ClO_2(g)$	24.5	28.8	61.36	10.03
$ClO_2(aq)$	17.9	28.7	39.4	
$ClO_3(g)$	37.			
$Cl_2O(g)$	19.2	23.4	63.60	10.85
$HCl(g)$	−22.062	−22.777	44.646	6.96
$HClO(aq)$	−28.9	−19.1	34.	
$ClO^-(aq)$	−25.6	−8.8	10.	
$HClO_2(aq)$	−12.4	1.4	45.0	
$ClO_2^-(aq)$	−15.9	4.1	24.2	
$ClO_3^-(aq)$	−23.7	−0.8	38.8	
$ClO_4^-(aq)$	−30.91	−2.06	43.5	
$ClF(g)$	−13.02	−13.37	52.05	7.66
$ClF_3(g)$	−37.0	−27.2	66.61	15.33
Bromine				
$Br_2(l)$	0.0	0.0	36.384	18.090
$Br(g)$	26.741	19.701	41.805	4.968
$Br^-(g)$	−52.36			
$Br^-(aq)$	−29.05	−24.85	19.7	
$Br_2(g)$	7.387	0.751	58.641	8.61
$Br_2(aq)$	−0.62	0.94	31.2	
$Br_3^-(aq)$	−31.17	−25.59	51.5	
$BrO(g)$	30.06	25.87	56.78	7.67
$BrO_2(s)$	11.6			
$HBrO(aq)$	−27.0	−19.7	34.	
$BrO^-(aq)$	−22.5	−8.0	10.	
$BrO_3^-(aq)$	−16.03	4.43	38.65	
$BrO_4^-(aq)$	3.1	28.2	47.7	
$HBr(g)$	−8.70	−12.77	47.463	6.965
$BrF(g)$	−13.97	−17.64	54.700	7.877
$BrF_3(g)$	−61.09	−54.84	69.89	15.92
$BrCl(g)$	3.50	−0.23	57.36	8.36
Iodine				
$I_2(s)$	0.0	0.0	27.757	13.011
$I(g)$	25.535	16.798	43.184	4.968
$I^-(g)$	−47.1			
$I^-(aq)$	−13.19	−12.33	26.6	
$I_2(g)$	14.923	4.627	62.28	8.82
$I_2(aq)$	5.4	3.92	32.8	
$I_3^-(aq)$	−12.3	−12.3	57.2	

Species	ΔH_f°	ΔG_f°	S°	$\overline{C}_P$
Iodine, continued				
$IO(g)$	41.84	35.80	58.68	7.86
$I_2O_5(s)$	−37.78			
$HIO(aq)$	−33.0	−23.7	22.8	
$IO^-(aq)$	−25.7	−9.2	−1.3	
$HIO_3(s)$	−55.0			
$HIO_3(aq)$	−50.5	−31.7	39.9	
$IO_3^-(aq)$	−52.9	−30.6	28.3	
$IO_4^-(aq)$	−36.2	−14.0	−53.	
$HI(g)$	6.33	0.41	49.351	6.969
$IF(g)$	−22.86	−28.32	56.42	7.99
$IF_5(l)$	−206.7			
$ICl(g)$	4.25	−1.30	59.140	8.50
$ICl_3(s)$	−21.4	−5.34	40.0	
$IBr(g)$	9.76	0.89	61.822	8.71
GROUP VIIIA (0)				
$He(g)$	0.0	0.0	30.1244	4.9679
$He^+(g)$	568.459			
$Ne(g)$	0.0	0.0	34.9471	4.9679
$Ne^+(g)$	498.77			
$Ar(g)$	0.0	0.0	36.9822	4.9679
$Ar^+(g)$	364.90			
$Kr(g)$	0.0	0.0	39.1905	4.9679
$Kr^+(g)$	324.32			
$Xe(g)$	0.0	0.0	40.5290	4.9679
$Xe^+(g)$	281.20			
$XeF_4(s)$	−62.5			
LANTHANIDES				
Cerium				
$Ce(s)$	0.0	0.0	16.6	6.44
$Ce(g)$	101.	92.	45.807	5.515
$Ce^{3+}(aq)$	−166.4	−160.6	−49.	
$Ce^{4+}(aq)$		−123.5		
$Ce_2O_3(s)$	−429.3	−407.8	36.0	27.4
$CeCl_3(s)$	−251.8	−233.7	36.	20.9
Neodymium				
$Nd(s)$	0.0	0.0	17.1	6.56
$Nd(g)$	78.3	69.9	45.243	5.280
$Nd^{3+}(aq)$	−166.4	−160.5	−49.4	
$Nd_2O_3(s)$	−432.1	−411.3	37.9	26.60
$NdCl_3 \cdot 6H_2O(s)$	−687.0	−588.1	99.7	86.25
Samarium				
$Sm(s)$	0.0	0.0	16.63	7.06
$Sm(g)$	49.4	41.3	43.722	7.255

Selected Values of Thermodynamic Properties at 298.15 K and 1 atm

Species	ΔH_f°	ΔG_f°	S°	$\overline{C}_P$
Samarium, continued				
$Sm^{2+}(aq)$		−118.9		
$Sm^{3+}(aq)$	−165.3	−159.3	−50.6	
$Sm_2O_3(s)$	−435.7	−414.6	36.1	27.37
$SmCl_3 \cdot 6H_2O(s)$	−686.0	−587.1	99.	86.4
Europium				
$Eu(s)$	0.0	0.0	18.60	6.61
$Eu(g)$	41.9	34.0	45.097	4.968
$Eu^{2+}(aq)$	−126.	−129.1	−2.	
$Eu^{3+}(aq)$	−144.6	−137.2	−53.	
$Eu_2O_3(s)$	−394.7	−372.1	35.	29.2
$EuCl_3 \cdot 6H_2O(s)$	−665.6	565.5	97.3	87.7

ACTINIDES
Thorium

Species	ΔH_f°	ΔG_f°	S°	$\overline{C}_P$
$Th(s)$	0.0	0.0	12.76	6.53
$Th(g)$	143.0	133.26	45.420	4.97
$Th^{4+}(aq)$	−183.8	−168.5	−101.0	
$ThO_2(s)$	−293.12	−279.35	15.590	14.76
$ThH_2(s)$	−33.4	−23.9	12.120	8.77
$ThF_4(s)$	−499.90	−477.30	33.950	26.420
$ThF_4(g)$	−418.0	−409.7	81.7	22.2
$ThCl_4(s)$	−283.6	−261.6	45.5	
$ThS_2(s)$	−149.7	−148.2	23.0	
$Th(NO_3)_4(s)$	−344.5			

Uranium

Species	ΔH_f°	ΔG_f°	S°	$\overline{C}_P$
$U(s)$	0.0	0.0	12.00	6.612
$U(g)$	128.0	117.4	47.72	5.663
$U^{4+}(aq)$	−141.3	−126.9	−98.	
$UO_2(s)$	−259.3	−246.6	18.41	15.20
$UO_2^{2+}(aq)$	−243.7	−227.9	−23.3	
$UO_3(s)$	−292.5	−273.9	22.97	19.52
$U_3O_8(s)$	−854.4	−805.4	67.54	56.97
$UF_4(s)$	−459.1	−437.4	36.25	27.73
$UF_4(g)$	−383.7	−377.5	88.	21.8
$UF_6(s)$	−525.1	−494.4	54.4	39.86
$UF_6(g)$	−464.	−452.	93.	26.2
$UCl_4(s)$	−243.6	−222.3	47.1	29.16
$US_2(s)$	−126.	−125.8	26.39	17.84
$UC(s)$	−23.5	−23.7	14.15	11.98
$USi(s)$	−19.2			

A Brief Look at Statistical Thermodynamics

How Dumb Molecules Naturally Account for the Laws of Thermodynamics

Introduction

The laws of thermodynamics owe nothing to any model based on molecules. For this reason the statistical molecular interpretation of thermodynamics proposed by Ludwig Boltzmann in the late 19th century (which we are about to describe) was strongly opposed by many of the older powers-that-were in physics and chemistry. They said a molecular theory was unnecessary, since thermodynamics forms a complete system for describing macroscopic events. There were many features of thermodynamics, however, that begged for interpretation, such as the temperature dependence of heat capacities, the widely varying magnitudes of heats of chemical reaction, and, most of all, the mysterious quantity called entropy. Nonetheless, strong opposition to the molecular model continued well into the 20th century. Boltzmann's suicide by hanging while on a picnic in 1906 is thought by many to have been an act of despair at the scientific community's intransigence toward this bold idea. Ironically, within only two years after his death, even Boltzmann's staunchest opponents had been won over to his views.

The Boltzmann Distribution

Any good theory is built on one or more reasonable postulates. Boltzmann assumed that *all matter is composed of tiny molecules* (quite acceptable to us!) and, further, that *these molecules are basically ignorant*. A large collection of *these "dumb molecules" will do whatever is most likely* in a statistical sense, since a molecule cannot decide to go off by itself and do whatever it wants. Boltzmann considered a system of N molecules, where N is very large, with a fixed total energy E. Each molecule has a particular energy ε; in the modern language of quantum mechanics, it occupies a certain energy level ε_j. Let's suppose there are N_1 molecules with energy ε_1, N_2 with energy ε_2, etc. Then the total energy is given by

$$E = \sum_j N_j \varepsilon_j \tag{D.1}$$

and the total number of molecules by

$$N = \sum_j N_j \tag{D.2}$$

with the sums running over all possible energy levels. *The object is to discover what the N_j are,* that is, to find a formula that allows you to calculate them. The N_j are clearly not susceptible to any thermodynamic measurement; only E and N are measurable in a thermodynamic sense. This fact was probably one of the sources of disgruntlement at Boltzmann's theory—the whole thing seemed superfluous!

The first step in finding the N_j is to write down the *number of possible arrangements W* for *N* molecules assigned to the various energy levels. Then we will ask that *this number be maximized*, subject to the constraints of Equations D.1 and D.2, and hope to find the N_j from the resulting equations. This maximization of *W* is basically the application of the "dumb molecule" postulate.

The number of ways of arranging *N* molecules so that each molecule occupies a *unique* energy level is just *N*! [*N*-factorial, $N(N-1)(N-2)\ldots 1$]. If several molecules N_j have the same energy ε_j, then this total number *N*! has to be divided by $N_j!$, because the possible *arrangements* of N_j molecules *within a given level ε_j are of no consequence to the distribution of energy*. The net number of arrangements corrected in this way for the population in each level ε_j is then

$$W = \frac{N!}{N_1! \, N_2! \, N_3! \, \ldots} = \frac{N!}{\prod_j N_j!} \tag{D.3}$$

where $\prod$ stands for "product of." You can verify this formula for *W* quite easily by writing down all the arrangements for a suitably small *N*. The factorials tend to become stupendously large even for modest *N* (70! will overflow most calculators and computers), so it is easier to deal with the natural logarithm of *W*:

$$\ln W = \ln N! - \sum_j \ln N_j! \tag{D.4}$$

For an Avogadro's number of molecules, the computation in Equation D.4 is still mind boggling; we therefore will invoke *Stirling's approximation* to *N*!, valid when *N* is large,

$$\ln N! \approx N \ln N - N \tag{D.5}$$

To maximize *W* with respect to all the N_j, we make use of the *calculus of variations* (see, for example, Courant's *Differential and Integral Calculus*, Vol. II, Chap. 7). In this case, applying the calculus of variations is similar to taking differentials in ordinary calculus. Instead of using *d*'s (for differentials) we use δ's (for variations, not necessarily infinitesimal). We ask that

$$\delta \ln W = 0 = \delta(\ln N! - \sum_j \ln N_j!) \tag{D.6}$$

$$= -\delta \sum_j \ln N_j!$$

as a condition for a maximum in *W*. The second line in Equation D.6 results because ln*N*! is a constant. If we plug in the Stirling formula, Equation D.5, for $\ln N_j!$, we get

$$\delta \sum_j (N_j \ln N_j - N_j) = 0 \tag{D.7}$$

Using the ordinary rules of calculus (assuming the δ is like a *d*), we can carry through the variation to obtain

$$\sum_j \ln N_j \, \delta N_j = 0 \tag{D.8}$$

We are almost there, but we mustn't forget about the constraints of Equations D.1 and D.2. These imply, respectively,

$$\delta E = \sum_j \varepsilon_j \, \delta N_j = 0 \qquad (D.9)$$

and

$$\delta N = \sum_j \delta N_j = 0 \qquad (D.10)$$

The way in which we can impose these constraints mathematically is by multiplying each of Equations D.9 and D.10 by arbitrary constants (called *Lagrange multipliers*) and adding them into Equation D.8:

$$\alpha \sum_j \delta N_j + \beta \sum_j \varepsilon_j \, \delta N_j + \sum_j \ln N_j \, \delta N_j = 0 \quad \text{or} \qquad (D.11)$$

$$\sum_j (\alpha + \beta \varepsilon_j + \ln N_j) \, \delta N_j = 0$$

where α and β are the Lagrange multipliers. Equation D.11 has to be true for *arbitrary* variations δN_j. This means that

$$\alpha + \beta \varepsilon_j + \ln N_j = 0 \qquad \text{for all } j \qquad (D.12)$$

since we could choose all the variations to be zero except the jth one. Solving Equation D.12 for N_j,

$$N_j = e^{-\alpha - \beta \varepsilon_j} \qquad (D.13)$$

This is the basic form of the Boltzmann distribution law; it remains only to find the unknown multipliers α and β. The condition of Equation D.2 leads to

$$e^{-\alpha} \sum_j e^{-\beta \varepsilon_j} = N \quad \text{or} \qquad (D.14)$$

$$e^{-\alpha} = \frac{N}{\sum_j e^{-\beta \varepsilon_j}}$$

and Equation D.13 becomes

$$\frac{N_j}{N} = \frac{e^{-\beta \varepsilon_j}}{\sum_j e^{-\beta \varepsilon_j}} \qquad (D.15)$$

If the energy of the lowest state is taken as the zero of energy, then we also have

$$\frac{N_j}{N_0} = e^{-\beta \varepsilon_j} \qquad (D.16)$$

The identification of $\beta = 1/(k_B T)$ follows from computing the average energy per molecule for one translational degree of freedom from

$$\bar{\varepsilon} = \frac{\sum \varepsilon_j N_j}{\sum N_j} = \frac{\sum \varepsilon_j e^{-\beta \varepsilon_j}}{\sum e^{-\beta \varepsilon_j}} \qquad (D.17)$$

and comparing it to the result of kinetic molecular theory, $\bar{\varepsilon} = \frac{1}{2}k_B T$. (We won't go through that here.)

If the energy levels ε_j possess degeneracy g_j (as in a 2p energy level, where $g = 3$), the most general form of the Boltzmann law is

$$\frac{N_j}{N} = \frac{g_j e^{-\varepsilon_j/k_B T}}{\sum_j g_j e^{-\varepsilon_j/k_B T}} \tag{D.18}$$

or

$$\frac{N_j}{N_0} = \frac{g_j}{g_0} e^{-\varepsilon_j/k_B T} \tag{D.19}$$

This prediction of state populations made by Boltzmann so long ago, based only on statistics, could not be verified by thermodynamics, since this would involve taking a "molecular survey" to find out what state each molecule is in. Modern optical spectroscopy has now accomplished just that for the molecular internal degrees of freedom (vibrations and rotations). And, of course, Boltzmann was right!

The sum in the denominator of Equation D.18 contains in its terms all the information about the molecules' energy levels and degeneracies. It is called the *molecular partition function q*, and plays a central role in finding thermodynamic functions. We'll examine q in the next section.

From Dumb Molecules to Thermodynamic Functions

Our low regard for the molecular intellect does not mean we do not respect the things molecules can do for us and the everyday facts they explain. Molecular stupidity allowed Boltzmann to figure out how the little fellows would arrange themselves among the available energy levels, as we saw in the last section. Now we can use this information to derive all of thermodynamics! This molecular miracle centers around the molecular partition function

$$q = \sum_j g_j e^{-\varepsilon_j/k_B T} \tag{D.20}$$

where q is a unitless number, and represents the *effective number of states* available to an individual molecule at a given temperature T. To the extent that the molecule's energy can be written as a sum of terms,

$$\varepsilon = \varepsilon_{trans} + \varepsilon_{rot} + \varepsilon_{vib} + \varepsilon_{elec} \tag{D.21}$$

it is readily shown that q becomes a product:

$$q = q_{trans} q_{rot} q_{vib} q_{elec} \tag{D.22}$$

This means that the results of spectroscopy, which tell us the energy-level spacings in the various degree of freedom, can be used to compute q for any temperature.

The importance of knowing how to find q will now become evident as we try to evaluate the thermodynamic functions. Let's start with the energy E. The average energy $\bar{\varepsilon}$ of one molecule is

$$\bar{\varepsilon} = \frac{\sum_j \varepsilon_j g_j e^{-\varepsilon_j/k_B T}}{\sum_j g_j e^{-\varepsilon_j/k_B T}} = \sum_j \varepsilon_j P_j \tag{D.23}$$

where P_j is the *probability* that the molecule will be found in level j, given by (see Equation D.18)

$$P_j = \frac{N_j}{N} = \frac{g_j e^{-\varepsilon_j/k_B T}}{q} \tag{D.24}$$

The sum in the numerator of Equation D.23 is just $-k_B \,\partial q/\partial(1/T) = k_B T^2 \,\partial q/\partial T$, so that $\bar{\varepsilon}$ becomes

$$\bar{\varepsilon} = \frac{k_B T^2}{q} \frac{\partial q}{\partial T} = k_B T^2 \frac{\partial \ln q}{\partial T} \tag{D.25}$$

If the intermolecular forces can be neglected, as in an ideal gas, then $E = N\bar{\varepsilon}$, where N is again the total number of molecules. Taking $N = N_A$, where N_A is Avogadro's number, and $N_A k_B = R$, the gas constant, we obtain the energy per mole:

$$E = RT^2 \left(\frac{\partial \ln q}{\partial T}\right)_V \tag{D.26}$$

We must hold the volume constant while taking $\partial \ln q/\partial T$ because q_{trans} changes with the volume of the container, owing to changes in the translational energy levels. (Recall the particle-in-a-box problem of Appendix B.) For a monatomic gas like Ar, $q = q_{\text{trans}}$ since there are no rotational or vibrational levels, and the electronic excitation energy is much too large to be accessible even at temperatures of 10^4 K. The result of putting particle-in-a-box energy levels into the formula for q (Equation D.20) and summing is

$$q_{\text{trans}} = \frac{(2\pi m k_B T)^{3/2} V}{h^3} \tag{D.27}$$

where m is the mass of the atom or molecule, h is Planck's constant, and other symbols are as usual. Equation D.26 requires $\ln q$, so

$$\ln q_{\text{trans}} = \ln \frac{(2\pi m k_B)^{3/2}}{h^3} + \tfrac{3}{2} \ln T + \ln V \tag{D.28}$$

The derivative in Equation D.26 is now straightforward, and the result for the energy is

$$E = RT^2 \left(\frac{3}{2T}\right) = \tfrac{3}{2} RT \tag{D.29}$$

just as anticipated from the kinetic molecular theory. The heat capacity at constant volume $C_V = (\partial E/\partial T)_V$ is $3R/2$, also as we've already seen, and in excellent agreement with experimental measurements on the noble gases.

The crucial test of Boltzmann's model comes in the prediction of the absolute entropy, S. Boltzmann postulated that entropy was a purely probabilistic quantity, depending only on the number of ways the molecules in the system could be arranged, W. Because entropies are additive while probabilities are multiplicative, it was natural to try a logarithmic relationship:

$$S = k \ln W \tag{D.30}$$

where the proportionality constant k turns out to be k_B, Boltzmann's constant. Equation D.30 is engraved on Boltzmann's tombstone.

Simply by substituting Equation D.4 for $\ln W$ into Equation D.30, using Stirling's approximation (Equation D.5) and the Boltzmann distribution law (Equation D.15), along with the definition of the partition function q (Equation D.20) written with the degeneracies omitted, you can show that the entropy per mole is

$$S = \frac{E}{T} + k_B \ln q^{N_A} = \frac{E}{T} + R \ln q \tag{D.31}$$

or, to put things entirely in terms of q using Equation D.26,

$$S = R \ln q + RT \left(\frac{\partial \ln q}{\partial T} \right)_V \tag{D.32}$$

The tomb of Ludwig Boltzmann (1844–1906), on the edge of the University of Vienna campus. His formula became his epitaph.

Equation D.32 is a pleasing result and gives accurate values for ΔS as well as the absolute entropies $S°_{298}$ of crystalline solids, but unfortunately the $S°$'s for gases are found to be far too large when compared with experiment. Why should the theory work for solids but fail for gases?

In Boltzmann's time, the only mechanics known for describing the motions of molecules was what we now call classical mechanics (Newtonian mechanics). In classical mechanics, we may (at least in our imagination) *label* each and every molecule (say, molecule A, molecule B, etc.) and follow its motion (position and momentum) for as long as we like, even though all the molecules are identical. The guiding principle of quantum mechanics, Heisenberg's uncertainty principle, says, however, that we cannot precisely follow the path of even one molecule, much less all 6×10^{23} of them! It was at the advent of quantum mechanics that the problem with Equation D.32 was discovered. Because we are *intrinsically* unable to "run a trace" on even one molecule, *identical molecules of a gas are therefore indistinguishable*. This means that *even if we had each molecule in a distinct energy level, we could not distinguish among the N! possible arrangements* and therefore the logarithmic term in Equation D.31 needs to be corrected for this indistinguishability, that is,

$$S = \frac{E}{T} + k_B \ln \left(\frac{q^{N_A}}{N_A!} \right) \tag{D.31$'$}$$

While this adjustment does not affect Boltzmann's argument leading to his distribution law, it does change the result for S. Using the Stirling formula (Equation D.5) for $\ln N_A!$ in Equation D.31$'$, we arrive at

$$S = R \ln\left(\frac{eq}{N_A}\right) + RT\left(\frac{\partial \ln q}{\partial T}\right)_V \tag{D.32'}$$

where e is the base of the natural logarithms, e = 2.718. When Equation D.32′ is used to calculate $S°$ for Ar, using $q = q_{trans}$ from Equation D.27, the result is $S° = 36.974$ cal K^{-1} mol^{-1}, while the third law experimental result is $S° = 36.95 \pm 0.05$ cal K^{-1} mol^{-1}, almost perfect agreement!

It is important to realize that, although the way in which $S°$ is obtained from the third law *makes it appear* that $S°$ depends on the properties of solid and liquid argon, $S°$ really is a state variable, dependent only on the gaseous properties of Ar. This is brought out by the statistical calculation of $S°$.

Once we have found the energy E and entropy S, all of the other thermodynamic functions follow from their definitions. For example, the Gibbs free energy $G = H - TS = E + PV - TS$ is given by

$$G = -RT \ln\left(\frac{q}{N_A}\right) \tag{D.33}$$

From Equation D.33 flows a lovely statistical interpretation of equilibrium that allows calculation of $\Delta G°$ and K_{eq} completely from the structure of the reagent and product molecules through their partition functions.

There are two aspects of statistical thermodynamics we haven't discussed, and will only briefly touch on here. The first is the neglect of intermolecular forces, which manifest themselves in deviations from ideal behavior in gases. This is most certainly not a good approximation in liquids or solids. The effect of these forces is to give the collection of molecules in the system an energy contribution, in the form of the intermolecular potential energy, over and above that from the individual, isolated molecules. This makes the energy and entropy deviate from predictions that neglect these forces. Proper inclusion of intermolecular energy is a difficult problem, still the subject of active research today.

The second aspect is the contribution from the internal degrees of freedom, rotation, vibration, and electronic. Here we are on pretty solid ground, since the relatively weak van der Waals forces do not influence these internal motions to a great extent in gases, even under very high pressure. The calculation of the partition functions q_{rot}, q_{vib}, and q_{elec} from the known energy levels is straightforward, and leads to some generalizations about the relative magnitudes of these quantities. At 298 K for most molecules $q_{trans} \approx 10^{30}$ V, where V is in liters (see Equation D.27), while $q_{rot} \approx 10$–1000 and $q_{vib} \approx 1$–10 per vibrational degree of freedom. Electronic energy levels are generally so widely spaced that only the ground state is populated, and $q_{elec} = g_0$, the degeneracy of the ground state. Recall that the q's count the number of states available to the molecules at a given T. This comparison makes it obvious that translation gives an overwhelming contribution to S, so that gases are by far preferred to condensed phases as far as probability goes. The functions q_{rot} and q_{vib} depend strongly on T, growing in magnitude at high temperature. They account extremely well for the temperature dependence of the heat capacity C_V or C_P we discussed earlier in the book. The accuracy of the statistical predictions for molecules is comparable to that we have already cited for Ar.

The marvelous union between classical thermodynamics and the molecular model fostered by Boltzmann and his students continues to provide the focus for modern-day research in many areas of chemistry. It has been of particular importance in formulating theories of *reaction rates,* which again draws on our ability to imagine events on a molecular level. Nowadays we think of the molecular model as existing quite apart from thermodynamic laws, bolstered as it is by spectroscopic, and now microscopic, evidence. You might say the molecular model proves that thermodynamics is a solid edifice, rather than vice versa. Boltzmann would have been pleased with this!

FURTHER READING
Dickerson, R. E., *Molecular Thermodynamics* (W. A. Benjamin, 1969).
Nash, L., *Chem Thermo: A Statistical Approach to Classical Chemical Thermodynamics* (Addison-Wesley, 1972).

EXERCISES
1. Verify Equation D.3 for a group of four molecules (label them a, b, c, and d) distributed among three energy levels (1, 2, and 3) by explicitly writing down all possible arrangements for
   a. three molecules in level 1, none in 2, one in 3.
   b. one molecule in level 1, two in 2, one in 3.

   Which configuration is more likely?

2. Derive Stirling's approximation, Equation D.5. (*Hint:* Use

   $$\ln N! = \sum_{m=1}^{N} \ln m$$

   and replace the sum by an integral.)

3. Show that Equation D.7 leads to Equation D.8.

4. Following the prescription outlined above Equation D.31, use Equation D.3 for W to derive Equation D.32.

5. Show that ΔS for the expansion of 1 mol of an ideal gas at constant temperature from volume V_1 to V_2 given by Equation D.32, using Equation D.27 for q, is $\Delta S = R \ln(V_2/V_1)$, in agreement with the result of classical thermodynamics.

6. Using the harmonic oscillator model for molecular vibrations, which yields $\varepsilon_v = hc\omega_e(v + \frac{1}{2})$, $v = 0,1,2,\ldots$ and $g_v = 1$ for all v, obtain an expression for q_{vib} by summing the terms in q_{vib} analytically (they form a geometric series). Calculate q_{vib} at 298 K for H_2 ($\omega_e = 4400$ cm^{-1}) and I_2 ($\omega_e = 215$ cm^{-1}). (In your derivation, choose the zero of energy to be the ground state, that is, take $\varepsilon_v = hc\omega_e v$, ignoring the zero-point energy.)

To see a World in a Grain of Sand,
    And a Heaven in a Wild Flower,
Hold Infinity in the palm of your hand,
    And Eternity in an hour . . .

—William Blake, *Auguries of Innocence,* 1789

Appendix E: Answers to Exercises

Chapter 1

1. $Hg_{0.92}O$, or HgO within experimental error
2. Formula: PO_3; change in parts oxygen $= -25$
3.
	m_{N_2}/m_{O_2}	m_{H_2}/m_{O_2}	m_{CO_2}/m_{O_2}
Lavoisier	0.8788	0.0700	1.2893
modern	0.8755	0.0630	1.3754

4. 136 in.^{3} or 2.22 L
5. $(m_C/m_H)_{ch} = 2.979$; $(m_C/m_H)_{og} = 5.958$; ratio og/ch = 2.000
6. a. $2CO + O_2 \longrightarrow 2CO_2$
   b. $3H_2 + N_2 \longrightarrow 2NH_3$
   c. $N_2 + O_2 \longrightarrow 2NO$
   d. $N_2 + 2O_2 \longrightarrow 2NO_2$; e.g., $N + 2O \longrightarrow NO_2$ predicts only one volume of NO_2
7. mole fraction of air $= 0.00827$
8. density $= 3.2$ mg cm^{-3}; molecular mass $= 72$ g/mol
9. $K = 360$ kJ; mass of $H_2 = 3.1$ g; $w = 360$ kJ
10. $V = 4400$ J; $K = 4400$ J; $v = 14$ m s^{-1}; $w = -4400$ J
11. a. $K = 0.5$ J; 1.0 J
    b. $v = 100$ cm/s; 140 cm/s
12. a. $k = 49$ eV Å^{-2}, $V = 0.245$ eV
    b. $K = 0.245$ eV
13. units: kg m^2 s^{-1} or J s; $K = L^2/(2mr^2)$
14. "Celsius 233"; $t(°C) = t(°F) = -40$; $-459.7°F$
15. average energy $= 4.12 \times 10^{-21}$ J $= 2480$ J/mol $= 592$ cal/mol
16. temperature rise $= 0.00283°C$
17. good for low Z; underestimates AM for high Z
18. mass defect $= -0.098940$ amu
19. $^{85}Rb = 77\%$; $^{87}Rb = 23\%$
20. AM(Fe) $= 55.85$ amu or g/mol
21. 70 : 72 : 74 = 0.574 : 0.367 : 0.059
22. AM(Ra) $= 226.45$
23. number of C atoms $= 3.0 \times 10^{23}$; 0.8×10^{23}; 1.3×10^{23}
24. more S atoms; Zn is limiting reagent; 51.0% S remains
25. more O atoms; Al is limiting reagent; 11.0% O remains
26. 69.9% Fe; 112.8 kg C; 9393 mol C; 699 kg Fe; 2.10×10^5 STP L CO_2
27. a. 5.59 L
    b. 66.9 mL
28. formula: MnO_2; $MnO_2 + 2H_2 \longrightarrow Mn + 2H_2O$; 0.414 g H_2O
29. empirical formula CH; molecular formula C_2H_2
30. a. 0.401 M
    b. 0.516 M
    c. 0.22 M
    d. 1.21 M
31. a. 6.00 g
    b. 12.5 mL
    c. 3.36 L
    d. 11.0 mL
32. 0.0150 M $BaCl_2$; 0.0075 M HCl (excess reagent); 1.25 g $BaCl_2$
33. 0.003047 mol HA; MM(HA) $= 58.1$ g/mol
34. 613 mg KHP; 0.0917 M NaOH
35. 16.01%
36. diameter $= 5.43$ Å
37. 7.56 cm; 2.32 g/cm^3
38. 1.25×10^{-22} J; 7.81×10^{-4} eV
39. 4.75×10^{-19} J; 2.96 eV
40. -2.33×10^{-3} dyne; F $= +2.31 \times 10^{-3}$ dyne
41. a. -93 eV
    b. $+19.5$ eV
    c. $+22.8$ MeV
    d. -43 eV
42. draw another r-axis raised by 27.2 eV
43. a. 4.67×10^{-4} Å
    b. 9.34×10^{-4} Å
44. $V = -6.79$ eV; $K = 3.39$ eV; $L = 2.11 \times 10^{-27}$ erg sec; $v = 1.09 \times 10^8$ cm s^{-1}; $\tau = 1.22 \times 10^{-15}$ s

Answers to Exercises

45. -14.40 eV; -43.2 eV; -50.4 eV if linear
46. 2.8926×10^{14} esu/mol; 96485 C/mol

Chapter 2

1. 1.552×10^{18} sec^{-1}; 1.028×10^{-15} J; 6417 eV
2. 320 cm; 6.21×10^{-26} J; 3.88×10^{-7} eV
3. 3.10 eV at 400 nm; 1.77 eV at 700 nm
4. 1×10^{19} photons/sec; 2×10^{-5} einsteins
5. 5400 K
6. 1.74 eV; 2.79×10^{-19} J; 713 nm
7. 0.7 eV
8. 37
9. 20.6 eV
10. No excited He in sample; form excited He in discharge
11. Li: 1.85 eV; 1.49×10^4 cm^{-1}
    Na: 2.10 eV; 1.69×10^4 cm^{-1}
    K: 1.62 eV; 1.31×10^4 cm^{-1}
12. a. ΔE/eV: 2.10443, 2.10230
    b. $\Delta E = 2.13$ meV, 17.2 cm^{-1}; $\lambda = 582$ μm, microwave
13. a. ΔE/eV: 2.03106, 2.02459, 2.01147
    b. ΔE/meV: 6.47, 13.12; ΔE/cm^{-1}: 52.2, 105.9; λ/μm: 192, 94.5
14. a. IE = 5.139 eV
    b. E/eV: -3.037, -3.035
15. a. 1.889 eV
16. 1.876×10^{-4} cm; 9.725×10^{-6} cm; 1.026×10^{-5} cm
17. 109,737 cm^{-1} (109,678 cm^{-1} using reduced mass)
18. 5290 Å; -1.36×10^{-3} eV
19. n_1^2/R_H; ionization
20. $1/\lambda = 4R_H(1/n_1^2 - 1/n_2^2)$; $4 \to 3$, $6 \to 4$, $7 \to 4$,... yield visible λ
22. 0.689 Å
23. 34 Å
24. 430 Å
25. $\lambda = h/\sqrt{2m(E-V)}$
26.

n	$v_n = 2\pi e^2/nh$	$\lambda_n = 2\pi n a_0$
1	2.19×10^8 cm/sec	3.33 Å
2	1.09×10^8 cm/sec	6.65 Å
100	2.19×10^6 cm/sec	333 Å

27. The virial theorem, $V = 2E$, may be substituted into the equation derived in Exercise 24 to give the formula for λ_n of Exercise 25.
28. r/a_0: 0   4.5   9.0   13.5
    λ/a_0: 0   10.88   18.85   32.65; as $r \to 18a_0$, $\lambda \to \infty$
29. Bohr's quantization of angular momentum, $L = mvr = nh/2\pi$, may be rearranged to give $2\pi r = n(h/mv) = n\lambda$
31. $n = 2.5 \times 10^{74}$; $\Delta n = 1$ is a negligible change
32. $E_{photon} = hc/\lambda = 7420$ eV; $2mE_{particle} = (E_{photon}/c)^2$

Chapter 3

1. a. $m = -2, -1, 0, 1, 2$
   b. $l = 0, 1, 2, 3, 4$
   c. $l = 2, 3$
   d. $n = 3, 4, 5, \ldots$
   e. $m = 0$
2. a. $3d$
   b. $5s, 5p, 5d, 5f, 5g$
   c. $4d, 4f$
   d. $3d, 4d, 5d, \ldots$
   e. $10s$
3. 200 and 210 are degenerate as are 322 and 31$-$1
4. IE($4p$) = 0.850 eV; IE($100s$) = 0.00136 eV
5. Work done = $-E$; make potential energy = 0 ($r \to \infty$)
6. a. 9: $3s, 3p_x, 3p_y, 3p_z, 3d_{z^2}, 3d_{xz}, 3d_{yz}, 3d_{xy}, 3d_{x^2-y^2}$
   b. 5: $4d_{z^2}, 4d_{xz}, 4d_{yz}, 4d_{xy}, 4d_{x^2-y^2}$
   c. 3: $5p_x, 5p_y, 5p_z$
   d. 2: $2p_x, 2p_y$
   e. 2: $3d_{xz}, 3d_{yz}$
8. a. 210
   b. 41 ± 1
   c. 400
   d. 32 ± 2
   e. 52 ± 1
9. $\Psi_{430} = R_{43}(r)Y_{30}(\theta,\phi) = (1/1024[5\pi]^{1/2})a_0^{-3/2}\rho^3(\frac{5}{3}\cos^3\theta - \cos\theta)e^{-\rho/4}$; $4f_{z^3}$; see Chapter 17
10.

Orbital:	$1s$	$6s$	$3d$	$5d$	$4f$	$2p$
Radial nodes:	0	5	0	2	0	0
Angular nodes:	0	0	2	2	3	1

11. $1s$: no nodes for finite r; $2s$: 1 node at $r = 2a_0 = 1.06$ Å; $3s$: two nodes at $r = 1.90a_0 = 1.01$ Å and $7.10a_0 = 3.76$ Å; all become nodal spheres
12. p_z: 1 node at $\theta = 90°$, nodal plane (xy); d_{z^2}: two nodes at $\theta = 54.7°, 125.3°$, nodal cones
13. Radial node occurs when the factor $(6 - \rho) = 0$; $r = 6a_0$; angular node occurs when $\cos\theta = 0$; $\theta = 90°$ (x-y plane).
14. See Figures 3.3 to 3.6 for help
15. See Figures 3.4 to 3.6; for $n = 4$, use nodal and size rules
16. ψ: length$^{-3/2}$; $|\psi|^2$: length^{-3} (probability per unit volume)
17. $(r,\theta,\phi) = (2a_0, 0, \phi)$ and $(2a_0, 180°, \phi)$ for any ϕ in $(0, 2\pi)$
18. RDF $= r^2 R^2 = a_0^{-1} 4\rho^2 e^{-2\rho}$

$\rho(r/a_0)$:	0.2	0.5	1.0	1.5	2.0	2.5	3.0
RDF(a_0^{-1}):	0.107	0.368	0.541	0.448	0.293	0.168	0.089

19. $P(r \geq 2a_0) = 0.238$
20. $2s$: $r = 0.764a_0 = 0.404$ Å and $5.24a_0 = 2.77$ Å; $2p$: $r = 4a_0 = 2.12$ Å
21. RDF $\propto \rho^{2n} e^{-2\rho/n}$
22. Watermelon: $\Delta p = 5 \times 10^{-33}$ kg m s^{-1}; $E = 3 \times 10^{-66}$ J
    $k_B T = 4.11 \times 10^{-21}$ J
    Be atom: $\Delta p = 1 \times 10^{-23}$ kg m s^{-1}; $E = 4 \times 10^{-21}$ J
23. $\Delta x = 100$ m (within a factor of 10)
24. For $\Delta r \approx a_0$, $K \approx (\Delta p)^2/2m = h^2/(32\pi^2 m a_0^2) = 3.4$ eV
25. $K_{vib} = K_{rot} = 1.70$ eV; $\Delta r = 0.75$ Å
26. $\lambda_n / r_n = 2\pi/n$
27. $\nu_{n,n-1} \longrightarrow \nu_n = 4\pi^2 m e^4 / h^3 n^3$
28. $4d \longrightarrow 3p \longrightarrow 1s$ or $4d \longrightarrow 2p \longrightarrow 1s$
29. $\Delta E = 4.95 \times 10^{-26}$ J $= 3.09 \times 10^{-7}$ eV; $\Delta\bar{\nu} = 2.49 \times 10^{-3}$ cm^{-1}; $\Delta\lambda/\lambda = 3.03 \times 10^{-8}$
30.

Δt (s)	ΔE (J)	ΔE (eV)	$\Delta\bar{\nu}$ (cm^{-1})	$\Delta\lambda$ (nm)
10^{-12}	1.05×10^{-22}	6.58×10^{-4}	5.31	0.150
10^{-15}	1.05×10^{-19}	0.658	5310	150.

Chapter 4

1. $r = \frac{4}{7} n^2 a_0$; $E = -83.3/n^2$ eV (cf. -79.00 eV)
2. $Z_{eff} = 2.38$; $r = a_0/2.38$; $E = -77.2$ eV
3. $E = -3.40$ eV
4. $E(3s) < E(3p) < E(3d)$ for many-electron atoms; orbitals are more penetrating the smaller the l value
5. $E = -108.8$ eV
6.
C	$1s^2 2s^2 2p^2$		Si	$1s^2 2s^2 2p^6 3s^2 3p^2$
Mg	$1s^2 2s^2 2p^6 3s^2$		S^{2-}	$1s^2 2s^2 2p^6 3s^2 3p^6$
S^{6+}	$1s^2 2s^2 2p^6$		C^-	$1s^2 2s^2 2p^3$
Cl^+	$1s^2 2s^2 2p^6 3s^2 3p^4$		Cl^-	$1s^2 2s^2 2p^6 3s^2 3p^6$
Mg^{2+}	$1s^2 2s^2 2p^6$		P^{3-}	$1s^2 2s^2 2p^6 3s^2 3p^6$
Kr	$1s^2 2s^2 2p^6 3s^2 3p^6 4s^2 3d^{10} 4p^6$			

7.
Be	[He] $2s^2$		As	[Ar] $4s^2 3d^{10} 4p^3$
Ba	[Xe] $6s^2$		Sn^{2+}	[Kr] $5s^2 4d^{10}$
Al	[Ne] $3s^2 3p^1$		Sn^{4+}	[Kr] $4d^{10}$
Mn	[Ar] $4s^2 3d^5$		Cu^+	[Ar] $3d^{10}$
Mn^{2+}	[Ar] $3d^5$		Nd	[Xe] $6s^2 4f^4$
Mn^{7+}	[Ar]		Ra	[Rn] $7s^2$
Ga	[Ar] $4s^2 3d^{10} 4p^1$		U	[Rn] $7s^2 5f^4$

8. Number of unpaired electrons: 1, 1, 2, 2, 6, 0
9. a. excited state
   b. ground state
   c. impossible
   d. excited state
   e. excited state
10. a. N
    b. [Ar]$4s^2$; 0
    c. 2; Ge
    d. [Ar]$3d^6$
    e. $+1$; 0
    f. [Ar]$3d^3$; $+3$
    g. $+2$; 0
11. 50 elements; ... $5s^2 5p^6 5d^{10} 5f^{14} 5g^{18}$; no
12. a. Ba<Ga<S<F<Ne
    b. Te<I<Br<Cl<Ar
    c. Cs<Al<Mg<Si<N
13. Pairing of electrons
14. Sr Mg Hg (lanthanide contraction) P (½-filled) N Zn Ca (filled) Cl Ar Kr
15. Extra electron is more destabilizing.
16. 1.65; 2.40; 2.79; 2.90; 4.00
17. a. (Ne) < F < S < Ga < Ba
    b. (Ar) < Cl < Br < I < Te
    c. N < Si < Al < Mg < Cs
18. Ba K Hg P Si Cu Ca I Br Rb
19. K^+, Cl^-, S^{2-}, P^{3-}; estimates of radii: 0.96, 0.79 Å
20. a. IE = 23.06 eV
    b. 0.314 Å

21. a. $Z_{eff} = 6.67$; 0.714 Å
    b. 16.5 eV
22. Li: $Z_{eff} = 1.28$, $r = 1.65$ Å; excited He: 1.14, 1.86 Å; excited He
23.

Atom:	C	N	O
Z_{eff}:	3.23	3.88	4.53
r/Å:	0.66	0.55	0.47
IE/eV:	11.2	12.6	13.8

24. Cl: $Z_{eff} = 6.02$, $r = 0.79$ Å; Cl$^-$: 5.67, 0.84 Å; EA = 5.3 eV
25. Na: $Z_{eff} = 2.12$, $r = 2.25$ Å; Na$^+$: 6.83, 0.31 Å; IE = 6.8 eV
26. 10, 595, 4186 terms, respectively

Chapter 5

1. a. RbBr
   b. Al$_2$S$_3$
   c. Li$_3$N
   d. CaO
   e. SrF$_2$
2. a. Ca$_3$P$_2$
   b. Li$_2$O
   c. BeCl$_2$
   d. Ga$_2$Se$_3$
   e. Mg$_3$N$_2$
3. 7/3, i.e., two charge states are present, +3, +2, and +2
4. Ba (AW = 137.34)
5.

	eV	kJ/mol	kcal/mol
a.	−9.54	−920	−220
b.	−4.38	−422	−101 (used 1.41 Å for Cs$^+$)
c.	−10.7	−1030	−247
d.	−21.9	−2110	−505 (1.23 Å for Ba^{2+})
e.	−25.6	−2470	−590 (0.99 Å for Ca^{2+})

6. $r_x = 7.24$ Å, $r_e = 1.51$ Å, $D_e = 7.55$ eV
7. $r_x = 53$ Å, $r_e = 2.91$ Å (1.41 Å for Cs$^+$), $D_e = 4.68$ eV
8. SiH$_4$, PH$_3$ (least polar), H$_2$S, HCl (most polar)
9. Lewis structures:

   H—Br: [:C≡N:]$^-$ H—C≡N: :S=C=S:

   H—Be—H   :N≡N—O: ↔ :N—N=O:

   :F—B—F: :O—N=O:$^-$:Cl—P—Cl:
     | ↕ 2 r.f. |
    :F: :O: :Cl:

   H—C≡C—H   H—N—N—H   :O:
                 | | ‖
                 H H   H—C—H

10.

Name	Lewis structure	Salt?
Sulfuric acid	H—O—S(=O)(=O)—O—H	no
Ammonia	H—N(H)—H	no
Calcium oxide	Ca^{2+}[:O:]$^{2-}$	yes
Oxygen	:O=O:	no
Nitrogen	:N≡N:	no
Ethylene	H$_2$C=CH$_2$	no
Sodium hydroxide	Na$^+$[:O—H]$^-$	yes
Chlorine	:Cl—Cl:	no
Phosphoric acid	H—O—P(=O)(O—H)—O—H	no
Ammonium nitrate	[H—N(H)(H)—H]$^+$:O—N=O:$^-$	yes

11. Lewis structures

    :Cl: :F—S—F: :F: :F:
    | :F: :F: \\ //
    :Cl—P—Cl: S
    :Cl: :Cl: // \\
                                       :F: :F:

    :F—Xe—F: :O—Xe—O: :O:
                                 ‖
                               :O—Xe—O:
                                 ‖
                                :O:

    [:I—I—I:]$^-$:F—Br—F:$^-$:F: :F:
                    :F: :F: \\ /
                                     I
                                    / \\
                                   :F: :F:
                                    :F: :F:

12. 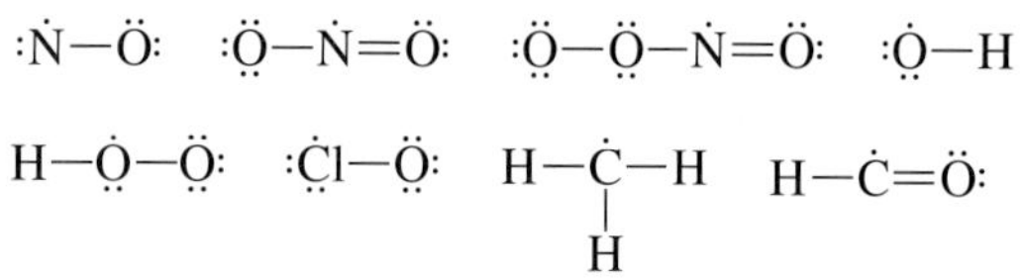 :N̈—Ö: :Ö—N̈=Ö: :Ö—Ö—N̈=Ö: :Ö—H

H—Ö—Ö: :C̈l—Ö: H—C̈—H   H—C̈=Ö:
                          |
                          H

13. | Name | Lewis structure |
|---|---|
| Sodium bicarbonate (baking soda) | Na⁺[H—Ö—C(=O:)—Ö:]⁻ |
| Carbonic acid (beverages) | H—Ö—C(=O:)—Ö—H |
| Acetic acid (vinegar) | H—C(H)(H)—C(=O:)—Ö—H |
| Sodium hypochlorite (bleach) | Na⁺[:C̈l—Ö:]⁻ |
| Hydrogen peroxide (disinfectant) | H—Ö—Ö—H |
| Ammonia (cleaning product) | H—N̈(H)—H |
| Hydrochloric acid (muriatic acid) | H—C̈l: |
| Sodium phosphate (detergent) | 3Na⁺ [:Ö—P(=O:)(—Ö:)—Ö:]³⁻ |
| Sodium nitrite (food preservative) | Na⁺[:Ö=N̈—Ö:]⁻ |
| Magnesium hydroxide (milk of magnesia) | Mg²⁺ 2[:Ö—H]⁻ |
| Methane (natural gas) | H—C(H)(H)—H |
| Methyl mercaptan (odiferous additive to natural gas) | H—C(H)(H)—S̈—H |
| Methanol (wood alcohol, solvent) | H—C(H)(H)—Ö—H |
| Ethanol (grain alcohol, beverage) | H—C(H)(H)—C(H)(H)—Ö—H |
| Isopropyl alcohol (rubbing alcohol) | H—C(H)(H)—C(H)(Ö:H)—C(H)(H)—H |
| n-Butane (fuel in lighters) | H—C(H)(H)—C(H)(H)—C(H)(H)—C(H)(H)—H |

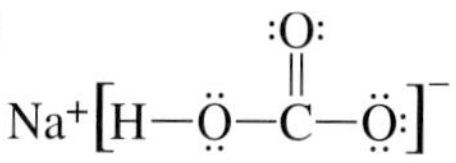

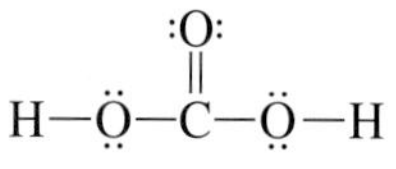

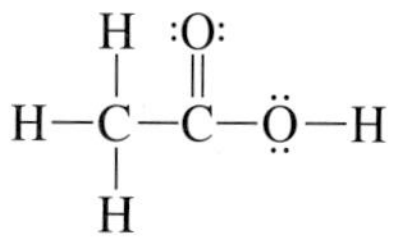

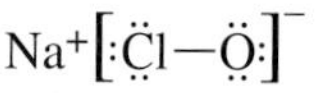

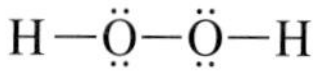

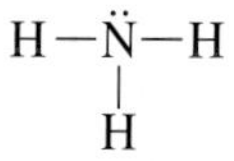

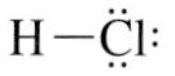

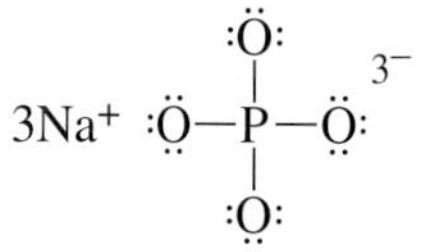

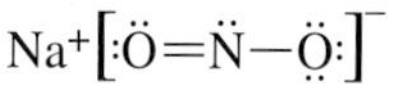

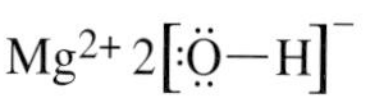

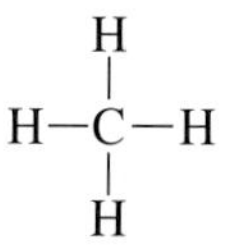

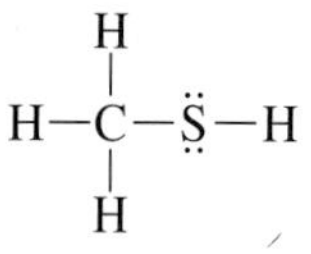

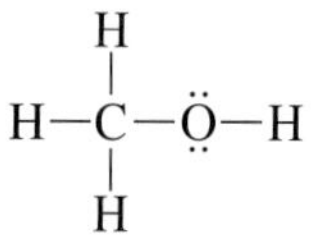

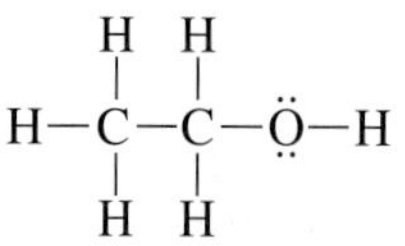

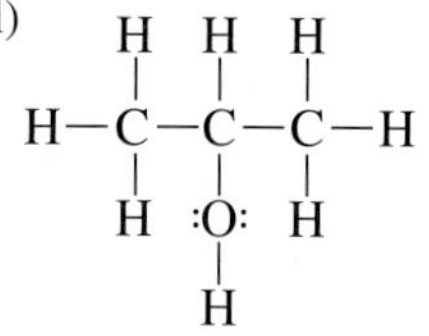

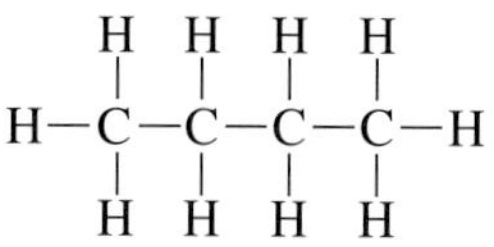

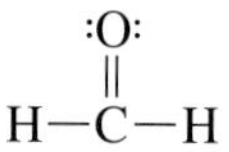

14.
Formaldehyde	H—C(=O:)—H
Acetone	H—C(H)(H)—C(=O:)—C(H)(H)—H
Nitric acid	
Sulfuric acid	
Phosphoric acid	
Hydrochloric acid	
Sodium hydroxide	
Ammonium hydroxide	
Hydrogen sulfide (hydrosulfuric acid)	H—S̈—H
Sodium thiosulfate	[:S̈—S(=O:)(—Ö:)—Ö:]²⁻
Chromium trioxide (chromium(VI) oxide)	:Ö—Cr(=O:)—Ö:

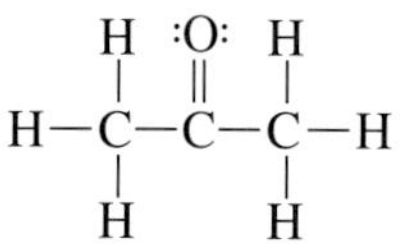

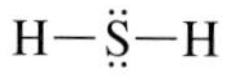

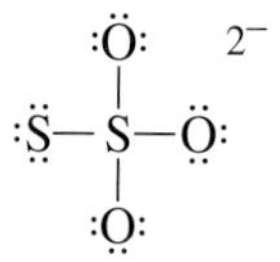

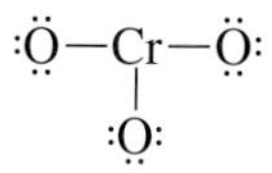

15. *Covalent:* P–H C–S Si–Te
Polar covalent: Ga–As N–H Sb–Br Al–I Ca–P B–O
Ionic: Rb–F

16. a. PC
    b. I
    c. I
    d. PC
    e. I
    f. I

17. 2.2

18. a. $2Rb + 2H_2O \longrightarrow 2RbOH + H_2$
    b. $Ba + 2H_2O \longrightarrow Ba(OH)_2 + H_2$
    c. $Be + 2H^+ \longrightarrow Be^{2+} + H_2$
    d. $Br_2 + H_2O \longrightarrow HBr + HOBr$
    e. $Br_2 + 2OH^- \longrightarrow BrO^- + Br^- + H_2O$

Answers to Exercises

f. NR
g. $2Al + 6H^+ \longrightarrow 2Al^{3+} + 3H_2$
h. $I_2 + 2OH^- \longrightarrow IO^- + I^- + H_2O$
i. $2Li + 2H^+ \longrightarrow 2Li^+ + H_2$
j. NR

19. a. basic
b. basic
d. acidic

20. See Eq. 5.10

21. See Eq. 5.12

22. c. $Be + 2HCl \longrightarrow BeCl_2 + H_2$;
$Be + H_2SO_4 \longrightarrow BeSO_4 + H_2$
g. $2Al + 6HCl \longrightarrow 2AlCl_3 + 3H_2$;
$2Al + 3H_2SO_4 \longrightarrow Al_2(SO_4)_3 + 3H_2$
i. $2Li + 2HCl \longrightarrow 2LiCl + H_2$;
$2Li + H_2SO_4 \longrightarrow Li_2SO_4 + H_2$

23. Cs ~ Rb > Ca ~ Sr > Al > Ga

24.
a. $Li_2O + H_2O \longrightarrow 2LiOH$ basic (B)
b. $SeO_3 + H_2O \longrightarrow H_2SeO_4$ acidic (A)
c. $CO_2 + H_2O \longrightarrow H_2CO_3$ (A)
d. $CaO + H_2O \longrightarrow Ca(OH)_2$ (B)
e. $SO_2 + H_2O \longrightarrow H_2SO_3$ (A)
f. $As_4O_{10} + 6H_2O \longrightarrow 4H_3AsO_4$ (A)
g. NR
h. $Cs_2O + H_2O \longrightarrow 2CsOH$ (B)
i. $N_2O_5 + H_2O \longrightarrow 2HNO_3$ (A)
j. $P_2O_3 + 3H_2O \longrightarrow 2H_3PO_3$ (A)

25. c. See Eqs. 5.23 and 5.25
d. and **h.** See Eq. 5.20

26. $Ca(OH)_2$ is a base since $\chi_{Ca} < \chi_H$; $SO_2(OH)_2$ is an acid since $\chi_S > \chi_H$

27.

Element	p only	s & p	
C	CO	CO_2	
P	P_2O_3	P_2O_5	
N	N_2O_3	N_2O_5	N has more oxides: N_2O, NO, NO_2, NO_3
S	SO_2	SO_3	
As	As_2O_3	As_2O_5	
Pb	PbO	PbO_2	
Se	SeO_2	SeO_3	

28. Peroxides, O_2^{2-} $[\ddot{\text{O}}-\ddot{\text{O}}]^{2-}$
Superoxides, O_2^- $[\dot{\text{O}}-\ddot{\text{O}}]^-$

29. a. $NaH + H_2O \longrightarrow NaOH + H_2$
b. $CaH_2 + 2H_2O \longrightarrow Ca(OH)_2 + 2H_2$
c. $AlH_3 + 3H_2O \longrightarrow Al(OH)_3 + 3H_2$
d. $LiAlH_4 + 4H_2O \longrightarrow LiOH + Al(OH)_3 + 4H_2$

30. See Eq. 5.28
M — electron attacks partial +H of water
MH — partial −H of hydride attacks partial +H of water
MO — lone pair on oxide attacks partial +H of water

31. $\chi_{Cl} > \chi_S$ makes HCl a stronger acid

32. Predict HF > HCl > HBr > HI from χ's (HI has weakest bond)

33. a. $2NaOH + H_2SO_4 \longrightarrow Na_2SO_4 + 2H_2O$
Arrhenius (A)
b. $NH_3 + HClO_4 \longrightarrow NH_4^+ + ClO_4^-$
Brønsted–Lowry (BL)
c. $CaO + CO_2 \longrightarrow CaCO_3$
Lewis (L)
d. $NaHCO_3 + HCl \longrightarrow NaCl + CO_2 + H_2O$
BL
e. $H_2O + HCl \longrightarrow H_3O^+ + Cl^-$
BL
f. $3Ca(OH)_2 + 2H_3PO_4 \longrightarrow Ca_3(PO_4)_2 + 6H_2O$
A
g. $H_2O + NH_3 \longrightarrow NH_4^+ + OH^-$
BL
h. $Na_2S + H_2O \longrightarrow NaOH + NaHS$
BL
i. $SO_2 + H_2O \longrightarrow H_2SO_3$
L
j. $LiH + AlH_3 \longrightarrow LiAlH_4$
L

34. a. $2OH^- + 2H^+ \longrightarrow 2H_2O$
b. $NH_3 + H^+ \longrightarrow NH_4^+$
c. $NaHCO_3 + H^+ \longrightarrow Na^+ + H_2O + CO_2$
d. $3OH^- + 3H^+ \longrightarrow 3H_2O$
e. $S^{2-} + H_2O \longrightarrow OH^- + HS^-$
f. $H^+ + OH^- \longrightarrow H_2O$, neutralization

35.

	Acid	Base	Conjugate acid	Conjugate base
b.	$HClO_4$	NH_3	NH_4^+	ClO_4^-
d.	HCl	HCO_3^-	H_2O	Cl^-
e.	HCl	H_2O	H_3O^+	Cl^-
g.	H_2O	NH_3	NH_4^+	OH^-
h.	H_2O	S^{2-}	HS^-	OH^-

36. $CaCO_3 + H_2SO_4 \longrightarrow CaSO_4 + CO_2 + H_2O$; BL

37.
	Acid	Base	Adduct
c.	CO_2	CaO	$CaCO_3$
i.	SO_2	H_2O	H_2SO_3
j.	AlH_3	LiH	$LiAlH_4$

38. a. BL
    b. L
    c. L
    d. BL
    e. L
    f. L
    g. NR
    h. BL
    i. L
    j. L

39. 0.4340 g HCl/g antacid

40. a. $K + NaCl \longrightarrow KCl + Na$ (1e)
    c. $H_2 + CuO \longrightarrow H_2O + Cu$ (2e)
    e. $Ba + Ag_2S \longrightarrow BaS + 2Ag$ (2e)
    g. $Au^{3+} + Cr \longrightarrow Cr^{3+} + Au$ (3e)
    i. $2Zn + PbO_2 \longrightarrow 2ZnO + Pb$ (4e)

41. Only coefficients are given
    a. 1,5 $\longrightarrow$ 3,4
    b. 1,6 $\longrightarrow$ 6,6
    c. 1,3 $\longrightarrow$ 2,3
    d. 2,13 $\longrightarrow$ 8,10
    e. 2,5 $\longrightarrow$ 4,2
    f. 4,5 $\longrightarrow$ 4,6
    g. 2,1 $\longrightarrow$ 2
    h. 4,9 $\longrightarrow$ 2,8,10

42. $C_8H_{18} + 12O_2 \longrightarrow 7CO_2 + CO + 9H_2O$; 31 g CO per mile

43. $2C_{20}H_{42} + 21O_2 \longrightarrow 40C + 42H_2O$

44. Left-to-right sequence of oxidation numbers is given for each species
    (1,6,−2) (1,5,−2) (1,5,−2) (1,−1) (4,−2) (5,−2) (2,−1)
    (1,1,4,−2)
    (5,−2) (3,−2) (−3,1) (−3,1) (−2,1,−2) (−4,1) (−2,1,−1)

45. For this exercise and the two that follow, unlabeled species are understood to be (aq)
    a. $3CuS(s) + 8HNO_3 \longrightarrow 3CuSO_4 + 8NO(g) + 4H_2O$
    b. $Br_2 + H_2C_2O_4 \longrightarrow 2HBr + 2CO_2(g)$
    c. $3H_2O_2 + CH_4O \longrightarrow 5H_2O + CO_2(g)$
    d. $3NO_2(g) + H_2O \longrightarrow 2HNO_3 + NO(g)$
    e. $3H_2S + K_2Cr_2O_7 + 8HCl \longrightarrow 3S(s) + 2CrCl_3 + 2KCl + 7H_2O$
    f. $2KMnO_4 + 16HBr \longrightarrow 2MnBr_2 + 5Br_2 + 8H_2O + 2KBr$
    g. $KIO_3 + 8KI + 3H_2SO_4 \longrightarrow 3KI_3 + 3H_2O + 3K_2SO_4$

46. a. $I_2 + 2S_2O_3^{2-} \longrightarrow 2I^- + S_4O_6^{2-}$
    b. $3CN^- + 2MnO_4^- + H_2O \longrightarrow 2MnO_2(s) + 3CNO^- + 2OH^-$
    c. $2CrO_4^{2-} + 3HSnO_2^- + H_2O \longrightarrow 3HSnO_3^- + 2CrO_2^- + 2OH^-$
    d. $2H_2O_2 + N_2H_4 \longrightarrow 4H_2O + N_2(g)$
    e. $6VO^{2+} + ClO_3^- + 6OH^- \longrightarrow 6VO_2^+ + Cl^- + 3H_2O$

47. 62.05% Fe in ore

48. $ClO^- + 2I^- + 2H^+ \longrightarrow Cl^- + I_2 + H_2O$;
    $2S_2O_3^{2-} + I_2 \longrightarrow S_4O_6^{2-} + 2I^-$; 0.8660 M ClO^-

49. $Cr_2O_7^{2-} + 6I^- + 14H^+ \longrightarrow 2Cr^{3+} + 3I_2 + 7H_2O$;
    0.12 g $K_2Cr_2O_7$

50. a. $4HCl + MnO_2(s) \longrightarrow MnCl_2 + Cl_2(g) + 2H_2O$
    b. $2NaI + 2H_2SO_4 \longrightarrow I_2 + SO_2 + Na_2SO_4 + 2H_2O$
    c. $Zn + 2HCl \longrightarrow ZnCl_2 + H_2(g)$
    d. $NH_4^+ + NO_2^- \longrightarrow N_2(g) + 2H_2O$
    e. $2KClO_3(s) \longrightarrow 2KCl(s) + 3O_2(g)$
    f. $8NH_3(g) + 3Br_2(l) \longrightarrow N_2(g) + 6NH_4Br(s)$
    g. $16H_2S(g) + 8SO_2(g) \longrightarrow 3S_8(s) + 16H_2O(l)$
    h. $2Ca_3(PO_4)_2(s) + 10CO(g) \longrightarrow 6CaO(s) + 10CO_2(g) + P_4(s)$
    i. $2KBr + 2H_2SO_4 + MnO_2(s) \longrightarrow Br_2 + K_2SO_4 + MnSO_4 + 2H_2O$
    j. $K_2SiF_6(s) + 4K(s) \longrightarrow Si(s) + 6KF(s)$
    k. $H_2SeO_3 + 2SO_2 + H_2O \longrightarrow Se(s) + 2H_2SO_4$

51. 15.5 g Zn

52. $S^{2-} + Cd^{2+} \longrightarrow CdS$
    LB   LA

53. a. no
    b. $Na_2CO_3 + BaCl_2 \longrightarrow BaCO_3(s) + 2NaCl$
    c. $Pb(C_2H_3O_2)_2 + 2NaOH \longrightarrow Pb(OH)_2(s) + 2NaC_2H_3O_2$
    d. no
    e. no
    f. $2Cr(NO_3)_3 + 3Na_2S \longrightarrow Cr_2S_3(s) + 6NaNO_3$

54. 1.83 g

55. a. acid–base Lewis
    b. redox

c. redox
   d. acid–base metathesis
   e. redox displ.
   f. acid–base B-L
   g. redox
   h. acid–base
   i. acid–base B-L
   j. acid–base Lewis
56. a. acid–base Lewis adduct decomp.
   b. redox displ.
   c. acid–base B–L
   d. redox addn.
   e. redox displ.

Chapter 6

1. a. Li_2 $\Psi = (1/\sqrt{2})\,[2s_A(\mathbf{r}_1)2s_B(\mathbf{r}_2) + 2s_A(\mathbf{r}_2)2s_B(\mathbf{r}_1)]$
    b. F_2 $\Psi = (1/\sqrt{2})\,[2p_A(\mathbf{r}_1)2p_B(\mathbf{r}_2) + 2p_A(\mathbf{r}_2)2p_B(\mathbf{r}_1)]$
    c. LiH $\Psi = (1/\sqrt{2})\,[2s_{Li}(\mathbf{r}_1)1s_H(\mathbf{r}_2) + 2s_{Li}(\mathbf{r}_2)1s_H(\mathbf{r}_1)]$
    Case c) omits an ionic term of the form $1s_H(\mathbf{r}_1)1s_H(\mathbf{r}_2)$

2. The integrand has equal magnitude but opposite signs above and below the plane in the $2p_\pi$ case, yielding $S = 0$

3. D_e's similar, but $r_e(H_2) < r_e(NaCl)$; NaCl shows Coulomb attraction at long range; exemplifies the difference between orbitals merging (H_2) or only touching (NaCl)

4. $r_e(HF) \approx 0.94$ Å; $r_e(H-O) \approx 0.96$ Å; $r_{orbital} > r_{atom}$ due to covalent overlap

5. $h_1 = sp_1 = (1/\sqrt{2})(2s + 2p_x) \propto (\rho - 2 + \rho\sin\theta\cos\phi)\,e^{-\rho/2}$
   $h_2 = sp_2 = (1/\sqrt{2})(2s - 2p_x) \propto (\rho - 2 - \rho\sin\theta\cos\phi)\,e^{-\rho/2}$

6. a. CH_3Cl sp^3 $109°$
    b. SiF_4 sp^3 $109°$
    c. AlH_4^- sp^3 $109°$
    d. MgI_2 sp $180°$
    e. BCl_3 sp^2 $120°$
    f. PH_3 sp^3 $\approx 109°$
    g. PO_4^{3-} sp^3 $109°$

7. a. CO_2 sp $180°$
    b. NO_2^- sp^2 $\approx 120°$
    c. NO_3^- sp^2 $120°$
    d. SO_3 sp^2 $120°$
    e. O_3 sp^2 $\approx 120°$
    f. ClO_3^- sp^3 $\approx 109°$

8. $\Psi_{C-H} = (1/\sqrt{2})\,[sp_C^3(\mathbf{r}_1)1s_H(\mathbf{r}_2) + sp_C^3(\mathbf{r}_2)1s_H(\mathbf{r}_1)]$
   $\Psi_{C-F} = (1/\sqrt{2})\,[sp_C^3(\mathbf{r}_1)2p_F(\mathbf{r}_2) + sp_C^3(\mathbf{r}_2)2p_F(\mathbf{r}_1)]$

9. $\Psi_{C-H} = (1/\sqrt{2})\,[sp_C^2(\mathbf{r}_1)1s_H(\mathbf{r}_2) + sp_C^2(\mathbf{r}_2)1s_H(\mathbf{r}_1)]$
   $\Psi_{\sigma,C=O} = (1/\sqrt{2})\,[sp_C^2(\mathbf{r}_1)2p_{\sigma O}(\mathbf{r}_2) + sp_C^2(\mathbf{r}_2)2p_{\sigma O}(\mathbf{r}_1)]$
   $\Psi_{\pi,C=O} = (1/\sqrt{2})\,[2p_{\pi C}(\mathbf{r}_1)2p_{\pi O}(\mathbf{r}_2) + 2p_{\pi C}(\mathbf{r}_2)2p_{\pi O}(\mathbf{r}_1)]$

10.   C   C   O
    $sp^3, 109°$ $sp^2, \approx 120°$ $sp^3, \approx 109°$
    $\Psi_{C-O} = (1/\sqrt{2})\,[sp_C^2(\mathbf{r}_1)sp_O^3(\mathbf{r}_2) + sp_C^2(\mathbf{r}_2)sp_O^3(\mathbf{r}_1)]$

    ```
         H   :O:
         |   ‖
    H — C — C — Ö — H
         |
         H
    ```

11. H_2SO_4: S sp^3, O sp^3; HNO_3: N sp^2, O sp^3; H_3PO_4: P sp^3, O sp^3

12. B is sp^2; its unused $2p$ orbital $\perp$ to molecular plane is a good e-pair acceptor

13. a. SCl_2 bent, $\approx 109°$ f. I_3^- linear, $180°$
    b. SO_4^{2-} tetrahedral, $109°$ g. SF_6 octahedral, $90°$
    c. NO_3^- trigonal planar, $120°$ h. PO_3^{3-} trigonal pyramidal, $\approx 109°$
    d. N_2O linear, $180°$ i. IF_5 square pyramidal, $90°$
    e. HCN linear, $180°$

14. a. GeH_3Cl tetrahedral, $109°$ f. NHF_2 trigonal pyramidal, $\approx 109°$
    b. CO_3^{2-} trigonal planar, $120°$ g. SF_4 sawhorse, $90°, 120°$
    c. S_2O bent, $120°$ h. HgI_2 linear, $180°$
    d. PCl_3 trigonal pyramidal, $\approx 109°$ i. XeF_4 square planar, $90°$
    e. PCl_5 trigonal bipyramidal, $90°, 120°$

15. a. $CHONH_2$: C sp^2, $120°$; N sp^3, $\approx 109°$
    b. CH_3OH: C sp^3, $109°$; O sp^3, $\approx 109°$
    c. CH_2CO: C_1 sp^2, $\approx 120°$; C_2 sp, $180°$
    d. HCOOH: C sp^2, $120°$; O sp^3, $\approx 109°$
    e. CH_2NH: C sp^2, $\approx 120°$; N sp^2, $120°$

16.
    ```
           H    H           H         sp², ≈120°
            \  /             \       ⋰
       cis   N=N     trans    N=N:
                                  \
                                   H
    ```

17. NO_2^+ sp, $180°$; NO_2^- sp^2, $\approx 120° \Rightarrow NO_2 \approx 150°$ (expt. $134°$)
    HO_2^+ sp^2, $\approx 120°$; HO_2^- sp^3, $\approx 109° \Rightarrow HO_2 \approx 114°$ (expt. $112°$)

18. NaCl $\delta = 0.79$ (Table 6.1); $\delta = 0.73$ (Table 5.1 and Eq. 6.9)
    HCl $\delta = 0.18$ (Table 6.1); $\delta = 0.20$ (Table 5.1 and Eq. 6.9)

19. KBr $\delta = 0.77$ (Table 6.1); $\delta = 0.70$ (Table 5.1 and Eq. 6.9)
    HBr $\delta = 0.12$ (Table 6.1); $\delta = 0.12$ (Table 5.1 and Eq. 6.9)

20. $\delta = 0.52$ for $\Delta\chi = 1.7$; $\delta = 0.02$ for $\Delta\chi = 0.4$
21. H_2O $\mu_{O-H} = 1.51$ D, $\delta = 0.33$ H_2S $\mu_{S-H} = 0.70$ D, $\delta = 0.11$
    NO_2 $\mu_{N-O} = 0.41$ D, $\delta = 0.07$ SO_2 $\mu_{S-O} = 1.61$ D, $\delta = 0.24$
22. PH_3 SF_4 $SeCl_2$ IF_5
23. $Cl_2O < HOCl < HOF < H_2O$
24. $\delta_{C-F} = 0.27$; $\delta_{C-I} = 0.19$
25.
$$\Psi_{HCl} = (a/\sqrt{2})\,[1s_H(\mathbf{r}_1)3p_{Cl}(\mathbf{r}_2) + 1s_H(\mathbf{r}_2)3p_{Cl}(\mathbf{r}_1)]\; \overset{covalent\;term}{} + \overset{ionic\;term}{b[3p_{Cl}(\mathbf{r}_1)3p_{Cl}(\mathbf{r}_2)]}$$
$b \approx \sqrt{\delta} = 0.42$ from Table 6.1, 0.45 from Eq. 6.9.

Chapter 7

1. $\oplus \cdot \oplus$ $V = -40.8$ eV
   $\oplus \cdot \oplus$ $V = -44.0$ eV (more stable)
2. 0.10 (10% increase)
3. $1 \le Z_{eff} \le 2$
4. $1 \le Z_{eff} \le 1.69$
5. a. $S = 0$
   b. $S = 0$
   c. $S = 0$
   d. $S \ne 0$
   e. $S \ne 0$
   f. $S \ne 0$
6. C_2 $(\sigma_{2s})^2(\sigma_{2s}*)^2(\pi_{2p})^4$ b.o. = 2
7. Be_2 $(\sigma_{2s})^2(\sigma_{2s}*)^2$ b.o. = 0
8. Al_2 $(\sigma_{3s})^2(\sigma_{3s}*)^2(\pi_{3p})^2$ b.o. = 1 $\pi_{3p}\uparrow\;\uparrow$
9. N_2^+ b.o. = 2½, $r_e(N_2^+) > r_e(N_2)$, $D_e(N_2^+) < D_e(N_2)$ (bonding e removed)
   O_2^+ b.o. = 2½, $r_e(O_2^+) < r_e(O_2)$, $D_e(O_2^+) > D_e(O_2)$ (antibonding e removed)
10. Ne_2 $(\sigma_{2s})^2(\sigma_{2s}*)^2(\sigma_{2p})^2(\pi_{2p})^4(\pi_{2p}*)^4(\sigma_{2p}*)^2$
    b.o. = 0, unstable
    Ne_2^+ $(\sigma_{2s})^2(\sigma_{2s}*)^2(\sigma_{2p})^2(\pi_{2p})^4(\pi_{2p}*)^4(\sigma_{2p}*)^1$
    b.o. = ½, stable
11. N_2: b.o. = 3 $\longrightarrow$ 2½; O_2 2 $\longrightarrow$ 1½; C_2: 2 $\longrightarrow$ 2½; only C_2 is stabilized
12. O_2^{2-} (as F_2) b.o. = 1; O_2^- b.o. = 1½ $\Rightarrow r_e(O_2^{2-}) > r_e(O_2^-)$
13. C_2^{2-} (as N_2) b.o. = 3
14. CN^- (as N_2) b.o. = 3
    CN $(\sigma_{2s})^2(\sigma_{2s}*)^2(\pi_{2p})^4(\sigma_{2p})^1$ b.o. = 2½
    $D_e(CN) < D_e(CN^-)$, $r_e(CN) > r_e(CN^-)$

15. ClO $(\sigma_s)^2(\sigma_s*)^2(\sigma_p)^2(\pi_p)^4(\pi_p*)^3$ b.o. = 1½ $D_e(ClO) > D_e(C-Cl)$
16. NO^+: b.o. = 3, strongest bond (see Exercise 9 answer)
17. BC $\longrightarrow$ BC^-: b.o. 1½ $\longrightarrow$ 2 = > $r_e\downarrow$ and $D_e\uparrow$
    CO $\longrightarrow$ CO^-: b.o. 3 $\longrightarrow$ 2½ = > $r_e\uparrow$ and $D_e\downarrow$
    ClO $\longrightarrow$ ClO^-: b.o. 1½ $\longrightarrow$ 1 = > $r_e\uparrow$ and $D_e\downarrow$
18. $OF^- < BC < BN < N_2^- < NO^+$
    b.o. 1   1½   2   2½   3
19. HCl $(\sigma_{3s})^2(\sigma_{3p})^2(n_{3p})^4$
20. NH $(\sigma_{2s})^2(\sigma_{2p})^2(n_{2p})^2$ b.o. = 1 $n_{2p}\uparrow\;\uparrow$, paramagnetic
21. OH $(\sigma_{2s})^2(\sigma_{2p})^2(n_{2p})^3$ b.o. = 1 $n_{2p}\uparrow\downarrow\;\uparrow$, paramagnetic
22. OH^- $(\sigma_{2s})^2(\sigma_{2p})^2(n_{2p})^4$ b.o. = 1 LUMO $\sigma_{2p}*$; HOMO n_{2p} (base)
23. LiH $(\sigma_{2s})^2$, $\sigma_{2s} = a\,1s_H + b\,2s_{Li}$, $a > b$
24. K HOMO 4s; ICl LUMO $\sigma_{5p3p}* = b\,5p_I - a\,3p_{Cl}$, b > a; expect preferential transfer to I end; more KI unless charge migrates
25. CH_2 is bent; HOMO as Figure 7.17, $a_1(n)$
26. H_2S acidity due to LUMO as Figure 7.17, $a_1*(\sigma)$
27. Three σ levels with PIB nodal structure; $(\sigma_1)^2(\sigma_2)^1$
28. Same as Exercise 27; $(\sigma_1)^2$; since $\sigma_1 \longrightarrow a_1(\sigma)$ is lowered by bending, molecule will be bent
29. Bond lengths equal within each ion; $r_e(NO_2^-) < r_e(NO_3^-)$
30. $r_e(SO_2) < r_e(SO_3)$

Chapter 8

1. He $T = 3$, $R = 0$, $V = 0$; F_2 321; NO 321; SO_2 333; C_2H_2 327
2. 2Ne $T = 6$, $R = 0$, $V = 0$; Ne_2 321; N_2O 324; CH_2CO 339; C_3H_8 3, 3, 27
3. 2OH $T = 6$, $R = 4$, $V = 2$; H_2O_2 336; $Cu(NH_3)_4^{2+}$ 3, 3, 45; $(H_2O)_n$ 3, 3, $9n - 6$; $H^+(H_2O)_n$ 3, 3, $9n - 3$.
6. H_2 0.148 ps; HI 0.450 ps; I_2 5.96 ps
7. 300 K 1.85 ps; 3000 K 0.584 ps; 30 K 5.84 ps
8. H_2 2.52 Å, 0.541; HI 1.79 Å, 0.178; I_2 0.224 Å, 0.0134; I_2
9. $K_{rot} = n^2h^2/(8\pi^2 m_e r_n^2)$, $n = 1, 2, 3, ...$
11. Both ν_{rot} and $\nu_{photon} \longrightarrow jh/(4\pi^2 I)$ or $J/(2\pi I)$
12. $\bar{j} = (2\pi/h)[2Ik_BT]^{\frac{1}{2}} - \frac{1}{2}$; $\bar{j} = 1, 5, 74$; large $\bar{j} \Rightarrow$ classical limit (small $\lambda/(2\pi r_e)$)
13. CsI $r_e = 3.32$ Å; LiH $r_e = 1.60$ Å
14. $\bar{\nu} = 13.17, 26.34, 39.51$ cm^{-1}
15. ClF, NO_2, CH_3Cl
17. $k = 2\beta^2 D_e$

18. H_2 7.59 fs, 19.5; I_2 0.155 ps, 38.4
19. $\varepsilon_0 \approx h\nu/(4\sqrt{2})$
20. H_2 0.124 Å, 0.33; I_2 0.050 Å, 0.037
21. 209 cm$^{-1} \ll \bar{\nu} \Rightarrow$ not enough to provide one vibrational quantum
23. All but Cl_2.
24. 660 nm (visible, red); overestimate, since real vibrational spacings decrease
25. $^2H^{35}Cl$ 2145 cm^{-1}, 4.54 eV; $^1H^{37}Cl$ 2988 cm^{-1}, 4.49 eV; $^{14}N^{15}N$ 2320 cm^{-1}, 9.76 eV
26. $\pi_{np}^* \longrightarrow \sigma_{np}^*$ in each case
27. $\sigma \longrightarrow \sigma^*$; 0.91 eV
28. $\pi \longrightarrow \pi^*$, since π is more strongly bonding; the $\pi \longrightarrow \pi^*$ band will be broader, since it populates more vibrational states of the excited electronic state
29. $\pi_4 \longrightarrow \pi_5$; 352 nm (19% error)
30. σ_{2s}^* −18.9 eV; π_{2p} −16.8; σ_{2p} −15.7; IE = 15.7
31. 2.89 σ_{2s}^*; 4.65 π_{2p}; 5.10 σ_{2p}; 11.67 π_{2p}^*; IE = 9.56 eV

Chapter 9

1. 10.3 m (34 ft)
2. **a.** 740. Torr, 0.973 atm, 9.86×10^4 Pa
   b. 2030 g
3. 8 km (5 mi)
4. h = 34 cm (13 in.); though short, they have deep pockets
5. 12.0 Torr (0.0158 atm)
6. 4970 N (1120 lb)
7. 0.40 atm at 298 K
8. 900 balloons
9. 56 L; 500 L
10. 3.3 L
11. 57 N (13 lb)
12. 66.9 mL; 2.98 mmol
13. 0.894 mmol; 25.8 mg
14. 14.4 atm
15. 24.5 L
16. 3.24×10^{11}
17. 12 mg
18. 2210 K
19. 188 g/mol
20. $\rho_N \equiv N/V = 3.24 \times 10^{16}\, P\,(\text{Torr})$; 2.46×10^{19} cm^{-3}; 324 cm^{-3}
21. 478 torr
22. 727 torr; 111 mg
23. P_{He} = 131 torr; P_{Ar} = 120 torr; P = 251 torr; X_{He} = 0.52; X_{Ar} = 0.48
24. $H_2 + Cl_2 \longrightarrow 2HCl$

Pressures (Torr)	H_2	Cl_2	HCl	Total
Before rxn	131	120	0	251
After rxn	11	0	240	251

25. 1.073 atm; 5.21 atm
26. 0.17 Torr HC; 9.1 Torr CO
27. 150 kg (330 lb)
28. M = 84.2 g/mol; C_6H_{12} (could be hexene or cyclohexane)
29. 743 L
30. 0.953 g
31. the lighter gas has greater average velocity, and hits the wall more often
32. one atom Ar: $P = 1.68 \times 10^{-19}$ Pa $= 1.66 \times 10^{-24}$ atm $= 1.26 \times 10^{-21}$ Torr
    one mole Ar: $P = 1.013 \times 10^5$ Pa = 1.00 atm = 760 Torr
33. H: 7.67×10^5 cm/s; I_2: 1.80×10^4 cm/s; C_8H_{18}: 2.96 Å/ps
34. $v_{esc} = [2gr]^{1/2}$ = 11.2 km/s (25,000 mph); T = 20,000 K (positively stellar!)
35. 10^{-379}; since there are only 2×10^{43} O_2 molecules in the atmosphere, not a single O_2 molecule can escape
36. Δv = 2.8, 4.9, 8.4 $\times 10^4$ cm/s; Δv vs. $\sqrt{T}$ is linear; energy spread $\propto T$
37. $f(E) = 4\pi\sqrt{2}\,[2\pi kT]^{-3/2}\,E^{1/2}\,e^{-E/kT}$
38. $\bar{v}_{air}$ = 1460 ft/s, comparable to $\bar{v}_{sound}$
39. 3.87 kJ, 0.924 kcal
40. **a.** 1.09×10^{24} cm^{-2}s^{-1}; 1.41×10^{10} s^{-1}; 1.26×10^{-5} cm
    b. 1.09×10^{17} cm^{-2}s^{-1}; 1410 s^{-1}; 126 cm
41. **a.** 2.0×10^{23}, 7.8×10^{23} collisions s^{-1}
    b. 830 Å, 220 Å
42. 50.7 Torr
43. 64 g/mol; SO_2
44. 3.79
45. 40.6 cm from HCl source; 26.7 cm from HI; 52.6 cm from HCl
46. 170 ns
47. 159 atm; 154 atm; a term dominates
48. 188 atm; a term much smaller, b more dominant
49. low V, high P
50. 0.335 mol
51. ideal 6.34 L/mol; van der Waals 6.32 L/mol; no (0.3% difference *cf.* 4% deviation in PV product from Table 9.1)

52. e.g., for propane C_3H_8, V_{STP} = 22.087 L/mol from the van der Waals equation; 2.2% ideal gas error, 0.7% van der Waals error

53. $d \approx (b/N_A)^{1/3}$ yields d = 3.70 Å; 4.16 Å; 4.30 Å
    A more rigorous relation between b and d is $b = \pi N_A d^3$; this yields d = 2.89 Å for H_2O; the value from Chapter 1 based on liquid density is d = 3.10 Å

Chapter 10

1. 61 J
2. −2390 J
3. -8.56×10^5 J
4. a. 1.89×10^5 J
   b. −5160 J
5. $h = c/g$ = 427 m (it is doubtful that such a high pop-up has ever been hit.)
6. a. 42,500 cal (1.78×10^5 J)
   b. 150 s (~2½ min)
   c. 18 m!
7. 0.394 cal/g°C (1.65 J/g°C)
8. 1.12×10^6 cal (4.67×10^6 J)
9. 9.6 servings
10. a. 37.8°C
    b. 50.6°C
    c. 24.9°C
11. 84.2°C
12. 31.8°C [6°C cooler than in Exercise 10a]
13. 0.095 cal/g°C; 63 g/mol
14. a. $q = +38.7$ cal (+162 J); $w = -38.7$ cal; $\Delta E = 0$
    b. $T_2 = 276$ K (3°C); $q = 0$; w = −27.3 cal (−115 J) = ΔE
    (What really happens is somewhere between these extremes.)
15. compression is partially adiabatic, so gas in cylinder is not able to throw off all the energy it acquires from work as heat transferred to the surroundings; ΔT = +580°C
16. q = 44.4 kJ; ΔE = 42.0 kJ; ΔH = 44.4 kJ
17. For 1 mol, $\Delta H = \Delta E + R\Delta T \geq \Delta E$; $\Delta H = \Delta E$ when $\Delta T = 0$; $R\Delta T = -w$, so ΔH corrects for work done; $\Delta H \leq \Delta E$ ($\Delta T < 0$)
18. 12.2×10^{-21} J; 4.07×10^{-21} J; 197 K; 590 cal (2450 J)
19. ΔE = −6230 cal; ΔH = −10380 cal; q and w cannot be determined without a knowledge of the path
20. −7810 cal/g; −93.8 kcal/mol
21. −6331 cal/g; −278.9 kcal/mol
22. −129 kcal/mol; 2%; 0.0046
23. −212.65 kcal/mol; 11%
24. +43.14 kcal/mol; +21.57 kcal/mol; −43.14 kcal/mol
25. $\Delta H_1 = \Delta H_2 + \Delta H_3$ or $\Delta H_1 - \Delta H_2 - \Delta H_3 = 0$
26. −18.04 kcal/mol
27. −196.536 kcal/mol
28. CH_3OH: 5.419 kcal/g; C_8H_{18}: 11.47 kcal/g; decrease
29. −68.22 kcal; −34.11 kcal/mol O_3; heat
30. −47.75 kcal/mol
31. +32.41 kcal/mol
32. $H^+(aq) + OH^-(aq) \longrightarrow H_2O(l)$ -13.34_5 kcal/mol
    $H^+(aq) + NH_3(aq) \longrightarrow NH_4^+(aq)$ −12.48 kcal/mol
    $NH_4^+(aq) + OH^-(aq) \longrightarrow NH_3(aq) + H_2O(l)$ -0.86_5 kcal/mol
33. −52.26 kcal/mol
34. +13.34 kcal/mol; Arrhenius acid–base
35. −148.1 kcal/mol
36. +3.12 kcal/mol; high T
37. −40 kcal/mol
38. 99.3 kcal/mol
39. +31 kcal/mol
40. $H + O_2 \longrightarrow OH + O$ +8 kcal/mol
    $O + H_2 \longrightarrow OH + H$ −6 kcal/mol, exothermic
41. From heats of formation (more accurate) −216.4 kcal/mol
    From bond energies   −214 kcal/mol
42. +27 kcal; no
43. HCl(g)—8.94 cal/mol K; CO_2(g)—14.90 cal/mol K; too high because quantized vibrations cannot hold the full RT per DF
44. N_2 — 6.96 cal/mol K; O_2 — 6.96 cal/mol K; CH_4 — 7.95 cal/mol K
45. 0.115 cal/g°C; Al, about twice as many atoms per unit mass, each holding $3k_BT$
46. 17.9 cal/mol K; 0.161 cal/g°C
47. q_P = 79.1 cal; q_V = 60.5 cal
48. 1.22×10^{16} cal
49. q_P = −20.8 kcal; q_V = −16.1 kcal
50. 6×10^{-10}; 2700 K
51.

v	$f(v)$	$E_v = N_A(\varepsilon_v - \varepsilon_0)$	$f(v)E_v$
0	0.9349	0.000 kcal	0.000 kcal
1	0.0611	1.615	0.099
2	0.0040	3.230	<u>0.013</u>

E_{vib} = 0.112 kcal/mol
RT = 0.592
Energy in vibration is 19% of classical

52. $\Delta H°_T = -22.04 - 0.01087(T - 298)$; -27.50 kcal
53. $\Delta H°_{f,230} = -17.38$ kcal/mol

Chapter 11

1. $w_{rev} = -1301$ cal $(-5440$ J$)$; $w_{rev}/w_{irrev} = 1.648$
2. $-w_2 = 316$ cal $(1324$ J$)$; $-w_1 < -w_2 < -w_{rev}$; $-w_n \longrightarrow -w_{rev}$
3. $w_1 = 543$ cal $(2270$ J$)$; $w_2 = 452$ cal $(1890$ J$)$; $w_{rev} = 376$ cal $(1573$ J$)$; $w_n \longrightarrow w_{rev}$
4. a. $V_1 = 7.385$ L; $V_2 = 13.567$ L; $P_2 = 1.814$ atm; $w = -447$ cal $(-1871$ J$)$
   b. $\gamma = C_P/C_V = (C_V+R)/C_V = 5/3$; $P_1V_1^\gamma = P_2V_2^\gamma = 140.0$
5. a. -298 cal $(-1250$ J$)$
   b. $T_2 = 270$ K, $P_2 = 1.20$ atm
   c. no; T_2 would be lower
6. $C_V \ln(T_c/T_h) = nR \ln(V_1/V_4)$
7. $\varepsilon = 0.311$
9. $-w_I = w_{III} = 596$ cal $(2490$ J$)$; $w_{II} = w_{IV} = 0$; hence, $\Sigma w_i = 0$ and $\varepsilon = 0$
10. $\Delta S = 4.37$ cal/K $(18.3$ J/K$)$ for both; $q/T = 4.37$ cal/K for rev, 2.65 cal/K for irrev; thus $\Delta S \geq q/T$
11. -0.890 cal/K $(-3.72$ J/K$)$
12. a. 1.82 cal/K $(7.60$ J/K$)$
    b. 3.03 cal/K $(12.7$ J/K$)$
13. -22.1 cal/K $(-92.3$ J/K$)$
14. -141 cal/K $(-590$ J/K$)$; a more precise value is -131 cal/K
15. 2.77 cal/K $(11.58$ J/K$)$
17. $\Delta V = 190$ mL; $\Delta S = 0.0595$ cal/K $(0.249$ J/K$)$
18. -2.77 cal/K; $+nR(X_{He} \ln X_{He} + X_{Ar} \ln X_{Ar})$; -0.0595 cal/K; all systems are isolated; hence, no processes can occur for which $\Delta S < 0$
19. $W_1 = 1$; $W_2 = 16$; 4L0R = 1 arr.; 3L1R = 4; 2L2R = 6; 1L3R = 4; 0L4R = 1; $\Delta S = +3.83 \times 10^{-23}$ J/K $(9.15 \times 10^{-24}$ cal/K$) > 0$; process is spontaneous
20. $W_1 = 1$, $W_2 = 729$; $\Delta S = 9.10 \times 10^{-23}$ J/K $(2.18 \times 10^{-23}$ cal/K$)$
21. most prob. config. 4 in ε_0; 2 in ε_1; 1 in ε_2
22. $SO_3(g)$; $Cl_2O(g)$; $CH_3OH(g)$; $Na(l)$
23. a. $-$
    b. $+$ (?)
    c. $-$
    d. $-$
    e. $+$
24. a. -43.29 cal/K
    b. -9.6
    c. -68.0
    d. -44.57
    e. $+5.077$
    For a)–d), $\Delta S°_{surr} > 0$ and $|\Delta S°_{surr}| > |\Delta S°_{sys}|$
25. $\Delta G \leq 0$ and path-independent; $\Delta G = 0$; $\Delta G° = \Sigma_P \nu_P \Delta G(P) - \Sigma_R \nu_R \Delta G(R)$
26. $\Delta G = -1301$ cal $(-5440$ J$)$; $\Delta G < 0$ spontaneous; $\Delta G = w_{rev}$ (maximum work)
27. 265 cal (1110 J); $\Delta G > 0$, system can now do more work
28. -17.7 cal $(-74.2$ J$)$; $\Delta G < 0$, dilution is spontaneous
29. $\Delta G°_{298} = -94.258$ kcal/mol $(-394.38$ kJ/mol$)$; spontaneous
30. $\Delta S° > 0$; $\Delta H° = +42.62$ kcal/mol; $\Delta S° = 38.4$ cal/mol K; $\Delta G°_{298} = +31.19$ kcal/mol $(+130.5$ kJ/mol$)$; no; yes; $T_{ta} = 1111$ K $(837°C)$
31. yes; 5452 K; it would spontaneously decompose into its elements—and, further, the elements would atomize! See Exercise 32.
32. $+97.16$ kcal/mol $(+406.6$ kJ/mol$)$; 4420 K; at 298 K, $\Delta G° \gg 0$, so no tendency to dissociate; but $T_{ta} < T_{sun}$, so $\Delta G° < 0$ at solar surface, and atomization occurs
33. $\Delta H° = -488.5$ kcal/mol; $\Delta S° = +24.07$ cal/K mol; $\Delta G_{98} = -495.7$ kcal/mol $(-2074$ kJ/mol$)$, spontaneous; since $\Delta H°$ and $\Delta S°$ have opposite signs, T_{ta} is not defined; $\Delta G°$ drops as T increases, and thus reaction remains spontaneous at all T

Chapter 12

1. $\Delta G = \Delta G° + RT \ln(P_{O_2}^3 / P_{O_3}^2)$; $\Delta G° = -78.0$ kcal $(-326$ kJ$)$; $\Delta G = -72.7$ kcal $(-304$ kJ$)$, spontaneous
2. $\Delta G = \Delta G° + RT \ln\{P_{NH_3}^2 / (P_{H_2}^3 P_{N_2})\} =$
   $= -7.88 + 0.5925 \ln\{P_{NH_3}^2 / (P_{H_2}^3 P_{N_2})\}$ kcal
   $= -32.97 + 2.478 \ln\{P_{NH_3}^2 / (P_{H_2}^3 P_{N_2})\}$ kJ

P_{H_2}	P_{N_2}	P_{NH_3}	ΔG, kcal (kJ)
00	1.00	1.00	-7.88 (-32.97)
00	1.00	0.00	$-\infty$
00	1.00	0.10	-12.56 (-52.56)
(equilibrium)			0

3. $Q = P_{SO_3}^2/(P_{SO_2}^2 P_{O_2})$; $\Delta G = \Delta G° + RT \ln Q$; $\Delta G° = -33.9$ kcal $(-142$ kJ$)$; $K = 6.8 \times 10^{24}$; $Q = 2.5 < K$, or $\Delta G = -33.3$ kcal $(-139$ kJ$) < 0 \Rightarrow$ spontaneous
4. $\Delta G° = +41.38$ kcal (173.1 kJ); $\Delta G = -\infty$, spontaneous!, but $K = 4.7 \times 10^{-31}$; $P_{NO} = 2.8 \times 10^{-16}$ atm; no

5. $K = 4.49 \times 10^{-5}$; $\Delta G° = +14.4$ kcal (60.2 kJ)

6. $K = 0.0105$; $\Delta G° = +12.45$ kcal (52.1 kJ)

7. $K = 2.17$; $P_{Cl_2} = P_{Br_2} = 115$ Torr, $P_{BrCl} = 170$ Torr

8. $K = 0.0502$; $P_{H_2} = 0.18$ atm

9.

	$\Delta H°$, kcal	incr T	incr P
a.	+25.9	to right	to left
b.	−99.5	left	right
c.	−34.3	left	left
d.	−23.5	left	no effect
e.	+119.0	right	left

10. $K = 0.068$; confirms LCP; $P_{NO_2} = 1.20$ atm, $P_{N_2O_4} = 0.098$ atm

11. a. increasing P decreases dissoc, but increasing T increases it

    b. $P_{Cl}/P = (K/(2P))\{[1 + 4P/K]^{1/2} - 1\}$; for $K \ll P$, $P_{Cl}/P \approx [K/P]^{1/2}$, while for $K \gg P$, $P_{Cl}/P \approx 1 - P/K$; in both limits, incr P decr P_{Cl}/P

    c. For $P = 1.00$ atm, $P_{Cl} = 0.107$ atm, $P_{Cl_2} = 0.893$ atm, $\alpha = 0.0565$

    For $P = 10.0$ atm, $P_{Cl} = 0.353$ atm, $P_{Cl_2} = 9.65$ atm, $\alpha = 0.0180$

    d. $\Delta H° = +59.7$ kcal (250. kJ); from App. B, $\Delta H° = 58.16$ kcal (243 kJ); T's differ and $\Delta C_P > 0$

12. $\Delta H° = -10.89$ kcal (-45.6 kJ) $\Rightarrow$ b) is most favorable

13. $\Delta H° = -22.04$ kcal (-92.2 kJ) $\Rightarrow$ high P, low T is best; no (industry uses high T to enhance reaction rate); $\Delta S° = -47.45$ cal/K (-198.5 J/K); $\Delta G°_{700} \approx +11.18$ kcal (46.8 kJ) and $K = 3.2 \times 10^{-4}$

14. $K(950\ K) = 11.4$; lower (see Exercise 3)

15. $\Delta H° = +22.07$ kcal (92.3 kJ); $K = 10.4$; 0.957; lower; decrease

16. $\ln(K/K_0) = -(\Delta H°_{T_0}/R)(1/T - 1/T_0) - (\Delta C_P/R)(\ln(T_0/T) - T_0/T + 1)$
    $K(198\ K) = 8.4 \times 10^5$, 5% larger

17. $\Delta G = -2.19$ kcal/mol (-9.14 kJ/mol)

18. a. 0.00, 2.60, 7.00

    b. 12.70, 11.19, 7.00

    c. 1.48

    d. 12.54

    e. 13.08

19. a. $[H^+] = 2.09 \times 10^{-11}$ M, $[OH^-] = 4.79 \times 10^{-4}$ M

    b. 3.98×10^{-8}, 2.51×10^{-7}

    c. 3.55×10^{-3}, 2.82×10^{-12}

    d. 2.0×10^{-6}, 5.0×10^{-9}

    e. 3.2×10^{-4}, 3.2×10^{-11}

    f. 3.55×10^{-12}, 2.82×10^{-3}

20. 6.0 M

21. 0°C: $pK_w = 14.90$, pH = 7.45; 37°C: $pK_w = 13.63$, pH = 6.81; 100°C: $pK_w = 12.04$, pH = 6.02

22. $\Delta G° = +4.55$ kcal/mol (19.0 kJ/mol); $\Delta H° = +3.5$ kcal/mol (15 kJ/mol) => T − dependence weak, ionization increases with increasing T

23. $\Delta G° \approx +5.5$ to $+13.6$ kcal/mol (23 to 57 kJ/mol)

24. 0.364 M; 0.356 M

25. 4.7×10^{-4}

26. Eq. 12.27: pH = 1.731; Eq. 12.26: 1.630; Eq. 12.28: 1.746

27. angle OCO = 120°; angle COH $\approx$ 109°; RCOO$^-$ has two res forms; VB σ bonds use sp^2 hybrids on central C; MO π bond delocalized over OCO in RCOO$^-$

    $\Delta H° = 0.00$ kcal $\Rightarrow D_0$(O—H) balanced by ion solvation + π resonance energy

    $\Delta S° = -22.0$ cal/K (-92 J/K) $\Rightarrow$ ionic products organize solvent

28. Since χ(Cl)>χ(Br)>χ(I), K_a(CH$_2$ICOOH) < K_a(CH$_2$BrCOOH) < K_a(CH$_2$ClCOOH)

29. $K_b = [NH_4^+][OH^-]/[NH_3]$; $K_w = [H^+][OH^-]$; $[H^+] + [NH_4^+] = [OH^-]$; $[NH_3]_0 = [NH_3] + [NH_4^+]$; $[H^+] \ll [OH^-]$

30. nitrogen lone pair is protonated; pH = 11.49

31. a. $K = 4.4 \times 10^{10}$; $\Delta G° = -14.5$ kcal/mol (-60.8 kJ/mol)

    b. $K = 4.6 \times 10^{10}$; $\Delta G° = -14.6$ kcal/mol (-60.9 kJ/mol)

    c. $K = 3.2 \times 10^5$; $\Delta G° = -7.5$ kcal/mol (-31 kJ/mol)

    d. $K = 3.1 \times 10^9$; $\Delta G° = -13.0$ kcal/mol (-54.2 kJ/mol)

    e. $K = 9.1 \times 10^3$; $\Delta G° = -5.4$ kcal/mol (-22.6 kJ/mol)

32. a. 11.15

    b. 5.48

    c. 8.18

    d. 10.70

33. $K_b = K_w/K_a = [OH^-][HA]/[A^-]$; $K_w = [H^+][OH^-]$; $[Na^+]_0 + [H^+] = [A^-] + [OH^-]$; $[A^-]_0 = [A^-] + [HA]$

34. $K_a = K_w/K_b = [B][H^+]/[BH^+]$; $I_w = [H^+][OH^-]$; $[BH^+] + [H^+] = [Cl^-]_0 + [OH^-]$; $[BH^+]_0 = [BH^+] + [B]$

35. pH = 1.60; [CH$_3$COO$^-$] = 8.8×10^{-5} M

36. [NH$_4^+$] should decrease; pure NH$_3$: [NH$_4^+$] = 1.5×10^{-3} M; NH$_3$/NaOH: [NH$_4^+$] = 1.8×10^{-5} M

37. pH = 4.755, 5.232; HH formula accurate to ~0.001 pH unit

38. 8.51 g KNO_2; pH = 3.52, increase of 0.18; pH = 12.60, increase of 9.26
39. a. $CH_2ClCOOH + NaOH \longrightarrow CH_2ClCOONa + H_2O$, $K = K_a/K_w = 1.4 \times 10^{11}$
    b. 2.52
    c. 11.25 mL
40. pH = $p(K_w/K_b) + \log_{10}([B]/[BH^+])$ (base always appears in numerator of log arg)
41. pH = 9.25, 8.55; pH decreases; 11.32, 11.13
42. for formic acid, $pK_a = 3.75$: $V_{HCOOH} = 305$ mL, $V_{NaOH} = 195$ mL
43. pH = 1.40; $[H_2PO_3^-] = 0.040$ M; $[HPO_3^{2-}] = 2.6 \times 10^{-7}$ M
44. $K_{a2}/K_{a1} \ll 1$, but $\gg K_{a2}$ and K_w/K_{a1}
45. a. pH = 4.02; $[HS^-] = 9.5 \times 10^{-5}$ M; $[S^{2-}] = 1.3 \times 10^{-14}$ M
    b. $[HS^-] = 9.1 \times 10^{-6}$ M; $[S^{2-}] = 1.2 \times 10^{-16}$ M
46. pH = 8.68
47. pH = 1.23; $[HSO_4^-] = 0.0415$ M; $[SO_4^{2-}] = 8.5 \times 10^{-3}$ M
48. $[H^+] \approx [H_2A^-] \approx \{K_{a1}[H_3A]_0\}^{1/2}$; $[HA^{2-}] \approx K_{a2}$
49. pH $\approx$ 10.68; $[N_2H_6^{2+}] \approx 1.0 \times 10^{-14}$ M
50. a. 0.3229 M
    b. pH = 2.12, 3.75, 8.32
    c. phenolphthalein
51. $K_a = 6.5 \times 10^{-5}$
52. pH = 11.13, 9.25, 5.28; methyl red
53. $[H_2SeO_3]_0 = 0.3140$ M; pH = 1.50, 2.46, 4.89, 7.31, 10.10
54. [NaOH] = 0.1041 M; pH = 4.20, 5.51, 9.08; phenolphthalein
55. a. $2H^+ + CO_3^{2-} \longrightarrow H_2O + CO_2(aq)$; $K = 4.8 \times 10^{16}$
    b. 0.2012 M
    c. pH = 11.41, 10.32, 8.34, 6.37, 3.99
    d. phenolphthalein
56. $[Ni^{2+}] = 1.7 \times 10^{-10}$ M, 2.6×10^{-12} M
57. $[NH_3] > 2.24$ M
58. a. $Q = [KOH]^2 P_{H_2}$
    b. P_{NH_3}
    c. $P_{NH_3}/[NH_3]$
    d. $1/P_{O_2}^3$
    e. $1/P_{O_2}^{3/2}$
    f. $P_{O_2}^3$
    g. $[H_2SO_3]/P_{SO_2}$
59. a. $P_{SiF_4}P_{H_2O}^2/P_{HF}^4$
    b. $1/(P_{NH_3}^2 P_{H_3PO_4})$
    c. $P_{CO_2}^3/P_{CO}^3$
    d. $P_{CO}P_{H_2}/P_{H_2O}$
    e. $[Na_2S]/(P_{H_2S}[NaOH]^2)$
    f. $[NaHCO_3]^2/(P_{CO_2}[Na_2CO_3])$
60. $K \approx 115$ atm = P_{CO_2}; need $T > T_{ta}$; explosive if not vented
61. $K = 4.0$; $P_{H_2} = 2.13$ atm, $P_{HI} = 0.73$ atm, $P = 2.86$ atm
62. a. to left, right, right, right
    b. $[CO_2]$ or $[H_2CO_3] = 0.035$ M
63. $[SO_2]_0$ or $[H_2SO_3]_0 = 8.1 \times 10^{-5}$ M; pH = 4.09
64. $K = 9.93$; $[H_2S] \le 0.101$ M
65. $2NaHCO_3(s) \rightleftharpoons Na_2CO_3(s) + CO_2(g) + H_2O(g)$; $K = 4.7 \times 10^{-7} = P_{CO_2}P_{H_2O}$
    $Q/K = 4.2 > 1 \Rightarrow$ stable; yes; $P_{H_2O} \le 0.71$ Torr
66. 7.35 g
67. $[Ca^{2+}] = 0.0106$ M; pH = 12.32
68. 0.0026 g
69. $Q_{ion} = [Cu^{2+}][OH^-]^2 = 2.5 \times 10^{-4} > K_{sp} \longrightarrow$ ppt will form; $[Cu^{2+}] = 2.2 \times 10^{-17}$ M; yes
70. pH $\le$ 1.47
71. pH = 5.05
72. $\Delta H° > 0$; $K = 7.2 \times 10^{-5}$ at 95°C
73. $0.60 < $ pH < 1.07
74. $[Ag^+] = 2.106 \times 10^{-8}$ M; $K_f = 1.77 \times 10^7$
75. $P = 1.00$, $\gamma = 1.001$; $P = 10.0$, $\gamma = 1.007$; $P = 1000.$, $\gamma = 1.973$; less ideal at high P; dominant force is repulsive
76. $w = 1367$ cal (5720 J) vs. ideal 1364 cal; extra work done by repulsive energy release
77. $\gamma = 0.995$ at 100°C, 0.997 at 200°C; more ideal at higher T; attraction dominates
78. $r_d = 30.4$ Å, $\gamma(K^+) = \gamma(Br^-) = 0.889$; $r_d = 15.2$ Å, $\gamma(Cu^{2+}) = \gamma(SO_4^{2-}) = 0.391$; $CuSO_4$ solution is less ideal because Coulomb forces are four times stronger; this also results in a factor-of-two reduction in Debye length for $CuSO_4$

Chapter 13

1. g Zn $\ge$ 1.63 g; $[Ag^+]_{eq} = 9.1 \times 10^{-28}$ M $\Rightarrow 2.7 \times 10^{-4}$ ions in 0.5 L $\Rightarrow$ no Ag^+ ions; $[Zn^{2+}] = 0.050$ M
2. $w = 360$ kJ = 86 kcal = power $\times$ time
3. $N_{e^-} = 1.9 \times 10^{22}$; $n_{e^-} = 0.031$ mol; 3.35 g Ag
4. $\mathscr{E} = 1.59$ V
5. $K = 6 \times 10^{52}$
6. $[Cu^{2+}] = 6.5 \times 10^{-38}$ M $\Rightarrow 8 \times 10^{-15}$ ions in 200 mL, not a single atom reacts

7. $Zn(-)$; $Ag(+)$
8. Ag; $Zn(NO_3)_2$; $Zn(NO_3)_2$, increase
9. Both $\mathscr{E}°$ and $(RT/nF) \ln Q$ are T-dependent; $\mathscr{E} = 1.533 - (0.06412/2) \log_{10}([Zn^{2+}]/[Ag^+]^2)$
10. $Q = 1.27 \times 10^5 C$; $I = 7.0$ A; $w = -139$ kJ $= -33.3$ kcal
11. $Zn + 2MnO_2 + 2NH_4^+ \longrightarrow Zn^{2+} + Mn_2O_3 + 2NH_3 + H_2O$; $\mathscr{E}°_{cath} = +0.74$ V (reduction); $Zn|Zn^{2+}|NH_4^+$, $MnO_2,NH_3|Mn_2O_3(C)$
12. $Zn + 2MnO_2 \longrightarrow ZnO + Mn_2O_3$; $\mathscr{E}°_{cath} = 0.146$ V; $\mathscr{E}°_{an} = -1.248$ V; $\mathscr{E}°_{cell} = 1.394$ V
    $Zn|ZnO,OH^-|OH^-,MnO_2|Mn_2O_3(C$ or $Cu)$
13. Reaction as in Exercise 12; $Q = 1.30 \times 10^4 C$; $w = -19.5$ kJ $= -4.66$ kcal; 3 days
14. For an n_c-cell battery, $\mathscr{E} = n_c\mathscr{E}_1$ and $w = Qn_c\mathscr{E}_1 = n_cw_1$; since every cell passes charge Q, n_c times as much reagent is consumed
15. $Zn + 2OH^- \longrightarrow ZnO + H_2O + 2e^-$ anode, $(-)$
    $HgO + H_2O + 2e^- \longrightarrow Hg + 2OH^-$ cathode, $(+)$
    $Zn + HgO \longrightarrow ZnO + Hg$ cell; $\mathscr{E}° =$
    $Zn|ZnO,OH^-||HgO,OH^-|Hg(?)$ $+1.35$ V
16. $Zn + 2OH^- \longrightarrow ZnO + H_2O + 2e^-$ anode, $(-)$
    $Ag_2O + H_2O + 2e^- \longrightarrow 2Ag + 2OH^-$ cathode, $(+)$
    $Zn + Ag_2O \longrightarrow ZnO + 2Ag$ cell; $\mathscr{E}° =$
    $Zn|ZnO,OH^-||Ag_2O,OH^-|Ag$ $+1.59$ V
17. $2Li + 2OH^- \longrightarrow Li_2O + H_2O + 2e^-$ anode, $(-)$
    $ZnO + H_2O + 2e^- \longrightarrow Zn + 2OH^-$ cathode, $(+)$
    $2Li + ZnO \longrightarrow Li_2O + Zn$ cell; $\mathscr{E}° =$
    $Li|Li_2O,OH^-||ZnO,OH^-|Zn$ $+1.26$ V
18. $\mathscr{E}_{batt} = 12.48$ V; 12.37 V; 11.77 V
19. $\mathscr{E}°_{298} = 2.042$ V; $\mathscr{E}°_{258} = 1.988$ V; for six cells, the voltage drops by $6(0.055) = 0.33$ V; this might impair starting
20. $Li \longrightarrow Li^+ + e^-$ anode
    $S + 2e^- \longrightarrow S^{2-}$ cathode
    $2Li + S \longrightarrow 2Li^+ + S^{2-}$ cell
    $\mathscr{E}° = +2.54$ V, vs. $\mathscr{E} = +2.28$ V in solid phase;
    $\mathscr{E}$ decreases with increasing T, since $\Delta S° < 0$
21. $Na \longrightarrow Na^+ + e^-$ anode
    $S + 2e^- \longrightarrow S^{2-}$ cathode
    $2Na + S \longrightarrow 2Na^+ + S^{2-}$ cell
    $\mathscr{E}° = +2.20$ V, vs. $\mathscr{E} = +1.81$ V in solid phase
22. $Cl_2|Cl^-$, $\Delta G° = -62.74$ kcal; $Al|Al^{3+}$, $\Delta G° = +116.0$ kcal; $\Delta G°_{cell} = -420.2$ kcal
23. a. $Ni \longrightarrow Ni^{2+} + 2e^-$; $Pb^{2+} + 2e^- \longrightarrow Pb$
       $\mathscr{E}° = +0.10$ V, spont. to right
    b. $Fe \longrightarrow Fe^{2+} + 2e^-$; $Ag^+ + e^- \longrightarrow Ag$
       $\mathscr{E}° = +1.21$ V, spont. to right
    c. $Mg \longrightarrow Mg^{2+} + 2e^-$; $2H^+ + 2e^- \longrightarrow H_2$
       $\mathscr{E}° = +2.36$ V, spont. to right
    d. $2Br^- \longrightarrow Br_2 + 2e^-$; $Sn^{2+} + 2e^- \longrightarrow Sn$
       $\mathscr{E}° = -1.23$ V, spont. to left
    e. $2I^- \longrightarrow I_2 + 2e^-$; $Cl_2 + 2e^- \longrightarrow 2Cl^-$
       $\mathscr{E}° = +0.825$ V, spont. to right
24. a. $Fe^{2+} \longrightarrow Fe^{3+} + e^-$
       $MnO_4^- + 8H^+ + 5e^- \longrightarrow Mn^{2+} + 4H_2O$
       $\mathscr{E}° = +0.72$ V, spont. to right
    b. $2Cr^{3+} + 7H_2O \longrightarrow Cr_2O_7^{2-} + 14H^+ + 6e^-$
       $Al^{3+} + 3e^- \longrightarrow Al$
       $\mathscr{E}° = -3.01$ V, spont. to left
    c. $Ag + Cl^- \longrightarrow AgCl + e^-$
       $Hg_2Cl_2 + 2e^- \longrightarrow 2Hg + 2Cl^-$
       $\mathscr{E}° = +0.05$ V, spont. to right
    d. $NO + 2H_2O \longrightarrow NO_3^- + 4H^+ + 3e^-$
       $Cu^{2+} + 2e^- \longrightarrow Cu$
       $\mathscr{E}° = -0.62$ V, spont. to left
    e. $H_2 \longrightarrow 2H^+ + 2e^-$
       $O_2 + 4H^+ + 4e^- \longrightarrow 2H_2O$
       $\mathscr{E}° = +1.229$ V, spont. to right
       or $H_2 + 2OH^- \longrightarrow 2H_2O + 2e^-$
       $O_2 + 2H_2O + 4e^- \longrightarrow 4OH^-$
       $\mathscr{E}° = +1.229$ V, spont. to right
25. Fe,Mn,Ni ($\mathscr{E}°>0$)
26. $\mathscr{E}°_{cath} = +0.77$ V; electrons cancel in cell reaction
27. $Ni^{2+} + 2e^- \longrightarrow Ni$ $\Delta G° = -2F\mathscr{E}° = +10.6$ kcal
    $Ni(OH)_2 \longrightarrow Ni^{2+} + 2OH^-$ $\Delta G° = -RT \ln K_{sp} = +20.8$ kcal
    $Ni(OH)_2 + 2e^- \longrightarrow Ni + 2OH^-$ $\Delta G° = +31.4$ kcal, $\mathscr{E}° = -0.68$ V
28. $Pb^{2+} + 2e^- \longrightarrow Pb$ $\Delta G° = -2F\mathscr{E}° = +6.0$ kcal
    $PbSO_4 \longrightarrow Pb^{2+} + SO_4^{2-}$ $\Delta G° = -RT \ln K_{sp} = +10.6$ kcal
    $PbSO_4 + 2e^- \longrightarrow Pb + SO_4^{2-}$ $\Delta G° = +16.6$ kcal, $\mathscr{E}° = -0.36$ V
29. $2H^+ + 2e^- \longrightarrow H_2$ $\Delta G° = -2F\mathscr{E}° = 0.0$ kcal
    $H_2O \longrightarrow H^+ + OH^-$ ($\times 2$) $\Delta G°_w = -RT \ln K_w = +19.1$ kcal
    $2H_2O + 2e^- \longrightarrow H_2 + 2OH^-$
    $\Delta G° = 2\Delta G°_w = 38.2$ kcal, $\mathscr{E}° = -0.828$ V

30. $Fe^{3+} + e^- \longrightarrow Fe^{2+}$ $\quad \Delta G°_1 = -1F\mathscr{E}°_1$
$\underline{Fe^{2+} + 2e^- \longrightarrow Fe \quad\quad \Delta G°_2 = -2F\mathscr{E}°_2}$
$Fe^{3+} + 3e^- \longrightarrow Fe$
$\Delta G° = -3F\mathscr{E}° = \Delta G°_1 + \Delta G°_2 = -F\mathscr{E}°_1 - 2F\mathscr{E}°_2$;
$\mathscr{E}° = (\mathscr{E}°_1 + 2\mathscr{E}°_2)/3 = -0.02$ V; electrons do not cancel

31. $O_2 + 4H^+ + 4e^- \longrightarrow 2H_2O \quad \Delta G°_1 = -4F\mathscr{E}°_1$
$\underline{2H^+ + 2e^- \longrightarrow H_2 \ (\times(-2)) \quad \Delta G°_2 = 0}$
$2H_2 + O_2 \longrightarrow 2H_2O$
$\Delta G° = \Delta G°_1 - 2\Delta G°_2 = \Delta G°_1$

32. $O_2 + 4H^+ + 4e^- \longrightarrow 2H_2O \quad \Delta G°_1 = -4F\mathscr{E}°_1$
$\underline{H_2O \longrightarrow H^+ + OH^- \ (\times 4) \quad \Delta G°_w = -RT \ln K_w}$
$O_2 + 2H_2O + 4e^- \longrightarrow 4OH^-$
$\Delta G° = \Delta G°_1 + 4\Delta G°_w \Rightarrow \mathscr{E}° = +0.401$ V

33. **a.** $\mathscr{E} = 0.391$ V
    b. $\mathscr{E} = 0.478$ V

34. anode: $CH_4 + 8OH^- \longrightarrow CO_2 + 6H_2O + 8e^-$;
$\mathscr{E}° = +1.06$ V; $\mathscr{E}°_{anode} = -0.659$ V; $\Delta G°_{reform} = +31$ kcal ($H_2O(l)$) or $+27$ kcal ($H_2O(g)$)

35. anode: $CH_3OH + 6OH^- \longrightarrow CO_2 + 5H_2O + 6e^-$;
$\mathscr{E}° = +1.21$ V; $\mathscr{E}°_{anode} = -0.81$ V

36. $H_2 + Br_2 \longrightarrow 2H^+ + 2Br^-$, $\mathscr{E}° = +1.087$ V

37. $Mg + 2H^+ \longrightarrow Mg^{2+} + H_2$, $\mathscr{E}° = +2.36$ V

38. anode: $Mg \longrightarrow Mg^{2+} + 2e^-$;
cathode: $Br_2 + 2e^- \longrightarrow 2Br^-$; $\mathscr{E}° = +3.45$ V

39. pH = 0; $\mathscr{E}° = 1.42, 1.77, 2.19$ V

40. $\mathscr{E} = (T/nF)\Delta S = -(RT/nF)\ln([Cu^{2+}]_{dil}/[Cu^{2+}]_{conc})$; statistics favor a uniform distribution of ions (equal concentrations) between the two cells, just as in the Joule expt

41. SHE is cathode; $(Pt)H_2(1\ atm)|H^+(c)\|H^+(1\ M)|H_2(1\ atm)(Pt)$;
$\mathscr{E} = 0.237, 0.414, 0.592$ V

42. $[Pb^{2+}]_{anode} = 1.8 \times 10^{-7}$ M; $K_{sp} = 1.8 \times 10^{-8}$

43. $\mathscr{E} = -0.59, -0.74, -0.77, -0.80, -0.95$; halfway

44. -1.229 V; $(-)$; $2H_2O + 2e^- \longrightarrow 2OH^- + H_2$ at $(-)$;
$2H_2O \longrightarrow 4H^+ + O_2 + 4e^-$ at $(+)$; current cannot flow without an electrolyte; get stoich. H_2/O_2 mixture at each electrode

45. $2H_2O + 2e^- \longrightarrow 2OH^- + H_2$ at cathode
$\underline{2Cl^- \longrightarrow Cl_2 + 2e^- \quad\quad\quad\quad\quad\text{at anode}}$
$2H_2O + 2Cl^- \longrightarrow 2OH^- + H_2 + Cl_2$ overall;
$\mathscr{E}° = -2.188$ V; $Q = 1.72 \times 10^6$ C

46. $Cu + Cu^{2+} \longrightarrow Cu^{2+} + Cu$; $Cu \longrightarrow Cu^{2+} + 2e^-$ at $(+)$;
Cu electrode will dissolve; water electrolysis requires more free energy

47. 0.129 g

48. 103.6 g/mol e^-; 207.2 g/mol

49. 5.6×10^6 A (5.6 MA)

50. 6.7×10^4 kg (67 metric tons)

51. $O_2 + 2H_2O + 4e^- \longrightarrow 4OH^-$
$\underline{Fe^{2+} + 3OH^- \longrightarrow Fe(OH)_3 + e^- \ (\times 4)}$
$4Fe^{2+} + O_2 + 2H_2O + 8OH^- \longrightarrow 4Fe(OH)_3$; $\mathscr{E}° = +1.91$ V

52. $Al + 3OH^- \longrightarrow Al(OH)_3 + 3e^-$ $(\times 4)$
$\underline{O_2 + 2H_2O + 4e^- \longrightarrow 4OH^- \quad\quad (\times 3)}$
$4Al + 3O_2 + 6H_2O \longrightarrow 4Al(OH)_3$ $\mathscr{E}° = +2.70$ V

Chapter 14

1. **a.** $n = 55.5$ mol; $q_P = \Delta H = -112.2$ kcal
   b. $\Delta S = -408$ cal/K; $\Delta G = +30.2$ kcal

2. $q_P = \Delta H = 6300$ cal

3. $n = 22.7$ mol; $q_P = \Delta H = 147$ kcal

4. $n = 0.358$ mol; $q_P = \Delta H = 4.4$ kcal; from Example 10.6, $\Delta H_{rxn} = -36.4$ kcal; yes

5. **a.** $n = 4350$ mol; $q_P = \Delta H = 4.82$ Mcal
   b. $\Delta S = 13.7$ kcal/K; $\Delta G = -4.12$ Mcal

6. $T_b \approx 6.7°C$

7. $T_b \approx 76.7°C$; 78.2 from table; T dependent of ΔH, ΔS neglected

8. $T_b \approx 41°C$; 40°C from table; T dependent of ΔH, ΔS neglected

9. **a.** $\Delta S = 26.83$ cal/K mol; T_b(rockies) $= 89.0°C$
   b. $89.0°C$

10. 13.4 atm

11. **a.** 0.0375 atm, 28.5 Torr
    b. no; $\Delta G = -215$ cal/mol
    c. 25.3°C

12. **a.** $\Delta H°_{sub} \approx 11.152$ kcal/mol; $P = 0.00056$ atm, 0.43 Torr
    b. $-28.3°C$
    c. 2.3%

13. $P < P_3 \Rightarrow$ sublimation; $-22.1°C$

14. 0.0154 atm, 11.7 Torr

15. 6.27 kcal/mol

16. 1.9 kcal/mol; $\Delta H°_{sub}$ and $\Delta H°_{vap}$ refer to different T's

17. Ave $T_c/T_b = 1.71$; $T_c(H_2O) = 638$K (647, err = 1.4%); $T_c(PH_3) = 317$K (324, 2.2%); $T_c(SiH_4) = 275$ K

18. $\ln(P/P°) = -(\Delta H°/R)(1/T - 1/T_b) - (\Delta C_P/R)(\ln(T_b/T) - T_b/T + 1)$; orig C–C, 27.9 Torr; mod C–C, 24.3 Torr

20. $\Delta V_{fus} = -1.6 \times 10^{-6}$ m³/mol; $\Delta H_{fus} = 6008$ J/mol; $P = 14.2$ atm

21. $G = H - TS = E + PV - TS$; $dG = dE + P\,dV + V\,dP - T\,dS - S\,dT$ since $dE = T\,dS - P\,dV$, $dG = V\,dP - S\,dT$; $dG_1 = dG_2 \Rightarrow V_1 dP - S_1 dT = V_2 dP - S_2 dT$, or $dP/dT = \Delta S/\Delta V = \Delta H/T\Delta V$ Q.E.D.

22. For Ge, $\Delta H°_{vap}/T_b = 25.7$; for Pb, 21.2; Si, $\Delta H°_{vap} \approx 77.8$ kcal/mol; SiO_2, 55.1

23. $CC + TR$, 242 Torr; CC alone, 251 Torr; CC should be more accurate, but isn't here

24. For CH_3CH_3, $\Delta S°_{vap} = 19.0$ cal/K mol; for CH_3F, 20.5; for CH_3OH, 24.9; order increases in this sequence, owing to orientations required for DD and HB IMFs

25. Na^+—OH_2, $V_{q\mu} = -23.1$ kcal/mol; Cl^-—H_2O, -14.2; Na^+, due to smaller ionic radius

26. $V_{\mu\mu} = -3.65$ kcal/mol; probably not; Pauli principle (overlap) repulsion

27. $V_{\mu\mu} = -2.35$ kcal/mol; smaller than but comparable to H_2O—H_2O

28. -7.89 kcal/mol; $\sim 1/3$ of leading term

29. a. 31.8 eV Å^6, -0.26 kcal/mol
    b. 19.9, -0.63
    c. 33.7, -0.45
    d. 71.6, -1.48
    e. 63.4, -0.40
    f. 25.7, -0.52
    g. 37.5, -0.96

30. $B = -(6C/R^7)/(\varepsilon - C/R^6)$; $A = (C/R^6 - \varepsilon)e^{BR}$; $B = 2.88$ Å^{-1}; $A = 1.36 \times 10^5$ kcal/mol $= 5910$ eV; 47%

31. $V = -3.65 - 0.63 = -4.28$ kcal/mol; close to H-bond; BMR omitted, would reduce net attraction

32. $V = +3.02$ kcal/mol; no

33.
Pair	DD	LD	Sum
H_2O—H_2O	-3.65	-0.63	-4.28
H_2O—NH_3	-2.35	-0.45	-2.80
H_2O—CH_4	0	-0.48	-0.48
CH_4—CH_4	0	-0.40	-0.40

The H_2O—H_2O attraction is comparable to H_2O—NH_3, but much greater than H_2O—CH_4; CH_4 is therefore excluded

34.
Substance	R_m	DD	LD	Sum	
HF	3.22	-2.86	-1.48	-4.34	34% LD
HCl	3.70	-0.66	-0.66	-1.32	
HBr	3.90	-0.33	-0.74	-1.07	
HI	4.20	-0.08	-0.93	-1.01	92% LD

35. 2.050 Å; yes, 92°

36. $2\theta = 52.2°, 123.4°$

37. $d = 2.083$ Å; $a = 3.608$ Å; $r = 1.276$ Å; yes, $n = 2$

38. $f_{bcc} = 0.680$; $f_{fcc} = 0.740$; fcc

39. For fcc, $N_A = 4M_{Ag}/(\rho a^3) = 6.0223 \times 10^{23}$ amu/g

40. For odd N, nodes in middle $= (N-1)/2$, and nodes in highest $= N - 1$

41. 4 to 7 nodes; bands will overlap (lowest $2p$ < highest $2s$) due small $2s-2p$ spacing and hybridization

42. a. bandwidth: for $N = 2$, -2β; for $N \longrightarrow \infty$, -4β
    b. level spacing: $\Delta E_{k,k+1} = 2\beta \sin(\pi/(N+1)) \longrightarrow 2\beta\pi/(N+1) \longrightarrow 0$ as $N \longrightarrow \infty$

43. $3s$, $3p$ bands must overlap

44. Other e^- are in π orbitals with nodes as in benzene

45. $r(Cl^-) = 1.994$ Å; $r(Na^+) = 0.826$ Å

46. $r(Cl^-) = 2.10$ Å; $r(Cs^+) = 1.54$ Å

47. -167.2 kcal/mol

48. $MgCl_2(s) \longrightarrow Mg^{2+}(aq) + 2Cl^-(aq)$: $\Delta H° = -38.2$ kcal/mol; $\Delta S° = -27.42$ cal/K mol; $\Delta G° = -30.00$ kcal/mol; exothermicity overcomes unfavorable entropy
$Mg(OH)_2(s) \longrightarrow Mg^{2+}(aq) + 2OH^-(aq)$: $\Delta H° = -0.55$ kcal/mol; $\Delta S° = -53.24$ cal/K mol; $\Delta G° = +15.32$ kcal/mol; dominated by unfavorable $\Delta S°$; lattice energy too great to be compensated by solvation

50. K_b(acetone) $= 1.80$

53. 100.75°C

54. -11.1°C

55. 670 g

56. $\Delta T_f(NaCl)/\Delta T_f(CaCl_2) = 1.27$; NaCl is better

57. 280. g/mol

58. $\Pi = 5.09$ atm; $h = 48.2$ m

59. $\Pi = 0.0048$ atm, 3.7 Torr; 35000 amu; 570

60. 47.3 atm

Chapter 15

1. a. $dx/dt = -\frac{1}{2}d[MnO_4^-]/dt = -\frac{1}{5}d[H_2C_2O_4]/dt = +\frac{1}{2}d[Mn^{2+}]/dt$

2. a. $d[BrO_3^-]/dt = -4.1 \times 10^{-5}$ mol L^{-1} s^{-1}; $[BrO_3^-] = 0.020 + (-4.1 \times 10^{-5})(48.2) = 0.018$ M
$d[Br^-]/dt = -5(4.1 \times 10^{-5}) = -2.05 \times 10^{-4}$ mol L^{-1} s^{-1}; $[Br^-] = 0.09_0$ M
$d[H^+]/dt = -2.46 \times 10^{-4}$ mol L^{-1} s^{-1}; $[H^+] = 0.08_8$ M

3. a. $dx/dt = -\frac{1}{2}d[O_3]/dt = +\frac{1}{3}d[O_2]/dt$
   b. $P = ([O_3]_0 + x)RT$

4. first order; $-d[N_2O_5]/dt = k[N_2O_5]$; $[N_2O_5] = [N_2O_5]_0 \exp(-2kt)$;
   $k = 0.045$ min$^{-1} = 7.5 \times 10^{-4}$ s^{-1}; $t_{1/2} = \ln2/(2k) = 7.6$ min $= 460$ s

5. a. first order; $-d[D]/dt = k[D]$; $k = 3.15 \times 10^{-4}$ s^{-1}; $t_{1/2} = 2200$ s.
   c. $t_{0.90} = 7310$ s

6. a. second order $(t_{0.10} \propto 1/c_0)$; $-d[HI]/dt = k[HI]^2$; $k = 5.8 \times 10^{-4}$ L mol^{-1} s^{-1}
   b. $t_{1/2} = 1/(2kc_0) = 5.4 \times 10^4$ s (15 hr)

7. 2nd wrt [NO], 1st wrt [H$_2$], 3rd overall; L^2 mol^{-2} s^{-1}; 4; $\frac{1}{2}$; 27

8. rate $= k[HCrO_4^-][HSO_3^-]^2[H^+]$; 1st, 2nd, 1st, 4th overall; L^3 mol^{-3} s^{-1}; not an elementary rxn

9. rate $= k[O_3]^2[O_2]^{-1}$; 2nd, $-$1st, 1st overall; s^{-1}

10. rate $= k[CH_3Cl][Cl_2]^{1/2}$; 1st, $\frac{1}{2}$, $\frac{3}{2}$ overall; L$^{1/2}$ mol$^{-1/2}$ s^{-1}

11. $dx/dt = k(c_0 - x) - k'x^2$; $x(t)$ would deviate from exp behavior; asymptote would be less than c_0

12. $k_{\text{eff}} = 1120$ s^{-1}; $k = k_{\text{eff}}[O_3]_0 = 2.8 \times 10^{-11}$ cm^3 molecule^{-1} s^{-1} (1.7×10^{10} L mol^{-1} s^{-1})

13. rate $= k[CH_3COCH_3][H^+]$; $k = 1.27 \times 10^{-5}$ L mol^{-1} s^{-1}

14. rate $= k[I^-][S_2O_8^{2-}]$; $k = 4.07 \times 10^{-3}$ L mol^{-1} s^{-1}; $t = 307$ s

15. a. 1. unimolecular, rate $= k_1[(CH_3)_3CCl]$
      2. bimolecular, rate $= k_2[(CH_3)_3C^+][H_2O]$; note that since [H$_2$O] $\approx$ const, this step will *appear* to be unimolecular
      3. bimolecular, rate $= k_3[H^+][OH^-]$
    b. rate $= k_1[(CH_3)_3CCl]$; yes; not unique as to fast steps

16. rate $= k_2[N_2O_2][H_2] = k_2K_1[NO]^2[H_2]$; $k = k_2K_1$

17. 1. unimolecular; $-$1. & 2. bimolecular;
    rate $= k_2[O_3][O] = k_2K_1[O_3]^2[O_2]^{-1}$; $k = k_2K_1$
    Since step 1 is at equil, can use LeChat: incr [O$_2$] shifts 1 to left, reduces [O] and rate of slow step

18. 1. $Cl_2 \rightleftharpoons 2Cl$, K_1
    2. $CHCl_3 + Cl \longrightarrow CCl_3 + HCl$, k_2, slow step
    3. $CCl_3 + Cl \longrightarrow CCl_4$, k_3
    rate $= k_2[CHCl_3][Cl] = k_2K[CHCl_3][Cl_2]^{1/2}$; consistent if $k = k_2K_1^{1/2}$

19. rate $= K_1K_2k_3[H_2][I_2]$, consistent if $k = K_1K_2k_3$

20. rate $= k_1k_2[AH][H^+][Br_2]/(k_{-1} + k_2[Br_2])$; not consistent; must have $k_{-1} \ll k_2[Br_2]$;
    step 1 equil $\Rightarrow$ rate $= K_1k_2[AH][H^+][Br_2]$, third–order rxn

21. rate $= k_1k_2[O_3]^2/(k_{-1}[O_2] + k_2[O_3])$; not consistent; must have $k_{-1}[O_2] \gg k_2[O_3]$; yes

22. rate $= k_1k_2[E][H^+]/(k_{-1} + k_2)$, second order; $k_{\text{obs}} = k_1k_2/(k_{-1} + k_2)$

23. $d[I]/dt = 2k_1[I_2] - 2k_{-1}[I]^2 - k_2[I][H_2] + k_{-2}[IH_2] - k_3[H_2I][I] = 0$
    $d[H_2I]/dt = k_2[I][H_2] - k_{-2}[IH_2] - k_3[H_2I][I] = 0$

25. a. $E_a = 62.7$ kcal/mol; $A = 1.3 \times 10^{15}$ s^{-1}
    b. $k = 1.4 \times 10^{-31}$ s^{-1}, too slow to be observed; $\exp(-E_a/RT) \approx 10^{-46}$, not a single molecule in a 1-mol sample will have enough energy to decompose at 25°C!
    c. $E_a' = 62.7 - 17.2 = 45.5$ kcal/mol

26. $k' = 5.6 \times 10^{-3}$ L mol^{-1} s^{-1}; $A = 1.3 \times 10^{10}$ L mol^{-1} s^{-1}

27. $E_a = 12.6$ kcal/mol

28. $E_a = 45.5$ kcal/mol; $A = 2 \times 10^{11}$ L mol^{-1} s^{-1}; $E_a' = 43.2$ kcal/mol

29. $O + CO_2 \rightleftharpoons CO + O_2$: from kinetics, $\Delta H° = E_a - E_a' = 3.2$ kcal/mol,
    $K = k/k' = 2.5 \times 10^{-2}$
    from thermodynamics $\Delta H° = 8.1$ kcal/mol, $K = 3.5 \times 10^{-5}$
    Considering the extrapolation to 298 K, agreement is not too bad

30. $v_{\text{rel}} = 6.5 \times 10^4$ cm s^{-1}; $\pi d_B^2 = 28$ Å^2 (2.8×10^{-15} cm^2); $k = 1.8 \times 10^{-10}$ cm^3 molecule^{-1} s^{-1} or 1.1×10^{11} L mol^{-1} s^{-1}; const $= 4.5713 \times 10^{-12}$ or 2.7529×10^9

31. $r_{\text{vdw}}(HI) = 2.10$ Å; $d_{AB} = 4.20$ Å; $A_{\text{ct}} = 7.9 \times 10^{10}$ L/mol s; $p \approx 1.8$, ~no steric effect; once collision occurs, H's will always find correct orientation; LD and DD attractions neglected

32. $p = 0.026$; O$\longrightarrow$O^O_O good, O$\longrightarrow$o^O_O bad, but more likely; $Q = 1.1 \times 10^{-16}$ cm^2 (1.1 Å^2)

33. const $= 1.263 \times 10^{10}$ (2.098×10^{-11}); $\Delta S°^{\ddagger} = -41.0$ cal K^{-1} mol^{-1}; $\Delta S° = -28.8$; activated complex is more constrained than ethane; perhaps C$_2$H$_2$ ring is involved

34. $\Delta S°^{\ddagger} = -24.6$ cal K^{-1} mol^{-1}; $\Delta S° = -2.61$; TS does not have a "loose" atom

35. a. $p_{\text{linear}} \approx 0.04$, $p_{\text{bent}} \approx 0.2$; most rxns in table seem to have bent TS's, except O + H$_2$ perhaps
    b. OH + H$_2$: $p_{\text{linear}} \approx 0.002$, $p_{\text{bent}} \approx 0.008$, $p_{\text{obs}} = 0.011$ => bent TS
       NO + O$_3$: $p_{\text{bent}} \approx 0.002$, $p_{\text{obs}} = 0.005$
       C$_2$H$_4$ + H$_2$: $p \approx 0.002$, $p_{\text{obs}} = 0.000002$, not well explained in this crude model

36. a. early
    b. late
    c. neither
    d. both

37. **a.** late
    b. early
    c. neither (symmetric)
38. 90 minutes
39. 74 ps
40. 2.90×10^{10} L mol^{-1} s^{-1}; better, no Coulomb attraction
41. $k = 7.1 \times 10^5$ s^{-1}; $t_{1/2} = 1$ μs
42. 14 ns
43. rate = $k_D k_{TST}[A][B]/(k_D' + k_{TST})$
44. NO + O$_3$ $\longrightarrow$ NO$_2$ + O$_2$; NO$_2$ + O $\longrightarrow$ NO + O$_2$; no, E_a is higher for step 1
45. $d[P]/dt \approx k_1[E]_0[S]/(1 + k_{-1}/k_2)$ famine; $d[P]/dt = k_2[E]_0$ feast
46. weak–binding limit: rate is controlled by adsorption rate, θ is small, in first-order regime

Chapter 16

1. 0.1015 amu fm^{-3}; 1.69×10^{14} g cm^{-3}; 1.69×10^8 tons cm^{-3}
2. **a.** 182 MeV; larger, due to finite outer wall of actual potential; no, since $\Delta E_{1 \longrightarrow 2} = 540$ MeV $\gg BE = 2$ MeV
   b. 0.022 zs
3. 37 MeV; no, since $\Delta E \gg BE$
4. $V(\text{MeV}) = 1.44\, Z_1 Z_2/r(\text{fm}) = 0.96$ MeV for p+p
5. 2.57 MeV; triangular, since this yields 3 bonds and a $D \approx 2.57$ MeV
6. 7.07 MeV; 6 bonds; no, since $D \approx 4.72$ MeV; ^{4}He is "doubly magic" with closed neutron and proton shells
7. 7.68 MeV; 66 bonds; $D \approx 1.40$ MeV, smaller than in d (suggests there are 40 bonds)
8. **a.** $\Delta m = -1.46 \times 10^{-8}$ amu
   b. $\Delta m = -4.81 \times 10^{-9}$ amu. Both are negligibly small
9. 0.141 amu ($256 m_e$) (measured Π meson mass is 0.150 amu); $a \longrightarrow \infty$
10. β^+: $^A_Z X \longrightarrow {}^{A}_{Z-1}Y + {}^{0}_{+1}\bar{e}$; EC: $^A_Z X + {}^{0}_{-1}e \longrightarrow {}^{A}_{Z-1}Y$; α: $^A_Z X \longrightarrow {}^{A-4}_{Z-2}Y + {}^4_2He$
11. ^{3}H: n-rich $\Rightarrow \beta$: $^3_1H \longrightarrow {}^3_2He + {}^0_{-1}e$
    ^{11}C: n-poor $\Rightarrow \beta^+$: $^{11}_6 C \longrightarrow {}^{11}_5 B + {}^0_1 \bar{e}$ or EC: $^{11}_6 C + {}^0_{-1}e \longrightarrow {}^{11}_5 B$; β^+ is observed
    ^{125}I: n-poor $\Rightarrow \beta^+$: $^{125}_{53}I \longrightarrow {}^{125}_{52}Te + {}^0_1 \bar{e}$ or EC: $^{125}_{53}I + {}^0_{-1}e \longrightarrow {}^{125}_{52}Te$; EC is observed
    ^{228}Th: overwt $\longrightarrow \alpha$: $^{228}_{90}Th \longrightarrow {}^{224}_{88}Ra + {}^4_2He$
    PET: ^{11}C
12. ${}^0_{-1}e + {}^0_1 \bar{e} \longrightarrow 2\gamma$; $E_\gamma = 0.511$ MeV; $\lambda = 0.0243$ Å; ${}^{13}_7 N \longrightarrow {}^{13}_6 C + {}^0_1 \bar{e}$; $\Delta E \approx K(\beta^+) = 1.20$ MeV; this energy is assumed lost before annihilation
13. -0.000219 amu; no, $\Delta m = +0.000878$ amu; no, $\Delta m = +0.00458$ amu
14. ${}^{99}_{43}Te \longrightarrow {}^{99}_{44}Ru + {}^0_{-1}e$; $\Delta m = -0.000315$ amu $< 0 \Rightarrow$ spontaneous. e^- are supplied by decay to reduce oxidized Al.
15. ${}^{40}_{19}K \longrightarrow {}^{40}_{20}Ca + {}^0_{-1}e$ $\Delta m = -0.001408 < 0$, spont.
    ${}^{40}_{19}K + {}^0_{-1}e \longrightarrow {}^{40}_{18}Ar$ $\Delta m = -0.001615 < 0$, spont.; yes, $\Delta m = -0.000518$
16. β^+ decay products must tunnel through a Coulomb barrier, making n-poor nuclides longer lived
17. -7.835 MeV
18. ${}^{197}_{79}Au \longrightarrow {}^{193}_{77}Ir + $ He; $\Delta m = -0.001024$, $\Delta E = -0.954$ MeV, spont.; lack of observed decay due to difficulty in tunneling through 22.6 MeV Coulomb barrier; rate too low to measure
19. $\lambda(\gamma_{1.17}) = 0.0106$ Å; $\lambda(\gamma_{1.33}) = 0.0093$ Å; $K_\beta \leq 0.32$ MeV
20. ^{205}Pb $\longrightarrow {}^{201}$Hg $+ \alpha$; $\Delta m = -0.00158$ amu
    ^{201}Hg $\longrightarrow {}^{197}$Pt $+ \alpha$; $\Delta m - 0.00036$ amu
    ^{197}Pt $\longrightarrow {}^{197}$Au $+ \beta$; $\Delta m - 0.000772$ amu

    all spontaneous

    Two other decay schemes can be written, but they are not consistent with stability trends, and contain one or more nonspontaneous steps
22. **a.** 2.30×10^{15} decays s^{-1} g^{-1}
    b. 1.61 ng
    c. 87 μCi
23. 3.8×10^5 decays s^{-1}; 8 min
24. 1.33×10^{-12}
25. 0.072 rem (72 mrem) yr^{-1}
26. 1.8×10^9 yr
27. $\geq 1.7 \times 10^8$ yr
28. 1.75×10^9 yr
29. ^{232}Th: $\Delta m = -0.00439$ amu $(-4.09$ MeV); ^{230}Th: -0.00512 amu $(-4.77$ MeV); ^{228}Th: -0.00593 amu $(-5.52$ MeV); expect $t_{1/2}$ to be inverse to $-\Delta m$, so $t_{1/2}(228) < t_{1/2}(230) < t_{1/2}(232)$; $V_c = 24.7$ MeV
    (Measured lifetimes are 2 yr, 1.4×10^4 yr, and 1.4×10^{10} yr)
30. ^{9}C: -0.01661 amu; ^{10}C: -0.01282 amu; ^{11}C: -0.00103 amu, implying $t_{1/2}(9) < t_{1/2}(10) < t_{1/2}(11)$ (measured half-lives are 0.13 s, 19.3 s, and 1220 s; β^+s, being lighter than αs, tunnel more easily, giving shorter lifetimes)
31. **a.** $7\alpha, 6\beta$
    b. 0.999896 Pb
    c. 0.233 atm

32. 180 d
33. 30.2 μg
34. 98 pg
35. 0.30
36. a. $4\,{}^{1}_{1}\text{p} \longrightarrow {}^{4}_{2}\text{He} + 2\,{}^{0}_{1}\bar{\text{e}} + 2\nu + 2\gamma$;
    b. -0.025407 amu, -23.667 MeV;
    c. 4.1×10^{35} protons s^{-1}; 2.0×10^{35} ν s^{-1}
37. $3\,{}^{4}_{2}\text{He} \longrightarrow {}^{12}_{6}\text{C}$; -0.007806 amu; 0.00171 Å
38. 2.41 MeV or 3.87×10^{-13} J; $2 \times 10^{-122} \longrightarrow$ rxn impossible without tunneling
39. a. $\begin{array}{lll} {}^{12}_{6}\text{C} + {}^{1}_{1}\text{H} \longrightarrow {}^{13}_{7}\text{N} + \gamma & (k_1) & -1.94 \text{ MeV} \\ {}^{13}_{7}\text{N} \longrightarrow {}^{13}_{6}\text{C} + {}^{0}_{1}\text{e} & (k_2) & -1.20 \text{ MeV} \\ {}^{13}_{6}\text{C} + {}^{1}_{1}\text{H} \longrightarrow {}^{14}_{7}\text{N} + \gamma' & (k_3) & -7.55 \text{ MeV} \\ \hline {}^{12}_{6}\text{C} + 2\,{}^{1}_{1}\text{H} \longrightarrow {}^{14}_{7}\text{N} + {}^{0}_{1}\text{e} + \gamma + \gamma' & & -10.69 \text{ MeV} \end{array}$
    b. $[{}^{13}\text{N}] \approx (k_1/k_2)[{}^{12}\text{C}][{}^{1}\text{H}]$; $[{}^{13}\text{C}] \approx (k_1/k_3)[{}^{12}\text{C}]$ (overall rate $= k_1[{}^{12}\text{C}][{}^{1}\text{H}]$)
40. -0.008711 amu, -8.11 MeV; both products are "doubly magic" nuclei and therefore especially strongly bound
41. 0.000890 Å
42. a. $+0.001280$ amu, $+1.192$ MeV; $E_{\text{prod}} \approx K_p = 6.643$ MeV; yes (single-humped E-profile)
    b. ${}^{14}\text{N} + {}^{4}\text{He} \longrightarrow {}^{18}\text{F}$, $\Delta E = -4.414$ MeV; $V_c' = 2.43$ MeV (double-humped E-profile)
43. 0.46 b; 0.51
44. a. $\Delta E = -5.70$ MeV; $V_c = 2.36$ MeV
    b. 0.52 b; 13.5 MeV
45. a. $\Delta E = +2.64$ MeV; $V_c = 6.14$ MeV; 0.25 b; 5.2 MeV
    b. ${}^{31}\text{P}$; $\Delta E = -9.67$ MeV
46. ${}^{210}\text{Po} \longrightarrow {}^{206}\text{Pb} + \alpha$; $\alpha + {}^{9}\text{Be} \longrightarrow {}^{12}\text{C} + \text{n}$; 11.1 MeV
47. ${}^{2}_{1}\text{d} + \gamma \longrightarrow {}^{1}_{1}\text{p} + {}^{1}_{0}\text{n}$ (${}^{2}_{1}\text{H} + \gamma \longrightarrow {}^{1}_{1}\text{H} + {}^{1}_{0}\text{n}$); 0.00558 Å
48. ${}^{98}_{42}\text{Mo} + {}^{1}_{0}\text{n} \longrightarrow {}^{99}_{42}\text{Mo}(+\gamma)$, -0.00636 amu; ${}^{99}_{42}\text{Mo} \longrightarrow {}^{99}_{43}\text{Tc} + {}^{0}_{-1}\text{e}$, -0.00146 amu
49. ${}^{235}_{92}\text{U} + {}^{1}_{0}\text{n} \longrightarrow 2\,{}^{116}_{46}\text{Pd} + 4\,{}^{1}_{0}\text{n}$; $\Delta E = -177$ MeV, $V_c = 235$ MeV; cf $\Delta E = -167$ MeV, $V_c = 224$ MeV; higher V_c
50. a. ${}^{244}_{94}\text{Pu} + {}^{48}_{20}\text{Ca} \longrightarrow {}^{289}_{114}\text{X} + 3\,{}^{1}_{0}\text{n}$; no
    b. 206 MeV; 288.99 amu
51. fission: 0.711 MeV/amu, 6.9×10^{13} J/kg, 1.6×10^{10} cal/g
    fusion: 0.812 MeV/amu, 7.8×10^{13} J/kg, 1.9×10^{10} cal/g
52. 511 MHz
53. ${}^{13}\text{C}$: 10.7 MHz; ${}^{19}\text{F}$: 40.1 MHz
54. 42.6 MHz: 0.9999931; 500 MHz: 0.9999195; the latter
55. $\Delta t \leq 23$ μs

Chapter 17

1. | Element or ion | Configuration | Unpaired electrons |
   |---|---|---|
   | Fe^{2+} | $[\text{Ar}]3d^6$ | 4 |
   | V^{3+} | $[\text{Ar}]3d^2$ | 2 |
   | Zn | $[\text{Ar}]4s^23d^{10}$ | 0 |
   | Sc^{3+} | $[\text{Ar}]$ | 0 |
   | Co^{2+} | $[\text{Ar}]3d^7$ | 3 |

2. | Element or ion | Configuration | Unpaired electrons |
   |---|---|---|
   | Ru^{3+} | $[\text{Kr}]4d^5$ | 5 |
   | Mo^{4+} | $[\text{Kr}]4d^2$ | 2 |
   | Pt^{2+} | $[\text{Xe}]4f^{14}5d^8$ | 2 |
   | Bh^{2+} | $[\text{Rn}]5f^{14}6d^5$ | 5 |
   | Au | $[\text{Xe}]4f^{14}5d^{10}6s^1$ | 1 |

3. $+3, +5, +4, +5,$ and $+3$
4. $+7, 0, +4, +6, +8/3, +2$
5. $\text{Fe}^{3+}, \text{Co}^{2+}, \text{Pd}^{4+}, \text{Sc}^{3+}, \text{Cr}^{3+}$
6. a. $2\,\text{KMnO}_4(aq) + \text{H}_2\text{SO}_4(aq) \longrightarrow \text{Mn}_2\text{O}_7(l) + \text{K}_2\text{SO}_4(aq) + \text{H}_2\text{O}(l)$; Brønsted–Lowry acid–base BLAB; $+7$
   b. $\text{MnO}_2(s) + 4\,\text{H}^+(aq) + 2\,\text{e}^- \longrightarrow \text{Mn}^{2+}(aq) + 2\,\text{H}_2\text{O}(l)$; redox; $+4, +2$
   c. $\text{Sc}_2\text{O}_3(s) + 6\,\text{H}^+(aq) \longrightarrow 2\,\text{Sc}^{3+}(aq) + 3\,\text{H}_2\text{O}(l)$; BLAB; $+3$
   d. $\text{V}_2\text{O}_5(s) + 2\,\text{H}^+(aq) \longrightarrow 2\,\text{VO}_2^+(aq) + \text{H}_2\text{O}(l)$; BLAB; $+5$
   e. $\text{V}_2\text{O}_5(s) + 6\,\text{OH}^-(aq) \longrightarrow 2\,\text{VO}_4^+(aq) + 3\,\text{H}_2\text{O}(l)$; Lewis acid–base LAB; $+5$
   f. $\text{VO}^{2+}(aq) + 2\text{H}^+(aq) + \text{e}^- \longrightarrow \text{V}^{3+}(aq) + \text{H}_2\text{O}(l)$; redox; $+4, +3$
7. $\text{MnO}_2 > \text{CrO}_2 > \text{VO}_2 > \text{TiO}_2$
8. $\varepsilon = \alpha\,/\,2.3026$; cm^2/mol
9. M energy levels drop as Z increases. Thus, ΔE_{CT} decreases from Sc to V; Sc_2O_3 and TiO_2 absorb in UV; V_2O_5 absorbs blue wavelengths
10. SiO_4^{4-}; colorless; absorb in UV; see Figure 17.6
11. green/blue; absorb around 710 nm; see Figure 17.6
12. a. $\text{Fe}_2\text{O}_3(s) + 6\,\text{H}^+(aq) \longrightarrow 2\,\text{Fe}^{3+}(aq) + 3\,\text{H}_2\text{O}(l)$; BLAB
    b. $\text{ZnO}(s) + 2\,\text{H}^+(aq) \longrightarrow 2\,\text{Zn}^{2+}(aq) + \text{H}_2\text{O}(l)$; BLAB
    c. $\text{ZnO} + 2\,\text{OH}^- + \text{H}_2\text{O} \longrightarrow \text{Zn(OH)}_4^{2-}$; BLAB & LAB
13. a. Co(II);
    b,c. Co(II) – 510 nm, 19,600 cm^{-1}; Ni(II) – 720 nm, 13,900 cm^{-1}

14. 0,1,3
15. Square−planar geometry leads to *cis* and *trans* isomers
16. a. pentacarbonylnickel(0)
    b. hexaamminecobalt(III) chloride
    c. hexaaquachromium(III) nitrate
    d. potassium hexacyanoferrate(III)
    e. potassium hexacyanoferrate(II)
    f. diamminebis(ethylenediamine)nickel(II) sulfate
    g. potassium tetrachlorocuprate(II)
    h. hexaaquazinc(II) tetracyanozincate(II)
17. a. $CuCl_4^{2-} < Cu(H_2O)_6^{2+} < Cu(NH_3)_4^{2+} < Cu(en)_2^{2+}$
    b. $CuCl_4^{2-} + 6 H_2O \longrightarrow Cu(H_2O)_6Cl_2 + 2 Cl^-$
18. H_2O—weak; CN^-—strong; high-spin compounds result from small Δ_o; small Δ_o implies long-λ absorption
19. Co^{3+}(4−high spin, 0−low spin); Cr^{2+}(4,2); Fe^{2+}(4,0); Co^{2+}(3,1); Fe^{3+}(5,1); Mn^{2+}(5,1)
20. tetrahedral: 2 unpaired electrons(ue); square planar: 0 ue; sq. pl. preferred
21. enhanced stability reflected in larger Δ_o for ammine comlex
22. Added NH_3 replaces H_2O ligands; NH_3 complex has larger Δ_o; $H_2O - 17{,}900$ cm^{-1}; $NH_3 - 20{,}000$ cm^{-1}
23. Cr^{3+} complex is more stable
24. Fe^{3+} complex has half-filled levels and is more stable
25. a. Fe^{2+}: −52.8 kcal/mol (−221 kJ/mol); Fe^{3+}: −86.9 kcal/mol (−364 kJ/mol)
    b. smaller radius leads to more overlap, therefore more covalent bonding, as reflected in greater Δ_0
26. Negative ions repel each other; F^- does not donate electrons very well
27. CO, N_2, CH_2O, NO, NO^+; see Figure 17.14
28. $d[ML_6-X]/dt = k_{OC}[ML6][X] - k_{-OC}[ML6-X] - k_{-L}[ML6-X] \approx 0$
29. a. $ML_5X + L \rightleftharpoons [ML_5X-L], K_{OC}; [ML_5X-L] \longrightarrow [ML_5-L] + X, k_L; [ML_5-L] \longrightarrow ML_6, k$
    b. $E_a' \approx D(M-X); E_a' = E_a - \Delta H°$
    c. reverse rate very small
30. 0.001 sec
31. $d[ML_6^{n+}-ML'_6^{m+}]/dt = k_{OC}[ML_6^{n+}][M'L'_6^{m+}] - k_{-OC}[ML_6^{n+}-M'L'_6^{m+}] - k_e[ML_6^{n+}-M'L'_6^{m+}] \approx 0$

33. Ion Configuration Unpaired electrons
    La^{3+} [Xe]   0
    Sm^{3+} [Xe]$4f^5$ 5
    Er^{3+} [Xe]$4f^{11}$ 3
    Ce^{4+} [Xe]   0
    Yb^{3+} [Xe]$4f^{13}$ 1
    Lu^{3+} [Xe]$4f^{14}$ 0
    U^{4+} [Xe]$5f^2$ 2

34. a. $2\ Dy(s) + 5\ H_2O(l) \longrightarrow 2\ Dy(OH)_3(aq) + H_2(g)$
    b. $Dy_2O_3(s) + 6\ HCl(aq) \longrightarrow 2\ DyCl_3(aq) + 3\ H_2O(l)$
    c. $3\ U_3O_8(s) + 20\ HNO_3(aq) \longrightarrow 9\ UO_2(NO_3)_2(aq) + 10\ H_2O(l) + 2\ NO(g)$
35. $[:\ddot{O}{=}U{=}\ddot{O}:]^{2+}$
36. 1.004298; 333; 692

Chapter 18

1. *n*−octane and isooctane (2,2,4−trimethylpentane) are given in Table 18.2. Some other structural isomers are:

   $CH_3CH(CH_3)(CH_2)_4CH_3$, (2−methylheptane)
   $CH_3CH_2CH(CH_3)(CH_2)_3CH_3$, (3−methylheptane)
   $CH_3(CH_2)_2CH(CH_3)(CH_2)_2CH_3$ (4−methylheptane)
   $CH_3CH(CH_3)CH(CH_3)(CH_2)_2CH_3$ (2,3−dimethylhexane)
   $CH_3CH_2CH(C_2H_5)(CH_2)_2CH_3$ (3−ethylhexane)

   plus 11 more. The longest chain determines the name; thus 2−ethylhexane is not a new isomer, being already represented by 3−methylheptane

2. between the zigzag and the fully−curled−up conformers there are four more

3. one C must bond to H, Me, Et, and Pr groups, so 3−methylhexane (with *n*−Pr) and 2,3−dimethylpentane (with *i*−Pr) are the smallest

4. K = [boat]/[chair] = 0.0016

5. $t_{1/2}$ = 1.8 μs

6. the intermediates are Cl· and ·CH_3, which cancel when adding the reactions

7. Rxn 18.3: $\Delta H°$ = −24 kcal/mol (−100 kJ/mol)
   Rxn 18.4b: $\Delta H°$ = −4 kcal/mol (−17 kJ/mol)
   Rxn 18.4c: $\Delta H°$ = −20 kcal/mol (−84 kJ/mol)

   the latter two add up to the first, in accord with Hess's law

8. $T_{ta} = 1136$ K
9.
$CH_2{=}CHCH_2CH_2CH_3$	1-pentene	straight chain
$CH_3CH{=}CHCH_2CH_3$	cis-, trans-2-pentene	straight chain
$CH_2{=}C(CH_3)CH_2CH_3$	2-methyl-1-butene	branched chain
$CH_2{=}CHCH(CH_3)_2$	3-methyl-1-butene	branched chain
$CH_3CH{=}C(CH_3)_2$	2-methyl-2-butene	branched chain

10. $CH_3{-}\overset{+}{C}H{-}CH_3$ is a secondary carbocation, more stable than $\overset{+}{C}H_2{-}CH_2CH_3$, a primary carbocation
11. key intermediates are $R{-}(HO)C{=}OH^+$ and $R{-}(HO)\overset{+}{C}{-}OH$; the latter is attacked by the alcohol O lone pair
12. $(CH_3)_3C{-}Cl$ and $Na^+{}^-{:}\ddot{O}{-}CH_3$; expected to be more facile, since Cl^- will depart a more stable carbocation more readily; S_N1 mechanism expected, analogous to that for 18.11
13. one pair, for CH_3CHFCl, 1,1-chlorofluoroethane
14. c,e
15. important resonance form shows $C{=}O$
16. yes, 6 π electrons, $n = 1$, as in benzene
17. $Br^+ \cdots 4p^4$; both are $4p$ orbitals; expect LUMO of Br^+ to interact with π HOMO of ethylene
18. $F_2C{=}CF_2$; $CF_2{-}CF_{2n}$
19. $HOCH_2CH_2{-}OH + HOOC{-}C_6H_4{-}COOH \longrightarrow$
    $HOCH_2CH_2{-}O{-}OC{-}C_6H_4{-}COOH + H_2O$
    glycol O lone pair attacks carboxyl C

Photo Credits

Chapter 1
CO: Taxi/Getty Images and Bettmann/CORBIS
Page 2 (*top*): *The Alchymist,* 1771 (oil on canvas), Joseph Wright of Derby (1734–1797), Derby Museum and Art Gallery, UK/Bridgeman Art Gallery
Page 2 (*bottom*): Peter Siska

Chapter 2
CO: Ernest Orlando, Lawrence Berkeley National Laboratory
Page 42: Photographie Benjamin Couprie, Institut International de Physique Solvay, courtesy AIP Emilio Segrè Visual Archives
Figure 2.10a-b: Millie LeBlanc

Chapter 3
CO: Courtesy John Oke, California Institute of Technology
Page 69: Bettman/Corbis

Chapter 4
CO: *Science,* vol. 185, pp 1163–1165, 1974
Page 99: AIP Emilio Segrè Visual Archives
Figure 4.15a: R. L. Tromp, *Physics Today,* January 1987, p. S-67
Figure 4.15b: R. L. Tromp, *Physics Today,* January 1987, p. S-68
Figure 4.16: M. J. Cardillio, *Physics Today,* January 1989, p. S-21
Figure 4.17: *Accounts of Chemical Research* 95 Arleen Courtney at (202) 872-4368
Figure 4.18a: D. Rugar & P. Hansma, *Physics Today,* October 1990, p. 26
Figure 4.18b: D. Rugar & P. Hansma, *Physics Today,* October 1990, p. 27

Chapter 5
CO: Richard Megna/Fundamental Photographs
Page 131: Courtesy of University Archives, The Bancroft Library, University of California, Berkeley
Figure 5.1a-b: Chip Clark
Figure 5.1c: Dorling Kindersley Media Library
Figure 5.1d: Tony Freeman/PhotoEdit
Figure 5.1e: Dorling Kindersley Media Library
Figure 5.1f: Charles D. Winters/Photo Researchers, Inc.
Figure 5.1g: Natural History Museum, London
Figure 5.1h: Erich Schrempp/Photo Researchers, Inc.
Figure 5.3a-c: Chip Clark
Figure 5.3c: General Electric Corporate Research & Development Center
Figure 5.3e: Chip Clark
Figure 5.3f: Dorling Kindersley Media Library
Figure 5.4a-b: Richard Megna/Fundamental Photographs
Figure 5.7: Chip Clark
Figure 5.8a-b: Richard Megna/Fundamental Photographs
Figure 5.9: Richard Megna/Fundamental Photographs
Figure 5.10: Richard Megna/Fundamental Photographs
Figure 5.11: Richard Megna/Fundamental Photographs
Figure 5.12a: Chip Clark
Figure 5.12b-c: Richard Megna/Fundamental Photographs
Figure 5.13: Richard Megna/Fundamental Photographs
Figure 5.14a-c: Chip Clark

Chapter 6
CO: Yoav Levy/Phototake
Page 178: Time Life Pictures/Getty Images
Figure 6.13a-b: Richard Megna/Fundamental Photographs

Chapter 7
CO: Clive Freeman, The Royal Institution/Photo Researchers, Inc.
Page 206: National Archives and Records Administration, courtesy AIP Emilio Segrè Visual Archives

Chapter 8
CO: SSPL/The Image Works
Page 238: AIP Emilio Segrè Visual Archives, W.F. Meggers Gallery of Nobel Laureates

Chapter 9
CO: Otto Von Guericke, *The New (So-Called) Magdenburg Experiments of Otto Von Guericke,* Trans. by Margaret Glover Foley Adams. Dordrecht/Boston/London: Kluwer Academic Publishers, 1994
Page 268: Sheila Terry/Photo Researchers, Inc.
Figure 9.1a-b: Richard Megna/Fundamental Photographs
Figure 9.9a-b: Richard Megna/Fundamental Photographs
Figure 9.10: Andrew Lambert/Photo Researchers, Inc.

Chapter 10
CO: Clive Streeter © Dorling Kindersley, Courtesy of The Science Museum, London
Page 302: SSPL/The Image Works
Figure 10.1a-b: Richard Megna/Fundamental Photographs

Chapter 11
CO: Ted Kinsman/Photo Researchers
Page 343: Library of Congress
Figure 11.1a-b: Chip Clark
Figure 11.2a-b: Richard Megna/Fundamental Photographs

Chapter 12
CO: Allphoto Images/Agefotostock
Page 376: AIP Emilio Segrè Visual Archives, Brittle Books Collection
Figure 12.1a-b: Richard Megna/Fundamental Photographs
Figure 12.6a: Richard Megna/Fundamental Photographs
Figure 12.7a-b: Richard Megna/Fundamental Photographs
Figure 12.8a-b: Richard Megna/Fundamental Photographs

Chapter 13
CO: Nathan S. Lewis/California Institute of Technology
Page 436: SSPL/The Image Works

Chapter 14
CO: Richard Walters/Visuals Unlimited
Page 473: Courtesy Newman Library, Virginia Tech
Figure 14.7: Richard Megna/Fundamental Photographs
Figure 14.10a: Prentice Hall, Inc.
Figure 14.10b: Dorling Kindersley Media Library
Figure 14.18c: A. Lessor, IBM Laboratories

Chapter 15
CO: AP Wide World Photos
Page 539: Hulton Archive/Getty Images
Figure 15.16: Corbis/Bettmann
Figure 15.17: NASA/Goddard Space Flight Center
Figure 15.20a-c: Richard Megna/Fundamental Photographs
Figure 15.21 a-d: Richard Megna/Fundamental Photographs

Chapter 16
CO: SPL/Photo Researchers, Inc.
Page 599: W. F. Meggers Collection/AIP/Photo Researchers
Figure 16.15: Joe Sohm/Chromosohm/The Stock Connection

Chapter 17
CO: Richard Megna/Fundamental Photographs
Page 644: Mary Evans Picture Library/The Image Works

Chapter 18
CO: Roger Ressmeyer/Corbis
Page 679: SSPL/The Image Works

Appendix A
Page 739: Elliot Plotkin, *Physics Today,* Vol. 42, No. 4, April 1989

Appendix D
Page 782: Central Library for Physics, Vienna, Austria

Index

A

Ab initio theory, 121
Abney, William, 256
Absolute entropies, 358–361
Absolute (Kelvin) temperature scale, 20
Absorption line, 47
Absorption spectroscopy, 376
Abstraction reactions, 585
Abundances of particles, 27
acac, 655
Acceleration, 14
Accuracy, 11–12
Acetalacetonato, 655
Acetaldehyde, 693
Acetamide, 693
Acetate, 146
Acetato, 655
Acetic acid, 146, 393, 397, 693
Acetone, 477, 483, 501, 554, 693
Acetyl CoA, 718
Acetylene, 188, 189, 657, 689
Acid, 148
Acid-base indicators, 417
Acid-base neutralization reaction, 38
Acid-base proton transfer reaction, 222
Acid-base reaction, 156–160, 168
    aqueous complexes, 416–418
    Arrhenius, 388–392
    Brønsted-Lowry, 400–403
    Lewis type, 416
Acid-base titrations, 411–416
Acid ionization constants, 398
Acid rain, 433, 465
Actinides, 107, 115, 672–674, 776
Activated complex, 566
Activation energy, 559
Active site, 720
Activity, 424, 612
Activity coefficient, 424
Activity series, 453
Addition polymers, 702
Addition reactions, 157, 691–692
Adduct, 155
Adduct decomposition, 169
Adenine, 399, 411, 712
Adenosine triphosphate (ATP), 711–713
Adiabatic compression, 349
Adiabatic expansion, 313, 349
Adsorption (sticking) coefficient, 582
Aerobic reduction pathway, 716
AFM, 125, 126
Ala-tRNA, 725
Alanine, 693, 695, 697, 708
Alchemy, 2
Alcohol, 692, 693
Aldehyde, 693, 694
Alder, K., 699
Aldohexose, 704, 705
Aldoses, 704
Alizarin, 417
Alizarin yellow, 417
Alkali halides, 518
Alkali metals, 115, 147
Alkali-sulfur cells, 447
Alkaline copper half-cell, 457
Alkaline Daniell cell, 457
Alkaline dry cell, 444
Alkaline earth metals, 147
Alkaline electrolytic cell, 463
Alkaline hydrogen-oxygen fuel cell, 448
Alkaline zinc-silver oxide cell, 468
Alkane, 499
Alkane reactions, 688–689
Alkanes, 685
Alkenes, 690
Alkynes, 690
Alkoxide, 694
Alkoxy radicals, 702
Alkyl amines, 400
Alkyl groups, 396, 687
Alkyl halide, 689
Alkyl radical, 688, 689
Alkyl side chains, 687
Allotrope, 141
Allotropic forms of carbon, 514
Allotropy, 513–514
α-amino acid, 397, 693, 707
α-decay, 609
α-helix, 710, 711
α particles, 23–24
Alternant hydrocarbons, 690
Alternating single and double carbon-carbon bonds, 230
Aluminum, 132, 447, 462, 764
Aluminum-air battery, 447, 454
Aluminum-chlorine cell, 468
AM, 26
Amide, 146, 693, 695
Amine, 693, 694
Amine groups, 397
Amines, 399, 656n
Amino acids, 400, 580, 707–711
Aminoacetic acid, 194
Aminoacyl tRNA synthetase, 724
Aminoacylating tRNA, 724–725
Ammine, 654, 655
Ammonia, 139, 146, 154, 155, 186, 393, 532, 534
Ammonia synthesis, 368, 424, 429
Ammonium chloride, 157
Ammonium ion, 145
Amphoteric, 648
Amplitude, 44
amu, 26
Anaerobic reduction pathway, 716
Analyte, 411
Angstrom, 30
Angular distributions, 573
Angular momentum, 37, 242
Angular momentum quantum number, 72, 91
Angular node, 79
Aniline, 399, 400, 699
Anion, 145, 146, 441
Annihilation, 608
Anode, 441
Anode sludge, 463
Anodized, 463
Anomalous configurations, 112
Antibonding molecular orbital, 208
Antibonding orbitals, 217
Antifreeze solution, 528, 537
Antihydrogen atom, 105n
Antimatter, 104
Antimony, 771
Antineutrin, 606
Antiparticle counterparts, 105n
Apollo lunar mission (1969-72), 439
Apse line, 570
Aqua, 654, 655
Aqua complex, 658
Aqueous acid-base chemistry, 387–418. *See also* Free energy and chemical equilibrium
Aqueous bimolecular reactions, 577
Aqueous complexes, 416–418
Aqueous oxidation of iodide by hydrogen peroxide, 546
Arginine, 708
Argon atmosphere, 313
Argon ion laser, 64
Aristotle, 302
Aromaticity, 698
Aromatics, 697–699
Arrhenius, Svante, 148, 528, 539, 559–561
Arrhenius acid, 156, 159
Arrhenius acid-base reaction, 156, 388–392
Arrhenius base, 156, 159
Arrhenius equation, 559
Arrhenius form of k, 560
Arrhenius parameters, 564
Arrhenius plot, 559
Arsenic, 771
Asimov, I., 704
Asparagine, 708
Aspartic acid, 708
Asphalt, 686
Aspirin, 699
Aston, Francis, 22
Aston's 1920 mass-spectrum of pure neon gas, 26
Asymmetric carbon, 696
Asymmetric carbon atom (enantiomers), 695–697
Asymmetric electron sharing, 195–199
atm, 269, 270
Atmospheric molecules and radicals, 550
Atom, 5, 201
Atom abstraction and dissociation, 625
Atom-molecule reaction, 570
Atomic bombs (WWII), 636
Atomic force microscope (AFM), 125, 126
Atomic hybridization, 183–187
Atomic mass (AM), 26
Atomic mass unit (amu), 26
Atomic number, 26, 598
Atomic spectroscopy, 623
Atomic theory, 5
Atomization energy, 325, 327
ATP, 711–713
atto-, 11
Attractive intermolecular forces, 496–499
Aufbau principle, 105–106
Aufbau valence electron configuration, 216
Autocatalytic, 591
Automotive catalytic converter, 578
Average energy per molecule, 282
Average relative velocity, 562
Average velocity, 280, 288
Avogadro, Amedeo, 6, 273
Avogadro's law, 273
Avogadro's number, 28, 30, 273
Azimuthal quantum number, 72

B

B_2, 216–217
B_{12} molecule, 520
^{139}Ba, 622
Back-bonding, 667
Back-titration, 174
Bacon, Francis, 301
Baekelund, L., 703
Bakelite, 703
Baking soda, 365
Ballhausen, C., 665
Balmer series, 48
Banana bond, 682
Band gap, 511
Band theory of solids, 508, 509
Barium, 755–756
Barium ions, 145
Barn, 620
Barometer, 268
Barometric formula, 297
Barometric pressure, 269n
Barometric principles, 268–270
Bartell, L., 123
Baryon, 602
Baryon numbers, 607
Base, 148
Battery, 436. *See also* Electrochemistry
bcc, 504
BE, 26, 600
BE/A versus A curve, 604
Becquerel, Henri, 598, 605
Beer-Lambert law, 675
BeH_2, 184, 187
Belousov, B. P., 587
Belousov-Zhabotinsky (BZ) reaction, 587, 588
Bending potential, 190
Bending vibrations, 240
Bent, 192
Benzene, 228, 229, 329, 657, 690, 697–698
Benzene derivatives, 697–699
Benzene ring, 699
Beratan, D., 697
Berthelot, M., 315, 343
Beryllium, 155, 510, 753
Beryllium hydride, 154
Berzelius, 1
Berzelius, J., 458, 470, 578
β-alumina, 468
β-decay, 607
β^+-decay, 608
β particle, 606
β^+ particle, 606
β-sheet, 710, 711
Bethe, H., 616, 661
BF_3, 186
Bicycle, 688
Bidentate, 656

Big Bang theory, 367
Big Crunch, 367
Bimolecular, 552
Bimolecular gas-phase reactions, 564
Binary oxides and hydrides, 150–156, 170
Binding energy (BE), 26, 600
Binnig, G., 123
Binomial coefficient, 356
Binuclear, 620
Biot, J., 696–697
bipy, 655
Bipyridyl, 655
Bismuth, 771
BL acid-baser reaction, 400–403
BL conjugate base, 401–402
Black, Joseph, 472, 473, 475
Black hole, 618n
Blackbody radiation, 49, 733–739
Bloch, F., 628
Boat form, 687, 688
Bodenstein, Max, 583
Body-centered cubic (bcc), 504
Bog pictures, 600
Bohr, Niels, 42, 53–56, 58–62, 102, 632, 728
Bohr-de Broglie model of hydrogen atom, 58–62
Bohr orbit, 85, 91
Bohr orbits/energy levels, 60
Bohr radius, 60
Bohr RDF, 85
Bohr's correspondence principle, 92
Bohr's stationary state hypothesis, 54
Boiling point, 292, 484
Boiling point elevation, 525
Boiling point elevation constant, 526, 527
Boiling water, 484
Boltzmann, Ludwig, 279, 287, 303, 331, 356–360, 735, 777–784
Boltzmann distribution, 331–333, 359, 735
Boltzmann distribution law, 779
Boltzmann energy distribution, 561
Boltzmann factor, 560, 735
Boltzmann's constant, 20, 282, 357
Boltzmann's formula, 332
Boltzmann's postulate, 356
Bomb building, 636
Bomb calorimeter, 315
Bond angle, 182
Bond angles, 192
Bond dipole, 196
Bond energies, 138, 255, 324–329
Bond length, 138, 180
Bond order, 139, 211, 215–217
Bond potential energy curve, 250
Bond-potential energy curves, 179
Bond potential forms, 251n
Bond rupture, 222
Bond-selective chemistry, 573
Bonding molecular orbital, 207
Boranes, 681
Borate, 146

Boric acid, 146
Born, Max, 83, 87
Born-Haber cycle, 517
Born-Mayer repulsions, 502, 516
Born-Oppenheimer approximation, 250n
Boron, 681–683, 763–764
Boron trichloride, 533
Boron trifluoride, 186
Bose-Einstein condensate, 66
Bottleneck, 552
Bound-bound transition, 261
Bound-free transition, 261
Bound state, 60
Boundary conditions, 71, 244, 743
Boundary surface diagram, 76
Boyle, Robert, 3, 270
Boyle's law, 270
Bragg, W. H., 504
Bragg, W. L., 504
Bragg law, 505, 506
Branched chain alkanes, 686
Branched chain hydrocarbons, 688
Bridging ligand, 671
Briggsian logarithms, 548
Brittleness, 517
Brockway, L., 123
Bromate, 147
Bromate/malonic acid redox reaction, 588
Bromcresol green, 417
Bromcresol purple, 417
Bromic acid, 147
Bromide, 146, 422
Bromination, 554
Bromine, 774
Bromo, 655
Bromphenol blue, 417
Bromthymol blue, 391, 417
Brønsted, Johannes, 157
Brønsted-Lowry acid-base, 157–158
Brønsted-Lowry acid-base reaction, 400–403
Brownian motion, 575
Buckminsterfullerene, 513, 514
Buckyball, 514
Buckybowl, 514
Buffer, 404–406
Building-up principle, 105–106
Bunch of grapes (bog) pictures, 600
Bunsen, Robert, 47, 175
Bunsen burner flame, 586
Buret, 38, 411
Butanoic acid, 397
Butene, 690
Button battery, 467, 468
Butylamine, 399
Butyric acid, 397
BZ reactions, 587, 588

C

C, 31
^{14}C, 607, 631
C_{60}, 514
C–C bond, 685, 686
C–F bond dipoles, 199

C–H bond, 685
Cadmium, 762
Calcium, 132, 754
Calculus, 13
Calculus of variations, 778
Calomel electrode, 451, 456
Caloric theory, 302
Calorie, 20, 303
Calorimeter, 314
Calorimetry, 314–317
Calvin, M., 716
Camphor, 527
Candy making, 537
Cannizzaro, Stanislao, 6
Carbocation, 691
Carbohydrates, 704–706
Carbohydrates metabolism, 717–720
Carbon, 513–514, 765–768
Carbon compounds, 680. *See also* Chemistry of carbon
Carbon dioxide, 487
Carbon solid, 134
Carbonate, 146, 407, 422
Carbonic acid, 146, 406
Carbonyl, 194, 654, 655
Carbonyl group, 695
Carboxyl group, 396
Carboxylate, 396
Carboxylate anion, 396
Carboxylic acid, 693, 694
Carlisle, Anthony, 436
Carnot, Sadi, 347
Carnot cycle, 349
Carnot ideal engine, 347–350
Carothers, W., 703
Cartesian coordinates, 70
Catalysis, 578–583
Catalyst, 578
Catalytic converter (automobile), 578
Catalytic cracking, 689, 690
Catalytic cycle, 579
Cathode, 441
Cathode ray, 20–21
Cation, 145, 147, 441
Caustic soda, 470
Cavendish, Henry, 4
ccp, 504
Cell, 704
Cell voltages, 470
Cellulose, 706
Celsius, 20
Celsius, Anders, 20
Center of mass, 238, 239
centi-, 11
Centigrade, 20
Central dogma of molecular genetics, 722
Centripetal acceleration, 14
Cerium, 775
Cesium, 752–753
cgs-esu system, 10, 11
CH, 220
CH_2, 235
Chadwick, James, 25, 621
Chain branching, 585, 595

Chain propagation, 584
Chain reaction, 583
Chair form, 687, 688
Charge, 11
Charge balance, 395
Charge-charge interaction, 496
Charge density, 84
Charge-dipole interaction, 496
Charge-induced dipole interaction, 497
Charge states, 646
Charge-to-mass ratio, 21
Charge-transfer transitions, 651
Charles, Jacques, 270, 298
Charles' law, 270, 272, 291
Chelating agent, 656–657
Chemical affinity, 295
Chemical dynamics of life
  carbohydrates metabolism, 717–720
  citric acid cycle, 718–719
  enzyme catalysis, 717
  future directions, 728
  genetic code, 722
  glycolysis, 717
  hydrolysis of peptide bond, 720–722
  molecular roots of life, 727–728
  mRNA/tRNA, 722
  photosynthesis, 715–716
  protein code, 722
  protein synthesis, 723–727
Chemical equilibria. *See* Equilibrium
Chemical gas law, 291
Chemical kinetics, 539. *See also* Rates and mechanisms of chemical reactions
Chemical periodicity, 150
Chemical potential, 383
Chemical reactions
  energy changes, 301–341. *See also* Energy changes in chemical reactions
  rates, 538–597. *See also* Rates and mechanisms of chemical reactions
  spontaneity, 342–374. *See also* Spontaneity of chemical reactions
Chemical shift, 628, 630
Chemiluminescent emissions, 256n
Chemiluminescent light, 573
Chemistry, 2–3
Chemistry of carbon, 678–731
  addition reactions, 691–692
  alkane reactions, 688–689
  alkanes, 685
  alkenes, 690
  alkyl side chains, 687
  aromatics, 697–699
  benzene/benzene derivatives, 697–699
  carbocation, 691
  chemical dynamics of life. *See* Chemical dynamics of life
  conformational isomers, 686

Diels-Alder reaction, 699–702
enantiomers, 695–697
formal charge, 692
hydrocarbon derivatives, 692–694
hydrocarbons, 684–692
molecules of life. *See* Molecules of life
orbitals, 699–702
paraffins, 685
periodic table, 680
polymers, 702–703
saturated hydrocarbons, 685
structural isomers, 686–688
substitution reactions, 694–695
unsaturated hydrocarbons, 689
Woodward-Hoffmann rules, 702
Chernobyl nuclear disaster, 634
Chiral, 696
Chlorate, 147
Chloric acid, 147
Chloride, 146, 422
Chlorine, 774
Chlorine-atom-catalyzed stratospheric ozone loss, 579
Chlorine gas, 134
Chlorine monofluoride, 141
Chlorite, 147
Chloro, 655
Chloroethane, 539–544
Chloromethanes, 198, 689
Chlorous acid, 147
Chromate, 147, 422
Chromic acid, 147, 648
Chromium, 757
Chromosome, 724
Churchill, Winston, 267
Chymotrypsin, 720–722
Ci, 613
ci-, 11
cis, 690
Cisplatin, 664
Citric acid cycle, 718–719
Classical equipartition theorem, 736
Classical mechanics, 782
Classical trajectories, 570
Classical wave equation, 67
Clausius, Rudolf, 303, 351, 493
Clausius equation, 493
Clausius-Clapeyron equation, 483
Clock reactions, 593
Cloning, 728
Closed system, 304
CN cycle (stellar nucleosynthesis), 617, 641
Cobalt, 759
Code of life (genetic code), 723
Codon, 723
Coenzymes, 716
Coffee cup calorimeter, 316–317
Cold (and heat). *See* Energy changes in chemical reactions
Collector, 512
Colligative properties, 524
Collision cross section, 289, 562
Collision diameter, 562

Collision rates, 289
Collision theory (CT), 562–566
Collision trajectories, 570
Collisional activation, 564
Collisional relaxation, 572
Colloids, 523–524
Combination bands, 257
Combustion, 688
Combustion reactions, 162
Common ion, 403
Competing equilibria, 395
Complex, 648
Complexation, 416–418
Compound, 3, 38
Comprehensive Test Ban Treaty, 636
Comstock Lode, 176
Concentration cell, 454–458
Condensation polymers, 703
Conduction band, 511
Conductor, 131
Conformational isomers, 686
Conformers, 194, 686
Conjugate acid, 158
Conjugate acids and bases, 401–402
Conjugate base, 158
Conjugated diene, 700
Conjugated hydrocarbons, 690
Conjugated systems, 328
Conservation of energy, 16–19, 303
Constructive interference, 45, 208
Continuous absorption spectrum, 261
Control rods, 633
Convergence, 394
Coordinate space, 279
Coordination complex, 652, 658–661
Coordination number, 504, 653
Copenhagen interpretation, 84
Copolymer, 703
Copper, 132, 451, 474, 644, 760–761
Core electrons, 107
Correlated motion, 99
Correspondence principle, 92
Corrosion, 464–465
Cosmic rays, 626, 631
Cosmos, 635
Coulomb, Charles Augustin de, 30
Coulomb (C), 31
Coulomb barrier, 601, 635
Coulomb force, 30–31
Coulomb hill, 210
Coulomb potential, 33
Coulomb repulsion, 601, 604, 661
Coulombic attraction, 576
Coulombic repulsion of electropositive hydrogens, 182
Coulomb's law, 30, 368
Coulometer, 470
Coulometric analysis, 463
Counter ions, 168
Counterion, 653
Covalent bond, 138, 209–210
Crick, F. C., 714, 722
Crim, F., 573
Critical point, 487
Critical temperature (T_c), 495

Crookes, William, 20
Crookes' tube, 21
Cross reaction, 670
Cross section (nuclear reactions), 620
Crude oil, 686
Cryogenics, 333
Cryolite, 462
Crystal, 474
Crystal field energy diagram, 662–664
Crystal field model, 661–663
Crystal lattice, 504
Crystal structure
  allotropic elements, 515
  metal-nonmetal ionic/polar covalent compounds, 516
  water ice, 520
Crystal violet, 417
Crystalline allotropy, 494
Crystallization, 420
CsCl, 516
CT, 562–566
Cubic closest-packed (ccp), 504
Cubic system, 504
Curie, Marie Sklodowska, 24, 37, 599, 605
Curie, Pierre, 598, 605
Curie (Ci), 613
Current amplifier, 512
Cyanic acid, 146
Cyanide, 146
Cyanide leaching, 176
Cyano, 655
Cycle, 310
Cyclic bases, 711
Cyclization, 699
Cyclohexane, 687, 688
Cyclohexatriene, 690
Cyclopentadienyl, 657
Cyclopropane, 688
Cyclotrons, 623
Cylindrical symmetry, 79
Cylindrically symmetric, 188
Cysteine, 708
Cytochromes, 720
Cytosine, 399, 712

D

D, 196
d-block metals, 645–647
D-line, 47
D-line doublet, 102
D_2O, 624
Dacron, 703
Dalton, John, 1, 5–6, 36, 302
Dalton's atomic model, 5
Dalton's law of definite proportions, 274
Dalton's law of partial pressures, 277
Daniell, John, 441
Daniell cell, 441–443
Dark reactions, 716
Daughter nucleus, 607, 610
Davisson, Clinton Joseph, 57
Davisson-Germer discovery of electron diffraction, 41, 57, 58

Davy, Sir Humphry, 458
DD interaction, 497–499
De Broglie, Louis, 42, 56
de Broglie wavelength, 57
Deactivation, 564
Dead cell, 457
Debye (D), 196
Debye, Peter, 62–63, 196, 425
Debye-Hückel model, 425
Debye unit, 196
Decalin, 688
Decay of the orbital, 76
Decay properties (radioisotope), 613
Decomposition, 168
Decomposition reactions, 540–544
Defined calorie, 20
Degeneracy, 73
Degree of freedom (DF), 238–240
Degrees of freedom, 783
Dehydration, 694
Dehydrogenase complex, 718
Dehydrogenation, 689
deka-, 11
Delocalization, 210, 328–329
Delocalized bonding, 227–230
Democritus, 5
Denatured, 711
Deoxyribonucleic acid (DNA), 713–714
Deoxyribonucleotide, 712
Descartes, René, 5
Destructive interference, 45, 208
Detailed balance, 557–559
Deuterium, 621
Deuteron, 600
Dew-point, 485
Dewar, James, 492
DF, 238–240
Diamagnetic, 657
Diamagnetism, 200
Diamminesilver(I), 418
Diamond, 134, 511, 513, 514
Diaphragm cell, 462
Diastereomers, 697
Diatomic hydrides, 219
Diatomic molecules, 197, 211–223
Dibasic nitrogen bases, 411
Diborane, 682
Diboron, 216–217
Dicarboxylic acids, 411
Dichloromethane, 533
Dichromate, 147, 648
Diels, O., 699
Diels-Alder reaction, 699–702
Diethyl ether, 693
Differential and Integral Calculus (Courant), 778
Differential calculus, 13
Differential operator, 69, 741
Differential volume element, 84
Diffraction experiments, 45
Diffraction grating, 46
Diffraction interference pattern, 45
Diffractograms, 507
Diffusion, 284, 573–576
Diffusion coefficient, 575

Diffusion-controlled reactions, 576
Diffusion limit, 597
Diffusion velocity, 576
Dihydrides, 224
Dihydrogen phosphate, 146
Diimine, 204
Dilution, 387
Dimerization, 377, 379, 381, 553
Dimethyl benzene, 699
Diol, 691
Dipolar molecules, 195
Dipole-dipole interaction, 497–499
Dipole moment, 195–199
Diprotic acid, 407
Dirac, Paul A. M., 98, 99, 104, 105, 606, 627
Dirac's relativistic electron, 104
Disaccharides, 706
Discharge-flow technique, 550
Discontinuous functions, 476n
Dispersion, 497
Disproportionation, 150
Dissociation, 363, 577
Dissociative interchange, 668
Distribution function, 285
Disulfide bonds, 709
DNA, 400, 713–714
DNA-directed RNA polymerate, 724
DNA double helix, 714
DNA polymer, 713
DNA polymerase, 726
DNA replication, 726–727
Double bond, 139
Double helix, 714
Doughnut repulsion, 661
Downs cell, 462
Dry air (sea level), 276
Dry cell, 443–445
Dry cell batteries, 445
Dry ice, 532
Dry-ice isopropanol, 377
Dual-beam infrared spectrometer, 257
Dumb molecule postulate, 777, 778
Dyne, 14

E

EA, 116–117
Early transition metals, 648–651
Earth, 368
Easiest proton, 415
EC, 608
Eckart, Johannes, 101n
Eclipsed form, 686
Economic predictions, 368
Eddington, Sir Arthur, 367
EDTA, 655, 656
EF, 726
Effective nuclear charge, 100
Efficiency, 350
Effusion, 283
Egg white lysozyme, 709, 711
Eigen, M., 577
einstein, 64
Einstein, Albert, 42, 51, 52, 63, 333, 575, 636

Electric charge, 10
Electric dipole, 93, 195
Electric dipole moment, 141, 195–199
Electric dipole transitions, 93
Electric monopole moment, 195
Electrical breakdown, 20
Electrical conductivity, 131
Electrical conductor, 131
Electrical insulators, 518
Electrical nature of matter, 20–23
Electrochemical cells, 436, 440–449
Electrochemistry, 435–471
    concentration cell, 454–458
    corrosion, 464–465
    Daniell cell, 441–443
    dry cell, 443–445
    electrode potentials, 449–454
    electrolysis, 458–464
    Faraday's laws, 460–461
    fuel cell, 447–448
    lead-acid battery, 445–447
    Nernst equation, 439, 440, 442
    pH meter, 455–457
    rust, 464–465
    SHE, 450–452
    Volta's piles, 436, 437
Electrode, 441
Electrode potentials, 449–454
Electrodeposition, 463
Electrolysis, 458–464
Electrolysis cell, 463
Electrolyte, 440
Electrolytic cell, 459
Electrolytic strike bath, 463
Electromagnetic force, 31
Electromagnetic radiation, 43–46
Electromagnetic spectrum, 46
Electron, 21
Electron affinity (EA), 116–117
Electron capacity, 107
Electron capture (EC), 608
Electron configuration, 100
Electron configurations, 108, 109–110
Electron deficient, 681
Electron delocalization, 328
Electron density, 84
Electron diffraction, 57
Electron-electron repulsion, 100
Electron-nucleus interaction, 510
Electron pair geometry, 192
Electron-pair repulsion, 190–195, 200–201
Electron pairing energy, 662
Electron sea, 508
Electron sharing, 138–145
Electron spectroscopy for chemical analysis (ESCA), 261
Electron spin, 101–105
Electron-transfer (redox) reaction, 223
Electron-volt (eV), 31
Electronegativity, 140–141
Electronegativity difference, 141
Electronegativity scale, 141
Electronic spectroscopy, 259–261
Electronic transitions, 93–94, 259–261

Electrophilic, 150
Electroplating, 463
Electrostatic (coulombic) interactions, 496
Electrostatic unit (esu), 31
Element, 3
Elementary reactions, 551–552
Eley-Rideal mechanism, 582
Elimination, 554
Elongation, 726
Elongation factor (EF), 726
Emission line, 47
Emitter, 512
Empirical formula, 38
en, 655
Enantiomers, 695–697
Endothermic but spontaneous process, 344
Endothermic reaction, 307
Endpoint, 411
Energy
    atomization, 325, 327
    binding, 26, 600
    bond, 324–329
    conservation of, 16–19, 303
    exchange, 181
    heat, 19
    hydration, 323
    internal, 309
    ionization, 113–116
    kinetic, 15
    lattice, 517
    potential, 17
    resonance, 329
    thermal, 20
    total, 17
    zero point, 90
Energy changes in chemical reactions, 301–341
    atomization energy, 325, 327
    Boltzmann distribution, 331–333
    bond energies, 324–329
    calorimetry, 314–317
    conceptual developments, 302–303
    enthalpy, 312
    enthalpy diagram, 318
    equipartition theorem, 330
    heat and work, 304–308
    heat capacities, 329–331
    heat of combustion, 314
    Hess's law, 317
    law of Dulong and Petit, 332
    molecular energy storage, 331–333
    Solvay process, 320
    standard heats of formation, 320–324
    state changes for ideal gas, 311–314
    state variables, 310
    systematics of heats of reaction, 317–324
    temperature dependence of enthalpy, 333–334
Energy density, 733
Energy gap, 511

Energy grain, 735
Energy levels, 60, 743, 744
Energy of internal motion, 240
Energy profile, 560–561
Energy splitting, 627
Enthalpy, 312
Enthalpy diagram, 318
Entropy, 351–352
    absolute, 358–361
    chemical reactions, 361–363
    dilution, 387
    evolution, and, 367–368
    large number of molecules, 357
    probability, and, 354–357
    society, and, 368
    time, and, 367
Entropy of activation, 567, 568
Entropy of mixing, 372
Enzyme, 580
Enzyme catalysis, 580–581, 717
Enzyme-substrate complex, 580
Enzymes, 707, 711, 720–722
Epoxide, 691
Equation of state, 274
Equatorial position, 191
Equilibrium, 307, 345, 364
    aqueous, 387–418. *See also* Free energy and chemical equilibrium
    attainment of, 376–383
    competitive, 395
    gas-phase reactions, 381–386
    macroscopically static/microscopically dynamic, 376
    metastable, 523
    multiphase systems, 418–423
    reversibility, 346
Equilibrium condition, 376
Equilibrium constant, 380–381
Equilibrium vapor pressure, 482
Equipartition theorem, 330
Equivalent weight, 470
Eriochrome black T, 417
Erlenmeyer flask, 304
Errors in measurement, 11–13
Erythrosin B, 417
ESCA, 261
Esson, William, 559, 591
Ester, 693, 695
Esterification, 624
Esters, 594
esu, 31
Ethane, 501, 685
Ethanoic acid, 397
Ethanol, 533, 630, 693
Ether, 693, 694
Ethyl, 687
Ethyl acetate, 693
Ethyl carbocation, 691
Ethylene, 657, 690, 700
Ethylene diamine, 400, 411, 655
Euler's formula, 79, 673
Europium, 672, 776
eV, 31

Evaporation, 482
Evaporation of water, 344
Evolution, 367
exa-, 11
Exchange, 180
Exchange energy, 181
Excited decay products, 610
Excluded volume, 294
Exothermic reaction, 307, 344, 363
Expanded octets, 141
Experimental potential curves, 179
Explosion hazards (chemical laboratory), 297
Explosions, 583–585
Exponential decay, 76
Extinction coefficient, 675
Eyeball spectroscopy, 652
Eyring, H., 566–568

F

f-block metals, 672–674
F_2, 216
Face-centered cubic (fcc), 504
FAD, 718
Fahrenheit, Gabriel, 19
Fahrenheit 451 (Bradbury), 37
Fahrenheit scale, 19
Far-infrared spectral region, 246
Faraday, Michael, 22, 439, 441, 460, 461, 470, 487, 523
Faraday constant, 439
Faraday's laws of electrolysis, 460–461
Fats, 706–707
Fatty acids, 397, 706–707
f-block metals, 672–674
fcc, 504
Feisher-Tropach process, 582
femto-, 11
Fermentation, 694
Fermi, 600
Fermi, Enrico, 105, 621, 631, 636
Fermi level, 510
Ferrocene, 657
Festing, E. R., 256
Feynman, Richard, 41, 474
Filament, 66
Fine-structure doublet, 65
First angle dependence, 79
First IE, 113
First law of thermodynamics, 303, 324, 335
First-order, 543
First-order differential equation, 543
First-order rate law, 543
Fischer, Emil Hermann, 679, 705
Fischer projection, 704–706
Fission, 608, 621–623
Flame, 585–586
Flame spectra, 47
Flame tests, 65
Flavin adenine dinucleotide (FAD), 718
Flavoproteins, 720
Flickering cluster model of liquid water, 522

Flow tube method for gas reactions, 550
Fluoride, 146, 422
Fluorine, 220, 773
Fluorite, 516
Fluoro, 655
Flux density, 620
FMO model, 700–702
Fock, V., 121
Force, 13
Force constant, 18, 255
Formal charge, 692
Formaldehyde, 693
Formation constant, 417, 658, 659
Formation reactions, 320
Formic acid, 397
4 + 2 cycloaddition, 700
4f orbitals, 673
Fourier transform NMR, 628
Fractional orders, 594
Francium, 753
Frank, H., 522
Franklin, Ben, 20
Franklin, R., 714
Fraunhofer, Josef, 46
Fraunhofer's D-line, 47
Fraunhofer's dark lines, 363
Fraunhofer's solar spectrum, 47
Free energy and chemical equilibrium, 375–434
 acid-base indicators, 417
 acid-base titrations, 411–416
 aqueous acid-base chemistry, 387–418
 aqueous complexes, 416–418
 Arrhenius acid-base reaction, 388–392
 attainment of equilibrium, 376–383
 Brønsted-Lowry acid-base reaction, 400–403
 buffer, 404–406
 common ion, 403
 Debye-Hückel model, 425
 deviations from ideal behavior, 423–426
 equilibrium constant, 380–381
 gas-phase reactions, 381–383
 Henderson-Hasselbalch equation, 404–406
 hydrohalic acids, 397
 ionization of polyprotic acids, 406–411
 LeChâtelier's principle, 384–387
 multiphase systems, 418–423
 oxyacids, 396, 406
 oxyhalic acids, 397
 pH scale, 391
 pH titration curves, 412–415
 qualitative metal ion analysis, 421, 423
 solubility and precipitation, 419–421
 standard free energy of formation, 377–378

 successive approximations, 394
 van't Hoff equation, 386
 weak acid, 392–397
 weak base, 398–400
 zwitterion, 403
Free gas-phase atoms, 325
Free internal rotation, 685
Free radical, 554, 688
Free-radical initiator, 702
Freezer unit (refrigerator), 532
Freezing point, 485
Freezing point depression, 526
Freezing point depression constant, 526, 527
Freon, 579
Frequency, 43
Frequency factor, 559
Frontier molecular orbital (FMO) model, 700–702
Frontier orbital, 221–223
Fuel cell, 447–448
Fuel rods, 633
Fugacity, 424
Fukui, Kenichi, 221
Fuller, R. Buckminster, 514
Fundamental theorem of the calculus, 17
Furanoses, 705
Fusion reaction, 600

G

Galena, 166
Gallium, 764
Galvani, Luigi, 436
Galvanic cell, 436
Galvani's animal electricity, 436
Galvani's frog legs, 436
Galvanizing, 465
γ-rays, 610
Gamov, G., 614, 615
Gamov's volcano potential energy model, 615
Gas dynamics, 288–291
Gas expansion, 353
Gas laser, 43
Gas mixtures, 276–278
Gas-phase collision chemistry, 289
Gas-phase reactions, 381–383
Gas thermometer, 275
Gases. *See* Properties of gases
Gasoline, 686
Gauss, Karl Friedrich, 286
Gaussian error distribution function, 574, 575
Gaussian function, 286
Gauss's error law, 286
Gay-Lussac, Joseph, 6, 268, 270, 272, 273
Geiger, Hans, 23, 24, 611
Geiger counter, 611–613
Gene, 714
Genetic code, 722, 723
Genetic engineering, 728
Genome, 724
Geometrical isomers, 654, 690

Germanium, 768
Germer, Lester Halbert, 57
Gibbs, Josiah Willard, 274, 343, 363, 364, 383, 479–480, 493
Gibbs free energy, 364, 481, 783
giga-, 11
Gillespie, R. J., 190
Glass, 494, 518
Glass electrode, 455
Globular, 711
Glucoside bond, 706
Gluon, 180
Glutamic acid, 708
Glutamine, 708
Glyceryl stearate, 706
Glycine, 194, 403, 693, 708
Glycolysis, 717
Glycolytic (anaerobic) pathway, 717
Goddard, W. A., 231
Gold, 761–762
Goudsmit, Samuel Abraham, 102
Grade distribution histograms, 285
Graham, Thomas, 283
Graham's 1846 experiment, 283
Graham's law, 283, 284
Graphite, 134, 514
Gravitational constant, 14
Gray tin (α, diamond), 515
Ground state, 54
Group I, 147
Group IA elements, 749–753
Group IB elements, 760–762
Group II, 147
Group IIA elements, 753–756
Group IIB elements, 762–763
Group IIIA elements, 763–765
Group IIIB elements, 756
Group IVA elements, 765–769
Group IVB elements, 756–757
Group VA elements, 769–773
Group VB elements, 757
Group VIA elements, 773
Group VIB elements, 757–758
Group VII elements, 773–775
Group VIIB elements, 758–760
Group VIII elements, 775
GTP, 725
Guanine, 399, 712
Guanosine triphosphate (GTP), 725
Guldberg, Cato Maximilian, 375

H

Haber-Bosch synthesis, 430
Hahn, O., 621
Hairpin, 724
Half-cell, 441
Half-filled band, 509
Half-life, 544, 612
Half-reaction, 437, 448, 449, 452
Half-reaction method, 164–165
Half-reaction time, 543
Halide, 692, 693
Hall, Charles, 462
Hall-Heroult process, 462, 463
Halogenation, 688–689
Hamilton, William, 69

Hamiltonian operator, 69, 99
Hamlet (Shakespeare), 577
Handbook of Chemistry and Physics, The, 321, 487, 602
Handedness, 104, 696
Haniwa, 632
Harcourt, A. Vernon, 559
Harmonic oscillator, 252
Harmonic oscillator (HO) potential, 19, 251–253
Hartree, D. R., 121
Hartree's iterative method, 121
Hawking, Stephen, 367
Haworth projection, 705, 706
HB, 222, 500–502
HCl, 247–248
$HClO_4$, 153
hcp, 504
HDPE, 703
Heat capacity, 307, 329–331
Heat capacity at constant pressure, 313
Heat capacity at constant volume, 312
Heat energy, 19
Heat of combustion, 314
Heating curves, 475–479
Heavy hydrogen, 621
Heavy water, 624
hecto-, 11
Heisenberg, Werner, 87
Heisenberg's uncertainty principle, 87–90, 741, 782
Heitler, W., 178–180
Heitler-London treatment of H_2, 178–180
Helicity, 104
Helium burning, 617
Helmholtz, Hermann von, 303
Heme group, 656
Hemoglobin, 711
Henderson-Hasselbalch equation, 404–406
Hermite, Charles, 253
Heroult, Paul, 462
Herschbach, Dudley, 130, 573
Hertz, Heinrich, 20, 51
Herzberg, Gerhard, 237, 238, 255, 260
Hess, G., 317
Hess's law of constant heat summation, 317
Heteroatoms, 193, 198
Heterogeneous catalyst, 578
Hexaaquanickel(II) complex, 658
Hexachloronickelate(II) complex, 658
Hexadentate, 656
Hexagonal closest-packed (hcp), 504
Hexagonal system, 504
Hexamminenickel(II) complex, 659
Hexane, 686
Hexose sugars, 704
HF, 221, 222
High-density polyethylene (HDPE), 703
High-energy/high-purity chemicals, 461
High-resolution spectrum, 630
High spin, 657

Highest energy occupied MO (HOMO), 221–223
Highest occupied atomic orbital (HOAO), 118
Highest occupied MO (HOMO), 700–702
Hindenburg, 538
Histidine, 708
Histograms, 285
HOAO, 118
Hodgkin, Dorothy, 520
Hoffmann, R., 221
HOMO, 221–223, 700–702
Homogeneous catalyst, 578
Homonuclear diatomic molecules, 217–221
Hooke's law of springs, 18
Hot atom chemistry, 625
Hotness and coldness. *See* Energy changes in chemical reactions
Hückel, E., 425, 698
Hückel model, 329
Human life. *See* Chemical dynamics of life; Molecules of life
Hund, Friedrich, 206
Hund's rule, 108–109
Huygens's wave construction, 45
Hybrid boundary surfaces, 186
Hybrid combustion/hydrogen-oxygen fuel cell automobile engine, 435
Hybrid orbital, 183, 184
Hybridization, 183–187
Hydration energies, 323
Hydrazine, 411, 683
Hydrazoic acid, 683
Hydride, 146
Hydrides, 154–156
Hydrobromic acid, 146
Hydrocarbon derivatives, 692–694
Hydrocarbons, 684–692
Hydrochloric acid, 146, 150
Hydrofluoric acid, 146
Hydrogen, 146, 749
Hydrogen atom, 76
Hydrogen atom orbitals, 81
Hydrogen atom spectrum, 47–49
Hydrogen bomb, 636
Hydrogen bonding (HB), 222, 500–502
Hydrogen bromide, 146
Hydrogen burning, 616
Hydrogen carbonate, 146
Hydrogen chloride, 146
Hydrogen dissociation reaction, 363, 374
Hydrogen fluoride, 139, 146, 154
Hydrogen iodide, 146
Hydrogen ion, 22
Hydrogen-like ions, 61
Hydrogen molecular ion, 206–207
Hydrogen molecule, 178–181
Hydrogen peroxide, 146
Hydrogen sulfate, 146
Hydrogen sulfide, 146, 235
Hydrogen sulfite, 146

Hydrohalic acids, 397
Hydroiodic acid, 146
Hydrolysis of peptide bond, 720–722
Hydrolyzing, 398
Hydronium, 156
Hydrophilic, 707
Hydrophobic, 707
Hydroselenic acid, 409
Hydrosulfuric acid, 146
Hydroxide, 146, 422
Hydroxide precipitation, 434
Hydroxo, 655
Hydroxyl (OH) compounds, 153
Hydroxyl hydrogen of phenol, 731
Hydroxyl radical (OH), 235
Hygroscopic, 39
Hyperbromite, 147
Hyperchlorite, 147
Hyperfine splittings, 249
Hypobromous acid, 147
Hypochlorous acid, 147, 150
Hypoiodite, 147
Hypoiodous acid, 147
Hysteresis, 494

I

Ice, 474, 519–521
Ice calorimeter, 314
Icosahedron, 683
Ideal gas constant, 274
Ideal gas law, 274
Ideal pressure, 293
IE, 600
IF-3, 725
Incomplete combustion, 162
Indicator, 38, 411
Indistinguishability, 105
Indium, 765
Infrared region of solar spectrum, 255
Infrared spectroscopy, 256
Inhibition, 593
Initial rate, 547
Initiation, 584, 726
Initiation factor 3 (IF-3), 725
Inner cone, 585
Inner-sphere mechanism, 671
Inner transition metals, 672
Insoluble chlorides, 421
Insoluble sulfides, 421
Instantaneous dipole-induced dipole interaction, 497–499
Integrated rate law, 543
Interconal region, 585
Intermolecular attraction, 293
Intermolecular distance, 474, 502
Intermolecular forces, 474, 496–503, 783. *See also* States of matter and intermolecular forces
Intermolecular potential energy, 474, 475, 783
Intermolecular repulsion, 294
Internal degrees of freedom, 331, 783
Internal energy, 309
Internal proton transfer, 403

International Nuclear Test Ban Treaty, 636
Inverse disproportionation, 587
Inverse-square intensity law, 613
Inverse-square law of force, 14, 30
Iodate, 147
Iodic acid, 147
Iodide, 146, 422
Iodine, 774–775
Iodo, 655
Iodometric analysis, 175
Ion-electron half-reaction method, 164
Ion-exchange chromatography, 673
Ion interaction in solution, 425, 426
Ion-selective electrodes, 456
Ionic bond, 137
Ionic bonding, 137–138
Ionic equation, 149
Ionic radii, 120, 647
Ionic solids, 333, 514–518
Ionic strength, 425
Ionic term, 199
Ionization, 21
Ionization continuum, 73
Ionization energy (IE), 73, 113–116, 600
Ionization equilibrium constants, 398
Ionization of ammonia, 431
Ionization of polytropic acids, 406–411
Ionization of weak acids and bases, 392–400
Iridium, 759–760
Iron, 132, 644, 758–759
Irreversible, 345, 346
Isentropic, 352
Isobaric, 312
Isobutane, 483, 484, 501, 685, 687
Isobutyl, 687
Isochoric, 312
Isoelectronic ions, 120
Isolated, 304, 307
Isolation method, 549
Isoleucine, 708
Isomer, 653
  conformational, 686
  geometric, 690
  optical, 697
  structural, 686–688
Isomerization, 554, 577
Isooctane, 685, 687
Isopropanol, 483, 501
Isopropyl, 687
Isopropyl amine, 693
Isopropyl chloride, 693
Isothermal, 311
Isothermal compression, 349
Isothermal expansion, 313, 349
Isothiocyanato, 655
Isotope, 26, 598, 603
Isotopic labeling, 624
Isotopic substitution, 259, 624
Isotopic chemistry, 624–625
Isotropic radiation, 613

Iteration, 121
Iterative method, 394

J

J, 16
J-tube, 270, 271
Joliot-Curie, Irène, 599, 621
Jordan, Pascual, 87
Jorgensen, W., 502
Joule, James, 301, 303, 353
Joule (J), 16
Joule heating, 462
Joule-Thomson effect, 353n

K

^{40}K, 639
K-shells, 505
Kamerlingh-Onnes, Heike, 492, 495
Kassel, L. S., 566
Kekulé, A., 698
Kelvin, Lord, 10, 20, 271, 302, 353
Kelvin temperature scale, 20
Kerosene, 686
Ketone, 693
Ketoses, 704
Khorana, H., 723
KHP, 699
kilo-, 11
Kinetic control, 582–583
Kinetic energy, 15
Kinetic energy of rotation, 240
Kinetic energy operator, 69
Kinetic molecular theory,
  278–291, 329
  collision rates, 289
  diffusion, 284
  effusion, 283
  gas dynamics, 288–291
  Gauss's error law, 286
  Graham's law, 283, 284
  ideal gas law, 282
  Maxwell-Boltzmann velocity
    distribution, 285–288
  predictions, 282–284
Kinetic theory interpretation of
  temperature, 281
Kinetics. *See* Rates and mechanisms of
  chemical reactions
Kirchhoff, Gustav, 47
Kjeldahl method, 39
Knudsen, B., 299
Knudsen cell, 299
Kossel, W., 135
^{94}Kr, 622
Krebs cycle, 718

L

La Châtelier's principle, 384–387
Lability, 657–658
Lagrange multipliers, 779
Laguerre-Legendre solutions, 71
Laguerre polynomial, 74
Langmuir, Irving, 581–582
Langmuir adsorption isotherm, 582

Langmuir adsorption theory, 582
Langmuir mechanism, 582
Lanthanides, 107, 115, 119, 672–673,
  775–776
Lanthanum, 756
Laplace, Pierre, 314
Larmor frequency, 628
Laser, 43
Laser-based Raman spectrometer, 249
Laser implosion, 635
Late barrier, 572
Late transition metals, 651
Latent heat, 476
Lattice energy, 517
Lattice parameter, 535
Lattice types, 515
Lavoisier, Antoine, 4, 5, 35, 302, 314
Law of combining volumes, 6
Law of conservation of energy, 303
Law of conservation of mass, 4
Law of definite proportions, 4
Law of Dulong and Petit, 332
Law of multiple proportions, 5
Law of octaves, 8
Laws of electrolysis, 461
LCAO approximation, 207–210
LCAO-MO, 208
LD interaction, 497–499
Lead, 132, 769
Lead-acid battery, 445–447
Least classical atoms, 570
LeChâtelier, Henri, 384
LeChâtelier's principle, 384
Leclanché, Georges, 443
Leclanché dry cell, 443–445
LED, 512
Lee, Y., 573
Legendre function, 74
Length, 11
Lepton, 602
Leucine, 708
Leveling effect, 160
Lewis, Gilbert Newton, 131, 135–136,
  138, 158–159, 423
Lewis acid, 158, 159
Lewis acid-base interactions, 158–159
Lewis base, 158, 159
Lewis covalent bond, 138–145
Lewis electron-dot structures, 136
Lewis noble gas rule, 135
Lewis structure, 139
Lewis structure description of reaction,
  144–145, 183, 190
L'Hôpital's rule, 738
Li_2, 215, 216
Libby, W., 631
Liborane, 154
Life energy. *See* Chemical dynamics of
  life; Molecules of life
Ligand, 417, 653, 654–657
Ligand exchange, 671
Ligand field MO energy-level
  diagram, 666
Ligand field model, 655–667

Ligand pecking order, 658
Ligand substitution (LS), 668–669
Light-emitting diode (LED), 512
Light reactions, 716
Lime, 433
Lime kilns, 433
Limewater, 434
Lind, Samuel Colville, 583
Lindemann, F., 563, 565
Lindemann mechanisms, 563
Lindemann rate law, 565
Line-of-centers model, 620
Line spectra of atoms, 46–49
Linear combination of atomic orbitals
  (LCAO), 207–210
Linear combinations, 79, 184
Lineberger, W. C., 116
Lipids, 706
Liquid sodium, 532
Liquids, 521–529
Liter, 29
Lithium, 468, 509, 749–750
Lithium aluminum hydride, 155
Lithium batteries, 468
Lithium crystal, 510
Lithium hydride, 154
Lithium-sulfur battery, 468
Loch Lomond, 538
Locke, John, 5
Lodestone, 651
Logarithmic time profiles, 551
Logarithms, 548
London, F., 178–180
London dispersion (LD) interaction,
  497–499
Lone pair electrons, 139
Lone pair substitutions, 192
Lone pairs, 190–195, 202
Long-range attractive intermolecular
  forces, 496–497
Los Alamos, NM (nuclear reactor), 633
Low resolution spectrum, 630
Low spin, 657
Low-spin complex, 662
Lowell, John, 739
Lowest energy unoccupied MO
  (LUMO), 221–223
Lowest unoccupied MO (LUMO),
  700–702
Lowry, T. M., 157
LS, 668–669
Lucretius, 5
Lummer, Otto, 49, 50, 736
LUMO, 221–223, 700–702
Lutetium, 672
Lyman-α, 48
Lyman series, 48
Lysine, 708, 723
Lysozyme, 709

M

$M(A,Z)$, 602
m-Xylene, 699
Maclaurin power series, 736

Macromolecules, 503, 702
Magic numbers, 602
Magnesium, 753–754
Magnetic field, 44
Magnetic moment, 626
Magnetic quantum number, 72
Magnetic resonance imaging
  (MRI), 629
Magnetism, 657
Magnetite, 651
Main group, 120
Malleability, 517
Manganese (Mn), 675, 758
Manhattan Project, 636
Many-electron problem, 120–122
Marsden, Ernest, 23, 24
Marcus, Rudolph A., 566
Markovnikov's rule, 731
Mass, 11, 603
Mass defect, 26, 600
Mass number, 26, 598
Mass spectrometry, 22
Mass spectrum, 26
Masses $M(A,Z)$, 602
Massive trajectory studies, 577
Matrix mechanics, 87
Maximum work, 345
Maxwell, James Clerk, 25, 43, 44, 49,
  278, 279, 286, 287, 303, 496
Maxwell-Boltzmann velocity
  distribution, 287
Maxwell's wave theory of light,
  43–46, 736
Mayer, Julius, 303
McMillan, E., 623
Mean free path, 289, 574
Mean square velocity, 281
Mean velocity, 280, 288
Measurement errors, 11–13
Mechanical energy, 237–238
Mechanical equivalent of heat, 303
Mechanical work, 14–16
Mechanistic resonance forms, 700
Medical diagnosis, 626
mega-, 11
Meitner, Lise, 621
MEK, 693
Melting, 521
Mendeleyev, Dmitri, 8
Mendeleyev's periodic law, 8
Menten, Maud, 580
Mercury, 132, 269, 495, 763
Meson, 180, 602
Mesosphere, 485
Messenger RNA (mRNA), 722
Metabolic energy production, 720
Metabolic rate, 720
Metabolism, 704, 717–718
Metal-acid reactions, 149–150
Metal crystals, 504–510
Metal ion analysis, 421, 423
Metal refining, 463
Metallic conduction, 131
Metallic hydrides, 155

Index

Metallic iron, 644
Metalloids, 132, 133
Metals, 131, 132
Metastable equilibrium, 523, 566
Metastable process, 494
Metathesis, 167
Methane, 139, 154, 183, 448, 501, 685
Methanoic acid, 397
Methanol, 501
Methionine, 708, 723
Method of initial rates, 547–549
Methyl, 687
Methyl amine, 693
Methyl chloride, 689
Methyl fluoride, 501
Methyl group, 685
Methyl orange, 417
Methyl red, 416, 417
Methyl salicylate, 698
Methylamine, 399
Methylene chloride, 533, 693
Methylene molecule (CH_2), 235
Methylpropane, 687
Methyltransferase, 727
Metric system, 10
Meyer, Lothar, 8, 538
Michaelis, Leonor, 580
Michaelis constant, 581
Michaelis-Menten rate law, 581
micro-, 11
Microscope, 123–126
Microscopic reversibility, 597
Microwave spectral region, 246
Microwave spectroscopy, 246
Midsummer Night's Dream, A (Shakespeare), 237
Milk of magnesia, 151
Miller, Stanley, 678, 727
Miller indices, 507
milli-, 11
Millikan, Robert A., 22, 23, 52
Millikan's oil drop experiment, 22, 23, 442
Mineral acids, 153
Minimization of potential energy, 190
Minimum work, 347
Mirror plane, 696
Miscible, 277
Mixture, 3
MO predicted bond order, 215–217
MO theory, 205–236. *See also* Molecular orbital (MO) theory
MO-VB hybrid description of molecules, 231–232
Moderator, 633
Mol, 28
Molality, 526
Molar heat capacities, 329
Molar kinetic energy, 282
Molarity, 29
Mole, 28
Mole balance, 395
Mole fraction, 276
Molecular beams, 573
Molecular bond lengths, 197

Molecular calisthenics (vibrational motion), 250–259
Molecular collision rate, 289
Molecular dissociation, 230
Molecular dynamics simulations, 502
Molecular electronic transitions, 260
Molecular energy levels, 201
Molecular energy storage, 331–333
Molecular formula, 38
Molecular liquids or solids, 333
Molecular mechanisms, 551
Molecular motion
  degrees of freedom, 238–240
  electronic transitions, 259–261
  rotational motion, 240–250
  vibrational motion, 250–259
Molecular orbital (MO) theory, 205–236
  bond order, 215–217
  delocalized bonding, 227–230
  diatomic molecules, 211–223
  frontier orbital, 221–223
  HOMO, 221–223
  homonuclear diatomic molecules, 217–221
  hydrogen molecular ion, 206–207
  LCAO approximation, 207–210
  LUMO, 221–223
  MO-VB hybrid description of molecules, 231–232
  molecular dissociation, 230
  resonance problem, 227–230
  shortcomings, 230
  water molecule, 224–227
Molecular oxygen, 200–201
Molecular partition function, 780
Molecular random walk, 574
Molecular reaction dynamics, 569–573
Molecular shapes, 190–195
Molecular solids, 519–520
Molecular spectroscopy, 238. *See also* Spectroscopy
Molecular stupidity, 780
Molecular trajectories, 569–572
Molecule, 6
Molecules of life, 704–715
  amino acids, 707–711
  carbohydrates, 704–706
  DNA, 713–714
  fats, 706–707
  nucleotides, 711–712
  proteins, 707–711
  purines, 712
  pyrimidines, 712
  RNA, 713
Molten Al, 462
Molten electrolyte, 462
Molten sodium-sulfur batteries, 468
Molybdenum, 757
Moment of inertia, 242
Momentum, 13
Monazite, 673
Monoclinic, 515
Monoclinic system, 494
Monohydrogen phosphate, 146

Monoprotic acid, 392, 396, 398
Monoprotic acid titration curves, 415
Monosaccharides, 704
Moore, Walter J., 205
Morse potential, 251n
Moseley, Henry, 8, 598, 599
Most probable velocity, 288
Mr. Tompkins in Wonderland (Gamov), 89
MRI, 629
mRNA, 722
mRNA codons, 723
mRNA-ribosome complex, 725
MTBE, 693, 695
Mulliken, Robert S., 74, 141, 206, 207, 212, 215
Mulliken's LCAO method, 207–210
Multidentate, 656
Multidentate ligands, 656
Multiphoton absorption, 64
Multiple-bonded molecules, 261
Multipole moments, 195
Mutation, 727
Myoglobin, 711

N

N, 14
n-Butane, 685
n-Butyl, 687
N-factorial, 778
n-Hexane, 685
n-Octane, 685
n-Propyl, 687
n-type semiconductor, 512
N_2, 216
N_2O_5, 541
NaCl, 516, 517
NAD, 717, 718
NADP, 715–716
NADPH, 716
NaI electrolysis cell, 459
nano-, 11
Nano-test tube, 514
Nanorope, 514
Naphtha, 686
Native, 711
Natural bond orbitals, 227n
Natural fission chain reactors, 637
Natural gas, 686
Natural logarithms, 548
Natural philosophy, 2
Nature of the Chemical Bond, The (Pauling), 178
Nearly reversible isothermal expansions, 346
Negative entropy changes, 363
Negative ions, 145
Negative total energy, 32
Neodymium, 775
Nernst, Walther, 439
Nernst equation, 439, 440, 442
Neutralization reaction, 156
Neutron, 25, 599
Neutron-induced nuclear fission, 622
Neutron star, 618n

New Ideas in Inorganic Chemistry (Werner), 644
Newcomen, Thomas, 348
Newlands, John, 7
Newman projection, 686
Newton, Sir Isaac, 5, 13, 342, 741
Newton (N), 14
Newtonian mechanics, 13, 570, 782
Newton's first law of motion, 238
Newton's gravity, 14
Newton's second law of motion, 13
Nicholson, Williams, 436
Nickel, 760
Nickel(II) chloride, 658
Nickel(II) complexes, 658–660
Nickel oxide-cadmium (nicad) battery, 445
Nicotinamide adenosine dinucleotide (NAD), 717, 718
Nicotinamide adenosine dinucleotide phosphate (NADP), 715–716
nido-pentaborane, 682
Nirenberg, M., 722
Nitrate, 146
Nitric acid, 146
Nitric oxide, 219
Nitride, 146
Nitrite, 146
Nitrito, 655
Nitro, 655
Nitrogen, 683, 769–770
Nitrogen gas, 134
Nitrogen ice, 474
Nitrogen weak bases, 400
Nitrogenase, 716
Nitrosyl, 655
Nitrous acid, 146, 396
NMR, 103n, 626–630
NMR spectra, 630
NMR spectrometers, 629
NO_2, 376n
NO_2 dimerization, 377, 381
Noble gas rule, 135
Nodal cones, 80
Nodal plane, 79
Nodal rule, 701
Nodal sphere, 77
Nodal surface, 79
Node, 77
Nonbonding, 666
Nonideal behavior, 291
Nonmetal oxides, 152
Nonmetallic network solids, 510–513
Nonmetals, 131, 134, 150
Normal alkanes, 686
Normal boiling point (T_b), 484
Normal distribution function, 286
Normal freezing point, 485
Normal mode, 240
Normal modes of vibration, 255
Normal temperature, 315
Normal temperature and pressure (NTP), 297
Normalization condition, 744
Normalization constants, 76, 84

Index

NTP, 297, 315
Nuclear applications, 623
   isotopic chemistry, 624–625
   MRI, 629
   NMR, 626–630
   nuclear energy sources, 632–635
   nuclear medicine, 625–626
   nuclear weapons, 636
   radioactive dating, 630–632
Nuclear atom, 23–25
Nuclear binding energies, 600
Nuclear chain reaction, 622
Nuclear chemistry, 618–623
Nuclear decay, 606–615
Nuclear energy sources, 632–635
Nuclear explosion (Alamogordo), 640
Nuclear fission, 621–623
Nuclear fission pressurized water power reactor (PWR), 633
Nuclear magnetic resonance (NMR), 103n, 626–630
Nuclear magneton, 628
Nuclear medicine, 625–626
Nuclear radiation, 612–613
Nuclear reaction cross section, 620
Nuclear reactions, 616–623
Nuclear reactors, 633–635
Nuclear spectroscopy, 610
Nuclear spin, 626, 627
Nuclear stability and radioactivity, 604–611
Nuclear stability diagram, 605–606
Nuclear structure and bonding, 599–604
Nuclear test ban treaties, 636
Nuclear weapons, 636
Nuclease, 724
Nucleon, 599
Nucleon-nucleon potential energy curves, 601
Nucleon spin, 626
Nucleophilic, 150
Nucleophilic substitution, 694
Nucleoside, 712
Nucleotides, 711–712
Nucleus, 25, 598, 599
Nuclide, 599
Number density, 289
Nylon-6, 703
Nylon-66, 703

O

O-H bond, 258
o-Phenanthroline, 655
o-Xylene, 699
O_2, 216
Objectives of modern chemistry, 3–9
Octahedral, 191, 192
Octahedral complex, 653, 654
Octahedron, 653
Octet rule, 137
Odd-electron molecules, 219
Odd oxygen, 578
OF cycle (stellar nucleosynthesis), 617, 641

OH, 235
OH compounds, 153
Oil droplets, 22, 23
On the Equilibrium of Heterogeneous Substances (Gibbs), 343, 480
"On the Motive Power of Heat" (Carnot), 347
On the Spring of the Air and Its Effects (Boyle), 270
One-dimension wavelength density, 734
One-electron orbitals, 206
One-molecule-at-a-time approach, 262
1-2-3 superconductor, 171, 495
1,3-Butadiene, 700
1,6-diaminohexane, 411
1-D orbitals, 744, 745
1-D radial wave function, 78
1s orbital, 76
Open system, 304
Operator, 68, 69
Oppenheimer, J. Robert, 636
Optical isomers, 656, 697
Optical rotation, 696
Orbital, 63, 91–92
   antibonding, 217
   Born's interpretation, 83
   characteristics, 76–82
   4f, 673
   frontier, 221–223
   H atom, 81
   hybrid, 183, 184
   organic reactions, in, 699–702
   overlap, 181
   screening, 106–108
   specification of wave functions, 74
   spherically symmetric, 76
Orbital hybridization, 183
Orbital model, 99
Orbital occupation diagrams, 108
Orbital overlap, 181
Orbital product, 121
Orbital screening, 106–109
Organic chemistry, 183, 679. *See also* Chemistry of carbon
Organic compounds, 678–679
Organic ligand derivatives, 655
Organic motifs in biology. *See* Molecules of life
Orientation, 519
Orlon, 702
Oscillating electric dipole, 246
Oscillating electric field, 44
Oscillating reactions, 586–589
Osmium, 759
Osmosis, 529
Osmotic pressure, 528
OsO_4, 647
Ostwald synthesis of nitric acid from ammonia, 340
Outer cone, 586
Outer-sphere complex, 668
Outer-sphere mechanism, 670
Outer turning point, 86, 614
Overall electron density, 226

Overall order, 546
Overlap repulsion, 502
Overtone transitions, 257
Overvoltage, 459
ox, 655
Oxalate, 146, 422
Oxalato, 655
Oxalic acid, 146
Oxidation number, 165–167
Oxidation-reduction reactions, 86, 133–137, 160–167, 168, 169, 669–671
Oxidation states, 646
Oxide, 146
Oxides, 151–154
Oxyacids, 153, 396, 406
Oxyanions, 146–162, 648
Oxygen, 772
Oxygen gas, 134
Oxyhalic acids, 397
Oxyions, 648
Ozone, 578
Ozone hole, 579
Ozonolysis, 691

P

p-n binding energy, 601
p-n junction, 512
p-type semiconductor, 512
p-Xylene, 699
Packing fraction, 535
Paddle-wheel experiment (Joule), 301, 303
Palladium, 760
PAN, 702
Pantothenic acid, 718
Paraffin waxes, 686
Paraffins, 499, 685
Paramagnetism, 200, 657
Partial charge separation, 141
Partial ionic character, 141
Partial pressures, 277
Particle-in-a-box model, 224, 508, 741–747
Particle-on-a-ring model, 229
Partition function, 341
Pascal, 270
Pascal, Blaise, 270
Pasteur, Louis, 696
Pathology, 728
Pauli, Wolfgang, 105, 626
Pauli repulsion, 502
Pauling, Linus, 140–142, 171–181, 636, 665, 710, 714
Pauling's valence bond model, 181–187
Pauli's exclusion principle, 105
Pentaborane-9, 682
Pentavalent carbon, 695
Pentose sugars, 704
Peptide bond, 708
Peptidyl site, 725
Perchlorate, 146
Perchloric acid, 146, 153
Periodic law, 8

Periodic properties of atoms of electrons, 113–120
   atomic radius, 118–119
   EA, 116–117
   IE, 113–116
   ionic configurations, 120
   isoelectronic sequences, 120
Periodic table, 107–110
Periodic table (short form), 133
Permanent gas, 487
Permanganate, 147, 648, 649
Peroxide, 146
PET scan, 626
peta-, 11
Petroleum, 686
Pfeffer, W., 529
pH, 390
pH electrode, 456
pH meter, 390, 391, 455–457
pH meter electrodes, 456
pH scale, 391
pH titration curves, 412–415
Phase changes, 479–495. *See also* States of matter and intermolecular forces
Phase diagram, 485–487
Phase-matching requirement, 702
Phase properties, 488–491
Phase rule, 493
phen, 655, 671
Phenanthroline (phen), 671
Phenol, 699
Phenol red, 417
Phenolic plastics, 703
Phenolphthalein, 417, 460
Phenyl, 702
Phenyl group, 699
Phenylalanine, 708
Phlogiston theory, 4
Phosphate, 146, 422
Phosphide, 146
Phosphine, 146
Phosphine gas, 300
Phosphite, 146
Phospholipids, 707
Phosphoric acid, 146, 152, 406
Phosphorus, 770–771
Phosphorus acid, 146
Phosphorus (V) oxide, 151
Phosphorus skeleton, 151
Phosphorus solid, 134
Photochemical reactor, 583
Photodissociation, 260, 584
Photoelectric effect, 51
Photoelectric equation, 52
Photoelectron spectroscopy, 54, 261, 262
Photoelectron spectrum, 262
Photoionization, 65, 261
Photon, 52, 642
Photosynthesis, 715–716, 716
Photosystem I, 716
Photosystem II, 716
pico-, 11
Piezoelectric effect, 125

Pimentel, G., 573
Pion, 602
pIon electrodes, 456
π bond, 188
π bonded molecules, 193
π-bonded molecules, 657
$\pi \longrightarrow \pi^*$ transition, 260
Pipet, 413
Pitchblende, 630, 674
Planck, Max, 42, 49–51, 733–737
Planck distribution law, 737
Planck's constant, 50, 736
Planck's hypothesis, 567
Planck's quantum hypothesis, 46, 736
Planck's quantum theory, 49–51
Planetary model, 25
Planté's lead-acid storage battery, 445–447
Plating of metals on other surfaces, 461
Platinum, 760
Pleated sheet, 710
^{210}Po, 640
Polanyi, J., 256n, 571–573
Polanyi rules, 596
Polar coordinates, 70
Polar covalent, 141
Polarizability α, 497
Polyacrylonitrile (PAN), 702
Polyamide, 703
Polyatomic molecules, 197
Polybasic acid, 396
Polyester, 703
Polyethylene, 702
Polymers, 524, 702–703
Polypeptide, 708
Polypropylene, 702
Polyprotic acid, 396, 406–411
Polysaccharides, 706
Polysilicate anions, 518
Polytetrafluoroethylene (PTFE), 703
Polyvinylchloride (PVC), 702
Polywater, 500n
Pope, Alexander, 375
Pople, J. A., 231
Porphyrin ring complex, 521
Porphyrins, 655, 656
Positive ions, 21, 145
Positron, 104, 606
Positron emission tomography (PET), 626
Potassium, 751–752
Potential difference, 438
Potential energy, 17
Potential energy barrier, 560
Potential energy contour map, 571
Potential energy minimization, 190
Potential energy surface, 258
Potentiometric titration, 457
Precipitation, 167, 419–421
Precipitation metathesis, 167
Precipitation reactions, 167–169, 597
Precision, 11–12
Predicting molecular shapes, 190–195
Premixed flame, 585
Pressure, 268

Pressurized water power reactor (PWR), 633
Priestley, Joseph, 4, 435
Primary carbocation, 691
Primary combustion zone, 585
Primary structure, 709
Principal quantum number, 72
Principle of detailed balance, 557–559
Pringsheim, Ernst, 49, 50, 256, 736
Probability amplitude, 84
Probability cloud, 84
Probability density, 84
Projection, 102
Projection quantum number, 72, 91
Proline, 708, 723
Propane, 685
Propanone, 554
Properties of gases, 267–300
　Avogadro's law, 273
　barometric principles, 268–270
　Boyle's law, 270
　Charles' law, 272
　collisions, 289
　Dalton's law of partial pressures, 277
　diffusion, 284
　effusion, 283
　gas dynamics, 288–291
　gas mixtures, 276–278
　Graham's law, 283, 284
　ideal gas law, 270–276
　kinetic theory of gases, 282–284
　Maxwell-Boltzmann velocity distribution, 287
　pressure of a gas, 268–270
　real gases, 291–295
　state changes for ideal gases, 311–314
　van der Waals equation of state, 294
Property tables. *See* Thermodynamic properties
Propionic acid, 397, 693
Proteases, 720
Protein, 580
Protein synthesis, 723–727
　step 1 (making RNA), 724
　step 2 (aminoacylating tRNA), 724–725
　step 3 (assembling ribosomes), 725
　step 4 (initiation), 726
　step 5 (elongation), 726
　step 6 (termination), 726–727
Proteins, 707–711
Proton, 22, 599
Proton acceptor, 157
Proton donor, 157
Proton Larmor frequency, 628
Proton-transfer reactions, 157
Protonation, 403
Pseudo-first-order, 550
PTFE, 703
^{239}Pu, 621, 633
Pulsed NMR, 642
Pump-probe experiment, 300
Punctuation, 722

Purcell, E., 628
Pure covalent, 141
Purines, 712
PVC, 702
PWR, 633
py, 655
Pyranoses, 705
Pyridine, 399, 400, 655
Pyrimidines, 712
Pyruvic acid, 717

Q

Qualitative metal ion analysis, 421, 423
Quanta, 50
Quantization of angular momentum, 55n, 67
Quantization of E, 59
Quantization of energy, 331
Quantization of rotational energy, 243–245
Quantization of vibration energy, 253–255
Quantum chemistry, 231
Quantum chemists, 121
Quantum chromodynamics, 602
Quantum corral, 125
Quantum dots, 518
Quantum electrodynamics, 180
Quantum hypothesis, 51, 737
Quantum mechanics, 62, 92n
Quantum number, 59, 72, 743
Quantum theory of radiation, 51
Quantum tunneling, 86
Quark, 180, 602
Quartz, 518
Quasibound state, 614
Quaternary structures, 711

R

Racemic acid, 696
Racemic mixtures, 697
rad, 625
Radial distribution function (RDF), 84–87
Radial electron distribution (Ar), 123
Radial node, 77
Radiation sickness, 625, 626
Radiation therapy, 623
Radiative lifetime, 93
Radio-frequency (RF), 627
Radioactive, 605
Radioactive dating, 630–632
Radioactive tracers, 624
Radioactivity, 23
Radioastronomy, 249
Radioisotopes, 613, 614, 623, 626
Radium, 756
Rail-welding thermite reaction, 322
Raisin pudding, 22
Raman effect, 249
Raman rotational spectra of homonuclear molecules, 627
Raman rotational spectrum, 249
Ramsay, W., 37

Ramsperger, Herman Carl, 566
Random coil, 711
Random walk, 574–575
Rankine scale, 37
Raoult, F., 525
Raoult's law, 525
Rare earths, 672
Rast method, 527
Rate constant, 543
Rate law, 542, 547–551
Rate-limiting step, 552–554
Rates and mechanisms of chemical reactions, 538–597
　Arrhenius plot, 559
　BZ reactions, 587, 588
　catalysis, 578–583
　chain reaction, 583
　collision theory (CT), 562–566
　determining the rate law, 547–551
　diffusion, 573–576
　elementary reactions, 551–552
　Eley-Rideal mechanism, 582
　energy profile, 560–561
　enzyme catalysis, 580–581
　explosions, 583–585
　Fischer-Tropsch process, 582
　flame, 585–586
　integrated rate law, 543
　isolation method, 549
　Langmuir mechanism, 582
　Lindemann mechanisms, 563
　method of initial rates, 547–549
　Michaelis-Menten rate law, 581
　molecular reaction dynamics, 569–573
　molecular theories of elementary reactions, 562–569
　oscillating reactions, 586–589
　principle of detailed balance, 557–559
　random walk, 574–575
　rate constant, 543
　rate law, 542, 547–551
　rate-limiting step, 552–554
　reaction mechanisms, 551–562
　reaction orders, 543–546, 594
　RRKM theory, 566
　rule-of-thumb reaction, 595
　solutions, reactions in, 573–577
　steady-state approximation (SSA), 555–557
　surface catalysis, 581–583
　temperature dependence of reaction rates, 559–562
　time profile, 540
　transition state theory (TST), 566–569, 572
Rayleigh, Lord, 256
Rayleigh-Jeans law, 736
RDF, 84–87
Reaction classification, 168–169
Reaction coordinate, 560
Reaction flame, 585
Reaction heats, 315
Reaction intermediate, 541, 552

Reaction mechanisms, 551–562
Reaction orders, 543–546, 594
Reaction quotient, 379
Reaction rate, 540. *See also* Rates and mechanisms of chemical reactions
Real gases, 291–295
Recharging a cell, 445
Recharging a storage battery, 461
Recreational wave pools, 45
Rectifier diode, 512
Red giants, 617
Red phosphorus, 152
Redox activity series, 161
Redox addition, 160
Redox chemistry. *See* Electrochemistry
Redox chemistry in reverse (electrolysis), 458–464
Redox displacement, 160–161, 436
Redox half-reaction, 164–165
Redox in aqueous solution, 162–165
Redox reactions, 133–137, 160–167, 168, 223, 669–671
Redox titration, 162–163, 457
Reduced mass, 242, 562
Reductionism, 3
Relative humidity (RH), 483
Release factor (RF), 726
rem, 625
Replication, 726–727
Representative elements, 135
Repulsive intermolecular forces, 502–503
Residue, 708
Resonance, 142
Resonance forms, 142, 328
Resonance problem, 227–230
Resonating double bonds, 230
Respiratory chain, 720
Reverse osmosis, 529
Reversibility, 346
Revolutionary council of physicists (1927), 42
RF, 627, 726
RH, 483
Rhenium, 758
Rhodium, 759
Rhombic form sulfur, 494
Rhombic system, 494
Rhombohedral, 515
Ribonucleic acid (RNA), 713, 724
Ribonucleotide, 712
Ribose, 712
Ribosome, 725
Ribozyme, 727
Rice, Oscar Knefler, 566
Right-handed amino acids, 707
Right-handed double helix, 714
Rigid-rotor approximation, 244
Ring, 687, 690
Ring-closing reactions, 699
rms velocity, 282
Rn decay product, 626
RNA, 713, 724
Road salt, 465
Rocket fuel, 299

Rohrer, H., 123
Röntgen, W., 605
Roosevelt, Franklin, 636
Root mean square displacement, 575
Root mean square (rms) velocity, 282
Roothaan, C. C. J., 231
Rotation, 239, 241
Rotational constant, 245, 246
Rotational energy, 244
Rotational motion, 91, 240–250
Rotational period, 243
Rotational selection rule, 247
Rotational spectroscopy, 246–250
Rotational wave function, 244
Rotationally resolved electronic band, 261
Rovibronic transitions, 261
RRKM theory, 566
Rubidium, 752
Rumford, Count, 302–303, 348
Rust, 464–465
Rust prevention, 465
Ruthenium, 759
Rutherford, Ernest, 22, 23–25, 53, 573, 599, 605, 606, 615, 616, 618, 619, 621, 623, 625, 639
Rutherford backscattering, 625
Rutherford's α-particle scattering apparatus, 24
Rutherford's α-particle scattering measurements, 599
Rutherford's planetary model, 25
Rutile, 516
Rydberg atoms, 66
Rydberg constant, 48
Rydberg formula, 48
Rydberg series, 66

S

Sagan, Carl, 635
Salt bridge, 441
Salt solubility rules, 167
Salts, 131, 169, 518
Samarium, 673, 775–776
Sand, 518
Sandwich compounds, 657
Saturated, 421
Saturated calomel electrode (SCE), 470
Saturated hydrocarbons, 499, 685
Saturated solution, 522
Sawhorse, 192
Scale height, 296
Scandium, 756
Scanning probe microscopy (SPM), 125–126
Scanning tunneling microscope (STM), 124, 125
Scattering light, 249
SCE, 470
SCF, 121
Schrödinger, Erwin, 42, 62–63, 69, 70–71, 244
Schrödinger equation, 63, 68–70, 742
Schrödinger RDF, 85
Schwinger, Julian, 3

Scientific notation, 12
Screening, 100, 106–109, 111, 115
Screening effect, 100
Seaborg, G., 623
sec-Butyl, 687
Second law of thermodynamics, 350–351, 354, 366–369
Second-order, 544
Second-period hydrides, 154
Secondary batteries, 445
Secondary carbocation, 691
Secondary structure, 710
Sedimentation coefficient, 725
"Seeing" atoms, 122–126
Selection rules, 93
Selective energy consumption, 572
Selenium, 773
Self-consistent-field (SCF), 121
Self-heating, 585
Self-reference, 726
Semiconductor, 511–513, 513
Semiempirical equation, 294n
Semipermeable membrane, 528
Separation of variables, 543
Separation-of-variables method, 544
Serine, 708
SF_4, 192
SHE, 450–452
Shell model of nuclear structure, 602
Short form of periodic table, 133
Shroud of Turin, 632
SI system, 10
SI unit prefixes, 11
σ bond, 188
σ bonded framework, 231
σ-bonded molecules, 190–193
$\sigma \longrightarrow \sigma^*$ transition, 260
Significant figures, 12
Silanes, 683
Silicate, 146, 518
Silicic acid, 146
Silicon, 511–512, 683–684, 768
Silicones, 684
Siloxanes, 684
Silver, 451, 761
Silver coulometer, 470
Simple harmonic motion, 19, 252
Simplest formula, 38
Simplicius, 1
Sine function, 44
Single bond, 139
Sinusoidal variation, 253
Slater, J. C., 121
Slow-step (bottleneck) rule, 669, 670
Small gas-phase molecules, 258
Smalley, R., 513
S_N1 reaction, 694
S_N2 reaction, 694
Snowflake, 472
Societal entropy, 368
Sodazide, 683
Soddy, Frederick, 605, 615, 616
Sodium, 132, 750–751
Sodium acetate, 401
Sodium borohydride, 155

Sodium D-line, 135
Sodium-water reaction, 148
Solar blackbody radiation, 363
Solar cells, 512
Solid
 allotropy, 513–514
 band theory, 508, 509
 ionic, 514–518
 metal crystals, 504–510
 molecular, 519–520
 nonmetallic network, 510–513
Solid-noble-gas matrices, 519
Solid-solid acid-base reaction, 343
Solid-solid transitions, 494
Solubility, 420–422
Solubility product, 420–422, 457
Solubility rules, 420
Solute, 522
Solutions, 521–529
Solutions, reactions in, 573–577
Solvay process, 320, 373
Solvent, 522
Solvent cage effect, 573
Sommerfeld, Arnold, 55
sp hybrids, 185
sp^2 hybrids, 185
sp^3 hybrids, 185, 187
Sparingly soluble salts, 167, 420–422
Spatial geometry, 192
Special theory of relativity, 600
Specific activity, 612
Specific energy disposal, 572, 597
Specific heat, 308
Specification of energy, 73
Specification of wave functions, 74
Specificity, 721
Spectator ions, 149
Spectral lines, 103
Spectrochemical ordering, 667
Spectrochemical series, 660
Spectroscopy, 43
 absorption, 376
 atomic, 623
 ESCA, 261
 eyeball, 652
 nuclear, 610
 electronic, 259–261
 infrared, 256
 microwave, 246
 photoelectron, 54, 261, 262
 rotational, 246–250
 vibrational, 255–259
Spectrum of electromagnetic radiation, 46
Spherical polar coordinates, 70
Spherically symmetric, 76
Spin, 101–105, 602, 626, 627
Spin quantum number, 102
SPM, 125–126
Spontaneity of chemical reactions, 342–374
 absolute entropies, 358–361
 Berthelot's hypothesis, 343–345
 Boltzmann's postulate, 356
 Carnot cycle, 347–351

Index

Spontaneity of chemical reactions—*continued*
  entropy, 351–352. *See also* Entropy
  entropy changes, 361–363
  Gibbs free energy, 364
  second law of thermodynamics, 350–351, 354, 366–369
  spontaneity and reversibility, 345–347
  temperature control of spontaneity, 365–366
  third law of thermodynamics, 357–358
Spread in energy, 288
Spring, 18–19
Square planar, 192
Square-planar complex, 664
Square pyramidal, 192
SSA, 555–557
Stability band, 606
Stability diagram, 605–606
Stabilization by delocalization, 329
Staggered form, 686
Stair-step appearance (stable isotopes), 606
Standard atmosphere, 269, 270
Standard deviation, 286
Standard free energy of formation, 377–378
Standard heats of formation, 320–324
Standard hydrogen electrode (SHE), 450–452
Standard pressure, 315
Standard reduction potentials, 449
Standard temperature and pressure (STP), 275
Standing-wave density, 736
Standing-wave vibration, 59
Starch, 706
Stark effect, 249
State, 310
State functions, 310
State variables, 310
States of matter and intermolecular forces, 472–537
  allotropy, 513–514
  attractive intermolecular forces, 496–499
  band theory of solids, 508, 509
  Born-Haber cycle, 517
  Bragg law, 505, 506
  Clausius-Clapeyron equation, 483
  Clausius equation, 493
  colloids, 523–524
  crystal structures. *See* Crystal structure
  evaporation, 482
  Fermi level, 510
  heating curves, 475–479
  hydrogen bonding, 500–502
  intermolecular forces, 474, 496–503
  ionic solids, 514–518
  liquids, 521–529
  metal crystals, 504–510
  molality, 526
  molecular solids, 519–520
  nonmetallic network solids, 510–513
  osmosis, 529
  phase changes, 479–495
  phase diagram, 485–487
  phase properties, 488–491
  phase rule, 493
  Raoult's law, 525
  rast method, 527
  repulsive intermolecular forces, 502–503
  semiconductor, 511–513
  solids, 503–520
  solutions, 521–529
  steric hindrance, 501
  temperature, 473
  Trouton's rule, 493
  Tyndall effect, 523
  x-ray crystallization, 508
  x-ray diffraction, 504–508
Stationary state, 54
Statistical thermodynamics, 568, 777–784
Steady-state approximation (SSA), 555–557
Steam engine, 348
Stearic acid, 706
Stefan, Josef, 737
Stefan-Boltzmann law, 737
Stellar nucleosynthesis, 616–618, 641
Stepwise equilibrium constants, 410, 411
Stereoisomers, 697
Steric effect, 690
Steric factor, 563
Steric hindrance, 501, 694
Stern, Otto, 287
Stern-Gerlach experiment, 103–104
Stimulated emission, 94
Stirling approximation, 778
Stoichiometric equivalence, 38
Stoichiometric methods, 28–29
Stoichiometry, 28
Stoney, G. Johnstone, 21
Storage batteries, 445
STP, 275
STP molar volume, 292
Straight-chain alkanes, 686
Straight-chain hydrocarbons, 499
Straight Dope (Adams), 747
Strained bond angles, 688
Strassman, F., 621
Stretching vibrations, 240
Strong acid, 160
Strong adsorption limit, 582
Strong field ligands, 662
Strong force, 599
Strontium, 755
Structural isomers, 199, 499, 686–688
Styrene, 699
Styrofoam, 317
Sublimation, 486
Substitution reactions, 689, 694–695
Substrate, 580
Successive approximations, 394
Sucrose, 706
Sulfate, 146, 422
Sulfide, 146, 422
Sulfide precipitation, 434
Sulfite, 146
Sulfur, 494, 772–773
Sulfur solid, 134
Sulfur trioxide, 153
Sulfuric acid, 146, 406
Sulfurous acid, 146
Sullivan, J., 594
Superconducting, 494–495
Superconductor, 495
Supercooling, 494
Superfluid, 492
Superheating, 494
Supernova, 618
Surface, 571
Surface catalysis, 581–583
Surroundings, 304
Surveyor 5, 625, 640
Symmetric tops, 245
Symmetry point groups, 661
Synthesis of ammonia, 368, 424, 429
Synthetic polymer, 702
System, 304
System temperature, 365
Szent-Györgi, Albert, 4

T

T_b, 484
T_c, 495
t-butyl group, 694–695
^{99}Tc, 639
Tantalum, 757
Tap water, 167
Tartaric acid, 696
Taylor series, 252n
Technetium-99, 639
Teflon, 703
Tellurium, 773
Temperature, 19
Temperature control of spontaneity, 365–366
Temperature dependence of enthalpy, 333–334
Temperature-jump method, 577
Temperature scales, 19–20
Temperature variation studies, 580
tera-, 11
Term values, 61
Terminal atoms, 194
Termination, 584, 726–727
Termolecular, 552
tert-Butyl, 687
Tertiary carbocation, 691
Tertiary structure, 711
Tetradentate ligands, 656
Tetragonal, 515
Tetrahedral, 183, 192
Tetrahedral complex, 664
Tetrahedral oxyanions, 650
Thallium, 765
Theoretical chemists, 2

Thermal cracking, 430, 690
Thermal energy, 20
Thermodynamic cycle, 319
Thermodynamic properties, 749–776
  aluminum, 764
  antimony, 771
  arsenic, 771
  barium, 755–756
  beryllium, 753
  bismuth, 771
  boron, 763–764
  bromine, 774
  cadmium, 762
  calcium, 754
  carbon, 765–768
  cerium, 775
  cesium, 752–753
  chlorine, 774
  chromium, 757
  cobalt, 759
  copper, 760–761
  europium, 776
  fluorine, 773
  francium, 753
  gallium, 764
  germanium, 768
  gold, 761–762
  hydrogen, 749
  indium, 765
  iodine, 774–775
  iridium, 759–760
  iron, 758–759
  lanthanum, 756
  lead, 769
  lithium, 749–750
  magnesium, 753–754
  manganese, 758
  mercury, 763
  molybdenum, 757
  neodymium, 775
  nickel, 760
  nitrogen, 769–770
  osmium, 759
  oxygen, 772
  palladium, 760
  phosphorus, 770–771
  platinum, 760
  potassium, 751–752
  radium, 756
  rhenium, 758
  rhodium, 759
  rubidium, 752
  ruthenium, 759
  samarium, 775–776
  scandium, 756
  selenium, 773
  silicon, 768
  silver, 761
  sodium, 750–751
  strontium, 755
  sulfur, 772–773
  tantalum, 757
  tellurium, 773
  thallium, 765
  thorium, 776

Index

tin, 768
titanium, 756
tungsten, 758
uranium, 776
vanadium, 757
yttrium, 756
zinc, 762
zirconium, 757
Thermodynamics, 302, 307
   first law, 303, 324, 335
   second law, 350–351, 354, 366–369
   statistical, 777–784
   third law, 357–358
Thiocyanate, 146
Thiocyanato, 655
Third law of thermodynamics, 357–358
Thisulfate, 146
Thompson, Benjamin, 302–303
Thomson, G. P., 57, 58, 123
Thomson, Joseph John, 21, 605, 627
Thomson, William. See Kelvin, Lord
Thorium, 776
Three-dimensional networks, 510
Three-dimensional wavelength density, 734
Three Mile Island nuclear meltdown, 634
3c2e bonding, 681, 683
3-D boundary surface diagrams, 77, 78
3-D cloud, 77
$3d$ transition metal, 645–651
Threonine, 708
Through-bond, 671
Thymine, 712
Thymol blue, 417
Thymolphthalein, 417
Time, 11, 367
Time-dependent Schrödinger equation, 93
Time profile, 540
Tin, 768
Titanium, 756
Titrant, 411
Titration, 38, 156, 162–163
   acid-base, 411–416
   pH titration curves, 412–415
   potentiometric, 457
   redox, 457
Tokamak, 635
Toluene, 699
Tomographies, 625
Torr, 269
Torricelli's barometer, 268, 269
Total energy, 17
Total kinetic energy, 281
Traité élémentaire de Chimie (Lavoisier), 35
Trajectory, 569–572
trans, 690
Trans-1,2-difluoroethylene, 199
Transcription bubble, 724
Transfer RNA (tRNA), 722, 724
Transient gas-phase species, 158n

Transistor, 512
Transition, 54
Transition element, 115
Transition metal, 133, 643–677
   actinides, 672–674
   back-bonding, 667
   charge-transfer transitions, 651
   color, 652
   coordination complex, 652, 658–661
   crystal field model, 661–663
   d-block metals, 645–647
   early transition metals, 648–651
   eyeball spectroscopy, 652
   f-block metals, 672–674
   geometry of complexes, 653–654
   group labels, 646
   lability, 657–658
   lanthanides, 672–673
   late transition metals, 651
   ligand field model, 655–667
   ligand substitution (LS), 668–669
   ligands, 654–657
   magnetism, 657
   nomenclature, 654
   octahedral complex, 653, 654
   outer-sphere complex, 668
   oxidation-reduction (redox) reactions, 669–671
   oxyions, 648
   spectrochemical series, 660
   square-planar complex, 664
   tetrahedral complex, 664
   $3d$ metals, 645–651
   what are they, 644
Transition metal complex, 417
Transition state, 566
Transition state theory (TST), 566–569, 572
Translation, 238, 241
Translational degrees of freedom, 570
Translocation, 726
Transuranic element, 621
Triatomic MO, 224
Triazane, 683
Triazene, 683
Tricarboxylic acid cycle, 718
Tricycle, 688
Tridentate ligands, 655n
Trigonal, 515
Trigonal bipyramidal, 191, 192
Trigonal planar, 186, 192
Trigonal pyramidal, 192
Triiodide, 147
Triose phosphate, 716
Triple bond, 139
Triple point, 486
tRNA, 722, 724
Trouton, F., 495
Trouton's rule, 493
Truncated icosahedron, 513
Tryptophan, 708, 723
TST, 566–569, 572
Tungsten, 758
Tunneling, 570, 614, 621

Tunnicle, 570
Turnaround temperatures, 480
Turning points, 252
Twain, Mark, 435
2,2,4-trimethylpentane, 687
2-D contour map, 77
2-D mountain, 77
Two-electron wave function, 180n
$2p$ orbital, 91
$2s$ orbital, 91
Two-slit experiment, 505
Tyndall, John, 737
Tyndall effect, 523
Tyrosine, 708

U
^{233}U, 633, 640
^{235}U, 621, 622, 633, 635
^{236}U, 621, 622
^{238}U, 612, 621, 633, 634
U_3O_8 oxide, 674
UF_6, 674
Uhlenbeck, George Eugene, 102
Ultraviolet catastrophe, 49, 736
Ultraviolet-to-visible (UV-vis) solution, 649
Uncertainty principle, 87–90
Unified mass scale, 26
Unimolecular, 552
Unit cell, 504
United atom, 206
Units of measure, 10–13
Universe, 367, 368
U_3O_8 oxide, 674
Unpaired, 109
Unpaired electron spins, 200
Unsaturated fats, 707
Unsaturated hydrocarbons, 689
Uracil, 712
Uranium, 626, 674, 776
Uranyl ion, 674
Urea, 679
Urease, 580
Urey, H. C., 621, 678, 727
U.S. Steel Building, 463
UV electronic spectra, 260
UV-vis solution absorption spectra, 649

V
Vacuum flange, 297
Valence, 138
Valence band, 511
Valence bond description of partial ionic character, 199
Valence bond σ–π picture, 188, 189
Valence bond (VB) theory, 181–187, 200–201
Valence bond wave function, 181
Valence electrons, 107
Valence-shell electron-pair repulsion (VSEPR), 190–195, 200–201
Valine, 708
van der Waals, Johannes, 291, 496
van der Waals constants, 292

van der Waals equation of state, 294, 424
van der Waals forces, 502
van der Waals radii, 503
van der Waals solids, 503
Vanadate, 649
Vanadium (V), 648, 675, 757
Vanillin, 698
van't Hoff, Jacobus H., 376, 528, 529, 697
van't Hoff equation, 386
van't Hoff factor, 528
van't Hoff plot, 559
van't Hoff plot/dissociation of fluorine, 387
Vapor pressure, 299
Variation theorem, 101n
VB description of partial ionic character, 199
VB hybridization model, 183–187
VB molecule, 201
VB overlap bond, 183
VB theory, 181–187, 200–201
VB-VSEPR model, 181–195
VB wave function, 181
Velocity distribution function, 286, 287
Velocity space, 279
Velocity-vector relations, 280
Vibration, 43, 240, 241
Vibrational band positions, 258
Vibrational frequencies, 255
Vibrational motion, 91, 250–259
Vibrational quantum number, 254
Vibrational selection rule, 256
Vibrational spectroscopy, 255–259
Vibrational transitions, 254
Virial theorem, 66
Vitamin B_{12}, 520
Van Vleck, J., 665
Volt, 438
Volta, Alessandro, 436, 437
Volta piles, 436, 437
Voltage ladder, 451
Voltaic cell, 436
Voltammetry, 463
von Guericke, Otto, 268
von Guericke's water barometer, 267
von Laue, Max, 504
von Weizsäcker, C., 616
VSEPR, 190–195, 200–201

W
Waage, Peter, 375
Wall collision rate, 289
Walsh, A. D., 225
Water, 139, 145, 154, 521
Water barometer, 268
Water hardness, 434
Water ice, 474, 519–521
Water molecule, 224–227
Water synthesis, 362
Watson, J. D., 714
Wave characteristics, 44
Wave equation, 63
Wave function, 63, 71, 742

Wave functions (hydrogen atom), 75
Wave mechanics, 62
Wave number, 43
Wave-particle duality, 53
Wave theory of light, 43–46
Waveguide microwave spectrometer, 248
Wavelength, 43
Wavelength density, 733, 734
Wavepackets, 570
Weak acid, 160, 392–397
Weak acid-weak base reaction, 402
Weak adsorption limit, 582
Weak base, 398–400
Weak base ionization constant, 399
Weak field ligands, 662
Weak interaction, 606
Werner, Alfred, 643, 644, 652, 653–654, 658

Whitetin (ß, metallic), 515
White dwarf, 618n
Whitehead, Alfred North, 68
Wien's law, 737
Wien's law of blackbody radiation, 64
Wilkins, M., 714
Williamson synthesis, 694
Wine barometer, 270
Wipf, P., 697
Witherite, 434
Wobble base, 723
Wöhler, Friedrich, 678, 679
Woodward, R. B., 221
Woodward-Hoffmann rules, 702
Work, 14–16, 305, 345
Work function, 52
World War II atomic bombs, 636
Wright, Thomas, 2
Wurtzite, 516

X
X-ray crystallography, 508
X-ray diffraction, 46, 504–508, 521

Y
yocto-, 11
yotta-, 11
Young, Thomas, 45
Young's two-slit experiment, 505
Ytterbium, 672
Yttrium, 756
Yukawa, H., 602, 639

Z
Z_{eff}, 114, 115
Zare, R., 573
zepto-, 11
Zero order, 545
Zero-order rate law, 545
Zero-order reaction, 545
Zero-point energy (ZPE), 90, 254, 282, 601, 737
Zero point vibrational energy, 253
zetta-, 11
Zewail, A., 573
Zhabotinsky, A. M., 587
Zinc, 132, 762
Zinc blende (cubic ZnS), 516
Zinc-mercuric oxide cell, 467
Zincate ion, 651
Zirconium, 757
ZPE, 90, 254, 282, 601, 737
Zwitterion, 194, 403

Greek alphabet

Alpha	A	α	Iota	I	ι	Rho	P	ρ
Beta	B	β	Kappa	K	κ	Sigma	Σ	σ
Gamma	Γ	γ	Lambda	Λ	λ	Tau	T	τ
Delta	Δ	δ	Mu	M	μ	Upsilon	Υ	υ
Epsilon	E	ε	Nu	N	ν	Phi	Φ	ϕ
Zeta	Z	ζ	Xi	Ξ	ξ	Chi	X	χ
Eta	H	η	Omicron	O	o	Psi	Ψ	ψ
Theta	Θ	θ	Pi	Π	π	Omega	Ω	ω

Conversion factors

Energy into joules (J)

4.184 J/cal 1.602×10^{-19} J/eV   101.325 J/(L atm) 10^{-7} J/erg 1.986×10^{-23} J/cm^{-1}

Length into meters (m)

10^{-10} m/Å   0.0254 m/in.   0.3048 m/ft   1609.3 m/mi

Volume into cubic meters (m^3)

10^{-3} m^3/L 3.7854×10^{-3} m^3/gal

Pressure into Pascals (Pa)

101,325 Pa/atm 10^5 Pa/bar   133.3 Pa/torr   6895 Pa/(lb in.$^{-2}$)

Mass into kilograms (kg)

0.4536 kg/lb 1.66054×10^{-27} kg/amu